KU-472-631

Basic and Practical
MICROBIOLOGY

BASIC AND PRACTICAL
Microbiology

Ronald M. Atlas *University of Louisville*

Macmillan Publishing Company
New York

Collier Macmillan Publishers
London

Printed in the United States of America

Macmillan Publishing Company
866 Third Avenue, New York, New York 10022

Collier Macmillan Canada, Inc.

Library of Congress Cataloging in Publication Data

Atlas, Ronald M.,
Basic and practical microbiology.

Includes index.
1. Microbiology. I. Title.
QR41.2.A83 1986 576 85-8821
ISBN 0-02-304350-4

Printing: 2 3 4 5 6 7 8 Year: 6 7 8 9 0 1 2 3 4 5

Cover photograph: This dermatophytic fungus, *Trichophyton tonsurans*, shown here growing on an agar plate and viewed under ultraviolet light, causes the so-called "black dot ringworm." In this particular type of tinea capitis infection, the hairs of the scalp break off near the scalp surface and exhibit a black dot appearance. Colonies of *T. tonsurans* have a powdery surface with wrinkled folds. Photograph © Carroll Weiss.

ISBN 0-02-304350-4

Acknowledgments

ANGUS & ROBERTSON, PUBLISHERS, A DIVISION OF BAY BOOKS PTY. LTD. A. D. Hope, "The Kings" (specified extracts) from *Poems* by A. D. Hope (London: Hamish Hamilton, 1960). Reprinted by permission of Angus & Robertson, Publishers.

THE ASSOCIATED PRESS. "56 shun measles vaccine, stay quarantined at camp" (8/19/84). "New drug is found to curb AIDS virus" (12/21/84). "Minister questions the ethics of 'bubble boy's' treatment" (1/5/85). "Boy who lacks immunity thriving" (11/25/84). "Disease risk found greater for early ages in day care" (11/10/84). "Drug produced on space flight is contaminated by microbe" (11/3/84). Reprinted by permission of The Associated Press.

THE COURIER JOURNAL. Eleanor Flagler, "The soap isn't squeaky clean" (1/12/84). Reprinted by permission of Eleanor Flagler and The Courier Journal.

CURTIS BROWN, LTD. Maxine Kumin, "The Microscope" from *The Wonderful Babies of 1809* by Maxine Kumin (New York: Putnam, 1968). Copyright © 1963, 1968 by Maxine W. Kumin. "The Microscope" originally appeared in *The Atlantic Monthly*. Ogden Nash, "Allergy Met a Bear," "The Germ," and "The Common Cold" from *I'm a Stranger Here Myself* and *Verses from 1929 On*, both published by Little, Brown and Company. Copyright © 1935, 1936, 1937, 1938 by Ogden Nash; Copyright © 1959 by Ogden Nash. Reprinted by permission of Curtis Brown Ltd.

GERALD DUCKWORTH & CO., LTD. Hilaire Belloc, "The Microbe" from *Cautionary Verses* by Hilaire Belloc. Reprinted by permission of Gerald Duckworth & Co., Ltd.

KATHERINE GALLAGHER. "Poison Ivy" by Katherine Gallagher. Reprinted by permission of the author. "Poison Ivy" originally appeared in *Jack and Jill.*

THE GLOBE AND MAIL. Stephen Strauss, "Death of a superbug: Nature is slicker oil eater" (1/17/85). Reprinted by permission of The Globe and Mail.

ESTATE OF ARTHUR GUITERMAN. Arthur Guiterman, "Ode to Amoeba" from *Gailey the Troubadour* (New York: Dutton, 1936) and "Strictly Germ-Proof" from *The Laughing Muse* (New York: Harper and Row, 1915), both by Arthur Guiterman. Reprinted by permission of Louise H. Sclove for the Estate of Arthur Guiterman.

HARCOURT BRACE JOVANOVICH, INC. T. S. Eliot, "The Dry Salvages" (one line) from *Four Quartets* reprinted from *Collected Poems,* 1909–1962. Copyright © 1943 by T. S. Eliot; renewed 1971 by Esme Valerie Eliot. Reprinted by permission of Harcourt Brace Jovanovich, Inc.

HARPER & ROW, PUBLISHERS, INC. Marie Louise Allen, "Sneezing" from *A Pocketful of Poems* by Marie Louise Allen. Copyright © 1939 by Harper & Row, Publishers, Inc. Reprinted by permission of Harper & Row, Publishers, Inc. Anonymous, "Little Willie" from *Shrieks at Midnight* edited by S. and J. Brewton (New York: Crowells, 1969). Robert Benchley, "How to Avoid Colds" from *From Bed to Worse* by Robert Benchley. Copyright © 1934 by Robert C. Benchley. Reprinted by permission of Harper & Row, Publishers, Inc. René Dubos, specified extracts (passim) from *Mirage of Health* by René Dubos, Volume Twenty-Two of *World Perspectives Series,* planned and edited by Ruth Nanda Anshen. Copyright © 1959 by René Dubos. Reprinted by permission of Harper & Row, Publishers, Inc. Gustav Eckstein, specified extracts (passim) from *The Body Has a Head* by Gustav Eckstein. Copyright © 1969, 1970 by Gustav Eckstein. Reprinted by permission of Harper & Row, Publishers, Inc. Shel Silverstein, "Sick" and "The Crocodile's Toothache" from *Where the Sidewalk Ends: The Poems and Drawings of Shel Silverstein.* Copyright © 1974 by Shel Silverstein. Reprinted by permission of Harper & Row, Publishers, Inc. Mark Twain, "Give an Irishman ..." from *Life on the Mississippi* by Mark Twain, Harper & Row, Publishers, Inc. Used by permission of the publisher.

THE HASTINGS CENTER. "Premarital blood tests: Mass screening for the wrong population" (12/8/78). Copyright © Institute of Society, Ethics and the Life Sciences, 360 Broadway, Hastings-on-Hudson, N.Y. 10706. Reprinted by permission of the Hastings Center Report.

HOUGHTON MIFFLIN COMPANY. Tom Parker, Facts 108, 212, 253, 265, 272, 302 from *In One Day* by Tom Parker. Copyright © 1984 by Tom Parker. Reprinted by permission of Houghton Mifflin Company.

EDITE KROLL. Shel Silverstein, "Sick" and "The Crocodile's Toothache" (text only) from *Where the Sidewalk Ends* by Shel Silverstein. Copyright © 1974 by Snake Eye Music, Inc. Reprinted by permission of the author in care of Edite Kroll.

LITTLE, BROWN AND COMPANY. Ogden Nash, "Allergy Met a Bear" and "The Common Cold" from *I'm a Stranger Here Myself* by Ogden Nash. Copyright © 1935, 1936, 1937, 1938 by Ogden Nash. "The Germ" from *Verses from 1929 On* by Odgen Nash. Copyright © 1959 by Ogden Nash. William T. Helmuth, "Oh, powerful bacillus ..." and Wallace Wilson, "He prayeth best ..." both quoted in *Familiar Medical Quotations* edited by Maurice B. Strauss, M.D. (1968). Hans Zinsser, short extract from *Rats, Lice and History* (1935). Emily Dickinson, "Faith is a fine invention," poem no. 185 of *The Complete Poems of Emily Dickinson* edited by Thomas H. Johnson (1960). All reprinted by permission of Little, Brown and Company.

CLARK MILLS MCBURNEY. Clark Mills, "Child With Malaria" from *Speech After Darkness: Five Young American Poets* (New York: New Directions, 1941). Reprinted by permission of the author.

NEWSWEEK. "Stinging a global killer" (8/13/84). Copyright © 1984 by Newsweek, Inc. All rights reserved. Reprinted by permission.

THE NEW YORK TIMES. Lawrence K. Altman, "How AIDS researchers strive for virus proof" (11/23/84). "Heart patient watched for infection" (12/12/84). "Applications of antibody discovery have broad impact on medicine" (10/16/84). "Mysterious form of Hepatitis seen as widespread threat" (5/28/84). "The mystery of Balanchine's death is solved" (5/8/84). "Publicity about airline incident leads to crucial diagnosis" (6/26/84). Harold M. Schmeck, "Agency reports genetic therapy is near" (12/18/84). "Odd virus now linked to major diseases" (11/20/84). "Gene related to immune system is purified by 3 research groups" (12/21/84). John Nobel Wilford, "Bacteria found to thrive in heat of volcanic vents on ocean floor" (6/3/83). Jo Thomas, "Big drop foreseen in infant deaths" (12/20/84). Perri Klass, "The age old fear of contagion arises when treating an AIDS patient" (10/25/84). Robert Hanley, "Eleven died of airborne fungus at hospital amid a delay over new air filter" (11/3/83). Andrew Pollack, "Japan's biotechnology effort" (8/28/84). Editorial Board, "The whooping cough crisis" (12/14/84). Bryan Miller, "From ale to stout: Guide to beer with food" (7/11/84). Philip Boffey, "U.S. panel issues first guide on coping with travellers' diarrhea" (1/31/85). Copyright © 1983, 1984, 1985 by The New York Times Company. Reprinted by permission of The New York Times.

OMNI INTERNATIONAL LTD. "Japanese drunkenness disease" (7/84). Copyright © 1984 by OPI. Reprinted by permission of OMNI International Ltd.

OXFORD UNIVERSITY PRESS. Henry Wadsworth Longfellow, extract from "Evangeline" reprinted from *Poetical Work, 1904, Oxford Standard Authors Series.* Oliver Goldsmith, "Elegy on the Death of a Mad Dog" from *The Collected Works of Oliver Goldsmith,* Volume 4 (London and New York). Susan Coolidge, "Measles in the Ark" reprinted from *The Oxford Book of Children's Verse* edited by Iona and Peter Opie. Reprinted with permission of Oxford University Press.

RAND MCNALLY. Anonymous, "Little Miss Muffet" from *The Real Mother Goose* (1944). Reprinted with permission of Rand McNally.

RANDOM HOUSE, INC./ALFRED A. KNOPF, INC. Hilaire Belloc, "The Microbe" from *Cautionary Verses* by Hilaire Belloc. Published 1941 by Alfred A. Knopf, Inc. Reprinted by permission of Alfred A. Knopf, Inc. John Updike, "Ode to Rot" from *Facing Nature* by John Updike. Copyright © 1985 by John Updike. Reprinted by permission of Alfred A. Knopf, Inc. Edward Fitzgerald, Extract from *The Rubaiyat of Omar Khayyam.* Reprinted with permission of Random House.

THE RICHMOND ORGANIZATION. Shel Silverstein, "Don't Give a Dose" from *Freakin' at the Freakers Ball,* words and music by Shel Silverstein. Copyright © 1972 by Evil Eye Music, New York, NY. Used by permission.

RUNNING PRESS. Frank Freudberg, specified extracts from *Herpes: A Complete Guide to Relief and Reassurance* by Frank Freudberg. Copyright © 1982 by Running Press. Reprinted by permission of Running Press, 125 S. 22nd St., Philadelphia, PA 19103.

SCIENCE NEWS. "Making snow the microbial way" (10/27/84). Copyright © 1984 by Science Service, Inc. Reprinted by permission of Science News, Inc.

THE SOCIETY OF AUTHORS. Walter de la Mare, extract from "Miss T." Reprinted by permission of The Literary Trustees of Walter de la Mare and The Society of Authors as their representative. George Bernard Shaw, extract from *The Doctor's Dilemma* (Copyright © Brentano's, New York, 1909). Reprinted by permission of The Bernard Shaw Estate and The Society of Authors as its representatives.

STERLING PUBLISHING COMPANY, INC. Specified extracts from *The Guinness Book of World Records* published by Sterling Publishing Co., Inc., New York, N.Y. Copyright © 1984 by Sterling Publishing Company Inc.

MRS. JAMES THURBER. James Thurber, "University Days" (specified extract), Copyright © 1933, 1961 by James Thurber. From *My Life and Hard Times* by James Thurber, published by Harper & Row. Reprinted by permission of Mrs. James Thurber.

TIME, INC. "Rush on Penicillin" (8/30/43). "The promise and peril of the new genetics" (4/19/71). J. D. Reed, "It's trendy, tasty and tofutti (7/9/84). Anastasla Toufexis, "Linking drugs to the dinner table" (9/24/84). "Chlamydia: The silent epidemic" (2/4/85). Copyright © 1943, 1971, 1984, 1985 by Time Inc. All rights reserved. Reprinted by permission from TIME.

UNITED PRESS INTERNATIONAL, INC. "Fungus kills koala at San Diego zoo" (8/19/84). "Maker's of aspirin agree" (1/11/85). "Vaccination drive for adults begins" (1/16/85). Reprinted with permission of United Press International, Inc.

UNIVERSITY OF UTAH PRESS. Adrien Stoutenberg, "V. D. Clinic" from *Greenwich Mean Time* published by the University of Utah Press. Reprinted by permission of the University of Utah Press. "V. D. Clinic" first appeared in *A Magazine of Poetry.*

UNIVERSITY PRESS OF KANSAS. Edmond R. Long, *A History of Therapy of Tuberculosis and the Case of Frederich Chopin* (specified extract, letter from Chopin to Fonatan), Copyright © 1956 by The University Press of Kansas. Reprinted by permission of the University Press of Kansas.

USA TODAY. Dan Sterling, "Four ways to fight hospital infection" (12/13/84). Reprinted by permission of USA TODAY.

VIKING PENGUIN INC. Elizabeth Madox Roberts, "Mumps" (extract) from *Under the Tree* by Elizabeth Madox Roberts. Copyright © 1922 by B. W. Huebsch, Inc. Copyright © 1930, 1958 by The Viking Press, Inc. Copyright © 1950 by Ivor S. Roberts. Reprinted by permission of Viking Penguin Inc. Theodore Rosebury, poem on p. 81 of *Microbes and Morals* by Theodore Rosebury. Copyright © 1971 by Theodore Rosebury. Reprinted by permission of Viking Penguin Inc. Lewis Thomas, *The Medusa and the Snail* (extract). Copyright © 1979 by Lewis Thomas. Reprinted by permission of Viking Penguin Inc.

THE WASHINGTON POST. Boyce Rensberger, "Researchers discover alga's primitive eyes aid in photosynthesis" (12/1/84). Christine Russell, "Scientists say common cold spread mainly by hand contact" (10/10/84). Victor Cohn, "Drug proves effective against genital herpes" (5/5/84). Philip J. Hilts, "U.S. approves dissemination of gene-engineered microbe" (9/14/83). Margaret Engel, "Toxic shock: A close brush with death" (10/21/84). Sally Squires, "Tests with humans could lead to vaccine against tooth decay and preventative measures include fluoride use" (1/3/85). Copyright © 1984, 1985 by The Washington Post. Reprinted by Permission.

YALE UNIVERSITY PRESS. Jonathan Swift, "On Poetry: A Rhapsody" (lines 351–356) from *The Complete Poems* edited by Pat Rogers (1983). Reprinted by permission of Yale University Press.

Preface

Basic and Practical Microbiology is intended for students who are not majoring in biology or microbiology but require a course in this subject as part of their program of study and for those without a science background who wish to learn about the part microorganisms play in life. The text does not assume prior courses in biology or chemistry. Students can read and understand the material without learning much chemistry, or they can use the appendix on chemical principles to develop a greater understanding of the basis for the relationship between biological structure and function.

In an age of "careerism," when most students are selecting courses that are directly related to their career goals, microbiology is an essential field of study for many students with diverse career interests. This text will be particularly useful for students in the health sciences, including those in allied health and nursing programs. It will also serve the needs of those interested in other areas of microbiology such as industrial and environmental microbiology.

The text covers both the basic and practical aspects of microbiology. Fundamental knowledge is developed within the context of applied relevance. Principles are supplemented with examples. This is done to maintain student interest, which is essential for learning any subject.

In a further attempt to maintain student interest, as well as to show the far-reaching relevance of microbiology, literary quotations have been included, placed throughout the book. These quotes reveal that microbiology reaches into all aspects of our daily lives, from the poetry our children read to the philosophical basis for many decisions that determine the course of history.

The text also includes numerous boxes that show microbiology in the news. These are complete reprints of actual news articles that have appeared in newspapers and magazines. New developments in microbiology appear in the news media on a daily basis. These boxes show some of the more recent developments and also show how the microbiological sciences are presented to the general public. The news headline boxes will reinforce the relevance of the textual material.

Several learning aids have been included in this text. Each chapter has a set of learning objectives. These objectives provide a prospectus for students so that they are prepared to cover the material in that chapter. The chapter objectives all begin with a list of key terms—these key terms are discussed within that chapter and also are defined in the glossary. The text has many figures to help illustrate the material being discussed. Each chapter has a summary to help place the material in perspective and a study outline to aid students in reviewing and learning the main points covered in that chapter. Students are provided with both short answer and thought-provoking study questions for each chapter to help them assess their own understanding of the material. For those interested in more information on a particular topic, each chapter has a list of annotated references that may prove of interest.

The presentation of material in this text, in terms of organization and level, is intended to aid students learn about microorganisms. The level is simple and straightforward. The organization of the text is from the basic to the practical—from the microscopic view of the cell to the macroscopic view of the importance of microbes in disease. The medical chapters are organized first according to the major anatomical site affected by a particular disease and then according to the taxonomy of the disease-causing organism. Many allied health and nursing students taking a microbiology course also take a course in human anatomy and physiology, and this organization permits these students to relate directly the material covered in these two courses.

I would particularly like to thank those colleagues at other institutions who gave so willingly of their time and expertise in reading the manuscript and providing suggestions for improvement: Ronald J. Downey, Ohio University; Denny O. Harris, University of Kentucky; Edward Kos, Rockhurst College; Gordon McFeters, Montana State University; Tom McQuistion, Milliken University; Harry E. Peery, Tompkins-Cortland Community College; Nancy Rapoport, Springfield Area Technical Community College; Pamela Tabery, Northampton County Area Community College; Kenneth J. Thomulka, Philadelphia College of Pharmacy; Patricia Warner, Palm Beach Junior College; and Mary E. Wise, Northern Virginia Community College. All read the manuscript at various stages, and it is much better for it. I am grateful to them all. I have not mentioned one reviewer, because I owe him special thanks: Jeffrey M. Libby, Miami University of Ohio, was especially helpful in helping the author steer a course appropriate for introductory students, and I want to publicly acknowledge his contribution to the present text. My thanks to the Macmillan staff: Greg Payne, editor; Dora Rizzuto, production supervisor; and Andy Zutis, designer.

R. M. A.

Contents

UNIT FIVE

Survey of Microorganisms

UNIT SIX

Microbial Growth

UNIT SEVEN

Host Defenses Against Disease-Causing Microorganisms

UNIT EIGHT

Microorganisms and Human Diseases

UNIT NINE

Applied Microbiology

Appendices

Basic and Practical MICROBIOLOGY

UNIT ONE
Introduction to Microbiology

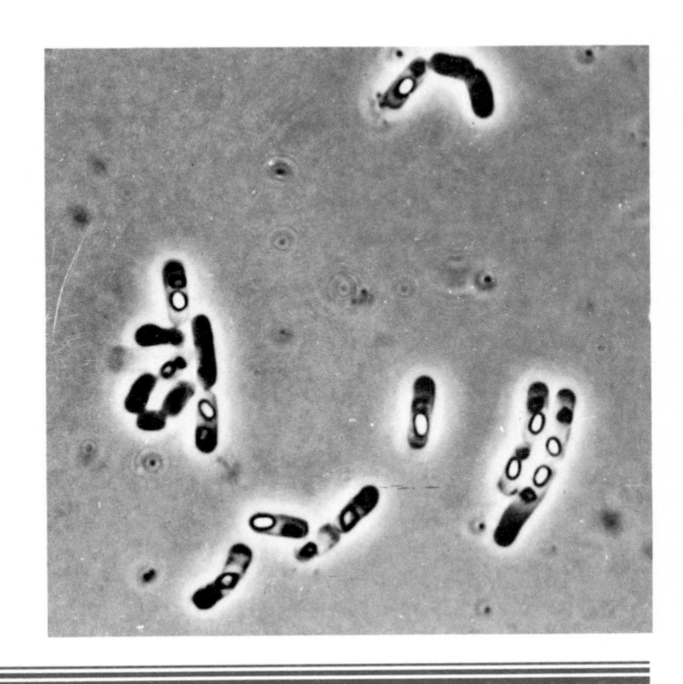

1 The Discovery of Microorganisms and Their Importance

KEY TERMS

agar
antibiotic
antiseptic
attenuated
autoclave
bacteria
bacteriological filter
chemotherapy
culture
etiology
fermentation
fomite
heat sterilization
immune
immunization
immunology
Koch's postulates
"magic bullets"
media (singular, medium)
microbiology
microorganism
microscope
pasteurization
petri dish
pure culture
putrefaction
spontaneous generation
sterile
tyndallization
vaccine
vaccination
virus

OBJECTIVES

After reading this chapter you should be able to

1. Define the key terms.
2. Discuss what each of the following individuals did to advance the field of microbiology and when their contributions were made.
   Paul Ehrlich
   Alexander Fleming
   Robert Hooke
   Edward Jenner
   Robert Koch
   Antony van Leeuwenhoek
   Joseph Lister
   Louis Pasteur
3. Discuss the importance of the microscope for the discovery of microorganisms.
4. Discuss how Pasteur discredited the theory of spontaneous generation and overcame objections that had been raised about previous experiments.
5. Discuss how elevated temperatures are used for sterilization.
6. Discuss the importance of culturing microorganisms on solid media for determining the relationship between microorganisms and disease.
7. List Koch's postulates.
8. Show how Koch's postulates are used today for the determination of the etiology of diseases such as Legionnaire's disease and AIDS.
9. Discuss the importance of chemotherapy for controlling infectious diseases.
10. Discuss the importance of vaccines for controlling infectious diseases.
11. Discuss the differences between chemotherapy and immunology for controlling disease.

Observation of Microorganisms

How many of us recognize that a glass of crystal clear water contains thousands of bacteria and that our bodies are covered with billions of bacteria? Our lack of awareness of the enormous numbers of diverse microorganisms all around us is not surprising. Microorganisms are extremely small (microscopic), and except for a few aggregated structures, microbes are invisible to the naked eye. For the most part only students who take courses in microbiology (the study of microorganisms) have the opportunity to peer into the microbial world and to appreciate the ubiquitous distribution of microorganisms.

Our ability to view microorganisms depends upon our use of a microscope. Before the advent of the microscope, no one could have imagined the existence of the enormous numbers of diverse microorganisms. The microscope was probably invented near the end of the sixteenth century by the Dutchman Hans Janssen. The advent of the microscope occurred at essentially the same time as the invention of the telescope. As Galileo looked outward toward the stars, others began to search inward to the never before seen, invisible microbial world. As some traveled on voyages of discovery to the Americas, others quietly carried out their private explorations into the fascinating realm of microbes.

Figure 1.1 Various species of fungi can clearly be identified in the drawings of Robert Hooke (1635–1703). This 1664 drawing shows the growth of blue mold on leather. Although macroscopic fungal structures had been described much earlier, Hooke's observations represent some of the earliest descriptions of many of the common fungi. (Reprinted by permission of the Royal Society, London, from R. Hooke, 1665, *Micrographia.*)

Where the telescope ends, the microscope begins. Which of the two has the grander view?

—VICTOR HUGO, *Les Misérable*

Among the first to discover the existence of microorganisms was Robert Hooke, an English experimental philosopher who also made significant contributions to physics, architecture, and clockmaking. Hooke made many microscopic observations of biological specimens and carefully recorded his findings. Hooke's detailed drawings in *Micrographia* (1665) reflect hours of tedious observation (Figure 1.1). Although Hooke clearly saw microbes, it was the amateur Dutch microscope maker Antony van Leeuwenhoek (Figure 1.2) who was the first to record observations of bacteria, yeasts, and protozoa.

Anton Leeuwenhoek was Dutch.
He sold pincushions, cloth, and such
The waiting townsfolk fumed and fussed
As Anton's dry goods gathered dust.

He worked, instead of tending store,
At grinding special lenses for
A microscope. Some of the things
He looked at were:
mosquitoes' wings,
the hairs of sheep, the legs of lice,
the skin of people, dogs, and mice;
ox eyes, spiders' spinning gear,
fishes' scales, a little smear
of his own blood,
and best of all
the unknown, busy, very small
bugs that swim and bump and hop
inside a single water drop.

Impossible! Most Dutchmen said.
This Anton's crazy in the head.
We ought to ship him off to Spain.
He says he's seen a housefly's brain.
He says the water that we drink
Is full of bugs. He's mad, we think!

They called him dumkopf, which means dope.
That's how we got the microscope.

—MAXINE KUMIN, *The Microscope*

Figure 1.2 Antony van Leeuwenhoek (1632–1723), here seen holding one of his microscopes, opened the doors to the hidden world of microbes when he described bacteria. Although only an amateur scientist, Leeuwenhoek had a keen interest in optics, and his diligence allowed him to make this important discovery. (Reprinted by permission of Russell and Russell Publishers, Dover Publications, and Mrs. Dobell, from C. E. Dobell, 1932, *Antony van Leeuwenhoek and his "Little Animals."*)

There is little resemblance between Leeuwenhoek's microscope (Figure 1.3) and those we use today. It is hard to imagine the persistence that must have been required to squint through this lens at a dimly illuminated specimen and to record the details of what he saw. But Leeuwenhoek had the necessary perseverance. A cloth maker and tailor by trade, as well as the official winetaster of Delft, Holland, Leeuwenhoek's interest in microscopes was probably related to the use of magnifying glasses by drapers to examine fabrics. He was a close friend of Christian Huygens, the noted Dutch physicist, mathematician, and astronomer, who undoubtedly encouraged Leeuwenhoek's interest in science and lenses.

Despite of his lack of training, Leeuwenhoek had an extraordinary talent for experimental design. After observing "animacules" in rainwater from his garden, Leeuwenhoek (being religious) decided to test whether the microbes were heavensent or (being practical) whether they were derived from earthly sources. He washed a porcelain bowl in fresh rainwater, set it out in his garden during a storm, and

Figure 1.3 Some early microscopes, used in the late seventeenth century. (A) This microscope used by Leeuwenhoek consists of a single biconvex lens (1) inserted into a small hole on the left side of the leather base. The specimen was placed on the small pointed wire attached to the screw, which moved the specimen back and forth to bring it into the focus of the fixed glass. (Reprinted by permission of Russell and Russell Publishers, Dover Publications, and Mrs. Dobell, from C. E. Dobell, 1932, *Antony van Leeuwenhoek and his "Little Animals."*) (B) Hooke's compound microscope of the same period more closely resembles the appearance of a modern microscope. Note the elaborate decoration of the body of the microscope, the candle illumination source and the separation of the eyepiece from the objective lens. (Courtesy American Optical.)

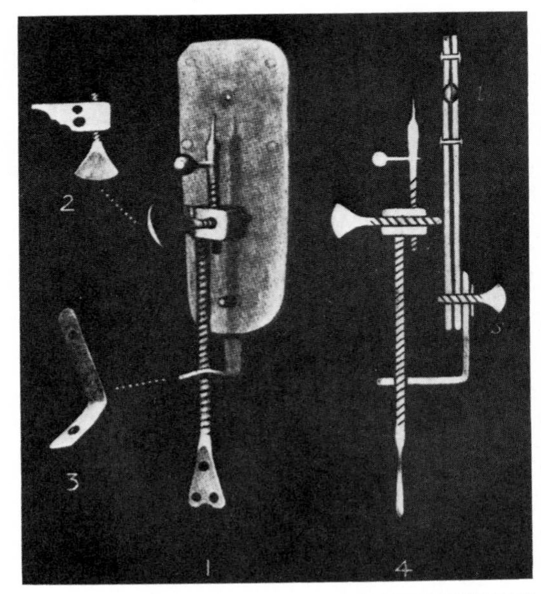

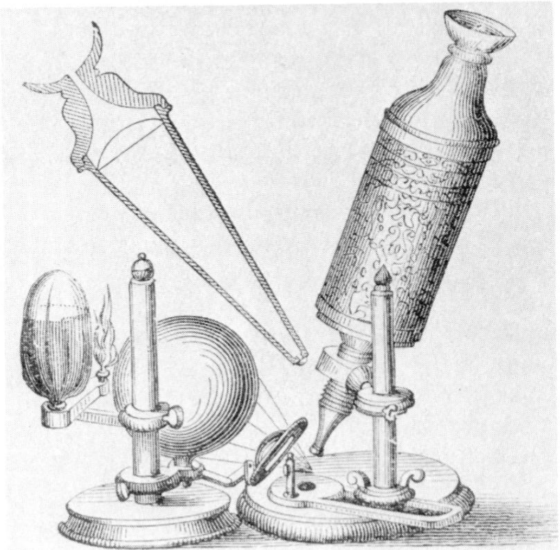

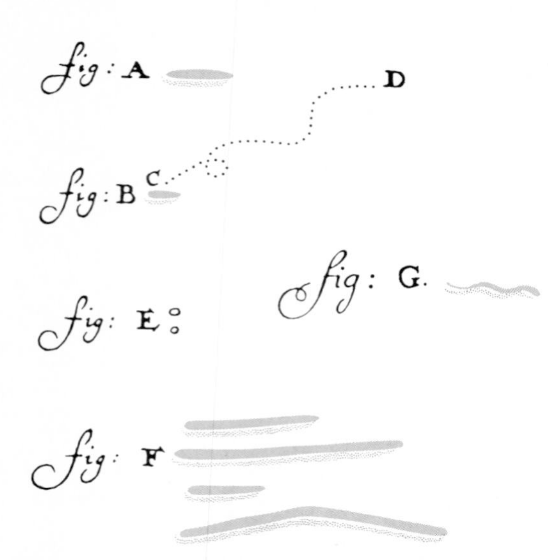

Figure 1.4 Leeuwenhoek's sketches of bacteria from the human mouth illustrate several common types of bacteria, including rods and cocci. (A) A motile *Bacillus*; (B) *Selenomonas sputigena*; (C) and (D) show the path of B's motion; (E) micrococci; (F) *Leptothrix buccalis*; and (G) a spirochete, probably *Spirochaeta buccalis*. (Reprinted by permission of the Royal Society, London, from Leeuwenhoek's Letter 39, 17 September 1683.)

observed no microbes in the freshly collected sample. After allowing the water to sit for a few days, he observed numerous microbes in the sample. He concluded that a few microbes in the sample had multiplied and that life begets life—even for animacules.

Leeuwenhoek transmitted his observations in a series of letters, from 1674 to 1723, to the Royal Society in London. During part of this period, Leeuwenhoek sent his communications to Robert Hooke, who was then secretary of the Royal Society. The letters contained detailed drawings, some of which clearly showed microorganisms (Figure 1.4). These letters to the Royal Society were widely disseminated, and thus the existence of microorganisms was made known to the scientific community of the early eigtheenth century.

By today's standards the early microscopes used by Hooke and Leeuwenhoek permitted only a limited view of the microbial world and only a superficial view of the organisms they were observing. Microscopic visualization of bacteria was greatly improved by changes in the design of the microscope and by Paul Ehrlich's introduction, in 1881, of vital staining with methylene blue. By using stains, the contrast between the microbe and the background is greatly enhanced, making the visualization of microorganisms much easier. Another major methodological improvement was the development of a differential staining method for bacteria by Hans Christian Gram in 1884. The Gram staining technique is basic to the taxonomic description of bacteria even today (see Chapter 3 and Figure 3.9 for a description of the Gram stain procedure).

When Leeuwenhoek discovered bacteria it was thought that these were the smallest living creatures. However, Dmitrii Ivanowski reported, in 1882 that the agent responsible for tobacco mosaic disease could pass through a bacteriological filter, which could trap all bacteria and larger microbes, thus making it apparent that the microbial world contained even smaller members than had been previously recognized. In 1898, Martinus Beijerinck, unaware of Ivanowski's work, ascribed tobacco mosaic disease to a "contagious living liquid." The visualization of these microbes, the viruses, did not occur, however, until the midtwentieth century. Viruses are too small to be viewed with the light microscope, and their observation depended upon the development of the electron microscope. Today the electron microscope permits us to observe the smallest of microbes and even to view individual molecules.

Cultivation of Microorganisms

Besides being able to view microorganisms, advances in microbiology depended upon the development of techniques for culturing (growing) and identifying microorganisms. These methods were a prerequisite for demonstrating the properties of individual microbial species and that particular diseases are caused by specific microorganisms. Many basic microbiological methods were developed in the late nineteenth century by Robert Koch and his disciples. Many of the methods developed in Koch's laboratories are still in use today with only minor modifications. It was Koch who, between 1881 and 1883, developed simple methods for the isolation and maintenance of pure cultures of microorganisms on chemically defined solid media. Prior to that time, microorganisms were grown in liquid broths. At first, Koch grew bacteria on solid fruits and vegetables, such as slices of boiled potato, but many bacteria cannot grow on such substances. Koch developed a way of solidifying liquid broths that could support the growth of a greater variety of microorganisms, initially using gelatin and later agar (an algal extract) as the solidifying agent. The suggestion for using agar originated with the New Jersey-born wife of one of the investigators at Koch's Institute, Frau Fanny Hesse, who had seen her mother using agar to make jellies. The use of these solidified media permitted the isolation and unequivocal identification of microorganisms.

The isolation and growth of microorganisms in pure culture has dominated most microbiological studies since the necessary techniques were developed by Koch. One of the modifications to Koch's original method was made by Richard J. Petri, who, in 1887, described the use of a new type of culture dish for growing bacteria on semisolid media. The basic design has become known as the petri dish and is used in virtually all microbiological laboratories in essentially the same design described by Petri.

The Importance of Microbial Metabolism

The ability to see and to culture (grow) microorganisms provided the necessary methodology for investigating the activities of microorganisms. The studies of Louis Pasteur in the late nineteenth century showed the importance of microorganisms and their metabolic activities. At the time of Pasteur the greatest scientific debates concerned the theory of spontaneous generation. A relationship had frequently been observed between the growth of infusoria (microorganisms) in organic broths and the onset of chemical changes, known as fermentation (souring of carbohydrates in grapes to produce wine) and putrefaction (decomposition of meat proteins), in the broth. Where did this infusoria come from? Did it arise spontaneously or did it arise from infectious agents in the air?

The experiments of Lazzarro Spallanzani, an eighteenth century priest who had an exceptionally inquiring mind and willingness to challenge the conventional wisdom of his time, demonstrated that the putrefaction of organic substances is caused by microorganisms, which multiply by reproductive divisions and that do not arise by spontaneous generation. He demonstrated that heating destroyed these organisms and that in sealed flasks spoilage was prevented indefinitely. However, nineteenth-century advocates of spontaneous generation claimed that the elimination of oxygen invalidated these experiments.

Chemists at that time held that changes in organic chemicals, such as the putrefaction of proteins and the transformation of sugar into alcohol, occurred by strictly chemical processes without the intervention of living organisms. This view was opposed in the 1830s by Charles Cagniard de Latour of France, and Theodor Schwann and Friedrich Küntzing, both of Germany, who separately proposed and conducted experiments to demonstrate that the products of fermentation, ethanol and carbon dioxide, were produced by microscopic forms of life. Schwann used a flame, and Latour and Küntzing used cotton plugs to prevent microorganisms from entering the heat-sterilized broth. Each of these experiments was criticized by the chemists because each destroyed or eliminated some essential component in air, which they claimed was needed for the spontaneous generation of the fermentation products.

Canning as a method of preserving foods has its roots in the works of Spallanzani, Schwann, and others, who steadfastly held that putrefaction was owing to living organisms. The process of canning was patented in 1810 by a French candy maker Nicolas Appert, and by 1820 the commercial production of canned foods had begun in the United States.

It was several decades later that Louis Pasteur (Figure 1.5) settled the dispute over spontaneous generation, finally discrediting this theory and establishing that living microorganisms are responsible for the chemical changes that occur during fermentation. His results were published in a report, *On the Organized Bodies which Exist in the Atmosphere; an Examination of the Doctrine of Spontaneous Generation* (1861). In his experiments, Pasteur demonstrated that liquids subjected to boiling remained sterile (free of living organisms) as long as microorganisms in the air were not allowed to contaminate the liquid. By using a swan-necked flask (Figure 1.6), Pasteur was able to leave a vessel containing a fermentable substance open to the air and show that fermentation did not occur. The shape of the flask prevented airborne microorganisms from entering the liquid, and the fact that air could enter the flask overcame the main argument that chemists had leveled against earlier studies using sealed flasks, namely, that oxygen was essential for the chemical reactions involved in the formation of alcohol.

No more shall spontaneous generation rear its ugly head!

—Louis Pasteur

Figure 1.5 Louis Pasteur (1822–1895), seen working in his laboratory in this 1885 woodcut, began as a chemist but soon became a pioneer microbiologist. Pasteur's work encompassed both pure research and many areas of applied science that produced many important practical discoveries. Among his many accomplishments, Pasteur discredited the theory of spontaneous generation, introduced vaccination to treat rabies, and solved industrial problems related to the production and spoilage of foods. (The Bettmann Archive, Inc.)

Pasteur was trained as a chemist, and this training had a marked influence on his approach to scientific questions. He followed the same investigative approach throughout his long scientific career: he identified the problem, sought out all the available information on the topic, formed a hypothesis, and devised experiments to test the validity of his theory. The problem of applied versus basic or pure research never bothered this aggressive, ambitious, argumentative, and highly patriotic Frenchman. Much of his work stemmed from the requests of local manufacturers to help solve the practical problems of their industrial processes. He loyally responded, attempting to solve these problems in order to improve the French economy and demonstrate French superiority. He was concerned with problems such as why French beer was inferior to German beer. The answer to this practical question eventually led him to the basic discovery of the existence of anaerobic life: life in the absence of air.

In 1854 Pasteur was appointed Dean of the Faculty of Science at the University of Lille. One of the first problems he then pursued was at the request of a local industrialist and concerned the souring of alcohol produced from beets. Pasteur's agreement to help solve this problem of the wine industry placed him directly on the main path of his scientific career, leading him from chemistry to microbiology. By comparing, with the aid of a microscope, samples taken from productive and nonproductive vats, Pasteur observed budding yeast cells in the productive vats and rod-shaped organisms in the nonproductive ones. He was able to demonstrate that these organisms determined the course of the chemical processes. The yeasts were responsible for the production of alcohol, and the rod-shaped bacteria produced the lactic acid that caused the wine to sour. Thus, Pasteur succeeded in demonstrating that different microorganisms are responsible for different fermentations. The work of Pasteur led to an understanding of the role of microorganisms in food spoilage and the use of heat to destroy microorganisms in food products. In 1857 Pasteur demonstrated that the souring of milk was also caused by the action of microorganisms, and in 1860 he showed that heating could be used to kill microorganisms in wine and beer.

Much of our economic life is under the control of microbial activities over which we have little if any control. Microorganisms contribute essential steps to many technological processes, but they also spoil or rot every kind of foodstuff and of goods. Except in a few situations, microorganisms are today as undisciplined a force of nature as they were centuries ago.

—René Dubos, *Mirage of Health*

Just as in the nineteenth century, the control of microorganisms today is very important. We rely on many processes to preserve foods and to steri-

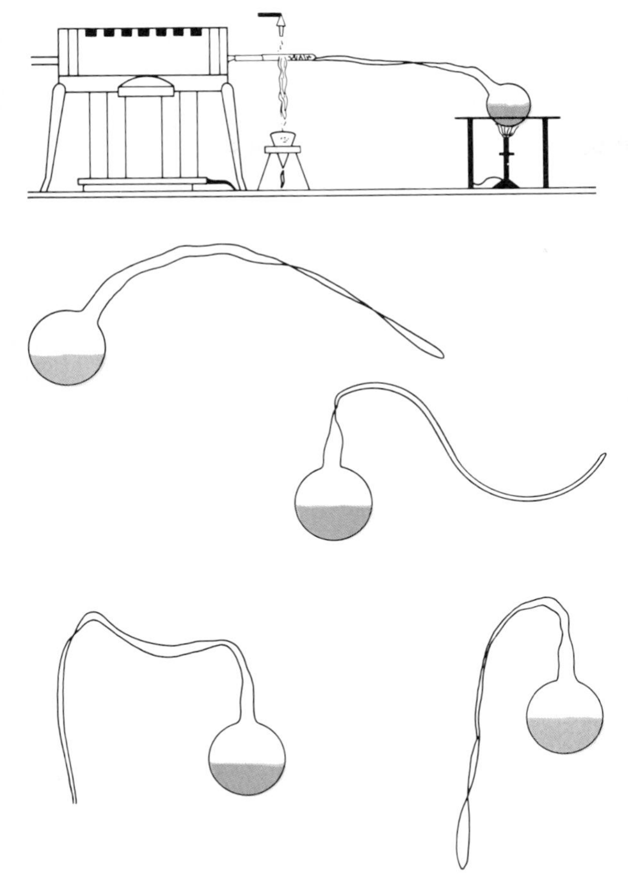

Figure 1.6 In his experiments aimed at settling the question of spontaneous generation, Pasteur used a variety of shapes in the design of his swan-necked flasks. Pasteur began his classic experiment in 1861 by placing yeast water, sugared yeast water, urine, sugar beet juice, and pepper water into ordinary flasks. He then reshaped the necks of the flasks under a flame so that there were several curves in each neck, hence the term swan-necked flasks. Next, Pasteur boiled the liquids until they steamed through the necks. By using curved ends, Pasteur could leave the flasks open to the air, thus overcoming a major criticism of previous experiments aimed at disproving spontaneous generation where air, an "essential life force," had been excluded. Dust and microbes settled out in the curved neck of the flask and thus, although exposed to air, the broth did not become contaminated with microorganisms. Contrary to the opinion of those who believed in spontaneous generation, no change or alteration appeared in the liquid. These flasks, later sealed, may be seen at the Pasteur Institute in Paris.

lize numerous medical solutions and instruments. The process of pasteurization, as it has come to be known, was introduced around 1867 when Pasteur pointed out that the application of a modification of Appert's canning method was a practical way of eliminating undesirable ferments.

Early work on heat sterilization (killing all microbes by exposure to high temperatures) was complicated by the occasional presence of endospore-forming bacteria. The English physicist, John Tyndall, trying to confirm the results of Pasteur's experiments refuting spontaneous generation, determined that the variability of results was owing to the ability of bacteria to exist in two forms—a heat-labile form that was killed by exposure to elevated temperatures, and a heat-resistant form that could survive at such high temperatures. He found that intermittent heating could eliminate viable microorganisms and sterilize solutions. Repeated heating on successive days, a process known as tyndallization, successfully sterilizes even solutions containing endospore-forming bacteria. In 1881, Robert Koch, Georg Gaffky, and Friedrich Loeffler published *Observations on the Effectiveness of Hot Steam for Disinfection*, in which they showed that it was necessary to achieve a temperature of 160°C for at least one hour in order to achieve sterility. The first laboratory autoclaves, which resembled large pressure cookers, were designed by Charles Chamberland, a colleague of Pasteur. The autoclave permits the heating of steam under pressure to achieve temperatures above 100°C, needed to kill bacterial endospores, and is an essential instrument in every bacteriology laboratory.

Recognizing the Relationship Between Microorganisms and Disease

Our earliest ancestors had to contend with disease. Major outbreaks of disease ran rampant through populated areas.

> The paleological records leave no doubt therefore that most of the known organic and microbial disorders of man and animals are extremely ancient. But it is also certain that the comparative prevalence and severity of various diseases have changed greatly in the course of historical times. . . . Influenza, yellow fever, smallpox, typhus, and cholera likewise conjure up the thought of visitations descending on mankind as the curse of some avenging deity. Other epidemics are now forgotten, but caused great terror in their days.
>
> —René Dubos, *Mirage of Health*

Figure 1.7 The devastating effects of plague in medieval Europe are shown in this painting. Plague was apparently introduced by sea-borne rats from Black Sea areas. The spread of the disease generally followed trade routes from the Near East to Europe, reaching Scandinavia by 1350 and perhaps spreading as far as Iceland and Greenland, whereas areas outside of the major trade routes remained virtually unaffected. In this engraving by Sabaleth, the plague of Florence in the fourteenth century is depicted as described by Bocaccio, a Florentine writer of that time. (The Bettmann Archive, Inc.)

The various epidemics of plague and other diseases constituted great destructive cataclysms (Figure 1.7). During the Justinian era, plague killed one half to two thirds of the inhabitants of certain cities of the Roman world. In Europe leprosy was prevalent in the fourteenth century, plague in the fifteenth, syphilis in the sixteenth, smallpox in the seventeenth and eighteenth centuries, scarlet fever, measles, and tuberculosis in the nineteenth century.

When, in the midfourteenth century more than 25 million people died in medieval Europe during a severe outbreak of plague, the populace turned to such unscientific and unrewarding practices as flagellation—beating themselves—to drive out the force causing "black death." A more relevant method of dealing with plague was that of removing individuals suspected of having the disease from contact with the general population. This practice dates back to Biblical times. At the height of the fourteenth century, when plague was epidemic, sea voyagers coming into Sicily had to wait 40 days before entering the city—a practice known as quarantine.

Two centuries after the major outbreaks of plague in Europe, Girolamo Fracastoro of Verona, a contemporary of Copernicus, published a work on contagious diseases and their treatment. *De Contagione* (1546) was largely philosophical and did not recognize the true nature of microorganisms. Nevertheless, Fracastoro did hypothesize that some diseases were caused by the passage of "germs" from one thing to another. He described three processes for the transmission of contagion: direct contact, transmission via **fomites** (inanimate objects, such as clothing, which may pick up and preserve germs), and transmission from a distance. Fracastoro recognized that germs causing disease exhibit specificity, which results in different diseases occurring in different hosts and different processes of transmission for different germs.

The germ theory of disease had little immediate influence on medicine and human health. The belief was that cleanliness and hygenic practices would control disease, and to a large degree sanitary practices introduced in Europe after the Industrial Revolution lowered the incidence of disease. The great microbial epidemics were brought under control not by treatment with drugs but largely by sanitation and by the general rise in living standards. Sanitary measures were accompanied by a decrease of typhus morbidity and mortality. These sanitary measures were implemented by boards of health that did not believe in contagion, let alone in the germ theory of disease. Faith in the healing power of pure air, with much contempt for the germ theory of disease, was the basis of Florence Nightingale's reforms of hospital sanitation in the mid 1850s (Figure 1.8).

> There are no specific diseases . . . there are specific disease conditions.
>
> —FLORENCE NIGHTINGALE

Figure 1.8 Florence Nightingale attempted to create sanitary conditions in the military hospitals during the Crimean War. This wood engraving, which appeared in the February 24, 1855 issue of *The Illustrated London News*, shows her in the hospital at Scutari. (From the collections of the Library of Congress.)

The last half of the nineteenth century marks a turning point in the epidemic history of the Western World. Transmissible diseases were, of course, still plentiful; and scarlet fever, diphtheria, meningitis, and measles—which had previously been masked to some extent by the more rapidly spreading and violent contagions—now attained greater prominence . . . But except for influenza, the pestilences which had, throughout preceding centuries, caused the most widespread destruction were distinctly declining and were becoming more limited in regional distribution.

—HANS ZINSSER, *Rats, Lice and History*

The studies of Robert Koch (Figure 1.9) changed the view of the cause of the disease and marked the beginning of modern medical microbiology. Koch, a German country physician, began his scientific studies isolated from any contact with the scientific community, working alone with primitive tools and materials. As a result of his medical practice, Koch was well aware of the diseases of man and other animals. From 1873 to 1876, Koch conducted experiments to show that the spores of anthrax bacilli isolated from pure cultures could infect animals. Koch demonstrated for the first time that germs grown outside the body could cause disease, and that specific microorganisms caused specific diseases. This was the beginning of Koch's illustrious career. He went on to determine the causative organisms for several other diseases, including tuberculosis and cholera.

Until late in the nineteenth century disease had been regarded as resulting from a lack of harmony between the sick person and his environment; as an upset of the proper balance between the yin and the yang, according to the Chinese, or among the four humors, according to Hippocrates. Louis Pasteur, Robert Koch, and their followers took a far simpler and more direct view of the problem. They showed by laboratory experiments that disease could be produced at will by the mere artifice of introducing a single specific factor—a virulent microorganism—into a healthy animal.

—RENÉ DUBOS, *Mirage of Health*

In describing the cause (etiology) of tuberculosis, Koch set forth the basic principles for establishing a cause and effect relationship between a given microorganism and a specific disease.

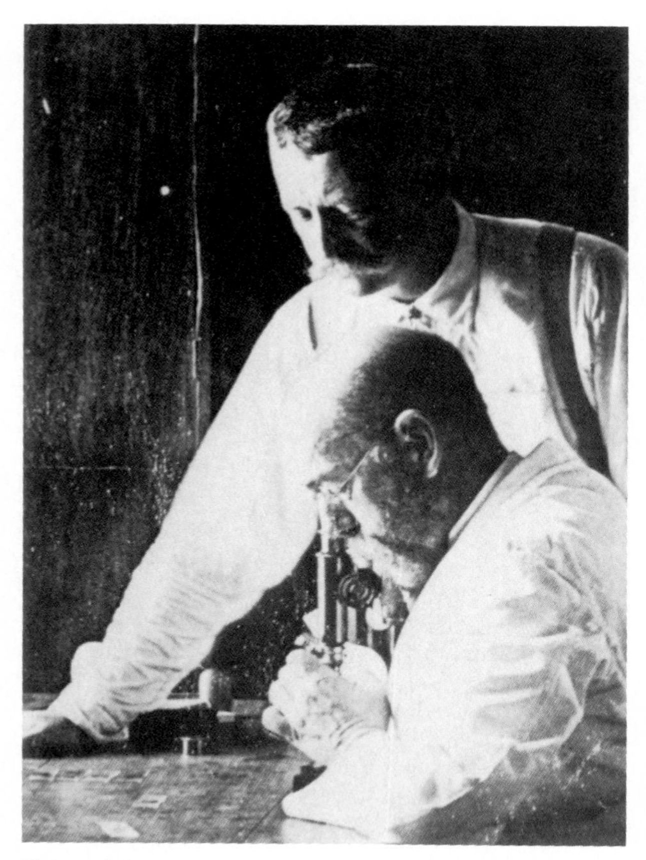

Figure 1.9 Robert Koch (1843–1910), seen here viewing a specimen while a disciple looks on, pioneered studies in medical microbiology and developed many of the basic methods essential for the study of microbiology. Koch's postulates for establishing the etiology of infectious diseases and the methodological techniques he developed still are used today in scientific investigations. Many of Koch's students also made significant contributions to the development of the field of microbiology. (The Bettmann Archive, Inc.)

Stated simply, Koch's four postulates for identifying the etiologic agent of a disease are

1. The organism should be present in all animals suffering from the disease and absent from all healthy animals.
2. The organism must be grown in pure culture outside the diseased animal host.
3. When such a culture is inoculated into a healthy susceptible host the animal must develop the symptoms of the disease.
4. The organism must be reisolated from the experimentally infected animal and shown to be identical with the original isolate.

> To obtain a complete proof of a causal relationship, rather than mere coexistence of a disease and a parasite, a complete sequence of proofs is necessary. This can only be accomplished by removing the parasites from the host, freeing them of all tissue elements to which a disease-inducing effect could be ascribed, and by introducing these isolated parasites into a healthy animal with the resulting reproduction of the disease with all its characteristic features. An example will clarify this type of approach. When one examines the blood of an animal that has died of anthrax one consistently observes countless colorless, non-motile, rodlike structures . . . When minute amounts of blood containing such rods were injected into normal animals, these consistently died of anthrax, and their blood in turn contained rods. This demonstration did not prove that the injection of the rods transmitted the disease because all other elements of the blood were also injected. To prove that the bacilli, rather than other components of blood produce anthrax, the bacilli must be isolated from the blood and injected alone. This isolation can be achieved by serial cultivation . . . The serial transfers can be continued for 3 or as many as 50 passages and in this manner the other blood components can be eliminated with certainty. Such pure bacilli produce fatal anthrax soon after injection into a healthy animal, and the course of the disease is the same as if produced with fresh anthrax blood or as in naturally occurring anthrax. These facts proved that anthrax bacilli are the unique cause of the disease.
>
> —Robert Koch, 1884, *The Etiology of Tuberculosis*

Koch's postulates, which are applicable to plant as well as animal diseases, still form the basis for determining that a particular disease is caused by a given microorganism. For example, the search for the cause of Legionnaire's disease in the 1970s followed Koch's 1890 postulates, resulting in the eventual identification of the bacterial etiologic agent. After many attempts, the bacterium *Legionella pneumophila* was isolated from patients with this disease, grown in the laboratory, inoculated into test animals in which the organism caused the onset of disease symptomology, and reisolated from the experimentally infected animals. Some modifications to Koch's postulates are required in cases when the disease is caused by opportunistic pathogens that are normally associated with healthy animals; when the experimental host is immune to the particular disease; when the disease process involves cooperation between multiple organisms; and when the causative agent cannot be grown in pure culture outside of host cells.

It is often difficult to find "volunteers" who are willing to be inoculated with presumed disease-causing microorganisms. Times have changed since the early 1900s when Pettenkoffer in Germany and Metchnikoff in France, together with several of their associates, drank tumblerfuls of cultures isolated from fatal cases of cholera. Fortunately these investigators were resistant to cholera; although high numbers of cholera vibrios could be recovered from their stools and some of the self-infected experimenters developed mild diarrhea, none of these microbiologists developed true cholera. Today, medical researchers try to avoid such experiments. Researchers have recently exposed large numbers of subjects to Rhinoviruses to study the transmission of the common cold but are not performing such experiments with more serious microbial pathogens. For example, although a virus has been isolated from individuals with AIDS (acquired immune deficiency syndrome) and grown in pure culture in the laboratory, it is not ethical to inject healthy individuals with the virus to establish a cause and effect relationship between the virus and AIDS. Despite these occasional difficulties, the philosophy of Koch's postulates for identifying the causes of infectious diseases remains the fundamental basis of modern epidemiology.

> *A mighty creature is the germ,*
> *Though smaller than the pachyderm.*
> *His customary dwelling place*
> *Is deep within the human race.*
> *His childish pride he often pleases*
> *By giving people strange diseases.*
> *Do you, my poppet, feel infirm?*
> *You probably contain a germ.*
>
> —Ogden Nash, *The Germ*

Control of Infectious Diseases

Chemotherapy

Responding to the growing awareness that microorganisms are associated with disease processes, Joseph Lister, an English Quaker and physician, revolutionized surgical practice in 1867 by introducing antiseptic practices (Figure 1.10). The discovery in the early 1850s of anaesthesia and its administration to patients made surgery much easier but of course did nothing to reduce the inci-

MICROBIOLOGY HEADLINES

How AIDS Researchers Strive for Virus Proof

By LAWRENCE K. ALTMAN, M.D.

WASHINGTON—Isolating a new microorganism among people afflicted with a disease of unknown origin is only the first step toward establishing a cause and effect relationship between the microbe and the disease. Such proof must come from additional laboratory tests that are usually done in conjunction with experiments on animals.

That process is known as fulfilling Koch's postulates, and doctors have been doing it over the last 100 years since Robert Koch, the German microbiologist, developed the scientific steps named for him.

Koch's postulates are often summarized in four steps. First, the microorganism is observed in all cases. Second, the microorganism is grown in pure culture. Third, inoculations of the pure culture in susceptible animals reproduce the condition. Fourth, the microorganism is recovered from the experimentally infected animal.

In the case of acquired immune deficiency syndrome, or AIDS, scientists have come tantalizingly close to proving that a retrovirus called HTLV-3/LAV is the cause of the usually fatal syndrome. However, scientists have not yet fulfilled Koch's postulates for AIDS, which has struck more than 6,250 people, according to the latest reports from the Centers for Disease Control in Atlanta.

Nevertheless, AIDS researchers have taken a major step toward fulfilling Koch's postulates by transmitting the HTLV-3/LAV retrovirus to chimpanzees.

The experiments have been done independently by researchers at the National Institutes of Health in Bethesda, Md., and at the Centers for Disease Control by different methods. In one, the injections were made with pure cultures of the AIDS virus. In the other, samples of blood from victims of AIDS were injected into animals.

At the National Institutes of Health, a team headed by Dr. Harvey Alter took plasma from AIDS patients and injected it into three chimpanzees housed at the Southwest Foundation for Biomedical Research in San Antonio. Two of the chimpanzees showed some evidence of AIDS.

A natural response to infections is to form antibodies, the immunological proteins that chemically fight off the invading microorganisms. One striking feature of the AIDS syndrome is the development of enlarged lymph nodes throughout the body. Another is the development of an immunological abnormality, depression of the so-called T4-T8 ratio. The T4 and T8 are types of white blood cells known as lymphocytes.

One chimpanzee injected with the AIDS virus developed enlarged lymph nodes as well as laboratory signs of the infection. The chimpanzee also developed antibodies to HTLV-3/LAV and also developed the T4-T8 immunological abnormality.

A second chimpanzee developed antibodies to the AIDS virus but not the two other abnormalities. The third has remained well and shown no laboratory abnormalities.

However, the incubation period is long in AIDS. The incubation period is the time it takes from exposure to the microorganism to the onset of symptoms of the condition it causes. The incubation period of AIDS can be up to five years in humans, and it took several months for the chimpanzees to develop the syndrome under the conditions of the experiment.

Thus, it is not clear whether animals injected with the AIDS virus will develop a more severe form of the disease. Sometimes microorganisms do not produce an illness of the same severity when injected across species barriers. But some researchers have accepted the experimental results as evidence toward fulfilling Koch's postulates.

However, the N.I.H. researchers have not yet succeeded in the final step, recovering the AIDS virus from the experimentally infected animals.

In the experiments that researchers at the disease control centers carried out with colleagues at Emory's Yerkes Regional Primate Research Center in Atlanta, two chimpanzees injected with the AIDS virus developed antibodies against it after four months. The animals developed swollen lymph glands but because they did not develop a clear case of AIDS, they did not fulfill Koch's postulates. However, the researchers did recover the virus from the infected animals.

According to Dr. Alfred S. Evans of Yale, Koch did not recommend the postulates as rigid criteria of causation and they should not be regarded as such today.

The postulates have been refined over the years to reflect advances in techniques and knowledge. Also, doctors have come to realize that the same syndrome can be produced by more than one agent and that one microorganism can produce different forms of illness.

Further, although doctors have fulfilled Koch's postulates for scores of infectious diseases, sometimes they have been unable to do so. Instead they have relied on immunological and other experimental evidence to support contentions that a microorganism causes a disease.

At a news conference last April to announce scientific advances concerning the HTLV-3 virus, Dr. Robert C. Gallo of the National Cancer Institute, a major figure in AIDS research, cited such problems. He said scientists might not be able to meet Koch's postulates in the case of AIDS.

In this connection, AIDS researchers have pointed to evidence from two humans who developed AIDS. One was a blood donor. The second received a transfusion of the first individual's blood, which was donated before the donor developed AIDS. The AIDS virus was isolated from both the donor and the recipient.

Although AIDS researchers have not yet fulfilled Koch's postulates, their success in transmitting the infection to chimpanzees is a crucial step toward the ultimate goal of developing a vaccine for AIDS.

Source: The New York Times, Oct. 23, 1984. Reprinted by permission.

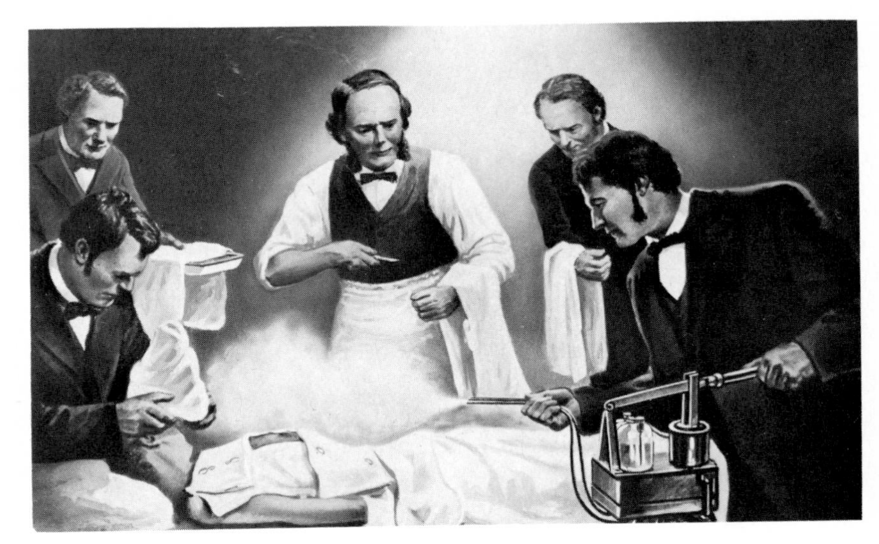

Figure 1.10 Joseph Lister (1827–1912) recognized the importance of preventing the contamination of wounds in order to curtail the development of infection. Here, Lister is shown spraying carbolic acid over a patient undergoing an operation, about 1867. (The Bettmann Archive, Inc.)

dence of postsurgical disease that often was as high as 90 percent, especially in military hospitals.

In the Nineteenth Century men lost their fear of God and acquired a fear of microbes.

—ANONYMOUS

Lister knew that in the 1840s, Ignaz Semmelweis, a Hungarian physician who worked in maternity wards in Vienna, had shown that physicians who went from one patient to another without washing their hands were responsible for transmitting childbed fever, a disease that killed many women after childbirth. He was also aware that Pasteur had demonstrated that microorganisms are present in the air. Lister used carbolic acid, phenol, as an antiseptic during surgery. He first used bandages soaked in carbolic acid to dress wounds from compound fractures in order to diminish the likelihood of infection. Later, he used a carbolic acid spray in addition to direct application of this compound during surgical procedures. Lister eventually discarded the practice of spraying after 17 years of trials as unnecessary, but he retained the use of direct application.

Various chemical formulations for preventing microbial growth and infection were described by Koch and his disciples, including Paul Ehrlich, who, like Pasteur, had been trained as a chemist. From 1880 to 1896, Ehrlich worked in Koch's laboratory, and in 1896 he became director of the first of his own institutes, which he dedicated to finding "substances which have their origin in the chemist's retort," that is, substances produced by chemical synthesis to cure infectious diseases. Ehrlich's research between 1880 and 1910 established the early basis for modern chemotherapy. He established the correct formula for atoxyl, an arsenical, which was being considered for use in treating sleeping sickness, and developed almost a thousand new derivatives of this compound. Compound 606, salvarsan, proved to be effective in treating syphilis. The drugs developed by Ehrlich and his coworkers became known as magic bullets and were portrayed as being able to find and kill disease-causing germs.

A major breakthrough in chemotherapy occurred in 1929, when the Scottish bacteriologist Alexander Fleming, working in a London teaching hospital, reported on the antibacterial action of cultures of a *Penicillium* species (Figure 1.11). Fleming observed that the mold *Penicillium notatum* killed his cultures of the bacterium *Staphylococcus aureus* when the fungus accidentally contaminated the culture dishes. It is likely that the fungal contaminant of Fleming's cultures, which was to bring medical practice into the modern era of drug therapy, blew into his laboratory from the floor below, where an Irish mycologist was working with strains of *Penicillium*. Such a serendipitous event can change history, but in science it takes a special individual, like Fleming, to recognize the significance of the observation. As Pasteur said, "Chance favors the prepared mind." After growing the fungus in a liquid medium and separating the fluid from the cells, Fleming discovered that the cell-free liquid was an inhibitor of many bacterial species. His publication on the active ingredient, which he called penicillin, was the first report of the production of an antibiotic. The first widespread use of penicillin

Figure 1.11 Sir Alexander Fleming (1881–1955), shown here working on penicillin development in his laboratory, had the insight to recognize the significance of the inhibition of bacterial growth in the vicinity of a fungal contaminant instead of simply discarding the contaminated plates. (Central Office of Information, London.)

did not occur until World War II. Since that time there has been a proliferation of new antibiotics, which have been discovered and put to use effectively in treating infectious diseases. We are continuously searching for new and better antibiotics that can be used for controlling human diseases.

Immunology

Whereas chemotherapy is effective in treating disease, immunization is used for preventing infectious diseases. The practice of immunization was used in the Far East for centuries before it was first introduced into England in 1718 by Lady Mary Montagu. A striking English beauty, Lady Mary contracted smallpox at age 26, leaving her scarred for life, mourning "My beauty is no more!" Accompanying her husband to Constantinople, where he had been appointed the British Ambassador in Turkey, Lady Mary learned about the practice of ingrafting.

I am going to tell you a thing that I am sure will make you wish your selfe here. The Small Pox, so fatal and general amongst us, is here entirely harmless by the invention of ingrafting.

—Lady Mary Montagu, *letter to her family*

Old Turkish women would collect material from the pustular sores of mild cases of smallpox and place it into a walnut shell. Using a needle, they would then place a small amount of the material into the vein of someone seeking to avoid smallpox. In about a week the recipient would become sick, feverish, and develop a number of sores and pustules. But in about another week the pustules would heal, and recovery would be complete, with no permanent scarring. After such inoculation with material from mild cases of smallpox, the individual became totally resistant (immune) to even the severe and deadly forms of this disease. Taking a personal interest in ingrafting, Lady Mary used her considerable influence in the court of King George I to gain publicity for the increased use of immunization. She had no explanation for how or why ingrafting worked. Nevertheless, she arranged for testing the effectiveness of ingrafting on prisoners and orphans, then a common practice.

Despite her efforts, immunization was not accepted by the scientists and physicians of the time as a useful practice for preventing disease. The failure to gain acceptance for immunization was due in part to the fact that some cases resulted in scarring and that there was a low, but significant, mortality rate associated with the practice. It was not until the report by Edward Jenner to the Royal Society in London 80 years later that credence was given to the practice of immunization. Jenner's 1798 report on the value of vaccination with cowpox as a means of protecting against smallpox established the basis for the immunological prevention of disease (Figure 1.12). Jenner was a middle class English country doctor, whose interest in science, like that of Leeuwenhoek, was typical of his class: scholarly but amateur. There was no hurry, no great impetus to discovery in his work, just the careful, methodical observation of outbreaks of disease. The work begun with Jenner's discovery of the effectiveness of vaccination for preventing smallpox culminated in the 1970s with the eradication of smallpox from the face of the earth.

Much work was needed beyond Jenner's report to achieve the elimination of smallpox and to use vaccines for preventing many other diseases. Pasteur greatly furthered the development of vaccines when, in 1880, he reported that attenuated microorganisms (microorganisms that had been modified to reduce their ability to cause disease) could be used to develop effective vaccines against chicken cholera. The production of attenuated vaccines that were effective against fowl cholera depended on prolonging the time between transfers of the cul-

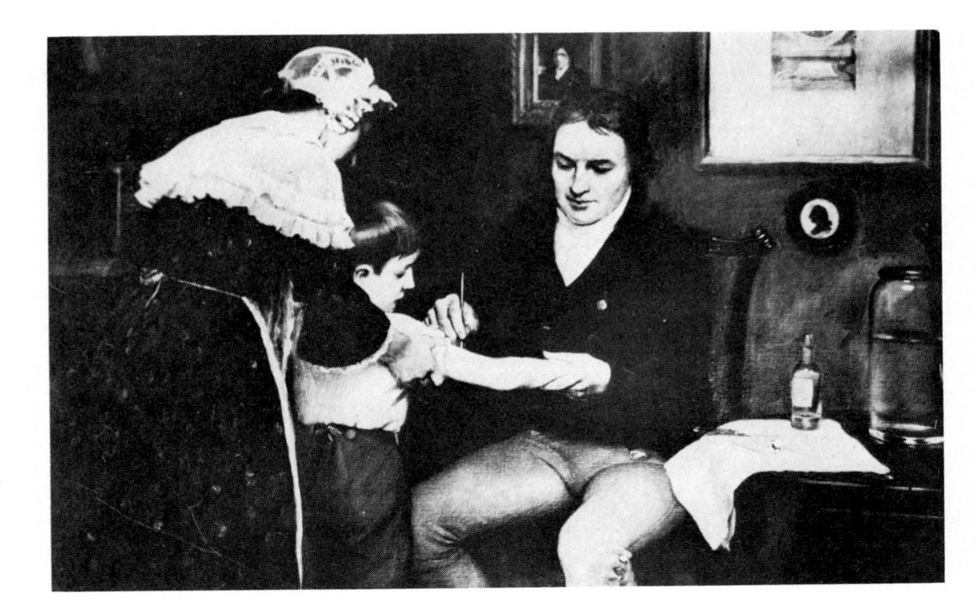

Figure 1.12 Edward Jenner (1749–1823) vaccinated James Phipps about 1800 with cowpox material, resulting in the development of resistance to smallpox infection by the boy and establishing the scientific credibility of vaccination to prevent disease. (Culver Pictures.)

MICROBIOLOGY HEADLINES

The Whooping Cough Crisis

Whooping cough once killed 7,000 American children a year and infected 265,000 more. It may again become a major killer if the problems now swirling around the vaccine are not speedily resolved.

Whooping cough vaccine is effective and quite safe. But not safe enough: It causes 1 or 2 deaths and about 32 cases of brain damage among every 10 million children immunized. As a result, parents in Britain lost confidence in the vaccine, and the proportion of children vaccinated steadily dropped through the mid-70's. The inevitable result was the return of whooping cough epidemics, with 100,000 cases and 28 deaths in three years. Japan has had a similar experience, losing 32 children a year for lack of immunity.

American parents have so far accepted the shots, which many states require for entering school. But parents of children injured by the vaccine have brought such a barrage of lawsuits that many vaccine makers have quit production.

Last week, Connaught Laboratories ceased making the vaccine rather than pay higher liability insurance, leaving only two American manufacturers, Lederle and Wyeth. The vaccine is in temporary shortage, forcing the Government to recommend that doctors delay booster shots. If Lederle or Wyeth should cease production, or if the vaccine's acceptability drops because of bad publicity, there would be serious risk of an epidemic.

Two steps would reduce the risk. One is a national program to compensate the inevitable victims of vaccines that benefit all. Such a program would properly shield vaccine makers from much, but not all, of their liability.

Another step is to hasten improvement of the vaccine. Despite the danger signals from Britain and Japan, there has been little sense of urgency, perhaps because responsibility for vaccines is divided among three Federal agencies. Whooping cough is strictly a human disease, which means that in theory it could, like smallpox, be eradicated worldwide. The solution is to find a wholly new kind of vaccine, not just improve the old one.

The vaccine now consists of the inactivated whole whooping cough bacterium. Side effects occur, perhaps because the bacterium's toxin, a necessary ingredient, is incompletely inactivated. Most experts agree that a better vaccine could be made with just the inactivated toxin and certain other components of the bacterium.

A promising component vaccine has been introduced in Japan, but it lacks the record of safety and efficacy required here. Also, experts still disagree as to which components are necessary. Uncertainty about how the bacterium works is a major reason for the lack of progress.

These disagreements need to be resolved and a clinical trial conducted in a country where whooping cough is prevalent. If Congress can focus resources on developing a better vaccine, it will avert the risk of whooping cough's returning to America, and lay the ground for eradicating the disease throughout the world.

Source: The New York Times, Dec. 21, 1984. Reprinted by permission.

tures, a fact accidently discovered through an error by Charles Chamberland, who used an old culture during one of the experiments he was conducting with Pasteur. Pasteur recognized that microorganisms could induce immunity even when the cultures had lost their ability to cause disease.

After his studies on fowl pox, Pasteur directed his attention to the study of anthrax. This irritated Koch, who considered anthrax to be his exclusive domain. There was an intense rivalry between Pasteur and Koch because of personal egotistical concerns and intense nationalistic pride. Because he enjoyed being the center of attention and controversy, Pasteur staged a very dramatic public demonstration to test the effectiveness of his anthrax vaccine. Witnesses to the demonstration were amazed to see that the 24 sheep, 1 goat, and 6 cows that had received the attenuated vaccine were in good health, but that all the animals that had not been vaccinated were dead of anthrax.

> The art of the experimenter is to create models in which he can observe some properties and activities of a factor in which he happens to be interested. Koch and Pasteur wanted to show that microorganisms could cause certain manifestations of disease. Their genius was to devise experimental situations that lent themselves to an unequivocal illustration of their hypothesis—situations in which it was sufficient to bring the host and the parasite together to reproduce the disease. By trial and error, they selected the species of animals, the dose of infectious agent, and the route of inoculation, which permitted the infection to evolve without fail into progressive disease.
>
> Pasteur, Koch, and their followers succeeded in minimizing in their tests the influence of factors that might have obscured the activity of the infectious agents they wanted to study. This experimental approach has been extremely effective for the discovery of agents of disease and for the study of some of their properties. But it has led by necessity to the neglect, and indeed has often delayed the recognition, of the many other factors that play a part in the causation of disease under conditions prevailing in the natural world—for example, the physiological status of the infected individual and the impact of the environment in which he lives.
>
> —René Dubos, *Mirage of Health*

Even considering the impressiveness of this display, Pasteur's greatest success in developing vaccines was yet to come. In 1885 Pasteur was able to announce to the French Academy of Sciences that he had developed a vaccine for preventing rabies. Although he did not understand the nature of the causative organism, Pasteur developed a vaccine that worked. Pasteur's motto was "seek the microbe," but the microorganism responsible for rabies is a virus, which could not be seen under the microscopes of the 1880s. He, nevertheless, was able to

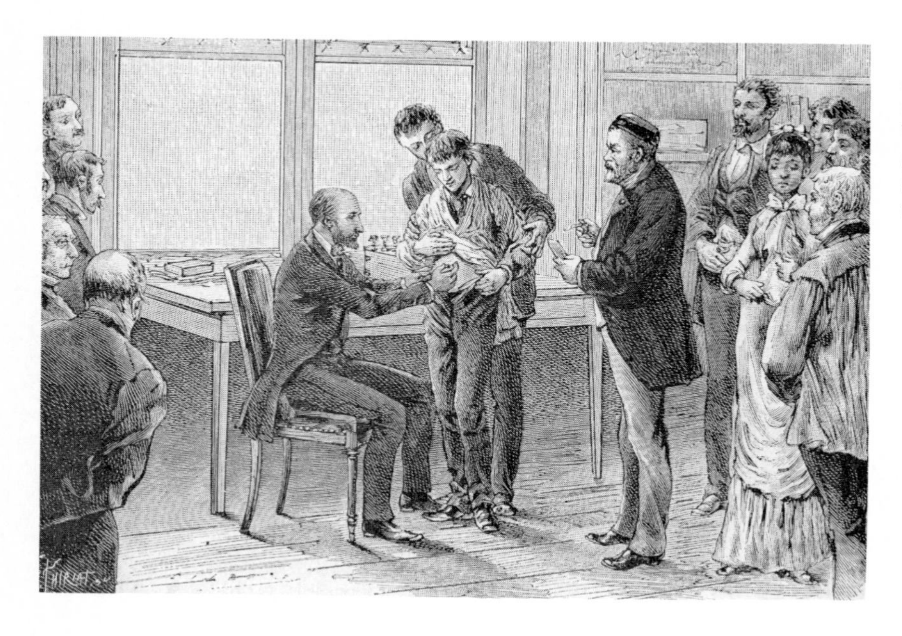

Figure 1.13 The vaccination of a child against rabies conducted under the direction of Louis Pasteur. With the successful development of a vaccine for preventing rabies, crowds flocked to Pasteur's laboratory. (The Bettmann Archive, Inc.)

weaken the rabies virus by drying the spinal cords of infected rabbits and allowing oxygen to penetrate the cords. Thirteen inoculations of successively more virulent pieces of rabbit spinal cord were injected over a period of two weeks during the summer of 1885 into Joseph Meister, a nine-year-old boy who had been bitten by a rabid dog (Figure 1.13).

> Since the death of the child was almost certain, I decided in spite of my deep concern to try on Joseph Meister the method which had served me so well with dogs . . . I decided to give a total of 13 inoculations in ten days. Fewer inoculations would have been sufficient, but one will understand that I was extremely cautious in this first case. Joseph Meister escaped not only the rabies that he might have received from his bites, but also the rabies which I inoculated into him.
>
> —LOUIS PASTEUR, 1885, *Compt. Rend. Acad. Sci.*

This treatment met with a highly publicized, personal success for Pasteur. The development of the rabies vaccine capped Pasteur's distinguished career. Donations sent to Pasteur as a consequence of his discoveries were used to erect l'Institute Pasteur, the first priority of which was to provide the proper facilities for the production of vaccines (Figure 1.14). Today vaccination is used to protect children around the world against many diseases.

Figure 1.14 Photograph of Pasteur's crypt in the lower level of the building at the Pasteur Institute where he performed his experiments. The crypt is lined with tile work depicting the great experiments of Pasteur.

> In a not-too-distant past children were expendable . . . it is now desired that all children survive. This ideal naturally makes it essential to discover ways that will substitute for the protective effects that childhood diseases used to exert in the past. Vaccination to elicit without danger the immunity formerly resulting from the disease itself and sanitary procedures designed to elimnate infectious agents from the environment are the technological procedures through which man attempts to substitute for the natural mechanisms of biological adaptation.
>
> —RENÉ DUBOS, *Mirage of Health*

Summary

The science of microbiology has evolved from many fields—from the contributions of numerous amateur scientific observers who, like Leeuwenhoek, took pleasure in searching for the very small and discovering the wonders of the microbial world, to professional scientists who, like Pasteur and Koch, brought their scientific background to the establishment of the new discipline of microbiology. As in most fields of science, the development of appropriate methodologies for advancements in one field often depend on developments in other areas of scientific endeavor. This is apparent in the historical development of microbiology (Table 1.1). Leeuwenhoek's observation of microorganisms depended upon the development of the microscope and lenses capable of magnifying while maintaining the detail of these organisms. Pasteur's discoveries came after major advancements in our understanding of chemistry, a prerequisite for investigating microbial metabolism.

Louis Pasteur and Robert Koch set the stage for the future development of this field of science, emphasizing the relationships between germs and disease (including host resistance) and microbes and fermentation (including metabolism and

Table 1.1 Time-line Showing Development of Microbiology Relative to Other Sciences

1200–1220	1221–1240	1241–1260	1261–1280	1281–1299
			Bacon describes laws of reflection and refraction of light and prototype magnifying glass	
1400–1420	**1421–1440**	**1441–1460**	**1461–1480**	**1481–1499**
		Gutenberg begins printing		Mercury used to treat venereal disease
1600–1620	**1621–1640**	**1641–1660**	**1661–1680**	**1681–1699**
			Hooke's *Micrographia;* van Leeuwenhoek observes bacteria and protozoa	
1800–1820	**1821–1840**	**1841–1860**	**1861–1880**	**1881–1899**
Appert develops canning	Schwann, Küntzing, and de Latour report that yeasts cause fermentation	Pasteur's studies on fermentation; Darwin's theory of natural selection	Lister begins practice of antiseptic surgery; Koch studies bacteria causing disease; Cohn discovers endospores of bacteria; development of autoclave—use of steam sterilization; Gram stain developed	Koch develops methods for pure culture of bacteria; Koch's postulates; Pasteur inoculates against rabies; discovery that mosquitoes carry malaria; Metchnikoff describes phagocytosis; Petri's culture dish; Beijerinck studies nitrogen fixation; Winogradsky studies autotrophy; discovery of antitoxins; field of virology begins with discovery of plant and animal viruses; complement discovered; Smith shows protozoan transmitted by tick vector

biochemistry). Louis Pasteur made major contributions to our understanding of fermentations; beginning as a chemist, Pasteur emerged as an outstanding microbiologist. Pasteur studied both lactic acid and alcoholic fermentations, describing the chemical changes that occurred during fermentation and the growth of yeast when air was present (aerobic conditions) and when it was absent (an-

1300–1320	1321–1340	1341–1360	1361–1380	1381–1399
		Plague in Europe	Quarantine used in Europe against plague	

1500–1520	1521–1540	1541–1560	1561–1580	1581–1599
Sewers begun in England		Fracastoro theory of contagious disease		Invention of compound microscope

1700–1720	1721–1740	1741–1760	1761–1780	1781–1799
Montague introduces smallpox vaccination			Linnaeus describes classification system for living organisms; Spallanzani disputes theory of spontaneous generation	Jenner introduces vaccination

1900–1920	1921–1940	1941–1960	1961–1980	
Ehrlich's studies on chemotherapy; discovery of bacteriophage; discovery that viruses can cause malignancy	Birdseye deep-freezes food; Fleming discovers penicillin; invention of electron microscope; first edition of *Bergey's Manual Of Determinative Bacteriology;* Kluyver and Van Niel work on comparative microbial metabolism	Discovery that actinomycetes produce antibiotics; beginning of molecular genetics; Avery shows bacterial transformation; Watson and Crick describe DNA double helix; use of antibiotics in medicine begins; antipolio vaccine	Discovery of interferon; breaking of the genetic code; DNA recombination studies lead to genetic engineering; elimination of smallpox	

aerobic conditions). Robert Koch extended his training as a physician to tackle the problems of treating and preventing infectious diseases. He and his disciples discovered the etiologic agents of several diseases and developed methods for their cure and prevention. The pure-culture techniques developed by Koch have dominated microbiological research up to the present day.

Pasteur's unique contribution was his conceptual ability, his ability to analyze and to bring together the pieces of the puzzle that form the grand scheme of microbiology. Koch's genius lay in his ability to devise the methods and techniques that allowed the science to progress. Both men were excited by the process of scientific discovery and by solving the problems they had set for themselves. Their inquiring minds, sense of adventure, egos, and competitiveness caused them to strive ahead at a furious pace. They set a pattern for the many great microbiologists who followed. The race to be first in making microbiological discoveries of basic and practical importance continues today. Both Koch and Pasteur trained students to carry on their work and to publish extensively to make available the knowledge they had acquired.

Improvements in scientific communication have allowed microbiologists to capitalize rapidly on scientific advances, hastening the rate of development in the field of microbiology. Today, there are numerous journals and publications through which worldwide distribution of microbiological information is made possible (Figure 1.15). The field of microbiology promises to continue its rapid development for many more years. Currently, studies in molecular genetics are producing new knowledge, enhancing our fundamental understanding of microorganisms and other living systems. Industrial and environmental problems provide many opportunities for the practical application of basic microbiological principles. We continuously face the challenge of infectious diseases, searching for better ways of preventing, diagnosing, and treating diseases caused by microorganisms. Microbes remain at the forefront of importance in today's news.

Figure 1.15 Some journals concerned primarily with microbiological studies. These journals contain articles that continuously increase our knowledge about microorganisms.

Study Outline

A. Observation of microorganisms.

1. Invention of the microscope in the mid-sixteenth century.
2. Robert Hooke published *Micrographia* (1665) in which the first drawings of microorganisms (fruiting bodies of fungi) appear.
3. Antony van Leeuwenhoek made his own microscopes and was the first to observe bacteria, yeasts, and protozoa. He transmitted his findings to the Royal Society in London between 1674 and 1723.
4. Paul Ehrlich (1881) introduced vital staining of bacteria with methylene blue.
5. Hans Christian Gram (1884) introduced the differential staining method now named in his honor, the Gram stain.
6. Dmitri Ivanowski (1882) reported that the agent causing tobacco mosaic disease (of tobacco plants) was a filterable agent (virus).

7. Martinus Beijerinck (1898) found the same results as Ivanowski.

B. Cultivation of microorganisms.

1. Robert Koch and his coworkers developed a method for isolation and maintenance of pure cultures on solid media (1881–1883).
2. At the suggestion of Frau Hesse, the use of agar as the solidifying agent of microbial media was introduced.
3. A contemporary of Koch, Richard Petri, introduced a new type of culture dish (Petri dish or plate) in 1887.

C. Importance of microbial metabolism.

1. Louis Pasteur (1856) discovered that yeasts produced the alcohol in wine and that rod-shaped organisms produced lactic acid, causing the wine to sour.
2. Louis Pasteur (1857) showed that the souring of milk is caused by microorganisms.
3. Louis Pasteur (1860) found that heating could be used to kill microorganisms in wine and beer; the process of pasteurization is named in his honor.
4. Louis Pasteur (1861) destroyed the theory of spontaneous generation with his experiments using a swan-necked flask. After boiling a liquid, he was able to leave a vessel containing a fermentable substrate open to the air and show that fermentation did not occur.
5. Autoclave (pressurized steam sterilization) designed for killing microorganisms.

D. Recognizing the relationship between microorganisms and disease.

1. The studies of Robert Koch greatly advanced the field of medical microbiology.
2. Robert Koch showed that microorganisms grown outside the body could cause disease.
3. Robert Koch (1873–1876) discovered that the spores of anthrax bacilli isolated from pure cultures could infect animals.
4. Robert Koch (1894) discovered the causative organisms of tuberculosis and cholera.
5. Robert Koch (1894) developed the postulates later named after him (Koch's postulates) for determining the cause-and-effect relationship between a given microorganism and a specific disease. They are
   - **(a)** The organisms should be present in all animals suffering from the disease and absent from all healthy animals.
   - **(b)** The organism must be grown in pure culture ouside the diseased animal host.
   - **(c)** When such a culture is inoculated into a healthy susceptible host, the animal must develop the symptoms of the disease.
   - **(d)** The organism must be reisolated from the experimentally infected animal and shown to be identical with the original isolate.

E. Chemotherapy.

1. Joseph Lister (England) introduced antiseptic surgery in 1867 by using carbolic acid (phenol) as an antiseptic on dressings, the surgical incision, and in the operating room air.
2. Paul Ehrlich discovered a wholly synthetic arsenical compound, compound 606, salvarsan, which was effective against syphilis. Salvarsan became known as "magic bullets" because it was portrayed as being able to find and kill disease-causing germs.
3. Alexander Fleming reported the antibacterial action of *Penicillium* cultures in 1929. He named this first antibiotic penicillin.

F. Control of microorgansisms—immunology.

1. Immunization introduced into England in 1718 by Lady Montagu.
2. Edward Jenner introduced smallpox immunization (vaccination) in 1798.
3. Louis Pasteur perfected the technique of attenuated (modified to reduce virulence) vaccines and developed vaccines against anthrax and rabies. He also coined the terms "vaccine" and "vaccination" in honor of Jenner, who developed the first vaccine from cowpox, or vaccinia.

Study Questions

1. Match these contributions with the names that appear in the list to the right.
   - **a.** First to see bacteria.
   - **b.** Introduced most widely used differential stain, named in his honor.
   - **c.** Developed microscope lenses.
   - **d.** Transmission of disease.
   - **e.** Invented the microscope.
   - **f.** First to see microorganisms.

   (1) Fracastoro
   (2) Hans Janssen
   (3) Robert Hooke
   (4) Antony van Leeuwenhoek
   (5) Hans Christian Gram
2. Describe the experiment of Louis Pasteur that destroyed the theory of spontaneous generation.
3. List Koch's postulates.
4. Match these contributions with the names that appear in the list below.
   - **a.** First reported the antibacterial action of *Penicillium* cultures.
   - **b.** Introduced the use of agar as a solidifying agent of microbial media.
   - **c.** Reported that the agent causing tobacco mosaic disease was a filterable agent (virus).
   - **d.** Suggested the use of agar as a solidifying agent of microbial media.
   - **e.** Introduced a new type of culture dish.
   - **f.** Discovered wholly synthetic arsenic compound, compound 606 (salvarsan).
   - **g.** Introduced antiseptic surgery by using carbolic acid.

   (1) Robert Koch
   (2) Frau Fanny Hesse
   (3) Richard Petri
   (4) Dmitri Ivanowski
   (5) Joseph Lister
   (6) Paul Ehrlich
   (7) Alexander Fleming
5. Indicate whether the following statements are true (T) or false (F).
   - **a.** Lady Montagu introduced immunization to England.
   - **b.** Pasteur introduced smallpox immunization.
   - **c.** Koch introduced a vaccine for rabies.
   - **d.** Pasteur introduced attenuated vaccines.
   - **e.** Ehrlich introduced a vaccine for syphilis.
   - **f.** Fleming introduced the first drug for treating infectious diseases.

Some Thought Questions

1. Why is the study of microbiology important to you?
2. Why did the plague follow medieval trade routes?
3. Is competition or cooperation better for the advancement of the science of microbiology?
4. Why weren't the discoveries of Pasteur and Koch made a century earlier?
5. How are microorganisms portrayed in literature?
6. Are microbes more important than statesmen? Explain.
7. How have microbes altered the course of history?
8. What would life be like without microbes?
9. Did we all evolve from a bacterial cell?
10. Why are there no descriptions of bacteria in the Bible?
11. How has the germ theory of disease changed medical practice?
12. Who should pay for public health care?

Additional Sources of Information

Brock, T. D. (ed.). 1979. *Milestones in Microbiology*. American Society for Microbiology, Washington, D.C. Contains a number of key papers by Louis Pasteur, Robert Koch, and others; reading these papers reveals how many of the major discoveries in microbiology were actually reported.

Cohen, I. B. 1984. Florence Nightingale. *Scientific American*, **250**(3):128–137. A well-written biographical account of the accomplishments of Florence Nightingale, including her role in the development of modern medical care.

De Kruif, P. 1926. *Microbe Hunters*. Harcourt, Brace and Co., New York. (1966. Harcourt Brace Javonovich, Inc., New York.) An inspiring narrative of the lives and works of several great historical figures in microbiology including the stories of Leeuwenhoek, Pasteur, and Koch.

Dobell, C. (ed.). 1932. *Antony van Leeuwenhoek and his "Little Animals."* Constable and Co., Ltd., London. (1960. Dover Publications, Inc., New York.) An interesting book describing the life and times of Antony van Leeuwenhoek; includes Leeuwenhoek's drawings of his microscopic observations.

Dubos, R. J. 1950. *Louis Pasteur: Free Lancer of Science*. Little Brown and Co., Boston. This work by a well-known microbiologist and philosopher provides an insightful view into the work of Louis Pasteur.

Fraser, D. W., and J. E. McDade. 1979. Legionellosis. *Scientific American* 241(4):82-99. A discussion of the investigations into the etiology of the mysterious Legionnaire's disease.

Hooke, R. 1665. *Micrographia*. Royal Society, London. (1961. Dover Publications, Inc., New York.) Contains the descriptions and drawings of Hooke's microscopic observations.

Lechevalier, H. A., and M. Solotorovsky. 1965. *Three Centuries of Microbiology*. McGraw-Hill Book Co., New York. (1974. Dover Publications, Inc., New York.) A comprehensive and authoritative review of the history of microbiology; many excerpts from original works of Pasteur, Koch, and others are included within a descriptive narrative.

Reid, R. 1975. *Microbes and Men*. Saturday Review Press, New York. An excellent, easy-to-read, script of a television series that documented several important historical events in microbiology including the discoveries of Pasteur, Koch, and Fleming.

2 The Scope of Microbiology and the Nature of Microorganisms

KEY TERMS

acellular
algae (singular, alga)
archaebacteria
bacteria (singular, bacterium)
cell
classification
cytoplasmic membrane
eubacteria
eukaryotic cell
fungi (singular, fungus)
genus
multicellular
mycology
nomenclature
nucleus
organelle
phycology
prions
prokaryotes
prokaryotic cell
protista
protozoa
species
taxonomic hierarchy
taxonomy
unicellular
urkaryote
viroids
viruses

OBJECTIVES

After reading this chapter you should be able to

1. Define the key terms.
2. Discuss how microorganisms impact our daily lives.
3. Discuss the relevance of microbiology to other basic and applied fields of science.
4. List the major groups of organisms that are recognized as microbes and the major fields of science that are within the realm of the microbiologist.
5. Discuss the essential features of living systems.
6. Compare the structural organization of microorganisms with that of plants and animals.
7. Differentiate between viruses and other microorganisms.
8. Differentiate between bacteria and other microorganisms.
9. Discuss the current view of the taxonomic position of microorganisms and compare this view with how microorganisms were classified in earlier taxonomic systems.
10. Write the names of microorganisms in proper format.

Scope of Microbiology

Despite their invisible nature, microorganisms affect our daily lives and influence the overall quality of life. Applications of microbiology are especially important in medicine, industry, agriculture, and environmental management. In the field of medical microbiology, for example, an understanding of the causative agents of disease has allowed us to develop preventive and treatment methods that have reduced morbidity (sickness) and mortality (death) arising from certain diseases. This has resulted in an increase in life expectancy. Because microorganisms are involved in causing infectious diseases, the field of microbiology has logically been extended to include the response of the diseased plant or animal. As such, the disciplines of immunology (the study of the immune defense systems of higher animals) and plant pathology are included in the study of microbiology. Maintaining our own health and the adequacy of our food supply depends upon using basic microbiological principles for achieving practical benefit.

The health of the people is really the foundation upon which all their happiness and all their powers as a state depend.

—Benjamin Disraeli

Microbiology is also an integral part of many industrial processes. Quality control in many industries is concerned with preventing microbial contamination of products that could lead to the spoilage of the products and thereby reduce their economic value. The proper handling of food products, for example, is based upon an understanding of the factors that affect microbial growth and death rates and the manipulation of those factors to control unwanted microbial growth. Various products, including many pharmaceuticals and food products such as beer, wine, and spirits, are produced by microbial fermentations. Without doubt microorganisms have great economic importance, both because of their destructive capacity and also their ability to produce useful products.

Microorganisms also play an important role in maintaining environmental quality. Although many of us equate microorganisms with germs and disease, the health of planet Earth and its inhabitants actually depends upon microorganisms. The unseen microbes are important in maintaining air, water, and soil quality. The maintenance of soil fertility, the role of microorganisms in causing plant diseases, the ability of some microbes to control pest populations, and the interactions of microorganisms with pesticides and fertilizers have an important influence on our agricultural practices. We rely upon microorganisms to degrade our waste materials. Without the biodegradative activities of microorganisms we would be buried in our own wastes and life on Earth could not persist for more than a few months.

In addition to the many practical applications of microbiology, the nature of microorganisms makes them suitable for use in many basic scientific studies. Our fundamental understanding of molecular genetics and the relationship between genetics and metabolism has been developed using microorganisms. Microorganisms reproduce rapidly, and large numbers of organisms with the same genetic composition can easily be grown for genetic and biochemical studies. Compared to higher organisms, microorganisms are simple, both genetically and biochemically. Besides, no one cares if bacteria are sacrificed for the advancement of science. Well, almost no one cares: Japanese microbiologists, feeling a subtle sense of guilt for killing microorganisms as part of their studies, have established a memorial to microorganisms in which they buried a Buddhist prayer scroll and the ashen remains of *Bacillus subtilis*. Today we are beginning to profit from the benefits of the basic understanding of genetics achieved using microorganisms, as evidenced by the rapidly developing field of genetic

Figure 2.1 The term **microorganism** encompasses five major groups of organisms: the viruses, bacteria, fungi, algae, and protozoa. These organisms exhibit diversity of form and size, but are similar in that a microscope is generally required for the observation of individual microorganisms. The micrographs in this figure show: **Viruses**—Respiratory syncytial virus (150,000×) (courtesy E. C. Ford and J. L. Gerin, Georgetown University Medical Center) and influenza A virus (200,000×) (courtesy Erskine Palmer, Centers for Disease Control, Atlanta); **Bacteria**—*Bacillus thuringiensis* (1500×) (courtesy B. N. Herbert, Shell Research, England) and *Caulobacter* (courtesy Jeanne Poindexter, City of New York Public Health Laboratories); **Fungi**—*Morchella esculata* (courtesy Orson K. Miller, Jr., Virginia Polytechnic Institute and State University, author of *Mushrooms of North America*, E. P. Dutton, Inc.) and Bakers' yeast, *Saccharomyces cerevisiae* (courtesy Robert P. Apkarian, University of Louisville); **Algae**—*Spirogyra elongans* and a marine diatom, *Actinoptychus* (courtesy Varley Wiedeman, University of Louisville); **Protozoa**—*Paramecium* sp. and *Amoeba* sp. (courtesy Robert P. Apkarian, University of Louisville).

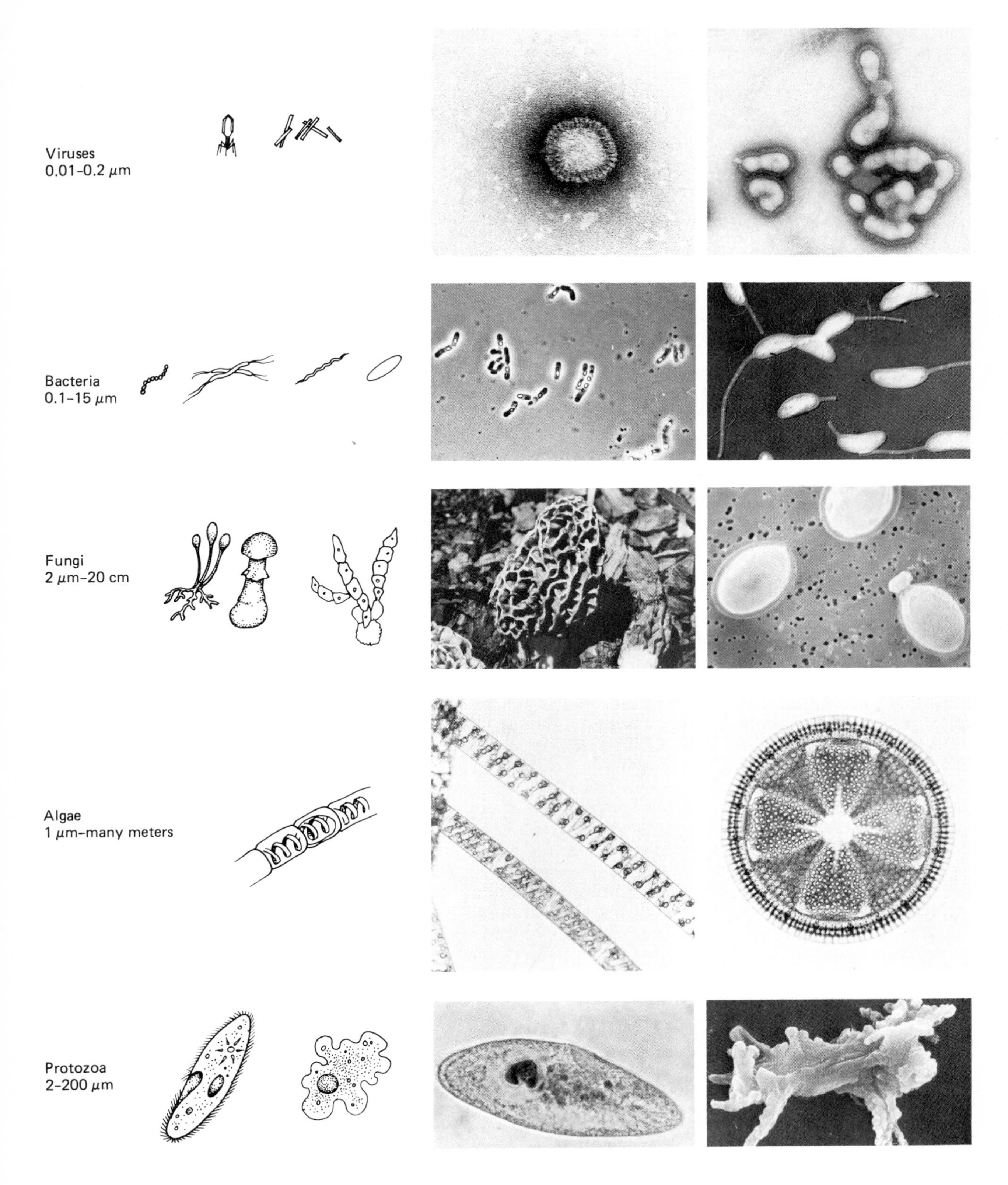
Viruses
0.01–0.2 μm
Bacteria
0.1–15 μm
Fungi
2 μm–20 cm
Algae
1 μm–many meters
Protozoa
2–200 μm

Table 2.1 Descriptions of the Major Subfields of Microbiology

Field	Description
Virology	Science that deals with viruses. Viruses are among the smallest of microorganisms, only a few billionths of a meter in size. They are incapable of independent existence and can only reproduce within the cells of other organisms. Outside host cells, viruses are effectively nonliving entities and are sometimes referred to as viral particles. Many animal and plant diseases, such as the common cold and the flu, are caused by viruses.
Bacteriology	Science that deals with bacteria. The bacteria are widely distributed in nature. A few bacteria, commonly referred to as germs, cause diseases of humans. Control of these diseases has been achieved through various practices that limit the growth of bacteria, including the use of antibiotics. A limited number bacteria are plant pathogens. Most bacteria, however, are beneficial. The metabolic activities of bacteria are essential for recycling of mineral resources. In lakes and oceans the growth of bacteria provides a food source for higher organisms. Bacteria are used for the decomposition of wastes, including the disposal of municipal refuse and the treatment of sewage. Many bacteria produce economically useful products, including antibiotics and other pharmaceuticals. A variety of foods, including cheeses and yogurt are produced by bacteria.
Mycology	Science that deals with fungi. The fungi include organisms that form long filaments (molds or filamentous fungi) and organisms that are typically single celled and reproduce by budding (yeasts). Many fungi are active decomposers. Decomposition of plant litter is largely carried out by fungi. Some fungi cause human disease and various plant diseases are caused by fungi. Numerous industrial products are produced by fungi, including beer, wine, and bread. Mushrooms, both poisonous and edible, are macroscopic fruiting bodies of some fungi.
Phycology	Science that deals with algae. The algae are photosynthetic and, like the higher plants, they are able to use sunlight, carbon dioxide, and water to generate organic matter and oxygen.
Protozoology	Science that deals with protozoa. The protozoa are single-celled microorganisms. A few human diseases, such as malaria, are caused by protozoa.

engineering. Biotechnology has the potential for improving many aspects of our daily lives.

It is thus apparent that microbiology encompasses a very broad field (Table 2.1). Some microbiologists are concerned with the basic sciences and the development of a fundamental understanding of living systems, and others are concerned with the application of basic scientific knowledge. Microbiology overlaps a number of other scientific disciplines, including biochemistry, genetics, zoology, botany, ecology, pharmacology, medicine, food science, agricultural science, industrial science, and environmental science. The broad scope of microbiology attests to the diversity of the microorganisms themselves, their ubiquitous distribution in nature and the importance of microorganisms in virtually all aspects of life. The unity of microbiology rests with its central subject matter—the organisms that are considered microbes (Figure 2.1).

Structural Organization of Microorganisms

Different kinds of microorganisms are distinguished largely on the basis of their structural organization. Compared to plants and animals, microorganisms have relatively simple structural organizations. In fact many microorganisms have only of a single cell (Figure 2.2). The cell is the fundamental organizational unit of all living systems. Each cell is separated from its surroundings by a cytoplasmic membrane. The cytoplasmic membrane acts as a boundary layer, selectively permitting material to move into and out of the cell, and thus allowing the organism to maintain its integrity and identity. When an organism no longer is able to sustain its cellular organization, it dies and the organism and its cellular components become disorganized.

Living cells have the capacity to reproduce their own organizational pattern independently. Cells come from the reproduction of preexisting cells. Many microorganisms are unicellular and, therefore, the reproduction of a single cell is synonymous with the reproduction of the entire organism. Other organisms are multicellular, consisting of aggregates of cells that are themselves organized. In some multicellular organisms, the cells are differentiated into specialized organizational units known as tissues and organs. Although microorganisms often form multicellular aggregations, they lack the high degree of specialized organization of cells into differentiated tissues exhibited by plants and animals.

Prokaryotic and Eukaryotic Cells

There are two basic types of cells: prokaryotic and eukaryotic. Eukaryotic cells are almost always larger than prokaryotic cells and are characterized by a greater degree of internal structural organization than is found within prokaryotic cells. The larger size of eukaryotic cells, compared to prokaryotic cells, depends upon the ability to form and manage subsystems within the eukaryotic cell (Figure 2.3). Eukaryotic cells contain membrane-bound organelles that effectively compartmentalize the specialized cellular functions.

Of prime importance is the separation of the genetic information from the rest of the cellular components in eukaryotic cells. The primary genetic information of a eukaryotic cell is contained within a membrane-bound organelle, the nucleus, and is separated from the rest of the cell by a membrane barrier (Figure 2.4). The DNA in prokaryotic cells is not separated by a membrane barrier from the rest of the cell constituents, as is the DNA within the nucleus of eukaryotic cells. By definition, all eukaryotic cells have a nucleus; prokaryotic cells never possess this organelle. The differences between prokaryotic and eukaryotic cells represent a fundamental division of living systems. Only bacteria have prokaryotic cells (Table 2.2). All organisms possessing prokaryotic cells (prokaryotes) are classified as bacteria. Plants, animals, protozoa, algae, and fungi all have eukaryotic cells.

In contrast with all other organisms, viruses, viroids, and prions are acellular and are not separated from their surroundings by a membrane. The lack of a cellular membrane precludes the ability of these microbes to carry out organized biochemical reactions outside a host cell. As such, they are obligate intracellular parasites and are only capable of acting as living systems within the cells of suitable host organisms. Within the confines of a host cell, there is a boundary membrane, provided by the host cell, that permits the genetic material of acellular microorganisms, such as viruses, to pro-

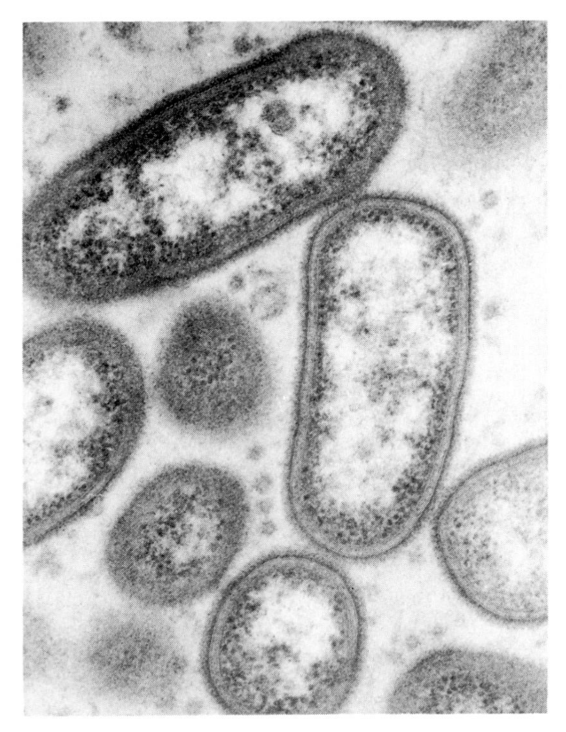

Figure 2.2 An electron micrograph of the bacterium *Bacteroides asaccharolyticus* isolated from the oral cavity where it causes gingivitis. The entire organism is one cell. The cytoplasmic membrane forms a complete circle and forms a barrier between the contents of the cell and the surroundings. (From BPS: Stanley C. Holt, University of Texas Health Science Center, San Antonio, Texas.)

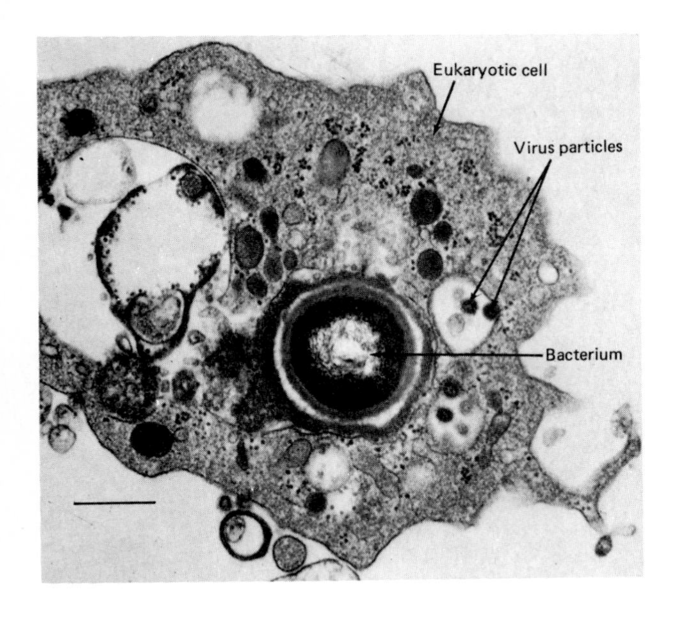

Figure 2.3 An electron micrograph showing a bacterium and several viruses within a eukaryotic cell. The relative sizes of these microorganisms is easily visualized in this micrograph. (Courtesy American Society for Microbiology. From J. S. Abramson, E. L. Mills, G. S. Gievink, and P. G. Quie, 1982, *Infection and Immunity* 35:350–355.)

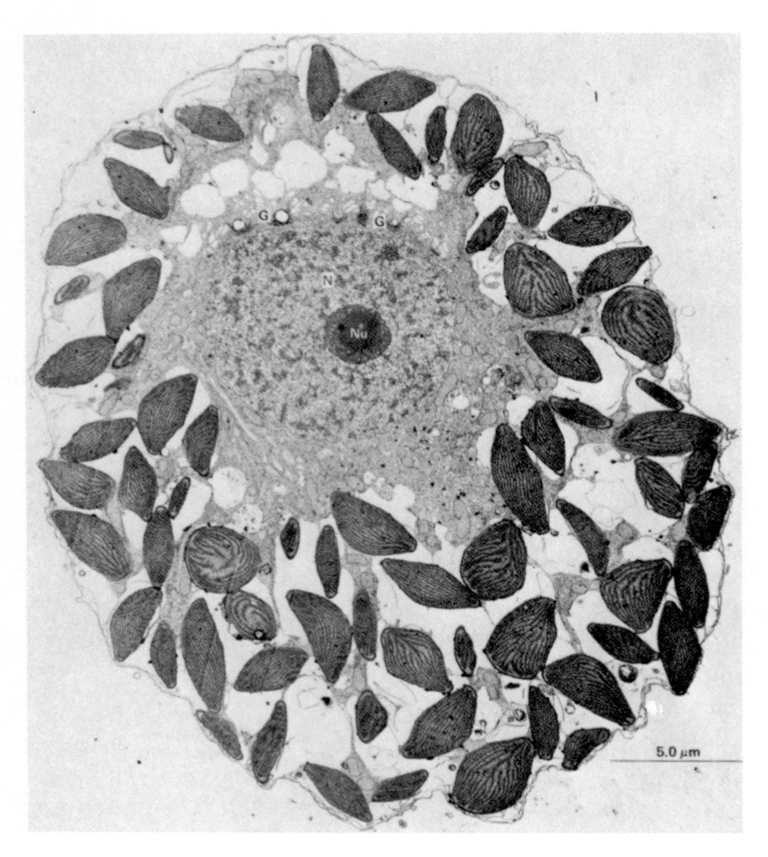

Figure 2.4 A photomicrograph of a eukaryotic cell. This cell comprises the entire organism of the alga *Vacuolaria virescens.* Several membrane-bound organelles can be seen within the cell, including the nucleus (Nu) that houses the genetic information of the cell. (Courtesy Peter Heywood, Brown University. Reprinted by permission from P. Heywood, *Journal of Phycology,* 13:69.)

gram the reproduction of its own macromolecules and structures. Outside such host cells, viruses, viroids, and prions are incapable of reproduction and act as nonliving entities.

The organizational differences between acellular, prokaryotic, and eukaryotic microorganisms have great practical importance. The use of antibiotics to treat human diseases caused by bacteria is dependent on the ability to target the mode of action of the antibiotic selectively against the prokaryotic

Table 2.2 Organizational Structure of the Major Groups of Microorganisms

Microbial group	Structural organization	Essential macromolecules
Prions	Acellular; infectious proteins, no nucleic acid	Protein
Viroids	Acellular; no protective coat around nucleic acid	RNA
Viruses	Acellular; protein coat surround nucleic acid	DNA or RNA, protein
Bacteria	Prokaryotic cell	Nucleic acids (DNA + RNA), protein, lipid, carbohydrate
Fungi	Eukaryotic cell	Nucleic acids (DNA + RNA), protein, lipid, carbohydrate
Algae	Eukaryotic cell	Nucleic acids (DNA + RNA), protein, lipid, carbohydrate
Protozoa	Eukaryotic cell	Nucleic acids (DNA + RNA), protein, lipid, carbohydrate

bacterial cells without at the same time killing the eukaryotic human cells. It is more difficult to target antibiotics against disease-causing fungi and protozoa because these organisms, like humans, have eukaryotic cells. It is also difficult to target antibiotics against viruses because they reproduce within the confines of the human host cell. Chemicals that inhibit viral reproduction generally do so by killing the host cells, eliminating their therapeutic value.

The Taxonomic Position of Microorganisms in the Living World

It is possible to create a taxonomic system (taxonomy), based partly on structural differences, that permits the unambiguous definition of individual groups of microorganisms and also shows the relationships among groups of microorganisms. Such a taxonomic system is essential for classification (ordering of microorganisms into groups based on their relationships), nomenclature (assigning names to the units described in a classification system), and identification (applying the system of classification and nomenclature to assign the proper name to an unknown microorganism and to place it in its proper position within the classification system).

Taxonomic Hierarchy

When classifying microorganisms, taxonomists use a hierarchy consisting of different organizational levels. The usual levels of a taxonomic hierarchy, from the highest to the lowest levels, are kingdoms, phyla, classes, orders, families, genera, and species (Table 2.3). Ideally, each level represents a different degree of genetic and evolutionary similarity.

> The general theory of evolution . . . assumes that in nature there is a great, unital, continuous and everlasting process of development, and that all natural phenomena without exception, from the motion of the celestial bodies and one fall of the rolling stone up to the growth of the plant and the consciousness of man, are subject to the same great law of causation—that they are ultimately to be reduced to atomic mechanics.
>
> —ERNST HAECKEL, *Freie Wissenschaft und Freie Lehre*

The species is the basic taxonomic unit of a classification system. Whereas species of higher organisms are readily recognized as a result of their reproductive isolation, microbial species are difficult to define objectively. A microbial species can be considered as a group of isolated strains that have an overall similarity and are different from other fundamental groups. The practical necessity of correctly distinguishing one microbial species from another is obvious in medical and industrial microbiology. Diagnosis of disease and patenting of biotechnology processes require unambiguous specification of microbial species.

Nomenclature

In order to communicate about specific microorganisms it is also critical to name microbial species in a formal unambiguous system. Many microorganisms have informal names such as pneumococcus, flu virus, and death angel mushroom. Microorganisms also have formal binomial names.

> *What's in a name?*
> *That which we call a rose*
> *By any other name would smell as sweet.*
>
> —WILLIAM SHAKESPSEARE, *Romeo and Juliet*

Table 2.3 Hierarchy of Taxonomic Organization

Level	Description	Examples	
		Fungus	Bacterium
Kingdom	A group of related phyla	Myceteae	Procaryotae
Phylum or division	A group of related classes	Amastigomycota	Eubacteria
Class	A group of related orders	Ascomycetes	—
Order	A group of related families	Xylariales	Spirochaetales
Family	A group of related tribes or genera	Sordariaceae	Spirochaetaceae
Tribe	A group of related genera	—	—
Genus	A group of related species	*Neurospora*	*Treponema*
Species	A group of organisms of the same kind	*N. crassa*	*T. pallidum*
Subspecies or type	Variants of a species	—	—

The formal binomial name has two parts, the genus name and the species epithet. The first letter of the genus name is capitalized. The species epithet is written in all lower case letters. The names of microorganisms and all other organisms are given in Latin. Latin is used because it was the classical

Table 2.4 The Names of Some Microbial Species

Alcaligenes faecalis	al-kah-LIH-jeh-knees fee-KAH-liss
Arthrobacter citreus	are-throw-BACK-ter SIT-ree-us
Bacillus anthracis	bah-SILL-lus an-THRAY-siss
B. cereus	bah-SILL-lus SEER-ee-us
B. subtilis	bah-SILL-lus SUH-till-us
Bordetella pertussis	bore-deh-TELL-ah per-TUSS-siss
Campylobacter fetus	kam-pee-lo-BACK-ter FEE-tus
Chlamydia psittaci	klah-MID-ee-ah SIT-tah-sigh
C. trachomatis	klah-MID-ee-ah tray-KOH-mat-iss
Clostridium botulinum	kloss-TRID-ee-um bah-chew-LINE-um
C. perfringens	kloss-TRID-ee-um per-FRINGE-ens
C. tetani	kloss-TRID-ee-um TET-an-ee
Corynebacterium diphtheriae	koh-rye-knee-back-TIER-ee-um dif-THER-ee-ee
Coxiella burnetii	koks-ee-EL-la ber-NET-ee-eye
Desulfovibrio desulfuricans	dee-sul-fo-VIH-bree-oh de-sul-FYUR-ee-kans
Enterobacter aerogenes	en-teh-row-BACK-ter air-AH-jeh-knees
Escherichia coli	esh-er-EEK-ee-ah KOH-lye
Francisella tularensis	fran-siss-EL-lah too-lah-REN-siss
Haemophilus influenzae	hee-MAH-fih-luss in-flew-EN-zee
Klebsiella pneumoniae	kleb-zee-EL-la noo-MOW-nee-ee
Lactobacillus acidophilus	lack-toh-bah-SILL-lus a-sih-DAH-fill-us
L. bulgaricus	lack-toh-bah-SILL-lus bull-GAIR-ih-kuss
Legionella pneumophila	lee-jon-EL-la noo-MAH-fill-ah
Micrococcus luteus	my-crow-KOK-kuss LOO-tee-us
Mycobacterium leprae	my-koh-back-TIER-ee-um LEP-ree
M. tuberculosis	my-koh-bak-TIER-ee-um too-berk-you-LOW-siss
Mycoplasma pneumoniae	my-koh-PLAZ-ma new-MOW-nee-ee
Neisseria gonorrhoeae	nye-SEER-ee-ah gon-or-REE-ee
N. meningitidis	nye-SEER-ee-ah men-in-jih-TIH-diss
Proteus mirabilis	PROH-tee-us mih-RAH-bih-liss
P. vulgaris	PROH-tee-us vul-GAIR-iss
Pseudomonas aeruginosa	soo-dough-MOW-nass eh-roo-jih-KNOW-sah
Rickettsia prowazekii	rih-KETT-see-ah proh-wah-ZEE-kee-eye
R. rickettsii	rih-KETT-see-ah rih-KETT-see-eye
R. typhi	rih-KETT-see-ah TIE-fee
Salmonella enteritidis	sal-moan-EL-la en-ter-IT-ih-diss
S. typhi	sal-moan-EL-la TIE-fee
Serratia marcescens	ser-RAH-tee-ah mar-SESS-sens
Shigella dysenteriae	shih-GELL-la diss-en-TEH-ree-ee
S. flexneri	shih-GELL-la FLEX-ner-eye
S. sonnei	shih-GELL-la SAHN-knee-eye
Spirillum volutans	spye-RIL-lum VOL-you-tans
Staphylococcus aureus	staff-ee-loh-KOK-kuss AWE-ree-us
S. epidermidis	staff-ee-loh-KOK-kuss eh-pee-DER-mih-diss
Streptococcus faecalis	strep-toh-KOK-kuss fee-KAL-iss
S. mutans	strep-toh-KOK-kuss MEW-tans
S. pneumoniae	strep-toh-KOK-us new-MOW-nee-ee
S. pyogenes	strep-toh-KOK-us pie-ah-JEH-knees
Thiobacillus thiooxidans	thigh-oh-bah-SILL-lus thigh-oh-OX-ee-dans
Treponema pallidum	treh-poh-KNEE-ma PAL-lih-dum
Vibrio cholerae	VIB-ree-oh KOHL-er-ee
Yersinia pestis	yer-SIN-ee-ah PES-tiss

language of science when early classification systems were developed and formal names were first given to organisms on a systematic basis. When typed or handwritten, bacterial species names are underlined to indicate that they are in Latin. In print, species names are italicized. When abbreviating the genus name will not create ambiguity, the genus is indicated by a single capital letter followed by a period. For example, in a list of the species in the genus *Streptococcus*, the species *Streptococcus pneumoniae* would be written as *S. pneumoniae*. The names of many microorganisms will become quite familiar to you as you continue through this book (Table 2.4).

Classification of Microorganisms

When living organisms were initially classified into taxonomic units, the existence of microorganisms was unknown. Early classification systems recognized the plant and animal kingdoms, with the primary criterion for separation being that animals were motile and plants were stationary. Classification systems, as expounded by Aristotle and other ancient philosophers, were developed centuries before the beginnings of evolution theory. Microorganisms were considered to be little plants and animals when they were discovered in the late seventeenth century. When the Swedish botanist Carl Linneaeus established the first comprehensive classification system he placed all microbes into a single genus, which he appropriately named *Chaos*.

The Microbe is so very small
You cannot make him out at all,
But many sanguine people hope
To see him through a microscope.
His jointed tongue that lies beneath
A hundred curious rows of teeth;
His seven tufted tails with lots
Of lovely pink and purple spots,
On each of which a pattern stands,
Composed of forty separate bands;
His eyebrows of a tender green,
All these have never yet been seen-
But scientists, who ought to know,
Assure us that they must be so . . .
Oh! let us never, never doubt
What nobody is sure about!

—HILAIRE BELLOC, *The Microbe*

Although a classification system should reflect the branching of groups that is a natural consequence of evolution, evolutionary affinities among microorganisms are difficult to discern because there is no fossil record of most microorganisms. The lack of a fossil record makes direct examination of the ancestors of today's microorganisms for the most part impossible. As a result, microbial taxonomy generally requires many subjective decisions, resulting in an artificial systematic classification scheme. Some taxonomists are "lumpers," tending to lump together many similar organisms into large taxonomic units. In contrast, other taxonimist are "splitters," favoring small taxonomic groups that emphasize even minor differences between organisms. Arguments can be quite heated over the "proper taxonomic position of a microorganism."

One of the earliest classification systems that attempted to reflect evolutionary relationships was developed by a German zoologist, Ernst Heinrich Haeckel, who was a disciple of Darwin and adopted his theories. Haeckel proposed a three-kingdom classification system that divided living organisms into animals, plants, and protists. The Protista were defined by a lack of tissue differentiation; therefore, in this system all microorganisms were classified as protists and, conversely, all protists as microorganisms. Haeckel's system has been superseded by other theoretical systems but still retains its practical appeal for microbiologists because Haeckel's kingdom Protista encompasses all the organisms (bacteria, fungi, algae, and protozoa) traditionally studied by microbiologists.
classification system proposed by Robert H. Whit-
Today, many biologists use the five-kingdom
taker in 1969. According to Whittaker's proposed classification, evolutionary divergence occurs along three principal modes of nutrition: synthetic to evolve plants; adsorptive to evolve fungi; and ingestive to evolve animals. In this system microorganisms exclusively comprise three of the five kingdoms: in addition to the kingdoms of Plants and Animals, there are the kingdoms Monera, which includes all bacteria, Protista, which includes the algae and protozoa, and Fungi, which is composed of the fungi (Figure 2.5). Whittaker's system assumes that the Monera (bacteria) represent the most primitive form of life and that evolutionary processes led to the development of the Protista (algae and protozoa), which then underwent further evolution to form the remaining three kingdoms.
that aim at reflecting the process of evolutionary
Determining the validity of classification systems
separation of species has been greatly bolstered by the development of an understanding of how genetic information is stored at the molecular level. Carl Woese in 1980 proposed a new classification system with three primary kingdoms—Archaebac-

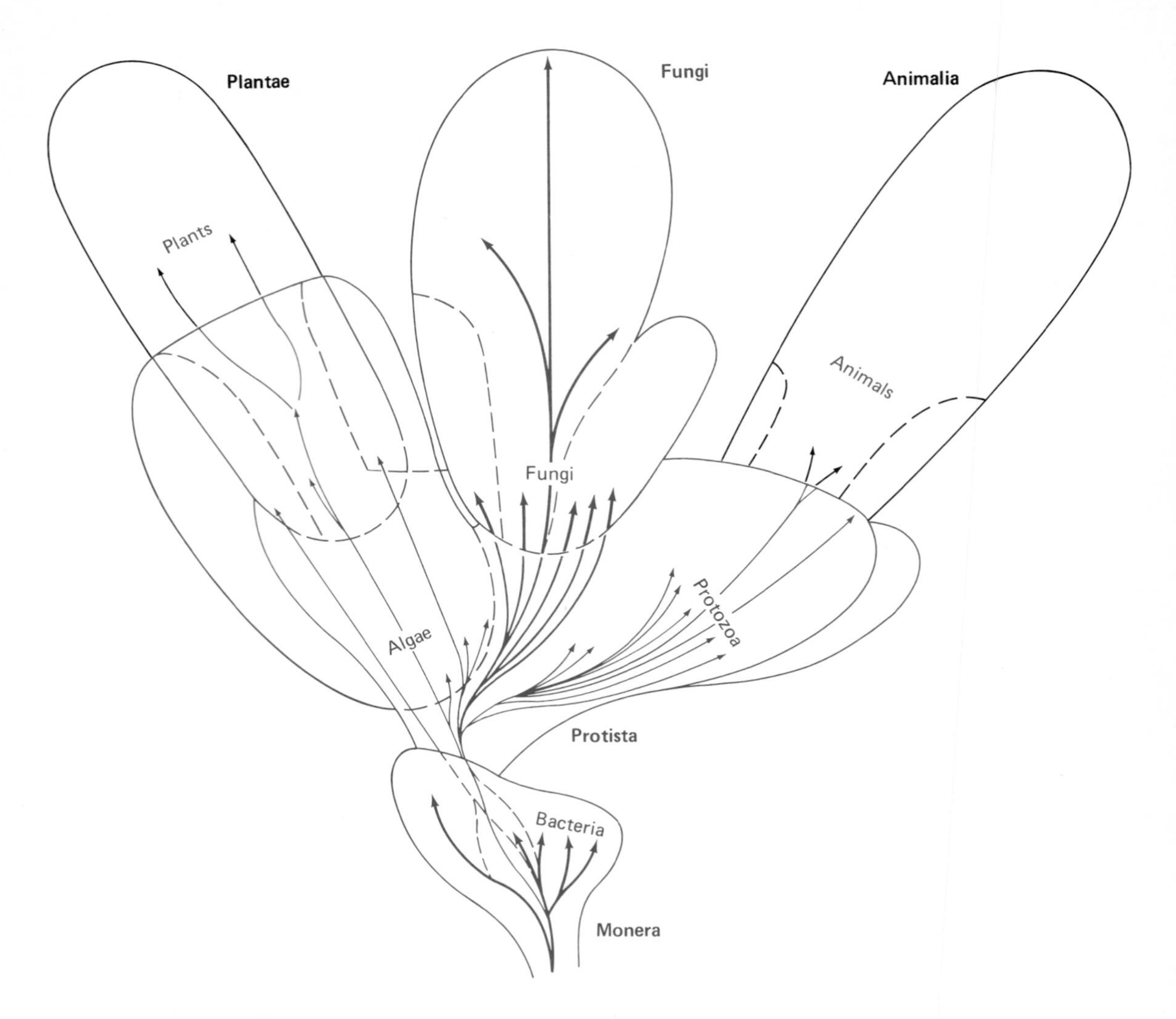

Figure 2.5 Whittaker's classification is a five-kingdom system based on three levels of organization, the prokaryotic (Kingdom Monera), eukaryotic unicellular (Kingdom Protista), and eukaryotic multicellular and multinucleate (Kingdom Fungi, Kingdom Animalia, and Kingdom Plantae).

teria, Eubacteria, and Eukaryotes—based upon molecular-level genetic analyses. Both eubacteria and archaebacteria have prokaryotic cells, which are simply organized, lack a nucleus, and are structurally quite different from eukaryotes, which have a nucleus and various other organelles. The archaebacteria and eubacteria, however, are genealogically as distantly related to each other as each is to eukaryotes. Woese proposed that the archaebacteria, eubacteria, and an urkaryote, which was the original eukaryotic cell (German *ur* = prototype), all evolved from a common simple ancestor, the progenote. According to Woese's theory, bacteria began to live within urkaryotes. Eventually these prokaryotic bacteria and their urkaryotic host cells became mutually dependent. Over time the prokaryotic partners in this mutually beneficial relationship developed into cellular organelles (mitochondria and chloroplasts), giving rise to the eukaryotic cell. This system is receiving a great deal of attention from microbiologists but has yet to gain wide support from general biologists.

You will notice that none of the classification systems deal with the acellular microorganisms. Effectively the viruses, viroids, and prions are treated as nonliving entities. The reproductive capacity of acellular microorganisms is integrally linked to the host organisms in which they multiply, and therefore viruses, viroids, and prions are normally considered along with their hosts. For example, viruses are classified as either animal, plant, fungal, algal, protozoan, or bacterial viruses.

Summary

The study of microbiology has great relevance. Microbiology is a broad, multidisciplinary field. It encompasses the fields of virology, bacteriology, mycology, phycology, and protozoology. Microbiology also overlaps many other fields, such as immunology, plant pathology, and cell biology.

Microorganisms impact our daily lives in many ways. The "balance of nature" depends upon microbes. Without essential microbial activities, life on Earth could not continue. Microbes also have great economic importance and many products such as cheese, wine, and antibiotics are produced by microorganisms. In some cases microbial growth, however, results in undesired biodeterioration or disease. Our fundamental understanding of microorganisms has allowed us to develop methods for controlling unwanted microbial growth. Today's reduced rates of sickness and death are a direct result of our ability to control the growth of microorganisms and our basic understanding of the involvement of microbes in infectious diseases. Many basic studies with microorganisms have added to our general understanding of genetics and metabolism, opening the door to new possibilities of applying this knowledge in practical ways. Today we are beginning to see the benefits of genetic engineering and a variety of microbial based biotechnology processes promise to have a major impact on our quality of life.

Even microorganisms, the smallest forms of life, are highly organized living systems. The structure of the cell provides for a separation of the living system from its surrounding environment, with the cytoplasmic membrane acting as a semipermeable barrier that controls the flow of material into and out of the cell.

The organizational patterns of viruses, bacteria, and other microorganisms are quite different. Viruses, viroids, and prions are acellular and are dependent upon host cells to provide a suitable environment for their reproduction. Bacteria are cellular organisms, but their cells lack a nucleus. Such cells are called prokaryotic. Fungi, algae, protozoa, plants, and animals (including humans) all have eukaryotic cells. In a eukaryotic cell the genetic information is contained within the nucleus and thus is separated from the rest of the cell. Eukaryotic cells also have a high degree of internal compartmentalization, with many functions being performed within specialized organelles.

Structural organization forms the basis, in part, for the taxonomic classification of microorganisms into major groups. For example, all bacteria have prokaryotic cells and conversely all prokaryotes are bacteria. The classification of microorganisms attempts to reflect evolutionary relationships. This is especially difficult because of the lack of a fossil record. Several classification systems have been proposed starting with the placement of all microbes into the single genus *Chaos*. The classification system of Haeckel recognized three kingdoms: plants, animals, and microbes (protists). The five-kingdom system of Whittaker places microorganisms into several different kingdoms. The recent system of Woese, which is based on genetic analyses at the molecular level, recognizes three primary kingdoms: Archaebacteria (primitive specialized prokaryotes), Eubacteria (common prokaryotes or "true bacteria"), and Eukaryotes (all other cellular organisms, including plants, animals, protozoa, algae, and fungi). The progression of systems for classifying organisms has gone from the proposal that the biological world contains two kingdoms, plants and animals, before the existence of microorganisms was suspected, to one in which microorganisms represent the majority of the primary kingdoms.

The ability to classify and name microorganisms is important in all fields of microbiology. Microorganisms are named using a binomial system. The name of a microorganism has two parts, the genus name and the species epithet. The formal naming of microorganisms permits one to refer to a particular organism unambiguouly. This is essential for communication regarding the properties of a particular microorganism. Learning how to properly specify the name of a microorganism is critical. In answer to the Shakespearean question "What's in a name?" the name of a microorganism collectively describes all the properties of that organism. Saying that you have isolated a culture of the bacterium *Yersinia pestis* from a patient specifies that you have found that the patient is infected with the organism that causes plague.

Study Outline

A. Scope of microbiology.

1. Microbiology is a very broad field.
2. Microorganisms affect our daily lives and influence the quality of life.
3. Some microorganisms cause disease. Our understanding of medical microbiology has led to decreased morbidity rates (ratio of sick to well persons) and to lower mortality rates (death rate).
4. Microorganisms have great economic importance. Some industries must limit microbial growth responsible for biodeterioration. Other industries grow microorganisms and harvest their metabolic products.
5. Microorganisms have a major impact on environmental quality. Agricultural management practices are based on an understanding of the roles of microorganisms in maintaining soil fertility, causing plant diseases, and transforming pesticides and fertilizers. Waste disposal relies upon the biodegradative power of microbes.
6. Our fundamental understanding of molecular biology has come from studies on the genetics and metabolism of microorganisms.
7. Microbiology includes the fields virology (the study of viruses), bacteriology (the study of bacteria), mycology (the study of fungi), phycology (the study of algae), and protozoology (the study of protozoa).

B. Structural organization of microorganisms.

1. Microorganisms have a simpler structural organization than higher plants and animals.
2. Microorganisms generally are unicellular (single-celled).
3. Microorganisms lack true tissue differentiation, which is characteristic of higher organisms.
4. Cells are the fundamental structural units of living systems.
5. Cells are separated from their surroundings by a cytoplasmic membrane which acts as a semipermeable barrier.
6. There are two types of cells, prokaryotic and eukaryotic.
7. Prokaryotic cells lack a nucleus, have a low degree of internal structural organization, and are relatively small.
8. Eukaryotic cells possess a nucleus, have membrane-bound organelles (compartmentalization), and are larger in size than prokaryotic cells.
9. All bacteria have prokaryotic cells and all prokaryotes are bacteria.
10. Plants, animals, protozoa, algae, and fungi all have eukaryotic cells.
11. Viruses, viroids, and prions are acellular and are not separated from their surroundings by a membrane.
12. Outside host cells, viruses, viroids, and prions act as nonliving entities; they are obligate intracellular parasites and act as living organisms only within host cells.

C. Taxonomic position of microorganisms in the living world.

1. Early classification systems did not include microorganisms.
2. Taxonomic systems try to show relatedness of organisms.
3. Haeckel proposed a three-kingdom classification system in which all microorganisms were placed into the Kingdom Protista.
4. The five-kingdom classification system proposed by Whittaker in 1969 places microorganisms into several kingdoms: bacteria in the Kingdom Monera; protozoa and some algae in Kingdom Protista; some algae in the Kingdom Plantae, and fungi in the Kingdom Fungi.
5. Woese in 1980 proposed a three-kingdom classification system based upon molecular-level genetic analyses that included the kingdoms Archaebacteria, Eubacteria, and Eukaryotes.
6. Based upon Woese's evidence it appears that all cells evolved from a simple ancestor (the progenote); the progenote evolved into an urkaryote and prokaryote cell line;

the urkaryote became a host for bacteria that evolved into mitochondria and chloroplasts, thereby giving rise to the eukaryotic cell.
7. Acellular microorganisms (viruses, viroids, and prions) are treated as nonliving entities and are, therefore, not included in any of these classification systems.
8. Microorganisms are classified using hierarchial levels of kingdom, phyla, class, orders, family, genus, and species.
9. Naming of microorganisms is important for communication.
10. Microorganisms have binomial (two part) names; the name of a species includes the genus name and the species epithet.
11. The names of microorganisms are italicized or underlined and the initial letter of the genus name is capitalized.

Study Questions

1. Briefly compare the five-kingdom classification system proposed by Whittaker with the three-kingdom classification system proposed by Woese.
2. Indicate whether the following statements are true (T) or false (F).
   a. Viruses are classified as living entities.
   b. Viruses are included in Whittaker's classification scheme.
   c. Viroids are composed of only RNA.
   d. Prions are composed of only protein.
   e. Viruses are composed of only DNA.
3. Match the statements with the appropriate terms in the list to the right.
   a. Plants, animals, protozoa, algae, and fungi all have this type of cell.
   b. Is smaller in size than the other cell type.
   c. Are acellular organisms.
   d. Greater degree of internal structure.
   e. More primitive.
   f. Membrane-bound organelles present.
   g. Never possesses a nucleus.
   h. More advanced cell type.
   i. Have a lesser degree of internal structural organization.
   j. Always possess a nucleus.
   k. Are obligate intracellular parasites.
   l. Only bacteria have this cell type.

   (1) Prokaryotic cell
   (2) Eukaryotic cell
   (3) Virus
   (4) Viroid
4. Name the organisms considered to be microbes and the taxonomic subfields of microbiology.
5. Write the names of 10 bacteria.
6. List the divisions in a taxonomic hierarchy from highest to lowest.

Some Thought Questions

1. What is a microorganism?
2. How should microorganisms properly be classified?
3. Is it morally right to kill bacteria in scientific studies?
4. Should bacteria be protected by the endangered-species act?
5. How can you distinguish living from nonliving microorganisms?
6. What is the answer to the Shakespearean question, what's in a name?
7. How would you define life for a visitor from another planet?
8. If you were going to mount an expedition to search for life on the moon, what would you take with you?

Additional Sources of Information

Dixon, B. 1979. *Magnificent Microbes*. Atheneum Publishers, New York. A popular work on the importance of microorganisms in our daily lives.

Holt, J. G. (ed.). 1977. *The Shorter Bergey's Manual of Determinative Bacteriology*. The Williams & Wilkins Company. The abbreviated version of the standard reference work for the taxonomy of bacteria. This work describes the recognized bacterial taxonomic groups and includes genus and species names.

Industrial Microbiology. 1981. *Scientific American*, **245**(3). A collection of articles describing the many industrial uses of microorganisms. The articles cover the importance of microbes in the food and pharmaceutical industries and the impact of genetic engineering on the field of biotechnology.

Postgate, J. 1975. *Microbes and Man*. Viking Penguin, New York. An easy-to-read book that discusses the impact of microoorganisms on our daily lives.

Rossmore, H. W. 1976. *The Microbes, Our Unseen Friends*. Wayne State University Press, Detroit. A light-reading book about the practical importance of microorganisms.

Whittaker, R. H. 1969. New concepts in kingdoms of organisms. *Science*, **163**:150–160. A description of the widely used five-kingdom classification system.

Woese, C. R. 1981. Archaebacteria. *Scientific American* **244**(6):98–122. A description of a revolutionary new three-kingdom classification system based upon genetic analyses. The basis for the new classification system is presented in an understandable manner.

UNIT TWO

Structure of Microorganisms

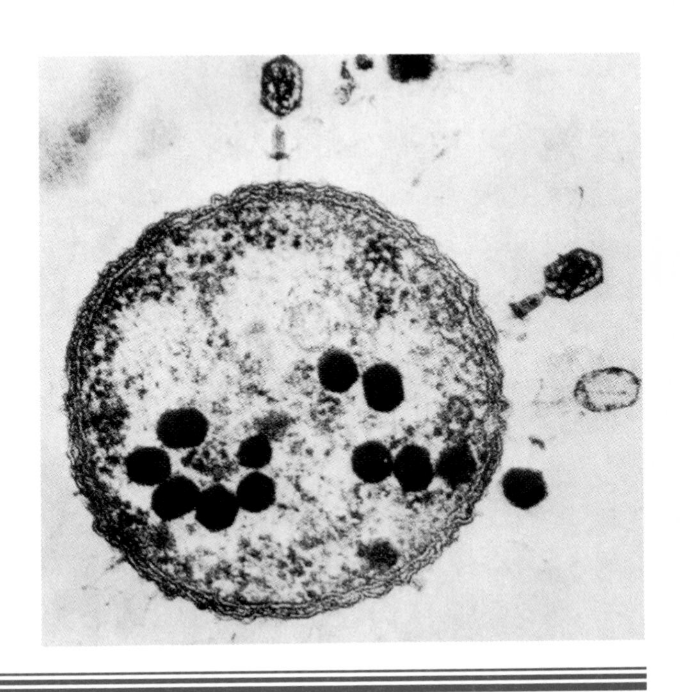

3 *The Observation of Microorganisms*

KEY TERMS

acid-fast stain
brightfield microscope
darkfield microscope
differential staining
fluorescence microscope
Gram stain
immunofluorescence
magnification
microscope
negative stain
numerical aperture
objective lens
ocular lens
oil immersion lens
phase contrast microscope
positive stain
resolution
resolving power
scanning electron microscope
simple stain
staining
transmission electron microscope

OBJECTIVES

After reading this chapter you should be able to

1. Define the key terms.
2. Describe the different types of microscopes, listing the advantages and limitations of each.
3. Discuss the factors that influence resolution and how optimal resolution can be achieved.
4. Calculate magnification.
5. Calculate the limit of resolution.
6. Describe how contrast can be increased for brightfield microscopy.
7. Compare darkfield and brightfield microscopy with respect to how contrast is achieved.
8. Describe the Gram stain procedure.
9. Compare simple and differential staining procedures.
10. Describe the appearances of microorganisms after different staining procedures.
11. Compare light microscopy with transmission electron microscopy with respect to useful magnification, resolution, sample preparation, and design of the microscope.
12. Discuss how samples are prepared for electron microscopy.
13. Compare the appearances of specimens viewed with transmission and scanning electron microscopy.

Light Microscope

The light microscope is the basic tool employed for the visualization of the microbial world. Microbes, although small, appear beautiful, diverse, and graceful in form when viewed through a microscope. The microscope truly opens our vistas into the magnificence of the microbial world.

By the assistance of microscopes have I seen animals of which many hundreds would not equal a grain of sand.
How Exquisite, how Stupendous, must the Structure of them Be!
The Whales . . . me thinks . . . are not such wonders, as these minute Fishes are.

—COTTON MATHER, *The Wonderful Words of God Commemorated: A Thanksgiving Sermon*, 1690

Virtually every microbiology laboratory has a light microscope capable of viewing bacteria and larger microorganisms. A simple magnifying glass permits some enlargement in the apparent size of the object and is useful in examining the details of macroscopic structures of microorganisms such as the mushroom fruiting bodies of certain fungi; however a simple magnifier does not normally permit the visualization of microorganisms the size of bacteria. The normal light microscope, which is used for viewing bacteria and other cellular organisms, is a compound microscope that has a light source, a condenser lens that focuses the light on the specimen, and two sets of lenses that contribute to the magnification of the image (Figure 3.1). Through the refraction (bending) of light rays by the system of microscope lenses, an image of the specimen (object) is formed that is larger than the object itself (magnified), permitting examination of the detailed structure of the specimen.

Magnification

The magnifying capability of a compound microscope is the product of the individual magnifying powers of the ocular lens (the one nearest the eye) and the objective lens (the lens nearest the specimen). A typical microscope used in microbiology has objective lenses with powers of 10×, 40×, and 100× and an ocular lens of 10× and thus is capable of magnifying the image of a specimen 100, 400, and 1000 times (Figure 3.2). If the microscope is parfocal, once the specimen is in focus with one lens, it remains in focus when switching to other

Figure 3.1 A light microscope and the path of light as it is modified by passage through the lenses.

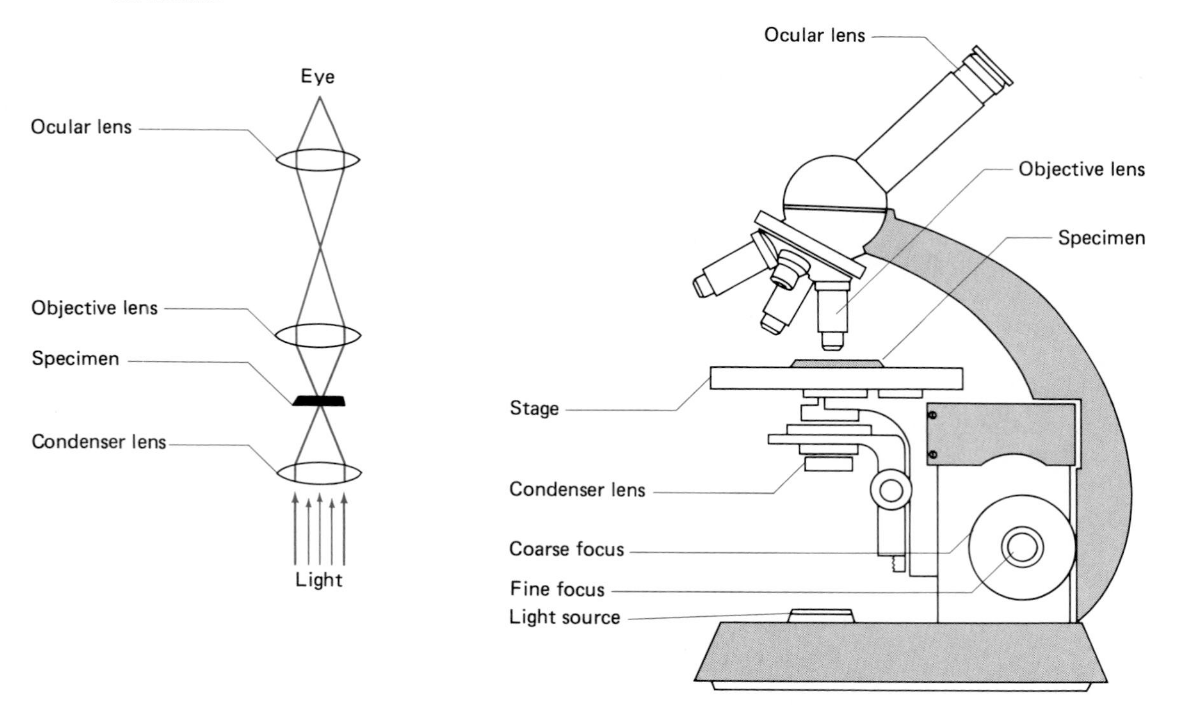

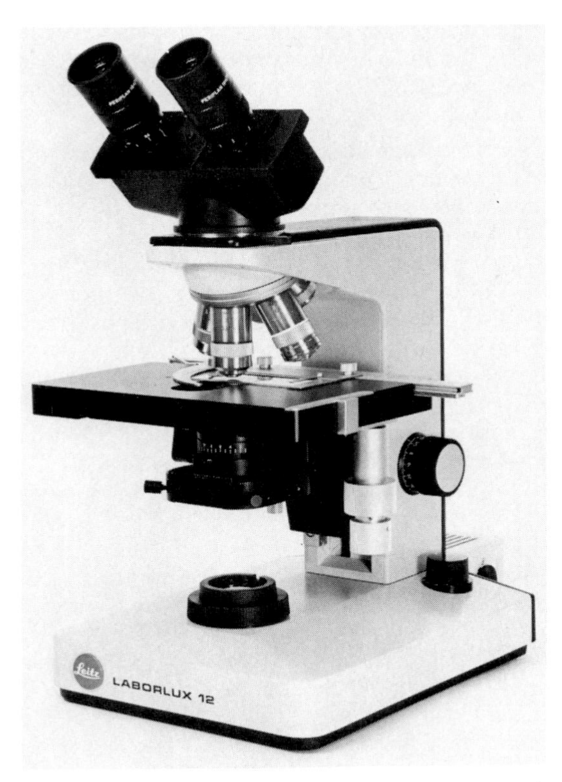

Figure 3.2 A brightfield microscope. (Courtesy Leitz Corporation.)

objective lenses. This permits focusing with a low power objective and changing to higher power objectives without refocusing. At a magnification of 1000×, bacteria and larger microorganisms can be visualized, but viruses and much of the fine structural detail of bacteria cannot be seen (Figure 3.3).

Of all the Inventions none there is Surpasses
the Noble Florentine's Dioptrich Glasses
For what a better, fitter guift Could bee
in this World's Aged Luciosity.
To help our Blindnesse so as to devize
a paire of new and Artificial eyes
By whose augmenting power wee now see more
than all the worls Has ever doun Before.

—HENRY POWERS, *In commendation of ye Microscope, 1664*

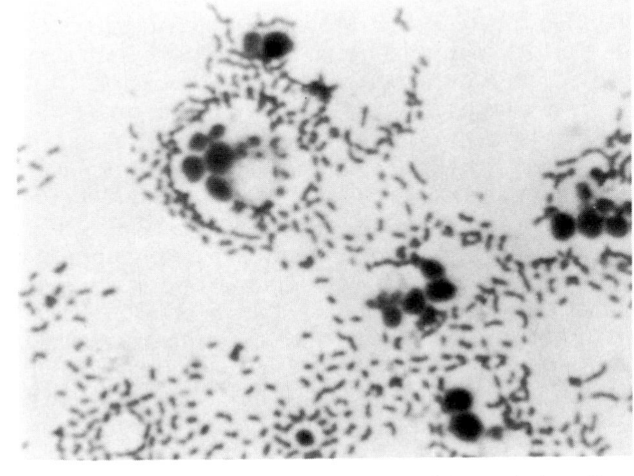

Figure 3.3 This photomicrograph of bacteria (small rods) and yeasts (larger oval-shaped cells) was produced by using brightfield microscopy. By using brightfield microscopy, the characteristic sizes, shapes, and arrangements of the microbial cells can be observed.

It is critical that the magnified image not be distorted and that the enlarged image retain the essential detail of the specimen. Seeing a large blur is of no use! The detail of the specimen should be seen clearly without distortion in the enlarged microscopic image. Modern microscopes contain complex lens systems that are designed to correct for various types of distortions that occur when light is bent (refracted) by passage through microscope lenses. The price of the microscope generally reflects the quality rather than the size of the image that is observed.

Resolution

In addition to magnifying the image of a specimen, the usefulness of a microscope is dependent on its resolving power, that is, the degree to which the detail present in the specimen is retained in the magnified image. Resolution is defined as the closest spacing between two points at which they still can clearly be seen as separate entities (Figure 3.4), and the resolving power of a microscope, therefore, is the distance between two structural entities of a specimen at which they can still be

Figure 3.4 Resolution of two points under a microscope. At low resolution structures blur together; the greater the resolution, the more detail can be observed.

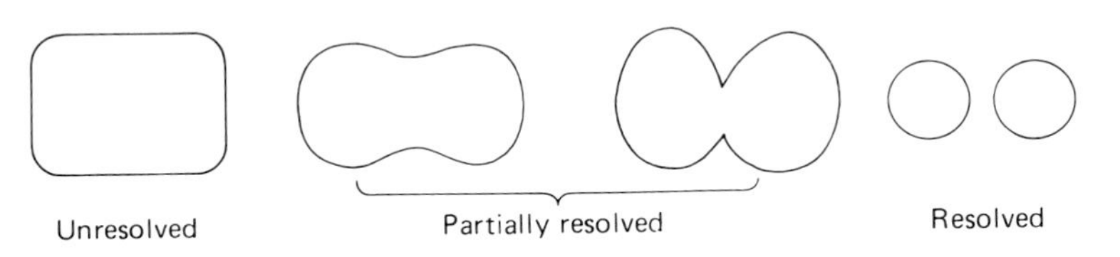

seen as individual structures in the magnified image. We measure the resolving power in billionths of a meter (nanometers). (See Appendix I for a review of the metric system of measurement.)

The resolving power of a microscope is dependent on the wavelength of light (λ) and on a property of the lens, which describes the amount of light that can be captured by that lens, known as the numerical aperture (N.A.). The limit of resolution of a microscope is approximately equal to $0.5\lambda/N.A.$, which for a light microscope is approximately 200 nanometers (nm), that is, objects smaller than 200 nm or closer than 200 nm can not be distinguished. The shorter the wavelength of light and the higher the numerical aperture of the lens, the better the resolving power of the microscope. It is thus apparent that the resolving power of a light microscope is restricted by the obtainable numerical apertures of the lens systems and the wavelengths of the visible light spectrum. Because blue light has a shorter wavelength than red light, greater resolution can be achieved by using a blue light source to illuminate the specimen. Along with using blue light for illumination, using a lens with a high numerical aperture permits high resolution. The numerical aperture of a lens is indicated on that lens.

Using immersion oil between the specimen and the objective also improves the resolving power of the microscope because more light can enter the lens. The observation of fungi, algae, and protozoa can be achieved with dry objectives, that is, where air occupies the space between the specimen and the objective, but the viewing of bacteria in sufficient detail to determine the shape and arrangement of cells normally requires the use of an oil immersion lens. As a rule of thumb, the useful magnification of a microscope is 1000 times the numerical aperture being used, so it is possible, using an oil immersion lens with a numerical aperture of 1.4, to achieve a useful magnification of approximately 1400×. At higher magnifications the quality of the image deteriorates, and the magnification is therefore considered to be empty.

A high numerical aperture lens has a short focal length, and therefore there is a short working distance between the lens and the object; that is, the lens and the specimen are very close to one another. Another consequence of the short focal length is a very shallow depth of field, so that only a very thin section can be in focus at any one time. Because of the short working distance and shallow depth of field of oil immersion lenses, many students at first have great difficulty trying to focus the microscope on a specimen of bacteria without breaking the slide and scratching the objective lens; however, with a little practice this problem is easily overcome.

I could never see through a microscope. I never once saw a cell through a microscope. This used to enrage my instructor. "I can't see anything," I would say. He would begin patiently enough, explaining how anybody can see through a microscope, but he would always end up in a fury, claiming that I could too see through a microscope but just pretended that I couldn't . . .

"Try it just once again," he'd say, and I would put my eye to the microscope and see nothing at all, except now and again a nebulous milky substance—a phenomenon of maladjustment . . .

Students to right of me and to left of me and in front of me were seeing cells; what's more, they were quietly drawing pictures of them in their notebooks. Of course, I didn't see anything. "We'll try it," the professor said to me, grimly, "with every adjustment of the microscope known to man. As God is my witness, I'll arrange this glass so that you can see cells through it or I'll give up teaching" . . .

So we tried it with every adjustment of the microscope known to man. With only one of them did I see anything but blackness or the familiar lacteal opacity, and that time I saw, to my pleasure and amazement, a variegated constellation of flecks, specks, and dots. These I hastily drew. "What's that?" he demanded, with a hint of a squeal in his voice. "That's what I saw," I said. "You didn't, you didn't you didn't!" he screamed, losing control of his temper instantly, and he bent over and squinted into the microscope. His head snapped up. "That's your eye!" he shouted. "You've fixed the lens so that it reflects! You've drawn your eye!"

—James Thurber, *University Days*

Contrast and Staining

Another factor that must be considered in light microscopy is contrast because, without adequate contrast, it is impossible to distinguish a structure from the surrounding background. Microorganisms are largely composed of water, as is the medium in which they are normally suspended, and simply viewing microorganisms with a light microscope without performing procedures to increase contrast can be likened to trying to see a white object on a white background.

Staining is used to increase the contrast between the specimen and the background, but, unfortunately, staining generally eliminates the possibility

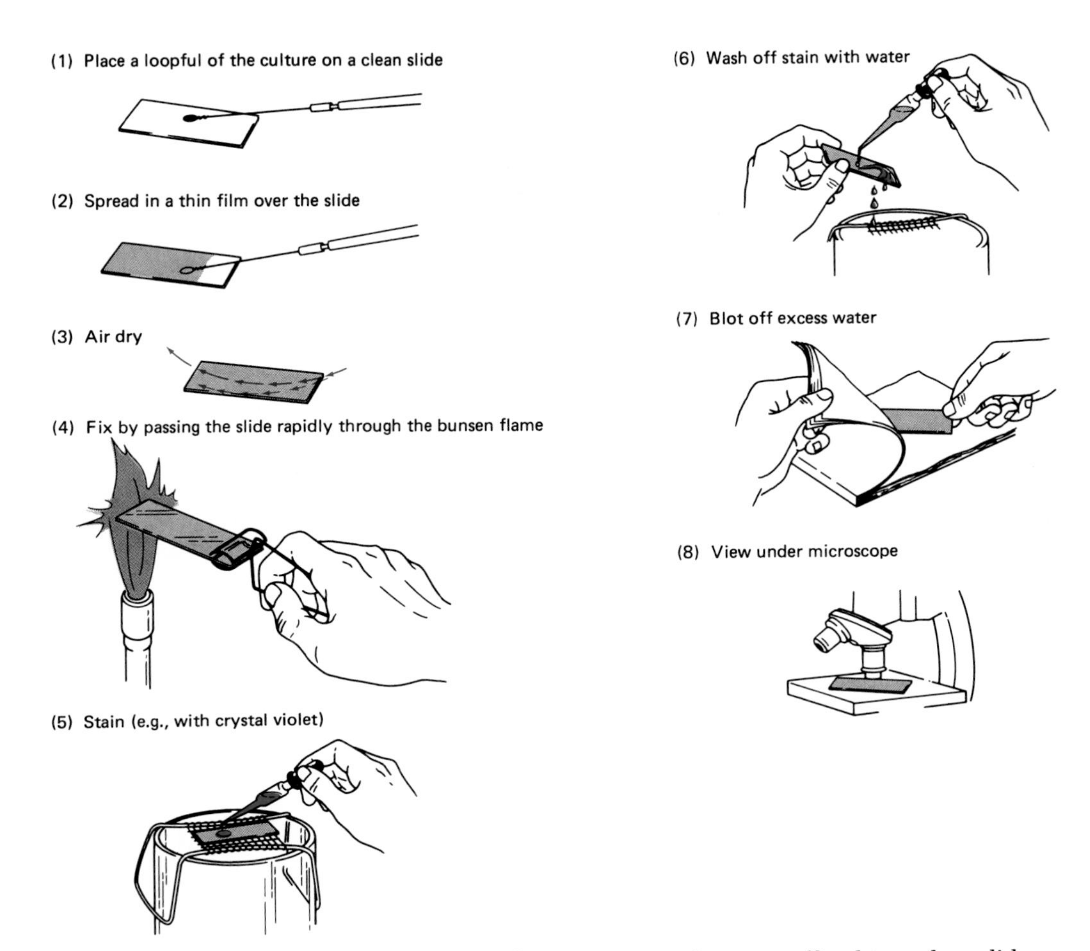

Figure 3.5 In a simple staining procedure, microorganisms are affixed to a glass slide and stained with an appropriate dye to increase the contrast between the cells and the background so that they can easily be seen using a light microscope.

of observing living microorganisms. In most staining procedures (Figure 3.5) a suspension of microorganisms is transferred to a glass microscope slide and allowed to dry. The slide is then quickly passed through a flame to heat fix the cells to the slide, that is, to affix the cells to the slide so that they cannot be washed off. A stain is then added to the slide, allowed to stand for a period of time, rinsed to remove excess stain, and viewed.

Simple Staining Procedures. In a simple staining procedure, a single stain reagent is used, and the procedure does not attempt to produce different staining reactions for different structures or types of microorganisms. Simple staining of microorganisms may be positive, where the stain is attracted to the microbial cells so that the cells appear dark or colored on a light or clear background, or may be negative, in which case the background is stained and the microorganisms appear bright on a dark background. Both of these staining procedures depend on the fact that bacteria and other cellular microorganisms have a negative charge associated with the outer surface of the cell, largely because of the PO_4^{3-} groups of the cell membrane (Figure 3.6).

In positive staining procedures, a stain that has a positively charged chromophore (colored portion of the stain molecule) is attracted to the negatively charged outer surface of the microbial cell. A stain such as methylene blue has a positively charged blue portion of the molecule that stains the microorganism (Figure 3.7). In negative staining procedures a negatively charged chromophore is repelled by the negatively charged microorganism,

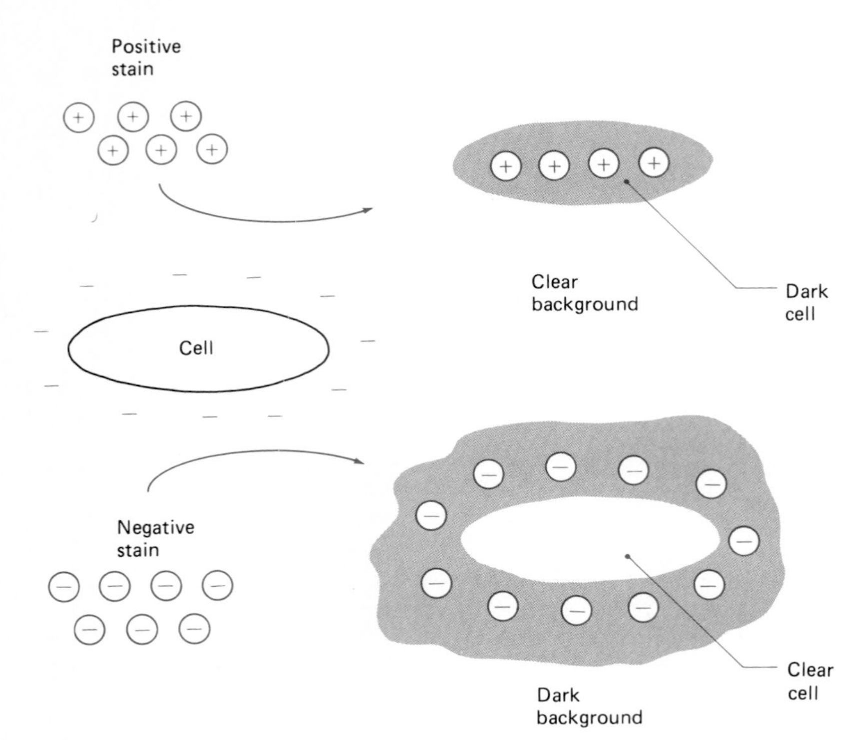

Figure 3.6 The interaction of a cell with negative and positive stain reagents. Because the outer layer of a cell is negatively charged, a positive stain is attracted to the cell, whereas a negative stain is repelled.

resulting in negative or indirect staining of the microbial cell. Nigrosin and India ink are frequently used for negative staining of microbial cells, and this type of staining is particularly useful for viewing some structures, such as, the capsules that surround some bacterial cells (Figure 3.8).

Differential Staining Procedures. In differential staining procedures, specific types of microorganisms and/or particular structures of a microorganism exhibit different staining reactions that readily can be distinguished. The development of a differential staining procedure by the Danish physician Hans Christian Gram remains one of the most important methodological contributions in bacteriology. At the time Gram published the description of his staining method, most bacteriologists were concerned with simply seeing difficult-to-detect bacteria and differentiating infecting bacteria from mammalian nuclei. Gram, working at the morgue of the City Hospital of Berlin, was trying to develop a method that would permit the visualization of bacteria within mammalian tissues. He failed at that

Figure 3.7 Positively stained bacteria. This photomicrograph shows a mixture of *E. coli* and *Staphylococcus aureus* (100×). (From BPS: Leon J. LeBeau, University of Illinois Medical Center, Chicago.)

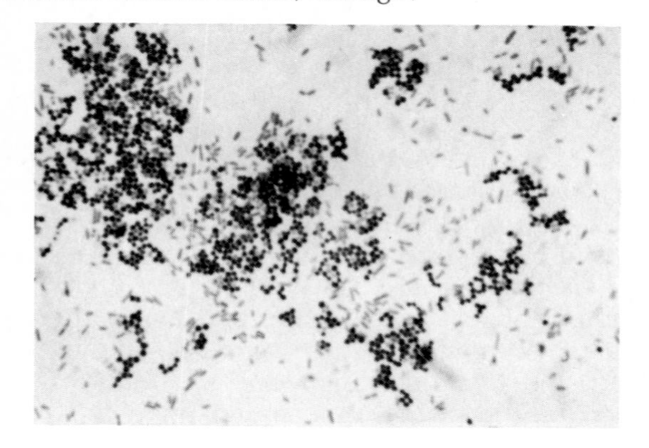

Figure 3.8 The bacterium *Thiocapsa floridiana* following negative staining with India ink. Note that the capsule surrounding this bacterium can be readily seen. (From BPS: Stanley C. Holt, University of Texas Health Science Center, San Antonio, Texas.)

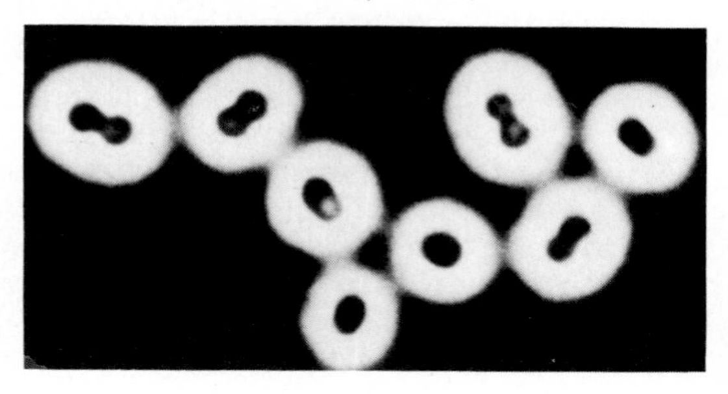

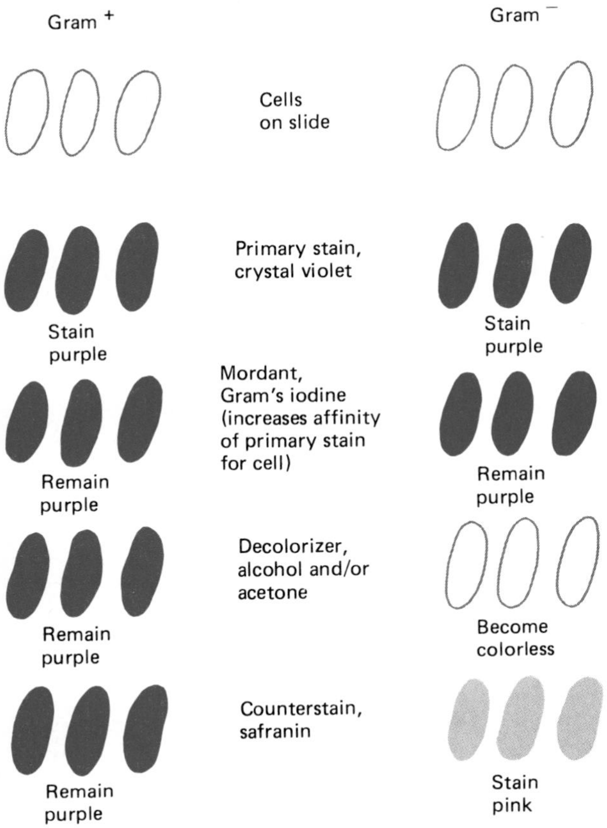

task but in doing so discovered a method for the diagnostic differentiation of bacterial species.

The **Gram stain** procedure undoubtedly is the most widely used differential staining procedure in bacteriology (Figure 3.9). This staining procedure includes primary staining with crystal violet, which stains all bacterial cells blue-purple, application of Gram's iodine, a mordant (a substance that increases the affinity of the primary stain for the bacterial cells), rinsing with acetone–alcohol or other decolorization agent, and application of a counterstain, which stains those bacteria that were decolorized in the previous step so that they can be easily seen. The decolorization step is the critical step that differentiates bacterial species based on their cell wall structure. About half of the species of bacteria are Gram positive and appear blue-purple at the end of this procedure; the remaining bacterial species are Gram negative and appear red-pink following Gram staining. The Gram stain procedure has great diagnostic value as a result of its ability to differentiate between different bacterial species and therefore is a key feature employed in many bacterial classification and identification systems.

Another differential staining procedure frequently used in bacteriology is acid-fast staining. The **acid-fast stain** procedure is especially useful in identifying members of the bacterial genus *Mycobacterium* and is important in identifying the causative organisms of tuberculosis (*M. tuberculosis*) and leprosy (*M. leprae*). The bacterial endospore, which is produced by members of relatively few bacterial genera but is very important because of its resistance to high temperatures, can also be visualized by using a differential staining procedure. In this procedure the bacterial endospore is stained one color and the rest of the bacterial cell another color, permitting differentiation of the endospore from the vegetative cell (Figure 3.10).

Fluorescence Microscopy

Other stains used in microbiology include fluorescent dyes, substances that when illuminated by light

Gram +
Gram −
Cells on slide
Primary stain, crystal violet
Stain purple
Stain purple
Mordant, Gram's iodine (increases affinity of primary stain for cell)
Remain purple
Remain purple
Decolorizer, alcohol and/or acetone
Remain purple
Become colorless
Counterstain, safranin
Remain purple
Stain pink

Figure 3.9 The Gram stain procedure is widely used to differentiate major groups of bacteria.

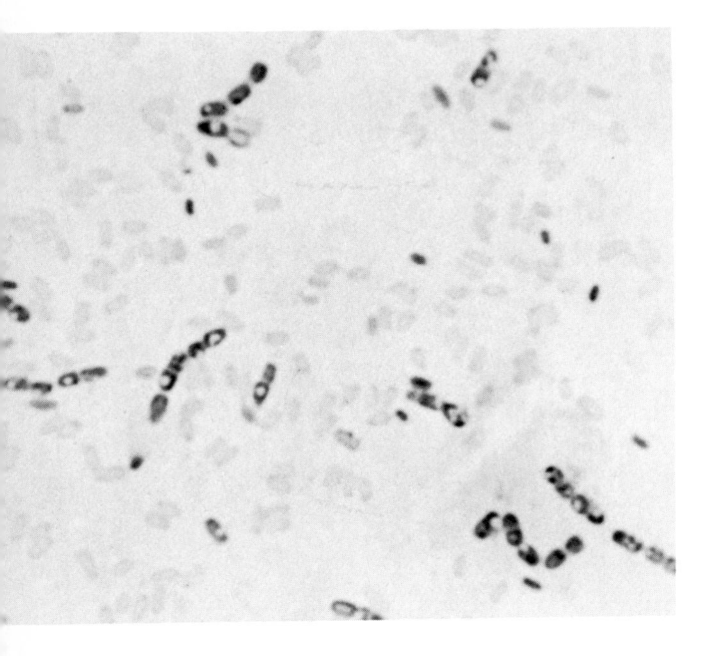

Figure 3.10 Photomicrograph of stained bacterial endospore, *Bacillus subtilis,* showing endospores (the light areas within cells) and free endospores (the lightly stained bodies in the background). (Courtesy R. A. Smucker, Chesapeake Biological Laboratory, Center for Environmental and Estuarine Studies, University of Maryland, Solomons, Maryland.)

of one wavelength, give off light at a different wavelength; for example, a dye illuminated with blue light will emit green or orange light. Microscopy that uses such fluorescent dyes is known as fluorescence microscopy, and the microscope used for viewing such specimens is referred to as a fluorescence microscope. If the light used to illuminate the stain is transmitted to the specimen through the objective lens, the system is referred to as "epifluorescence." If the light is transmitted from below the specimen, the procedure is termed "transmitted fluorescence." The principles of magnifying power and resolution are no different for a fluorescence microscope than for a normal light microscope. The wavelength of the light used to excite the dye may be in the ultraviolet range, but the emitted light viewed must be in the visible range. Fluorescent dyes can be conjugated (linked) with antibodies for immunofluorescent microscopy, which provides great specificity in staining procedures, because of the specificity of immunological reactions. Immunofluorescence staining is one of the applications that make fluoresence microscopy an important method for viewing microorganisms.

Darkfield Microscopy

There are several alternative microscope designs that enhance the contrast between the specimen and the background without the use of staining. In the simplest of these, the darkfield microscope, the normal condenser of the light microscope is replaced with a darkfield condenser that does not permit light to be transmitted directly through the specimen and into the objective lens (Figure 3.11). The darkfield condenser focuses light on the specimen at an oblique angle, such that any light that does not reflect off an object does not enter the objective lens. Thus, only light that reflects off the specimen will be seen, and in the absence of a specimen the entire field will appear dark. Bacteria viewed with a darkfield microscope appear very bright on a dark (black) background (Figure 3.12). The contrast between the specimen and the background is sufficient to permit the visualization of even small bacteria and large viruses, but it is not generally possible with darkfield microscopy to distinguish the internal structures of the microorganisms being viewed.

Phase Contrast Microscopy

The phase contrast microscope is useful for visualizing living microorganisms because it permits viewing of microbial structures without the necessity of staining. Light passing through a cell of higher refractive index (one that has a greater ability to change the direction of a ray of light) than the surrounding medium is slowed down relative to the light that passes directly through the less dense

Figure 3.11 Diagram of a darkfield microscope showing the path of light passing through the background and light striking the specimen.

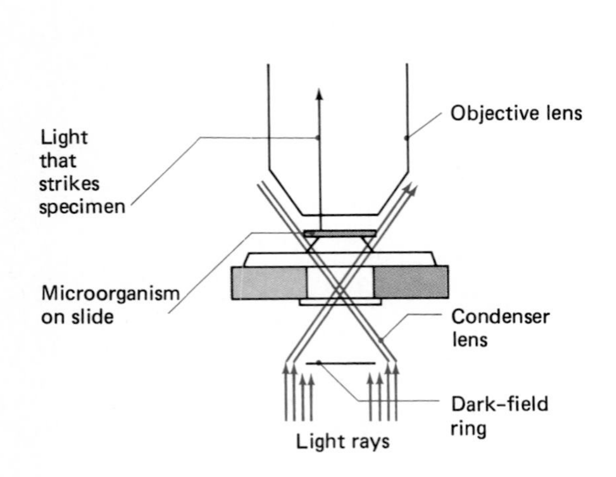

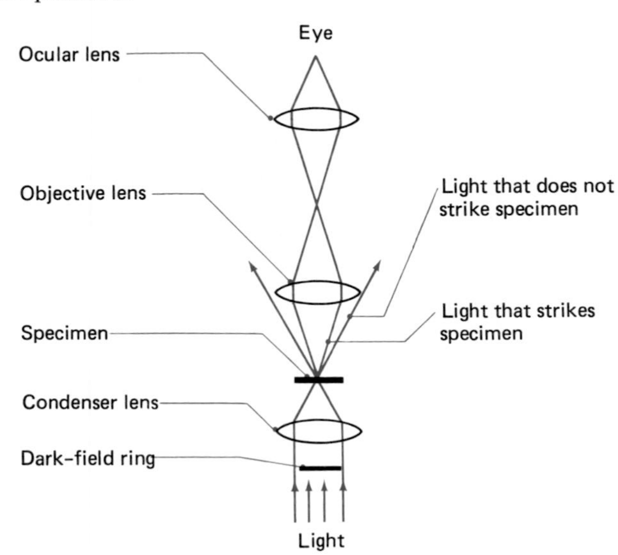

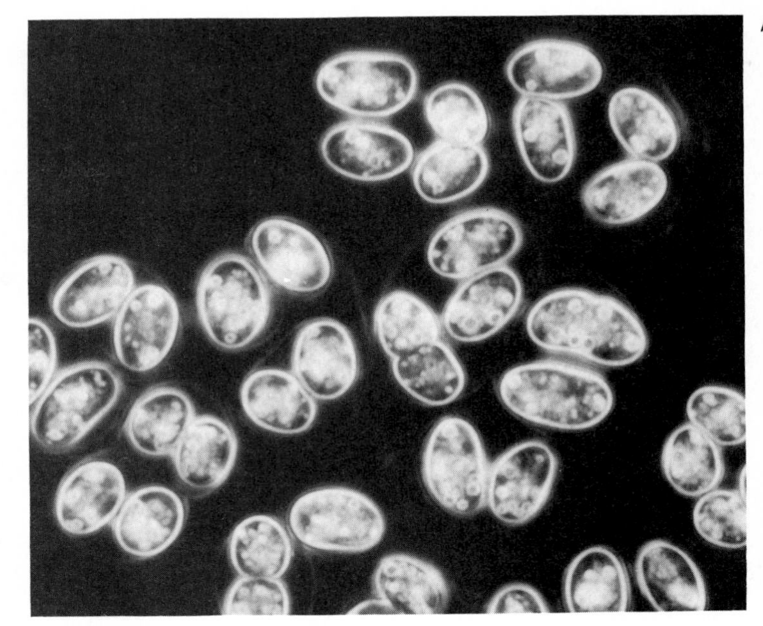

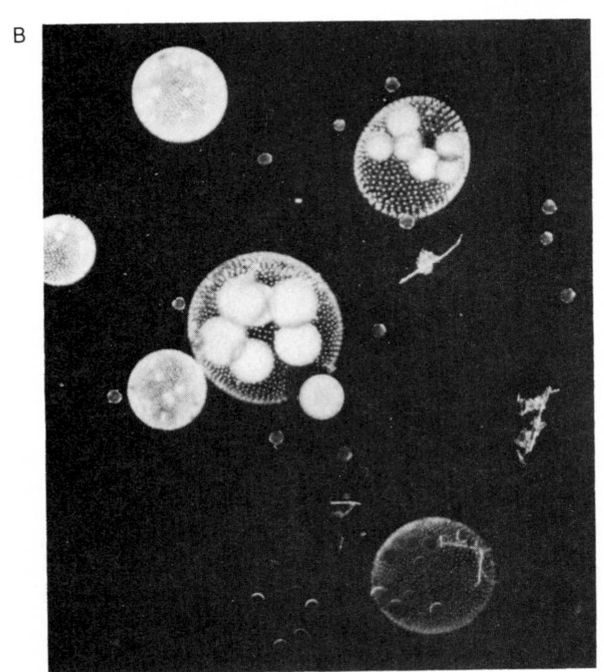

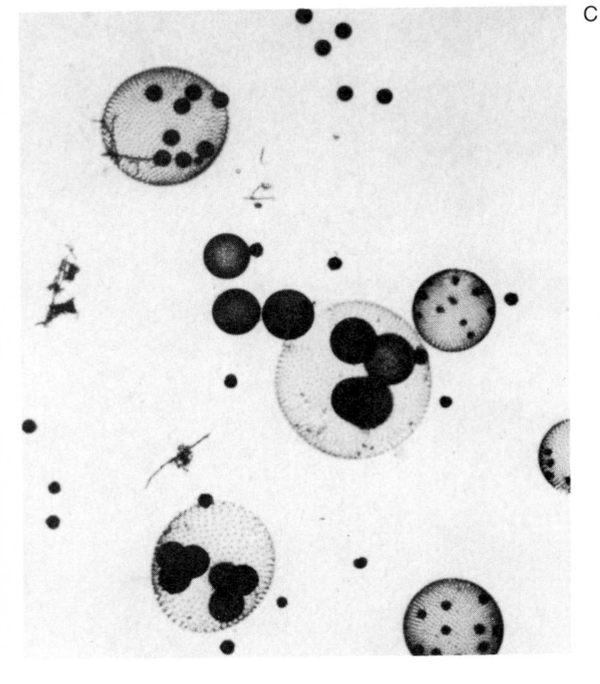

Figure 3.12 (A) This micrograph shows the appearance of the cyanobacterium *Gleocapsa* (1400×) using darkfield microscopy. In this type of microscopy the cells appear bright against a dark background. Darkfield microscopy is useful for visualizing bacterial cells without the need for staining to enhance contrast. For a comparison of how these cells of *Gleocapsa* appear in other types of microscopy, see Figure 3.13. (From BPS: J. Robert Waaland, University of Washington.) (B) Micrograph of a colony of the green alga *Volvox* as it appears in darkfield microscopy. (C) Micrograph of a colony of the green alga *Volvox* as it appears in light microscopy is shown for comparison. (B and C courtesy Gary B. Collins, U.S. Environmental Protection Agency, Cincinnati.)

background medium. The greater the refractive index of the cell or cellular structure, the greater the retardation of the light wave. Thus, when light passes through a microorganism, there is a slight alteration in the phase (stage of cyclic movement) of the light wave. The phase contrast microscope is designed to take advantage of the differences in refractive indexes between structures of a microbial cell, translating differences in the phase of light into changes in light intensity that are visible to the eye. Even difficult-to-stain structures often are conspicuous under a phase contrast microscope because small phase changes give rise to high contrast images (Figure 3.13). Thus, with the phase contrast microscope, living organisms can be clearly observed in great detail without staining them, permitting the study of their movements in the medium in which they are growing.

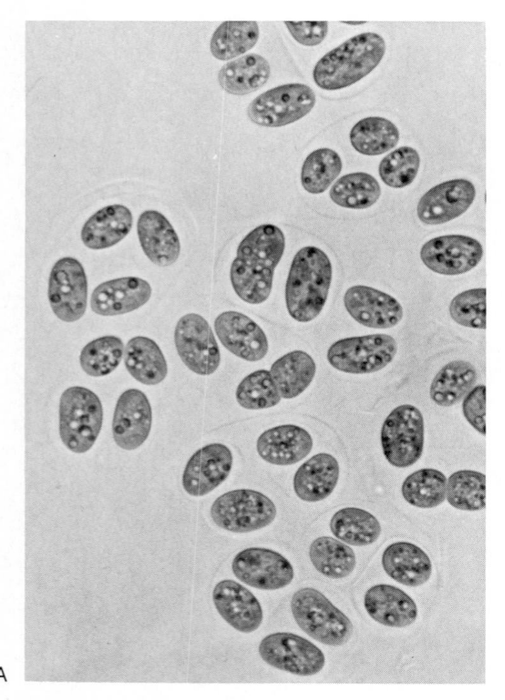

Figure 3.13 These photomicrographs of the cyanobacterium *Gleocapsa* compare its appearance by using (A) brightfield and (B) phase contrast microscopy. The cell structures are easier to see in the micrograph taken with the phase contrast microscope. (From BPS: J. Robert Waaland, University of Washington.)

Electron Microscopy

The advent of the electron microscope marked a significant improvement over light microscopes for the visualization of microorganisms. Many of the fine structures of microorganisms have been elucidated using electron microscopy. The electron microscope permits better resolution, and therefore, higher useful magnifications than can be achieved with light microscopy. There are two basic types of electron microscopes: the transmission electron microscope (TEM) and the scanning electron microscope (SEM).

Transmission Electron Microscope

The design of the transmission electron microscope is similar to that of the compound light microscope, except that an electron beam is substituted for the light source and a series of electromagnets are substituted for the glass lenses (Figure 3.14). The transmission electron microscope permits much greater resolution and thus much higher useful magnifications than the light microscope because the wavelength of an electron beam, generated at a high accelerating voltage, is much shorter than that of light in the visible range of the electromagnetic spectrum. The actual wavelength of the electron beam depends on the accelerating voltage of the microscope. At 60,000 volts, a typical accelerating voltage used in a transmission electron microscope, the wavelength of the electron beam is approximately 0.005 nm, permitting a theoretical resolution of approximately 0.2 nm, which is about a thousand times better than can be achieved when using light microscopy. The useful magnification for an electron microscope, consequently, is in excess of 100,000×, which permits the visualization of all microorganisms, including viruses (Figure 3.15).

There are several problems in viewing biological specimens, including microorganisms, with the transmission electron microscope. There is a great potential for creating artifacts that could be mistakenly viewed as real structures in electron micrographs. An artifact is the appearance of something in an image or micrograph that is due to causes

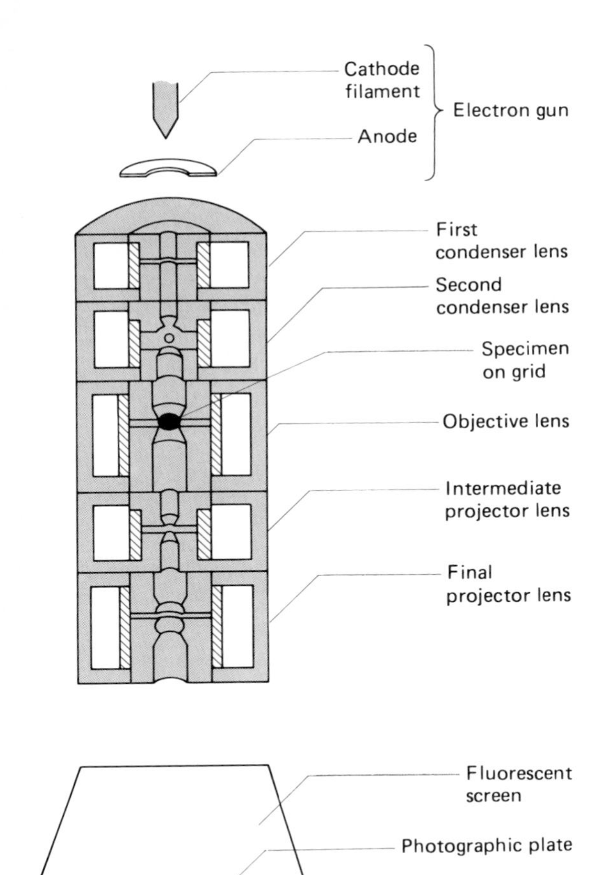

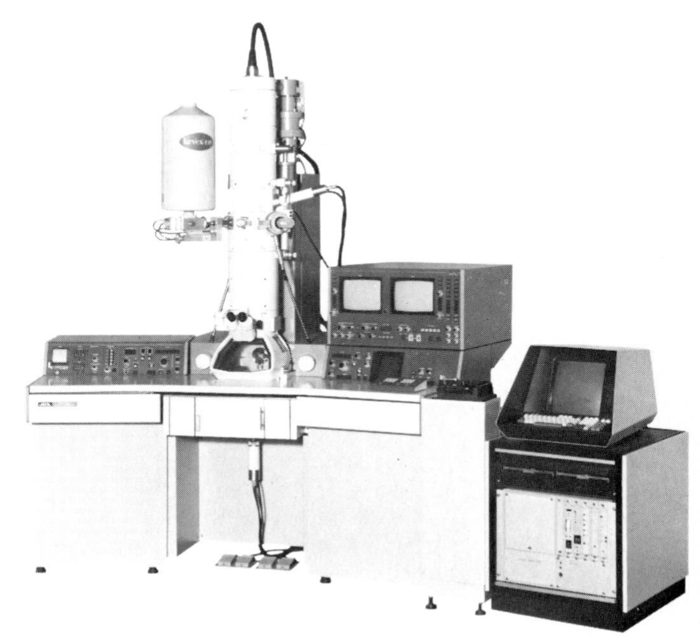

Figure 3.14 The transmission electron microscope (TEM) allows the visualization of the fine detail of the microbial cell. (A) Diagram of a TEM and (B) photograph of a high resolution TEM. (Courtesy JEOL, Peabody, Massachusetts.)

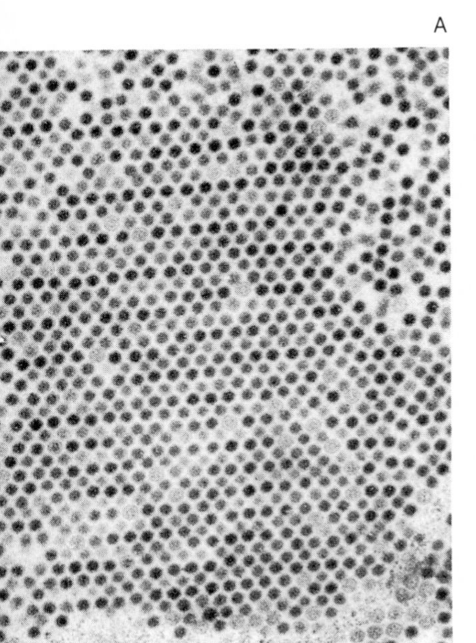

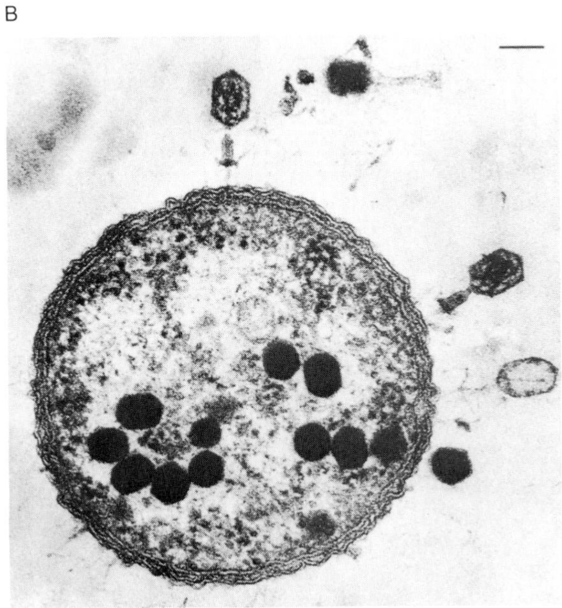

Figure 3.15 To view viruses an electron microscope is needed. (A) Here an electron micrograph shows an intracellular crystalline array of adenoviruses viewed with a TEM (44,000 ×). (From BPS: S. Dales and S. L. Wilton, University of Western Ontario.) (B) This electron micrograph shows phage infecting a bacterium. Some viruses are outside of the cell while new viruses are being made within the bacterium (200,000 ×). (Courtesy Lee O. Simon, Rutgers, The State University, New Brunswick, New Jersey.)

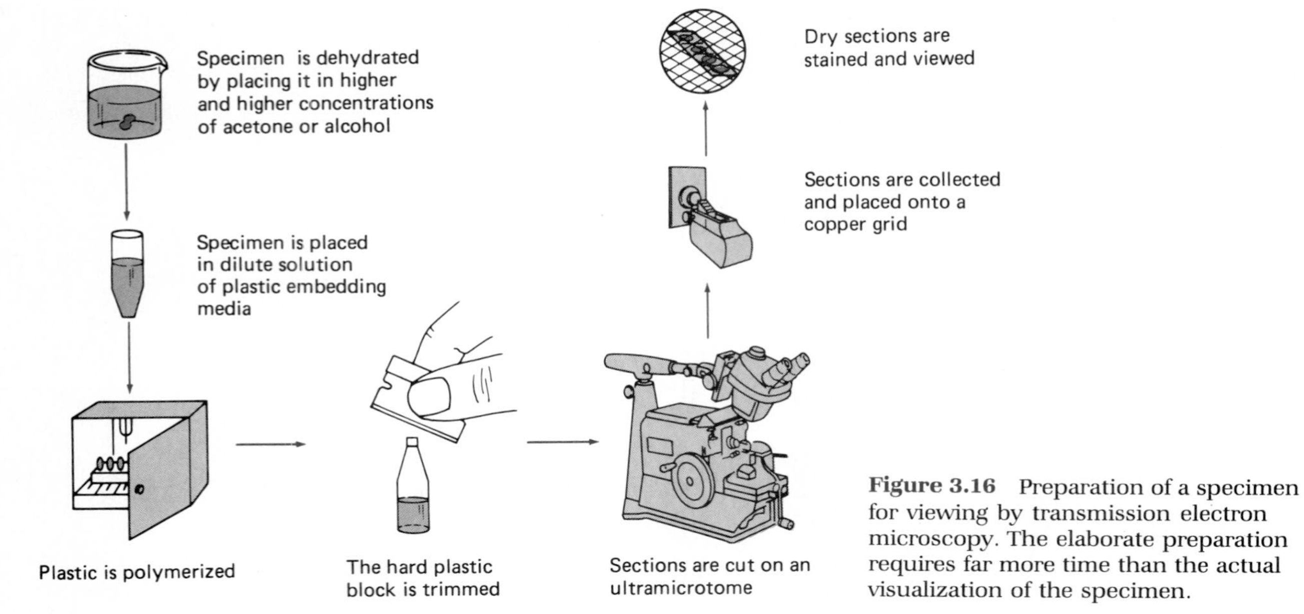

Figure 3.16 Preparation of a specimen for viewing by transmission electron microscopy. The elaborate preparation requires far more time than the actual visualization of the specimen.

within the optical system and not a true representation of the features of the specimen on view. This is a problem common to all microscopes but is particularly troublesome in electron microscopy because of the high magnifications that are used, the need to dehydrate the specimen, and the fact that the specimen must be placed in a high vacuum chamber.

> *Faith is a fine invention*
> *When Gentlemen can see—*
> *But Microscopes are prudent*
> *In an Emergency.*
>
> —Emily Dickinson

Before they can be viewed with a transmission electron microscope, biological specimens, including microorganisms, must be prepared in order to maximize the amount of detail that can be visualized and to avoid the formation of artifacts (Figure 3.16). Biological specimens containing water cannot simply be placed under high vacuum because the water would boil, destroying the integrity of the organisms. Therefore, before viewing a microorganism with a transmission electron microscope, it is necessary to preserve (fix) and dehydrate (remove water) the specimen. Additionally, microorganisms are too large (thick) to view with a transmission electron microscope and see the maximum amount of detail that is possible with the high resolving power of the TEM. Therefore, it is normally necessary to slice the microorganisms into thin sec-

Figure 3.17 The detailed structure of a bacterium is seen in this electron micrograph of a thin section of *Pseudomonas aeruginosa* (120,000×) viewed with a transmission electron microscope. (From BPS: John J. Cardamone, Jr., University of Pittsburgh.)

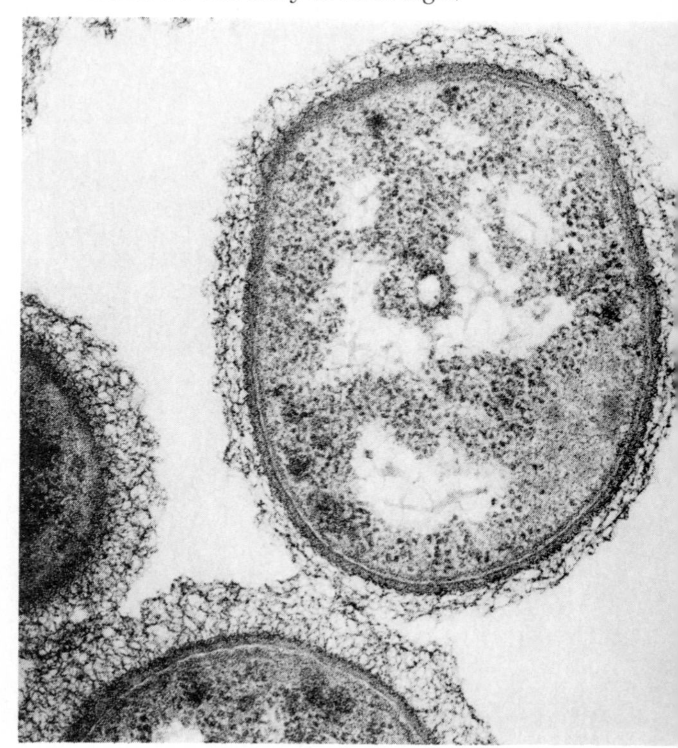

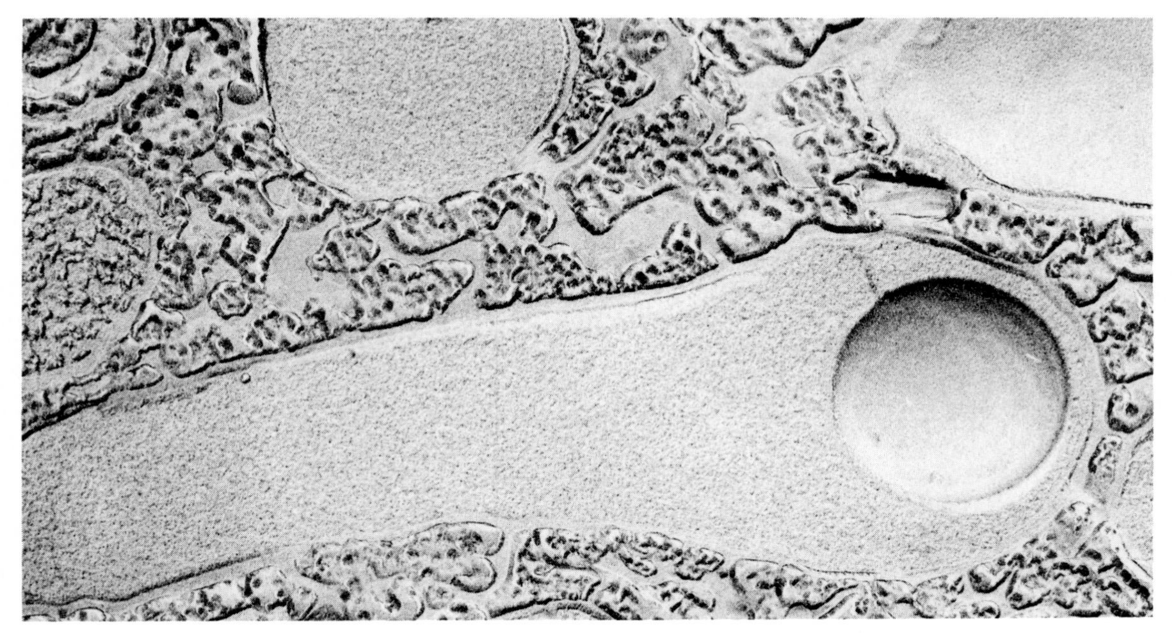

Figure 3.18 Electron micrograph of the endospore-forming bacterium *Clostridium botulinum* (type A) following freeze-etching, as viewed under a TEM (55,000×). Note the prominence and topographical relief of the terminal endospore. (From BPS: T. J. Beveridge, University of Guelph, Ontario.)

tions (Figure 3.17) or to fracture them into pieces to observe their detailed structures (Figure 3.18).

Scanning Electron Microscope

The scanning electron microscope is used primarily for viewing surface details rather than the internal structures of microorganisms that are revealed by using the transmission electron microscope. The operational principles and design of the SEM are quite different from those of the TEM (Figure 3.19). In scanning electron microscopy, a beam of electrons is rapidly scanned across the surface of the specimen, forming a raster (a set of straight lines, such as those used in a television set).

The primary electron beam knocks electrons out of the specimen surface, and the secondary electrons produced in this process are transmitted to a collector, amplified, and used to generate an image on a cathode ray tube (CRT) screen. The cathode ray tube is scanned across its screen by another beam of electrons, and the intensity of the electron beam used to illuminate the cathode ray tube is controlled by the number of electrons collected by the detector. The image shown on the CRT screen shows a shadowing effect that gives a three-dimensional appearance to the image, and the topography of the specimen surface is easily seen with the SEM (Figure 3.20).

Magnification in a scanning electron microscope is not achieved by using "lenses," as is the case in both light microscope and transmission electron microscopy. Magnification in the SEM is achieved by having different lengths of scan for the specimen and the CRT display and is determined by the ratio of the length of the scan across the specimen surface to the length of the scan of the CRT. If the electron beam scans a set of lines of 100 nm × 100 nm on the specimen and the image is displayed on a CRT screen of 100 mm × 100 mm, the magnification will be 100,000×. The primary beam that scans across the specimen is synchronized with the raster in the CRT display so that the image on the CRT screen is an accurate reproduction of the scanning image.

As with other forms of microscopy, resolution is essential for achieving useful magnification. Resolution of the SEM depends on the spot size of the screen and the size (diameter) of the primary electron beam. Most scanning electron microscopes are capable of achieving resolution in the 1–10 nm range, and thus, useful magnification of the SEM is of the range 10,000–100,000×.

Figure 3.19 The scanning electron microscope (SEM) is used for viewing surface structures and their three-dimensional spatial relationships. (A) Diagram of SEM; (B) photograph of SEM.

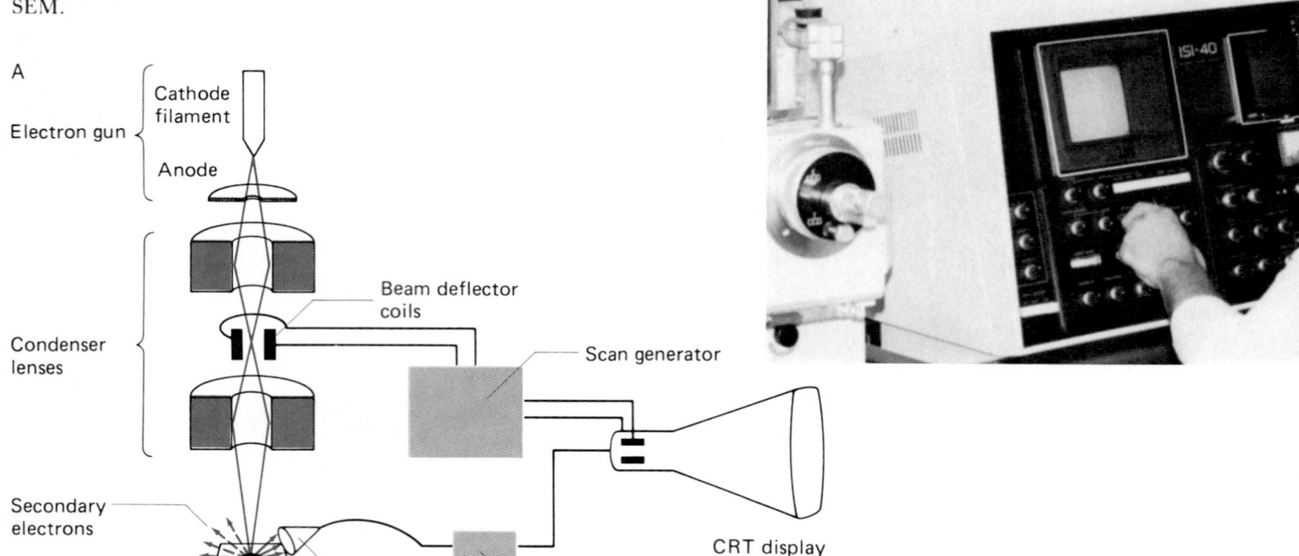

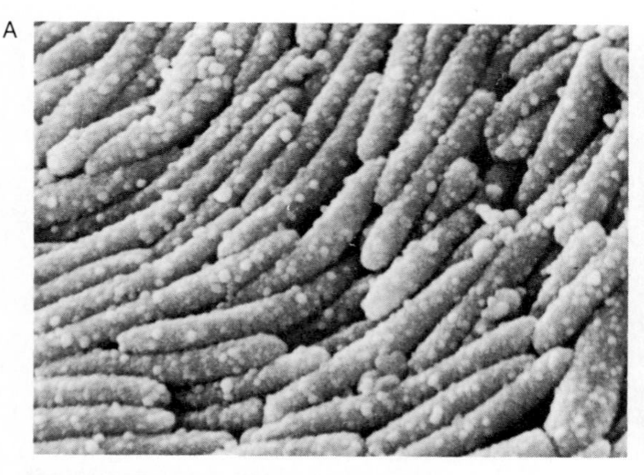

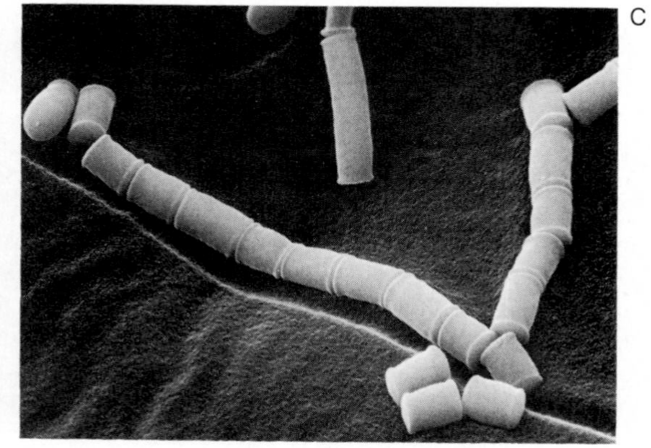

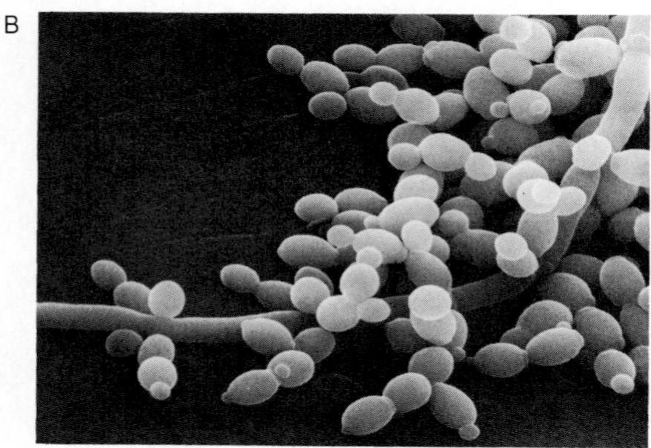

Figure 3.20 Scanning electron micrographs of microorganisms showing their three-dimensional appearance. (A) Note the bumps on the outer membranes of these rod-shaped bacteria; (B) *Candida albicans* yeast and hyphal phases (3400×); and (C) *Geotrichum candidum,* a fungus associated with food spoilage, forming arthrospores (3250×). (A from BPS: Z. Skobe, Forsythe Dental Center; B and C from BPS: Garry T. Cole, University of Texas, Austin.)

Summary

A variety of microscopes are used for viewing different microorganisms (Figure 3.21). The usefulness of a particular type of microscopy depends upon the ability to produce a magnified image of a microorganism that (1) is large enough to be seen; (2) has not been distorted by the method of sample preparation or viewing; (3) retains sufficient detail for resolution of the structures of interest; and (4) has sufficient contrast so that the organisms and the structures of interest can be distinguished from the surrounding background. Each type of microscope has its advantages and limitations (Table 3.1). Magnification is useful only as long as detail can still be resolved. Resolution is the shortest distance between two points that still permits visualization of the distinct points.

The light microscope is the basic tool of the microbiologist. This microscope has a series of ground glass lenses that focus light onto a specimen and bend the light that passes through the specimen to produce a magnified image. With the light microscope, images can be magnified 1000 times and structures larger than 200 nm can be resolved. The resolving power of the light microscope depends on the wavelength of the illuminating light source and the numerical aperature of the lens; the shorter the wavelength of light and the higher the numerical aperture, the better the resolving power. Modern microscopes are designed to minimize optical distortions such as spherical aberration, chromatic aberration, and curvature of field.

To enhance the contrast between the specimen and the surrounding background a variety of stains are employed. The Gram stain procedure is the most widely used staining procedure in bacteriology. This procedure differentially stains some taxonomic groups red-pink (Gram negative) and other groups blue-purple (Gram positive). Other differential

Figure 3.21 A diagrammatic representation of the types of specimens and their size ranges that can be viewed with different types of microscopes.

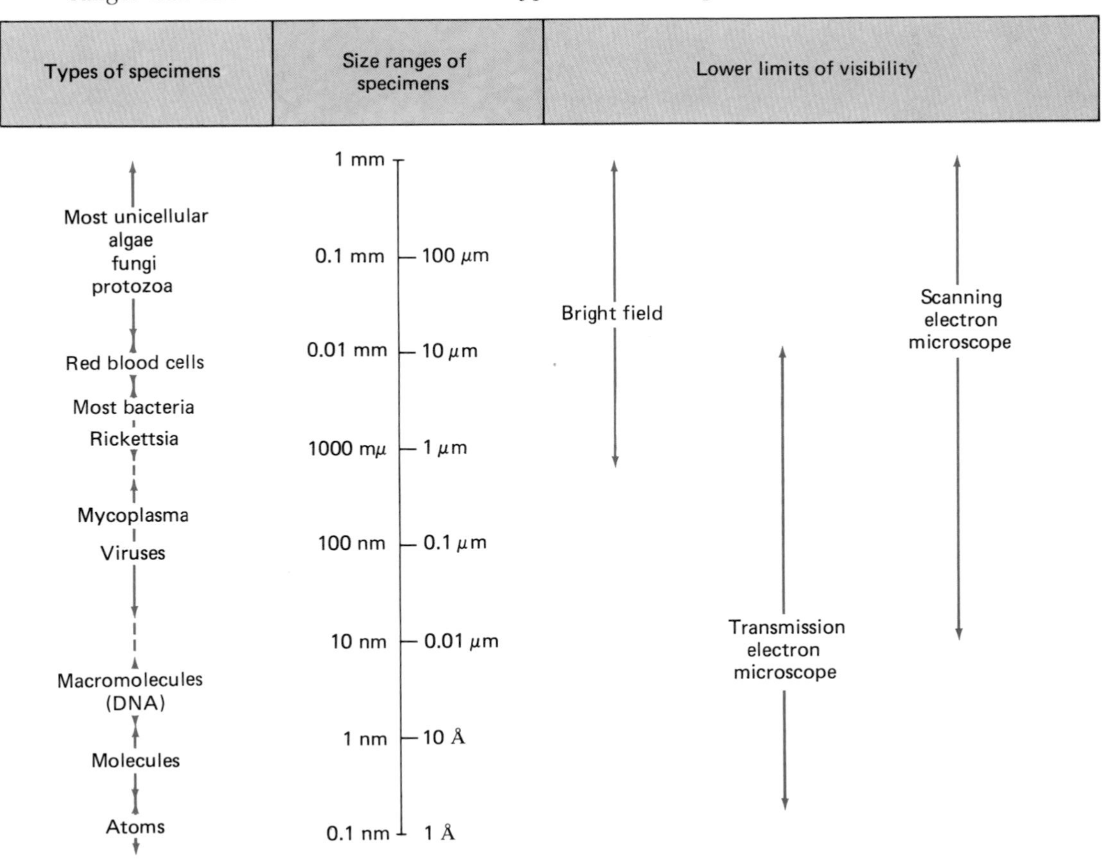

Table 3.1 **Comparison of Various Types of Microscopes**

Type of Microscope	Maximum Useful Magnification	Resolution	Comments
Brightfield	1500×	200 nm[a]	Extensively used for the visualization of microorganisms; usually necessary to stain specimens for viewing
Darkfield	1500×	200 nm	Used for viewing live microorganisms, particularly those with characteristic morphology; staining not required; specimen appears bright on dark background
Ultraviolet	2500×	100 nm	Improved resolution over normal light microscope; largely replaced by electron microscopes
Fluorescence	1500×	200 nm	Uses fluorescent staining; useful in many diagnostic procedures for identifying microorganism
Phase contrast	1500×	200 nm	Used to examine structures of living microorganisms; does not require staining
Interference	1500×	200 nm	Used to examine structures of microorganisms, particularly those near edges; produces sharp multicolored image with three-dimensional appearance
Transmission electron microscope (TEM)	500000–1000000×	1 nm	Used to view ultrastructure of microorganisms, including viruses; much greater resolving power and useful magnification than can be achieved with light microscopy
Scanning electron microscope (SEM)	10000–100000×	10 nm	Used for showing detailed surface structures of microorganisms; produces three-dimensional image

[a]nm = nanometer, 10^{-9} meter.

staining procedures have specific diagnostic value; for example, the acid-fast staining procedure is used to identify *Mycobacterium* species, including those that cause tuberculosis and leprosy.

For some applications fluorescent dyes are used to stain microorganisms and a fluorescence microscope is used for their visualization. The fluorescence microscope permits light of one wavelength to be focused onto the specimen and light of a different wavelength to be viewed. Certain immunofluorescent procedures, in which fluorescent dyes are coupled with specific components of the body's immune defense system, are widely used for the specific diagnosis of infectious diseases.

Some specialized microscopes are designed to eliminate the need for staining. The darkfield microscope relies upon the reflection of light striking the specimen at an oblique angle for visualization. In darkfield microscopy the specimen appears bright against a dark background. The phase contrast microscope is another type of microscope that is designed to achieve enhanced contrast without the need to stain the specimen. Living microorganisms can easily be seen swimming through a drop of water when viewed with the phase contrast microscope.

The smallest microorganisms and greatest amount of detail about microbial structure can be seen using the electron microscope. The short wavelength of an electron beam permits better resolution and hence higher useful magnifications than can be achieved with the light microscope. Viruses and even molecules can be seen with electron microscopes. There are two types of electron microscopes: transmission (TEM) and scanning (SEM). Both scanning and transmission electron microscopy require extensive preparatory procedures to prevent the production of artifacts. The transmission electron microscope is used to see finely detailed structures. The scanning electron microscope is most useful for observing surface structures. The many micrographs in this text reveal the information that can be obtained with these microscopes.

Study Outline

A. Light microscopy.

1. The microscope is the basic tool employed for the visualization of the microbial world.
2. Magnification is the product of magnifying power of ocular lens times the magnifying power of objective lens.
3. The typical light microscope used in microbiology magnifies 1000 times.
4. Modern microscope lenses correct for optical defects.
5. Resolving power is the degree to which the detail present in the specimen is retained in the magnified image.
6. Resolution is dependent upon the wavelength of light and the numerical aperture of the lens.
7. The best resolution for a light microscope is 200 nm.
8. The shorter the wavelength of the illuminating source, the better the resolving power.
9. The greater the numerical aperture of the lens, the better the resolving power.
10. Contrast is necessary to discern a structure from the surrounding background.
11. Staining is used to increase contrast.
12. In a positive simple stain procedure the dye is attracted to the microorganism, leaving a clear background.
13. In a negative simple stain procedure the dye is repelled by the microorganism, leaving the microbe clear against a colored background.
14. A differential stain procedure distinguishes (differentiates) one microorganism or structure from another or one microbial structure from the rest.
15. The Gram stain procedure is the most widely used differential staining procedure in bacteriology. The steps of this procedure are (a) primary staining with crystal violet; (b) applying a mordant, Gram's iodine, which increases the affinity of the primary stain for the bacterial cells; (c) rinsing with acetone–alcohol or other decolorization agent; (d) counterstaining with a red stain (usually safranin).
16. Acid-fast staining is useful in identifying members of the genus *Mycobacterium*.
17. In brightfield microscopy, the object is dark and the background is bright.
18. In darkfield microscopy, the object appears bright on a dark background.
19. The darkfield condenser focuses light on the specimen at an oblique angle so that only light that is reflected off the specimen will be seen.
20. In fluorescence microscopy objects are stained with dyes that fluoresce, that is, dyes that, when illuminated by light of one wavelength, give off light at a different wavelength; flurescent dyes can be conjugated with antibodies to produce immunofluorescent stains for immunofluorescent microscopy.
21. In phase contrast microscopy contrast is enhanced by shifts in the phase of light passing through the specimen.

B. Electron microscopy.

1. Electron microscopes use an electron beam instead of visible light.
2. Better resolution and higher useful magnification can be achieved with the electron microscope than with the light microscope.
3. There are two basic types of electron microscopes, the transmission electron microscope (TEM) and the scanning electron microscope (SEM).
4. The design of a transmission electron microscope is similar to the compound light microscope, but electromagnets are used for the lens systems instead of glass.
5. In a transmission electron microscope the electron beam passes through the specimen; the beam is enlarged by electromagnetic objective and projector lenses and projected onto a fluorescent screen or photographic film.
6. The theoretical resolution of the transmission electron microscope is about 0.2 nm, a thousand times better than a light microscope.

7. The useful magnification of most transmission electron microscopes is over 100,000x so that even the smallest microbes can be visualized.
8. Extensive sample preparation is necessary for electron microscopy in order to avoid artifacts and maximize observation of detail.
9. In a scanning electron microscope the electron beam is scanned across the surface of the specimen. The electrons emitted from the surface of the specimen are used to generate an image on a cathode ray tube (CRT).
10. The intensity of the image seen with scanning electron microscopy reflects the composition and topography of the specimen surface.
11. The useful magnification of a scanning electron microscope generally is 10,000–100,000x.
12. The resolution obtained with an SEM is 1–10 nm.

Study Questions

1. Find the *incorrect* definition among the following statements.
   a. Resolving power is the degree to which the detail present in the specimen is retained in the magnified image.
   b. Resolution is the closest space between two points at which they still can clearly be seen as separate points.
   c. Resolution is dependent upon magnification.
   d. Resolution is 200 nm for a light microscope.
2. Indicate whether the following statements are true (T) or false (F).
   a. Positive simple stain darkens the background, leaving a bright, clear microorganism.
   b. Methylene blue is an example of a positive simple stain.
   c. A differential stain distinguishes one microorganism or structure from another.
   d. In a negative stain, a basic stain is repelled by the negatively charged microorganism.
   e. Nigrosin and India ink are examples of differential stains.
   f. The Gram stain is the most widely used differential stain.
   g. The spore stain and acid-fast stain are examples of simple stains.
3. List the steps of the Gram stain.
4. Identify the type of microscopy described in each of the following statements.
   a. Object is dark, field is bright.
   b. Only light that is reflected off the specimen will be seen.
   c. Objects are stained with fluorescent dyes.
   d. Objects are living cells whose details are enhanced by contrast obtained by shifts in light phase.
   e. Usually do not see internal structures.
   f. Fluorescent dyes can be conjugated with antibodies.
   g. The most common type of compound microscope.
   h. Produces psuedo-three-dimensional images.
5. Indicate whether the following statements are true (T) or false (F).
   a. Scanning electron microscopy has a greater resolution than transmission electron microscopy.
   b. A fixative kills the specimen in such a way that its natural state is not affected.
   c. For most types of transmission electron microscopy, the specimen must be thin-sectioned.
   d. In scanning electron microscopy, the useful magnification is around 10,000–100,000x.
   e. In scanning electron microscopy, resolution usually is 10–100 nm.

Some Thought Questions

1. If laid end on end how many bacteria would be needed to stretch from the earth to the moon?
2. If Noah included two of every species in the ark, how much room was needed to house the bacteria?
3. How can you catch a bacterium?
4. If you were going to buy a microscope, what criteria would you use in its selection?
5. Why don't we use ultraviolet light rather than blue light to illuminate specimens for brightfield microscopy?
6. How would our view of microorganisms have been altered if Leeuwenhoek had developed an electron microscope?
7. How could you use fluorescent antibody technique to confirm that an oyster contained *Vibrio cholerae*?
8. How far would the average bacterium (2 micrometers, μm, length) have to swim to be equivalent to a 5-ft tall woman running a 5-km race?

Additional Sources of Information

Collins, C. H., and P. M. Lyne. 1984. *Microbiological Methods*. Thornton Butterworth, Ltd., Woburn, Massachusetts. Provides simple descriptions of the methods used for the preparation of specimens for microscopy.

Gabriel, B. L. 1982. *Biological Electron Microscopy*. Van Nostrand Reinhold Company, New York. Thorough well-explained discussion of principles, design, applications, and methods for transmission electron microscopy. Includes a handy compilation of protocols for preparation of biological specimens for examination by transmission electron microscopy.

Gerhardt, P. (ed.). 1981. *Manual of Methods for General Bacteriology*. American Society for Microbiology, Washington, D.C. Extensive descriptions of procedures used for different microscopical applications.

Gray, P. 1981. *Encyclopedia of Microscopy and Micro-Technique*. Robert E. Krieger Publishing Co., Melbourne, Florida. A well-illustrated work describing the various types of microscopes and microscopical methods.

Hayat, M. A. 1978. *Introduction to Biological Scanning Electron Microscopy*. University Park Press, Baltimore. A comprehensive work on the design, principles, and applications of scanning electron microscopy.

James, J. 1976. *Light Microscopic Techniques in Biology and Medicine*. Nijhoff Medical Division, The Hague. Higham, Massachusetts. A good discussion of the applications of light microscopy.

Kessel, R. G., and C. Y. Shih. 1976. *Scanning Electron Microscopy in Biology*. Springer Verlag, Berlin. A stunning collection of scanning electron micrographs of biological specimens including microbes. Includes an easy-to-read discussion of the principles of scanning electron microscopy and the methods used for sample preparation.

Meek, G. A. 1976. *Practical Electron Microscopy for Biologists*. John Wiley & Sons, Inc., London. A comprehensive technical discussion of electron microscopy.

Sieburth, J. M. 1975. *Microbial Seascapes*. University Park Press, Baltimore. A magnificent collection of micrographs of microorganisms showing the beauty of the microbial world. Includes a discussion of the principles of microscopy.

Wischnitzer, S. 1981. *Introduction to Electron Microscopy*. Pergamon Press, Inc., New York. A thorough primer on the principles of electron microscopy.

Yoshii, Z., J. Tokunaga, and J. Tawara. 1976. *Atlas of Scanning Electron Microscopy*. The Williams & Wilkins Company, Baltimore. A beautiful collection of scanning electron micrographs of microorganisms. Includes a discussion of scanning electron microscopy.

4 *The Structure of Prokaryotic (Bacterial) Cells*

KEY TERMS

active transport
ATP (adenosine triphosphate)
bacterial chromosome
bilipid structure
capsule
cell wall
chemiosmosis
chemotaxis
differential permeability
diffusion
endospores
endotoxin
envelope
F pilus
facilitated diffusion
flagella (singular, flagellum)
fluid mosaic model
gas vacuoles
genome
glycocalyx
LPS (lipopolysaccharide)
lyse
magnetosomes
magnetotaxis
metachromatic granules
murein
nitrifying bacteria
nucleoid region
osmosis
osmotic shock
peptidoglycan
periplasm
peritrichous flagella
permeability
phagocytosis
phospholipid
phototaxis
pilus
plasmids
polar flagella
protoplast
ribosomes
mRNA (messenger RNA)
rRNA (ribosomal RNA)
slime layer
spheroplast
Svedberg unit (S)
volutin

OBJECTIVES

After reading this chapter you should be able to

1. Define the key terms.
2. Describe the structure and functions of the cytoplasmic membrane.
3. Describe the structure and function of the bacterial cell wall.
4. Compare the cell-wall structures of Gram negative and Gram positive bacteria.
5. Describe how bacteria move.
6. Describe the properties of the bacterial endospore and its importance for establishing effective sterilization procedures.
7. Describe the structures involved in bacterial attachment.
8. Discuss the role of the bacterial capsule and its importance related to infectious diseases.
9. Describe the function and structure of the ribosome.
10. Discuss the practical importance of the differences between the ribosomes of prokaryotic and eukaryotic cells.
11. Describe the specialized functions of internal membrane structures of prokaryotic cells.

Prokaryotic Cells

The prokaryotic or bacterial cell is relatively simple structurally. The lack of separation of the bacterial genome (complete set of genetic information) within a specialized organelle, the nucleus, distinguishes the prokaryotic bacterial cell from the eukaryotic cells of all other organisms. However, as with other cellular organisms, the essential metabolic reactions of the living system occur within the confines of the cytoplasmic membrane of a prokaryotic cell.

Long ago it became evident that the key to every biological problem must finally be sought in the cell; for every living organisms is, or at some time has been, a cell.

—E. B. WILSON

Cytoplasmic Membrane

The cytoplasmic membrane, the boundary layer of the cell, is a differentially permeable barrier; that is, the movement of molecules across the cytoplasmic membrane is selectively restricted. Transport of materials into and out of the cell is regulated by the cytoplasmic membrane. Small molecules such as water move across the membrane quite readily; the passage of medium sized molecules such as glucose across the membrane is restricted; and very large molecules such as cellulose cannot pass through the membrane barrier. The membranes themselves mediate the selective transport process, and the specific biochemical structure of the membrane determines which molecules can enter and leave the cell.

The Structure of the Cytoplasmic Membrane. The cytoplasmic membrane is composed largely of phospholipid. As the name implies, a phospholipid molecule contains two parts, a lipid or fat-like portion and a phosphate portion. The lipid portion is hydrophobic (water-fearing) and, like an oil, does not mix with water. The phosphate portion of the molecule, on the other hand, is hydrophilic (water-loving) and mixes freely with water. The fact that phospholipid molecules have polar (hydrophilic) and nonpolar (hydrophobic) ends establishes the basis for the orientation of the membrane around the cell.

The normal cytoplasmic membrane has a bilipid structure; that is, there are two layers of phospholipid (Figure 4.1). The hydrophobic ends of the phospholipids orient toward each other and form the internal matrix of the membrane; the hydrophilic ends point away from each other, with one layer of polar ends pointing away from the cell and the other layer of polar ends pointing to the cell's cytoplasm—the fluid contents within the cell. When viewed with the transmission electron microscope, the cytoplasmic membrane has a railroad track ap-

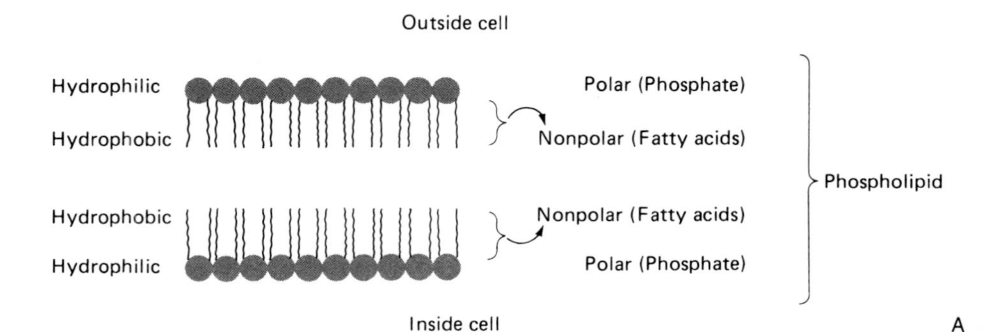

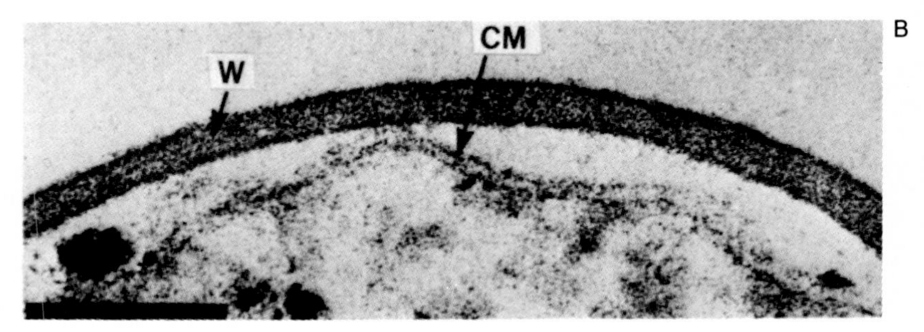

Figure 4.1 (A) The typical cytoplasmic membrane is a bilipid membrane. This illustration shows the orientations of the hydrophilic and hydrophobic ends of the phospholipids that make up this structure. (B) The railroad-track-like appearance of the cytoplasmic membrane of *Bacillus subtilis* is seen in this electron micrograph: bar = 100 nm, W = cell wall outside of membrane, CM = cytoplasmic membrane. (From BPS: T. J. Beveridge, University of Guelph, Ontario.)

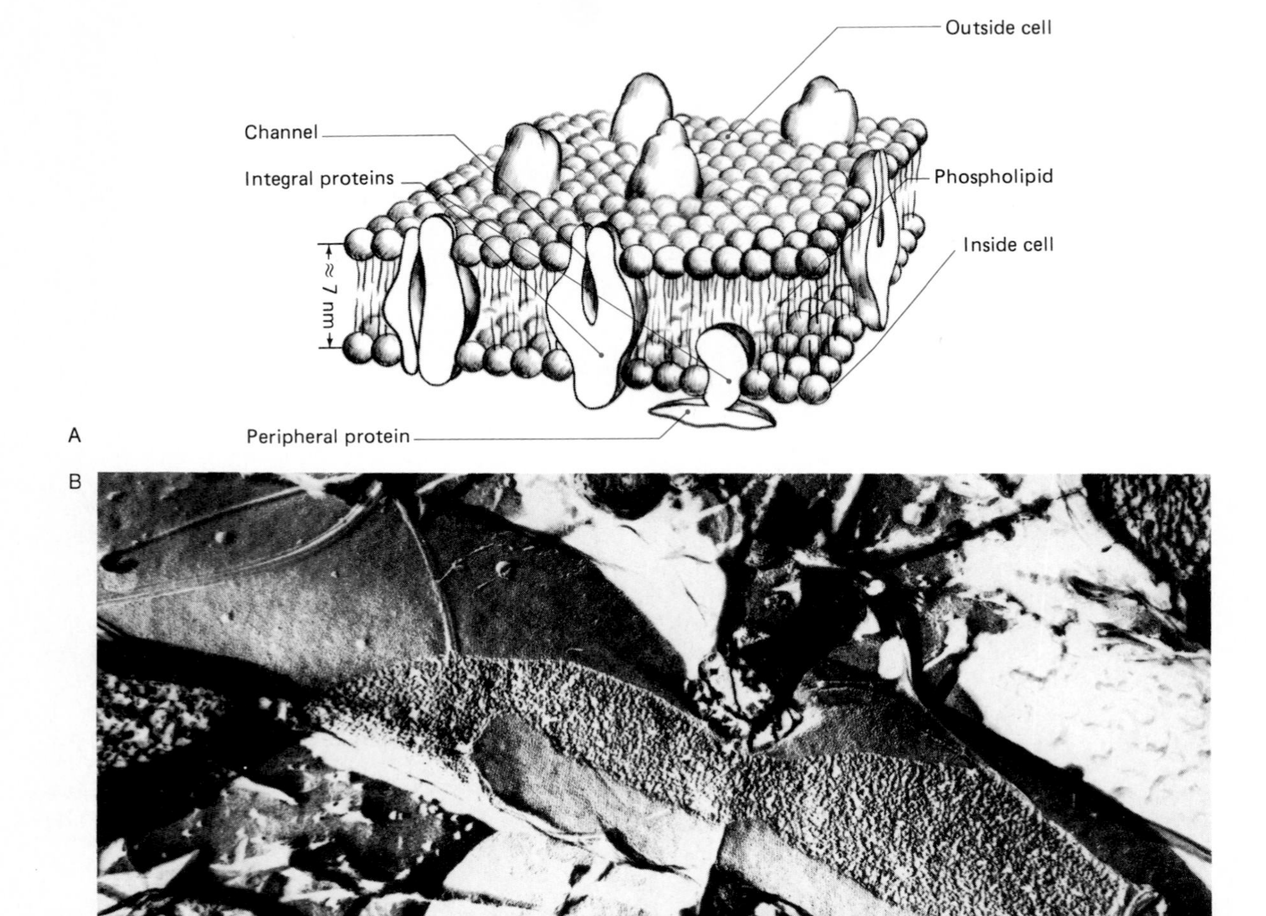

Figure 4.2 (A) The fluid mosaic model of membrane structure accounts for the fact that proteins as well as phospholipids form an integral part of membranes and that the structure is dynamic as opposed to static in nature. (B) This electron micrograph of a freeze-fractured and etched *Aquaspirillum putridiconchylium* shows visible particles (proteins), which were exposed by the freeze-fracturing, arrayed in a mosaic pattern protruding from the membrane; width of cell = 0.6 μm. (From BPS: T. J. Beveridge, University of Guelph, Ontario.)

pearance; the dark rail-like portions of the membrane correspond to the electron-dense hydrophobic portions of the phospholipid molecule.

In addition to the lipid constituents of the cytoplasmic membrane, there are other biochemicals, including proteins, that are integrated into or associated with the basic membrane structure. The protein molecules are used for transport of nutrients and wastes across the membrane, as well as for other essential physiological functions.

Several models have been proposed to explain the relationship between membrane structure and function. The currently accepted fluid mosaic model of membrane structure allows for the movement of proteins within the phospholipid matrix of the membrane. According to this model, the membrane is a bilipid layer with the proteins associated with the membrane distributed in a mosaic pattern, both on the surfaces and in the interior of the membrane (Figure 4.2). The structure of the membrane is not viewed as static. Lipids held together by weak bonds can move laterally through the fluid membrane matrix. Proteins can also move laterally, but to a lesser extent than the phospholipid molecules. The fluid mosaic model accounts for the biochemical and microscopical analyses of the membrane which have revealed the integral and dynamic relation between protein and phospholipid molecules in the membrane matrix, and also explains the permeability properties of the cytoplasmic membrane.

Movement of Substances across the Cytoplasmic Membrane. The primary function of the cytoplasmic membrane is to regulate the flow of material into and out of the cell. There are several different ways by which molecules can enter or leave a cell, including diffusion and active transport (Figure 4.3). It is the properties of the membrane that

Figure 4.3 This figure illustrates several ways for substances to cross a cell membrane and enter a cell. In cytosis the substance is transported into the cell without actually passing through the membrane. Diffusion across a membrane occurs when substances can pass through the pores of the membrane and when there is a favorable concentration gradient; this type of transport represents the downhill flow of a substance along a concentration gradient. In contrast to diffusion, active transport can occur along an unfavorable concentration gradient but requires the input of energy. In some cases the required energy is supplied by hydrogen ion pumping (chemiosmosis), in other cases the hydrolysis of ATP is used to drive the pumping substances across the membrane. A special form of active transport, **group translocation**, occurs exclusively in prokaryotes; in group transport the substrate is chemically modified during transport; that is, the substance is phosphorylated during transport across the membrane with the energy derived from phosphoenolpyruvate.

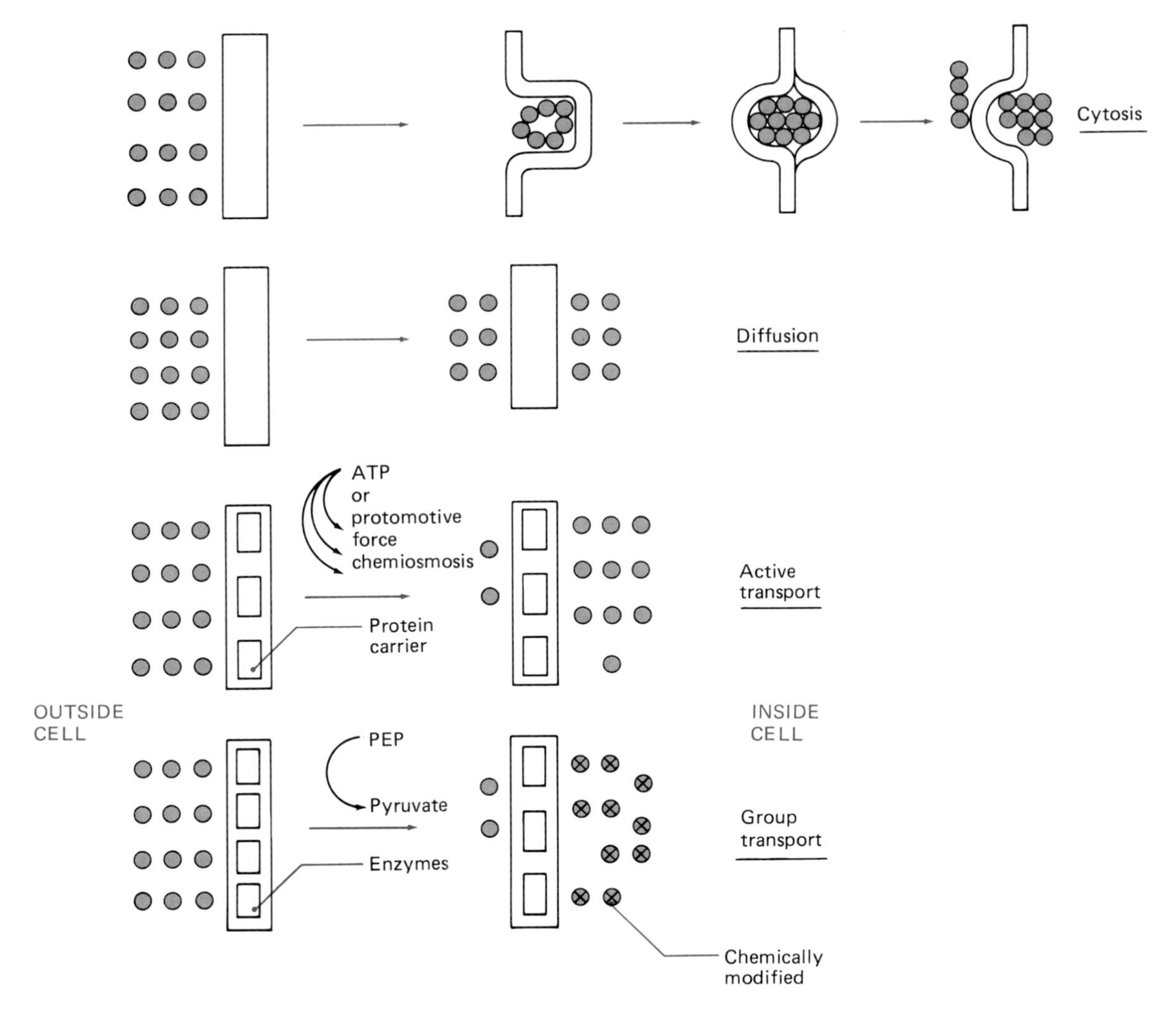

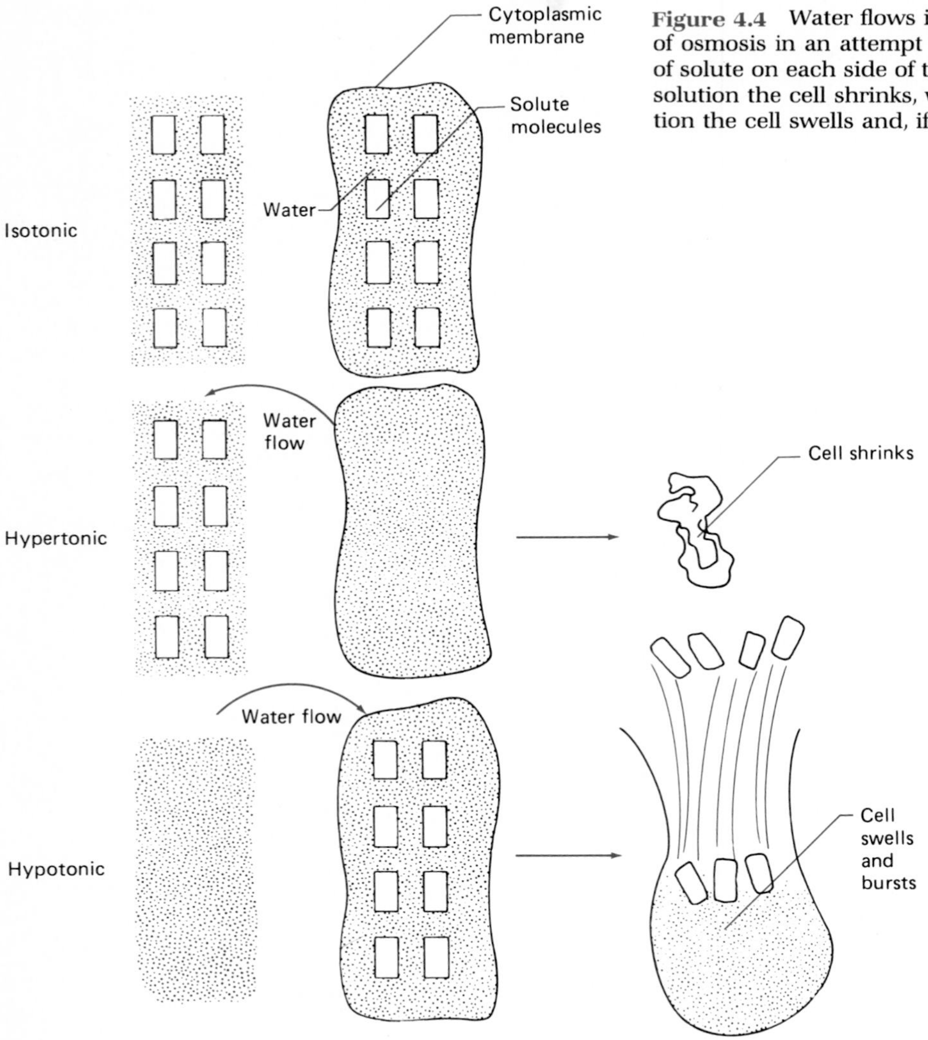

Figure 4.4 Water flows into and out of cells because of osmosis in an attempt to balance the concentration of solute on each side of the membrane. In a hypertonic solution the cell shrinks, whereas in a hypotonic solution the cell swells and, if unprotected, bursts.

regulate these different transport mechanisms, establishing which materials can be translocated across the membrane. Organisms use different transport mechanisms for moving various biochemicals into and out of the cell and have different capabilities for transporting molecules across the cytoplasmic membrane.

When there is a difference in the concentrations of a solute on either side of a membrane, water will move across the membrane from the region of lower to the region of higher concentration of solute until the concentrations of the solute are equalized on both sides of the membrane, or until a pressure force prevents further flow of the water. The process by which the water crosses the membrane in response to the concentration gradient of the solute is known as osmosis. Osmosis may be viewed as a form of passive diffusion in which a small water molecule can cross a membrane, but the larger solute molecules cannot freely move across this barrier. This process exerts a pressure known as the osmotic pressure on the membrane, which represents the force that must be exerted to maintain the concentration differences between solutions on opposite sides of the membrane.

Now, let us examine the consequences of osmosis for a microbial cell. In a medium where the solute concentration inside the cell is equal to the solute concentration outside the cell, water will flow equally in both directions across the membrane (Figure 4.4). However, in a medium where the solute concentration is higher (hypertonic) outside the cell than inside the cell, water will tend to flow out of the cell and the cell will tend to shrink. The reverse is true if the cell is in a medium where the solute concentration is lower (hypotonic) outside of the membrane than inside of the cell, in which case water will tend to flow into the cell, causing the cell to expand and—if unrestricted—the cytoplasmic membrane to burst. Microorganisms usually find themselves in this latter situation because the various biochemicals of the cell normally are in considerably higher concentrations than the solutes in the dilute aqueous medium in which most microorganisms exist.

In addition to water, various small solute molecules can diffuse passively across the membrane (see Figure 4.3). Passive diffusion occurs when a solute molecule moves from an area of higher concentration on one side of the membrane to an area of lower concentration on the other side of the membrane. The rates of passive diffusion are determined by the concentration gradient; the greater the gradient, the more rapid the movement. The rates of simple diffusion across the cytoplasmic membrane are not rapid enough for many of the exchanges between biochemicals and surroundings that must be accomplished by a cell, and microorganisms have developed additional membrane transport mechanisms.

Some molecules can move across a membrane at higher rates than would occur by simple diffusion in a process referred to as facilitated diffusion. Facilitated diffusion involves some of the proteins associated with the cytoplasmic membrane. Some of the membrane proteins—called carrier proteins or permeases—selectively increase the permeability of the membrane for specific substances. In both simple and facilitated diffusion movement is from a region of higher concentration to a region of lower concentration.

Substances may also move across the membrane against a concentration gradient by active transport. Active transport can be likened to a pump that requires energy to move water uphill. The cell must actively work to move substances up a free energy gradient and therefore, unlike diffusion, active transport requires that the cell expend energy to move biochemicals across the membrane against a concentration gradient (Figure 4.5). As with most energy-requiring activities of living systems, the energy for active transport may come from the conversion of ATP (adenosine triphosphate) to ADP (adenosine diphosphate). ATP, a universal carrier of energy in living systems, is one of several energy sources used by different organisms to drive active transport. The active transport capabilities of a cell determine in large part what substrates can move into a cell and thus help define the metabolic potential of that cell.

Additional Functions of the Bacterial Cytoplasmic Membrane. Because the prokaryotic cell lacks internal membrane-bound organelles, several functions associated with such organelles in eukaryotic organisms are associated with the cytoplasmic membrane in prokaryotes. For example, the cytoplasmic membrane of bacteria plays a critical role in the generation of ATP during respiration. Specific proteins associated with the cytoplasmic membrane of the prokaryotic cell establish channels through which hydrogen ions are pumped to establish a concentration gradient. In accordance with fundamental chemical principles, hydrogen ions will move by diffusion toward a region of lower concentration, which in this case would be to move back across the membrane into the cell. The return flow of ions into the cell is channelled through protein pores associated with the enzyme ATPase. This enzyme catalyzes the formation of the energy-rich compound ATP.

The process that drives the formation of ATP is known as chemiosmosis (see Chapter 7 and Figure 7.8 for a discussion of ATP generation and chemiosmosis). It was first hypothesized by Peter Mitchell in 1965. Chemiosmosis is now recognized as fundamental to the generation of ATP by both cellular respiration and photosynthesis. In eukaryotic cells, the chemiosmotic generation of ATP during oxidative phosphorylation—a process in cellular respiration—occurs within a specialized organelle—the mitochondrion—during respiration and the chemiosmotic generation of ATP during oxidative photophosphorylation—another process for generating ATP—occurs within the chloroplast during photosynthesis. The involvement of membranes in these ATP-generating processes is essential, as it provides the necessary semipermeable boundary across which an electrochemical gradient can be established. By performing work to pump hydrogen ions across the membrane, the return flow can be efficiently coupled with the synthesis of ATP. The process works like an electricity generating station, where water is pumped up across a dam and the return gravitational flow of water is routed through specific channels so that it turns a turbine. The work of turning the turbine generates electricity that can be used elsewhere. In the case of a cell, the ATP formed at the membrane by chemiosmosis can then be an energy source within the cell.

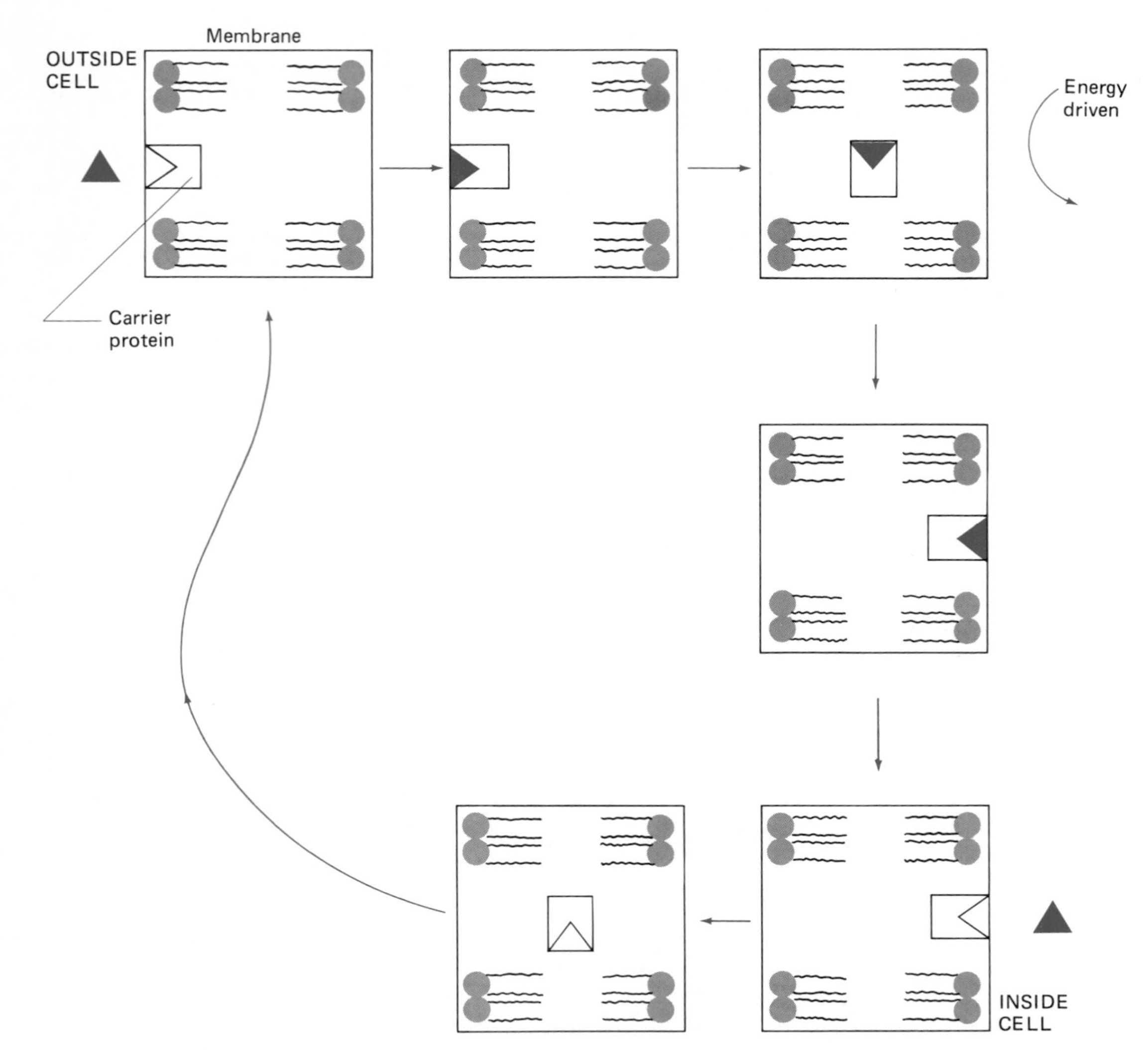

Figure 4.5 Active transport requires temporary binding of the substance with a carrier protein during transfer across the membrane and energy activation of the membrane from chemiosmosis or the hydrolysis of ATP.

Figure 4.6 The bacterial cell wall is found outside the cytoplasmic membrane, protecting the cell from various types of damage, such as bursting owing to osmotic pressure. The periplasmic space between the cell wall and cytoplasmic membrane is an area of active enzymatic activity.

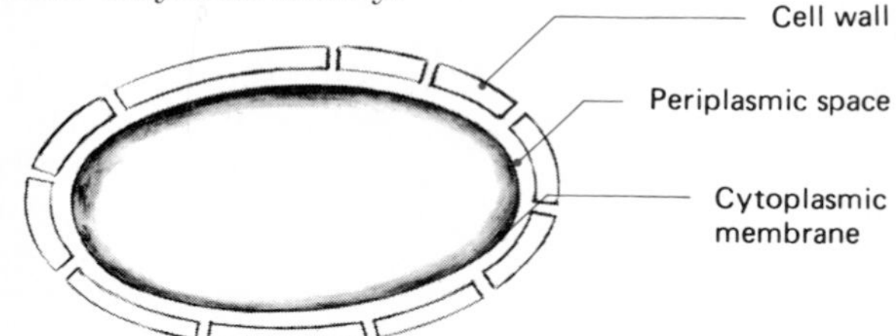

Bacterial Cell Wall

A cell wall surrounds the cytoplasmic membrane of almost all prokaryotic cells (Figure 4.6). The cell wall is a very important structure because it protects the cell against osmotic shock. Without a cell wall the cells of most microorganisms would burst from the osmotic pressure exerted on their cytoplasmic membranes, because these organisms normally exist in dilute aqueous environments.

The cell wall is generally a relatively porous structure that does not restrict the flow of small molecules to or from the cytoplasmic membrane, although very large polymers generally are unable to pass across the cell wall. Some enzymes secreted by the cell can be trapped in the periplasmic space, the region between the cell wall and the cytoplasmic membrane; the periplasm, therefore, can be a region of intense enzyme activity. Cell-wall structures are normally rigid. This rigidity of the prokaryotic cell wall is responsible for maintaining the specific shape of a bacterium. Bacteria owe their shapes, which are characteristic of particular species, to their cell-wall structure. Commonly, bacteria occur as spheres called cocci and cylinders called rods, although, as can be seen in the micrographs throughout this book, many diverse forms typify different bacterial species.

The bacterial cell wall is a biochemically unique structure. The cell wall of all but a few bacteria contains murein, which is also known as peptidoglycan or mucopeptide. This peptidoglycan layer is not found in any eukaryotic organism. As the name peptidoglycan implies, the biochemical contains two parts: a peptide portion made up of amino acids, and a glycan portion composed of carbohydrates. Murein is composed of a backbone of alternately repeating units of two carbohydrate derivatives, the amino sugars N-acetylglucosamine and N-acetylmuramic acid (Figure 4.7). These repeating amino sugars form the glycan portion of the molecule. Attached to some of the N-acetylmuramic acid units is a short peptide chain, consisting of four amino acids. Some of the amino acids occurring in the peptide portion of the molecule are relatively unusual in biological systems. In protein we only find 20 amino acids, all of which have an L configuration (analogous to all being left handed). Both L- and D-amino acids occur in murein (analogous to having both left- and right-handed molecules).

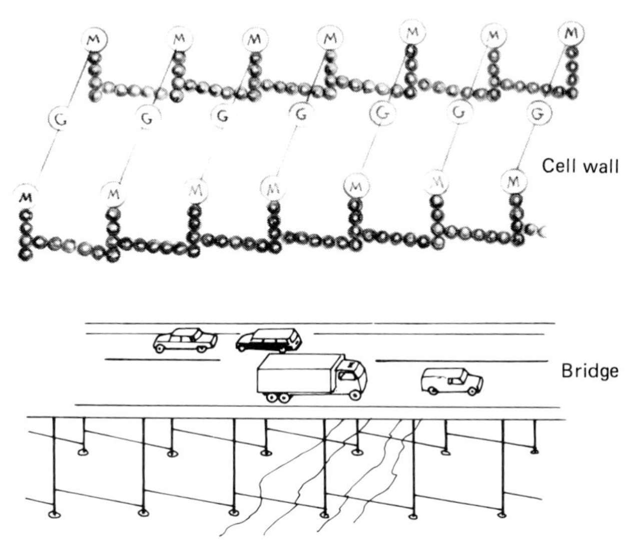

Figure 4.7 Murein or peptidoglycan is the backbone biochemcial of the bacterial cell wall; it is composed of repeating alternating units of N-acetylglucosamine and N-acetylmuramic acid and has crosslinked short peptide chains, some of which have unusual amino acids.

The short peptide chains that are attached to the N-acetylmuramic acid molecules are themselves interlinked by peptide bridges. This crosslinked peptidoglycan layer can be likened to a girder-supported bridge and in fact, the crosslinkages are referred to as bridges. Without the crosslinkage of the peptide chains, the murein layer would not be rigid and would not protect the cell against osmotic shock. When this crosslinkage is disrupted, the cell wall is defective and cannot adequately protect the bacterial cell against osmotic shock.

Some antibiotics act by disrupting the underlying biochemical structure that gives the bacterial cell wall its strength and rigidity. The reason that the antibiotic penicillin is effective in controlling bacterial infections is because penicillin prevents the formation of crosslinkages between the peptides. This results in the production of defective cell walls and the death of growing bacteria. Penicillin, however, does not destroy existing crosslinkages of the peptidoglycan layer and thus has no effect on bacteria that are not growing.

One bacterial genus, *Mycoplasma*, lacks a cell wall entirely. This particular bacterial genus is also unusual in that it incorporates sterols, which it normally acquires from a eukaryotic host cell, as a component of its cytoplasmic membrane. Members of the genus *Mycoplasma* will not be inhibited by penicillin because they lack a cell wall and thus lack the biochemical that this antibiotic affects. *Mycoplasma pneumoniae* causes atypical pneumonia, a disease that requires antibiotics other than penicillin for effective treatment.

In contrast to penicillin, the enzyme lysozyme will degrade a preformed peptidoglycan molecule. Lysozyme acts by breaking the glycan rather than the peptide portion of the peptidoglycan layer. This enzyme is produced by various organisms that consume bacterial cells, aiding in the digestion of the bacteria. Lysozyme also occurs as part of various normal body secretions, such as tears, providing protection against would-be bacterial invaders. By using lysozyme, it is possible to remove all or part of the cell-wall structure. If the bacterial cell remains intact following the partial removal of the

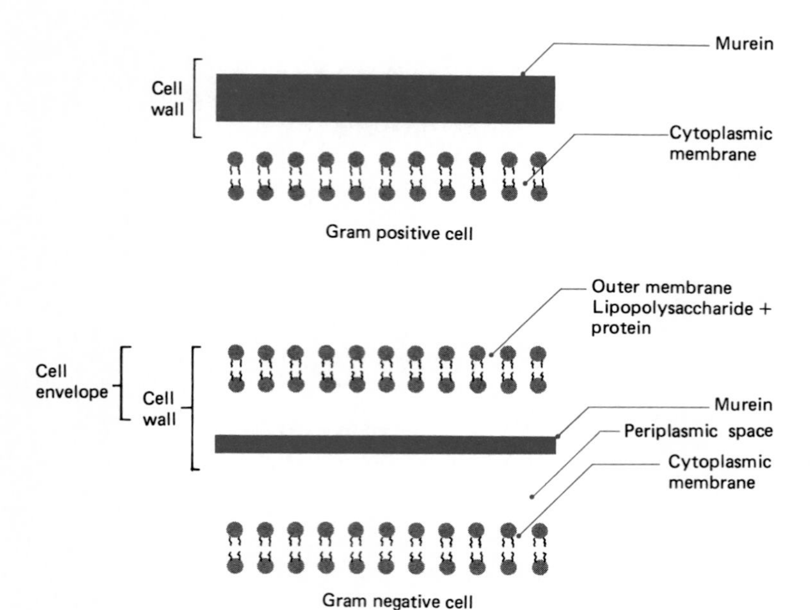

Figure 4.8 Gram positive and Gram negative bacteria differ in the structure of their cell walls. The Gram positive cell wall has a relatively thick murein layer. The Gram negative cell wall has a thin murein layer but also has an outer membrane and additional lipopolysaccharides and proteins not present in Gram positive cell walls.

cell wall, it is referred to as a spheroplast; if the wall has been completely removed, such an intact bacterial cell is called a protoplast. Both protoplasts and spheroplasts can exist in a supporting medium of high solute concentration in which the osmotic pressure is not high enough to **lyse**, or burst, the cell.

The Gram Negative and Gram Positive Bacterial Cell Walls. As discussed in Chapter 3, bacteria can be differentiated by using the Gram stain reaction. The difference in staining reaction reflects differences in cell-wall structure. Regardless of the staining reaction, which sometimes gives anomalous results depending on the condition and age of the bacteria, the difference between Gram negative and Gram positive bacteria is defined by the inherently different structure of the cell wall in these two groups (Figure 4.8). Both Gram positive and Gram negative cell-wall structures contain a peptidoglycan layer, but they have different amounts of peptidoglycan; further, they differ in which biochemicals, in addition to murein, form the cell-wall complex.

The Gram positive cell wall has a peptidoglycan layer that is relatively thick and comprises approximately 90 percent of the cell wall (Figure 4.9). The murein layer acts to fulfill the primary protective function of the cell wall. It is this thick murein layer that is believed to trap the primary stain within the periplasmic space during the Gram stain procedure, preventing the primary stain from being washed out when the cells are treated with a decolorizing agent.

In contrast to the relatively simple Gram positive bacterial cell wall, which is composed mostly of murein, the Gram negative cell wall is biochemically far more complex (Figure 4.10). The peptidoglycan layer of the Gram negative cell wall is very thin and often comprises only 10 percent of the cell wall. The cell wall of Gram negative bacteria contains lipids, polysaccharides, and proteins, which form an outer layer of the wall. This outer layer is sometimes referred to as the cell envelope

The Gram negative bacterial cell wall effectively forms a complex with an outer membrane layer. In reality Gram negative bacteria possess two mem-

Figure 4.9 Electron micrograph of the cell wall of *Bacillus subtilis*, a typical Gram positive cell wall with a thick murein layer (227,000 ×). (From BPS: T. J. Beveridge, University of Guelph, Ontario.)

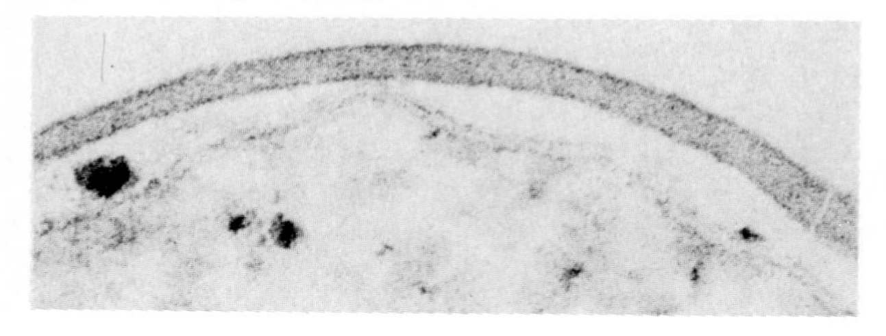

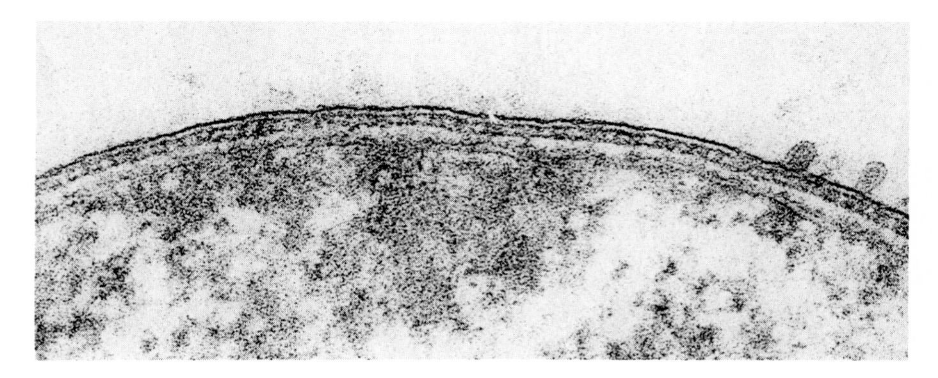

Figure 4.10 Electron micrograph of the cell wall of *Escherichia coli*, a typical Gram negative cell wall with a thin murein layer and an outer membrane (220,000×). (From BPS: T. J. Beveridge, University of Guelph, Ontario.)

branes—the normal cytoplasmic membrane and the additional membrane of the cell envelope layer surrounding the cell wall—that regulate the flow of material into and away from the cell. In addition to the normal membrane phospholipids, the outer membrane of Gram negative bacteria contains lipopolysaccharides (LPS). Lipopolysaccharides are molecules containing both lipid and carbohydrate components. The LPS of the Gram negative outer wall is known as endotoxin and is associated with disease symptoms such as those that characterize traveler's diarrhea.

Functionally, the outer membrane of the Gram negative bacterial cell is a coarse molecular sieve. Despite its permeability to small molecules, the outer membrane is less permeable than the cytoplasmic membrane to certain molecules. Therefore, Gram negative bacteria are less sensitive to some antibiotics than are Gram positive bacteria. Both because of the endotoxin and the increased resistance to antibiotics resulting from the biochemical nature of their cell walls, Gram negative organisms are more prominent than Gram positive bacteria in human infections.

Bacterial Capsule

The cell wall is not always the outermost layer of a cell. Some bacteria form a capsule external to the cell wall (Figure 4.11). The capsule is composed of polysaccharides and/or proteins. The capsule is especially important in protecting bacterial cells against phagocytosis, that is, against engulfment and digestion by white blood cells. The presence of a capsule can be a major factor in determining the pathogenicity of a bacterium, that is, the ability of a bacterium to cause disease in the organism that it infects. In some cases a bacterial species will have two variants, one that forms a capsule and is a viru-

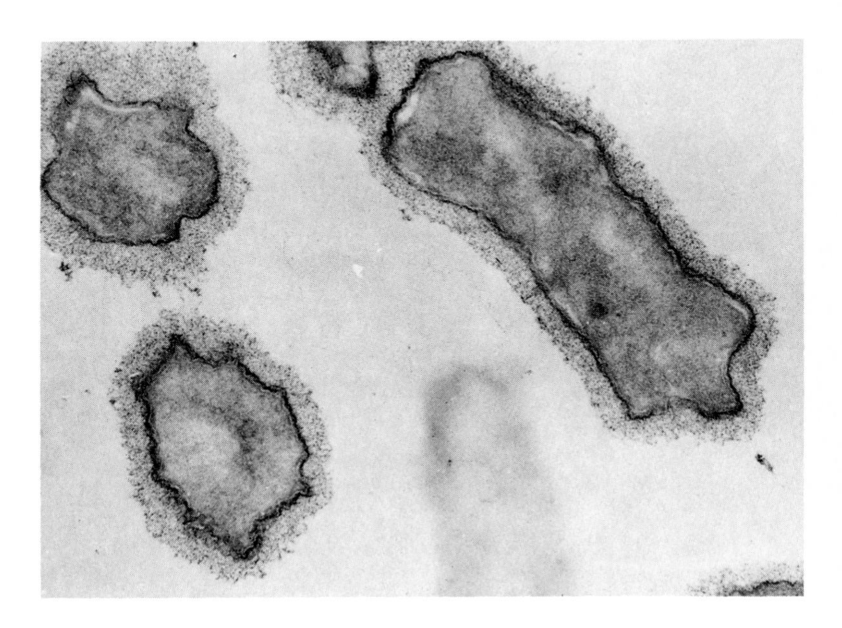

Figure 4.11 Photomicrograph showing the thick capsule of *Klebsiella aerogenes* (58,000×). (From BPS: Stanley C. Holt, University of Texas Health Science Center, San Antonio, Texas.)

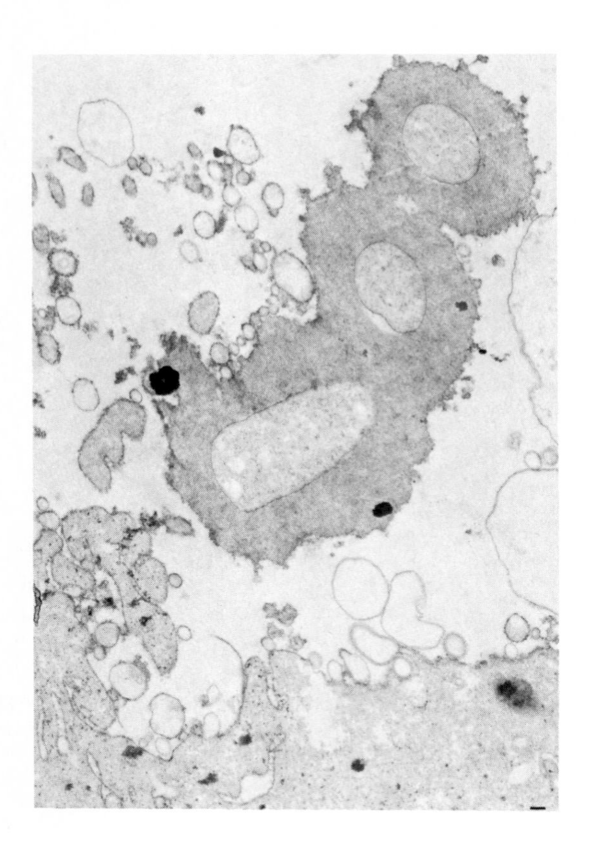

Figure 4.12 The bacterial glycocalyx is involved in the attachment of bacteria to solid surfaces; in this electron micrograph the stabilized glycocalyx surrounds enteropathogenic *E. coli* cells. (Courtesy William Costerton, University of Calgary, Alberta.)

lent (disease-causing) pathogen, and a nonencapsulated form that is avirulent and does not cause disease. The reason for this is that the nonencapsulated bacteria are subject to phagocytosis by blood cells involved in the host defense response of the infected organism. On the other hand, phagocytizing blood cells involved in the host defense response are unable or less able to adhere to, to engulf, and to digest those bacteria that have a capsule.

Slime Layers and the Glycocalyx

Slime layers are similar to capsules but are not as tightly bound to the cell. The outer slime layer is also known as a glycocalyx. The glycocalyx is a mass of tangled fibers of polysaccharides or branching sugar molecules surrounding an individual cell or a colony of cells. This external layer may protect the cell against dehydration (loss of water) and loss of nutrients. In some cases the slime layer appears to act as a trap. The viscosity of the slime restricts the movement of substrates away from the cell. The glycocalyx may also act to bind

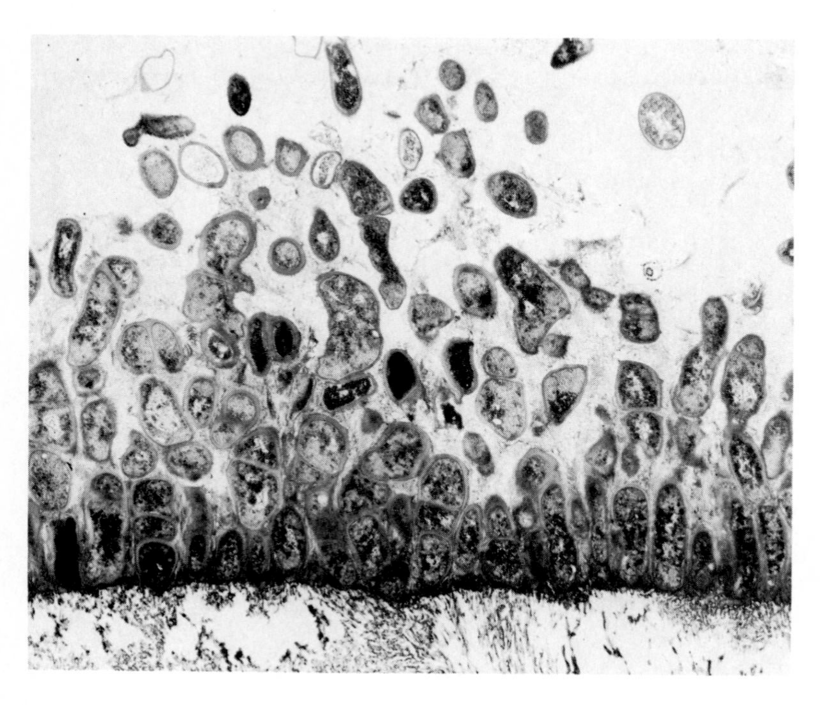

Figure 4.13 The slime layer of *Streptococcus* species permits the attachment of bacterial cells to tooth surfaces and the formation of dental plaque. This electron micrograph, showing bacteria in plaque and actinomycetes in the slime (plaque) layer, exemplifies the normal microbiota associated with human tooth enamel (13,600 ×). (From BPS: Max Listgarten, School of Dental Medicine, University of Pennsylvania.)

cells together, forming multicellular aggregates. Additionally, the glycocalyx of some bacteria are involved in attachment to solid surfaces (Figure 4.12). Some bacteria in aquatic habitats, for example, appear to be held to rocks through the slime layers they secrete. Bacteria occurring in the oral cavity on the surfaces of teeth form a polysaccharide slime, dental plaque, which enables them to adhere to the tooth (Figure 4.13). This adherence to the tooth surface is important in the formation of dental caries.

Pili

In addition to the glycocalyx, pili appear to be involved in attachment processes (Figure 4.14). Pili are short hair-like projections, composed primarily of protein, that emanate from the surface of some bacteria. There appear to be several different types of pili that may be associated with the bacterial surface, each serving a different function, but attachment is central to these different functions. The F or sex pilus is involved in bacterial mating and is found exclusively on the donor or male cells. Mating pairs cannot form in the absence of an F pilus or if the bridge established by the F pilus between the donor and recipient cell is interrupted. Pili also act as the receptor sites for bacteriophage, providing these bacterial viruses with a site of attachment to the bacterial cell. The phage attach to the pili and subsequently transfer their genetic information to the bacterial cell. Pili further have been implicated in the ability of bacteria to recognize specific receptor sites on host cell membranes. The pili appear to allow the bacteria to attach to and colonize the host cells, sometimes leading to disease in the host organism. *Neisseria gonorrhoeae*, for example, appears to use pili in order to attach to the cells of the human genitourinary tract, leading to the sexually transmissible disease gonorrhea. In all of these processes the pili act as a point of specific contact and attachment between the bacterial cell and another surface.

Flagella

In addition to pili, flagella project out from the cell surfaces of many bacterial cells. Flagella are the most common means of motility of a bacterial cell, permitting the bacterium to move from place to place so that it can obtain nutrition, grow, and reproduce. In contrast to pili, which are short projections, flagella are relatively long projections extending outward from the cytoplasmic membrane.

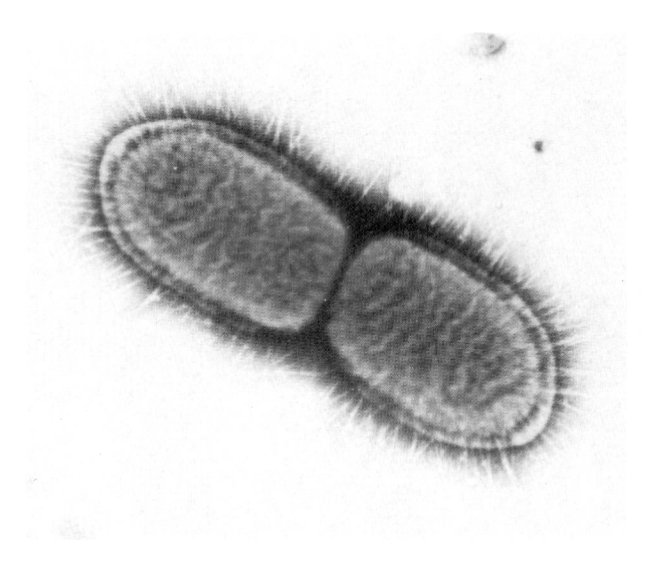

Figure 4.14 *Proteus mirabilis*, negatively stained with phosphotungstic acid and surrounded by pili (36,000×). (Courtesy J. F. M. Hoeniger, University of Toronto. Reprinted by permission of the Society for General Microbiology, from J. F. M. Hoeniger, 1965, *Journal of General Microbiology*, 40:29–42.)

The rod-shaped bacillus *Bdellovibrio bacteriovorans* by means of a polar flagellum rotating 100 times per second can move 50 times its own length of 2 μm per second. This would be equivalent of a human sprinter reaching 200 miles per hour.

— *1985 Guinness Book of World Records*

Two basically different types of arrangements of flagella occur in bacteria (Figure 4.15). In some bacteria, such as *Pseudomonas*, the flagella emanate from an end of the bacterial cell; such flagella are known as polar flagella because they originate from the pole of the cell. In contrast to polar flagella, some bacteria, such as those in the genus *Proteus*, have peritrichous flagella that surround the cell. The arrangement of the flagella is characteristic of a bacterial genus and is an important characteristic used in classifying bacteria.

The structure of the bacterial flagellum allows it to spin like a propeller and thereby propel the bacterial cell. Effectively, the structure allows the flagellum to spin like the shaft of an electric motor. The bacterial flagellum provides the bacterium with a mechanism for swimming toward or away from chemical stimuli, a behavior known as chemotaxis (Figure 4.16). Bacteria move toward certain chemicals, known as attractants, and away from others,

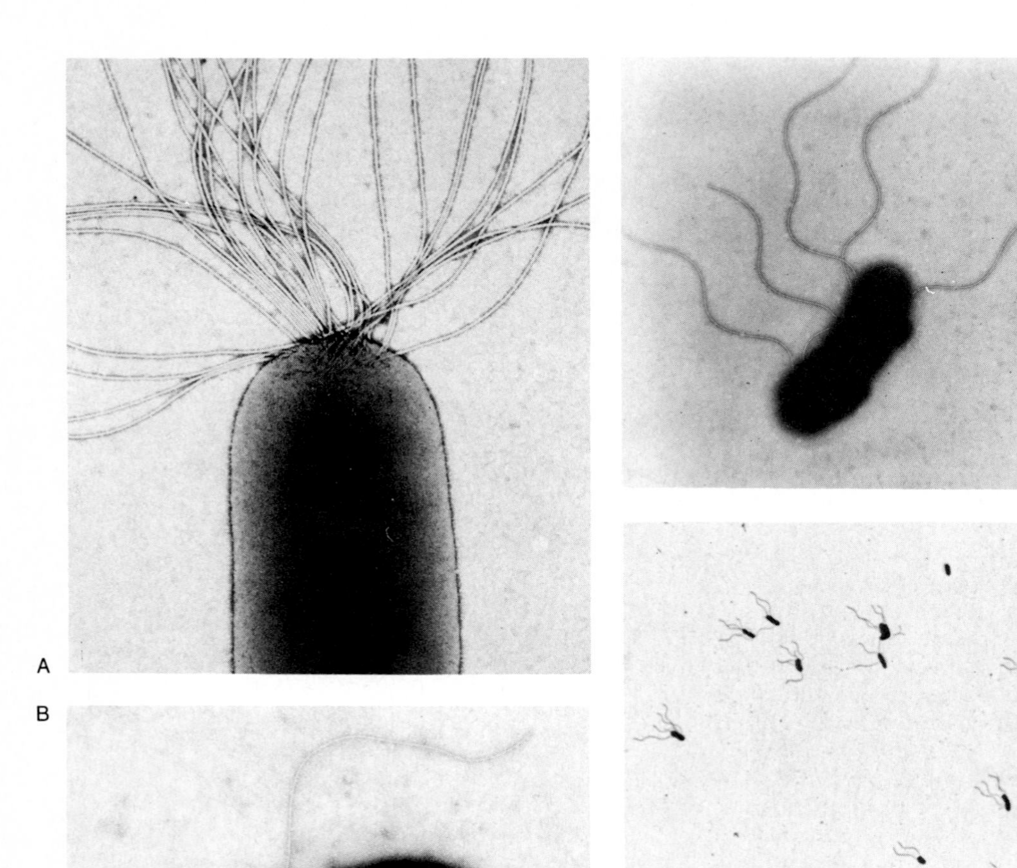

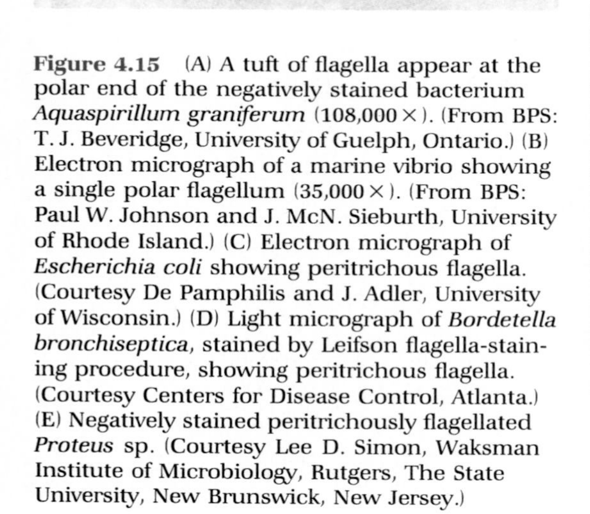

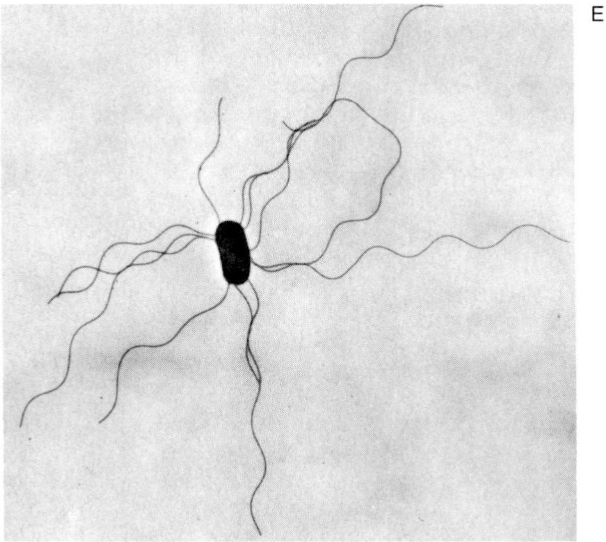

Figure 4.15 (A) A tuft of flagella appear at the polar end of the negatively stained bacterium *Aquaspirillum graniferum* (108,000 ×). (From BPS: T. J. Beveridge, University of Guelph, Ontario.) (B) Electron micrograph of a marine vibrio showing a single polar flagellum (35,000 ×). (From BPS: Paul W. Johnson and J. McN. Sieburth, University of Rhode Island.) (C) Electron micrograph of *Escherichia coli* showing peritrichous flagella. (Courtesy De Pamphilis and J. Adler, University of Wisconsin.) (D) Light micrograph of *Bordetella bronchiseptica*, stained by Leifson flagella-staining procedure, showing peritrichous flagella. (Courtesy Centers for Disease Control, Atlanta.) (E) Negatively stained peritrichously flagellated *Proteus* sp. (Courtesy Lee D. Simon, Waksman Institute of Microbiology, Rutgers, The State University, New Brunswick, New Jersey.)

Figure 4.16 A demonstration of a chemotactic response to an attractant. That bacteria respond to their chemical surroundings by chemotaxis was an important finding, showing that even the "simplest" organisms can exhibit approach–avoidance behavior. Chemotactic behavior is readily demonstrated and measured by placing the tip of a thin capillary tube containing an attractant solution in a suspension of motile *E. coli* bacteria. The suspension is placed on a slide in a chamber created by a U-tube and a coverslip. (A) At first, the bacteria are distributed at random throughout the suspension. (B) After 20 minutes they have congregated at the mouth of the capillary. (C) After about an hour many cells have moved up into the capillary tube. If the capillary had contained a repellant, few bacteria, if any, would have entered the tube. Using this technique, it is possible to show which chemicals attract bacteria and which substances bacteria choose to avoid.

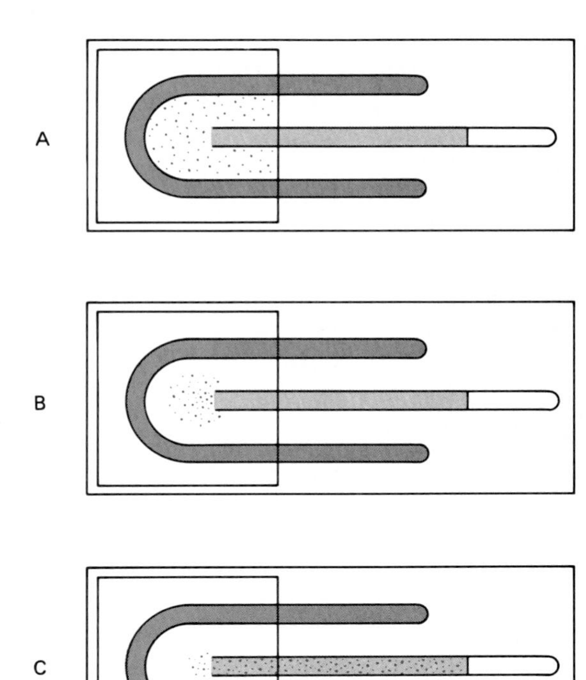

known as repellents. When bacteria move, they periodically change direction rather than reaching their destination by swimming in one straight line. The straight-line movements of bacteria are known as runs, and the turns are called tumbles or twiddles. The basis for this movement rests with the ability of the bacterium to reverse the direction of rotation of the flagellum, depending on whether they are moving toward or away from a chemical stimulus. At least in bacteria with peritrichous flagella, the counterclockwise rotation of the flagella results in a run and the clockwise rotation in a twiddle.

Storage of Genetic Information

Bacterial Chromosome. The storage of genetic information is of paramount importance as it specifies the potential structures and functions of the cell. The genetic information of all cells is housed in DNA molecules. Most of the genetic information of the bacterial cell is contained within a bacterial chromosome composed of a single DNA macromolecule (Figure 4.17). Bacteria have only one bacterial chromosome. The area occupied by the DNA is sometimes referred to as the nucleoid region although the DNA is not contained within a separate membrane-bound organelle. The DNA in the nucleoid region occurs as a single, large, circular macromolecule of a DNA double helix. The lack of a membrane surrounding the bacterial chromosome differentiates the prokaryotic from the eukaryotic cell.

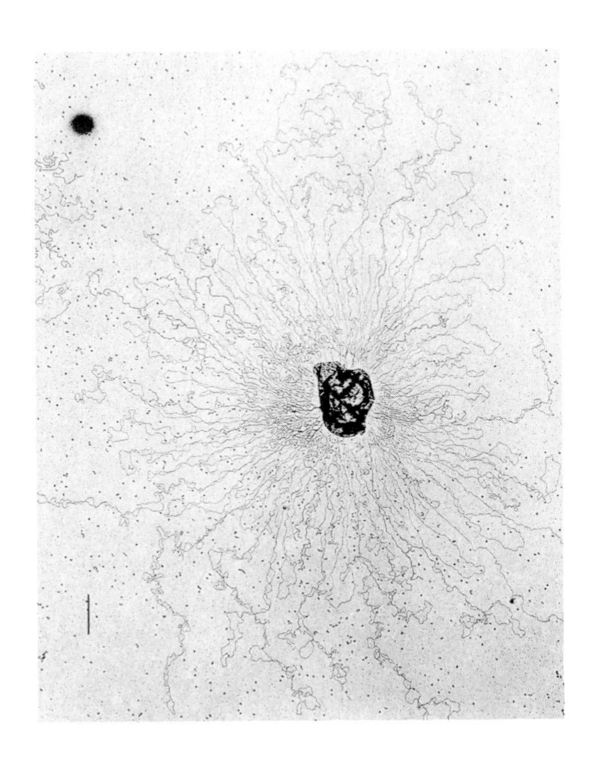

Figure 4.17 The mass of DNA contained within the circular loop of the bacterial chromosome is shown in this micrograph. (Courtesy Ruth Kavenoff, Designergenes Posters Ltd. Reprinted by permission of Springer-Verlag from R. Kavenoff and O. Ryder, 1976, *Chromosoma*, 55:23.)

Plasmids. In addition to the bacterial chromosome, bacteria may contain one or more small circular macromolecules of DNA—known as plasmids All bacterial cells contain a bacterial chromosome, but not all bacteria contain plasmids. Plasmids contain a limited amount of specific genetic information that supplements the essential genetic information contained in the bacterial chromosome. This supplemental information can be quite important, establishing mating capabilities, resistance to antibiotics, and tolerance of toxic metals. Such supplemental genetic capability can permit the survival of the bacterium under conditions that are normally unfavorable for growth and survival.

Pathogenic bacteria that contain plasmids that code for multiple drug resistance have become a particular problem in treating some infectious human diseases because such bacteria are resistant to many antibiotics and can continue to grow in the body despite antibiotic treatment. Plasmids are also quite useful in research and are employed in genetic engineering as carriers of genetic information from a variety of sources. Because they are relatively small, plasmids are relatively easy to manipulate. They can be isolated, genetic information from other sources can be spliced into them, and they can be implanted into viable bacterial cells, permitting expression of the genetic information they contain. Genetic engineering appears to have many industrial and medical applications that will be discussed later.

Ribosomes

The expression of the genetic information of a cell requires that the information stored in the DNA macromolecules be used to direct the synthesis of functional proteins. Proteins, acting as enzymes, then catalyze the metabolic functions of the cell. The ribosomes are the sites where protein synthesis occurs within the cell. This is an essential function without which the cell can not long survive.

A typical prokaryotic cell may have 10,000 or more ribosomes; eukaryotic cells contain considerably more. Ribosomes are composed of ribosomal ribonucleic acid (rRNA) and protein. During protein synthesis the information stored in the DNA is transferred to a messenger RNA molecule (mRNA), which as the name implies, acts as a messenger carrying the transcribed information to the ribosomes located in the cell's cytoplasm, where the information is used to direct the synthesis of the protein.

The prokaryotic cell has 70S ribosomes. The S refers to Svedberg units and is a measure of the "size" of the ribosome. Svedberg units (S) are a measure of how far substances move in density gradient ultracentrifugation. The functional ribosome is composed of two structural subunits of different sizes. The 70S bacterial ribosome is composed of 50S and 30S subunits (Figure 4.18). Svedberg units are nonadditive, and one cannot simply add the sizes of the subunits to determine the size of the intact ribosome. Thus, attaching a 30S subunit and a 50S subunit produces a 70S rather than an 80S ribosome.

There are significant differences between the ribosomes that occur in prokaryotic cells and those that are found in the cytoplasm of eukaryotic cells. Eukaryotic cells have 80S ribosomes composed of 60S and 40S subunits. The differences in the structural composition of prokaryotic and eukaryotic ribosomes forms an important basis for using antibiotics in the treatment of animal and plant diseases caused by bacteria. Protein synthesis that occurs at the ribosomes is essential for cells to carry out life-supporting metabolism, and any disruption of the ribosomal conformation can disrupt this essential process. Many antibiotics, such as erythromycin and streptomycin, are effective because they bind to and alter the shape of 70S ribosomes. Such

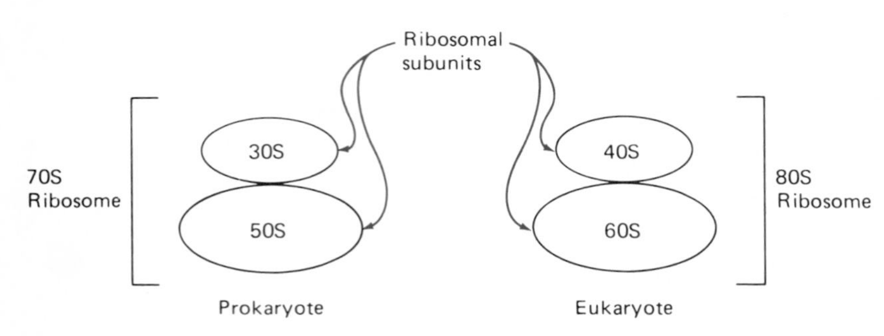

Figure 4.18 Diagram of the structure of a ribosome.

antibiotics are useful therapeutically because they selectively attach to 70S ribosomes and hence disrupt protein synthesis in bacteria; they do not exhibit any affinity for 80S ribosomes, and therefore do not disrupt protein synthesis in eukaryotic human cells. Here we can see the practical application of a fundamental difference in the cellular structure of eukaryotes and prokaryotes.

Reserve Materials

Cells often store various biochemicals, which they have synthesized or accumulated, to act as nutrient reserves to be used in times of need. Many bacteria accumulate granules of polyphosphate, which are reserves of inorganic phosphate that can be used in the synthesis of ATP. Polyphosphate granules can be seen in light microscopy after staining and are sometimes termed volutin or metachromatic granules.

Cells may also accumulate reserves of organic carbon molecules that can be metabolized at a later time for the generation of ATP and cell constituents. In eukaryotic cells such reserve materials normally accumulate in membrane-bound vacuoles. In bacteria the reserve materials, most commonly the lipid-like material poly-beta-hydroxybutyric acid (PHB), accumulate as cytoplasmic inclusions not separated by a boundary membrane from the rest of the cytoplasm. Rather, the separation of the reserve inclusions is generally based on differential solubility.

Spores

Of the many types of spores produced by microorganisms, one specific type—the bacterial endospore—has special importance (Figure 4.19). The endospore is a complex seven-layered structure containing murein within its complex spore coat and calcium dipicolinate within its core. The endospore is highly refractory and resistant to desiccation (drying), retaining its viability over extended periods of time under conditions that do not permit growth of the organism. The major importance of the bacterial endospore for humans rests with its resistance to high temperatures. Endospores can survive exposure to high temperatures for extended periods, whereas normal bacterial vegetative cells are killed by brief exposures to such high temperatures. The mechanism of heat resistance must protect the cell's macromolecules from denaturation.

Figure 4.19 Electron micrographs of bacterial endospores (A) *Bacillus sphaericus* (125,000 ×) in thin section; (B) freeze-etched preparation of a *Bacillus polymyxa* spore. (From BPS: Stanley C. Holt, University of Texas Health Science Center, San Antonio, Texas.)

A

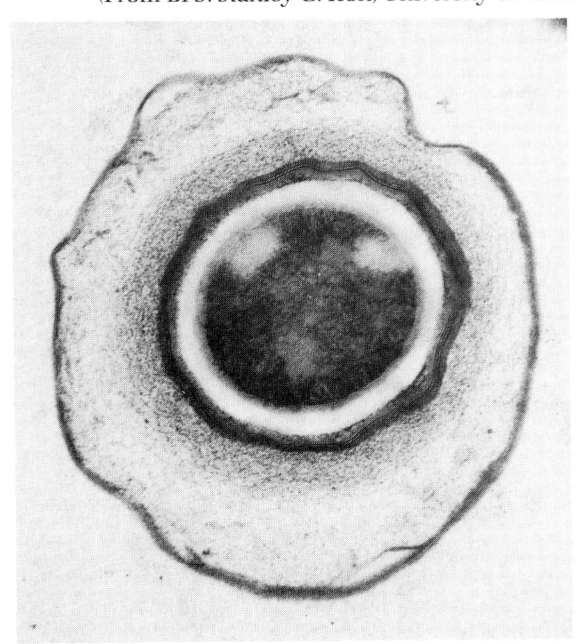

B

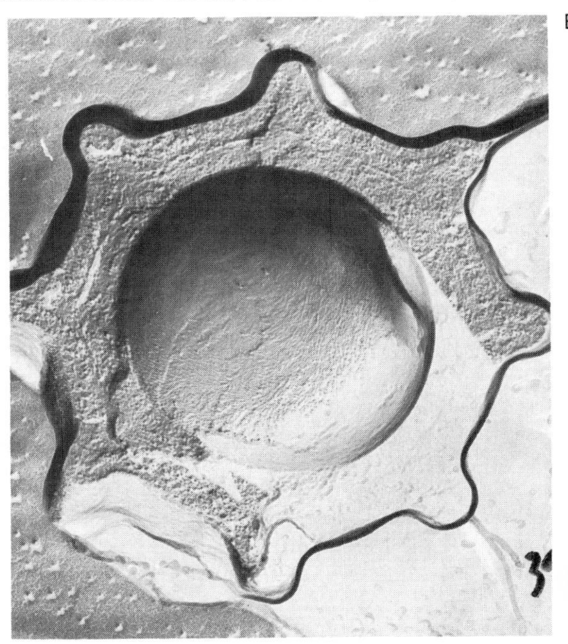

The resistance of the swollen *Bacillus* spores to at least 15 minutes and even 1–2 hours boiling water is probably due to the fat-like contents of the spores . . . The longer the boiling proceeds, the fewer the spores which remain alive . . . Finally, by heating over 100°C, all of the spores are killed and the liquid is completely sterilized . . . I have found only spore-forming bacilli . . . the one that is present in the blood of animals and humans with anthrax has an especially important place, since it is without doubt important in the pathology of the disease.

—FERDINAND COHN, *Studies on the Biology of the Bacilli*

Few bacterial genera are capable of forming endospores; the most important endospore formers are members of the genera *Bacillus* and *Clostridium*. Both *Bacillus* and *Clostridium* are defined as Gram positive rods that form endospores; *Bacillus* is aerobic, growing in the presence of oxygen, and *Clostridium* is obligately anaerobic, growing only in the absence of oxygen. Endospores are resistant to harsh environments including elevated temperatures. The heat resistance of the endospore has great practical importance. Sterilization procedures used in medicine and industry, which are aimed at killing all microorganisms, are designed to kill even endospore-producing bacteria. The canning industry in particular uses temperatures and exposure times that will kill both vegetative cells and endospores of endospore-producing bacteria, including *Clostridium botulinum*—the bacterium that causes deadly botulism. In bacteriology laboratories we generally sterilize media in autoclaves using 15 minutes exposure at 121°C. Clearly our understanding of the basic structure of a bacterial cell helps us design practical processes for controlling bacterial populations.

Membrane Structures Within the Cell

Some electron micrographs of prokaryotic cells reveal extensively invaginated extensions of the cytoplasmic membrane which have been termed mesosomes. Establishing the existence and role of mesosomes has proven difficult. Some of the microbiologists who originally described mesosomes now feel they may be artifacts. If they are real structures mesosomes probably have different functions in different cells. Among the proposed functions, mesosomes may play a role in cell-divisional processes, various metabolic processes—including the generation of ATP—and secretion of enzymes from the cell.

Although prokaryotic cells are generally characterized by a lack of internal membrane-bound organelles, certain specialized groups of bacteria do contain extensive internal membranes. Such groups of bacteria include some nitrifying bacteria (bacteria that convert nitrate to nitrite ions or ammonium to nitrate ions) and the photosynthetic bacteria. In these nitrifying and photosynthetic bacteria, the cells literally may appear to be filled with membranes. In the photosynthetic bacteria these membranes are the anatomical sites of the energy-generating reactions of photosynthesis. In the nitrifying bacteria the internal membranes similarly are involved in the generation of ATP, only in this case through the oxidation of inorganic nitrogen compounds.

In addition to these membranous networks, some bacteria form "membrane-bound" gas vacuoles. The formation of gas vacuoles by aquatic bacteria provides a mechanism for adjusting the buoyancy of the cell and thus the height of the bacterium in the water. Many aquatic cyanobacteria, for example, use their gas vacuoles to move up and down in the water column. Depending upon light intensity levels, they are able to move to, and remain at, a depth where conditions are best for carrying out their photosynthetic metabolism. The presence of gas vacuoles thus provides the necessary structural units for phototaxis (movement in response to light) in aquatic bacteria.

Some bacteria also contain membrane-bound iron granules that permit them to navigate along the earth's magnetic field. The response to a magnetic field is known as magnetotaxis. For some bacteria that live in aquatic habitats, magnetotaxis allows them to orient the cells pointed downward into the sediment. This is important because it points the bacteria toward potential sources of nutrients. The direction of orientation depends upon the polarity of the magnetosome. Some bacteria move predominantly north, and others move south. When a magnetotactic bacterium from the northern hemisphere is transported to the southern hemisphere, it begins to swim with its cell pointed upward instead of the normal downward orientation. Similarly, when bacteria from the southern hemisphere are moved to the northern hemisphere, they reverse the orientation of swimming.

Thus, even bacteria are able to sense and respond to environmental stimuli in adaptive ways. The structures of the prokaryotic cell permit the survival and reproductive success of bacteria. Bacteria with specialized cell structures have selective advantages, under certain conditions, over bacteria that lack such structures.

Summary

The bacteria all have prokaryotic cells. The prokaryotic cell is surrounded by a cytoplasmic membrane that regulates the flow into and out of the cell. Substances can enter a bacterial cell by simple diffusion, facilitated diffusion, and active transport. In diffusion, substances move from a region of high concentration to a region of lower concentration. Substances move across the membrane more rapidly in facilitated diffusion than in simple diffusion. The increased rate of movement involves permeases that are associated with the movement. Unlike diffusion, active transport requires an expenditure of energy and can result in the movement of substances against a concentration gradient. ATP often supplies the energy for active transport.

In addition to its primary function in regulating the movement of material into and out of the cell, the cytoplasmic membrane of the prokaryotic cell can have an important role in the generation of cellular energy. During respiration, bacteria pump hydrogen ions across the cytoplasmic membrane, thereby establishing a hydrogen ion gradient. The movement of hydrogen ions back into the cell by diffusion is used to generate ATP. This process of ATP generation, called chemiosmosis, is very important for the generation of cellular energy.

The movement of water across the cytoplasmic membrane, known as osmosis, also is important for the survival of the cell. Excessive movement of water can result in swelling or shrinking of the cell. Most bacterial cells are surrounded by a rigid cell wall that protects them from bursting because of osmotic pressure. The bacterial cell wall is composed of murein, which is also known as peptidoglycan or mucopolysaccharide. Murein is a unique biochemical structure found only in bacterial cell walls. The biochemical composition of murein provides for its rigidity and permits it to meet its primary function of protecting the cell. Disruption of the cell wall usually results in death of the bacterial cell. Penicillin prevents the formation of intact cell walls, and this is the basis for using this antibiotic to treat bacterial infections. Lysozyme, a substance found in tears and other body fluids, also disrupts the bacterial cell wall. Members of the genus *Mycoplasma* do not have a cell wall and therefore are not affected by penicillin and other cell-wall inhibitors. Atypical pneumonia caused by *Mycoplasma pneumoniae* does not respond to penicillin.

There is a primary difference in the construction of the cell wall between Gram positive and Gram negative bacteria. The Gram positive cell wall is composed almost entirely of murein. The Gram negative cell wall has a relatively thin layer of murein and also contains several additional biochemicals that constitute the cell envelope. The biochemically complex wall of Gram negative bacteria contains lipopolysaccharide (LPS), which is known as endotoxin. Endotoxin can cause adverse reactions in humans, including gastrointestinal disorders. The Gram negative bacterial cell wall also has an outer membrane that restricts the flow of some materials, including some antibiotics, to the cell. The nature of the Gram negative cell wall contributes to the fact that Gram negative bacteria are more prevalent than Gram positive bacteria in human infections.

Some bacteria are surrounded by a capsule, which also can contribute to the virulence of disease-causing bacteria. The capsule appears to protect the cell against destruction by components of the host defense system. Sometimes a strain of a particular species that has a capsule is a deadly pathogen, whereas a strain of the same species that lacks a capsule is a harmless nonpathogen.

In addition to protective layers, the ability to attach to surfaces is important in bacterial pathogenicity. Several structures, including pili and the glycocalyx, are involved in the attachment of different bacteria to surfaces. Pili are short hair-like projection that are involved in various attachment process, such as the mating of bacteria and the infection of bacteria by bacteriophage. The glycocalyx surrounds the cell and is responsible for the ability of some bacteria to attach to inanimate surfaces, such as submerged rocks and living tissues like those lining the human intestine.

Several bacteria produce specialized structures that provide important adaptations for their survival. A number of bacteria store reserve materials, such as polyphosphate (volutin or metachromatic granules) and poly-beta-hydroxybutyrate, that permit their survival under adverse conditions. These reserve materials are not contained within membrane-bound vacuoles. Although internal membrane structures are generally lacking in prokaryotic cells, some specialized bacteria produce internal membrane structures. Both the photosynthetic and nitrifying bacteria have extensive internal membrane networks involved in ATP generation. Some bacteria have gas vacuoles that allow them to adjust their buoyancy in a water column. Other newly discovered bacteria produce magnetosomes that act as "magnetic compasses" and allow them to navigate along the earth's magnetic field.

Members of a few genera, including species of *Bacillus* and *Clostridium* produce a specialized structure—the endospore—that has great survival value. The endospores formed by members of these genera present special problems for the food industry, which employs processes that rely on heat to sterilize products. Some endospores can withstand boiling for more than 1 hour. Endospores are formed when conditions are unfavorable for continued growth of the bacterium. Once formed, endospores can retain viability for millennia. Under favorable conditions, the endospore can germinate and give rise to an active vegetative cell of the bacterium. In order to ensure that endospores are killed, it is necessary to heat liquids to a temperature greater than 120°C and hold that temperature for at least 15 minutes; dry materials require several hours at this temperature to ensure sterilization.

Study Outline

A. Prokaryotic cells.
- **1.** All bacteria have prokaryotic cells.
- **2.** All organisms with prokaryotic cells are bacteria.
- **3.** Prokaryotic cells lack internal compartmentalization.
- **4.** Cytoplasmic membrane of the prokaryotic cell.
  - **a.** The cytoplasmic membrane is a semipermeable barrier and regulates flow of material into and out of the cell.
  - **b.** The cytoplasmic membrane is composed of phospholipid and protein.
  - **c.** The cytoplasmic membrane is a bilipid layer that gives a railroad track appearance when viewed by electron microscopy.
  - **d.** The structure and function of the cytoplasmic membrane is explained by the fluid mosaic model.
  - **e.** Substances move across the cytoplasmic membrane by simple diffusion, facilitated diffusion, and active transport.
  - **f.** Facilitated diffusion involves proteins called permeases.
  - **g.** Active transport requires energy, which may be supplied by ATP.
  - **h.** The cytoplasmic membranes of bacteria may serve additional functions such as energy (ATP) generation.
  - **i.** During respiration, hydrogen ions are pumped across the cytoplasmic membrane and the return flow is used to generate ATP by chemiosmosis.
  - **j.** Chemiosmosis was proposed by Peter Mitchell and is now known to be very important for the generation of cellular energy.
- **5.** Bacterial cell wall.
  - **a.** A cell wall surrounds the cytoplasmic membrane of most prokaryotic cells.
  - **b.** The bacterial cell wall is relatively rigid and gives bacteria their shape.
  - **c.** The cell wall protects bacteria against osmotic shock
  - **d.** The bacterial cell wall contains murein. Murein, or peptidoglygan, is composed of a backbone of alternately repeating units of N-acetylglucosamine and N-acetylmuramic acid. Attached to some of the N-acetylmuramic acid units is a short chain of amino acids. The short chains of amino acids are crosslinked, making the wall rigid.
  - **e.** Penicillin prevents formation of crosslinkages in a growing cell, causing cell wall to be defective and resulting in the death of the cell.
  - **f.** Lysozyme, which is found in tears, breaks the glycan portion of the peptidoglycan rather than the peptide portion, as penicillin does.
  - **g.** A spheroplast is a bacterium whose cell wall is only partially removed.
  - **h.** A protoplast is a bacterium whose cell wall is completely removed.
  - **i.** The Gram-positive bacterial cell wall has a thick peptidoglycan layer comprising 90 percent of cell wall.
  - **j.** The Gram negative bacterial cell wall is more complex than Gram positive cell walls. It has a thin peptidoglycan layer comprising only 20 percent of the cell

wall and forms a complex with an outer membrane and lipopolysaccharides (LPS).
   - **k.** The LPS portion of the cell wall is known as endotoxin.

6. Pili.
   - **a.** Pili are short, hair-like projections composed of protein subunits emanating from the bacterial surface.
   - **b.** Pili are important in attachment processes.
   - **c.** The F or sex pilus is involved in mating and is found on the donor cells.
   - **d.** Pili also act as the receptor sites for phages.

7. Flagella.
   - **a.** Bacterial flagella have a different structure from eukaryotic flagella.
   - **b.** Bacteria exibit two basic types of flagella arrangements: polar flagella, which originate from the pole, and peritrichous flagella, which surround the cell.
   - **c.** Bacterial flagella spin and can reverse direction.
   - **d.** Flagella permit bacteria to move toward and away from chemical stimuli.
   - **e.** Bacterial movement in response to chemical stimuli is known as chemotaxis.
   - **f.** Straight-line movements are known as runs; turns are called tumbles or twiddles.

8. Storage of genetic information.
   - **a.** In the prokaryotic cell, the DNA is not segregated from the rest of the cell by a membrane barrier. This is the feature that differentiates prokaryotic from eukaryotic cells.
   - **b.** The bacterial chromosome is composed of a double-stranded DNA helix in the form of a single circular macromolecule.
   - **c.** The region occupied by the bacterial chromosome is called the nucleoid region.
   - **d.** Plasmids are small circular DNA molecules that contain supplemental genetic information.

9. Ribosomes.
   - **a.** Ribosomes are the sites where protein synthesis occurs.
   - **b.** Ribosomes are composed of rRNA (ribosomal RNA) and protein.
   - **c.** The ribosome is two structural subunits.
   - **d.** The prokaryotic ribosome is a 70S ribosome composed of 50S and 30S subunits. (The S refers to Svedberg units, which are a measure of size and cannot be simply added.)

10. Spores.
   - **a.** The bacterial endospore is a heat-resistant structure.
   - **b.** Bacterial genera that form endospores include *Bacillus*, a Gram positive aerobe, and *Clostridium*, a Gram positive anaerobe.

11. Internal membrane structures.
   - **a.** Usually, prokaryotic cells do not have internal membrane-bound organelles.
   - **b.** Specialized bacteria, such as photosynthetic bacteria and nitrifying bacteria, do contain extensive internal membranes.
   - **c.** Some bacteria form gas vacuoles, which allow them to adjust buoyancy in aquatic habitats.
   - **d.** Some bacteria contain membrane-bound iron granules that allow them to navigate along the earth's magnetic field, a movement called magnetotaxis.

Study Questions

1. Indicate whether the following statements are true (T) or false (F).
   - **a.** The cytoplasmic membrane is composed of three layers of phospholipids.
   - **b.** In the cytoplasmic membrane, the nonpolar ends are oriented toward each other and the polar ends are oriented toward the outside and inside surfaces of the membrane.
   - **c.** The modern concept of the cytoplasmic membrane structure is called the fluid mosaic model.

2. Describe the primary function of the cytoplasmic membrane.
3. List three ways material is moved across the membrane.
4. Fill in the blanks.
   a. The passive movement of molecules from regions of higher concentration to regions of lower concentration is called ________ .
   b. The movement of water molecules across the membrane in response to a concentration gradient of solute is called ________ .
   c. An environment in which the cell shrinks as water osmotically moves out in response to a higher extracellular concentration of solute is called ________ .
   d. An environment in which the cell swells as water osmotically moves into the cell in response to a higher intracellular concentration of solute is called ________ .
   e. A type of passive transport involving a carrier protein is called ________ .
   f. A type of transport where energy is required in the form of ________ is called ________ .
   g. In active transport, movement is across a membrane from an area ________ in concentration to one that is ________ in concentration.
   h. The form of energy generation that involves a difference in concentration of hydrogen ions across the cell membrane is called ________ .
5. Match the following phrases with the terms in the list to the right.
   a. Peritrichous or polar in arrangement.
   b. Projections involved in attachment, shorter than flagella.
   c. Lies external to cell wall; relatively tightly bound to cell wall and a major factor in pathogenicity.
   d. Surrounds the cytoplasmic membrane of most prokaryotic cells and many eukaryotic microorganisms.
   e. In bacteria, contains a backbone of murein.
   f. Lies external to cell wall; is involved in attachment; important in formation of dental plaque.
   g. Short, hair-like projections composed of protein subunits emanating from the bacterial surface.

   (1) Cell wall
   (2) Capsule
   (3) Glycocalyx
   (4) Pili
   (5) Flagella
6. Indicate whether the following statements are true (T) or false (F).
   a. In bacterial cell walls, murein (peptidoglycan) is composed of alternately repeating units of N-acetylglucosamine and N-acetylmuramic acid.
   b. Penicillin prevents the formation of the crosslinkages in murein.
   c. Lysozyme is found in tears.
   d. A spheroplast is a bacterium whose cell wall is completely removed.
   e. Peritrichous flagella originate from the pole of the cell.
   f. Tumbles and twiddles are technical terms for straight line movements.
7. Fill in the blanks.
   a. Bacterial genome is not surrounded by a ________ as eukaryotic cell genomes are.
   b. In a prokaryotic cell, the region occupied by the bacterial chromosome is called the ________ .
   c. The number of bacterial chromosomes is (are) ________ and is (are) in the form of ________ .
   d. In bacteria, small circular DNA molecules that contain supplemental genetic information are called ________ .
   e. The site where protein synthesis occurs is the ________ .
   f. The prokaryotic ribosome is ________ S and is made up of ________ S and ________ S subunits.
   g. Special structures involved in the reproduction, dispersal, or survival of an organism are called ________ and ________ confer heat resistance on bacteria that produce them.
   h. The two bacterial genera that produce endospores are ________ and ________ .

Some Thought Questions

1. Why are cells the smallest units of life?
2. Why are cell membranes important?
3. How do bacteria move?
4. How was the mechanism for the movement of bacterial flagella determined?
5. When a bacterium reaches the corner of Oxford and Divinity Streets how does it choose which way to go?
6. Why is there no fossil record for bacteria?
7. Can X rays be used to see bacterial structures?

Additional Sources of Information

Adler, J. 1976. The sensing of chemicals by bacteria. *Scientific American* **234**(4):40–47. This interesting article describes how bacteria respond to chemical stimuli, moving toward some chemicals and away from others.

Berg, H. C. 1975. How bacteria swim. *Scientific American*, **233**(2):36-44. This fascinating article describes how cells move, the way in which flagella propel bacteria, and how the mode of flagella movement was discovered.

Blakemore, R. P., and R. B. Frankel. 1981. Magnetic navigation in bacteria. *Scientific American*, **245**(6):58–67. This article discusses the surprising discovery that some bacteria respond to magnetic fields and can navigate by magnetotaxis.

Costerton, J. W., G. G. Geesey, and K.-J. Cheng. 1978. How bacteria stick. *Scientific American*, **238**(1):86–95. This article considers new revelations about how bacteria attach to substances. It includes some spectacular micrographs.

Costerton, J. W., and R. T. Irvine. 1981. The bacterial glycocalyx in nature and disease. *Annual Reviews of Microbiology*, **35**:299–324. This technical article discusses the functions of the glycocalyx and the importance of attachment by the glycocalyx in disease and ecological processes.

Jensen, W. A., and R. B. Park. 1967. *Cell Ultrastructure*. Wadsworth Publishing Co., Inc., Belmont, California. An excellent primer on the structures of a cell.

Loewy, A. G., and P. Siekevitz. 1970. *Cell Structure and Function*. Holt, Rinehart and Winston, New York. A classic and simple presentation on the structures of a cell.

Molecules to Living Cells: Readings from Scientific American. 1980. W. H. Freeman and Company, Publishers, San Francisco. An excellent work on various levels of organization of biological systems, including articles on the macromolecules and structural units that comprise living systems.

Walsby, A. E. 1977. The gas vacuoles of blue-green algae. *Scientific American*, **237**(2):90–97. An interesting article on the structures that allow cyanobacteria to adjust their buoyancy.

Woese, C. R. 1981. Archaebacteria. *Scientific American*, **244**(6):98–122. An easy to understand presentation about the structural and other differences between archaebacteria and other organisms.

5 The Structure of Eukaryotic Microorganisms: Algae, Fungi, and Protozoa

KEY TERMS

ascospores
ascus
basidium
basidiospores
chlorophyll
chloroplasts
chromosomes
cilia
coenocytic hypha
contractile vacuoles
cytoplast
cytoskeleton
cytosis
digestive vacuole
endocytosis
endoplasmic reticulum
exocytosis
flagella
frustule
Golgi apparatus
histones
hyphae (singular, hypha)
lysosomes
microbodies
microtubules
mitochondria (singular, mitochondrion)
mycelium
nuclear membrane
nucleolus
nucleosome
nucleus
pellicle
peroxisomes
phagocytosis
phagosome
photosynthesis
pseudopodia
ribosomes
storage vacuole
vesicle

OBJECTIVES

After reading this chapter you should be able to

1. Define the key terms.
2. List the structural differences between eukaryotic and prokaryotic cells.
3. Discuss the similarities and differences between the cytoplasmic membranes of eukaryotic and prokaryotic cells.
4. Describe how substances enter and leave eukaryotic cells.
5. Discuss how genetic information is stored in eukaryotic cells.
6. Describe the cell wall structures of eukaryotic microorganisms.
7. Describe the structure and function of the flagella and cilia of eukaryotic cells.
8. Discuss the structure and functions of mitochondria and chloroplasts.
9. Compare the mitochondria and chloroplasts of eukaryotic cells with the prokaryotic cell.
10. Discuss the functions of the endoplasmic reticulum.
11. Discuss the functions of the Golgi apparatus.
12. Compare vacuoles, lysosomes, and microbodies.
13. Discuss the types of ribosomes found within eukaryotic cells.
14. Discuss the types of spores produced by eukaryotic microorganisms.
15. Describe the multicellular arrangements of algae and fungi.

Eukaryotic Cell Structure

Except for the bacteria all other living organisms have eukaryotic cells. Among the microbes this includes the fungi, algae, and protozoa. The eukaryotic cell also is the fundamental structural unit of plants and animals, including humans.

The body is a state in which each cell is a citizen.

—RUDOLPH VIRCHOW

Eukaryotic cells have a far more complex internal organization than prokaryotic cells (Figure 5.1). There are major differences between the prokaryotic bacterial cell and the eukaryotic human cell that provide several sites against which substances can be targeted for the treatment of human diseases caused by bacteria. The similarities between the eukaryotic cells of fungi, protozoa, and humans makes it much more difficult to find substances that will inhibit eukaryotic microorganisms that cause human diseases without at the same time adversely affecting human cells, with serious side effects to the patient.

Each bodily function, and even the life of the organism as a whole, may thus in one sense be regarded as a resultant arising through the integration of a vast number of cell activities; and it cannot be adequately investigated without the study of the individual cell activities that lie at its root.

—E. B. WILSON

Cytoplasmic Membrane

As with the prokaryotic cell, the eukaryotic cell is separated from its surroundings by a cytoplasmic membrane, the primary function of which is to regulate the flow of materials into and out of the cell. Like its prokaryotic counterpart, the eukaryotic cell's cytoplasmic membrane is composed largely of phospholipid with interspersed proteins. The same fluid mosaic model explains the structure–function relationships of cytoplasmic mem-

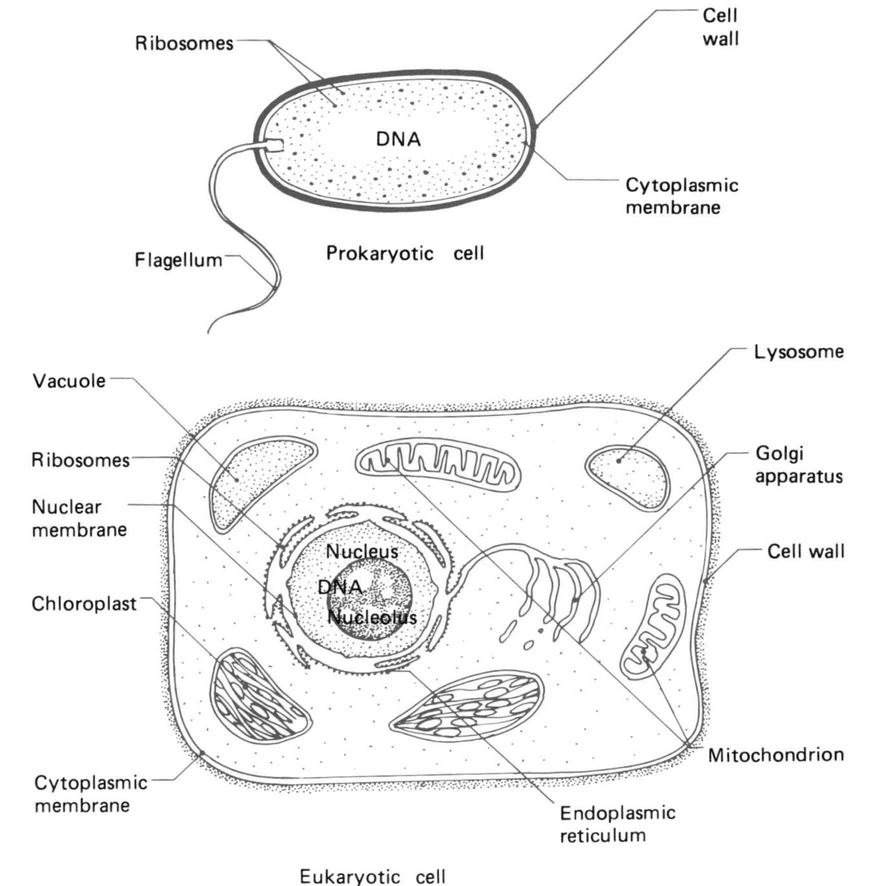

Figure 5.1 A comparison of the structural organization reveals that the eukaryotic cell has far more internal organization than the prokaryotic cell; many of the organelles found in eukaryotic cells do not occur in prokaryotic cells.

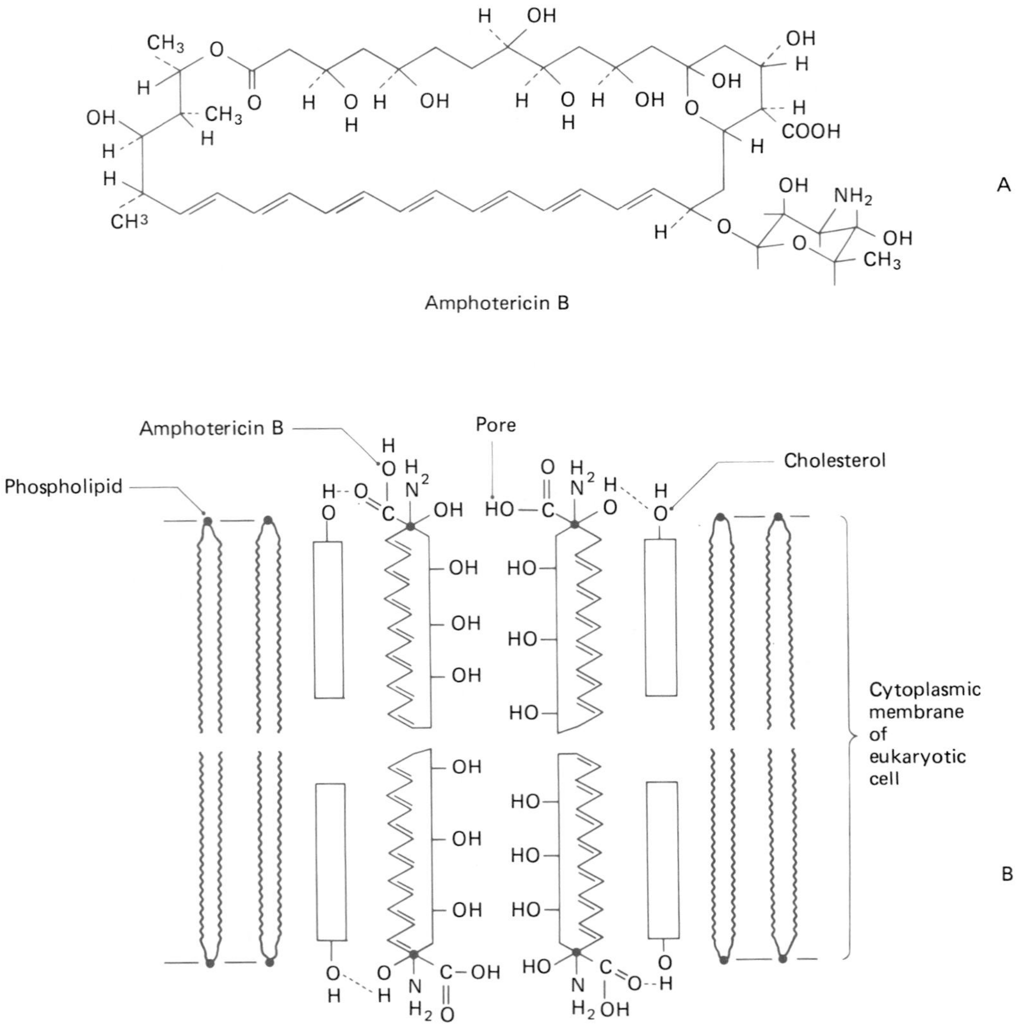

Amphotericin B

A

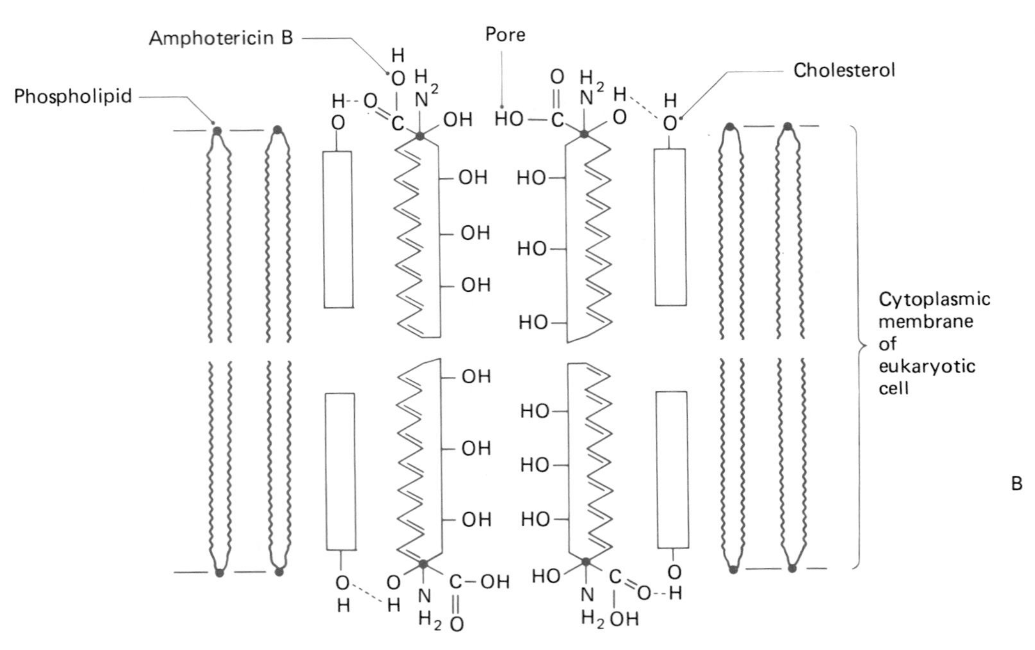

Figure 5.2 Diagram showing the interaction of the antibiotic amphotericin B with the cytoplasmic membrane of a eukaryotic cell. The action of amphotericin B is directed at disrupting the semipermeable nature of the cytoplasmic membrane through interactions with the sterols associated with that structure.

branes of all cells. Viewed with an electron microscope, the cytoplasmic membrane of a eukaryotic cell has the same railroad track appearance as that of the prokaryotic cell, reflecting the same basic bilipid structure of the membrane.

There are some important differences, however, between the cytoplasmic membranes of prokaryotic and eukaryotic cells. Sterols, such as cholesterol, occur in eukaryotic cell membranes but generally are absent from the cytoplasmic membranes of prokaryotic cells. The inclusion of sterols makes the cytoplasmic membrane of eukaryotic cells somewhat more rigid than the cytoplasmic membrane of the prokaryotic cell. The presence of sterols in the membrane provides a site of action for certain antibiotics, such as amphotericin B, making such antibiotics useful in treating infections by eukaryotic microorganisms (Figure 5.2).

An additional difference between eukaryotic and prokaryotic cells is how substances may enter and leave the cell. Substances may pass through the cytoplasmic membranes of both prokaryotic and eukaryotic cells by simple diffusion, facilitated diffusion, and active transport. These three processes

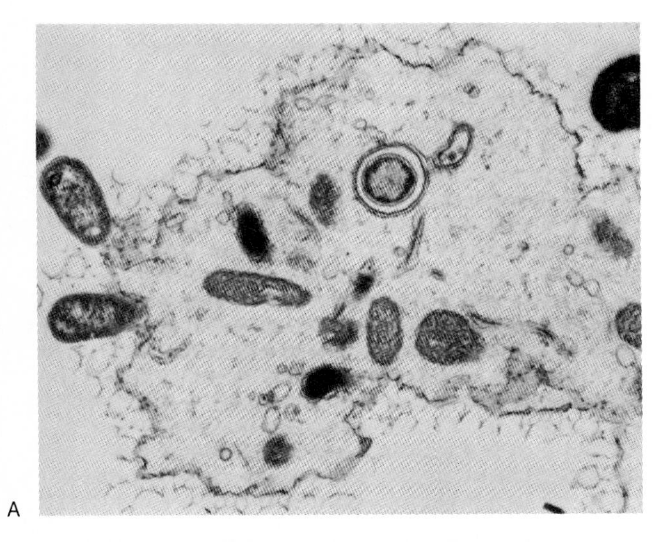

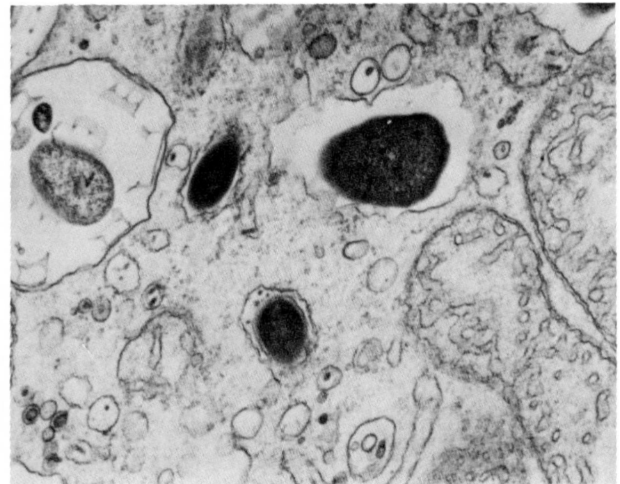

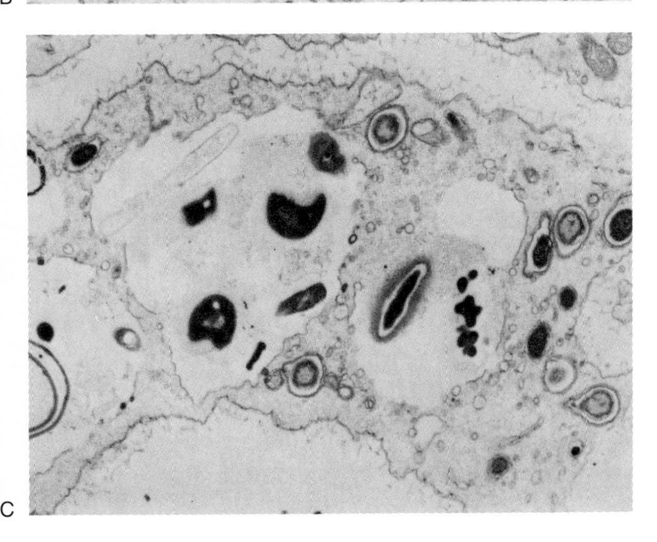

Figure 5.3 Phagocytosis of rod-shaped bacteria by an ameboid protozoan. (A) In the early stages of bacterial engulfment, one bacterium is attached to the surface scales while the other is already in the process of being surrounded by the cytoplasmic membrane of the ameba (20,000 ×). (B) This phagosome (a) enclosed a bacterium shortly after engulfment; there are still scales on the vacuolar membranes. A later stage (b) shows advanced digestion and no scales on the vacuolar membrane (44,000 ×). (C) Here, digestive vacuoles containing bacteria and other food particles are in advanced stages of digestion (14,000 ×). (Courtesy O. Roger Anderson, Columbia University.)

do not permit the transport of very large sustances into and out of the cell. Eukaryotic cells, however, possess another mechanism—cytosis—by which substances may enter or leave a cell. Instead of transporting substances through the membrane, cytosis involves the engulfment of a substance by the membrane to form a vesicle. The membrane-bound vesicle then opens, releasing the substance on the opposite side of the cytoplasmic membrane. If the substance moves into the cell, the process is called endocytosis. Movement of materials out of the cell by this mechanism is called exocytosis

Phagocytosis is a specific example of this transport mechanism in which a eukaryotic cell engulfs a smaller cell or other particle (Figure 5.3). This transport mechanism is particularly important in some protozoa that feed on bacteria. Through phagocytosis, the protozoan is able to ingest the bacteria as its food source and to derive its needed nutrition. Phagocytosis is also an essential line of defense of the human body against microbial infections. Certain white blood cells can engulf and ingest bacteria by this process. Within the white blood cell, the vesicle containing the ingested bacteria—called a phagosome—merges with a membrane-bound vesicle containing digestive enzymes. The bacterial cells are killed when they are mixed with the digestive enzymes. The debris from the digested bacterial cells, still contained within a membrane-bound vesicle, is moved to the cytoplasmic membrane where it is released by exocytosis. The ability to ingest potentially pathogenic bacteria and expel their degraded remains is a major part of the integrated immune defense network that is responsible for the maintainance of a healthy condition.

Nucleus and Chromosomes of Eukaryotic Cells

The genetic information or genome of a eukaryotic cell, like that of the prokaryotic cell, is stored within molecules of DNA. As already indicated, the way in which the genetic information is stored is the prime distinction between prokaryotic and eukaryotic cells. In the eukaryotic cell the genome is separated from the rest of the cell and contained within the nucleus, whereas in the prokaryotic cell the DNA is not segregated from the rest of the cell constituents by a membrane barrier. Additionally, it is now clear that there are fundamental differences in the specific way that the genetic information is stored within the DNA molecules of eukaryotic and prokaryotic cells and that extensive processing of the genetic information occurs within the nucleus.

Do bacteria contain a nucleus? This question has agitated many workers; but at the present moment it appears no more meaningful than the everlasting controversies about the authorship of Shakespeare's plays which was finally resolved by the recognition that another fellow with the same name must have been the author. The important property of the nucleus of animal and plant cells is not so much its appearance or shape as the orderly changes, connected with definite biological tasks, which it can be seen to undergo.

—EDWIN CHARGAFF

A

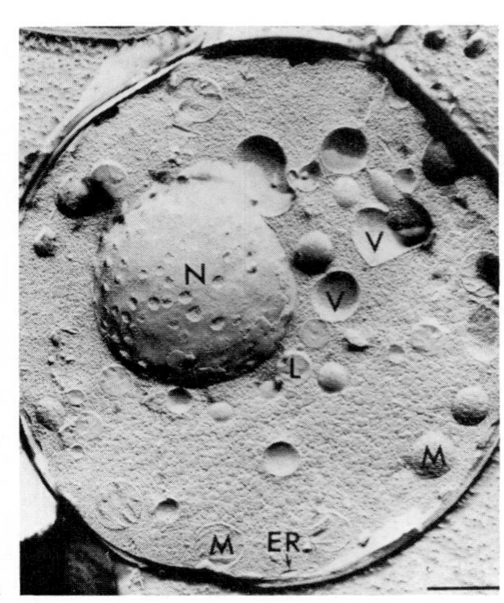

B

C

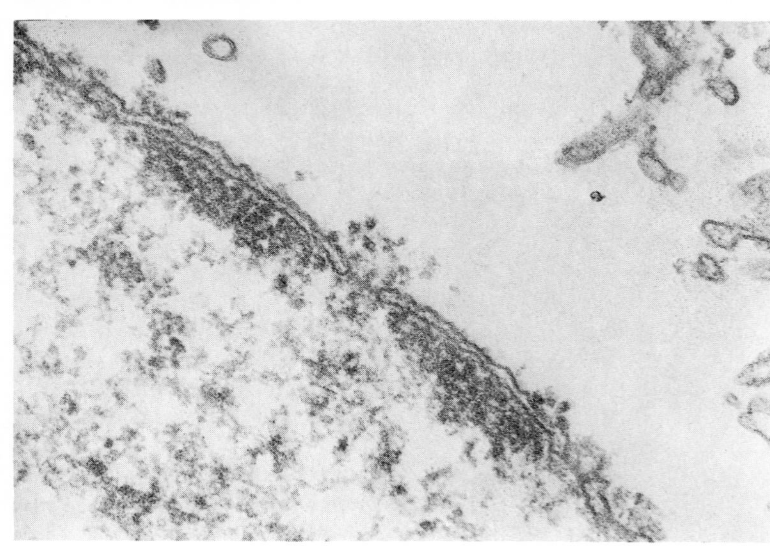

Figure 5.4 The double membrane structure and pores of the nucleus are visible in these electron micrographs. (A) An electron micrograph of a freeze-etched preparation of *Saccharomyces cerevisiae*, showing the surface of a nucleus (16,650 ×). Note the large number of pores in the membrane of the nucleus (N), making it look rather like a golf ball. ER = endoplasmic reticulum; L = lipid granules; M = mitochondrion; and V = vacuole. (Courtesy S. F. Conti, University of Massachusetts. Reprinted by permission of American Society for Microbiology, from E. Guth, T. Hashimoto, and S. F. Conti, 1972, *Journal of Bacteriology*, 109:869–880.) (B) High magnification freeze-etched preparation of nuclear pores (200,000 ×). (Courtesy Daniel Branton, Harvard University.) (C) Thin section of nuclear envelope (160,000 ×), showing pore through which granular material is escaping. (From BPS: W. Rosenberg, Iona College.)

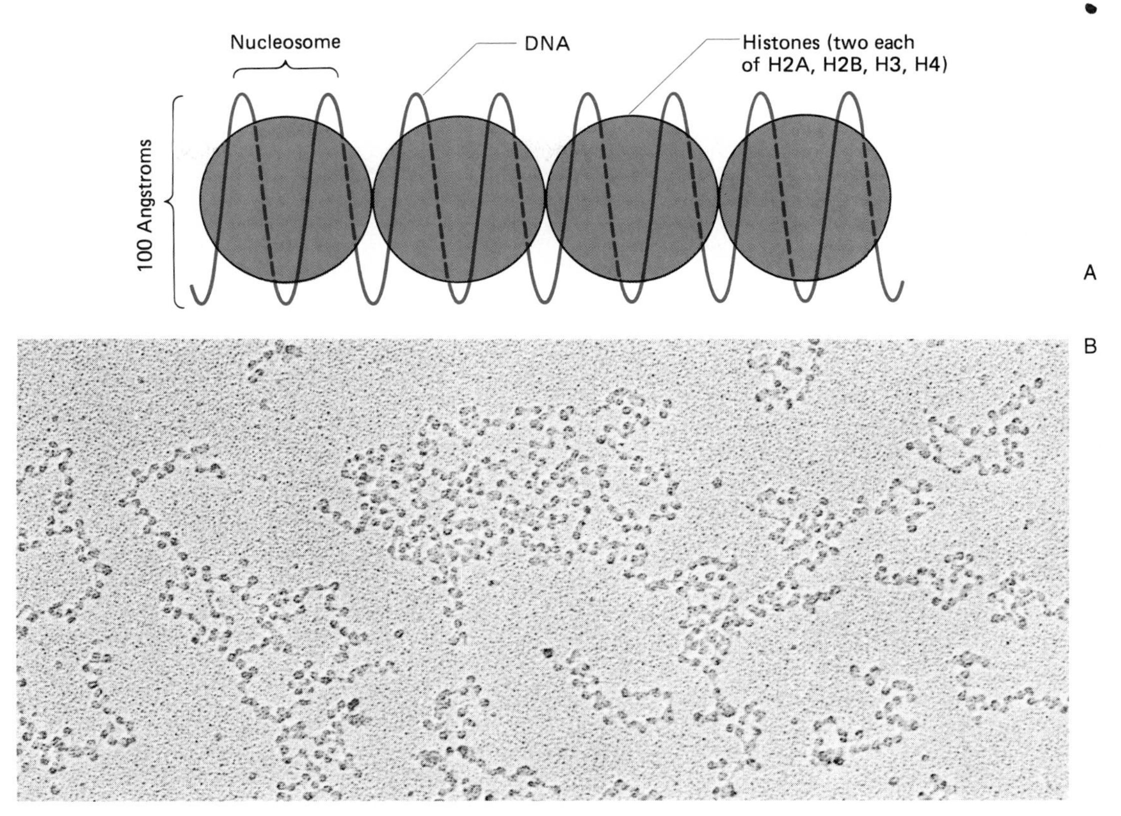

Figure 5.5 (A) This illustration of a nucleosome shows how histones establish the configuration of DNA in eukaryotic cells. (Based on R.D. Kornberg and A. Klug, 1981, "The nucleosome," *Scientific American,* **244**(2):55.) (B) In this electron micrograph of chromatin fibers, the nucleosomes look like beads on a string. The elucidation of the histone-DNA organization of chromosomes was among several accomplishments cited in awarding the 1982 Nobel Prize in Chemistry to Aaron Klug. (Courtesy J. T. Finch and A. Klug, MRC Laboratory for Molecular Biology, Cambridge, England.)

The genome of eukaryotic cells, contained within the cell's nucleus, is separated from the rest of the cell by an inner and an outer membrane (Figure 5.4). In contrast to the cytoplasmic membrane, the nuclear membrane is a double layer with a distinct space between the two layers. Further, the nuclear membrane has pores that permit the exchange of relatively large molecules between the nucleus and the cytoplasm of the cell. This nuclear membrane nevertheless is selective and controls which molecules pass into and out of the nucleus.

Within the nucleus the genetic information is stored within chromosomes, which are composed of chromatin consisting of DNA and protein. The chromosomes are only visible with the light microscope when the cell is undergoing division, and the DNA is in a highly condensed form. At other times the chromosomes are not condensed and are not visible as distinct thread-like structures when the nucleus is viewed with the light microscope. All the genetic information resides in the DNA, although the protein component of chromatin is more abundant than the DNA. Unlike the bacterial chromosome where the DNA forms a circular double helical macromolecule, the chromosomes of eukaryotic cells contain linear DNA macromolecules arranged as a double helix.

The chromatin proteins consist primarily of five histone proteins. These basic proteins bind to the DNA and determine the three-dimensional configuration of the chromatin. The DNA coils around the histones to form subunits of the chromatin known as nucleosomes (Figure 5.5). Each nucleosome is composed of about 200 nucleotides of DNA coiled around the histones. The resulting structures appear as spherical particles or beads on a string when viewed by electron microscopy. The nucleosomes, which establish the structural configuration of eukaryotic chromosomes, are a fundamental unit of the eukaryotic genetic material. Such nucleosomes are absent in the bacterial chromosome, which is composed almost exclusively of DNA and lacks associated histones.

In calling the structure of the chromosome fibers a code-script, we mean that the all-penetrating mind could tell from their structure whether the egg would develop, under suitable conditions, into a black cock or into a speckled hen, into a fly or a maize plant, a beetle, a mouse or a woman . . . But the term code-script is, of course, too narrow. The chromosome structures are at the same time instrumental in bringing about the development they foreshadow. They are law-code and executive power—or, to use another simile, they are architect's plan and builder's craft—in one.

—ERWIN SCHRÖDINGER

The highly structured arrangement of the DNA within chromosomes is critical because the genetic information coding for the synthesis of specific proteins is split in eukaryotic cells and must be extensively processed before it can be properly expressed. Before the information contained within the genome of a eukaryotic cell can be interpreted, it must be edited, a process that involves cutting and splicing of the informational molecules transcribed from the DNA. The editing of the genetic information of a eukaryotic cell to form a readable message, in the form of a messenger RNA (mRNA) molecule, occurs within the nucleus. Compartmentalization of the genetic information within the nucleus, that is, separation from the rest of the cell, is necessary for this editing process to occur. The editing process is analogous to the production of a motion picture, where the order of scenes can be rearranged, scenes can be deleted, and special effects can be added before viewing by the general public is permitted. In prokaryotic cells, the information in the bacterial chromosome is not split; instead, the information coding for specific proteins occurs as a contiguous sequence of nucleotides within the bacterial chromosomes. Hence, the genetic information of prokaryotes does not have to be extensively edited to produce a functional mRNA molecule. In the eukaryotic cell, the mRNA molecule leaves the nucleus and migrates to a ribosome where it is translated into a protein that can carry out the metabolic functions of the cell.

Ribosomes

We mentioned in Chapter 4 that there are significant differences between the ribosomes of prokaryotic cells and those that are found in the cytoplasm of eukaryotic cells. The ribosomes of eukaryotic cells are larger and contain different size rRNA molecules than the ribosomes of prokaryotic cells. The eukaryotic cell has 80S ribosomes in its cytoplasm, composed of 60S and 40S subunits. (Remember Svedberg units are nonadditive.) In eukaryotic cells the ribosomal subunits are synthesized within the nucleus in a region known as the nucleolus and are transported through the pores of the nuclear membrane to the cytoplasm, where the assembly of the 80S ribosome occurs. Antibiotics such as erythromycin, streptomycin, neomycin, chloramphenicol, and numerous others have little or no effect on eukaryotic cells because they react with 70S and not with 80S ribosomes. Only a few antibiotics, such as Actinomycin D, disrupt the functioning of 80S ribosomes. Such 80S ribosomal inhibitors are not useful in treating infectious diseases because they do not inhibit bacteria. Although they would inhibit infecting eukaryotic microorganisms (fungi or protozoa), they also block protein synthesis in human cells, which are eukaryotic. Actinomycin D and similar compounds have been used as chemotherapeutic agents in treating cancer where the aim is to block protein synthesis and active cell reproduction.

Mitochondria

The mitochondria are the sites of extensive ATP synthesis in eukaryotic cells. An organelle carrying out such a function should have a large membrane surface area. The mitochondria have an outer unit membrane, which acts as the boundary between this organelle and the cell cytoplasm, and an independent interior membrane, which exhibits extensive folding (Figure 5.6). The convolutions of the inner membrane, which extend into the interior of the mitochondrion, are called cristae. This inner membrane has a higher proportion of protein associated with it than the outer mitochondrial membrane. Many of these proteins are involved in energy-transferring metabolic reactions. The establishment of a gradient of hydrogen ions across the inner membrane of the mitochondrion is used to drive the chemiosmotic generation of ATP in eukaryotic cells.

In addition to considering the relationship between structure and function of the mitochondrion, several aspects of the structure of the mitochondrion are of interest in evolutionary theory. These organelles of eukaryotic cells show a marked resemblance to the prokaryotic cell. Mitochondria are approximately the same size as a bacterial cell,

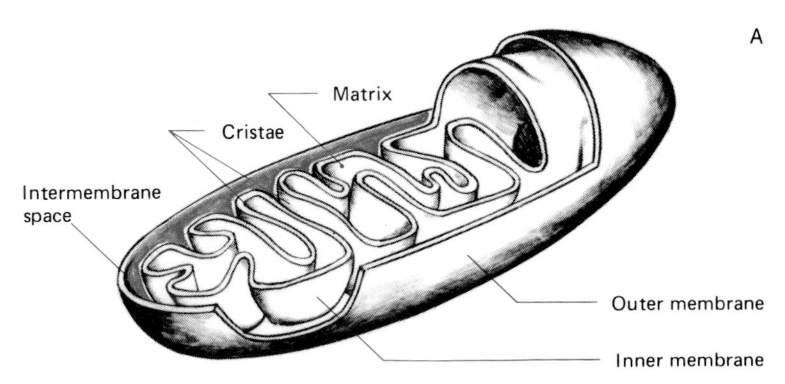

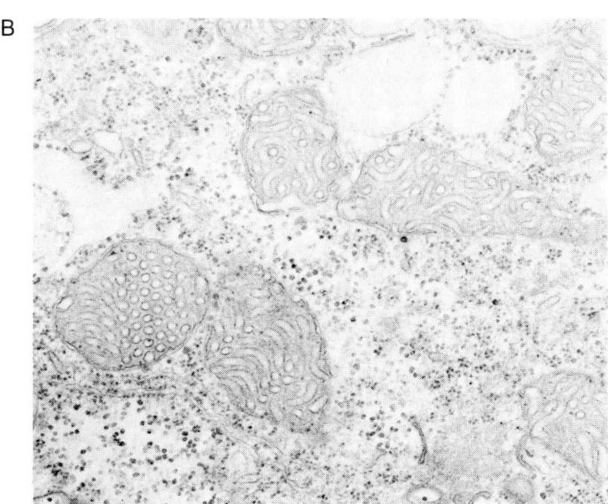

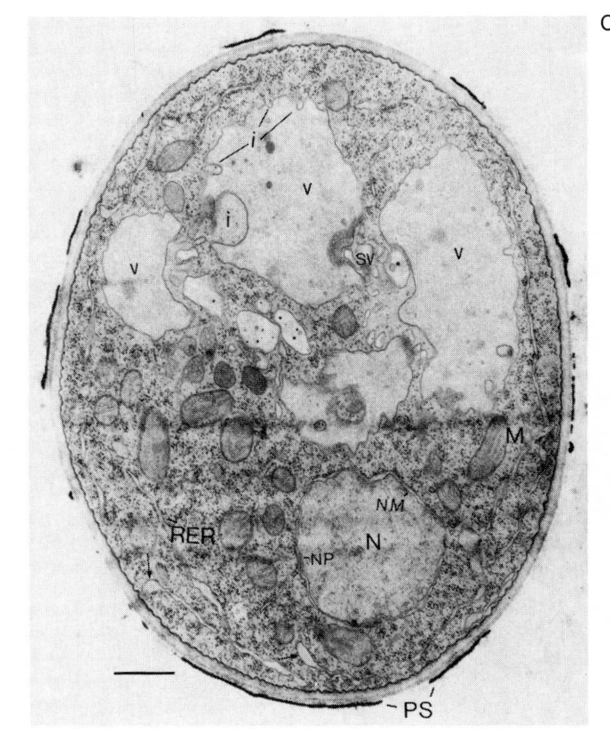

Figure 5.6 (A) Drawing of the structure of a mitochondrion, showing its two distinct membranes and the extensive folding of the internal membrane; (B) electron micrograph of *Tetrahymena rostrata*, showing the mitochondria. As shown here the inner membranes of the mitochondrion of many microorganisms appear as tubular structures. (Courtesy Eugene W. McArdle, Northwestern Illinois University.) (C) Cross section of a growing hypha of the fungus *Trichoderma reesei*: M = mitochondrion with distinct cristae, N = nucleus, NM = nuclear membrane, NP = nuclear pore, RER = rough endoplasmic reticulum, V = vacuole, i = invagination common to vacuoles, SV = smooth surface vesicles, Ps = polysaccharide layer on the outside surface of the cell wall. (Courtesy Bijan Ghosh and Arati Ghosh, Rutgers, The State University.)

a fact that probably explains why one never finds mitochondria within prokaryotic cells. Within the mitochondrion there are 70S ribosomes and a circular strand of DNA arranged in a manner similar to the bacterial chromosome. The mitochondrial membranes appear to lack the sterols found in the cytoplasmic membranes of eukaryotic cells. Further, some of the RNA molecules of the 70S mitochondrial ribosomes show a high degree of similarity to the rRNA of prokaryotic cells. These similarities have led some to propose that mitochondria evolved from prokaryotic cellular organisms. Based on these facts it would appear that the mitochondria have descended from prokaryotic cells that became entrapped in larger eukaryotic cells, eventually forming a stable relationship and evolving into the present mitochondrial structure. However, the DNA contained in mitochondria is not sufficient to direct the synthesis of the entire structure, and it has been recently found that the

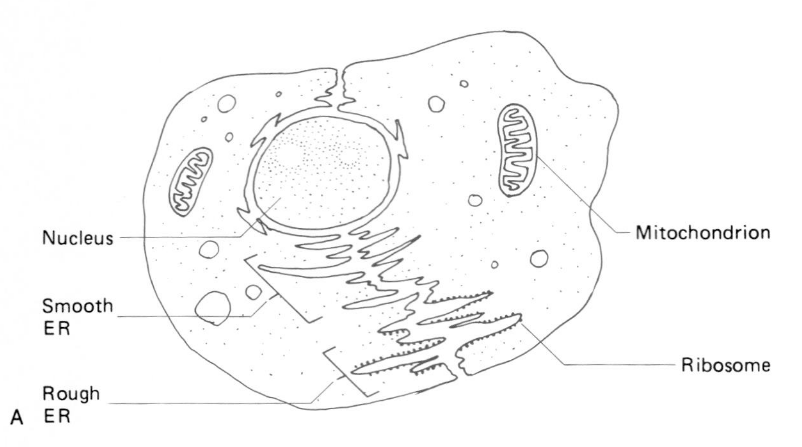

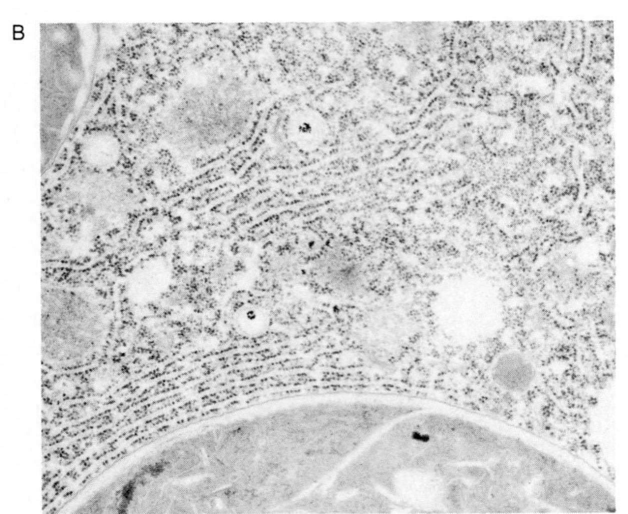

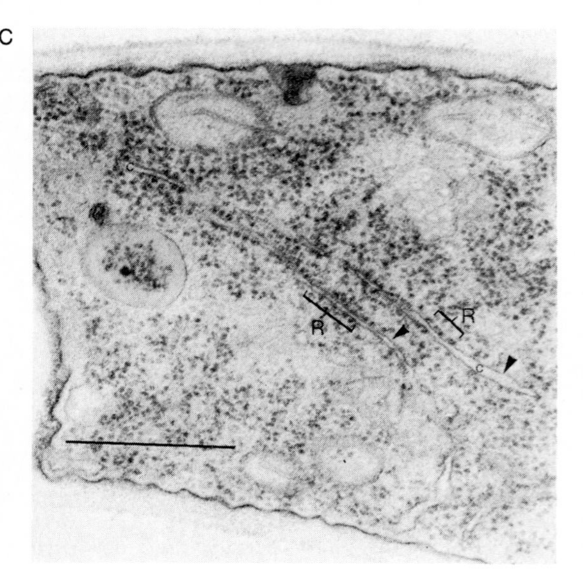

genetic coding of information is different for mitochondrial DNA than for DNA from other sources, fueling the debate over the relatedness of the mitochondria of eukaryotes to the prokaryotic cell.

Chloroplasts

Chloroplasts are quite similar in many ways to mitochondria, as they are composed of extensively invaginated membranes, contain 70S ribosomes, and have a circular DNA macromolecule. Chloroplasts are large cytoplasmic organelles occurring within the cytoplasm of photosynthetic eukaryotic organisms. They are the sites of ATP synthesis and carbon dioxide fixation, which occur in photosynthesis. Among the microorganisms, chloroplasts occur exclusively in the algae. Chloroplasts also occur in plant cells. The chloroplasts contain chlorophylls and the proteins responsible for the conversion of light energy to chemical energy in the form of ATP. The process of ATP generation in the chloroplast is analogous to the synthesis of ATP in the mitochondria, except that the flow of hydrogen ions is inward in the case of the chloroplast and outward in the case of mitochondria.

Figure 5.7 (A) Drawing of rough and smooth endoplasmic reticulum. The terminology is derived from the fact that, when viewed by electron microscopy, the endoplasmic reticulum with attached ribosomes appears bumpy or rough, whereas the endoplasmic reticulum lacking attached ribosomes appears smooth. (B) Electron micrograph of the ribosomes arranged along the rough endoplasmic reticulum of the protozoan *Tetrahymena rostrata*. (Courtesy Eugene W. McArdle, Northeastern Illinois University.) (C) Electron micrograph of a thin section of the fungus *Trichoderma reesei* showing large numbers of ribosomes (R) attached to the endoplasmic reticulum membrane and the cisternal space (c) of the endoplasmic reticulum. (Courtesy Bijan K. Ghosh and Arati Ghosh, Rutgers, The State University.)

Endoplasmic Reticulum

In addition to the chloroplasts and mitochondria, which are membranous organelles involved in energy conversions, eukaryotic cells contain an extensive membranous network known as the endoplasmic reticulum. The appearance of the endoplasmic reticulum varies greatly among different eukaryotic cells but always forms a system of fluid-filled sacs enclosed by the membrane network. This large membranous network appears to serve several functions involved with coordination of the metabolic activities of the cell within the eukaryotic cell.

The endoplasmic reticulum shows two distinct appearances when examined by electron microscopy: in one case the endoplasmic reticulum (ER) appears rough and has attached ribosomes; in the other case the ER appears smooth and is not associated with ribosomes (Figure 5.7). The attachment of ribosomes to the endoplasmic reticulum allows for coordinated activity, and the channels of the endoplasmic reticulum provide a system through which the synthesized protein can be transported. In fact, it appears that many of the proteins synthesized by ribosomes attached to the endoplasmic reticulum are destined to be transported out of the cell or for incorporation into membranes. Proteins synthesized on free ribosomes not associated with the endoplasmic reticulum are not transported through the channels of this membranous network and appear to be destined for use within the cytoplasm of the cell. Additionally, many enzymes are associated with both the smooth and rough endoplasmic reticulum, and this large membrane network appears to provide a large surface for enzymatic activities, including the synthesis of lipids of membrane structures.

The Golgi Apparatus

The Golgi apparatus is closely associated with the rough endoplasmic reticulum and appears to interact with the rough ER to establish an integrated function. Normally, four to eight Golgi bodies, which are flattened membranous sacs, are stacked to form the Golgi apparatus (Figure 5.8). The Golgi apparatus is sometimes referred to as the Golgi complex and the individual stacks of membranes as dictysomes. Golgi bodies are the sites of various synthetic activities through which polysaccharides and lipids can be added to proteins to form lipoproteins and glycoproteins. These complex biochemicals are essential for the synthesis of various cell constituents. The Golgi apparatus also

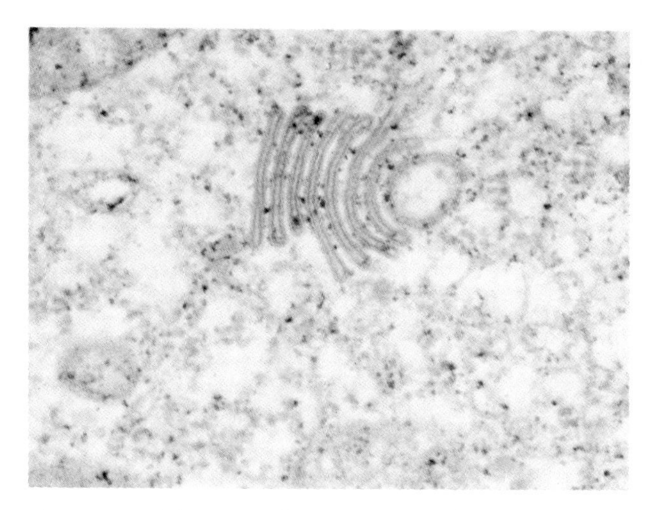

Figure 5.8 Electron micrograph of Golgi apparatus of *Tetrahymena rostrata*, showing stacked Golgi bodies (60,000 ×). (Courtesy Eugene W. McArdle, Northeastern Illinois University.)

plays an important role in the excretion of material from the cell.

Membrane sacs from the endoplasmic reticulum carry protein and lipids to the Golgi apparatus, where repackaging into secretory vesicles occurs. The secretory vesicles, which are formed by the Golgi apparatus, then move to the cytoplasmic membrane, where they release their contents through exocytosis. Such a process is important for the construction of cellular structures that occur externally to the cytoplasmic membrane, such as the cell wall. Taken together, the outer nuclear membrane, the endoplasmic reticulum, the Golgi bodies, and the secretory vesicles form a sequence of coordinated activities that move and process biochemicals associated with protein synthesis through the cytoplasm of the cell to the exterior of the cytoplasmic membrane.

Lysosomes

Lysosomes are specialized membrane-bound organelles that are probably produced in the Golgi apparatus. They contain various enzymes, and various digestive activities of the cell occur within the lysosome. The digestive enzymes must be packaged rapidly into the lysosome to prevent their release into the cytoplasm, and the coordination between the endoplasmic reticulum and the Golgi apparatus establishes the mechanism by which the containment of the enzymes can be achieved. The lysosome membrane is impermeable to the outward movement of these digestive enzymes and is also resistant to their action. This segregation of certain enzymes within the lysosome is necessary to prevent the degradation of essential biochemical components of the eukaryotic cell because the enzymes inside the lysosomes often are capable of digesting many of the cell's structural components. Indeed, one of the functions of the enzymes within the lysosomes is to digest prokaryotic cells that have been ingested by phagocytosis.

Microbodies

Microbodies have been found in eukaryotic cells and appear to be similar in function to lysosomes in that they isolate specialized enzyme functions within the cell. Microbodies are smaller than lysosomes and appear to isolate metabolic reactions that involve hydrogen peroxide. Hydrogen peroxide is produced in various biochemical reactions. It is quite toxic to cells, a fact that we utilize when we apply dilute hydrogen peroxide to a wound as an antiseptic in order to kill contaminating microbes.

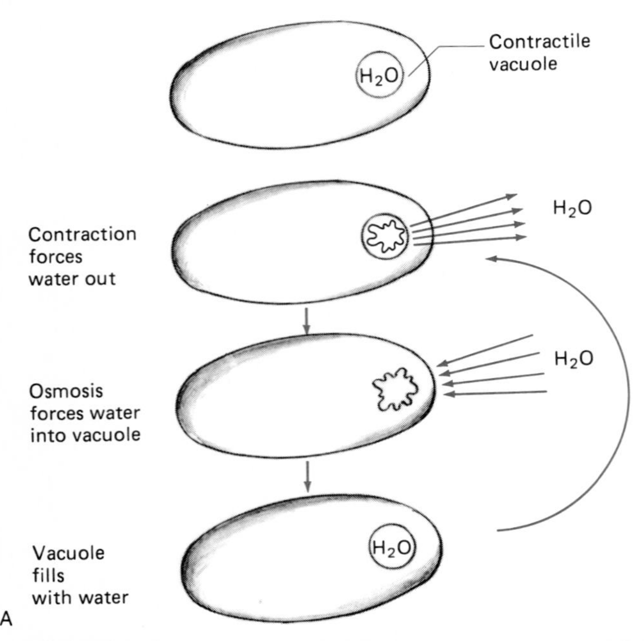

Figure 5.9 (A) Diagram showing the action of the contractile vacuole. (B and C) Electron micrographs showing the contractile vacuole system of *Paramecium aurelia*. (Courtesy Eugene W. McArdle, Northeastern Illinois University. Reprinted by permission from *Paramecium: A Current Survey*, 1974, W. J. Van Wagtendonk, ed., Elsevier Biomedical Press, B.V., Amsterdam.) (B) The nephridial canals (NC) connect directly through openings with surrounding tubules. (C) The discharge channel (dc) is surrounded by microtubules that run in two directions; the contractile vacuole (cv) opens to the outside of the cell. (D) Electron micrograph of a contractile vacuole pore in *Stentor coeruleus* (1,200 ×); CVP = contractile vacuole pore, MB = membranellar band of cilia, FF = frontal field of cilia. (Reprinted by permission of Springer-Verlag, Heidelberg, from R. G. Kessel and C. Y. Shih, 1976, *Scanning Electron Microscopy in Biology: A Students' Atlas on Biological Organization* [Springer-Verlag, Berlin, Heidelberg, New York].)

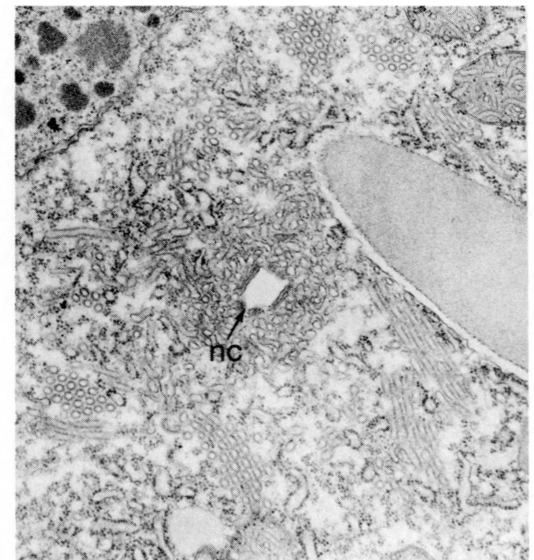

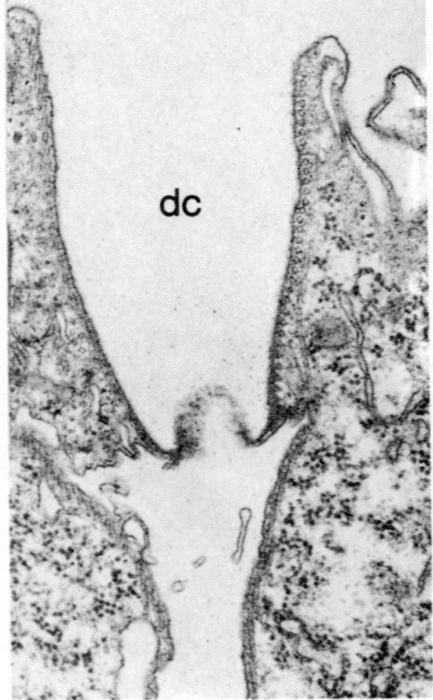

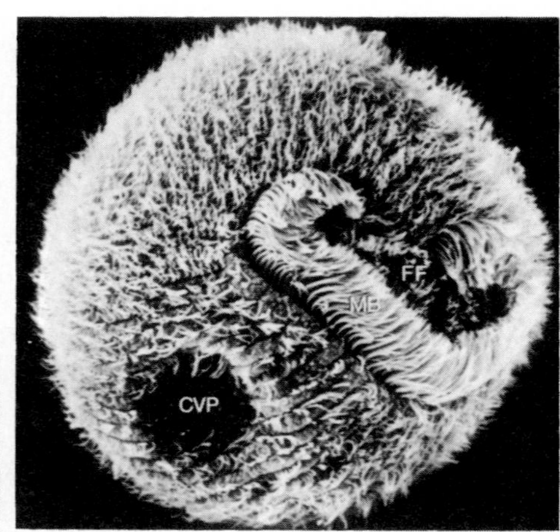

Microbodies contain catalase, an enzyme that immediately breaks down hydrogen peroxide to oxygen and water.

The peroxisome is a special type of microbody that has been found in some eukaryotic microorganisms. Peroxisomes are involved in the oxidation of amino acids, a process that produces hydrogen peroxide. If the peroxides formed in these reactions were not contained or destroyed, they could oxidize a number of biochemicals within the cell, resulting in the death of the cell.

Vacuoles

Various types of membrane-bound vacuoles, serving differing purposes, occur within eukaryotic microorganisms. One type of vacuole, the storage vacuole, is involved in maintaining accumulated reserve materials segregated from the cytoplasm within eukaryotic cells. For example, yeast vacuoles can store polyphosphate, amino acids, and uric acid as reserve materials. Other eukaryotic cells store other forms of organic carbon, nitrogen, and phosphate reserves for times of need. In some cases a vacuole formed when the cell engulfs a food source fuses with lysosomes, establishing a digestive vacuole that permits digestion of the contents.

Other vacuoles are involved in the movement of materials out of the cell. These vacuoles can unite with the cytoplasmic membrane during endo- and exocytosis. A specialized type of vacuole, the contractile vacuole, occurs in some protozoa and acts to pump water and wastes out of the cell, affording protection for the organism against osmotic shock and a mechanism for removing some toxic materials from the cell (Figure 5.9). As water enters the cell because of differences in water potential, it fills the contractile vacuole. The contractile vacuole then suddenly contracts, forcing the water out of the cell. This is an interesting adaptation for protecting the cell against osmotic shock in organisms lacking a rigid cell-wall structure.

Flagella and Cilia

Unlike the bacterial flagella that rotate, the cilia and flagella of eukaryotic cells undulate in a wavelike motion to propel the cell. Eukaryotic flagella emanate from the polar region of the cell, whereas cilia, which are somewhat shorter than flagella, surround the cell. Both cilia and flagella are generally involved in cell locomotion, but cilia may also be involved in moving materials past the cell surface while the organism or cell remains stationary.

The protozoan *Monas stigmata* has been measured to move a distance equivalent to 40 times its own length in a second. No human can cover even seven times his own length in a second.

— *1985 Guinness Book of World Records*

In contrast to the rather simple structure of the bacterial flagellum, the flagella and cilia of eukaryotic microorganisms are thicker and far more complex (Figure 5.10). Both the flagella and cilia of eukaryotic cells have the same basic structure and biochemical composition, consisting of a series of microtubules (hollow cylinders composed of proteins) surrounded by a membrane. The tubular proteins are known as tubulin. The normal arrangement of microtubules in eukaryotic flagella and cilia is termed the "9 + 2" system because it consists of nine peripheral pairs of microtubules surrounding two single central microtubules. The nine pairs of microtubules form a circle surrounding the central microtubules. The peripheral microtubule doublets are linked to the central microtubules by radial spokes of protein microfilaments. The peripheral microtubule doublets are also similarly linked to each other to form a circular network surrounding the spokes. This movement of the microtubules requires energy in the form of ATP. The peripheral spokes of the microtubular network contain the protein dynein, which is involved in coupling ATP utilization to movement of the flagella or cilia. The movement of flagella or cilia appears to be based on a sliding microtubule mechanism in which the peripheral doublet microtubules slide past each other, resulting in bending of the flagella or cilia (Figure 5.11).

Cytoskeleton

In addition to numerous membrane-bound organelles, the eukaryotic cell has a cytoskeletal network. The cytoskeleton consists of microtubules and microfilaments. This cytoskeletal network appears to link the various components of the cytoplasm into a unified structure, the cytoplast, providing the rigidity needed to hold the various structures in their appropriate locations (Figure 5.12). The cytoskeleton is implicated in the ability of the cell to move and to maintain its shape.

The microtubular–microfilament arrangement of the cytoskeleton runs throughout the eukaryotic cell, connecting membrane-bound organelles with the cytoplasmic membrane. This cytoskeletal structure appears to be involved in both the support and movement of membrane-bound structures, includ-

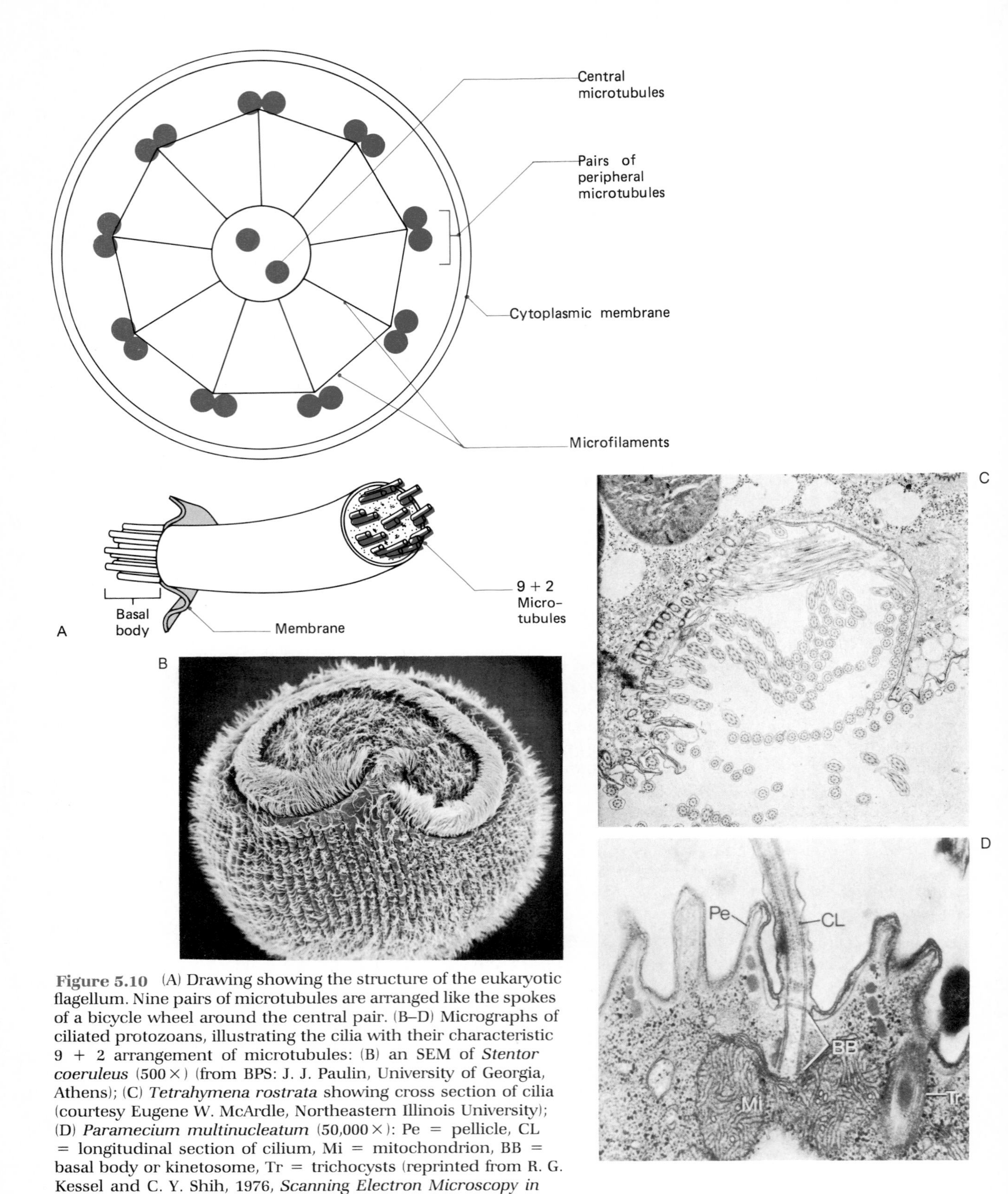

Figure 5.10 (A) Drawing showing the structure of the eukaryotic flagellum. Nine pairs of microtubules are arranged like the spokes of a bicycle wheel around the central pair. (B–D) Micrographs of ciliated protozoans, illustrating the cilia with their characteristic 9 + 2 arrangement of microtubules: (B) an SEM of *Stentor coeruleus* (500×) (from BPS: J. J. Paulin, University of Georgia, Athens); (C) *Tetrahymena rostrata* showing cross section of cilia (courtesy Eugene W. McArdle, Northeastern Illinois University); (D) *Paramecium multinucleatum* (50,000×): Pe = pellicle, CL = longitudinal section of cilium, Mi = mitochondrion, BB = basal body or kinetosome, Tr = trichocysts (reprinted from R. G. Kessel and C. Y. Shih, 1976, *Scanning Electron Microscopy in Biology: A Students' Atlas on Biological Organization* [Springer-Verlag, Berlin, Heidelberg, New York]).

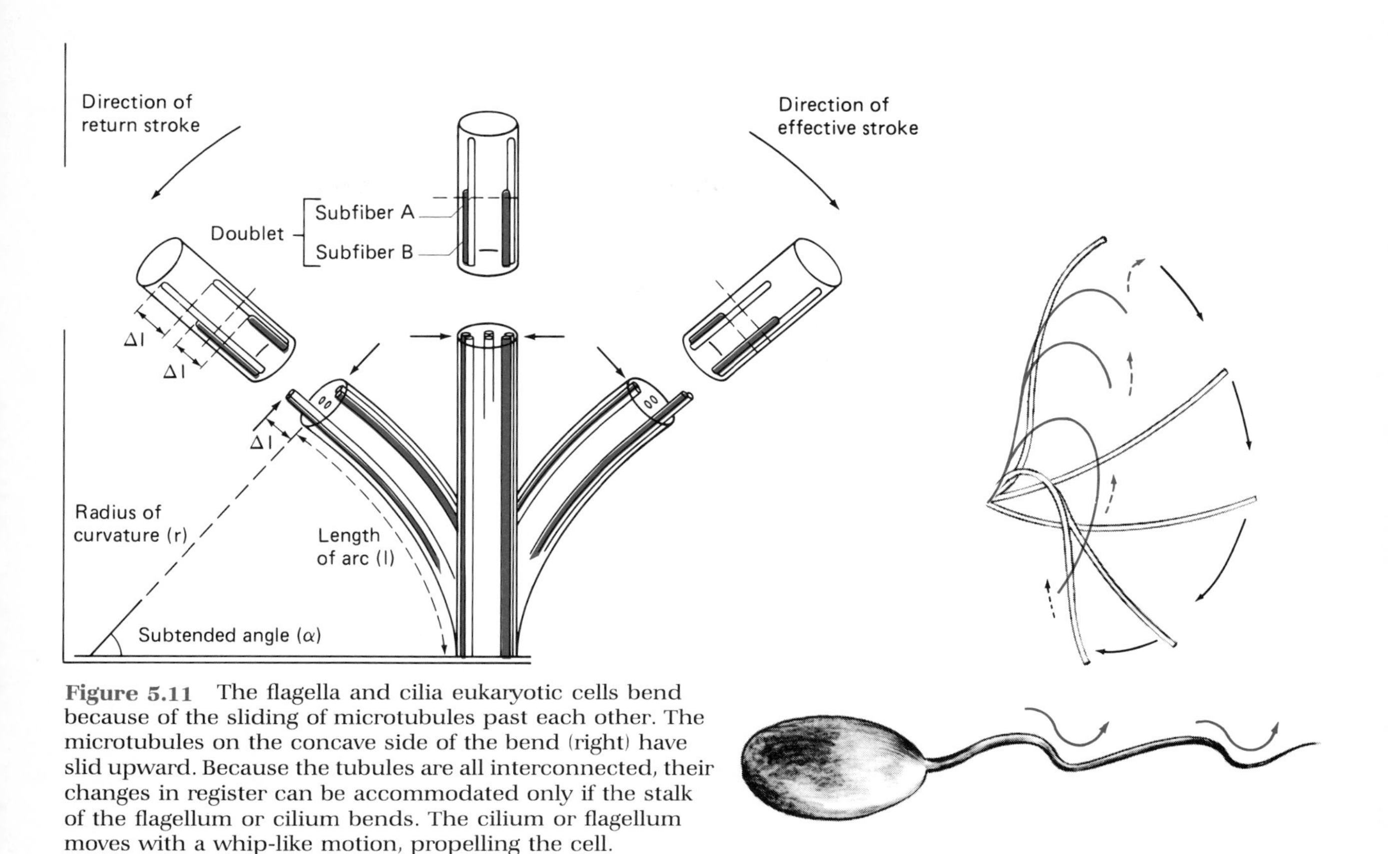

Figure 5.11 The flagella and cilia eukaryotic cells bend because of the sliding of microtubules past each other. The microtubules on the concave side of the bend (right) have slid upward. Because the tubules are all interconnected, their changes in register can be accommodated only if the stalk of the flagellum or cilium bends. The cilium or flagellum moves with a whip-like motion, propelling the cell.

Figure 5.12 Drawing showing the complex cytoskeleton of the eukaryotic cell.

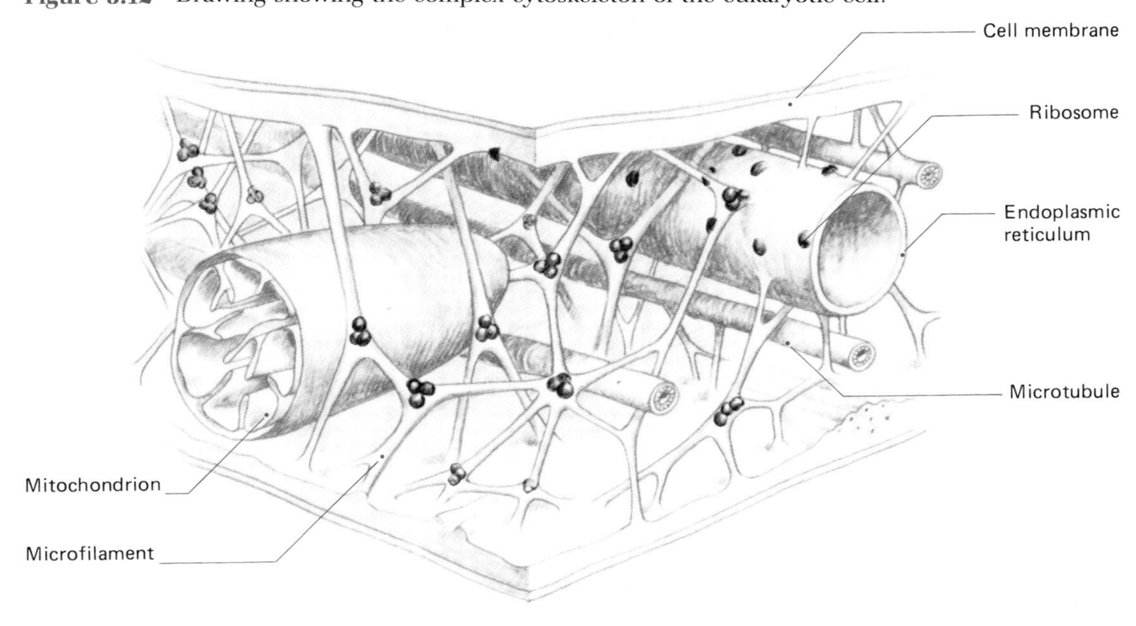

ing the cytoplasmic membrane and the various organelles of the eukaryotic cell. The sliding of microtubules and microfilaments in eukaryotic cilia and flagella provides the basis for locomotion by many eukaryotic cells. The movement of the microtubules also permits the extension of the cytoplasmic membrane and cytoplasm, providing the basis for locomotion through the formation of false-feet (pseudopodia). This means of locomotion occurs in some protozoa, such as *Amoeba*.

Cell Wall

Many eukaryotic microorganisms have a cell-wall structure, the primary function of which is to protect the cell. There is a great diversity of biochemical structures found in the cell walls of eukaryotic microorganisms. In fungi the cell walls are normally composed of cellulose or chitin or a combination of these biochemicals. (Wood is composed largely of cellulose and the shells of crabs are primarily made of chitin.) Detailed analyses of the cell walls of fungi are useful taxonomic criteria for classifying these organisms.

Cell-wall analyses are also used in the classification of the algae. Many algae have a cell wall made up primarily of cellulose, but various other polysaccharides are found as major components of some algae. The cell walls of some algal species contain elements not commonly found in biological structures, for example, calcium or silicon. Such cell-wall structures are sometimes called the test or frustule. The diatoms, for example, have a frustule that contains silicon dioxide. The frustule has two overlapping halves and distinctive markings that give these organisms their characteristically symmetrical and beautiful shapes (Figure 5.13). Large accumulations of diatoms sometimes occur, producing deposits of diatomaceous earth. The coral algae deposit calcium carbonate in their cell-wall structures, forming the basis for coral reefs. These structures protect the cell against physical damage rather than against osmotic shock.

The protozoa usually do not have a true cell wall surrounding the membrane, and many protozoa have developed alternative mechanisms for protection against osmotic shock. Many protozoa do have a thin pellicle surrounding the cell that maintains the shape of the organisms. If the pellicle of a ciliate protozoan such as *Paramecium* is removed, the cell becomes spherical. Some protozoa form an outer wall or shell. For example, the foraminifera produce a wall composed of calcium carbonate, and the radiolaria a wall containing silicon dioxide or strontium sulfate (Figure 5.14). These distinctive structures are preserved long after the organisms die and form part of the fossil record.

Spores

The diversity of cell-wall composition of the cells of eukaryotic microorganisms is matched by the variety of spores that they produce. Spores typically are involved in the reproduction, dispersal, or survival of the organism. Numerous different types of spores are produced by eukaryotic microorganisms. Often a microorganism produces several different spores, each type of spore serving a distinctive function. None of the eukaryotic spores exhibit the heat resistance of the bacterial endospore. Those spores involved in reproduction are metabolically quite active, and those involved in dispersal or survival of the microorganism often are metabolically dormant. Spores involved in the dispersal of microorganisms usually are quite resistant to desiccation, and the production of such spores is an important adaptive feature that permits the survival of microorganisms for long periods of time during transport in the air. The spread of many fungi depends on the successful transport of fungal spores

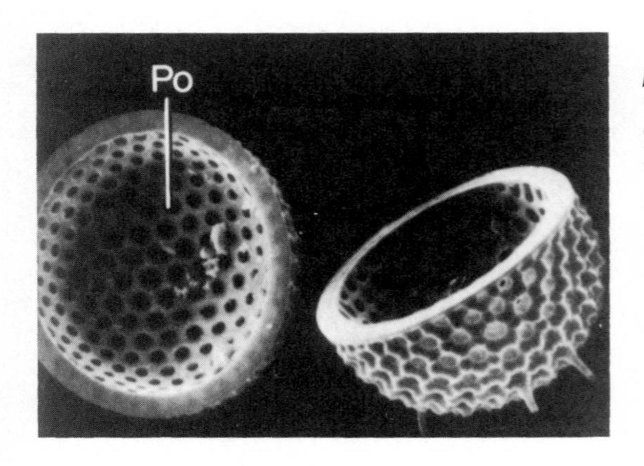

Figure 5.13 Electron micrograph of diatom frustule of *Xanthopyxis* (780×); Po = pore. (Reprinted from R. G. Kessel and C. Y. Shih, 1976, *Scanning Electron Microscopy in Biology: A Students' Atlas on Biological Organization* [Springer-Verlag, Berlin, Heidelberg, New York].)

Figure 5.14 Micrograph of radiolarian shells (400×). (Reprinted from R. G. Kessel and C. Y. Shih, 1976, *Scanning Electron Microscopy in Biology: A Students' Atlas on Biological Organization* [Springer-Verlag, Berlin, Heidelberg, New York].)

from one place to another. Unfortunately, many fungi that are plant pathogens produce large numbers of aerial spores. These fungi cause widespread agricultural damage as a result of their ability to move effectively from field to field. Tracking the aerial dispersal of fungal spores can aid in the prediction of the spread of plant pathogens and allow farmers to take the necessary steps to minimize crop damage.

In addition to the economic importance of the spores of plant pathogens, spores are important in the taxonomic classification of eukaryotic microorganisms. The primary taxonomic separation of the fungi is based upon the production of sexual spores involved in reproduction (Figure 5.15). One of the major divisions of the fungi is defined based upon the formation of sexual spores, called ascospores, within a specialized structure called the ascus. Many yeasts such as *Saccharomyces cereviseae*, the yeast commonly used to make bread, wine, and beer, produce ascospores. Identification of yeasts is based in part upon the shape of the ascospore. Another division of the fungi produces sexual spores, called basidiospores, on a specialized spore-bearing body called the basidium. We see the fruiting bodies of the basidiomycetes, upon which basidiospores are borne, as mushrooms. The identification of wild mushrooms depends upon an accurate description of the fruiting body and the basidiospores. Eating wild mushrooms is very dangerous unless they are positively identified because many mushrooms produce potent poisons.

The yellow-olive death cup (*Amanita phalloides*) is regarded as the world's most poisonous fungus. From 6 to 15 hours after tasting, the effects are vomiting, delirium, collapse, and death. Among its victims was Cardinal Giulio de' Medici, Pope Clement VII (1478–1534).

— *1985 Guinness Book of World Records*

Multicellular Structures

Eukaryotic cells often form unified structures. In plants and animals we find differentiated tissues and organs serving specialized functions. Although they do not exhibit true tissue differentiation, many algae and fungi form characteristic multicellular structures. In some cases these multicellular struc-

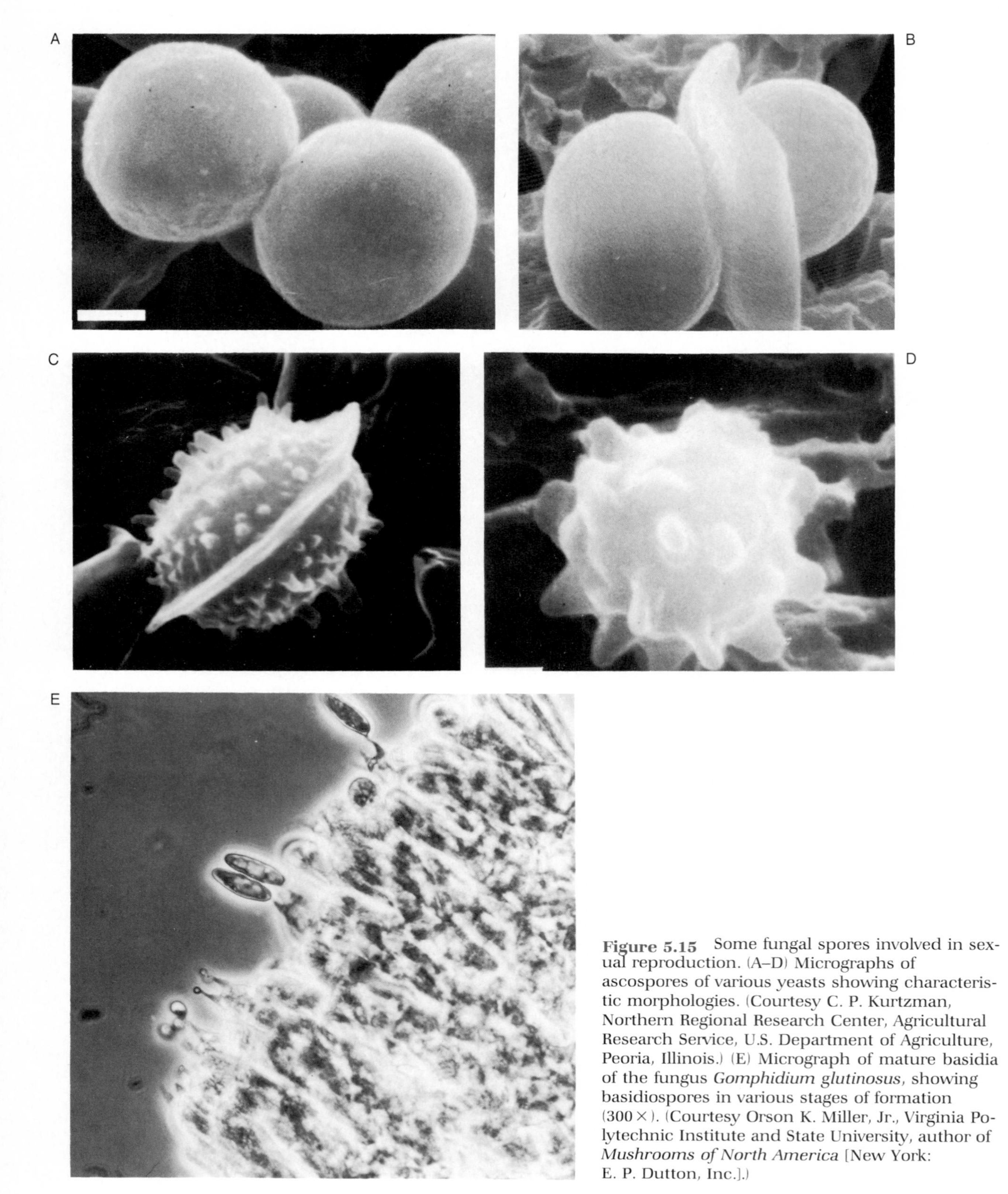

Figure 5.15 Some fungal spores involved in sexual reproduction. (A–D) Micrographs of ascospores of various yeasts showing characteristic morphologies. (Courtesy C. P. Kurtzman, Northern Regional Research Center, Agricultural Research Service, U.S. Department of Agriculture, Peoria, Illinois.) (E) Micrograph of mature basidia of the fungus *Gomphidium glutinosus*, showing basidiospores in various stages of formation (300×). (Courtesy Orson K. Miller, Jr., Virginia Polytechnic Institute and State University, author of *Mushrooms of North America* [New York: E. P. Dutton, Inc.].)

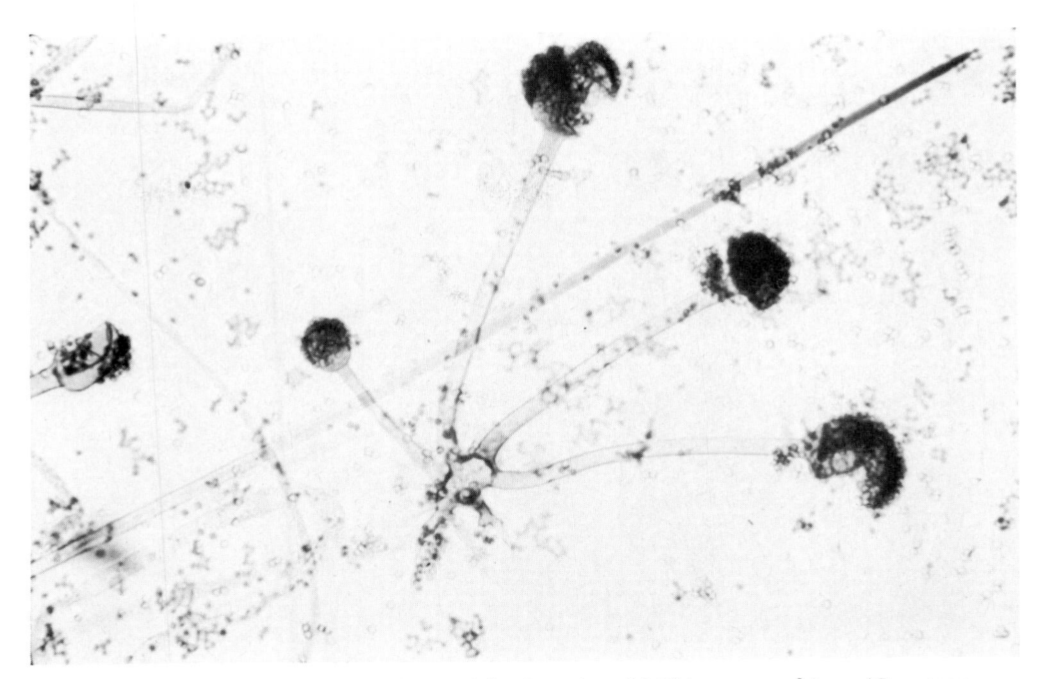

Figure 5.16 Micrograph of mycelium of the bread mold *Rhizopus arrhizus*. (Courtesy C. W. Emmons, Greensboro, North Carolina)

tures form the normal body of the organism. In other cases they are only formed at specific times during the life cycle of the organism.

Human body and human head are recognized as having their billions of cells, and in the cells, billions and billions of molecules, in the molecules, billions and billions and billions of atoms . . . each molecule, each atom is no more an individual than a Roman soldier in a Roman phalanx. Each is the result of change. Each survives because it fits the design, Friend. Everything under the stars must fit the design.

—GUSTAV ECKSTEIN, *The Body Has a Head*

The filamentous fungi form long chains of cells known as hyphae. These fungi are also known as molds. The hyphae form an intertwined mass known as a fungal mycelium (Figure 5.16). We often observe the fluffy mycelia of filamentous fungi growing on surfaces such as bread. The hyphae of some fungal species lack cross walls so that the hypha is really one long multinucleate cell. Such hyphae are termed coenocytic. Many of the fungi exhibit complex life cycles, part of which involves a phase of sexual reproduction. The structures that are produced during such reproduction are quite different from the structures that are formed during normal asexual growth. For example, some fungi produce complex multicellular reproductive structures that we see as mushrooms; the mushrooms are the fruiting bodies of the basidiomycetes.

Like the fungi, many algae produce multicellular arrangements (Figure 5.17). Some algae are filamentous, forming long hyphae; some algae are colonial and form complex integrated units of differentiated cells. The red and brown algae have even more complex multicellular forms; they form complex multicellular bodies of differentiated cells that can be many meters long. The complexity of these structures makes them very much like plants.

Summary

The differences between prokaryotic and eukaryotic cells represent a true split in the architectural strategies of organizing living systems. These different strategies give rise to the formation of distinct structures in prokaryotic and eukaryotic organisms, with some structures occurring exclusively in prokaryotes and others occurring exclusively in

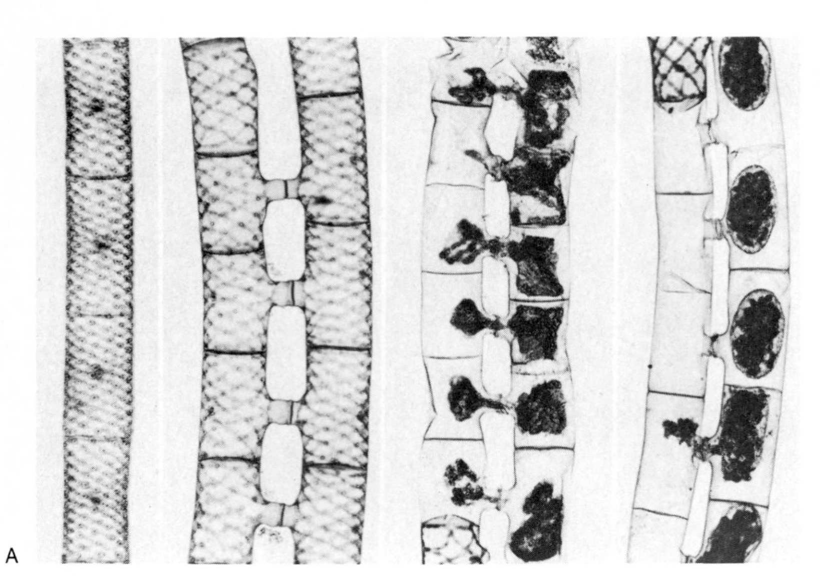

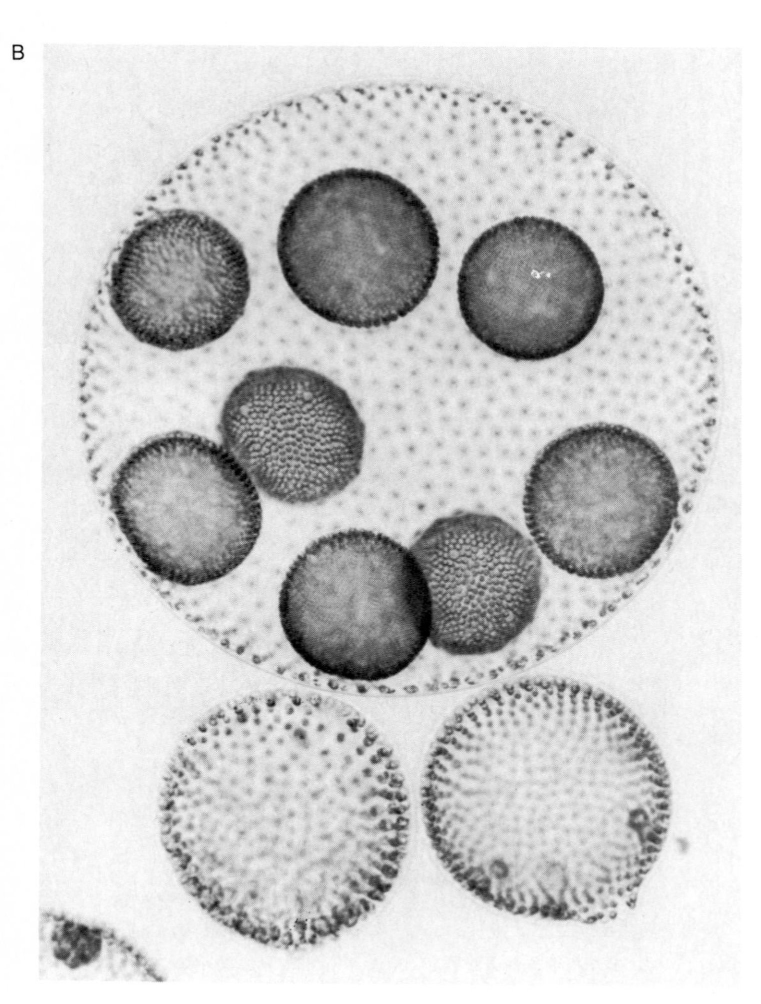

Figure 5.17 (A) Micrograph of the filamentous alga *Spirogyra*; (B) colony of the alga *Volvox aureus*. (Courtesy Carolina Biological Supply Co.)

eukaryotes (Table 5.1). Although the genetic information of both prokaryotic and eukaryotic cells is encoded within DNA, the storage of the information is inherently different. In eukaryotic cells the DNA occurs in linear chromosomes. The chromosomes also contain basic proteins called histones that maintain the specific structure of the chromosome. The DNA is highly coiled and wrapped around histones to form functional units called nucleosomes. The genetic information of eukaryotes often is stored as split sections, whereas in prokaryotic cells functional units occur as contiguous segments of the genome. The nucleus of the eukaryotic cell allows for the processing of the genetic information before it is translated into proteins within a protected organelle. In bacteria such processing cannot occur because the DNA is not segregated within a membrane-bound organelle.

In marked contrast to the prokaryotic cell, the eukaryotic cell is filled with membranous organelles. We have already discussed the importance of the nucleus in separating the genome of the eukaryotic cell from the rest of the cell's contents. The extensive internal membrane systems of eukaryotic microorganisms permit the efficient segregation of function, adding versatility to the metabolic functioning of the eukaryotic cell and increasing the need to coordinate and manage the functions of the cell's subunit organelles. The mitochondria and chloroplasts are specially designed organelles that are involved in energy generation. The extensive membrane surfaces of these two organelles allow the pumping of hydrogen ions for the chemiosmotic generation of ATP. Both are double membrane structures so that the necessary hydrogen ion gradient can be established for chemiosmotic ATP generation.

Other specialized organelles of the eukaryotic cell are involved in nutrition, movement, and reproduction. Various types of vacuoles are involved in the digestion and storage of various nutrients. Enzymes packaged into vesicles at the Golgi apparatus are responsible for some of the specialized digestive activities of the cell. Some of these specialized digestive enzymes occur in lysosomes and microbodies. By packaging these digestive enzymes into such membrane-bound organelles the eukaryotic cell is able to degrade ingested material, including foreign cells, without digesting itself.

Flagella and cilia are responsible for the locomotion of many eukaryotic microorganisms. The flagella and cilia have a "9 + 2" arrangement of microtubules. The movement of the flagella and cilia of eukaryotic cells is very different than the movement of the bacterial flagellum. The sliding of microtubules of the eukaryotic flagellum results in bending that produces a wave-like motion.

Many of the organelles of the eukaryotic cell are linked together so that they can function in a coordinated manner. The eukaryotic cell has an extensive support and communication network that permits the coordination of complex cellular activ-

Table 5.1 Comparison of Eukaryotic and Prokaryotic Cell Structure

Structure	Prokaryotic Cell	Eukaryotic Cell
Cytoplasmic membrane	+	+
Sterols in membranes	−	+
Nucleus	−	+
DNA arranged as true chromosomes with associated histone proteins	−	+
Ribosomes	70S	80S
Ribosomal subunits	50S + 30S	60S + 40S
Cell wall	Contains murein	Several types none with murein
Internal membranous organelles	±	+
Chloroplasts	−	+
Mitochondria	−	+
Endoplasmic reticulum	−	+
Golgi apparatus	−	+
Vacuoles	±	+
Flagella	+	+
9 + 2 microtubular arrangement	−	+
Cytoskeleton	−	+

± denotes general absence although membranous structures occur in some species of prokaryotes

ities. In particular the endoplasmic reticulum connects several organelles so that their activities can be coordinated. Ribosomes on the rough ER are the sites of protein synthesis. The proteins made at these sites can be modified, transported directly to the Golgi apparatus, packaged into vesicles, moved to the cytoplasmic membrane, and exported from the cell. The cytoskeleton also is involved in coordination of cell functions. This structure provides the functional framework to support various cell functions. Many of the organelles of the eukaryotic cell are connected to the cytoskeleton and processes such as cytosis depend upon this support. The ability of the eukaryotic cell to carry out cytosis and the inability of prokaryotic cell to perform this function can be attributed to the presence of the cytoskeleton only in eukaryotic cells.

The differences in structural elements of acellular, prokaryotic, and eukaryotic organisms forms the fundamental basis for many applied aspects of microbiology, including the use of particular antibiotics to control various diseases of humans, other animals, and plants, which will be examined in later chapters.

Study Outline

A. Eukaryotic cell structure.

- **1.** Cytoplasmic membrane.
  - **a.** The cytoplasmic membrane of the eukaryotic cell has the same basic structure as the cytoplasmic membrane of the prokaryotic cell.
  - **b.** The cytoplasmic membrane of a eukaryotic cell contains sterols that are not found in the cytoplasmic membranes of most prokaryotic cells.
  - **c.** The bacterium *Mycoplasma* does have sterols in its cytoplasmic membrane that it obtains from the eukaryotic cells in which it reproduces.
  - **d.** The action of the antibiotic amphotericin B is directed at the cytoplasmic membrane of a eukaryotic cell and works because of the presence of sterols.
  - **e.** Large substances can enter and leave the eukaryotic cell by endocytosis and exocytosis.
  - **f.** Phagocytosis of bacterial cells is important in the nutrition of some protozoa and because of its role in immune defense network of the body; phagocytosis by white blood cells is an important line of defense against bacterial pathogens.
- **2.** Nucleus and chromosomes.
  - **a.** The genetic information of a eukaryotic cell is stored in chromosomes within the nucleus.
  - **b.** The nucleus has a double membrane that contains many pores.
  - **c.** The chromosomes are composed of DNA, which actually encodes the genetic information, and associated proteins called histones, which establish the three-dimensional shape of the chromosome.
  - **d.** Chromosomes are linear and contain double-stranded helical DNA.
  - **e.** The DNA is coiled around histones to form subunits called nucleosomes.
- **3.** Ribosome.
  - **a.** The ribosome is the site of protein synthesis.
  - **b.** The eukaryotic ribosomes found in the cytoplasm are 80S.
  - **c.** The 80S ribosome is composed of 60S and 40S subunits.
- **4.** Mitochondria.
  - **a.** Mitochondria are the sites of extensive ATP synthesis.
  - **b.** A mitochondrion is composed of an outer and an inner membrane; the invaginations of the inner membrane are called cristae.
  - **c.** A hydrogen ion gradient across the inner membrane is used to drive the production of ATP by chemiosmosis.
  - **d.** Mitochondria appear to be related to prokaryotic cells because they have 70S ribosomes, circular chromosomes, no sterols in their membranes, and are the same size and shape as bacteria.

5. Chloroplasts.
    a. The chloroplasts are the sites of light-driven ATP generation in photosynthetic eukaryotes.
    b. The chloroplast has many similarities to both mitochondria and prokaryotic cells.
    c. Chlorophyll in the chloroplasts traps light energy, initiating the biochemical reactions of photosynthesis.
6. Endoplasmic reticulum.
    a. The endoplasmic reticulum (ER) is an extensive membranous network within the eukaryotic cell.
    b. The ER appears to be a communications network and is involved in several coordinated activities.
    c. Smooth ER has no attached ribosomes; rough ER has attached ribosomes.
7. Golgi apparatus.
    a. The Golgi apparatus consists of four to eight stacked, flattened, membranous sacs called Golgi bodies or dictysomes.
    b. Proteins formed on rough ER are transferred to the Golgi apparatus.
    c. The Golgi apparatus is the site where polysaccharides and lipids can be added to proteins to form lipoproteins, glycoproteins, and polysaccharide derivatives.
    d. Substances formed in the Golgi apparatus are packaged in secretory vesicles, which are generally transported to the cytoplasmic membrane where their contents are released.
8. Lysosomes.
    a. Lysosomes, which are probably formed by the Golgi apparatus, contain hydrolytic enzymes.
    b. Lysosomes function by digesting ingested bacteria or other macromolecule nutrients, cell organelles, or stored cellular material.
9. Microbodies.
    a. Microbodies are similar in function and origin to lysosomes.
    b. Microbodies are smaller than lysosomes.
    c. Microbodies are also called peroxisomes.
    d. Amino acids are oxidized within microbodies.
    e. Hydrogen peroxide, a product of metabolism that is lethal to cells, is then further broken down by catalase to water and oxygen within microbodies.
10. Vacuoles.
    a. Vacuoles are membrane-bound structures that serve several different specific functions.
    b. A storage vacuole stores reserve materials.
    c. A digestive vacuole digests material.
    d. A contractile vacuole acts to pump water and wastes of the cells of certain protozoa, thereby preventing osmotic shock.
11. Flagella and cilia.
    a. Flagella and cilia are involved in movement.
    b. Flagella always emanate from the end of a cell; cilia cover the cell surface.
    c. Cilia and flagella have a "9 + 2" arrangement that is composed of a series of nine peripheral pairs of microtubules and two central pairs of microtubules.
    d. The sliding of microtubles past each other produces a bending of the flagellum or cillium and results in a wave-like motion.
12. Cytoskeleton.
    a. The cytoskeleton consists of microtubules and microfilaments.
    b. The cytoskeleton functions to maintain cell shape and aid in cellular movement.
13. Cell wall.
    a. Many, but not all, eukaryotic cells are surrounded by a cell wall that protects the cell against mechanical damage.

b. Fungi often have cellulose or chitin in their cell walls.
c. The cell walls of algae are composed of a variety of biochemicals; diatoms have silicon in their cell walls (frustules), which can result in the formation of deposits of diatomaceous earth; some algae deposit calcium carbonate in their cell walls, which can result in the formation of coral reefs.
d. Protozoa do not have a true cell wall but are surrounded by a thin pellicle.

14. Spores.
a. Eukaryotic microorganisms produce a variety of spores that are involved in survival, dispersal, and reproduction.
b. The primary taxonomic separation of the fungi is based on the production of sexual spores.
c. Ascospores are sexual fungal spores produced within a specialized structure known as the ascus.
d. Basidiospores are sexual fungal spores produced on the surface of a specialized structure known as the basidium. The sexual fruiting bodies of basidiomycetes are seen as macroscopic structures, which we recognize as mushrooms.

15. Multicellular structures.
a. Many algae and fungi form characteristic multicellular structures.
b. The filamentous fungi, which are also known as molds, form long chains of cells known as hyphae. The hyphae form an intertwined mass known as a fungal mycelium.
c. The hyphae of some fungal species lack cross walls so that the hypha is really one long multinucleate cell. Such hyphae are termed coenocytic.
d. Some fungi produce complex multicellular reproductive structures such as the mushrooms, which are the fruiting bodies of the basidiomycetes.
e. Some algae are filamentous, forming long hyphae.
f. Some algae are colonial and form complex integrated units of differentiated cells.
g. The brown algae form complex multicellular bodies of differentiated cells that can be many meters long. The complexity of these structures makes them very much like plants.

Study Questions

1. Fill in the blanks.
a. The eukaryotic nuclear envelope (membrane) contains many ________ that allow substances to enter and leave the nucleus.
b. The genetic material within the nucleus is called ________ and is composed of ________ and ________. Each unit of ________ is called a ________.
c. A type of active transport that occurs only in eukaryotes is called ________.
d. The engulfment of a bacterial cell by a white blood cell is called ________.
e. The site where protein synthesis occurs in the cytoplasm of eukaryotic cells is the ________. It is composed of ________ S and ________ S subunits.
f. Special structures involved in the reproduction, dispersal, or survival of an organism are called ________.
g. In eukaryotic cells, such as fungal and human cells, an organelle composed of an outer membrane and an inner membrane with extensive invaginations is the ________. It is the site of extensive ________ formation. The invaginations are called ________ and are the sites of ________. The space between these is called the ________ and is the site of the ________.
h. An organelle that is responsible for converting light to ATP is the ________.
i. An extensive membranous network in eukaryotic cells is the ________. There are two forms: ________ and ________. Most proteins synthesized on the ________ are for use ________ the cell, and those synthesized on the ________ are for use inside the cell.
j. A stacked structure associated with the endoplasmic reticulum is ________. The "stack" is composed of flattened membranous sacs called ________ or ________.
k. Membrane-bound structures that contain digestive enzymes are called ________.
l. A structure that functions to maintain cell shape and aids in movement is the ________. It consists of ________ and ________.
m. A multicellular mass of hyphae that characterizes filamentous fungi is called ________.
o. A hypha lacking septa is called ________.

p. Spores produced by mushrooms are called __________.

2. Describe three ways eukaryotic chromosomes differ from prokaryotic ones.

3. Name three types of vacuoles found in eukaryotic cells and briefly describe the functions.

4. Describe the functions of the cytoskeleton.

5. Discuss the functions of spores produced by eukaryotic microoranisms.

6. Describe the types of multicellular arrangements formed by eukaryotic microorganisms.

Some Thought Questions

1. Why do eukaryotic cells have organelles?
2. Why is the endoplasmic reticulum considered a communications network?
3. Are plants more similar to humans than bacteria are to fungi?
4. Did mitochondria and chloroplasts come from bacteria?
5. Why don't most protozoa need cell walls?
6. Why are eukaryotic but not prokaryotic cells capable of endocytosis?

Additional Sources of Information

Alexopoulos, C. J., and C. W. Mims. 1979. *Introductory Mycology*. John Wiley & Sons, Inc., New York. A comprehensive text on the fungi that is largely taxonomically oriented but includes a discussion of fungal structure.

Bold, H. C., and M. J. Wynne. 1985. *Introduction to the Algae: Structure and Reproduction*. Prentice-Hall, Inc., Englewood Cliffs, New Jersey. A well written and extensively illustrated text on the algae that emphasizes structure.

Dyson, R. D. 1978. *Cell Biology: A Molecular Approach*. Allyn & Bacon, Inc., Boston. An excellent, comprehensive text on cell structure and function.

Farmer, J. N. 1980. *The Protozoa: Introduction to Protozoology*. The C. V. Mosby Company, St. Louis. An introductory text on the protozoa, including discussions of the structures found in various protozoa.

Karp, G. 1984. *Cell Biology*. McGraw-Hill Book Company, Inc., New York. A well presented text covering all aspects of cell structure and function.

Kornberg, R. D., and A. Klug. 1981. The nucleosome. *Scientific American*, **244**(2):55–72. A detailed article describing the structure of the nucleosome.

Loewy, A. G., and P. Siekevitz. 1970. *Cell Structure and Function*. Holt, Rinehart and Winston, New York. A classical presentation of cell structure.

Miller, O. K. 1979. *Mushrooms of North America*. E. P. Dutton & Co., Inc., New York. A former book of the month club selection, this superbly illustrated work shows the diversity of mushrooms produced by fungi found in North America.

Molecules to Living Cells: Readings from Scientific American. 1980. W. H. Freeman and Company, Publishers, San Francisco. Articles include detailed discussions of the various structures of eukaryotic cells.

Satir, P. 1974. How cilia move. *Scientific American*, **231**(4):44–63. An interesting article on the structure of cilia with a thorough discussion of the basis of movement of cilia.

Wolfe, S. L. 1981. *Biology of the Cell*. Wadsworth Publishing Co., Inc., Belmont, California. A comprehensive discussion of cell structure and function.

UNIT THREE
Microbial Metabolism

6 *The Cultivation of Microorganisms*

KEY TERMS

aerobic
anaerobic
agar
aseptic transfer technique
autoclave
autotrophic
bacteriological filter
colony
culture
dilution gradient
enrichment culture
fastidious
heterotrophic
incubation
in vitro
in vivo
isolation
media (singular, medium)
pour plate
pure culture
selective culture medium
spread plate
sterilization
streak plate
substrate
tissue culture

OBJECTIVES

After reading this chapter you should be able to

1. Define the key terms.

2. Describe the steps that are necessary for establishing a pure culture.

3. Describe the differences between the spread plate, pour plate, and streak plate methods.

4. Describe the general nutritional requirements for the cultivation of bacteria.

5. Describe the differences in nutritional requirements for the cultivation of an autotrophic, heterotrophic, and a fastidious bacterium.

6. Discuss the principles of the enrichment culture method and how it can be applied to the isolation of a bacterium with a particular feature.

Pure Culture Methods

The ability to determine the characteristics of a microorganism depends in large part on being able to grow pure cultures (cultures containing *only* that organism) of that microbe for study. To cultivate, or culture microorganisms it is necessary to establish a suitable environment, one in which the particular microbe can survive and reproduce. For each type of microorganism there are minimal nutritional requirements, tolerance limits for a variety of environmental factors, and optimal conditions for growth. By understanding the growth requirements of a given microbial species, it usually is possible to establish the necessary conditions in vitro (within glass or plastic culture vessels) to support the optimal growth of that microorganism. Cultures are routinely grown in the clinical microbiology laboratory to aid in the determination of the cause of a patient's disease. Water quality testing laboratories culture microorganisms to determine the safety of the water supply. Various industries grow pure cultures of microorganisms in huge vessels called fermentors to produce numerous products of economic value.

Give me matter and motion and I will make the world.

—RENE DESCARTES

The growth of a microorganism in pure culture requires that all other microbial species be eliminated. Obtaining and maintaining pure cultures requires the elimination of other microorganisms from the growth medium (sterilization), the separation of the microorganism being cultured from a mixture of microbes (isolation), the movement of the microorganism from one place to another without contamination (aseptic transfer), and the maintenance of the culture in a favorable enviroment (incubation).

Sterilization

Sterilization procedures eliminate all viable microörganisms from a specified region. Culture dishes, test tubes, flasks, pipettes, transfer loops, and media must be free of viable microorganisms before they can be used for establishing pure cultures of microorganisms. There are various ways of sterilizing liquids, containers, and instruments used

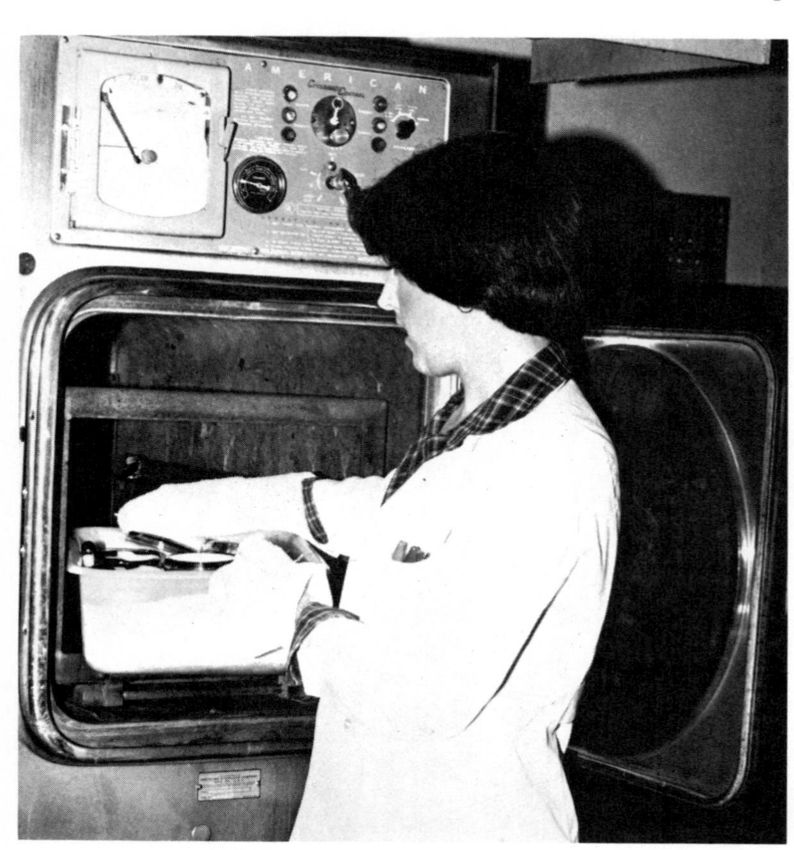

Figure 6.1 Sterilization of materials using an autoclave. (Danae Associates, courtesy Danbury Hospital, Danbury, Connecticut.)

in pure culture procedures. These methods include exposure to elevated temperatures, toxic chemicals, or ionizing radiation to kill microorganisms, and filtration to remove microorganisms from solutions. Sterilization by filtration is accomplished by passage through a 0.2μm or smaller pore size bacteriological filter. In such procedures, most bacteria are trapped on the filter, but viruses and some very small bacteria pass through.

Media preparation for the microbiology laboratory typically involves exposure to high temperatures for a specified period of time. Generally, a temperature of 121°C (achieved by using steam at 15 pounds per square inch) for 15 minutes is used to heat sterilize bacteriological media. An autoclave, which is like a large automated pressure cooker, is used for these sterilization procedures (Figure 6.1). Much time is spent in the preparation of media for the bacteriology laboratory. Mixing and sterilizing the growth media in suitable sterile culture vessels usually takes several hours.

Other methods are used for sterilizing materials that cannot withstand elevated temperatures. Hospitals routinely use ethylene oxide sterilizers for elimiating microorganisms from various materials, ranging from bedding to plastics. Ethylene oxide is toxic to microorganisms. The ethylene oxide sterilizer permits exposure of materials to this toxic substance in an isolated chamber. Slow and thorough venting of the ethylene oxide is necessary because this substance is also very toxic to humans. Exposure to gamma radiation sometimes is used for the sterilization of perishable substances. The product does not come in direct contact with the radiation source and does not acquire any radioactivity. Exposure to gamma radiation, however, is an effective means of sterilizing the product. Most plastic petri plates and culture vessels that are routinely used in microbiology laboratories are sterilized by exposure to radiation.

Aseptic Transfer Technique

Many of the petri dishes and tissue culture plates used for growing pure cultures of microorganisms now are plastic and come presterilized from the manufacturer. Filling these vessels with a sterile medium requires the use of aseptic technique (Figure 6.2). Aseptic technique is also required to transfer the pure culture from one vessel to another.

By aseptic technique the microbiologist means taking all prudent precautions to prevent contamination of the culture. Proper aseptic transfer technique also protects the microbiologist from contamination with the culture, which should always be treated as a potential pathogen. Aseptic technique involves avoiding any contact of the pure culture, sterile medium, and sterile surfaces of the growth vessel with contaminating microorganisms.

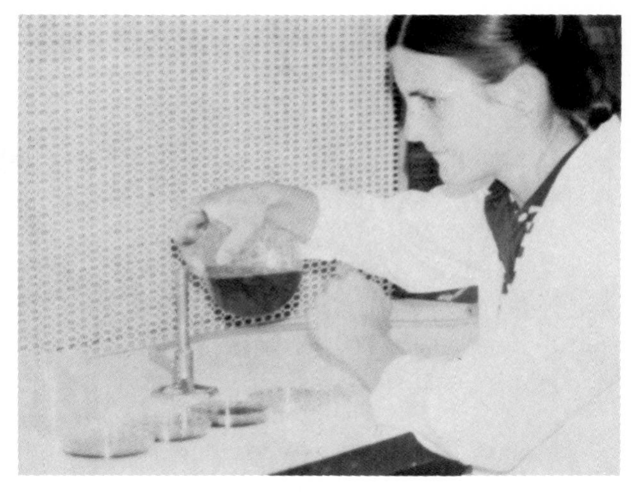

Figure 6.2 A laboratory technician flaming the mouth of a flask during aseptic pouring of medium into a petri dish. It is critical that microorganisms from the environment do not contaminate the medium as it is being poured; flaming the mouth of the flask eliminates a source of potential contaminating microorganisms.

To accomplish this task: (1) the work area is cleansed with an antiseptic to reduce the numbers of potential contaminants; (2) the transfer instruments are sterilized, for example, the transfer loop is sterilized by heating in a bunsen burner flame before and after transferring (Figure 6.3); and (3) the

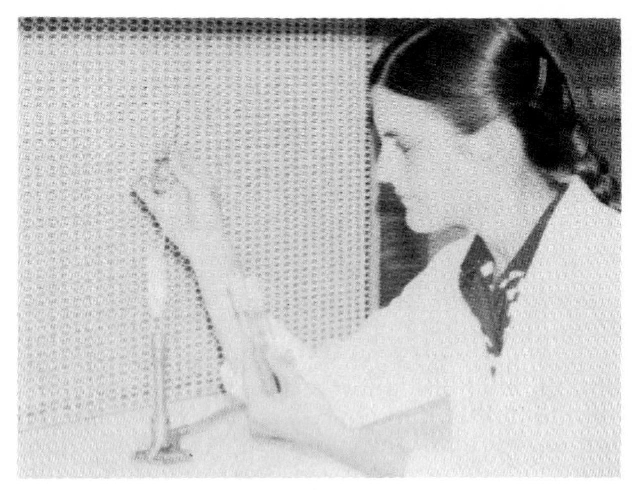

Figure 6.3 A laboratory technician is shown here flaming an inoculating loop. To prevent contamination and spread of bacteria, the loop is flamed before and after transferring bacterial cultures.

Figure 6.4 Steps in the aseptic transfer of bacteria.

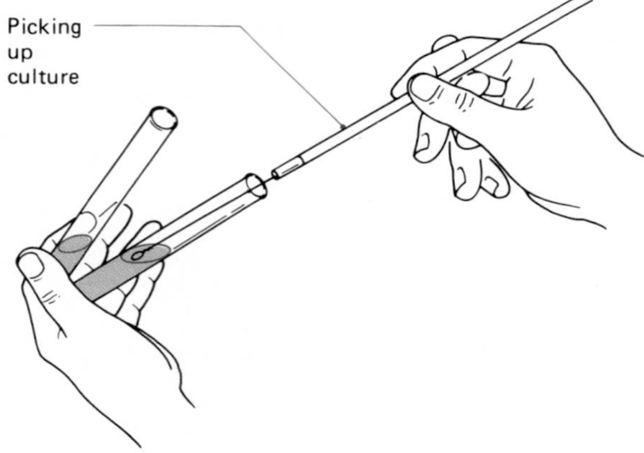

work is accomplished quickly and efficiently to minimize the time of exposure during which contamination of the culture or laboratory worker can occur.

The normal steps for transferring a culture from one vessel to another are (1) flame the transfer loop; (2) open and flame the mouths of the culture tubes; (3) pick up some of the culture growth and transfer it to the fresh medium; (4) flame the mouths of the culture vessels and reseal them; and (5) reflame the inoculating loop (Figure 6.4). Essentially the same technique is used for inoculating petri dishes and for transferring microorganisms from a culture vessel to a microscope slide. We do not however, flame the mouth of the petri plate.

Isolation

Several different methods are used for the isolation of pure cultures of microorganisms. These methods involve separating microorganisms on a solid medium into individual cells that are then allowed to reproduce to form a colony. A colony is

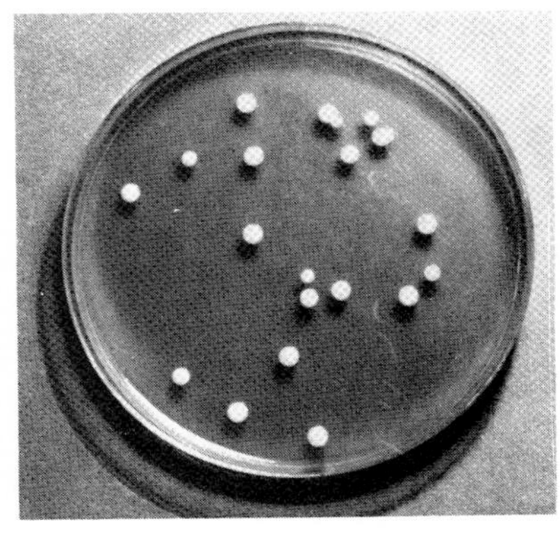

Figure 6.5 Photograph of bacterial colonies growing on two types of agar media in petri dishes. The larger colonies are *Staphylococcus aureus* and the smaller colonies *Micrococcus luteus*. (From K.-H. Schleifer and E. Kramer, 1980, *Zbl. Bakt. Hyg. I Abt. Orig. C.*, 1:270–280. Reprinted by permission of Gustav Fischer Verlag.)

a visible clone of microorganisms (Figure 6.5). The most commonly used solidifying agent for preparing media is agar, a relatively nonreactive algal extract. Agar is particularly useful because it melts at 100°C but does not resolidify until it cools to 42°C. Often the agar is contained within a petri plate. Both the use of agar and the design of the petri plate were methodologies introduced in the nineteenth century by the laboratory of Robert Koch.

Generally a colony that is well separated from others (isolated) arises from a single bacterial cell. Isolated colonies, thus, usually are pure cultures of a single microbial type. Isolation is achieved by the physical separation of the microorganisms, but the success of an isolation method also involves the maintenance of the viability and growth of a pure culture of the microorganism. Care must be taken to ensure that the microorganisms are not killed during the isolation procedure, which can easily occur by exposing the microorganisms to conditions they cannot tolerate, such as air in the case of obligately anaerobic microorganisms that are sensitive to oxygen. The success of an isolation method also depends on being able to grow the microorganism, that is, to define the growth medium and to establish the appropriate incubation conditions that permit the growth of the microorganism.

Streak Plate. Several different streaking patterns can be used to achieve separation of individual bacterial cells on the agar surface (Figure 6.6). In the streak plate technique, a loopful of bacterial cells is streaked across the agar-solidified surface of a nutrient medium. The plates are then incubated under favorable conditions to permit the growth of the microorganisms. The key to this method is that, by streaking, a dilution gradient (a decreasing concentration of bacterial cells) is established across the face of the plate, so that while confluent growth

Figure 6.6 Streaking for the isolation of pure cultures, showing two different streaking patterns. In this procedure a culture is diluted by drawing a loopful of the organism across a medium until only single cells are deposited at a given location. The multiplication of each isolated cell results in the formation of a discrete colony.

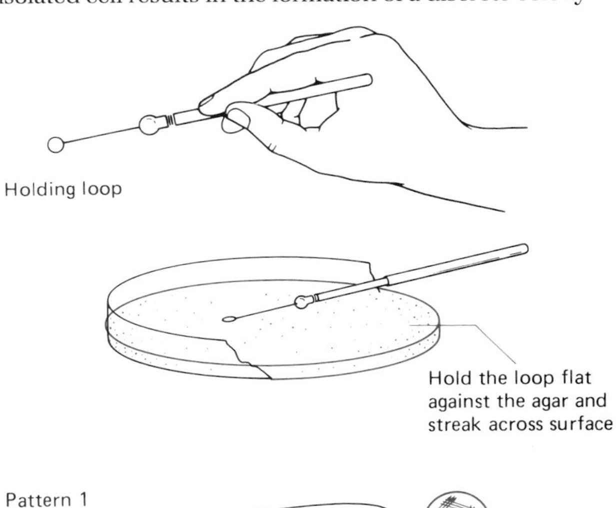

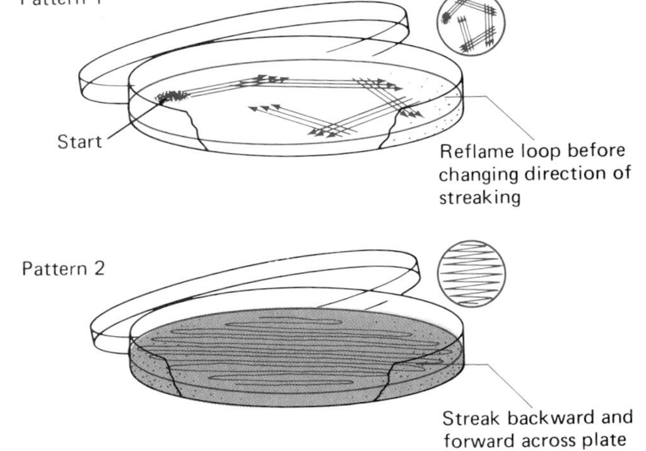

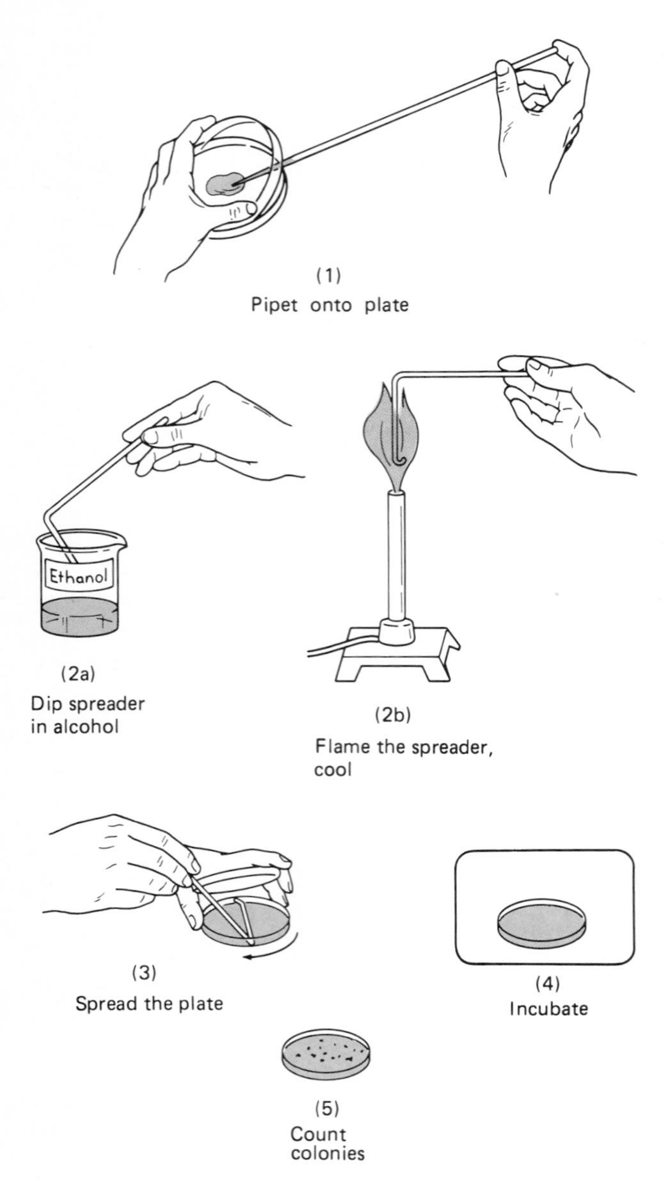

Figure 6.7 The spread plate technique for enumerating microorganisms. (1) Aseptically apply a known volume of a suspension to a suitable solid medium; (2) sterilize a spreading rod by dipping in alcohol and flaming; (3) use the sterile rod to spread the suspension over the surface of the medium; (4) incubate; (5) count colonies and calculate number of microorganisms in the original suspension.

(overlapping colonies) occurs on part of the plate where the bacterial cells are not sufficiently separated, isolated colonies do develop in another region of the plate. The colonies are macroscopic and easily seen with the naked eye. In theory, each discrete (well-isolated) colony arises from a single bacterium and represents a clone of a pure culture. The isolated colonies can then be picked, using a sterile inoculating loop and restreaked onto a fresh medium to ensure purity. A new colony is again picked and transferred to an agar slant or other suitable medium for maintenance of the pure culture.

Spread Plate. In the spread plate method a small volume of a suspension of microorganisms is placed on the center of an agar plate and spread over the surface of the agar by using a sterile glass rod (Figure 6.7). The glass rod is normally sterilized by dipping in alcohol and flaming to burn off the alcohol. By spreading the suspension over the plate, an even layer of cells is established so that individual microorganisms are separated from the other organisms in the suspension and deposited at a discrete location. In order to accomplish this, it is often necessary to dilute the suspension before application to the agar plate to prevent overcrowding and the formation of confluent growth rather than the desired development of isolated colonies. After incubation isolated colonies are picked and streaked onto a fresh medium to ensure purity.

Pour Plate. In the pour plate technique, suspensions of microorganisms are added to melted agar tubes that have been cooled to approximately 42–45°C (Figure 6.8). The bacteria and agar medium are mixed well and the suspensions are poured into sterile petri dishes using aseptic technique. The agar is allowed to solidify, trapping the bacteria at separate discrete positions within the matrix of the medium. Although the medium holds bacteria in place, it is soft enough to permit growth of bacteria and the formation of discrete isolated colonies both within the fluid and on the surface of the agar. As with the other isolation methods, individual colonies are then picked and streaked onto another plate for purification. In addition to its use in isolating pure cultures, the pour plate technique is used for the quantification of numbers of viable bacteria. The facts that agar solidifies below 42°C and that many bacteria survive at these temperatures ensure the success of this isolation technique. However, in some cases significant numbers of bacteria can be killed by the heat shocking so that this method cannot always be used.

Incubation

After the agar plates are appropriately inoculated, the plates must be incubated under suitable environmental conditions. If conditions are favorable the microorganisms will reproduce until they

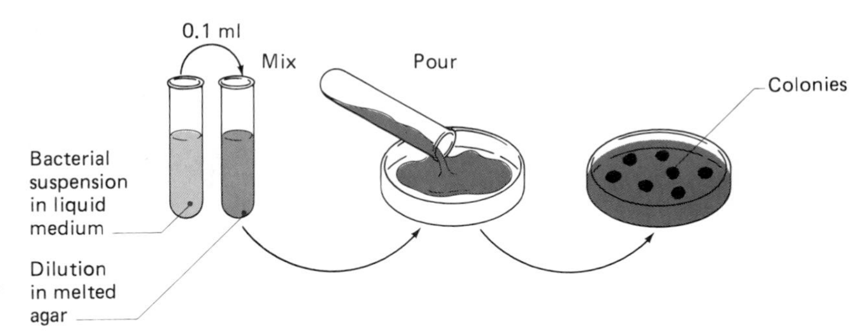

Figure 6.8 The pour plate technique for enumerating microorganisms. A known volume of a microbial suspension is mixed with a liquefied agar medium and poured into a petri plate. After incubation the numbers of colonies that develop are counted, and the concentration of microorganisms in the original suspension is calculated.

form a colony. Various environmental parameters are controlled in the laboratory for cultivating microorganisms. Among the most critical parameters that influence microbial growth are temperature and oxygen concentration.

During incubation cultures are routinely maintained within a temperature controlled incubator (Figure 6.9). Each microorganism has an optimal growth temperature. For most bacteria of medical importance the optimal growth temperature is approximately 37°C, the normal human body temperature. Thus the incubators of the clinical microbiology laboratory typically are set at 37°C.

The concentration of oxygen also is adjusted for the cultivation of microorganisms. Some microorganisms, called obligate aerobes, require oxygen; others, termed obligate anaerobes, can only grow in the absence of oxygen. To grow aerobic microorganisms we simply incubate in the presence of air. The overlapping lid of the petri plate permits the exchange of gases with the surrounding atmosphere without permitting contaminating microorganisms to enter the plate. To cultivate anaerobic microorganisms we can use specialized incubaton chambers that remove oxygen from the atmosphere under which the plates are incubated. In the widely used GasPak system oxygen is replaced with carbon dioxide and hydrogen (Figure 6.10). Other more elaborate incubation systems are also used to permit the cultivation of microorganisms in an environment that is free from molecular oxygen.

Other environmental factors can also be controlled for culturing microorganisms. For example, some pathogenic microorganisms grow best under an atmosphere with elevated carbon dioxide. Special incubators, which permit control of the carbon dioxide concentration, are used for the cultivation of such microbes. Other microorganisms grow best at particular humidity levels or light intensities. Again these are parameters that can be easily controlled to optimize growth conditions.

Assuming optimal growth conditions most microorganisms reproduce very rapidly. Visible colonies of most microorganisms develop in less than 24 hours if the culture is incubated under optimal conditions. The ability to obtain pure cultures within a matter of hours is especially important in the clinical microbiology laboratory where speed is essential. Many microbial identifications can be completed in less than a day, permitting the physician to begin appropriate treatment in a timely fashion.

Figure 6.9 Photograph of a technician loading an incubator. (Danae Associates, courtesy Danbury Hospital, Danbury, Connecticut.)

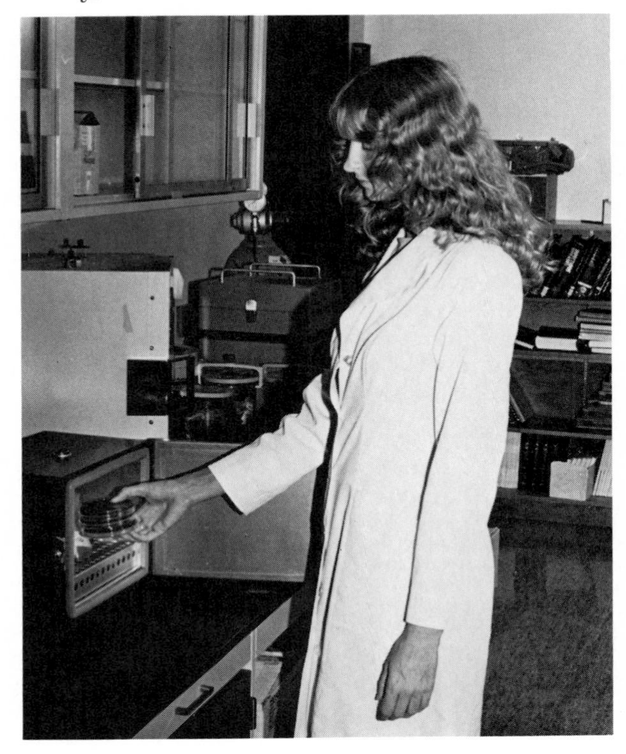

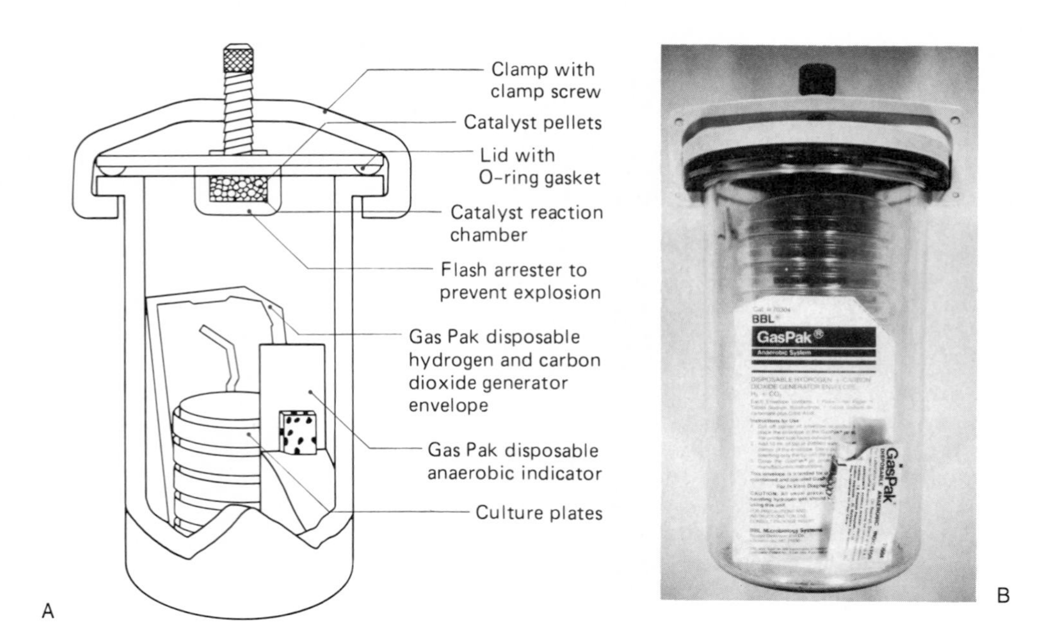

Figure 6.10 Anaerobic jars are used to maintain cultures under strict anaerobic conditions. (A) Diagram and (B) photograph of the anaerobic incubation of microbial cultures using the Gas Pak system.

Microbial Nutrition

To achieve optimal growth, the composition of the culture medium also must be suitable for the cultivation of the particular microorganism. A variety of organic and inorganic nutrients are required for microbial metabolism. All organisms require carbon, nitrogen, oxygen, hydrogen, phosphorus, sulfur, and various other elements. These elements must be available in a usable chemical form to meet the nutritional requirements of a microorganism for that organism to grow. In nature the distribution of microorganisms is determined in part by the availability of specific nutrients in a given environment needed to support the growth of particular microbial species.

It's a very odd thing—
As odd as can be—
That whatever Miss T. eats
Turns into Miss T.

—WALTER DE LA MARE, *Miss T*

The specific nutritional requirements for different microorganisms vary greatly. The substrates (growth substances) that a microorganism can utilize and the specific growth factors it requires for growth are determined by the enzymatic capabilities of that microorganism as specified by its genetic information. For example, heterotrophic microorganisms require organic matter, but autotrophic (self-feeding) microorganisms do not.

Autotrophic organisms obtain their cellular carbon from inorganic carbon, generally by the fixation of carbon dioxide. They obtain their energy from light or inorganic compounds. Often autotrophic microorganisms are grown in a medium containing simple inorganic compounds. Plants, algae, and several ecologically important groups of bacteria are autotrophic.

In contrast to autotrophs, heterotrophic organisms require organic compounds as a source of cellular carbon and energy. Some heterotrophic microorganisms have simple nutritional requirements and are able to reproduce by using a single carbon source as their growth substrate. Such organisms have the metabolic capability of converting a particular organic carbon compound, such as glucose, into all the macromolecular constituents that compose the structures of those organisms. Other heterotrophic microorganisms have more complex growth requirements which must be met before they can be grown in vitro. The nutritionally most demanding microorganisms are termed fastidious; some such fastidious microorganisms are important human pathogens, and their culture in the clinical microbiology laboratory requires the use of various specialized media. Fastidious microorganisms are able to reproduce only under very restricted conditions. In some cases we are unable to grow such microorganisms in pure culture because we have been unable to define and to create the necessary growth conditions.

For the growth of heterotrophic microorganisms, an organic substrate serves as the source of essential carbon and energy. Normal laboratory culture media often contain proteins and/or carbohydrates as growth substrates (Table 6.1). For the growth of specific heterotrophic microorganisms, individual growth substrates may be included in the culture medium.

A culture medium must also include a variety of required inorganic chemicals to support the growth of either heterotrophic or autotrophic microorganisms. A general growth medium normally contains a source of nitrogen, such as ammonium nitrate, phosphate in the form of phosphate buffer, sulfate, magnesium, sodium, potassium, and chloride ions. These inorganic chemicals are required for the biosynthesis of a variety of cellular biochemicals and for the maintenance of transport activities across the cytoplasmic membrane. Microorganisms generally have many other specific inorganic nutritional requirements. For example, various metals, like zinc, manganese, and copper among others, are generally required as trace elements. Some growth factors, such as vitamins and amino acids, may also be included in a medium designed to support the growth of microorganisms.

Not all microorganisms can be grown by using defined media. The nutritional requirements of many microorganisms are simply not known. These microorganisms are able to reproduce in nature where their nutritional needs are met, but we do not understand their growth requirements well enough to define the appropriate laboratory conditions needed for their growth. For example, we are typically able to culture less than 1 percent of the microorganisms that are present in a natural water or soil sample.

In some cases the natural environment in which the microorganism reproduces can be established in the laboratory without actually defining the growth requirements. For example, soil extract is often added to growth media to support the reproduction of soil microorganisms. Some microorganisms, such as viruses or other obligate intracellular microbial species can reproduce only within cells. Generally such microorganisms have limited metabolic capabilities and rely on the metabolism of the host cell to meet their nutritional needs. Such organisms can be grown in tissue culture or by using other methods that provide the necessary cells within which these microorganisms can reproduce. In these methods the host cells are cultivated in vitro (within glass) and the microorganisms are grown in vivo (within a living system) inside of the host cells. For example, some viruses are cultivated by inoculating them into eggs containing developing chicken embryos (Figure 6.11). Other viruses are cultivated in human or other animal cells that are grown as a tissue culture.

Often, the task of defining the proper medium for growing such organisms is tedious and taxes the creativity of the microbiologist. In many cases, rather than defining the specific constituents, complex media are used. For example, many media contain beef extract (a complex mixture of proteins, carbohydrates, lipids, and other biochemical constituents obtained by extracting the water soluble components from beef tissue), peptones (an enzymatic digest of protein-containing amino acids and other nitrogen-containing compounds, as well as

Table 6.1 The Composition of Some Culture Media for the Growth of Heterotrophic Bacteria

Nutrient Broth (general medium for cultivation of heterotrophic bacteria)	
Beef extract	3 g
Peptone	5 g
Water	1000 g
Glucose–Inorganic Salts Medium (for growth of *E. coli*)	
Glucose	5 g
K_2HPO_4 (potassium dibasic phosphate)	7 g
KH_2PO_4 (potassium monobasic phosphate)	2 g
$MgSO_4$ (magnesium phosphate)	0.1 g
$(NH_4)_2SO_4$ (ammonium sulfate)	1 g
Water	1000 g
Glucose–Yeast Extract Medium (for growth of bacteria with growth factor requirements)	
Glucose	5 g
K_2HPO_4 (potassium dibasic phosphate)	7 g
KH_2PO_4 (potasssium monobasic phosphate)	2 g
$MgSO_4$ (magnesium phosphate)	0.1 g
$(NH_4)_2SO_4$ (ammonium sulfate)	1 g
$FeSO_4$ (ferrous sulfate)	0.1 g
Yeast extract (contains numerous vitamins, amino acids, and other growth factors)	5 g
Water	1000 g

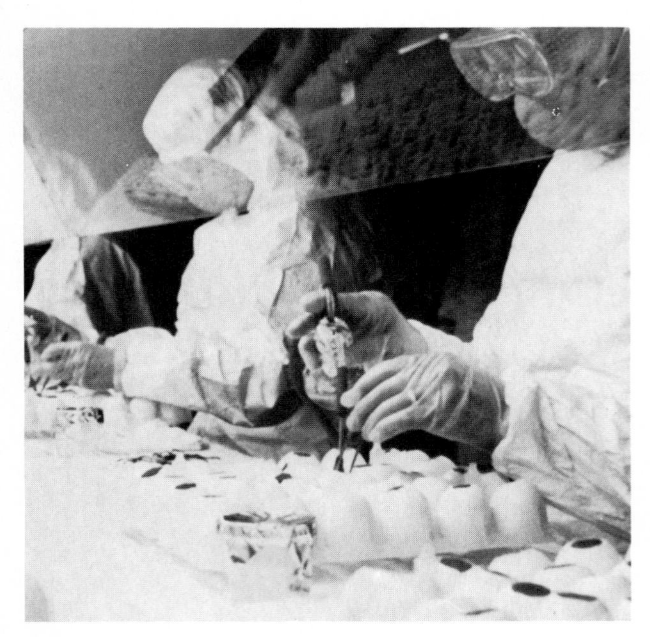

Figure 6.11 Photograph showing inoculation of chick embryos for the cultivation of viruses. (Courtesy Wyeth Laboratories, Philadelphia.)

vitamins and other compounds), and yeast extract (an aqueous extract of yeast cells containing vitamins and other growth factors). In the clinical laboratory we often incorporate blood into media that are designed for the culture of human pathogens. Such complex culture media will support the growth of many different types of heterotrophs but not those fastidious organisms that have very specific growth requirements or those organisms that cannot tolerate particular components in the undefined medium.

Enrichment Culture Technique

By taking into account the metabolic capabilities of specific microorganisms, it is possible to design growth media that will select for the growth of particular microorganisms based on their nutritional requirements. In the enrichment culture technique employed to isolate specific groups of microorganisms, the culture medium and incubation conditions are designed to favor the growth of a particular microorganism (Table 6.2). For example, in order to isolate microorganisms capable of metabolizing petroleum hydrocarbons, one can design a culture medium containing a hydrocarbon as the sole source of carbon and energy. By doing so, one establishes conditions whereby only microorganisms capable of metabolizing hydrocarbons can grow. Because other microorganisms cannot reproduce in this medium, one thereby selects for hydrocarbon-utilizing microorganisms.

Similarly, a culture medium that favors the growth of a particular pathogen can be used to increase the numbers of that organism before attempting to

Table 6.2 Some Culture Conditions That Are Employed for Enrichment Cultures

Carbon Source	Oxygen	Nitrogen Source	Inorganic Sulfur Source	Light	Organism Enriched for
Nonfermentable organic carbon	+	N_2	–	–	*Azotobacter*
Nonfermentable organic carbon	–	NO_3^-	–	–	Denitrifying bacteria
Nonfermentable organic carbon	–	–	SO_4^{2-}	–	*Desulfovibrio*
Hydrocarbon	+	NH_4^+	SO_4^{2-}	–	Hydrocarbon oxidizers
–	+	NH_4^+	SO_4^{2-}	–	*Nitrosomonas*
–	+	NO_2^-	SO_4^{2-}	–	*Nitrobacter*
–	+	–	S	–	*Thiobacillus*
–	+	N_2	SO_4^{2-}	+	Cyanobacteria
–	–	NH_4^+	H_2S	+	Green sulfur bacteria

isolate it in pure culture. Such enrichment culture techniques are sometimes used in the clinical microbiology laboratory to increase the likelihood of isolating a particular disease-causing microorganism, which might otherwise be obscured by the occurrence of larger numbers of nonpathogenic microorganisms in a sample. The enrichment culture technique models many natural situations in which the growth of a particular microbial population is favored by the chemical composition of the system and by environmental conditions.

Selective Culture Media

Microbial reproduction is not only affected by the qualitative composition of the growth medium but also by the concentrations of the various components in the medium. Although a nutrient may be essential for growth in low concentrations, it may be toxic at higher concentrations. For example, some autotrophic microorganisms require trace amounts of growth factors but cannot grow in the presence of high amounts of organic matter. Similarly, many microorganisms may use hydrogen sul-

Table 6.3 Selective Media for Isolation of Enterobacteriaceae

MacConkey Agar

MacConkey agar is a differential plating medium for the selection and recovery of Enterobacteriaceae and related enteric Gram negative rods. Bile salts and crystal violet are included to inhibit the growth of Gram positive bacteria and some fastidious Gram negative bacteria. Lactose is the sole carbohydrate. Lactose-fermenting bacteria produce colonies that are varying shades of red because of the conversion of the neutral red indicator dye (red below pH 6.8) from the production of mixed acids. Colonies of non-lactose-fermenting bacteria appear colorless or transparent.

Eosin Methylene Blue (EMB) Agar

Eosin methylene blue (EMB) agar is a differential plating medium that can be used in place of MacConkey agar in the isolation and detection of the Enterobacteriaceae and related coliform rods from specimens with mixed bacteria. The aniline dyes (eosin and methylene blue) in this medium inhibit Gram positive and fastidious Gram negative bacteria. They also combine to form a precipitate at acid pH, thus also serving as indicators of acid production.

Desoxycholate–Citrate Agar (DCA)

Desoxycholate–citrate agar is a differential plating medium used for the isolation of members of the Enterobacteriaceae from mixed cultures. The medium contains about three times the concentration of bile salts (sodium desoxycholate) as MacConkey agar, making it most useful in selecting species of *Salmonella* from specimens overgrown or heavily contaminated with coliform bacteria or Gram positive organisms. Sodium and ferric citrate salts in the medium retard the growth of *Escherichia coli*. Lactose is the sole carbohydrate, and neutral red is the pH indicator and detector of acid production.

Endo Agar

Endo agar is a solid plating medium used to recover coliform and other enteric organisms from clinical specimens. The medium contains sodium sulfite and basic fuchsin, which serve to inhibit the growth of Gram positive bacteria. Acid production from lactose is not detected by a pH change, rather from the reaction of the intermediate product, acetaldehyde, which is fixed by the sodium sulfite.

***Salmonella–Shigella* Agar (SS)**

SS agar is a highly selective medium formulated to inhibit the growth of most coliform organisms and permit the growth of species of *Salmonella* and *Shigella* from clinical specimens. The medium contains high bile salts concentration and sodium citrate, which inhibit all Gram positive bacteria and many Gram negative organisms, including coliforms. Lactose is the sole carbohydrate and neutral red the indicator for acid detection. Sodium thiosulfate is a source of sulfur, and any bacteria that produce hydrogen sulfide gas are detected by the black precipitate formed with ferric citrate.

Hektoen Enteric Agar (HE)

HE agar is devised as a direct plating medium for fecal specimens to increase the yield of species of *Salmonella* and *Shigella* from the heavy numbers of normal microbiota. The high bile salt concentration of this medium inhibits the growth of all Gram positive bacteria and retards the growth of many strains of coliforms. Acids may be produced from three carbohydrates, and acid fuchsin reacting with thymol blue produces a yellow color when the pH is lowered. Sodium thiosulfate is a sulfur source, and hydrogen sulfide gas is detected by ferric ammonium citrate, producing a black precipitate.

Xylose lysine desoxycholate agar (XLD)

XLD agar is less inhibitory to growth of coliform bacteria than HE and was designed to detect *Shigella* species in feces after enrichment in broth. Bile salts in relatively low concentrations make this medium less selective than the other media included in this table. Three carbohydrates are available for acid production and phenol red is the pH indicator. Lysine-positive organisms, such as most *Salmonella enteriditis* strains, produce initial yellow colonies from xylose utilization and delayed red colonies from lysine decarboxylation. The hydrogen sulfide detection system is similar to HE agar.

fide at low concentrations, although at higher concentrations hydrogen sulfide is very toxic to most microorganisms. Accordingly, in order to achieve maximal growth rates, it is necessary to define carefully the growth medium.

Inhibitory substances may intentionally be added to a culture medium to prevent the growth of particular microorganisms. For example, the stain methylene blue is more toxic to Gram positive bacteria than to Gram negative bacteria. By incorporating appropriate concentrations of methylene blue (about 0.5 percent) into a culture medium, the growth of Gram positive bacteria may be inhibited, while not interfering with the growth of Gram negative bacteria. The medium eosin methylene blue agar is frequently used for the selective culture of Gram negative bacteria, such as *E. coli*. The methylene blue in this medium inhibits the growth of Gram positive bacteria, and the other constituents of the medium favor the growth of microorganisms found in association with animal intestinal tracts.

Various antibiotics can be similarly incorporated into a medium to prevent the growth of specific types of microorganisms. Antibiotics are substances produced by microorganisms that inhibit or kill other microorganisms; these substances are used in medicine to inhibit the growth of pathogenic microorganisms. Such culture media are frequently used in microbiological laboratories in order to select for the growth of specific microbial populations. For example, by incorporating an antibacterial antibiotic into a medium, fungi can be grown free of bacterial contamination. By incorporating the right inhibitory compound into a growth medium, the selectivity of that medium for a particular type of microorganism is established. Also, lowering the pH of the medium by adding acid inhibits bacterial growth, permitting the selective culture of fungi. Selective media are used in many areas of microbiology, including clinical microbiology where such media are employed for the isolation of pathogenic microorganisms (Table 6.3).

Summary

A culture medium must contain the nutrients that are required by the organism for growth. In some cases it is necessary to include a variety of growth factors, such as vitamins and amino acids. In other cases, microorganisms are capable of growing on simple media. In general, microorganisms require a source of carbon, nitrogen, phosphorus, iron, magnesium, sulfur, sodium, potassium, and chloride. Oxygen is also required by some microorganisms, but oxygen is toxic to others. Parameters that have a great influence on microbial growth and death rates include organic and inorganic nutrient concentrations, concentrations of organic and inorganic inhibitory substances, temperature, oxygen concentrations, water availability, and pH. The combined effects of these factors generally determine the ability of a microorganism to survive and grow in a natural system. It is also these parameters that are manipulated in controlled situations to regulate the rates of microbial growth. The modification of environmental parameters to control the rates of microbial growth is applied in many areas of microbiology. Understanding the factors that control the growth of a microorganism provides the basis for cultivating that microorganism.

Most microorganisms can be grown in pure culture on an appropriate medium or within cultivated cells. Microorganisms are transferred from one medium to another using aseptic technique, that is, methods that prevent contamination of the microbial culture and also protect against the uncontained spread of the organism being transferred. Mastering aseptic transfer technique is essential for all microbiologists.

Microorganisms can be isolated using several different techniques. Streak plate, spread plate, and pour plate methods are all designed so that individual microbial cells are sufficiently separated so that when they reproduce they develop well-isolated colonies. Colonies are macroscopic aggregations of clones of microbial cells. In some cases specialized selective or differential media are employed to isolate specific microbial species. Enrichment cultures that favor the selective multiplication of specific types of microorganisms are also used to isolate microorganisms. These methods have many applications, including the isolation and culture of pathogens that is essential for the diagnosis and initiation of appropriate treatment of various diseases.

Study Outline

A. Pure culture methods.

   1. The ability to examine and study the characteristics of microorganisms depends in large part on being able to grow the organisms in pure culture.

   2. For each type of microorganism there are optimal conditions for growth.

   3. The growth of microorganisms in pure culture requires that all other microbial species be eliminated.

   4. Sterilization eliminates all viable microorganisms from a specified region.

      a. Exposure to elevated temperatures kills microorganisms. Often exposure to 121°C for 15 minutes in an autoclave is used for the sterilization of microbiological media.

      b. Exposure to certain radiation levels can be used to sterilize heat-labile materials such as plastics.

      c. Filtration is used to sterilize solutions by removing microorganisms.

      d. Various chemicals, such as ethylene oxide, are used for sterilizing materials that cannot be autoclaved.

   5. Aseptic transfer technique is used to prevent contamination of the culture and the microbiologist. This method involves avoiding any contact of the pure culture, sterile medium, and sterile surfaces of the growth vessel with contaminating microorganisms.

      a. The steps for transferring a culture from one vessel to another are (1) flame the transfer loop; (2) open and flame the mouths of the culture tubes; (3) pick up some culture growth and transfer it to fresh medium; (4) flame mouths of culture vessels and reseal them; (5) reflame the inoculating loop.

   6. The isolation of pure cultures involves the separation of microorganisms into individual cells, each of which grows, giving rise to a colony of like cells. The desired colony is removed and grown in fresh medium.

      a. In the streak plate technique a loopful of bacterial cells is streaked in a pattern across the surface of a solidified nutrient agar plate; the plate is then incubated under favorable conditions to permit microbial growth; the desired isolated colonies are picked and restreaked onto fresh medium to ensure purity; a new colony from this second plate is picked and transferred to an agar slant.

      b. In the spread plate method a dilute suspension drop of microorganisms is placed on the center of an agar plate and, with a glass rod, is spread over the plate surface; the plate is incubated and the desired isolated colonies are picked and restreaked.

      c. In the pour plate technique suspensions of microorganisms are added to melted agar tubes that have been cooled to approximately 42–45°C; the agar is allowed to solidify, trapping the separated bacteria within the medium and on its surface; after incubation, the desired colonies are picked and streaked onto another plate to establish purity.

      d. Incubation conditions are adjusted to support optimal growth. Temperature is controlled by using incubators; oxygen can be eliminated by using special incubation chambers; other factors such as carbon dioxide concentration and humidity also can be controlled to achieve optimal growth.

B. Microbial nutrition.

   1. Microorganisms require specific nutrients to grow.

   2. Autotrophic microorganisms obtain their carbon from carbon dioxide and their energy from light or inorganic chemicals.

   3. Heterotrophic microorganisms obtain their carbon and energy from organic compounds.

4. Media are designed to meet the nutritional requirements of the microorganisms that are cultivated with those media.
5. Some microorganisms can be grown on defined media in which all the constituents are known.
6. Some microorganisms can only be cultivated on complex media in which some constituents are not fully defined.
7. Fastidious microorganisms have very demanding growth requirements.
8. Tissue culture can be used to cultivate viruses and other microorganisms that can only reproduce within host cells. In tissue culture the host cells are cultivated in vitro and the microbes then are grown in vivo within the host cells.

C. Enrichment culture technique.
1. Media and incubation conditions can be adjusted to favor the growth of a particular type of microorganism.
2. Enrichment culture can be used to increase the relative numbers of a particular microbial species that is originally present in low numbers in a mixed population; this method is used for the recovery of some pathogens.
3. Selective culture media are designed to permit the growth of only microorganisms with particular metabolic features; a variety of selective media are used in the clinical microbiology laboratory for the recovery and cultivation of pathogenic microorganisms.

Study Questions

1. Name the steps in obtaining and maintaining a pure culture.
2. Match the procedures or steps of a procedure with the terms in the list to the right.
   - a. Accomplish the work quickly and efficiently to minimize exposure time for contamination.
   - b. Exposing microorganisms to elevated temperatures.
   - c. Elimination of all viable microorganisms from a specified region.
   - d. Streak plate technique.
   - e. Filtration.
   - f. Avoid any contact of the pure culture, sterile medium, and sterile surfaces of the growth vessel with contaminating microorganisms.
   - g. Spread plate method.
   - h. Flame mouths of culture vessels and reseal them.

   (1) Sterilization
   (2) Aseptic transfer technique
   (3) Isolation of pure cultures

Some Thought Questions

1. How would you isolate a dioxin-degrading microbe?
2. How does the medium used for the culture of human cells differ from the medium used to cultivate *E. coli*?
3. Why are we unable to cultivate viruses in a cell-free medium?
4. Why are we unable to cultivate *Treponema pallidum*, the bacterium that causes syphilis, in a cell-free medium?
5. How would you determine the best medium for cultivating a newly discovered bacterium?
6. How was the medium for culturing *Legionella pneumophila* (causative organism of Legionairre's disease) developed?

Additional Sources of Information

Collins, C. H., and P. M. Lyne. 1976. *Microbiological Methods*. Butterworth Publishers, Ltd., London. Easy-to-follow descriptions of basic methods used for cultivating bacteria.

Gerhardt, P. (ed.). 1981. *Manual of Methods for General Bacteriology*. American Society for Microbiology, Washington, D.C. An extensive manual of methods used in microbiology, including procedures for the cultivation of various microorganisms.

Laskin, A., and H. A. Lechevalier. 1977–1981. *Handbook of Microbiology* (four volumes). CRC Press, Inc., Boca Raton, Florida. An extensive manual of relevant microbiological information, including nutritional characteristics of different microorganisms.

Norris, J. R., and D. W. Ribbons (eds.). 1969–. *Methods in Microbiology*. Academic Press, Inc., New York. A series of articles on microbiological methods, including chapters on the cultivation of specific groups of microorganisms.

Sirockin, G., and S. Cullimore. 1969. *Practical Microbiology*. McGraw-Hill Publishing Co., Ltd., London. A simple text describing fundamental microbiological procedures.

7 *Microbial Metabolism*

KEY TERMS

aerobic respiration
anabolic pathway
anabolism
anaerobic respiration
autotrophic metabolism
autotrophs
butanediol fermentation
butyric acid fermentation
Calvin cycle
catabolic pathway
catabolism
chemoautotrophic
chemolithotrophic
electron transport chain
Embden–Meyerhof pathway
enzymes
ethanolic fermentation
facultative anaerobes
fermentation
glycolysis
heterolactic fermentation
heterotrophic metabolism
heterotrophs
homolactic fermentation
Krebs cycle (tricarboxylic acid cycle, citric acid cycle)
lactic acid fermentation
metabolism
mixed acid fermentation
obligate anaerobes
oxidation
oxidative phosphorylation
oxidative photophosphorylation
photoautotrophic
propionic acid fermentation
reducing power
reduction
respiration

OBJECTIVES

After reading this chapter you should be able to

1. Define the key terms.
2. Discuss the importance of carbon flow, reducing power, and ATP generation in cellular metabolism.
3. Discuss the roles of enzymes, coenzymes, and ATP in microbial metabolism.
4. Compare autotrophic and heterotrophic metabolism.
5. Describe the metabolism of a chemolithotrophic microorganism.
6. Describe the metabolism of a photoautotrophic microorganism.
7. Compare respiration and fermentation.
8. Describe the three stages of respiratory metabolism and the key metabolic events that occur during each stage.
9. Describe the different fermentation pathways.

Chemical Principles of Metabolism

The transformation of substrates into the biochemical components of microbial cells is dependent on a complex integrated network of biochemical reactions that collectively constitutes the metabolism of the organism.

A seed grows to a sapling, a zygote to a baby, a baby to a man, the man falls in love with a woman, stubbornly stays a man, as his riding horse his riding horse, the innumerable shifts of energy that make those possibilities possible are metabolism. Man, woman, horse, horsefly that bit the horse, bacteria and viruses that infected the horsefly, toxins manufactured in the bacteria and the viruses, toxins coming out and killing a physician or a politician, all are metabolism. The infinitude of change forever taking place in us, and in those who inhabit air and earth and sea with us, is metabolism.

—GUSTAV ECKSTEIN, *The Body Has a Head*

The process of microbial metabolism can be viewed from two perspectives: the flow of energy through the cell, which is needed for biosynthesis and the maintainence of the organizational structure of a living system; and the flow of carbon through the cell, which is needed for incorporation into the macromolecules of the cell and for the biosynthesis of the structural components of the microorganism.

There is no good evidence that in any of its manifestations life evades the second law of thermodynamics, but in the downward course of the energy-flow it interposes a barrier and dams up a reservoir which provides potential for its own remarkable activities.

—SIR FREDERICK GOWLAND HOPKINS

A chain of enzymes remakes that glucose molecule into intermediate molecules, these combine with others, combine with phosphorus, produce high-energy phosphorus, and this when called on can accomplish at the low temperature of the body the high-temperature work that must be done.

—GUSTAV ECKSTEIN, *The Body Has a Head*

Virtually every step in the metabolism of a cell involves an enzyme. (See appendix I for a review of enzymes and biochemistry.) Enzymes are biological catalysts. They lower the amount of energy necessary to initiate a chemical reaction. This lowering of the activation energy is critical because it permits reactions to occur at temperatures where life can exist. The alternative, which would be to use elevated temperatures—as we do when we use a bunsen burner in the laboratory to initiate chemical reactions—is not a viable option. Because enzymes are involved in virtually every step of cellular metabolism, microbial metabolism occurs rapidly at temperatures that are not disruptive to the three-dimensional configurations of the organism's macromolecules.

Enzymes exhibit great specificity in the reactions that they catalyze. Different enzymes are needed to bring about reactions that transform even very similar chemical compounds. Thousands of enzymes, involved in the metabolic reactions of each cell, are necessary for microbial growth and reproduction. These enzymatically mediated metabolic reactions proceed via a series of small discrete steps that establish a metabolic pathway. The sequential steps between the starting substrate molecule(s) and the end product(s) constitute the intermediary metabolism of the cell.

Intermediary metabolism can be likened to moving between the first and second floors of a building. Most of us lack the energy necessary to jump the required 10 feet to reach the next floor, and the energy released by jumping down from a height of greater than one story can be quite disruptive to the organization of the human body. In order to move up and down efficiently between the floors of a building, therefore, we normally use a staircase by which we are able to alter energy levels in small discrete steps. Likewise, the microbial cell does not convert a substrate into a product in one large step but rather carries out a series of smaller intermediary metabolic reactions.

A central focus of the intermediary metabolism of a microorganism concerns the flow of energy through the cell. The energetics of a cell are based on the molecule ATP (adenosine triphosphate). During metabolism energy is transferred to and stored within molecules of ATP. ATP can then be used to drive the energy-requiring reactions of biosynthesis. A growing cell of the bacterium *Escherichia coli*, as a typical example, must synthesize approximately 2.5 million molecules of ATP per second to support its energy needs. Approximately 7300 calories are needed for every mole of ATP that is formed from ADP (adenosine diphosphate).

In many cases the energy for the generation of ATP comes from oxidation reactions. In an oxidation reaction there is a loss of electrons. Such oxidation reactions are always coupled with reduction reactions, that is, with chemical reactions in which there is a gain of electrons. For example, some cells obtain their energy for generating ATP from the oxidation of glucose. If the oxidation of glucose is tied to the reduction of oxygen, the end products of metabolism are carbon dioxide (product of glucose oxidation) and water (product of oxygen reduction). As we will see, the complete oxidation of a glucose molecule by respiratory metabolism yields 6 carbon dioxide, 6 water, and 38 ATP molecules.

In addition to requiring energy in the form of ATP, microorganisms must carry out metabolic activities that transform the starting substrate molecule into the many different macromolecules of the cell, including among others, proteins for enzymes, lipids for membranes, carbohydrates for various structures such as cell walls, and nucleic acids for the storage and expression of genetic information. The basic strategy of the cell, in terms of carbon flow, is to form relatively small biochemical molecules that can act as central building blocks for establishing the carbon skeletons of a variety of large macromolecules. When a heterotrophic microorganism starts with an organic substrate, like glucose, it carries out a catabolic pathway to break that molecule down into smaller compounds; these smaller compounds then act as precursors for the biosynthesis of macromolecules (catabolism means degradative process). Then the microorganism uses an anabolic pathway in which small molecules are transformed into larger molecules (anabolism means biosynthetic process).

Microbial metabolism must also balance reactions in which molecules gain electrons (reduction reactions) with those in which molecules lose electrons (oxidation reactions). If oxidation and reduction reactions were not balanced the cell would become electrically charged. Because the macromolecular components of the cell generally are in a more reduced oxidation state than their precursors, microbial metabolism must generate reducing power that can be used to convert the substrate into the more reduced molecules of the cell. This is accomplished by coupling oxidation reactions with the reduction of coenzymes. In these reactions electrons are transferred to coenzymes, such as NAD (nicotinamide adenine dinucleotide), to form reduced coenzymes, such as NADH (reduced nicotinamide adenine dinucleotide). The coenzyme acts as a temporary holder of the electrons. In subsequent reactions the reduced form of the coenzyme supplies electrons for reduction reactions; hence the term reducing power. In the course of supplying electrons for a reduction reaction the coenzyme is reoxidized. Both oxidation–reduction reactions and the transfer of energy are central to the metabolic reactions of a cell.

Autotrophic Metabolism

Microorganisms have developed several strategies of metabolism for meeting their common needs of synthesizing ATP, forming reduced coenzymes, and transforming carbon-containing molecules into the macromolecules that constitute the microorganism. Two distinct modes of microbial metabolism have evolved for accomplishing these tasks: autotrophy and heterotrophy. Microorganisms with autotrophic (literally self-feeding) metabolism (called autotrophs) do not require preformed organic matter to generate ATP or as a source of carbon for the biosynthesis of the macromolecules of the cell. Rather, autotrophic microorganisms are able to generate ATP either from the oxidation of inorganic compounds or through the conversion of light energy to chemical energy. The carbon for the macromolecules of autotrophic microorganisms originates from inorganic carbon dioxide.

There are two separate components to the metabolism of autotrophic organisms. One portion of the metabolic activities involves the conversion of carbon dioxide to organic matter. An independent part of the metabolic sequence is devoted to the generation of ATP. Photoautotrophic (photosynthetic) microorganisms use light energy and chemoautotrophic (chemolithotrophic) microorganisms use the energy derived from the oxidation of inorganic compounds to supply the energy needed for the synthesis of ATP.

Autotrophic microorganisms carry out a metabolic sequence of reactions known as the Calvin cycle (Figure 7.1). In the Calvin cycle carbon dioxide is reduced to form organic matter. Because organic carbon is in a more reduced oxidation state than carbon dioxide, the autotrophic conversion of carbon dioxide to organic carbon requires reducing power in the form of reduced coenzyme NADPH. Additionally the pathway of carbon dioxide fixation requires energy in the form of ATP.

The Calvin cycle is a complex series of reactions that really represents three slightly different but fully integrated metabolic sequences. It effectively takes three turns of the Calvin cycle to synthesize one molecule of the organic product of this metabolic pathway, which is glyceraldehyde 3-phosphate. Because glyceraldehyde 3-phosphate contains three carbon atoms, the Calvin cycle is sometimes referred to as a C3 pathway. The formation of one

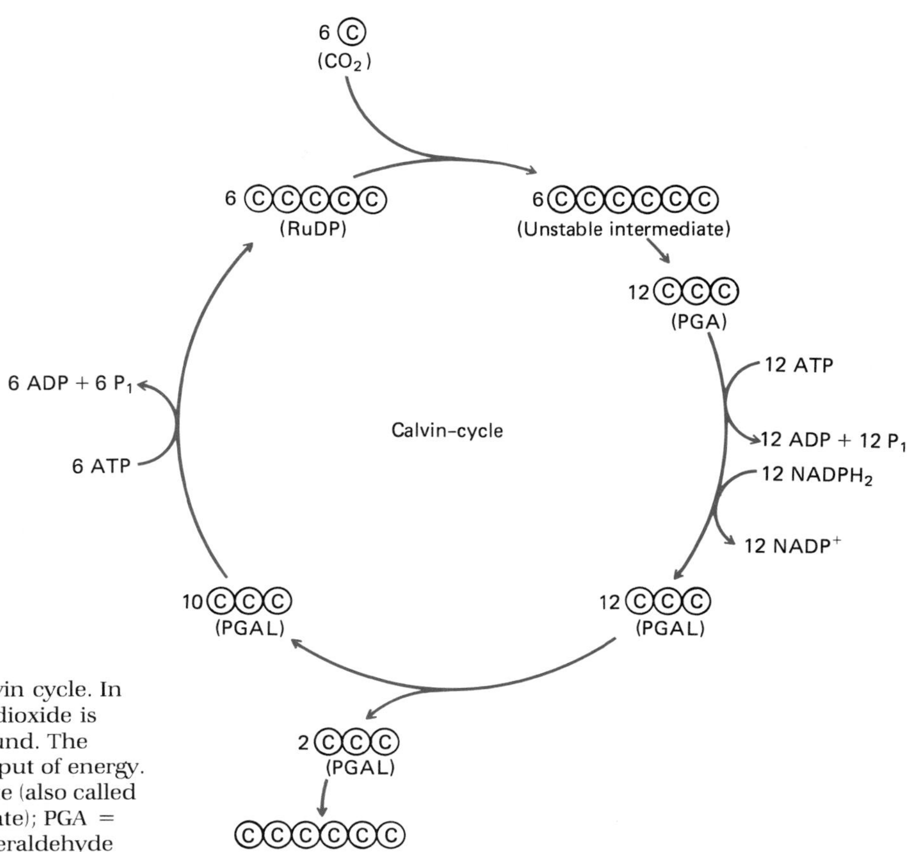

Figure 7.1 Diagram of the Calvin cycle. In this metabolic pathway carbon dioxide is converted to an organic compound. The process requires a substantial input of energy. RuDP = ribulose-1,5-diphosphate (also called RuBP = ribulose-1,5-bisphosphate); PGA = phosphoglycerate; PGAL = glyceraldehyde phosphate.

molecule of glyceraldehyde 3-phosphate requires three molecules of carbon dioxide, nine molecules of ATP, and six molecules of NADPH. The glyceraldehyde 3-phosphate molecules that are formed during the Calvin cycle can further react to form glucose and polysaccharides of glucose, such as starch and cellulose. It takes six turns of the Calvin cycle to form a six-carbon carbohydrate, such as glucose. The net input of energy, as ATP, and reducing power—as NADPH—required for the conversion of carbon dioxide to glucose is 18 ATP and 12 NADPH molecules.

Clearly the Calvin cycle requires a great deal of energy and reducing power. In photoautotrophs the ATP and NADPH (energy and reducing power) to drive the Calvin cycle come from the light reactions of photosynthesis. In chemolithotrophs the needed ATP and reduced coenzymes come from the oxidation of inorganic compounds. The Calvin cycle itself is known as a dark reaction because, although it requires ATP and NADPH, it does not require any light reactions (photo acts) as an integral part of the metabolic reactions of this cycle. As long as there is an adequate supply of ATP and NADPH, the Calvin cycle continues in the absence of light.

In my hunt for the secret of life, I started my research in histology. Unsatisfied by the information that cellular morphology could give me about life, I turned to physiology. Finding physiology too complex I took up pharmacology. Still finding the situation too complicated I turned to bacteriology. But bacteria were even too complex, so I descended to the molecular level, studying chemistry and physical chemistry. After twenty years' work, I was led to conclude that to understand life we have to descend to the electronic level, and to the world of wave mechanics. But electrons are just electrons, and have no life at all. Evidently on the way I lost life; it had run out between my fingers.

—ALBERT SZENT-GYORGYI

Photoautotrophic ATP Generation

Plants, algae, and photosynthetic bacteria are able to convert light energy to chemical energy in a process known as oxidative photophosphorylation. These photosynthetic organisms contain various types of chlorophyll molecules. When a chlorophyll molecule absorbs light energy, it becomes energetically excited and emits an electron. Within specialized membranes of the photosynthetic organisms, the electrons emitted from the chlorophyll are transferred through a series of carrier molecules. The transfer of electrons through the carriers of this electron transport chain establishes a hydrogen ion gradient across the membrane. The resulting electrochemical gradient supplies sufficient energy to drive the chemiosmotic synthesis of ATP.

The basic chemiosmotic mechanism driving the synthesis of ATP during oxidative photophosphorylation is the same in all photoautotrophs. However, different pathways of electron transfer (photosystems) are employed by different organisms to transfer the energy absorbed from light irradiation to the synthesis of ATP (Figure 7.2). In the cyanobacteria, algae, and higher plants, the electron released from chlorophyll when it is excited by light flows through a series of electron carriers and is eventually used to reduce the coenzyme NADP. The resulting reduced coenzyme (NADPH) represents the reducing power that is needed for the Calvin cycle. An electron donor, namely water, is used to balance the positive charge left on the chlorophyll molecule when it loses an electron. The electron flow is thus from the electron donor, water (H_2O), to the electron acceptor, NADP. When water donates an electron, it is transformed to oxygen. The molecular oxygen produced by these oxygen-evolving photosynthetic organisms provides the oxygen needed by ourselves and other organisms that require molecular oxygen to carry out their metabolism. Not all photosynthetic microorganisms pro-

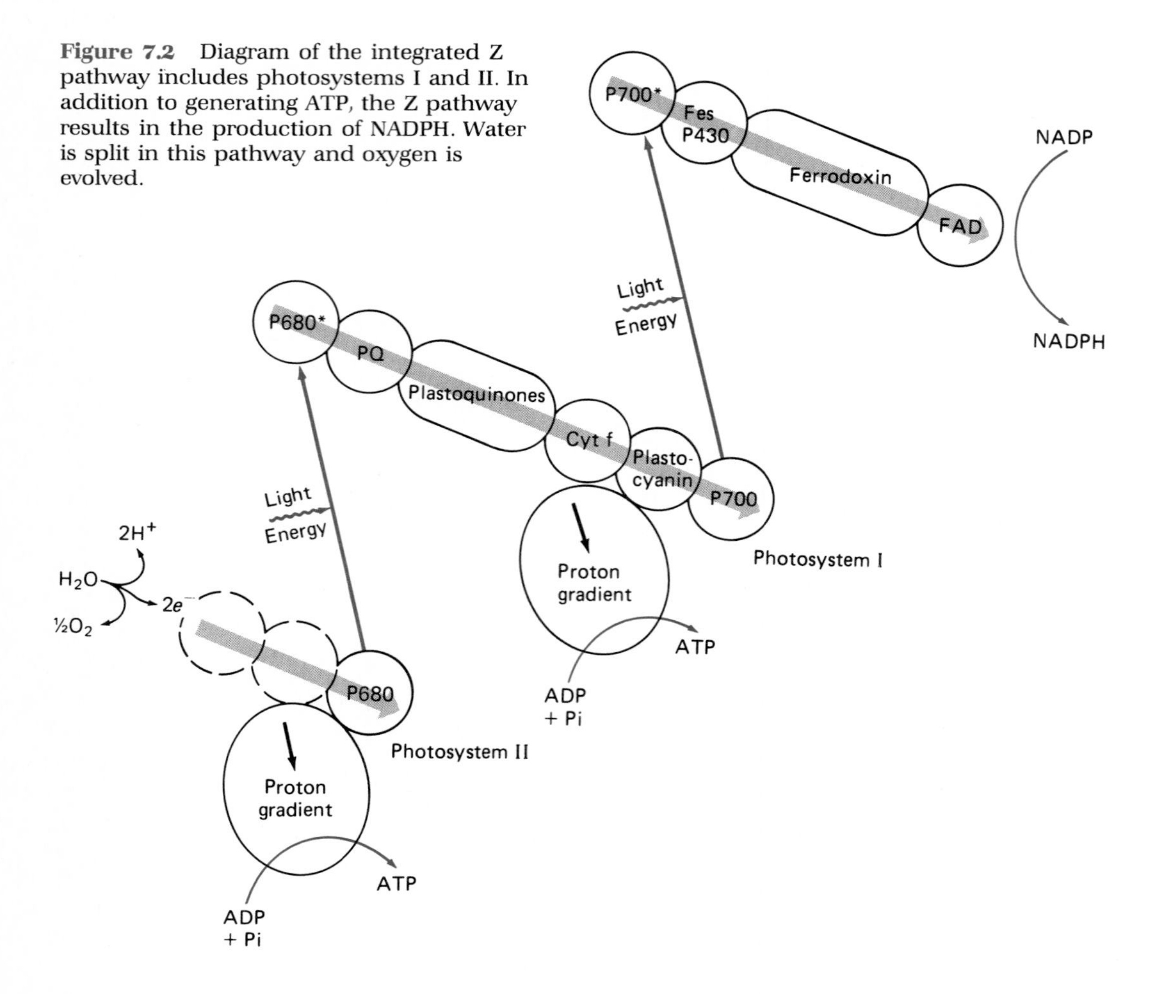

Figure 7.2 Diagram of the integrated Z pathway includes photosystems I and II. In addition to generating ATP, the Z pathway results in the production of NADPH. Water is split in this pathway and oxygen is evolved.

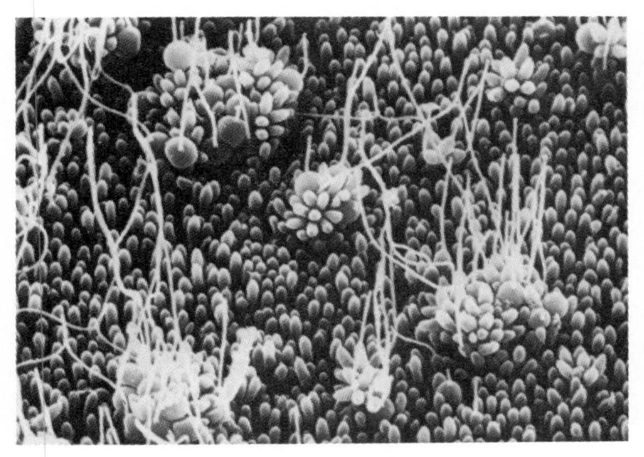

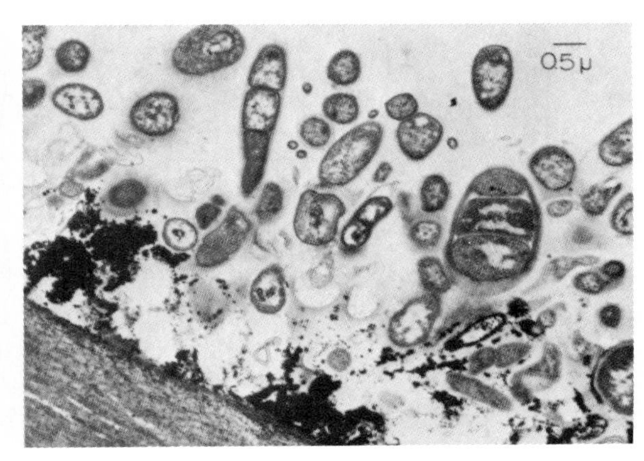

Figure 7.3 Areas of high biological productivity have been discovered near thermal vents in the deep reaches of the oceans. These regions are supported by chemolithotrophic bacteria. (A) Scanning electron micrograph of bacteria growing at a depth of 2250 meters in the thermal rift area off the Galapagos Islands. (B) Transmission electron micrograph of chemolithotrophic bacteria collected from deep ocean thermal vents. These bacteria are able to obtain energy from the oxidation of hydrogen sulfide, and some bacteria in this region may grow at temperatures above 100°C. (Courtesy Holger W. Jannasch, Woods Hole Oceanographic Institution, Woods Hole, Massachusetts. Reprinted by permission of the American Society for Microbiology, Washington, D.C., from H. W. Jannasch and G. O. Wirsen, 1981, *Applied and Environmental Microbiology,* 41:528–538.)

duce oxygen. The anaerobic green and purple photosynthetic bacteria, for example, utilize hydrogen sulfide (H_2S) as an electron donor instead of water and produce sulfur instead of oxygen.

Chemolithotrophic Generation of ATP

Chemolithotrophic microorganisms couple the oxidation of an inorganic compound with the reduction of a suitable coenzyme. Various sulfur compounds, for example, can be oxidized by chemolithotrophs to meet their energy needs. The chemolithotrophic activities of sulfur-oxidizing microorganisms have received considerable attention as a result of the finding that a highly productive submarine area off the Galapagos Islands is supported by the productivity of chemolithotrophs growing on reduced sulfur released from thermal vents in the ocean floor (Figure 7.3). It is quite unusual to find an ecological system driven by chemolithotrophic metabolism. Some sulfur-oxidizing chemolithotrophic bacteria, such as *Thiobacillus thiooxidans,* can oxidize large amounts of reduced sulfur compounds with the formation of sulfate. The sulfur-oxidizing activities of this bacterium are important because of their involvement in the formation of acid mine drainage and mineral recovery processes.

The nitrifying bacteria oxidize either ammonium or nitrite ions. Bacteria, such as *Nitrosomonas,* oxidize ammonia to nitrite. Other bacteria, such as *Nitrobacter,* oxidize nitrite to nitrate. Because the chemolithotrophic oxidation of reduced nitrogen compounds yields relatively little energy, chemolithotrophic bacteria carry out extensive transformations of nitrogen in soil and aquatic habitats in order to synthesize their required ATP. The activities of these bacteria are quite important because they have a marked influence on soil fertility.

Heterotrophic Metabolism

Let us now examine some of the central metabolic pathways of energy and carbon flow in heterotrophic microorganisms. In heterotrophic metabolism generation of ATP involves the conversion of an organic substrate molecule to end products via a metabolic pathway that releases sufficient energy to be coupled with the synthesis of ATP. This process involves the breakdown of an organic molecule to smaller molecules, a process called catabolism. A catabolic pathway is one in which larger molecules are split into smaller ones. The biochemical reactions in such a pathway which liberate sufficient energy to drive the conversion of ADP to ATP, are oxidation reactions. For such reactions to occur, they must be coupled with simultaneous reduction reactions, often the reduction of the coenzyme NAD to its reduced form, NADH. To sustain its metabolism, the cell must reoxidize the reduced coenzyme in subsequent biochemical reactions. The reoxidation of NADH ensures the continuous supply of NAD required for use as an oxidizing agent in metabolic pathways aimed at generating ATP. Thus, the heterotrophic generation of ATP is integrally tied to the cell's ability to balance its oxidation–reduction reactions.

Heterotrophs exhibit two basic strategies for oxidizing organic compounds to drive the formation of ATP, both of which maintain the required balance between oxidation and reduction reactions. These two strategies are fermentation and respiration.

Respiration

In respiratory metabolism some molecule other than a product derived from the organic substrate must be involved in reoxidizing the reduced coenzyme formed during the oxidation of the organic substrate molecule. Stated another way, a respiration pathway requires an external terminal electron acceptor. By reducing this terminal electron acceptor, the cell is able to balance the change in the oxidation state of the metabolic products relative to the starting substrate. When molecular oxygen serves as the terminal electron acceptor, the pathway is known as aerobic respiration. When other molecules, such as nitrate or sulfate, serve as the terminal electron acceptor, the metabolic pathway is called anaerobic respiration (literally meaning respiration not requiring the presence of air).

There are three distinct phases to a complete respiration pathway (Figure 7.4). In the first phase, the organic substrate is converted to small organic molecules. The products of this catabolic pathway then enter the Krebs cycle, during which organic carbon is oxidized to inorganic carbon dioxide and reduced coenzyme is generated. Finally, oxidative phosphorylation occurs, during which reduced coenzyme molecules are reoxidized. Oxidative phosphorylation involves the transport of electrons through a series of membrane-bound carriers, establishing a hydrogen ion gradient across a membrane that is used to generate ATP by chemiosmosis.

In the case of carbohydrates, the substrate molecule is initially broken down to pyruvate via a metabolic pathway called glycolysis. Glycolysis is a

Figure 7.4 This drawing shows a simplified view of the interrelationships of several major pathways involved in respiratory metabolism: glycolysis, the Krebs cycle, and oxidative phosphorylation.

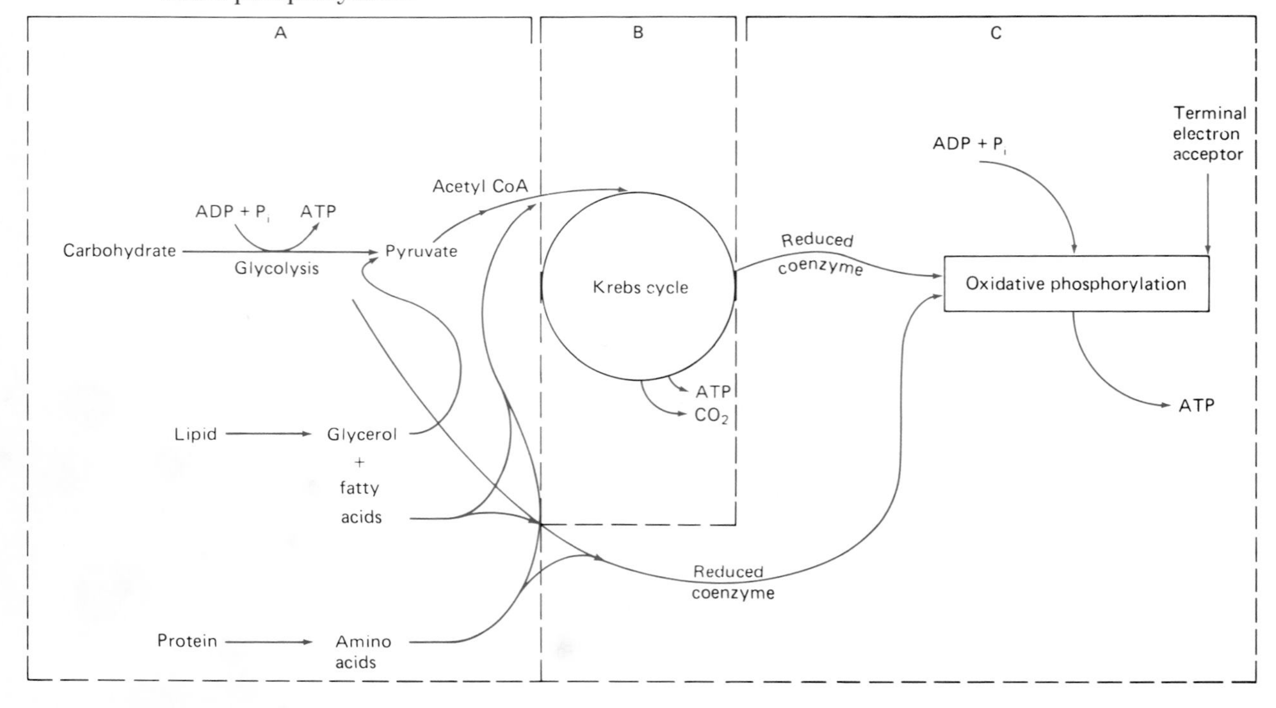

central pathway in the metabolism of carbohydrates by microorganisms. Many different carbohydrates or their derivatives can enter into the metabolism of a cell via a glycolytic pathway.

Different metabolic pathways are involved in the conversion of other classes of biochemicals into small organic molecules that can enter the Krebs cycle during respiration. For example, lipids are converted to fatty acids, which are broken down by beta-oxidation to products that can enter the Krebs cycle. Proteins are broken down into amino acids, which are converted to form various small organic acids that can enter the Krebs cycle.

The overall respiration pathway results in the formation of carbon dioxide from the organic substrate molecule. In aerobic respiration water is also produced as a result of the reduction of oxygen, whereas in anaerobic respiration molecular nitrogen, hydrogen sulfide, or other reduced compounds are produced in addition to carbon dioxide, depending on the specific terminal electron acceptor. The classic equation for aerobic respiration of glucose is

$$C_6H_{12}O_6 + 6\ O_2 \rightarrow 6\ CO_2 + 6\ H_2O$$

Glycolysis

The most common of pathway of glycolysis is the Embden–Meyerhof pathway (Figure 7.5). This pathway results in the conversion of the six-carbon molecule glucose to two molecules of the three-carbon molecule pyruvate. It also results in the net production of two molecules of reduced coenzyme (NADH) and the net synthesis of two ATP molecules. The overall equation for glycolysis by the Embden–Meyerhof pathway can be written as

$$\text{glucose} + 2\ \text{ADP} + 2\ P_i + 2\ \text{NAD} \rightarrow 2\ \text{pyruvate} + 2\ \text{NADH} + 2\ \text{ATP}$$

where P_i stands for inorganic phosphate.

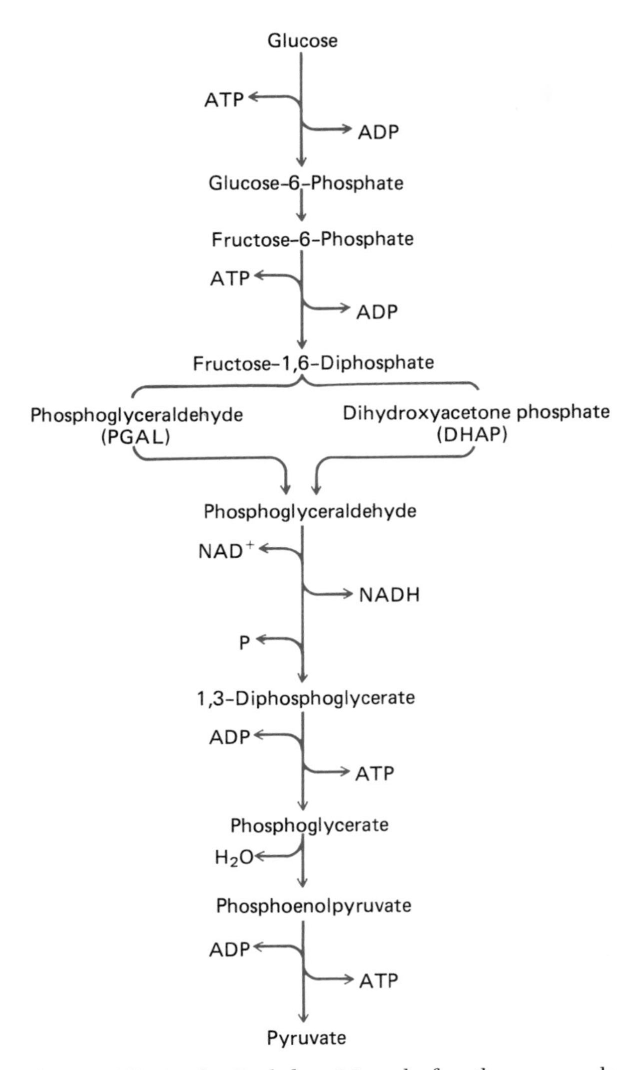

Figure 7.5 In the Embden–Meyerhof pathway a molecule of glucose is converted to two molecules of pyruvate with the net production of two molecules of ATP and two molecules of reduced coenzyme NADH.

The initial step in the Embden–Myerhof pathway of glycolysis actually uses ATP rather than producing it. In this reaction glucose is converted to glucose-6-phosphate. As the pathway continues, the reactive glucose-6-phosphate is modified and converted to fructose-1,6-diphosphate. Here again ATP is consumed rather than produced so that at this point in the pathway the net ATP balance is minus two. The fructose-1,6-diphosphate molecule is enzymatically split into phosphate-containing molecules: phosphoglyceraldehyde-3-phosphate (PGA) and dihydroxyacetone phosphate (DHAP). The DHAP is converted into PGA so that there are two molecules of PGA, and beyond this point in the pathway each step occurs twice for each molecule of glucose that is metabolized. In the next step the PGA molecule loses an electron, thereby becoming more oxidized. The electron is transferred to a coenzyme, forming reduced nicotinamide adenine dinucleotide (NADH). An additional phosphate group is added in this reaction to form 1,3-diphosphoglycerate; this reaction, which does not require the expenditure of ATP, is termed a substrate level phosphorylation. It is after this step that net ATP production begins. The 1,3-diphosphoglycerate is used to convert ADP to ATP, thus balancing the original energy investment of ATP. The resulting molecule undergoes rearrangements to produce phosphoenolpyruvate (PEP). The PEP is transformed into pyruvate, the end product of the glycolytic pathway. In this last reaction, a phosphate group is donated to ADP to form additional ATP, accounting for the net gain of two ATP molecules via this pathway.

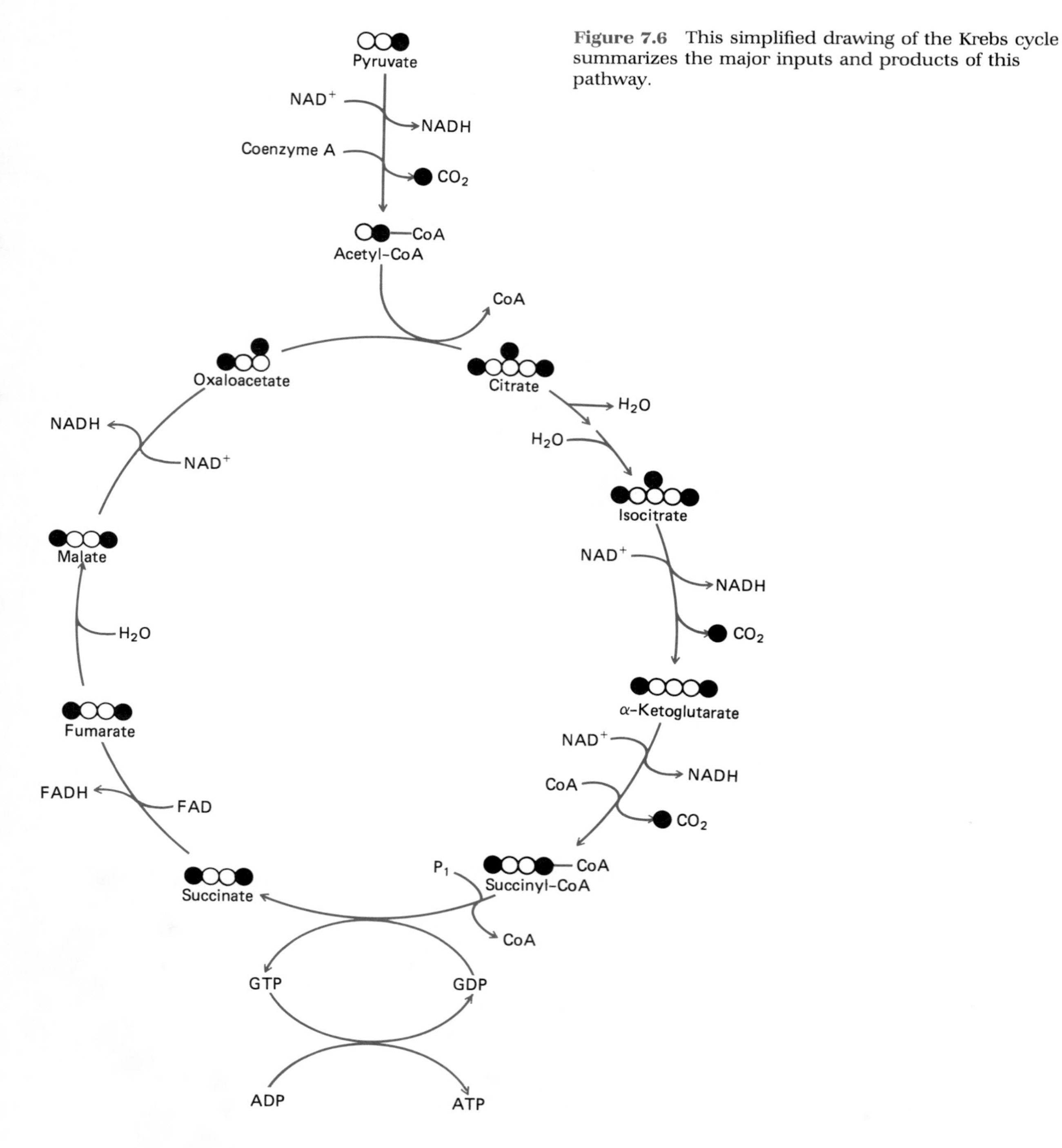

Figure 7.6 This simplified drawing of the Krebs cycle summarizes the major inputs and products of this pathway.

Krebs Cycle

The pyruvate generated by glycolysis can feed into the Krebs cycle (Figure 7.6). The Krebs cycle is also known as the tricarboxylic acid cycle or the citric acid cycle. As a result of the reactions of the Krebs cycle the pyruvate molecules formed during glycolysis are converted to carbon dioxide. Thus, at the end of the Krebs cycle six carbon dioxide molecules are produced for each six-carbon glucose molecule that is metabolized.

Reduced coenzyme NADH is generated during several reactions of the Krebs cycle. Additionally, another coenzyme, flavin adenine dinucleotide (FAD), is reduced to $FADH_2$. One of the reactions of the Krebs cycle is also directly coupled with the generation of a high-energy phosphate-containing compound. In this reaction, instead of the normal generation of ATP, guanidine triphosphate (GTP) is synthesized from guanidine diphosphate (GDP) and inorganic phosphate (P_i). The GTP can be converted

Figure 7.7 The Krebs cycle is a central metabolic pathway to respiratory metabolism and provides a critical link between the metabolism of the different classes of macromolecules.

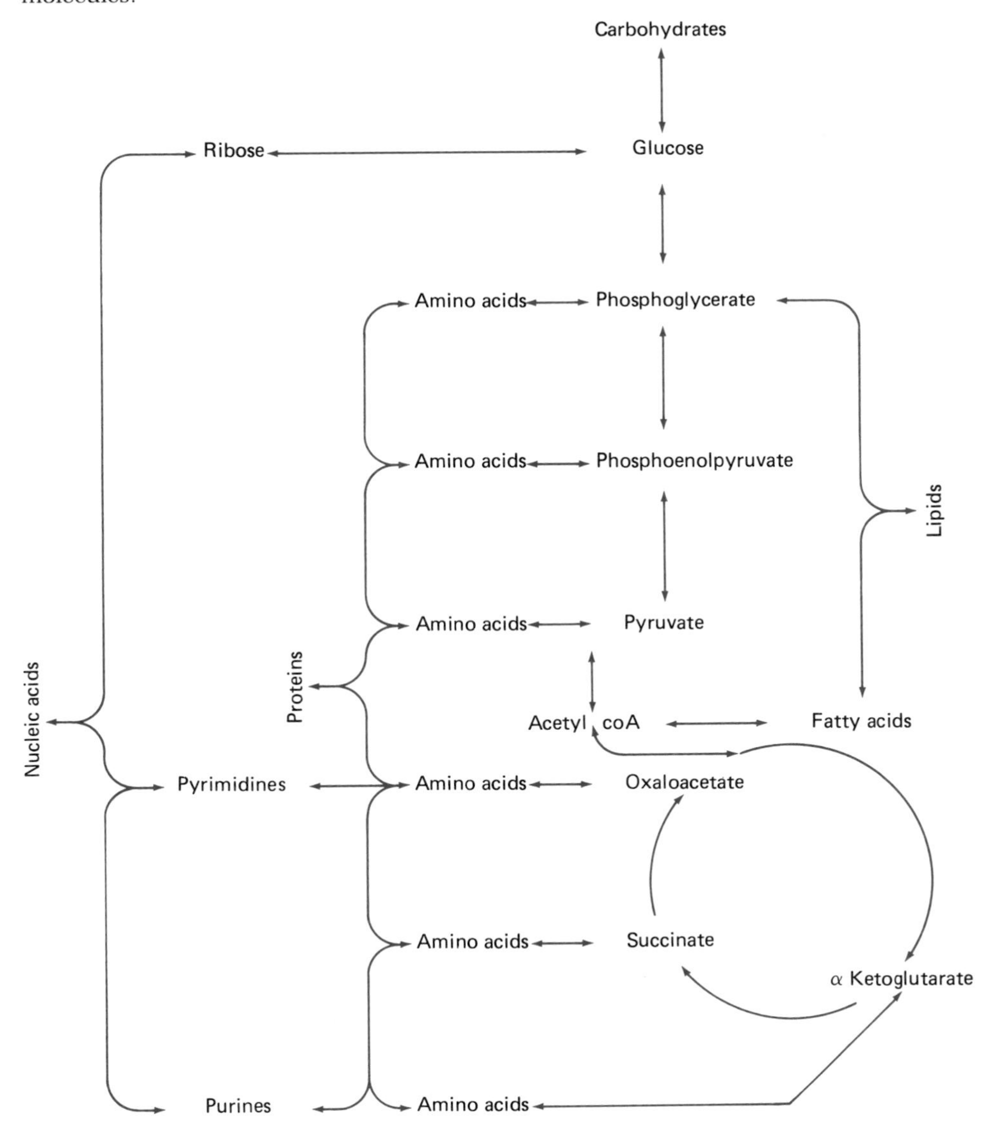

to ATP, and for accounting purposes, the GTP generated in this reaction will be treated as equivalent to ATP in determining the net synthesis of ATP during respiration. The net reaction of the Krebs cycle, starting with the pyruvate generated from glucose can be written as

$$2 \text{ pyruvate} + 2 \text{ ADP} + 2 \text{ FAD} + 8 \text{ NAD} \rightarrow 6 \text{ CO}_2 + 2 \text{ ATP} + 2 \text{ FADH}_2 + 8 \text{ NADH}$$

At the end of the Krebs cycle, the cell has managed to convert all of the substrate carbon of the glucose molecule to carbon dioxide. There also has been a net synthesis of four ATP molecules—the production of ten reduced coenzyme molecules as NADH and the generation of two reduced coenzyme molecules as $FADH_2$.

In addition to their roles within the overall respiratory generation of ATP, the Krebs cycle and glycolytic pathways occupy a central place in the flow of carbon through the cell (Figure 7.7). As a result of its function of supplying small biochemical molecules for biosynthetic pathways, the Krebs cycle rarely acts as a complete cycle. Because some of the intermediates are siphoned off out of the cycle, some of the intermediary metabolites of this pathway must be resynthesized to maintain an active Krebs cycle. In many microorganisms only part of the substrate is completely oxidized for driving the synthesis of ATP, and the remainder is used for biosynthesis. Similarly, the reduced coenzymes generated in this pathway can be used for generating ATP or for the synthesis of the reduced coenzyme NADPH (reduced nicotinamide adenine dinucleotide phosphate) for use in biosynthesis.

Oxidative Phosphorylation

When we consider the respiration pathway from the standpoint of the synthesis of ATP, the reduced coenzyme molecules (generated both during glycolysis and the Krebs cycle) can be reoxidized during oxidative phosphorylation with the generation of additional ATP. During oxidative phosphorylation electrons from NADH and $FADH_2$ are transferred through a series of steps, known as the electron transport chain This transfer of electrons involves a series of oxidation–reduction reactions of membrane-bound carrier molecules and the eventual reduction of a terminal electron acceptor (Figure 7.8).

Figure 7.8 The electron transport chain is a membrane-bound series of reactions that results in the reoxidation of reduced coenzymes. The transport of electrons through the cytochrome chain of this pathway results in the pumping of hydrogen ions across the membrane, and the return flow of hydrogen ions resulting from this proton gradient drives the generation of ATP. The electron chain terminates with reduction of a terminal electron acceptor. The requirement for such a terminal electron acceptor differentiates respiration from fermentation pathways. In aerobic respiration oxygen is the terminal electron acceptor; in the absence of oxygen some microorganisms can carry out anaerobic respiration by using the reduction of nitrate or sulfate to terminate the reaction sequence of this pathway.

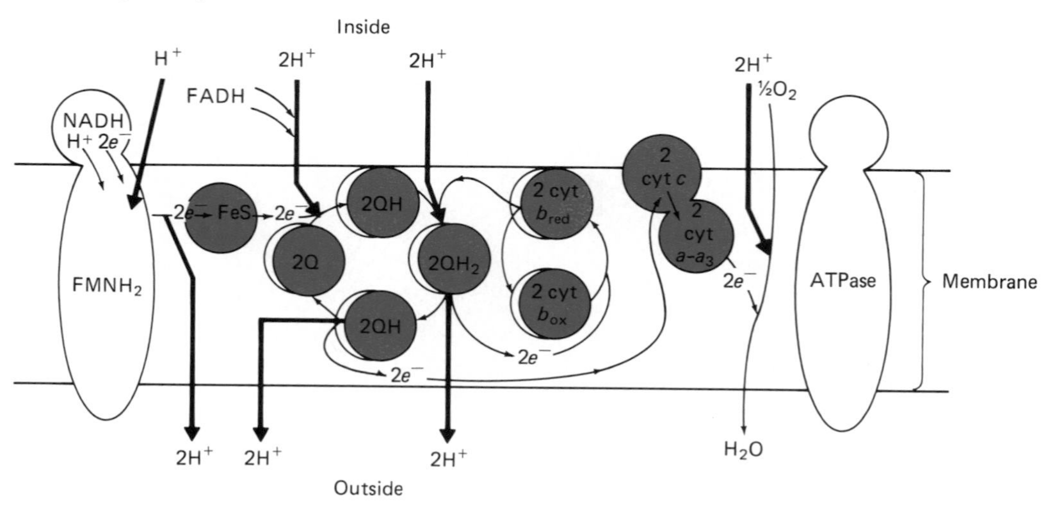

The terminal electron acceptor during aerobic respiration is oxygen, which is reduced to form water in this process. In the absence of molecular oxygen, nitrate can serve as terminal electron acceptor, with the production of molecular nitrogen and water. In the absence of both oxygen and nitrate, sulfate can serve as terminal electron acceptor, becoming reduced to hydrogen sulfide and water in the process. Phosphate may also act as a terminal electron acceptor, forming phosphine, which is a green-glowing gas; this process may occur around cemeteries where extensive decomposition occurs.

Bacteria keep us from heaven and put us there.

—Martin H. Fisher

As a consequence of the electron transport chain, the energy contained in the reduced coenzyme molecule is used to drive the synthesis of ATP. This generation of ATP is based on the establishment of a hydrogen ion gradient across a membrane and chemiosmosis to drive the formation of ATP. The reduced coenzyme NADH contains more stored chemical energy than the reduced coenzyme $FADH_2$. For each NADH molecule three ATP molecules can be synthesized during oxidative phosphorylation, compared to only two ATP molecules for each $FADH_2$. The ten NADH molecules generated during glycolysis and the Krebs cycle, therefore, can be converted to 30 ATP molecules during oxidative phosphorylation. The two $FADH_2$ molecules generated during the Krebs cycle can generate four ATP molecules. This synthesis of ATP is in addition to the ATP formed during glycolysis and the Krebs cycle. Thus, the overall reaction for the respiratory metabolism of glucose using the Embden–Meyerhof pathway of glycolysis can be written as

$$\text{glucose} + 6\ O_2 + 38\ \text{ADP} + 38\ P_i \rightarrow 6\ CO_2 + 6\ H_2O + 38\ \text{ATP}$$

Fermentation

In fermentation pathways the organic substrate acts as the internal electron donor (reducing agent), and a product of that substrate acts as an internal electron acceptor (oxidizing agent). There is no net change in the oxidation state of the products relative to the starting substrate molecule in fermentation pathways. The oxidized products are exactly counterbalanced by reduced products, and thus the required oxidation–reduction balance is achieved. Such a metabolic pathway can occur in the absence of air because there is no requirement for oxygen or other electron acceptor to balance a change in the oxidation state of the organic molecule.

Unlike respiration, fermentation does not involve oxidative phosphorylation for generating ATP. Rather, the synthesis of ATP in fermentation is largely restricted to the amount formed during glycolysis.

Fermentation yields far less ATP per substrate molecule than respiration. This is because the organic substrate molecule must serve as both the internal electron donor and electron acceptor during a fermentation pathway. Thus the substrate cannot be completely oxidized to carbon dioxide. As a result, not as much energy can be released from the substrate molecule to drive the synthesis of ATP as can be obtained during a respiration pathway.

The energy that can be obtained from the complete oxidation of glucose to carbon dioxide and water by respiration is more than 10 times greater than can be realized when glucose is metabolized by fermentation. Because more ATP can be generated per molecule of substrate by a respiration pathway than by a fermentation pathway, fewer substrate molecules must be metabolized during respiration than during fermentation to achieve equivalent growth, that is, to support the metabolic requirements of an equivalent number of cells. From both the viewpoints of bioenergetics and conservation of available organic nutrient resources, respiration is more favorable than fermentation. Organisms that have the metabolic capability to carry out both types of metabolism will generally use the energetically more favorable respiration pathway, when conditions permit, and will rely on fermentation only when there is no available external electron acceptor that can be used.

Because they do not require oxygen, all fermentation pathways are anaerobic, and microorganisms that generate their energy by fermentation are carrying out anaerobic metabolism, regardless of whether or not the organism is growing in the presence of molecular oxygen. Some microorganisms, known as facultative anaerobes, are capable of carrying out both respiration and fermentation pathways. Other microorganisms, which are restricted to fermentative pathways, are metabolically obligate anaerobes. The use of the terms "obligate anaerobe" and "obligate aerobe" is often confusing. Some fermentative microorganisms are inhibited by oxygen and grow only in the absence of air; in such cases the designation obligate anaerobe can be used without ambiguity. However, many microorganisms that are metabolically obligate anaerobes, because they are restricted to fermentative metabolism, can grow in the presence of air even though they do not use molecular oxygen. Similarly some organisms that are referred to as obligate aerobes, because they are restricted to respiratory metabolism, can grow in the absence of air using terminal electron acceptors other than oxygen.

In the fermentation of a carbohydrate, the initial metabolic steps are identical to those in respiration, and the metabolic pathway for carbohydrate fermentation begins with glycolysis. If the microorganism uses the Embden–Meyerhof glycolytic pathway, it generates two pyruvate molecules, two reduced coenzyme NADH molecules, and two ATP molecules for each molecule of glucose that goes through glycolysis. In general, the two ATP molecules generated during glycolysis represent the total energy conversion of the fermentation pathway, as measured by the number of ATP molecules synthesized.

The remainder of the fermentation pathway is really concerned with reoxidizing the coenzyme. In respiration the reoxidation of the coenzymes occurs in the electron transport chain and requires an external electron acceptor, whereas in fermentation the reoxidation of NADH to NAD depends on the reduction of the pyruvate molecules formed during glycolysis to balance the oxidation–reduction reactions. Different microorganisms have developed different pathways for utilizing the pyruvate to reoxidize the reduced coenzyme with the different terminal sequences of the various fermentation pathways, resulting in the formation of various end products (Figure 7.9). The complete fermentation pathway begins with the substrate, includes glycolysis, and terminates with the formation of end products. There is no net change in the oxidative state of the coenzymes during the overall fermentation pathway, and thus, the coenzyme does not appear in the overall equation for the fermentation. The different fermentation pathways generally are named for the characteristic end products that are formed.

In the ethanolic or alcoholic fermentation pyruvate is converted to ethanol and carbon dioxide. The ethanolic fermentation is carried out by many yeasts, such as *Saccharomyces cerevisiae*, but by relatively few bacteria. The ethanolic fermentation pathway can be written as

$$\text{glucose} + 2\ \text{ADP} + 2\ \text{P}_i \rightarrow 2\ \text{ethanol} + 2\ \text{CO}_2 + 2\ \text{ATP}$$

The lactic acid fermentation pathway is carried out by bacteria, which by virtue of their metabolic reactions are classified as the lactic acid bacteria. When the Embden–Meyerhof scheme of glycolysis is used in the lactic acid fermentation pathway, the overall pathway is a homolactic fermentation because the only end product formed is lactic acid. Homolactic fermentation is carried out by *Streptococcus* (Gram positive cocci that tend to form

Figure 7.9 Various fermentation pathways, each branching from pyruvate, are carried out by different microorganisms. Compared to respiration, these fermentation pathways are energetically less favorable. In each complete fermentation there is a balance of oxidation–reduction reactions so that there is no net production of reduced coenzyme. Each pathway results in the formation of characteristic products, many of which are of industrial importance.

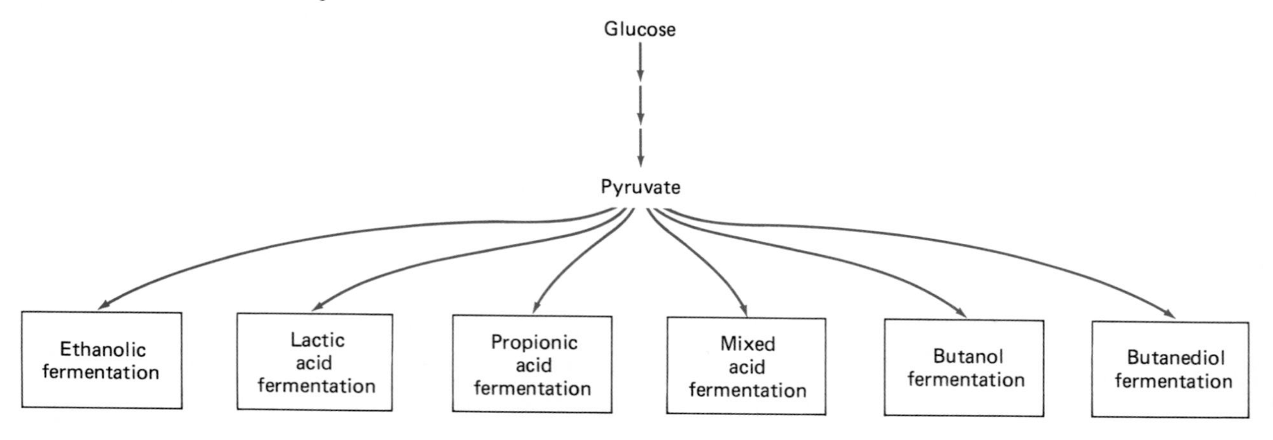

MICROBIOLOGY HEADLINES

Japanese Drunkenness Disease

For Charlie Swaart, the nightmare began right after World War II, when he was a soldier in charge of an information unit in occupied Japan. On a visit to Tokyo, Charlie suddenly discovered that he was roaring drunk. "I had to be poured into bed," he says, "and I spent the whole weekend with a monumental hangover. The only trouble was, I hadn't had a drop to drink in three days."

Swaart was an unknowing victim of *meitei-sho*—Japanese drunkenness disease—a rare and mysterious malady that makes even the strictest of teetotalers act and feel as if he were in his cups. There were years of unsuccessful attempts to convince American doctors that Swaart was not merely a closet drunk, until the nature of his ailment was finally explained by microbiologist Kazuo Iwata, of the University of Tokyo School of Medicine. Iwata had studied 30 cases with symptoms similar to Swaart's and had discovered that the blame lay with an intestinal yeast called *Candida albicans*.

The yeast is ubiquitous, and in humans it constitutes a normal part of the population of intestinal flora. In *meitei-sho* victims, though, the yeast has run wild. For example, Swaart's *C. albicans* count was about 6,000 times as high as it should have been. The net effect was that the yeast acted like an internal moonshine still, transforming any carbohydrate in Swaart's diet into pure alcohol.

Iwata and Swaart (a former medical writer) think that *meitei-sho* is caused by a mutant strain of *C. albicans* that might have been created by radioactivity surrounding the atomic blasts in Hiroshima and Nagasaki. Since the Japanese use human feces as fertilizer, the *Candida* could have been passed on through the food chain.

Although this interpretation is controversial, some American researchers—like Dr. Alan Levin, of the University of California at San Francisco, and Dr. Sidney Baker, of Yale—are blaming overabundance of *C. albicans* for everything from depression to simple colitis to such immune-system diseases as AIDS. In fact, Baker calls *C. albicans* nothing less than "the most important issue in our medical culture today."—Bill Lawren

Source: Omni, July 1984. Reprinted by permission.

chains) and various *Lactobacillus* (Gram positive rods that tend to form chains) species. The homolactic acid fermentation pathway can be written as

$$\text{glucose} + 2\ \text{ADP} + 2\ \text{P}_i \rightarrow 2\ \text{lactic acid} + 2\ \text{ATP}$$

In some bacteria, however, a different pathway of glycolysis is used. Such bacteria are termed heterolactic because ethanol and carbon dioxide are produced in addition to lactic acid. The ethanol and carbon dioxide come from the glycolytic portion of the pathway. This fermentative pathway is carried out by *Leuconostoc* and various *Lactobacillus* species. The overall reaction for the heterolactic fermentation can be written as

$$\text{glucose} + \text{ADP} + \text{P}_i \rightarrow \text{lactic acid} + \text{ethanol} + CO_2 + \text{ATP}$$

Another fermentation pathway of some interest is the propionic acid fermentation pathway. This metabolic sequence is carried out by the propionic acid bacteria. The bacterial genus *Propionibacterium* is defined as Gram positive rods that produce propionic acid from the metabolism of carbohydrates. These bacteria are especially interesting because they have the ability to carry out this fermentation pathway beginning with lactic acid as the substrate. The ability to utilize the end product of another fermentation pathway is quite unusual.

In contrast to the propionic acid fermentation, which is restricted to a few bacterial species, the mixed acid fermentation pathway is relatively common. The mixed acid fermentation is carried out by members of the family Enterobacteriaceae. The Enterobacteriaceae includes *Escherichia coli* as well as hundreds of other bacterial species. In the mixed acid fermentation pathway the pyruvate formed during glycolysis is converted to a variety of products, including ethanol, acetate, formate, molecular hydrogen, and carbon dioxide. The proportions of the products vary, depending on the bacterial species.

In addition to the mixed acid fermentation, some bacteria, such as members of the bacterial genus *Klebsiella*, can carry out a butanediol fermentation pathway. An intermediary product in the butanediol fermentation pathway is acetoin (acetyl methyl carbinol). One of the classic diagnostic tests used for separating *E. coli* from *Enterobacter aerogenes* is the Vogues–Proskauer test, which detects the presence of acetoin. *E. coli* does not carry out a butanediol pathway, whereas *Ent. aerogenes* does. Thus *Ent. aerogenes* is Vogues–Proskauer positive and *E. coli* is Vogues–Proskauer negative.

Another test used for separating *Ent. aerogenes* from *E. coli* is the methyl red test, which detects very low pH resulting from high amounts of acid production. Because *Ent. aerogenes* channels part of its substrate into the neutral fermentation end product, butanediol, it does not produce as much acid and thus does not lower the pH as much as *E. coli*, which channels all of its substrate into the mixed acid fermentation pathway. Thus, *E. coli* shows a positive methyl red test, whereas *Ent. aerogenes* yields the opposite reaction.

In yet another pathway, members of the genus *Clostridium* carry out a butanol fermentation. Different *Clostridium* form a variety of end products via this fermentation pathway with pyruvate being converted either to acetone and carbon dioxide, propanol and carbon dioxide, butyrate, or butanol. Several of these products are good solvents; acetone, for example, is used as nail polish remover. Regardless of the particular products, this pathway is usually referred to as the butyric acid fermentation

Summary

It is clear that microorganisms exhibit a variety of strategies for converting chemical and light energy into the energy stored within ATP, the central currency of energy of the cell. The synthesis of ATP can be achieved autotrophically—either through the oxidation of inorganic substrates or through the conversion of light energy to chemical energy—or may be generated heterotrophically through the utilization of organic substrates. The amount of ATP that can be synthesized in these processes varies greatly, and microorganisms accordingly show great variation in the efficiency with which they can synthesize sufficient ATP to meet their energy requirements. Chemolithotrophic metabolism, for example, is relatively inefficient energetically; thus, chemolithotrophic microorganisms must metabolize a large number of inorganic substrate molecules to synthesize sufficient ATP for their metabolic needs during growth and reproduction.

Respiration is energetically far more favorable than fermentation. To take a representative number, many microorganisms are capable of generating 38 ATP molecules per molecule of glucose by respiration, compared to only two molecules of ATP per molecule of glucose by fermentation. As a consequence of the "inefficiency" of the fermentation pathway, more substrate must be utilized during fermentation than during respiration to achieve similar amounts of growth. Although fermentation does not yield as much ATP per molecule of substrate as does respiration, the end products of fermentation are of more practical interest than are those of respiration. Respiration normally results in the formation of carbon dioxide and water. The various fermentation pathways carried out by different organisms produce numerous end products of economic value. For example, the ethanolic and lactic acid fermentation pathways are of particular importance in the food industry because they are directly involved in the production of bread, beer, wine, and cheese, among other foods and beverages.

Although microorganisms show great metabolic versatility for generating ATP and reducing power, several central pathways play key roles in the metabolism of microbial cells. The metabolic pathways utilized in ATP generation involve various intermediary metabolites linked together in a series of small steps to form unified biochemical pathways.

There are various ways in which a microorganism can drive the synthesis of ATP. Heterotrophic metabolism of an organic substrate may occur either by respiration or fermentation. In respiration an external electron acceptor is required to complete the metabolic pathway. Oxygen serves as the terminal electron acceptor in aerobic respiration, and nitrate or sulfate act as possible terminal electron acceptors in anaerobic respiration. In fermentation the organic substrate molecules serve as internal electron donors, and products of their metabolism serve as acceptors.

Autotrophic metabolism may involve the oxidation of inorganic compounds or the conversion of light energy. Chemolithotrophic microorganisms are able to oxidize inorganic compounds to generate ATP. Photoautotrophic microorganisms are able to trap light energy, in some cases involving one cyclic photosystem and in others involving two different photosystems linked in a united pathway.

With the exception of fermentation, all of these pathways, which are aimed at the synthesis of ATP, involve an electron transport through a series of membrane-bound carriers and the establishment of a hydrogen ion gradient across a membrane. For example, in photosynthetic microorganisms the flow of electrons—initiated when a chlorophyll molecule is energetically excited by absorbing light energy—establishes an electrochemical gradient across a membrane during the process of oxidative photophosphorylation. Similarly, a hydrogen ion gradient across a membrane is formed during oxidative phosphorylation in both respiration and chemolithotrophic metabolism. The counterflow of hydrogen ions, channeled through membrane-bound ATPase, drives the synthesis of ATP. This process of chemiosmosis is central to many of the ATP-synthesizing pathways of microorganisms.

As a consequence of their metabolic activities, microorganisms are capable of channeling energy into the synthesis of ATP. They are also capable of channeling reducing power into the synthesis of the reduced coenzymes NADH and NADPH. The reducing power and the ATP generated by microbial metabolism are used by microorganisms for their growth and reproduction. The particular pathways that a microorganism can use for generating ATP and reducing power depend on the enzymes that the organism possesses. The additional end products of metabolism that are produced is also a function of the cell's enzymatic potentialities, which are genetically determined. Regardless of the mode of metabolism the strategies are the same: synthesize ATP, reduce coenzyme (NADPH) and small precursor molecules to serve as the building blocks of macromolecules, and then use the energy, reducing power, and precursor molecules to synthesize the macromolecular constituents of the organism.

Study Outline

A. Microbial metabolism.
- **1.** All metabolic reactions are enzymatic.
- **2.** Intermediary metabolism is the sequential steps between the starting substrate molecules and the end products.
- **3.** Cycling of ADP and ATP is fundamental to the bioenergetics of microbial cells.
- **4.** The central themes of metabolism are generating and using ATP; generating and using reducing power; and synthesizing and using macromolecular precursors.
- **5.** Catabolic pathways are involved in the breakdown of organic molecules; biosynthetic or anabolic pathways are involved in the synthesis of larger molecules.

B. Autotrophic metabolism.
- **1.** Autotrophic microorganisms do not require preformed organic matter.
- **2.** Autotrophs obtain their cellular carbon by the fixation of carbon dioxide; that is, autotrophs can convert inorganic carbon dioxide to organic compounds.
- **3.** Carbon dioxide is fixed by autotrophs via the Calvin cycle.
- **4.** Photosynthetic organisms use light energy to generate ATP.
- **5.** Chemolithotrophic bacteria oxidize inorganic compounds to obtain needed energy for ATP generation.

C. Heterotrophic microorganisms.
- **1.** Heterotrophic generation of ATP involves the conversion of an organic substrate molecule to end products with the release of sufficient energy to form ATP.
- **2.** Respiration.
  - **a.** Respiratory metabolism requires an external electron acceptor, which is necessary to terminate the flow of electrons; hence its name, terminal electron acceptor.
  - **b.** Molecular oxygen is the most common terminal electron acceptor, in which case the process is called aerobic respiration; when nitrate or sulfate act as the terminal electron acceptor the process is called anaerobic respiration.
  - **c.** There are three phases in the respiratory pathway: glycolysis (carbohydrate broken down to pyruvate), Krebs cycle (organic carbon oxidized to carbon dioxide and coenzyme NADH is generated), and oxidative phosphorylation (coenzyme NADH is oxidized, a terminal electron acceptor is reduced, and ATP is synthesized).
  - **d.** In glycolysis, glucose is transformed to pyruvate with the production of reduced coenzyme and the generation of ATP.

e. The Embden–Meyerhof pathway is the most common glycolytic pathway; this pathway produces two pyruvate, two ATP, and two NADH molecules per molecule of glucose.
f. In the Krebs cycle pyruvate is transformed into carbon dioxide with the generation of additional ATP and reduced coenzymes.
g. Starting with glucose, the products of glycolysis and Krebs cycle are: four ATP, ten NADH, two $FADH_2$, and six CO_2
h. In oxidative phosphorylation, reduced coenzyme molecules are reoxidized with the generation of ATP; electrons from NADH and $FADH_2$ are transferred through a series of steps known as the electron transport chain; three ATP molecules are generated for each NADH and two ATP molecules are formed for each $FADH_2$ molecule; chemiosmosis is the mechanism of ATP generation.
i. The combined reactions of Embden–Meyerhof glycolytic pathway, Krebs cycle, and oxidative phosphorylation are represented as

$$\text{glucose} + 6\ O_2 + 38\ \text{ADP} + 38\ P_i \rightarrow 6\ CO_2 + 6\ H_2O + 38\ \text{ATP}$$

j. The simplified overall equation for the repiratory metabolism of glucose is

$$C_6H_{12}O_6 + 6\ O_2 \rightarrow 6\ CO_2 + 6\ H_2O$$

3. Fermentation.
a. In a fermentation pathway an organic substrate acts as both electron donor and, later in the pathway, electron acceptor.
b. Fermentation can occur in absence of air because there is no need for molecular oxygen as an electron acceptor.
c. In fermentation, a substrate cannot be completely oxidized to carbon dioxide because it must act as the electron acceptor; hence, the fermentation yields less energy than respiration.
d. In fermentation pathways the energy yield is generally restricted to the two ATP molecules produced during glycolysis.
e. In addition to glycolysis, a fermentation pathway includes a series of terminal reactions in which the reduced coenzyme formed during glycolysis is reoxidized.
f. Different pathways are named for the end products formed from the remainder of the fermentation pathway.
g. In the ethanolic fermentation pathway, ethanol and carbon dioxide are the end products.
h. In the homolactic acid fermentation pathway, lactic acid is the end product; two important genera of lactic acid bacteria are *Streptococcus* and *Lactobacillus*.
i. In the heterolactic acid fermentation, the products are lactic acid, ethanol, and carbon dioxide.
j. In the propionic acid fermentation pathway, which is carried out by propionic acid bacteria (*Propionibacterium*), the end products are propionic acid and carbon dioxide.
k. The mixed acid fermentation yields ethanol, acetic acid, formic acid, hydrogen, and carbon dioxide. This pathway is carried out by members of Enterobacteriaceae, including *Escherichia coli*.
l. In the butanediol fermentation pathway, butanediol is the end product. An intermediary metabolite in this pathway, acetoin, can be detected by the Vogues–Proskauer test.
m. The Vogues–Proskauer (VP) and methyl red (MR) tests are used to distinguish between *Escherichia coli* and *Enterobacter aerogenes. E. coli* is MR + and VP −; *Ent. aerogenes* is MR − and VP +.
n. The butanol fermentation pathway is carried out by members of the genus *Clostridium*; the end products of this pathway can be acetone and carbon dioxide, propanol and carbon dioxide, butyrate, and/or butanol.

Study Questions

1. Briefly compare autotrophic to heterotrophic metabolism.
2. Match the following phrases with the terms in the list to the right.

   (1) Catabolism
   (2) Fermentation
   (3) Respiration

   a. Substrate is not completely oxidized to carbon dioxide.
   b. Process can be either aerobic or anaerobic.
   c. Process can occur in absence of air because there is no need for oxygen to be the electron acceptor.
   d. Requires an external electron acceptor.
   e. Molecular oxygen is the most common electron acceptor.
   f. An organic substrate acts as both electron donor and acceptor.
   g. Yields less energy than the other type of ATP generation.
   h. The breakdown of an organic molecule to smaller molecules.
3. Fill in the blanks.
   a. The three phases in the respiratory pathway are: ________ , in which ________ is broken down to ________ . ________ , in which is oxidized to ________ and ________ is generated; and ________ , in which ________ is ________ , a terminal electron acceptor is ________ , and ________ synthesized. The overall equation for respiratory metabolism of glucose is ________ .
   b. The ________ pathway is the most common glycolytic pathway. It begins with ________ glucose, ________ ADP, ________ Pi, and ________ coenzyme NAD; and ends with ________ pyruvate, ________ ATP, and ________ coenzyme NADH.
   c. The second phase of respiratory metabolism is the ________ . In this phase, the end product of glycolysis ________ , reacts with ________ to form ________ and ________ . In the first step of the process, coenzyme NAD is ________ . In the second step, ________ combines with ________ from the cyclic portion of the Krebs cycle to form ________ and ________ . This is then oxidized to ________ . In the process, ________ is produced, coenzyme NAD is ________ in three places, coenzyme FAD is ________ in one place, and ________ is synthesized in one place. The summary of the events in the Krebs cycle is ________ .
   d. The final phase of the respiratory metabolism of glucose is ________ . In this phase, reduced ________ is ________ with the generation of ________ . During this phase, electrons from ________ and ________ are transferred through a series of steps known as the ________ . Finally, the electron is accepted by a ________ , which can be ________ or ________ .
   e. From NADH, ________ ATP are generated; from $FADH_2$, ________ ATP are generated. The driving force of ATP generation is ________ . In this process, the transfer of electrons from reduced coenzyme to the terminal acceptor establishes a ________ across a membrane. As it diffuses, it ________ .
   f. A metabolic pathway that also generates ATP, but does not involve oxidative phosphorylation, is ________ . All such processes are ________ , regardless of whether molecular ________ is present or not. This process always begins with ________ . The ________ gives the particular type of pathway its name.
   g. Ethanol is the end product in the ________ pathway. In this pathway, ________ is converted to ________ and ________ . It is carried out by many ________ , but few ________ .
   h. In the ________ pathway, pyruvate is reduced to lactic acid. There are two types of this fermentation pathway. In ________ fermenation, only ________ is produced.
   i. In the ________ pathway, the final products are propionic acid and ________ .
   j. The ________ fermentation is carried out by species of *Clostridium*. The end products include ________ , ________ , and other organic solvents.
4. Match the the folowing phrases with the terms in the list to the right. Be as specific as possible.

   (1) Autotroph
   (2) Photoautotroph
   (3) Chemoautotroph
   (4) Chemolithotroph

   a. Important because metabolic activities are essential for normal biogeochemical cycling of various elements.
   b. Uses hydrogen, iron, sulfur, or nitrogen as substrates.
   c. Another term for chemoautotroph.
   d. Uses energy derived from oxidation of inorganic compounds for ATP synthesis.
   e. Microorganisms that do not require preformed organic matter to synthesize ATP.
   f. Photosynthetic microorganisms.

Some Thought Questions

1. Why do microbiologists have to learn chemistry?
2. Which bacteria would you call workaholics?
3. Why is ATP, rather than a more energy-rich compound, the central focus of microbial metabolism?
4. Why doesn't a bacterial cell ever really produce 38 ATP molecule per molecule of glucose it metabolizes?
5. Why does glucose disappear from a medium in the absence of air faster than when oxygen is present?
6. Considering the chemiosmotic mechanism for ATP generation, why do bacteria living in acid lakes still need to metabolize substrates?
7. Considering the chemiosmotic mechanism for ATP generation, how can respiratory bacteria live in alkaline lakes?
8. How can methanogenic bacteria explain the observation of fire breathing dragons?

Additional Sources of Information

Dawes, I. W., and I. W. Sutherland. 1976. *Microbial Physiology*. Blackwell Scientific Publications, Oxford, England. An overview of microbial metabolism, including coverage of the flow of energy and carbon through different microorganisms.

Hinkle, P. C., and R. E. McCarthy. 1978. How cells make ATP. *Scientific American*, **238**(3):104–123. An interesting and thorough article on the strategies of ATP production.

Lehninger, A. L. 1971. *Bioenergetics*. Benjamin/Cummings Publishing Co., Menlo Park, California. An advanced work on the flow of energy through biological systems.

Lehninger, A. L. 1982. *Principles of Biochemistry*. Worth Publishers, Inc., New York. An advanced text covering the field of biochemistry, including the pathways of cellular metabolism.

Mandelstam, J., K. McQuillen, and I. W. Dawes. 1982. *Biochemistry of Bacterial Growth*. Blackwell Scientific Publications, Oxford, England. A treatise on microbial metabolism including the details of the reactions that occur in different metabolic pathways.

Mitchell, R. 1979. Keilin's respiratory chain concept and its chemiosmotic consequences. *Science*, **206**:1148–1159. An article describing the chemiosmotic basis for ATP generation and the involvement of the electron transport chain.

Parson, W. W. 1974. Bacterial photosynthesis. *Annual Reviews of Microbiology*, **28**:41–59. An advanced work describing the details of bacterial photosynthesis.

Stryer, L. 1981. *Biochemistry*. W. H. Freeman Company, Publishers, San Francisco. An advanced, well-illustrated biochemistry text covering the roles of enzymes and coenzymes and the major metabolic reactions of cellular metabolism.

8 *Industrial Uses of Microbial Metabolism*

KEY TERMS

adjuncts
ale
amylases
antibiotic
beer
cheese
cortisone
distilled liquors
fermentor
gasohol
industrial fermentation
leavening
malt
mashing
must
pitching
processed cheeses
proteases
rennin
ripened cheeses
starter cultures
steroids
substrates
unripened cheeses
vitamins
whey
wine
wort
yogurt

OBJECTIVES

After reading this chapter you should be able to

1. Define the key terms.
2. List the food products produced by the lactic acid fermentation.
3. Describe how the same metabolic pathway can be used for the formation of different commercial products.
4. Describe the differences between ripened and unripened cheeses.
5. List the products produced by the ethanolic fermentation.
6. Describe the production of beer, wine, and spirits.
7. Describe the uses of microbial fermentations in the pharmaceutical industry.
8. List the commercial uses of organic acids produced by microorganisms.
9. List the commercial uses of amino acids produced by microorganisms.
10. List the commercial products produced by the butanol fermentation and their uses.
11. List the microbial enzymes that have commercial uses and their applications.

The Fermentation Industry

The metabolic activities of microorganisms are used to produce numerous commercial products of important economic value (Table 8.1). When referring to industrial processes the term fermentation need not refer to metabolic pathways that proceed by fermentation as compared to respiration. Rather an industrial fermentation process can refer to any chemical transformation of organic compounds carried out by using microorganisms and their enzymes. Industrial processes using microorganisms exploit the enzymatic activities of the microbe to produce substances of commercial value. Production methods in industrial microbiology bring together the raw materials (substrates), microorganisms (specific strains of organisms, called starter cultures, or microbial enzymes, such as amylases and proteases), and a controlled favorable environment (created in a fermentor) to produce the desired substance. The essence of an industrial process is to combine the right organism, inexpensive substrate, and proper environment to produce high yields of a desired product.

Production of Foods and Beverages

The controlled growth of microorganisms on or in foods is used to produce numerous foods and beverages that are considered gastronomic delights. Many of the foods and beverages we commonly enjoy, such as wine and cheese, are examples of the products of microbial enzymatic activity.

Here with a Loaf of Bread beneath the Bough,
A flask of Wine, a Book of Verse—and Thou
Beside me singing in the Wilderness—
And Wilderness is Paradise enow.

—EDWARD FITZGERALD, *The Rubaiyat of Omar Khayyam*

For the most part, it is the fermentative metabolism of microorganisms that is exploited in the production of food products. The accumulation of fermentation products, such as ethanol and lactic acid, is desirable because of their characteristic flavors and other properties. Only a few processes, such as the production of vinegar, make use of microbial oxidative metabolism. The microbial production of foods can be viewed as an exercise in harnessing microbial biochemistry to produce desired changes in food products. The microbial processes in the production of food traditionally employ microbial

Table 8.1 Some Microbial Species Used for Producing Commercial Products

Industrial Chemicals	
Saccharomyces cerevisiae	Ethanol (from glucose)
Kluyveromyces fragilis	Ethanol (from lactose)
Clostridium acetobutylicum	Acetone and Butanol
Aspergillus niger	Citric acid
Xanthomonas campestris	Polysaccharides
Amino Acids and Flavor-enhancing Nucleotides	
Corynebacterium glutamicum	L-Lysine
Corynebacterium glutamicum	5′-Inosinic acid and 5′-guanylic acid
Corynebacterium glutamicum	Monosodium glutamate
Vitamins	
Ashbya gosspii	Riboflavin
Eremothecium ashbyi	Riboflavin
Pseudomonas denitrificans	Vitamin B_{12}
Propionibacterium shermanii	Vitamin B_{12}
Enzymes	
Aspergillus oryzae	Amylases
Aspergillus niger	Glucamylase
Trichoderma reesii	Cellulase
Saccharomyces cerevisiae	Invertase
Kluyveromyces fragilis	Lactase
Saccharomycopsis lipolytica	Lipase
Aspergillus sp.	Pectinases and proteases
Bacillus sp.	Proteases
Mucor pussilus	Microbial rennet
Mucor meihei	Microbial rennet
Polysaccharides	
Leuconostoc mesenteroides	Dextran
Xanthomonas campestris	Xanthan gum
Pharmaceuticals	
Penicillum chrysogenum	Penicillins
Cephalosporium acremonium	Cephalosporins
Streptomyces species	Amphotericin B, kanamycins, neomycins, streptomycin, tetracyclines and others
Bacillus brevis	Gramicidin S
Bacillus subtilis	Bacitracin
Bacillus polymyxa	Polymyxin B
Rhizopus nigricans	Steroid transformation
Arthrobacter simplex	Steroid transformation
Mycobacterium sp.	Steroid transformation
Escherichia coli (via recombinant-DNA technology)	Insulin, human growth hormone, somatostatin, interferon

enzymatic activities to transform one food into another, with the microbially produced food product having properties vastly different from the starting material.

Alcoholic Beverages

Microorganisms, principally yeasts in the genus *Saccharomyces*, are used to produce various types of alcoholic beverages. The production of alcoholic beverages relies on the alcoholic fermentation, that is, the conversion of sugar to alcohol by microbial enzymes (Figure 8.1). The flavor and other characteristic differences among various types of alcoholic beverages reflect differences in the starting substrates and the production process, rather than differences in the microbial culture or the primary fermentation pathways employed in the production of alcoholic beverages.

> *Give an Irishman lager for a month*
> *and he's a dead man.*
> *An Irishman is lined with copper,*
> *and the beer corrodes it.*
> *But whiskey polishes the copper*
> *and is the saving of him*
>
> —MARK TWAIN, *Life on the Mississippi*

Beer and Ale. Beer is a very popular beverage. The worldwide production of beer is over 18 billion gallons per year. Beer and ale are malt beverages, so-named because the initial preparation of the substrate for microbial fermentation involves barley

Figure 8.1 The steps in the brewing of beer. The production of beer begins with the malting of barley. The barley is induced to sprout and to produce enzymes that will catalyze the breakdown of starch. The malt is ground and mixed with warm water, and often other cereals such as corn, before going to the mash tun, where over a period of a few hours enzymes break down the long chains of starch into smaller molecules of carbohydrate. The aqueous extract called wort is separated from the mix and boiled with hops in a brew kettle. The boiling extracts flavors from the hops and stops the enzyme action in the wort. The hops are removed and the wort is put in a fermenting vessel, where it is pitched, or seeded, with yeast. After fermentation the beer may go to a lagering tank to mature, then it is pasteurized and bottled.

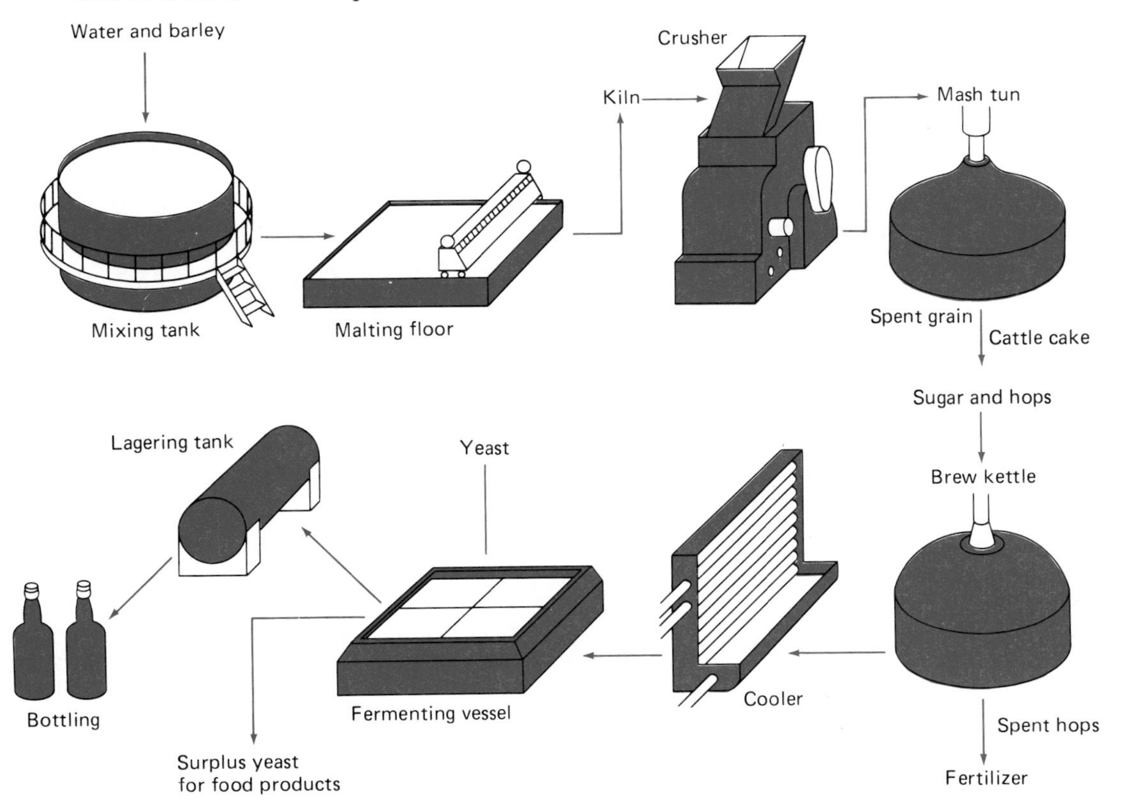

No Longer Just for Quaffing, the Bourgeois Beverage Is Being Fitted to the Food

By BRYAN MILLER

NEW YORK—For generations of Americans, beer has been a quaffing beverage, simple and straightforward, brewed for those who could put away a six-pack after a softball game or at a backyard picnic. The notion of pairing specific beers with certain foods was rarely a serious issue.

Today, beer is taking on new sophistication with the flood of imported brands on the American market—more than 200—and the rise of small, distinctive breweries from coast to coast. There are light-bodied Pilseners, dark German bocks, English and Canadian ales, hearty Irish stouts and porters, steam beer and more.

With such a varied selection of styles and flavors, it may be time to consider beer, like wine, as a complement to food.

"I feel very strongly that beer should be thought of in this light," says Michael Jackson, editor of "The World Guide to Beer" (Exeter Books, $9.95) and "The Pocket Guide to Beer" (Perigee Books, $5.95).

"One reason this has not happened in the United States until recently is that beer there has had sort of an Archie Bunker image. But now you can find all kinds of beer, some of which cost $6 a bottle. Archie Bunker sure doesn't buy that."

Another reason is that the major American brewers—Anheuser-Busch, Miller, Schlitz, G. Heileman, Pabst, Coors, Stroh and Olympia—make similar products. This is not out of any lack of competence or imagination, the experts point out, but because American beer drinkers have shown an overwhelming preference for mildly flavored, light-bodied beers.

Beer connoisseurs use a lexicon remarkably similar to that of wine lovers, describing brews as having bouquet, astringency, bitterness, body, yeastiness and aftertaste. It makes perfect sense, they contend, to think about how these characteristics marry with particular foods.

"When you eat food that is full-flavored, such as red meat, you need a beer with lots of flavor," says Matthew Reich, owner of the fledgling Old New York Beer Co., which produces New Amsterdam Amber.

Reich, who taught courses in beer appreciation before getting into the business, said beer with a good balance of body, which comes from malt, and bitterness, which comes from hops, can enhance a tasty meat dish far better than "bland quaffing beers."

Of course, Reich's prize-winning New Amsterdam, an amber beer with a red tint, flowery aroma and faintly sweet flavor, falls into that category, but he is quick to add that many others do as well.

Those could include products of other small breweries, called "microbreweries," such as William S. Newman in Albany, N.Y., Anchor Steam in San Francisco and Sierra Nevada in Chico, Calif., as well as some of the more flavorful Mexican imports, such as Carta Blanca and Dos Equis, both of which are slightly flowery with a touch of caramel in the aroma.

"To tell you the truth," Reich said, "if I were eating a really spicy meal, say an Indian curry or Sichuan food, I probably would go with Budweiser or Miller because they are so thirst-quenching."

Beer, like wine, can be characterized by isolating its key components: body, which can be felt on the palate as well as in the stomach—the "filling" sensation; and astringency or bitterness, similar to the tannins in wine—intensity of flavor. Once you recognize these qualities in your favorite beers it is easier to match them with foods.

"With rich foods, such as those with sauces that tend to coat your tongue, you need a beer with lots of astringency to cleanse the palate," said Joseph Owades, director for the Center for Brewing Studies, an independent organization in San Francisco, and an international consultant on brewing techniques.

He suggests the German Pilseners or English ales or beers. Some of the more widely available include Dinkelacker, DAB and Spaten from Germany and Whitbread, Watney, Courage and Bass from Britain.

Jackson concurs, giving an edge to the ales. "English ales are the cabernet sauvignons of the beer world," he said in a telephone interview from his London home. "They are full in flavor without being too heavy, and at the same time they are fruity."

Even a simple meat dish such as a hamburger can be enhanced by a complementary beer.

At the American Festival Cafe in Rockefeller Center, part of a new three-restaurant complex surrounding the skating rink, 13 American beers are offered. Andrew Young, director of restaurant development for the complex, says waiters are trained to explain characteristics of various beers to customers.

"If someone asks for a hamburger and a Heineken, we might ask if the customer would like to try something a little different to drink, such as a Ballantine India Pale Ale or a Fred Koch Jubilee Porter," Young said.

"With a light dish such as a pasta salad we might suggest a Rolling Rock, Cold Spring Export or a Lone Star, which are lighter." These three regional beers, from Pennsylvania, Minnesota and Texas, are typical light American-style lagers.

Young said customers had been enthusiastic about experimenting. "We have been open only four weeks and we've sold 10 cases of Prior Double Dark, which is a lot for an unknown beer," he said. Prior Double Dark, made by C. Schmidt & Sons of Philadelphia, is a full-bodied, copper-colored beer with what some describe as a "malty" flavor and smooth aftertaste.

Seafood and shellfish prepared without heavy sauces, particularly boiled lobster, go well with lighter beers—most American brands or the lighter Canadian ales such as Moosehead and Molsen. The exception might be oysters, which many beer lovers say go beautifully with dry English or Irish stout.

"I never could explain it very well," Jackson said, "but there is something about a dry stout that brings out the best in oysters. It has that sort of dry, tangy sensation that you might expect would drown out oysters, but it doesn't."

Howard Hillman, author of "The Gourmet Guide to Beer," calls Guinness Stout and oysters "a marriage that was made in heaven."

The consumption of Japanese and Chinese beers with Oriental food may have more to do with the "hot dogs taste better at the ball park" syndrome than any natural affinity, some of the beer experts say.

Oriental beers are, by and large, brewed in a German style established under the tutelage of German brewmasters. Kirin is a malty, filling beer while Sapporo is slightly more bitter; Suntory is one of the lightest. Most Oriental beers are made with a combination of malt and rice (large American breweries use malt and corn).

There is a lack of consensus about beer and spicy Oriental food. Some say a quaffing beer is what is required, while others, such as Jackson, favor a dark German beer in the Munich style such as Spaten.

The same might be said for Southern Italian food, with its acidic tomato sauces and sharp cheeses. "You really need a powerful tasting beer to stand up to all that," said Owades, of the Center for Brewing Studies. "I think Anchor Porter would be great, or maybe Bass Ale or New Amsterdam."

As for aperitifs and digestifs, most experts feel that a light-style beer, even the "Lite" brews made by American companies, work best.

"You want something that will not fill you up but will give you a slight warm feeling," Owades said. Jackson added that the foreign counterpart would be a Pilsen Urquell from Czechoslovakia or a lighter Pilsener-style German beer.

As for an after-dinner brew, a sweet stout such as Mackeson seems to be the choice. Another could be Newcastle Brown, a bittersweet ale from northern England with a creamy head. Dry stouts such as Guinness from Ireland are considered by some too heavy and cloying to sip after a meal.

"If you could get it down, Guinness would be good because of its burnt caramel flavor," Owades said. "But I really prefer sweet sherry."

Here's a Class Way to Get a Hop on Beer Basics

NEW YORK—Beer, theoretically, is a simple product: a fermented beverage made from water, barley malt (which is sprouted barley), hops and yeast. However, hundreds of variations are used to yield distinctive flavors.

Brewers in some countries, including the United States, use less malt and add corn, which makes a lighter product; Japan does the same with rice. Styles of fermentation, aging and clarifying also account for differences.

Among the principal categories of beer are these:

Ale

Technically the term refers to any beer made from a yeast that floats to the top during fermentation. Ales, in general, are relatively full-flavored and slightly higher in alcoholic content than beer, which is usually just under 5 percent by volume. Top-fermenting yeasts tend to produce a fruitier, more distinctive aroma, although brewers can overcome that if they want a milder product. In addition, ales are not aged like some beers.

Lager

Any bottom-fermented beer that has been aged, usually from one to six weeks. All leading American beers are lagers.

Porter

A dark lager. The color and extra flavor come from toasting the malt before brewing. Porters are normally stronger in flavor and alcoholic content than regular lager.

Stout

A dark ale made with toasted malt. Normally stronger than regular ale.

Pilsner

A generic term based on a style of Bohemian brewing developed in Pilsen, now in Czechoslovakia, in the 19th century. Pilseners are made with water that is hard but not alkaline. The term has little meaning today other than to indicate a pale golden beer.

All the principal American beers are Pilsener-style lagers. The "Lite" beers, which account for 20 percent of domestic consumption, are light Pilsener-style lagers with about a third fewer calories and at least 20 percent less alcohol than regular beer.

Bock

Traditionally a strong German dark beer, although there are now a few domestic bocks. In Germany bocks are brewed in the spring to launch the new beer season. Usually extremely dense, they are made with roasted malts, although some pale bocks can be found. They are almost always rich and with a distinct malty flavor.

Source: New York Times News Service, July 18, 1984. Reprinted by permission.

malt. Malt is a mixture of amylases and proteinases prepared by germinating barley grains for about a week and crushing the grains to release the plant enzymes. In the production of most beers, the malt is added to adjuncts in a process known as mashing The malt adjuncts, such as corn, rice, and wheat, provide carbohydrate substrates for ethanol production. During the mashing process the amylases from the barley malt hydrolyze the starches in the malt adjunct. The mash is heated to reach temperatures of about 70°C, which favors the rapid enzymatic conversion of starch to sugars. The clear liquid that is produced in this process is called wort

The wort is then cooked with the dried flowers of the hop plant. This cooking concentrates the mixture, inactivates the enzymes, extracts soluble compounds from the hops, and greatly reduces the numbers of microorganisms prior to the fermentation process. Additionally, compounds in the hops extract have antibacterial properties and protect the wort from the undesirable growth of Gram positive bacteria that could sour the beer.

The fermentation of the wort to produce beer in most countries is carried out by the yeast *Saccharomyces carlsbergensis*, a bottom fermenter. This means that at a late stage in fermentation, the yeasts floculate or aggregate and settle, partially clarifying the beer. *Saccharomyces cerevisiae* is also sometimes used in beer production, particularly in Great Britain and parts of the United States, but it is a top yeast and rises to the surface during fermentation. The inoculation of the yeast into the cooled wort, known as pitching, uses a heavy inoculum, of about 1 pound of yeast per barrel of beer. During fermentation the yeasts convert the sugars in the wort to alcohol and carbon dioxide and also produce small amounts of glycerol and acetic acid from the fermentation of the carbohydrates. Proteins and lipids are converted to small amounts of higher alcohols, acids and esters, which contribute to the flavor of beer. The active fermentation process is accompanied by extensive foaming of the mixture because of the production of carbon dioxide (Figure 8.2). Foaming decreases as the fermentation nears completion. The product is then known as a green beer and requires aging to achieve the characteristic flavor and aroma of the finished product.

Figure 8.2 Photograph showing beer being produced. The foaming is due to the rapid action of the yeast, which produces carbon dioxide as well as alcohol. In the traditional European manner, the beer is carefully inspected by the brewmaster at all stages of the fermentation to ensure production of a quality product. (Reprinted by permission of Quarto Ltd. from M. Jackson, 1977, *The World Guide to Beer*.)

During the aging process precipitation of proteins, yeasts, and resins occurs, resulting in a mellowing of the flavor. In the commercial production of beer, the carbon dioxide is normally collected during the fermentation phase and reinjected during the finishing process. In home production of beer, a small additional amount of sugar is usually added to each bottle to permit limited additional fermentative production of carbon dioxide to provide carbonation within the bottle. Commercially produced beer has an alcohol content of about 3.8 percent.

In addition to normal beer, there are a number of other malt beverages. Light or low carbohydrate beers are produced by using a wort prehydrolyzed with fungal enzymes. This pretreatment converts various carbohydrates in the wort to maltose and glucose. This permits the yeasts to ferment the carbohydrates completely to alcohol and carbon dioxide, greatly reducing the concentration of residual carbohydrates in the beer. These low calorie beers are particularly popular today for those who consume large amounts of beer and don't wish to develop a "beer belly."

Distilled Liquors. Distilled liquors or spirits are produced in a manner similar to beer, except that after the fermentation process the alcohol is collected by distillation, permitting the production of beverages with much higher alcohol concentrations than could accumulate during the fermentation process (Figure 8.3). The initial steps in the production of distilled spirits are analogous to those in beer production, beginning with a mashing process in which the polysaccharides and proteins in a starting plant material are converted to sugars and other simple organic compounds that can be readily fermented by yeasts to form alcohol.

Various starting plant materials are used for the production of different distilled liquor products. Rum

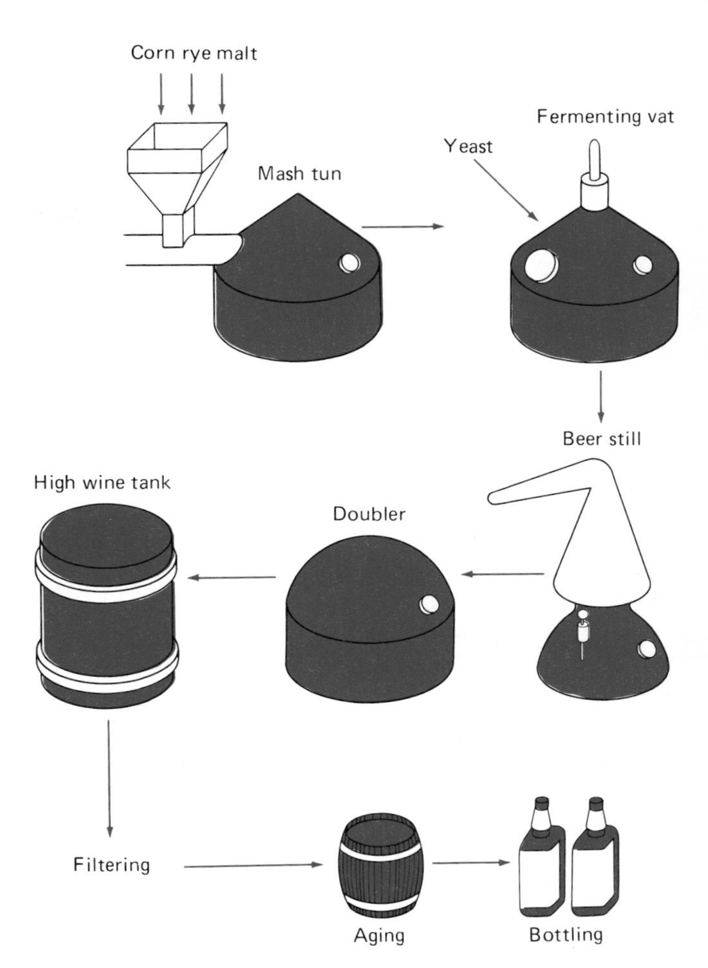

Figure 8.3 The process for producing distilled spirits. Distilled spirits, such as whiskey, are made in a process much like the brewing of beer. Bourbon production is illustrated here. Grains of corn are mixed with smaller amounts of rye and malted barley, crushed and mixed with warm water. The wort that emerges from the mash is transferred to a fermentor and pitched with yeast. After fermentation the beer is conveyed to a unit consisting of a beer still and doubler. The condensate is collected in a high wine tank and then matured for several years in oak casks before bottling.

is produced by using sugar cane syrup or molasses as the initial substrate, rye whiskey is produced from the fermentation of a rye mash, bourbon or corn whiskey uses corn mash, and brandy comes from the fermentation of grapes. The yeasts used in the production of distilled liquors typically are special distiller strains of *Saccharomyces cerevisiae*, which yield relatively high concentrations of alcohol.

The alcoholic product formed from the fermentation of the wort, known as a beer or wine, is heated in a still and the alcohol is collected. In addition to the alcohol, various volatile organic compounds, fusel oils, are collected with the distillate and contribute to the characteristic flavors of the different distilled liquor products. The distilled alcohol product is normally aged to yield a mellow tasting alcoholic beverage.

Wines. Wine is fermented primarily from grapes. Red wines are produced by using red grapes, whereas white wines are made from white grapes or from grapes that have had their skins removed.

And Noah the husbandsman began and planted a vineyard and he drank of the wine and was drunken

—GENESIS 9:21

Drink no longer water but use a little wine for thy stomach's sake and thine other infirmities

—I TIMOTHY 5:23

The production of wine begins when the grapes are crushed to form a juice or must. In the classical European method of wine making, wild yeasts from the surface skins of the grapes are the only inoculum for the fermentation. In modern wine production, however, the naturally occurring organisms (microbiota) associated with the grapes are removed by sulfur dioxide fumigation or by the addition of metabisulfite. The grape must is then inoculated with a specific strain of yeast, normally a variety of *Saccharomyces cerevisiae*. By using specific yeast strains and controlled fermentation conditions, a consistent quality product can be made. Initially, the grape must and yeasts are stirred to increase aeration and permit the proliferation of the yeasts. The mixing is later discontinued, permitting anaerobic conditions to occur that favor the production of alcohol. The sugar content of the grapes determines the final ethanol concentration. At the end of fermentation, wines typically have an alcohol content of 11–16 percent by volume.

Upon the first goblet he read this inscription, monkey wine; upon the second, lion wine; upon the third, sheep wine; upon the fourth, swine wine. These four inscriptions expressed the four descending degrees of drunkenness; the first, that which enlivens; the second, that which irritates; the third, that which stupefies; finally the last, that which brutalizes.

—VICTOR HUGO, *Les Misérables*

Wines are aged after the fermentation process to achieve their final bouquet and essence of flavor. During aging some fermentation of the malic acid of grape juice is carried out by lactobacilli (malolactic fermentation), reducing the acidity of the wine. Normally, the carbon dioxide produced during the alcoholic fermentation is allowed to escape and the wine is therefore still. In the case of champagne and other sparkling wines, however, the carbonation is essential. In some commercially produced champagne, carbon dioxide is reinjected into the wine after fermentation. In the classic French method of producing champagne, the wine is fermented in the bottle. After fermentation is complete, the bottles are inverted, and the yeast sediments into the neck of the specially shaped champagne bottles. The yeasts are frozen and removed as a plug without excessive loss of carbon dioxide.

Not only does one drink wine, but one inhales it, one looks at it, one tastes it, one swallows it . . . and one talks about it.

—KING EDWARD VII OF ENGLAND

Leavening of Bread

Yeasts are added to bread dough to ferment the sugar. The production of carbon dioxide leavens the dough and causes it to rise. The principal yeast used in bread baking is *Saccharomyces cerevisiae*, which is known as Bakers' yeast. In the baking process, yeast is used strictly as a source of enzymes to carry out an alcoholic fermentation. No growth of the yeast occurs during the baking process. Amylases in the dough convert starch to sugars, and the yeasts metabolize the sugars that are formed, producing carbon dioxide and ethanol. The ethanol is removed during baking.

The bread I eat in London is a deleterious paste, mixed up with chalk, alum and bone ashes, insipid to the taste and destructive to the constitution. The good people are not ignorant of the adulteration, but they prefer it to wholesome bread because it is whiter . . . and the miller or the baker is obliged to poison them.

—TOBIAS SMOLLET

In addition to leavening bread, microbes produce the characteristic flavors of some breads. For example, the production of San Francisco sour dough bread utilizes the yeast *Torulopsis holmii* and a heterofermentative *Lactobacillus* species to sour the dough and give this bread its characteristic sour flavor. Rye bread is also produced by initially souring the dough; cultures of *Lactobacillus plantarum*, *L. brevis*, *L. bulgaricus*, *Leuconostoc mesenteriodes*, and *Streptococcus thermophilus* are employed as starter cultures in making different rye breads. The action of heterofermentative lactic acid bacteria produces the bread's characteristic flavor.

Fermented Dairy Products

Numerous products are made by the microbial fermentation of milk, which is primarily carried out by lactic acid bacteria. The lactic acid fermentation pathway and the accumulation of lactic acid, from the metabolism of the milk sugar lactose, is common to the production of fermented dairy products. The accumulation of lactic acid in these products acts as a natural preservative. The differences in flavor and aroma of the various fermented dairy products are due to additional fermentation products that may be present in only relatively low concentrations.

Different fermented dairy products are produced by using different strains of lactic acid bacteria as starter cultures and different fractions of whole milk as the starting substrate (Table 8.2). Sour cream, for example, uses *Streptococcus cremoris* or *S. lactis* for the production of lactic acid, and *Leuconostoc cremoris* or *S. lactis diacetilactis* for production of

Table 8.2 Some Foods Produced from Fermented Milks

Fermented Product	Microorganisms Responsible for Fermentation	Description
Sour cream	*Streptococcus* sp. *Leuconostoc* sp.	Cream is inoculated and incubated until the desired acidity develops
Cultured buttermilk	*Streptococcus* sp. *Leuconostoc* sp.	Made with skimmed or partly skimmed pasteurized milk
Bulgarian buttermilk	*Lactobacillus bulgaricus*	Product differs from commercial buttermilk in having higher acidity and lacking aroma
Acidophilus milk	*Lactobacillus acidophilus*	Milk for propagation of *L. acidophilus* and the milk to be fermented is sterilized and then inoculated with *L. acidophilus*; this milk product is used for its medicinal therapeutic value
Yogurt	*Streptococcus thermophilus* *Lactobacillus bulgaricus*	Made from milk in which solids are concentrated by evaporation of some water and addition of skim milk solids; product has consistency resembling custard
Kefir	*Streptococcus lactis* *Lactobacillus bulgaricus* Yeasts	A mixed lactic acid and alcoholic fermentation; bacteria product acid, and yeasts produce alcohol

the characteristic flavor compounds. Cream is used as the starting substrate for this product.

If skim milk is used as the starting material, cultured buttermilk is produced. Bulgarian buttermilk is produced by using *Lactobacillus bulgaricus* for the production of both lactic acid and flavor compounds.

Butter is normally made by churning cream that has been soured by lactic acid bacteria. *S. cremoris* or *S. lactis* are used to produce lactic acid rapidly, and *Leuconostoc citrovorum* produce the necessary flavor compounds. The *Leuconostoc* enzymes attack citrate in milk, producing diacetyl, which gives butter its characteristic flavor and aroma.

Yogurt is produced by fermenting milk with a mixture of *L. bulgaricus* and *S. thermophilus*. The characteristic flavor of yogurt is due to the accumulation of lactic acid and acetaldehyde produced by *L. bulgaricus*. Because of the tart taste of acetaldehyde, most yogurt produced in the United States is flavored by adding fruit. Over 550,000 pounds of yogurt are produced annually in the United States.

Cheese. A wide variety of cheeses are produced by microbial fermentations. Cheeses consist of milk curds that have been separated from the liquid portion of the milk (whey).

Little Miss Muffet
Sat on a tuffet
Eating her curds and Whey
Along came a spider
That sat down beside her
And Frightened Miss Muffet away

—MOTHER GOOSE

The curdling of milk is accomplished by using the enzyme rennin and lactic acid bacterial starter cultures. Cheeses are classified as (1) soft, if they have a high water content (50–80 percent); (2) semihard, if the water content is about 45 percent; and (3) hard, if they have a low water content (less than 40 percent). Cheeses are also classified as unripened cheeses if they are produced by a single-step fermentation, or as ripened cheeses if additional microbial growth is required during maturation of the cheese to achieve the desired taste, texture, and aroma (Table 8.3). Processed cheeses are made by blending various cheeses to achieve a desired product. If the water content is elevated during processing, thereby diluting the nutritive content of the product, the product is called a "processed food" rather than a cheese.

Table 8.3 The Microorganisms Used To Produce Various Cheeses

Cheese	Microorganisms		
	Primary Fermenter		Secondary Ripening
Soft, unripened			
Cottage	*Streptococcus lactis*	*Leuconostoc citrovorum*	
Cream	*Streptococcus cremoris*		
Neufchatel	*Streptococcus diacetilactis*		
Soft, ripened, 1–5 months			
Brie	*Streptococcus lactis* *Streptococcus cremoris*	*Penicillium camemberti* *Penicillium candidum*	*Brevibacterium linens*
Camembert	*Streptococcus lactis* *Streptococcus cremoris*	*Penicillium camemberti* *Penicillium candidum*	
Limburger	*Streptococcus lactis* *Streptococcus cremoris*	*Brevibacterium linens*	
Semisoft, ripened, 1–12 months			
Blue	*Streptococcus lactis* *Streptococcus cremoris*	*Penicillium roqueforti* or *Penicillium glaucum*	
Brick	*Streptococcus lactis* *Streptococcus cremoris*	*Brevibacterium linens*	
Gorgonzola	*Streptococcus lactis* *Streptococcus cremoris*	*Penicillium roqueforti* or *Penicillium glaucum*	
Monterey	*Streptococcus lactis* *Streptococcus cremoris*		
Muenster	*Streptococcus lactis* *Streptococcus cremoris*	*Brevibacterium linens*	
Roquefort	*Streptococcus lactis* *Streptococcus cremoris*	*Penicillium roqueforti* or *Penicillium glaucum*	
Hard, ripened 3–12 months			
Cheddar	*Streptococcus lactis* *Streptococcus cremoris* *Streptococcus durans*	*Lactobacillus casei*	

Table 8.3 The Microorganisms Used To Produce Various Cheeses *(Continued)*

Cheese	Microorganisms		
	Primary Fermenter		Secondary Ripening
Colby	*Streptococcus lactis* *Streptococcus cremoris* *Streptococcus durans*	*Lactobacillus casei*	
Edam	*Streptococcus lactis* *Streptococcus cremoris*		
Gouda	*Streptococcus lactis* *Streptococcus cremoris*		
Gruyere	*Streptococcus lactis* *Streptococcus thermophilus*	*Lactobacillus helveticus*	*Propionibacterium shermanii* or *Lactobacillus bulgaricus* and *Propionibacterium freudenreichi*
Swiss	*Streptococcus lactis* *Streptococcus thermophilus*	*Lactobacillus helveticus*	*Propionibacterium shermanii* or *Lactobacillus bulgaricus* and *Propionibacterium freudenreichi*
Very hard, ripened, 12–16 months			
Parmesan	*Streptococcus lactis* *Streptococcus cremoris* *Streptococcus thermophilus*	*Lactobacillus bulgaricus*	
Romano	*Lactobacillus bulgaricus*	*Streptococcus thermophilus*	

Many's the long night I've dreamed of cheese

—Robert Louis Stevenson, *Treasure Island*

The natural production of cheeses involves a lactic acid fermentation with various mixtures of *Streptococcus* and *Lactobacillus* species used as starter cultures to initiate the fermentation. The flavors of different cheeses result from the use of different microbial starter cultures, varying incubation times and conditions, and the inclusion or omission of secondary microbial species late in the fermentation process. Ripening of cheeses involves additional enzymatic transformations after the formation of the cheese curd.

Figure 8.4 Photograph showing cheese production room with fungi growing on the surface of the ripening Camembert cheese. (Courtesy of Foods and Wines of France, New York.)

Swiss cheese formation involves a late propionic acid fermentation, with ripening accomplished by *Propionibacterium shermanii* and *P. freudenreichii*. The propionic acid yields the characteristic aroma and flavor, and the carbon dioxide produced during this late fermentation forms the holes in Swiss cheese. Various fungi are also used in the ripening of different cheeses. The unripened cheese is normally inoculated with fungal spores and incubated in a warm moist room to favor the growth of filamentous fungi (Figure 8.4). For example, blue cheeses are produced by using *Penicillium* species, roquefort cheese is produced by using *P. roqueforti*, and camembert and brie are produced by using *P. camemberti* and *P. candidum*.

Have Roquefort with a fine Chambertin wine, and you'll be ready for love

—CASANOVA

Fermented Meats

Several types of sausage, such as Lebanon bologna, the salamis, and the dry and semidry summer sausages, are produced by allowing the meat to undergo a heterolactic acid fermentation during curing. The fermentation has both a preservative effect and also adds a tangy flavor to the meat. Various lactic acid bacteria are normally involved in the fermentation.

Fermented Vegetables

Vegetables are fermented by using lactic acid bacteria to produce various foods. Sauerkraut is produced from a lactic acid fermentation of wilted, shredded cabbage. Coliform bacteria, such as *Enterobacter cloacae*, are prominent in the initial mixed community and produce gas and volatile acids as well as some lactic acid. The accumulating lactic acid exerts a selective pressure on the microbial community, causing population shifts and continued succession. As a result, after the initial fermentation there is a shift in the microbial community and *Leuconostoc mesenteroides* becomes the dominant microbial population. The continuing succession of bacterial populations next favors the development of *Lactobacillus plantarum*, which produces acid but no gas. During this phase of the fermentation the concentration of lactic acid reaches 1.5–2 percent. Growth of *L plantarum* also removes mannitol, which is produced by *Leuconostoc* and has an undesirable bitter flavor.

The traditional method for producing pickles by fermentation uses the natural microbiota associated with the cucumber and controlled temperature and salt concentrations to regulate the fermentation process. As the lactic acid and salt concentrations increase, *Lactobacillus plantarum* becomes the dominant bacterium, beginning several days after the fermentation and continuing until the salt concentration surpasses 10 percent. The completion of the fermentation process involves yeasts that grow at high salt concentrations. During the final yeast fermentation stage, some carbohydrates are converted to alcohol. The growth of film-forming yeasts lowers the lactic acid concentration. The sourness of the pickle reflects the amount of lactic acid that accumulates during the fermentation.

The production of green olives involves a lactic acid fermentation. The harvested olives are washed with a solution of sodium hydroxide that removes most of the oleuropein, a very bitter phenolic glucoside, which gives unfermented olives a very undesirable flavor. The olives are next placed in a brine solution and a lactic acid fermentation of 6–10 months is permitted to occur.

Soy sauce. Several Oriental foods are prepared by fermenting soybeans or rice. Soy sauce, a brown, salty tangy sauce, which in Japanese is called shoyu, is produced from a mash consisting of soybeans, wheat, and wheat bran. Soy sauce is used as a condiment or as an ingredient in other sauces (Figure 8.5). The starter culture for the production of soy sauce is produced by a koji fermentation, a dry fermentation, in which a mixture of soybeans and wheat is inoculated with spores of *Aspergillus oryzae*. The fungi grow on the surface of the soybeans and wheat, accumulating various enzymes, including proteinases and amylases. Various bacterial populations, normally dominated by lactic acid bacteria, also develop during this koji fermentation.

The cultures are mixed with a mash consisting of autoclaved soybeans, autoclaved and crushed wheat, and steamed wheat bran. The mash is then incubated from 10 weeks to over a year, depending on the incubation temperature. During this incubation period the proteinases, amylases, and other enzymes of the koji are active, and there is a succession of microbial populations. The maturation begins with lactic acid bacteria, and later involves alcoholic fermentations by yeasts. The most important organisms during the fermentation process are *Aspergillus oryzae*, which produce proteinases and amylases; *Lactobacillus* species, which produce sufficient amounts of lactic acid to prevent spoilage by other microorganisms; and yeasts, which produce sufficient alcohol to increase the flavor content.

Poi is a fermented food product from the Ha-

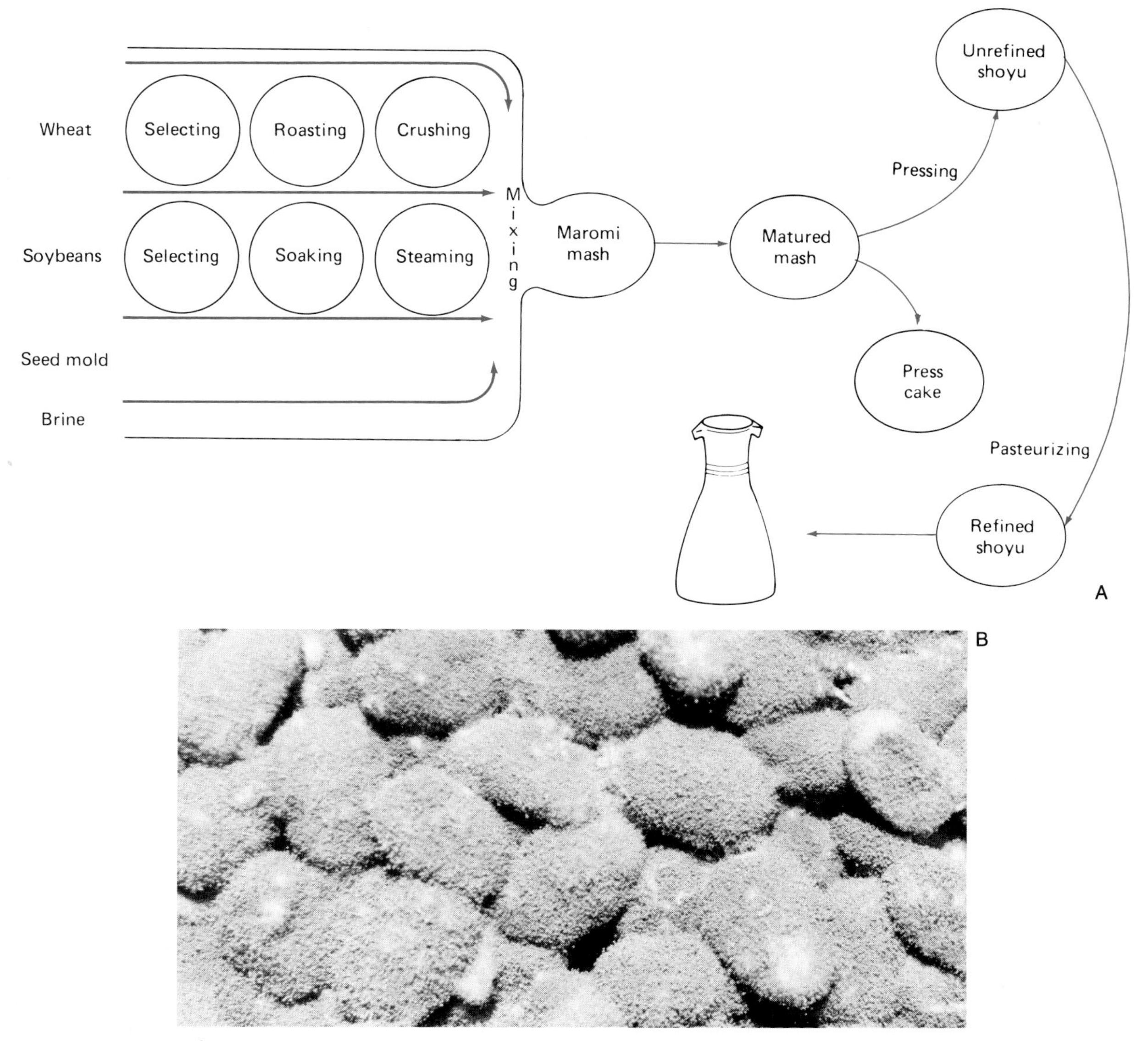

Figure 8.5 (A) As shown in this diagram, soy sauce production is a two-stage fermentation process, involving a dry fermentation followed by a submerged fermentation. (B) Photograph of koji fermentation tray, showing growth of *Aspergillus* on a mixture of soybeans and wheat. (Courtesy Kikkoman Foods, Walsworth, Wisconsin.)

waiian Islands. In the production of poi, the stems of the taro plant are steamed, ground, and subjected to fermentation for 1–6 days. There is a succession of microbial populations involved in poi production that includes coliforms, *Pseudomonas, Lactobacillus, Streptococcus, Leuconostoc,* and yeasts. The fermentation products, principally lactic acid, acetic acid, formic acid, ethanol, and carbon dioxide, contribute to the characteristic texture, flavor, and aroma of poi.

Vinegar

In the production of vinegar, an initial anaerobic fermentation by *Saccharomyces cerevisiae* converts sugars to a sufficient concentration of alcohol. After production of the alcoholic liquid, acetic bacteria,

MICROBIOLOGY HEADLINES

It's Trendy, Tasty and Tofutti

The summer's hot frozen dessert is no-cream ice cream

By J. D. REED

Sure, once upon a time, they all used to scream for ice cream. But many Americans today are more likely to pine for the impossible: a tasty, low-cal no-cholesterol, nondairy frozen delight—and make it all natural too, please. Only a dream, say cynics. Tofutti, says David Mintz. Bless you, say many converts who believe Mintz's Tofutti may be the answer. Indeed, Americans are licking up 40,000 gal. a week of his chilly concoction. Production has nearly doubled in the past month. Move over Frusen Glädjé and frozen yogurt; this is the summer of Tofutti.

At prices ranging from 75¢ for a 4-oz. soft-whipped portion to $2.49 for a hard-packed pint, Tofutti is selling briskly from pushcarts and at custard shops and specialty stores almost all the way across the country. Strangely, it has not caught on yet in Southern California, normally a hotbed of food fads. But Tofutti is moving smoothly at chic Neiman-Marcus in Dallas and at Bloomingdale's in New York City. Häagen-Dazs, the designer-ice-cream folks, will begin nationwide distribution to retailers later this month. On Wall Street, stock in Mintz's Tofu Time, Inc., which went public in December, has tripled in value and split.

This epicurean success rests on an improbable ingredient: a bland, gelatinous, soybean derivative called tofu, which many consider an affront to the taste buds. "Tofu is like eating your pillow," pronounces Washington-based Researcher Lisa Frangos. But she likes Tofutti. Creator Mintz, 53, is not surprised. A former caterer and a devout Orthodox Jew, Mintz suspected eight years ago that tofu could be the milk substitute he needed to make an ice-cream-like dessert that would not violate religious prohibitions against the mixing of meat and dairy products at the same meal."It didn't come out right at the outset," he admits. But he kept trying: "I am a guinea pig with a hell of a palate." He was finally satisfied in 1980. Mintz keeps his formula a secret but admits to contemplating odd flavors like mango and even pineapple-yam. Says he: "My wife Rachel and I live and breathe Tofutti."

While available in plain vanilla, Tofutti best masks the aftertaste of its novel ingredient with strident flavors such as banana-pecan and forest maple. In Texas, the bestseller is made by swirling peanut butter and chocolate flavors together into something that tastes like a Reese's Cup. Aficionados swear Tofutti is better than ice cream. Says one: "If you eat it too fast, you even get the headache."

That is about the only unwelcome side effect. Tofutti is a dieter's and a cardiologist's dream. Blended with high-fructose corn sweeteners, it contains 128 calories in four ounces, about the same as frozen yogurt, but fewer than the 250 or more for premium ice cream. NutraSweet, the sugar substitute, may soon be used to further reduce calories. Because Tofutti is both butterfat- and cholesterol-free, Mount Sinai Medical Center in Miami Beach even serves the confection to heart patients. It contains no lactose, good news for the millions of Americans who suffer an adverse reaction to milk-based products. At the Schnelli Deli in Detroit, blacks, who have a higher incidence of lactose sensitivity than whites, are among the best customers.

There are other frozen soy-based competitors. Ice Bean, for example, has been sold in health-food emporiums since 1976, and a new effort is reported to be in the works in Atlanta. But Tofutti seems to have the edge, and Mintz is looking to new horizons. He is experimenting with tofu-based drinks and a vegetable salad. His next obsession is a nonmeat burger. Says the tycoon of tofu: "I am close to this tremendous breakthrough. It is made of tofu and has the taste and smell of a beefburger." Given his marketing skills, it may not be long before Tofutti addicts sit down to the main course: a Big MacFutti.

such as *Acetobacter,* are added and allowed to convert the alcohol slowly to acetic acid. The type of vinegar is determined by the starting material. For example, wine vinegar comes from grapes and cider vinegar from other fruits. The rate of vinegar production is limited by the availability of oxygen. Many industrial producers of vinegar use forced aeration to maximize the rate of acetic acid production. Using a 10 percent alcohol solution as substrate, the acetic acid yield can be 13 percent.

And Boaz said unto her, At mealtime come thou hither, and eat the bread, and dip thy morsel in the vinegar

—RUTH 2:14

Production of Pharmaceuticals

The pharmaceutical manufacturing industry—a major source of employment for industrial microbiologists—is primarily concerned with disease processes (some of which are caused by microorganisms) and with making drugs (many of which are produced by microorganisms) to control disease processes. The world's supply of pharmaceuticals, including many antibiotics, steroids, and vitamins are produced in substantial part by microorganisms. The microbial production of pharmaceuticals is a major industry. Antibiotic production alone accounted for approximately five billion dollars in worldwide sales in 1980. The role of microorganisms in producing these pharmaceuticals is economically important for industry and is essential for making these compounds available at a cost low enough to permit their wide use in preventing and treating numerous diseases.

Antibiotics

Antibiotics are secondary metabolites produced by microorganisms that inhibit or kill other microorganisms. A secondary metabolite is not directly involved in the central energy-generating metabolic pathways. Rather, a secondary metabolite is made only under a special set of favorable conditions. Of the thousands of different antibiotic compounds made in nature by various microorganisms, relatively few are produced commercially. Several of the major antibiotics used in medicine and the microorganisms used for producing these antibiotics are shown in Table 8.4.

Antibiotics are widely used in medicine for treating infectious diseases. Although it is hard to imagine medical practice without antibiotics, these compounds have only been in use since the end of World War II. Penicillin was the first antibiotic that was discovered. The strain of *Penicillium* that is used today for the production of penicillin is very different from the strain first observed by Alexander Fleming. The organism has been modified and efficient strains selected for industrial use. Penicillin and other antibiotics are typically produced in large fermentors (Figure 8.6). The pH, temperature, nutrient levels, and oxygen concentration are carefully controlled to maximize the yield of the desired antibiotic.

Table 8.4 Some Antibiotics Produced by Microorganisms

Antibiotic	Produced by
Amphotericin-B	*Streptomyces nodosus*
Bacitracin	*Bacillus licheniformis*
Carbomycin	*Streptomyces halstedii*
Chlorotetracycline	*Streptomyces aureofaciens*
Chloramphenicol	*Streptomyces venezuelae* or total chemical synthesis
Erythromycin	*Streptomyces erythreus*
Fumagillin	*Aspergillus fumigatus*
Griseofulvin	*Penicillium griseofulvin*
	Penicillium nigricans *Penicillium urticae*
Kanamycin	*Streptomyces kanamyceticus*
Neomycin	*Streptomyces fradiae*
Novobiocin	*Streptomyces niveus*
	Streptomyces spheroides
Nystatin	*Streptomyces noursei*
Oleandomycin	*Streptomyces antibioticus*
Oxytetracycline	*Streptomyces rimosus*
Penicillin	*Penicillium chrysogenum*
Polymyxin-B	*Bacillus polymyxa*
Streptomycin	*Streptomyces griseus*
Tetracycline	Dechlorination and hydrogenation of chlorotetracycline; direct fermentation in dechlorinated medium
Vira-A (adenine arabinoside)	*Streptomyces antibioticus*

Figure 8.6 (A) Intermediate-size fermentors are used in a pilot plant to scale up the fermentation. (B) Large-scale fermentors are used for commercial antibiotic production. (Courtesy Marvin Hoehn and Bernard Abbott, Eli Lily Research Laboratories, Indianapolis, Indiana.)

A

B

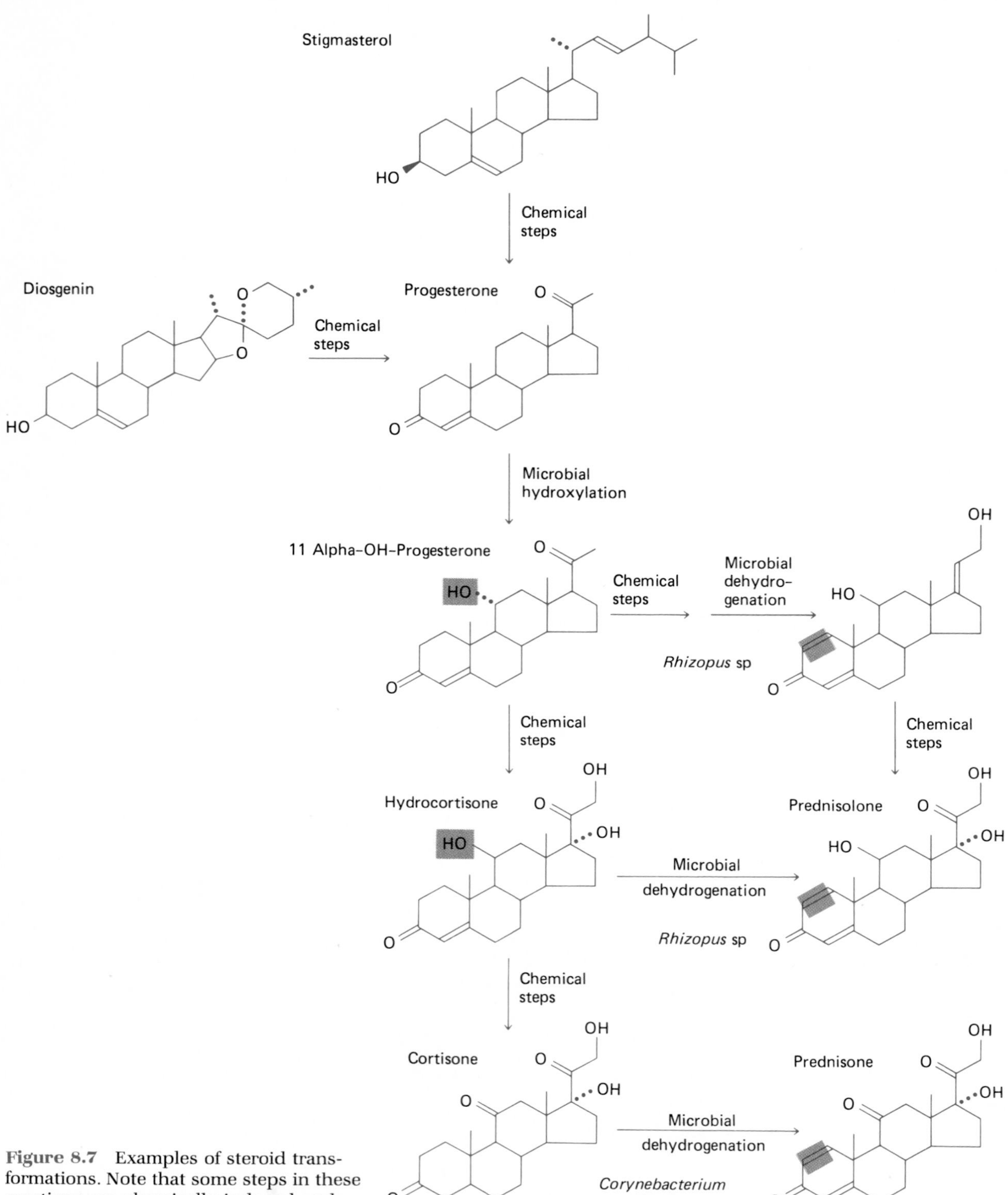

Figure 8.7 Examples of steroid transformations. Note that some steps in these reactions are chemically induced and that others are microbial transformations.

Microorganisms can also be used to modify the antibiotics produced in these processes. For example, the penicillin G produced by *Penicillium* can be modified to form various penicillin derivatives. The modification of penicillin may be accomplished chemically or by using microbial enzymes. Transformations of the basic penicillin structure can be used to produce penicillin derivatives with differing properties. For example, in 1981 piperacillin was approved as a broad-spectrum antibiotic, and in 1982 azlocillin was also introduced for use against strains that are resistant to earlier generation penicillins.

Similar semisynthetic approaches can be used for manufacturing other antibiotics. For example, cephalosporin C is made as the fermentation product of *Cephalosporium acremonium*. Although this form of the antibiotic is not potent enough for clinical use, the cephalosporin C molecule can be transformed. We are now into the so-called third generation cephalosporins, such as moxlactam, which have been developed to combat bacteria that produce penicillin-destroying enzymes.

Other antibiotics are produced by using different microbial strains, media, incubation conditions, and recovery methods. Many antibiotics, such as streptomycin, are produced using strains of *Streptomyces griseus*.

Steroids

The use of microorganisms to carry out biotransformations of steroids is very important in the pharmaceutical industry (Figure 8.7). Steroid hormones regulate various aspects of metabolism in animals, including humans. One such hormone, cortisone, has been found to relieve the pain associated with rheumatoid arthritis. Various cortisone derivatives are also useful in alleviating the symptoms associated with allergic and other undesired inflammatory responses of the human body. Additionally, various steroid hormones regulate human sexuality, and some of these steroids are manufactured as oral contraceptives. Thus, there is a great demand for steroids. The physiological properties of a steroid depend on the nature and the exact position of the chemical constituents on the basic steroid ring structure. The chemical synthesis of steroids is very complex because of the necessity of locating substituents so precisely.

For example, cortisone can be synthesized chemically from desoxycholic acid, but the process requires 37 steps, many of which must be carried out under extreme conditions. Chemical production of cortisome results in a product costing over $200 per gram. The major difficulty in chemically synthesizing cortisone is the need to introduce an oxygen atom at the number 11 position of the steroid ring, but this can be accomplished by microorganisms. The fungus *Rhizopus arrhizus*, for example, hydroxylates progesterone, forming another steroid with the introduction of oxygen at the number 11 position. The fungus *Cunninghamella blakesleeana* similarly can hydroxylate the steroid cortexolone to form hydrocortisone with the introduction of oxygen at the number 11 position. The use of such microbial transformations in the formation of cortisone has lowered the original cost over 400 fold, so that in 1980 the price of cortisone in the United States was less than 50 cents per gram, compared to the original $200.

Vitamins

Various vitamins, which are essential animal nutritional factors, can be produced by microbial fermentations (Table 8.5). Vitamin B_{12} can also be produced commercially by direct fermentation, using *Propionibacterium shermanii* or *Pseudomonas denitrificans*, and these are the organisms used today for the production of this vitamin.

Riboflavin can also be produced as a fermentation product by using various microorganisms. Riboflavin is a byproduct of the acetone butanol fermentation and is produced by various *Clostridium* species. Commercial production of riboflavin by direct fermentation often uses the fungal species *Eremothecium ashbyii* or *Ashbya gossypii*.

Production of Organic Acids

Several organic acids, including acetic, citric, lactic, itaconic, gluconic, and gibberellic acids can be produced by microbial fermentation (Table 8.6). Acetic acid can be used for commercial purposes, for example, as a stop bath in photographic processing, though, for economic reasons most production for such purposes today is accomplished by using chemical syntheses.

Gluconic acid also has a variety of commercial uses. Calcium gluconate, for example, is used as a pharmaceutical to supply calcium to the body; ferrous gluconate similarly is used to supply iron in the treatment of anemia; and gluconic acid in dishwasher detergents prevents spotting of glass surfaces owing to the precipitation of calcium and magnesium salts. Gluconic acid is produced by

Japan's Biotechnology Effort

By ANDREW POLLACK

TOKYO, Aug. 27—For centuries, soy sauce has been made by fermenting soybeans and other ingredients for six to eight months. Now, the Kikkoman Corporation, the world's largest soy sauce producer, is experimenting with a process that can reduce the time to one week.

The soy sauce "bioreactor" is one example of Japan's distinctive push into biotechnology. Indeed, one of Japan's major advantages in this emerging industry is what might be called the soy sauce factor.

Hundreds of years of making soy sauce, tofu, sake and other traditional food and drink, have given Japan great expertise in fermentation—the ability to change substances through the use of microorganisms. And fermentation is becoming a key technique in the biotechnology industry.

While small entrepreneurial companies have led the move into biotechnology in the United States and Europe, food and liquor companies are in the vanguard here.

The Office of Technology Assessment, a research branch of Congress, said in a report in January that Japan might be in a position to surpass the United States in the commercialization of biotechnology.

"If we can combine the old technology with the new technology, we will have great power in this area," said Akio Sato, an official at the government-run Fermentation Research Institute, which has been studying fermentation since 1940.

Biotechnology usually brings to mind advanced techniques, such as genetic engineering. The best-known example involves recombinant DNA, in which a gene to produce, say insulin, is implanted into bacteria, which then start making insulin.

But that is only part of the story. Once a few bacteria are induced to make a substance, the bacteria must be grown and multiplied so that the substance can be mass-produced. It is in this mass-production stage that Japan's expertise could be of great advantage.

Still, many researchers here say that fermentation is not enough to put Japan in the forefront of biotechnology. They say that Japan must also catch up to the West in genetic research.

"The Office of Technology Assessment study very much overestimates the Japanese situation in biotechnology," said Masami Tanaka, director of the bioindustry office at the Ministry of International Trade and Industry.

Japan was a late starter in genetic engineering. Only after Western companies demonstrated the principles and potential of gene-splicing did industry and Government here take much interest.

The Japanese, typically, attacked the problem vigorously. More than 150 companies have biotechnology projects. Several Government agencies have also started programs.

Corporate research and development spending on biotechnology totaled about $220 million in 1982 and has been growing at 20 percent a year. Government expenditures directly on biotechnology amounted to about one-fifth that level. In addition, the Government spends about $435 million a year on basic research in biology and medicine, according to Mr. Tanaka of the trade ministry.

One example of the Japanese approach is the Kyowa Hakko Kogyo Company, which specializes in fermentation. The company started by making alcohol for sake and shochu, two popular Japanese drinks. Then, in the 1950's, it began applying its fermentation expertise to pharmaceutical products such as streptomycin.

In 1956, Shukuo Kinoshita, now the company's chairman, discovered microorganisms that could be used to make L-Glutamic acid. That substance, in turn, is used to make monosodium glutamate—better-known as MSG, a flavor enhancer—as well as amino acids used in seasonings and intravenous fluids.

Now, Kyowa Hakko wants to apply its expertise to produce interferon, a potential cancer-fighting substance, and other products of gene-splicing. Last September, it announced a process for mass-producing interferon.

To get started in genetic engineering, many Japanese companies initially teamed up with American or European concerns.

Shionogi & Company, a pharmaceutical concern, recently completed clinical testing of human insulin produced through recombinant DNA techniques. The technology was licensed from Eli Lilly & Company and Genentech Inc., both United States concerns.

But Japanese companies and universities have recently begun producing original work and transferring technology abroad.

Suntory Ltd., Japan's major liquor company, was the first concern to use a synthetic gene, rather than one isolated from a cell, to produce gamma-interferon, the type of interferon viewed as most promising as an anti-cancer agent.

In December, Suntory and the Schering-Plough Corporation, the United States pharmaceutical company, announced an agreement in which Schering-Plough would get access to Suntory's technology. In return, Suntory would get access to some of Schering-Plough's technology and help in running clinical tests, an area in which the Japanese company has little experience.

"For the first time in Japan, we transferred our technology," said Teruhisa Noguchi, head of biotechnology research for Suntory.

Last weekend it was announced that researchers at the Dai-Ichi Seiyaku Company, a pharmaceutical manufacturer, and Tottori University had produced interferon by genetically altering a virus that lives in silkworms. The researchers say the silkworms might be more efficient interferon producers than the bacteria usually used.

While pharmaceuticals have been the initial focus of biotechnology research, attention is shifting to agriculture and the chemical industry.

Kyowa Hakko used to make solvents using fermentation. But in 1961, when oil prices were low, it decided to switch to petrochemical technology. Now, with oil prices higher and Japan worried about its dependence on foreign petroleum, some scientists hope production methods will change again.

Petrochemical plants operate at high temperatures and pressures, whereas biological factories would operate best at the low temperatures and pressures at which living things thrive. That could save huge amounts of energy.

Patent filings indicate the gathering strength of Japanese industry in biotechnology. A study by the international trade ministry found that, from 1973 through 1981, foreigners accounted for 19 percent of all patent applications filed in Japan, but they filed 76 percent of the applications involving recombinant DNA. In the two years after that, however, foreigners accounted for only 55 percent of patent applications involving recombinant DNA.

However, few here think that Japan will catch up to the West in basic research.

One problem is a shortage of researchers trained in biotechnology. When Suntory decided to start a biotechnology program, it recruited about 20 Japanese researchers working abroad. It also now employs 40 women researchers, one-third of its total—an extraordinarily high proportion for Japan.

Source: The New York Times, August 28, 1984. Reprinted by permission.

Table 8.5 Production of Some Vitamins by Using Microorganisms

Vitamin	Culture	Medium	Fermentation Conditions	Yield
Riboflavin	*Ashbya gossypii*	Glucose, collagen, soya oil, glycine	6 days 36°C aerobic	4.25 g/L
L-Sorbose (in vitamin C synthesis)	*Gluconobacter oxidans* subsp. *suboxidans*	D-sorbitol, 30% corn steep	45 h at 30°C aerobic	70% based on substrate used
5-Keto gluconic acid (in vitamin C synthesis)	*Gluconobacter oxidans* subsp. *suboxidans*	Glucose, $CaCO_3$, corn steep	33 h at 30°C aerobic	100% based on substrate used
Vitamin B_{12}	*Propionibacterium shermanii*	Glucose, corn steep, ammonia, cobalt pH 7.0	30°C anaerobic (3 days) + aerobic (4 days)	23 mg/L

Abbreviations: h = hours, g/L = grams per liter, mg/L = milligrams per liter, $CaCO_3$ = calcium carbonate

Table 8.6 Some Organic Acids Produced by Fermentation

Product	Culture	Substrate (yield, %)	Process
Acetic acid	*Acetobacter* sp.	Ethanol (98–99)	Continuous aerated process using an alcoholic solution containing (%): glucose 0.9, ammonium phosphate 0.4, magnesium sulfate 0.1, potassium citrate 0.1, pantothenic acid 0.00005; extraction by filtration
Lactic acid	*Lactobacillus delbrueckii*	Milk whey, molasses, pure, sugars (90)	10–15% glucose, 5–6 days, 50°C in corrosion resistant fermentor, pH 5.5–6.0 buffered with $CaCO_3$; no aeration, growth factors provided by malt; extraction by precipitation after heating to 80°C and the addition of chalk (calcium lactate is formed); extraction with solvents; esterification with methanol followed by distillation
Fumaric acid	*Rhizopus* sp.	Glucose (60)	3 days at 30°C with aeration; pH 5–6 maintained by the addition of NaOH; extraction by acidification of media and crystallization.
Gluconic acid	*Aspergillus niger*	Glucose and corn steep liquor (90)	36 h at 30°C with aeration; pH 6.5; extraction by filtration and purification using cation exchange column

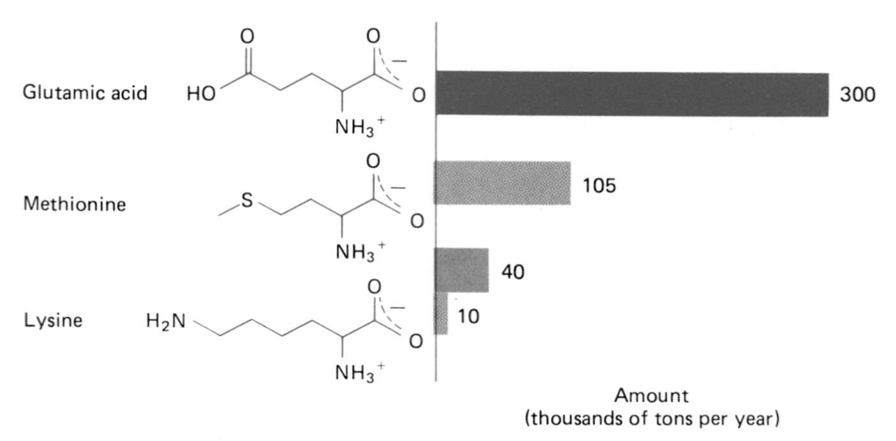

Figure 8.8 Amino acids are commercially important products. Worldwide sales of amino acids were $1.7 billion in 1980. Glutamic acid, methionine, and lysine are the ones now made in the largest quantities. Methionine and lysine are nutritionally essential amino acids used as animal feed additives. Glutamic acid and 80 percent of the lysine is made by fermentation. Methionine is manufactured by chemical synthesis. All amino acids have two isomers, only one of which can participate in biological reactions. Fermentation yields only the biologically active isomer. In chemical synthesis half of the yield is the inactive one. This specificity makes the biological means more efficient, but it has not always been possible to exploit it. With increased understanding of cellular metabolism, all industrially valuable amino acids may soon be made by fermentation.

various bacteria, including *Acetobacter* species, and by several fungi, including *Penicillium* and *Aspergillus* species.

Citric acid is also produced by cultures of *Aspergillus niger*. Commercially produced citric acid is used in various ways, including as a food additive, especially in the production of soft drinks, as a metal chelating and sequestering agent, and as a plasticizer.

Itaconic acid is used as a resin in detergents. The transformation of citric acid by *Aspergillus terreus* can be used for the fermentative production of itaconic acid, although chemical procedures for producing this compound are also available.

Gibberellic acid and related gibberellins are plant hormones and are extensively used as growth-promoting substances to stimulate plant growth, flowering, and seed germination and to induce the formation of seedless fruit. The commercial production of gibberelins can be used to enhance agricultural productivity. Gibberellic acid is formed by the fungus *Gibberella fujikuroi* (= *Fusarium moniliforme*) and can be produced commercially.

Lactic acid has various commercial uses. It is used in foods as a preservative, in leather production for deliming hides, and in the textile industry for fabric treatment. Various forms of lactic acid are also used for other purposes, in resins as polylactic acid, in plastics as various derivatives, in electroplating as copper lactate, and in baking powder and animal feed supplements as calcium lactate. *Lactobacillus delbrueckii* is widely used in the commercial production of lactic acid, but various other *Lactobacillus*, *Streptococcus*, and *Leuconostoc* species also are of industrial importance for the production of this compound.

Production of Amino Acids

Microbial production of the amino acids lysine and glutamic acid presently accounts for over $1 billion in annual worldwide sales (Figure 8.8). Animals require various amino acids in their diets. Lysine and methionine are essential amino acids but are not present in sufficient concentrations in grains to meet animal nutritional needs. Lysine produced by microbial fermentation and methionine produced synthetically are used as animal feed supplements and as additives in cereals. Glutamic acid is principally made for use as monosodium glutamate (MSG), a widely used ingredient in soup produc-

Table 8.7 **Important Uses for Enzymes Produced by Microorganisms**

Industry	Application	Enzyme	Source
Analytical	Sugar determination	Glucose oxidase	Fungi
	Glycogen determination	Galactose oxidase	Fungi
Food	Uric acid determination	Urate oxidase	Fungi
Baking	Bread baking	Amylase	Fungi
		Protease	Fungi
Brewing	Mashing in beer making	Amylase	Bacteria
		Glucamylase	Fungi
Carbonated beverages	Oxygen removal	Glucose oxidase	Fungi
Cereals	Breakfast foods	Amylase	Fungi
Chocolate, cocoa	Syrups	Amylase	Fungi, bacteria
Coffee	Coffee bean fermentation	Pectinase	Fungi
Confectionery	Soft-center candies	Invertase	Bacteria, fungi
Dairy	Cheese production	Rennin	Fungi
Dry cleaning	Spot removal	Protease, amylase	Bacteria, fungi
Eggs	Glucose removal	Glucose oxidase	Fungi
Fruit juices	Clarification	Pectinases	Fungi
	Oxygen removal	Glucose oxidase	Fungi
	Debittering of citrus	Naringinase	Fungi
Laundry	Spot removal	Protease, amylase	Bacteria
	Cold-soluble laundry starch	Amylase	Bacteria
Leather	Bating	Protease	Bacteria, fungi
Meat	Meat tenderizing	Protease	Fungi, bacteria
Mayonnaise and salad dressings	Oxygen removal	Glucose oxidase	Fungi
Paper	Starch modification for paper coating	Amylase	Bacteria
Pharmaceutical and clinical	Digestive aids	Amylase	Fungi, bacteria
		Protease	Fungi, bacteria
		Lipase	Fungi
		Cellulase	Fungi
	Wound debridement (tissue removal)	Streptokinase–streptodornase, protease	Bacteria
Photographic	Recovery of silver from spent film	Protease	Bacteria
Starch and syrup	Corn syrups	Amylase, dextrinase	Fungi
		Glucose isomerase	Fungi, bacteria
	Production of glucose	Glucamylase, amylase	Fungi, bacteria
Textile	Desizing of fabrics	Amylase protease	Bacteria
Wine	Clarification	Pectinases	Fungi

tion. The flavoring industry in the United States consumed more than 30,000 tons of MSG in 1980, some of which was imported from Japan, a major producer of amino acids by fermentation, as well as some from Taiwan and South Korea.

Production of Enzymes

Enzymes have a variety of commercial applications, some of which are shown in Table 8.7. Enzymes for industrial applications can be obtained from plants, animals, and microorganisms. Microbial production of useful industrial enzymes is advantageous because of the large number of different enzymes and the virtually unlimited supply of enzymes that can be produced by microorganisms. Enzymes produced for industrial processes include proteases, amylases, glucose isomerase, glucose oxidase, rennin, pectinases, and lipases. The four extensively produced microbial enzymes are protease, glucamylase, alpha-amylase, and glucose isomerase (Figure 8.9).

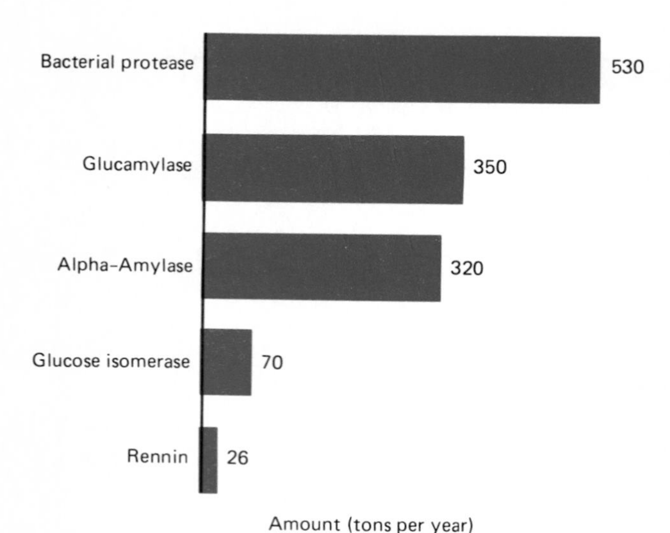

Figure 8.9 Commercial sales of enzymes were $300 million in 1980. The five enzymes illustrated here are produced by microbiological methods. Bacterial protease, which degrades proteins by cleaving peptide bonds, is used as a cleaning aid. Glucamylase, alpha-amylase, and glucose isomerase convert starch into high-fructose corn syrup sweetener, which is replacing sucrose in soft drinks. Amylases break down starch to yield glucose; glucose isomerase converts glucose into fructose. Renin, employed in making cheese, can be extracted from the fourth stomach of a cow or can be made microbiologically.

Proteases

Proteases represent a class of enzymes that attack the peptide bonds of protein molecules, forming small peptides. Proteases produced by different bacterial species are used for different industrial purposes. The largest commercial application of bacterial alkaline proteases is in the laundry industry, principally in modern detergent formulations. In the cleaning industry, proteases are used as spot removers in dry cleaning, as presoak treatments in laundering, and in laundry detergents. The action of the enzyme degrades various proteinaceous materials, such as milk and eggs, forming small peptide fragments that can be washed out readily. In dry cleaning, proteases are effective spot removers and are even effective in removing blood spots. These protease enzymes are relatively heat stable and are able to remain active in warm–hot water long enough for them to degrade the proteinaceous materials contaminating the fabric being washed. When used as a presoak, protease enzymes have sufficient time to act to degrade insoluble proteinaceous materials staining the fabric. Proteases for detergents are largely produced by *Bacillus licheniformis*.

Another major use of microbial proteases is in the baking industry. Protease enzymes are used to alter the properties of the gluten proteins of flour. Fungal protease is added in the manufacture of most commercial bread in the United States to reduce the mixing time and improve the quality of the loaf. Either fungal or bacterial protease is used in the manufacture of crackers, biscuits, and cookies. Fungal proteases are principally obtained from *Aspergillus* species, and bacterial proteases are primarily produced using *Bacillus* species.

Proteases are also used for various other purposes, including as digestive aids. Adding protease enzymes to beef can soften or tenderize the meat, making the meat more edible. A typical meat tenderizer contains 5 percent fungal protease as well as monosodium glutamate and other ingredients. In the leather industry microbial proteases are used for bating of hides, which improves the quality by softening the leather. In the textile industry proteases are used for removing proteinaceous sizing and freeing silk fibers from the proteinaceous material in which they are imbedded.

Amylases

Amylases are used for the preparation of sizing agents and removal of starch sizing from woven cloth, preparation of starch sizing pastes for use in paper coatings, liquefaction of heavy starch pastes that form during heating steps in the manufacture of corn and chocolate syrups, production of bread, and removal of food spots in the dry cleaning industry, where amylase functions in conjunction with protease enzymes. Amylases are also sometimes used to replace or augment malt for starch hydrolysis in the brewing industry, as in the production of low calorie beers.

Other Enzymes

Various other microbial enzymes are produced for industrial applications. Rennin is used in the production of cheese, and *Mucor pussilus* or *M. meihei* can be used for the commercial production of rennin for curdling milk in cheese production. Fungal pectinase enzymes are used in the clarification of fruit juices. Glucose oxidase, produced by fungi, is used for removing glucose from eggs prior to drying, since powdered dried eggs brown because of the chemical reaction of proteins with glucose and removing the glucose stabilizes and prevents deterioration of the dried egg product. Glucose oxidase is also used to remove oxygen from various products, such as soft drinks, mayonnaise, and salad dressings, preventing oxidative color and flavor changes.

Production of Solvents

Several organic solvents can be produced by microbial fermentation, but organic solvents are produced today by chemical synthesis. For example, although ethanol is produced by fermentation for beverages, industrial alcohol is produced by chemical synthesis. Fermentation processes have been important in the past in the industrial production of organic solvents and, as economic conditions change, will likely be used again in the future. The process for producing acetone and butanol was discovered by Chaim Weizmann in England. During World War I the microbial production of acetone was very important in the manufacture of the explosive cordite, and microbially produced butanol was converted to butadiene for use in making synthetic rubber. After the war the demand for acetone declined, but the need for *n*-butanol increased. Butanol is used in brake fluids, urea–formaldehyde resins, and in producing protective coatings, such as lacquers used on automobiles.

The microbial production of acetone and butanol employs anaerobic *Clostridium* species. The fermentation discovered by Weizmann was based on the conversion of starch to acetone by *C. acetobutylicum*. Other species, such as *C. saccharoacetobutylicum*, are able to convert the carbohydrates in molasses to acetone and butanol. These *Clostridium* species synthesize butyric and acetic acids, which are then converted to butanol and acetone. The yields of these neutral solvents are typically low, approximately 2 percent by weight of the fermentation broth, representing a 30 percent conversion of carbohydrates to neutral solvents. In South Africa, because of the scarcity of petroleum and the abundance of plant residues as substrates for fermentation, butanol and acetone are produced by microbial fermentation employing this process. Elsewhere, these solvents are produced from petroleum; however, as the costs of petrochemicals increase, the feasibility of more extensive use of fermentation for producing *n*-butanol is enhanced.

Today, glycerol production is also primarily by chemical synthesis, but in the past microorganisms have been used to produce glycerol. Glycerol is used as (1) a solvent in flavorings and food coloring agents; (2) a lubricant in the manufacture of pet food, candy, cake icings, toothpaste, glue, cellophane, and other products; (3) an emollient and demulcent in pharmaceuticals and cosmetics; and (4) in the production of explosives and propellants. The production of glycerol by fermentation in Germany was an important factor during World War I, because the glycerol was used for the production of munitions. The microbial production of glycerol is accomplished by adding sodium sulfite to a yeast-mediated ethanol fermentation process. The sodium sulfite reacts with the carbon dioxide to produce sodium bisulfite, which prevents the reduction of acetaldehyde ethanol. This blockage results in a divergence of the metabolic pathway with the accumulation of glycerol. Glycerol can be produced by using yeasts, such as *Saccharomyces cerevisiae*, and bacteria, such as *Bacillus subtilis*. The microbial production of glycerol may be renewed as a result of the finding that some yeasts can synthesize glycerol without the need for adding sodium sulfite, thus making the process economically competitive with chemical methods of glycerol production.

Production of Fuels

Limited petroleum resources are forcing many industrialized nations to seek alternative fuel resources. Microbial production of synthetic fuels has the potential for helping meet world energy demands. Useful fuels produced by microorganisms include ethanol, methane, hydrogen, and hydrocarbons. The use of microorganisms to produce commercially valuable fuels depends on finding the right strains of microorganisms to produce the desired fuel efficiently and having an inexpensive supply of substrates available for the fermentation process. It is obviously imperative that the production of synthetic fuels not consume more natural fuel resources than are produced. Microbial production of fuels can be a particularly attractive process when waste materials, such as sewage and municipal garbage, are used as the fermentation substrate.

Ethanol

The microbial production of ethanol has become an important source of a valuable fuel, particularly in regions of the world having abundant resource supplies of plant residues. Brazil produces and uses large amounts of ethanol as an automotive fuel and plans to replace gasoline with ethanol by the 1990s. Gasohol, a 9:1 blend of gasoline and ethanol, has become a popular fuel in the midwestern United

States. At present about 100 million gallons of ethanol per year are used as a fuel in the United States, but 12 billion gallons of ethanol per year would be required to replace completely gasoline use in the United States. There are several major limitations to the successful production of sufficient quantities of ethanol to serve as a major fuel source, including (1) ethanol is relatively toxic to microorganisms, and therefore, only limited concentrations of ethanol can accumulate in a fermentation process; (2) carbohydrate substrates normally used for the production of ethanol in the food industry are relatively expensive, making the cost of fuel produced by fermentation high; and (3) distillation to recover ethanol requires a substantial energy input, reducing the net gain of fuel as an energy resource produced in this process.

Despite the problems with the economics of ethanol production, several processes can be employed for the commercial production of ethanol as a fuel. The finding that the bacterium *Zymomonas mobilis* ferments carbohydrates, forming alcohol twice as fast as yeasts, appears to represent a significant advance in the search for a microbial strain for producing ethanol as a fuel. *Thermoanaerobacter ethanolicus*, a bacterium that grows at high temperatures, may be even more efficient than the organisms currently used for the fermentative production of ethanol. Corn sugar and plant starches are currently used as substrates for the production of ethanol, but the prices of these substrates vary greatly, depending on plant harvests, and they are also needed as food resources. Biomass produced by growing photosynthetic microorganisms is a potential source of an inexpensive substrate for ethanol production, but cellulose from wood and other plant materials is probably the most promising substrate. A two-step fermentation process can be used for the conversion of cellulose to ethanol, in which cellulose is first converted to sugars, generally by *Clostridium* species, and the carbohydrates are then converted to ethanol by yeasts, *Zymomonas* or *Thermoanaerobacter* species. It is very likely that genetic engineering can create a microbial species that can efficiently convert cellulose directly to ethanol, and also tolerate high concentrations of ethanol. Such an organism should permit the commercial production of ethanol as a fuel.

Methane

Methane produced by methanogenic bacteria is another important potential energy source. Methane can be used for the generation of mechanical, electrical, and heat energy. Large amounts of methane can be produced by anaerobic decomposition of waste materials. Many sewage treatment plants are able to meet their own energy needs from the production of methane in their anaerobic sludge digesters. Excess methane produced in such facilities is sufficient to supply power for some municipalities. Efficient generation of methane can be achieved by using algal biomass grown in pond cultures, sewage sludge, municipal refuse, plant residue, and animal waste. Methanogenic bacteria are members of the archaebacteria; they are obligate anaerobes and produce methane from the reduction of acetate and/or carbon dioxide. The production of methane generally requires a mixed microbial community, with some bacterial populations converting the available organic carbon into low molecular weight fatty acids that are substrates for methanogens.

Summary

Many foods and industrial chemicals are produced by microbial fermentations. The use of microorganisms to produce food products is important to meet the food demands of the world's expanding population. Microbial fermentation products, such as cheeses, are part of the normal diet of most individuals. Various beverages, such as beer and champagne, are also frequently consumed. Many of these microbially produced foods are considered delicacies, adding enjoyment to eating.

The lactic acid and ethanolic fermentation pathways are especially important in the production of foods. Cheeses, yogurt, sausages, sauerkraut, soy sauce, pickles, and various other foods are produced using lactic acid bacteria. Wine, beer, spirits, and bread are made using yeasts that carry out an ethanolic fermentation. The accumulation of the fermentation end products helps preserve the food and gives it a distinctive flavor.

Industrial fermentations are also important in the production of pharmaceuticals. Antibiotics, steroids, and vitamins are made by microbial fermentations. These drugs are the cornerstone of modern health maintenance.

Numerous other commercial products are made by microorganisms, including enzymes, organic acids, alcohols, amino acids, solvents, and fuels. Tiny microbes produce millions of dollars worth of products that we use in our daily lives. The products of the microbiological fermentation industry are found in many of our goods.

Study Outline

A. Industrial fermentation.
  1. Many economically valuable products are made by using microorganisms.
  2. Industrial fermentations include all chemical transformations of organic compounds carried out by using microorganisms or their enzymes.
  3. An industrial fermentation combines the right organism(s), substrate(s), and proper environment to bring about the formation of high yields of the desired product(s).

B. Alcoholic beverages.
  1. Microorganisms, principally yeasts in the genus *Saccharomyces*, are used to produce various alcoholic beverages.
  2. Beer.
    a. The initial preparation of substrate for fermentation involves barley malt (mixture of amylases and proteinases).
    b. The malt is added to adjuncts (corn, rice, wheat), which provide carbohydrate substrate for ethanol production.
    c. The supernatant (wort) is then cooked with hops, which concentrates the mixture, inactivates the enzymes, extracts soluble compounds from the hops, and greatly reduces the numbers of microorganisms prior to fermentation.
    d. The fermentation of the wort to produce beer is carried out by yeasts *Saccharomyces carlsbergensis* or *S. cerevisiae*.
    e. Light beers are produced by prehydrolysis of the complex carbohydrates in the malt so that the fermentation removes most of the carbohydrates and results in the production of a low calorie beer.
  3. Distilled liquors are produced in a manner similar to beer, except that after the fermentation process, the alcohol is collected by distillation.
  4. Wines are primarily produced from grapes by an ethanolic fermentation carried out by yeasts.
    a. In the classic European method, yeasts from the grape surface are used.
    b. In modern wine production methods the surface microbiota are removed from the grape by sulfur fumigation, the grapes are crushed to form a juice (must), and the grapes are inoculated with a specific strain of yeast (normally a variety of *Saccharomyces cerevisiae*).
    c. The sugar content of the grapes determines the final ethanol concentration (11–16 percent).

C. Leavening of bread.
  1. Yeasts (principally *Saccharomyces cerevisiae*, baker's yeast) are added to bread dough to ferment the sugar, producing carbon dioxide, which causes the dough to rise.

D. Fermented dairy products.
  1. Various foods, including buttermilk and sour cream are produced by using different strains of lactic acid bacteria as starter cultures and different fractions of whole milk as the starting substrate.
  2. Yogurt is produced by fermenting milk with a mixture of *Lactobacillus bulgaricus* and *Streptococcus thermophilus*.
  3. A wide variety of cheeses are produced by microbial fermentations. Cheese consists of milk curds that have been separated from the whey; curdling is accomplished by using rennin and lactic acid bacterial starter cultures. Unripened cheeses are produced by a single-step fermentation. Ripening of cheeses requires additional microbial growth during maturation of the cheese to achieve desired taste, texture, and aroma; for example, Swiss cheese is formed by a late propionic acid fermentation after the cheese curd is formed by a lactic acid fermentation.

E. Fermented meats.
  1. Several types of sausage are produced by allowing meat to undergo heterolactic acid fermentation during curing.

F. Fermented vegetables.

1. Sauerkraut is produced by a lactic acid fermentation of wilted, shredded cabbage.
2. Pickles can be made from cucumbers by a lactic acid fermentation carried out by the natural microbiota associated with the cucumber or by pure cultures of *Lactobacillus plantarum* and *Pediococcus cerevisiae*; the sourness of pickles reflects the amount of lactic acid accumulated during fermentation.
3. Green olives are produced by a lactic acid fermentation.
4. Soy sauce is produced from a mash consisting of soybeans, wheat, and wheat bran; the most important organisms during fermentation process are *Aspergillus oryzae, Lactobacillus* species, and yeasts.
5. Poi is a food product produced by lactic acid fermentation of the taro plant by a succession of organisms.

G. Vinegar production involves an initial anaerobic fermentation to convert carbohydrates to alcohol by *Saccharomyces cerevisiae*. A secondary oxidative transformation of the alcohol forms acetic acid.

H. Pharmaceuticals.

1. Antibiotics.
   - **a.** Antibiotics are produced by microorganisms and are widely used in medicine.
   - **b.** Specific strains of fungi and bacteria are used to produce a wide variety of antibiotics.
   - **c.** Antibiotics are secondary metabolites that have antimicrobial activities.
2. Steroids.
   - **a.** The chemical synthesis of steroids is very complex and expensive because of precision of substituent location on the basic steroid ring structure; it can be accomplished economically by microorganisms.
3. Vitamins, which are essential animal nutrition factors, can be produced by microbial fermentations.

I. Organic acids.

1. Gluconic acid.
   - **a.** Gluconic acid has a variety of commercial uses as calcium gluconate (dietary calcium supplement), ferrous gluconate (iron supplement for treatment of anemia), and gluconic acid in dishwasher detergents prevents glass spotting.
   - **b.** Gluconic acid can be produced by various bacteria, including *Acetobacter* species, and by several fungi, including *Penicillium* and *Aspergillus* species.
2. Citric acid.
   - **a.** Citric acid can be produced by cultures of *Aspergillus niger*.
   - **b.** Commercially produced citric acid is used in various ways: food additive, metal chelating and sequestering agent, and as a plasticizer.
3. Itaconic acid.
   - **a.** Itaconic acid is used as a resin in detergents.
   - **b.** Itaconic acid is produced by the transformation of citric acid by *Aspergillus terreus*.
4. Gibberellic acid.
   - **a.** Gibberellic acid and related gibberellins are plant hormones used extensively as growth-promoting substances to stimulate plant growth, flowering, and seed germination, and to induce seedless fruit formation.
   - **b.** Gibberellic acid is formed by the fungus *Gibberella fujikuroi*.
5. Lactic acid.
   - **a.** Lactic acid has various commercial uses: food preservation, fabric treatment, hide preparation; various forms of lactic acid also used in resins, plastics, electroplating, baking powder, animal feed supplements.
   - **b.** *Lactobacillus delbrueckii* is widely used in commercial production of lactic acid.

J. Amino acids.
  1. The direct production of lysine from carbohydrates uses mutant strains of *Corynebacterium glutamicum*.
  2. Glutamic acid and monosodium glutamate (MSG) can be produced by direct fermentation using strains of *Brevibacterium, Arthrobacter*, and *Corynebacterium*.

K. Production of enzymes.
  1. The four extensively produced microbial enzymes are protease, glucaylase, alpha-amylase, and glucose isomerase.
  2. Proteases.
    a. Proteases are a class of enzymes that attack peptide bonds of protein molecules, forming small peptides.
    b. The largest commercial application of bacterial alkaline proteases (largely produced by *Bacillus licheniformis*) is in the laundry industry, especially in modern detergents.
    c. The baking industry is the major user of microbial proteases.
    d. Additional uses of proteases include digestive aids, tenderizing beef, softening hides, and fabric treatment.
  3. Amylases.
    a. Amylases are used for preparation of sizing agents and removal of starch sizing from woven cloth, preparation of starch sizing pastes for paper coatings, liquefaction of starch pastes that form in corn and chocolate syrup manufacture, production of bread, and, with proteases, removal of food spots in the cleaning industry.
  4. Other enzymes.
    a. Renin, used in cheese production, is made with *Mucor pussilus* or *M. meihei*.
    b. Fungal pectinases are used to clarify fruit juices.
    c. Glucose oxidase, used for removing glucose and oxygen from food products, is produced by fungi.

L. Solvents.
  1. Several organic solvents can be produced today by chemical synthesis.
  2. Microbial production of acetone and butanol uses anaerobic *Clostridium* species.
  3. Glycerol can be produced by using yeasts such as *Saccharomyces cerevisiae* and bacteria such as *Bacillus subtilis*.

M. Production of fuels.
  1. Ethanol
    a. Microbial production of ethanol is important in regions having abundant resource supplies of plant residues.
    b. Gasohol is a 9:1 blend of gasoline and ethanol.
    c. Several processes can be used for commercial production of alcohol as a fuel.
  2. Methane can be produced by methanogenic bacteria.

Study Questions

1. Identify the microorganisms and metabolic processes associated with each of the following food products.
  a. Buttermilk
  b. Sour cream
  c. Butter
  d. Yogurt
  e. Swiss cheese
  f. White bread
  g. Sour dough bread
  h. Beer
  i. Wine
  j. Scotch
  k. Sauerkraut
  l. Soy sauce

2. Describe the steps involved in the production of beer.

3. Compare the classical and modern methods of wine production.

4. Describe the metabolic processes involved in the production of vinegar.

5. List examples of five different types of pharmaceutical products produced by microorganisms.

6. List four organic acids produced by microorganisms and briefly describe their uses.

7. List two amino acids produced by microorganisms and briefly describe their uses.

8. List three enzymes produced by microorganisms and briefly describe their uses.

Some Thought Questions

1. How would life be different if there were no microbes?
2. Would you eat bacteria for dinner?
3. Why is spoilage of wine more of a problem than spoilage of beer?
4. Do you have to kill all microorganisms before you eat them?
5. If you were organizing a picnic for the microbiology class, what foods or drink could you make in the laboratory?
6. If all cheeses start from the same materials, why do they taste so different?
7. If you wanted to make your own yogurt, would it be necessary to buy a starter culture or could you use a spoonful of commercial yogurt?
8. If the yeast *Saccharomyces cerevisiae* can metabolize 3000 times its own weight of glucose per hour, would it be a suitable candidate for the industrial production of automobile fuel?
9. Why isn't pure glucose used as a substrate for industrial processes?
10. Why do draft and canned beers taste differently?
11. How do light and low alcohol (LA) beers differ from regular beers?
12. If yeast cells were not seen until the eighteenth century, how could beer and wine have been produced in ancient times?
13. Why is gasohol particularly suitable as a fuel in tropical countries?
14. Why are microbes better than chemists at making steroid drugs?

Additional Sources of Information

Ayres, J. C., J. O. Mundt, and W. E. Sandine. 1980. *Microbiology of Foods*. W. H. Freeman and Company, Publishers, San Francisco. A text covering the field of food microbiology, including the microbial production of foods.

Crueger, W. and A. Crueger. 1984. *Biotechnology: A Textbook of Industrial Microbiology*. Science Tech Inc., Madison, Wisconsin. An up-to-date text describing paractical uses of microbial metabolism by industry.

Frazier, W. C., and D. C. Westhoff. 1978. *Food Microbiology*. McGraw-Hill Book Company, New York. A popular text on all aspects of food microbiology.

Industrial Microbiology. 1981. *Scientific American*, **245**(3). A collection of articles dealing with industrial microbiology. Individual articles describe the uses of microorganisms for the production of foods, pharmaceuticals, and industrial chemicals.

Jay, J. M. 1970. *Modern Food Microbiology*. Van Nostrand Reinhold Company, New York. A widely-read text on the microbiology of foods.

Pederson, C. S. 1979. *Microbiology of Food Fermentations*. AVI Publishing Co., Westport, Connecticut. Part of a series of books on food microbiology dealing with the microbial production of foods.

9 Ecological Aspects of Microbial Metabolism

KEY TERMS

ammonification
bacteroids
biomagnification
biogeochemical cycling
carbon cycle
denitrification
energy transfer
eutrophication
food web
hydrogen cycle
grazers
leaching
manganese nodule
methanogenic archaebacteria
nitrification
nitrogen cycle
nitrogen fixation
nitrogenase
nodules
oxygen cycle
predators
primary producers
recalcitrant
sulfur cycle
trophic level

OBJECTIVES

After reading this chapter you should be able to

1. Define the key terms.
2. Describe the major ecological transformations of elements that are mediated by microorganisms and their importance.
3. Discuss the importance of microorganisms in food webs.
4. Discuss the major processes involved in the nitrogen cycle, indicating the organisms responsible for each process.
5. Discuss the importance of nitrogen fixation, nitrification, and denitrification for agricultural productivity.
6. Describe which portions of the sulfur cycle occur under anaerobic conditions and which occur when oxygen is available.
7. Discuss the role of the sulfur cycle in the formation of acid mine drainage.
8. Discuss the role of microbial biogeochemical cycling in coral reef formation.
9. Discuss the importance of microbial transformation of heavy metals.

Biogeochemical Cycling

All living organisms carry out chemical transformations that influence their surrounding environment. The metabolic activities of microorganisms, in particular, have an ecological impact that influences the quality of human life. Without the activities of microorganisms, higher forms of life, including humans, could not exist.

A thing . . . never returns to nothing, but all things after disruption go back into the first bodies of matter . . . None of the things therefore which seem to be lost is utterly lost, since nature replenishes one thing out of another and does not suffer any thing to be begotten, before she has been recruited by the death of some other.

—LUCRETIUS

Just as transformations of compounds occur via central metabolic pathways within cells, there are central pathways through which materials flow within ecosystems. Changes in the chemical forms of various elements can lead to the physical movement of materials, sometimes mediating transfers between the atmosphere (air), hydrosphere (water), and lithosphere (land). The exchange of elements between the biotic and abiotic portions of the biosphere occurs through characteristic cyclic pathways known as biogeochemical cycles

Because of their ubiquitous distribution and diverse enzymatic activities, microorganisms play a major role in biogeochemical cycling. The major elements of living biomass, carbon, hydrogen, oxygen, nitrogen, phosphorus, and sulfur are the most intensively cycled by microorganisms. Other elements are generally cycled to a lesser extent. The biogeochemical cycling activities of microorganisms are essential for the survival of plant and animal populations and determine in large part the potential productivity level of a given habitat. Human activities, such as those that result in the release of pollutants, can have a major influence on the rates of microbial cycling activities, leading to significant changes in the biochemical characteristics of a given habitat.

Little causes, great effects. Nature's balance depends on a straw.

—ROGER HEIM, *Un Naturaliste Autour du Monde*

Our land, compared with what it was, is like the skeleton of a body wasted by disease. The plump soft parts have vanished, and all that remains is the bare carcass.

—PLATO, *Critias* III

The relationship between modern man and the planet . . . has been that not of symbiotic partners, but of the tapeworm and the infested dog, of fungus and the blighted potato.

—ALDOUS HUXLEY, *Ape and Essence*

The world is being carried to the brink of ecological disaster not by a singular fault, which some clever scheme can correct, but by the phalanx of powerful economic, political, and social forces that constitute the march of history. Anyone who proposes to cure the environmental crisis undertakes thereby to change the course of history.

—BARRY COMMONER

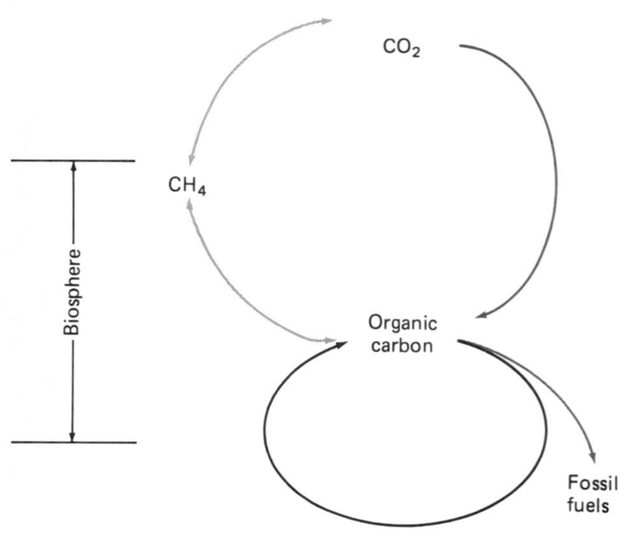

Figure 9.1 A simplified diagram of the carbon cycle.

The Carbon Cycle

Carbon is actively cycled between inorganic carbon dioxide and the variety of organic compounds that compose living organisms (Figure 9.1). The carbon cycle begins with autotrophic metabolism of photosynthetic and chemolithotrophic organisms, which is responsible for primary production, the conversion of inorganic carbon dioxide to organic carbon. Once carbon is fixed, that is, reduced into organic compounds, it can be transferred from population to population within the biological community. The organic compounds made by autotrophs support the growth of a wide variety of heterotrophic organisms. The respiratory and fermentative metabolism of heterotrophic organisms returns inorganic carbon dioxide to the atmosphere, completing the carbon cycle. The production of methane by a specialized group of methanogenic archaebacteria represents a shunt to the normal cycling of carbon because the methane that is produced cannot be used by most heterotrophic organisms and thus is lost from the biological community to the atmosphere.

The carbon and energy in organic compounds, formed by primary producers, move through the biological community of an ecosystem. The transfers of energy stored in organic compounds (energy transfer) between the organisms in the community forms a food web (Figure 9.2). The feeding relationships between organisms establish the trophic structure of an ecosystem. Energy, stored in organic compounds, is transferred through an ecosystem occurring in steps from one trophic level to another. Grazers feed upon primary producers. Grazers in turn are preyed upon by predators, establishing a pyramid of biological populations in the food web. Only a small portion of the energy stored in any trophic level is transferred to the next higher trophic level in foodwebs. Normally, 85–90

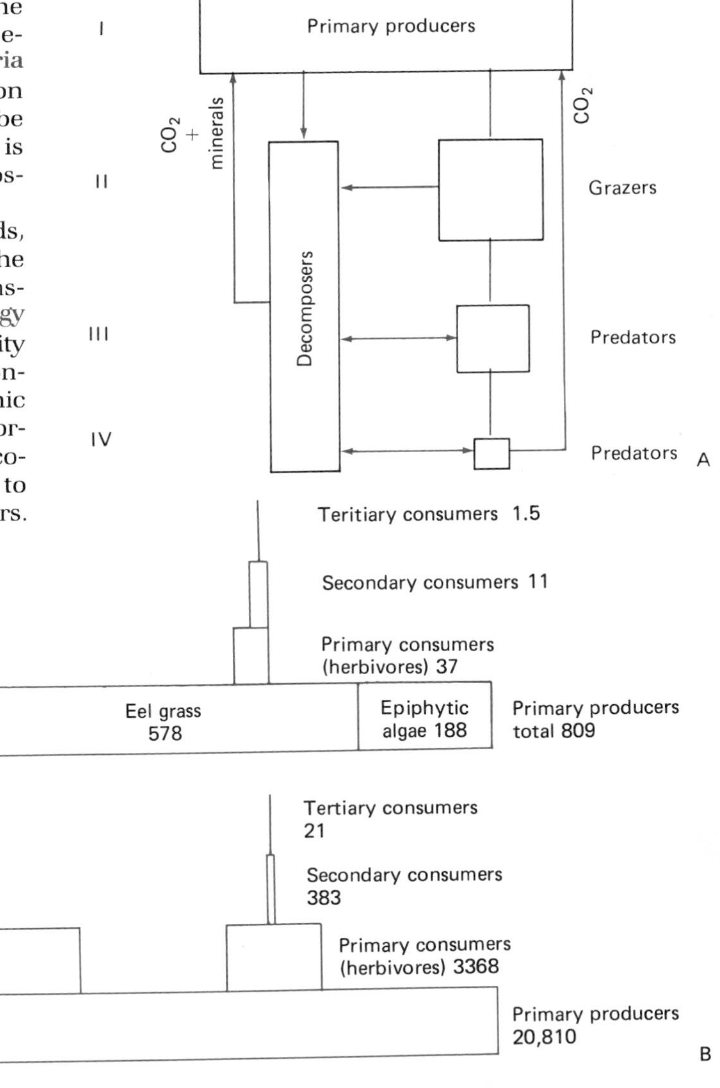

Figure 9.2 (A) This idealized food web shows the transfers between trophic levels. Organic carbon formed by primary producers is transferred to grazers and predators. Carbon dioxide is returned to primary producers by decomposers and by respiration of grazers and predators. The population sizes that can be supported decline at higher trophic levels. (B) Food webs are structured in the shape of a pyramid.

Death of a Superbug: Nature Is Slicker Oil Eater

By STEPHEN STRAUSS

The "superbug" is dead. The oil-gobbling microbe, which made bioengineering history in the United States in 1980 when it became the first living organism to be patented, has quietly been allowed to die.

The problem was that the superbug was not very super after all.

Experts, including the superbug's creator, University of Illinois geneticist Ananda Chakrabarty, now say it and other similarly engineered microbes hold little practical application for cleaning up the giant oil spills on which they were designed to feed. Nature has proved to be more efficient, more economical and, most important, less potentially lethal than the man-made microbes.

"The general feeling right at the moment is that it is unlikely that one would knowingly release an engineered organism into the open environment. If you are wrong (about health hazards), it turns out to be a horrendous problem. To be effective, you are not talking about using grams or micrograms, you are talking about tonnes of the stuff," said Herman Finkbeiner, manager of biological science at the General Electric Research and Development Centre in Schenectady, N.Y., which owns the superbug patent.

The failure of the superbug's promise is so profound that neither the General Electric Co. nor Dr. Chakrabarty was aware it no longer existed. Dr. Finkbeiner said that his company no longer had any of the microbes, but assumed Dr. Chakrabarty had some. Dr. Chakrabarty said he didn't have any, but assumed G.E. did.

The superbug concept has died with the growing appreciation of how talented nature is at breaking down oil by itself. Some 40 to 50 per cent of the tens of thousands of species of natural bacteria have an ability to break down one element or another in oil.

Large oil spills and well blowouts aroused the public's interest in the superbugs. The Amoco Cadiz tanker accident dumped 7 million barrels of oil off the coast of Brittany in 1978 and the Mexican Ixtoc I oil well spewed 14 million barrels into the Gulf of Mexico in 1979. However, these spills did not appear to have had a clear, long-term negative effect on the environment because of the natural degradation process.

"The artificial emulsifiers people poured on the oil to disperse it after the Torrey Canyon disaster killed more sea life than the oil itself," said Jacques Berger, a microbial-zoologist at the University of Toronto.

In addition, although the superbug—created in 1972 by genetically combining the oil-eating talents of five separate pseudomonas microbes—was efficient in the laboratory, it flunked its "real world" tests: it is of little use in cold water or in sea water.

Dr. Finkbeiner said scientists now know that the naturally appearing species of the pseudomonas bacteria from which the superbug's genetic properties were taken work together. Jointly, they are as efficient as the bioengineered microbe.

Then there are many concerns about biological hazards. The most efficient oil-eating microbes are highly mutable and closely related to microbes which cause respiratory infections, tuberculosis and other diseases in man.

Other scientists worry that the bioengineered microbes will massively displace the naturally appearing microbes and cause dramatic and unforeseen catastrophes in the food chain.

Some examples of the dangers which loom when the most basic level of the environment is tampered with are seen in recent experiments carried out by Prof. Berger and his associate, Andrew Rogerson. They examined how a one-celled protozoa would react after eating a microbe which had dined extensively on oil. They found that the protozoa readily ate the microbe—apparently without any ill effects. However, the protozoa grew 20 per cent larger than usual.

The added weight turned out to be bits of hydrocarbon deposited in fatty sacs all over the protozoa's body. "In a crude analogy, it is like the protozoa's body puts the hydrocarbons in a garbage bag and just leaves them there," Prof. Berger said.

However, when they looked at the microscopic worm-shaped rotifers which fed on the swollen protozoa, they found sick creatures. The rotifers vomited, ate less, moved erratically, and reproduced much less than normal.

In turn, the rotifers are eaten by fish larvae which in turn are eaten by fish.

It seems the more microbiologists learn about the complicated checks and balances which operate at the microscopic level, the less clear they are that even a super-super microbe could prove effective.

If the original superbug idea has literally died, the concept of a super microbe specially designed for breaking down dangerous substances in the environment hasn't. The attention of Prof. Chakrabarty and others has shifted to bioengineering microbes which break down highly toxic and long-lasting dioxin, polychlorinated biphenyls (PCBs), and insecticides into less dangerous components. These substances are attacked by few natural organisms.

John Loper, a molecular geneticist at the University of Cincinnati and scientists from the University of Nebraska are hoping to place a gene or genes from a mouse or chicken liver into a yeast cell.

The liver genes produce a natural enzyme which breaks down dioxin and other substances in the body.

Prof. Chakrabarty and his researchers are hoping to evolve new genes among microbes which would completely break down dangerous substances to harmless chemicals.

Source: The Globe and Mail, Toronto, May 13, 1983. Reprinted by permission.

percent of the energy stored in the organic matter of a trophic level is consumed by respiration during transfer to the next trophic level. Consequently, the higher the trophic level the smaller its biomass.

Microbial decomposition of dead plants and animals and partially digested organic matter is largely responsible for the conversion of organic matter to carbon dioxide and the reinjection of inorganic carbon dioxide into the atmosphere. Some natural organic compounds, such as lignin, cellulose, and humic acids are relatively resistant to attack and decay only slowly. Various synthetic compounds, such as DDT (dichlorodiphenyltrichloroethane), may be completely resistant to enzymatic degradation (recalcitrant). We depend on the activities of microorganisms to decompose organic wastes, and when microbial decomposition is ineffective, organic compounds accumulate. This is evidenced by the environmental accumulation of plastic materials that are recalcitrant to microbial attack. Many modern problems relating to the accumulation of environmental pollutants reflect the inability of microorganisms to degrade rapidly enough the concentrated wastes of industrialized societies.

Der gut Herr Gott
said, "Let there be rot,"
and hence bacteria and fungi sprang
into existence to dissolve the knot
of carbohydrates photosynthesis
achieves in plants, in living plants.
Forget the parasitic smuts,
the rusts, the scabs, the blights, the wilts, the spots,
the mildews and aspergillosis–
the fungi gone amok,
attacking living tissue,
another instance, did Nature need another,
of predatory heartlessness.
Pure rot
is not
but benign; without it, how
would the forest digest its fallen timber,
the woodchuck corpse
vanish to leave behind a poem?
Dead matter else would hold the elements in thrall—
nitrogen, phosphorus, gallium
forever locked into the slot
where once they chemically triggered
the lion's eye, the lily's relaxing leaf.

All sparks dispersed
to that bad memory where the dream of life
fails of recall, let rot
proclaim its revolution:
the microscopic hyphae sink
the fangs of enzyme into the rosy peach
and turn its blush a yielding brown,
a mud of melting glucose:
once-staunch committees of chemicals now vote
to join the invading union,
the former monarch and constitution routed
by the riot of rhizoids,
the thalloid consensus.

The world, reshuffled, rolls to renewed fullness;
the oranges forgot
in the refrigerator "produce" drawer
turn green and oblate
and altogether other than edible,
yet loom as planets of bliss to the ants at the dump.
The banana peel tossed from the Volvo
blackens and rises as roadside chicory.
Bodies loathsome with their maggotry of ghosts resolve
to earth and air,
their fire spent, and water present
as a minister must be, to pronounce the words.
All process is reprocessing;
give thanks for gradual ceaseless rot
gnawing gross Creation fine while we sleep,
the lightning-forged organic conspiracy's
merciful counterplot.

—JOHN UPDIKE, *Ode to Rot*

Hydrogen and Oxygen Cycles

Hydrogen and oxygen are also cycled by microbial activities (hydrogen cycle and oxygen cycle). The cycling of hydrogen and oxygen through the biosphere is closely tied to the carbon cycle (Figure 9.3). Hydrogen is an essential component of all organic compounds. Oxygen occurs in all organic compounds except hydrocarbons. When water acts as an electron donor during photosynthesis, molecular oxygen is liberated and hydrogen and electrons are transferred to form reduced coenzymes, which are used in biosynthesis. The oxygen liberated by photolysis, the splitting of water that occurs during photosynthesis, is the major source of molecular oxygen upon which aerobic respiration depends. During respiration both oxygen and hydrogen are recycled with the formation of water and carbon dioxide.

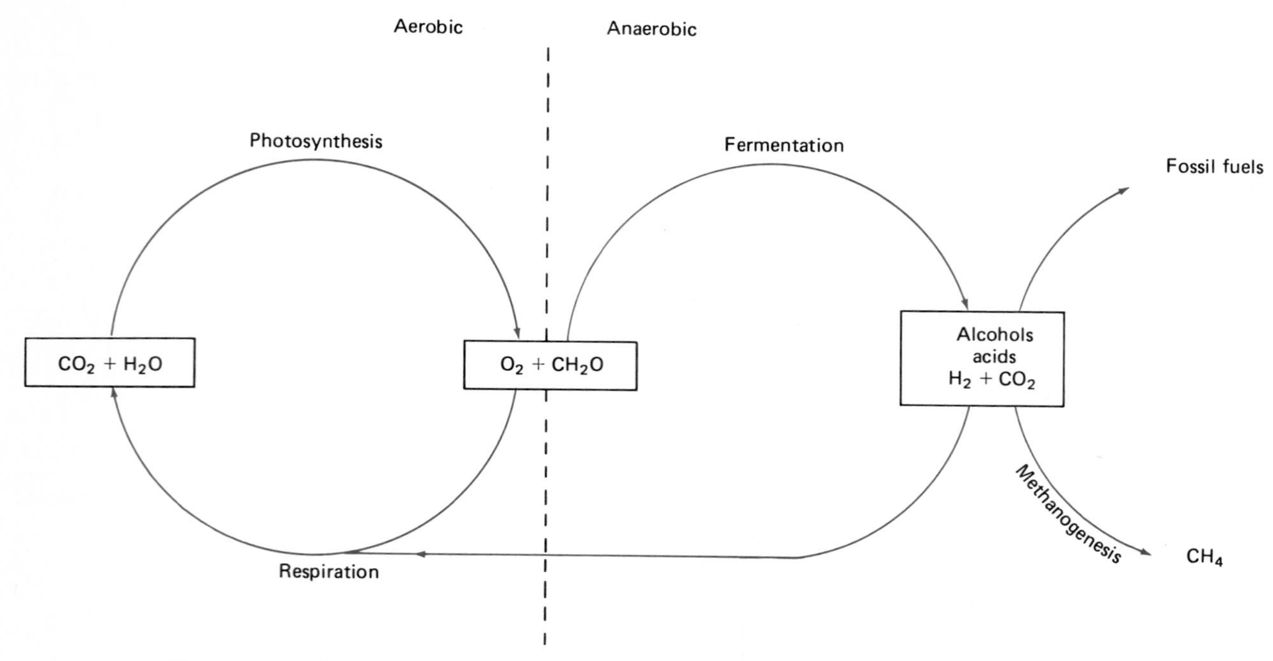

Figure 9.3 This diagram of integrated carbon, hydrogen, and oxygen cycles shows the involvement of oxygen and hydrogen in the aerobic and anaerobic oxidation of organic carbon and in the reduction of carbon dioxide.

The Nitrogen Cycle

Nitrogen Fixation

The biogeochemical cycling of nitrogen (nitrogen cycle) is largely dependent on the metabolic activities of microorganisms (Figure 9.4). Aside from the chemical fixation of molecular nitrogen by human beings to form nitrogen fertilizers, the ability to fix atmospheric nitrogen (the conversion of molecular nitrogen, N_2, to ammonia or organic nitrogen) is restricted to a limited number of bacterial species. Ammonia is the first detectable product of nitrogen fixation. It is assimilated into amino acids and subsequently synthesized into proteins and nucleic acids. Proteins, amino acids, and inorganic ammonium ions are used as a source of nitrogen by many organisms that are unable to assimilate atmospheric nitrogen directly. Animals, plants, and most microorganisms are unable to use atmospheric nitrogen directly and depend on the availability of fixed forms of nitrogen for incorporation into their cellular biomass. Other than one exceptional case, where a green alga has been shown to fix atmospheric nitrogen, this process is carried out strictly by bacteria. The productivity of many ecosystems is limited by the supply of fixed forms of nitrogen.

Figure 9.4 The nitrogen cycle: the chemical forms and key processes in the biogeochemical cycling of nitrogen. The critical steps of nitrogen fixation, nitrification, and denitrification are all mediated by bacteria.

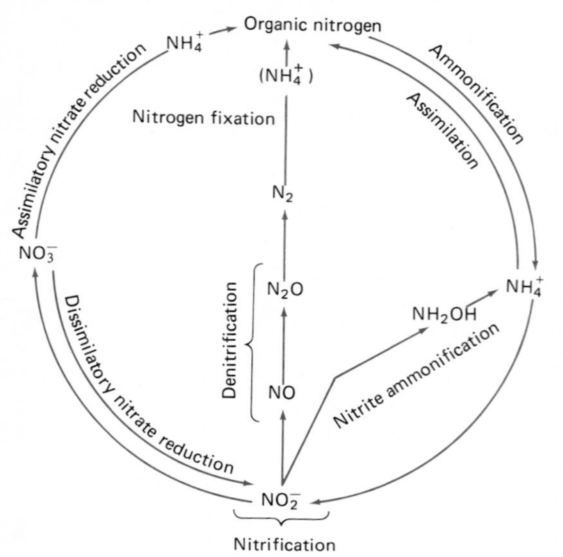

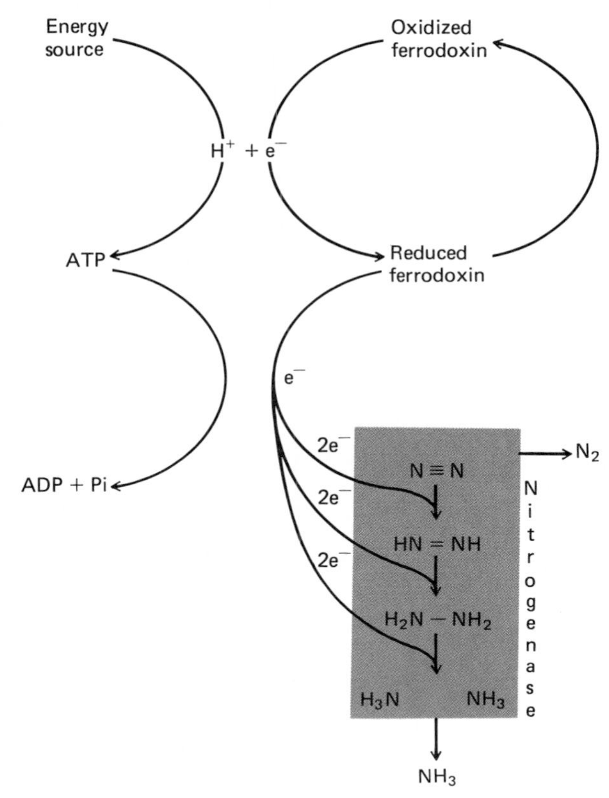

Figure 9.5 A generalized scheme for the action of the nitrogenase enzyme.

It is estimated that microorganisms convert approximately 200 million metric tons of nitrogen to fixed forms of nitrogen per year compared to about 30 million metric tons produced by industrial production of nitrogen fertilizers. The fixation of atmospheric nitrogen depends on the nitrogenase enzyme system (Figure 9.5). Nitrogenase is very sensitive to oxygen, and nitrogen fixation is restricted to regions with appropriately low levels of free oxygen. ATP is required to drive the reactions catalyzed by the nitrogenase enzyme system. Nitrogen fixation is largely dependent on the availability of relatively high concentrations of organic matter for use in generating ATP. The fixation of atmospheric nitrogen requires a high energy input (approximately 30 ATP/N_2 fixed).

In soil, the microbial fixation of atmospheric nitrogen is carried out by free-living bacteria and by bacteria living in symbiotic association with plants. Symbiotic nitrogen fixation by *Rhizobium* is most important in agricultural fields, where this bacterium lives in association with various crop plants. The symbiotic relationship between *Rhizobium* and leguminous plants is of extreme importance in the maintenance of soil fertility. *Rhizobium* species are able to invade the roots of suitable host plants, leading to the formation of nodules. Within root nodules the *Rhizobium* bacteria are able to fix atmospheric nitrogen (Figure 9.6). The establishment of a symbiotic association between a *Rhizobium* species and a plant is very specific. There is a mutual recognition between the bacteria and the binding sites on the surfaces of the plant roots. Within the infected plant tissue the *Rhizobium* cells multiply, forming unusually shaped pleomorphic cells called bacteroids (Figure 9.7).

The bacteroid cells contain active nitrogenase enzymes, not found in free-living *Rhizobium* cells, that allow them to fix molecular nitrogen and provide their symbiotic plant partner with an available source of fixed nitrogen for growth. The biochemicals of the nodule supply oxygen to the bacteroids

Figure 9.6 Nodules occur as tumorous growths on the roots of plants infected with nitrogen-fixing bacteria. It is within these tumor-like growths that *Rhizobium* fixes atmospheric nitrogen. Only a few types of plants can enter into a symbiotic relationship with nitrogen-fixing bacteria. Shown here is the extensive nodulation of the root system of a soybean plant. (Courtesy Nitragin Co., Milwaukee, Wisconsin.)

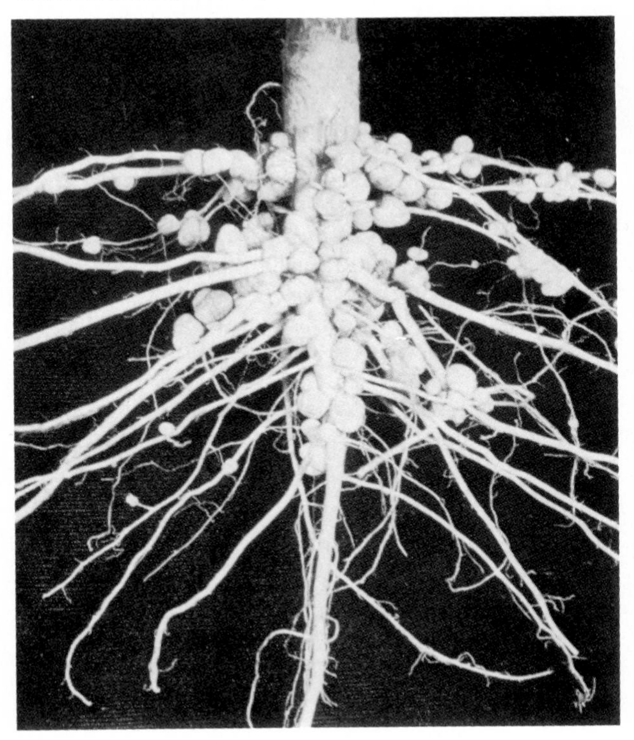

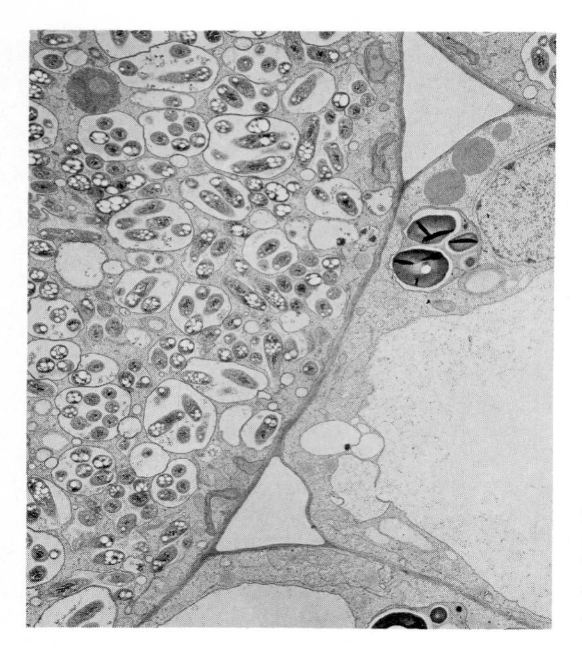

Figure 9.7 Electron micrograph of *Rhizobium japonicum* bacteroids in an infected cell (right) of a root nodule of soybean. The left cell is uninfected and shows the appearance of the cell without bacteroids (13,000×). (From BPS: E. H. Newcomb, University of Wisconsin.)

for their respiratory metabolism but maintain a sufficiently low concentration of free oxygen so as not to inactivate the nitrogenase enzymes. In particular, leghemoglobin produced by the plant supplies oxygen to the bacteroids for production of ATP. The control of oxygen is critical because oxygen is both required and inhibitory for the nitrogen fixation process. When growing in association with plants, symbiotic nitrogen-fixing *Rhizobium* species generally exhibit rates of nitrogen fixation that are two to three orders of magnitude higher than are accomplished by free-living nitrogen-fixing soil bacteria. *Rhizobium* species associated with alfalfa, for example, can account for an input of 250 kilograms of nitrogen fixed per hectare per year, as compared to 2.5 kilograms of nitrogen fixed per hectare per year for free-living nitrogen-fixing *Azotobacter* species.

Through the use of the acetylene reduction assay, a method for detecting nitrogen fixation based on the fact that the nitrogenase enzyme system also catalyzes the reduction of acetylene to ethylene, many free-living bacteria have now been shown to be capable of fixing atmospheric nitrogen. Most of these free-living nitrogen-fixing bacteria exhibit nitrogen-fixing activities only at oxygen levels below 0.2 atmospheres. Such conditions frequently occur in subsoil and sediment environments. Although the amount of nitrogen fixed per hectare by free-living soil bacteria is considerably lower than the amount fixed by symbiotic nitrogen-fixing species, the widespread distribution of the free-living bacteria in soil means that they make a significant contribution to the input of nitrogen to terrestrial habitats.

In aquatic habitats cyanobacteria, such as *Anabaena* and *Nostoc*, are very important in determining the rates of conversion of atmospheric nitrogen to fixed forms of nitrogen. Cyanobacteria, capable of nitrogen fixation, are distributed in both marine and freshwater habitats. These cyanobacteria are able to couple the ability to generate ATP (through the conversion of light energy) and organic matter (through the reduction of carbon dioxide) with the ability to fix atmospheric nitrogen to efficiently form nitrogen-containing organic compounds. In low nutrient aquatic environments the use of light energy for generating ATP is critical to supplying sufficient ATP to drive the nitrogen-fixation reactions. Rates of nitrogen fixation by cyanobacteria are typically ten times higher than shown by free-living bacteria. As such, cyanobacteria form a very important component of aquatic food webs.

Ammonification

Microorganisms also perform important transformations of organic and inorganic fixed forms of nitrogen that alter the ecological availability of nitrogen. Nitrogen in organic matter is present predominantly in the reduced amino form, such as occurs in amino acids. Many microorganisms as well as plants and animals are capable of converting organic amino nitrogen to ammonia (ammonification). Deaminase enzymes play an important role in this process of ammonification, which transfers nitrogen from organic to inorganic forms of nitrogen. The microbial decomposition of urea, for example, results in the release of ammonia. Ammonium ions can be assimilated by a number of organisms, continuing the transfer of nitrogen within the nitrogen cycle.

Nitrification

Although many organisms are capable of ammonification, relatively few are capable of nitrification. Nitrification is a two–step process in which ammonium ions are initially oxidized to nitrite ions and subsequently to nitrate ions. Both steps of nitrification are energy-yielding processes from which chemolithotrophic bacteria are able to derive needed energy. Relatively low amounts of ATP, however, are generated by the oxidation of inorganic nitrogen compounds. Therefore, large amounts of inorganic nitrogen compounds must be transformed in order to generate sufficient ATP to support the growth of these chemolithotrophic bacteria. The oxidation of approximately 35 moles of ammonia is required to support the fixation of 1 mole of carbon dioxide, and approximately 100 moles of nitrite must be oxidized to support the fixation of 1 mole of carbon dioxide. As a consequence of the high amounts of nitrogen that must be transformed to support the growth of chemolithotrophic bacterial populations, the magnitude of the nitrification process is typically very high, whereas the growth rates of nitrifiers are generally relatively low compared to other bacteria.

Table 9.1 Genera of Nitrifying Bacteria

Genus	Converts	Habitat
Nitrosomonas	Ammonia to nitrite	Soils, freshwater, marine
Nitrosospira	Ammonia to nitrite	Soils
Nitrosococcus	Ammonia to nitrite	Soils, freshwater, marine
Nitrosolobus	Ammonia to nitrite	Soils
Nitrobacter	Nitrite to nitrate	Soils, freshwater, marine
Nitrospina	Nitrite to nitrate	Marine
Nitrococcus	Nitrite to nitrate	Marine

The two steps of nitrification, the formation of nitrite from ammonium and the formation of nitrate from nitrite, are carried out by different microbial populations (Table 9.1). For the most part, the oxidative transformations of inorganic nitrogen compounds in the nitrification process are restricted to a few species of autotrophic bacteria. In soils *Nitrosomonas* is the dominant bacterial genus involved in the oxidation of ammonia to nitrite, and *Nitrobacter* is the dominant genus involved in the oxidation of nitrite to nitrate (Figure 9.8).

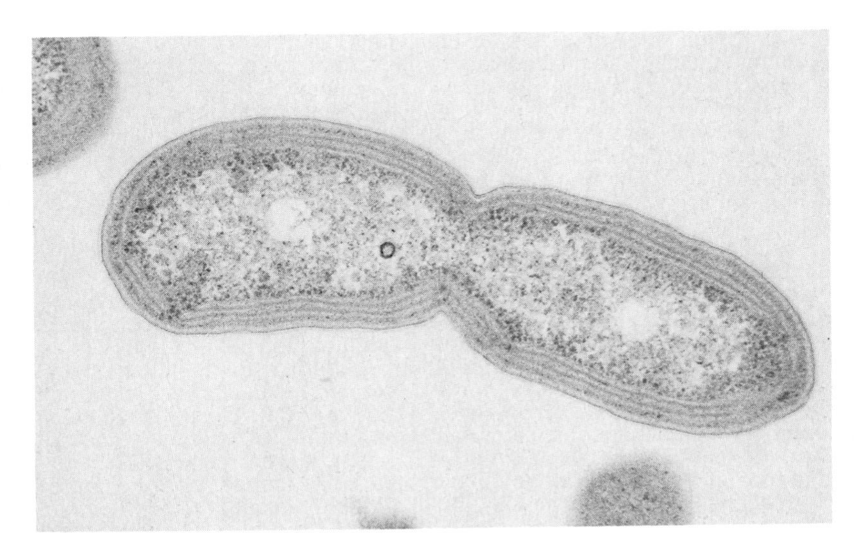

Figure 9.8 Micrograph of the internal membranes of the nitrifying bacteria *Nitrosomonas europaea* (77,750 ×). (Courtesy Stan W. Watson, Woods Hole Oceanographic Institution, Woods Hole, Massachusetts. Reprinted by permission of the American Society for Microbiology from S. W. Watson and M. Mandel, 1971, *Journal of Bacteriology,* 107:563–569.)

Plans to Release New Organisms Into Nature Spur Concern

By PHILIP M. BOFFEY

WASHINGTON—The rapidly approaching day when American scientists and industry will begin releasing new genetically engineered organisms into the environment is sending shivers of apprehension through many prominent ecologists.

They are coming to the conclusion that a revolution in genetic engineering is under way whose potential ecological consequences have not yet been thoroughly analyzed.

Their rising chorus of concern has prompted governmental and scientific organizations to take a new and harder look at how to assess whether microorganisms, plants and animals whose genes are altered for scientific or commercial purposes might somehow get out of control and cause unexpected damage to other life forms or to atmospheric and ecological processes.

Both Congress and the courts have already begun to respond to these concerns. In February the staff of a House science subcommittee issued a report warning, "While there is only a small possibility that damage could occur, the damage that could occur is great." And last month, Federal District Judge John J. Sirica delayed the first proposed outdoor experiment using genetically engineered organisms and indicated that he would order the Federal Government to conduct a full-scale assessment of environmental effects that are likely to result from releasing such organisms out of doors.

At the same time, scientific agencies and organizations have begun their own assessments of the problem. On Friday, the National Academy of Sciences decided to seek financing for a study of the environmental and public health consequences of releasing genetically engineered organisms into the open; the study will attempt to set forth the underlying biological principles governing how organisms evolve, compete, survive and move from one place to another so that Federal regulators will have some basis on which to judge the risks in particular experiments or commercial activities. The Environmental Protection Agency is preparing its own draft document on the environmental implications of genetic engineering; it has commissioned academic studies of the key issues and how to deal with them. The White House Office of Science and Technology Policy is coordinating a Cabinet-level review of the best regulatory mechanisms for dealing with genetic engineering.

Most or all of these reviews might well end up concluding that the dangers are minimal and that scare-talk should not be allowed to interfere with attaining the benefits promised by genetic engineers, ranging from more nutritional crops to novel ways of cleaning up pollutants or recovering scarce oil. But the new assessments respond to rising complaints that, in the pell-mell rush to achieve the benefits of genetic engineering, the potential side effects have been given far too little rigorous scientific scrutiny.

"There are a number of ecological changes that could be very serious," warns Martin Alexander, a professor of soil science at Cornell University who is a former chairman of a genetic engineering study group established by the science advisory board to the E.P.A.

"In general, I think the incidence of problems will be very low," Dr. Alexander said in a telephone interview. "But I don't think it will be zero." If a rare mistake does occur, he warned in Congressional testimony, "the consequences of this low-probability event may be enormous."

The most visible prophet of ecological harm from genetic engineering has been a layman, Jeremy Rifkin, an author and self-described activist, who has filed suit in Judge Sirica's court to block Federal approval of the first outdoor experiments with genetically engineered organisms. Although many prominent scientists have challenged Mr. Rifkin's statements as unduly alarmist, several respected ecologists have also begun to raise warning flags, although in more moderate and tentative terms.

"The real danger is that Jeremy Rifkin's association with the cause can lead people to say there is absolutely nothing there, so let's blow that view away and get on with the job," says Peter H. Raven, director of the Missouri Botanical Garden. "But the point is that there is something there. It's not a fairy tale."

As part of his legal filings, Mr. Rifkin gathered affidavits from nine scientists who expressed concern about potential dangers and inadequate screening of genetic engineering work for possible ecological consequences. The list included Dr. Raven; Eugene P. Odum, director of the Institute of Ecology at the University of Georgia; Ralph Baker, professor of botany and plant pathology at Colorado State University; Liebe F. Cavalieri, of the Memorial Sloan-Kettering Institute; Stephen R. Gliessman, director of the agroecology program at the University of California at Santa Cruz, and Michael W. Fox, scientific director of The Humane Society of the United States.

The Government collected more than 30 affadavits from prominent scientists contending that there was little danger in the initial experiments planned and that all had been subjected to thorough review. Most of those affidavits were from molecular biologists, geneticists or other scientists who wish to use the techniques of genetic engineering. Dr. Raven signed affidavits for both sides, concluding that the initial experiment planned seemed quite benign, although his general fears about future activities remained unanswered.

Few ecologists are willing to describe worst-case scenarios they can envisage, partly because they do not want to be branded crackpot alarmists, and partly because they themselves have not thought long enough about the possibilities.

But to give some flavor of the kind of mishap they are worried about, some ecologists noted in interviews that proposals have been floating around for years to produce a genetically engineered enzyme that could destroy lignin, an organic substance that gives wood its rigidity. Such an enzyme might be useful in cleaning up the effluent from paper mills or in decomposing biological material for energy or agricultural purposes.

But what if the enzyme somehow

got out of control and invaded living trees in the forests, destroying the substance that gives trees their rigidity? "You might harvest the trees with scissors rather than with saws," jokes Arthur Stern, a senior E.P.A. scientist.

Various ecologists also raise other "what if" possibilities. For example, what if an organism designed to clean up oil spills got out of control and started eating its way through the world's gas tanks; or what if an organism designed to consume toxic wastes suddenly mutated and began eating more valuable substances, or what if scientists trying to enhance the nitrogen-fixing capabilities of bacteria in soil inadvertently introduced genes that depleted soil of nitrogen needed by plants for growth?

Genetic engineers complain such speculation is unsupported by any credible evidence and is therefore impossible to answer. Some also note that organisms have been mutating spontaneously in nature since the beginning of life without gobbling up all the world's oil or chemicals or trees. And virtually all genetic engineers contend their new techniques are more precise and controllable than cruder laboratory techniques that have been causing genetic alterations for years without visible harm.

In the absence of any hard data on what harm genetically engineered organisms might do, ecologists draw analogies with other "exotic organisms" which move from one environment, where they are held in check by natural enemies, to another, where they sometimes run wild and cause immense damage. Thus an Asian fungus that came to be known as the "chestnut blight" virtually wiped out the American chestnut tree; another foreign fungus is destroying the Dutch elm trees; starlings from Europe have become major agricultural pests, and gypsy moths are eating their way through American forests.

Only a minority of the tens of thousands of exotic species that have invaded new environments have caused major damage, ecologists acknowledge, but there is no way to predict which organisms will be harmful. It is a "game of chance" or "biological roulette," according to ecological opinions cited by Frances E. Sharples, an ecologist at the Oak Ridge National Laboratory in Tennessee who recently surveyed scientific literature on the subject for the E.P.A.

Genetic engineers retort that analogies between their products and exotic species are irrelevant and misleading. They say that genetically engineered organisms are not "exotic" or "novel" in the sense that the Japanese beetle or chestnut blight were, but are only slightly altered versions of organisms already present in the environment, much as a new rose strain developed by classical plant breeding techniques is a harmless variation of the parent organism.

However, Dr. Sharples concluded that the analogy between minor genetic alterations and exotic species was valid because even very slight genetic alterations have made various organisms resistant to antibiotics or to chemical pesticides, thus indicating that "very small changes" in genetic information can sometimes produce "large changes" in environmental effects and "extremely large" economic consequences.

The most thoroughly analyzed of all proposed genetic engineering experiments is probably the one that was originally scheduled to occur first, until it was blocked by Mr. Rifkin's suit. The experiment, proposed by Steven Lindow and Nikolas Panopoulos of the University of California at Berkeley, seeks to use genetically altered bacteria to retard frost damage on plants. Ordinarily, certain strains of wild bacteria found on plants serve as nucleation points for the formation of ice crystals, but if a bacterial gene is deleted, the ice crystals do not form and the plants can tolerate much lower temperatures. Thus Dr. Lindow proposed to saturate a row of potato plants with genetically altered bacteria, thereby crowding out the natural bacteria that would promote ice formation.

In opposing the experiment, Mr. Rifkin contended that if the modified bacteria were to multiply in large numbers, they could "drastically alter" the region's plant and insect population, allowing those that had previously been killed by frost to become more prevalent and jeopardizing survival of the traditional species.

However, even many ecologists consider this experiment relatively safe because it simply involves deleting a gene; nothing potentially dangerous is being added to the genes. Moreover, plenty of the ice-resistant bacteria already exist in nature, and the same ice-resistance bacteria have already been produced and disseminated artifically, without ill effects, by standard laboratory techniques requiring no Government approval.

In seeking closer scrutiny of possible dangers, most ecologists stress that they have no desire to block constructive progress in genetic engineering. They say they simply want to avoid the belated discovery of such unexpected side effects as those that undercut support for nuclear energy and chemical pesticides.

Because relatively few microbial genera make significant contributions to the rates of nitrification, it is not surprising that this process is particularly sensitive to environmental stress. Toxic chemicals can block the nitrification process. Nitrification is an obligately aerobic process, and under anaerobic conditions, such as may exist when high concentrations of organic matter are added to soil or aquatic ecosystems, the nitrification process may cease. The process of nitrification is very important in soil habitats because the transformation of ammonium ions to nitrite and nitrate ions results in a change from a positively charged to a negatively charged ion. Positively charged ions are bound by negatively charged soil clay particles and thus are retained in soils. Negatively charged anions, such as nitrate, are not absorbed by soil particles and are readily leached from the soil (Figure 9.9). Nitrification, therefore, results in the transfer of inorganic fixed forms of nitrogen from surface soils to subsurface

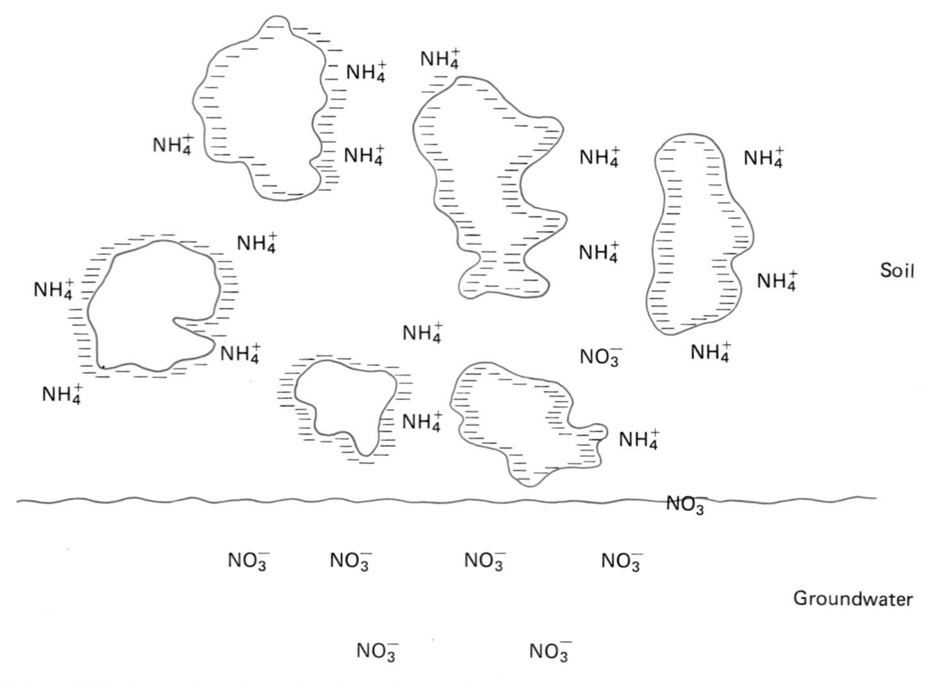

Figure 9.9 Drawing showing the relationship of ammonium and nitrate ions to soil particles; the positively charged ammonium ions are retained and the negatively charged nitrate ions are leached.

groundwater reservoirs by leaching (dissolving out by percolation of water). In agriculture, inhibitors of nitrification, such as nitrapyrin, sometimes are intentionally added to soils to prevent the transformation of ammonium to nitrate, ensuring better fertilization of crops.

The transfer of nitrate and nitrite ions from surface soil to groundwater supplies is critical for two reasons: (1) it represents an important loss of nitrogen from the soil, where it is needed to support the growth of higher plants; and (2) high concentrations of nitrate and nitrite in drinking water supplies pose a serious human health hazard. Nitrite is toxic to humans because it can combine with blood hemoglobin to block the normal gas exchange with oxygen. Additionally, nitrites can react with amino compounds to form highly carcinogenic nitrosamines. Further, nitrate, although not highly toxic itself, can be microbially reduced in the gastrointestinal tracts of human infants to form nitrite, causing "blue baby syndrome"; this reduction of nitrate does not occur in adults because of the low pH of the normal adult gastrointestinal tract. The presence of nitrate and nitrite in groundwater is a particular problem in agricultural areas, such as the "cornbelt" of the midwestern United States, where high concentrations of nitrogen fertilizers are applied to soil. The use of nitrification inhibitors in combination with the application of ammonium nitrogen fertilizers can minimize the nitrate leaching problem, and at the same time supporting better soil fertility and increased plant productivity.

Denitrification

Denitrification, the conversion of fixed forms of nitrogen to molecular nitrogen, is another important process in the biogeochemical cycling of nitrogen that is mediated by microorganisms. Denitrification occurs when nitrate ions serve as terminal electron acceptors in anaerobic respiration. The reduction of nitrate leads to the production of gaseous forms of nitrogen, including nitrous oxide and molecular nitrogen. The return of nitrogen to the atmosphere by the denitrification process completes the nitrogen cycle.

Sulfur Cycle

Sulfur can exist in a variety of oxidation states within organic and inorganic compounds. The sulfur cycle is established by microbial oxidation–reduction reactions that change the oxidation states of sulfur within various compounds (Figure 9.10). Under aerobic conditions the removal of sulfur (**desulfurization**) of organic compounds results in the formation of sulfate, whereas under anaerobic conditions hydrogen sulfide is normally produced from the microbial metabolism of organic sulfur compounds. Hydrogen sulfide may also be formed by sulfate-reducing bacteria, which utilize sulfate as the terminal electron acceptor during anaerobic respiration. Hydrogen sulfide can accumulate in toxic concentrations in areas of rapid protein decomposition; the gas is highly reactive and is very toxic to most biological systems.

Although hydrogen sulfide is toxic to many microorganisms, the photosynthetic sulfur bacteria use hydrogen sulfide as an electron donor for generating reduced coenzymes during their photosynthetic metabolism. The anaerobic photosynthetic bacteria often occur on the surface of sediments where there is available light to support their photosynthetic activities and a supply of hydrogen sulfide from dissimilatory sulfate reduction and anaerobic degradation of organic sulfur-containing compounds. Some photosynthetic bacteria deposit elemental sulfur as an oxidation product, whereas others form sulfate.

Some bacteria are capable of generating ATP by oxidizing hydrogen sulfide. Chemolithotrophic members of the genus *Thiobacillus* oxidize sulfur as their source of energy. Some *Thiobacillus* species are acidophilic and grow well at pH 2–3, and the growth of such *Thiobacillus* species can produce sulfate from the oxidation of elemental sulfur, leading to the environmental accumulation of sulfuric acid.

Acid mine drainage is a consequence of the metabolism of sulfur- and iron-oxidizing bacteria. Coal, in geological deposits, is often associated with pyrite (FeS_2), and when coal mining activities expose pyrite ores to atmospheric oxygen, the combination of autoxidation and microbial sulfur and iron oxidation produces large amounts of sulfuric acid. Anytime pyrites are mined as part of an ore recovery operation, oxidation may produce large amounts of acid. The acid draining from mines kills aquatic life and renders the water it contaminates unsuitable as a drinking water supply or for recreational uses. At present, approximately 10,000 miles of

Figure 9.10 The sulfur cycle, showing various transformations of organic and inorganic compounds.

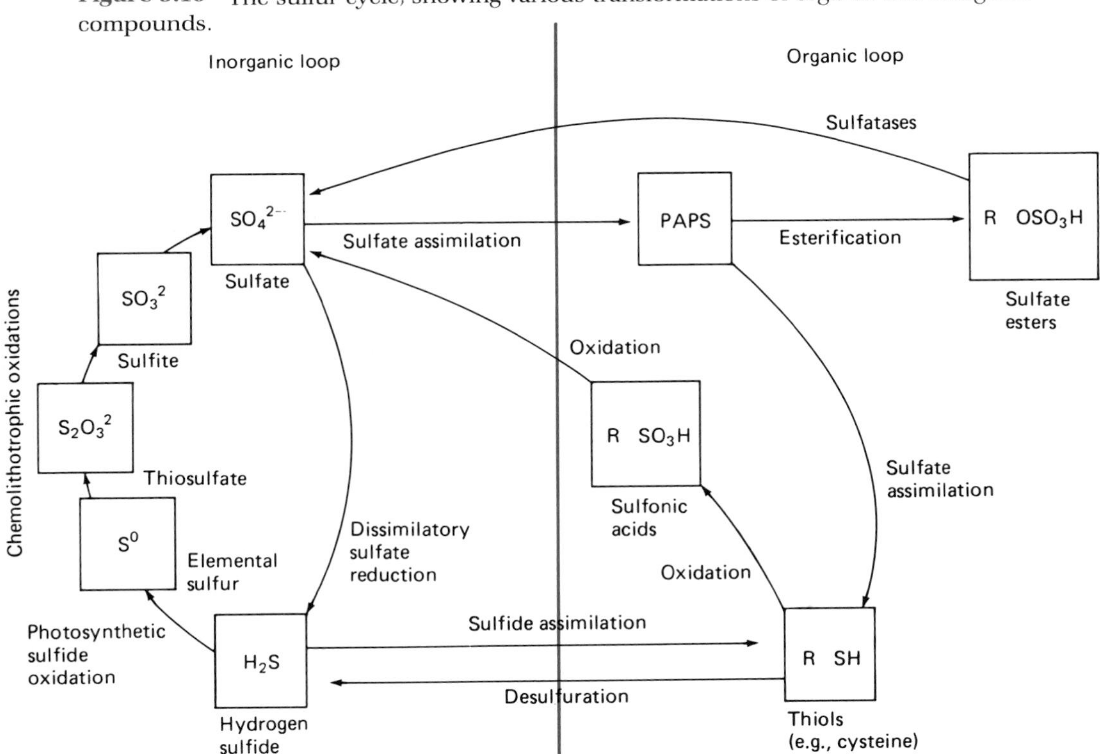

Figure 9.11 An algal bloom in a small lake resulting from eutrophication. (Courtesy Ronald F. Lewis, U.S. Environmental Protection Agency, Cincinnati.)

United States waterways are affected in this manner, predominantly in the states of Pennsylvania, Virginia, Ohio, Kentucky, and Indiana. Strip mining is a particular problem with respect to acid mine drainage because this method of coal recovery removes the overburden, leaving a porous rubble of tailings exposed to oxygen and percolating water. Oxidation of the reduced iron and sulfur in the tailings (refuse material) produces acidic products, causing the pH to drop rapidly and preventing the reestablishment of vegetation and a soil cover that would seal the rubble from oxygen. A strip-mined piece of land continues to give rise to acid mine drainage until most of the sulfide is oxidized and leached out; recovery of this land may take 50–150 years.

Figure 9.12 The iron cycle, showing interconversion of ferrous and ferric iron.

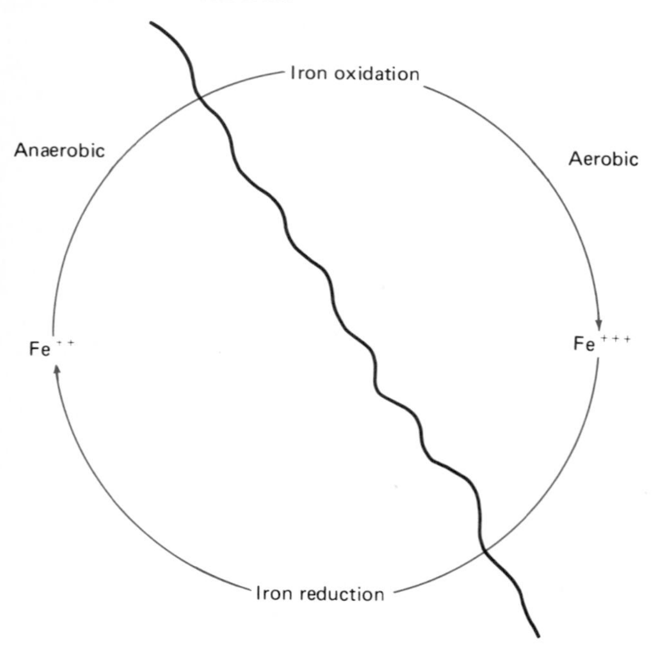

Other Element Cycles

Phosphorus

Phosphorus normally occurs as phosphates in both inorganic and organic compounds. Microorganisms assimilate inorganic phosphate and mineralize organic phosphorus compounds; microbial activities are involved in the solubilization or mobilization of phosphate compounds. Unlike the other elements discussed, microorganisms normally do not oxidize or reduce phosphorus. In many habitats phosphates are combined with calcium, rendering them insoluble and unavailable to most organisms. Various heterotrophic microorganisms are capable of solubilizing phosphates primarily through the production of organic acids. These actions of microorganisms mobilize phosphate, and activities of other microorganisms act to immobilize phosphorus. For example, microorganisms compete with plants for available phosphate resources because the assimilation of phosphates by microorganisms removes phosphates from the available nutrient pool required by plants.

In many habitats productivity is limited by the availability of phosphate. When excess concentrations of phosphate enter aquatic habitats, such as when phosphate detergents are added to lakes, there can be a sudden increase in productivity, a process called eutrophication (Figure 9.11). The blooms of algae and cyanobacteria associated with eutrophication can greatly increase the concentrations of organic matter in bodies of water. During the subsequent decomposition of this organic matter, oxygen can be severely depleted from the water column, causing major fish kills. The introduction of

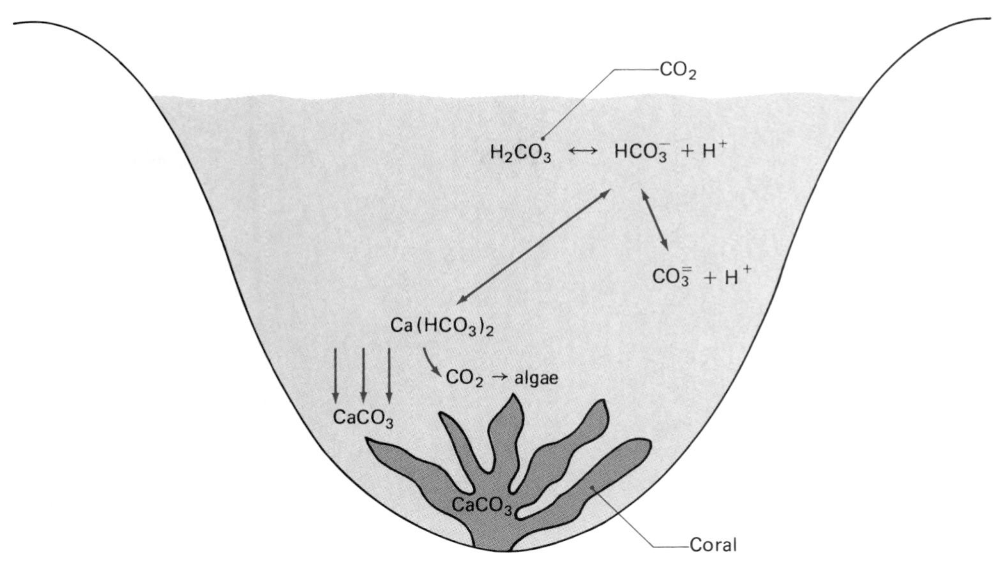

Figure 9.13 The reactions of calcium in seawater lead to the formation of coral reefs.

high concentrations of phosphate from phosphate laundry detergents created such eutrophication problems in many water bodies that some municipalities banned the use of such detergents.

Iron

The cycling of iron compounds has a marked effect on the availability of this essential element for other organisms. Iron is transformed between the ferrous (Fe^{2+}) and ferric (Fe^{3+}) oxidation states by microorganisms (Figure 9.12). Ferric and ferrous ions have very different solubility properties, ferric compounds tending to be less soluble than ferrous compounds. The microbial oxidation of ferrous ions, therefore, tends to result in the precipitation of ferric iron. Various bacteria, including members of the genera *Thiobacillus, Gallionella,* and *Leptothrix* are capable of oxidizing iron compounds.

Calcium

Calcium also exhibits biogeochemical cycling between soluble and insoluble forms. Calcium bicarbonate is soluble, but calcium carbonate is much less soluble. The microbial production of acidic compounds solubilizes precipitated and immobilized calcium compounds. There is an interesting cycling of calcium in marine habitats, where dissolved carbon dioxide reacts with available calcium, forming calcium bicarbonate and calcium carbonate (Figure 9.13). During the formation of coral, calcium carbonate precipitates when carbon dioxide held in solution as calcium bicarbonate is removed by algal cells of the coral. This process results in the deposition of calcium carbonate and the formation of coral reefs. Calcium carbonate is also precipitated by various algae to form an outer frustule. Accumulation of algal frustules can lead to the formation of major limestone deposits.

Silicon

Various algae, most notably the diatoms, form silicon impregnated structures (Figure 9.14). These al-

Figure 9.14 These micrographs of diatoms show the delicate beauty of their frustules. Exterior view of *Stephanodiscus niagarae* (3740 ×). (Courtesy Edward Theriot, Great Lakes Research Division, University of Michigan, Ann Arbor, Michigan.)

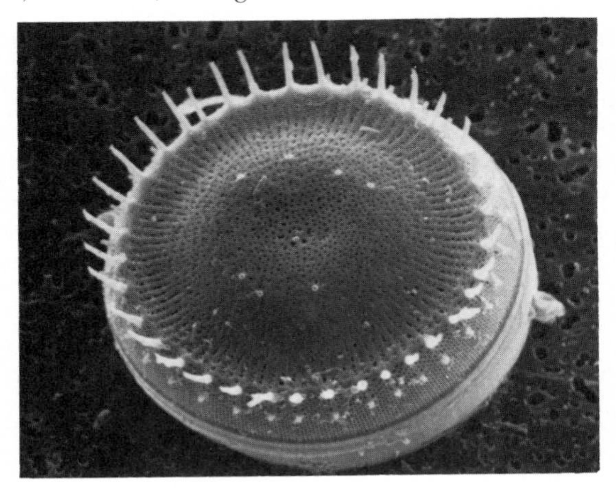

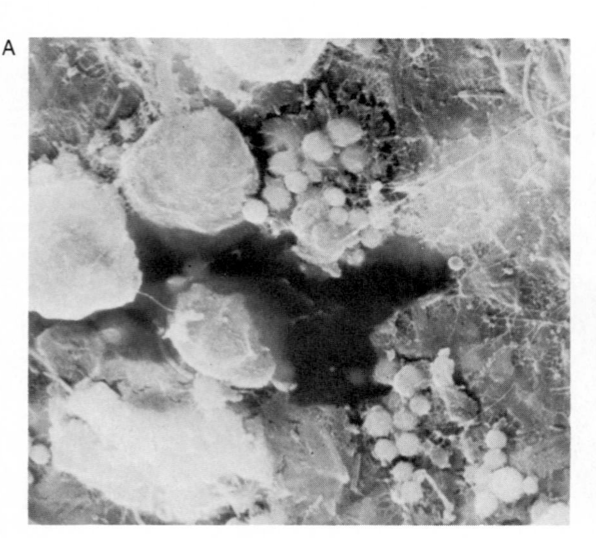

Figure 9.15 A manganese nodule showing associated bacteria and algae. (A) Scanning electron micrograph showing microcolonies of bacteria. (B) Scanning electron micrograph of a region of the nodule covered with debris of cocolith algae. (Courtesy Paul LaRock, Florida State University, Tallahassee.)

gae precipitate silicon dioxide to build their delicate and decorative shells. As much as 10 billion metric tons of silicon dioxide is precipitated by microorganisms in the oceans each year. The shells of these dead microorganisms accumulate and form silicon-rich oozes that later develop into deposits of diatomaceous earth, which is also known as Fuller's earth.

Manganese

Manganese exists as a water soluble divalent manganous ion (Mn^{2+}) and as a relatively insoluble tetravalent manganic ion (Mn^{4+}). The microbial oxidation of manganous ions forms manganese oxides, which produce characteristic manganese nodules (Figure 9.15). The manganese for the nodules originates in anaerobic sediments. When the manganese enters aerobic habitats it is oxidized and precipitates to form the nodules. The farming of manganese nodules in deep ocean sediments is considered a possible source for obtaining manganese for industrial usage.

Heavy Metals

Mercury, arsenic, and other heavy metals are also subject to microbial biogeochemical cycling (Figure 9.16). These transformations are important because they alter both the mobility and toxicity of the metals. For example, mercury is released into the environment largely as a consequence of its widespread use in industry and the burning of fossil fuels, although some mercury also leaches from rocks. Mercury salts, though fairly toxic, are excreted efficiently. Therefore, the release of mercury salts into the environment was not originally viewed with much concern. However, in anaerobic sediments, some microorganisms are capable of methylating mercury.

As crude a weapon as the cave man's club, the chemical barrage has been hurled against the fabric of life.

—RACHEL LOUISE CARSON, *Silent Spring*

The methylation of mercury causes increased danger because the resultant product, methylmercury, is highly neurotoxic. Methylmercury is readily concentrated in filter-feeding shellfish. The increase in concentration in the tissues of higher members of the food web, originating from the ingestion of shellfish, is called biomagnification. Methylmercury also accumulates in humans. Unlike inorganic and phenylmercury compounds, methylmercury is excreted by humans only very slowly. In Japan in the 1950s, the ingestion of shellfish containing methylmercury led to outbreaks of Minamata disease, a severe disturbance of the central nervous system associated with mercury poisoning. The buildup of methylmercury compounds in Scandanavian freshwater lakes and the North American Great Lakes forced large areas to be condemned for fishing.

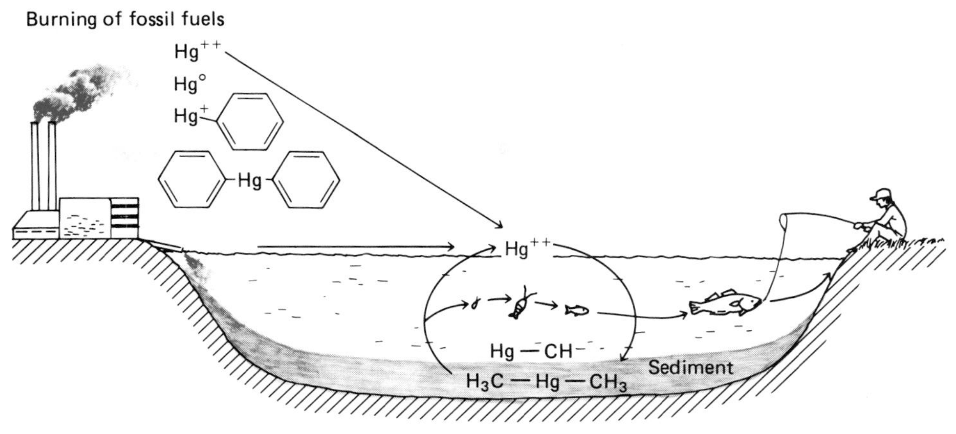

Figure 9.16 The mercury cycle, showing the biological magnification of methylmercury in the aquatic food chain.

Summary

Microorganisms, because of their diversity and ubiquitous distribution, are extremely important in ecological processes. Critical steps of the major global biogeochemical cycles are mediated by microorganisms; for example, virtually all steps in the biogeochemical cycling of nitrogen involve microorganisms. The growth of higher organisms in various habitats relies on the biogeochemical cycling activities of microorganisms, in many cases forming crucial links in the food webs of an ecosystem, that permit the flow of energy and organic compounds to reach higher trophic levels.

Microorganisms play essential roles in the biogeochemical cycling of carbon, oxygen, hydrogen, nitrogen, sulfur, and various other elements. Carbon is cycled between organic compounds and carbon dioxide. Molecular nitrogen in the atmosphere is fixed by a restricted number of bacteria. Microbial biomass is important at the base of food webs that support higher organisms.

The nitrogen cycle, and nitrogen fixation in particular, is extremely important because plants require fixed forms of nitrogen for growth. Agricultural productivity depends upon the availability of fixed forms of nitrogen. Symbiotic nitrogen–fixing bacteria that form root nodules are particularly important in this regard. In addition to nitrogen fixation, nitrification and denitrification determine the availability of nitrogen in an ecosystem. Nitrification, which is carried out by chemolithotrophic bacteria, transforms ammonium ions into nitrate ions; the change in electronic charge associated with this transformation leads to leaching of nitrogen from soils. Denitrification returns molecular nitrogen to the atmosphere, thereby also removing useable forms of nitrogen from soil and aquatic ecosystems.

Microorganisms are also important in environmental pollution. In some cases, such as sulfur oxidation by *Thiobacillus* which results in acid mine drainage and bacterial methylation of mercury, microbial activities are responsible for serious deterioration of environmental quality. In most cases, however, microorganisms act as biological incinerators and prevent the harmful accumulation of environmental pollutants. When microorganisms fail, pollutants accumulate and environmental quality deteriorates.

Study Outline

A. The metabolic activities of microorganisms transform the chemical forms of elements and establish global biogeochemical cycles.

B. The carbon cycle.

1. The carbon cycle primarily involves the transfer of carbon dioxide and organic carbon.
2. Carbon and energy in organic compounds formed by primary producers move through the biological community of an ecosystem, forming a food web: grazers feed on primary producers, predators feed on grazers and smaller predators. Only 10–15 percent

of energy stored in any trophic level is transferred to next level; the higher the trophic level, the smaller the biomass.

C. The hydrogen and oxygen cycles.

1. The biogeochemical cycling of hydrogen and oxygen is closely tied to carbon cycle.

2. Water acts as an electron donor during photosynthesis: molecular oxygen is liberated (major source of molecular oxygen upon which aerobic respiration depends), and hydrogen and electrons are transferred to form reduced coenzymes for redox reactions for organic compounds.

3. During respiration, oxygen and hydrogen are recycled with the formation of molecular hydrogen and carbon dioxide.

D. The nitrogen cycle.

1. Fixation of atmospheric nitrogen.

a. Nitrogen fixation is the conversion of molecular nitrogen to ammonia or organic nitrogen that can be assimilated into biomass.

b. Nitrogen fixation is restricted to a limited number of microbial species.

c. Nitrogen fixation depends upon the nitrogenase enzyme system; the process requires a high energy input.

d. Nitrogen fixation is carried out by free-living bacteria and by bacteria living in symbiotic association with plants.

2. Ammonification.

a. Ammonification is the conversion of organic amino nitrogen to ammonia.

b. Many microorganisms as well as plants and animals carry out ammonification.

3. Nitrification.

a. Relatively few organisms carry out nitrification.

b. Nitrification is a two-step, energy-yielding process carried out by different microbial populations; step one is the oxidation of ammonium ions to nitrite ions and is carried out by *Nitrosomonas, Nitrosospira, Nitrosococcus*, and *Nitrosolobus*; step two is the oxidation of nitrite ions to nitrate ions and is carried out by *Nitrobacter, Nitrospira*, and *Nitrococcus*.

c. Nitrification is sensitive to environmental stress.

d. In soil, nitrification results in the transfer of inorganic fixed forms of nitrogen from surface soils, where ammonium ions are bound to clay particles, to subsurface groundwater reservoirs because the negatively charged nitrite and nitrate ions are not absorbed by soil particles.

e. Nitrification causes an important loss of nitrogen from soil.

f. In regions of extensive nitrification, high concentrations of nitrite and nitrate in drinking water often pose a serious human health hazard.

g. Nitrification inhibitors can be added to soils to block nitrification.

4. Denitrification.

a. Denitrification is the conversion of fixed forms of nitrogen to molecular nitrogen.

b. Denitrification returns nitrogen to the atmosphere, completing the nitrogen cycle.

E. The sulfur cycle.

1. Oxidation–reduction reactions, mediated by microorganisms, change the oxidation states of sulfur within various compounds.

2. Desulfurization (removal of sulfur) of organic compounds results in the formation of sulfate under aerobic conditions and hydrogen sulfide under anaerobic conditions.

3. Anaerobic photosynthetic sulfur bacteria use hydrogen sulfide as an electron donor for generating reduced coenzymes.

4. Chemolithotrophic members of genus *Thiobacillus* oxidize sulfur as their energy source.

5. Acid mine drainage is a consequence of metabolism of sulfur- and iron-oxidizing bacteria (sulfuric acid is produced).

F. Other element cycles.

1. Microorganisms assimilate inorganic phosphate and mineralize organic phosphorus compounds; they normally do not oxidize or reduce phosphorus.

2. Iron is transformed between ferrous and ferric oxidation states by microorganisms (ferric compounds are less soluble than ferrous compounds).
3. Calcium is cycled between soluble (bicarbonate) and insoluble (carbonate) forms.
   a. Microbial production of acidic compounds solubilizes precipitated and immobilized calcium compounds.
   b. Algal cells remove calcium bicarbonate from the surrounding water, resulting in precipitation of calcium carbonate and the formation of coral reefs.
4. Various algae (primarily diatoms) form silicon-impregnated structures; the shells of these dead microorganisms accumulate to form silicon-rich oozes that later develop into deposits of diatomaceous or Fuller's earth.
5. Microbial oxidation of manganous ions forms manganese oxides, which produce characteristic manganese nodules; manganese nodules may be recovered from deep ocean sediments for industrial use.
6. Some heavy metals are transformed by microorganisms.
   a. Biogeochemical cycling of heavy metals is important because it alters their mobility and toxicity.
   b. In anaerobic sediments, some microorganisms methylate mercury; methylmercury is readily concentrated in filter-feeding shellfish, is excreted very slowly by humans, and is highly neurotoxic.

Study Questions

1. Indicate whether the following statements are true (T) or false (F).
   a. The hydrogen and oxygen cycles are closely tied to the carbon cycle.
   b. Water acts as an electron acceptor during photosynthesis, during which molecular oxygen is liberated.
   c. During respiration, oxygen and hydrogen are recycled with the formation of molecular hydrogen and carbon dioxide.
   d. The fixation of atmospheric nitrogen is the conversion of molecular nitrogen to ammonia or organic nitrogen that can be assimilated into biomass.
   e. Many bacterial species can fix nitrogen.
   f. Nitrogen fixation in soil is carried out by free-living bacteria and bacteria living in symbiotic association with plants.
   g. *Rhizobium* is a nitrogen-fixing, free-living microorganism.
   h. Agricultural productivity depends upon the availability of fixed forms of nitrogen.
   i. Under aerobic conditions decomposition of proteins results in the formation of hydrogen sulfide.
   j. Chemolithotrophic bacteria are responsible for acid mine drainage.
   k. The incorporation of calcium into algal cell walls results in the destruction of coral reefs.
   l. Microbial methylation of mercury lowers toxicity.
2. List the steps involved in nitrification. For each step indicate the microorganisms involved.
3. Describe how nitrification mobilizes fixed forms of nitrogen in soil.
4. Fill in the blanks.
   a. The carbon cycle primarily involves the transfer of ________ and ________ between the ________ and the ________ and ________ .
   b. In primary production, ________ is converted into ________ by ________ in an ecosystem.
   c. The relationship established is such that the grazers feed on the ________ , and the predators feed on the ________ and smaller ________ .
   d. The higher the ________ level, the smaller the ________ .

Some Thought Questions

1. Could human life survive in a sterile world?
2. Do microbes make the world go round?
3. When searching for life on Mars, why did NASA first look for microorganisms?
4. If there is a heaven would you expect to find bacteria there?
5. Are microbes fallible? Are there really recalcitrant molecules?
6. Why isn't the garbage piled over our heads?
7. Why are there so few big fierce tigers?
8. Why shouldn't you build a house with a basement too close to an active landfill?
9. How do microorganisms affect the price of potatoes? oranges?
10. Do microorganisms heighten or dampen the greenhouse effect?
11. Some years ago the U. S. Environmental Protection Agency reported that burping cows were a major source of air pollutants. What gas does a cow belch and where does it come from?

Additional Sources of Information

Alexander, M. 1977. *Introduction to Soil Microbiology*. John Wiley and Sons, Inc., New York. An excellent text on the microbiology of soils with several chapters devoted to elemental cycling.

Atlas, R. M., and R. Bartha. 1981. *Microbial Ecology: Fundamentals and Applications*. Addison-Wesley Publishing Co., Inc., Reading, Massachusetts. A text covering all aspects of microbial ecology including the biogeochemical cycling of elements.

Campbell, R. E. 1984. *Microbial Ecology*. Blackwell Scientific Publications, Oxford, England. A short text that includes coverage of biogeochemical cycling.

Codd, G. A. (ed.) 1984. *Aspects of Microbial Metabolism and Ecology*. Academic Press, Inc., Orlando, Florida. This book contains a wealth of information on recent advances in selected areas of microbial metabolism, genetics, and ecology.

Kosikowski, F. V. 1985. Cheese. *Scientific American*, 252(5):88–99. A thorough discussion of the many varieties of cheese and the microbiology of their production.

Lynch, J. M., and N. J. Poole (eds.). 1979. *Microbial Ecology: A Conceptual Approach*. Blackwell Scientific Publications, Oxford, England. An advanced text that covers the major elemental cycles.

Ourisson, G., P. Albrecht, and M. Rohrer. 1984. The microbial origin of fossil fuels. *Scientific American*, 251(2):44–51. This interesting article discusses the role of microorganisms in the formation of fossil fuel deposits.

UNIT FOUR

Microbial Genetics

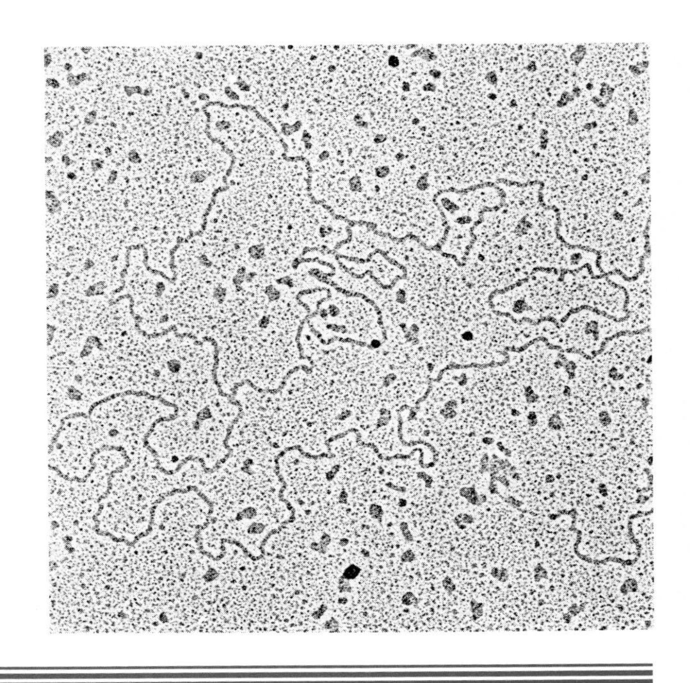

10 *Storage and Expression of Genetic Information*

KEY TERMS

Ames test
anticodon
auxotroph
carcinogen
catabolite repression
cistron
codon
constitutive enzyme
degenerate
deoxyribose
deoxyribonucleic acid (DNA)
diploid
fertility (F) plasmid
frame shift mutation
gene
genome
genotype
haploid
heterogeneous RNA (hnRNA)
3′ hydroxyl end (3′-OH)
5′ hydroxyl end (5′-OH)
inducer
inducible enzymes
intron
messenger RNA (mRNA)
mutagen
mutation
nonsense codon
nuclotide
nutritional mutant
operator region
operon
phenotype
phosphate diester linkage
plasmids
polymerase
promoter region
prototroph
repressor protein
resistance (R) plasmid
ribonucleic acid (RNA)
ribosomal RNA (rRNA)
silent mutation
structural gene
transcription
transfer RNA (tRNA)
translation
translocation
wobble hypothesis

OBJECTIVES

After reading this chapter you should be able to

1. Define the key terms.
2. Describe how genetic information is stored within the DNA macromolecule.
3. Describe the process of transcription.
4. Describe the process of translation.
5. Discuss the consequences of changes in the sequence of bases in DNA.
6. Discuss the factors that influence the rates of mutations.
7. Describe the Ames test for detecting carcinogens.
8. Discuss how the expression of genetic information can be regulated.

Storage of Genetic Information

The metabolic and structural properties of a microorganism are determined by its genetic information. Whether the microbe can produce a deadly poison or whether it has the potential for producing substances that can be used in the treatment of disease are specified within the DNA molecule(s) that encode(s) the genetic information of the cell(s) of that organism (Figure 10.1). The discovery of the structure of DNA and how genetic information is encoded within an organism has revolutionized our understanding of genetics and has opened up many new vistas for controlling harmful microorganisms and reaping benefit from others.

> Since scientists studied in 1944 a substance which they called DNA, and since this undoubtedly revolutionized the life sciences with full breakthrough force, I feel confident that—if we survive—the year 2004 will see molecular biology introducing triumphs that can now barely be imagined. Many of us will survive to see that. And if we reach 2004 in safety, man may then be knowlegeable enough to guarantee his own safety against the possibility of self-destruction.
>
> —ISAAC ASIMOV, *The Genetic Code*

The DNA Macromolecule

DNA (deoxyribonucleic acid) is a large macromolecule composed of nucleotides linked together to form a long chain. The information within the DNA defines the organism's genotype. A single copy of the genetic information of a cell constitutes the genome of that organism. The genome is divided into segments, known as genes, that have specific functions. Genes that code for the synthesis of RNA and proteins are known as structural genes or cistrons. Other genes have regulatory functions and act to control the activities of the cell. The order or sequence of the nucleotides determines the functions specified in that region of the DNA. The nucleotide subunits of the DNA are composed of a nitrogenous base (a heterocyclic ring structure containing nitrogen as well as carbon, sometimes referred to as a nucleic acid base or simply as the base), a five-carbon sugar (deoxyribose), and a phosphate group (Figure 10.2).

Four nitrogenous bases occur in DNA: adenine, guanine, cytosine, and thymine. Adenine (A) and guanine (G) are substituted purine bases that are two-ringed structures. Cytosine (C) and thymine (T) are substituted pyrimidine bases and have only one ring. The nucleotides are linked by 3′-5′phosphate diester linkages (Figure 10.3). Consequently, at one end of the nucleic acid molecule, there is no phosphate diester bond to the carbon 3 of the deoxyribose, and that end of the molecule is called the 3′-hydroxyl free end. The other end of the molecule, where the carbon 5 is not involved in forming a phosphate diester linkage, is called the 5′-hydroxyl free end. The fact that the ends of the nucleic acid macromolecule differ is extremely important because this permits directional recognition at the biochemical level in the same sense that we can recognize left and right. The directional nature of nucleic acid molecules is critical for establishing the necessary direction of reading the genetic information.

Figure 10.1 James Watson (left), age 23, and Francis Crick, age 34, posed with their DNA model in the Cavendish Laboratory at Cambridge University, England, in 1953, when they announced their discovery of the molecular structure of deoxyribonucleic acid. They shared the Nobel Prize for Medicine in 1962 with Maurice Wilkins. (Photograph by Barr-Brown Camera Press Ltd., London.)

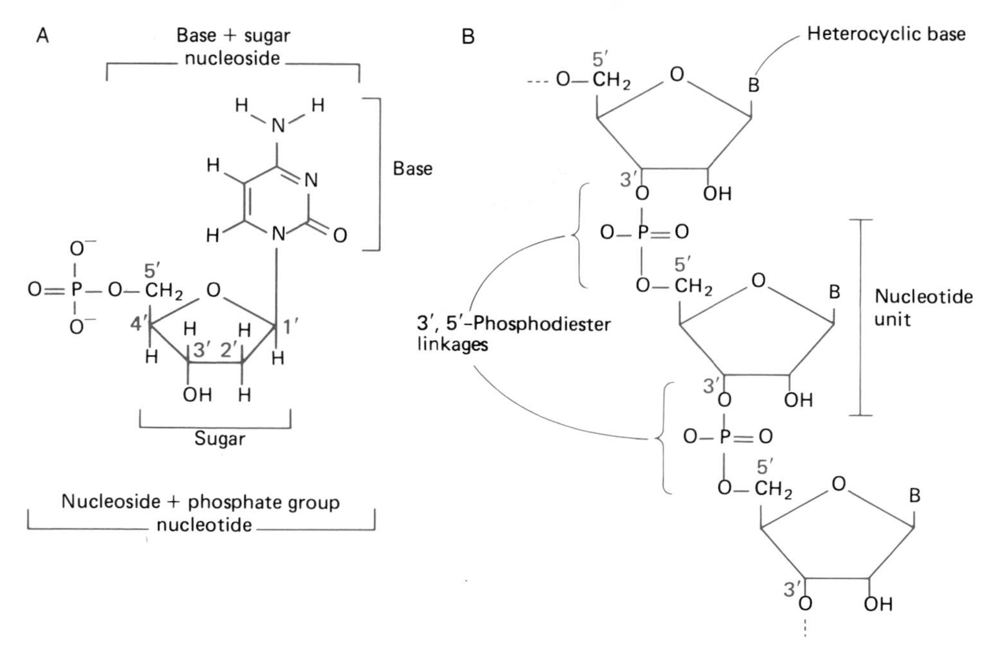

Figure 10.2 (A) The structural formula of a nucleotide, the basic unit of the informational macromolecules of living systems. The five carbon atoms of the deoxyribose are numbered 1 through 5 so that the positions can be specified. The nucleic acid base is linked to the carbon 1 of the deoxyribose. The nitrogenous base plus deoxyribose is called a nucleoside. (B) Nucleotides are linked by phosphate diester bonds to form dimers and polymeric units. The nucleosides are linked together by a phosphate group between carbon 3 of one deoxyribose and carbon 5 of another deoxyribose.

The DNA macromolecule normally occurs as a double helical molecule (Figure 10.4). The DNA double helix is composed of two primary polynucleotide chains held together by hydrogen bonding between complementary nucleotide bases (see Appendix for discussion of chemical bonding). Within the double helical DNA, adenine always pairs with thymine, and guanine always pairs with cytosine. There are two hydrogen bonds established between the adenine–thymine base pairs and three hydrogen bonds established between the guanine–cytosine base pairs.

Within the double helical DNA macromolecule, the two polynucleotide chains run in opposite directions. One chain runs from the 3′-hydroxyl to the 5′-hydroxyl free end and the complementary chain runs in an antiparallel direction from the 5′-hydroxyl to the 3′-hydroxyl free end. A consequence of the antiparallel nature of the DNA molecule is that different information is stored within each of the chains. The genetic code, based on only the few "letters" (nucleotides) in its "alphabet," provides the necessary biochemical basis for encoding the genetic information of the great diversity of living organisms.

Genome Structure

Prokaryotic and eukaryotic cells differ in their genome structure. Bacterial cells have a single genome contained in the bacterial chromosome. Because there is a single set of genes, bacteria are **haploid**. In contrast, eukaryotic microorganisms generally have pairs of matching chromosomes and are **diploid**, having two copies of each gene during at least part of their life cycles. When both copies of the gene are identical, the microorganism is **homozygous**. When the corresponding copies of the gene differ, the microorganism is **heterozygous**. The information encoded in one gene may be dominant and the other recessive. In some cases alleles may exhibit codominance, producing a hybrid state with an intermediate appearance. For example, microorganisms with homozygous genes for a particular feature may appear red or colorless, and those with heterozygous genes may appear pink.

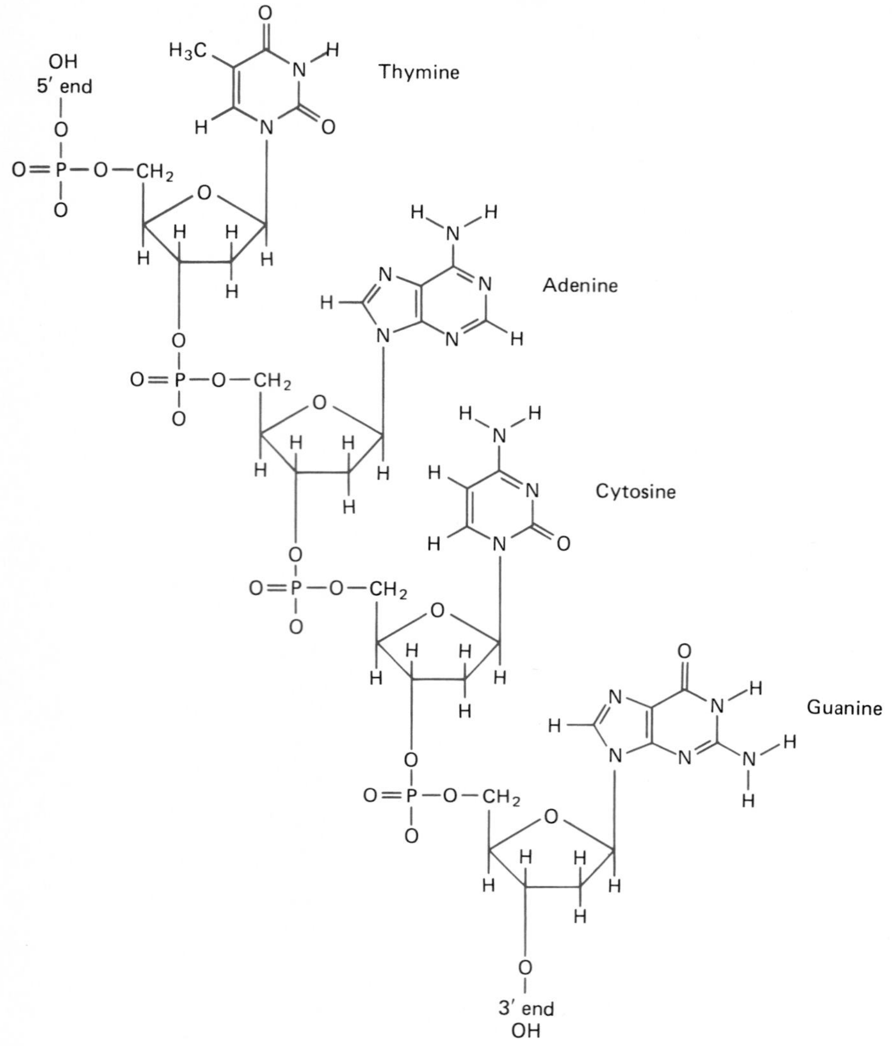

Figure 10.3 Because nucleic acids are linked by phosphodiester bonds between carbon 3 of one nucleic acid base and the carbon 5 of the other, nucleic acids have 3′-hydroxyl and 5′-hydroxyl free ends at opposite ends of the polymeric molecules.

We are a spectacular, splendid manifestation of life.
We have language . . . We have affection.
We have genes for usefulness,
and usefulness is about as close to a
"common goal" of nature as I can guess at.

—LEWIS THOMAS, *The Medusa and the Snail*

In addition to the genetic information contained within the bacterial chromosomes of prokaryotic cells and the chromosomes of eukaryotic cells, some microorganisms contain plasmids (Figure 10.5). Plasmids are small extrachromosomal genetic elements that permit microorganisms to store additional genetic information. The acquisition or loss of plasmids alters the genomes of microorganisms because individuals of a microbial population that possess plasmids contain different genetic information than those individuals in the population that lack plasmids. Plasmids do not normally contain the genetic information for the essential metabolic activities of the microorganism; generally, they contain genetic information for specialized features. Plasmids are capable of self-replication and can be exchanged between bacteria, permitting the transfer of the information encoded within the plasmid DNA from one microbial population to another.

Although plasmids contain only a very small portion of the microbial genome, they are important. Plasmids can contain: the genetic information that determines the ability of bacteria to mate and

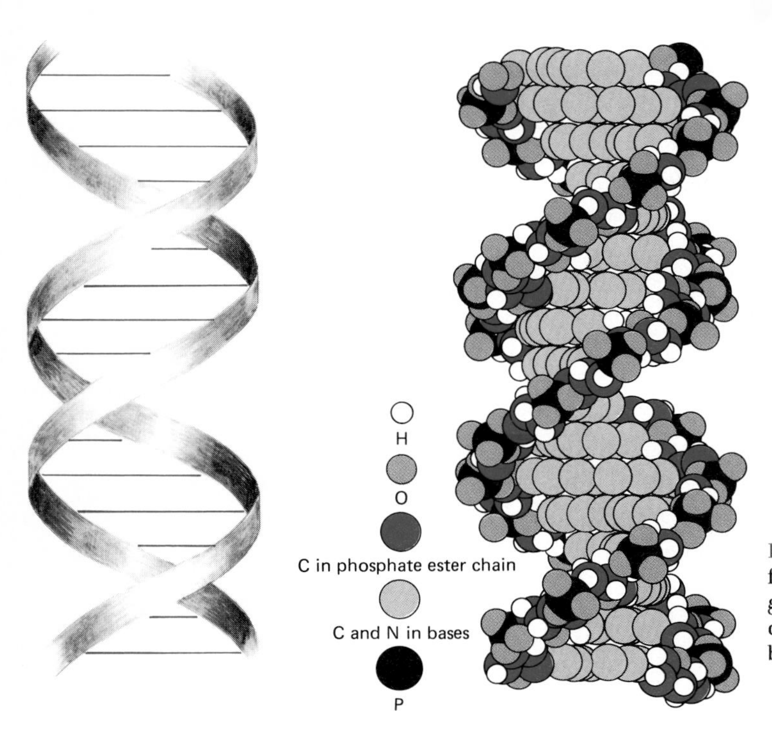

Figure 10.4 The double helix is the fundamental structure of DNA in which a cell's genetic information is stored. The two strands of DNA are held together by hydrogen bonding between complementary base pairs.

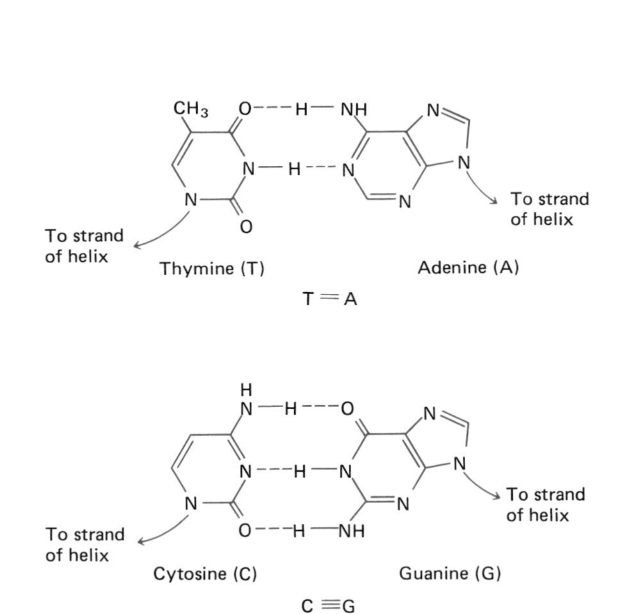

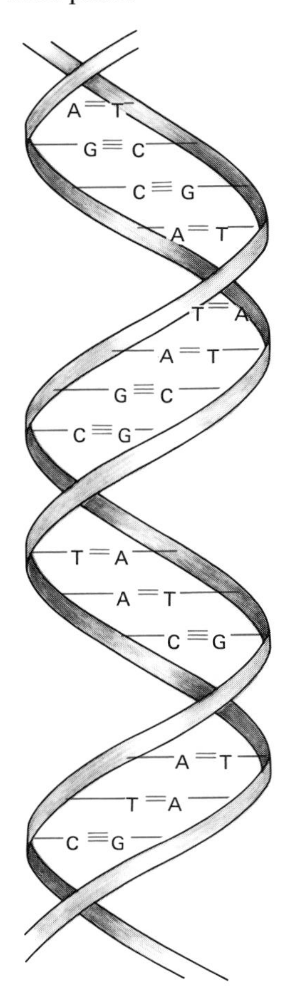

MICROBIOLOGY HEADLINES

The Promise and Peril of the New Genetics

The Cell: Unraveling the Double Helix and the Secret of Life

Wildly excited, two men dashed out of a side door of Cambridge University's Cavendish Laboratory, cut across Free School Lane and ducked into the Eagle, a pub where generations of Cambridge scientists have met to gossip about experiments and celebrate triumphs. Over drinks, James D. Watson, then 24, and Francis Crick, 36, talked excitedly, Crick's booming voice damping out conversations among other Eagle patrons. When friends stopped to ask what the commotion was all about, Crick did not mince words. "We," he announced exultantly, "have discovered the secret of life!"

Brave words—and in a sense, incredibly true. On that late winter day in 1953, the two unknown scientists had finally worked out the double-helical shape of deoxyribonucleic acid, or DNA. In DNA's famed spiral-staircase structure are hidden the mysteries of heredity, of growth, of disease and aging—and in higher creatures like man, perhaps intelligence and memory. As the basic ingredient of the genes in the cells of all living organisms, DNA is truly the master molecule of life.

The unraveling of the DNA double helix was one of the great events in science, comparable to the splitting of the atom or the publication of Darwin's *Origin of Species*. It also marked the maturation of a bold new science: molecular biology. Under this probing discipline, man could at last explore—and understand—living things at their most fundamental level: that of their atoms and molecules.

Using laboratory skills that were unheard of a generation ago, scientists have isolated, put together and manipulated genes, and have come close to creating life itself. In 1967 Stanford University's Arthur Kornberg synthesized in a test tube a single strand of DNA that was actually able to make a duplicate of itself. Kornberg's "creation" was only a copy of a virus, a coated bit of genetic material that occupies a twilight zone between the living and inanimate. But many scientists have become convinced that they may eventually be able to create functioning, living cells.

Molecular biology, in part, is rooted in the science of genetics. Ever since Cro-Magnon man, parents have probably wondered why their children resemble them. But not until an obscure Austrian monk named Gregor Mendel began planting peas in his monastery's garden in the mid-19th century were the universal laws of heredity worked out. By tallying up the variations in the offspring peas, Mendel determined that traits are passed from generation to generation with mathematical precision in small, separate packets, which became known as genes (from the Greek word for race).

Mendel's ideas were so unorthodox that they were ignored for 35 years. But by the time the Mendelian concept was rediscovered at the turn of the century, scientists were better prepared for it. They already suspected that genetic information was hidden inside pairs of tiny threadlike strands in cell nuclei called chromosomes, or colored bodies (for their ability to pick up dyes). During cell division they always split lengthwise, giving each daughter cell a full share of what was presumed to be hereditary material.

By the 1940s, the molecular biologists had come on the scene, and they insisted that fundamental life processes could be fully understood only on the molecular level. In their investigations, some used the electron microscope, which revealed details of structure invisible to ordinary optical instruments. Others specialized in X-ray crystallography, a technique for deducing a crystallized molecule's structure by taking X-ray photographs of it from different angles.

Inspired by these experiments, Watson, then a young Ph.D. in biology from Indiana University, decided to take a crack at the complex structure of DNA itself. The same thought struck Crick, a physicist turned biologist who was preparing for his doctorate at Cambridge. Neither man was particularly well equipped to undertake so formidable a task. Watson was deficient in chemistry, crystallography and mathematics. Crick, on the other hand, was almost totally ignorant of genetics. But together, in less than two years of work at Cambridge, these two spirited young scientists showed how it is possible to win a Nobel Prize without really trying.

In 1968 Watson himself produced a highly irreverent, gossipy bestseller, *The Double Helix,* which revealed the human story behind the discovery of DNA's structure: the bickering, the academic rivalries, even the deceits that were practiced to win the great prize. From the X-ray crystallography laboratory at King's College in London, where Biochemist Maurice Wilkins was also investigating the molecule's structure, they quietly obtained unpublished X-ray data on DNA. Relying as much on luck as logic, they constructed Tinkertoy-like molecular models out of wire and other metal parts. To everyone's astonishment, they suddenly produced a DNA model that not only satisfied the crystallographic evidence but also conformed to the chemical rules for fitting its many atoms together.

Source: Time, April 19, 1971. Reprinted by permission.

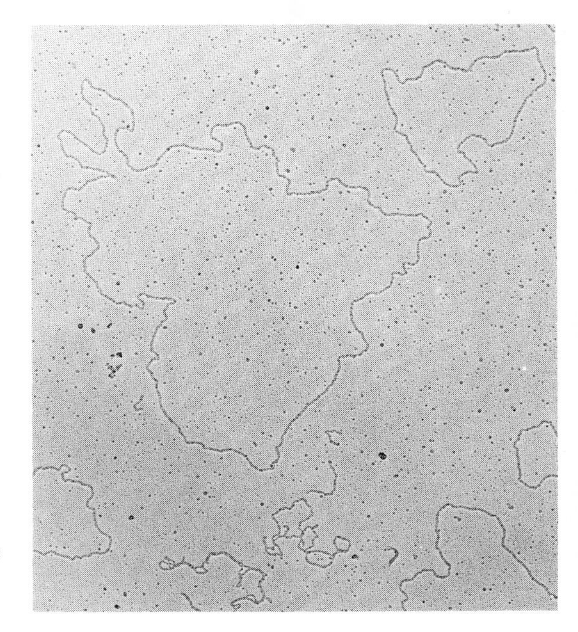

Figure 10.5 This electron micrograph shows two different plasmid DNA molecules. The larger loop is plasmid pBF4 and was isolated from *Bacteroides fragilis;* it encodes resistance to the antibiotic clindamycin. The smaller DNA loop is plasmid pSC101 from *Escherichia coli;* it encodes resistance to tetracycline (30,000 ×). (From BPS: Rod Welch, School of Medicine, University of Wisconsin, Madison.)

whether a bacterial strain acts as a donor during mating; the information that codes for resistance to antibiotics and other chemicals, such as heavy metals, which are normally toxic to microorganisms; the genetic information for the degradation of various complex organic compounds, such as the aromatic hydrocarbons found in petroleum; and the genetic information for toxin production that renders some bacteria pathogenic for man.

There are several different types of plasmids that serve different functions. For example, the F (fertility) plasmid codes for mating behavior. Strains of *E. coli* that have the F plasmid are donor or male strains; those that lack the F plasmid are recipient strains. The R (resistance) plasmids carry genes that code for antibiotic resistance. R plasmids can be passed from one bacterial species to another, such as from *E. coli* to pathogenic strains of *Shigella* or *Salmonella*. Antibiotic-resistant strains of bacteria have become a serious health problem because R plasmids can occur in pathogenic bacteria, and the treatment of bacterial diseases of man has been complicated by the occurrence of these pathogens, which are resistant to multiple antibiotics.

Expression of Genetic Information

The expression of genetic information involves using the information encoded within the genome (DNA) to control the synthesis of proteins. By controlling protein synthesis, the genetic information macromolecules exert control over the metabolic capabilities of microorganisms. The sequence of bases within the genome of the cell determines the sequence of amino acids within protein molecules and thus the functional properties of enzymes that are produced. The expression of the genetic information of the organisms, through the activities of the enzymes that are produced, is reflected in the phenotypic (observable) features that distinguish one microorganism from another.

Transferring the information contained in DNA to form a functional enzyme occurs through protein synthesis, a process accomplished in two stages. The information in the DNA molecule is initially transcribed to form RNA molecules. One type of RNA molecule, messenger RNA (mRNA), carries the information from the DNA molecule to the ribosomes, the anatomical sites of protein synthesis. The information encoded in the mRNA molecule is then translated into the sequence of amino acids that comprise the protein. The specific polypeptide sequences are specified by the sequences of nucleotide bases within the mRNA. In addition to mRNA, the process of translation requires transfer RNA (tRNA) molecules that help align the amino acids during the translational process. Further, the ribosomes themselves are largely comprised of structural (ribosomal) RNA (rRNA) molecules. Thus, in most microorganisms, the nucleic acids of RNA act as informational mediators between the DNA, where the genetic information is stored, and the proteins that functionally express that information.

The central dogma states that once "information" has passed into protein it cannot get out again. The transfer of information from nucleic acid to nucleic acid, or from nucleic acid to protein, may be possible, but transfer from protein to protein, or from protein to nucleic acid, is impossible. Information means here the precise determination of sequence, either of bases in the nucleic acid or of amino acid residues in the protein.

—FRANCIS CRICK

RNA Synthesis—Transcription

Transcription is the process in which the information stored in the DNA molecule is used to code for the synthesis of RNA. RNA (ribonucleic acid) does not contain thymine, but instead has uracil. Also the sugar in RNA is ribose instead of the deoxyribose that occurs in DNA. In transcription the DNA serves as a template for the synthesis of RNA, accomplishing a critical transfer of information for the eventual expression of genetic information (Figure 10.6).

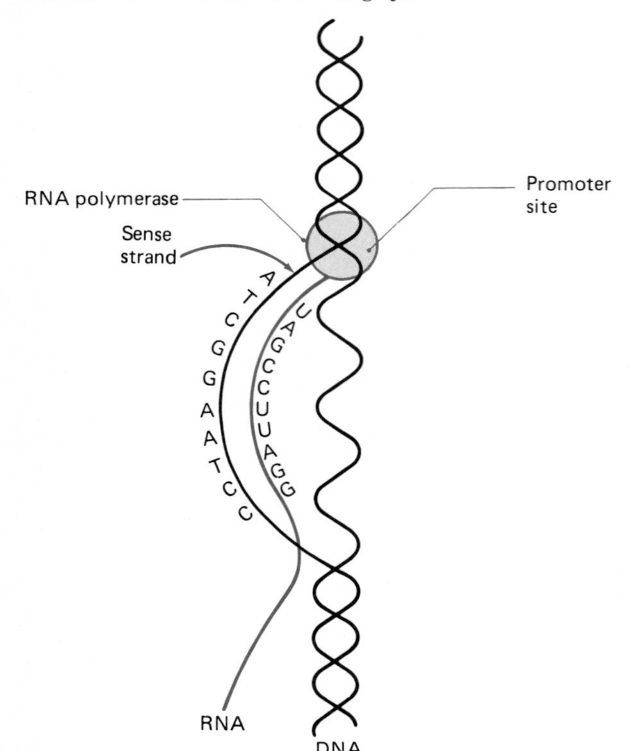

Figure 10.6 During transcription one strand of the DNA codes for the synthesis of RNA. The process of transcription begins at the promoter site where RNA polymerase begins its job of linking the aligned nucleotides of the RNA molecule being synthesized.

During transcription, the RNA nucleotides pair with complementary DNA bases. The base pairs between DNA and RNA are: thymine (DNA) and adenine (RNA); adenine (DNA) and uracil (RNA); guanine (DNA) and cytosine (RNA); and cytosine (DNA) and guanine (RNA). Once the RNA bases are properly aligned in the order specified by the DNA molecule, an enzyme links the bases together. This enzyme is called RNA polymerase. Just as in DNA, the nucleotides are joined by a phosphate group between the 3′ and 5′ positions of the sugar. The molecule of RNA that is synthesized is antiparallel to the strand of DNA that serves as the template. The RNA that is synthesized is single-stranded, and thus, for a given region only one strand of the DNA serves as a template, the strand of DNA coding for the synthesis of RNA is known as the sense strand. Both strands of the DNA can serve as sense strands in different regions, and the term sense strand is applied only to the specific region of the DNA that is being transcribed.

The transfer of the information from DNA to RNA requires that transcription begin at precise locations. There are multiple initiation sites for transcription along the DNA molecule in both prokaryotes and eukaryotes (Figure 10.7). The promoter region of the DNA is the site where RNA polymerase binds to initiate transcription. The presence of the promoter region specifies both the site of transcription initiation and which of the two DNA strands is to serve as the sense strand for transcription in that region.

There is a fundamental difference in how the mRNA is formed in prokaryotic and eukaryotic cells. In prokaryotic cells transcription leads to the direct formation of mRNA. In eukaryotic cells additional processing is necessary after transcription to form functional mRNA. The sequence of nucleotides in the mRNA molecule of prokaryotic cells corresponds exactly with the sequence of nucleotides in the DNA (Figure 10.8).

In contrast to the formation of mRNA in prokaryotic cells, the RNA molecules of eukaryotic microorganisms are generally extensively modified after transcription from the DNA to form mRNA (Figure 10.9). The precursor of mRNA in eukaryotes, known as hnRNA (heterogeneous nuclear RNA) is subjected to substantial posttranscriptional modification within the nucleus to form the mRNA. The processing of the hnRNA involves removing, add-

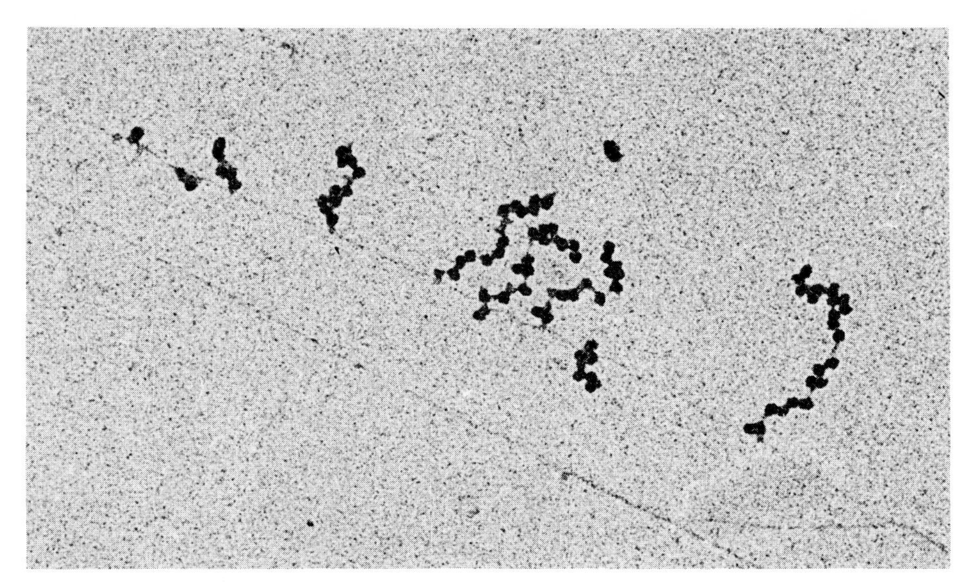

Figure 10.7 This autoradiograph shows transcription occurring at multiple sites along a region of the genome of *E. coli* (165,000 ×). The double-stranded DNA is visible as thin parallel tracks and the RNA as dark chains growing away from the DNA. (Courtesy of O. L. Miller, Jr., Biology Division, Oak Ridge National Laboratory. Reprinted by permission of the American Association for the Advancement of Science, from O. L. Miller, Jr., B. A. Hankalo, and C. A. Thomas, Jr., 1970, *Science* 169:392–395, Fig. 3, ©1970 AAAS.)

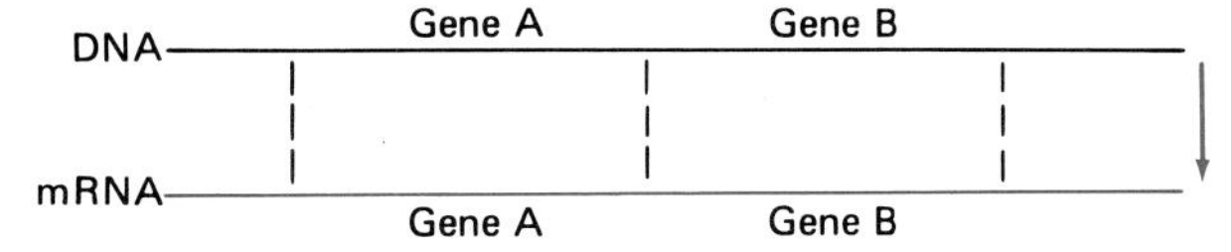

Figure 10.8 In prokaryotes mRNA is colinear with the region of the DNA that is transcribed; that is, there is an exact match in the order of complementary bases between mRNA and DNA molecules.

ing, and rearranging sequences of nucleotides. Intervening sequences of nucleotides, known as introns, are removed in this process. As a result, the mRNA molecules of eukaryotic cells generally are not colinear with the DNA molecule; that is, the sequence of nucleotide bases in the mRNA is not complementary to the specific contiguous linear sequence of bases in the DNA molecule.

If hnRNA molecules much larger than mRNA act as messenger precursors, it is tempting to speculate that at least some repetitive sequences act as elements that control processing . . . One approach to the critical problem of defining the relationship between mRNA and genome, that is how structural genes are expressed, may be to define the structure of the precursor hnRNAs comprising the primary transcripts, so that the unit of transcription can be compared both with the unit of translation and with the organization of the genome.

—Benjamin Lewin

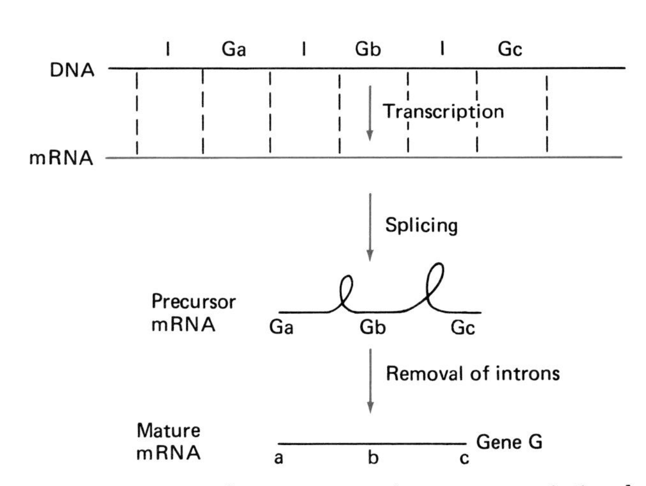

Figure 10.9 In eukaryotes extensive posttranscriptional modification of the primary RNA transcript, including the removal of introns, is needed to produce the mRNA. As a result of these modifications, the eukaryotic mRNA is not colinear with the DNA.

Protein Synthesis—the Translation of the Genetic Code

The translation of the genetic code into protein molecules, which can functionally express the genetic information, occurs at the ribosomes. During translation, the information in mRNA specifies the sequence of amino acids in a protein. The ribosomes provide the spatial framework and structural support for aligning the translational process of protein synthesis. Distortion of the proper configuration of the ribosome can prevent the proper informational exchange and expression of the genetic information, and this forms the basis for the action of many antibiotics.

Within the mRNA three sequential nucleotides are used to code for a given amino acid, and the genetic code, therefore, is termed a triplet code (Figure 10.10). Each of the triplet nucleotide sequences is known as a codon. Because there are four different nucleotides there are 64 possible codons; that is, there are 64 possible three-base combinations of the four different nucleotides. The genetic language, which is almost universal, can therefore be said to have four letters in the alphabet and 64 words in the dictionary, each word containing three letters.

Although there are 64 possible codons, proteins in biological systems normally contain only 20 L-amino acids. (Some proteins do contain other amino acids, but in such cases the unusual amino acids are usually formed by posttranslational modification.) Therefore, there are many more codons than are strictly needed for the translation of genetic information into functional proteins. More than one codon can code for the same amino acid, and therefore, the genetic code is said to be degenerate. Stated another way, the genetic code is redundant with several codons coding for the insertion of the same amino acid into the polypeptide chain. Additionally, there are three codons that do not code for any amino acid. These codons have been referred to as nonsense codons. In actuality, the nonsense codons serve a very important function acting as punctuators that signal the termination of the synthesis of a polypeptide chain.

Figure 10.10 The codons of the genetic code. Each three-base codon calls for a specific amino acid or acts as a stop signal.

First letter	Second letter: U	C	A	G	Third letter
U	UUU Phe UUC Phe UUA Leu UUG Leu	UCU Ser UCC Ser UCA Ser UCG Ser	UAU Tyr UAC Tyr UAA Stop UAG Stop	UGU Cys UGC Cys UGA Stop UGG Trp	U C A G
C	CUU Leu CUC Leu CUA Leu CUG Leu	CCU Pro CCC Pro CCA Pro CCG Pro	CAU His CAC His CAA Gln CAG Gln	CGU Arg CGC Arg CGA Arg CGG Arg	U C A G
A	AUU Ile AUC Ile AUA Ile AUG Met	ACU Thr ACC Thr ACA Thr ACG Thr	AAU Asn AAC Asn AAA Lys AAG Lys	AGU Ser AGC Ser AGA Arg AGG Arg	U C A G
G	GUU Val GUC Val GUA Val GUG Val	GCU Ala GCC Ala GCA Ala GCG Ala	GAU Asp GAC Asp GAA Glu GAG Glu	GGU Gly GGC Gly GGA Gly GGG Gly	U C A G

> As biological studies proceed to a molecular and a genetic level, the parallel between mammalian and microbial cells has become increasingly prominent.
>
> —BERNARD D. DAVIS

The translation of the information in the mRNA molecule, that is, reading the codons, is a directional process (Figure 10.11). The mRNA is read in a 5′-hydroxyl to 3′-hydroxyl direction. The polypeptide that is made is synthesized from the amino terminal to the carboxyl terminal end. The mRNA molecule is read one codon at a time. In other words, the three nucleotides that specify a single amino acid are read together.

There are no spaces between the codons that are read. Therefore, establishing a reading frame is critical for extracting the proper information. Here, we see the importance of having a mechanism for recognizing direction in the informational macro-

Figure 10.11 Protein synthesis (translation) is a directional process. The mRNA is read in the 5′-hydroxyl to 3′-hydroxyl direction, and the peptide chain is synthesized from the amino end to the carboxylic acid end.

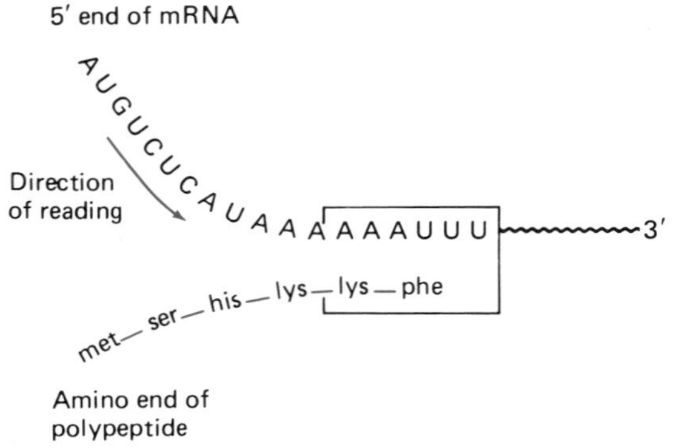

5′ end of mRNA
AUGUCUCAUAAAAAAUUU 3′
Direction of reading
met—ser—his—lys—lys—phe
Amino end of polypeptide

molecules. Just as we establish a convention for reading the English language from left to right, the correct interpretation of the information stored in the mRNA molecule requires that it be read from the 5′-hydroxyl to the 3′-hydroxyl free end.

Before going on we should consider the role of transfer RNA (tRNA) in the translation process (Figure 10.12). The tRNA brings the amino acids to the ribosomes and properly aligns them during translation. Each tRNA molecule contains a different anticodon region, and the anticodon contains three nucleotides that are complementary to the three-based nucleotide sequence of the codon. The pairing of the codons of the mRNA molecules and the anticodons of the tRNA molecules determines the order of amino acid sequence in the polypeptide chain. The third base of the anticodon does not always properly recognize the third base of the mRNA codon. As a result the tRNA molecules may wobble, resulting in the recognition by a tRNA anticodon of more than one codon.

During translation, tRNA molecules bring individual amino acids to be sequentially inserted into the polypeptide chain. When tRNA molecules arrive at the ribosome, the proper anticodon pairs with its matching codon of the mRNA (Figure 10.13). After a peptide bond is established between amino acids that have been properly aligned in this process, the mRNA moves along the ribosome by three nucleotides. The movement of mRNA, tRNA, and the growing polypeptide chain along the ribosome is known as translocation. The process is repeated over and over, resulting in the elongation of the polypeptide chain. Eventually, one of the nonsense codons appears, and because no charged tRNA molecule pairs with the nonsense codon, the translational process is terminated.

Mutations

Any change in the sequence of nucleotide bases in the DNA can be reflected in the expression of the genetic information. Because the genetic code consists of triplet nucleotide sequences that determine the biochemical structure of proteins, changing the order of bases can alter the ability of the cell to produce enzymes that function properly. Even a simple change, such as the deletion or addition of a single nucletotide, can have a major effect; once the reading frame is altered, all of the three-base codons read after that may be different. Also, because part of the microbial genome regulates the synthesis of proteins, mutations may alter the capacity of the cell to control properly its metabolic activity.

Physics tells us that—save at absolute zero, an inaccessible limit—no microscopic entity can fail to undergo quantum perturbations, whose accumulation within a macroscopic system will slowly but surely alter its structure . . . A mutation is in itself a microscopic event, a quantum event, to which the principle of uncertainty consequently applies. An event which is hence and by its very nature essentially unpredictable.

—JAQUES MONOD, *Chance and Necessity*

There are various classes of mutations. One type of mutation, base substitution, occurs when one pair of nucleotide bases in the DNA is replaced by another pair of nucleotides. Some base substitutions, known as missense mutations, result in a change in the amino acid inserted into the polypeptide chain. Although missense mutations can result in the production of an inactive enzyme, changes in a single amino acid within a polypep-

Figure 10.12 The transfer RNA (tRNA) molecule has a characteristic four lobe structure. Each lobe has a distinct function. The lower lobe contains the anticodon region which is complementary to the codon of the messenger RNA that specifies the amino acid to be incorporated at that point in the synthesis of a protein.

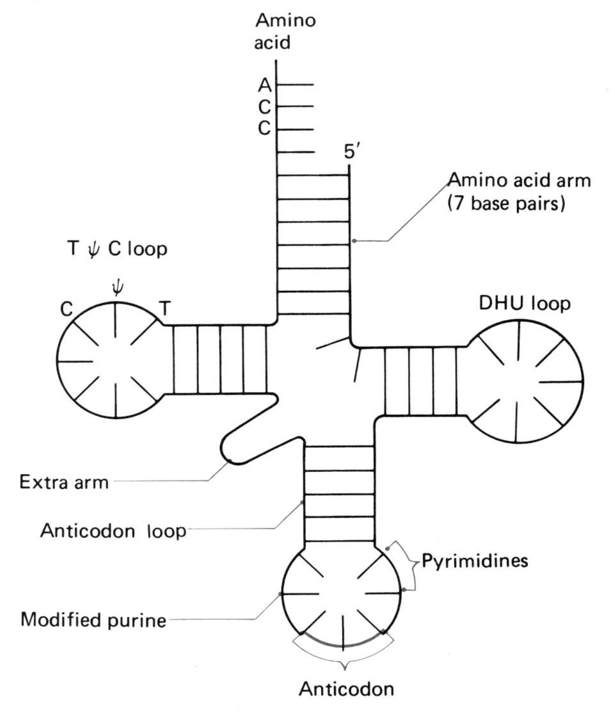

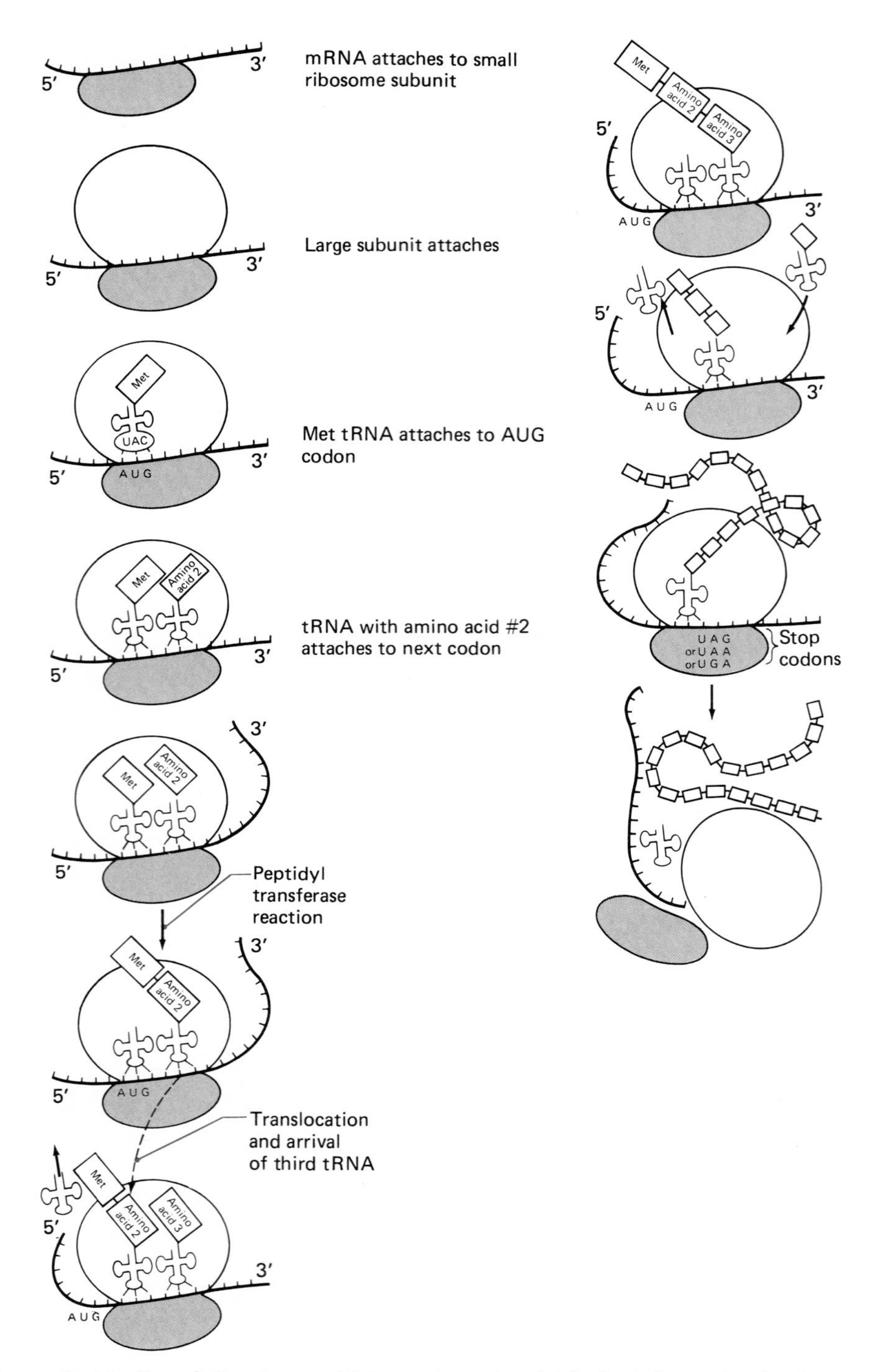

Figure 10.13 Translation is a multistep process in which the information in the mRNA is used to produce a polypeptide. This process occurs at the ribosomes, which provide the structural support for the proper alignment of the macromolecules involved in the translational process. During translation the anticodon of the charged tRNA pairs with the codon of the mRNA. Amino acids are transferred to the growing peptide chain as the mRNA is translocated along the ribosome, moving in three base jumps.

tide often do not drastically reduce the activity of an enzyme and are rarely fatal to the organism.

The substitution of one nucleotide base for another may not even change the amino acid specified by the codon. This is because the genetic code is degenerate. Changes in the nucleotide sequence that alter the third base of codon are most likely to produce such silent mutations because this is where most of the redundancy in the genetic code occurs. This process is expressed by the wobble hypothesis, which states that changes in the third position of the codon often do not alter the amino acid sequence of the polypeptide. Silent mutations do not alter the phenotype of the organism and go undetected.

Changing even a single nucleotide base, however, can radically change the genetic information of a cell. One special type of mutation that often has a major effect on the expression of the genetic information occurs when the alteration in the base sequence of the DNA results in the formation of a nonsense codon. Because the nonsense codons act as terminator signals during protein synthesis, the formation of a nonsense codon often signals the premature termination of a polypeptide chain, preventing the formation of a functional enzyme molecule.

Deletions or additions of nucleotide bases also can have serious consequences. A deletion mutation involves removal of one or more nucleotide base pairs from the DNA, and an insertion mutation involves the addition of one or more base pairs. Adding or deleting a single base pair changes the reading frame of the transcribed mRNA. Such frame shift mutations can result in the misreading of large numbers of codons.

One way of viewing mutations is to consider their effect on the phenotype of the organism. When the mutation results in the death of the microorganism or its inability to reproduce, the mutation is said to be lethal. Such mutations may be conditional, causing the death of the organism only under certain environmental conditions, or may be unconditional, being lethal to the organism regardless of the environmental conditions. A conditionally lethal mutation causes a loss of viability only under some specified conditions where the organism would normally survive. Temperature-sensitive mutations, for example, alter the range of temperatures over which the microorganism may grow.

Nutritional mutations occur when a mutation alters the nutritional requirements for the progeny of a microorganism. Often, nutritional mutants will be unable to synthesize essential biochemicals, such as amino acids. Nutritional mutants (auxotrophs) require growth factors that are not needed by the parent (prototroph) strain. Nutritional mutants are often detected using the replica plating technique (Figure 10.14). In this method, microorganisms can be repeatedly stamped onto media of differing composition. The distribution of microbial colonies should be exactly replicated on each plate, and a colony that develops on a complete medium but not on a minimal medium (lacking a specific growth factor) indicates the occurrence of a nutritional mutation. The microorganisms that do not grow on the minimal medium represent auxotrophic strains.

Various chemicals can modify the nucleotide bases and act as chemical mutagens. The fact that microorganisms are susceptible to chemical mutagens can be used to determine the mutagenicity (the ability to increase the rate of mutation) of various chemicals. In the Ames test procedure, a strain of the bacterium *Salmonella typhimurium* is used as a test organism for determining chemical mutagenicity (Figure 10.15). The bacterial strain used in the Ames test procedure is an auxotroph that requires the amino acid histidine. The organisms are exposed to a gradient of the chemical being tested on a solid growth medium that lacks histidine. Normally, the bacteria cannot grow, and in the absence of a chemical mutagen no colonies can develop. If the chemical is a mutagen, lethal mutations will occur in the areas of high chemical concentration, and no growth will occur in these areas. At lower chemical concentrations along the concentration gradient, however, fewer mutations will occur, and some of the cells may revert, because of mutagenic action, producing histidine prototrophs that will be able to grow and produce visible bacterial colonies on the medium. The appearance of bacterial colonies demonstrates that the chemical has mutagenic properties.

The Ames test is also useful in determining if a chemical is a potential carcinogen (cancer-causing agent). There is a high correlation between a chemical that is a mutagen and whether that chemical also is a carcinogen. Thus, determining whether a chemical has mutagenic activity is useful in screening large numbers of chemicals for potential mutagens, although the Ames test does not positively establish whether a chemical causes cancer. In testing for potential carcinogenicity, the chemical is incubated with a preparation of rat liver enzymes to simulate what normally occurs in the liver, where many chemicals are inadvertently transformed into carcinogens in an apparent effort by the body to detoxify the chemical. Following this activation step,

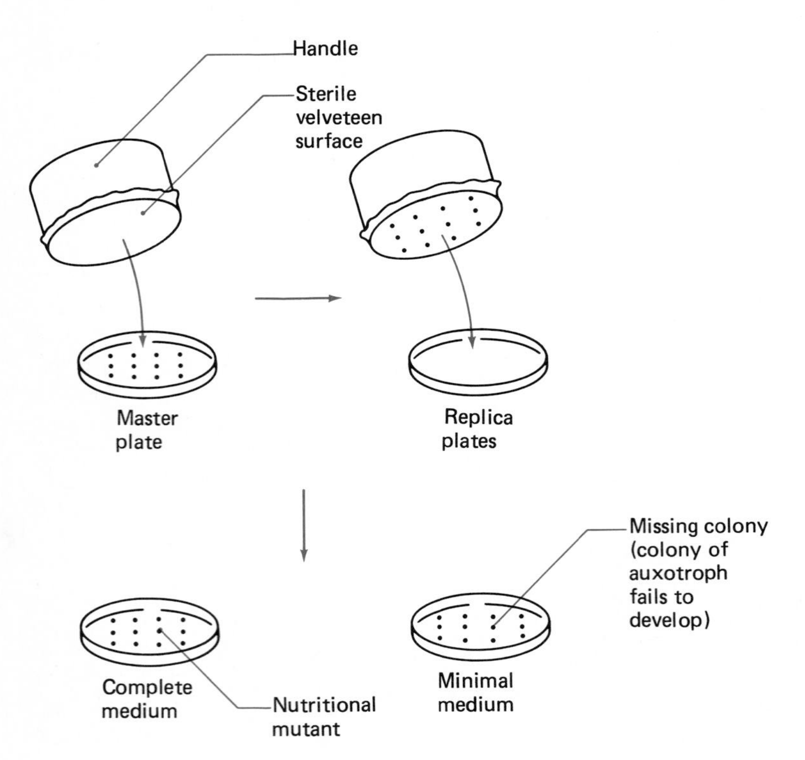

Figure 10.14 Replica plating is used to identify mutants by transferring identical colonies to different types of media and comparing which colonies develop on respective plates. This method is critical in identifying auxotrophic mutants.

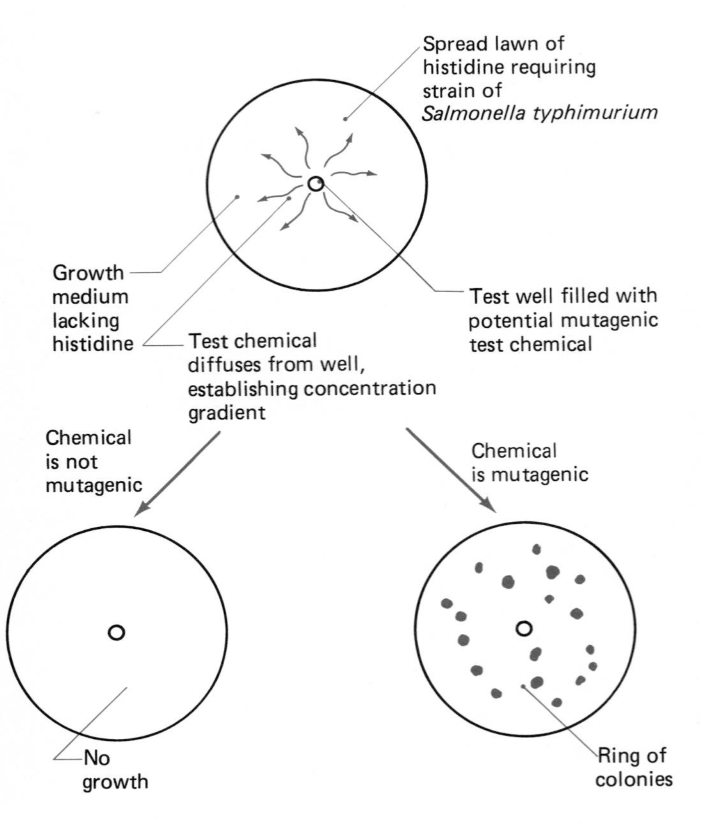

Figure 10.15 The Ames test procedure is used to screen for mutants and potential carcinogens. The theoretical basis of the Ames test is that nearly all proven carcinogens that act directly, that is, by attacking the DNA as opposed to indirect hormonal action, are also mutagens. So rather than screening chemicals directly for carcinogenicity, Bruce Ames and his coworkers thought it would be simpler to test them for mutagenicity: those chemicals not shown to be mutagenic would be assumed to be noncarcinogenic or indirect acting, and those shown to be mutagenic would be subjected to further testing. They wanted a convenient bacterial genetic system that involved a positive selection for the desired event, so they chose a reversion mutation, the reacquisition of a genetically controlled trait; specifically, the reversion of the histidine mutations in *Salmonella typhimurium.* They assembled a collection of 13 mutant strains that revert to the prototype very rarely and that represent different types of mutagenic events, including frame shift, base substitution, and nonsense mutations. Today, the use of this bacterial assay greatly simplifies the task of screening many potentially dangerous chemicals, allowing us to recognize potentially carcinogenic compounds.

various concentrations of the transformed chemical are incubated with the *Salmonella* auxotroph to determine whether it causes mutations and is a potential carcinogen.

The difficulties seem almost insurmountable. But considering the incredible progress made in developing techniques for the manipulation of genes, of chromosomes, of DNA and RNA, for the identification and determination of the structures of tremendously complex molecules, and even the synthesis of genes, hope returns.

—THOMAS CARNEY, *Instant Evolution*

In addition to chemical mutagens, exposure to radiation can increase the rate of mutation. Exposure to high energy radiation such as X rays, can cause mutations because such high-energy ionizing radiation produces breaks in the DNA molecule. Exposure to gamma radiation, such as emitted by cobalt-60, can be used for sterilizing objects, including plastic petri plates, because sufficient exposure to ionizing radiation results in lethal mutations and the death of all exposed microorganisms. The time and intensity of exposure determines the number of lethal mutations that occur, and thus establishes the required exposure when ionizing radiation is employed in sterilization processes.

The bacterium *Micrococcus radiodurans* can withstand atomic radiation of 6,500,000 Roentgens or 10,000 times greater than the radiation that is fatal to the average human.

—1985 GUINNESS BOOK OF WORLD RECORDS

Exposure to ultraviolet light also can result in base substitutions by creating covalent linkages between pyrimidine bases on the same strand of the DNA. A thymine dimer cannot act as a template for DNA polymerase, and the occurrence of such dimers therefore prevents the proper functioning of polymerase enzymes. Exposure to ultraviolet light can cause lethal mutations and is sometimes used to kill microorganisms in sterilization procedures.

Regulation of Gene Expression

In addition to encoding the information for the specific polypeptide sequences of proteins, the genome of the cell codes the information that regulates its own expression; some sequences of the DNA are involved in regulatory functions rather than coding for specific polypeptide sequences. By controlling which of the genes of the organism are to be translated into functional enzymes, the cell is able to regulate its metabolic activities. Some regions of the DNA are specifically involved in regulating transcription, and these regulatory genes can control the synthesis of specific enzymes.

In some cases gene expression is not subject to specific genetic regulatory control, and the enzymes coded for by such regions of the DNA are continuously (constitutively) synthesized at a constant rate by the organism. In contrast to **constitutive enzymes**, some enzymes are sythesized only when the cell requires the enzymes, and those are either **inducible enzymes** or **repressible enzymes**. Often several enzymes that have related functions are controlled by the same regulatory genes. The control genes and those genes under their control contitute the **operon**.

One of the best studied regulatory systems concerns the enzymes produced by *E. coli* for the metabolism of lactose (Figure 10.16). In this integrated system, a regulatory gene codes for a **repressor protein**, which in the absence of lactose binds to the operator region of the DNA. The binding of the repressor protein at the **operator region** blocks the transcription of the structural genes under the control of that operator region. In the case of the **lac operon**, this means that in the absence of lactose, the three structural lac genes, which code for the three enzymes involved in lactose metabolism, are not transcribed. The operator region is adjacent to or overlaps the promoter region (site where RNA polymerase binds to initiate transcription). The binding of the repressor protein at the operator region interferes with the binding of RNA polymerase at the promoter region. An **inducer** substance, which is a derivative of lactose, binds to the repressor protein and alters the properties of the repressor protein so that it is unable to interact with and bind at the operator region. Thus, in the presence of an inducer that binds with the repressor protein, transcription of the lac operon is derepressed. When this occurs the synthesis of the three structural proteins needed for lactose metabolism proceeds. As the lactose is metabolized and its concentration diminishes, the concentration of the inducer also declines. Therefore, active repressor protein molecules are again available for binding at the operator region, and the transcription of the lac operon is repressed.

In addition to the specific operator-mediated regulation of transcription, there is a generalized type of repression known as **catabolite repression**. In the presence of an adequate concentration of

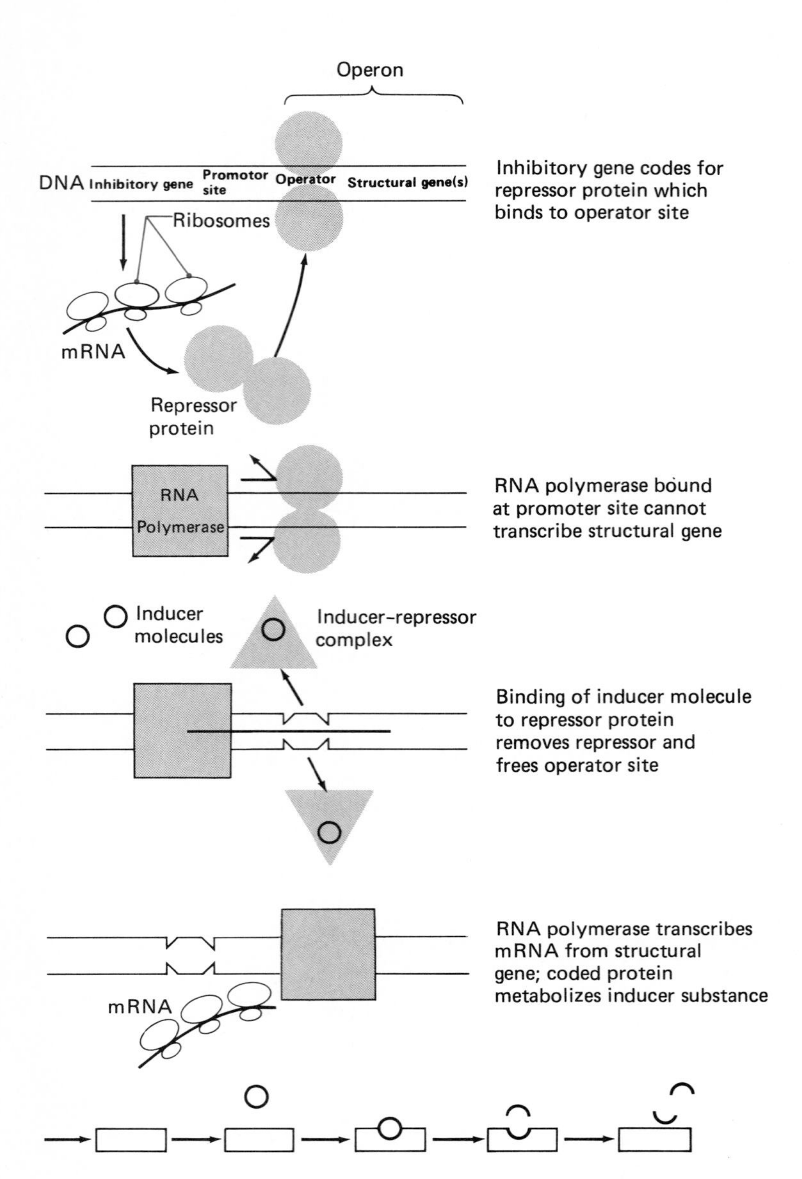

Figure 10.16 The operation of the lac operon permits the turning on of the synthesis of the enzymes needed for lactose utilization when lactose is present. Three enzymes are synthesized by *E. coli* for the metabolism of lactose. The control of the expression of these structural genes for lactose metabolism is explained in part by the operon model. The regulation of these enzymes is controlled by the promoter region (p), a regulatory gene (i) that codes for the synthesis of a repressor protein, and an operator region (o) that occurs between the promoter and the structural genes involved in lactose metabolism.

glucose, a number of catabolic pathways are repressed, including those involved in the metabolism of lactose, arabinose, and galactose. When glucose is available for catabolism in the glycolytic pathway, disaccharides and polysaccharides need not be hydrolyzed to supply monosaccharides for the metabolic activities of the cell. By blocking the metabolism of these more complex carbohydrates, the cell is able to conserve its metabolic resources.

Catabolite repression acts via the promoter region of DNA, and by doing so it complements the control exerted by the operator region. The efficient binding of RNA polymerase to promoter regions subject to catabolite repression requires the presence of a **catabolite activator protein** (Figure 10.17). In the absence of the catabolite activator protein, the RNA polymerase enzyme has a greatly decreased affinity to bind to the promoter region. The catabolite activator protein in turn cannot bind to the promoter region unless it is bound to cyclic AMP. In the presence of glucose, cyclic AMP levels are greatly reduced, and thus, the catabolite activator protein is unable to bind at the promoter region. Consequently, the RNA polymerase enzymes are unable to bind to catabolite repressible promoters, and transcription at a number of coordinated regulated structural genes ceases. In the absence of glucose, cyclic AMP is synthesized from

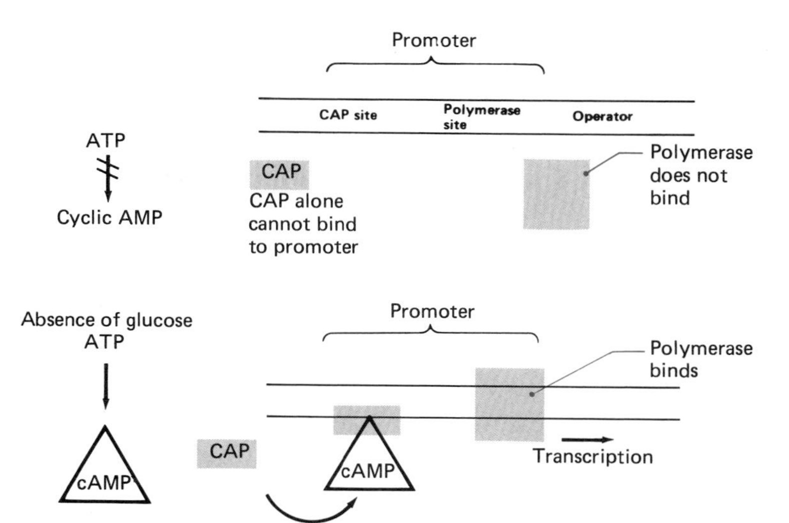

Figure 10.17 Catabolite repression explains why in the presence of glucose several catabolic pathways are shut off. As shown in this illustration, cyclic AMP plays a critical role in permitting RNA polymerase to bind to some promoters; thus the concentration of cyclic AMP can regulate metabolism by controlling the synthesis of enzymes in operon regions adjacent to such promoters.

ATP, and there is an adequate supply of cyclic AMP to permit the binding of RNA polymerase to the promoter region. Thus, when glucose levels are low, cyclic AMP stimulates the initiation of many inducible enzymes.

Summary

In prokaryotic and eukaryotic microorganisms DNA is the molecule that stores the genetic information. The DNA molecule is double helical. It contains four nucleotide bases, adenine, thymine, guanine, and cytosine, which are attached to the sugar deoxyribose. The deoxyribose sugars are linked by phosphate diester bonds between the carbon 3 and carbon 5. The individual chains of the DNA molecule are directional, one running from 3′-hydroxyl to 5′-hydroxyl and the other running from 5′-hydroxyl to 3′-hydroxyl.

A single copy of the genetic information constitutes the genome of the cell. The genome contains segments called genes. Some genes code for the synthesis of RNA and proteins; other genes have regulatory functions.

The expression of genetic information involves transferring the information encoded in the genome to functional enzymes. This is accomplished by having the genome direct the process of protein synthesis, a two-stage process in which information encoded in the DNA is first transcribed to RNA molecules and then translated into the amino acid sequence of the polypeptide. Three types of RNA molecules are involved in protein synthesis, ribosomal RNA (rRNA), transfer RNA (tRNA), and messenger (mRNA). Transcription to form RNA uses the DNA as a template, but only one strand of the DNA, the sense strand, serves as the template for the transcription of any given RNA molecule. Transcription requires that RNA polymerase bind to the promoter region of the DNA to initiate RNA synthesis; the promoter region contains a sequence of nucleotides recognized by the RNA polymerase enzyme as the site of transcriptional initiation.

A critical function of transcription is the formation of mRNA because the information for the sequencing of the amino acids in a polypeptide is encoded within mRNA molecules. In prokaryotic microorganisms, mRNA is colinear with the DNA molecule. In contrast, eukaryotes have split genes, and an hnRNA molecule is initially synthesized and then extensively processed after transcription to produce mRNA molecules. Part of the processing of a hnRNA molecule involves the removal of intron regions, which occur between the nucleotides that actually code for the amino acids of a polypeptide.

The translation of the mRNA molecule occurs at the ribosomes. During translation the genetic code, consisting of 64 triplet nucleotide sequences known as codons, specifies the amino acids that are to be inserted into a polypeptide chain. Three of the codons do not specify any of the 20 amino acids found in proteins; instead these nonsense codons act as termination signals for the translational process. Because more than one codon can specify the same amino acid, the genetic code is said to be degenerate. The genetic code appears to be almost universal in both prokaryotic and eukaryotic microorganisms, although some exceptions have been found in mitochondria where different nucleotide sequences are used to code for particular amino acids. The direction of reading of the messenger RNA molecule is from the 5′-hydroxyl to the 3′-hydroxyl free end, and the direction of synthesis of the polypeptide chain is from the amino to the carboxyl end. In the process of translation, the ribosome supplies the structural framework for the proper alignment required for protein synthesis, the mRNA specifies the order of the amino acid sequence for the polypeptide chain, and the tRNA molecule brings the amino acid to the ribosome and aligns it in its proper sequence.

Various types of mutations can occur that modify the DNA molecule. These mutations produce multiple allelic forms of the same gene. Mutations can alter the phenotype (appearance) as well as the genotype (genetic composition) of the organism. Even a simple additional deletion of the base can have a major effect because of the alteration of the reading frame. Some mutations are lethal; other mutations alter nutritional requirements, producing an auxotroph. Some chemicals, called mutagens, increase the rate of mutation. Mutagens and potential carcinogens can be detected by the Ames test. Ionizing and ultraviolet radiation also increase the rate of mutation.

Both prokaryotic and eukaryotic microorganisms can regulate the expression of genetic information in ways that permit them to carry out specialized activities during specific periods of time. The expression of genetic information can be controlled at several levels. In prokaryotic microorganisms some genes are clustered together, and their transcription is under the control of a regulatory operator region. The operator acts as a switch that can regulate transcription and thus can turn on and off protein synthesis, permitting the coordinated expression of genetic information. The functioning of the operon can be likened to a building with multiple light fixtures and electrical switches. When a light switch is turned on, the light bulbs function and emit light in a specified area of the building. In some rooms the light switch is turned on only when someone enters the room, a case that is analogous to the lac operon where the system is induced (derepressed) only in the presence of an inducer. In other areas of the building, such as the entrances, the lights are normally left on but may be shut off when they are not needed. Thus, the genome of the cell directs the synthesis of proteins, determining the metabolic capabilities and potential phenotype of the organism. It contains regulatory genes for controlling genetic expression that allow microorganisms to finely regulate their metabolic activities.

Study Outline

A. Storage of genetic information.

  1. The genetic information of a cell is stored within its DNA macromolecules.

  2. Structure of DNA.

    a. The DNA is composed of nucleotides that are linked together.

    b. A nucleotide consists of a nucleic acid base, a deoxyribose sugar, and a phosphate group.

    c. Four nucleic acid bases occur in DNA: cytosine, guanine, adenine, and thymine.

    d. The deoxyribose sugars are linked by phosphate diester bonds. The linkage establishes a directional orientation to the DNA. Nucleotides are linked by 3′-5′ phosphate diester linkages. At the ends of the DNA strand, there are no linkages and free hydroxyl groups are present. One end has a free hydroxyl group at the carbon 3 of the monosaccharide (3′-hydroxyl free end); the other end of the strand has free hydroxyl group at the carbon 5 of the monosaccharide (5′-hydroxyl free end).

    e. DNA is a double-helix molecule composed of two primary polynucleotide chains. The chains are held together by hydrogen bonding between complimentary nucleotide bases.

    f. The complementary base pairs are adenine and thymine, which are held together by two hydrogen bonds, and guanine and cytosine, which are held together by three hydrogen bonds.

    g. The complimentary strands are antiparallel; where one strand has the 3′-hydroxyl free end, the complimentary strand will have the 5′-hydroxyl free end.

3. A genome is a single copy of the genetic information of the cell.
4. The coded information within the genome is used to control the synthesis of proteins.
5. The sequence of bases determines the sequence of amino acids in proteins.
6. A gene is a segment of a genome.
7. Alleles are different forms of the same gene.
8. Prokaryotic cells have a single genome and therefore are haploid.
9. Eukaryotic cells generally have pairs of matching chromosomes, making them diploid.
10. Plasmids are small extrachromosomal genetic elements that permit microorganisms to store additional genetic information.
11. Plasmids contain genetic information for specialized features rather than for essential metabolic activities.
    a. F (fertility) plasmid codes for mating behavior in *E. coli*.
    b. R (resistance) plasmids carry genes that code for antibiotic resistance.

B. Expression of genetic information.
1. The genetic information that is contained within the genome represents the genotype; not all of the genotype is expressed at all times.
2. The genetic information that is expressed appears as the phenotype, the discernible characteristics of an organism.
3. Protein synthesis occurs in two stages: transcription and translation.
4. Transcription.
    a. In transcription the information in the DNA is transferred to RNA.
    b. During transcription DNA serves as a template that determines the order of the bases in the RNA.
    c. The RNA that is formed by transcription is complementary to the DNA.
    d. In prokaryotes transcription produces mRNA that is not extensively modified after synthesis; the mRNA is colinear with the bases of the DNA.
    e. In eukaryotes transcription produces hnRNA which is extensively modified to form the mature mRNA; the mRNA is not colinear with the DNA because of excision of introns and other modifications during posttranscriptional processing.
5. Translation.
    a. In translation the mRNA is used to establish a sequence of amino acids that make up the protein.
    b. Translation occurs at the ribosomes.
    c. The genetic code has 64 possible codons; each codon is a triplet containing three nucleotides.
    d. There is more than one codon for each amino acid; therefore the genetic code is said to be degenerate because different codons can specify the same amino acid.
    e. Nonsense codons are those for which there are no amino acids; the nonsense codons signal termination of synthesis of polypeptide chains.
    f. The ribosome moves along the mRNA exposing one codon at a time.
    g. As each triplet is exposed by the ribosome, a tRNA brings the specified amino acid to the ribosome; the tRNA has an anticodon region that is complementary to the codon and is responsible for bringing the correct amino acid specifed by the codon.
    h. The ribosome moves to the next triplet and the process is repeated.
    i. The amino acid of the second tRNA bonds to the amino acid carried by the first tRNA.
    j. As soon as the bonding between amino acids takes place, the bonds between the first tRNA and its amino acid and between it and the mRNA are broken and it leaves the ribosome.

C. Mutations.

1. A mutation is a change in the sequence of bases in the DNA.

2. Types of mutations.

a. A lethal mutation results in the death of a microorganism or in its inability to reproduce; a conditionally lethal mutation exerts an effect only under certain environmental conditions; an unconditionally lethal mutation is lethal regardless of environmental conditions.

b. Temperature-sensitive mutations alter the range of temperatures over which the microorganisms may grow.

c. Nutritional mutations alter the nutritional requirements for the progeny; nutritional mutants (auxotrophs) require growth factors not needed by the parental (prototrophic) strain.

3. Factors influencing rates of mutation.

a. Some chemicals, called mutagens, increase the rate of mutation.

b. Ames test procedure utilizes *Salmonella typhimurium* as test organism for determining chemical mutagenicity and potential carcinogenicity.

c. High-energy ionizing radiation causes mutation and can be used for sterilizing objects.

d. Ultraviolet light can cause mutations.

D. Regulation of genetic expression.

1. The expression of genetic information can be regulated at the level of transcription.

2. Constitutive enzymes are continuously synthesized at a constant rate and are not regulated.

3. Inducible enzymes are made only at appropriate times, for example, when synthesis is induced by appropriate factors.

4. Operon model of gene control explains the basis of control of transcription.

5. An operon consists of structural genes that contain the code for making proteins; an operator region, which is the site where repressor protein binds and prevents RNA transcription; a promoter region, which is the site where RNA polymerase binds; and a regulatory gene, which codes for the repressor protein.

6. The lac operon regulates the utilization of lactose.

a. The regulator gene codes for repressor protein.

b. The repressor protein binds to operator region and prevents structural genes from transcribing the mRNA needed to synthesize the enzymes for lactose catabolism.

c. In the presence of lactose, an inducer binds to the repressor protein preventing it from binding to the operator region of the operon; this results in derepression of lac operon and structural genes needed for the utilization of lactose are transcribed until the lactose has been broken down.

d. When lactose is used up, no inducer is present and the repressor protein is free to bind to the operator region; when this happens the lac operon is repressed and the structural genes are no longer transcribed.

7. Catabolite repression is a generalized type of repression.

a. Catabolite repression supersedes the control exerted by the operator region.

b. Catabolite repression acts via a promoter region of DNA by blocking the normal attachment of RNA polymerase; a catabolite activator protein is needed to bind RNA polymerase to the promoter region and cyclic AMP is required for efficient binding to occur.

c. In the presence of glucose, the amount of cyclic AMP is reduced; therefore the catabolite activator protein cannot bind to the promoter and transcription is unable to occur.

Study Questions

1. Match the following phrases with the terms in the list to the right.
   - **a.** Single copy of the genetic information of the cell.
   - **b.** A sequence of nucleotides.
   - **c.** The linear sequence of triplets on the mRNA.
   - **d.** Encoding mRNA with information from DNA.
   - **e.** Genetic information that has the potential of being expressed.
   - **f.** Expression of genetic information.
   - **g.** Composed of structural genes.
   - **h.** Translating the information on mRNA into a sequence of amino acids.
   - **i.** A segment of a genome.

   (1) Gene
   (2) Genome
   (3) Cistron
   (4) Phenotype
   (5) Genotype
   (6) Transcription
   (7) Translation
   (8) Codon

2. Fill in the blanks.
   - **a.** The synthesis of RNA using DNA as a ________ is called ________ . Three classes of RNA are synthesized: ________ , ________ , and ________ .
   - **b.** Synthesis of RNA begins opposite the ________ strand of DNA. The ________ enzyme links RNA together.
   - **c.** The formation of the protein from the codon on mRNA is called ________ . The genetic code is composed of ________ nucleotides, ________ nucleotides per codon. This gives ________ possible codons. There are ________ amino acids, and thus, more than one ________ for each amino acid. Some codons do not code for an amino acid. These are called ________ codons and their function is to ________ synthesis of the polypeptide chain.
   - **d.** ________ brings the amino acids to the ribosomes. One lobe of the RNA contains the ________ , which will match the ________ on the mRNA. This matching is called the ________ .

3. Match the following phrases with the terms in the list to the right.
   - **a.** Results in the death of the microorganism or inability to reproduce.
   - **b.** Results in a change in the amino acid inserted into the polypeptide chain.
   - **c.** Is not always changed as a result of mutation.
   - **d.** Provide additional site for storage of genetic information.
   - **e.** Is always changed as a result of mutation.

   (1) Plasmids
   (2) Variability
   (3) Phenotype
   (4) Genotype
   (5) Lethal mutation

4. Identify the two factors that increase the relatively low natural mutation rates.
5. Name the test used to determine chemical mutagenicity and the organism used in this test.
6. Indicate whether the following statements are true (T) or false (F).
   - **a.** In the absence of lactose, the regulator gene codes for the repressor protein.
   - **b.** An inducer binds to the regulator gene.
   - **c.** An inducible enzyme is made all the time.
   - **d.** When an inducer is present the lac operon is derepressed and the structural genes begin to transcribe mRNA.
   - **e.** Structural genes will continue to transcribe mRNA until no more inducer is formed.
   - **f.** The lac operon is derepressed when no more inhibitor is present.
   - **g.** Catabolite repression affects the promoter region of the DNA.
   - **h.** In catabolite repression, RNA polymerase cannot bind to the promoter because of a lack of cyclic AMP.

Some Thought Questions

1. What structural characteristics of nucleic acids make them suitable to serve for the storage of genetic information?
2. Why do cells have more genetic information than they need?
3. Can microorganisms be used as computer memory chips?
4. Why are there no four-letter words in the language of genes?
5. Why is feedback inhibition important, and how does it work?
6. Why are Gram negative bacteria called promiscuous?
7. How can recombination account for evolutionary spurts?
8. What is the difference between reproduction and recombination?
9. How can recombination explain regular outbreaks of influenza?

Additional Sources of Information

Ayala, F. J., and J. A. Kiger, Jr. 1984. *Modern Genetics*. Benjamin/Cummings Publishing Co., Menlo Park, California. A well written text on genetics including sections on the molecular basis of genetics.

Chambon, P. 1981. Split genes. *Scientific American*, 244(5):60–71. An excellent article describing how genetic information is stored in eukaryotic cells.

Freifelder, D. 1978. *The DNA Molecule: Structure and Properties*. W. H. Freeman and Company, Publishers, San Francisco. A thorough description of the DNA molecule and how genetic information is stored in this molecule.

Genetics: Readings from Scientific American. 1981. W. H. Freeman and Company, Publishers, San Francisco. A series of articles that includes discussion of gene structure and function.

Mays, L. L. 1981. *Genetics: A Molecular Approach*. Macmillan Publishing Co., Inc., New York. An introductory text that presents the basis for storage and expression of genetic information in an easily understood manner.

Strickberger, M. W. 1985. *Genetics*, 3rd ed. Macmillan Publishing Co., Inc., New York. An excellent text on genetics including coverage of microorganisms.

Suzuki, D. T., A. J. F. Griffiths, and R. C. Lewontin. 1981. *An Introduction to Genetic Analysis*. W. H. Freeman and Company, Publishers, San Francisco. An excellent text on genetics including coverage of microorganisms.

Watson, J. D. 1976. *Molecular Biology of the Gene*. Benjamin/Cummings Publishing Co., Menlo Park, California. A superb work describing the molecular basis of genetics.

11 *Transmission of Genetic Information*

KEY TERMS

clone
complementary strands
conjugation
continuous strand
crossing over
discontinuous strand
donor strain
endonuclease
F+ strain
F− strain
fertility pilus
genetic engineering
Hfr strain
high frequency recombinant
lagging strand
leading strand
ligase
mating
meiosis
pallindromic sequence
plasmids
recipient strain
recombinant DNA technology
recombination
replication
replication fork
reverse transcriptase
restriction enzyme
semiconservative replication
sexual reproduction
syngamy
transduction
transformation

OBJECTIVES

After reading this chapter you should be able to

1. Define the key terms.
2. Describe the replication of DNA.
3. Describe the differences between prokaryotic and eukaryotic DNA replication.
4. Describe the mechanisms of genetic exchange in bacteria.
5. Describe recombination and discuss how it forms the basis for genetic engineering.
6. Describe the steps involved in creating a new organism by genetic engineering.
7. Discuss the importance of plasmids for genetic engineering.
8. Describe the use of the Ti plasmid for transferring plant genes.

DNA Replication

Having discussed how genetic information is stored and how it controls the properties of an organism, we now turn to the question of how genetic information is passed from one generation to the next. The transmission of hereditary information necessitates the faithful replication of the DNA macromolecules that encode the genetic information. The process of DNA replication must ensure that the sequence of nucleotides in the genome is accurately reproduced. This is accomplished by using the original DNA as a template to determine the order of nucleotides in the newly synthesized DNA. The replication of the genome is called semiconservative replication, because when a DNA molecule is replicated to form two double helical DNA molecules, each of the new daughter DNA molecules consists of one intact (conserved) strand from the parental double helical DNA and one newly synthesized complementary strand (Figure 11.1).

In the genetic programme, therefore, is written the result of all past reproductions, the collection of successes, since all traces of failures have disappeared. The genetic message, the programme of the present-day organism, therefore, resembles a text without an author, that a proof-reader has been correcting for more than two billion years, continually improving, refining and completing it, gradually eliminating all imperfections.

—FRANCOIS JACOB

During replication there is a localized separation of the DNA double helix (Figure 11.2). The localized unwinding of the DNA double helix establishes a replication fork, which is the site of DNA synthesis. At the replication fork, free nucleotide bases are aligned opposite their base pairs in the parental DNA molecules. The complementary base pairs are adenine opposite thymine and guanine opposite cytosine. The parental DNA molecule acts as a template that directs the order of bases in the synthe-

Figure 11.1 Bands of ultracentrifuged labelled DNA. The experiments of Meselson and Stahl relied on the ability to specifically label DNA in one generation by the incorporation of heavy nitrogen (^{15}N) into the DNA and to follow the fate of this tagged DNA from one location to the next using density gradient ultracentrifugation. The location of the bands obtained by the ultracentrifugation, that is, the distance that the DNA moves, is a function of the molecular weight of the DNA. The difference in molecular weight permitted the tracking of the fate of the heavy DNA when the cells were then grown in the presence of normal-light nitrogen (^{14}N). Of several hypotheses, the only logical explanation for the banding pattern obtained in these experiments was that DNA replication occurs by a semiconservative method. The design of the experiments was eloquently simple and the results so clear-cut that an unambiguous answer to the fundamental question of how DNA is replicated was clearly achieved.

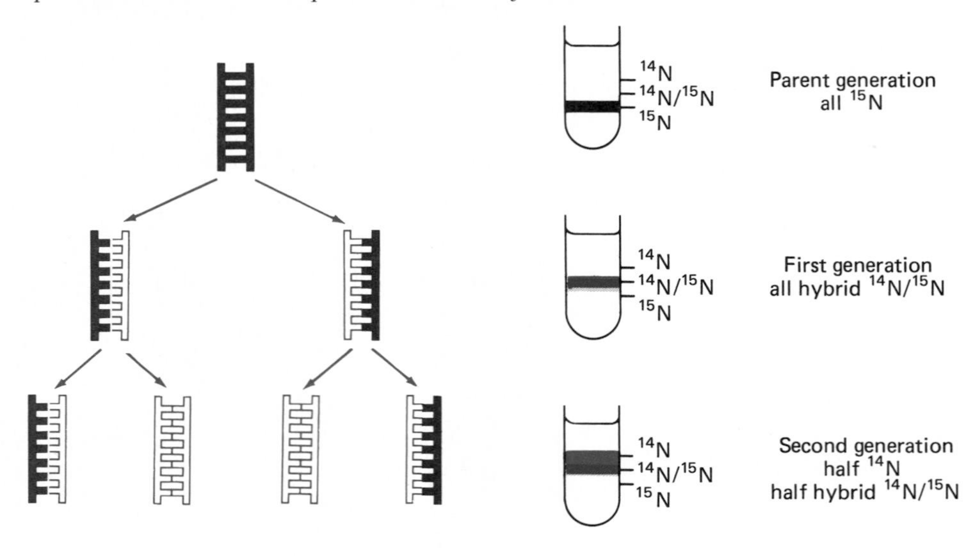

sized DNA. The newly synthesized strands of DNA are formed by linking the nucleotides together with phosphodiester bonds through the action of DNA polymerases. The action of a DNA polymerase enzyme can be likened to a zipper where the teeth of the zipper are initially aligned and progressively linked together in a continuous motion. The DNA polymerase can add nucleotides only to a 3′-hydroxyl free end, and thus the direction of DNA synthesis is 5′-hydroxyl to 3′-hydroxyl.

The fact that DNA polymerase enzymes can add deoxynucleotides only to the 3′-hydroxyl free end of a nucleic acid primer creates a paradox for our understanding of the replication of DNA. The two strands of the double helical DNA molecule are antiparallel. One strand runs from the 5′-hydroxyl to the 3′-hydroxyl free end, and the other complementary strand runs from the 3′-hydroxyl to the 5′-hydroxyl free end. Therefore, synthesis of complementary strands requires that DNA synthesis proceed in opposite directions. However, the double DNA helix is progressively unwound and replicated in one direction.

One of the DNA strands can be continuously synthesized because it runs in the appropriate direction for the continuous addition of new free nucleotide bases to the free 3′-hydroxyl end of the primer molecule (see Figure 11.2). Synthesis of this **continuous** or **leading strand** of DNA occurs simultaneously with the unwinding of the double helical molecule and progresses towards the replication fork. The other strand of the DNA, however, must be synthesized discontinuously. The initiation of the synthesis of the **discontinuous strand** of DNA begins only after some unwinding of the double helix has occurred and therefore lags behind the synthesis of the continuous strand. It is therefore, referred to as the **lagging strand**. Short segments of DNA can be formed by the DNA polymerase enzyme running opposite to the direction of unwinding of the parental DNA molecule.

The fragments are joined together by the action of **ligases**, which are enzymes that establish phosphodiester bonds between the 3′-hydroxyl and 5′-hydroxyl ends of chains of nucleotides. The ligase enzymes are not involved in chain elongation, but rather they act as repair enzymes for sealing "nicks" within the DNA molecule. Thus, through the combined actions of DNA polymerase and DNA ligase enzymes, both complementary strands of the DNA can be synthesized during DNA replication.

Although DNA is the universal macromolecule for storing genetic information and DNA replication

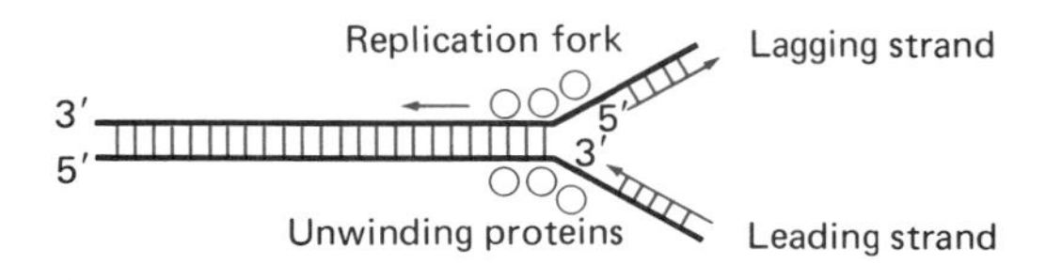

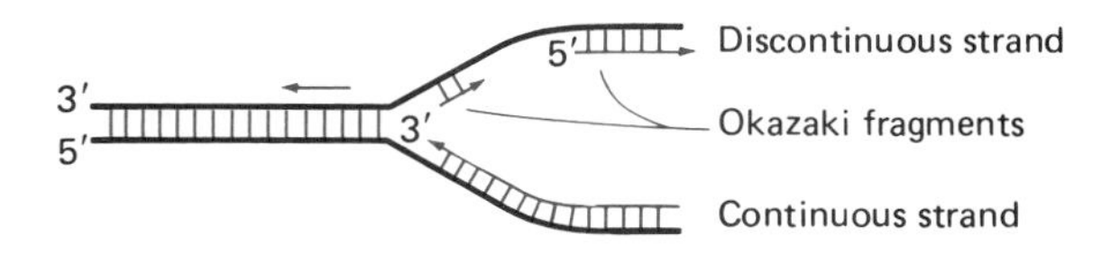

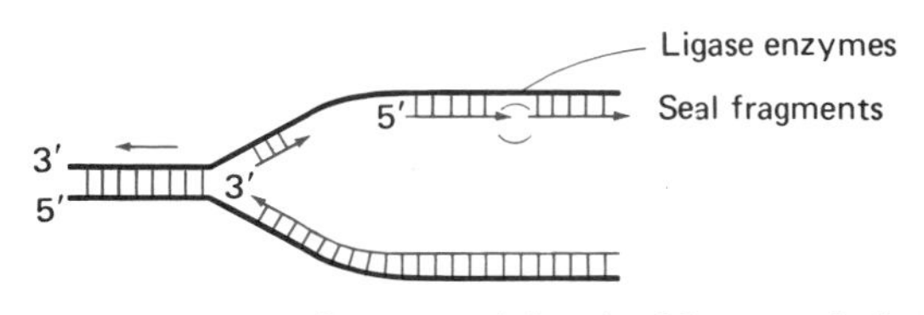

Figure 11.2 Replication of the double-stranded DNA molecule requires localized unwinding for the establishment of a replication fork. The unidirectional nature of the DNA polymerase and the antiparallel nature of the double-stranded DNA molecule means that one strand of DNA is synthesized continuously and the other DNA strand discontinuously.

is semiconservative in both eukaryotic and prokaryotic cells, there are differences in how DNA is replicated in prokaryotic and eukaryotic organisms. One major difference between prokaryotic and eukaryotic DNA replication concerns the number of sites at which DNA replication can begin. Bacteria have a single point of origin of DNA replication. DNA replication moves bidirectionally from the origin so that there are two replicating forks moving in opposite directions. The bidirectional movements of the replication forks produce a loop of DNA that extends out of the plane of the parental bacterial chromosome (Figure 11.3).

In contrast, the replication of eukaryotic DNA begins at multiple points of origin (Figure 11.4). Consequently, even though there is much more DNA in a eukaryotic chromosome than in a bacterial chromosome, the eukaryotic genome can be replicated much faster than the bacterial genome because of the multiple initiation points for DNA syn-

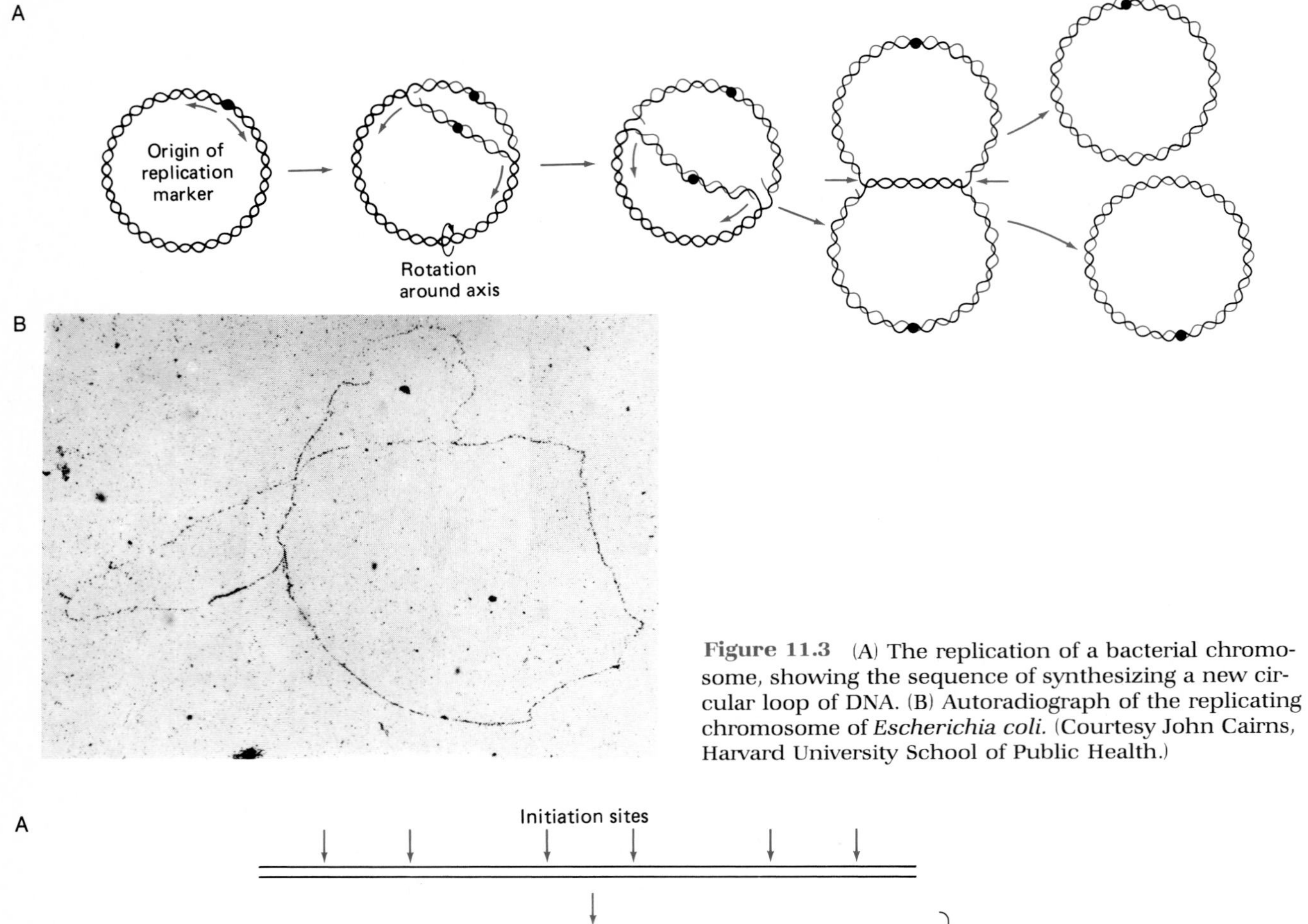

Figure 11.3 (A) The replication of a bacterial chromosome, showing the sequence of synthesizing a new circular loop of DNA. (B) Autoradiograph of the replicating chromosome of *Escherichia coli.* (Courtesy John Cairns, Harvard University School of Public Health.)

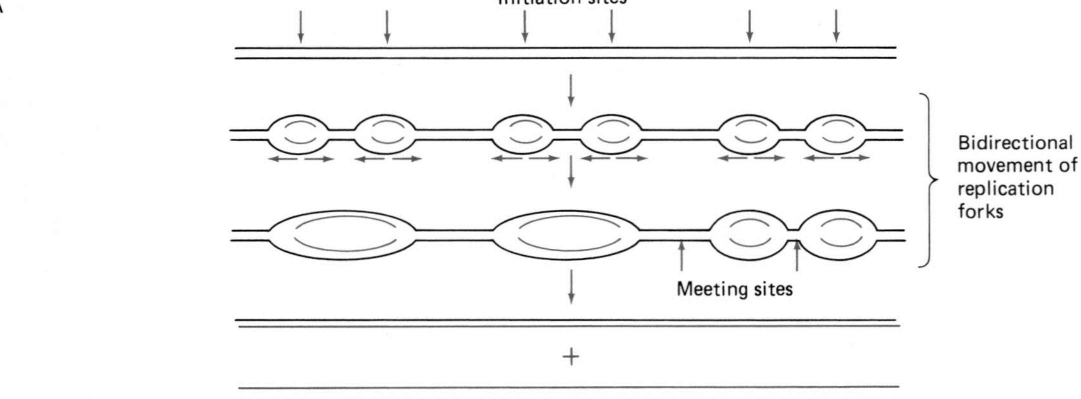

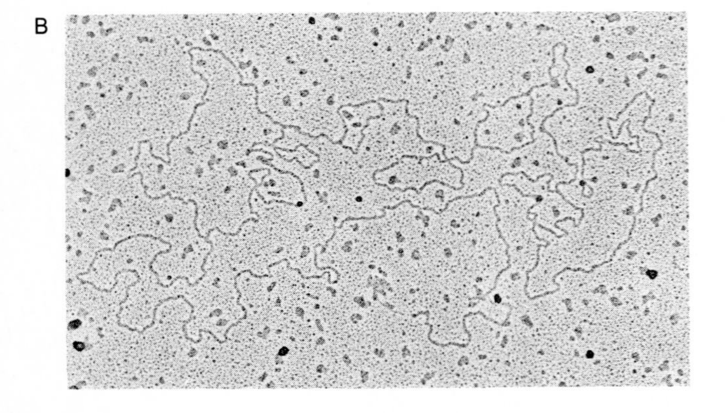

Figure 11.4 (A) As shown in this illustration, DNA replication begins at multiple points in eukaryotes. (B) The multiple replication forks of a eukaryotic chromosome are visible in this electron micrograph of a replicating DNA molecule. (Courtesy D. R. Wolstenholme, University of Utah. Reprinted by permission of Springer-Verlag, Heidelburg, from D. R. Wolstenholme, 1973, *Chromosoma,* 43:1–18.)

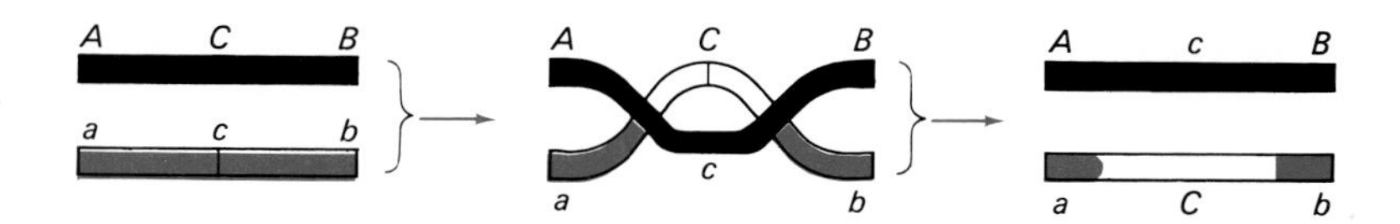

Figure 11.5 Chromosomal crossing over results in the recombination of genes.

thesis. This distinction between single and multiple origins of DNA synthesis appears to represent a fundamental difference between prokaryotic and eukaryotic cells.

Recombination

Although the process of DNA replication assures with great precision the production of a faithful copy of the parental genome, the transmission of genetic information from one generation to the next also permits change. Recombination is a process that results in a reshuffling of genes. This process provides a mechanism for redistributing the information within the genome. It can produce numerous new combinations of genetic information. In some cases recombination involves the movement of segments of the genome to new locations within the genome. These changes in the locations of genes can alter which regulatory genes control particular segments of information and thereby the expression of genetic information. In other cases recombination involves the exchange and combining of genetic information from two different sources. Such recombinational events result in an exchange of different forms of genes that can produce new combinations of genes.

The new genetics is revolutionizing practices and investigative approaches in fields as diverse as medicine and the use of atomic power; therefore, every taxpayer and contributor to research institutions should understand the purposes for which his money is being sought or spent.

—George and Muriel Beadle, *The Language of Life*

In the classical type of recombinational process, there is an exchange of homologous regions of DNA molecules. This is seen in the crossing over of chromosomes, where pairs of chromosomes containing the same gene loci pair and exchange corresponding portions of the same chromosomes (Figure 11.5). The term homologous indicates that the exchange is between alleles of the same gene but does not imply that the DNA segments that are exchanged have the same nucleotide sequences. In eukaryotic microorganisms this often occurs during the process of meiosis, which results in the reduction of the number of chromosomes and the conversion of a diploid cell into a haploid cell. A similar homologous alignment of DNA molecules can occur when a bacterial chromosome, or portion thereof, is transferred from a donor to a recipient bacterium.

DNA Transfer in Prokaryotes

Genetic exchange and recombination in bacteria can occur by three different mechanisms: transformation, transduction, and conjugation (Figure 11.6). These mechanisms differ in the way the DNA is transferred between a donor and a recipient cell.

In transformation a free (naked) DNA molecule is transferred from a donor to a recipient bacterium. The donor bacterium leaks its DNA, generally as a result of lysis of the bacterium, and the recipient bacterium is able to take up the DNA. A classic example of transformation involves the bacterium *Streptococcus pneumoniae* (Figure 11.7). One strain of this bacterium produces a capsule and is a virulent (disease-causing) pathogen. Another strain of the same bacterial species lacks the genetic information for capsule production and is avirulent (not disease-causing). When dead cells of the virulent strain are mixed with avirulent live bacteria, transformation can occur, producing a mixture of avirulent and virulent bacteria. What occurs in this process is that the DNA containing the genes for capsule production leaks out of the dead bacteria and is taken up by the living bacteria, which normally lack the genetic information for capsule production. Recombination can occur, and the progeny of the transformed bacteria become capable of producing capsules. In this manner nonpathogenic strains can be transformed into deadly pathogens.

Whereas in transformation free DNA is transferred, in transduction the DNA is transferred from a donor to a recipient cell by a viral carrier. For transduction a virus must acquire a portion of the genome of the host cell in which it reproduces (Figure 11.8). Phage can carry bacterial DNA from a donor cell and inject it into a recipient bacterial cell.

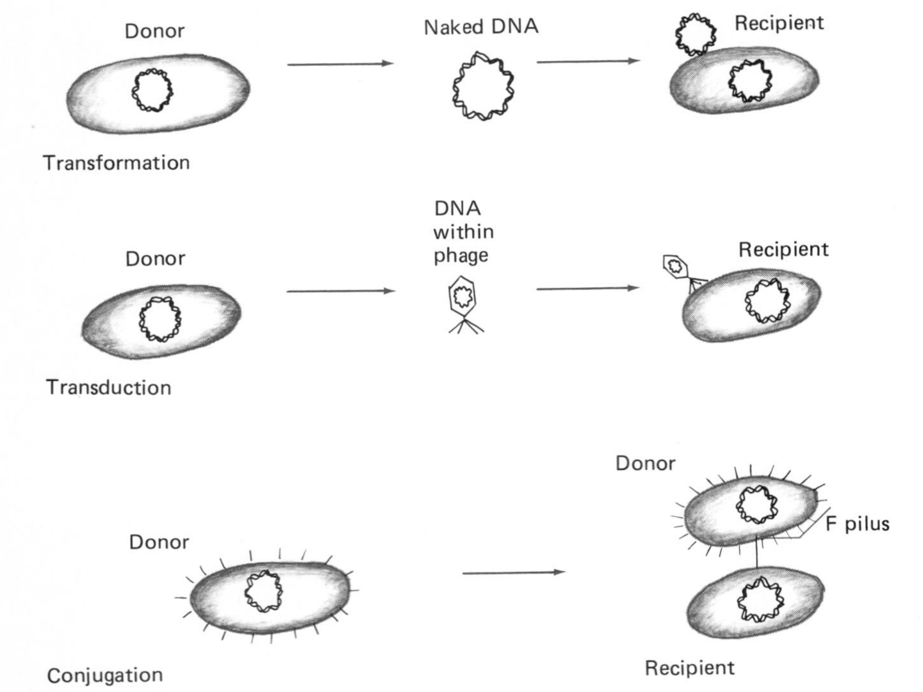

Figure 11.6 Genetic transfer in bacteria occurs by three distinct mechanisms: transformation, involving transfer of naked DNA; transduction, involving transfer of DNA by a phage; and conjugation, involving direct contact of bacteria during transfer of DNA.

Once inside the recipient cell, the DNA may be degraded by nuclease enzymes, in which case genetic exchange does not occur, or the injected DNA may undergo recombination. If recombination occurs, the transduced recipients may possess new combinations of genes.

There are two different types of transduction, generalized and specialized, which as the names imply differ in whether they can bring about the general transfer of genes or only the transfer of specific genes. In generalized transduction, pieces of bacterial DNA are accidentally acquired by developing phage during their normal lytic growth cycle. (The normal lytic growth cycle of phage involves the invasion of a host cell, the reproduction of the virus within the host, and the lysis or bursting of the host cell to release the newly formed phage.) If a phage carries bacterial instead of viral DNA, it cannot reestablish a lytic growth cycle and cannot cause lysis in a recipient bacterium. Such a phage, however, can attach to and inject DNA into a recipient bacterium, permitting it to carry bacterial DNA from a donor cell and inject it into a recipient bacterial cell. The injected DNA may undergo recombination and the transduced recipients may possess new combinations of genes. In specialized transduction temperate phage act as carriers of specific bacterial genes. (Temperate phage insert DNA into the bacterial chromosome at specific sites and do not normally bring about the lysis and death of the bacterial host cell.) Because there is a specific site for incorporation of the phage DNA into the bacterial chromosome only the genes that are adjacent to the site of insertion of the viral genome may be transferred by specialized transduction.

The third exchange process, conjugation (mating), requires the establishment of physical contact between the donor and recipient bacteria. The physical contact between mating bacteria is established through the F fertility pilus (Figure 11.9). Bacterial strains that produce F pili act as donors during conjugation, and are designated F+ strains if the F plasmid, which codes for F pilus production, is independent (free in the cytoplasm) or Hfr strains (high frequency recombinant strains) if the F plasmid DNA is incorporated into the bacterial strains (high frequency recombinant strains) if the F plasmid DNA is incorporated into the bacterial chromosome. Those strains lacking F pili are recipient strains, and are designated F− strains. During bacterial mating, DNA from the donor strain is replicated and transferred into the recipient cell. The precise portion of the DNA that is transferred depends on the time of mating, that is, how long the

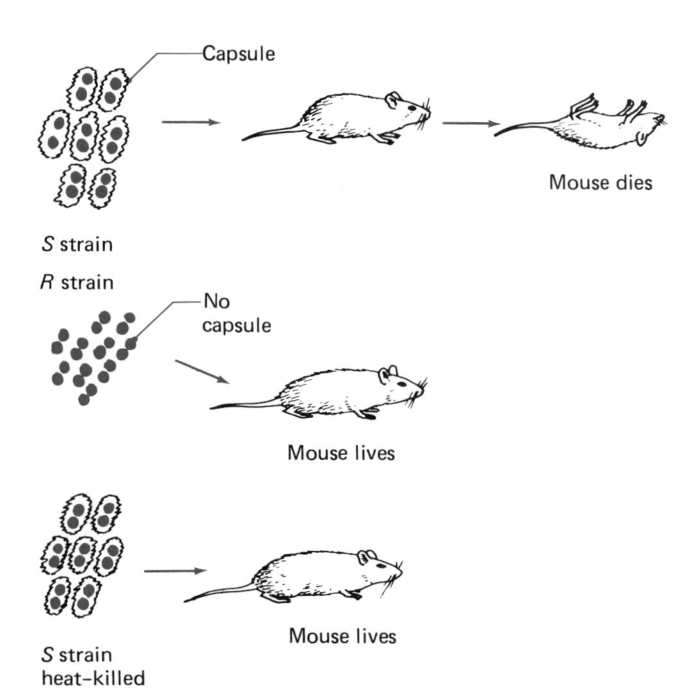

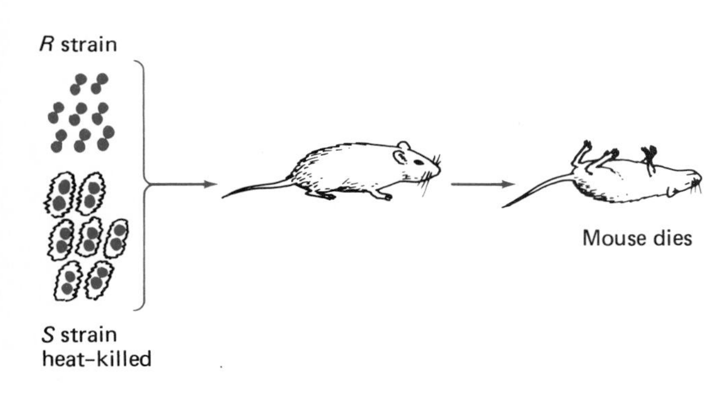

Figure 11.7 The transformation of *Streptococcus pneumoniae* shows how the properties of a bacterial strain can be altered by this recombination process. The discovery of the transformation process provided the first direct proof that DNA is the genetic material. (*Streptococcus pneumoniae,* a causative organism of pneumonia, produces a polysaccharide capsule. Avirulent (not disease-causing) mutants, called *R* strains because they appeared as rough colonies on agar, lack a capsule and are unable to cause disease. Mice infected with even minimal doses of *S* (nonmutant, capsulated, smooth colony-producing) strains die a few days after exposure to the bacteria, whereas injection of even massive doses of *R* strains does not cause death. When mixtures of heat-killed *S* strains and live *R* strains are injected into mice, the mice surprisingly die and *S* strains can be isolated from their corpses. The *R* strains had been transformed into a new encapsulated, pathogenic strain by what appeared to be a genetic process, later called transformation.

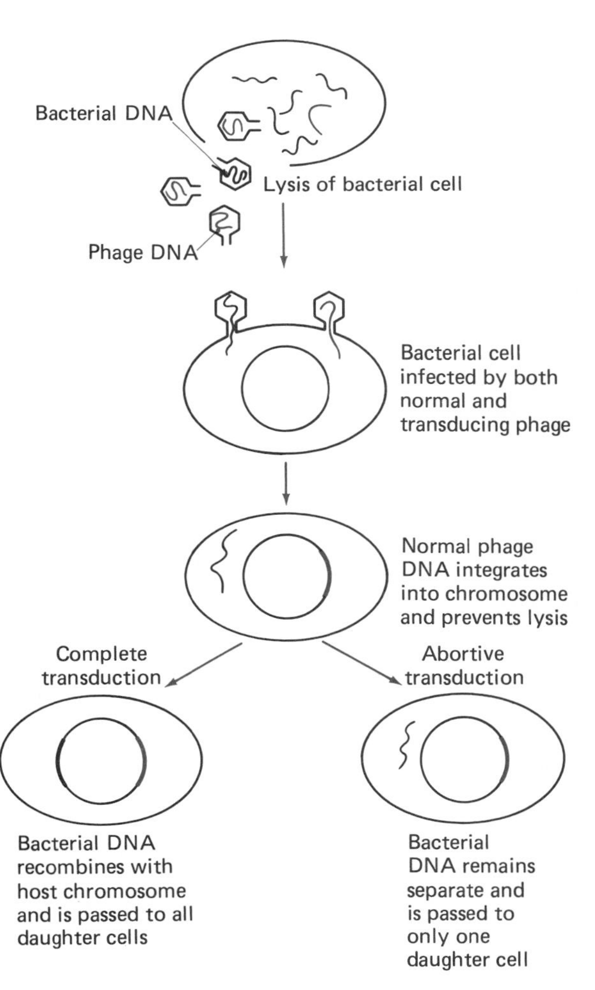

Figure 11.8 (*right*) Diagram showing how phage can acquire bacterial DNA for generalized transduction.

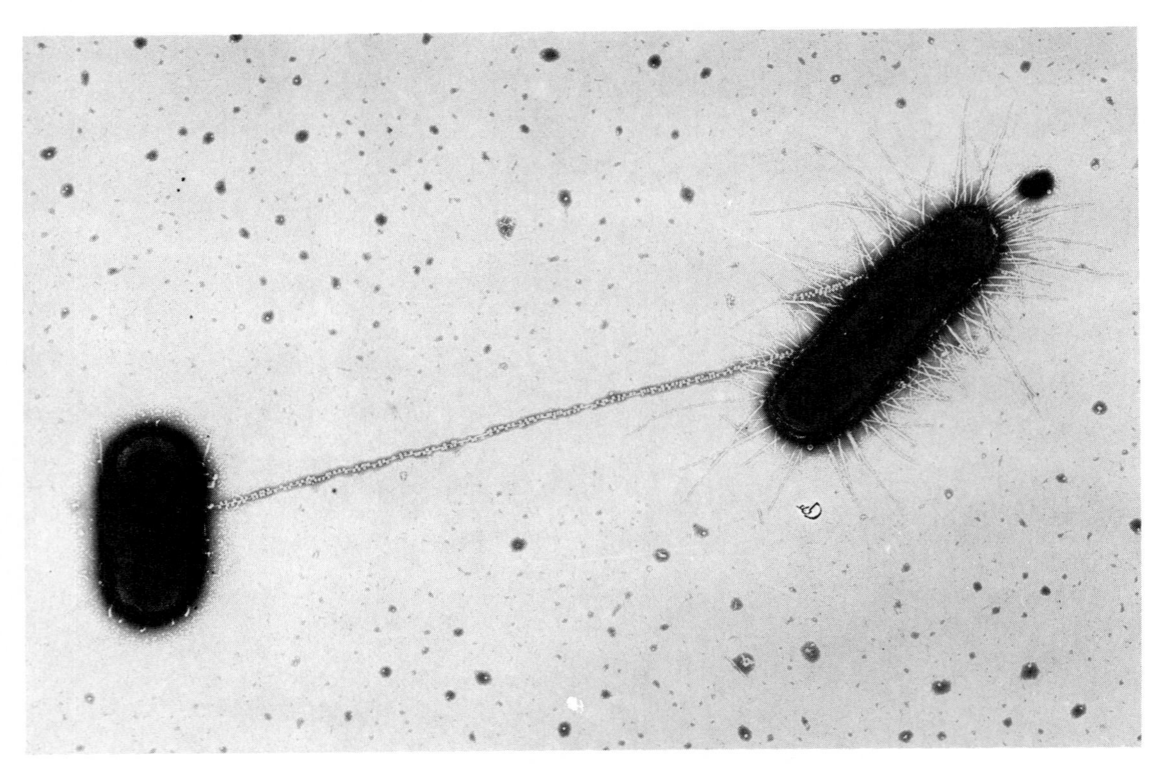

Figure 11.9 Electron micrograph showing the attachment of mating *Escherichia coli* cells by an F pilus. The bacterial cell covered with numerous appendages is the genetic donor and the cell without appendages is the recipient. The F pilus is necessary for the transfer of genes from the donor to the recipient. (Courtesy Charles C. Brinton, Jr. and Judith Carnahan, University of Pittsburgh.)

F pilus maintains contact between the mating cells.

When an F+ cell is mated with an F− cell, the F plasmid DNA is usually transferred from the donor to the recipient (Figure 11.10). The F plasmid confers the genetic information for acting as a donor strain, and the offspring of the recipient of such a cross, therefore, are mostly donor strains. When an Hfr strain is mated with an F− strain, the bacterial chromosome with the integrated F plasmid gradually is transferred to the recipient strain. Because the F plasmid is often not near the beginning of the DNA that is transferred and only rarely is there sufficient mating time to accomplish the complete transfer of the genome, the recipient cell normally remains F−. However, a relatively large portion of the bacterial chromosome is transferred from the donor to the recipient, and thus, there is a relatively high frequency of recombination of genes of the bacterial chromosome when Hfr strains are mated with F− strains.

Mating of different strains of microorganisms that have different forms of genes can be used for genetic mapping. The frequency of recombination that results from mating is used to map the order and thus determine the relative locations (loci) of genes. If genes are close together, the chances that a break will occur between them are less than if they are far apart. Genes that are located near each other on a chromosome should segregate together with a high frequency, and the rates of recombination between such closely linked genes should be low. By permitting different times for mating, the order of genes on the bacterial chromosome can be determined by examining the times at which recombinants for given genes are found. If a gene of unknown location shows a high frequency of recombination along with the marker gene, it is likely that the marker and unknown genes are closely associated in the chromosome. If, however, the genes are far apart, it is unlikely that recombinants of both the marker gene and the gene of unknown location will occur in the progeny.

Genetic Exchange in Eukaryotes

In eukaryotic microorganisms genetic exchange normally occurs as a result of a sexual reproduction phase of the life cycle. The vegetative cells of many eukaryotic organisms are diploid, but in order to exchange genetic information these organisms normally form specialized reproduction gametes or haploid spores. The conversion of a diploid to a haploid state occurs in the process of meiosis, which is also known as reductive division. Meiosis begins after DNA replication, so that the starting cell is actually tetraploid (Figure 11.11). During this tetraploid state the chromosomes are aligned side by side, and recombination can occur between homologous chromosomes by crossing over. The chromosomes are pulled apart by spindle fibers to reestablish a diploid state. A second meiotic division then occurs without further DNA replication, forming four nuclei, each of which contains a haploid number of chromosomes. The haploid nuclei of these reproductive cells can later fuse with the nuclei of reproductive cells of an appropriate mating type. The union of the haploid nuclei during

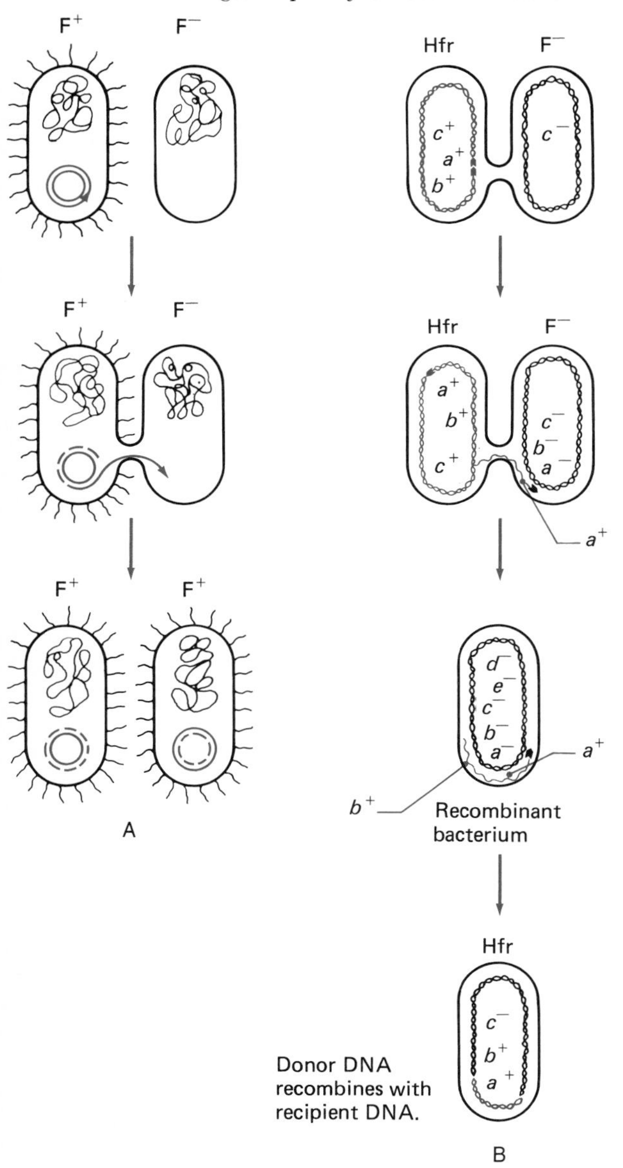

Figure 11.10 (A) Diagram showing conjugation of F^+ and F^- strains. (B) Conjugation of Hfr × F^- results in a high frequency of recombination.

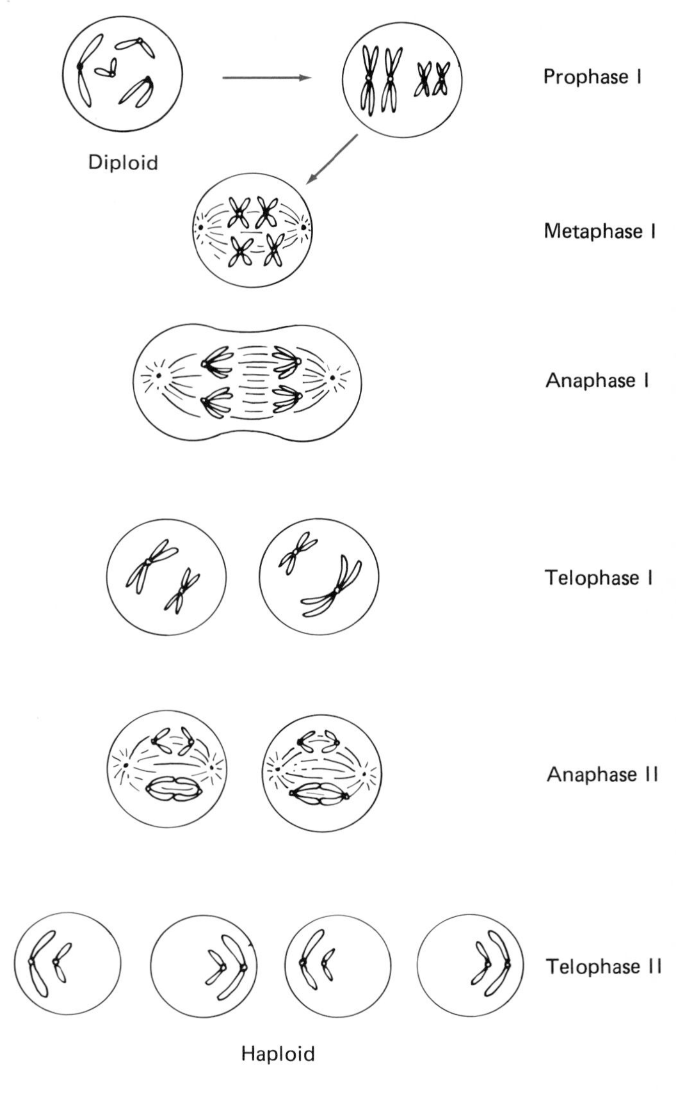

Figure 11.11 Stages of meiosis. Meiotic division results in a reduction of the chromosome number.

Agency Reports Genetic Therapy Is Near

By HAROLD M. SCHMECK JR.

The policy question concerning human gene therapy is no longer whether to attempt such treatments, but when trials in patients should begin, according to a report released yesterday by the Congressional Office of Technology Assessment.

Gene therapy designed to treat individual patients without causing changes that could be inherited is just an extension of conventional medical science and technology and does not raise new ethical issues, the report said.

"The factor that most distinguishes it from other medical technologies is its conspicuousness in the public eye," the report said. Some critics have asserted that such treatments would be the beginning of misguided attempts to redesign the human race.

The type of gene therapy under most active research is the introduction of a normal gene, produced in the laboratory, to serve a function lacking in the patient's comparable but defective gene. For example, cells containing a transplanted gene for a vital enzyme might cure a disease caused by the patient's natural lack of that enzyme.

The document made public yesterday made a sharp distinction between treating individual patients to cope with their own genetically based disorders and treatment that could cause changes transmissible to the patients' children and succeeding generations. Inheritable changes presumably could be made only by transplanting genes into the patient's reproductive cells.

Such treatment is not practical today, may or may not prove possible in the future, and would require much further evaluation and public discussion before it is ever attempted, the report said.

The 105-page document was described as a background paper prepared to help Congress in considering future policy.

A few attempts at human gene therapy were made in 1980 and earlier, but none were successful. Many experts called the attempts premature. Since then, the Office of Technology Assessment report said, knowledge and technology have advanced rapidly.

Experts say noninheritable gene therapy will be attempted soon, possibly as early as next year. The first such attempts are likely to be done by removing a sample of the patient's bone marrow, treating it to insert normal copies of a gene that is defective in the patient, and then returning the treated bone marrow to that patient.

Two Rare Diseases Are Targeted

Such attempts are expected first in two rare genetic diseases that leave their victims immunologically crippled. The diseases are called adenosine deaminase deficiency and purine nucleoside phosphorylase deficiency.

Only about 50 cases of the first disease, known as ADA deficiency, have been reported in the world, the report said. The other disease is even more rare. No conventional cures are known, and the patients usually die young.

In each disease, however, the gene that is defective in the patients is known and the equivalent normal genes have been grown in the laboratory.

The Office of Technology Assessment was set up by Congress to aid the Senate and House of Representatives in dealing with complex technical issues that confront society. The background paper on gene therapy was requested by Senator-elect Albert Gore Jr., Democrat of Tennessee, when he was chairman of an investigations subcommittee of the House Science and Technology Committee.

Within the past few years, interest in human gene therapy has intensified as a result of improved methods for the actual transplantation of genes into living cells.

The most promising technique is the use of genetically engineered viruses to carry these genes. The viruses are modified to incorporate the foreign genes and to make the virus particles harmless to the cells they penetrate.

The modified viruses are little more than biological guided missiles capable of penetrating certain types of cells. Presumably the modified viruses have no effects, other than that of delivering the new genes they carry.

It is expected that gene therapy will first be performed on patients who have no other prospects for treatment and who suffer from devastating diseases that are likely to be rapidly fatal.

For the foreseeable future, only diseases that are known to result from the failure of single genes will be the subject of such treatment. Genetic diseases that result from defects in more than one gene appear to be beyond the capabilities of gene transplantation today.

Whether gene therapy of the kind envisioned today will indeed prove curative in many cases is still unknown, but experts are hopeful that some carefully chosen patients may benefit greatly from the treatments.

In addition to the two immune deficiency diseases that are the most likely candidates for early trial, there are many other single-gene disorders that might be attacked by gene therapy.

One candidate is likely to be a disorder called Lesch-Nyhan disease that has serious ill effects on the central nervous system as well as the rest of the body. The normal counterpart of the gene that is defective in this disorder has been identified and grown in the laboratory.

Another promising case is phenylketonuria, a cause of serious mental retardation in patients who are not put on rigorous restrictive diets very early in infancy.

Source: The New York Times, Dec. 18, 1984. Reprinted by permission.

syngamy (the fusion of reproductive cells) reestablishes the diploid state. One of the haploid sets of chromosomes comes from a donor strain (male) and the other haploid set from a recipient strain (female). In most cases there are differences between the mating types, so that the progeny are heterozygous.

Genetic Engineering

The recognition that genetic information can be exchanged between organisms with different genetic characteristics and that the underlying genetic code is the same in all organisms has led to the ability artificially to direct recombinant processes for beneficial purposes, a technique called recombinant DNA technology or genetic engineering. DNA recombination can be employed to engineer the evolution of new microorganisms (Figure 11.12). This recombinant DNA technology promises to be a powerful tool for understanding basic genetic processes and has tremendous potential industrial applications. The organisms formed in this manner have a combination of genes from a donor source and from a recipient. The source can be any DNA: human, plant, bacterial, or viral. The recipient is often a bacterium but could be any organism.

Genetic engineering has opened up many new possibilities for employing microorganisms to produce economically important substances. The use of recombinant DNA technology permits the purposeful manipulation of genetic information to engineer microorganisms that have the capacity to produce high yields of a great variety of useful products. Until the development of recombinant DNA technology, bacteria could produce only substances from their own genomes. Now, however, there are bacteria that have had plant and animal genes added to their genomes by genetic engineers (Figure 11.13). These bacteria can produce plant and animal products such as insulin, inteferon, and human growth hormone. Several new vaccines have been developed by using recombinant DNA technology for preventing human diseases. These genetically engineered bacteria are revolutionizing the economics of the pharmaceutical industry. Genetic engineering also has great potential for improving agricultural productivity. For example, a bacterium has been genetically engineered that can reduce frost damage to agricultural crops. This bacterium has the potential for saving millions of dollars of crops that are destroyed annually by frost. Normally bacteria living on the surfaces of plants produce proteins that serve as ice condensation nuclei, initiating the formation of frost. The genetically engineered bacteria lack the genes for producing the proteins that initiate frost formation. If the normal microbial populations on the plant surface can

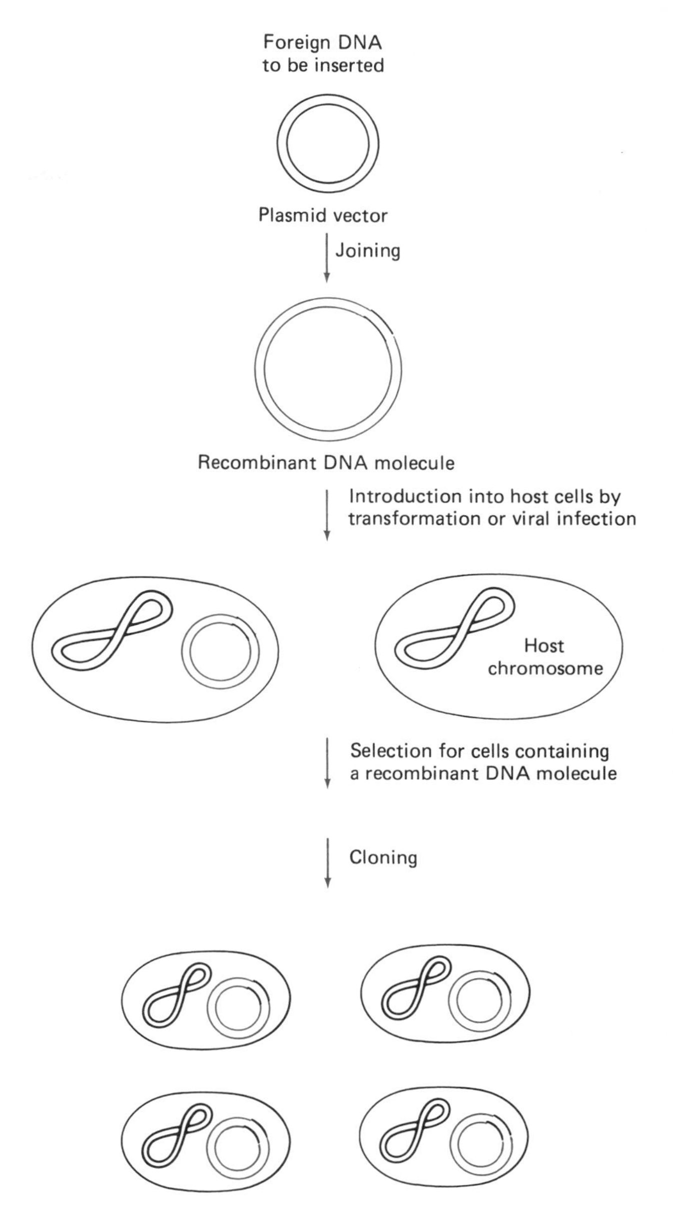

Figure 11.12 Diagram showing a generalized outline for genetically engineering a new organism. The ability to create new organisms in this way holds promise for improving the quality of life by using genetic engineering in medicine and industry to produce useful products.

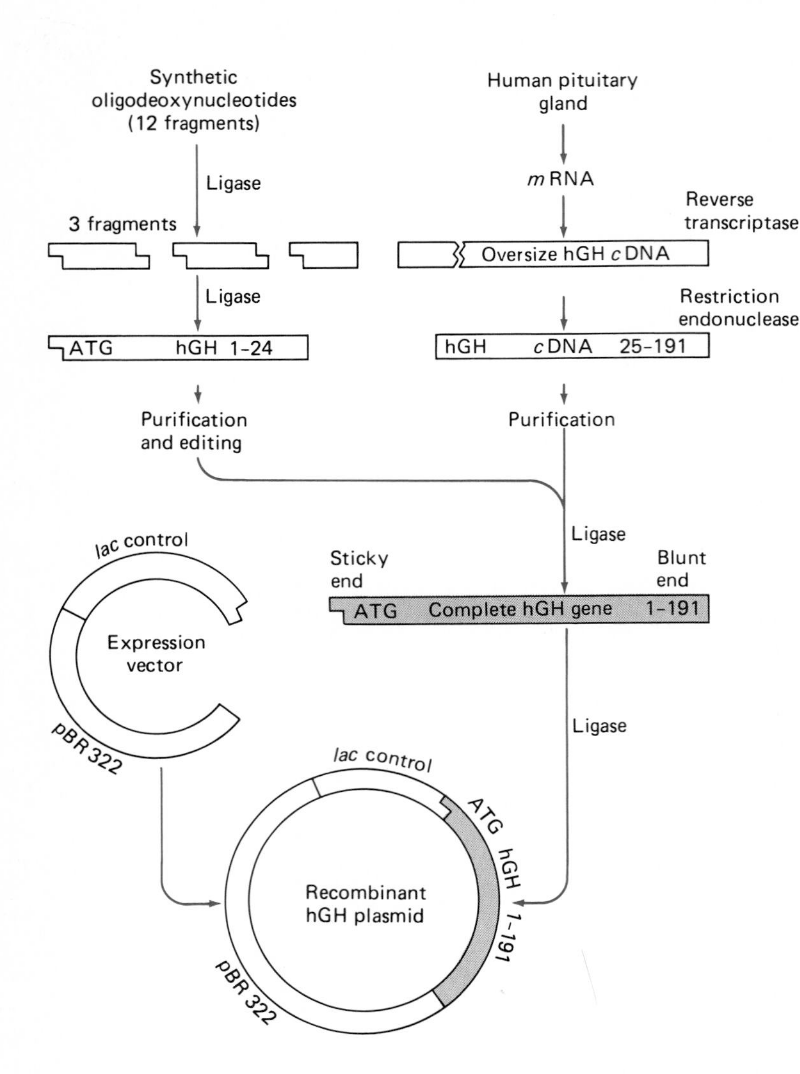

Figure 11.13 A genetic engineering process for producing a bacterium capable of synthesizing human growth hormone. The recognition that genes are split in eukaryotic cells has forced the genetic engineers to develop special techniques for incorporating human genes into bacteria in order to produce useful gene products. Through a series of novel procedures, involving a combination of chemical synthesis and isolation of the natural molecules, a gene for human growth has been created and placed into *Escherichia coli.* Human growth hormone is naturally produced by tissues of the pituitary gland; its absence leads to a form of dwarfism that can be cured by administration of the hormone. Clinical trials of human hormone produced by genetically engineered bacteria are quite promising; children receiving injections of bacterially produced human growth hormone for a few months are approaching normal heights for their age group with no significant side effects.

be replaced by the genetically engineered bacteria, the temperature would have to be several degrees lower before frost began to form. This would provide the necessary margin for the survival of many crops.

The suggestion that knowledge resulting from recombinant DNA, genetics, or any other research should not be obtained because it is too dangerous is repulsive to me. I do not think there is any knowledge too dangerous to possess . . . Knowledge is what gives us freedom. And if we want freedom we must pay for it. Part of that payment is the risk that we must take . . . Whether we are free or enslaved because of knowledge does not depend on what knowledge we have. It depends on how the knowledge is used.

—THOMAS CARNEY, *Instant Evolution*

Genetic engineering often uses plasmids as carriers of unrelated DNA and the enzymes involved in the normal recombination and replication of DNA for splicing foreign DNA into the plasmid carriers. Plasmids can be isolated, and the plasmid DNA can be cut open by using a site-specific endonuclease, commonly called a restriction enzyme, to create a site where foreign donor DNA can be inserted. Some restriction endonucleases cut the DNA at a palindromic sequence of bases (a sequence of nucleotide bases that can be read identically in both the 3′-hydroxyl to 5′-hydroxyl and 5′-hydroxyl to 3′-hydroxyl direction), producing DNA with staggered single-stranded ends. These ends of the cut DNA can act as cohesive or sticky ends during recombination, making them amenable for splicing with segments of DNA from a different source that has been excised by using the same endonuclease.

MICROBIOLOGY HEADLINES

U.S. Approves Dissemination of Gene-Engineered Microbe

By PHILIP J. HILTS

A government board has approved the first release into the environment of a gene-engineered microbe—one designed to retard the formation of frost on plants—but four organizations are expected to file suit today to halt the experiment.

The gene is in a bacterium that is found in fields and on plants and normally helps start frost crystals when the temperature falls below 32 degrees. A National Institutes of Health board has approved field tests of an altered version that does not produce frost even when temperatures fall as low as 23. Since most frost damage occurs between 25 and 30 degrees, the new bacterium has enormous commercial potential, researchers say.

But, the four organizations contend in their court suit that the experiment could be hazardous. In the worst case, opponents speculate that the frost-preventing bacterium might rise into the upper atmosphere and disrupt the natural formation of ice crystals there, which could effect the global climate.

They also claim that the NIH board is unqualified to assess the environmental risk of such an experiment since it has no ecologists or botanists, and has failed to carry out an environmental assessment of the experiment's dangers.

Such environmental assessments or impact statements are required by the National Environmental Protection Act on "major federal actions." The suit's plaintiffs, which include the Humane Society of the United States, say approval of the research constitutes "major" action.

The NIH offered no comment on the suit yesterday, but scientists familiar with the experiment said it would be hard to argue that it was risky, since very similar experiments with frost-stopping bacteria already have been carried out successfully with no apparent risk. Those experiments altered the bacterium in similar ways, but by methods other than gene-engineering and so were beyond NIH regulation.

The experiment—to be carried out by Steven Lindow of the University of California at Berkeley and partly funded by Advanced Genetic Sciences Company of Berkeley—was approved by the National Institutes of Health under federal rules governing gene engineering.

Until research was delayed recently, Lindow had planned to spray a field of potatoes this month with an altered version of the extremely common organism called *Pseudomonas syringae*. The natural organism lives on the outside of plants, and makes a chemical that encourages the formation of frost and ice a degree or so below freezing.

Lindow and his colleagues have identified the genes in the organism that trigger this quick-freezing action, and have excised them from the bacterium.

These neutered organisms are sprayed on the plants as soon as they sprout. Once the bacteria are in place, they take over the niche usually occupied by their frost-causing relations, and so crowd out any of the frost-causing variety before they can take hold.

The altered bacteria protect plants down to about 23 degrees, and possibly lower. Department of Agriculture officials said successful results could save millions of dollars in crops.

Besides the Humane Society, plaintiffs in the suit are the Environmental Task Force Inc., Environmental Action Inc. and the Foundation on Economic Trends. Leading them is Jeremy Rifkin, author of "Algeny," a book on genes, ecology and the future of man.

In the suit, expected to be filed in U.S. District Court here, the plaintiffs said introducing gene-engineered bacteria into the environment is much like introducing a foreign plant into the Country.

"Some of our most significant problems are of that nature," suit documents said. "These include the kudzu weed, the chestnut blight, the gypsy moth, Dutch elm disease, and starlings and house sparrows, both of which are agricultural pests."

The suit was bolstered by affidavits from several scientists, including Eugene P. Odum of the University of Georgia and David Pimentel of Cornell University, two of the country's most eminent ecologists. Odum and Pimentel did not criticize the Lindow experiment, but said they were worried about releasing newly engineered organisms into the environment without first testing to see their effects.

NIH defenders say the frost-stopping bacteria already exists on virtually all plants. The experiments do not add new genes or new powers to the bacterium, but simply take away some of its destructive powers.

Source: The Washington Post, Sept. 14, 1983. Reprinted by permission.

Having been banished from the court of Queen Elizabeth on suspicion of too great familiarity with a nobleman in high favor, a lady adopted the following pallindrome for a motto: ABLATA ATALBA—Secluded but pure.

—FRANK STAUFFER, *The Queer, The Quaint, The Quizzical*

If two different endonuclease enzymes are used, one to open the plasmid ring and another one to form a segment of donor DNA, it often is necessary to establish an artificial homology at the terminal ends of the donor and plasmid DNA molecules (Figure 11.14). This can be accomplished by adding polyA (poly-adenine) tails to the plasmid and polyT (poly-thymine) tails to the donor DNA. A transferase

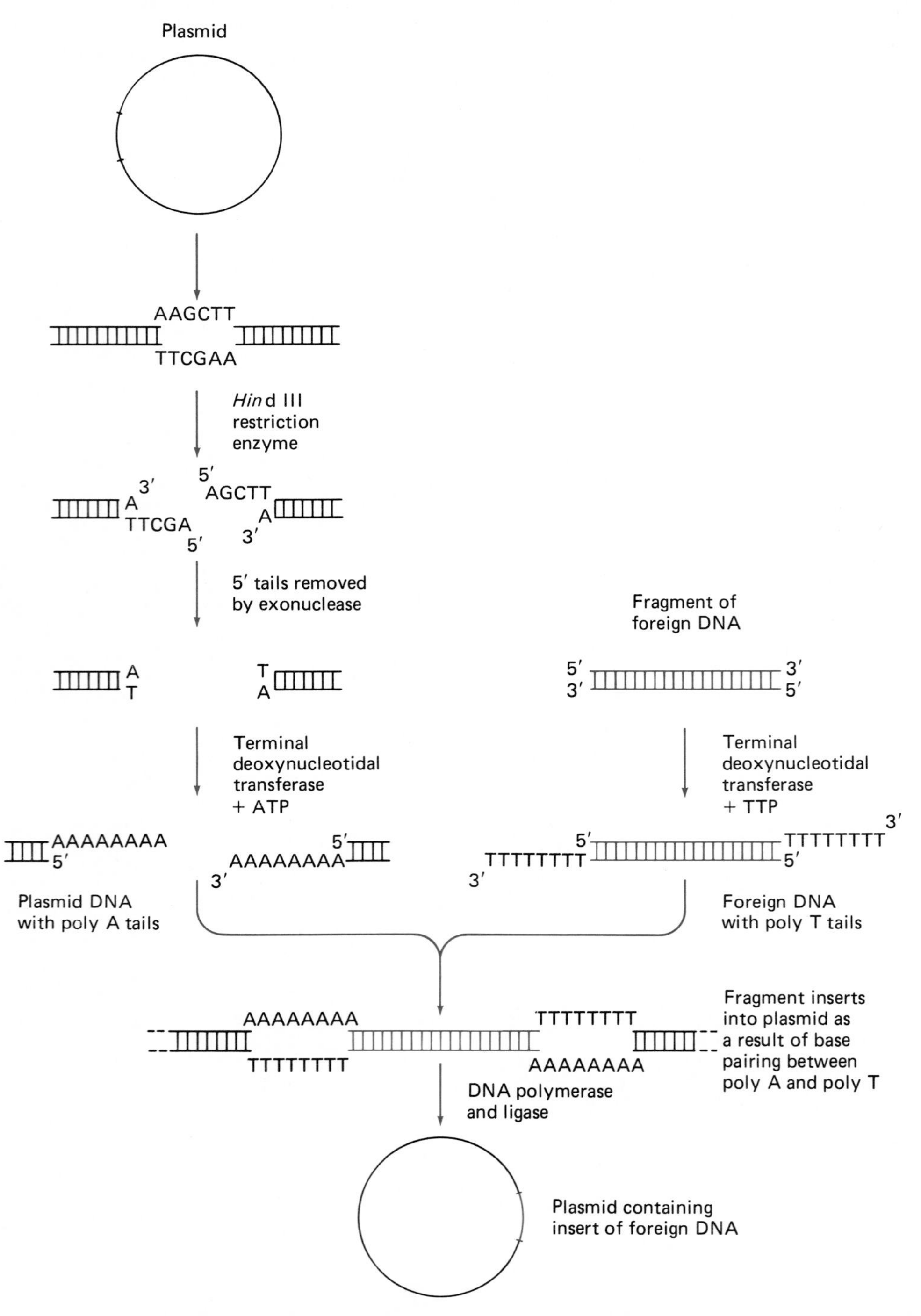

Figure 11.14 Diagram showing the establishment of artificial homology when different endonuclease enzymes are employed.

enzyme can then be employed to add polyA tails to the 3′-hydroxyl ends of the DNA molecule. Pairing occurs between homologous regions of complementary bases, and ligase enzymes are used to seal the circular plasmid. The tails left by the action of the endonuclease are cleaved in vitro, using exonuclease enzymes. Virtually any source of DNA can be used as a donor, including human DNA. By adding a polyT tail to the donor DNA after its excision with an endonuclease, the donor DNA can be made complementary to the polyA tails of the plasmid DNA, permitting the formation of a circular plasmid molecule.

If the same restriction endonuclease is used to cut both the donor and recipient plasmid DNA, the strands will have homologous ends and it will be unnecessary to add polyA and polyT tails (Figure 11.15). The sealing of the ends of the DNA molecules is accomplished by using ligases, and in this manner a plasmid is created that contains a foreign segment of DNA. Once the plasmid containing the desired additional DNA segments is formed, it can be added to a culture of a suitable recipient bacterium that will incorporate the plasmid. The plasmid is taken up by the bacterium, and then, regardless of its source, it comprises part of the

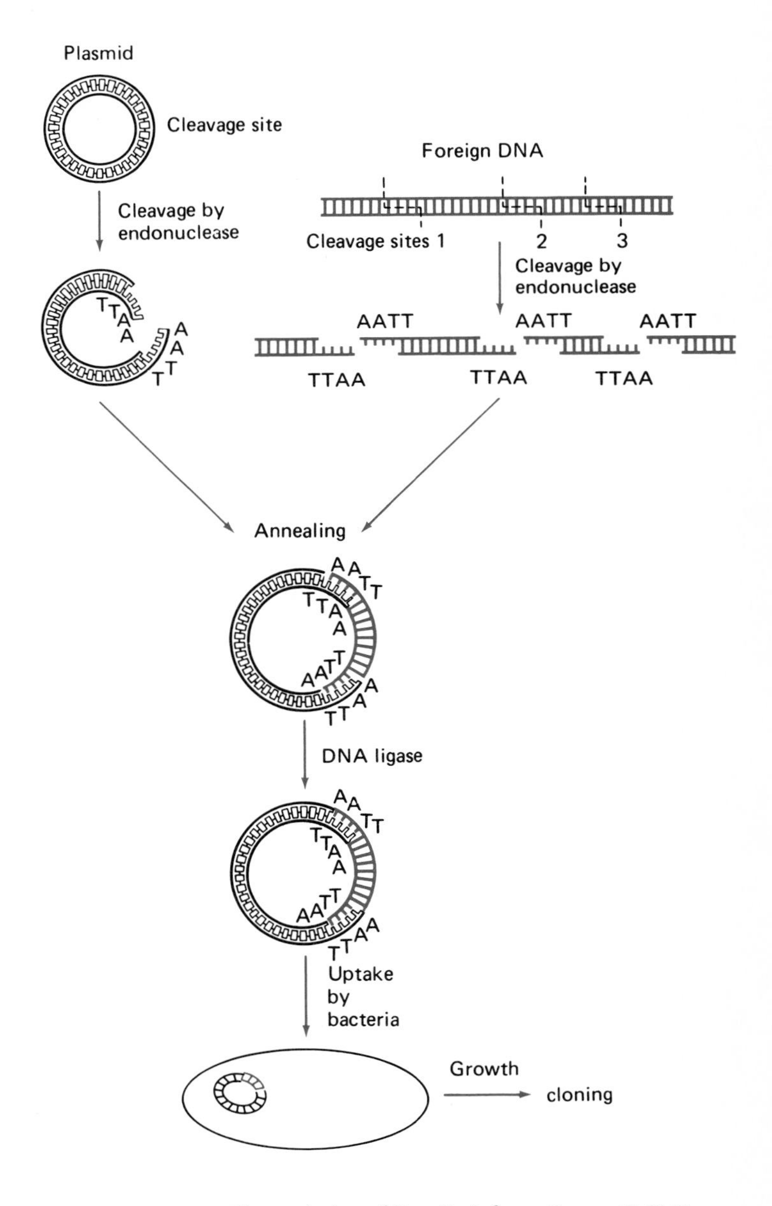

Figure 11.15 Diagram showing the formation of a recombinant DNA molecule when the same endonuclease enzyme is used for both the donor and recipient DNA.

bacterial genome. The plasmid DNA, including the foreign DNA segments, can be replicated and passed from one generation to another. Genes can be **cloned** (asexually reproduced) in this manner, with bacteria acting as factories to produce multiple copies of identical genes. Because the genetic code is essentially universal, the information encoded in the DNA sequence can be expressed, and the polypeptide chain specified by the foreign DNA segment can be transcribed and translated to form an active protein molecule. The expression of the foreign genetic information, however, requires that the appropriate reading frame be established and that the transcriptional and translational control mechanisms be turned on to permit the expression of the DNA.

The construction of bacterial plasmids containing complete gene sequences, derived from eukaryotic organisms, is complicated by the fact that eukaryotic genes are generally split. In eukaryotic organisms posttranscriptional modification of the hnRNA is required in order to produce a mature mRNA molecule that can be properly translated; bacteria, however, do not possess the capacity to remove introns from eukaryotic DNA to form the mRNA needed to produce a functional protein molecule. Therefore, it is necessary artificially to cut and splice eukaryotic DNA, or to use an alternative procedure, to establish a contiguous sequence of nucleotide bases that will define the protein to be expressed (Figure 11.16). The problem of the discontinuity of the eukaryotic gene can be overcome by using an mRNA molecule and a **reverse transcriptase** enzyme to produce a DNA molecule with a contiguous sequence of nucleotide bases containing the complete gene. The single-stranded DNA molecule formed in this procedure is complementary to the complete mRNA molecule. The RNA can

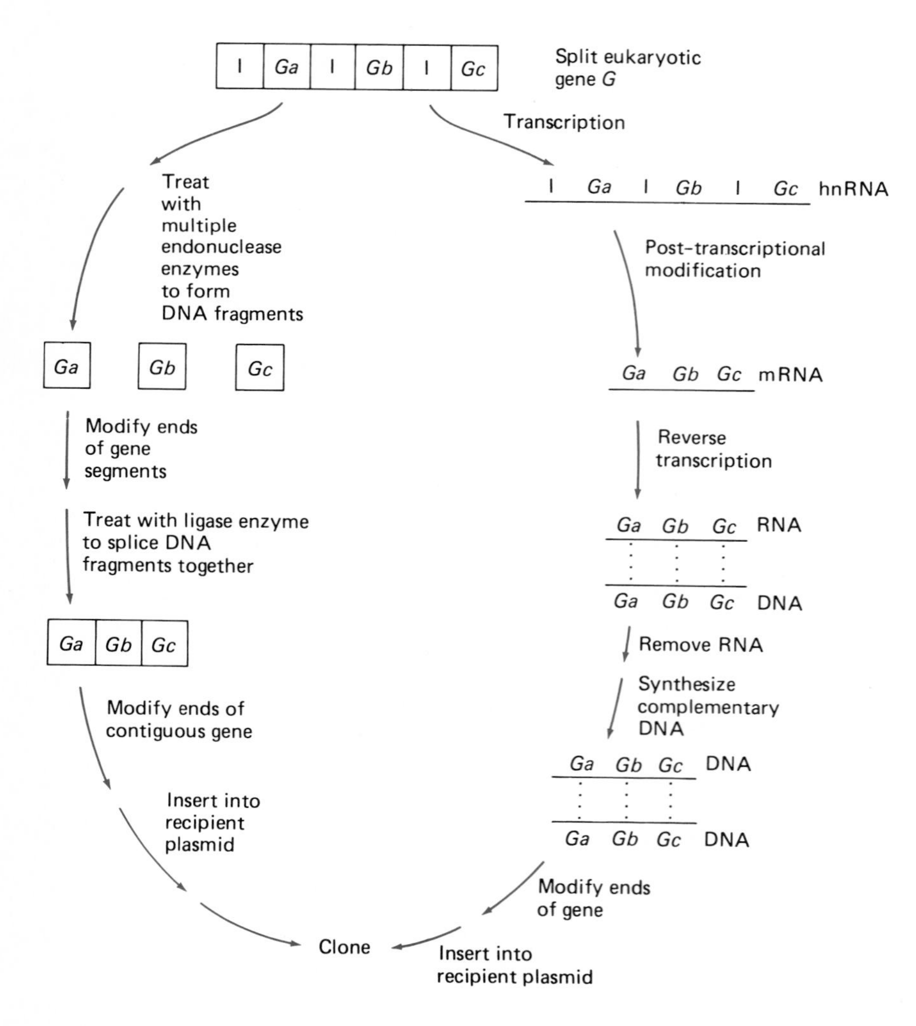

Figure 11.16 The establishment of a contiguous sequence of nucleotides is needed for the cloning of eukaryotic genes in bacterial cells. The need to remove intervening sequences is an unexpected complication for employing bacteria to produce eukaryotic gene products.

be removed by using a nuclease and a complementary strand of DNA synthesized by the reverse transcriptase enzyme. The double-stranded DNA molecule formed in this manner can then be inserted into a carrier plasmid. In this way many eukaryotic genes can be cloned in prokaryotic cells.

One of the greatest benefits that may be realized through genetic engineering is the introduction of the capacity to fix nitrogen into plants, such as wheat and rice, that are not able to utilize atmospheric nitrogen (Figure 11.17). For years scientists have been exploring the relationships between *Rhizobium* and the plants with which this nitrogen-fixing bacterium can establish symbiotic relationships. Winston Brill and colleagues at the University of Wisconsin have tried to apply the screening procedures of the pharmaceutical industry in an effort to find especially effective nitrogen-fixing strains of *Rhizobium* that could increase crop yields. Using mutagens and screening procedures, they isolated strains of *Rhizobium* capable of very high rates of nitrogen fixation. However, field tests with these efficient *Rhizobium* strains did not increase crop yields; the superior nitrogen-fixing strains could not successfully compete with indigenous strains. Because of the inefficiency and lack of success with the mutation-screening approach, microbiologists are studying the genetics and biochemistry of infection by *Rhizobium* with the aim of employing recombinant DNA techniques to genetically engineer plants containing the bacterial genes for nitrogen fixation.

In one series of studies, the genes for nitrogen fixation are first inserted into the genome of a yeast. Plasmids from *Escherichia coli* and from a yeast cell are cleaved and then fused to form a single hybrid plasmid, which can be recognized by the yeast cell and integrated into its chromosomal DNA. In the next step, the genes that will be introduced into the yeast are isolated from the chromosome of *Klebsiella pneumoniae*, a nitrogen-fixer. The genes, collectively designated *nif*, code for some 17 proteins. Another *E. coli* plasmid is cleaved, and the isolated *nif* genes are introduced to form a second hybrid plasmid. Because of the bacterial DNA already inserted into one of the yeast chromosomes, the yeast cell recognizes the hybrid *E. coli* plasmid. The plasmid is then integrated into the yeast chromosome. Although the insertion of the prokaryotic nif genes into the eukaryotic yeast cell demonstrates that genetic material can be transferred between different biological systems, the nitrogen-fixing proteins are not expressed in the yeast. More studies are needed to elucidate the factors controlling expression of the nif genes before success is obtained.

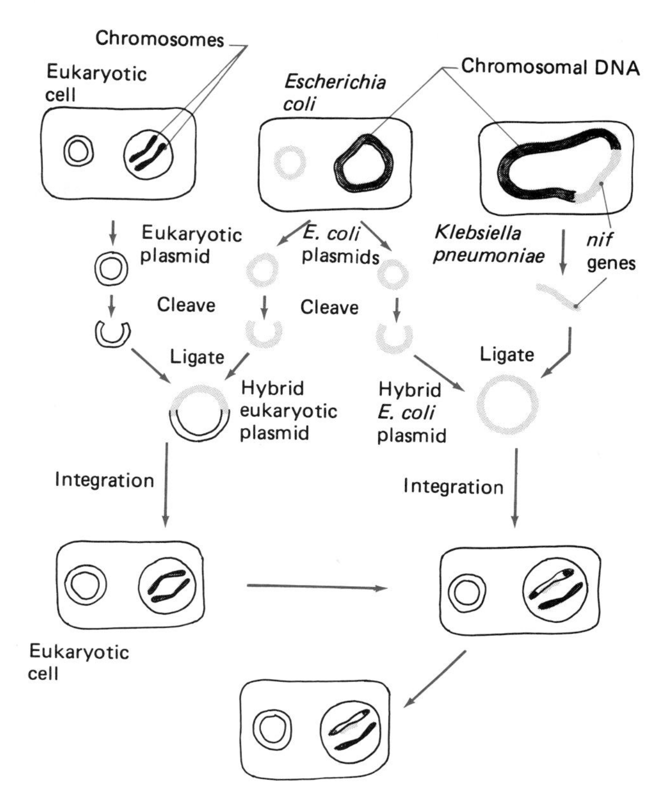

Figure 11.17 Genetic engineering techniques can be used to create new plants that contain the bacterial nif gene for nitrogen fixation.

It is increasingly apparent that the ability to engineer organisms, such as eukaryotic plant cells that can fix atmospheric nitrogen, depends on developing a thorough understanding of the molecular biology of gene expression. Once we understand the mechanisms of gene regulation in eukaryotes, we will be able to apply this knowledge through genetic engeineering to create organisms with novel properties.

One approach for improving agricultural crops that holds great promise is to use bacterial plasmids as vectors to move genetic information from one plant to another. The Ti (tumor-inducing) plasmid of *Agrobacterium tumefaciens* appears to be suitable for such a process (Figure 11.18). *A. tumefaciens* causes crown gall tumors in most dicotyledonous plants. The Ti plasmid of *A. tumefaciens* induces the infected plant cells to synthesize nitrogen compounds called opines. During the normal infectious process, a section of the plasmid

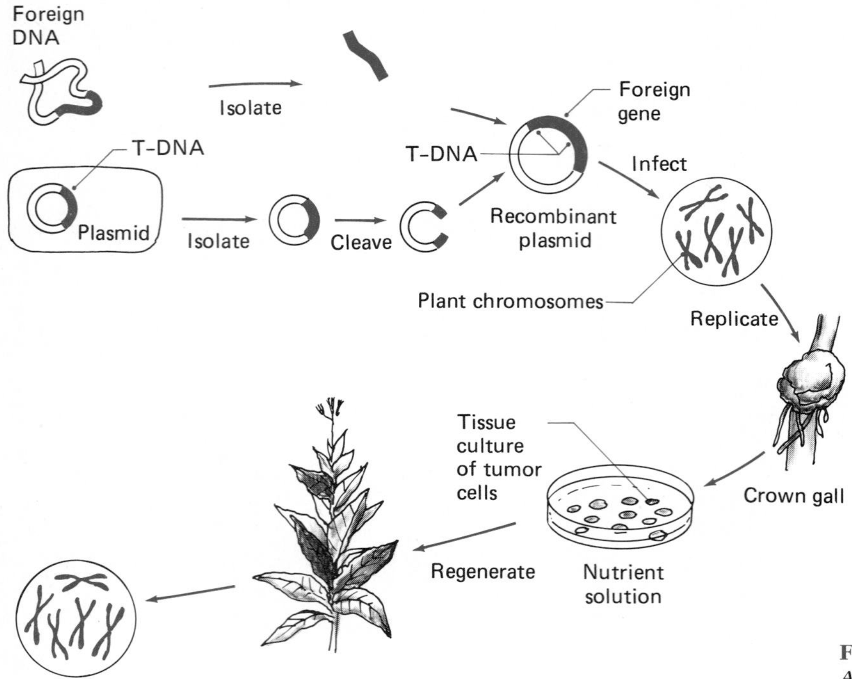

Figure 11.18 The Ti plasmid of *Agrobacterium tumefaciens* is a useful vector for introducing foreign DNA into the genome of a plant.

combines with chromosomal DNA in the nucleus of the plant cell. It may therefore be possible to employ the Ti plasmid as a vector for inserting foreign DNA into the genomes of plant cells. To do so the plasmid would be cut open at a site within the Ti plasmid DNA, and the foreign gene would be spliced into it. When the tumor cells are grown in tissue culture, they continue to carry and replicate Ti-DNA during the normal divisional process. If foreign genes inserted into Ti-DNA are also transmitted to plant progeny, new plant strains can be genetically engineered.

The potential of genetic engineering to short-circuit evolution has raised numerous ethical, legal, and safety questions. The Supreme Court of the United States has ruled, in a landmark decision, that genetic engineering can "create" novel living systems that can be patented as inventions. This ruling has established the precedent for future genetic engineering efforts. The safety question was temporarily solved by using a fail-safe strain of *E. coli* that is unable to grow outside of a carefully defined culture medium. Nevertheless, the question remains: what if a genetically engineered plasmid were to enter another microorganism and spread? Could this represent a serious biological hazard? The debate also continues as to whether it is ethical to clone DNA and whether recombinant DNA technology will permit the cloning of higher eukaryotic organisms, including plants and animals. One must weigh these ethical questions against the benefits that can be derived through genetic engineering. There is little question that through genetic engineering the quality of human life can be improved. Scientists have a responsibility to see that the public is informed about genetic engineering, to see that research remains within acceptable guidelines, to use this powerful technique for examining basic questions about molecular level genetics, and to develop genetically engineered organisms of agricultural, ecological, medical, and economic importance.

Summary

The transmission of genetic information depends upon the faithful replication of DNA. DNA replication is a semiconservative process. The DNA unwinds to form a replication fork where new strands of DNA are synthesized. The original DNA serves as a template that specifies the order of bases that are synthesized. Nucleotides are aligned opposite complementary bases (adenine opposite thymine and guanine opposite cytosine). Once the bases are aligned they are linked by DNA polymerases. One strand can be synthesized continuously, but synthesis of the opposite strand is discontinuous. The short segements of DNA that are formed by copying the lagging strand are linked by ligases. After replication, each double helical DNA molecule contains one original (conserved) strand and one newly synthesized strand.

Recombination involves a restructuring of DNA molecules so that new genomes are formed containing information from different DNA sources. A variety of mechanisms permit the transfer of genetic information from a donor to a recipient microorganism so that recombination can occur. In eukaryotes sexual reproduction results in the exchange of genetic information between closely related organisms. In bacteria, DNA may be transferred by conjugation, which involves direct contact between the donor and recipient strain; transduction, which utilizes a virus as a vector for carrying the DNA from the donor to the recipient strain; and transformation, which involves the uptake of naked DNA by a competent recipient strain.

Recombination involving plasmids provides a mechanism for the particularly rapid dissemination of genes through a population. Plasmids are quite useful as carriers of foreign genetic information in genetic engineering. Recombinant DNA technology can be employed to create organisms that contain combinations of genetic information that do not occur naturally. For example, bacteria containing genetically engineered plasmids can synthesize proteins that are normally produced only in eukaryotic organisms. Human genes have been successfully transferred to bacteria so that now bacteria exist that make human insulin, human interferon, and human growth hormone. In theory, recombinant DNA technology can be used to engineer organisms that can produce any desired combination of proteins. As a result genetic engineering holds great promise in industry, agriculture, and medicine because various proteins of economic importance or those of use in curing disease may be produced.

Study Outline

A. DNA replication.

   1. Transmission of hereditary information necessitates the faithful replication of DNA.

   2. DNA is replicated by a semiconservative process; after replication, half of the double helix (that is, one strand) has been retained (conserved) from the original DNA, and the other strand (called the complementary strand) has been synthesized.

   3. In DNA replication, one strand acts as a template for the complementary strand (which contains the complements of the bases on the template); adenine is aligned opposite thymine and cytosine is aligned opposite guanine.

   4. The site of replication is called the replication fork.

   5. Replication proceeds slightly differently on each of the parent strands; one strand of the DNA is called the leading (continuous) strand because its formation is in the direction of the movement of the replication fork; the other strand is called the lagging or discontinuous strand because DNA polymerase synthesizes small segments of DNA; ligases attach the short segments of DNA.

   6. In prokaryotes, DNA replication starts at only one site; in eukaryotes DNA replication is initiated at multiple sites.

B. Recombination.

   1. Recombination involves a reshuffling of genes.

   2. In some recombination processes there is a transposition in the locations of genes.

   3. In many recombination processes there is an exchange of genetic information from different sources that results in new combinations of genetic information.

C. Genetic exchange.

   1. Genetic exchange in prokaryotes occurs by transformation, transduction, and conjugation.

   2. In transformation, a free DNA molecule is transferred from donor to recipient bacterium.

3. In transduction, DNA is transferred from donor to recipient cells by a viral carrier; the virus acquires a portion of the genome of the host (donor) cell and transfers it to the recipient bacterium.
4. Conjugation (mating) requires physical contact between donor and recipient bacteria; there are different mating pairs that differ in the extent of successful genetic exchange (recombination) and whether the progeny are donor or recipient strains.
5. In eukaryotic microorganisms, genetic exchange normally occurs through sexual reproduction.
6. The vegetative cells of eukaryotic microorganisms typically are diploid, and sexual reproduction normally involves formation of specialized reproduction gametes or spores that are haploid; conversion of diploid to haploid states occurs during meiosis (reductive division); haploid nuclei of reproductive cells later fuse with nuclei of reproductive cells of an appropriate mating type (syngamy), reestablishing the diploid state.

D. Genetic engineering.

1. Recombinant DNA technology can be used to engineer the creation of new microorganisms.
2. Genetic engineering often uses plasmids as carriers of unrelated DNA; it employs enzymes involved in normal recombination and replication of DNA for splicing foreign DNA into plasmid carriers.
3. A typical procedure for genetically engineering a new organism is to isolate plasmids or another vector; to cut open the plasmid DNA by using site-specific endonuclease (restriction enzyme) to create an insertion site for foreign donor DNA; to use endonucleases to form a segment of donor DNA; to pair the plasmid and donor DNA; to use ligases to seal the circular plasmid; and to add the plasmid to a culture of recipient bacterium, which incorporates it as part of the bacterial genome; to clone the new organism.
4. Many economically beneficial products can be formed by genetically engineered organisms.
5. Genetic engineering permits the exchange of genetic information across species lines, including the intermixing of microbial, plant, and animal genes; bacteria have been created by recombinant DNA technology that produce human proteins, and attempts are underway to create plants containing bacterial genes.

Study Questions

1. DNA replication is ________ , which means that in each double-stranded DNA molecule, ________ strand(s) has(have) been ________ from the original DNA while the other strand is ________ . The enzymes that are responsible for DNA synthesis are called ________ . One strand that is synthesized is called the ________ or ________ strand; the other strand that is synthesized is called the ________ or ______ strand. On this strand, DNA is synthesized ________ in small segments that are then joined together by the enzyme ________ .
2. List the three ways DNA can be transferred from one prokaryotic cell to another.
3. Match the following phrases with the terms in the list to the right. Use each term as often as necessary.
   - **a.** Pieces of bacterial DNA are accidentally acquired by developing phage.
   - **b.** Requires physical contact between donor and recipient bacteria.
   - **c.** DNA is transferred from donor to recipient cell by viral carrier; the virus acquires portion of genome of host (donor) cell.
   - **d.** Only a portion of donor chromosome is normally transferred, depending on the time of mating.
   - **e.** Free DNA molecule is transferred from donor to recipient bacterium.
   - **f.** Lysis of recipient cell may not occur due to replacement of viral DNA with bacterial DNA.

   (1) Transformation
   (2) Transduction
   (3) Conjugation
4. List the steps involved in creating a bacterium that will synthesize human growth hormone.

Some Thought Questions

1. Should genetically engineered bacteria be released into the environment?
2. Is bacterial sex comparable to human sex?
3. If you were an investment counselor would you advise your clients to invest in a genetic engineering company?
4. How would you determine whether gene exchange and recombination occurred by conjugation, transduction, or transformation?
5. Why would *Time* magazine label genetic engineering as the most powerful and awesome skill acquired by man since the splitting of the atom?
7. What are some of the possible untoward effects of recombinant DNA technology?
8. What regulations would you consider necessary to protect the public and environment from experiments in genetic engineering?

Additional Sources of Information

Chakrabarty, A. M. 1978. *Genetic Engineering*. CRC Press, Inc., Boca Raton, Florida. A useful introduction to the techniques and potential applications of recombinant DNA technology.

Cohen, S. N., and J. A. Shapiro. 1980. Transposable genetic elements. *Scientific American*, **242**(2):40–49. An interesting article on a form of recombination that results in rearrangements of the locations of genes within the genome.

Genetics: Readings from Scientific American. 1981. W. H. Freeman and Company, Publishers, San Francisco. A well-worth-reading collection of articles on genetics, including coverage of genetic exchange mechanisms, recombination, and genetic engineering.

Gilbert, W., and L. Villa-Komanoff. 1980. Useful proteins from recombinant bacteria. *Scientific American*, **242**(4):74–94. A very interesting discussion of the potential uses of genetic engineering.

Grobstein, C. 1977. The recombination debate. *Scientific American*, **237**(7):22–33. A fascinating article on the potential benefits and risks of genetic engineering.

Novick, R. 1980. Plasmids. *Scientific American*, **243**(6):103–123. A discussion of the properties of plasmids including their potential as vectors for use in the transfer of genetic information between eukaryotic and prokaryotic cells.

Old, R. M., and S. B. Primrose. 1981. *Principles of Gene Manipulation: An Introduction to Genetic Engineering*. Blackwell Scientific Publications Ltd., Oxford, England. An informative introduction to the field of genetic engineering.

UNIT FIVE
Survey of Microorganisms

12 *Replication and Structure of Viruses and Other Acellular Microorganisms*

KEY TERMS

adsorption
animal virus
bacteriophage
budding
burst size
capsid
capsomeres
cytopathic effect
double-stranded DNA virus
double-stranded RNA virus
eclipse period
latent period
lysis
lysozyme
lytic phage
lysogeny
phage
plant virus
plaque
prion
prophage
retroviruses
reverse transcriptase
scrapie
single-stranded DNA virus
single-stranded RNA virus
temperate phage
transformation
viroid
virus

OBJECTIVES

After reading this chapter you should be able to

1. Define the key terms.
2. Describe the structures of acellular microorganisms.
3. Discuss why viruses are incapable of independent existence.
4. Discuss how genetic information is stored in viroids and viruses.
5. Compare the structural differences between viruses and bacteria.
6. Describe the stages in the replication of a lytic bacteriophage.
7. Describe the steps in the growth curve of a lytic virus.
8. Describe the replication of a temperate phage.
9. Describe the replication of an animal virus.
10. Discuss the differences in the replication of RNA and DNA viruses.
11. Describe how viruses transform animal cells.
12. Discuss the criteria that are used for the classification of viruses.

Acellular Microorganisms

The acellular microorganisms range from individual macromolecules to highly organized complex assemblages. They have in common their ability to reproduce their own structure within the confines of the cell of a compatible organism. It is this reproductive capacity that gives the acellular microorganisms their lifelike character and distinguishes them from other chemical combinations of molecules. Clearly the acellular microbe represents a borderline case between a living organism and a nonliving entity. The lack of a cytoplasmic membrane precludes an acellular microorganism from acting independently as a living system. Without a suitable barrier layer it is not possible to carry out the metabolic reactions needed to provide the energy and structural molecules of life's dynamic processes.

> Now, however, systems are being discovered and studied which are neither obviously living nor dead and it is necessary to define these words or else give up using them and coin others. When one is asked whether a filter-passing virus is living or dead the only sensible answer is, "I don't know; we know a number of things it will do and a number of things it won't and if some commission will define the word 'living' I will try to see how the virus fits into the definition." This answer does not as a rule satisfy the questioner who generally has strong but unfortunate opinions about what he means by the words living and dead.
>
> —N. V. PIRIE, *The Meaninglessness of the Terms Life and Living*

Even if you consider the acellular microbes as nonliving because they do not have all the properties we associate with living systems, the importance of acellular microorganisms cannot be overlooked. When acellular microbes reproduce within the cells of other organisms, they disrupt the normal cellular functions, often producing diseases in those organisms. Each acellular microorganism produces characteristic symptoms when it reproduces within a host organism. For example, tobacco mosaic virus produces a characteristic mosaic pattern on the leaves of a tobacco plant when it infects the cells of that plant, and we easily recognize the red rash of the measles when the measles virus reproduces within human cells.

There is great specificity between the acellular microbe and the host cell within which it can reproduce. For example, viruses that reproduce within a bacterial cell do not reproduce within human cells. The specificity of the relationship between an acellular microorganism and its host cell together with other characteristics of acellular microorganisms suggest that they evolved from host cells rather than as primitive or independent entities. The viruses appear to serve a useful function in genetic exchange between the host cells they infect. They also unfortunately cause abnormalities, which may include malignancies, in the host organisms they infect.

Prions

The most recently discovered and least understood microbes are the prions. The discovery of prions in 1983 was quite unexpected, and many scientists still find their existence difficult to accept. What is so unusual about prions is that they are composed only of protein. These "organisms" are nothing more than specific infectious protein molecules that contain the information that codes for their own replication.

In the discovery of prions we have found an exception to what appeared to be a universal characteristic of living systems—the storage of genetic information in nucleic acid molecules. All other organisms store their genetic information in nucleic acids, DNA (deoxyribonucleic acid) or, less commonly, RNA (ribonucleic acid). We know how the genetic information in these nucleic acid molecules is replicated so that it can be passed from one generation to the next. We also have broken the genetic code and know how the information contained within nucleic acid molecules is expressed, that is, how it is processed to determine the specific characteristics of each organism. We do not know how a protein can direct its own replication, and thus we do not understand how prions replicate.

Although we do not know exactly how prions reproduce nor how prevalent they are, we recognize their potential importance. Prions were discovered during the search for the cause of scrapie. Scrapie is an infectious and usually fatal disease of sheep. It is characterized by a wild facial expression, nervousness, twitching of the neck and head, grinding of the teeth, and scraping of portions of the skin against rocks with subsequent loss of wool. This disease was known to be caused by an agent that could pass through a bacteriological filter and was believed to be caused by a virus. However, no virus could be found, and eventually scrapie was shown to be caused by a prion. It is likely that other diseases, including some human diseases, result from the reproduction of prions within host cells. Prions and the consequences of their reproduction within the cells of the organisms they infect may explain several diseases that up until now have no known cause. Some scientists have suggested that Alzheimer's disease, a degenerative disease of the nervous system that afflicts a large number of people over 40 years old, is caused by a prion. This is one of several hypotheses to explain the etiology of degenerative nervous disorders. Much more research is needed to reveal the importance of the discovery of prions.

Viroids

Like prions, viroids are simply molecules that contain the information for their own reproduction. However, whereas prions are protein molecules, viroids are composed exclusively of RNA. They have no other structures, and their RNA genomes (complete sets of genetic information) are quite small (Figure 12.1).

Inside a suitable host cell the RNA of a viroid is capable of initiating its own replication. The replication of viroids sometimes manifests itself as disease symptoms in the host organism, and certain plant and animal diseases have been identified as caused by viroids. The viroids were discovered in the early 1970s to cause potato spindle tuber disease, chrysanthemum stunt, and citrus exocortis. It is also suspected that some human diseases are caused by viroids, but this has yet to be confirmed. In essence, viroids are simply molecules that can be preserved and transmitted to cells, where they are reproduced.

The RNA molecule of the viroid genome contains the information needed for directing its own replication. Viroids, however, are totally dependent on the metabolic activities of a host cell for accomplishing the task of replication. It is not yet clear how these molecules survive outside of host cells, and how they are transported from one cell to another. The lack of protective structures surrounding the RNA genome makes these "organisms" unique in the biological world. That such naked RNA molecules can be transmitted and cause infectious diseases of higher organisms is a relatively new finding the ramifications of which have yet to be fully appreciated.

Viruses

Viruses are more highly structured than viroids and prions, having an outer protective layer surrounding their genetic material. A central genetic nucleic acid core composed of DNA or RNA surrounded by a protective coat composed of protein normally constitutes the entire structure of the vi-

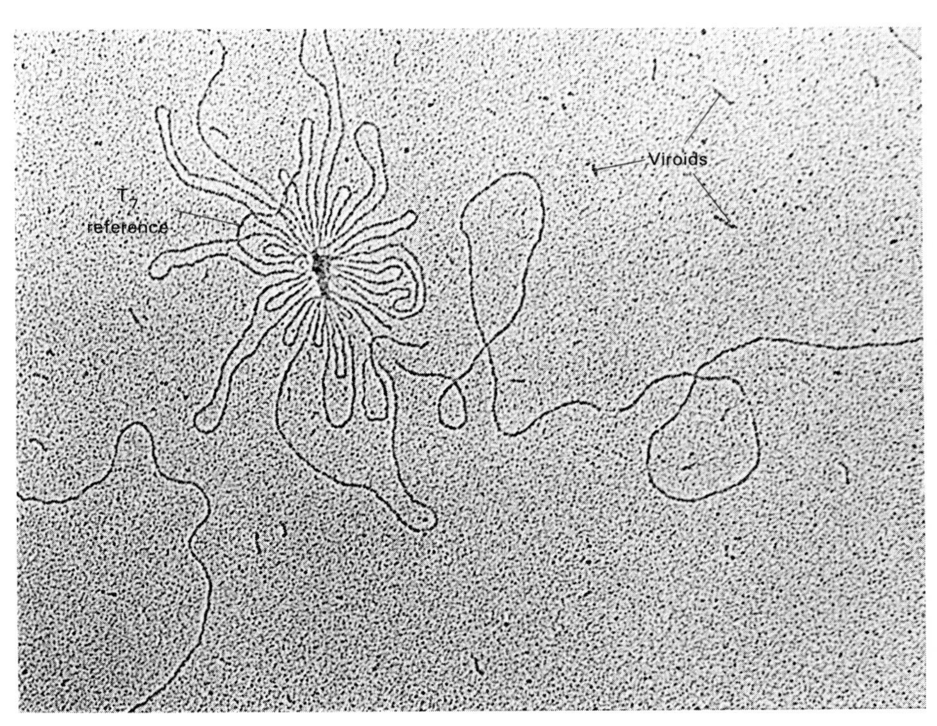

Figure 12.1 Photomicrograph of a viroid. Note the size difference between the reference bacteriophage T7 DNA and the RNA viroids. (From BPS: T. O. Diener, U.S. Department of Agriculture, and T. Koller and J. S. Sago, Swiss Federal Institute of Technology, Zurich.)

rus. The viral coat structure surrounding the nucleic acid, known as the capsid, is composed of protein subunits called capsomeres. The capsid protects the nucleic acids from enzymatic digestion and promotes attachment to susceptible host cells. Whereas the capsid of some viruses is composed of a single type of protein, many other viruses have more complex capsids containing several different proteins. Unlike prokaryotic and eukaryotic cells, viruses only contain one type of nucleic acid, either RNA or DNA. The genome of the virus may consist of double-stranded DNA, single-stranded DNA, single-stranded RNA, or double-stranded RNA. The ability of nucleic acid molecules other than double-stranded DNA to store the genetic information of the organism is unique to the acellular microbes.

The capsids of some viruses, such as tobacco mosaic virus, form a helical coil around the nucleic acid (Figure 12.2). The elucidation of the structure of tobacco mosaic virus (TMV) was accomplished in large part by work at the Medical Research Council's Laboratory of Molecular Biology in Cambridge, England, under the direction of Aaron Klug. It took 20 years to determine the complete details of the structure and assembly of TMV. Klug developed electron microscopic methods for the visualization of three-dimensional structures, permitting the elucidation of the details of how viruses are constructed. Klug's work showed how the components of TMV are put together; starting with one disk of protein surrounding a strand of RNA, the initial complex serves as a center for the assembly of a stack of 100 such disks composed of a total of 2200 identical protein molecules that make up the complete virus. Structure is important in learning how biological entities function.

Unlike the helical structure of tobacco mosaic virus, some viruses form a geometric structure, known as an icosahedron, which resembles a pyramid. The capsids of icosahedral viruses have 20 triangular faces. In yet other viruses that infect bacteria, such as the T-even bacteriophage, the capsid structures are relatively complex. These bacteriophage or phage (viruses that infect bacteria) have a head and tail structure that have quite different structural appearances (Figure 12.3).

Besides the essential nucleic acid and capsid components, some viruses have an envelope that surrounds the virus particle. Some of the proteins of the envelope are involved in the binding of the virus to a host cell, and others cause the host cell to rupture (Figure 12.4). Part of the envelope is specified by the viral genome, but some components of the viral envelope are obtained from the host cell. When some viruses leave a host cell, they pick up a portion of the nuclear or cytoplasmic membrane; that piece of host cell membrane can

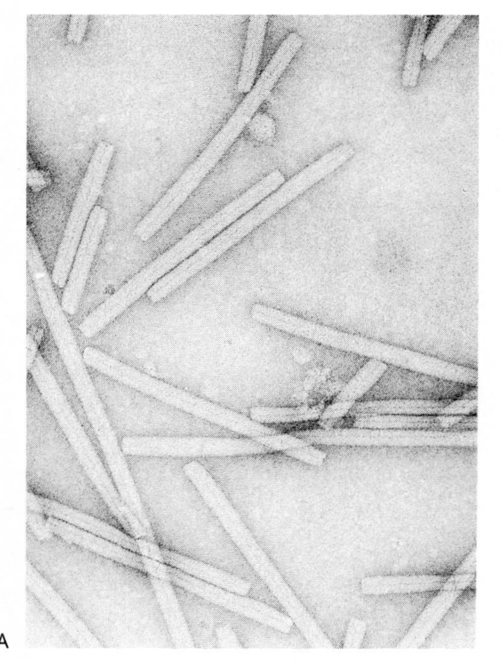

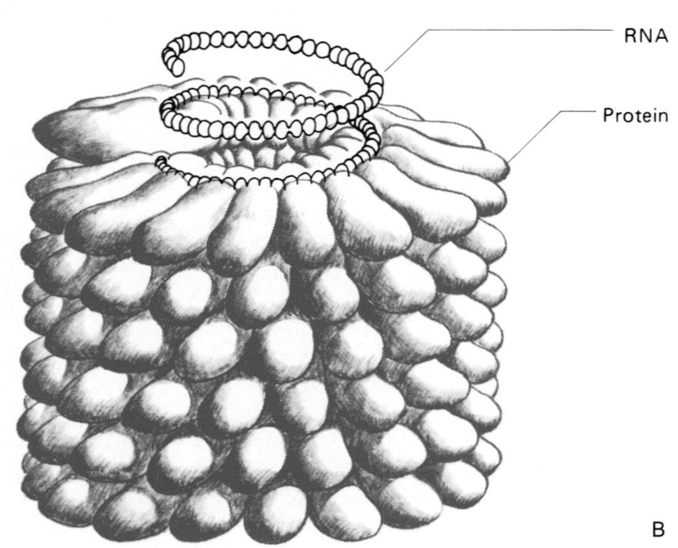

Figure 12.2 In tobacco mosaic virus (TMV) the subunits of the capsid are helically coiled around the virus's RNA nucleic acid. (A) An electron micrograph of TMV. (Courtesy Lee Simon, Waksman Institute of Microbiology, Rutgers, the State University, New Brunswick, N.J.) (B) A drawing showing the detailed structure of TMV.

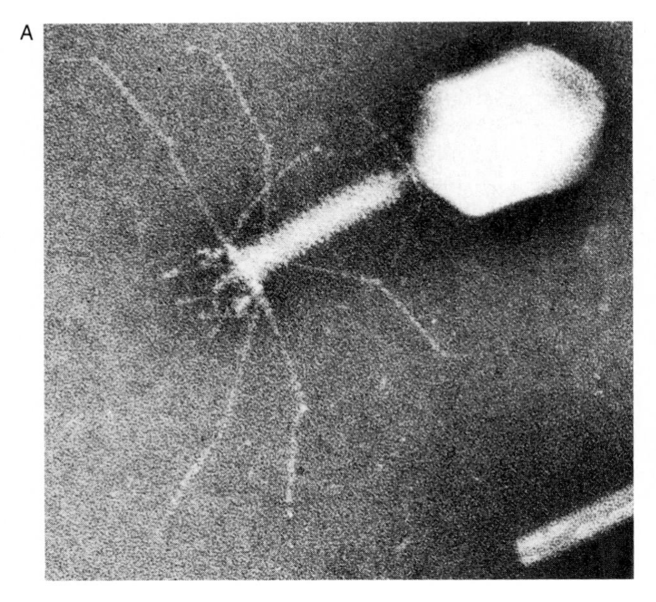

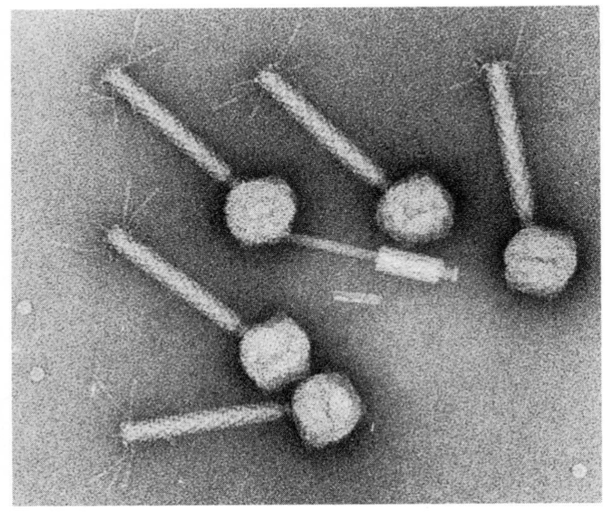

Figure 12.3 Electron micrographs of some bacteriophage with differing morphologies: (A) T4 phage (280,000 ×); (B) P2 phage (240,000 ×). (Courtesy Robley C. Williams, University of California, Berkeley.)

surround the viral capsid, forming part of the envelope. The presence of host cell membranes surrounding a viral particle can help the virus evade the normal host defense mechanisms designed to recognize and destroy foreign substances by making the outer layer of the virus look like a normal host cell. The viruses that cause herpes have such a host-cell derived envelope surrounding the virus, which in part contributes to the persistence of herpes infections.

Figure 12.4 (A) Micrograph of influenza A virus, strain USSR. (Courtesy E. Palmer, Centers for Disease Control, Atlanta.) (B) Micrograph of a rabies virus particle (280,000 ×). (From BPS: Alyne K. Harrison, Centers for Disease Control, Atlanta.)

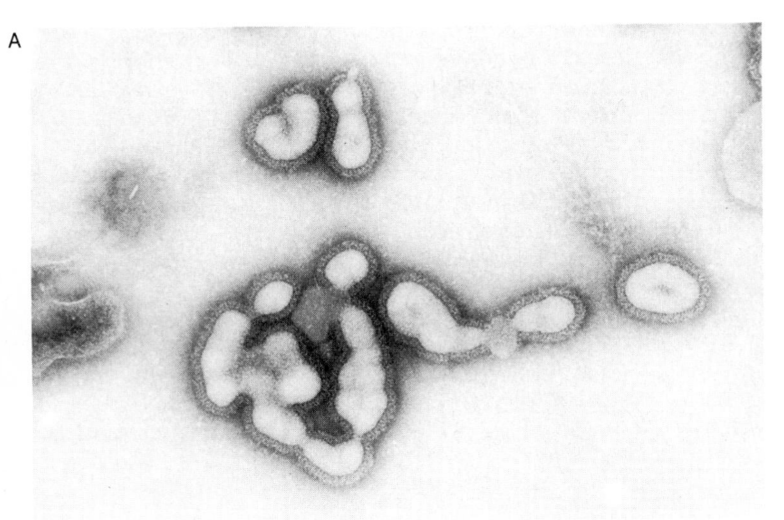
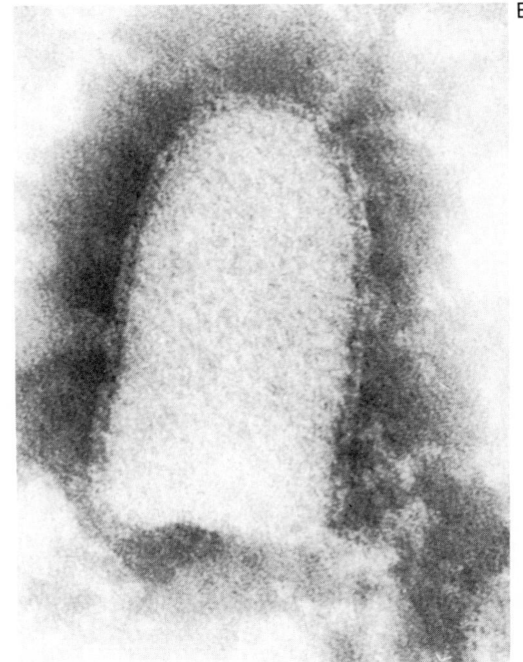

Throughout most of the life of the infected individual the herpes simplex virus lies latent in the body, without causing any symptom or obvious pathology . . . herpetic blisters thus provide a striking example of an infectious disease of man in which, contrary to the original tenets of the germ theory, the living agent of the disease can be present all the time in the host—be intrinsic, so to speak . . .

—RENE DUBOIS, *Mirage of Health*

Replication of Viruses

Although the specific details of viral replication vary from one virus to another, the general strategy for reproduction is the same for most viruses. The virus initially attaches to the outer surface of a suitable host cell. Generally, the attachment of a virus to a host cell involves adsorption to specific binding sites on the cell surface, which explains in part the high degree of specificity between the virus and the host cell. The nucleic acid of the virus then penetrates the cytoplasmic membrane of the host cell. Unlike the genomes of cellular organisms, which are always double helical DNA, the viral nucleic acid genome may be single-stranded RNA, double-stranded RNA, single-stranded DNA, or double-stranded DNA. Within the host cell the viral genome achieves control of the cell's metabolic activities. In many cases the viral genome actually codes for the shutdown of those metabolic activities normally involved in host cell's reproduction. The virus then uses the needed biochemical components and anatomical structures of the host cell for the production of new viruses. In particular, the virus employs the host cell's ribosomes for producing viral proteins and the cell's ATP and reduced coenzymes for carrying out biosynthesis. The nucleic acid and protein capsid structures of the virus (the essential structural components of a virus) are synthesized separately and then assembled prior to release from the host cell. Generally, many virus particles are produced within a single host cell and are released together. The replication of a virus results in changes in the host cell, often causing the death of the cell within which the virus reproduces. The cycle of viral adsorption to a host cell, invasion, synthesis of viral nucleic acid and proteins, assembly of the virus structure, and release of viral progeny is repeated when a virus encounters another suitable host cell.

Reproduction of Lytic Phage

The normal reproductive cycle of bacteriophage (viruses that infect bacteria) results in the lysis of the host cell at the completion of the viral replication cycle; thus, these bacteriophage are referred to as lytic phage. The lytic phage exhibit characteristic developmental sequences or stages in their reproductive cycles. The general sequence of events in the lytic reproductive cycle of a T-even phage is depicted in Figure 12.5. The reproduction of T-even phages within cells of bacteria such as *Escherichia coli* begins with the attachment of a T-even phage to the bacterial cell. There appear to be specific receptor sites on the bacterial cell surface where the phage may attach (Figure 12.6). The entire phage particle does not penetrate into the bacterial cell, but rather, the phage, using its tail structure like a syringe, injects its DNA into the bacterium. The contraction of the tail structure drives a tube through a pore in the outer layers of the bacterium, providing an entry tube for the phage DNA. Within the host cell, the T-even phage genome codes for a nuclease enzyme that degrades the host cell DNA, and the deoxynucleotides released by the degradation of the bacterial chromosome can be used as precursors for the synthesis of viral DNA.

The entire sequence of penetration, shutting off host cell transcription and translation, and the degradation of the bacterial chromosome takes only a few minutes. After these early events in the phage reproduction cycle, the host cell begins to synthesize proteins involved in the replication of the phage genome and for the various proteins that make up the capsid structure of the phage. The tail, tail fiber, and head structures of the phage are made up of proteins coded for by different phage genes, with at least 32 genes involved in the formation of the tail structure and at least 55 genes involved in the formation of the head structure of the phage.

The transition from early to late gene transcription involves an interesting shift in which one of the two DNA strands of the phage genome serves as the sense strand. The early genes are transcribed in a counterclockwise direction, whereas the late genes are transcribed in a clockwise direction. Because transcription proceeds in the 5′-hydroxyl to 3′-hydroxyl direction, the change from counterclockwise to clockwise must mean that the opposing strands of the viral DNA code for the early and late proteins. By altering reading frames and by changing which DNA strand serves as the sense strand, the relatively small phage genome can en-

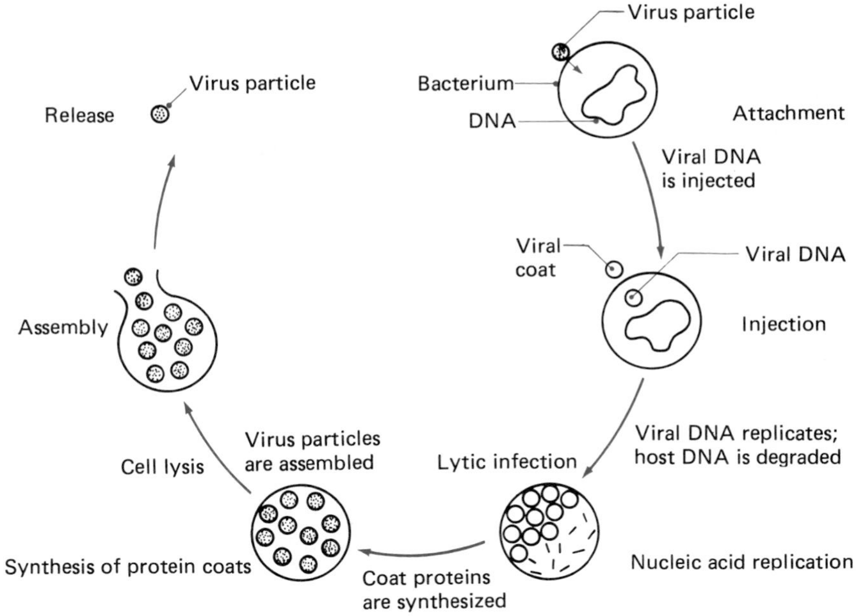

Virus particle

Bacterium

DNA

Attachment

Viral DNA is injected

Viral coat

Viral DNA

Injection

Viral DNA replicates; host DNA is degraded

Lytic infection

Nucleic acid replication

Coat proteins are synthesized

Synthesis of protein coats

Virus particles are assembled

Cell lysis

Assembly

Release

Virus particle

Figure 12.5 In the lytic phage reproduction cycle, the bacteriophage attaches to specific chemical sites on the surface of the bacterium and the viral nucleic acid penetrates the host cell. Once the viral nucleic acid is inside the bacterium, bacterial metabolism is halted by the destruction of the host genome. First viral DNA, then viral-coat proteins appear; they then self-assemble, and the host cell lyses. The free virions may infect other susceptible host bacteria.

code the almost 150 genes involved in the replication of T4 phage.

After the production of the individual components of the virus, the virus is assembled with packaging of the nucleic acid genome within the protein capsid. The assembly of the T-even phage capsid is a complex process. Assembly of the head and tail structures requires several enzymes that are coded for by the phage genome. The head, tail, and tail fiber units of the T-even phage capsid are assembled separately, and the tail fibers are added after the head and tail structures are combined. The small size of the viral particle means that the DNA must be tightly packed within the phage head assembly. Thus, packaging DNA into the head structure involves stuffing the head with DNA and cutting away

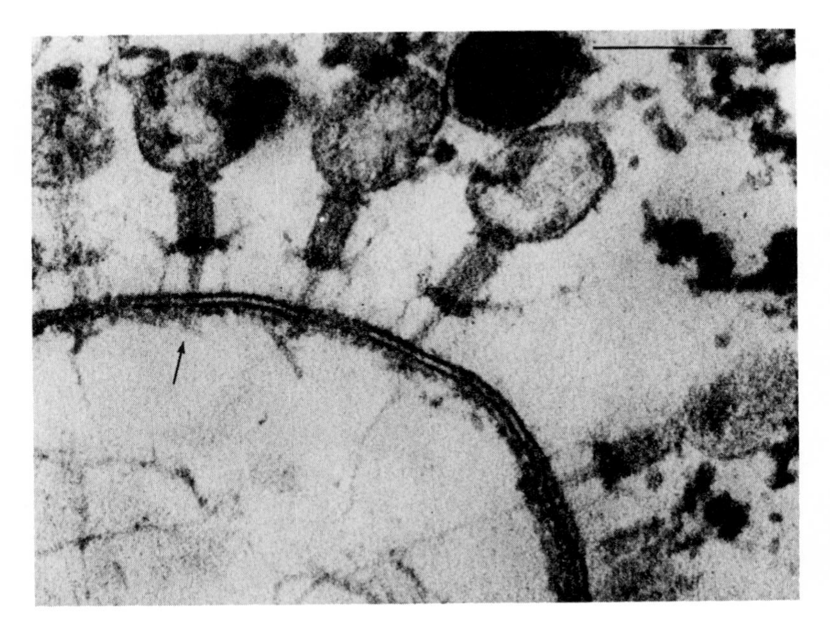

Figure 12.6 Electron micrograph showing the attachment of T-even phage to *Escherichia coli.* Different viruses attach at different receptor sites on the host cell surface. The tail structure penetrates the bacterial cell (arrow), and the phage DNA is injected into the bacterium as the sheath contracts. (Courtesy Lee Simon, Waksman Institute of Microbiology, Rutgers, The State University, New Brunswick, N.J.)

the excess. When the head structure is completely filled with DNA, any extra DNA is cleaved by a nuclease enzyme.

Once the phages are assembled, they must be released from the host cell and encounter another host cell for further reproduction. One of the late proteins coded for by the phage genome is lysozyme, which catalyzes the breakdown of the peptidoglycan wall structure of bacteria. The action of the lysozyme results in sufficient damage to the cell wall so that the wall is unable to protect the cell against osmotic shock, which results in the lysis of the bacterial cell and the release of the phage particles into the surrounding medium. The action of the lysozyme enzyme appears to be subject to phage-directed regulation, which ensures that the wall is not degraded prematurely before a sufficient number of phage particles has been completely assembled.

The lysis of the bacterial cell releases a large number of phage simultaneously, and consequently, the lytic reproduction cycle exhibits a one-step growth curve (Figure 12.7). The growth curve begins with an eclipse period during which there are no complete infective phage particles, and the naked DNA within the host cell is unable to infect other cells to initiate a new reproductive cycle. The end of the eclipse period is taken as the time at which an average of one infectious unit has been produced for each productive cell. The eclipse period is part of a longer period, the latent period. The latent period, which for a T-even phage often is about 15 minutes, begins when the phage injects DNA into a host cell and ends when one viral infectious unit per productive cell, on average, has appeared extracellularly. Because only complete phage particles are able to initiate the lytic replication cycle, the latent period does not end until assembled phages begin to accumulate within the bacterial cell.

Completely assembled phage particles continue to accumulate within the bacterial cell until, at the time of lysis, they reach a particular number known as the burst size (Figure 12.8). The burst size, which varies from cell to cell, represents the average number of infectious viral units produced per cell; a typical burst size for a T-even phage is 200. As a result of the simultaneous release of a number of infective phages, the number of phages that can initiate a lytic reproduction cycle increases greatly in a single step. The entire lytic growth cycle for some T-even phage can occur in less than 20 minutes under optimal conditions.

The growth curve for a bacteriophage can be de-

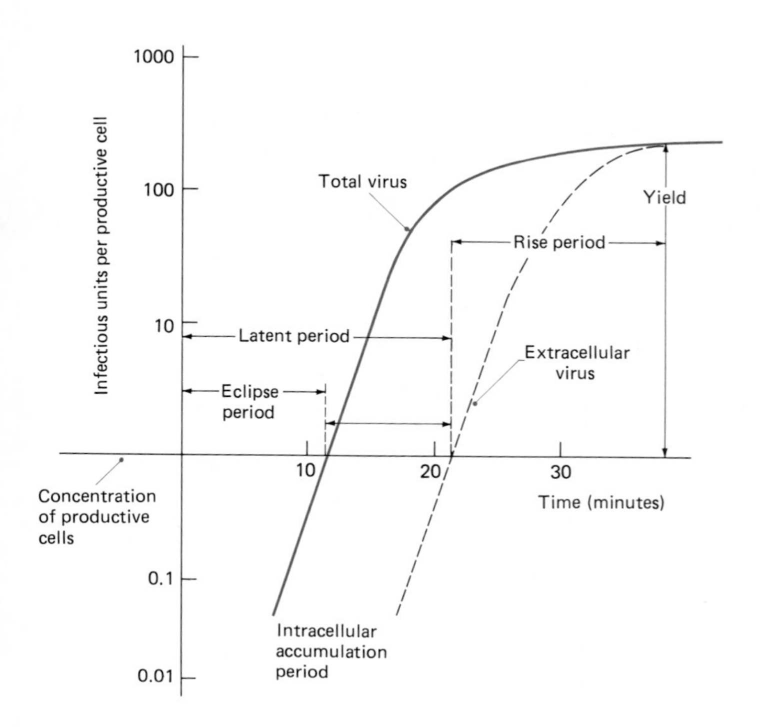

Figure 12.7 Detailed growth curve for bacteriophage T2, showing the eclipse, latent, and rise periods. To determine this viral growth curve, bacteria and phage were mixed for 2 minutes. The bacteria were recovered by centrifugation and plated to determine the number of productive bacteria. Other samples periodically were collected and used to assay for total phage after chloroform treatment to disrupt bacterial cells and extracellular phage in the supernatant.

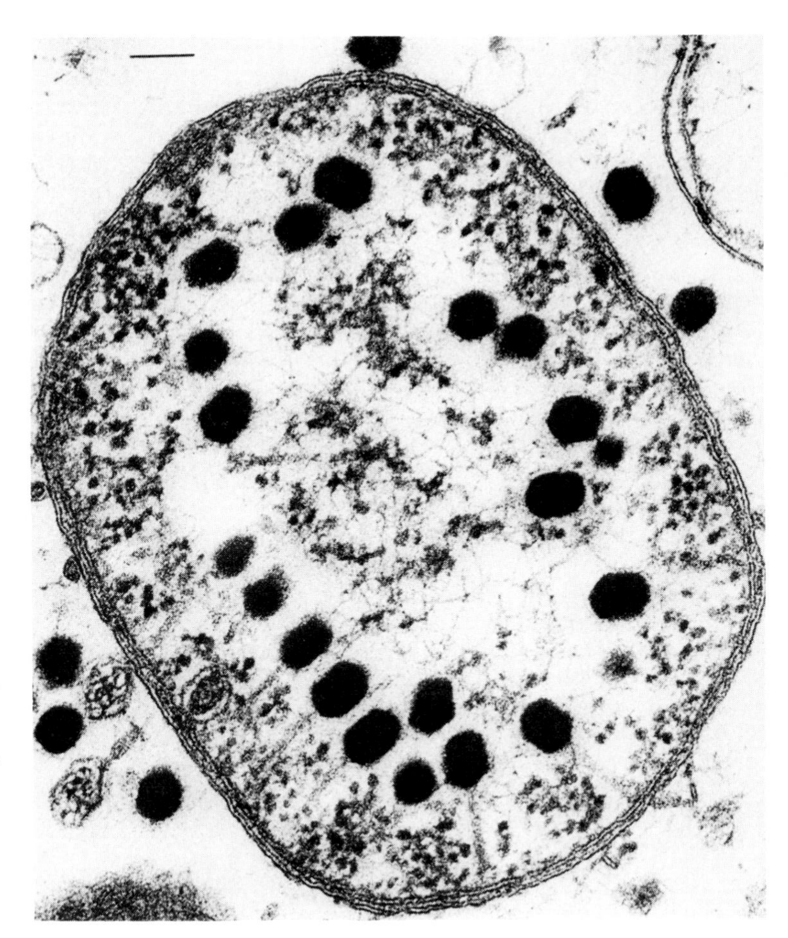

Figure 12.8 Assembled phage, T2, within the bacterial cell, *E. coli* strain B. *Note:* cell envelopes of the wall and cytoplasmic membrane; the clear area of the phage DNA pool contains many condensed phage DNA cores; and three phages attached to the cell surface, one is empty and two still partially filled. The phage at the top shows the long tail fibers and the spikes of the tail plate in contact with the cell wall. The tail sheath is contracted, and the tail core has apparently reached the cell surface but has not penetrated it. (Courtesy Lee Simon, Waksman Institute of Microbiology, Rutgers, The State University, New Brunswick N.J.)

Figure 12.9 The lytic replication of phage on a lawn of bacteria growing on a solid medium produced zones of clearing known as plaques. Here, the plaques of bacteriophage λ growing on *E. coli* are clearly visible. (From BPS: Richard Humbert, Stanford University.)

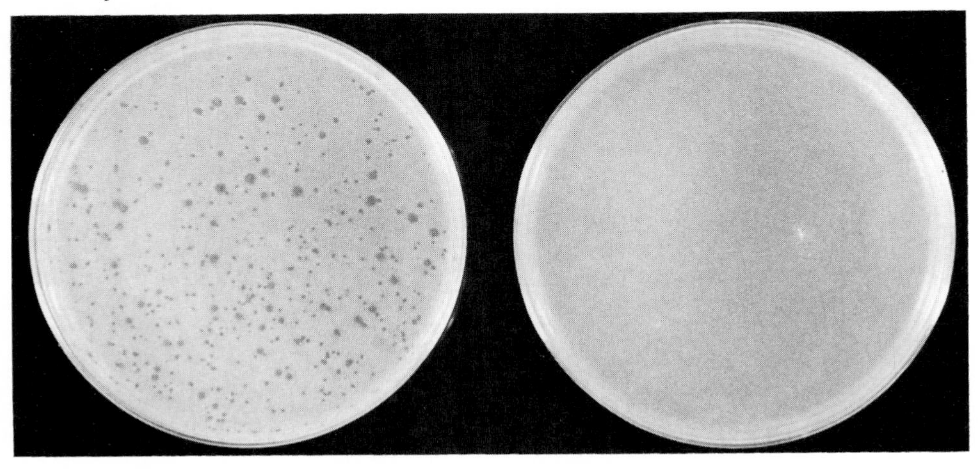

termined by inoculating a suspension of host bacterial cells with the phage and assaying for the number of infective phage particles at various time intervals. In one assay for infective bacteriophage, a lawn of bacteria is prepared on a suitable solid nutrient medium, and dilutions of the phage suspension are then spread over the same surface (Figure 12.9). In the absence of lytic bacteriophage, the bacteria form a confluent lawn of growth. Lysis by bacteriophage is indicated by the formation of a zone of clearing or plaque within the lawn of bacteria. Each plaque corresponds to the site where a single bacteriophage initiated its lytic reproductive cycle.

Temperate Phage Replication—Lysogeny

In addition to the normal lytic cycle, some bacteriophage, called temperate phage, are capable of carrying out a lysogenic replication cycle (lysogeny) (Figure 12.10). During the lysogenic replication of a virus, only the integrated genome of the virus is reproduced, and the virus does not code for the production of complete viral particles nor for the lysis of the host cell. The lysogenic reproduction cycle consists of three phases: (1) establishment of the prophage (incorporation of the phage genome into the bacterial genome, generally into the bacterial chromosome but in some cases into a bacterial plasmid); (2) maintenance of the prophage (replication of the phage DNA along with the normal replication of the DNA of the bacterium); and (3) release of the phage genome (excision of the phage DNA and initiation of the normal lytic reproduction cycle). Once incorporated into the bacterial chromosome, the viral genome (referred to as a prophage) is replicated along with the bacterial DNA during normal host cell DNA replication. At a later time the prophage can be excised from the bacterial chromosome or plasmid DNA, reestablishing a normal lytic reproductive cycle. The release of the phage genome from the bacterial chromosome completes the lysogenic reproduction cycle.

Plant Viruses

Many plant viruses exhibit a reproductive cycle analogous to the lytic reproduction cycle of bacteriophage, involving adsorption of the virus onto a susceptible plant cell, penetration of the viral nucleic acid into the plant cell, assumption by the viral genome of control of the synthetic activities of the host cell, coding by the viral genome for the

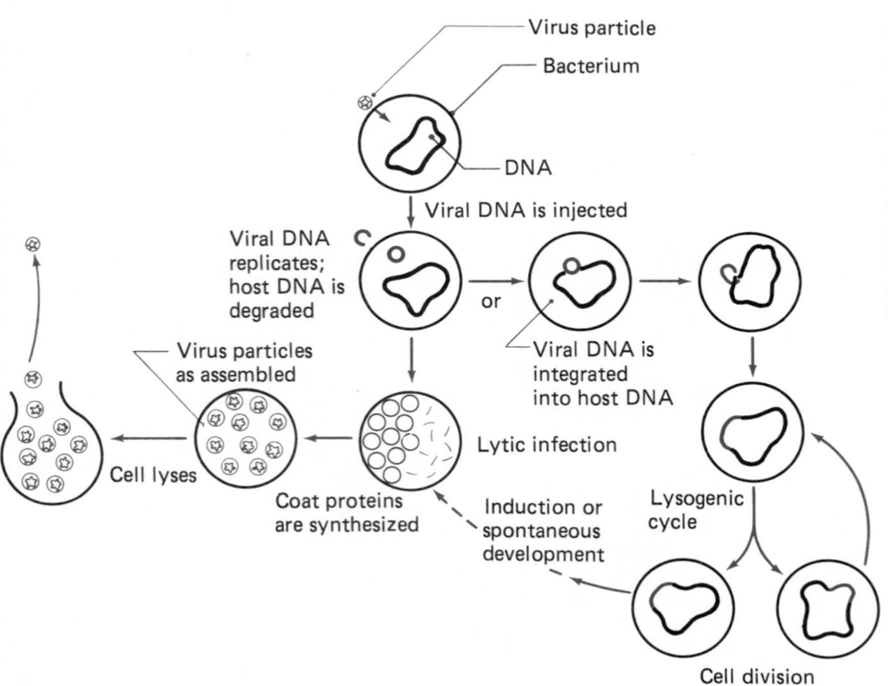

Figure 12.10 When a temperate bacteriophage infects, the result is usually a lysogenic cycle. The viral genome and the bacterial genome are fused enzymatically. The lysogenic cell continues growth and division normally through binary fission. When the cell is induced, the viral DNA deintegrates, and a productive or lytic infection is carried out.

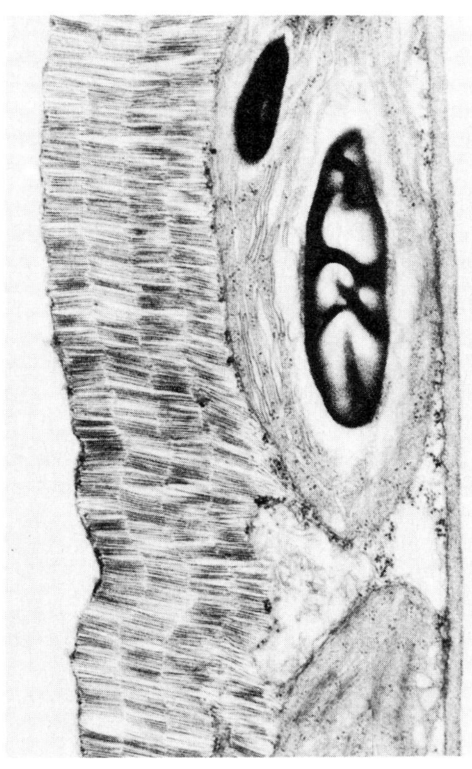
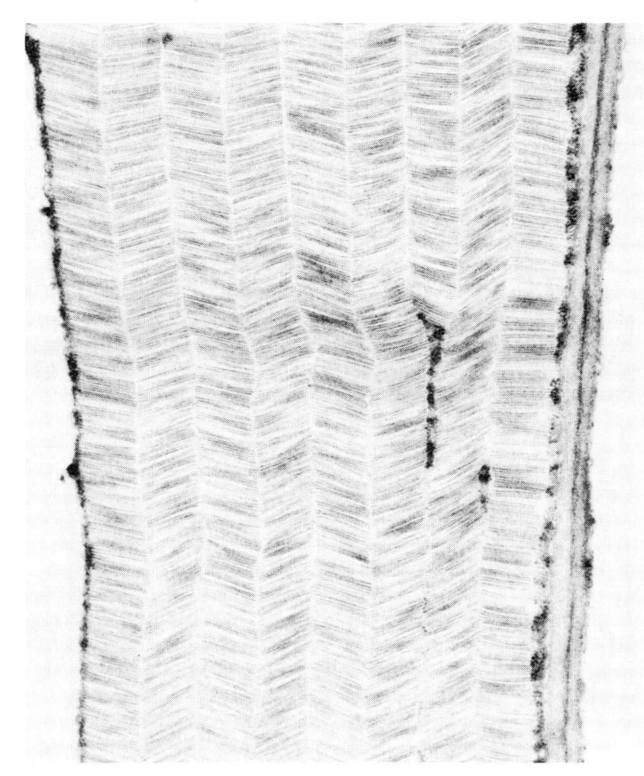

Figure 12.11 Electron micrograph of a thin section of a tobacco leaf infected with TMV. (A) Note chloroplast at bottom and section of viral crystal showing alignment of multiple viral particles. (B) Note herringbone pattern of three-dimensional array of viral particles. (Courtesy Harry E. Warmke, University of Florida, Gainesville.)

synthesis of viral nucleic acid and capsid components, assembly of the viral particles within the host cell, and finally, release of the complete viral particles from the host plant cell. In contrast to bacteriophage, both the capsid and nucleic acid core of viruses infecting eukaryotic cells may cross the cytoplasmic membrane by endocytosis, with release of the nucleic acid from the capsid occurring within the host cell.

As an example of the reproductive cycle of a plant virus, let us examine tobacco mosaic virus (TMV). TMV has a single-stranded RNA genome contained within a helical array of protein subunits that comprise the viral capsid. Replication of TMV occurs within the cytoplasm of the infected cell. The RNA genome of TMV codes for an RNA-dependent RNA replicase enzyme that is used for the synthesis of a complementary RNA strand to serve as a template for the synthesis of the RNA genome of TMV. The RNA genome acts as a template for the synthesis of mRNA, which is subsequently translated at the plant cell ribosomes for the production of the protein coat subunits. Once the RNA and protein components of tobacco mosaic virus are synthesized, the assembly of the protein coat around the central RNA core can proceed spontaneously; that is, tobacco mosaic virus is self-assembled.

Within infected plant cells, the replicated tobacco mosaic viral particles form cytoplasmic inclusions (Figure 12.11). These viral inclusions are crystalline in nature. The chloroplast of a tobacco mosaic virus infected leaf becomes chlorotic, leading to the death of the cell. The death of the plant cell releases completely assembled TMV particles and viral nucleic acid that has not been packaged with the protein subunits. Within plants both completely assembled viral particles and viral RNA can move from one cell to another, establishing new sites of infection. As a consequence of the reproduction of the viruses within the plant cells, the plant develops characteristic disease symptoms, which include the appearance of a mosaic pattern of chlorotic spots on the leaves that gives both the disease and the virus their names.

Animal Viruses

There are many types of animal viruses and many variations in the specific details of animal virus reproduction. In some cases the reproductive cycle of animal viruses closely resembles that of lytic bacteriophage; in such instances there is a stepwise growth curve, with a burst of a large number of viruses released simultaneously. Unlike bacteriophage, however, the single-step growth curve for viruses occurs in hours rather than minutes (Figure 12.12). Though many viruses exhibit single-step growth curves characterized by the death of the host cell and the simultaneous release of a large number of viruses, some animal viruses characteristically do not kill the host cell and reproduce with a gradual slow release of intact viruses. Additionally, some animal viruses transform the host cells, resulting in tumor formation, rather than death of the host cell. A change within animal cells grown in tissue culture due to a viral infection can be observed as a characteristic cytopathic effect (CPE).

The essential steps of the reproductive cycle for animal viruses are (1) attachment (adsorption of the virus to the surface of the animal cell); (2) penetration (entry of the intact virus or the viral genome into the host cell); (3) uncoating (release of the viral genome from the capsid); (4) transcription to form viral mRNA; (5) translation using viral-coded mRNA to form early proteins; (6) replication of viral nucleic acid to form new viral genomes; (7) transcription of mRNA to form late proteins needed for structural and other functions; (8) assembly of complete viral particles; and (9) release of new virions (Figure 12.13).

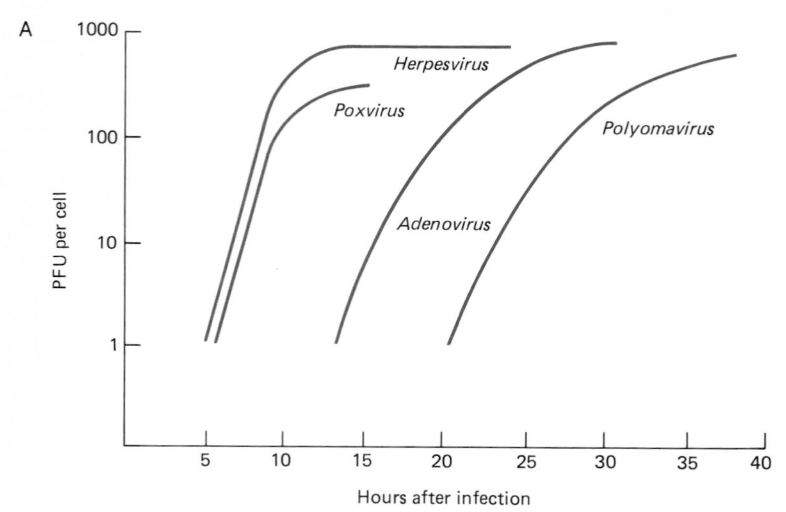

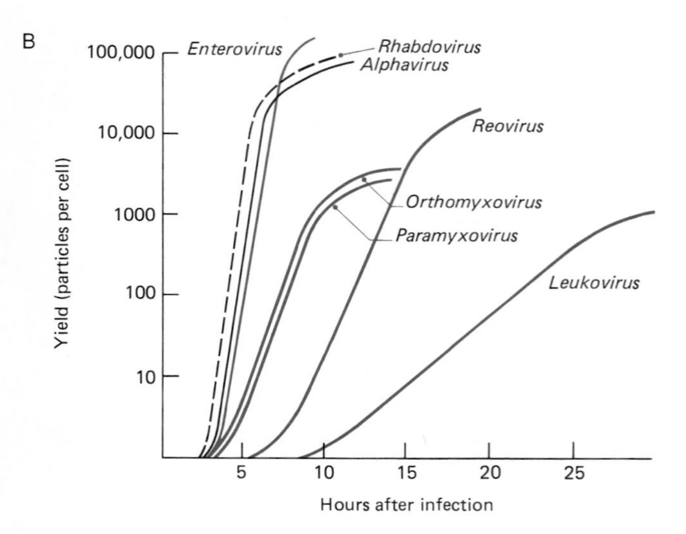

Figure 12.12 Growth curves for various DNA and RNA viruses. (A) Idealized multiplication curves (total infectious virus particles per cell at intervals after infection at high multiplicity) of viruses representing the major genera of DNA viruses. (B) Idealized multiplication curves (total virus particles per cell at intervals after infection at high multiplicity) of viruses representing the major genera of RNA viruses (Enterovirus, poliovirus, type 1; Alphavirus, Sindbis virus; Rhabdovirus, vesicular stomatitis virus; Orthomyxovirus, influenza type A; Paramyxovirus, Newcastle disease virus; Reovirus, reovirus type 3; Leukovirus, RAV-1). Latent periods, multiplication rates, and final yields are all affected by species and strain of virus and by cell strain and the kind of culture; latent period and multiplication rate, but not yield, by multiplicity of infection.

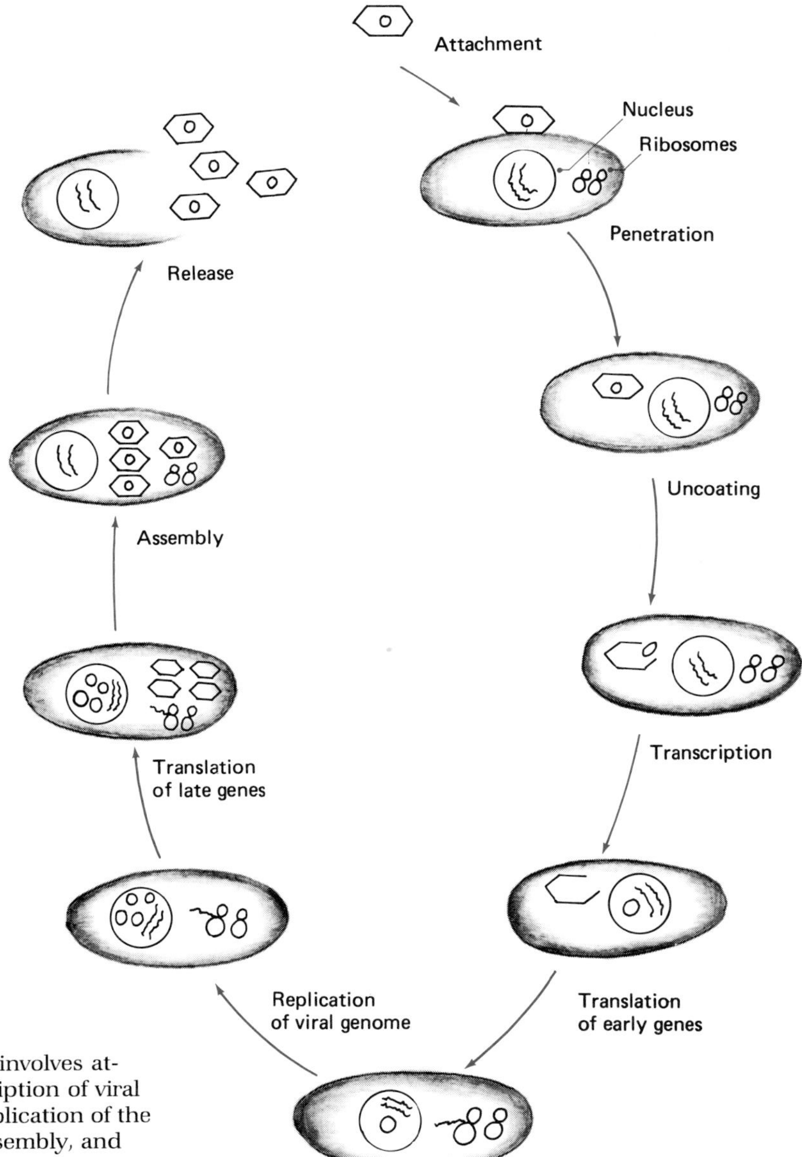

Figure 12.13 Animal viral reproduction involves attachment, penetration, uncoating, transcription of viral nucleic acid, translation of early genes, replication of the viral genome, translation of late genes, assembly, and release.

Viruses appear to adsorb onto specific receptor sites on animal cell surfaces, and as a rule the entire viral particle enters the cell, often by endocytosis.

Within the host cell, uncoating of animal viruses varies from one virus to another (Figure 12.14). The viral nucleic acids may be released at the cytoplasmic membrane, as occurs in enteroviruses, which are single-stranded RNA viruses; the virus may be uncoated in a series of complex steps within the host cell, as occurs in poxviruses, which are large double-stranded DNA viruses; or the virus may never be completely uncoated, as occurs in reoviruses, which are large double-stranded RNA viruses. After uncoating, the genome of a DNA animal virus generally enters the nucleus, where it is replicated, whereas the genome of most RNA animal viruses need only enter the cytoplasm of the animal cell to be replicated. The diversity and complexity of animal virus reproduction make it impossible to go into further detail. Rather, a few representative examples of the reproduction of different types of animal viruses will be discussed.

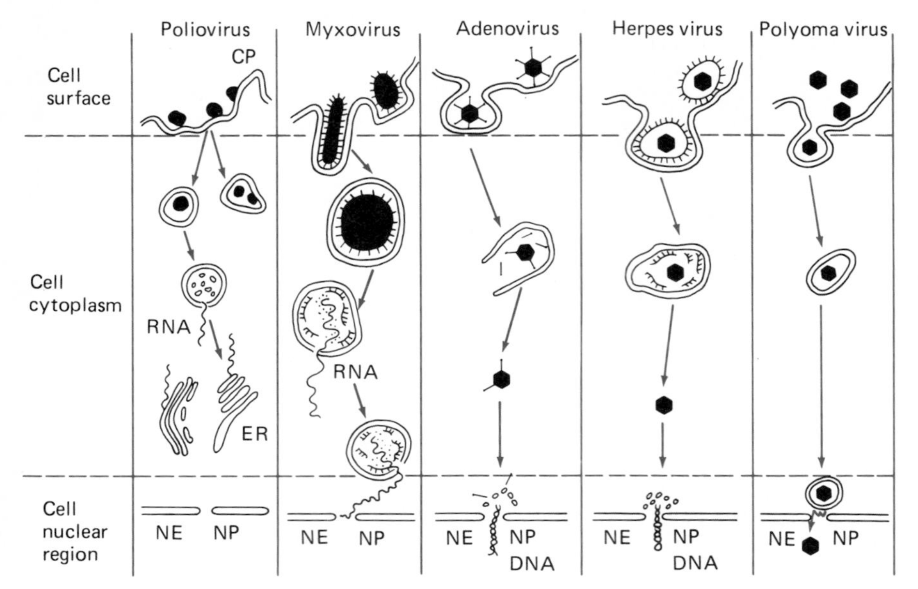

Figure 12.14 Some of the mechanisms associated with the attachment, penetration, and uncoating of certain viruses as observed in the electron microscope from thin sections of infected cells sampled over known time courses. NE = nuclear envelope; NP = nuclear pore; CP = cell plasmalemma; and ER = endoplasmic reticulum. (Based on S. Dales, 1973, *Bacteriological Reviews*, **37**:103–135.)

Reproduction of DNA Viruses

In the replication of adenovirus, a representative DNA animal virus, the host cell continues its normal metabolic activities for a short period of time after the entry of the virus. Uncoating the virus takes several hours, and during this period the viral nucleic acid is released from the capsid, entering the nucleus possibly through a nuclear pore. Within the nucleus, the viral genome codes for the inhibition of normal host cell synthesis of macromolecules. The viral genome also acts as a template for its own replication. Viral genes are transcribed, the resulting mRNA translated, and the proteins and enzymes needed for the assembly of the viral capsid are produced, with synthesis of viral proteins occurring at the ribosomes within the cytoplasm. The assembly of the adenovirus particles occurs within the nucleus, and therefore, the nucleus of an infected animal cell contains inclusion bodies consisting of crystalline arrays of densely packed adenovirus particles (Figure 12.15). Upon lysis of the host cell, numerous adenovirus particles are released.

Reproduction of RNA Viruses

The RNA animal viruses exhibit many diverse strategies for reproduction. In the case of poliovirus, the RNA genome of the virus acts as an mRNA

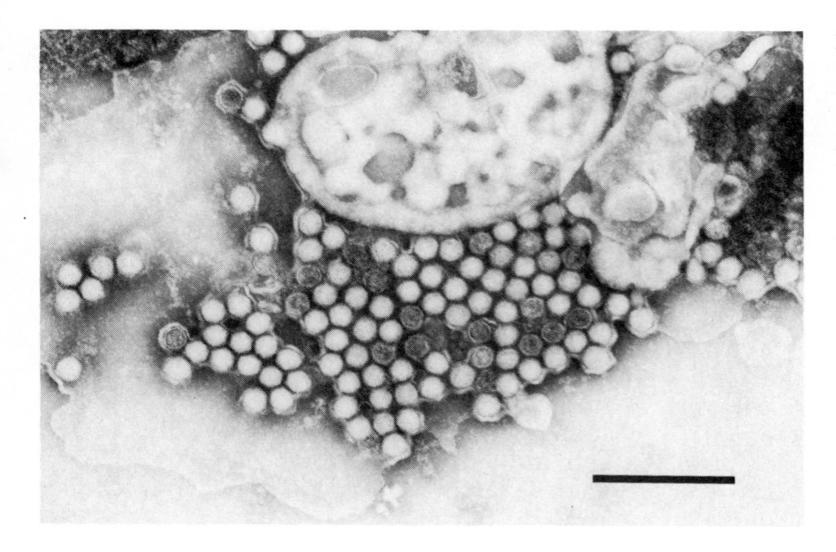

Figure 12.15 Electron micrograph of densely packed adenovirus particles within the nucleus of an infected cell. (Courtesy F. Williams, U.S. Environmental Protection Agency, Cincinnati.)

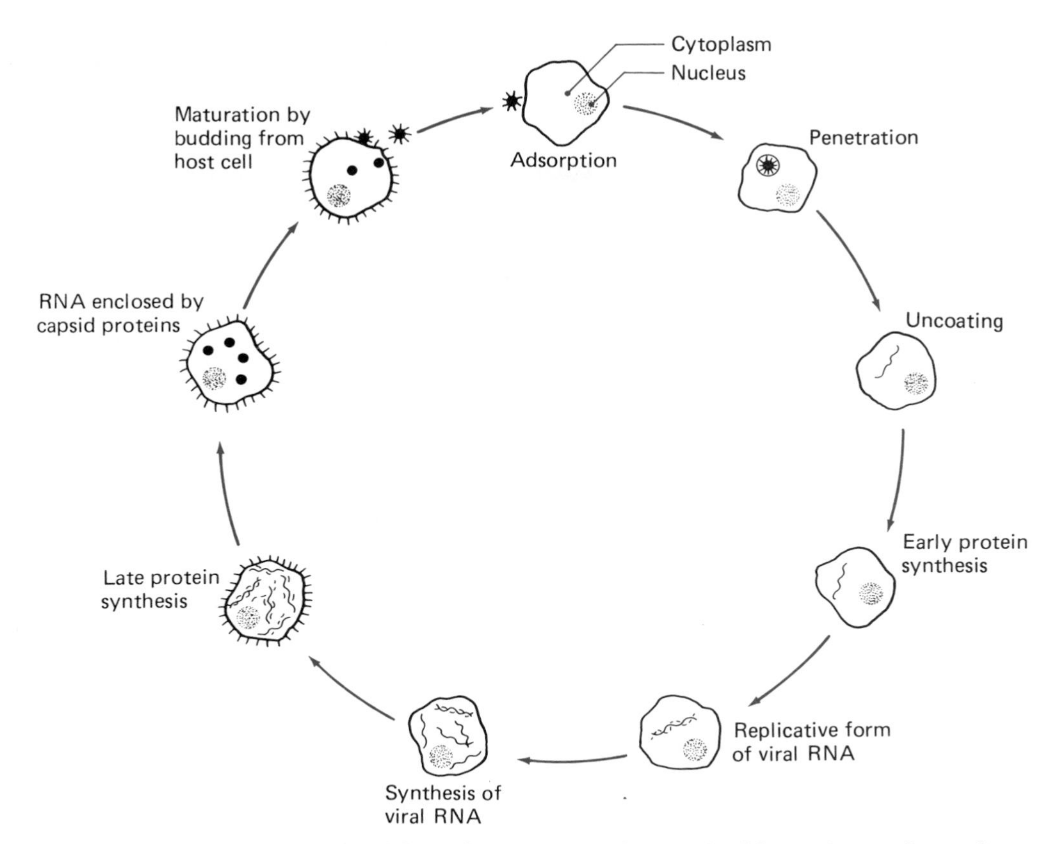

Figure 12.16 The life cycle of the influenza virus, showing budding release. The replication of the influenza virus illustrates many of the features that characterize animal virus interaction with host cells. The virion is taken into the cell intact; it is then uncoated, which releases the single-stranded RNA genome in seven segments (only one of which is shown here). A replicative form (Rf = shaded area) is made for each segment. The Rf in turn serves as the template for the synthesis of new genomes. As the late proteins are synthesized, they come to lie just beneath the cell membrane. Virions exit by budding through the membrane. A segment of the host cell's membrane becomes their envelope. It is chemically modified by the inclusion of the viral proteins hemagglutin and neuraminidase. In vitro virus production is eventually brought to a halt by interferon. Virus replication does not lyse the cell.

on entering the host cell, coding for the production of capsid proteins and an RNA-dependent RNA polymerase enzyme. Interestingly, the poliovirus RNA codes for a very large polypeptide chain, which is cleaved by protease enzymes, probably of the host cell, to form multiple different proteins, including the RNA-dependent RNA polymerase enzyme and four proteins of the viral capsid. The RNA polymerase enzyme is used to produce a complementary replicative RNA strand that can act as a template for the synthesis of new viral genomes. The morphogenesis of the viral particle, that is, the assembly of the capsid and insertion of the RNA genome, is followed by the release of a large number of viral particles. The poliovirus is believed to code for enzymes that are involved in the cleavage of the cytoplasmic membrane of the host cell, and thus, the mechanism for release of poliovirus from the host cell resembles the release of lytic bacteriophage.

In the case of influenza viruses, the viral RNA genome serves as a template for the transcription of mRNA molecules within the nucleus of the host cell rather than acting as the messenger RNA, as occurs in poliovirus. The genome of the influenza viruses is segmented, consisting of eight different RNA molecules, which code for different monocistronic mRNA molecules (RNA carrying information

Odd Virus Now Linked to Major Diseases

By HAROLD M. SCHMECK JR.

An unusual group of viruses, never implicated in human illness during more than a half century of study, have now been linked to three important diseases: cancer, the acquired immune deficiency syndrome and, most recently, hepatitis.

These viruses are called retroviruses. Some of them have been known since the early 1900's as causes of cancers in chickens and other animals. They have been much-studied over the years because of the clues they offer to the nature of cancer and even to the basic organization of life.

In addition to their newly discovered role in some human diseases, retroviruses seem to be the most promising prospect for use as vehicles to transplant genes from one species to another. As such, genetically engineered retroviruses may become agents of future attempts at human gene therapy. Some experts believe the first such attempts may come within two years.

In the past few years some of the retroviruses have been analyzed down to the ultimate molecules of their genes. Important clues have been collected on the specific means by which they cause cancer and why some of them do this with terrible swiftness while others take a large part of a lifetime. But the viruses continue to cause surprises.

One of the greatest surprises was the discovery that some retroviruses have been able to kidnap genes from animal cells and reintroduce them through infections, setting up the circumstances that can lead to cancer.

Only in recent years has the name retrovirus been applied to the viruses. The name comes from the virus's most characteristic and unusual trait, the possession of an enzyme, called reverse transcriptase, that reverses the flow of genetic information normal to all living things. In all animals, plants and microbes, DNA (deoxyribonucleic acid) is normally the archive of genetic instructions for every cell. It determines what each cell can become and every product that it will be able to make.

The messages of this genetic archive are then translated into the form of another nucleic acid, RNA (ribonucleic acid), which actually serves as the blueprint from which the cell makes its products. This progression from DNA to RNA to the manufacture of proteins has been called the central dogma of molecular biology.

Reverse transcriptase, as the name implies, reverses an important element of the process and translates the message of RNA into the form of DNA. The core of each retrovirus is composed of two identical strands of RNA. The enzyme makes it possible for the genetic blueprints embodied in these strands to be translated into DNA in cells infected by the virus. After this, the cell can be induced to manufacture new virus particles or their parts. Perhaps even more important, the DNA copy of the virus's genes can become a permanent part of the cell's genetic endowment.

The discovery, about 15 years ago, of reverse transcriptase and the chemical process the enzyme carries out has had pervasive effects on modern research in molecular biology. For this discovery, Dr. David Baltimore of the Massachusetts Institute of Technology and Dr. Howard Temin of the University of Wisconsin won a Nobel Prize in 1975. Since then, the reverse transcriptase has become a key tool in many aspects of biological research, and the retroviruses themselves have become central to much of the study of cancer. Retroviruses have been a key to the discovery and understanding of oncogenes, the particular genes of animals and humans that appear to play a role in the process by which cancer originates. These genes are thought to have normal functions in regulating cell growth and development, but they seem to contribute to the cancer process when activated too much or too little, when they're put in the wrong place within the genetic apparatus, or when a mutation has changed such a gene's chemistry.

"Basic cancer research with viruses has proved directly relevant to human cancers by identifying genes which can lead to cancer when several of them are altered in one cell," Dr. Temin said earlier this year in testimony before a Congressional subcommittee. "This identification provides targets for therapy and means for diagnosis and perhaps prevention."

Many Species Are Affected

Beginning in the early years of this century, scientists have identified more and more viruses that can cause cancers in such diverse species as fowl, mice, cats and nonhuman primates. More recently it has been discovered that many of these viruses are equipped with reverse transcriptases.

The inner core of an ordinary retrovirus contains three genes called pol, gag and env, as well as flanking sequences of RNA that are crucial to the process by which the virus genetic material is activated to subvert the cell it invades. Pol is the gene for the reverse transcriptase enzyme, gag is the gene for the virus's internal proteins, and env is the genetic code messages for production of the virus's outer envelope.

Some of the retroviruses also carry an oncogene picked up originally from an animal cell and incorporated permanently in the virus. Sometimes these genes replace part or all of one of the other genes. Sometimes the oncogene is simply an addition.

Scientists have discovered that those retroviruses that have oncogenes are usually the ones that produce deadly cancers in animals most rapidly, sometimes within only a few weeks. Some of the other retroviruses also cause cancer, but usually at a far slower rate.

Suspected in Human Disease

Because of the wealth of evidence linking various retroviruses with cancers in animals, many scientists thought one or another such virus must be a factor in some human cancers, too. Many different viruses were suspected, but none stood the test of thorough research. Meanwhile, other viruses totally unrelated to retroviruses were shown to be probable factors in some human cancers, although probably not the direct and sole cause.

Notable among these were some herpes viruses linked to cervical cancer; a herpes virus called the Epstein Barr virus linked to a kind of lymphoma especially in parts of Africa and to nasophryngeal carcinoma in Asia; and the hepatitis B virus, linked to liver cancer in parts of southern China and Southeast Asia.

But, the class of retroviruses, which included the first known cancer viruses of animals, still could not be tied clearly to any human cancer.

The break came about 1980 with the publication of reports by Dr. Robert C. Gallo's group at the National Cancer Institute linking some leukemia and lymphoma patients with what appeared to be a retrovirus. These were cancers of important white blood cells called T-lymphocytes. The virus was named human T-cell leukemia-lymphoma virus. It is now known as HTLV-1 and has been linked particularly to cancers in a region of southern Japan. The virus has been found in many areas of the world including the United States, the Caribbean region and Africa. It is generally regarded as the first cancer-causing human retrovirus.

A closely related virus called HTLV-2 has also been found in one cancer patient and in one chronic drug abuser, but has not yet been linked conclusively to human disease, Dr. Gallo and Dr. Samuel Broder of the National Cancer Institute said in an article in the Nov. 15 issue of The New England Journal of Medicine. The scientists said the genes of both viruses have been cloned and analyzed, and they do not have oncogenes. Nevertheless, they can transform cells growing in the laboratory into a cancerlike state.

A third member of that family of retroviruses, HTLV-3, is considered to be the cause of AIDS. A virus linked to AIDS discovered by scientists in France and named LAV is now generally believed to be identical to HTLV-3.

One of the greatest recent surprises in retrovirus research was the report a few weeks ago implicating something resembling one, in a puzzling category of hepatitis infections.

A group led by Dr. Robert J. Gerety of the Food and Drug Administration found evidence of reverse transcriptase activity in blood serum samples known to have transmitted what is called non-A non-B hepatitis. This hepatitis includes most of the liver disease transmitted by blood transfusion. It has been called non-A non-B because the two hepatitis viruses, hepatitis A and hepatitis B, have been ruled out as causes. Neither virus resembles retroviruses.

As directed by Dr. Gerety, the discovery of reverse transcriptase activity in non-A non-B hepatitis was to some extent a consequence of two earlier discoveries, the link between HTLV-1 and cancer, and, particularly, the relationship of HTLV-3 to AIDS.

The scientists at the F.D.A. were concerned over the possible risk of transmitting the AIDS virus with material used to test for the presence of antibodies against that virus. In testing human material for such use the scientists found clear evidence of reverse transcriptase activity. They also found related evidence including particles visible under the electron microscope that looked like retroviruses.

The link to hepatitis still remains to be confirmed.

Nevertheless, Dr. Gerety, as well as many other scientists, thinks we have by no means heard the last of retroviruses in human disease.

"That should be obvious," Dr. Gerety said. "I think there are a lot of chronic human diseases for which there are no known causes and for which one or another of the ominous retroviruses should make excellent candidates.

Source: The New York Times, Nov. 20, 1984. Reprinted by permission.

---

for single proteins), with each of the segments coding for separate mRNA molecules. The RNA-dependent RNA polymerase enzyme required for transcription of the viral genome is coded for by one of the RNA genome segments. The replication of the viral RNA genome involves the production of a complementary RNA strand that then serves as a template for the synthesis of new viral RNA genome molecules. The maturation of some viruses, such as influenza virus, occurs by budding, a process in which the viral particles are wrapped within a piece of the cytoplasmic membrane of the host cell (Figure 12.16). As a result, budding releases encapsulated (lipid-enveloped) influenza viruses slowly from infected host cells.

Reovirus, a double-stranded RNA virus, carries an RNA polymerase enzyme that is used for the synthesis of new viral genome molecules. The reovirus genome is segmented, containing ten different double-stranded RNA molecules. Each of the ten RNA genome molecules codes for the production of a different protein. The proteins are then assembled into the viral capsid, and the RNA genome is inserted prior to release of the completed reoviruses.

Retroviruses are RNA viruses that use a reverse transcriptase to produce a DNA molecule within the host cell. The production of the DNA molecule using RNA as a template is the reverse of the normal transcription process. It is the DNA molecule "transcribed" from the viral RNA genome that actually codes for viral replication within the host cell. Retroviruses are released from host cells by budding, and viral replication is nondestructive to the host cell because it does not result in lysis and death of the infected cell. Thus, these viruses can be released slowly and continuously from infected host cells.

It is of biological advantage for parasites not to kill their hosts, since disappearance of the host jeopardizes the parasite's survival. The most successful parasite, in other words, is the one that allows to its victim as much life as is compatible with its own needs.

—THEOBALD SMITH, *Parasitism and Disease*

Transformation of Animal Cells

The DNA produced during the reproduction of retroviruses can also be incorporated into the host cell's chromosomes. Within the chromosomes of the host cell the viral genome can be transcribed, resulting in the production of viral-specific RNA and viral proteins. The DNA coded for by the virus, which is incorporated into the host cell genome, can be passed from one generation of animal cells to another. It is, therefore, possible for animals to inherit a viral genome. The presence of viral-derived DNA within the host cell can transform the animal cell. Transformed cells have altered surface properties and continue to grow even when they contact a neighboring cell, resulting in the formation of a tumor. Viruses that transform cells and cause cancerous growth are called oncogenic viruses. Several different retroviruses produce malignancies within infected cells. Rous sarcoma virus, for example, is an RNA tumor virus that causes malignancies in chickens. Similarly, some DNA viruses, such as simian virus (SV40) and polyoma virus, are capable of transforming host cells and producing malignant tumors. These oncogenic viruses reproduce in permissive hosts (cells in which the virus can fully replicate) by using a normal lytic reproduction cycle. Viral reproduction does not occur in nonpermis-

Table 12.1 The Charcteristics of Various Families of Viruses

Viridae	Nucleic acid	Symmetry	Nucleocapsid	Number of capsomeres	Diameter of the nucleocapsid (nm)	Diameter of the envelope (nm)	Diameter and length of the nucleocapsid (nm)	Dimensions of the enveloped virions (nm)	Mol wt ($\times 10^6$) of nucleic acid	Number of nucleic acid strands
Ino-	D	H	N				0.5 × 85		1.7–3	1
Pox-	D	H	E					250 × 160 300 × 230	160–240	2
Micro-	D	C	N	12	25				1.7	1
Parvo-	D	C	N	32	22				1.8	1
Denso-	D	C	N	42	20				160–240	1
Papilloma- (papova)	D	C	N	72	45–55				3–5	2
Adeno-	D	C	N	252	70				20–25	2
Irido-	D	C	N	812	130				126	2
Herpes	D	C	E	162	77	150–200			54–92	2
Uro-	D	BC	N							2
Rhabdo-	R	H	N				2 × 13 1 × 125			1
Myxo-	R	H	E				9 × ?	100	2–3	1
Paramyxo-	R	H	E				18 × ?	120	7.5	1
Stomato- (rhabdo)	R	H	E				18 × ?	175 × 68	6	1
Thylaxo-	R	H	E					1000	10	1
Napo-	R	C	N	32	22–27				1.1–2	1
Reo-	R	C	N	92	70				10	2
Cyano-	R	C	N	32 or 42	54					2
Encephalo-	R	C	E				60–80		2–3	1

Code: D = DNA; R = RNA; H = helical; C = cubic; B = binal; N = naked; E = enveloped.

Based on A. Lwoff and P. Tournier. 1971. Remarks on the classification of viruses. In *Comparative Virology*, K. Maramorosch and E. Kurstak (Eds.), Academic Press, New York.

sive host cells, but rather, part of the viral genome is incorporated into the host cell genome resulting in the transformation of the host cell.

Classification of Viruses

Viruses have generally been classified nonsystematically; the common name of the virus generally describes the disease caused by that virus. Formal systems for the classification and nomenclature of viruses are relatively new, and most viruses are still referred to by their common names, as opposed to using a name assigned in a formal classification system. For example, we commonly refer to the measles virus, the poliovirus, and others, even though these viruses have been classified and given other formal names.

Viruses are often classified into large groups according to the type of host cell they infect. Three groups of viruses have usually been recognized: animal viruses, plant viruses, and bacteriophage. Animal viruses traditionally refer to those viruses that infect only vertebrate animals. Viruses that infect invertebrate animals and microorganisms other than bacteria are not covered in classification systems that rely on these three lines of division. As a result separate taxonomic consideration has been given to the insect viruses and to the viruses that infect fungi, protozoa, and algae.

Viral taxonomists working with viruses that infect different types of host cells have developed separate taxonomic schemes for classifying the viruses. Formal classification systems are largely based on the nature of the nucleic acid molecule and the arrangement of the capsid. Critical questions include whether the virus is a DNA or RNA virus, whether the nucleic acid is single-stranded or double-stranded, and whether the virus is naked or enveloped. Important features used in classifying viruses include the molecular weight of the nucleic acid, the symmetry of the capsid, the number of capsomeres, and the site of capsid assembly (Table 12.1).

Summary

Acellular microorganisms lack a cytoplasmic membrane. They are obligately dependent upon cellular organisms to reproduce and carry out any of the dynamic processes of living systems. These organisms include the prions, viroids, and viruses. The prions and viroids are simply molecules that can direct their own replication within the confines of a suitable host cell. Prions are infectious proteins. Viroids are naked RNA molecules. Both prions and viroids cause diseases of the host organisms in whose cells they replicate.

Viruses are more highly structured than prions and viroids. A virus has two essential components, a genome and a surrounding coat. The protective coat of a virus, called a capsid, is composed of protein. The capsid is constructed from capsomere subunits. The capsid is symmetrically wound around the nucleic acid core. Some viruses exhibit helical symmetry, some form complex geometric forms. The nucleic acid core of a virus houses the genome containing the genetic information of the virus. The genome of a virus can be composed of double-stranded DNA, single-stranded DNA, double-stranded RNA, or single-stranded RNA. In addition to the essential genome and capsid structures, some viruses have an envelope. The presence of an envelope often increases the ability of the virus to cause disease.

Viral reproduction can occur only intracellularly. Viruses lack the necessary metabolic capability and cellular structures to accomplish independent reproduction and therefore are dependent on their compatibility with host cells for continued existence. The key steps in the reproduction of viruses are (1) adsorption of the virus onto the host cell; (2) penetration of the viral nucleic acid across the host cell cytoplasmic membrane; (3) separation of the nucleic acid from the viral capsid; (4) shutdown of normal host cell synthesis of macromolecules; (5) production of proteins coded for by the viral genome; (6) replication of the viral genome; (7) assembly of the viral particles; and (8) release of the viral progeny. RNA and DNA viruses use different modes of reproduction for the expression and replication of the viral genome. In many viruses different reading frames and different directions of reading are used during transcription, enabling a small viral genome efficiently to code for the variety of proteins that are required for viral reproduction.

Often, the reproduction of a virus within a host cell results in the death of that cell. The lytic reproductive cycle of bacteriophage, for example, results in a one-step growth curve, with a large number of viruses released simultaneously from the lysed bacterial cell. Similarly, some plant and animal viruses reproduce with the simultaneous release of a large number of progeny from the host cell. Other viruses reproduce without killing the host cell, exhibiting a slow continuous release of progeny from the infected host cell, as occurs in the budding mode of reproduction. Some bacteriophage are temperate and are able to establish a prophage state that does not result in the death of the host cell. During lysogeny the viral DNA is incorporated into the bacterial chromosome and is replicated along with the bacterial DNA. At some later time the viral DNA can be excised from the bacterial chromosome, and a lytic reproduction cycle can be reestablished.

The viruses are classified both systematically, based largely on their molecular properties (modern approach), and nonsystematically, based largely on the host cells they reproduce within and the diseases they cause (classical approach). Different classification systems are used for vertebrate animal viruses, insect viruses, plant viruses, bacteriophages, and viruses that infect other microorganisms. The nature of the genome and capsid structure are important characteristics used in the formal classification of viruses. In the case of the viruses, it seems appropriate to consider the host cell as a primary basis for classification because viruses may have evolved as genetic extensions of prokaryotic and eukaryotic cells rather than as a direct evolutionary line from one viral species to another.

Study Outline

A. Acellular microorganisms.

1. Acellular microorganisms do not have a cytoplasmic membrane and are incapable of independent existence.
2. Prions, viroids, and viruses are acellular microorganisms.
3. Prions are infectious proteins. They are unlike any other organisms in that in addition to being acellular they do not have genetic information stored in nucleic acids (DNA or RNA).
4. Viroids are composed exclusively of RNA. They can reproduce within a suitable host cell and can cause disease.
5. Viruses have a nucleic acid core that houses the genetic information and a surrounding protein coat called a capsid.
6. Some viral capsids are helical, some are organized in a geometrical pattern called an icosahedron, and others have more complex symmetry.
7. The genetic information of a virus may be composed of double-stranded DNA, single-stranded DNA, double-stranded RNA, or single-stranded RNA.
8. Some viruses have an outer covering called an envelope that surrounds the virus. The presence of an envelope can help a virus evade normal host defense mechanisms.

B. Viral reproduction.

1. In order to reproduce, a virus must invade a suitable host cell; there is a high degree of specificity between virus and host.
2. The stages of viral reproduction generally include adsorption of virus to specific binding sites on cell surface; penetration of the cytoplasmic membrane by viral nucleic acid; control of the cell's metabolic activities by the viral genome; use of biochemical components and anatomical structures of the host cell for production of new viruses; and release of viral progeny.
3. The reproductive cycle of lytic phage results in host-cell lysis at completion of viral replication cycle.
4. The general developmental sequence for lytic phage involves: attachment of phage to specific receptor sites on the bacterial cell surface; injection of phage DNA into the bacterial cell, using its tail structure like a syringe; coding by phage genome for DNA polymerases; stoppage of transcription of bacterial genes; degradation of host DNA; production of phage DNA and capsid proteins; assembly of virus by packing nucleic acid genome into the capsid; and lysis of host cell, resulting in release of viruses.

5. The average number of viruses released when the infected cells lyse is known as the burst size.
6. The lytic phage exhibit a one-step growth curve.
    a. The eclipse period begins the growth curve, during which there are no complete infective phage particles.
    b. The latent period begins with injection of viral DNA into the host cell and ends with the extracellular appearance of one viral infectious unit per productive cell.
7. A bacteriophage growth curve is determined by inoculating a host bacterial cell suspension with phage and assaying for the number of infective phage particles at various time intervals by plaque formation (a plaque is a zone in which all bacteria have been lysed).

C. Lysogeny—temperate phage replication.
1. In lysogeny, only the integrated genome of a virus is reproduced; the virus does not code for production of complete viral particles nor for host cell lysis.
2. Lysogenic reproduction cycle consists of three phases:
    a. Establishment of prophage: incorporation of phage genome into the bacterial genome.
    b. Maintainence of prophage: replication of phage DNA along with the normal replication of bacterial DNA.
    c. Release of phage genome: excision of phage DNA and initiation of normal lytic reproduction cycle.

D. Plant viruses.
1. The reproductive cycle of many plant viruses is analogous to the lytic reproductive cycle of bacteriophage.
2. The cycle begins with the adsorption of the virus onto a susceptible plant cell; next the viral nucleic acid penetrates into the plant cell; then the viral genome asssumes control of the host cell's synthetic activities; the viral genome codes for synthesis of viral nucleic acid and capsid components; the viral particles then assemble within host cell, and the complete viral particles are released from the host plant cell.

E. Animal viruses.
1. There are many types and variations in animal virus reproduction.
    a. Some have a reproductive cycle that closely resembles that of lytic bacteriophage (although it takes much longer).
    b. Some characteristically do not kill the host cell, but reproduce with gradual slow release of intact viruses.
    c. Some transform host cells, resulting in tumor formation rather than host-cell death.
2. The essential steps in the animal virus reproductive cycle are attachment (adsorption of virus to surface of animal cell); penetration (entry of the intact virus or viral genome into the host cell); uncoating (release of viral genome from the capsid); transcription (formation of viral mRNA); translation of early proteins (using viral-coded mRNA to form early proteins); replication (formation of viral nucleic acid to form new viral genomes); transcription and translation of late proteins (for viral structure and other functions); assembly of complete viral particles; and release of new virons.
3. Reproduction of different animal viruses.
    a. Reproduction of DNA viruses uses the viral genome to direct the shutdown of host cell activities and the synthesis of new viruses; within the nucleus the viral genome codes for inhibition of normal host-cell synthesis of macromolecules and acts as template for its own replication.
    b. The reproduction of some RNA viruses uses the viral RNA genome as mRNA on entering host cell, coding for production of capsid proteins and RNA polymerase; the RNA polymerase then is used to produce a complementary replicative RNA strand that can act as a template for new viral genome synthesis.

c. The reproduction of other RNA viruses uses the RNA genome as a template for transcription of mRNA molecules within the host cell nucleus and for the synthesis of a DNA molecule that serves as the template for viral genome production; the DNA is synthesized by reverse transcription and the viral RNA genome is then made by normal transcription; this type of reproduction occurs in the retroviruses.

4. Some viruses are able to transform animal cells; DNA produced during reproduction of retroviruses can be incorporated into the host cell's chromosomes, where viral genome can be transcribed, resulting in production of viral-specific RNA and viral proteins; the viral genetic informtaion can be passed from one generation to another, making it possible for animals to the inherit viral genome; the presence of viral DNA in a host cell can transform an animal cell; transformed animal cells have altered surface properties and can continue to grow even when in contact with neighboring cells, resulting in tumor formation; viruses that transform cells and cause cancerous growth are called oncogenic viruses.

Study Questions

1. Match the following statements with the terms in the list to the right.
   - **a.** Acellular.
   - **b.** Has no membrane-bound organelles.
   - **c.** Cannot reproduce outside of host cell.
   - **d.** Contain DNA, RNA, and proteins.
   - **e.** Contain DNA and protein.
   - **f.** Contain RNA and protein.
   - **g.** Contain only RNA.
   - **h.** Contain only protein.

   (1) Prions
   (2) Viroids
   (3) Viruses
   (4) Bacteria
2. List the steps in the reproduction of a lytic bacteriophage.
3. List the steps in the reproduction of a temperate bacteriophage.
4. List the steps in the reproduction of a DNA animal virus.
5. List the steps in the reproduction of a RNA animal virus that does not produce reverse transcriptase.
6. List the steps in the reproduction of a RNA animal retrovirus.
7. Describe how a virus can transform an animal cell.
8. Match the following statements with the terms in the list to the right.
   - **a.** Characterized by temperate phage.
   - **b.** Lasts about 15 minutes for T-even phage.
   - **c.** Virus does not produce new virus particles.
   - **d.** Results in cell lysis at the completion of the viral reproductive cycle.
   - **e.** Viral genome is replicated along with that of the host cell.
   - **f.** Characterized by an eclipse and latent period.
   - **g.** Begins with the injection of viral DNA into the host cell and ends with the extracellular appearance of the first infectious unit from that cell.

   (1) Lytic phage
   (2) Latent period
   (3) Lysogeny

Some Thought Questions

1. Are viruses alive or not?
2. Can you send your enemies viruses through the mail?
3. If prions are really made of only protein, how can they reproduce?
4. Which came first, the virus or the host cell?
5. Is cancer caused by viruses?
6. When is it appropriate to give penicillin to treat a viral disease?
7. How do retroviruses differ from other living organisms?

Additional Sources of Information

Dalton, A. J., and F. Haguenau (eds.). 1973. *Ultrastructure of Animal Viruses and Bacteriophages: An Atlas*. Academic Press, Inc., New York. A pictorial view of the viruses, containing superb micrographs of the different animal viruses.

Diener, T. O. 1981. Viroids. *Scientific American*, **224**:66–73. A general description of the viroids and the role of these "organisms" in disease processes.

Fenner, F., B. R. McAuslan, C. A. Mims, J. Sambrook, and D. O. White. 1974. *The Biology of Animal Viruses*. Academic Press, Inc., New York. A text describing the different types of animal viruses, how they reproduce, and their significance.

Fraenkel-Conrat, H., and P. C. Kimball. 1982. *Virology*. Prentice-Hall Inc., Englewood Cliffs, New Jersey. A well-written introductory text describing the various aspects of virology.

Luria, S. E., J. E. Darnell Jr., D. Baltimore, and A. Campbell. 1978. *General Virology*. John Wiley & Sons, Inc., New York. A well-written introductory text describing the various aspects of virology.

Maramorosch, K. (ed.). 1977. *Insect and Plant Viruses: An Atlas*. Academic Press, Inc., New York. A collection of excellent micrographs showing the diversity of viruses that infect plants and insects.

Palmer, E. L., and M. L. Martin. 1982. *An Atlas of Mammalian Viruses*. CRC Press, Inc., Boca Raton, Florida. A series of micrographs showing the structural properties of the viruses that infect mammalian cells.

Prusner, S. B. 1984. Prions. *Scientific American*, **251**(4):50–60. This article reports the startling discovery of prions, which are infectious proteins that appear to lack genetic material (nucleic acids).

Wildy, P. 1971. Classification and nomenclature of viruses. *Monographs in Virology*, **5**:1–81. A comprehensive discussion of the classification of viruses.

Williams, R. C., and H. W. Fisher. 1974. *An Electron Micrographic Atlas of Viruses*. Charles C. Thomas, Publisher, Springfield, Illinois. A superb collection of electron micrographs showing the structures of viruses.

13 *Prokaryotic Microorganisms: Classification and Diversity of Bacteria*

KEY TERMS

actinomycetes
anoxyphotobacteria
appendaged bacteria
archaebacteria
asporogenous
bacteroids
budding bacteria
chemolithotrophic bacteria
"Chinese letter" formation
chlamydias
cocci
coccobacilli
computer-assisted identification
coryneform group
cyanobacteria
diagnostic table
dichotomous key
diplococci
endospore-forming bacteria
endosymbionts
fruiting bodies
galls
gliding bacteria
identification key
methanogens
mole% G + C
mycoplasmas
numerical profile
oxyphotobacteria
phenotypic characteristics
phototrophic bacteria
pleomorphic
probabilistic identification matrix
prosthecae
rickettsias
sheath
sheathed bacteria
snapping division
spiral bacteria
spirochetes

OBJECTIVES

After reading this chapter you should be able to

1. Define the key terms.
2. Describe the criteria for classifying bacteria.
3. Discuss why genetic analyses (for example, mole% G + C and DNA hybridization) are better than phenotypic characterization for establishing classification systems and why phenotypic characteristics are preferable for identification purposes.
4. Describe the major groups of bacteria including: distinguishing features, representative genera, ecological habitats and importance, economic importance, and medical importance.

Classification of Bacteria

Criteria for Classifying Bacteria

For practical reasons, phenotypic characteristics (observable characteristics that are actually expressed) are employed in most microbial classification and identification systems (Table 13.1). The morphological, physiological, biochemical, and nutritional features normally examined include mode of reproduction, morphology (form and structure), staining reactions, ability to form spores, motility, specific metabolic activities, and life cycle stages. Analysis of the biochemical constituents of the cell and particular structures, such as cell walls and membranes among others, are also valuable in revealing major differentiating features of groups of microorganisms.

In addition to phenotypic analyses, it is also possible to include genetic analyses in the development of classification schemes for microorganisms. One of the genetic analyses used for classifying microorganisms is the determination of the relative proportion of guanine (G) and cytosine (C) base pairs compared to adenine (A) and thymine (T) base pairs in the DNA. Because there is pairing between complementary bases in DNA, specifying G + C also specifies A + T. Therefore, the composition of the DNA is normally described by specifying the mole% G + C. Measuring the proportion of G + C in the DNA is a crude analysis of the microbial genome. Closely related organisms should have similar proportions of G + C in their DNA, and organisms with vastly differing proportions of G + C in their DNA can safely be said to be unrelated. Microorganisms that are otherwise similar, but differ by 2–3% G + C content in their genomes, represent different species. However, a similarity of G + C ratios does not necessarily establish the relatedness of organisms because the sequence of genes may be different even when the mole% G + C content is the same. A more precise measure of relatedness is the DNA homology between two organisms. Measuring the DNA homology accounts for the order of nucleotides as well as the overall composition of the genome.

Table 13.1 Some Criteria for Classifying Bacteria

- Microscopic characteristics
  - Morphology
    - cell shape
    - cell size
    - arrangement of cells
    - arrangement of flagella
    - capsule
    - endospores
  - Staining reactions
    - Gram stain
    - acid fast stain
- Growth characteristics
  - Appearance in liquid culture
  - Colonial morphology
  - Pigmentation
- Biochemical characteristics
  - Cell-wall constituents
  - Pigment biochemicals
  - Storage inclusions
  - Antigens
  - RNA molecules
- Physiological characteristics
  - Temperature range and optimum
  - Oxygen relationships
  - pH tolerance range
  - Osmotic tolerance
  - Salt requirement and tolerance
  - Antibiotic sensitivity
- Nutritional characteristics
  - Energy sources
  - Carbon sources
  - Nitrogen sources
  - Fermentation products
  - Mode of metabolism (autotrophic, heterotrophic, fermentative, respiratory)
- Genetic characteristics
  - DNA mole% G + C
  - DNA hybridization

Identification Keys and Diagnostic Tables

The classical approach to microbial identification involves the development of keys or diagnostic tables. An identification key consists of a series of questions that lead through a classification system to the determination of the identity of the organism (Table 13.2). In a dichotomous key a series of yes–no questions is asked that leads through the branches of a flow chart to the identification of a microorganism as a member of a specified microbial group (Figure 13.1). The path to an identification in a true dichotomous key is unidirectional, and a single atypical feature or error in determining a feature will result in a misidentification. Most students in introductory microbiology laboratory courses use dichotomous keys for the identification of an unknown organism.

Besides identification keys, diagnostic tables can be developed to aid in microbial identification. Such tables summarize the characteristics of the taxonomic groups but do not indicate a hierarchical separation of the taxa. Diagnostic tables generally give the appearance of being far more complicated than keys for the identification of microorganisms

Table 13.2 Dichotomous Key for the Diagnosis of Species of the Genus *Pseudomonas*

	Characteristic	Species
1.	Oxidase negative (go to 2)	
	Oxidase positive (go to 4)	
2.	Lysine positive	*P. maltophilia*
	Lysine negative (go to 3)	
3.	Motile	*P. paucimobilis*
	Nonmotile	*P. malleii*
4.	Fluorescent (go to 5)	
	Nonfluorescent (go to 8)	
5.	Pyocyanin and pyorubin positive	*P. aeruginosa*
	Pyocyanin and pyorubin negative (go to 6)	
6.	Growth at 42°C	*P. aeruginosa*
	No growth at 42°C (go to 7)	
7.	Gelatinase positive	*P. fluorescens*
	Gelatinase negative	*P. putida*
8.	Nonmotile	probably not a pseudomonad
	Motile (go to 9)	
9.	Peritrichous	not a pseudomonad
	Polar (go to 10)	
10.	Glucose oxidation negative (go to 11)	
	Glucose oxidation positive (go to 13)	
11.	2 or more flagella	*P. diminuta*
	1 flagellum (go to 12)	
12.	PHB positive	*P. testosteroni*
	PHB negative	*P. alcaligenes*
13.	Mannose negative (go to 14)	
	Mannose positive (go to 16)	
14.	Ornithine positive	*P. putrefaciens*
	Ornithine negative (go to 15)	
15.	Mannitol positive	*P. acidovorans*
	Mannitol negative	*P. pseudoalcaligenes*
16.	Arginine positive (go to 17)	
	Arginine negative (go to 19)	
17.	6.5% NaCl positive	*P. mendocina*
	6.5% NaCl positive (go to 18)	
18.	Gelatinase positive	*P. fluorescens*
	Gelatinase negative	*P. putida*
19.	Galactose negative (go to 20)	
	Galactose positive (go to 21)	
20.	Mannose positive	*P. maltophilia*
	Mannose negative	*P. vesicularis*
21.	Lactose negative (go to 22)	
	Lactose positive (go to 23)	
22.	6.5% NaCl positive	*P. stutzeri*
	6.5% NaCl negative	*P. pickettii*
23.	Nitrogen production	*P. pseudomallei*
	No nitrogen production (go to 24)	
24.	Citrate positive	*P. cepacia*
	Citrate negative	*P. paucimobilis*

Figure 13.1 A dichotomous key for classification, showing branching based on critical features.

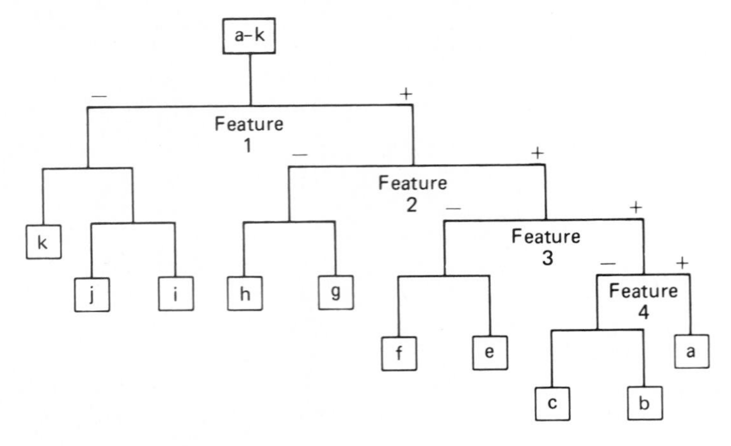

because they contain more information. However, in cases where some features are variable for different groups, diagnostic tables are better than keys for the successful identification of an unknown microorganism.

Computer-assisted Identification

Computers greatly facilitate the identification of microorganisms. When computers are used, the data gathered on an unknown microorganism can rapidly be compared to a data bank containing information on the characteristics of defined taxa. Keys and diagnostic tables can readily be programmed for computer-assisted identification of unknown isolates. Because computers rapidly perform large numbers of calculations and comparisons, computerized identification systems have also been developed to assess the statistical probability of correctly identifying a microorganism. In these methods the results of a series of phenotypic tests are scored and compared to the test results of organisms that have been classified as belonging to a particular taxonomic group. Unlike keys, where individual tests are critical in reaching proper diagnostic identification, these identification systems assess the statistical likelihood of obtaining a particular pattern of test results.

Such identification systems often involve the development and use of probabilistic identification matrices (Figure 13.2). These probability matrices are developed by characterizing large numbers of strains belonging to each taxonomic group. In this way the variability of the group for a particular feature can be determined. Many of the commercial identification systems used in clinical laboratories are based on such probabilistic identification matrices. Several of the commercial systems simplify the process of identification by calculating a numerical profile to describe unambiguously the pattern of test results (Figure 13.3). The numerical profile of an unknown organism can be compared with the test pattern of a defined group to determine the probability that the test results could represent a member of that taxon. Because of the critical nature of making correct identifications in medical microbiology, a positive identification in a clinical identification system generally requires that the unknown organism be far more similar to one group, to which it is identified as belonging, than to any other group. For example, in some identification systems an unknown microorganism must be a thousand times more similar to one group than to all other groups in the system in order to establish a positive identification.

Probability matrix

TAXA	Probability of positive results in test: a	b	c
A	0.99	0.99	0.99
B	0.99	0.01	0.10
C	0.01	0.95	0.02

Unknown test results

	Test: a	b	c
Organism X	+	+	+

Probability scores comparing X with taxa A, B, C

$P_A = (.99)\,(.99)\,(.99) = 0.9703$

$P_B = (.99)\,(.01)\,(.10) = 0.0001$

$P_C = (.01)\,(.95)\,(.02) = 0.0002$

Sum $= 0.9706$

Normalized identification scores

$I_A = \frac{0.9703}{0.9706} = 0.9997$

$I_B = \frac{0.0001}{0.9706} = 0.0001$

$I_C = \frac{0.0002}{0.9706} = 0.0002$

Organism X identified as belonging to TAXON A

Figure 13.2 The probabilistic matrix approach to organism identification allows organisms of unknown affiliation to be identified as members of established taxa.

Bacterial Systematics

The bacteria encompass many diverse groups of organisms exhibiting widely differing morphological, ecological, and physiological properties.

My daughter Alice is a student in High School. One of the prescribed courses is General Science. The section on Bacteria left her with the vague impression of a world teeming with deadly germs awaiting an opportunity to infect mankind. It seems probable that this malignant conception of bacteria is very generally held. In reality civilisation owes much to the microbe.

—A. I. KENDALL, *Civilisation and the Microbe*

The unifying feature of the bacteria is the fact that they all have prokaryotic cells. Bacterial systematics is always in a state of flux, with frequent descriptions of new bacterial genera and species and revisions to older taxonomic classifications. Pe-

	Digit 1			Digit 2			Digit 3			
	A	B	C	D	E	F	G	H	I	
	1	2	4	1	2	4	1	2	4	
Unknown	+	−	−	−	−	+	+	−	−	Test results

Profile 141

A – ornithine decarboxylase (ODC)
B – citrate utilization (CIT)
C – H_2S production (H_2S)
D – Voges Proskauer test (VP)
E – gelatin hydrolysis (GEL)
F – glucose utilization (GLU)
G – mannose utilization (MAN)
H – inositol utilization (INO)
I – sorbose utilization (SOR)

Probability matrix showing percent positive for all strains which have been classified

		ODC	CIT	H_2S	VP	GEL	GLU	MAN	INO	SOR	Profile
Escherichieae	*Escherichia coli*	75.7	0	1.0	0	0	99.9	99.0	0.5	94.4	145
	Shigella dysenteriae	0	0	0	0	0	94.1	2.0	0	18.6	040
	Sh. flexneri	0.9	0	0	0	0	99.0	91.8	0	24.0	041
	Sh. boydii	3.4	0	0	0	0	99.1	94.1	0	55.9	041
	Sh. sonnei	92.9	0	0	0	0	99.9	99.0	0	2.2	141

Unknown matches profile of *Shigella sonnei*

Figure 13.3 A numerical profile can be created to identify an unknown organism. This approach is used in several widely used commercial systems for the identification of clinical isolates.

riodically, the status of bacterial taxonomy is summarized in *Bergey's Manual of Determinative Bacteriology*. The eighth edition of *Bergey's Manual*, published in 1974, was the last comprehensive volume of *Bergey's Manual*. Newer volumes of *Bergey's Manual* describe the taxonomy of particular groups of bacteria. The major groups of bacteria are defined primarily on physiological and morphological criteria. An examination of the volumes describing the taxonomy of the bacteria reveals that most bacteria are not associated with disease processes. Most bacterial genera described in *Bergey's Manual* are important for their ecological rather than medical activities.

The history of warfare always proves more glamorous than accounts of co-operation. Plague, cholera, and yellow jack have found their way into the novel, the stage, and the screen, but no one has made a success story of the useful role played by microbes in the intestine or the stomach.

—RENE DUBOS, *Mirage of Health*

Phototrophic Bacteria

The phototrophic bacteria are distinguished from other bacterial groups by their ability use light energy to drive the synthesis of ATP (Table 13.3). Most of the organisms included in this group are autotrophs, capable of using carbon dioxide as the source of cellular carbon. Some of the phototrophic bacteria use water as an electron donor and liberate oxygen. These bacteria can be considered as belonging to oxyphotobacteria. The remainder of the photobacteria do not evolve oxygen and with one exception can be classified as belonging to anoxyphotobacteria

The cyanobacteria, or blue-green bacteria, are the most diverse and widely distributed group of photosynthetic bacteria. Over 1000 species of cyanobacteria have been reported. Unlike the cyanobacteria, the anoxyphotobacteria do not evolve oxygen. The anaerobic photosynthetic bacteria typically occur in aquatic habitats, often growing at the sediment water interface of shallow lakes where (1) there is sufficient light penetration to permit photosynthetic activity; (2) anaerobic conditions are sufficient to permit the existence of these organisms; and (3) there is a source of reduced sulfur or organic compounds to act as electron donors for the generation of reduced coenzymes. The phototrophic bacteria include the Rhodospirillaceae

Table 13.3 **Characteristics of the Major Groups of Phototrophic Bacteria**

Metabolism	Taxonomic Group	Photosynthetic Pigments	Electron Donors	Carbon Source	Comments
Anoxygenic photosynthesis	Purple bacteria	Bacteriochlorophyll a or b, carotenoids	H_2,H_2S,S	Organic C or CO_2	Rhodospirillaceae (purple nonsulfur bacteria) use organic substrates; most unable to grow with only hydrogen sulfide—termed photoheterotrophs; Chromatiaceae (purple sulfur bacteria) cells able to grow with sulfide and sulfur as the sole electron donor—sulfur deposited inside or outside of cell
Anoxygenic photosynthesis	Green bacteria	Bacteriochlorophyll a or b, carotenoids	H_2,H_2S,S	CO_2	Chlorobiaceae (green sulfur bacteria) cells able to grow with sulfide and sulfur as the sole electron donor—sulfur deposited outside of cell; Chloroflexeae (green flexibacteria) cells have flexible walls, gliding motility, form filaments, use organic carbon sources
Oxygenic photosynthesis[a]	Cyanobacteria	Chlorophyll a, phycobiliproteins	H_2O	CO_2	
Oxygenic photosynthesis	Prochlorobacteria	Chlorophyll a + b, β carotenes	H_2O	CO_2	
Purple membrane mediated	*Halobacterium*	Bacteriorhodopsin	—	Organic C	Unusual form of photosynthesis based upon excitation of purple membrane

[a]Under some conditions photosynthesis is anoxygenic, and H_2S serves as the electron donor.

(purple nonsulfur bacteria), Chromatiaceae (purple sulfur bacteria), Chlorobiaceae (green sulfur bacteria), and Chloroflexaceae (green flexibacteria) (Figure 13.4).

The Gliding Bacteria

The Myxobacterales (fruiting myxobacteria) and the Cytophagales are grouped together based on their gliding motility on solid surfaces. The myxobacteria are small rods that are normally embedded in a slime layer. They lack flagella but are capable of gliding movement. A unique feature of the myxobacteria is that under appropriate conditions they aggregate to form fruiting bodies (Figure 13.5). The fruiting bodies of myxobacteria occur on decaying plant material, on the bark of living trees, or on animal dung, appearing as highly colored slimy growths that may extend above the surface of the substrate. In contrast to the Myxobacterales, the Cytophagales do not produce fruiting bodies. Members of both groups, however, do exhibit gliding motion.

The Sheathed Bacteria

The sheathed bacteria comprise those bacteria whose cells occur within a filamentous structure known as a sheath (Figure 13.6). The formation of a sheath enables these bacteria to attach themselves to solid surfaces. This is important to the ecology of these bacteria because many sheathed bacteria live in low nutrient aquatic habitats. By absorbing nutrients from the water that flows by the attached cells, these bacteria are able to conserve their limited energy resources. Additionally, the sheaths afford protection against predators and parasites. *Sphaerotilus natans* is a sheathed bacterium that is often referred to as the sewage fungus. This organism normally occurs in polluted flowing

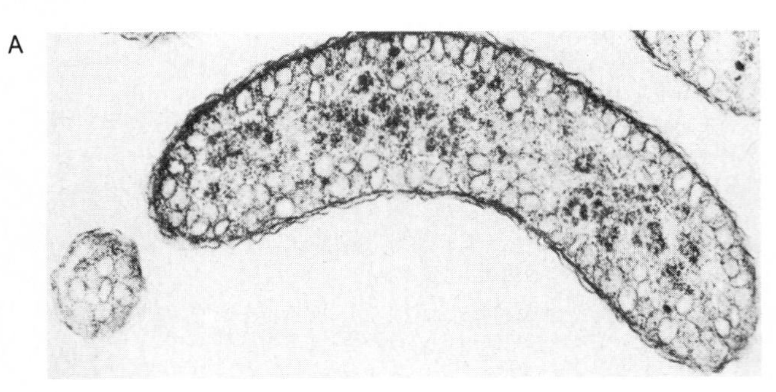

Figure 13.4 Morphologies of anaerobic photosynthetic bacteria. (A) Purple sulfur bacteria *Rhodospirillum rubrum* (32,900 ×); (B) green sulfur bacteria *Pelodictyon clathratiforme* (1,600 ×), note gas vacuoles; (C) purple nonsulfur bacteria *Chromatium vinosum* (1,600 ×). (From N. Pfenning and H. Trüper, reprinted by permission of the Bergey's Trust, John Holt, executor, from *Bergey's Manual of Determinative Bacteriology, eighth edition,* Williams & Wilkins Company, Baltimore, Maryland.)

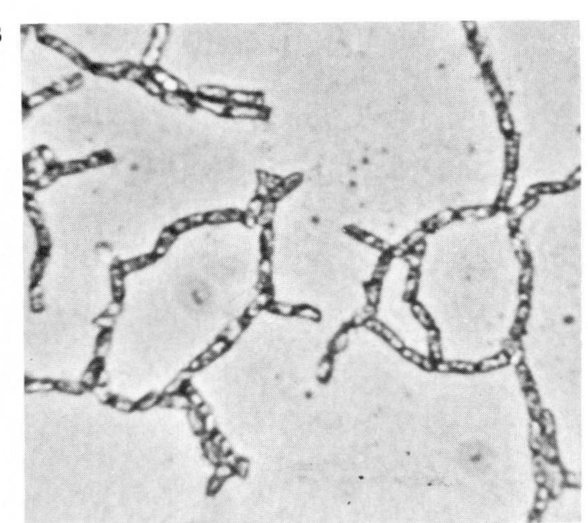

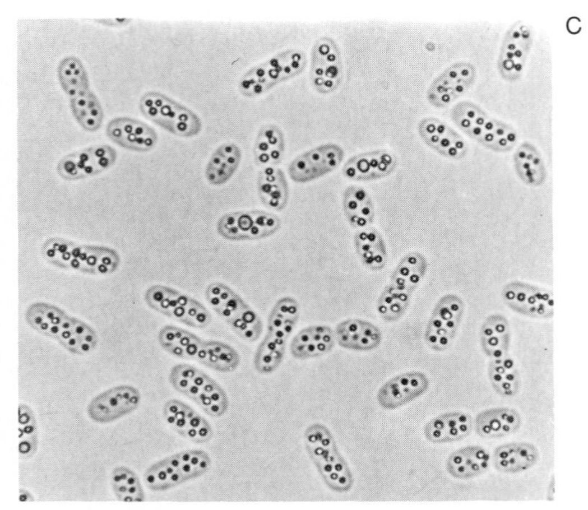

Figure 13.5 Micrograph showing morphologies and fruiting bodies of myxobacteria. The stages of fruiting body formation of the myxobacterium *Stigmatella aurantiaca* are shown in (A) late stalks and (B) mature fruiting body. (C) *Cystobacter fuscus* fruiting bodies (170 ×); (D) *Chondromyces crocatus* (125 ×). (A and B from BPS: Karen Stephens, Stanford Medical Center; C and D courtesy Hans Reichenbach, Gesellschaft für Biotechnologische Forschung, Braunschweig, Federal Republic of Germany.)

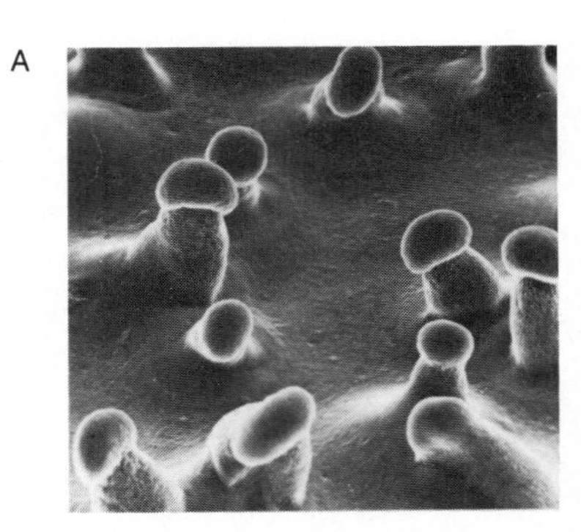

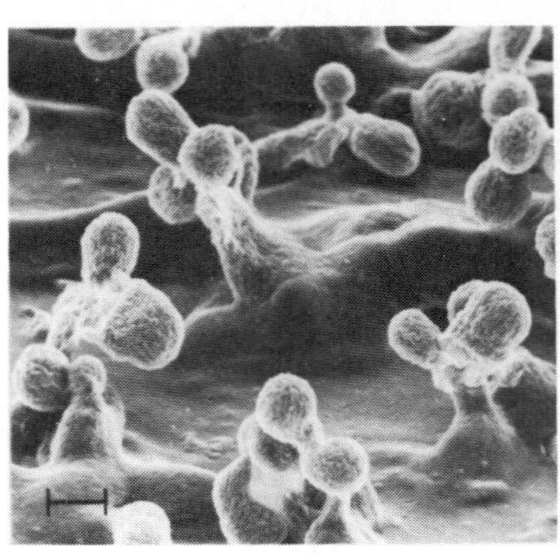

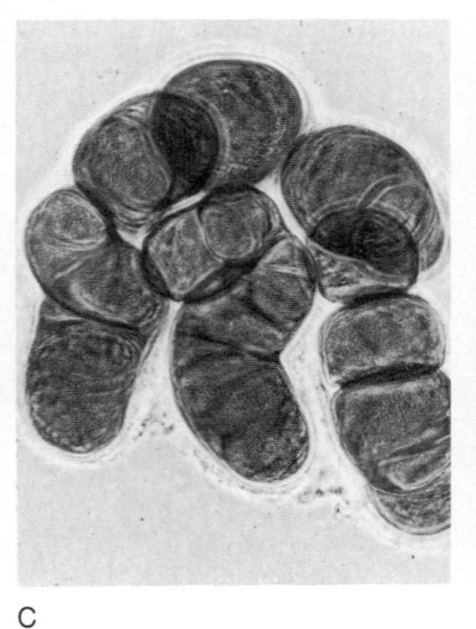

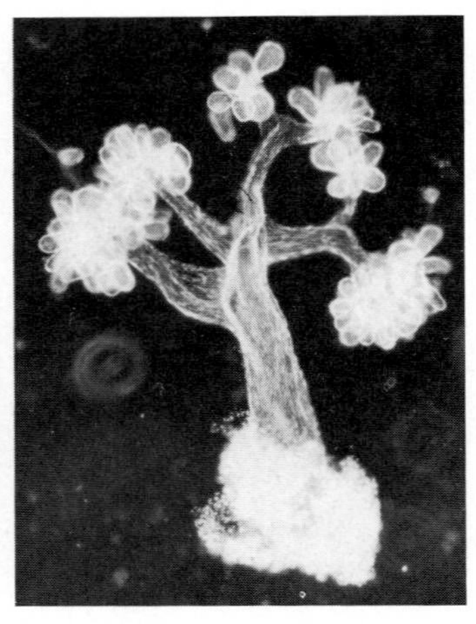

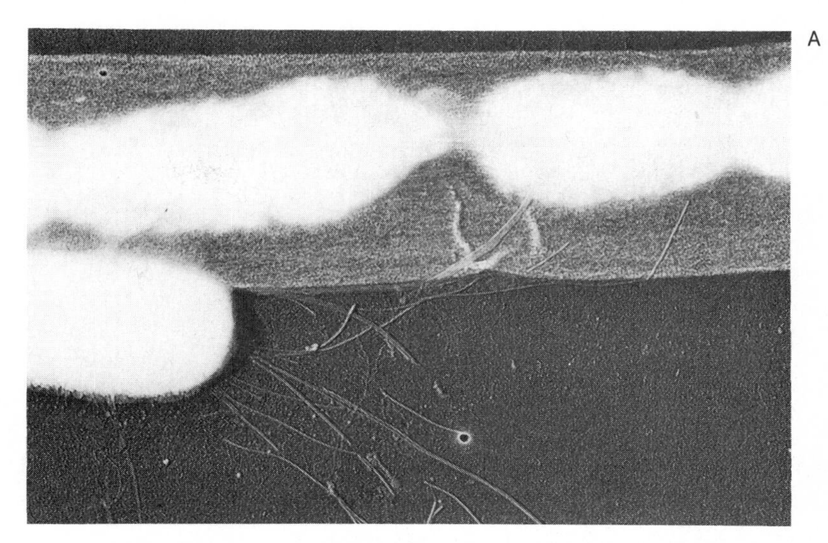

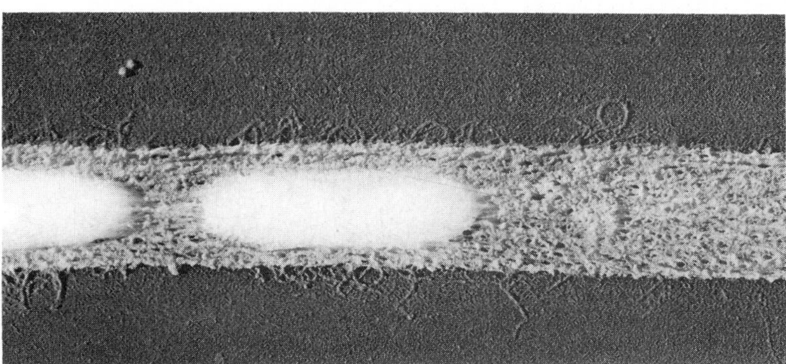

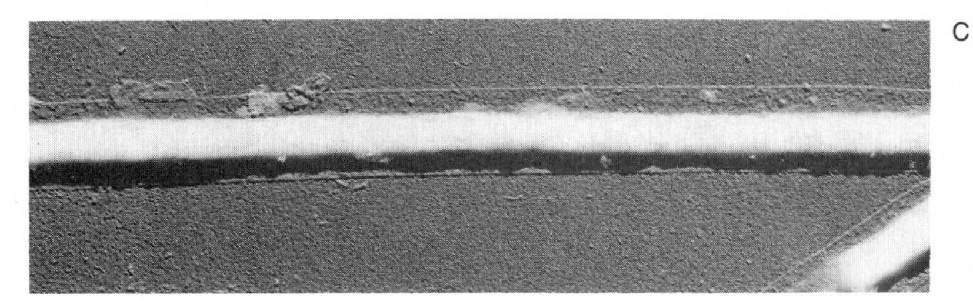

Figure 13.6 Micrographs showing morphologies of sheathed bacteria. (A) *Sphaerotilus natans* (19,000×). (B) *Leptothrix cholodnii* (19,000×). (C) *Haliscomenobacter hyrossis* (19,000×). (A, C courtesy M. H. Deinema, Landbouwhogeschool Wageningen, Netherlands. A: Reprinted by permission of the American Society for Microbiology, Washington, D.C., from W. L. vanVeen, E. G. Mulder, and M. H. Deinema, 1978, *Bacteriological Reviews*, **42**:329–356. B: Reprinted by permission of Springer-Verlag, Heidelberg, from E. G. Mulder and M. H. Deinema, 1981, "The sheathed bacteria" in *The Prokaryotes*, M. P. Starr, H. Stolp, H. G. Truper, A. Balows, and H. G. Schlegel, eds.)

waters, such as sewage effluents, where it may be present in high concentrations just below sewage outfalls.

Budding and/or Appendaged Bacteria

Like the sheathed bacteria, the budding and/or appendaged bacteria represent a heterogeneous group based on a particular morphological feature. These bacteria reproduce by budding or binary fission. The cell appendages of the bacteria in this group, known as prosthecae, afford the cell greater efficiency in concentrating available nutrients. Many of the appendaged bacteria grow well at low nutrient concentrations. The appendages provide sufficient membrane surface to transport adequate nutrients into the cell to support the metabolic requirements of the organism. *Caulobacter*, for example, is able to grow in very dilute concentrations of organic matter in lakes and even is able to grow in distilled water. The isolation of various new types of appendaged bacteria has greatly increased our view of the morphological diversity among the bacteria and the relationship between morphology and nutritional status.

A

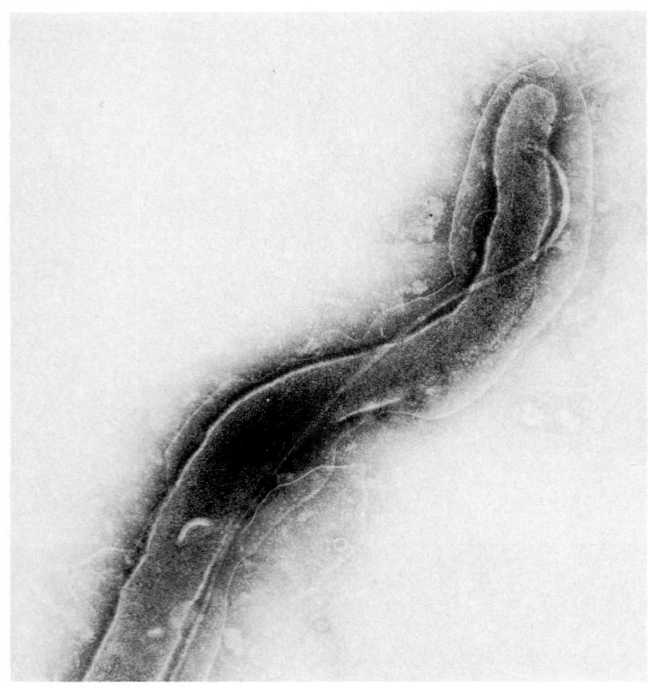

B

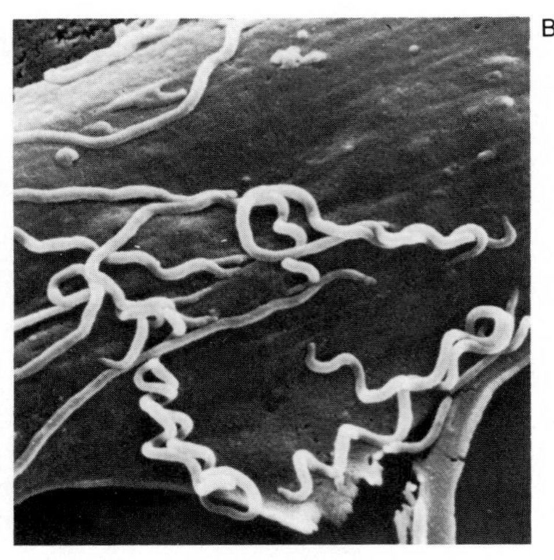

Figure 13.7 (A) Spirochete showing central axial filament (99,560 ×). (From BPS: Stanley C. Holt, University of Texas Health Science Center, San Antonio, Texas.) (B) Scanning electron micrograph of *Treponema pallidum* (8,000 ×). (Reprinted by permission of the American Society of Microbiology, Washington, D.C., from N. S. Hayes, K. E. Muse, A. M. Collier, and J. B. Baseman, 1977, *Infection and Immunity,* 17:174.)

The Spirochetes

The spirochetes are another group that has a distinct morphology. Spirochetes are helically coiled rods (Figure 13.7). The cells are wound around one or more central axial fibrils. In addition to their characteristic morphology, the spirochetes exhibit a unique mode of motility. These bacteria move by a flexing motion of the cell, exhibiting the greatest velocities in very viscous solutions where motility by bacteria with external flagella is slowest. Many spirochetes are human pathogens. Several members of the genus *Treponema,* for example, are human pathogens, with *Treponema pallidum* causing syphilis and *Treponema pertenue* causing yaws.

Spiral and Curved Bacteria

Members of the spiral and curved bacteria group are helically curved rods that may have less than one complete turn (comma shaped) to many turns (helical), but unlike the spirochetes the cells are not wound around a central axial filament. Members of *Spirillum,* for example, have multiple polar flagella, usually at both ends. Some members of the genus *Campylobacter* are important pathogens of humans and other animals. *Bdellovibrio,* a genus of uncertain affiliation within this group, has the outstanding characteristic of being able to penetrate and reproduce within prokaryotic cells (Figure 13.8). All naturally occurring strains of *Bdellovibrio* have been found to be bacterial parasites. Unlike viruses, *Bdellovibrio* reproduce by binary fission and do not lose their integrity within host cells. After reproduction of *Bdellovibrio* within a host cell, the host cell lyses, releasing the *Bdellovibrio* progeny.

Gram Negative Aerobic Rods and Cocci

The Gram negative aerobic rods and cocci encompass a large number of taxonomic units. Several major families are included in this group: Pseudomonadaceae, Azotobacteraceae, Rhizobiaceae, and Methylomonadaceae. The Pseudomonadaceae are motile rods with polar flagella; they are obligately respiratory. Many *Pseudomonas* species are nutritionally versatile and are capable of degrading many natural and man-made organic compounds. Some *Pseudomonas* species produce characteristic fluorescent pigments, but others do not. For example, *Pseudomonas aeruginosa* produces yellow-green diffusible pigments that fluoresce. Some *Pseudomonas* species are plant and animal pathogens. *P. aeruginosa,* for example, can be a human pathogen and is commonly isolated from wound, burn, and urinary tract infections.

The family Azotobacteraceae is characterized by its capacity to fix molecular nitrogen. The genera

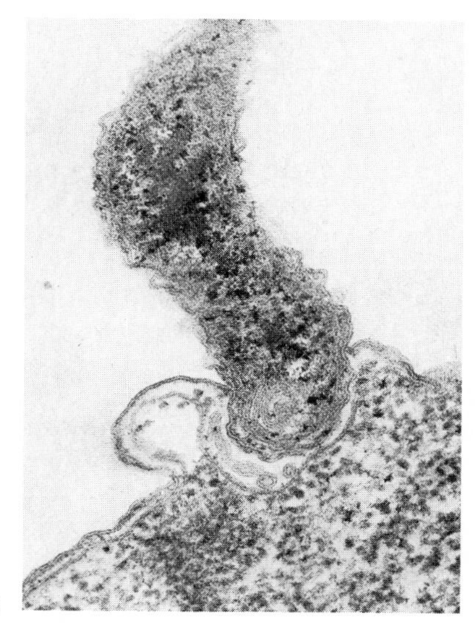

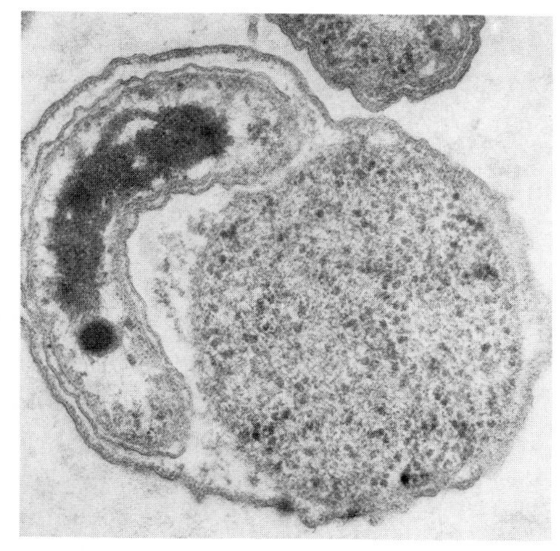

Figure 13.8 *Bdellovibrio bacteriovorus* entering and within an *Escherichia coli* host cell. (Courtesy Jeffrey C. Burnham, Medical College of Ohio, and Sam F. Conti, University of Massachusetts. Reprinted by permission of the American Society for Microbiology from J. C. Burnham and S. F. Conti, 1968, *Journal of Bacteriology*, 96:1374.)

Azotobacter and *Beijerinckia* are particularly important free-living nitrogen-fixing bacteria. The Rhizobiaceae are also capable of fixing atmospheric nitrogen. *Rhizobium* species are able to infect leguminous plant roots, causing the formation of tumorous growths called nodules. Free-living cells of *Rhizobium* are rod-shaped, but within the nodules they occur as pleomorphic (irregularly-shaped) cells termed bacteroids. *Rhizobium* can fix atmospheric nitrogen only within root nodules and thus is considered an obligately symbiotic nitrogen fixer. Unlike *Rhizobium*, *Agrobacterium* species do not fix molecular nitrogen. *Agrobacterium* species, however, produce tumorous growths on infected plants, known as galls. *Agrobacterium tumefaciens* causes galls of many different plants and is an extremely important plant pathogen, causing large economic losses in agriculture.

Gram Negative Facultatively Anaerobic Rods

There are two major families of Gram negative facultatively anaerobic rods: the Enterobacteriaceae, (motile by means of peritrichous flagella) and the Vibrionaceae (motile by means of polar flagella). The Enterobacteriaceae are divided further into five tribes: the Escherichieae, Klebsielleae, Proteeae, Yersinieae, and Erwinieae.

The family Enterobacteriaceae includes the genera *Citrobacter, Edwardsiella, Enterobacter, Erwinia, Escherichia, Hafnia, Klebsiella, Proteus, Salmonella, Serratia, Shigella*, and *Yersinia*. Members of the genus *Escherichia* occur in the human intestinal tract. *E. coli* has achieved a special place in microbiology. It has been used as the test organism in many metabolic and genetic studies, and much of what we know about bacterial metabolism and bacterial genetics has been elucidated in studies using *E. coli*. Additionally, *E. coli* is employed as an indicator of fecal contamination in environmental microbiology. The genera *Salmonella* and *Shigella* contain numerous different species, many of which are important human pathogens. In particular, typhoid fever and various gastrointestinal upsets are caused by *Salmonella* species, and bacterial dysentery is caused by *Shigella*.

The family Vibrionaceae includes the genera *Vibrio, Aeromonas, Plesiomonas, Photobacterium*, and *Lucibacterium*. Many of the *Vibrio* have curved rod-shaped cells. The habitat of *Vibrio* species is generally aquatic. *Vibrio cholerae* is an important human pathogen that causes cholera.

Gram Negative Anaerobic Bacteria

There is only one family, Bacteroidaceae, and relatively few genera in the Gram negative anaerobic bacteria group. *Bacteroides* are important members of the normal microbiota of humans. They may be the dominant microorganisms occurring in the human intestinal tract. *Bacteroides* species characteristically form pleomorphic rods. *Fusobacter-*

ium and *Leptotrichia* are important pathogens in this family.

Members of the genus *Desulfovibrio* are curved rods capable of reducing sulfates, or other reducible sulfur compounds, to hydrogen sulfide. *Desulfovibrio desulfuricans* is normally found in bogs and anaerobic sediments, where it plays an important role in the biogeochemical cycling of sulfur.

Gram Negative Cocci and Coccobacilli

The family Neisseriaceae includes the genera *Neisseria, Branhamella, Moraxella,* and *Acinetobacter.* The cells of *Neisseria* and *Branhamella* are cocci, whereas the cells of *Moraxella* and *Acinetobacter* are coccobacilli (oval-shaped). Members of the genera *Neisseria, Branhamella,* and *Moraxella* are parasitic, and some are important human pathogens. For example, *Neisseria gonorrhoeae* causes gonorrhea (Figure 13.9), and *Neisseria meningitidis* causes meningitis.

Gram Negative Anaerobic Cocci

The Gram negative anaerobic cocci include only four genera, *Veillonella, Acidaminococcus, Megasphaera,* and *Gemmiger*. Each of these genera contains very few species. The cells of species in all of these genera typically occur as diplococci (pairs of cocci). *Veillonella* species have complex nutritional requirements and are unable to grow on individual organic substrates. They also require carbon dioxide for growth. Although these organisms are fastidious in their nutritional requirements, they comprise part of the normal human microbiota, representing, for example, 5–16 percent of the bacteria found in the oral cavity. Some *Veillonella* species are human pathogens causing infections in the oral cavity and intestinal and respiratory tracts.

Figure 13.9 Electron micrograph of *Neisseria gonorrhoeae,* showing diplococci (100,000×). (From BPS: Centers for Disease Control, Atlanta.)

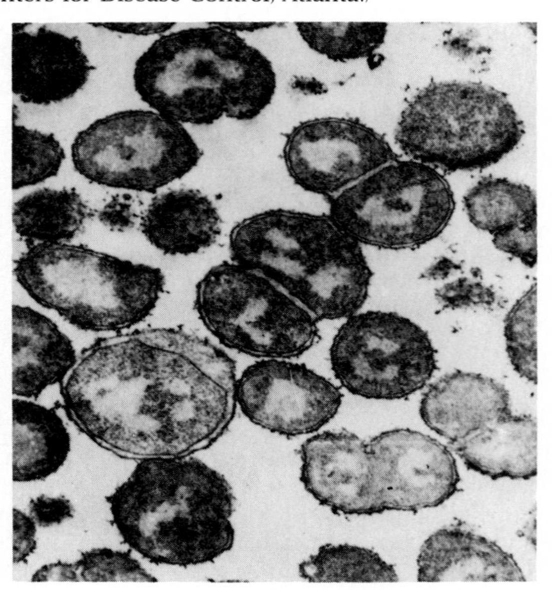

Gram Negative Chemolithotrophic Bacteria

The metabolic activities of the Gram negative chemolithotrophic bacteria are extremely important in biogeochemical cycling reactions (Table 13.4). These bacteria oxidize inorganic compounds in order to generate ATP. Because their ATP-generating metabolism is inefficient, they metabolize large amounts of substrate to meet their energy requirements. The metabolic transformations of inorganic compounds mediated by these organisms cause global-scale cycling of various elements between the air, water, and soil.

The family Nitrobacteraceae oxidize ammonia or nitrite in order to generate ATP. Organisms in this family, usually referred to as nitrifying bacteria, are commonly found in soil, freshwater, and seawater. Many of the nitrifying bacteria have extensive internal membrane systems (Figure 13.10). There are two physiological groups in the family Nitrobacteraceae; the first group oxidizes nitrite to nitrate, and the second group oxidizes ammonia to nitrite. *Nitrobacter* species are extremely important nitri-

Table 13.4 The Gram Negative Chemolithotrophic Bacteria

Oxidize ammonia or nitrite	Family Nitrobacteraceae
Oxidize nitrite to nitrate	*Nitrobacter*
	Nitrospina
	Nitrococcus
Oxidize ammonia to nitrite	*Nitrosomonas*
	Nitrosospira
	Nitrosococcus
	Nitrosolobus
Oxidize sulfur and sulfur compounds	*Thiobacillus*
	Sulfolobus
	Thiobacterium
	Macromonas
	Thiovulum
	Thiospira
Oxidize iron or manganese	Family Siderocapsaceae
Iron or manganese oxides deposited	*Siderocapsa*
	Naumanniella
	Ochrobium
Iron but not manganese deposited	*Siderococcus*

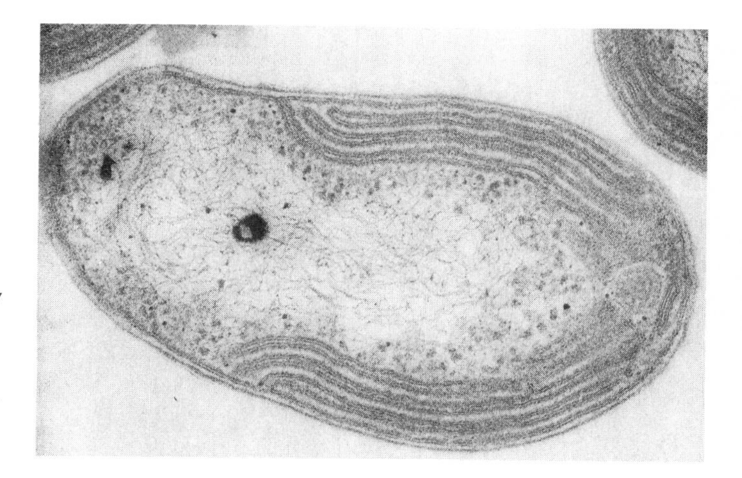

Figure 13.10 Micrograph showing the extensive internal membrane system of the nitrifying bacterium *Nitrobacter winogradsky* (165,759×). (Courtesy Stan Watson, Woods Hole Oceanographic Institute, Woods Hole, Massachusetts. Reprinted by permission of the American Society for Microbiology, Washington, D.C., from S. W. Watson, 1971, *International Journal of Systematic Bacteriology*, 21:254–270.)

fiers in soil, oxidizing nitrite to nitrate. *Nitrosomonas* species, likewise, are important nitrifiers in soil, oxidizing ammonia to nitrite. The combined actions of the members of the genera *Nitrosomonas* and *Nitrobacter* permit the conversion of ammonia to nitrate. The change in electronic charge between NH_4^+ and NO_3^- alters the mobility of these nitrogenous ions in soil and has a major influence on soil fertility.

Several different genera of chemolithotrophic bacteria metabolize sulfur and sulfur-containing inorganic compounds. The genus *Thiobacillus* derives energy from the oxidation of reduced sulfur compounds; they are Gram negative rods, motile by means of polar flagella. Some members of the genus *Thiobacillus* oxidize only sulfur compounds, whereas others, such as *Thiobacillus ferrooxidans*, can also oxidize ferrous iron to ferric iron in order to generate ATP. *Thiobacillus* species can be used in the recovery of minerals, including uranium, and its oxidation of reduced iron and sulfur compounds mobilizes various metals so that they can be extracted from even low-grade ores. *Thiobacillus thiooxidans* is often found in association with waste coal heaps. The metabolic activities of this organism produce acid mine drainage, a serious ecological problem associated with some coal mining operations.

Gram Positive Cocci

The Gram positive cocci include three families: the Micrococcaceae, the Streptococcaceae, and the Peptococcaceae. The coccoid cells of the Micrococcaceae may occur singly or as irregular clusters (Figure 13.11). For example, the genus *Staphylococcus* typically forms grape-like clusters. Most strains of *Staphylococcus* can grow in the presence of 15 percent sodium chloride. Species of *Staphylococcus* commonly occur on skin surfaces. *Staphylococcus aureus* is a potential human pathogen, infecting wounds and also causing food poisoning.

In the family Streptococcaceae, Gram positive cocci occur as pairs or chains (Figure 13.12). The metabolism of the Streptococcaceae is fermentative. Even though their metabolism is anaerobic, the Streptococcaceae are listed as being facultatively anaerobic because they grow in the presence of air. Indeed, many species *Streptococcus* occur in the oral cavity, where they are continuously ex-

Figure 13.11 Scanning electron micrograph of *Staphylococcus aureus*, showing grape-like clusters (6000×). (Courtesy Robert P. Apkarian, University of Louisville.)

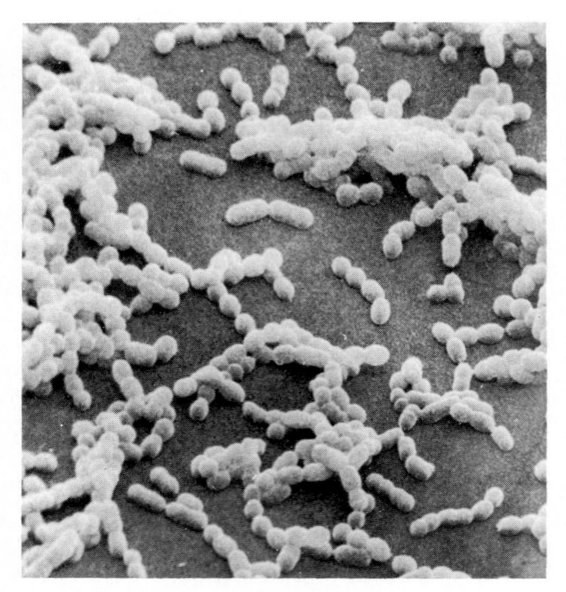

Figure 13.12 Micrograph of *Streptococcus*. Note chains of coccal-shaped bacteria. This particular strain, *S. mutans*, occurs on tooth surfaces where it produces acid that erodes the enamel, leading to dental caries. (From BPS: Z. Skobe, Forsythe Dental Center.)

posed to air. Some members of *Streptococcus* are human pathogens. For example, rheumatic fever is caused by *Streptococcus pyogenes*. Several *Streptococcus* species are also responsible for the formation of dental caries.

Endospore-forming Rods and Cocci

The two most important genera of endospore-forming bacteria are the genera *Bacillus* and *Clostridium*. Bacillus species are strict aerobes or facultative anaerobes. *Clostridium* species are obligately anaerobic. The endospore-forming bacteria are extremely important in food, industrial, and medical microbiology. Food spoilage by *Bacillus* and *Clostridium* species is of great economic importance. Several *Clostridium* species are important human pathogens. For example, *Clostridium botulinum* is the causative agent of botulism, *Clostridium tetani* causes tetanus, and *Clostridium perfringens* causes gas gangrene.

Oh, powerful bacillus,
With wonder how you fill us,
Every day!
While medical detectives
With powerful objectives,
Watch your play.

—William T. Helmuth, *Ode to Bacillus*

Gram Positive, Asporogenous, Rod-shaped Bacteria

The Gram positive, asporogenous (nonsporulating), rod-shaped bacteria include the family Lactobacillaceae. Members of this family produce lactic acid as the major fermentation product; they occur in fermenting plant and animal products that have available carbohydrate substrates. They are also found as part of the normal human microbiota, in the oral cavity, vaginal tract, and intestinal tract. *Lactobacillus* is the only genus within the family Lactobacillaceae. *Lactobacillus* is extremely important in the dairy industry; cheese, yogurt, and many other fermented products are made by the metabolic activities of *Lactobacillus* species.

Actinomycetes and Related Organisms

The coryneform group of bacteria is a heterogeneous group defined by the characteristic irregular morphology of the cells and the tendency of the cells to show incomplete separation following cell division. This group includes the genera *Corynebacterium, Arthrobacter, Brevibacterium, Cellulomonas*, and *Kurthia*. Many species of *Corynebacterium* are plant or animal pathogens. For example, *Corynebacterium diphtheriae* is the causative agent of diphtheria. Cells of *Corynebacterium* exhibit snapping division; that is, after binary fission the cells do not completely separate (Figure 13.13) and appear to form groups resembling "Chinese letters"

Figure 13.13 This micrograph of *Corynebacterium diphtheriae* shows the snapping division typical of the species. (From Centers for Disease Control, Atlanta.)

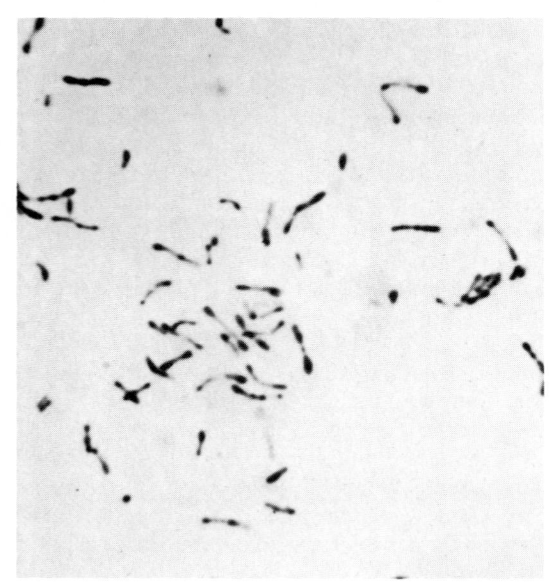

when viewed under the microscope. The genus *Arthrobacter*, which is widely distributed in soils, is interesting because it exhibits a simple life cycle, in which there is a change from rod-shaped cells to coccoid cells.

The order Actinomycetales contains bacteria characterized by the formation of branching filaments. Many of the actinomycetes resemble the fungi in appearance, but their cells are prokaryotic, and they are clearly bacteria. The various families of the order Actinomycetales are distinguished from one another by the nature of their mycelia and spores. Various types of spores are produced by actinomycete species. Many of these spores are involved in the dispersal of actinomycetes. Spore production is an important diagnostic characteristic for the identification of actinomycetes.

The actinomycetes are widely distributed in nature. Some actinomycetes are human pathogens. *Mycobacterium tuberculosis* is the causative agent for tuberculosis. The genus *Mycobacterium* is considered an actinomycete, although the formation of mycelia is rudimentary. *Mycobacterium* is acid fast; that is, stained cells resist decolorization with acid alcohol.

The production of antibiotics by actinomycetes, such as *Streptomyces griseus*, is extremely important in the pharmaceutical industry. The availability of such antibiotics has revolutionized medical practice. Many previously fatal diseases are now easily controlled by using antibiotics produced by actinomycetes.

Rickettsias

The rickettsias are intracellular parasites. They lack the enzymatic capability of producing sufficient amounts of ATP to support their reproduction; they are able to obtain the ATP from the host cells in which they grow. Many rickettsia cause disease in humans and other animals. Many members of the genus *Rickettsia* are carried by insect vectors and cause diseases in humans (Figure 13.14). For example, *Rickettsia rickettsii* is transmitted by ticks and causes Rocky Mountain spotted fever.

The chlamydias are obligate intracellular parasites whose reproduction is characterized by the change of the small, rigid-walled infectious form of the organism (elementary body) into a larger, thin-walled noninfectious form (initial body) that divides by fission (Figure 13.15). These bacteria are unable to generate sufficient ATP to support their reproduction. The reproductive cycle for these organisms takes about 40 hours. *Chlamydia* cause human respiratory and urogenital tract diseases, and in birds they cause respiratory diseases and generalized infections. For example, the disease psittacosis, parrot fever, is caused by *Chlamydia psitaci*.

Mycoplasmas

The mycoplasmas lack a cell wall. They are the smallest organisms capable of self-reproduction. When growing on artificial media, mycoplasmas form small colonies that have a characteristic "fried egg" appearance (Figure 13.16). Members of the genus *Mycoplasma* require sterols for growth. Several

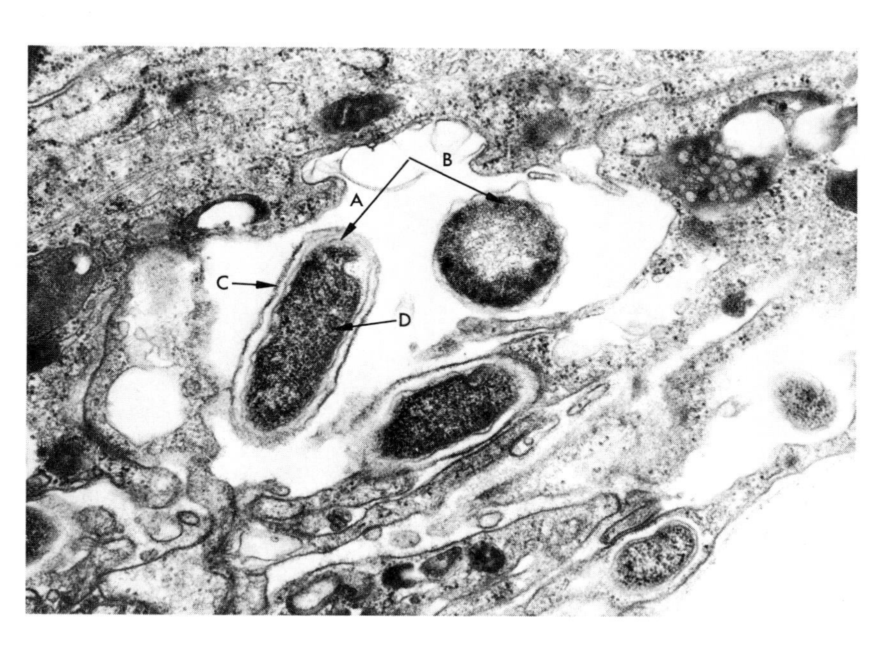

Figure 13.14 Electron micrograph of rickettsiae within cells of their wood tick vector (28,000×). (A) Bacillary form; (B) coccal form; (C) cell wall; (D) ribosomes. (Courtesy L. P. Brinton and W. Burgdorfer, Rocky Mountain Laboratory, U.S. Public Health Service, Hamilton, Montana.)

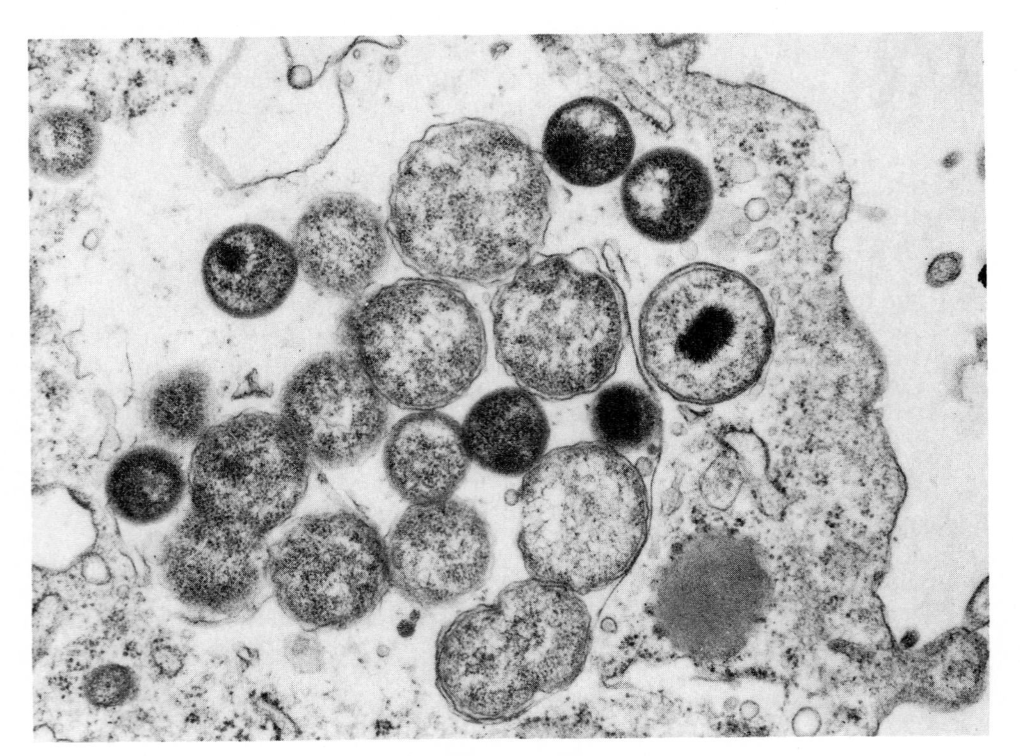

Figure 13.15 Micrograph of *Chlamydia psittaci* in the cytoplasm of a cell (47,500 ×). (From R. C. Cutlip, National Animal Disease Center, Ames, Iowa.)

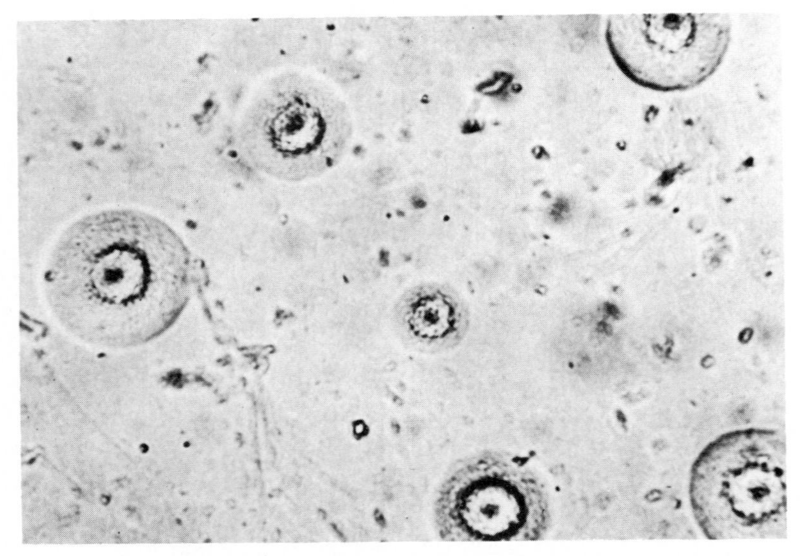

Figure 13.16 Micrograph of colonies of *Mycoplasma,* showing "fried egg" appearance. (By permission of Leonard Hayflick, University of Florida, Gainesville, Florida.)

members of this genus cause diseases in humans. For example, some forms of pneumonia are caused by *Mycoplasma* species.

Endosymbionts

Several bacterial genera have been recognized that are obligate endosymbionts of invertebrates; that is, they live within the cells of invertebrate animals without adversely affecting the animal. For example, the protozoan *Paramecium aurelia* can harbor a variety of endosymbiotic bacteria. Additionally, new genera of endosymbionts have recently been described for other protozoa, insects, and various other invertebrates.

The Archaebacteria

The archaebacteria appear to be "primitive" bacteria that are distantly related to other prokaryotes. They share in common several morphological and physiological features that make them distinct from other bacteria, including the lack of murein in their cell walls and the unusual ether linkage that occurs in their phospholipid molecules. Genetic analyses have shown that members of this group are phylogenetically related and appear to represent an evolutionary line that is distinct from other bacteria. Many of the archaebacteria have unusual physiological properties, some of which would have made them compatible with conditions on earth at the time that life first evolved. Some are thermophiles and other extreme halophiles.

The methane-producing bacteria (methanogens) represent a highly specialized physiological group of archaebacteria. The methanogenic bacteria are very strict obligate anaerobes that form methane by the reduction of carbon dioxide. In order to produce methane, these organisms utilize electrons generated in the oxidation of hydrogen or simple organic compounds, such as acetate and methanol. Microorganisms associated with the methanogens maintain the low oxygen tensions and provide the carbon dioxide and fatty acids required by the methanogenic bacteria. Such associations are extremely important in the rumen of animals, such as cows. A major source of atmospheric methane comes from the rumen of such animals. The U. S. Environmental Protection Agency once issued an indelicate report stating that the burping cow was the major source of atmospheric hydrocarbon pollutants. Clearly though, in urban areas hydrocarbon pollutants originate primarily from automotive exhausts and not cows.

Summary

Bacteria are classified and identified using a variety of morphological and physiological features. Most commonly, phenotypic characteristics that can be readily determined are used for identification systems. Modern classification systems also employ genetic analysis such as determination of mole% G + C and DNA homology. Genetic analyses show the true evolutionary relatedness but are not as convenient as phenotypic analyses.

Classical classification and identification systems use keys and diagnostic tables. These systems permit a stepwise diagnosis of a particular microbial species. Dichotomous keys ask a series of yes–no questions that lead to the identification of a bacterial species. Another approach to microbial identification employs probabilistic identification matrices. This approach is less dependent on individual key tests than dichotomous keys and provides a statistical probability for the correctness of an identification. The results of such numerical approaches are simply presented in several commercial identification systems as a numerical profile that describes the test result pattern. Many commercial systems also employ computer-assisted probabilistic identifications and the physician now commonly receives from the clinical microbiology laboratory a list of possible pathogens associated with the disease condition in order of their probability.

Only a small portion of the total bacterial taxa are ever seen in the clinical laboratory or ever discussed in introductory microbiology laboratory courses. The major groups, organized based on unifying morphological and/or physiological properties, represent the great diversity of microbial morphological forms and physiological functions. As seen in *Bergey's Manual*, the standard reference volume describing the taxonomy of bacteria, there is great diversity among the bacteria. The major groups of bacteria include: phototrophic bacteria (photosynthetic bacteria: cyanobacteria, prochlorales, purple sulfur, purple nonsulfur, green sulfur, and green flexibacteria); gliding bacteria (bacteria that move by gliding motion: myxobacteria, cytophagales); sheathed bacteria (bacteria that form filaments within a sheath); budding and/or appendaged bacteria (bacteria that reproduce by budding or form prosthecae); spirochetes (rods that are helically coiled around an axial filament: *Treponema*); spiral and curved bacteria (bacteria that are curved but not coiled around a central filament: *Spirillum, Campylobacter, Bdellovibrio*); Gram negative aerobic rods and cocci (Pseudomonadaceae, Azotobacteraceae, Rhizobiaceae, and Methylomonadaceae); Gram negative facultatively anaerobic rods (Enterobacteriaceae and Vibrionaceae); Gram negative anaerobic bacteria (Bacteroidaceae); Gram negative cocci and coccobacilli (Neisseriaceae); Gram negative anaerobic cocci (anaerobic cocci: *Veillonella*);

Gram negative chemolithotrophic bacteria (chemolithotrophs: nitrifying bacteria, sulfur oxidizers); Gram positive cocci (Micrococcaceae, Streptococcaceae, and Peptococcaceae); endospore-forming rods and cocci (bacteria that form endospores: *Clostridium, Bacillus*); Gram positive, asporogenous, rod-shaped bacteria (*Lactobacillus*); actinomycetes and related organisms (coryneforms, mycobacteria, and branching filamentous bacteria); rickettsias (obligate intracellular parasites: *Rickettsia, Chlamydia*); mycoplasmas (bacteria lacking a cell wall); endosymbionts (bacteria living exclusively within invertebrates); and archaebacteria (physiologically unique bacteria distantly related to other prokaryotes).

Study Outline

A. Classification of bacteria.
  1. Phenotypic characteristics have been traditionally used for classifying and identifying bacteria.
  2. Genetic analyses, such as the mole% G + C and DNA hybridization are used in modern classification systems.
  3. The classical approach to identification involves the use of keys or diagnostic tables.
  4. Computer-assisted identification systems are based upon probabilistic matrices.
  5. In clinical laboratories, the identification of pathogens is aided by the use of commercial systems that employ computers or numerical profiles.

B. Bacterial systematics.
  1. Phototrophic bacteria use light energy to drive ATP synthesis and most are autotrophs capable of using carbon dioxide as the source of cellular carbon.
    a. Oxyphotobacteria use water as an electron donor and liberate oxygen in their process of photosynthetic metabolism; the primary photosynthetic pigment is chlorophyll a; include Cyanobacteria (blue-green bacteria) and Prochlorales.
    b. Anoxyphotobacteria do not evolve oxygen, carry out photosynthesis anaerobically; typically occur in aquatic habitats; include Chromatiaceae (purple sulfur bacteria), Chlorobiaceae (green sulfur bacteria), Rhodospirillaceae (purple nonsulfur bacteria), and Chloroflexaceae (green flexibacteria).
  2. Gliding bacteria.
    a. Myxobacteriales are small rods embedded in a slime layer, exhibit gliding motility, lack flagella, decompose organic matter and aggregate to form fruiting bodies.
    b. Cytophagales exhibit gliding motility, produce hydrolytic enzymes, but do not produce fruiting bodies.
  3. Sheathed bacteria are those bacteria whose cells occur within a filamentous sheath, the formation of which permits the bacteria to attach to solid surfaces, conserve energy, live in low-nutrient environments; it also protects them from predators and parasites.
  4. Budding and/or appendaged bacteria form extensions from the cell; they reproduce by budding or binary fission; they occur in all nutritrional categories; their cell appendages (prosthecae) concentrate nutrients efficiently.
  5. Spirochetes are helically coiled rods that move by a flexing motion.
  6. Spiral and curved bacteria are helically curved rods having less than one complete turn not wound around a central axial filament; they are motile by means of polar flagella.
  7. Gram negative aerobic rods and cocci.
    a. Pseudomonadaceae are straight or curved rods with polar flagella; metabolism is respiratory, some are obligately aerobic others are anaerobic.
    b. Azotobacteriaceae can fix nitrogen and exhibit pleomorphic morphology.
    c. Rhizobiaceae can fix atmospheric nitrogen.

8. Gram negative facultatively anaerobic rods.
   a. Enterobacteriaceae are motile by means of peritrichous flagella and include genera *Escherichia, Citrobacter, Salmonella, Shigella, Enterobacter,* and *Yersinia*.
   b. Vibrionaceae are motile by means of polar flagella and include genera *Vibrio, Photobacterium,* and *Lucibacterium*.
9. Gram negative anaerobic bacteria include only one family, Bacteroidaceae, which include the genera *Bacteroides, Fusobacterium,* and *Leptotrichia*; *Desulfovibrio* reduces sulfates.
10. Gram negative aerobic cocci and coccobacilli include only the Neisseriaceae family.
    a. Cocci genera include *Neisseria* and *Branhamella*.
    b. Coccobacilli (oval-shaped) genera include *Moraxella* and *Acinetobacter*.
11. Gram negative anaerobic cocci typically occur as diplococci and include the genera *Veillonella, Acidaminococcus, Megasphaera,* and *Gemmiger*.
12. Gram negative chemolithotrophic bacteria are important in biogeochemical cycling reactions.
    a. Nitrobacteraceae are the nitrifying bacteria, oxidizing ammonia or nitrite to generate ATP.
    b. Several genera metabolize sulfur and sulfur-containing compounds, for example, *Thiobacillus*.
    c. Members of the family Siderocapsaceae oxidize iron or manganese.
13. Gram positive cocci
    a. Micrococcaceae may occur singly or as irregular clusters.
    b. Streptococcaceae occur as pairs or chains, metabolize fermentatively, and are facultative anaerobes.
    c. Peptococcaceae have complex nutritional requirements; may occur singly, in pairs or in regular or irregular masses; are obligately anaerobic and produce low molecular weight volatile fatty acids, carbon dioxide, hydrogen, and ammonia as main products of amino acid metabolism.
14. Endopsore-forming rods and cocci are extemely important because of the heat resistance of the endospore structure; include *Bacillus* and *Clostridium*.
15. Gram positive asporogenous, rod-shaped bacteria include the family Lactobacillaceae, which produce lactic acid as a major fermentation product (*Lactobacillus* genus important in dairy industry).
16. Actinomycetes and related organisms.
    a. Coryneform bacteria are a heterogeneous group having a characteristic irregular cellular morphology, cells tend to show incomplete separation following cell division and do not form true filaments.
    b. Order Actinomycetales form branching filaments; many produce antibiotics.
17. Rickettsias.
    a. Rickettsiales are obligate intracellular parasites; Gram negative; reproduce by binary fission.
    b. Chlamydiales are obligate intracellular parasites; reproduction characterized by change of small, rigid-walled infectious form into larger, thin-walled noninfectious form.
18. Mycoplasmas lack a cell wall; they are bounded by a single triple-layered membrane and are the smallest organisms capable of self-reproduction.
19. Endosymbionts of invertebrates live within animal cells without adversely affecting the host organism.
20. Archaebacteria are primitive bacteria that are not closely related to other prokaryotes.
    a. The archaebacteria do not have murein in their cell walls, have ether linkages in the membrane phospholipids, and have distinct ribosomal RNA.
    b. The archaebacteria include physiologically unusual bacteria such as the methanogens (strictly anaerobic methane-evolving bacteria), halophiles, and acidophilic thermophiles.

Study Questions

1. Identify the phenotypic characteristics normally examined in classification.
2. List two types of genetic analyses used in the development of classification schemes.
3. Match the following statements with the terms in the list to the right.
   - **a.** Include *Treponema pallidum*.
   - **b.** Do not evolve oxygen; carry out photosynthesis anaerobically.
   - **c.** Myxobacteriales and Cytophagales.
   - **d.** The phosynthetic pigments are chlorophyll a and chlorophyll b.
   - **e.** The most diversified and widely distributed group of photosynthetic bacteria.
   - **f.** Use water as an electron donor and liberate oxygen in the process as part of their photosynthetic metabolism.
   - **g.** Comprise bacteria whose cells occur within a filamentous structure
   - **h.** The photosynthetic pigment is chlorophyll a.
   - **i.** Form extensions or protrusions from the cell.
   - **j.** Helically curved rods not wound around a central axial filament (have less than one complete turn).
   - **k.** Use light energy to drive ATP synthesis.
   - **l.** Helically coiled rods.
   - **m.** Comma-shaped.

   (1) Phototrophic bacteria
   (2) Oxyphotobacteria
   (3) Cyanobacteria
   (4) Prochlorales
   (5) Anoxyphotobacteria
   (6) The gliding bacteria
   (7) The sheathed bacteria
   (8) Budding/appendaged bacteria
   (9) Spirochetes
   (10) Spiral and curved bacteria

4. Match the following statements with the terms in the list to the right.
   - **a.** A family of Gram negative facultative anaerobic rods that are motile by means of peritrichous flagella.
   - **b.** A Gram negative anaerobic bacteria that is the dominant microorganism in the intestinal tract.
   - **c.** Has one family, *Neisseriaceae*.
   - **d.** Include genera *Escherichia, Salmonella, Proteus*, and *Shigella*.
   - **e.** Include families Enterobacteriaceae and Vibrionaceae.
   - **f.** Include genera *Vibrio* and *Photobacterium*.
   - **g.** Includes the species *Neisseria gonorrhoeae*.
   - **h.** Pseudomonadaceae, Azotobacteraceae, Rhizobiaceae, and Methylomonadaceae.

   (1) Gram negative aerobic rods and cocci
   (2) Pseudomonadaceae
   (3) Gram negative facultatively anaerobic rods
   (4) Enterobacteriaceae
   (5) Vibrionaceae
   (6) Gram negative anaerobic bacteria
   (7) Gram negative aerobic cocci
   (8) Gram negative anaerobic cocci
   (9) Bacteroides

5. Match the following statements with the terms in the list to the right.
   - **a.** Lack a cell wall.
   - **b.** Important group of bacteria in biogeochemical cycling reactions.
   - **c.** Oxidize nitrite to nitrate.
   - **d.** Include the genus *Staphylococcus*.
   - **e.** Include genera *Nitrosomonas* and *Nitrobacter*.
   - **f.** Oxidize ammonia to nitrite.
   - **g.** Include the families Nitrobacteraceae and Siderocapsaceae.
   - **h.** Include genera *Bacillus* and *Clostridium*.
   - **i.** Gram positive cocci occurring as pairs or chains.
   - **j.** Oxidize ammonia or nitrite to generate ATP.
   - **k.** Include the families Micrococcaceae, Streptococcaceae, and Peptococcaceae.
   - **l.** Include methane-producing and halophilic bacteria.
   - **m.** Include genus *Corynebacterium*.
   - **n.** Include *Chlamydia*.
   - **o.** As a group, they oxidize inorganic compounds to generate ATP.
   - **p.** Include genus *Lactobacillus*.

   (1) Gram negative chemolithotrophs
   (2) Nitrobacteraceae
   (3) *Nitrosomonas* and *Nitrospora*
   (4) *Nitrobacter* and *Nitrococcus*
   (5) Gram positive cocci
   (6) Streptococcaceae
   (7) Micrococcaceae
   (8) Endospore-forming rods and cocci
   (9) Gram positive, nonsporulating rods
   (10) Actinomycetes and related organisms
   (11) Rickettsias
   (12) Mycoplasmas
   (13) Archaebacteria

Some Thought Questions

1. What is a bacterial species?
2. What would Humpty Dumpty say about bacterial nomenclature?
3. If Noah collected two of every species, how many bacteria were in the ark?
4. Why do microbiologists argue about bacterial names?
5. If you discovered a new bacterium what would you name it and how would you go about doing this?
6. Do bacteria ever get sick?
7. Are the "blue greens" bacteria or algae?
8. Which came first, *Bacillus* or its endospore?
9. Why are methanogens archaebacteria?
10. What are the differences between classification and identification?
11. How have computers changed the way diseases are diagnosed?

Additional Sources of Information

Buchanan, R. E., and N. E. Gibbons (eds.). 1974. *Bergey's Manual of Determinative Bacteriology*, 8th ed. Williams & Wilkins Company, Baltimore, Maryland. The last comprehensive volume of the standard taxonomic reference covering all groups of bacteria.

Cowan, S. T., and L. R. Hill (eds.). 1978. *A Dictionary of Microbial Taxonomy*. Cambridge University Press, New York. A very useful dictionary providing definitions of terms used in classification of microorganisms and descriptions of the bacterial taxa.

Goodfellow, M. (ed.) 1980. *Microbiological Classification and Identification*. Academic Press, Inc., Orlando, Florida. A valuable reference work for all taxonomists, microbiologists, and biochemists.

Holt, J. (ed.). 1984. *Bergey's Manual of Determinative Bacteriology*, 9th ed. Part I. Williams & Wilkins Company, Baltimore, Maryland. The first part of the newest edition of the standard taxonomic reference work covering the bacteria. This part describes the Gram negative facultatively anaerobic bacteria.

Laskin, A. I., and H. A. Lechevalier. 1977. *Handbook of Microbiology: Bacteria*. CRC Press, Inc., Boca Raton, Florida. A multivolume set describing the properties of microoganisms, including the various bacterial taxa.

Skerman, V. B. D. 1967. *A Guide to the Identification of the Genera of Bacteria*. Williams & Wilkins Company, Baltimore, Maryland. A reference work describing the characteristics of the major bacterial genera.

Skinner, F. A., and D. W. Lovelock. 1980. *Identification Methods for Microbiologists*. Academic Press, Inc., New York. A reference work describing the specific methodologies that are employed for the identification of the bacteria.

Stanier, R. Y., and G. Cohen-Bazire. 1977. Phototrophic prokaryotes: the cyanobacteria. *Annual Review of Microbiology*, 31:225–274. A detailed discussion of the taxonomy of the cyanobacteria.

Starr, M. P., H. Stolp, H. G. Truper, A. Ballows, and H. G. Schlegel (eds.). 1981. *The Prokaryotes: A Handbook on Habits, Isolation, and Identification of Bacteria*. Springer-Verlag, Berlin. A superb, well-illustrated, two-volume work describing the numerous taxonomic groups of bacteria.

Woese, C. R. 1981. Archaebacteria. *Scientific American*, 244(6):98–122. An excellent, easy-to-read description of the arachaebacteria.

14 *Eukaryotic Microorganisms: Classification of Fungi, Algae, and Protozoa*

KEY TERMS

algae
arthrospores
ascomycetes
ascospores
ascus
asexual spores
axopodia
basidia
basidiomycetes
basidiospores
bioluminescence
brown algae
budding
chlamydospores
chlorophylls
ciliates
Ciliophora
clamp cell connections
club fungi
coenocytic
conidia
conidiophores
deuteromycetes
diatoms
dinoflagellates
dolipore septa
euglenoids
filamentous fungi
fire algae
foraminiferans
fruiting bodies
frustules
fungi
fungi imperfecti
green algae
hyphae
kelps
Mastigophora
molds
multilateral budding
mycelia
myxamoebae
pellicle
plasmodium
polar budding
protozoa
pseudoplasmodium
pseudopodia
radiolarians
red algae
red tide
sac fungi
Sarcodina
Sarcomastigophora
sexual spores
slime molds
sporangiospores
sporangium
Sporozoa
sporozoites
swarm cells
teliospores
tests
theca
trophozoites
xanthophylls
zoospores
zygospores

OBJECTIVES

After reading this chapter you should be able to

1. Define the key terms.
2. Describe the modes of reproduction exhibited by the fungi.
3. Describe the characteristics of the major taxonomic groups of fungi.
4. Describe the characteristics of the major taxonomic groups of algae.
5. Describe the characteristics of the major taxonomic groups of protozoa.
6. Compare the main criteria used to classify the major groups of fungi, algae, and protozoa.

Classification of Fungi

As with other microorganisms, the classification of the fungi is difficult, sometimes ambiguous, vehemently debated, and subject to constant revision. The diversity of the fungi and the interrelationships between the fungi and other groups of eukaryotic microorganisms make fungal taxonomy particularly difficult. The fungi are eukaryotic, heterotrophic microorganisms. They are nonphotosynthetic and typically form reproductive spores.

By a concise definition: the fungi are achlorophyllous, saprophytic or parasitic, with unicellular or more typically filamentous vegetative structures usually surrounded by cell walls composed of chitin or other polysaccharides, propagating with spores and normally exhibiting both asexual and sexual reproduction. If this definition seems complex and vague, it is; of necessity, a broad definition is needed to accommodate all the morphological and physiological anomalies that occur among the fungi.

The classification of fungi is based largely on the means of reproduction, including the nature of the life cycle, reproductive structures, and reproductive spores. Many fungi exhibit both sexual and asexual forms of reproduction. During the asexual phase of fungal reproduction a variety of spores may be produced. These include various conidia that are asexual spores borne externally on hyphae or specialized conidiophore structures. One type of conidia, the arthrospore, represents fragmented hyphae. The fragmentation of multicellular eukaryotic microorganisms constitutes a form of reproduction because the individual fragments are each capable of reproducing the original organism. Other asexual fungal spores include sporangiospores, which are produced within a specialized structure known as the sporangium, and chlamydospores, thick-walled spores that occur within hyphal segments. These fungal spores can be dispersed from the fungal hyphae and later germinate to form new vegetative structures.

Fungi also can produce various types of sexual reproductive spores. The primary taxonomic groupings of the fungi are based on the sexual reproductive spores. Some sexual spores of fungi, called ascospores, are formed within a specialized structure known as the ascus. The ability to produce ascospores distinguishes the ascomycete fungal group, which includes both yeasts and filamentous fungi, from other major fungal groups. In another major group of fungi, the basidiomycetes,

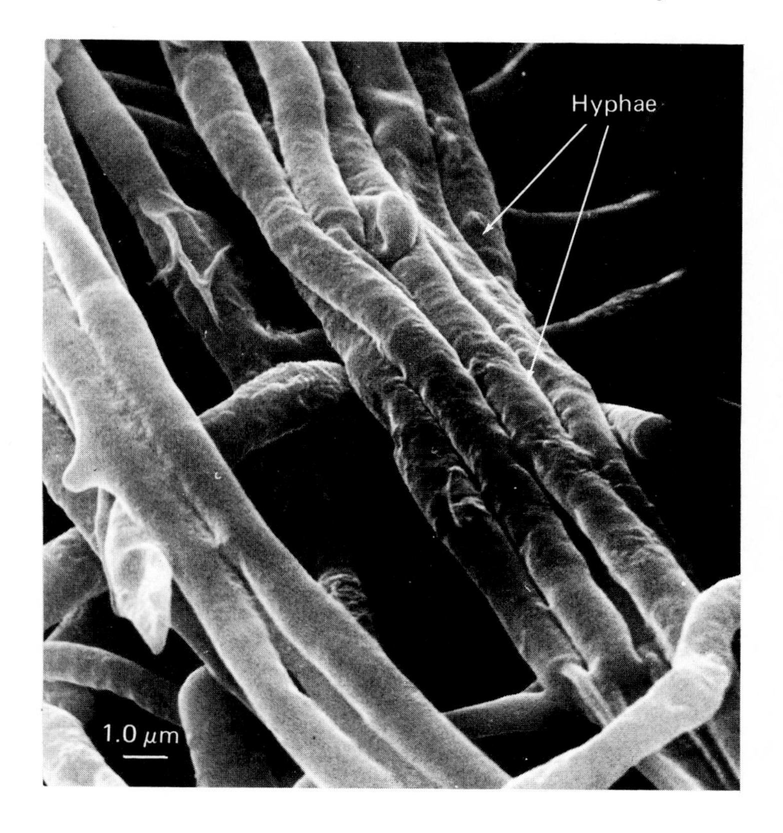

Figure 14.1 Micrograph of filamentous fungi showing tubular hyphae that form threads of mycelia. (From D. H. Ellis and D. A. Griffiths, *Canadian Journal of Microbiology*, **21**:442–452. Reprinted by permission of the National Research Council of Canada.)

the sexual spores are produced on a specialized structure known as the **basidium**. The sexual reproduction of basidiomycetes usually involves the fusion of hyphal cells. In several fungal groups sexual reproduction involves fusion of specialized gametes. These gametes are haploid (having a single set of homologous chromosomes) and their fusion reestablishes a diploid state (having the normal

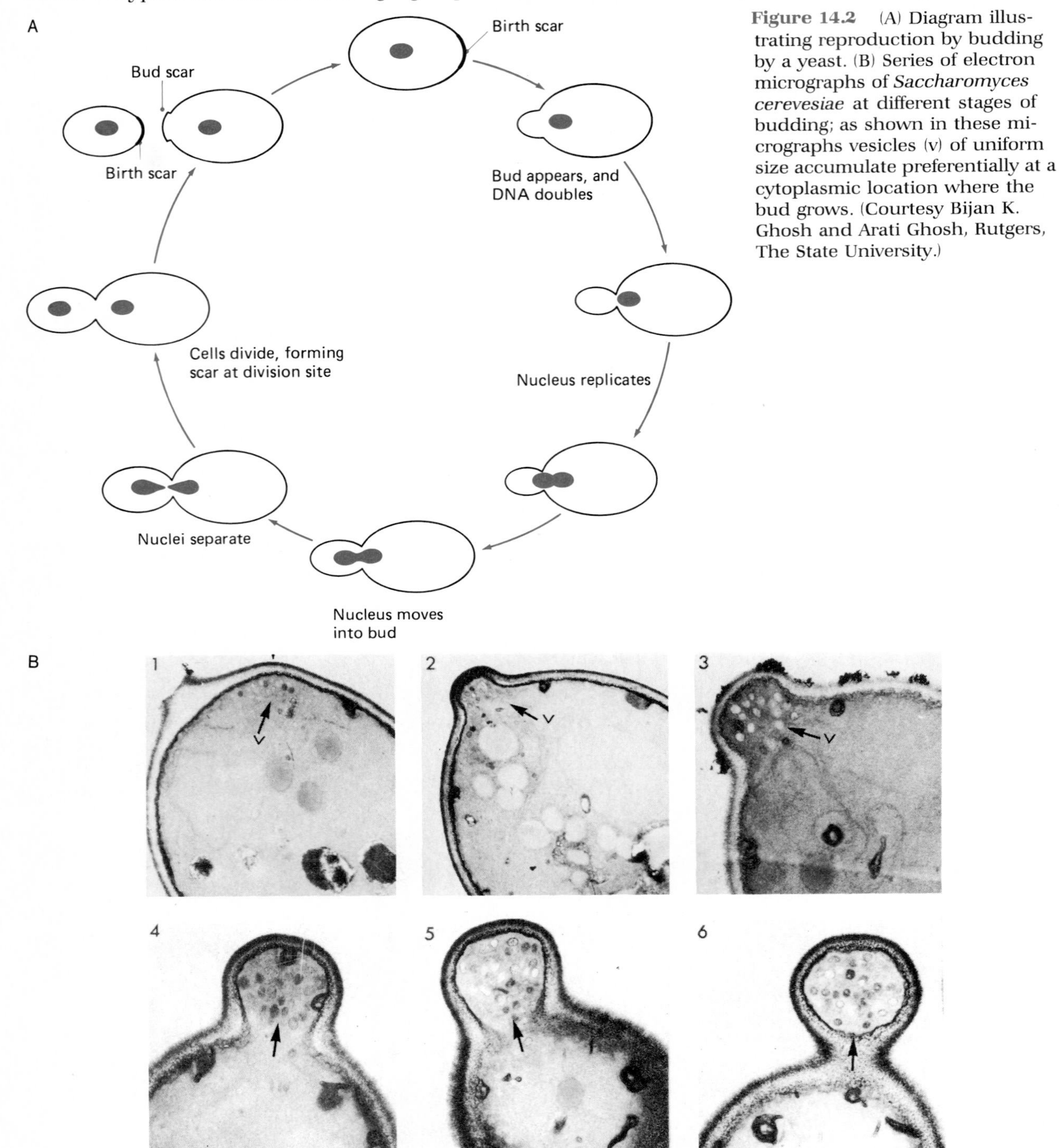

Figure 14.2 (A) Diagram illustrating reproduction by budding by a yeast. (B) Series of electron micrographs of *Saccharomyces cerevesiae* at different stages of budding; as shown in these micrographs vesicles (v) of uniform size accumulate preferentially at a cytoplasmic location where the bud grows. (Courtesy Bijan K. Ghosh and Arati Ghosh, Rutgers, The State University.)

number of chromosomes). If the gametes are motile they are called zoospores. Not all fungi produce sexual spores. The deuteromycetes, or fungi imperfecti, have no known sexual reproductive phase, and as far as we know they are restricted to asexual means of reproduction. If a sexual stage is discovered for a fungus that has been classified among the fungi imperfecti, it is reclassified into one of the other major fungal groups based on the type of sexual spores that are produced.

Many fungi form filaments of vegetative cells known as mycelia (Figure 14.1). These fungi are known as the molds or filamentous fungi. The filamentous fungi normally develop multicellular structures known as hyphae. The growth of the fungus involves the elongation of the hyphae, generally with the formation of branches and crosswalls separating individual cells of the fungus. Some fungi, however, form coenocytic mycelia that lack crosswalls; therefore, these fungi are really one-celled multinucleate organisms. Mycelia are integrated masses of individual tube-like filaments of hyphae. The mycelia usually exhibit branching and are typically surrounded by cell walls containing chitin and/or cellulose.

Some fungi are unicellular. This is characteristic of the yeasts. In a formal systematic sense, yeasts are not recognized as being separate from the rest of the fungi and are classified along with their filamentous counterparts. In practice, however, the yeasts are typically treated separately from the filamentous fungi for purposes of identification. The most common mode of reproduction for yeasts is by budding (Figure 14.2). The budding process involves the formation of a crosswall that separates the bud from the mother cell. Budding can occur all around the mother cell (multilateral budding) or may be restricted to the end of the mother cell (polar budding). The budding process leaves a bud scar on the mother cell, and consequently, only a limited number of progeny may be derived from an individual yeast cell. Although budding is the most common form of reproduction, various other reproductive strategies exist among the yeasts, including sexual reproduction and fission.

The Slime Molds

The slime molds represent a borderline case between the fungi and the protozoa. They can just as well be classified with the protozoa, but traditionally they have been studied by mycologists. The vegetative cells of the slime molds lack cell walls. Their nutrition is phagotrophic; that is, they engulf and ingest nutrients.

All slime molds exhibit characteristic life cycles, a feature used to subdivide the classification of these organisms (Figure 14.3). The Acrasiales or cellular slime molds often feed upon bacteria. These fungi form a fruiting body (spore-bearing body) known as a sporocarp. The sporocarps are generally stalked structures. The stalks normally consist of walled cells, and this characteristic forms the basis for designating these organisms as the cellular slime molds. The sporocarp releases spores that germinate, forming myxamoebae (amoeboid cells that form pseudopodia, or false feet). The myxamoebae swarm together or aggregate to form a pseudoplasmodium. Within the pseudoplasmodium, the cells do not lose their integrity. The pseudoplasmodium undergoes a developmental sequence (differentiation), culminating in the formation of a sporocarp, which is a special type of fruiting body that bears a mucoid droplet at the tip of each branch, containing spores with cell walls.

The plasmodium of the myxomycetes or true slime molds is a multinucleate protoplasmic mass that is devoid of cell walls and is enveloped in a gelatinous slime sheath. The plasmodium gives rise to brilliantly colored fruiting bodies. The classification of the myxomycetes is based largely on the structure of the fruiting body. In many species of myxomycetes spores are formed inside the fruiting body. These spores are sometimes referred to as endospores, but they do not bear any resemblance to the endospores of bacteria. The spores of myxomycetes generally have a definite thick wall. In the life cycle of a myxomycete the spores are released from the sporangia and disseminated. At a later time they germinate, producing myxamoebae and/or swarm cells. The myxamoebae and/or swarm cells later unite in a sexual fusion process to initiate formation of the plasmodium. The brightly colored fruiting bodies of myxomycetes are often seen on decaying logs or other moist areas of decaying organic matter (Figure 14.4). Myxomycetes are often conspicuous on grass lawns, appearing as large blue-green colonies. To remove these unsightly blemishes from an otherwise luxuriant lawn, simply mow the grass.

Mastigomycota

The Mastigomycota comprise the second major taxonomic division of the fungi (Table 14.1). Most members of this division form extensive filamentous coenocytic mycelia. Mastigomycota typically produce motile cells with flagella during part of their life cycle. Asexual reproduction by species in this division normally involves motile spores called

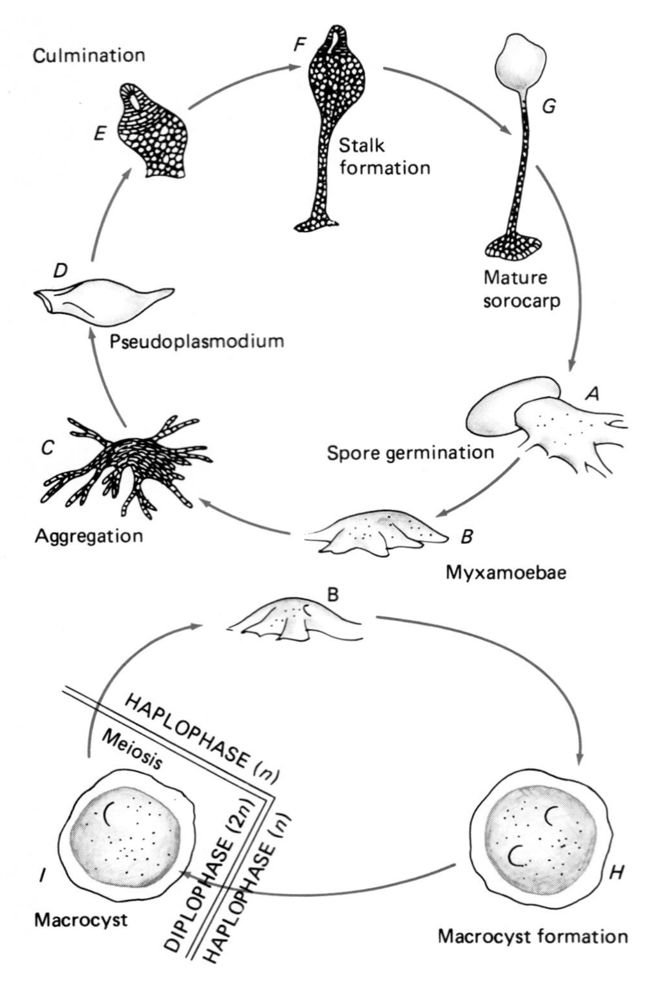

Figure 14.3 The life cycle of a slime mold, *Dictyostelium discoideum.* (A) Germination of a spore with a single myxamoeba issuing from it. (B) Myxamoebae. (C) Streams of aggregating myxamoebae. (D) Pseudopodium, grex, slug. (E) Beginning of culmination, myxamoebae at the front end of the grex pushing down the middle to stalk cylinder. (F) Later stage, showing stalk formation. (G) Mature sorocarp with cellular stalk. (H) Young macrocyst in which keryogamy occurs. (I) Mature macrocyst where meiosis takes place. Myxamoebae form upon germination of the mature macrocyst. The pseudoplasmodium formation of the Acrasiomycetes is of special interest because of the biochemical communication involved in initiating swarming activity. The swarming behavior of *D. discoideum* has been extensively studied. Under appropriate conditions, when food sources become limiting, the myxamoebae cease their feeding activity and swarm to an aggregation center. The swarming activity is initiated when one or more cells at the aggregation center release cyclic AMP (acrasin). At the biochemical level, cyclic AMP is responsible for communication between the myxamoebae. The myxamoebae move along the concentration gradient of cyclic AMP until they reach the center of aggregation. When the myxamoebae reach the center of aggregation, they mass together to form a pseudoplasmodium. Swarming occurs as a pulsating wave motion in which the chemical stimulus, cyclic AMP, is transmitted from cells that are near the aggregation center to distant cells.

Figure 14.4 (A) The slime mold *Lycogala epidendrum* growing on a log. (Courtesy Varley Wiedeman, University of Louisville.) (B) *Lycogala epidendrum* (*aethalium*). (Courtesy Orson K. Miller, Jr., author of *Mushrooms of North America,* Virginia Polytechnic Institute and State University.)

A

B

Table 14.1 **The Four Classes in the Division Mastigomycota**

Chytridiomycetes	Vegetative form varied, produces posteriorly uniflagellate motile cells with whiplash flagella
Hyphochytridiomycetes	A small group of fungi, produces motile anteriorly uniflagellate cells with tinsel flagella
Plasmodiophoromycetes	Parasitic fungi with multinucleate thalli (plasmodia) within the cells of their hosts; resting cells (cysts) produced in masses but not in distinct sporophores; motile cells with two anterior whiplash flagella
Oomycetes	Vegetative form varied, usually filamentous, with a coenocytic, walled mycelium; hyphal wall, produces zoospores, each with one whiplash and one tinsel flagellum; sexual reproduction oogamous, resulting in the formation of oospores

zoospores. As opposed to the phagotrophic mode of nutrition exhibited by the slime molds, members of this division absorb their nutrients. Many members of this division are parasitic on other fungi, algae, and plants. For example, *Phytophthora infestans* causes potato blight and was responsible for the great Irish potato famine of 1845 and 1846, which resulted in the great wave of immigration from Ireland to the United States.

Since the Kennedy family rose to political power, there has been greater recognition of the tremendous part the Irish have played in forming our country. The role in America would have been a lesser one had it not been for the actions of a fungus . . .

—LUCY KAVALER, *Mushrooms, Molds, and Miracles*

Amastigomycota

The vegetative cells of the Amastigomycota, the third division of the fungi, vary from single cells to mycelia that may be coenocytic (multinucleate) or may have extensive septation (crosswalls separating cells within mycelia). Unlike slime molds and Mastigomycota, the Amastigomycota do not produce motile cells. There are four subdivisions in the Amastigomycota: Zygomycotina, Ascomycotina, Basidiomycotina, and Deuteromycotina (Table 14.2).

Zygomycotina. The Zygomycotina typically have coenocytic mycelia and are characterized by the formation of a zygospore, a sexual spore that results from the fusion of specialized structures called gametangia. For sexual reproduction some species require gametangia of two different mating types (genetically distinct sexual partners), whereas others are homothallic, requiring only one type of gametangia for zygote formation. In addition to sexual reproduction, the Zygomycotina characteristically produce asexual sporangiospores within a sporangium (Figure 14.5). Many members of the Zygomycotina are plant or animal pathogens. For example, *Mucor* species are important opportunistic human pathogens and can cause serious infections in burn wounds.

Table 14.2 **The Four Subdivisions of the Amastigomycota**

Zygomycotina	Saprophytic, parasitic, or predatory fungi, coenocytic mycelium; asexual reproduction usually by sporangiospores; sexual reproduction, where known, by fusion of equal or unequal gametangia resulting in the formation of zygosporangia containing zygospores
Ascomycotina	Saprophytic, symbiotic, or parasitic fungi; unicellular or with a septate mycelium, producing ascospores in sac-like cells (asci)
Basidiomycotina	Saprophytic, symbiotic, or parasitic fungi; unicellular or, more typically, with a septate mycelium, producing basidiospores on the surface of various types of basidia
Deuteromycotina	Saprophytic, symbiotic, parasitic, or predatory fungi; unicellular or, more typically, with a septate mycelium, usually producing conidia from various types of conidiogenous cells; sexual reproduction unknown

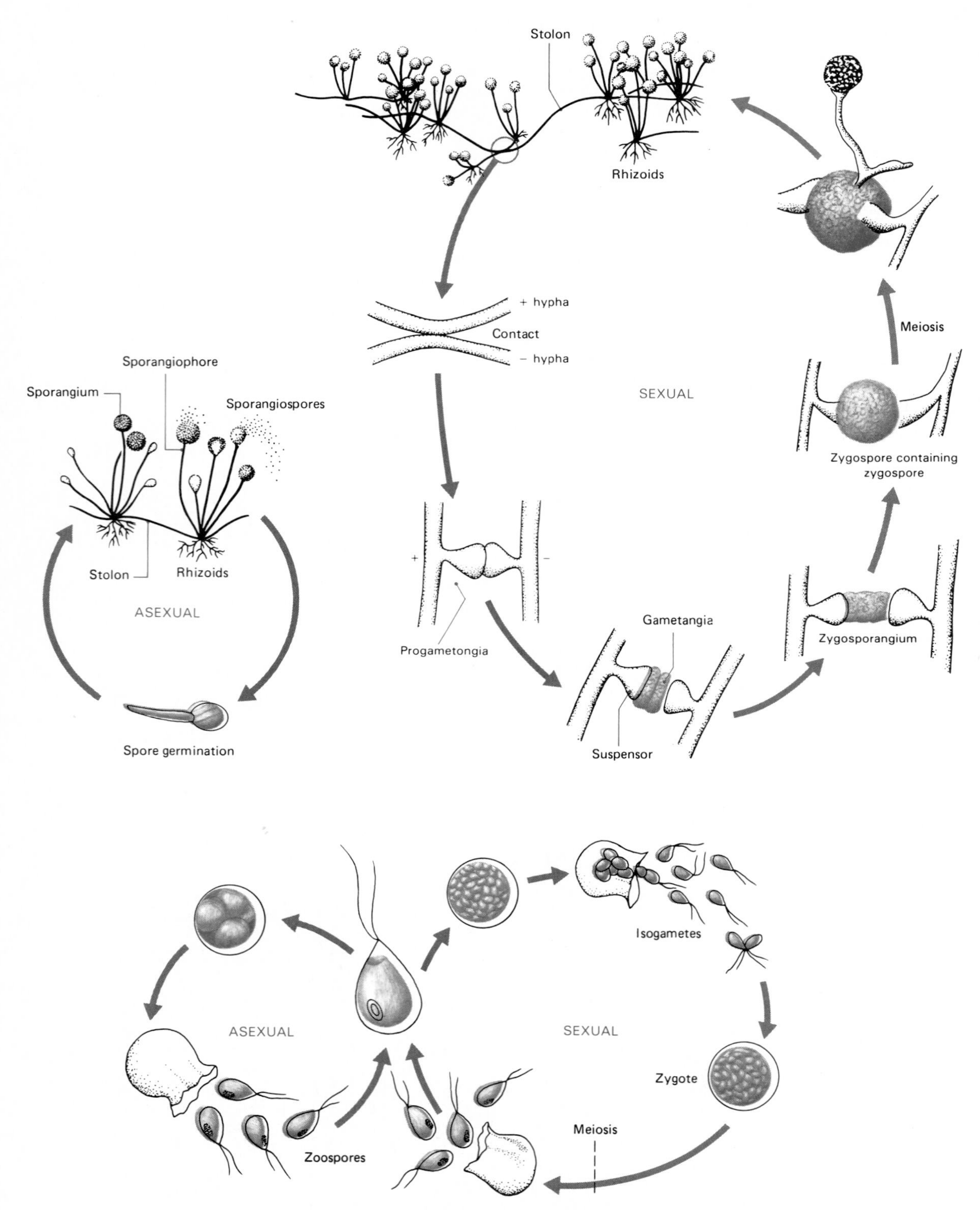

Figure 14.5 The life cycle of the bread-mold fungus *Rhizopus stolonifer*, showing the asexual and sexual phases. The asexual phase involves the production, release, and germination of sporangiospores.

Ascomycotina. The ascomycetes, which are sometimes called the sac fungi, produce sexual spores within a specialized sac-like structure known as the ascus. The mycelia of ascomycetes is composed of septate hyphae and the cell walls of the hyphae of most ascomycetes contain chitin. Asexual reproduction in the ascomycetes may be carried out by fission, fragmentation of the hyphae, formation of chlamydospores (thick-walled spores within the hyphal filaments), and production of conidia (nonmotile spores produced on a specialized spore-bearing cell). During sexual reproduction, the ascomycetes normally exhibit a short-lived **dikaryotic stage** (having cells containing two nuclei) between the time of fusion of gametes (plasmogamy) and the time of fusion of the two nucleii (karyogamy).

Many yeasts are ascomycetes (Figure 14.6). The morphology of the ascospore is a critical taxonomic feature for classifying yeasts to the genus level. Classification of the yeasts to the species level normally employs numerous biochemical and physiological characteristics as well as morphological features. The metabolic activities of the ascosporogenous yeasts have many industrial applications. *Saccharomyces cerevisiae* is used as baker's yeast. Many fermented beverages are also produced by using members of the genus *Saccharomyces*. Most commonly, *S. carlsbergensis* and *S. cerevisiae* are used for the production of beer, wine, and spirits. The morels and truffles, which are gastronomical delights, are also in this taxonomic group.

Not presume to dictate, but broiled fowl and mushrooms—capital thing

—CHARLES DICKENS, *Pickwick Papers*

Some ascomycetes are important plant and animal pathogens. The powdery and black mildews occur in this taxonomic group. *Endothia parasitica* is the causative agent of chestnut blight. This organism was introduced into North America from eastern Asia in the early twentieth century and quickly devastated the chestnut trees of the United States and Canada. *Ceratocystis ulmi* is the causative agent of Dutch elm disease, a great threat to elm trees in North America.

Claviceps purpurea causes the ergot of rye. Cattle and other such animals are poisoned when grazing on grasses contaminated with the resting bodies (sclerotia) of the fungus. Sclerotia are hard resting bodies that are resistant to unfavorable conditions and may remain dormant for prolonged periods. Various alkaloid biochemicals are produced by *Claviceps purpurea*. Some of these have hallucinogenic properties, whereas others, such as those used to induce labor for childbirth, are useful medicinals. Recent evidence suggests that ergoticism, caused by ingestion of grains in which *C. purpurea* had grown, was responsible for the Salem witch hunts of the 1690s.

Basidiomycotina. The basidiomycetes are the most complex fungi. This taxonomic group includes the smuts, rusts, jelly fungi, shelf fungi, stinkhorns, bird's nest fungi, puffballs, and mushrooms. The basidi-

Figure 14.6 Micrograph of the yeast *Schizosaccharomyces* containing ascospores. (From BPS: J. Robert Waaland, University of Washington.)

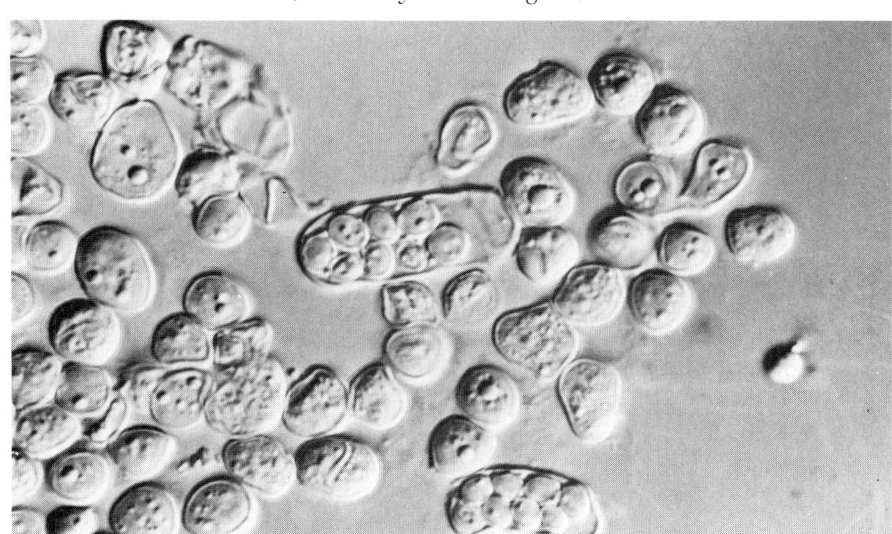

omycetes are distinguished from other classes of fungi by the fact that they produce sexual spores, known as basidiospores, on the surfaces of specialized spore-producing structures, known as basidia. This group is also known as the club fungi based on the typical shape of the basidia.

There is something absolutely fascinating to me about being present at the exact moment when a mushroom is mushrooming.

—JOHN CAGE

The mycelia of basidiomycetes typically form clamp connections between cells. Additionally, the mycelia of many basidiomycetes are characterized by specialized crosswalls between connecting cells, known as dolipore septa. The dolipore septum has a central pore surrounded by a barrel-shaped swelling of the crosswall. Effectively, the clamp cell connections and the dolipore septa permit enhanced communication of chemicals through the mycelia of the organism.

The shelf or bracket fungi are some of the most conspicuous fungi. These basidiomycetes are often seen growing on trees. Mushrooms, which we also often see, are the fruiting bodies (basidiocarps) of basidiomycetes (Figure 14.7). The spore-bearing structures (basidia) of mushrooms are borne on the surface of the gills of the basidiocarp. Some mushrooms are edible, but others are extremely poisonous. The proper identification of mushrooms is critical lest one become the victim of mushroom poisoning. It is sometimes easy to confuse an edible species with one that is deadly poisonous. Many *Amanita* species are quite beautiful, but most are deadly poisonous. *Amanita phalloides* is known as the death cap because most deaths due to mushroom poisoning have been attributed to the ingestion of this species.

She stretched herself up on tiptoe and peeped over the edge of the mushroom and her eyes immediately met those of a large blue caterpillar.

—LEWIS CAROLL, *Alice's Adventures in Wonderland*

The numerous species of rust and smut fungi are the most serious fungal plant pathogens. There are over 20,000 species of rust fungi and over 1000 species of smut fungi. Rusts and smuts are characterized by the production of a resting spore known as a teliospore, which is thick-walled and binucleate.

A

B

C

Figure 14.7 Mushrooms are visible macroscopic fruiting bodies of basidiomycetes. Some mushrooms are edible, but others are quite poisonous. (A) This *Pleurotus ostreatus* is edible and good. (B) This *Amanita virosa* is deadly poisonous. (Courtesy Orson K. Miller, Jr., author of *Mushrooms of North America,* Virginia Polytechnic Institute and State University.) (C) This *Amanita pantherina* is also a poisonous mushroom. (From BPS: J. Robert Waaland, University of Washington.)

All rust fungi are plant pathogens. These fungi require two unrelated hosts for the completion of their normal life cycle (Figure 14.8). Important plant diseases are caused by rust and smut fungi, and these fungal plant pathogens cause great economic losses in agriculture.

For 300 years after the birth of Christ rainfall was abnormally heavy . . . The crops fluorished and so did the host of tiny living organisms—the rusts, rots, mildews, molds, smuts, and blights that were to ultimately destroy the grain . . . Field after field was laid waste. History is made by a multiplicity of little things, as well as major revolutions, wars, earthquakes and floods. And so it was that the fungi, by bringing hunger and unrest, contributed to the decline of the great Roman Empire.

—LUCY KAVALER, *Mushrooms, Molds, and Miracles*

Deuteromycotina. The deuteromycetes or fungi imperfecti have no known sexual reproductive phase. There are about 15,000 species in the fungi imperfecti. The vegetative structures of most of the fungi in this class resemble ascomycetes, although the vegetative structures of a few members of the fungi imperfecti resemble basidiomycetes.

The fungi imperfecti are classified largely on the basis of the morphological structure of the vegetative phase and on the types of asexual spores produced (Figure 14.9). The deuteromycetes include many economically and medically important genera of filamentous fungi (Table 14.3). Some deuteromycetes cause diseases in plants and animals. For example, *Candida albicans* can cause serious human infections. Some species of fungi imperfecti are important in food and industrial microbiology. The antibiotic penicillin, for example, is produced

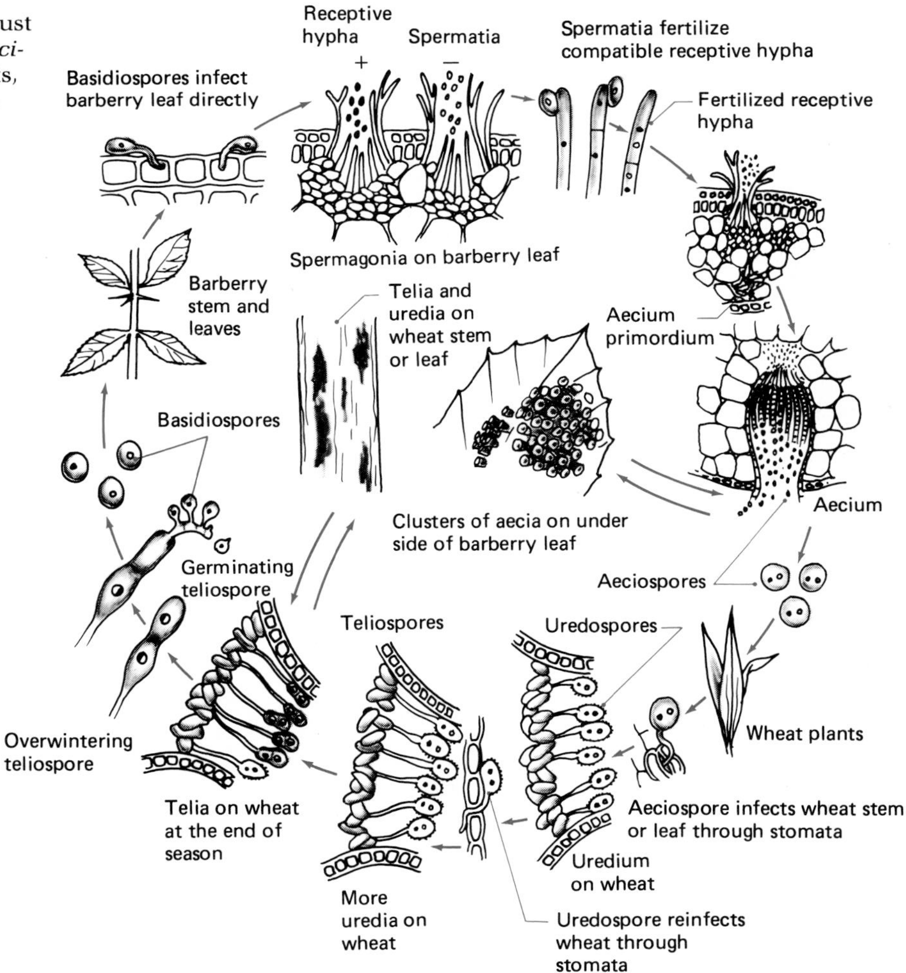

Figure 14.8 Life cycle of stem rust of wheat, which is caused by *Puccinia gramminis.* As with other rusts, *Puccinia* requires alternate hosts.

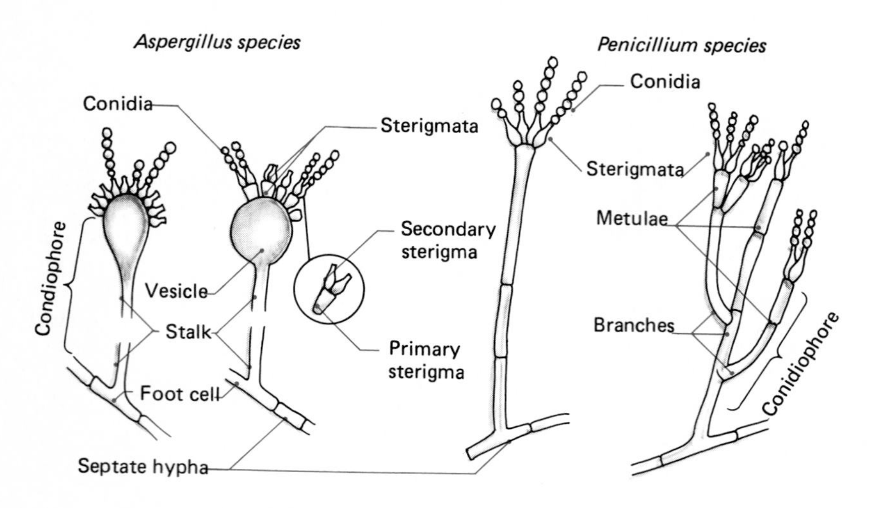

Figure 14.9 Diagram of asexual spores in *Penicillium* and *Aspergillus* spp. These asexual spores are important diagnostic characteristics of the fungi imperfecti.

Table 14.3 Some Representative Genera of Deuteromycetes

Alternaria	Soil saprophytes and plant pathogens, muriform spores fit together like bricks of a wall
Arthrobotrys	Soil saprophytes, some form organelles for capture of nematodes
Aspergillis	Common molds, radially arranged colored, often black, conidiospores
Aureobasidium (Pulullaria)	Short mycelial filaments, lateral blastospores; often damage painted surfaces
Candida	Common yeast; some cause mycoses; some species able to grow in concentrated sugar solutions; some species able to grow on hydrocarbons
Coccidioides	*C. immitis* causes mycotic infections in humans and animals
Cryptococcus	Yeasts, saprophytic in soil but some may cause mycoses in animals and humans
Geotrichum	Common soil fungus; older mycelial filaments break up into arthrospores
Helminthosporium	Cylindrical, multiseptate spores; many are economically significant plant pathogens
Penicillium	Common mold with colored, often green, conidiospores arranged in brush shape
Trichoderma	Common soil saprophyte with highly branched conidiophores

by *Penicillium* species and both *Aspergillus* and *Penicillium* species are used in the production of various foods, such as blue cheese and soy sauce.

Classification of Algae

Whereas the fungi are nonphotosynthetic, the algae are eukaryotic photosynthetic organisms. As such, they are differentiated from all other microorganisms. The algae are separated from the plants by their lack of tissue differentiation. They contain chlorophylls and convert light energy to ATP. In Whittaker's five kingdom classification system (see Figure 2.5), some of the algae are placed in the Kingdom Protista along with the protozoa, and other algae exhibiting more extensive organizational development are placed in the Kingdom Plantae. Indeed, some organisms that are classified as algae are borderline cases with higher plants, and others are borderline cases with protozoa.

I suspect, a fish—a little fish anyway—generally regards many of the algae as great stationary or oscillating forests of intertwining, variously colored, slippery, slimy strands through which it must swim in hunting for its breakfast . . . At some time during its life the average fish must regard algae as a rabbit considers lettuce, as a horse enjoys green pastures: in other words, as something to eat.

—Lewis H. Tiffany, *Algae*

MICROBIOLOGY HEADLINES

Fungus Kills Koala at San Diego's Zoo

SAN DIEGO—A common fungus has killed one rare koala at the San Diego Zoo and may have claimed the life of a second, a zoo spokesman said Saturday.

Officials said they did not fear an epidemic that would harm the zoo's other 24 koalas—bearlike marsupials native to Australia—but they said steps were being taken to make certain the eucalyptus leaves they eat are clean.

A koala, named Tully, died April 16 of cryptococcosis, a fairly common soil-borne fungus that affects the central nervous system, said zoo spokesman Jeff Jouett.

A second koala, known as both Pavo and Gumdrop, is believed to have died of the same fungus Aug. 2.

Tissue samples were sent for testing but no results have been released, Jouett said.

"Swabs of fluid taken from the other koalas at the zoo following Gumdrop's demise turned up evidence of the fungus in one other animal," said Jouett.

"We are concerned enough to take some preventive measures because the koalas are very special animals to San Diego," Jouett said. "For the longest time, until 1982, San Diego had the only koalas outside Australia."

Jouett said officials did not think there was an epidemic because the deaths were four months apart and only one other animal had the fungus.

Keepers have started washing down the eucalyptus leaves that are the only food koalas eat. The trees in the exhibit areas will be changed.

The movement of the koalas in the nearby koala barn also was being restricted to keep the stress levels of the animals down and give their immune systems a better chance of fighting the fungus.

"We haven't positively tracked down the source of the fungus because it is very common in the environment," said zoo veterinarian Jane Meier.

"The main staple of the koala diet is eucalyptus leaves, and the fungus could have been spread from pigeon droppings or dust on the eucalyptus trees."

San Diego is one of two zoos in the United States with resident koalas. The other is the Los Angeles Zoo.

There are some algae that lose their ability to carry out photosynthetic metabolism, rendering them indistinguishable from the protozoa. Some motile unicellular algae have traditionally been studied by both protozoologists and phycologists. This has led to an inevitable confusion in the literature because zoologists and botanists typically use different features and criteria for establishing classification systems. Most traditional algal classification systems include the blue-green algae, but these organisms are properly considered as cyanobacteria because of their prokaryotic cells. The reclassification of the

Table 14.4 The Major Divisions of Algae

Chlorophycophyta (green algae)
Photosynthetic pigments: chlorophylls a and b, carotenes, several xanthophylls. Storage product: starch. Cell wall: cellulose, xylans, mannans, absent in some, calcified in some. Flagella: 1, 2–8, many, equal, apical.

Chrysophycophyta (golden and yellow-green algae, including diatoms)
Photosynthetic pigments: chlorophylls a and c, carotenes, fucoxanthin, and several other xanthophylls. Storage product: chrysolaminaran. Cell wall: cellulose, silica, calcium carbonate. Flagella: 1–2, unequal or equal, apical.

Cryptophycophyta (cryptomonads)
Photosynthetic pigments: chlorophylls a and c, carotenes, xanthophylls (alloxanthin, crocoxanthin, monadoxanthin), phycobilins. Storage product: starch. Cell wall: absent. Flagella: 2, unequal, subapical.

Euglenophycophyta (euglenoids)
Photosynthetic pigments: chlorophylls a and b, carotenes, several xanthophylls. Storage product: paramylon. Cell wall: absent. Flagella: 1–3, apical, subapical.

Phaeophycophyta (brown algae)
Photosynthetic pigments: chlorophylls a and c, carotenes, fucoxanthin and several other xanthophylls. Storage product: laminaran. Cell wall: cellulose, alginic acid, sulfated mucopolysaccharides. Flagella: 2, unequal, lateral.

Pyrrophycophyta (dinoflagellates)
Photosynthetic pigments: chlorophylls a and c, carotenes, several xanthophylls. Storage product: starch. Cell wall: cellulose or absent. Flagella: 2, one trailing, one girdling.

Rhodophycophyta (red algae)
Photosynthetic pigments: chlorophylls a (also d in some), phycocyanin, phycoerythrin, carotenes, several xanthophylls. Storage product: Floridean starch. Cell wall: cellulose, xylans, galactans. Flagella: absent.

"blue-greens" as cyanobacteria is still considered controversial and is opposed by many phycologists.

The algae are classified into seven major divisions, based largely on the types of photosynthetic pigments they produce, the types of reserve materials that are stored intracellularly, and the morphological characteristics of the cell (Table 14.4). The relative concentrations of photosynthetic pigments give the algae characteristic colors. Many of the major algal divisions have common names based on these characteristic colors, such as the green algae, red algae, and brown algae.

Chlorophycophyta

The Chlorophycophyta are known as the green algae. These algae contain chlorophylls a and b and typically appear green in color. Starch is stored as a reserve material. They are widely distributed in aquatic ecosystems. The green alage may be unicellular, filamentous, or colonial.

Probably the best known genus of green algae is *Spirogyra*, a filamentous green alga. The chloroplasts of *Spirogyra* form a spiral within the filaments. Another well-known algal genus, *Ulva*, is commonly known as sea lettuce. *Ulva* grows in marine habitats attached to rocks and other surfaces. Another marine form, the genus *Acetabularia* is a tubular green alga; this organism is known as the mermaid's wine goblet (Figure 14.10). Members of the genus *Volvox* form spheroidal colonies. The cells within a colony of *Volvox* act in a cooperative fashion so that the entire colony behaves as a "superorganism." The flagella of the vegetative cells face outward and are able to move the entire colony in a unified manner. The colonies act as reproductive individuals, some colonies producing male gametes and others producing female gametes. The colonies of *Volvox* approach the level of tissue differentiation.

Euglenophycophyta

The Euglenophycophyta or euglenoids are similar to the Chlorophycophyta in that they contain chlorophylls a and b and typically appear green in color. The Euglenophycophyta differ from the Chlorophycophyta with respect to their cellular organization, and their intracellular reserve storage products. The Euglenophycophyta are unicellular. The Euglenophycophyta do not store starch but rather store paramylon. They lack a cell wall but normally are surrounded by an outer layer, known as a pellicle, composed of lipid and protein.

The Euglenophycophyta appear to be closely related to the protozoa. Members of this division that lose their photosynthetic apparatus are indistinguishable from protozoa. Reproduction in the Euglenophycophyta is normally by longitudinal division. These algae are widely distributed in aquatic and soil habitats. *Euglena* is the best-known genus of Euglenophycophyta. *Euglena* species have two flagella for locomotion and normally contain a contractile vacuole to protect them against osmotic shock.

Chrysophycophyta

The Chrysophycophyta include the Xanthophyceae (yellow-green algae), Chrysophyceae (golden algae), and Bacillariophyceae (the diatoms). Typically these algae produce carotenoid and xantho-

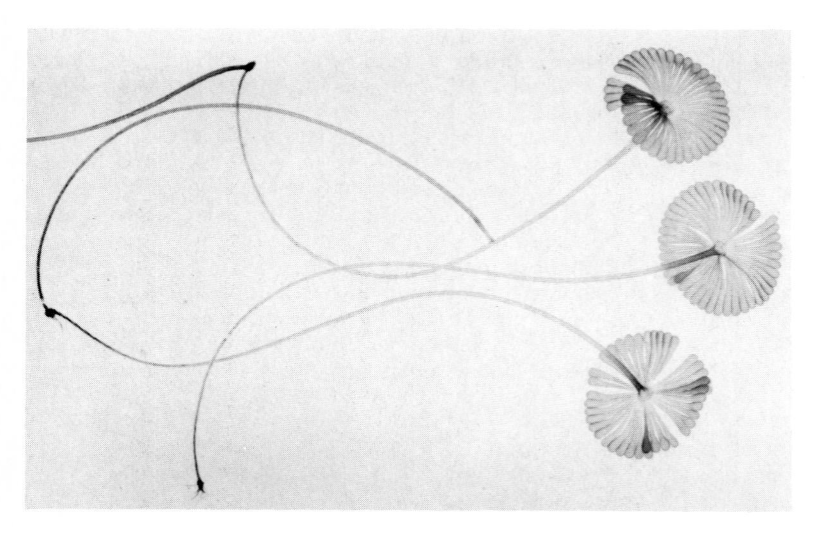

Figure 14.10 Photograph of *Acetabularia*, showing its delicate shape. (Courtesy Carolina Biological Supply Co.)

MICROBIOLOGY HEADLINES

Researchers Discover Alga's Primitive Eyes Aid in Photosynthesis

By BOYCE RENSBERGER

WASHINGTON—Algae have eyes.

Not exactly the way human beings have eyes, perhaps; but a team of scientists has discovered that a common one-celled alga has a visual system that works rather like the human eye's retina, converting light into electrical signals that govern the organism's behavior.

It has long been known that the alga, a pond-dwelling species called *Chlamydomonas*, has a light-sensitive "eyespot." What has been discovered recently is that the eyespot uses the same light-sensitive chemical—a visual pigment called rhodopsin—that is in the retinas of the eyes of humans and other higher animals.

Although it is a green plant, *Chlamydomonas* has two thread-like flagella that lash about to pull it through the water, like a human swimmer doing the breast stroke. The organism senses the amount of light coming through the water, and steers itself up or down to find the optimal level of light it needs for photosynthesis, the solar-powered process that all green plants use to manufacture sugar.

Because algae are among the earliest forms of life to have appeared on Earth, the discovery suggests that rhodopsin was "invented" quite early in the process of evolution. The eyes of higher organisms simply employ more sophisticated ways of making use of rhodopsin's remarkable ability to produce an electrical signal when struck by light.

In humans and other animals, the eyes have millions of light-sensitive cells, each of which sends a different signal to the brain, where the pattern of simultaneous signals is interpreted as a picture. *Chlamydomonas*, with just one light-sensitive spot and no brain, sees no picture, but still uses the signal to guide its movement through its watery world.

The algal eye can detect light only 1 percent as strong as necessary for photosynthesis. This directs the organism to swim toward the light. However, if the light is too bright, photosynthesis stops, and the alga swims away.

Nobody knows how it happens, but the signal—a burst of electrically charged atoms—apparently sets in motion a chain of chemical reactions that determines which of the alga's two flagella strokes more strongly through the water. That determines whether the little cell swims up or down.

The discovery was made by Kenneth W. Foster of Syracuse University and a team of six others at Columbia University and the City University of New York, Foster's former school.

"In a way," Foster said, "this sort of vindicates (Gottfried) Ehrenberg, who named *Chlamydomonas* back in 1831. He wrote that they had an eye and could see—but everybody thought he was crazy."

Foster said that, although the rhodopsin of humans and algae appear similar in molecular structure, it is likely that slight differences will be identified as they are compared more closely. "In any case," he said, "it seems quite likely that we share a common ancestry with *Chlamydomonas*. We both inherited versions of the rhodopsin molecule."

Source: The Washington Post, December 1, 1984. Reprinted by permission.

phyll pigments that tend to dominate over the chlorophyll pigments, making them golden-brown in color. All members of the Chrysophycophyta produce the storage product chrysolaminarin.

The diatoms are particularly interesting because they produce distinctive cell walls known as frustules. The frustules of diatoms, which are also known as valves, have two overlapping halves; the larger portion is referred to as the epitheca and the smaller as the hypotheca (Figure 14.11). The halves of the frustule fit together like a petri dish. The geometric appearance of diatoms renders them aesthetically attractive. The growth of diatoms is dependent on the concentrations of available silica, because the cell walls of diatoms are impregnated with silica. Holes in the silica walls, called puntae, allow exchange of nutrients and metabolic wastes between the cell and its surroundings. The frustules of diatoms are resistant to natural degradation and accumulate over geologic periods. As a result diatoms are preserved in fossil records dating back to the Cretaceous period, 65 million years ago. There are significant deposits of diatom frustules in the world. Such deposits are known as diatomaceous earth and are mined for a number of commercial uses. Diatomaceous earth is sometimes used as an abrasive in toothpaste and metal polish. The most extensive use of diatomaceous earth is in the filtration of liquids, especially those from sugar refineries.

Pyrrophycophyta

The Pyrrophycophyta or fire algae are generally brown or red in color because of the presence of xanthophyll pigments. The Pyrrophycophyta are unicellular, biflagellate, and store starch or oils as their reserve material. The dinoflagellates, which are members of the Pyrrophycophyta, are characterized by the presence of a transverse groove that divides the cell into two semicells (Figure 14.12). The two flagella of the dinoflagellates emerge from

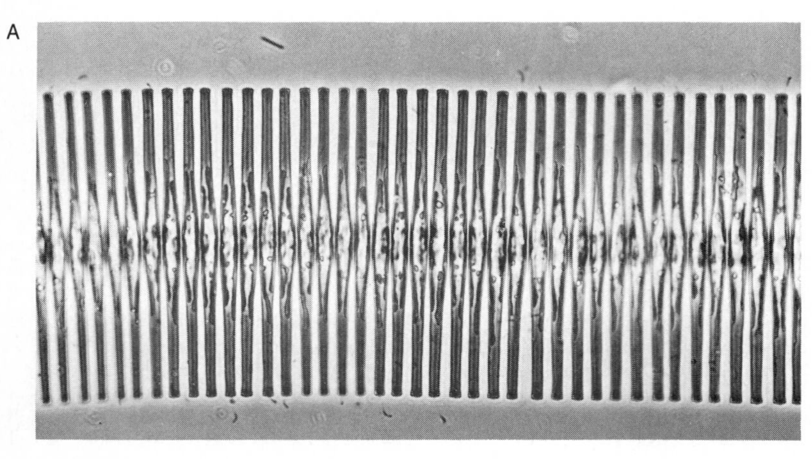

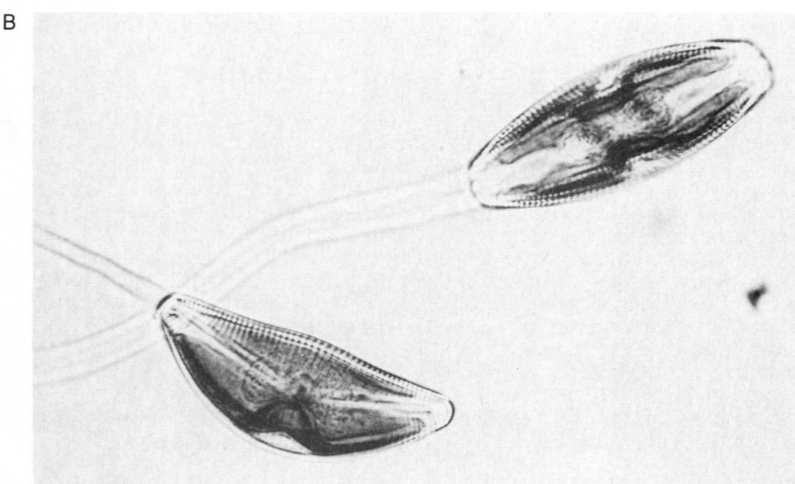

Figure 14.11 (A) Micrograph of *Fragilaria*, a colonial diatom (1120×); (B) front and side view of *Cymbella*, a stalked diatom (1280×). (From BPS: J. Robert Waaland, University of Washington.)

Figure 14.12 Scanning electron micrograph of the dinoflagellate *Gonyaulax tamerensis*, the organism that causes red tide (4100×). (From BPS: P. W. Johnson and J. McN. Sieburth, University of Rhode Island.)

an opening in the groove. The cell walls of Pyrrophycophyta contain cellulose and sometimes form structured plates, called theca. Because of their cell-wall structures, some dinoflagellates that produce thecal plates are referred to as armored dinoflagellates.

Several dinoflagellates exhibit bioluminescence, the characteristic on which the designation fire algae is based. Some water bodies glow from the luminescence of these algae. Species of *Gonyaulax* and other dinoflagellates are extremely important because they produce the toxic blooms known as red tides, which tend to color water red or red-brown in the vicinity of the bloom. During such blooms the toxins of dinoflagellates may kill invertebrate organisms. Although the blooms kill relatively few marine organisms, their toxins are concentrated in the tissues of filter-feeding molluscs, such as clams and oysters. Ingestion of shellfish containing dinoflagellates results in paralytic shellfish poisoning. This is a serious form of food poisoning, and in order to prevent outbreaks of paralytic shellfish poisoning in humans, the collection of shellfish is banned during outbreaks of red tide.

Rhodophycophyta

The Rhodophycophyta or red algae contain phycocyanin and phycoerythrin in addition to chlorophyll pigments, which gives them their red color. The primary reserve material in the Rhodophycophyta is Floridean starch, a polysaccharide similar to one found in higher plants. The red algae form complex, plant-like structures. They exhibit a specialized type of sexual reproduction involving female cells called carpogonia and male cells called spermatia. Most red algae occur in marine habitats.

Various biochemicals, including agar and carrageenin, occur in the cell walls of red algae. The agar and carrageenin of red algae are widely used as thickening agents and binders in various food products. Agar is also used as a solidifying agent in culture media, upon which the cultivation of bacteria largely depends. The carrageenin of *Chondrus crispus* is utilized in puddings. The red alga *Porphyra* is cultivated and harvested by the Japanese as a source of food.

Phaeophycophyta

The division Phaeophycophyta or brown algae includes over 200 genera and 1500 species. The Phaeophycophyta produce xanthophylls that dominate over the carotenoid and chlorophyll pigments and impart a brown color to these organisms. The brown algae are almost exclusively marine organisms and are found primarily in coastal zones.

The kelps, which are brown algae, can form macroscopic structures up to 50 meters in length (Figure 14.13). It is difficult to consider organisms of that size as members of the microbial world. These are complex organisms that exhibit a degree of cellular differentiation. Most kelps have vegetative structures consisting of a holdfast, stem, and blade. The genera *Fucus* and *Sargassum* are important representatives of the Phaeophycophyta. Large populations of *Sargassum natans* occur in the Atlantic Ocean in the region known as the Sargasso Sea. Species of *Fucus* commonly occur along rocky shores, attached to the rocks by disc-like holdfasts. These brown algae clearly are the most complex organisms classified as algae or for that matter as microorganisms, representing a borderline case between algae and plants.

Figure 14.13 Photograph of kelp, *Lammanaria saccharina*. (From BPS: J. Robert Waaland, University of Washington.)

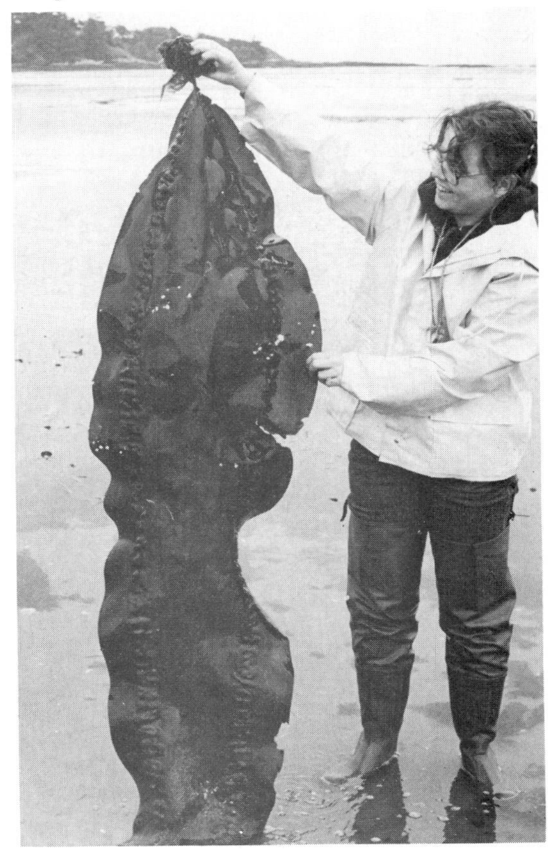

Classification of Protozoa

Protozoa are unicellular, nonphotosynthetic organisms that occur in the kingdom Protista. This subkingdom includes over 65,000 species. There has been great controversy over which organisms to include in the protozoa. The traditional classification system for the protozoa, which is primarily based on the means of motility, recognizes four phyla: the Mastigophora (flagellates); Sarcodina (pseudopodia formers); Ciliophora (cilliates); and Sporozoa (spore formers). In 1980, a new classification system was introduced by protozoologists (Table 14.5). The new system encompasses the classification of several groups claimed by other disciplines, including most of the algae and the lower fungi. For purposes of overall classification of microorganisms within this text, the slime molds are still considered as fungi and photosynthetic microorganisms as algae.

There are several noteworthy aspects to the new protozoa classification system. First, the Sarcodina and the Mastigophora are included in one phylum, the Sarcomastigophora. Within this phylum the Mastigophora and Sarcodina are treated as separate subphyla. The other major change in this classification system is the division of the Sporozoa into four separate phyla: the Apicomplexa, Microspora, Acetospora, and Myxospora. The last three have spores, but many members of the Apicomplexa do not; the Apicomplexa have an apical complex visible by electron microscopy.

Table 14.5 The Classification of the Protozoa

Old System

Ciliophora. Locomotion: cilia. Reproduction: asexual, transverse fission; sexual, conjugation. Nutrition: ingestive.

Mastigophora. Locomotion: usually paired flagella. Reproduction: asexual, longitudinal fission. Nutrition: heterotrophic, absorptive.

Sarcodina. Locomotion: pseudopodia (false feet). Reproduction: asexual, binary fission. Nutrition: phagocytic.

Sporozoa. Locomotion: usually none, some stages with flagella. Reproduction: asexual, multiple fission; sexual, within host; spores formed. Nutrition: absorptive.

New System

Sarcomastigophora. Locomotion: flagella, pseudopodia, or both. Reproduction: sexuality, when present, essentially syngamy. Representative genera: *Monosiga, Bodo, Leishmania, Trypanosoma, Giardia, Opalina, Amoeba, Entamoeba, Difflugia.*

Labyrinthomorpha. Synonymous with the net slime molds. Produce ectoplasmic network with spindle-shaped or spherical nonamoeboid cells; in some genera amoeboid cells move within network by gliding.

Apicomplexa. Produce apical complex visible with electron microscope; all species parasitic. Representative genera: *Eimeria, Toxoplasma, Babesia, Theileria.*

Microspora. Unicellular spores, each with imperforate wall; obligate intracellular parasites. Representative genus: *Metchnikovella.*

Ascetospora. Spore multicellular; no polar capsules or polar filaments; all species parasitic. Representative genus: *Paramyxa.*

Myxospora. Spores of multicellular origin with one or more polar capsules; all species parasitic. Representative genera: *Myxidium, Kudoa.*

Ciliophora. Cilia produced at some stage in life cycle. Reproduce by binary transverse fission, budding and multiple fission also occur. Sexuality involving conjugation, autogamy, and cytogamy; most free-living heterotrophs. Representative genera: *Didinium, Tetrahymena, Paramecium, Stentor.*

Sarcodina

The Sarcodina are motile by means of pseudopodia. Pseudopodia are cytoplasmic extensions, sometimes referred to as false feet or rhizopods.

Figure 14.14 Illustrations of members of the Sarcodina with different types of false feet. (A) *Elphidium crispa* with rhizopodia; (B) *Actinospaerium eichhorni* with axopodia; (C) *Chaos carolinensis* with lobopodia; and (D) *Euglypha alveolata* with filopodia.

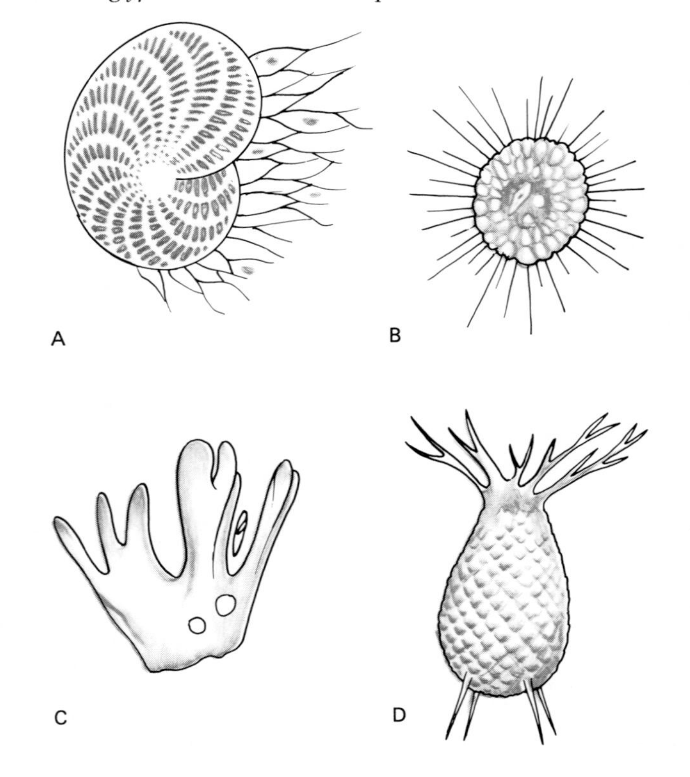

The false feet may occur in a variety of forms, including extensions of the ectoplasm that include flow of endoplasm (lobopodia); filamentous projections composed entirely of ectoplasm (filopodia); filamentous projections with branching (rhizopodia); and axial rods within a cytoplasmic envelope (axopodia) (Figure 14.14). The pseudopodia are used for engulfing and ingesting food as well as for locomotion. Members of the Sarcodina may move at rates of 2–3 centimeters per hour under optimal conditions. Some members of the Sarcodina are important because they cause human diseases. For example, *Entamoeba histolytica* causes amoebic dysentery, a serious debilitating human disease.

Members of the genus *Amoeba* have no distinct shape because the flow of cytoplasm continuously changes the shape of true amoeba (Figure 14.15). *Amoeba* feed on a number of smaller organisms, including bacteria and other protozoa. For example, *Amoeba proteus* can ingest the protozoa *Tetrahymena* and *Paramecium*.

Recall from Time's abysmal chasm
That piece of primal protoplasm
The First Amoeba, strangely splendid,
From whom we're all of us descended.
That First Amoeba, weirdly clever,
Exists today and shall forever,
Because he reproduced by fission;
He split himself, and each division
And subdivision deemed it fitting
To keep on splitting, splitting, splitting;
So, whatsoe'er their billions be,
All, all amoebas still are he.
Zoologists discern his features
In every sort of breathing creatures,
Since all of every living species,
No matter how their breed increases
Or how their ranks have been recruited,
From him alone were evoluted.
King Solomon, the Queen of Sheba
And Hoover sprang from that amoeba;
Columbus, Shakespeare, Darwin, Shelley
Derived from that same bit of jelly.
So framed he is and well-connected,
His statue ought to be erected,
For you and I and William Beebe
Are undeniably amoebae!

—ARTHUR GUITERMAN, *Ode to the Amoeba*

Radiolarians typically have axopodia, with a skeleton of silicon or strontium sulfate. The radiolarians occur in marine ecosystems. The silica-containing exoskeletons of the radiolarians are quite attractive when viewed microscopically. Foraminiferans are also marine members of the Sarcodina. Foraminiferans form one or many chambers composed of siliceous or calcareous tests (Figure 14.16). A test is a skeletal or shell-like structure. The tests of the Foraminferida accumulate in marine sediments and are preserved in the geological record. The White Cliffs of Dover are composed largely of the test structures of foraminiferans. Many of the foraminiferans are recognized in fossil records, whereas there is no fossil record for many microorganisms.

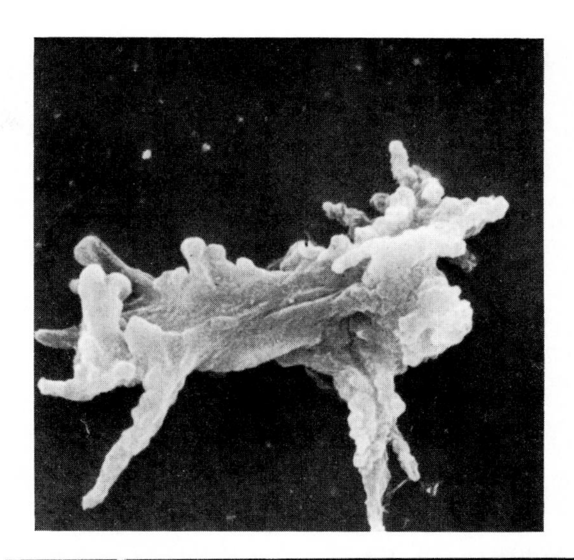

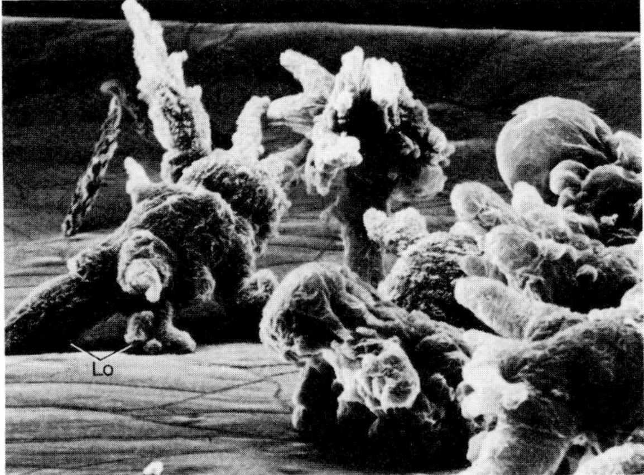

Figure 14.15 Micrographs of *Amoeba*, showing extensions of pseudopodia. (A) *Amoeba proteus*. (Courtesy Robert Apkarian, University of Louisville.) (B) *Pelomyxa carolinensis*, Lo = lobopodia (210 ×). (Reprinted by permission of Springer-Verlag, Heidelberg, from R. G. Kessel and C. Y. Shih, 1976, *Scanning Electron Microscopy in Biology: A Students' Atlas on Biological Organization*, Springer-Verlag, Berlin, Heidelberg, New York.)

Figure 14.16 The larger foraminiferan, *Heterostegina depressa,* showing the chambers of the test (test size = 2 millimeters). The calcareous test consists of many chambers coiled in a plane. Each chamber is subdivided into chamberlets. As in all larger foraminiferans, it harbors unicellular symbiotic algae in its protoplasm, which give a yellow coloration to the living specimen. These larger foraminiferans occur in tropical and subtropical shallow seas. (Courtesy Rudolf Röttger, Institute für Allgemeine Mikrobiologie der Universität Kiel, Kiel, Federal Republic of Germany.)

Figure 14.17 Micrograph of *Trypanosoma cruzi,* the causative agent of Chagas' disease. (From BPS: Stephen Baum, Albert Einstein College of Medicine, New York.)

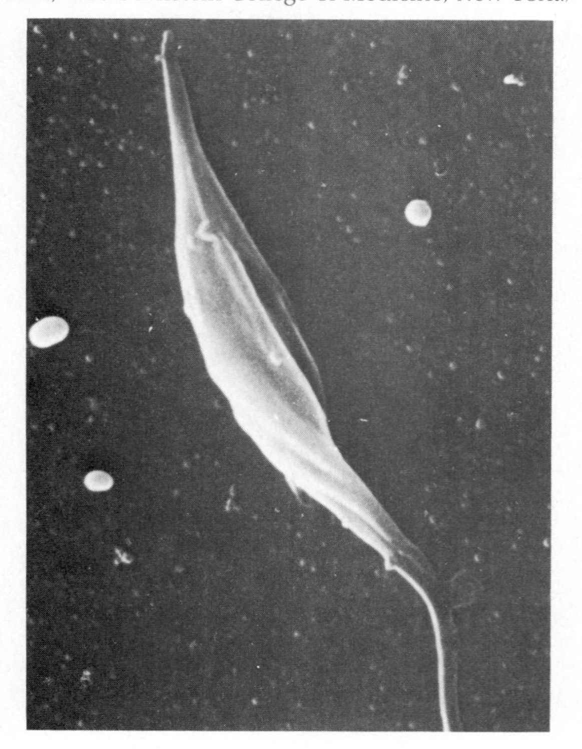

Mastigophora

The Mastigophora are the flagellate protozoa. Because some members of the Mastigophora are able to produce pseudopodia in addition to possessing flagella, these organisms are now classified together with the Sarcodina. It is in this subphylum that protozoologists place the dinoflagellates, euglenoids, and other algae.

Many members of the Mastigophora are plant and animal parasites. The genera *Trypanosoma* and *Leishmania* contain species that produce serious human diseases (Figure 14.17). *Trypanosoma gambiense,* for example, causes African sleeping sickness, and *T. cruzi* is the causative agent of Chagas' disease. Infections with the flagellate protozoan *Giardia* can cause severe diarrhea. *Leishmania donovani* is the causative agent for Kala-azar disease, also known as dum dum fever. Human diseases caused by flagellate protozoa are normally transmitted by arthropods, and control of many of these diseases rests with controlling the carrier rather than eliminating the disease-causing protozoa.

Ciliophora

Ciliophora are motile by means of cilia. The ciliate protozoa reproduce by various asexual and sexual means. Asexual reproduction is often by binary fission, and sexual reproduction is usually by conjugation (Figure 14.18). The Ciliophora normally contain two nuclei, a macronucleus and a micronucleus, both of which are diploid. The macronu-

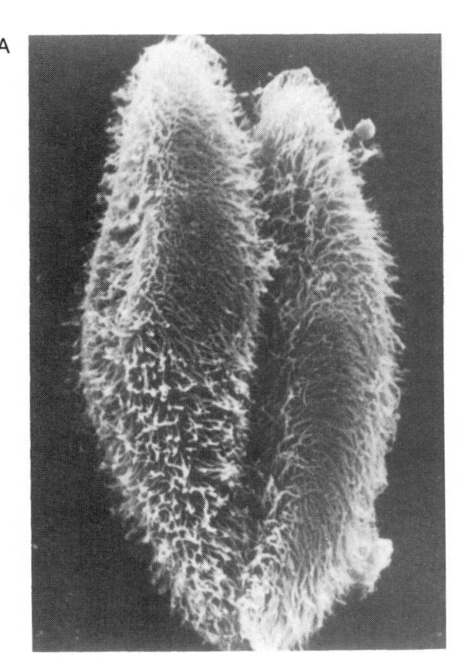

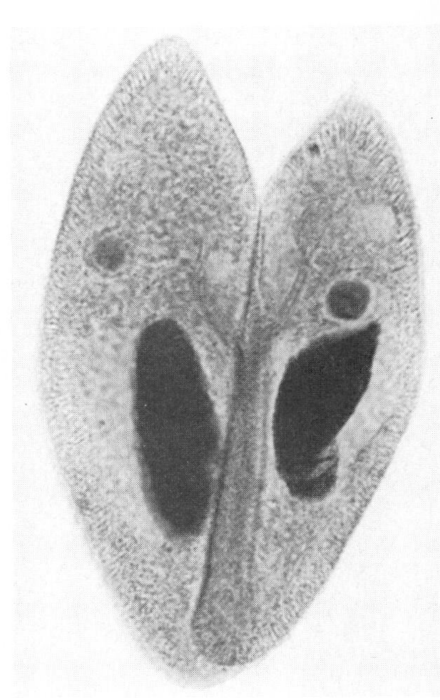

Figure 14.18 Micrographs of conjugating *Paramecium.* (A) (Reprinted with permission of Springer-Verlag, Heidelberg, from R. G. Kessel and C. Y. Shih, 1976, *Scanning Electron Microscopy in Biology: A Students' Manual on Biological Organization,* Springer-Verlag, Berlin, Heidelberg, New York.) (B) From Carolina Biological Supply Co.)

cleus is involved in asexual reproduction and the micronucleus in sexual reproduction. *Paramecium* is perhaps the best-known genus of ciliate protozoa. Other genera of Ciliophora include *Stentor, Vorticella, Tetrahymena,* and *Didinium*.

The protozoan *Glaucoma*, which reproduces by binary fission, divides as frequently as every 3 hours. Thus in the course of a day it could become a "six greats grandparent" and the progenitor of 510 descendants.

—1985 GUINNESS BOOK OF WORLD RECORDS

Ciliate protozoa consume other microorganisms, including other protozoa as their food source. For example, the genus *Didinium* can consume *Paramecium* species, providing a dramatic picture of the microbial world when viewed by scanning electron microscopy (Figure 14.19).

Sporozoa

The Sporozoa are parasites, exhibiting complex life cycles. The adult forms are nonmotile, but immature forms and gametes may be motile. The sporozoans derive nutrition by absorption of nutrients

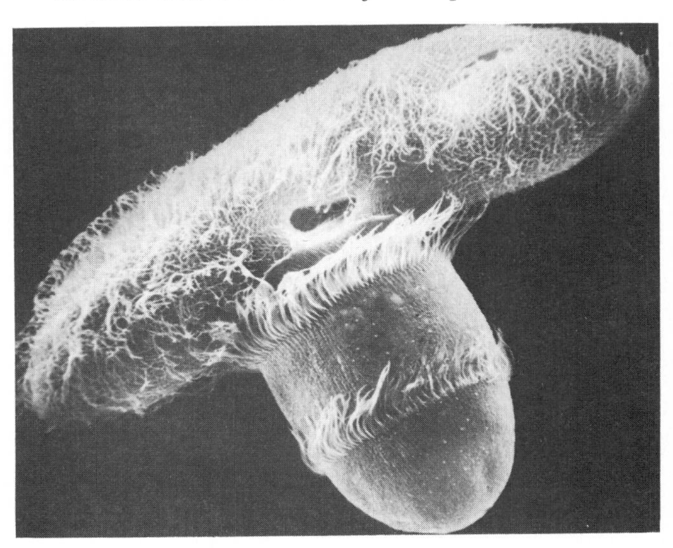

Figure 14.19 Micrograph of one ciliate protozoan, *Didinium nasutum,* consuming another ciliate, *Paramecium multimicronucleatum,* at an early stage of the ingestion (1400×). (From BPS: H. S. Wessenberg and G. A. Antipa, 1970, *Journal of Protozoology,* **17**:250–270.)

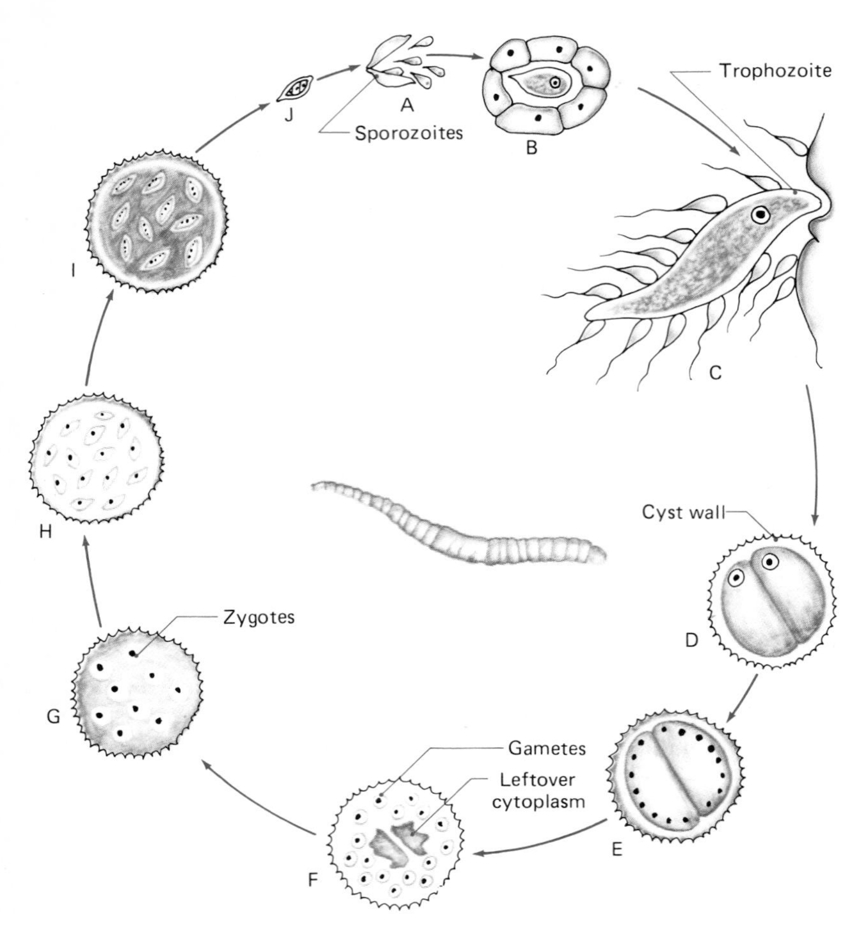

Figure 14.20 Diagram of life cycle of the sporozoan, *Monocystis. Monocystis* lives in the common earthworm. (A) Spores are eaten by the earthworm and sporozoites are released. (B) Young trophozoite in sperm morula. (C) Trophozoite grows into worm, earthworm sperm develops from morula cells and attaches to trophozoite. (D) Trophozoites associate in pairs and form a cyst wall (gametocytes). (E) Nuclei of gametocytes divide. (F) Gametes are produced. (G) Gametes fuse, that is, fertilization occurs, to form zygotes. (H) Zygotes secrete spore walls, sporocysts. (I) Nuclei divide to form eight sporozoites in each spore. (J) Spores are liberated when earthworm dies. Spores may also be transferred to another worm during intercourse and infect testes of other worms. They may also be transferred to the worm cocoon, entering the ground when the cocoon disintegrates.

from the host cells they inhabit. The immature stages are referred to as **sporozoites**. Reproduction of the **trophozoite**, the adult stage of a sporozoan, occurs asexually by multiple fission (Figure 14.20). The cells produced by multiple fission eventually mature into gametes that can be involved in sexual reproduction. The multiple fission process can result in the production of thousands of spores. *Plasmodium vivax* and various other species of *Plasmodium* cause malaria. During the course of a malarial infection, life cycles of *Plasmodium* involve reproduction within human red blood cells, with the periodic release of large numbers of protozoa.

Summary

The fungi exhibit a great variety of reproductive strategies, often exhibiting both asexual and sexual reproductive modes. Having several strategies for reproduction enables them to establish relatively complex life cycles, in many cases involving an alternation between haploid and diploid phases. As part of their reproductive strategies, fungi produce various sexual and asexual spores.

The fungi are classified largely on the basis of their modes of reproduction. The sexual spores of fungi are the most important features that are used in their classification and identification. Additionally, the asexual spores and vegetative structures of the fungi are used for the finer definition of taxa. Although fungi typically form filamentous mycelia, one group, the yeasts, are characteristically unicellular. The fungi include the slime molds, a taxonomic group having interesting life cycles. During these life cycles, amoeboid cells migrate and unite to form a fruiting body. The fruiting body produces spores that are disseminated.

The ascomycetes produce sexual spores within an ascus. Many yeasts are ascosporogenous. Another major group of fungi, the basidiomycetes, is characterized by the production of sexual spores on a basidium. The fruiting bodies of the basidiomycetes, which are often conspicuous, include the mushrooms. Some mushrooms are edible but others are deadly poisonous. Many basidiomycetes, particularly the rusts and smuts, are plant pathogens that cause great agricultural losses. One group of fungi, the deuteromycetes or fungi imperfecti, is defined by the lack of observed sexual spores.

The algae are classified into seven groups based largely on pigment production and the biochemical nature of the storage reserve materials. The seven groups of algae are the green algae, euglenoids, brown algae, golden and yellow-green algae, dinoflagellates, cryptomonads, and red algae. Some of these groups, such as the euglenoids, are closely related to protozoa. Other algae, such as the brown algae, are closely related to plants. Many algae produce complex macroscopic structures; the kelps, for example, are often 50 meters in length. The algae include the diatoms, organisms that produce frustules containing silica, which are highly symmetric and quite beautiful.

The protozoa are classified into groups based largely on their means of locomotion. Some protozoa form extensions of the cytoplasm known as pseudopodia or false feet. This group is known as the Sarcodina and includes the genus *Amoeba*. The pseudopodia are involved in both locomotion and the ingestion of food. The Ciliophora are protozoa that are motile by means of cilia. The genus *Paramecium* is an example of a ciliate protozoan. The Mastigophora are protozoa that are motile by means of flagella. The Sporozoa, all of which are parasites, are generally nonmotile. Members of this group produce spores during their life cycles.

Study Outline

A. Fungi.

- **1.** The fungi are eukaryotic, heterotrophic, nonphotosynthetic microorganisms.
- **2.** The fungi typically form reproductive spores; many fungi exhibit both sexual and asexual forms of reproduction.
- **3.** Many fungi form filaments of vegetative cells (mycelia); mycelia are integrated masses of individual tube-like filaments of hyphae that usually exhibit branching.
- **4.** The classification of the fungi is based largely on the means of reproduction (nature of life cycle, reproductive structures, reproductive spores); the primary taxonomic groupings are based on sexual reproductive spores.
- **5.** Slime molds.
  - **a.** The slime molds represent a borderline case between fungi and protozoa.
  - **b.** Slime molds exhibit characteristic life cycles.
  - **c.** The cellular slime molds form a stalked fruting body (sporocarp) consisting of walled cells; the sporocarp releases spores that germinate to form myxamoebae (amoeboid cells that form pseudopodia); the myxamoebae aggregate to form a pseudoplasmodium; the pseudoplasmodium differentiates, culminating in formation of sporocarp.
  - **d.** The true slime molds form myxamoebae or swarm cells that fuse together to form a true plasmodium (multinucleate protoplasmic mass devoid of cell walls and enveloped in gelatinous slime sheath); the plasmodium gives rise to fruiting bodies.
- **6.** Mastigomycota.
  - **a.** Some are unicellular; most form extensive filamentous coenocytic mycelia.
  - **b.** The mastigomycota typically produce motile cells with flagella during part of the life cycle.
  - **c.** Asexual reproduction of the mastigomycota normally involves motile spores (zoospores).
  - **d.** The nutrition of the mastigomycota is by absorption of nutrients.
- **7.** Amastigomycota.
  - **a.** The vegetative cells of the amastigomycota vary from single cells to mycelia that may be coenocytic or may have extensive septation.

b. There are four phyla: Zygomycotina, Ascomycotina, Basidiomycotina, and Deuteromycotina.
c. The Zygomycotina have coenocytic mycelia and produce sexual spores from fusion of gametangia called zygospores.
d. The Ascomycotina (sac fungi) produce sexual spores within specialized sac-like structures (ascus).
e. The Basidiomycotina (basidiomycetes or club fungi) are the most complex fungi.
f. The basidiomycetes are distinguished from other classes of fungi by production of sexual spores (basidiospores) on surfaces of specialized club-shaped spore-producing structures (basidia).
g. The mushrooms are basidiomycetes.

8. Deuteromycotina (deuteromycetes or fungi imperfecti).
a. The deuteromycetes are characterized by lack of observed sexual spores.
b. The deuteromycetes are classified on the basis of morphological structure of vegetative phase and types of asexual spores produced.
c. The fungi imperfecti include many important genera of filamentous fungi, such as *Penicillium, Aspergillus*, and *Candida*.

B. Algae.
1. The algae are eukaryotic, photosynthetic microorganisms; they are separated from plants by their lack of tissue differentiation.
2. The algae are classified into divisions based on types of photosynthetic pigments produced, the relative concentrations of pigments give the algae their characteristic colors; types of reserve materials stored intracellularly; and morphological characteristics of the cells.
3. The seven major divisions of algae are
a. Chlorophycophyta (green algae): are widely distributed in aquatic ecosystems; contain chlorophylls a and b; most cells are uninucleate; chloroplasts of many unicellular green algae contain pigmented region (stigma or red eyespot); some contain osmoregulatory contractile vacuoles; normally store starch; cell walls composed of cellulose, mannans, or xylans; high proportion of protein may also be present; cellular organization may be unicellular, colonial, or filamentous.
b. Euglenophycophyta: contain chlorophylls a and b; appear green in color; surrounded by an outer layer (pellicle); store paramylon; reproduction normally by longitudinal division; widely distributed in aquatic and soil habitats.
c. Chrysophycophyta: produce diversity of pigment biochemicals, cell wall biochemicals, and cell types; produce chrysolaminarin and carotenoid and xanthophyll pigments that dominate the chlorophyll pigments and confer golden-brown color; most are unicellular; some are colonial; include the Xanthophyceae (yellow-green algae), Chrysophyceae (golden algae), and Bacillariophyceae (diatoms); diatoms produce distinctive cell walls (frustules or valves), impregnated with silica; growth dependent on availability of silica; tests are resistant to natural degradation and accumulate over geologic periods.
d. Pyrrophycophyta (fire algae): generally brown or red due to xanthophyll pigments; unicellular; biflagellate; store starch or oils; reproduction primarily by cell division; cell walls contain cellulose; include the dinoflagellates, which are characterized by transverse groove dividing cells into two semicells with flagella emerging from opening in groove, some exhibit bioluminescence and/or circadian rhythm, produce toxic blooms (red tides).
e. Rhodophycophyta (red algae): contain phycocyanin and phycoerythrin in addition to chlorophyll pigments; primary reserve material is Floridean starch; exhibit specialized type of oogamous sexual reproduction involving specialized female (carpogonia) and male (spermatia) cells; most occur in marine habitats; typically have bilayered cell wall.

f. Phaeophycophyta (brown algae); produce xanthrophylls that dominate over carotenoid and chlorophyll pigments; main reserve materials are laminarin and mannitol; have bilayered cell walls.

C. Protozoa.

1. The protozoa are unicellular, heterotrophic microorganisms.
2. The Sarcodina form pseudopodia (false feet), which occur in a variety of forms and are used for engulfing and ingesting food and locomotion.
3. The Mastigophora are the flagellate protozoa.
4. The Ciliophora are motile by means of cilia.
5. The Sporozoa include the Apicomplexa, Microspora, Acetospora, and Myxospora; these protozoa are parasitic, produce spores, and have complex life cycles; immature forms (sporozoites) may be motile; mature forms (trophozoites) are nonmotile; reproduction may occur by multiple fission.

Study Questions

1. Match the following statements with the terms in the list to the right.

a. Include the classes Chytridiomycetes, Hyphochytridiomycetes, Plasmodiophoromycetes, and Oomycetes.
b. Have four subdivisions: Zygomycotina, Ascomycotina, Basidiomycotina, and Deuteromycotina.
c. Produce a sexual spore.
d. Bread molds belong to this subdivision.
e. Also called the sac fungi.
f. Reproduce either by sexual spores in an ascus or by asexual reproduction.
g. Also called the club fungi.
h. Include the mushrooms.
i. Also called the fungi imperfecti.
j. Do not have sexual spores.
k. Include the genera *Penicillium* and *Candida*.

(1) Mastigomycota
(2) Amastigomycota
(3) Zygomycotina
(4) Ascomycotina
(5) Basidiomycotina
(6) Deuteromycotina

2. Match the following statements with the terms in the list to the right.

a. Are also known as red algae.
b. Contain chlorophyll a and b.
c. Include the classes Xanthophyceae, Chrysophyceae, and Bacillariophyceae.
d. *Euglena* is the best known genus.
e. Are also known as fire algae.
f. Are also known as brown algae.
g. Include the diatoms.
h. *Chylamydomonas, Volvox*, and *Ulothrix* are examples of genera.
i. Closely related to protozoa.
j. Produce toxic blooms known as red tides.
k. Also known as green algae.
l. Species of this division form agar and carageenin, the source of laboratory agar and thickening in foods.
m. Lack a cell wall.
n. Include the kelps.
o. Organisms of this division are the most complex of those called microorganisms.
p. Include dinoflagellates.
q. In the colonial form, cells act cooperatively; the colony behaves as a "superorganism."

(1) Chlorophycophyta
(2) Euglenophycophyta
(3) Chrysophycophyta
(4) Pyrrophycophyta
(5) Cryptophycophyta
(6) Rhodophycophyta
(7) Phaeophycophyta

3. Match the following statements with the terms in the list to the right.

a. Include *Plasmodium*.
b. Form pseudopodia.

(1) Arcomastigphora
(2) Sarcodina

c. Include the genus *Trypanosoma*.
d. Some members form tests (skeletal or shell-like structures), which are responsible for the White Cliffs of Dover.
e. Include the Foraminiferida.
f. Human pathogens of this subphylum are normally transmitted by arthropods.
g. The best-known genus is *Paramecium*.
h. Include the subphylum Sarcodina and Mastigophora.

(3) Mastigophora
(4) Ciliophora
(5) Sporozoa

Some Thought Questions

1. If you were going to a costume party as a fungus, what would you wear?
2. If you could be an alga or a fungus, which would you be?
3. If you were a slime mold, what would you say to another slime mold and how would you say it?
4. What are the signs of aging in a yeast?
5. Why are no old bold mushroom hunters?
6. Why aren't all algae green?
7. Why are some organisms considerd algae by phycologists and protozoa by protozoologists?

Additional Sources of Information

Ainsworth, G. C., and A. S. Sussman (eds.). 1965–1973. *The Fungi: An Advanced Treatise* (4 volumes). Academic Press, Inc., New York. An extensive work describing the fungi, including detailed taxonomic descriptions.

Alexopoulos, C. J., and C. W. Mims. 1979. *Introductory Mycology*. John Wiley & Sons, Inc., New York. A taxonomically oriented introductory text describing the fungi.

Bold, H. C., and M. J. Wynne. 1985. *Introduction to the Algae: Structure and Reproduction*. Prentice-Hall, Inc., Englewood Cliffs, New Jersey. A well-written and extensively illustrated introductory text covering the algae.

Corliss, J. O. 1979. *The Ciliated Protozoa: Characterization, Classification and Guide to the Literature*. Pergamon Press, Inc., New York. A detailed guide to the ciliates written by a leading protozoan taxonomist.

Farmer, J. N. 1980. *The Protozoa: Introduction to Protozoology*. The C. V. Mosby Company, St. Louis. A basic introductory text covering the protozoa.

Jahn, T. L., E. C. Bovee, and F. F. Jahn. 1979. *How to Know the Protozoa*. William C. Brown Company, Publishers, Dubuque, Iowa. A widely used guide that provides a taxonomic key for the identification of the common protozoa.

Kudo, R. R. 1977. *Protozoology*. Charles C. Thomas, Publisher, Springfield, Illinois. An extensive compilation of taxonomic descriptions of the protozoa.

Lee, R. E. 1980. *Phycology*. Cambridge University Press, Cambridge, England. A well-written introduction to the algae.

Levine, N. D., J. O. Corliss, F. E. G. Cox, G. Deroux, J. Grain, B. M. Honigberg, G. F. Leedale, A. R. Loeblich, J. Lom, D. Lynn, E. G. Meringeld, F. C. Page, G. Poljansky, V. Sprague, J. Vavra, and F. G. Wallace. 1980. A newly revised classification of the Protozoa. *Journal of Protozoology*, **27**:37–58. A description of the newly introduced classification system for the protozoa.

Lodder, J., and N. Kreger-van Rij. 1970. *The Yeasts: A Taxonomic Study*. North Holland Publications, Amsterdam. A massive compilation describing the taxonomic features of the yeasts.

Miller, O. K. 1979. *Mushrooms of North America*. E. P. Dutton & Co., Inc., New York. A beautifully illustrated guide to the mushrooms.

Moore-Landecker, E. 1982. *Fundamentals of the Fungi*. Prentice-Hall Inc., Englewood Cliffs, New Jersey. A comprehensive introductory text about the fungi.

Phaff, H. J., M. W. Miller, and E. M. Mrak. 1978. *The Life of Yeasts*. Harvard University Press, Cambridge, Massachusetts. A concise introductory book about the yeasts.

Rose, A. H., and J. S. Harrison (eds.). 1969. *The Yeasts*. Academic Press, Inc., New York. An extensive reference work devoted to the yeasts.

Trainor, F. R. 1978. *Introductory Phycology*. John Wiley & Sons, Inc., New York. An introductory text covering the algae.

Westphal, A. 1976. *Protozoa*. Blackie & Sons, Ltd., Glasgow, United Kingdom. An introductory text covering the protozoa.

UNIT SIX
Microbial Growth

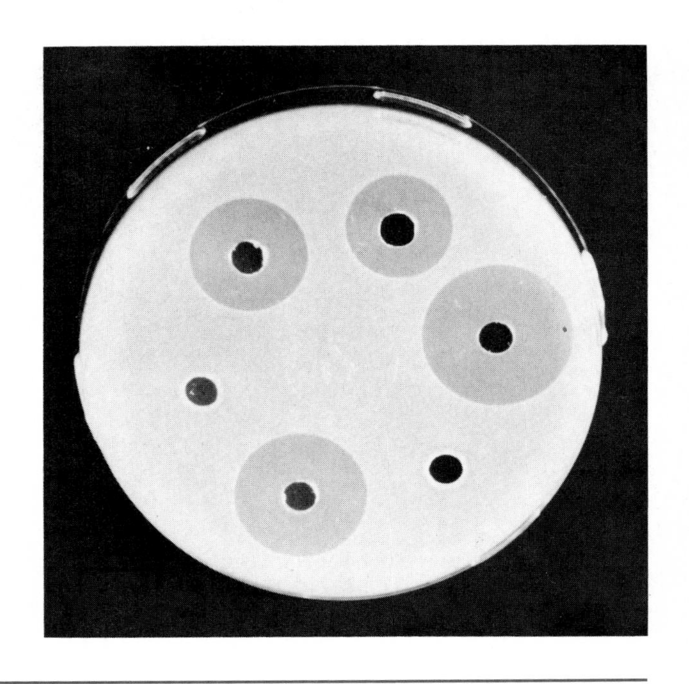

15 *Growth of Microorganisms: Environmental Influences*

KEY TERMS

acidophiles
aerobes
anaerobes
barophiles
barotolerant
batch culture
binary fission
catalase
chemostat
continuous culture
death phase
doubling time
exponential growth
facultative anaerobes
generation time
growth curve
halophiles
hydrostatic pressure
incubators
lag phase
log phase
mesophiles
microaerophiles
normal growth curve
optimal growth temperature
osmophiles
osmotolerant
oxyduric anaerobes
oxylabile anaerobes
pH
preservation
psychrophiles
psychrotrophs
salt tolerant
stationary phase
superoxide dismutase
temperature growth range
thermophiles
water activity
xerotolerant

OBJECTIVES

After reading this chapter you should be able to

1. Define the key terms.
2. Describe the ways by which bacteria reproduce.
3. Discuss the events that occur when bacteria reproduce by binary fission.
4. Describe the phases of the normal bacterial growth curve.
5. Compare batch and continuous culture.
6. Discuss the difference between optimal growth temperature and growth range.
7. Discuss the difference between a salt tolerant and a halophilic bacterium.
8. Discuss the effect of lowering water activity on microbial growth.
9. Discuss how environmental conditions can be adjusted to achieve maximal growth rates.

Bacterial Reproduction

Bacteria normally reproduce by binary fission with the formation of two equal-sized progeny cells (Figure 15.1). Reproduction requires the replication of the bacterial chromosome so that each daughter cell receives a complete genome. During cell division the bacterial chromosome appears to be attached to the cytoplasmic membrane and cell wall. The inward movement of the cytoplasmic membrane and cell wall, septa formation, pinches off and separates the two complete bacterial chromosomes, providing each of the progeny cells with a complete genome. The formation of septae physically cuts apart the bacterial chromosomes and distributes them to the two daughter cells. Completion of the replication of the bacterial chromosome is a prerequisite for cell division, and if the termination of DNA replication is blocked, the next cell division also is prevented. The control of the cell division cycle is complex, and there appear to be several different regulatory mechanisms that control cell division.

> If it be true that the essence of life is the accumulation of experience through the generations, then one may perhaps suspect that the key problem of biology, from the physicist's point of view, is how living matter manages to record and perpetuate its experiences. Look at a single bacterium in a large volume of fluid of suitable chemical composition. It assimilates substance, grows in length, divides in two. The two daughters do the same, like the broomstick of the Sorcerer's apprentice. Occasionally the replica will be slightly faulty and an individual arises with somewhat different properties, and it perpetuates itself in this modified form. It is quite easy to believe that the variety of types will be multiplied indefinitely.
>
> —MAX DELBRUCK

Figure 15.1 In binary fission the ingrowth of the new wall material, septa formation, progresses to separate the two daughter cells, assuring the equal distribution of genetic information between the cells. (A) Schematic of the binary fission process. (B) Micrographs showing stages of cell division for synchronously grown *Erwinia* showing invagination of cytoplasmic membrane and formation of crosswall septum. (Reprinted by permission of the American Society for Microbiology, Washington, D.C., from E. A. Grula and G. L. Smith, 1965, *Journal of Bacteriology*, 90:1054–1058.)

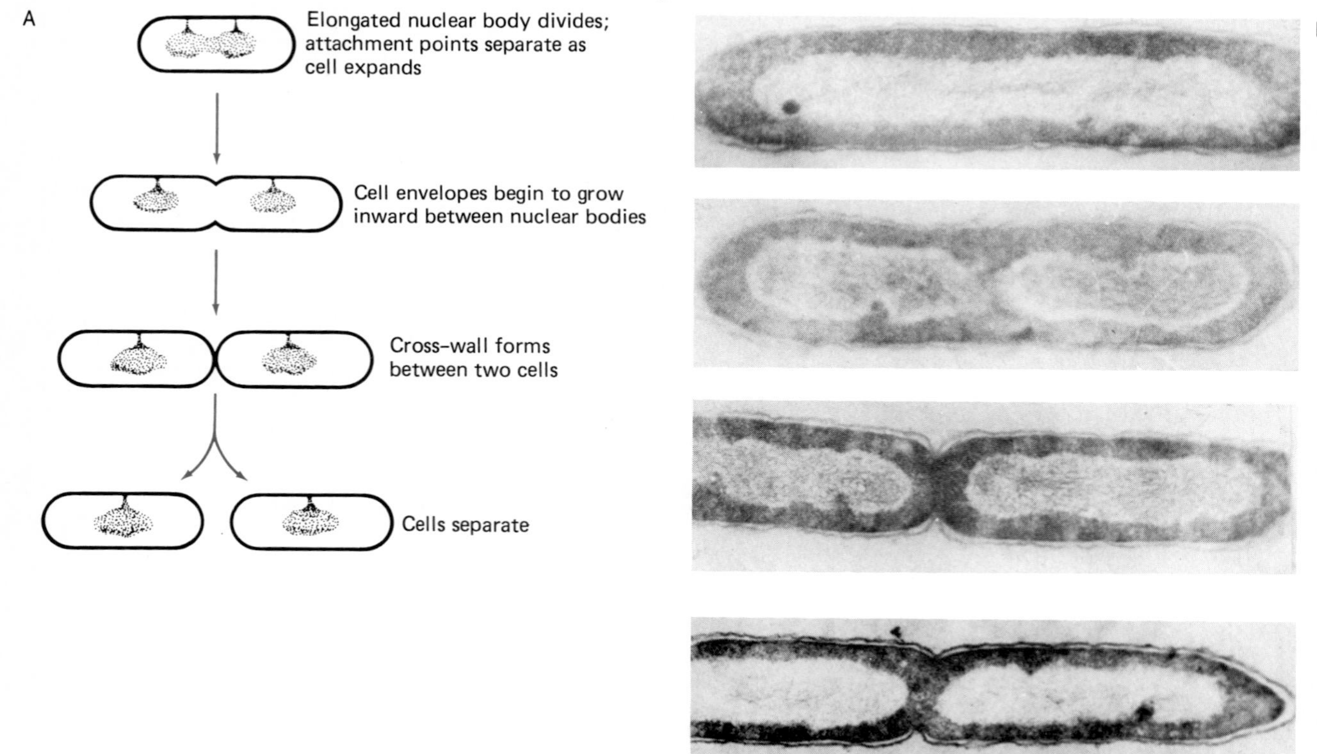

In addition to requiring duplicate bacterial chromosomes, cell division necessitates the synthesis of new cell wall and cytoplasmic membrane structures to surround and protect the daughter cells. During the reproductive cycle of bacteria, the parent cell elongates, and the cell wall grows inward from opposing sides of the bacterial cell, establishing a crosswall (Figure 15.2). The cross septum that divides the cells consists both of cell wall and cytoplasmic membrane. Upon completion of the crosswall formation, there are two equal-sized cells that can separate. Repeating the process results in the multiplication of the bacterial population.

Each organic being is striving to multiply . . . each has to struggle for life . . . the vigorous survive and multiply.

—CHARLES DARWIN, *The Origin of Species*

Although binary fission is the most common mode of bacterial reproduction, some bacteria reproduce by other means. The predominant modes of reproduction in all bacteria are asexual. The various reproductive modes differ in how the cellular material is apportioned between the daughter cells and whether the cells separate or remain together as part of a multicellular aggregation. For example, *Hyphomicrobium* reproduces by budding, a type of division characterized by an unequal division of cellular material. The daughter cell develops when a crosswall forms, segregating a portion of the cytoplasm containing a duplicate genome. In the case of the actinomycetes, such as *Streptomyces*, reproduction involves the formation of hyphae. In this mode of reproduction the cell elongates, forming a relatively long and generally branched filament or hypha. Regardless of the mode of reproduction, bacterial multiplication requires replication of the genome and synthesis of new boundary layers, including cell wall and cytoplasmic membrane structures.

In addition to binary fission and other modes of vegetative cell reproduction, some bacteria produce spores (specialized resistant resting cells) to facilitate their survival. The production of spores represents an interesting variation from the normal binary fission mode of replication. Various types of spores, including endospores (heat resistant spores formed within the cell), myxospores (resting cells of the myxobacteria formed within a fruiting body), cysts (resting or dormant cells sometimes enclosed in a sheath), and arthrospores (spores formed by the fragmentation of hyphae), are produced by different bacteria as part of their reproductive cycles. For example, under certain conditions *Bacillus* and *Clostridium* do not divide by binary fission to form two equal cells but rather sporulate to form endospores (Figure 15.3). The endospore is a highly resistant structure that can remain metabolically dormant for a prolonged period of time and that can germinate at a later time (Figure 15.4).

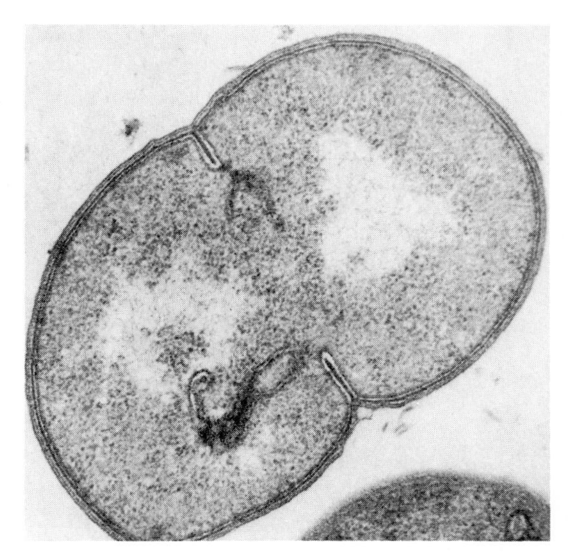

Figure 15.2 Electron micrograph showing crosswall formation. In this thin section of *Sporosarcina urea* both the ingrowing septum and the nucleoid regions of the bacterial chromosome can be clearly seen (63,500 ×). (From BPS: T. J. Beveridge, University of Guelph, Ontario.)

Normal Bacterial Growth Curve

By its very nature, bacterial reproduction by binary fission results in doubling of the number of viable bacterial cells. Therefore, during active bacterial growth, the size of the microbial population is continuously doubling. Once cell division begins it proceeds exponentially as long as growth conditions permit, with one cell dividing to form two, each of these cells dividing so that four cells form, and so forth in a geometric progression (Figure 15.5). The time required to achieve a doubling of the population size, known as the generation time or doubling time, is the unit of measure of microbial growth rate (Table 15.1). The generation time for bacteria can be expressed mathematically as

$$g = t/3.3(\log B_t - \log B_0)$$

where g is the generation time, log B_t is the loga-

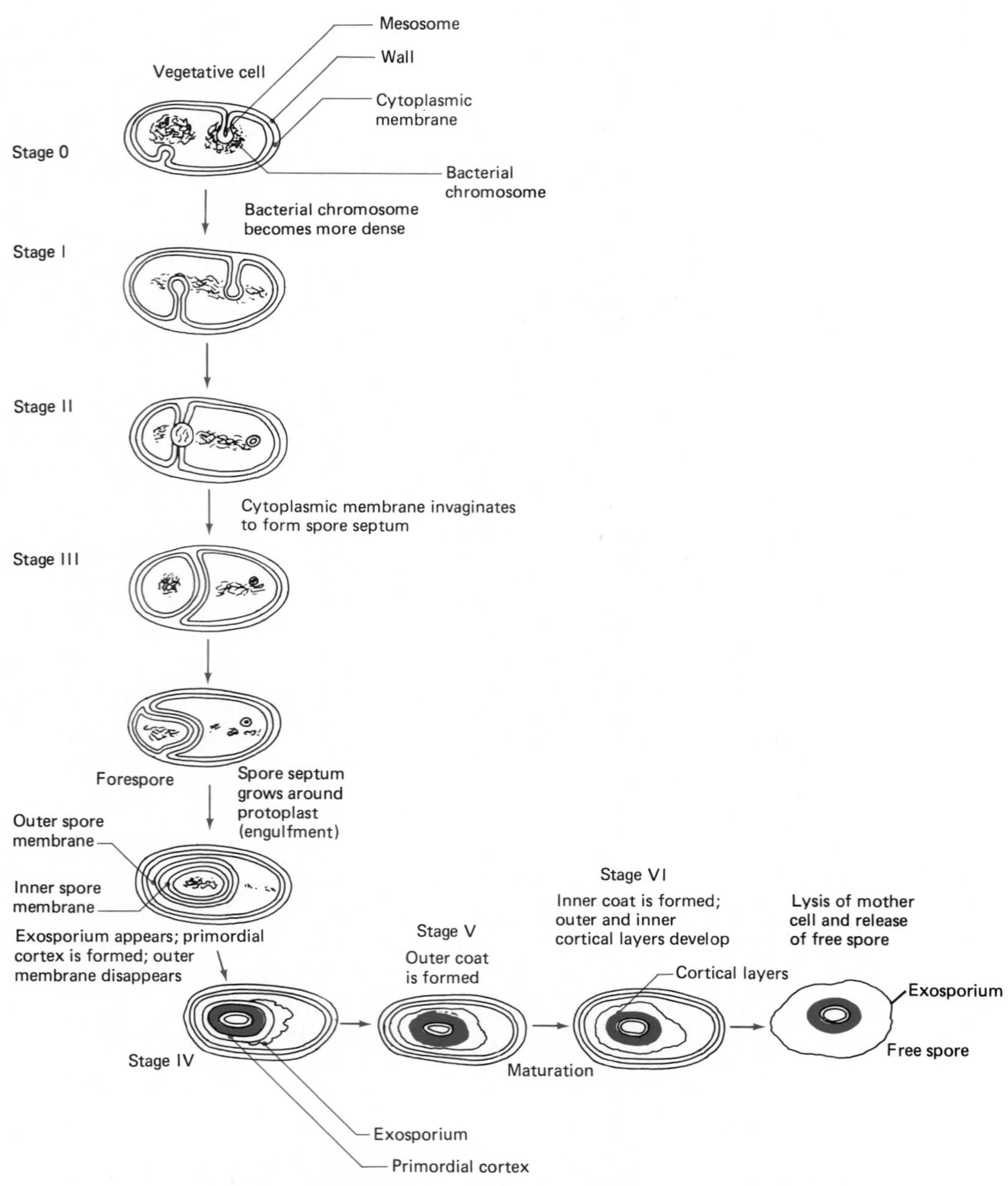

Figure 15.3 The stages of endospore formation. A typical vegetative cell is shown as stage 0. In stage I the bacterial chromosome material becomes more dense; in stage II the cytoplasmic membrane invaginates to form a spore septum; in stage III engulfment occurs, in which a spore septum grows around the protoplast; in stage IV the exosporium appears, the primordial cortex is formed, and the outer membrane disappears; in stage V the outer coat is formed; in stage VI maturation occurs and the inner coat is formed as the outer and inner cortical layers develop; in stage VII the mother cell lyses and releases a free spore.

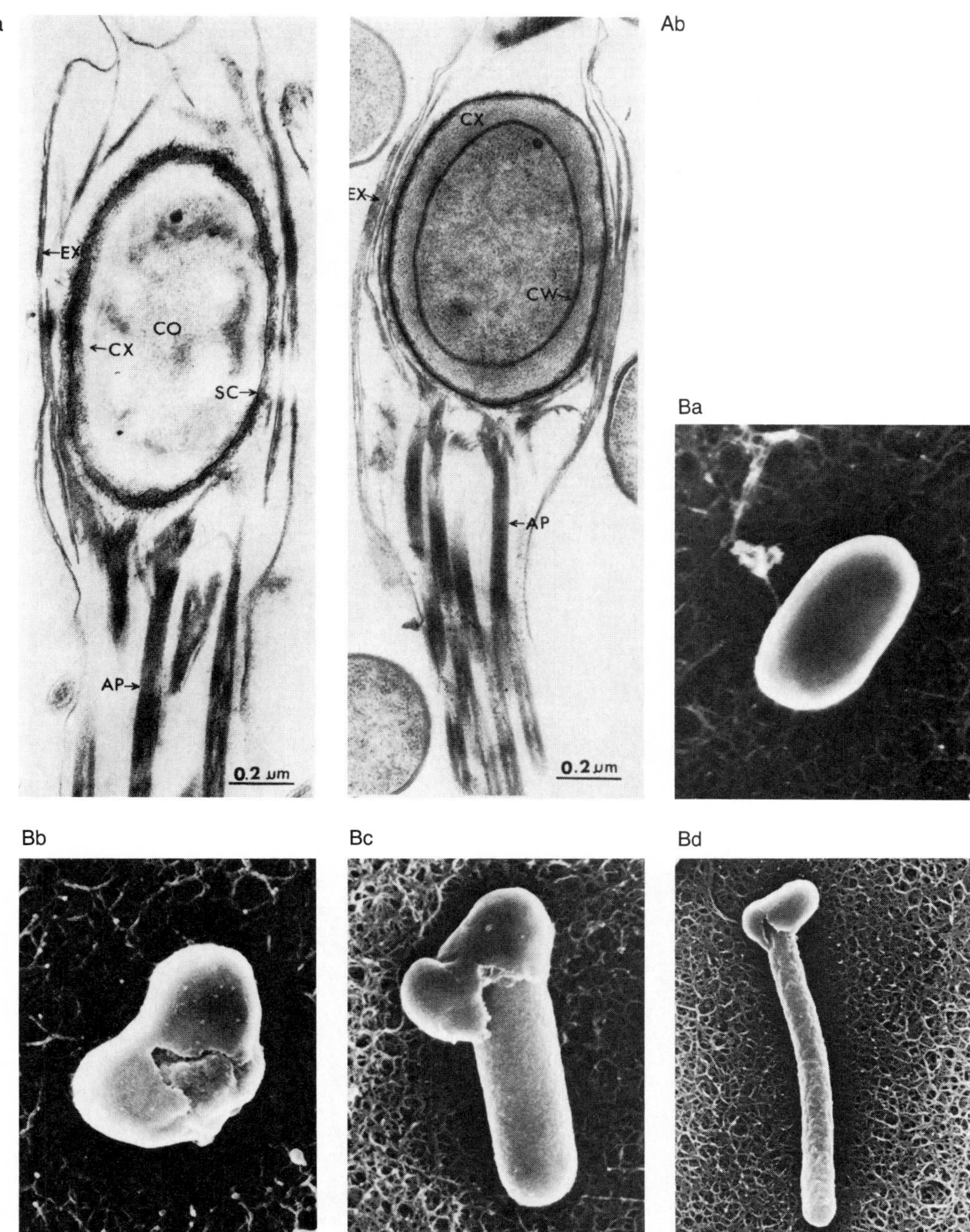

Figure 15.4 Micrographs showing endospore germination. (A) (a) A thin section of a dormant spore of *Clostridium bifermentans,* the components include appendages (AP), core (CO), cortex (CX), spore coat layers (SC), and exosporium (EX); after the stages of elongation and outgrowth, (b) the bacterial cell is constricted by the spore coat as it emerges, and the cortex material (arrows) is squeezed from the remains of the spore. (Reprinted by permission American Society of Microbiology from W. A. Samsonoff *et al.,* 1970, *Journal of Bacteriology,* 101:1038–1045.) (B) (a) Scanning electron micrograph of a spore of *Bacillus subtilis* strain (30,000×) 0 minutes following heat activation at 60°C for one hour; (b) after 90 min. incubation on nutrient agar at 37°C, a crack can be seen in an equatorial position (30,000×); (c) after 120 minutes on an agar surface, the vegetative bacterium has started to grow (22,500×); (d) after 150 minutes incubation on agar, the vegetative bacterium has grown extensively, the bacterium retains an empty spore shell on one or both ends, and the cell is about 4–5 µm across. (Courtesy of Akiko Umeda and Kazunobu Amako, Fukoaka University.)

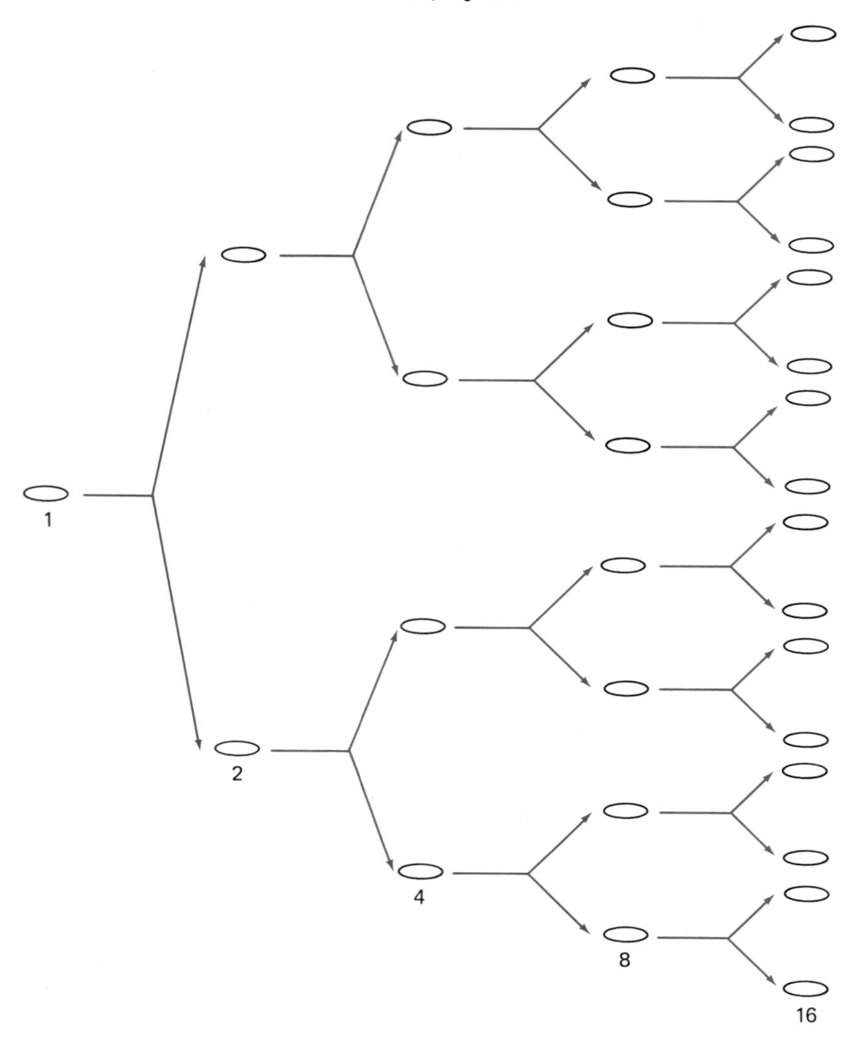

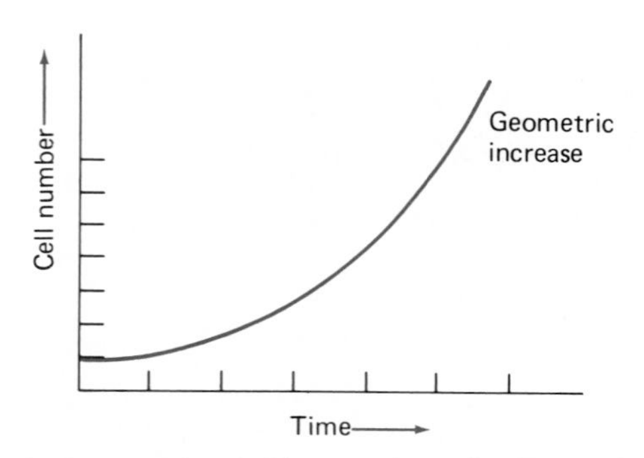

Figure 15.5 Geometric progression in the number of cells resulting from binary fission.

Table 15.1 Growth Rates for Some Representative Bacteria under Optimal Conditions

Organism	Temperature (°C)	Generation Time (min)
Bacillus stearothermophilus	60	11
Escherichia coli	37	20
Bacillus subtilis	37	27
Bacillus mycoides	37	28
Staphylococcus aureus	37	28
Streptococcus lactis	37	30
Pseudomonas putida	30	45
Lactobacillus acidophilus	37	75
Vibrio marinus	15	80
Mycobacterium tuberculosis	37	360
Rhizobium japonicum	25	400
Nostoc japonicum	25	570
Anabaena cylindrica	25	840
Treponema pallidum	37	1980

rithm to the base 10 of the number of bacteria at time t, log B_0 is the logarithm to the base 10 of the number of bacteria at the starting time, and t is the time period of growth. By determining cell numbers during the period of active cell division, the generation time can be estimated (Figure 15.6). A bacterium such as *E. coli* can have a generation time as short as 20 minutes under optimal conditions. Considering a bacterium with a 20-minute generation time, one cell would multiply to 1000 cells in 3.3 hours and to 1,000,000 cells in 6.6 hours.

When a bacterium is inoculated into a new culture medium, it exhibits a characteristic growth curve (Figure 15.7). The normal growth curve of bacteria has four phases, the lag phase, the log or exponential growth phase, the stationary phase, and the death phase. During the lag phase there is no increase in cell numbers. Rather, during this phase the bacteria are preparing for reproduction, synthesizing DNA and various enzymes needed for cell division. During the log phase of growth, so-named because the logarithm of the bacterial biomass increases linearly with time, bacterial reproduction occurs at a maximal rate for the specific set of growth conditions. It is during this period that the generation time of the bacterium is determined.

If a bacterial culture in the exponential growth phase is inoculated into an identical fresh medium, the lag phase is bypassed, and exponential growth continues. This occurs because bacteria are already actively carrying out the metabolism necessary for continued growth. If, however, the chemical composition of the new medium differs significantly from the original growth medium, the bacteria do go through a lag phase before entering the logarithmic growth phase when they synthesize the enzymes needed for growth in the new medium.

However, if the bacterium is not transferred to a new medium and no fresh nutrients are added, the stationary growth phase eventually is reached, and

Figure 15.6 The estimation of the generation time of a bacterium based on observed cell numbers during a period of active division.

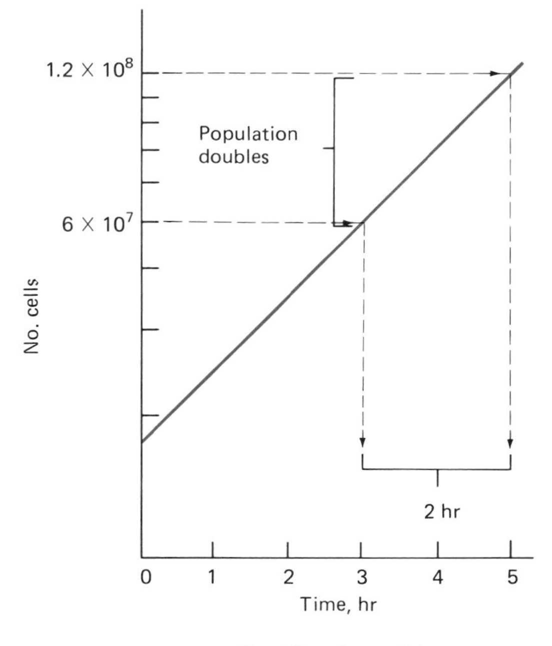

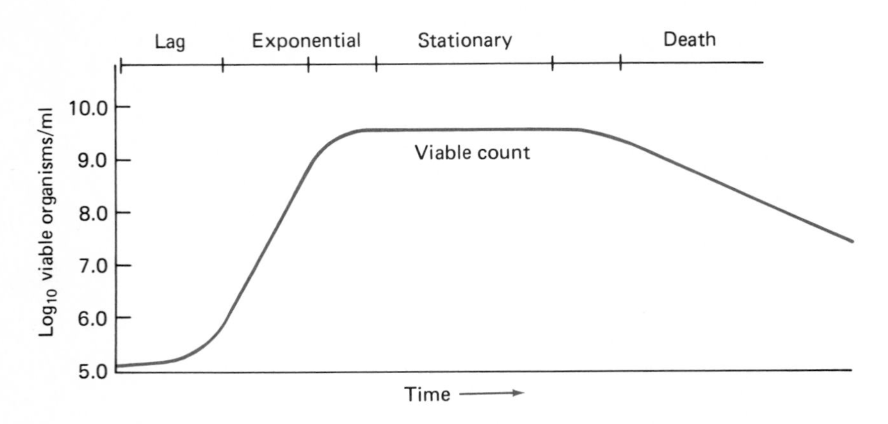

Figure 15.7 The normal bacterial growth curve has four stages: lag, exponential, stationary, and death.

there is no further net increase in bacterial cell numbers. During the stationary phase the growth rate is exactly equal to the death rate. A bacterial population may reach stationary growth when a required nutrient is exhausted, when inhibitory end products accumulate, or when physical conditions are restricting. The duration of the stationary phase varies; some bacteria exhibit a very long stationary phase.

Eventually the number of viable bacterial cells begins to decline, signalling the onset of the death phase. The kinetics of bacterial death follow the same exponential kinetics as the logarithmic growth phase. This is true because the death phase really represents the result of the inability of the bacteria to carry out further reproduction. The rate of the death phase need not, however, be equal to the rate of growth during the exponential phase. The rate of death is proportional to the number of the survivors and can be expressed as

$$dt = 3.3 \log B_0 - \log B_t$$

where d is the death rate, t is the time period, log B_0 is the logarithm of the number of bacteria at the starting time, and log B_t is the logarithm of the survivors at time t.

Figure 15.8 A chemostat is a continuous culture device. In such a device the population density is controlled by the concentration of the limiting nutrient, and the growth rate is controlled by the flow rate, which can be arbitrarily set.

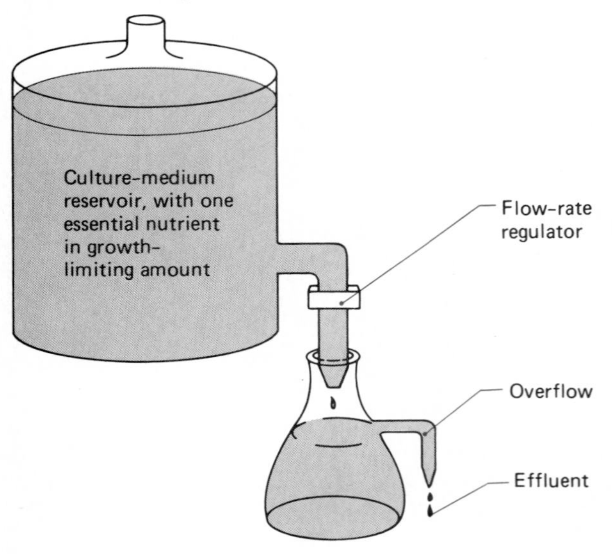

Batch and Continuous Growth

The normal bacterial growth curve is characteristic of bacteria in batch culture, that is, under conditions when a fresh medium is simply inoculated with a bacterium. A flask containing a liquid nutrient medium inoculated with a bacterium such as *E. coli* is an example of a batch culture. In batch culture growth nutrients are expended and metabolic products accumulate in the closed environment. The batch culture models situations such as occur when a canned food product is contaminated with a bacterium.

Bacteria may also be grown in continuous culture in a chemostat, where nutrients are supplied and end products continuously removed so that the exponential growth phase is maintained. Because end products do not accumulate and nutrients are not completely expended, the bacteria never reach stationary phase. The chemostat is a device in which a liquid medium is continuously fed into the bacterial culture (Figure 15.8). The liquid medium contains some nutrient in growth limiting concentration, and the concentration of the limiting nutrient in the growth medium determines the rate of bacterial growth. Even though bacteria are continuously reproducing, a number of bacterial cells are continuously being washed out and removed from the culture vessel.

Bacterial Growth on Solid Media

The development of bacterial colonies on solid growth media follows the basic normal growth curve. However, the dividing cells do not disperse; hence, the population is densely packed. Under these conditions nutrients rapidly become limiting at the center of the colony, and microorganisms in this area rapidly reach stationary phase. At the periphery of the colony, cells can continue to grow exponentially even while those at the center of the colony are in the death phase. Bacterial colonies generally do not extend indefinitely across the surface of the media but have a well-defined edge. Therefore, individual well-isolated colonies develop from the growth of individual bacterial cells. The fact that the bacteria have reproduced asexually by binary fission means that, barring mutation, all the bacteria in the colony should be genetically identical; that is, each colony contains a clone of cells derived from a single parental cell.

Factors Influencing Microbial Growth

The rates of microbial growth and death are greatly influenced by a number of environmental parameters, some environmental conditions favoring rapid microbial reproduction and others not permitting any microbial growth. Conditions permitting the growth of one microorganism may preclude the growth of another. Not all microorganisms can grow under identical conditions. Each microorganism has a specific tolerance range for specific environmental parameters. Outside the range of environmental conditions under which a given microorganism can reproduce, it may either survive in a relatively dormant state or may lose viability; that is, it may lose the ability to reproduce and consequently die.

In both laboratory and natural situations some environmental parameter or interaction of environmental parameters controls the rate of growth or death of a given microbial species. In nature, where conditions cannot be controlled and many species coexist, fluctuating environmental conditions favor population shifts because of the varying growth rates of individual microbial populations within the community of a given location. In the laboratory it is possible to adjust conditions to achieve optimal growth rates for a given microorganism. Similarly, in industrial fermentors conditions can be adjusted to optimize microbial growth rates, thereby maximizing the accumulation of desired microbial metabolic products. Many laboratory and industrial applications use pure cultures of microorganisms, facilitating the adjustment of the growth conditions so that they favor optimal growth of the particular microbial species.

Effects of Temperature on Microbial Growth Rates

Temperature is one of the most important of factors affecting both the rates of microbial growth and death. The minimum and maximum temperatures at which a microorganism can grow establish the temperature growth range for that microorganism (Figure 15.9). Most microorganisms are killed when exposed to temperatures above the maximal growth temperature. At very low temperatures, however, microbial metabolism ceases, preserving the microorganisms in a dormant state. Some strains of microorganisms in culture collections are maintained indefinitely by freezing them at very low temperatures, such as in liquid nitrogen; such preservation of reference type cultures is important because it provides necessary standards.

The oldest deposits from which living bacteria are claimed to have been extracted are salt layers near Irkutsk, USSR, dating from about 600 million years ago. The discovery was not accepted internationally. The United States Dry Valley Drilling Project in Antarctica claimed resuscitated rod-shaped bacteria from caves up to a million years old.

—1985 GUINNESS BOOK OF WORLD RECORDS

Figure 15.9 Effect of temperature on growth rate showing growth ranges and optimal temperatures for microorganisms.

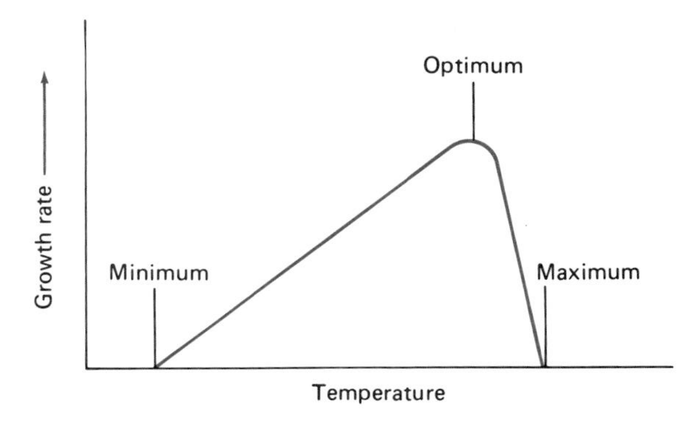

MICROBIOLOGY HEADLINES

Making Snow the Microbial Way

By C. Mlot

Earlier this year, a federal court blocked the University of California's release of a genetically engineered microbe aimed at preventing frost damage to crops (SN: 5/26/84, p. 325). Now, a Berkeley, Calif., company is marketing the natural form of the microbe—which triggers, rather than blocks, ice formation—as a snowmaking device at ski resorts. The microbial snow will demonstrate its fluff at a number of Rocky Mountain ski slopes this season after three years of field-testing at Copper Mountain in Colorado.

The frost-forming organism that damages $1 billion worth of crops annually in the United States is a natural for snowmaking; it may even play a role in cloud formation. The bacterium, *Pseudomonas syringae,* makes a protein complex in its outer membrane that serves as a nucleus for ice crystal formation. On leaves, where the bacteria naturally exist, the protein complex induces ice crystals at temperatures just below freezing that a plant could otherwise withstand. Added as a freeze-dried powder in a snowgun, the bacteria make snow more efficiently by triggering freezing at higher temperatures. Water is normally supercooled with compressed air—the expensive part of snowmaking—to about −10°C before it will crystallize, but the bacteria can form ice at about −3°C. "Basically, you don't need as much compressed air to cool the water," says Thomas Dyott, president of Advanced Genetic Sciences (AGS), which markets the bacteria as Snomax.

The company purchased rights to the bacterial snowmaking method from the University of Wisconsin in Madison, where Steven Lindow, now working on bacterial frost damage at the University of California at Berkeley, first struck upon the use of the ice-forming organism to make snow. The company, unaffected by the ruling against the University of California experiment, is also preparing to test the altered form of the bacteria in field crops (SN: 6/9/84, p. 356). Although Snomax is a naturally occurring strain, the company's researchers have developed genetic variants with improved ice-nucleating ability for potential future use. Right now, the company is increasing production of the natural strain for large-scale marketing next year. Dyott estimates a ski resort could use 200 to 1,000 pounds of bacteria to ice up about 200 million gallons of water in one season.

Are there any dangers to microbial snow? AGS bacteriologist Trevor Suslow thinks not, since the cells are debilitated with gamma-radiation so they can't reproduce in the environment. Because the snowmaking bacterium is not genetically altered, it was not included in the suit filed by the Foundation on Economic Trends against the University of California, though foundation Director Jeremy Rifkin says the introduction of the bacteria could "change the balance" of the ecosystem. Bacteriologist John Ingraham of UC Davis says the snow-related use is probably benign.

Source: Science News, Oct. 27, 1984. Reprinted by permission.

Figure 15.10 Temperature growth ranges and optima for psychrophilic, mesophilic, and thermophilic organisms.

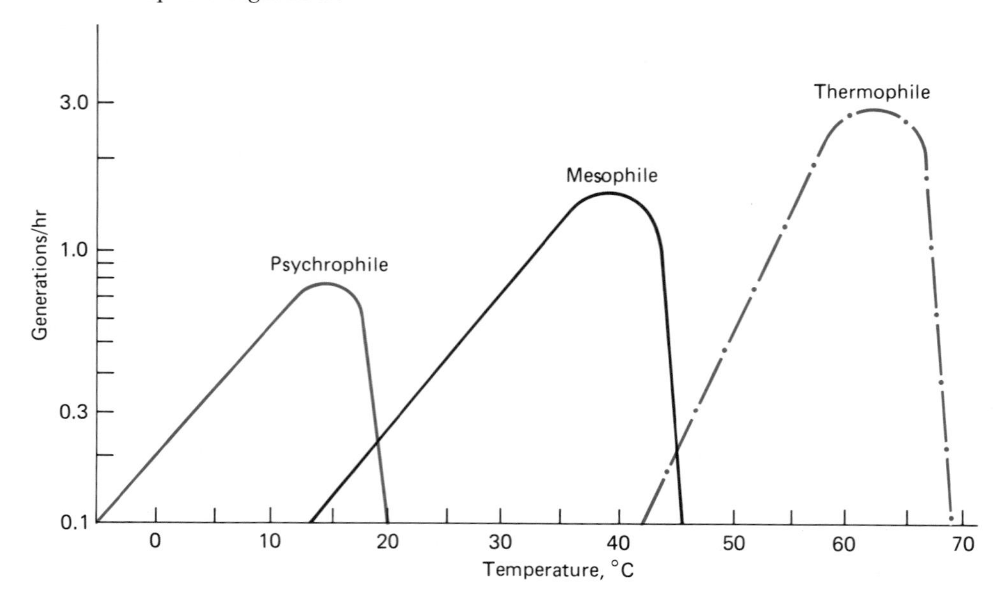

Within the growth range each microorganism has an optimal growth temperature at which the highest rate of reproduction occurs. The optimal growth temperature is defined by the maximal growth rate (shortest generation time), not the maximal cell yield. Sometimes, greater cell or product yields are achieved at lower or higher temperatures. In the laboratory, incubators (controlled temperature chambers) are normally used to establish conditions that permit the growth of a microbial culture at temperatures favoring optimal growth rates. Different microorganisms have different optimal growth temperatures (Figure 15.10).

Some microorganisms grow best at low temperatures. Such organisms, known as psychrophiles, have optimal growth temperatures of under 20°C. Psychrophilic microorganisms have enzymes that are inactivated at even moderate temperatures, about 25°C. Some psychrophilic microorganisms are capable of growing below 0°C as long as liquid water is available. Psychrophilic microorganisms are commonly found in the world's oceans and are also capable of growing in a household refrigerator, where they are important agents of food spoilage.

Other microorganisms, known as mesophiles, have optimal growth temperatures in the middle temperature range between 20 and 50°C. Most of the bacteria grown in introductory microbiology laboratory courses are mesophilic. Many mesophiles have an optimal temperature of about 37°C, which corresponds to human body temperature. Many of the normal resident microorganisms of the human body, such as *E. coli*, are mesophiles. Similarly, most human pathogenic microorganisms are mesophiles and thus are able to grow rapidly and establish an infection within the human body. Although some microorganisms are restricted to growth near the optimal growth temperature, others can grow over a wide range of temperatures. For example, psychrotrophic microorganisms have an optimal growth temperature between 20 and 50°C, but they are capable of growing at low temperatures (about 5°C), as inside a refrigerator.

Whereas mesophiles grow in the middle temperature range, thermophilic microorganisms grow at higher temperatures, often growing only above 40°C. Many thermophilic microorganisms have optimal growth temperatures of about 55–60°C. One finds thermophilic microorganisms in such exotic places as hot springs and effluents from laundromats. Recently bacteria have been found that can grow above 100°C. These bacteria live at great ocean depths, near thermal rifts, where the water is hot and still liquid because of the very high pressure.

Oxygen

Another factor that greatly influences microbial growth rates is the concentration of molecular oxygen. Microorganisms can be classified as aerobes, anaerobes, facultative anaerobes, or microaerophiles based on their oxygen requirements and tolerances (Figure 15.11). Aerobic microorganisms (aerobes)grow only when oxygen is available to support their respiratory metabolism, whereas anaerobic microorganisms (anaerobes), such as *Clostridium*, grow in the absence of molecular oxygen. Anaerobic microorganisms may carry out fermentation or anaerobic respiration to generate ATP. Some anaerobes have very high death rates in the presence of oxygen, and such organisms are termed oxylabile anaerobes. Other anaerobic microorganisms, known as oxyduric anaerobes, although unable to grow, have low death rates in the presence of oxygen. In contrast to obligate anaerobes, which

Figure 15.11 These test tubes show the relationship of the growth of various types of microorganisms to the presence of oxygen. (A) Aerobes grow in the presence of oxygen on the surface; (B) facultative anaerobes grow throughout the tube; (C) microaerophiles grow in a narrow band where the oxygen tension is reduced; (D) obligate anaerobes grow at the bottom of the tube where there is no free oxygen.

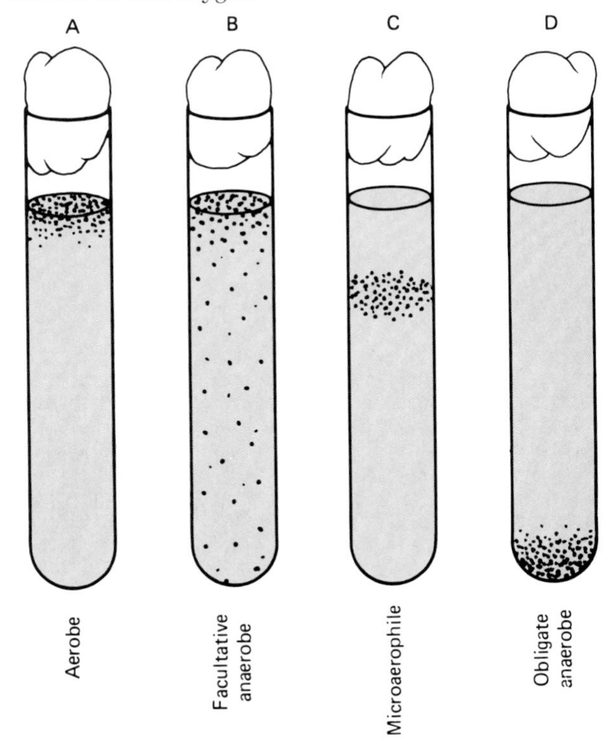

MICROBIOLOGY HEADLINES

Bacteria Found to Thrive in Heat of Volcanic Vents on Ocean Floor

By JOHN NOBLE WILFORD

Bacteria found at volcanic vents on the ocean floor are capable of living and multiplying at temperatures as high as 482 degrees Fahrenheit, more than twice as hot as conditions in which life had previously been known to be possible, according to a report published yesterday by two American biologists.

Scientists said the discovery established that life, either on Earth or elsewhere in the universe, could exist under environmental conditions far more hostile than was generally suspected. It also suggested new approaches to research on the origins of life on Earth. These bacteria, one of the discoverers said, may be physiologically similar to the earliest forms of life on Earth.

Results of experiments with the bacteria were reported in the British science journal Nature by Dr. John Baross, a microbiologist at Oregon State University, and Dr. Jody Deming, a research scientist at the Chesapeake Bay Institute of Johns Hopkins University. The experiments were conducted in a high-temperature, high-pressure laboratory chamber at the Scripps Institution of Oceanography at La Jolla, Calif.

Environment Under Pacific

The bacteria, unlike any other known microbes, were discovered in 1979 in submarine hot springs, or hydrothermal vents, at the bottom of the Pacific Ocean south of Baja California. They lived just above sulfide chimneys, or "black smokers," which leak sulfurous gases from inside the Earth's crust and create an environment supporting exotic clams, crabs, fish and giant tube worms that live on inorganic chemicals and without sunlight. Inside the vents, the water reaches 700 degrees.

In a statement issued by Johns Hopkins, Dr. Deming said: "Our studies of the black smoker bacteria suggest that microbial life is limited not by some maximum temperature or the availability of sunlight, but by the existence of liquid water that contains an appropriate supply of inorganic chemicals or gases. Under conditions of high temperature and high pressure, then, life could exist at the unicellular level on another planet or deep in the Earth itself."

Dr. Baross and Dr. Deming, with the assistance of Dr. Art Yayanos of Scripps, grew the bacteria under various temperature and pressure conditions, including those conditions simulating the area around the hydrothermal vents. They reported that the microbes did not grow at all below 167 degrees. Most plants and animals die if their temperatures exceed 104 degrees for some length of time. Most bacteria die when heated above 150 degrees, and until now only one strain of bacteria was known to endure up to 221 degrees.

In a sea-level atmosphere of 14.7 pounds per square inch, they further said, the bacteria grow quite rapidly just below the boiling point of water, which is 212 degrees. At pressures of 265 atmospheres, approximately 3,900 pounds per square inch, which exist at the depths of the hydrothermal vents, the bacteria continued to survive and grow at temperatures of 482 degrees. As pressure increases, the boiling point of water rises; at 265 atmospheres, seawater boils when its temperature exceeds 860 degrees.

When Bacteria Can Thrive

Not only did the bacteria survive at 482 degrees, the biologists said, but they also increased their number a hundred-fold in a few hours. The bacteria could even exist at up to 572 degrees, but without multiplying.

"This greatly increases the number of environments and conditions both on Earth and elsewhere in the universe where life can exist," Dr. Baross and Dr. Deming concluded in their article.

In a telephone interview, Dr. Deming said the research was "the first demonstration that these bacteria not only are present at the hydrothermal vents, but are viable and growing and actually prefer those temperatures and pressures."

These bacteria are also fascinating to scientists because they thrive on inorganic chemicals, primarily sulfur, manganese and iron, for energy. Their process of converting energy from inorganic matter independent of sunlight is known as chemosynthesis. All other organisms draw their energy directly or indirectly from the products of photosynthesis, which is the conversion of sunlight energy into organic carbon.

This raises the possibility, the biologists said, that life on Earth could have originated in the many hydrothermal vents that were presumably common in the primordial ocean. Scientists like Dr. Cyril Ponnamperuma of the University of Maryland, who have been frustrated thus far in the effort to understand how living matter could spring from nonliving matter, now suggest the hydrothermal vents are a promising natural laboratory for studying the conditions out of which the first microorganisms emerged on Earth some four billion years ago.

The discovery also reopens the question of possible life on a planet like Venus, which has extremely high temperatures and atmospheric pressures.

Source: The New York Times, June 3, 1983.

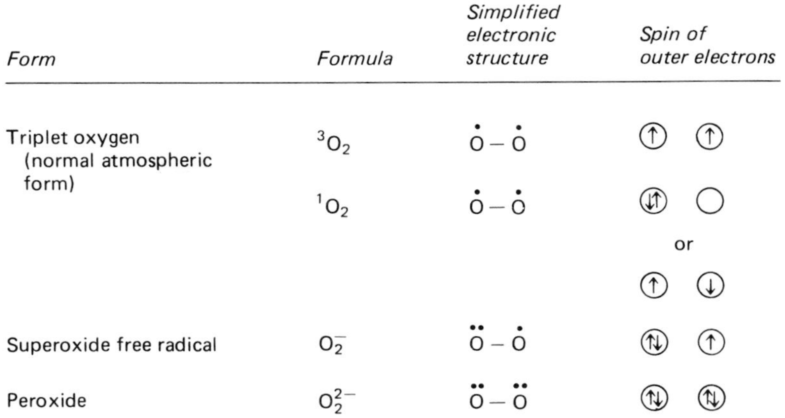

Form	Formula	Simplified electronic structure	Spin of outer electrons
Triplet oxygen (normal atmospheric form)	$^3O_2$	Ȯ – Ȯ	(↑) (↑)
	$^1O_2$	Ȯ – Ȯ	(⇅) ( ) or (↑) (↓)
Superoxide free radical	O_2^-	Ö – Ȯ	(⇅) (↑)
Peroxide	O_2^{2-}	Ö – Ö	(⇅) (⇅)

Figure 15.12 Oxygen occurs in a variety of electronic states. Some forms of oxygen are particularly toxic to microorganisms, and microorganisms have evolved various enzymes for removal of such toxic forms of oxygen.

grow only in the absence of molecular oxygen, facultative anaerobes, such as *E. coli*, can grow with or without oxygen. As a rule, facultative anaerobes are capable of both fermentative and respiratory metabolism.

Although oxygen is required for the growth of many microorganisms, it can also be toxic. Some microorganisms, known as microaerophiles, grow only over a very narrow range of oxygen concentrations. Microaerophiles require oxygen but exhibit maximal growth rates at reduced oxygen concentrations because higher oxygen concentrations are toxic to these organisms. As a consequence of the toxicity of oxygen to microorganisms, even aerobic microorganisms generally possess enzyme systems for detoxifying various forms of oxygen.

Oxygen can exist in a number of energetic states, some of which are more toxic than others (Figure 15.12). The common state of oxygen is the triplet form in which two of the electrons in the valance shell are unpaired and spin in parallel directions. In singlet oxygen, which has a higher energy level than the triplet form, these two electrons have antiparallel spins. Various enzymes are capable of catalyzing the conversion of triplet oxygen to singlet oxygen. The singlet oxygen form is chemically reactive and is extremely toxic to living organisms. Phospholipids can be oxidized by singlet oxygen, leading to a disruption of membrane function and the death of microorganisms. Peroxidase enzymes in saliva and phagocyte cells (blood cells involved in the defense mechanism of the human body against invading microorganisms) generate singlet oxygen, accounting in part for the antibacterial activity of saliva and the ability of phagocytic blood cells to kill invading microorganisms.

Additionally, the conversion of oxygen to water, which occurs when oxygen serves as a terminal electron acceptor in respiration pathways, involves the formation of an intermediary form of oxygen known as the superoxide anion (O_2^-), in addition to forming singlet oxygen. Hydrogen peroxide and free hydroxyl radicals are also generated as intermediates in this process. These are extremely reactive oxidative chemicals that can irreversibly denature many biochemicals. Hydrogen peroxide is frequently utilized as an antiseptic because it is toxic and kills microorganisms. Microorganisms generally contain enzymes that protect them against

Figure 15.13 A positive test for catalase shows bubbles arising from a *Bacillus* bacterial colony upon addition of hydrogen peroxide.

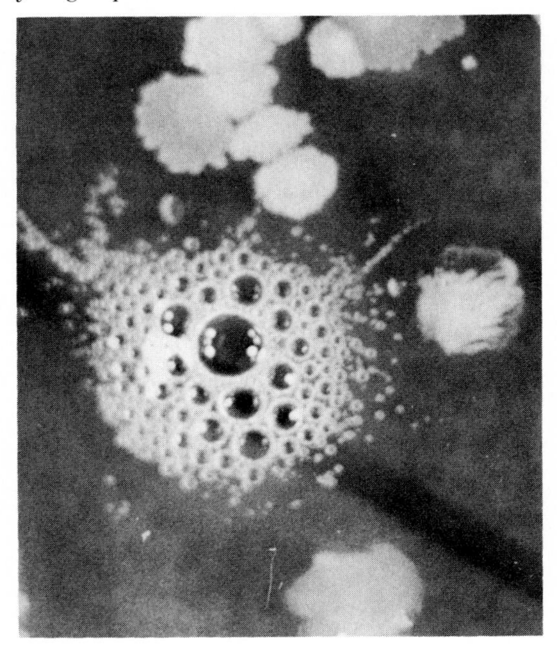

	Catalase ($2 H_2O_2 \rightarrow 2 H_2O + O_2$)	Superoxide dismutase ($2 O_2^- + 2 H^+ \rightarrow H_2O_2 + O_2$)
Aerobe	+	+
Facultative anaerobe	+	+
Oxyduric anaerobe	—	+
Oxylabile anaerobe	—	—

Figure 15.14 Enzymes produced by aerobes and anaerobes that remove toxic forms of oxygen.

the toxicity of these forms of oxygen. Both catalase and peroxidase enzymes are involved in the destruction of hydrogen peroxide. Catalase converts hydrogen peroxide to water and triplet state oxygen. Microbial production of catalase can be demonstrated by adding a loopful of a microbial culture to a 3 percent solution of hydrogen peroxide (Figure 15.13). The evolution of gas bubbles, oxygen, is evidence for the action of the catalase enzymes. Peroxidase uses the coenzyme NADH to convert hydrogen peroxide to water. In addition to these enzymes, the superoxide radical is converted to hydrogen peroxide and water by the action of the enzyme superoxide dismutase.

Both obligate aerobes and facultative anaerobes usually produce both catalase and superoxide dismutase enzymes (Figure 15.14). These enzymes permit such microorganisms to use oxygen and continue growing without accumulating toxic forms of oxygen that would kill the organism. In contrast, obligate oxylabile anaerobes generally lack both catalase and superoxide dismutase enzymes. The inability of these organisms enzymatically to remove toxic forms of oxygen probably accounts for the fact that they are obligately anaerobic and sensitive to oxygen. The oxyduric anaerobes generally produce superoxide dismutase but not catalase, evidence that they can survive in the presence of oxygen.

Figure 15.15 The effect of water activity (A_w) on the growth rate of microorganisms. Normally, microorganisms grow best at high water activities and are severely inhibited by reduced activity.

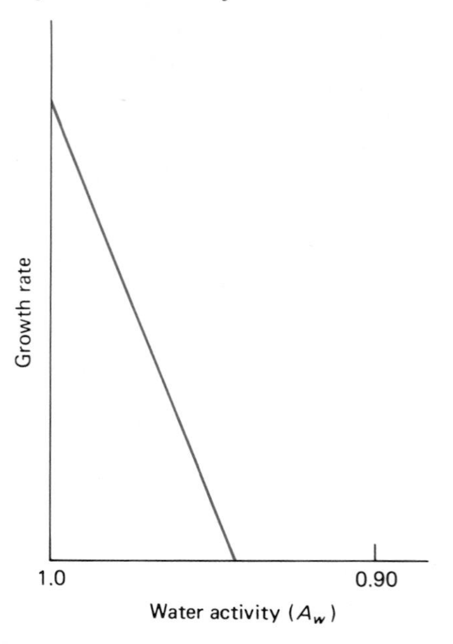

Water Activity

All microorganisms require water for growth and reproduction. Water is essential for most biochemical reactions in living systems, and the availability of water has a marked influence on microbial growth rates (Figure 15.15).

Water activity is an index of the water that is actually available for utilization by microorganisms. In the atmosphere, availability of water is expressed as relative humidity (RH). RH $= 100 \times A_w$; thus a relative humidity of 90 percent corresponds to an A_w of 0.90. Pure distilled water has a water activity (A_w) of 1.0. Adsorption and solution factors, however, can reduce the availability of water and thus lower the water activity. Water, for example, may be bound by a solute and hence be unavailable to the microorganisms. Adding high concentrations of sugars, such as sucrose, to a solution lowers the availability of water. For example, maple syrup has a water activity of 0.9. Similarly, adding salt (NaCl) to a solution can lower the availability of water. A saturated solution of sodium chloride has an A_w of 0.8.

Most microorganisms require a water activity above 0.9 for active metabolism (Table 15.2). The relatively low availability of water in the atmosphere accounts for the inability of microorganisms to grow in the air. Microorganisms likewise are unable to grow on dry surfaces, except when there is a relatively high humidity. Microbial growth on surfaces is a problem in tropical zones, where the available water in the atmosphere can support microbial growth, permitting microorganisms to grow on clothing, tents, and numerous other surfaces where microbial growth normally does not occur in temperate regions.

Some microorganisms, known as xerotolerant organisms, can grow at much lower water activities. Some yeasts grow on concentrated sugar solutions with an A_w of 0.60. As a rule, fungi are able to grow at lower water activities than other microorganisms, such as bacteria. Fungi, therefore, grow on many surfaces where the available water will not support bacterial growth. This is why fungal growth, but not bacterial growth, is commonly observed on the surface of bread.

Table 15.2 Approximate Limiting Water Activities for Microbial Growth

Water Activity (A_w)	Bacteria	Fungi	Algae
1.00	*Caulobacter*		
	Spirillum		
0.90	*Lactobacillus*	*Fusarium*	
	Bacillus	*Mucor*	
0.85	*Staphylococcus*	*Debaromyces*	
0.80		*Penicillium*	
0.75	*Halobacterium*	*Aspergillus*	*Dunaliella*
	Halococcus	*Chrysosporum*	
0.60		*Saccharomyces rouxii*	
		Xeromyces bisporus	

Pressure

Osmotic Pressure

The solute concentration of a solution effects the osmotic pressure that is exerted across the cytoplasmic membrane of a microorganism. The cell-wall structures of bacteria and other microorganisms make them relatively resistant to changes in osmotic pressure, but extreme osmotic pressures can result in the death of microorganisms. In hypertonic solutions, microorganisms may shrink and become desiccated; and in hypotonic solutions, the cell may burst. Organisms that can grow in solutions with high solute concentrations are called osmotolerant (Figure 15.16). These organisms are able to withstand high osmotic pressures and also grow at low water activities. Some microorganisms are actually osmophilic, requiring a high solute concentration for growth. For example, the fungus *Xeromyces* is an osmophile, and the optimum A_w for *Xeromyces* is approximately 0.9.

Hydrostatic Pressure

In addition to osmotic pressure, hydrostatic pressure can influence microbial growth rates.

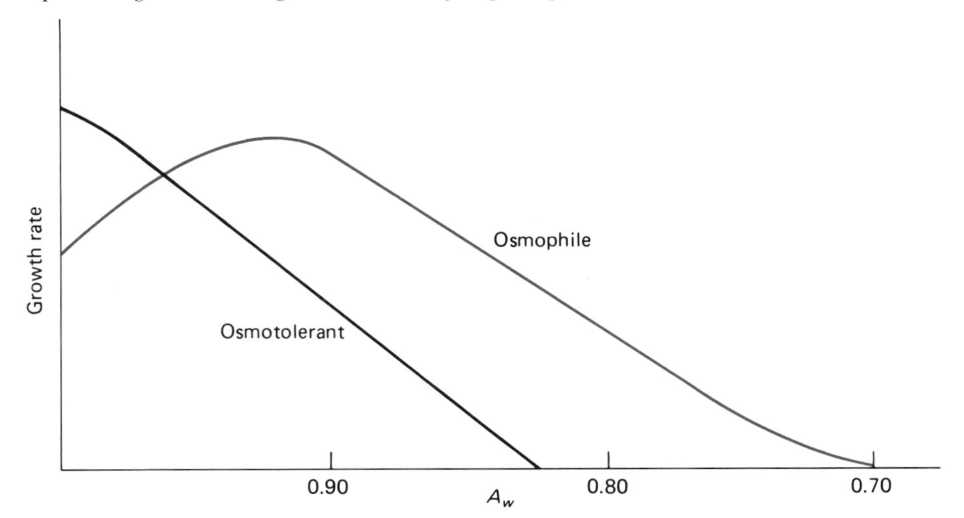

Figure 15.16 The growth of most microorganisms is restricted by high osmotic pressure, such as those produced by high sugar concentrations. However, osmotolerant and osmophilic organisms can grow at relatively high sugar concentrations.

Hydrostatic pressure refers to the pressure exerted by a water column as a result of the weight of the water column. Each 10 meters of water depth is equivalent to approximately 1 atmosphere pressure. Most microorganisms are relatively tolerant to the hydrostatic pressures in most natural systems but cannot tolerate the extremely high hydrostatic pressures that characterize deep ocean regions. High hydrostatic pressures of greater than 200 atmospheres generally inactivate enzymes and disrupt membrane transport processes. However, some microorganisms—referred to as barotolerant—can grow at high hydrostatic pressures and there even appear to be some microorganisms—referred to as barophiles—that grow best at high hydrostatic pressures.

Salinity

A special terminology is used to describe the salt tolerance and requirements of microorganisms. Some microorganisms, known as halophiles, specifically require sodium chloride for growth (Figure 15.17). Moderate halophiles, which include many marine bacteria, grow best at salt (NaCl) concentrations of about 3 percent NaCl. Extreme halophiles exhibit maximal growth rates in saturated brine solutions. These organisms grow quite well in salt concentrations of greater than 15 percent NaCl and can grow in places like salt lakes and pickle barrels. High salt concentrations normally disrupt membrane transport systems and denature proteins, and the extreme halophiles must possess physiological mechanisms for tolerating high salt concentrations. For example, the extreme halophilic bacterium, *Halobacterium*, possesses an unusual cytoplasmic membrane and many unusual enzymes that require a high salt concentration for activity.

Figure 15.17 The effect of salt concentration (sodium ion concentration) on the growth of microorganisms of varying salt tolerances. Halophiles require salt for growth.

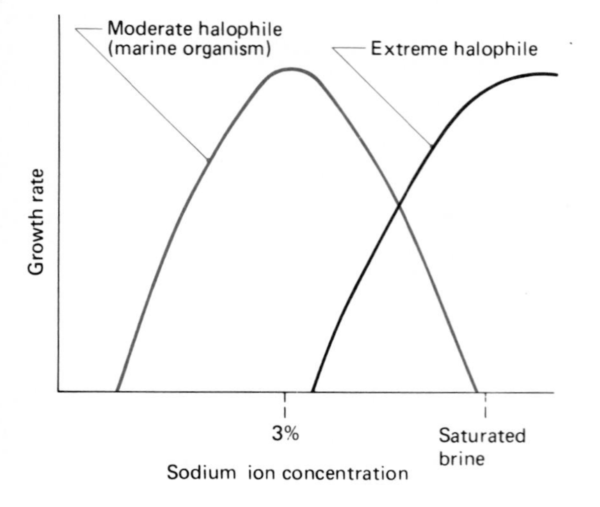

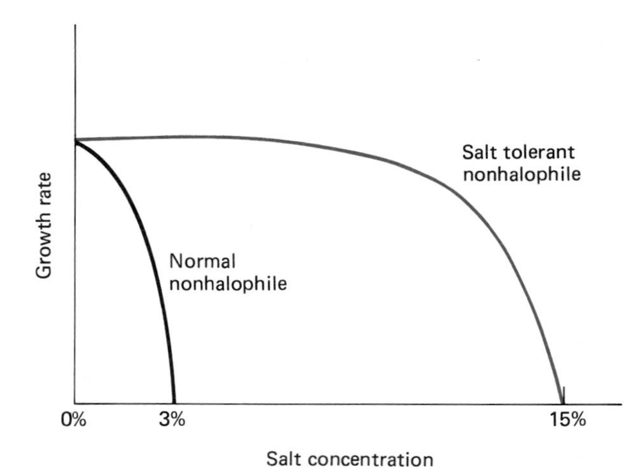

Figure 15.18 Diagram showing relative salt tolerance. Salt-tolerant microorganisms can grow on the skin and in brines where the growth of most microorganisms is restricted.

Most microorganisms, however, do not possess these physiological adaptations and are not tolerant of high salt concentrations. The degree of sensitivity to salt varies for different microbial species (Figure 15.18). Many bacteria will not grow at a salt concentration of 3 percent. Some strains of *Staphylococcus*, however, are salt tolerant and grow at concentrations greater than 10 percent NaCl. This physiological adaptation in *Staphylococcus* is important, as some members of this genus grow on skin surfaces where salt concentrations can be relatively high.

Acidity and pH

The pH of a solution describes the hydrogen ion concentration $[H^+]$. The pH is equal to $-\log [H^+]$ or $1/\log [H^+]$. A neutral solution has a pH of 7.0; acidic solutions have pH values less than 7; and alkaline or basic solutions have pH values greater than 7.

Microorganisms vary in their pH tolerance ranges (Table 15.3). Fungi generally exhibit a wider pH range, growing well over a pH range of 5–9, compared to most bacteria, which grow well over a pH range of

Table 15.3 Table of pH Tolerances of Various Bacteria

Organism	pH		
	Minimum	Optimum	Maximum
Thiobacillus thiooxidans	1.0	2.0–2.8	4.0–6.0
Lactobacillus acidophilus	4.0–4.6	5.8–6.6	6.8
Escherichia coli	4.4	6.0–7.0	9.0
Proteus vulgaris	4.4	6.0–7.0	8.4
Enterobacter aerogenes	4.4	6.0–7.0	9.0
Clostridium sporogenes	5.0–5.8	6.0–7.6	8.5–9.0
Pseudomonas aeruginosa	5.6	6.6–7.0	8.0
Erwinia cartovora	5.6	7.1	9.3
Nitrobacter spp.	6.6	7.6–8.6	10.0
Nitrosomonas spp.	7.0–7.6	8.0–8.8	9.4

6–9 (Figure 15.19). Similarly, some fungi grow well at lower pH values, as low as 0. Some other eukaryotic microorganisms, including protozoa and algae, are able to grow at low pH values; the lower limit for growth of some protozoa is approximately 2 and for some algae approximately 1. Although most bacteria are unable to grow at low pH, there are some exceptional cases. Some bacteria tolerate pH values as low as 0.8. There are even some bacteria, called acidophiles, that are restricted to growth at low pH values. Some members of the genus *Thiobacillus* are acidophilic and grow only at pH values near 2. Acidophilic bacteria possess physiological adaptations that permit enzymatic and membrane transport activities. The cytoplasmic membrane of an acidophilic bacterium breaks down and cannot function at neutral pH values.

Light Radiation

Although exposure to high intensities of light can be lethal to some microorganisms, photosynthetic microorganisms require light in the visible spectrum to carry out photosynthesis. The rate of photosynthesis is a function of light intensity. At some light intensities, rates of photosynthesis reach a maximum, and although light intensities above this level do not result in further increases in the rates of photosynthesis, light intensities below the optimal level result in lower rates of photosynthesis. The wavelength of light also has a marked effect on the rates of photosynthesis. Different photosynthetic microorganisms are capable of using light of different wavelengths. For example, anaerobic photosynthetic bacteria use light of longer wavelengths than eukaryotic algae are capable of using. Many photosynthetic microorganisms have accessory pigments that enable them to use light of wavelengths other than the absorption wavelength for the primary photosynthetic pigments. The distribution of photosynthetic microorganisms in nature reflects the variations in the ability to use light of different wavelengths and the differential penetration of different colors of light into aquatic habitats.

Exposure to visible light can also cause the death of microorganisms. Visible light may be absorbed by various biochemicals of microbial cells, resulting in changes in the energetic states of those biochemicals. In some cases this can lead to interference with metabolic activities or damage to cellular structures. Additionally, exposure to visible light can lead to the formation of singlet oxygen. As indi-

Figure 15.19 The pH ranges for the growth of fungi and bacteria. As a rule fungi are more tolerant of low pH than bacteria, but some acidophilic bacteria are restricted to growth at very low pH.

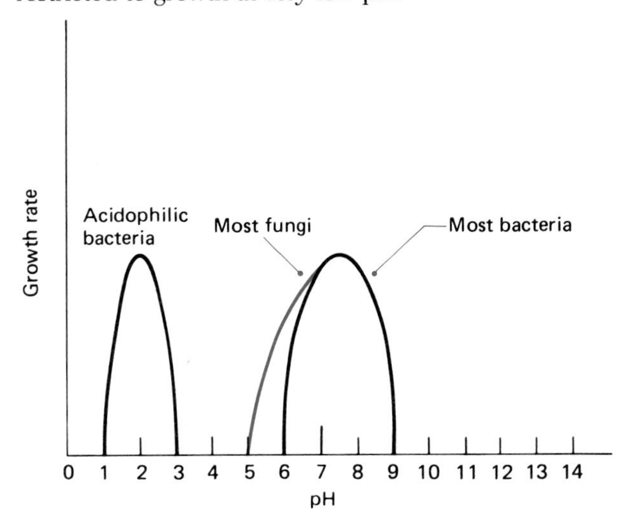

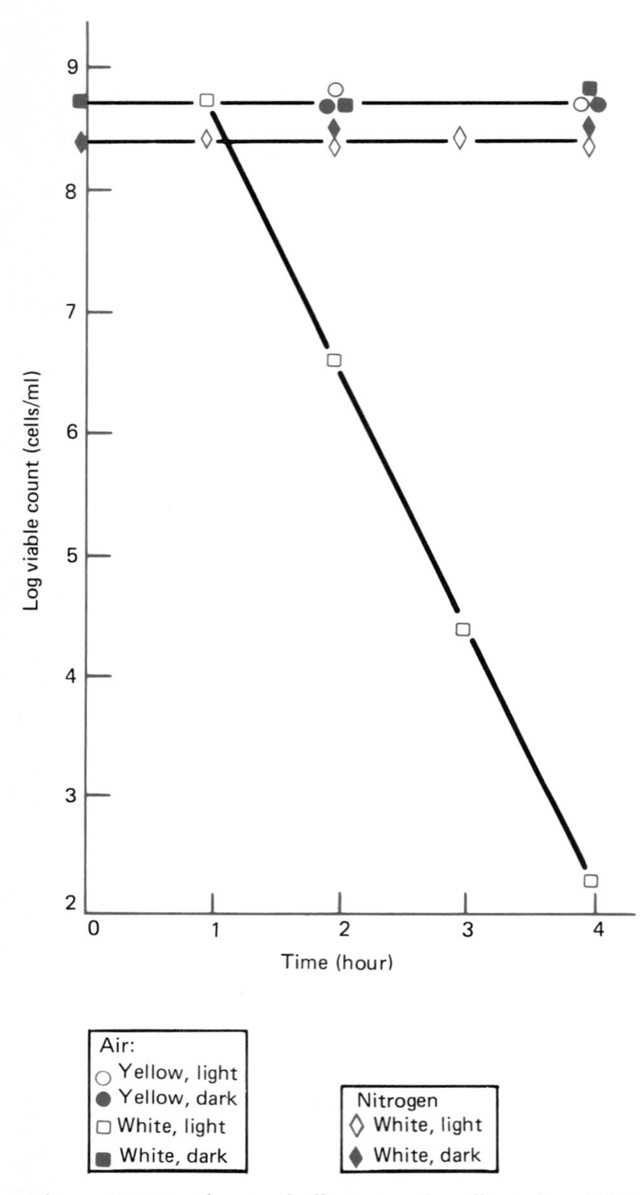

Figure 15.20 The graph illustrates the effect of sunlight on the viability of colorless and yellow-pigmented *Micrococcus lutea.* Note that death occurs in the white strain but not in the pigmented strain. (Based on M. M. Matthews, and W. R. Sistrom, 1959, *Nature,* 184:1892.)

cated earlier in this chapter, singlet oxygen is chemically reactive and can result in the death of a microorganism. Some microorganisms produce pigments that protect them against the lethal effects of exposure to light (Figure 15.20). For example, yellow, orange, or red carotenoid pigments interfere with the formation and action of singlet oxygen, preventing its lethal action. Microorganisms possessing carotenoid pigments can tolerate much higher levels of exposure to sunlight than nonpigmented microorganisms. Pigmented microorganisms are often found on surfaces that are exposed to direct sunlight, such as on leaves of trees. Many viable microorganisms found in the air produce colored pigments. Likewise, some microbial spores that are principally used for aerial dispersal are pigmented.

Summary

Bacteria normally reproduce by binary fission. This process results in doubling of the number of bacteria. Binary fission requires that cells replicate their genome and synthesize new cell wall and cytoplasmic membrane components. The replication of the genome actually can take longer than the normal division process. This is accomplished by synchronizing DNA replication and cell division such that each time a cell divides it begins to make a new copy of the genome.

Bacteria may also reproduce by other means, including budding and formation of hyphae. One alternative to binary fission is sporulation. Under certain conditions some bacteria are able to form endospores. This normally occurs when conditions are not favorable for further reproduction by binary fission. The endospore is a dormant resting stage that is highly resistant to adverse environmental conditions. Then, under appropriate conditions the endospore can germinate, reestablishing the normal binary fission reproduction cycle.

There are many factors that influence microbial growth and death rates. Microorganisms exhibit ranges of tolerance to many abiotic factors. When a given parameter exceeds the tolerance range of a microorganism, the microorganism ceases to grow. In some cases exceeding a tolerance range increases the death rate, and the microorganism is unable to survive; whereas in other cases the inability to grow does not result in the death of the microorganism but rather in its preservation. Some microorganisms tolerate a wide range of values for a given parameter, whereas other microorganisms are restricted to growth near their optimal value for that parameter.

For each of the parameters affecting microbial growth and death rates, microorganisms exhibit optimal growth rates at specific values of the given parameter. In laboratory and industrial situations it is possible to adjust conditions to favor optimal growth rates or to preclude microbial growth. For example, cultures can be incubated at the optimal growth temperature if maximal growth rates are desired. If microbial growth is not desired, temperatures can be lowered to prevent microbial reproduction. Temperatures can also be raised to increase the death rate of microorganisms and sterilize various materials.

Parameters that have a great influence on microbial growth and death rates include temperature, oxygen concentrations, water availability, radiation intensity, pressure, and pH. The combined effects of these factors generally determine the ability of a microorganism to survive and grow in a natural system. It is also these parameters that are manipulated in controlled situations to regulate the rates of microbial growth and death.

Study Outline

A. Bacterial Reproduction.
  1. Binary fission is the normal form of bacterial reproduction.
    a. Binary fission produces two equal-sized daughter cells.
    b. Replication of the bacterial chromosome is required in order that each daughter cell receives a complete genome.
    c. During cell division the bacterial chromosome appears to be attached to the cytoplasmic membrane and cell wall.
    d. A septa (crosswall) is formed by the inward movement of the cytoplasmic membrane, separating the two complete bacterial chromosomes, physically cutting the chromosomes apart and distributing them to the two daughter cells.
    e. Cell division is synchronized with chromosome replication.
  2. Other modes of bacterial reproduction are predominantly asexual, differing in how the cellular material is apportioned between the daughter cells and whether the cells separate or remain together as part of a multicellular aggregation.
    a. Budding is characterized by the unequal division of cellular material.
    b. In hyphae formation the cell elongates, forming relatively long, generally branched filaments; crosswall formation results in individual cells containing complete genomes.
  3. Sporulation results in the formation by bacteria of specialized resistant resting cells called spores.
    a. Types of spores: endospores (heat resistant spores formed within the cell); myxospores (resting cells of myxobacteria formed within a fruiting body); cysts (dormant cells sometimes enclosed in a sheath); and arthrospores (spores formed by hyphae fragmentation).
    b. During sporulation the cytoplasmic membrane invaginates within the cell to establish an endospore formation site; no crosswall forms; the endospore forms within the parent cell; a copy of the bacterial chromosome is incorporated into the endospore and various layers of endospore are synthesized around the DNA; finally the completed endospore is released by the lysis of the parent cell.
    c. Spore germination occurs upon suitable stimulation; the spore swells, breaks out of its spore coat; the germ cell elongates; the metabolic rate shifts from dormancy to a high level; enzymes are activated; and after germination the organism renews normal vegetative growth.

B. Binary fission leads to a doubling of a bacterial population at regular intervals; the generation or doubling time is a measure of the growth rate.

C. The normal growth curve for bacteria has four phases: lag, log or exponential, stationary, and death.
  1. During the lag phase bacteria are preparing for reproduction, synthesizing DNA and various enzymes needed for cell division; there is no increase in cell numbers during this time.

2. The log or exponential phase is so-named because the logarithm of bacterial biomass increases linearly with time; during log growth the instantaneous growth rate is proportional to bacterial biomass; this phase of growth is used to determine the generation or doubling time.
3. Stationary phase occurs if bacteria are not transferred to new medium and no fresh nutrients are added; during the stationary phase there is no further net increase in bacterial cell numbers and the growth rate is exactly equal to death rate.
4. The death phase begins when the number of viable bacterial cells begins to decline; the kinetics of the death phase are exponential.

D. Rates of microbial growth are greatly influenced by a number of environmental parameters.
1. Each microorganism has a specific tolerance range for specific environmental parameters; outside the tolerance range it may either survive in a dormant state or lose its viability.
2. In nature, fluctuating environmental conditions favor population shifts; laboratory and industrial growth conditions are adjusted to favor optimal growth of a particular microbial species.
3. Temperature has a major effect on microbial growth rates.
   a. There are minimum and maximum temperatures at which microorganisms can grow; these extremes of temperature at which growth occurs establish the temperature growth range.
   b. Above the maximal growth temperature, microbes generally die.
   c. Many microorganisms survive in a dormant state at temperatures below the minimal growth temperature; freezing can be used to preserve cultures of microorganisms.
   d. Microorganisms have optimal growth temperatures at which the maximal growth rate occurs; the highest rate of reproduction (shortest generation time) occurs at the optimal growth temperature.
   e. Several categories of bacteria are defined based upon optimal growth temperatures; psychrophiles have optimal growth temperatures of under 20°C; mesophiles have optimal growth temperatures in the middle range (20–50°C); psychrotrophs have optimal temperatures in the mesophile range but growth occurs at low temperatures; thermophiles grow optimally at higher temperatures, above 40°C.
4. Oxygen influences the survival and growth of microorganisms.
   a. Aerobic microorganisms grow only when oxygen is available (respiratory metabolism).
   b. Anaerobic microorganisms grow in the absence of molecular oxygen by fermentation or anaerobic respiration; obligate anaerobes grow only in the absence of molecular oxygen; oxylabile anaerobes have very high death rates in the presence of oxygen; oxyduric anaerobes, although unable to grow, have low death rates in the presence of oxygen.
   c. Facultative anaerobes can grow with or without oxygen, and usually are capable of both fermentative and respiratory metabolism.
   d. Microaerophiles grow over very narrow range of oxygen concentrations; they require oxygen, but higher concentrations are toxic to these organisms.
   e. Microorganisms possess enzyme systems for detoxifying various forms of oxygen; catalase and peroxidase enzymes are involved in destruction of hydrogen peroxide; superoxide dismutase destroys the toxic superoxide radical.
5. Availability of water has a marked influence on microbial growth rates.
   a. Water activity (A_w) is a measure of the availability of water; pure distilled water has an A_w of 1.0.
   b. Relative humidity (RH) is an expression of the availability of water in the atmosphere; RH = 100 x A_w; relatively low availability of water in the atmosphere accounts for inability of microorganisms to survive in the air.

c. Most microorganisms require A_w above 0.9 for active metabolism.
d. Fungi generally are able to grow at lower A_w values than bacteria.
e. Xerotolerant organisms can grow at very low water activities.

6. Pressure sometimes influences microbial growth rates.
   a. Extreme osmotic pressures can result in microbial death; cells shrink and become dessicated in hypertonic solutions; cells may burst in hypotonic solutions.
   b. Organisms that can grow in solutions with high solute concentrations are osmotolerant; those that require high solute concentrations are osmophilic.
   c. Hydrostatic pressure refers to pressure exerted by a water column as a result of the weight of the water column (10 meters water = 1 atmosphere pressure); most microorganisms are relatively tolerant to hydrostatic pressures in most natural systems except in deep ocean regions.
7. Salinity effects microbial growth rates.
   a. Most microorganisms are not tolerant of high salt concentrations, but some salt tolerant microorganisms, such as *Staphylococcus*, will grow at high salt concentrations.
   b. Halophiles require sodium chloride for growth; obligate halophiles can grow at very high salt concentrations.
8. The pH of a solution describes the hydrogen ion concentration.
   a. Microorganisms vary in their pH tolerance ranges; fungi generally exhibit wider pH range (pH 5–9) than bacteria (pH 6–9).
   b. Acidophiles are restricted to growth at low pH values.
9. Light radiation can influence microbial growth rates.
   a. Exposure to visible light can cause death of some microorganisms; some microorganisms produce pigments that protect them against the lethal action of light radiation.
   b. Photosynthetic microorganisms require visible light to carry out their metabolism; the rate of photosynthesis is a function of light intensity.

Study Questions

1. List four ways in which bacteria can reproduce.
2. Match the following statements with the terms in list to the right.
   a. Growth rate exactly equal to death rate.
   b. Bacteria are preparing for reproduction, synthesizing DNA and various enzymes needed for cell division.
   c. Begins when number of viable bacterial cells beings to decline.
   d. No further net increase in bacterial cell numbers.
   e. Instantaneous growth rate proportional to bacterial biomass.
   f. Generation time determined during this phase.
   g. Occurs in bacteria not transferred to new medium and no fresh nutrients are added.
   h. Bacterial reproduction occurs at maximal rate for specific growth conditions.
   i. No increase in cell numbers.
   j. Named because the logarithm of bacterial biomass increases linearly with time.
   k. Kinetics are exponential as in log phase.

   (1) Lag phase
   (2) Log or exponential phase
   (3) Stationary phase
   (4) Death phase
3. Match the following phrases with the terms in the list to the right.
   a. Grow only in the absence of molecular oxygen.
   b. Have optimal growth temperatures in the range of 20–50°C
   c. Have optimal growth temperature of under 20°C.
   d. Grow over a wide range of temperature.
   e. Grow with or without oxygen.

   (1) Optimal growth temperature
   (2) Eurythermal organisms
   (3) Psychrophiles
   (4) Mesophiles
   (5) Thermophiles
   (6) Oxylabile anaerobes
   (7) Oxyduric anaerobes
   (8) Obligate anaerobes
   (9) Facultative anaerobes

f. Require oxygen, but higher concentrations are toxic.
g. Grow only when oxygen is available.
h. The temperature at which the highest rate of reproduction occurs, as defined by the maximum growth rate.
i. Grow best at higher temperatures.
j. Have very high death rates in the presence of oxygen.
k. Have low death rates in the presence of oxygen, although they will be unable to grow.

(10) Microaerophiles
(11) Aerobes

4. Indicate whether the following statements are true (T) or false (F).
   a. Water is an essential solvent and is needed for most biochemical reactions in living systems.
   b. Pure distilled water has an A_w of 1.0.
   c. RH $= 100 \times 10\ A_w$.
   d. Relative humidity expresses the availability of water in the atmosphere.
   e. Water activity is an index of the water that is not available for utilization by microorganisms.
   f. Lack of available water always accelerates the death rate.
   g. Extreme osmotic pressures can result in microbial death.
   h. Organisms that can grow in solutions with high solute concentrations are osmotolerant.
   i. Barotolerant microorganisms can grow at high hydrostatic pressure; barophiles grow best at a low hydrostatic pressure.
   j. Acidophiles are bacteria restricted to growth at low pH.

Some Thought Questions

1. If it takes 60 minutes to replicate a bacterial chromosome, how can bacteria reproduce every 20 minutes?
2. How old is an old bacterium?
3. Where does the oldest bacterium live?
4. If bacteria can live in a freezer, why do we put food there?
5. How long would it take a single *E. coli* cell to divide into a sufficient number of cells to equal the volume of the Earth?
6. How can anaerobic bacteria live in your open mouth?
7. Are there more or less bacteria in a cigar and cigarette smoke-filled room than in a room where smoking is not permitted?
8. What kinds of differences are there between microorganisms in the Antarctic dry valleys and the Sahara desert?
9. If we use steam to kill bacteria, including their endospores, how can bacteria live at 300°C?
10. When the submersible Alvin was flooded and sank, it carried with it the lunches (bologna sandwiches) of the crew to the bottom of the ocean (1500 meter depth). Why were the sandwiches still unspoiled when Alvin was recovered 10 months later?
11. Why can the shirt literally rot off your back in the tropics but not in temperate zones?
12. Why don't bacteria exhibit circadian rhythms?
13. Do alcoholics have fewer bacteria in their guts than teetotalers?

Additional Sources of Information

Brock, T. D. 1978. *Thermophilic Microorganisms and Life at High Temperatures*. Springer-Verlag, New York. An advanced treatise describing the growth of microorganisms at elevated temperatures.

Brown, A. D. 1976. Microbial water stress. *Bacteriological Reviews*, **40**:803–846. A thorough discussion of the effect of water activity on microbial growth and survival.

Fridovich, I. 1977. Oxygen is toxic! *BioScience*, **27**:462–466. A very interesting discussion about the toxicity of various forms of oxygen and the enzymatic protective mechanisms that permit the continued growth of microorganisms in the presence of oxygen.

Friedman, E. I. 1982. Endolithic microorganisms in the Antarctic cold desert. *Science*, **215**:1045–1053. A fascinating discussion of the growth of microorganisms within rocks found in the cold deserts of the Antarctic.

Gould, G. W., and J. E. Corry. 1980. *Microbial Growth and Survival in Extremes of Environment*. Academic Press, Inc., New York. A detailed examination of the ability of microorganisms to withstand extremes of a variety of environmental factors.

Jannasch, H. W., and C. W. Wirsen. 1977. Microbial life in the deep sea. *Scientific American*, **236**(6):42–52. An easy-to-read article about the growth of microorganisms in regions of the oceans that are characterized by very high hydrostatic pressures.

Kushner, D. J. 1968. Halophilic bacteria. *Advances in Applied Microbiology*, **10**:73–97. An advanced discussion of the physiological basis of halophilic growth.

Kushner, D. J. (ed.). 1978. *Microbial Life in Extreme Environments*. Academic Press, Inc., London. A series of articles concerning the growth of microorganisms at extremes of a variety of environmental parameters.

Morita, R. Y. 1976. Psychrophilic bacteria. *Bacteriological Reviews*, **39**:144–167. An in-depth discussion of the growth of bacteria at low temperatures.

Morris, J. G. 1975. The physiology of obligate anaerobiosis. *Advances in Microbial Physiology*, **12**:169–246. A discussion of the physiological basis underlying the inability to grow in the presence of air.

16 Physical Control of Microbial Growth

KEY TERMS

autoclave
canning
D value
decimal reduction time (*D*)
desiccation
dry heat sterilization
filtration
freeze-drying
gamma radiation
HTST (high temperature–short time) process
infrared radiation
ionizing radiation
isolation
laminar flow
LTH (low temperature–hold) process
microwave radiation
packaging
pasteurization
quarantine
radappertization
radurization
shelf life
sterilization
ultraviolet radiation
X rays

OBJECTIVES

After reading this chapter you should be able to

1. Define the key terms.
2. Describe how microorganisms can be removed from a liquid.
3. Describe how microorganisms can be removed from the air.
4. Describe how temperature can be used to prevent the growth of microorganisms.
5. Describe how temperature can be used to reduce the number of viable microorganisms and the limitations of using elevated temperatures for sterilization of products.
6. Describe how the *D* value can be used to establish a process for canning or pasteurizing of foods.
7. Describe how desiccation can be used to preserve foods.
8. Describe how radiation can be used to limit microbial populations.
9. Select several food products and describe the physical methods used to preserve each of them.
10. Discuss how physical methods are used to control disease-causing microorganisms.

Physical Removal and Containment of Microorganisms

Several environmental conditions can be adjusted to kill microorganisms or to limit microbial growth. These approaches to the control of microorganisms rely upon physically removing microorganisms, altering conditions to lower the growth rate or to prevent growth entirely, and establishing conditions that increase the death rates of microorganisms. The ability to limit microbial growth rates and to eliminate viable microorganisms is especially important when trying to preserve products and/or totally eliminate viable microorganisms. Sterilizing products is very important in many areas, including medical procedures, the pharmaceutical industry, and food processing. In some cases sterilization of the product is essential for its intended use. For example, it is necessary to eliminate all viable microorganisms from surgical instruments, suturing material, and pharmaceutical solutions. In other cases the complete elimination of microorganisms is not essential and limiting microbial growth will suffice.

Physically removing microorganisms and preventing microorganisms from coming in contact with particular substances is an effective way of controlling microbial populations. Many modern sanitary practices are aimed at reducing the incidence of diseases by preventing the spread of pathogenic microorganisms or by reducing their numbers to concentrations that are insufficient to cause disease. For example, washing one's hands before eating is practiced to avoid the accidental contamination of one's food with soil or other substances that may harbor populations of disease-causing microorganisms. Similarly we wash fruits and vegetables before they are consumed in order to remove soil.

Quarantine and Isolation

We often try to avoid exposure to pathogenic microorganisms in order to preclude contracting infectious diseases. Quarantine is an old procedure that was introduced to contain physically in one area humans potentially carrying pathogenic microorganisms, thereby preventing the spread of dread diseases such as plague. Historically, the isolation of leprosy patients in remote colonies is an example of the extreme steps that have been taken to prevent contact of such individuals with the general population. Isolation of individuals with contagious microbial diseases, in which the infectious agent is airborne, is often practiced. For example, children with measles, chicken pox, or mumps are often kept away (isolated) from other children who are not immune to these diseases. Such practices decrease the probability of exposure to pathogenic organisms and prevent, or at least reduce, the transmission of disease.

When it is necessary to come in contact with persons harboring infectious microorganisms other precautions can be taken to prevent physical transmission of the disease. Masks should be worn in the presence of patients with tuberculosis or other pulmonary diseases or else other precautions, such as maintaining a safe distance from the infected individual, should be taken to minimize exposure to the pathogenic microorganisms. Likewise, great care is taken during surgical procedures to prevent contamination by accidental introduction of microorganisms into the exposed tissues. In operating theaters, walls are disinfected, instruments sterilized, and staff appropriately garbed to prevent infection of exposed tissues (Figure 16.1). Surgical staff wear hats, masks, gloves, gowns, and even shoe covers and extensively scrub (wash) their skin before entering the operating room.

Preservation of Food

Preventing the contamination of food products with pathogenic and spoilage microorganisms is an important part of food preservation methods. There are a variety of food preservation methods (Table 16.1), all of which attempt to eliminate microorganisms from foods and/or limit their growth rates. The use of sanitary methods minimizes microbial contamination of foods and is critical whenever food products are handled. This includes both the commercial handling of foods during processing and the handling of foods in the home during preparation. Washing of utensils, such as cutting instruments that may contaminate foods minimizes the possibility of microbial contamination. It should be remembered that anything that contacts the food, including hands and work surfaces, may be a source of microbial contamination and that once the food is inoculated, the microorganisms may reproduce rapidly under favorable conditions.

Packaging. The packaging of food products is extremely important in preventing microbial contamination during transport and storage. Many foods have a natural protective covering that prevents or delays microbial decomposition of the sterile inner tissues, such as the outer skins of fruits and vegetables and the shells of eggs. Processing of foods, however, often removes or disrupts these protective layers, exposing the product to microbial spoil-

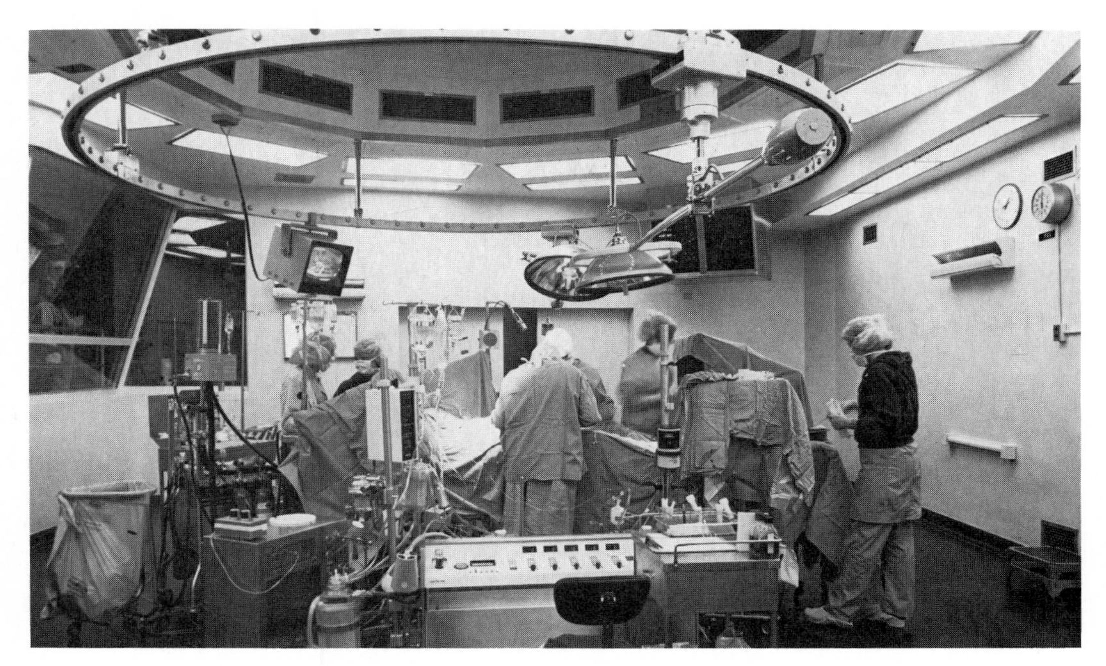

Figure 16.1 In operating theaters the surgical staff are attired so as to minimize possible contamination of the exposed surgical wound to microorganisms. Masks are used to filter aerosols from the breaths of the surgical staff. In some operating facilities the air entering the room is filtered to remove airborne microorganisms. (Courtesy National Institutes of Health, Bethesda, Maryland.)

age. Most processed food products are wrapped or placed in sealed containers to preclude microbial contamination and spoilage of the food. Quality control in packaging is an important aspect of the food processing operation.

Filtration. In some cases it is necessary to remove microorganisms from a product rather than simply to prevent them from entering the product. Passing a solution through a 0.2–0.45 micrometer pore size filter will remove bacteria and larger microorganisms. Filtration is widely used for some industrial applications to sterilize liquids that cannot tolerate exposure to high temperatures. For example, filtration is used to eliminate completely (sterilize) or to reduce the numbers of microorganisms in many pharmaceuticals and beverages (Figure 16.2). Industrial filtration methods employ various types of bacteriological filters, including sintered glass, cellulose, and nitrocellulose with pore sizes small enough to trap bacteria. Because of problems with clogging of the filters, only clear liquids can effectively be processed by filtration.

In addition to the filtration of liquids, air can be filtered to remove microorganisms (Figure 16.3). Thorough decontamination of air is accomplished by using high efficiency particulate air (HEPA) fil-

Table 16.1 **Principles and Principal Methods of Food Preservation**

Principles of Food Preservation

In accomplishing the preservation of foods by the various methods, the following principles are involved

1. Prevention or delay of microbial decomposition
   - **a.** by keeping out microorganisms (asepsis)
   - **b.** by removal of microorganisms, e.g., filtration
   - **c.** by reducing rate of microbial growth, e.g., by low temperature, drying, anaerobic conditions, and chemical inhibitors
   - **d.** by killing microorganisms, e.g., by heat or radiation.

2. Prevention or delay of self-decomposition of the food
   - **a.** by inactivation of food enzymes, e.g., blanching
   - **b.** by prevention of chemical reactions, e.g., by using antioxidants

Principal Methods of Food Preservation

1. Asepsis, or keeping out of microorganisms
2. Removal of microorganisms
3. Maintenance of anaerobic conditions
4. Use of high temperatures
5. Use of low temperatures
6. Drying
7. Irradiation
8. Use of chemical preservatives

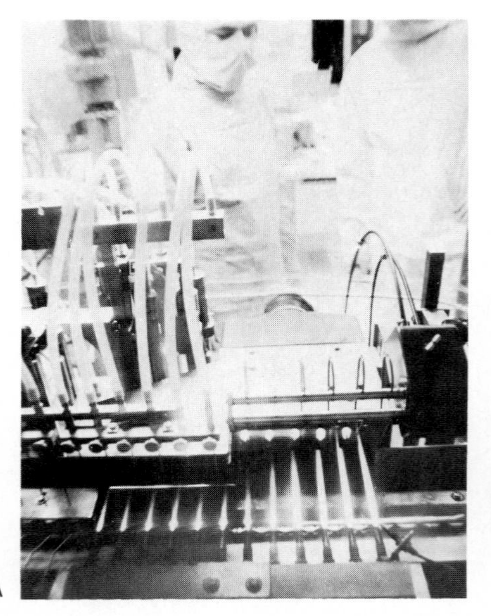

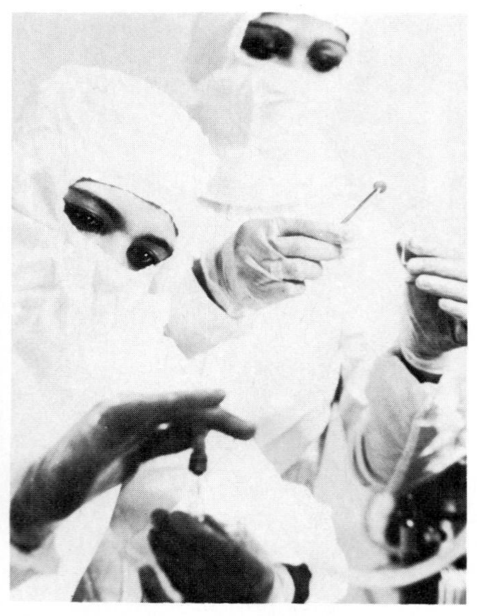

Figure 16.2 Photographs of sterile drug filling operations for producing pharmaceuticals in ampules. (A) Workers at Giza, Egypt, manufacturing plant produce pharmaceuticals in an area completely isolated and supplied with a change of filtered, sterile air each minute. (B) Modern high-speed equipment at a pharmaceutical manufacturing plant in Regensburg, West Germany, washes, sterilizes, and fills ampules for hypodermic solutions. (Courtesy E. R. Squibb & Sons, Inc.)

Figure 16.3 Diagram of a laminar flow hood used for aseptic transfer procedures.

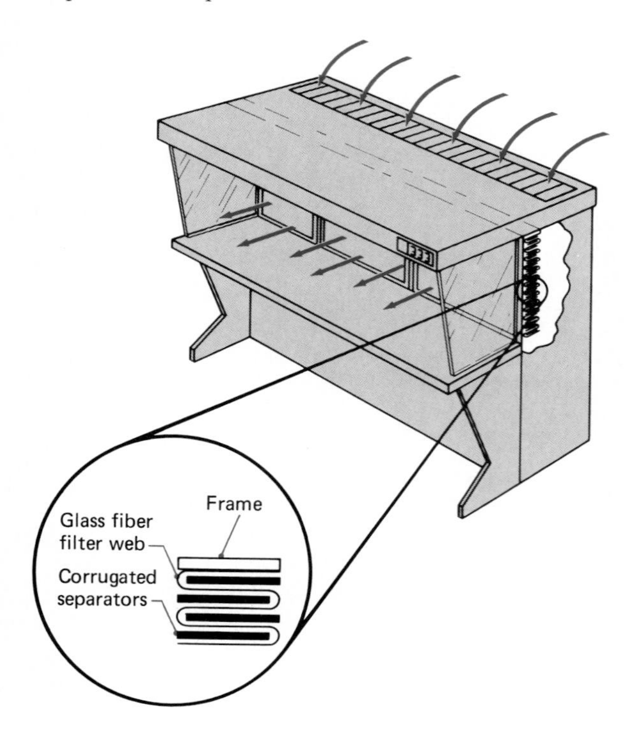

ters. These filters remove 99.97 percent of all particles, including microbes that are 0.3 micrometer in diameter or larger. In some operating theaters and specialized treatment facilities, such as burn wards, the air entering the room is passed through a filter to prevent microorganisms from entering the room. Similarly in some infectious disease wards that specialize in respiratory diseases, the air is circulated through filters to trap airborne pathogens and prevent their spread. Many laboratories use laminar flow hoods, in which filtered air is unidirectionally blown across a surface, to transfer cultures safely. Likewise many vials of pharmaceuticals are filled in laminar flow rooms where the air is sterilized by passage through filters.

Temperature

High Temperatures

As we discussed in Chapter 15, temperature has a major influence on the rates of microbial growth and death. At temperatures above the maximal growth temperature, the death rate exceeds the growth rate. The higher the temperature above the maximal growth temperature, the higher the death rate for that microorganism (Figure 16.4). Consequently, high temperatures can be used to kill microorganisms in order to control their proliferation.

The heat killing of microorganisms can be described by the decimal reduction time (*D value*) (Figure 16.5). *D* is defined as the time required for a tenfold reduction in the number of viable cells at a given temperature, that is, the time required for a log reduction in the number of microorganisms. As the temperature is increased above the maximal growth temperature for a microorganism, the decimal reduction time is shortened. The decimal reduction time varies for different microorganisms (Table 16.2). In the food industry, the decimal reduction time is important in establishing appropriate processing times for sterilizing food products.

Time the destroyer is time the preserver

—T. S. Eliot, *The Dry Salvages*

Pasteurization. Exposure to high temperatures is used to reduce the numbers of microorganisms associated with certain foods without trying to eliminate all viable microorganisms. Such procedures prolong shelf life and ensure the safety of the food as long as human pathogens are eliminated. Pasteurization is a process that uses relatively brief exposures to moderately high temperatures to reduce the numbers of viable microorganisms and to eliminate human pathogens. A pasteurized food, however, retains viable microorganisms, which means that additional preservation methods are needed to extend the shelf life of the product. Such processing of foods is normally followed by other preservation methods, often refrigeration, to reduce the growth rates of the surviving microorganisms.

The temperatures and exposure times required for the effective preservation of foods depend on the nature of the food and the heat resistance properties of the microorganisms associated with that food. Pasteurization of milk, for example, is largely aimed at eliminating a limited number of non-spore-forming pathogenic bacteria, namely *Brucella* sp., *Coxiella burnetii*, and *Mycobacterium tuberculosis*, which are associated with the transmission of disease via contaminated milk. These microorganisms are relatively sensitive to elevated temperatures, and the pasteurization of milk is therefore normally achieved by exposure to 62.8°C for 30 minutes (low temperature–hold or LTH process) or 71.7°C for 15 seconds (high temperature–short time or HTST process). Recently, there

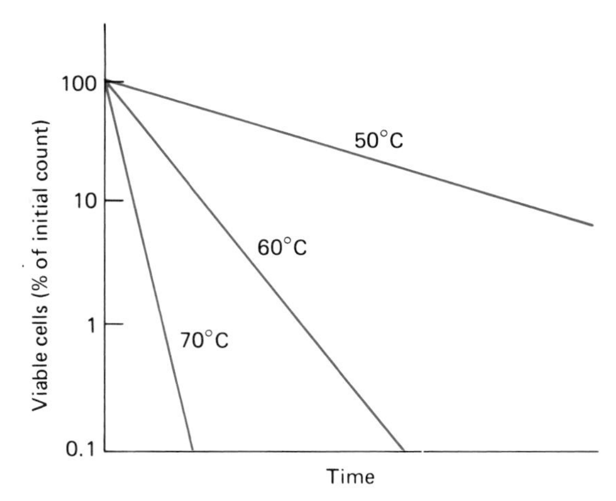

Figure 16.4 Effect of varying temperature on the death rate of a mesophilic bacterium. Above the maximal tolerance of the organism, increasing the temperature greatly accelerates the death rate.

have been several outbreaks of foodborne disease associated with the inadequate pasteurization of milk, including a widespread outbreak of salmonellosis in the Chicago area associated with one dairy and a number of deaths in California from a *Listeria* infection of Mexican-style cheese. Although milk in the United States is normally preserved by pasteurization and requires refrigeration to extend its shelf life, exposure to 141°C for 2 seconds can

Figure 16.5 This is a hypothetical survivor curve for a decimal reduction time (*D* value) = 10 with an initial concentration of 10^4 cells per milliliter. After 10 minutes there is a log reduction in numbers of surviving cells. Increased process times are needed to sterilize highly contaminated materials. (Based on W. C. Frazier and D. C. Westhoff, 1978, *Food Microbiology*, McGraw-Hill Book Company, New York.)

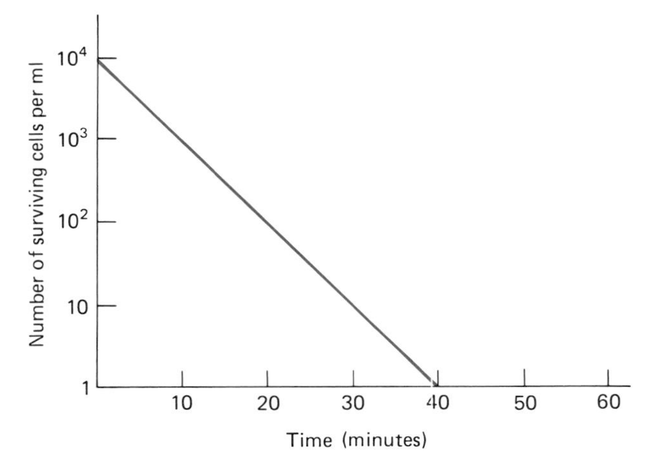

Table 16.2 Effects of Temperature on Various Microorganisms

Organism	Temperature Growth Range (°C)	Thermal Death (°C)	Thermal Death (min)
Acetobacter roseum	10–41	50	5
Bacillus subtilis	15–55	100	14
Clostridium botulinum	18–55	120	5
Clostridium tetani	14–50	105	10
Corynebacterium diphtheriae	15–40	54	10
Diplococcus pneumoniae	18–42	56	6
Enterobacter aerogenes	12–44	60	30
Escherichia coli	10–45	60	10
Haemophilus influenzae	26–43	52	30
Klebsiella pneumoniae	12–43	55	30
Mycobacterium tuberculosis	30–42	65	15
Neisseria gonorrhoeae	25–40	55	<5
Neisseria meningitidis	25–42	50	<5
Proteus vulgaris	10–43	55	60
Pseudomonas aeruginosa	0–42	62	10
Salmonella typhi	4–46	60	2
Shigella dysenteriae	10–40	60	10
Streptococcus pyogenes	15–40	60	15
Vibrio cholerae	14–42	55	2
Yersinia pestis	0–45	55	5

be used to sterilize the milk (ultra high temperature or UHT process). Sterilized milk is currently marketed in several European countries and is now being introduced in the United States; it has an indefinite shelf life provided that the container remains sealed.

Sterilization. In many microbiological procedures high temperatures are used to kill all viable

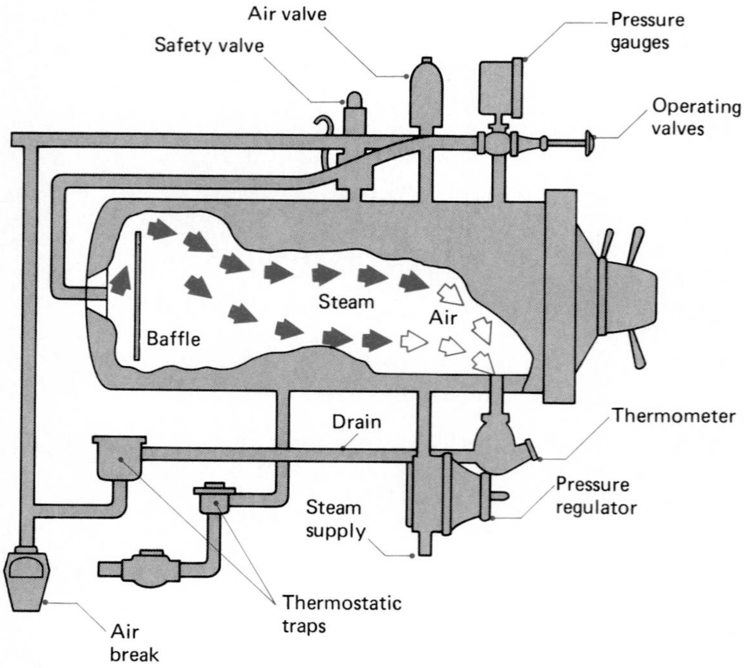

Figure 16.6 Diagram of an autoclave. This instrument is routinely used for sterilization of media and other items in the microbiology laboratory.

MICROBIOLOGY HEADLINES

Drug Produced on Space Flight Is Contaminated by Microbe

WASHINGTON—A drug created in space by America's first industry-sponsored astronaut two months ago has been contaminated and may be destroyed, the McDonnell Douglas Corp. said yesterday.

The firm, which hopes to market the drug to treat a disease affecting millions of people, did not explain the contamination but said it will use "different procedures for sterilization of the equipment before flight."

The drug, a hormone, was manufactured aboard the space shuttle Discovery in late August and September by Charles D. Walker, chief test engineer for the McDonnell Douglas project.

The contamination is so bad that the material returned from space is unsuitable for testing in laboratory animals.

A source in the National Aeronautics and Space Administration said it appeared "a microbe contaminated the material" after the shuttle landed at Edwards Air Force Base in California.

Walker's tests during the flight confirmed that the hormone was present and being collected, said McDonnell Douglas spokeswoman Susan Flowers in St. Louis.

"However, no hormone activity was detected when the material was returned to St. Louis" after the shuttle landed in California, a McDonnell Douglas statement said. "It is presumed that the contamination is either masking the presence of or has destroyed the hormone."

The firm said it will ask NASA soon to allow Walker and the equipment to fly on one additional shuttle mission to recover from the loss of the material.

Source: Associated Press, Nov. 11, 1984. Reprinted by permission.

microorganisms, that is, to sterilize, materials. As long as there are no endospore-forming bacteria, boiling at 100°C for 10 minutes is adequate to eliminate microorganisms from water. When water supplies are potentially contaminated with enteric pathogens, such boiling is used to ensure the bacteriological safety of the water. In many cases higher temperatures are needed to ensure that all microorganisms, including endospore producers, are killed.

Autoclaving. Culture media in bacteriological laboratories are normally prepared by heat sterilization, killing all microorganisms, including even viruses and resting stages of other microorganisms. In the normal heat sterilization process, the medium is exposed to steam at a temperature of 121°C, which corresponds to 15 pounds per square inch pressure, for 15 minutes in an autoclave (Figure 16.6). The autoclave is basically a chamber that can withstand pressures of greater than 2 atmospheres. The materials to be sterilized are placed into a chamber, and the chamber is sealed; steam then is transferred from a jacket into the chamber, forcing out all the air. The steam is held in the chamber for the necessary time and then vented from the chamber. Steam has great penetrating power and is able to kill microorganisms efficiently; thus autoclaving for 15 minutes at 121°C kills all microorganisms, including endospores.

Dry heat. Dry heat sterilization requires higher temperatures for much longer exposure periods, in order to kill all the microorganisms in a sample. We routinely sterilize transfer loops by flaming them red hot before aseptically transferring a culture from one site to another. In this case we use a very high temperature for a short time. Exposure in an oven for 2 hours at 170°C (328°F) is generally used for the dry heat sterilization of glassware and other items. There are many practical applications of heat sterilization procedures. For example, heat sterilization is very important in medical microbiology where sterile instruments are required for surgical procedures, where sterile culture media are required for diagnostic purposes, and where contaminated materials must be sterilized before they can be reused or safely discarded. Incineration is an effective means of disinfection for the safe disposal of many contaminated materials.

Canning. Many foods are exposed to a given elevated temperature for a period of time sufficient to kill all the microorganisms associated with that food, followed by asepsis to prevent recontamination of the food. Canning, for example, uses heat for sterilizing the food and hermetic sealing under anaerobic conditions to prevent food spoilage. In this preservation method the high temperature exposure kills all the microorganisms in the product, the can or jar acts as a physical barrier to prevent recontamination of the product, and the anaerobic conditions prevent oxidation of the biochemicals in the food (Figure 16.7).

Canning to preserve foods has its origin in 1795

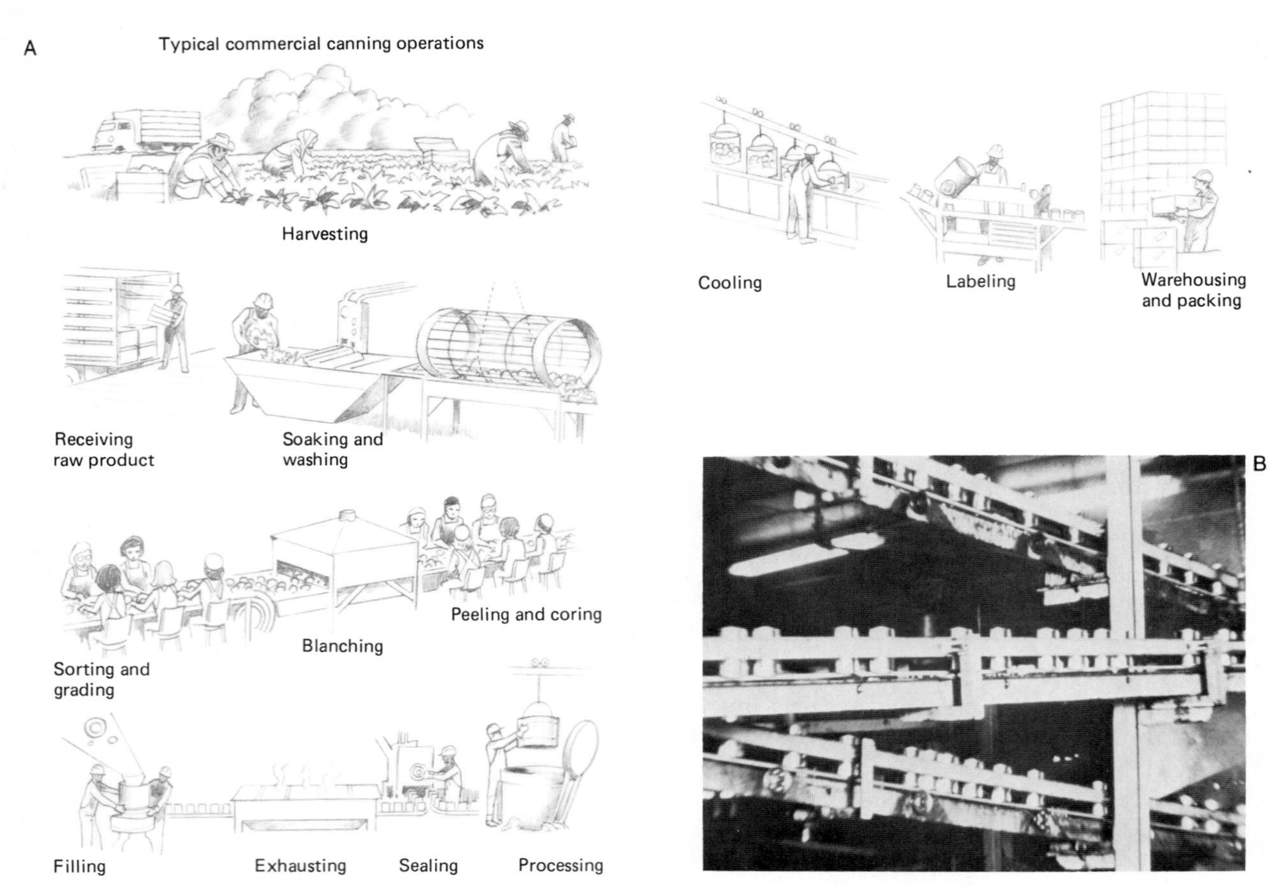

Figure 16.7 (A) Illustration showing the steps in a commercial canning procedure. (B) Photographs of a commercial canning operation; the cans are carried along a conveyor belt through the filling and sealing operations and into the retort; in modern automated systems the speed of the conveyor determines the exposure time of the cans within the retort. (Courtesy American Can Company, Barrington, Illinois.)

when the French government offered a prize of 12,000 francs for the development of a practical method of food preservation. In 1809 Francois (Nicolas) Appert succeeded in preserving meats in glass bottles that had been kept in boiling water for varying periods of time. Appert was issued a patent for his process in 1810. As early as 1820 the commercial production of canned foods was begun in the United States by W. Underwood and T. Kensett. Appert's discovery that foods could be preserved for prolonged periods of time when they were heated and stored in the absence of oxygen came at the same time that the questions of spontaneous generation and the role of microorganisms in fermentation and putrefaction were being debated by the premier scientists of the day. Spallanzani in 1765 had shown that beef broth that had been boiled for an hour and sealed did not spoil. Appert applied to home economics the results of Spallanzani's experiments. Not being a scientist, Appert probably did not understand either why his method worked or its long-range significance. It was not until almost a half century later that Pasteur, in disproving the theory of spontaneous generation, provided the scientific basis for understanding why Appert's canning method works. Pasteur pointed out that Appert's method, even when modified by using temperatures below 100°C and relatively short incubation times, was a practical method for preventing undesirable ferments. Today, the method of canning, begun over a century ago without the knowledge of why it worked, that is, its scientific rationale, is one of the most widely used methods for preserving foods.

Table 16.3 **Classification of Canned Foods and Their Processing Requirements**

Acidity Class	pH	Representative Foods	Spoilage Organisms	Processing
Low acid	7	Ripe olives, eggs, milk, poultry, beef, oysters	Mesophilic *Clostridium* species	High temperature, 121°C
	6	Beans, peas, carrots, beets, asparagus, potatoes	Thermophiles, plant enzymes	High temperature, 121°C
Medium acid	5	Figs, tomato soup, ravioli	Limit of growth of *C. botulinum*	High temperature, 121°C
Acid	4	Potato salad, pears, peaches, oranges, apricots, tomatoes,	Aciduric bacteria	Boiling water, 100°C
		Sauerkraut, apple, pineapple, grapefruit, strawberry	Plant enzymes	Boiling water, 100°C
Highly acid	3	Pickles, relish, lemon juice, lime juice	Yeasts, fungi	Boiling water, 100°C

Exposure to 115°C for 15 minutes is generally considered necessary in home canning to ensure killing of endospore formers in medium acid (pH 4.5–5.3) to low acid (pH greater than 5.3) foods. Particular concern must be given to ensuring sterility of such foods because of the possible growth of *Clostridium botulinum* and the seriousness of the disease botulism. Somewhat lower temperatures, for example, exposure to 100°C for 10 minutes, are often employed in the home canning of acidic (pH less than 4.5) foods, in part because of the lowered thermal resistance of microorganisms under acidic conditions, and because of the fact that *C. botulinum* is unable to grow at low pH values (Table 16.3). The canning industry employs a 12*D* process for nonacid foods. Heating the food at 121°C for 2.52 minutes, 12 times the *D* value, reduces the probability of the survival of *Clostridium botulinum* endospores to 10^{-12}. Thus, if there were one spore in every can, the probability of contamination remaining after processing should be reduced to one in every million million cans. Heating at 121°C for 2.52 minutes therefore should ensure the safety of canned foods with respect to possible contamination with *Clostridium botulinum*. Several commercial canning operations that have not adhered to the necessary standards and whose operations resulted in outbreaks of botulism have been put out of business. Most problems occur with home canning, where a lack of care and effort can result in insufficient heating.

Sterility Testing. Testing of sterility is an important quality control procedure. Autoclaves have pressure gauges and thermometers that monitor the sterilization operation. These monitors are supplemented with internal monitors. Chemicals that darken when exposed to specific heat treatments are sometimes included with each sterilization batch. These chemicals are often impregnated into a tape or wrapping material so that, upon adequate exposure to elevated temperature, the darkened chemical spells out the word STERILE. In this way the proper operation of the sterilizer is monitored and, simultaneously, the sterilized material is properly labelled.

Another method for monitoring the operation of the sterilizer is to use a biological indicator. This is the best method for assuring the adequacy of the sterilization procedure because the method actually monitors the death of heat-resistant endospores. Typically spores of *Bacillus staerothermophilus* are used for monitoring the effectiveness of steam sterilizers. The indicator consists of a strip of filter paper impregnated with spores of this bacterium. After exposure of the test strip to the sterilization procedure, a nutrient solution is added to the spore strip. After incubation for 24 hours the tube containing the spore strip and nutrient solution is examined for growth. The lack of growth indicates successful sterilization; the occurrence of any growth indicates a failure of the sterilization system.

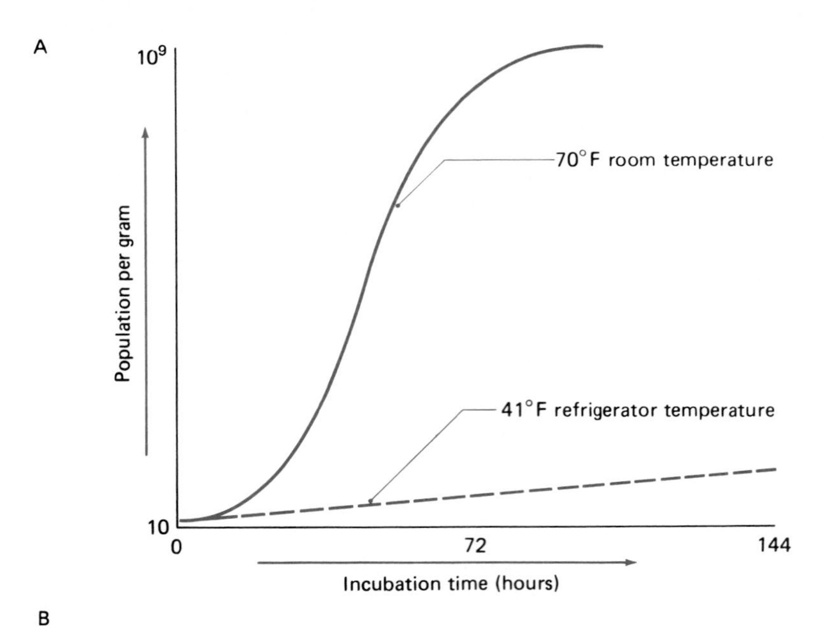

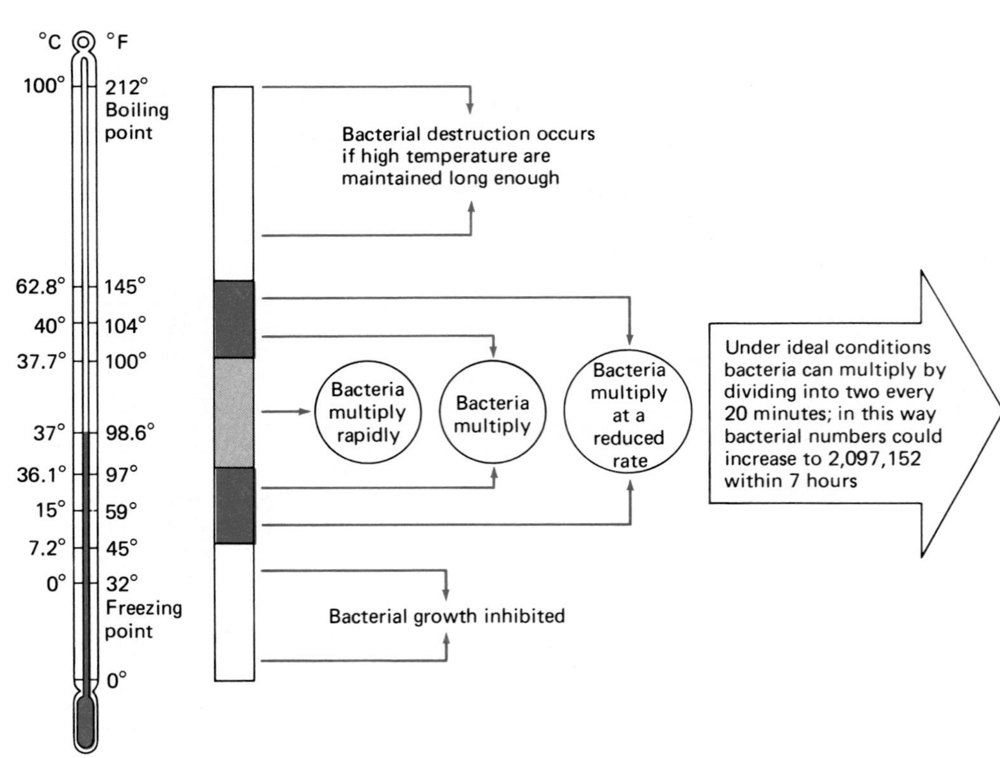

Figure 16.8 Temperature–growth relationships of microorganisms. (A) Diagram shows that the growth rate of microorganisms is much lower at refrigerator than at room temperatures. (B) Shows that microorganisms grow over a wide range of temperatures. Knowing the temperature growth ranges for organisms, such as those capable of causing spoilage or disease, is important to selecting the proper temperature for the storage of different foods. The safe and growth supportive temperatures for bacteria are indicated.

Low Temperatures

Whereas exposure to high temperatures accelerates the death rate, and thus can be used to kill microorganisms, low temperatures limit the rates of microbial reproduction, and thus can be used to prevent or to limit microbial growth. Refrigeration and freezing are widely used to restrict microbial growth. Samples collected in hospital wards for microbiological analysis sometimes are chilled to prevent microbial growth prior to reaching the clinical microbiologist. Many foods are kept in the refrigerator or freezer to prevent microbial spoilage. Most mesophilic microorganisms are unable to grow at refrigerator temperatures (5°C). However, though most pathogenic microorganisms are unable to grow in refrigerated foods, *Clostridium botulinum* type E will grow and produce toxin (Figure 16.8). Psychrophilic and psychrotrophic microorganisms are also able to grow slowly at 5°C, and thus, although refrigeration extends the shelf life of the product, it does not do so indefinitely.

Freezing at temperatures of −20°C or lower precludes microbial growth entirely. Not all food products can be preserved by freezing because of the damage that may occur to the food as a result of ice formation. Desiccation of frozen foods (freezer burn), although not a microbial spoilage process, causes serious quality defects. Freezing does not kill most microorganisms, although some microbial death may occur during freezing and during thawing as a result of ice crystal damage to microbial membranes. In fact, freezing at extremely low temperatures is routinely used for preserving microorganisms in type culture collections. Therefore, when food is thawed, the microorganisms associated with that food can grow, leading to food spoilage and potential accumulation of microbial pathogens and toxins if the food is not promptly prepared or consumed. Once frozen, it is generally not advisable to thaw and refreeze food products. Refreezing of food products also alters the texture, flavor, and nutritional qualities of the food; and, even though these changes are not related to microbial spoilage, the changes lower the acceptability of the food. Freezing, thawing, and refreezing generally disrupts the texture of the food and in addition permits invasion of the food by microorganisms that are normally restricted to the food surfaces. When thawed a second time, refrozen food products are more prone to microbial spoilage than are foods that are only allowed to thaw once. This problem can be avoided if proper care is taken during handling to prevent excessive microbial contamination and growth.

Desiccation

In addition to freezing, many foods are preserved by **desiccation**. Water is required for microbial growth. Although lack of available water prevents microbial growth, it does not necessarily accelerate the death rate of microorganisms. Some microorganisms, therefore, can be preserved by drying. One can readily purchase active dried yeast for baking purposes and after the addition of water, the yeasts begin to carry out active metabolism. Freeze-drying or lyophilization is a common means of removing water that can be used for preserving microbial cultures (Figure 16.9). During freeze-drying, water is removed by sublimation. This process generally eliminates damage to microbial cells from the expansion of ice crystals.

Whereas some microorganisms are relatively resistant to drying, other microorganisms are unable to survive desiccating conditions for even a short period of time. For example, *Treponema pallidum*, the bacterium that causes syphilis, is extremely sensitive to drying and dies almost instantly in the air or on a dry surface.

The fact that microorganisms are unable to grow at low water activities can be used for the preservation of many products. Salting was one of the early means of preserving foods and is still employed today. By adding high concentrations of salt, the A_w is lowered sufficiently to prevent the growth of most microorganisms.

Canvas and other textiles are preserved in temperate zones by the lack of water in the air, but in tropical zones these same materials are subject to biodeterioration because the humidity is sufficiently high to permit microbial growth. Exposed wood surfaces are often painted to keep the wood dry enough to preclude microbial growth. Many food products are also preserved by drying. This preservation method depends on maintaining the product in a dry state, and exposure to high humidity can negate the factor limiting microbial growth and promote microbial spoilage of food products preserved in this manner. If food can be maintained at an A_w value of 0.65 or less, spoilage is unlikely for several years. Products preserved by drying include fruits, vegetables, eggs, cereals, grains, meat, and milk.

Ionizing Radiation

High-energy, short wavelength radiation disrupts DNA molecules, and exposure to short wavelength

A

B

C

Figure 16.9 (A) This freeze-dryer has a 6-liter capacity and is mobile. (Courtesy Virtis Co.) (B) This photograph shows the storage of freeze-dried cultures in a culture collection. (Courtesy American Type Culture Collection.) (C) This commercial freeze-dryer has a 500-pound condensate capacity. Such commercial units are used in the production of freeze-dried coffee and other freeze-dried foods. (Courtesy Virtis Co.)

radiations may cause mutations, many of which are lethal. Exposure to gamma radiation (short wavelengths of 10^{-3}–10^{-1} nanometers), X ray (wavelengths of 10^{-3}–10^{2} nanometers), and ultraviolet radiation (ultraviolet light with wavelengths of 100–400 nanometers) increases the death rate of microorganisms and is used in various sterilization procedures to kill microorganisms. Viruses as well as other microorganisms are inactivated by exposure to ionizing radiation (Figure 16.10).

Sensitivities to ionizing radiation vary (Table 16.4). Resistance to ionizing radiation is based on the biochemical constituents of a given microorganism. Nonreproducing (dormant) stages of microorganisms tend to be more resistant to radiation than are growing organisms. For example, endospores are more resistant than are the vegetative cells of many bacterial species. Exposure to 0.3–0.4 Mrads (million units of radiation) is necessary to cause a tenfold reduction in the number of viable bacterial endospores. An exception is the bacterium *Micrococcus radiodurans*, which is particularly resistant to exposure to ionizing radiation. Vegetative cells of *M. radiodurans* tolerate as much as 1 Mrad of exposure to ionizing radiation with no reduction in viable count. It appears that efficient DNA repair mechanisms are responsible for the high degree of resistance to radiation exhibited by this bacterium.

Ionizing radiation is used to pasteurize or sterilize some products. Some commercially produced plastic petri plates are sterilized by exposure to gamma radiation. Most sterilization procedures involving exposure to radiation employ gamma radiation from cobalt-60 or cesium-137. Bacon, for example, can be sterilized by **radappertization**, a process of sterilizing foods by exposure to radiation, using radiation doses of 4.5–5.6 Mrads. **Radurization**, functionally equivalent to pasteurization, is used to kill nonspore-forming human pathogens that may be present in food. Radurization can be used to increase the shelf life of seafoods, vegetables, and fruits (Table 16.5).

Unlike gamma radiation, ultraviolet light does not have high penetrating power and is useful for killing microorganisms only on or near the surface of clear solutions (Figure 16.11). The strongest germicidal wavelength of 260 nanometers coincides with the absorption maxima of DNA, suggesting that the principle mechanism by which ultraviolet light exerts its lethal effect is through the disruption of

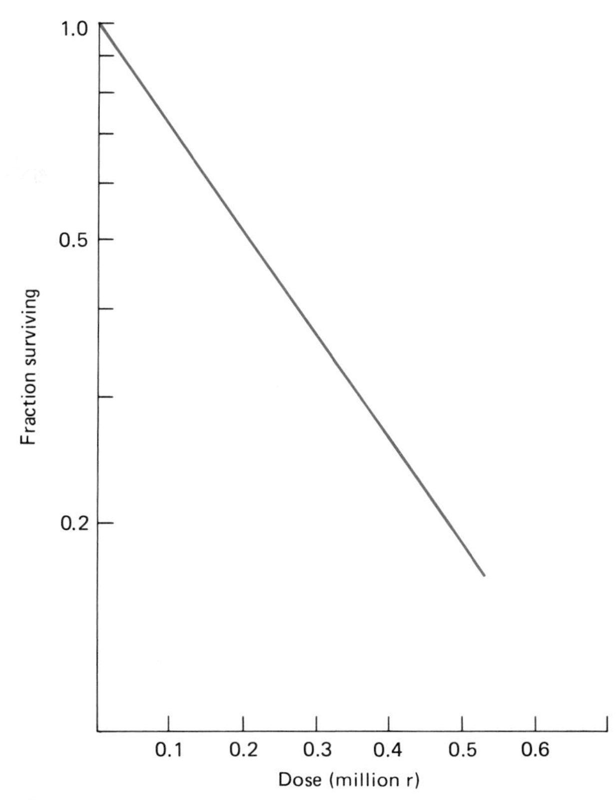

Figure 16.10 Viral inactivation by exposure to ionizing radiation. The graph summarizes the results of three experiments exposing tobacco mosaic virus to X rays. (Based on W. Ginoza and A. Norman, 1957, *Nature*, 179:520–521.)

Table 16.4 Radiation Tolerances for Various Microorganisms

Organism	Dose (Mrad)
Bacteria	
Clostridium botulinum (type E)	1.5
Enterobacter aerogenes	0.16
Escherichia coli	0.18
Micrococcus radiodurans	6.0
Mycobacterium tuberculosis	0.14
Salmonella typhimurium	0.33
Staphylococcus aureus	0.35
Streptococcus faecalis	0.38
Viruses	
Polio virus	3.8
Vaccinia virus	2.5

Table 16.5 Potential Useful Radiation Dosages for Extending the Shelf Life of Food Products

Product	Dose (krads)	Shelf life (days)
Fishery Products		
Atlantic haddock fillets	100–250	30–37
Fresh shrimp	100–200	21–28
Pacific cod fillets	100	16–18
Pacific oysters	100	31
King crab meat	100	21
Meats and Poultry		
Fresh meat and poultry	50–100	21
Fruits		
Cherries	250	14–20
Oranges	200	90
Peaches, nectarines	200	14
Pineapples	300	14
Strawberries	200	14–18

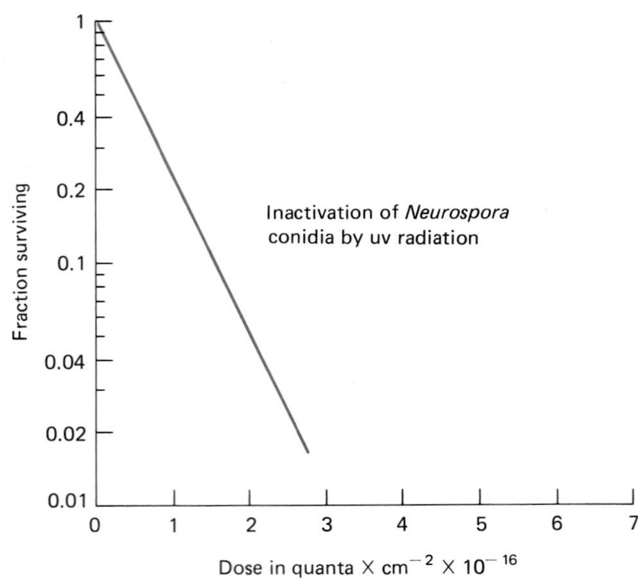

Figure 16.11 The effect of ultraviolet (UV) radiation on *Neurospora* conidia. Conidia with higher numbers of nuclei are less susceptible to UV inactivation, probably because UV acts by interference with DNA replication and the presence of redundant DNA information prevents lethal damage. (Based on A. Norman, 1954, *Journal of Cellular Comparative Physiology,* 44:1–10.)

the DNA. In fact, ultraviolet light causes the formation of covalently linked thymine dimers within the DNA in place of the normal thymine-adenine hydrogen bonded base pairs. Microorganisms have enzymes that can repair the alterations in the DNA caused by exposure to ultraviolet light. The photoreactivation enzymes require exposure to light in the visible spectrum. Exposure to ultraviolet light sometimes is used to maintain the sterility of some surfaces. In some hospitals benchtops are maintained bacteria-free when not in use by using an ultraviolet lamp. The dangers involved in human exposure to excess ultraviolet radiation, which include blindness if an ultraviolet light is viewed directly, have led to the use of alternative methods for maintaining the sterility of such areas.

Like ultraviolet radiation, long wavelength infrared radiation (103–105 nanometers) and microwave radiations (wavelenghts greater than 106 nanometers) have poor penetrating power. Infrared and microwave radiations do not appear to kill microorganisms directly. Absorption of such long wavelength radiation, however, results in increased temperature. Exposure to infrared or microwave radiations can thus indirectly kill microorganisms by exposing them to temperatures that are higher than their maximal growth temperatures. Because microwaves generally do not kill microorganisms directly, there is some concern in the food industry that cooking with microwave ovens may not adequately kill microorganisms contaminating food products.

Summary

A number of factors can be adjusted to increase microbial death rates and kill microorganisms. These include exposure to high heat and high irradiation levels. The modification of environmental parameters to control the rates of microbial growth and death is applied in many areas of microbiology and many of these applications will be discussed in later chapters. Understanding the environmental factors controlling microbial growth gives insight into the natural distribution of microorganisms in nature and provides the basis for developing methods to control microbial growth.

High temperatures are used to kill microorganisms in a variety of pasteurization and heat sterilization processes. The use of elevated temperatures aims at reducing or totally eliminating viable microorganisms. The decimal reduction time (*D* value) describes the rate at which microorganisms are killed at a particular temperature. The *D* value is used in designing processes for the preservation of foods. Ionizing radiation is also used for killing microorganisms. In contrast to factors that kill microorganisms, some factors simply prevent microbial growth. Low temperature (refrigeration and freezing) and lack of water (desiccation) prevent microbial growth but without necessarily killing microorganisms.

Various food preservation methods are employed to increase the shelf life of numerous food products. Several modern preservation methods, including canning and freezing, can extend the shelf life of the product indefinitely. Increasing the shelf life of products is extremely important because much of the world's population now lives in urban centers, requiring increased storage and transport, and because many foods are grown seasonally but desired year-round. The shipping and distribution of food to urban population centers, which are distant from the agricultural areas of food production, necessitates increased storage times for foods, compared to agricultural societies where the food can be consumed quickly after harvesting. Proper preservation of foods is important for public health reasons as well as for saving needed resources for meeting human nutritional needs.

Study Outline

A. Physical removal of microorganisms.

  1. Control of microbial populations can be effected by physically removing and thus preventing microorganisms from coming into contact with objects.

  2. Disease transmission can be interrupted by avoiding physical contact with pathogenic microorganisms.

    a. Quarantine removes potential carriers of disease from the general population.

    b. Washing food and hands removes substances that may contain pathogens.

    c. Filtration can be used to sterilize pharmaceuticals and to remove microorganisms from the air in hospital wards and operating rooms.

  3. Food preservation methods attempt to eliminate microorganisms from food and/or limit their growth rates.

    a. Sanitary methods minimize microbial contamination of food.

    b. Packaging of food products is important in preventing contamination during transport and storage.

    c. Filtration can be used to sterilize liquids.

B. High temperatures can be used to kill all microorganisms in a sample (heat sterilization).

  1. Decimal reduction time (D) describes the heat killing of microorganisms; it is defined as the time required for tenfold reduction in the number of viable cells at a given temperature.

    a. The D value is shortened as the temperature is raised above maximal.

    b. The D value for heat-resistant, endospore-producing microorganisms is used for designing processes in the canning industry.

  2. Pasteurization uses relatively brief exposures to moderately high temperatures to reduce the numbers of viable microorganisms and to eliminate human pathogens.

  3. Sterilization uses high temperatures to kill all viable microorganisms.

    a. Autoclaving, using steam under pressure, is routinely used to sterilize materials; the normal process involves exposure at 121°C (15 pounds per square inch pressure) for 15 minutes.

    b. Dry heat sterilization procedures require higher temperatures and/or longer times than steam sterilization.

C. Low temperatures limit the rates of microbial reproduction and can be used to prevent or limit microbial growth.

  1. Freezing at -20°C or lower precludes microbial activity entirely.

  2. Refrigeration at 5°C limits the rate of microbial growth.

D. Desiccation removes the water required for microbial growth; it does not necessarily increase the death rate.

  1. Some microorganisms can be preserved by drying, for example, freeze-dried yeasts; in freeze-drying, water is removed by sublimation.

  2. The fact that microorganisms are unable to grow at low water activities can be used in the preservation of foods.

    a. Adding salt lowers the A_w, limiting microbial growth and increasing the shelf life of some foods.

    b. Drying is used to preserve many foods, such as powdered milk, dried fruits, cereals, and so on.

E. Exposure to certain forms of radiation is another way of killing microorganisms.

  1. Gamma radiation and X radiation have high penetrating power and kill microorganisms.

  2. Ultraviolet light does not have high penetrating power; it is useful for killing microorganisms on surfaces.

  3. Infrared and microwave radiations have poor penetrating power and do not appear to kill microorganisms directly; the absorption of long wavelength radiation results in increased temperature, killing microorganisms.

Study Questions

1. What are the two principal goals of food preservation?
2. What precautions to prevent infection can be taken when coming into contact with an infected individual?
3. How can viable microorganisms be totally eliminated from a liquid product without using elevated temperatures?
4. What is a laminar flow hood?
5. For each of the following products describe the principle physical method(s) of preservation.
   - **a.** Condensed milk
   - **b.** Pasteurized milk
   - **c.** Hamburger meat
   - **d.** Potato chips
   - **e.** Canned peas
   - **f.** Breakfast cereals
   - **g.** Vitamins
6. List five types of radiation that can kill microorganisms.
7. True or false
   - **a.** The higher the temperature above maximal for growth, the higher the death rate will be.
   - **b.** The decimal reduction time is lengthened as the temperature is raised above maximal.
   - **c.** Lack of available water always accelerates the death rate.
   - **d.** The decimal reduction time is the time required for a tenfold reduction in the number of viable cells at a given temperature.
   - **e.** *D* value equals death rate.
   - **f.** Ionizing radiation is a method of food preservation.
   - **g.** Pasteurization uses long periods of exposure at high temperatures to ensure food safety.
   - **h.** All food products can be sterilized to prevent microbial spoilage.
   - **i.** The time and temperature required for processing canned foods at home is determined by the acidity of the food.
   - **j.** Freeze-drying can be used to preserve microbial cultures.

Some Thought Questions

1. Should bacteria fear a nuclear holocaust?
2. When you shower, do your bacteria get clean?
3. To paraphrase Shakespeare: Can these bacteria ever be washed out?
4. Why is eating forbidden in clinical microbiology laboratories?
5. Why isn't it necessary to refrigerate the new sterile milk?
6. Why are different processing times used for heat sterilization of different foods?
7. How could you use filters to separate a mixture of fungi, bacteria, and viruses?
8. Why should ultraviolet light germicidal lamps be used with caution?
9. Why does it take longer to sterilize glassware in an oven than it does to sterilize media in an autoclave?
10. Why do we paint wooden houses?

Additional Sources of Information

Block, S. S. (ed.). 1977. *Disinfection, Sterilization, and Preservation*. Lea & Febiger, Philadelphia. An extensive work covering various methods for killing and preventing the growth of microorganisms.

Castle, M. 1980. *Hospital Infection Control*. John Wiley & Sons, Inc., New York. A practical discussion of the maintenance of sterility and asepsis in the hospital environment.

Desrosier, N. W. 1977. *The Technology of Food Preservation*. AVI Publishing Company, Westport, Connecticut. A text describing the principles of preserving foods from microbial spoilage.

Frazier, W. C., and D. C. Westhoff. 1978. *Food Microbiology*. McGraw-Hill Book Co., New York. A basic text describing all aspects of food microbiology including the physical means of preventing microbial spoilage.

Phillips, G. B., and W. S. Miller (eds.). 1973. *Industrial Sterilization*. Duke University Press, Durham, North Carolina. Discussions of the methods used in industrial processes for achieving sterility.

Russell, A. 1982. *Destruction of Bacterial Spores*. Academic Press, Inc., Orlando, Florida. Factors affecting the activity of sporicidal agents together with their possible practical uses are the main focus of this book.

Stumbo, C. R. 1973. *Thermobacteriology in Food Processing*. Academic Press, Inc., New York. A discussion of the use of elevated temperatures for controlling microorganisms in foods.

17 *Chemical Control of Microorganisms*

KEY TERMS

algicides
antimicrobial agents
antiseptics
bactericidal
bacteriostatic
biocides
chlorination
detergents
disinfectants
fungicides
germicides
microbicidal
microbiostatic
oligodynamic action
ozonation
phenol coefficient
phenolics
preservatives
quaternary ammonium compounds
quats
sanitizers
spectrum of action
sporicidal
sterilizing agents
toxicity index
use-dilution method
virucides

OBJECTIVES

After reading this chapter you should be able to

1. Define the key terms.
2. Describe the types of preservatives that are used and for each give examples of the products that it is used to preserve.
3. Describe the test procedures that are used to determine the effectiveness of disinfectants.
4. Discuss the types of disinfectants, their advantages, limitations, and applications.
5. Describe the test procedures that are used to determine the effectiveness of antiseptics.
6. Discuss the types of antiseptics, their advantages, limitations, and applications.

Control of Microbial Growth by Antimicrobial Agents

Chemical inhibitors are widely used to prevent the spread of disease-causing microorganisms and to preclude the growth of microbes that would cause spoilage of foods or biodeterioration of industrial products. Such chemicals that kill microorganisms or prevent the growth of microorganisms are called antimicrobial agents.

The Antiseptic Baby and the Prophylactic Pup
Were playing in the garden when the Bunny gamboled up;
They looked upon the Creature with a loathing undisguised;—
It wasn't Disinfected and it wasn't Sterilized.

They said it was a Microbe and a Hotbed of Disease;
They steamed it in a vapor of a thousand-odd degrees;
They froze it in a freezer that was cold as Banished Hope
And washed it in permanganate with carbolated soap.

In sulphureted hydrogen they steeped its wiggly ears;
They trimmed its frisky whiskers with a pair of hard-boiled shears;
The donned their rubber mittens and they took it by the hand;
And 'lected it a member of the Fumigated Band.

There's not a Micrococcus in the garden where they play;
They bathe in pure iodoform a dozen times a day;
And each imbibes his rations from a Hygienic Cup—
The Bunny and the Baby and the Prophylactic Pup.

—ARTHUR GUITERMAN, *Strictly Germ-Proof*

There are many different types of antimicrobial agents employed for the control of microbial growth (Table 17.1). Microorganisms vary in their sensitivity to particular antimicrobial agents. Generally, growing microorganisms are more sensitive than dormant stages, such as spores. Many antimicrobial agents are aimed at blocking active metabolism and preventing the organism from generating the macromolecular constituents needed for reproduction. Because resting stages are metabolically dormant and are not reproducing, they are not affected by such antimicrobial agents. Similarly, viruses are more resistant than other microorganisms to antimicrobial agents because they are metabolically dormant outside host cells.

Terminology Used to Describe Antimicrobial Agents

The terminology used to describe antimicrobial agents is designed to indicate the intended usage of that product. Antimicrobial agents are classified according to their application and spectrum of action. Biocides are agents that kill living organisms. Germicides are chemical agents that specifically kill microorganisms. Such chemicals may exhibit selective toxicity and, depending on their spectrum of action, may act as virucides (killing viruses), bactericides (killing bacteria), algicides (killing algae), or fungicides (killing fungi). The suffix -cidal is generally added to indicate that the particular agent kills microorganisms. A sporocidal agent, for example, kills bacterial spores.

The term disinfectant refers to antimicrobial substances that kill microorganisms. Specifically its use should refer to agents that are applied to inanimate objects. Under appropriate conditions, a disinfectant may produce complete sterilization, that is, disinfectants may kill all microorganisms, in which case they are termed sterilizing agents. A sanitizer represents a particular kind of disinfectant. Sanitizing agents are used to reduce numbers of bacteria to levels judged safe by public health requirements. Antiseptics kill or prevent the growth of microorganisms. The term antiseptic literally means a substance that opposes sepsis, that is, a substance that works against putrefaction or decay. The proper use of this term is restricted to agents that are applied to living tissue.

Whereas germicides are microbicidal and kill microorganisms, microbiostatic agents inhibit the growth of microorganisms but do not kill them (Figure 17.1). The suffix -static is added to indicate that a substance inhibits microbial growth but does not kill microbes. For example, a bacteriostatic agent will prevent the growth of bacteria but will not lower the number of bacteria. When a microbiostatic agent is removed, microorganisms resume their growth. Therefore, to prevent growth it is necessary to maintain an adequate concentration of the microbiostatic agent. Microbiostatic agents are frequently used to prevent the growth of microorganisms un-

Table 17.1 Some Common Antimicrobial Agents

Compound	Concentration[a]	Application
Acids		
Acetic acid	1%	Food preservative
Benzoic acid	0.01%	Food preservative
Lactic acid	—	Food preservative
Propionic acid	0.1%	Food preservative
Sorbic acid	0.01–0.1%	Food preservative
Sulfurous acid (added as SO_2)	0.05%	Food preservative
Alcohols		
Ethanol	75%	Topical antiseptic
Lysol	2%	Disinfectant
Dichlorophene	1%	Fungal inhibitor
Hexachlorophene	3%	Topical antiseptic
Aldehydes		
Formaldehyde	5%	Preservative
Halogens		
Chlorine	0.2–0.4 ppm	Water purification
Iodine	2%	Topical antiseptic
Heavy metals		
Copper sulfate	0.8 ppm	Algal inhibitor
Copper-8-hydroxy quinolate	1%	Fungal/bacterial inhibitor
Mercuric chloride	0.1%	Disinfectant
Phenylmercuric nitrate	0.002%	Fungal inhibitor
Mercuric naphthenate	0.01%	Fungal/bacterial inhibitor
Phenylmercuric acetate	10 ppm	Inhibitor of slime-producing microorganisms

[a]% = percent; ppm = parts per million

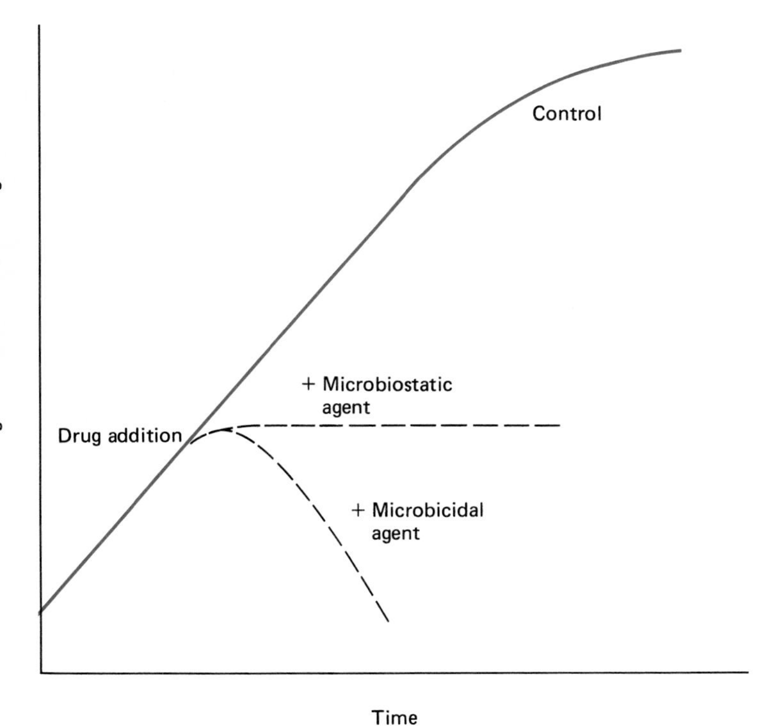

Figure 17.1 This graph shows the effects of microbicidal and microbiostatic agents on numbers of viable microorganisms.

der a given set of conditions, thereby preventing the establishment of an infection or preserving a product against spoilage or biodeterioration. The term preservative refers to microbiostatic agents that are used for these purposes.

Preservatives

The use of chemical additives is widespread and represents an important means of preserving foods and other products (Table 17.2). The choice of the chemical preservative used depends on the nature of the food and the likely spoilage microorganisms. Although there is great concern today over the addition of any chemicals to foods because of the finding that some chemicals that have been used as food additives, such as red dye number 2, are potential carcinogens, it must remembered that the effective preservation of food prevents spoilage and the transmission of food-borne diseases. In the United States the federal Food and Drug Administration is responsible for determining and certifying the safety of food additives and must approve any chemicals that are added to foods as preservatives.

Salt and Sugar

The addition of salt or sugar to a food reduces the amount of available water and alters the osmotic pressure. High salt concentrations, such as occur in saturated brine solutions, are bacteriostatic and the shrinkage of microorganisms in brine solutions can cause loss of viability. Salting is effectively used for the preservation of fish, meat, and other foods. However, because of the association of high levels of salt in the diet with high blood pressure and heart disease, there is currently great interest in lowering the salt contents of foods. Sugars, such as sucrose, also act as preservatives and are effective in preserving fruits, candies, condensed milk, and other foods. Some foods, including maple syrup and honey, are preserved naturally by their high sugar content.

Acids

Various low molecular weight carboxylic acids are inhibitors of microbial growth. Lactic, acetic, propionic, citric, benzoic, and sorbic acids or their salts are effective food preservatives. An examination of the lists of food additives in the various foods in your pantry will rapidly convince you of the wide use of organic acids as preservatives. The effectiveness of a particular organic acid depends upon the pH of the food. For example, at the same pH the citric acid is less effective than lactic acid, which in turn is less effective than acetic acid.

Propionates are primarily effective against filamentous fungi. The calcium and sodium salts of propionic acid are used as preservatives in bread, cake, and various cheeses, and because propionates are effective inhibitors of rope formation, they are added to bread dough and milk. Besides their

Table 17.2 Some Representative Chemical Food Preservatives

Preservatives	Maximum[a] Concentration	Target Organisms	Foods
Proprionic acid and propionates	0.32%	Fungi	Bread, cakes, some cheeses
Sorbic acid and sorbates	0.2%	Fungi	Cheeses, syrups, jellies, cakes
Benzoic acid and benzoates	0.1%	Fungi	Margarine, cider, relishes, soft drinks, catsup
Sulfur dioxide, sulfites, bisulfites, metabisulfites	200–300 ppm	Microorganisms	Dried fruits, grapes, molasses
Ethylene and propylene oxides	700 ppm	Fungi	Spices
Sodium diacetate	0.32%	Fungi	Bread
Sodium nitrite	200 ppm	Bacteria	Cured meats, fish
Sodium chloride	None	Microorganisms	Meats, fish
Sugar	None	Microorganisms	Preserves, jellies
Wood smoke	None	Microorganisms	Meats, fish

[a]% = percent; ppm = parts per million

MICROBIOLOGY HEADLINES

Scientists Report New Tools for Curbing Colds and Diarrhea

By PHILIP M. BOFFEY

WASHINGTON, Oct. 9—New medical tools that can help prevent the spread of common colds in Americans and the onset of devastating diarrhea in third world infants were described here today at a meeting of the American Society for Microbiology.

Dr. Elliot C. Dick, professor of preventive medicine at the University of Wisconsin at Madison, reported that among participants in a small, scientifically controlled study, facial tissues treated with virus-killing chemicals appeared to block completely the spread of rhinoviruses, which are considered the leading cause of common colds.

Dr. Dick described two sets of experiments in which 16 people who had been deliberately infected with rhinoviruses played poker for 12 hours with 24 healthy people who were initially free of the viruses.

In the first set of experiments, when no special precautions were taken to prevent colds, the 16 infected individuals passed their virus on to 14 of the 24 originally healthy individuals.

But in the second set of experiments, when the infected individuals used the special virucidal tissues to smother coughs and sneezes and wipe their noses, none of the 24 healthy individuals caught the virus.

The three-ply tissues had been treated with citric acid and malic acid, chemicals found in fruit, and with sodium lauryl sulfate, a substance often used in shampoos and toothpaste. Each square inch of the treated tissues destroyed thousands of virus particles in less than a minute, Dr. Dick said.

Dr. Dick previously used tissues treated with iodine in an apparently successful attempt to interrupt an outbreak of respiratory illness in Antarctica in 1979, but he said that that experiment had no control group to prove that the tissues were effective.

The Kimberly-Clark Corporation of Neenah, Wis., which collaborated in developing the tissues, said that the tissues, under the brand name Avert, were being put through market tests in Albany, Buffalo and Rochester to gauge consumer response.

In a related study, Dr. J. Owen Hendley, of the University of Virginia School of Medicine, reported that two-ply tissues treated with the same chemicals were "extremely effective" in preventing the spread of rhinoviruses through handshakes. Such spreading can occur if an infected individual blows his nose, shakes hands with someone else and that person touches his own nasal passages.

Dr. Hendley's experiments found that when infected volunteers used placebo tissues to blow their noses and then shook hands with healthy individuals, they passed the virus on to 3 out of 25 recipients. When they used no tissues at all, they spread the virus even more, infecting 4 out of 8 recipients. But when they used the specially treated virucidal tissues to blow their noses, they did not infect any of the 24 individuals who shook their hands.

Elsewhere at the conference, an investigator from Finland reported new clinical trials confirming the effectiveness of an oral vaccine to prevent much of the acute diarrhea that kills 4.5 million children in developing countries each year.

The vaccine is designed to combat the rotaviruses that cause 20 to 40 percent of all acute diarrhea cases in children around the world as well as half of the pediatric hospital admissions for acute diarrhea in the United States. Theoretically an effective rotavirus vaccine could prevent 1 to 2 million deaths each year, according to estimates presented here.

The vaccine under study is made by Smith Kline-RIT, a Belgian subsidiary of the SmithKline Beckman Corporation. It is derived from a calf rotavirus and stimulates the body to develop defenses against the human rotavirus.

Dr. Timo Vesikari, associate professor of pediatrics at the University of Tampere in Finland, reported that his latest trial, the largest yet conducted, had found that the vaccine was "a very promising candidate" for general use.

Only five cases of severe diarrhea, lasting more than 24 hours and requiring medical attention, occurred among 168 vaccine recipients, he said, as against 26 severe diarrhea cases among 160 children who were given a placebo. Dr. Vesikari said that the "take rate" for the vaccine could be improved by feeding the children milk at the time of vaccination to neutralize gastric acid that might destroy the vaccine. "By this simple procedure," he said, "it may be possible to induce over 90 percent protection against rotavirus diarrhea."

Source: The New York Times, Oct. 10, 1984. Reprinted by permission.

---

intentional addition to various foods, propionates form naturally during the production of Swiss cheese and act as a natural preservative.

Lactic and acetic acids also are effective preservatives that form naturally in some food products. Cheeses, pickles, and sauerkraut contain concentrations of lactic acid that normally protect the food against spoilage. Vinegar contains acetic acid, an effective inhibitor of bacterial and fungal growth. Acetic acid is used to pickle meat products and is added as a preservative to various other products, including mayonnaise and catsup. Both of these preservatives, however, will prevent surface fungal growth on a food only if molecular oxygen is excluded.

Benzoates, including sodium benzoate, methyl *p*-hydroxybenzoate (methylparaben), and propyl-*p*-hydroxybenzoate (propylparaben) are extensively used as food preservatives. Benzoates are used as preservatives in such products as fruit juices, jams, jellies, soft drinks, salad dressings, fruit salads, relishes, tomato catsup, and margarine. They are also used as preservatives in a great variety of pharmaceutical preparations.

Sorbic acid, used primarily as calcium, sodium, or potassium salts (for example, sodium sorbate) is more effective as a preservative at pH 4–6 than the benzoates. Sorbates inhibit fungi and bacteria, such as *Salmonella, Staphylococcus*, and *Steptococcus* species. Sorbates are frequently added as preservatives to cheeses, baked goods, soft drinks, fruit juices, syrups, jellies, jams, dried fruits, margarine, and various other products.

Boric acid is used as a preservative in eyewash and other products. The limited toxicity of boric acid to human tissues makes it suitable for such applications. Boric acid is also used in urine collection jars to prevent bacterial growth between the time of collection and analysis.

Nitrates and Nitrites

Nitrates and nitrites are added to cured meats to preserve the red meat color and protect against the growth of food spoilage and poisoning microorganisms. Nitrates are effective inhibitors of *Clostridium botulinum* in meat products such as bacon, ham, and sausages. Recently, however, there has been great concern over the addition of nitrates and nitrites to meats because these salts can react with secondary and tertiary amines to form nitrosamines, which are highly carcinogenic.

Disinfectants

Disinfectants are widely used to kill or prevent the growth of microorganisms. These agents are used for reducing the numbers of microorganisms on the surfaces of inanimate objects, such as floors and walls. Many household cleaning agents contain disinfectants. Disinfectants are also used to limit microbial populations within liquids; for example in swimming pool water. Disinfectants, however, are not considered safe for application to human tissues or for internal consumption. If they are used to remove microorganisms from a consumable product, such as drinking water, their concentration must be reduced before the product is consumed.

Disinfectants obviously should have high germicidal activity. They should rapidly kill a wide range of microorganisms including spores. The agent should be chemically stable and effective in the presence of organic compounds and metals. The ability to penetrate into crevices is desirable. It is essential that a disinfectant should not destroy the materials to which it is applied. Furthermore it should be inexpensive and aesthetically acceptable.

Evaluation of the Effectiveness of Disinfectants

Concentration and contact time are critical factors that determine the effectiveness of an antimicrobial agent against a particular microorganism (Figure 17.2). Several standardized test procedures have been employed for evaluating the effectiveness of disinfectants. The classic test procedure used until a few decades ago is the phenol coefficient. Phenol, which is also known as carbolic acid, was the antimicrobial chemical used by Joseph Lister for preventing infections following compound fractures and for antiseptic surgical practice. The phenol coefficient test compares the activity of a given product with the killing power of phenol under the same test conditions.

To determine the phenol coefficient, dilutions of phenol and the test product are added separately to test cultures of *Staphylococcus aureus* or *Salmonella typhi*. The tests are run in liquid culture. After exposure for 5, 10, and 15 minutes, a sample from each tube is collected and transferred to a nutrient broth medium. After incubation for 2 days the tubes from the different disinfectant dilutions are examined for visible evidence of growth. The phenol coefficient is defined as the ratio of the highest dilution of a test germicide that kills the test bacteria in 10 minutes but not in 5 minutes to the dilution of phenol that has the same killing effect. For example, if the greatest dilution of a test disinfectant producing a killing effect was 1:100, and the greatest dilution of phenol showing the same result was 1:50, the phenol coefficient would be 100/50 or 2.0.

The phenol coefficient indicates the relative toxicity of various disinfectants but does not establish the appropriate concentration that should be used for disinfecting surfaces. The Association of Official Analytical Chemists (AOAC) use-dilution method, which has replaced the phenol coefficient as the standard method for evaluating the effectiveness of disinfectants, establishes appropriate dilutions of a germicide for actual conditions. In this procedure, disinfectants are tested against *Staphylococcus aureus* strain ATCC 6538, , *Salmonella cholerasuis* strain ATCC 10708, and *Pseudomonas aeruginosa* strain ATCC 15442. Small stainless steel cylinders are con-

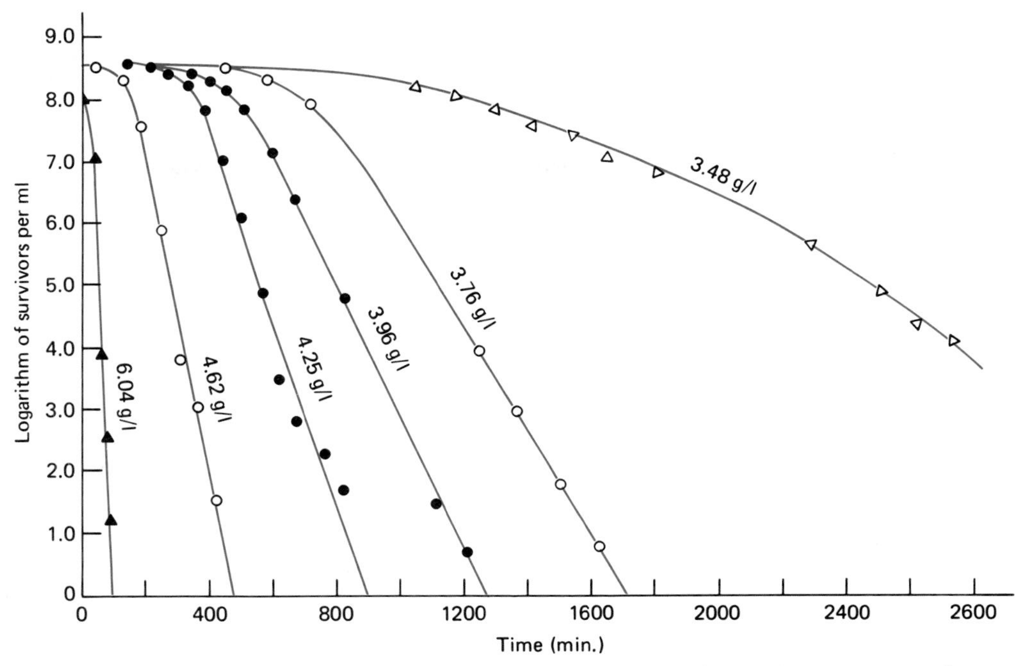

Figure 17.2 This graph charts the results of exposure of *E. coli* to various concentrations of phenol at 35°C. The number of survivors, expressed logarithmically, is plotted against time. The concentrations of phenol used were (A) 5.04 grams per liter (g/L); (B) 4.62 g/L; (C) 4.26 g/L; (D) 3.96 g/L; (E) 3.76 g/L; and (F) 3.48 g/L. Increasing the concentration of disinfectant increases the speed with which bacterial cells are killed. (Based on R. C. Jordan and S. E. Jacobs, 1945, *Journal of Hygiene,* 44:210.)

taminated with specified numbers of the test bacteria. After drying, the cylinders are placed into a series of specified dilutions of the test disinfectant. At least ten replicates of each organism at the test dilutions of the disinfectant are used. The cylinders are exposed to the disinfectant for 10 minutes, allowed to drain, transferred to appropriate culture media, and incubated for 2 days. After incubation the tubes are examined for growth of the test bacteria. No growth should occur if the disinfectant was effective at the test concentration. An acceptable use-dilution is one that kills all test organisms at least 95% of the time.

Types of Disinfectants

Chlorine. Chlorine kills microorganisms by disrupting membranes and inactivating enzymes. A variety of inorganic and organic forms of chlorine are used for disinfection purposes. The germicidal action of chlorine is based upon the formation of hypochlorous acid when it is added to water. The hypochlorous acid releases an active form of oxygen that reacts with cellular biochemicals. Chlorine gas condensed into liquid form is widely used for such disinfection. Chlorination is the standard treatment for disinfecting drinking water in most communities. It is also used to disinfect the effluents from sewage treatment plants in order to minimize the spread of pathogenic microorganisms. Because most forms of chlorine are inactivated in the presence of organic matter, this disinfection process does not completely eliminate microorganisms from the water.

The disinfection of water is very important because the same water supplies that are used as sources of drinking water often are also used for the disposal of human wastes. The outfall from one city's sewage treatment plant flows downstream to the drinking water intake of another city. When the disinfection of drinking water is inadequate, other measures, such as boiling water before use, must be used to ensure the safety of the water supply. For example, when flooding washes untreated sewage directly into the water supply, such alternative disinfection procedures become necessary. Outbreaks of diseases such as cholera and typhoid fever often occur if disinfection of drinking water supplies is not achieved.

Campers often use organic compounds containing chlorine to disinfect water that may be contaminated with fecal matter and associated human pathogens. Halazone (parasulfone dichloramidobenzoic acid) and succinchlorimide are examples of such chlorine-containing compounds. These organic chlorides, which are quite stable in tablet form, become active when placed in water. A halazone concentration of 4–8 milligrams per liter (mg/L) safely disinfects water containing *Salmonella typhi*, the bacterium that causes typhoid fever, within 30 minutes. Succinchlorimide at a concentration of 12 mg/L will disinfect water within 20 minutes. NASA uses iodine on space vehicles to treat the potable water.

In addition to drinking water supplies, chlorine is also used to disinfect swimming pools. Liquid chlorine and hypochlorite solutions are frequently used for such purposes. A residual chlorine level of 0.5 mg/L will achieve control of microbial populations and prevent the multiplication of enteric pathogens in swimming pools. Such levels of chlorine are relatively harmless to human tissues, although prolonged exposure can cause irritation to the eyes and will bleach bathing attire.

Chlorination of water in air conditioning cooling towers is important to control populations of *Legionella pneumophila*, the bacterium that causes Legionnaire's disease. Outbreaks of Legionnaire's disease have frequently been traced to aerosols released from cooling towers that are dispersed through the air. It has also been suggested that disinfection of home water heaters may be necessary to prevent the multiplication of *L. pneumophila* and its subsequent release in aerosols produced by shower heads. When excessive levels of *L. pneumophila* are detected, shock treatment with calcium hypochlorite at a dose of 50 mg/L per day will lower the concentration of this disease-causing bacterium to acceptable levels (Figure 17.3).

Figure 17.3 Graph showing reduction in counts of *Legionella* in a cooling tower as a result of chlorination. (Reprinted with permission from C. B. Fliermans *et al.* 1981. "Continuous high levels of *Legionella*," *Water Research* **16**:903–906, © 1981, Pergamon Press, Ltd.)

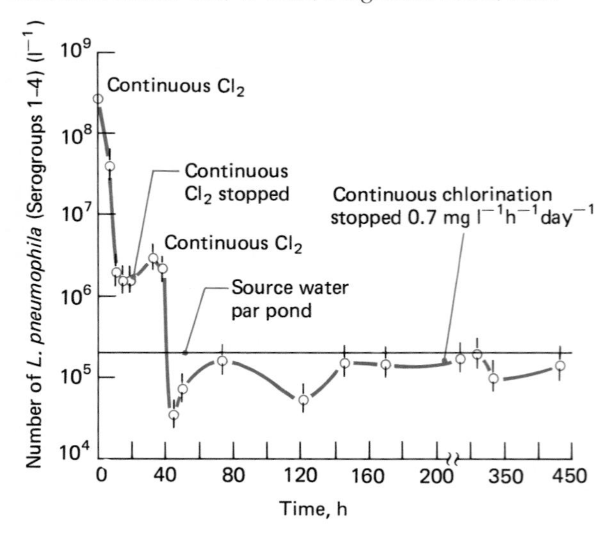

Hypochlorite solutions are also commonly used in other disinfecting and deodorizing procedures. Sodium hypochlorite (Chlorox bleach) is widely used as a household disinfectant. Food processing plants and restaurants also use calcium and sodium hypochlorite solutions to disinfect utensils. In some hospitals hypochlorite is used to disinfect rooms, surfaces, and nonsurgical instruments.

Phenolics. Phenolics are more commonly used than chorinated compounds for the disinfection of hospital floors and walls. Phenol (carbolic acid) is probably the oldest recognized disinfectant; its use as a germicide in operating rooms was introduced by Joseph Lister in 1867. Phenol and its chemical derivatives (phenolics) disrupt cytoplasmic membranes, inactivate enzymes, and denature proteins, thereby exerting antimicrobial activities. Phenolics are particularly useful as disinfectants because they are very stable when heated or dried and retain activity in the presence of organic material.

Cresol is one of the phenolics that is a good disinfectant. Cresols are derived from coal tars. Although only moderately effective against bacterial spores and disruptive to human tissues, cresols are useful for disinfecting inanimate objects. The active ingredient in Lysol, a commonly used household disinfectant, is the cresol *o*-phenylphenol. In hospitals the commercial phenolics One-Stroke Vesthene and Wexide are used to disinfect wards and operating theaters. The distinctive aroma of these phenolics gives many hospitals their characteristic smell. In addition to their use for disinfecting floors and walls, phenolics are incorporated into telephone poles, railroad ties, and other wood products to prevent microbial deterioration of the wood.

Detergents. Detergents, particularly quaternary ammonium compounds, also are effective disinfectants used for removing microorganisms from floors and walls. Detergent molecules have one end that is hydrophilic and mixes well with water and another end that is hydrophobic and is attracted to nonpolar organic molecules. If the detergents are electrically charged, they are termed ionic. The negatively charged anionic detergents are only mildly bactericidal. Anionic detergents are used as laundry detergents to remove soil and debris, thereby reducing numbers of microorganisms associated with the product being washed. In contrast to the limited antimicrobial activities of anionic detergents, the positively charged cationic detergents are highly bactericidal. In particular cationic detergents are effective against *Staphylococcus* and various viruses, making them excellent candidates for disinfecting agents for hospital use.

The most widely used cationic detergents are quaternary ammonium compounds (quats). These compounds have four organic groups bonded to a nitrogen atom. Some examples of commonly used quats are Ceepryn (cetylpyridinium chloride), Phemerol (benzethonium chloride), and Zephiran (benzylalkonium chloride). Their bactericidal action appears to be based upon the disruption of the functioning of cellular membranes and enzymes. The quaternary ammonium compounds that are effective antimicrobial agents are used in concentrations that are not irritating to human tissues.

Figure 17.4 Photograph of an ethylene oxide sterilizer. This unit is used to sterilize materials that cannot tolerate the high temperatures of an autoclave. (From BPS: The Surgery Center, Middleburg Heights, Ohio.)

Many mouthwash formulations contain quats, as do storage solutions for contact lenses. Many hospitals also use quats for disinfecting floors and walls.

However, there are some problems associated with the use of quats as disinfectants. Their antimicrobial activity is lowered if they are absorbed by porous or fibrous materials such as gauze bandages. Hard water, containing calcium or magnesium ions, interferes with their action. Also, they can cause metal objects to rust. More importantly, rather than killing *Pseudomonas* species, quats actually support the growth of these bacteria. As such, quats are not used in operating theaters, because of the danger that they will permit *Pseudomonas* to survive and infect surgical wounds.

Alcohols. Alcohols are among the most effective and heavily relied upon agents for sterilization and disinfection. Alcohols denature proteins and disrupt membrane structure. Methanol, ethanol, and isopropanol are commonly used for disinfection. Of the three, isopropyl alcohol has the highest bactericidal activity, and therefore is the most widely used. In practice, a solution of 70 to 80 percent alcohol in water is generally employed although isopropyl alcohol is effective in solutions of up to 99 percent. A 10-minute exposure is sufficient to kill vegetative cells but not spores. Surfaces of cabinets are frequently disinfected by wiping with alcohol. Some instruments are left in alchohol to maintain their sterility.

Aldehydes. Formaldehyde and glutaraldehyde are useful for disinfecting medical instruments. These aldehydes kill microorganisms by denaturing proteins. Instruments can be sterilized by placing them into a 20 percent solution of formaldehyde in 70 percent alcohol for 18 hours. Formaldehyde, however, leaves a residue and it is necessary to rinse the instruments before use. A solution of glutaraldehyde at pH 7.5 kills *Staphylococcus* within 5 minutes and *Mycobacterium tuberculosis* within 10 minutes, but endospores may survive for up to 12 hours. The use of glutaraldehyde is limited by its expense. It is used for sterilizing specialized medical instruments such as a bronchoscope.

Ethylene oxide. Ethylene oxide has a number of applications as a sterilizing agent. The ethylene portion of the molecule reacts with proteins and nucleic acids. A special autoclave-type sterilizer is used for ethylene oxide sterilization (Figure 17.4). Several hours of exposure to 12 percent ethylene oxide at 60°C is used for sterilization. In hospitals ethylene oxide sterilization is used to disinfect materials that cannot withstand steam sterilization. Linens, suturing material, and plastic items are often sterilized in this manner. Additionally some foods, such as nuts and spices are sterilized by exposure to ethylene oxide.

Ozone. Ozone is a strong oxidizing agent that kills microorganisms by oxidizing cellular biochemicals. In some communities ozone has replaced chlorination as the primary means of disinfecting drinking water. Unlike chlorination, ozonation leaves no residue. Economic considerations currently limit the widespread introduction of ozonation for disinfection purposes.

Heavy Metals. Microorganisms are inhibited by heavy metals such as mercury, silver, and copper. Mercuric chloride has been widely used as a disinfectant solution, but because it is inactivated by organic matter, it is not widely used anymore. Copper sulfate is effective as an algicide. This compound is frequently added to swimming pools and aquaria to control algal growth.

Antiseptics

Antiseptics are similar to disinfectants but may be applied safely to biological tissues. These substances are used for topical (surface) applications, but are not necessarily safe for consumption. A check of your pharmacy's shelves will illustrate the number of different agents and specific chemical formulations that are marketed as antiseptics.

Evalutation of the Effectiveness of Antiseptics

Two factors must be evaluated in determining the effectiveness of antiseptics: the antimicrobial activity of the agent and the lack of toxicity to living tissues. Potential antimicrobial activity can be evaluated by adding pieces of filter paper that have been saturated with the test agent to agar media that have been seeded with test bacterial cultures (Figure 17.5). The lack of bacterial growth in the vicinity of the filter paper containing the test agent indicates the antimicrobial effect of the test agent. Standardized test procedures, such as the use-dilution test, are also useful for comparing the effectiveness of different antiseptics.

A particularly useful approach for comparing antiseptics that encompasses both the antimicrobial activity and the toxicity to tissues is the generation of a toxicity index. In the tissue toxicity test germicides are tested for their ability to kill bacteria and their toxicity to chick-heart tissue cells. The toxicity index is defined as the ratio of the greatest dilution of the product that can kill the animal cells in 10 minutes to the dilution that can kill the bacterial cells in the same period of time and under

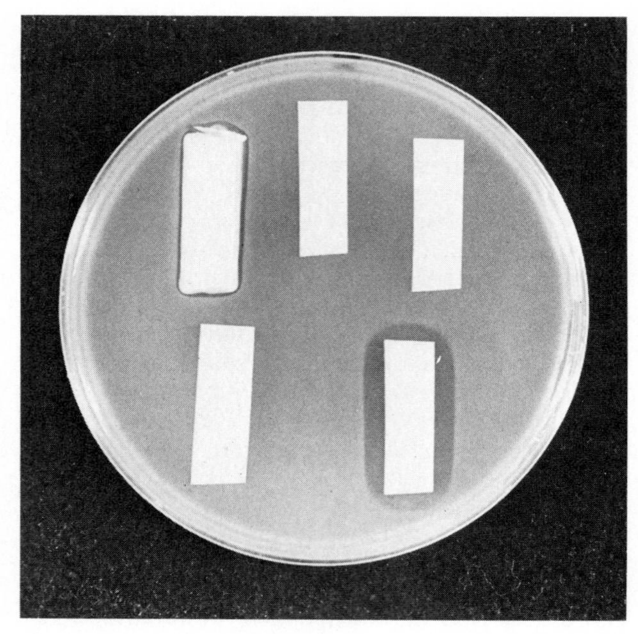

Figure 17.5 Photograph showing filter paper saturation test for evaluating the effectiveness of antiseptics. (From BPS: Richard Humbert, Stanford University.)

identical conditions. For example, if a substance is toxic to chick-heart tissue at a dilution of 1:1000 and bactericidal for *Staphylococcus aureus* at a dilution of 1:10,000, the toxicity index would be 1000/10,000 or 0.1. Typical toxicity values for tincture of iodine solution and tincture of merthiolate are 0.2 and 3.3, respectively. Ideally an antiseptic should have a toxicity index of less than 1.0, that is, it should be more toxic to bacteria than to human tissue.

Commonly used Antiseptics

Alcohols. Alcohols, such as isopropanol and ethanol, probably are the most widely used antiseptics. Alcohols are used to reduce the numbers of microorganisms on the skin surface in the area of a wound, as well as for the disinfection of various contaminated objects. Alcohol denatures proteins, extracts membrane lipids, and acts as a dehydrating agent, all of which contribute to its effectiveness as an antiseptic. Even viruses are inactivated by alcohol.

Alcohol is far more effective than soap and water in reducing the numbers of microorganisms on the skin surface (Figure 17.6). On the skin, 70 percent ethanol kills nearly 90 percent of the cutaneous bacterial population within 2 minutes. Before puncturing the skin with a hypodermic syringe, the

MICROBIOLOGY HEADLINES

The Washroom Bar Soap Isn't Squeaky Clean

By ELEANOR FLAGLER

Nothing is cleaner than soap, right? Not necessarily.

The bar soap you use to wash your hands may be loaded with disease-causing germs. That is what researchers from the medical school at Michigan State University reported recently.

They studied bar and liquid soaps in 26 public bathrooms. Of the 84 samples collected from bar soaps, all were contaminated with microorganisms.

Dr. Jon J. Kabara, a microbiologist and former dean of the medical school, said more than 100 strains of bacteria, yeast and fungi were found on bar soaps used by healthy people, including some organisms associated with serious infection and botulism.

The findings have serious implications for hospitals, nursing homes and other institutions that care for sick people. The findings also are significant for people caring for newborn babies, the aged, cancer patients, post-operative patients and others with compromised immune systems.

The study called bar soap "a potential reservoir for the spread of cross-infection in public institutions."

Kabara recommends liquid soap in disposable containers. Such soaps never became contaminated during the study, although some organisms were found on the outside of the soap dispensers.

Source: The Courier-Journal, Jan. 12, 1984. Reprinted by permission.

area is generally wiped with alcohol. Oral thermometers also are often wiped with alcohol to kill disease-causing microorganisms that otherwise might be transmitted from one patient to another. Even though brief exposure to alcohol is not sufficient to achieve sterility, it does reduce the numbers of microorganisms to levels the make infection unlikely.

Figure 17.6 This graph illustrates the comparative effectiveness of various antiseptic agents. In each test the bacterial microbiota before antiseptic application was considered as 100 percent. The residual biota immediately after use of the antiseptic is shown as a proportion of the original one. The steeper the curve, the greater the effect. (Based on P. B. Price, 1957, "Skin antisepsis," in *Lectures on Sterilization*, J. H. Brewer, ed., Duke University Press, Durham, North Carolina.)

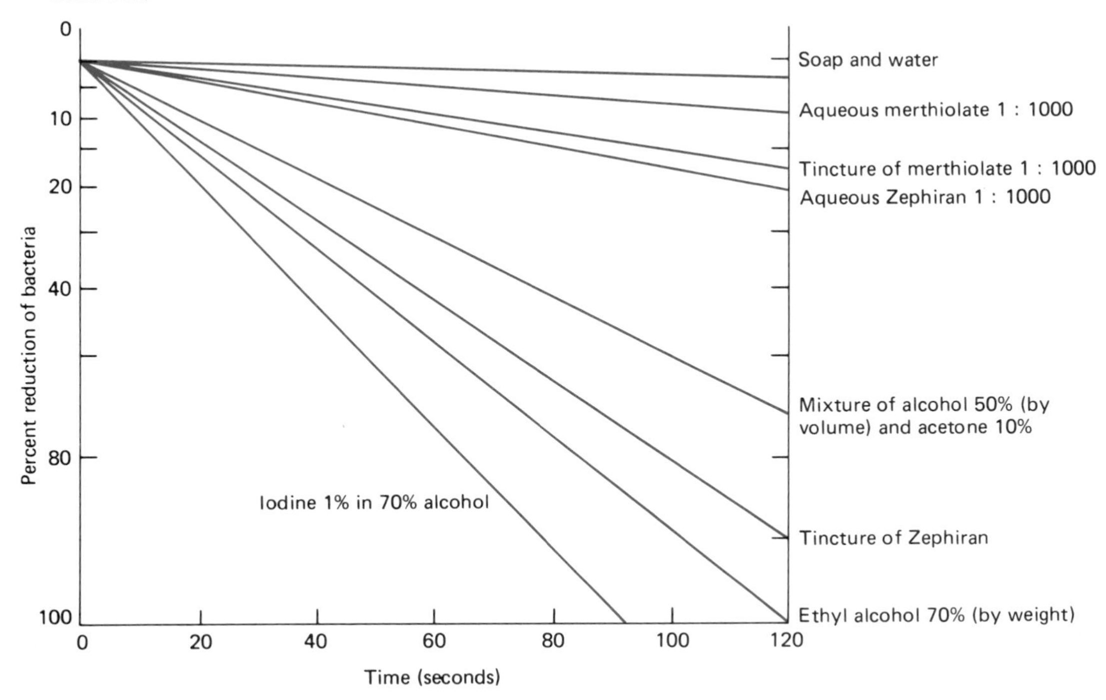

A

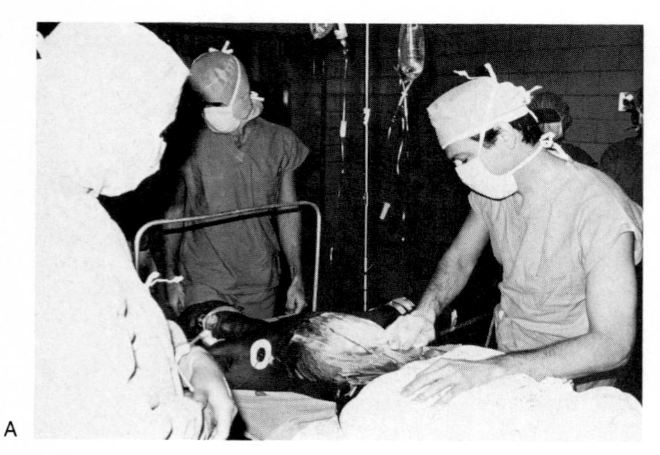

B

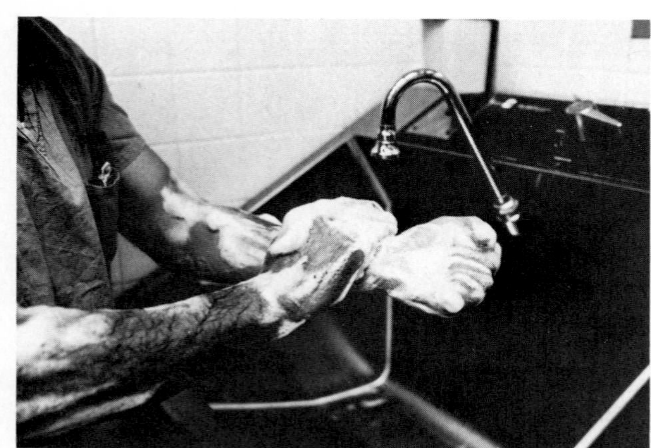

Figure 17.7 (A) Photograph showing use of iodophors (betadine) to prepare a patient for a surgical incision. (From Medichrome/Stock Shop: L. L. T. Rhodes.) (B) Photograph of surgical staff scrubbing with providine iodine, the scrub form of betadine. (From Medichrome/ Stock Shop: David York.)

Iodine. Iodine is a very effective antiseptic agent. It is bactericidal and sporicidal; that is, it kills all types of bacteria including endospores. It is also effective against many protozoa. Iodine is used in alcohol solution (tincture of iodine) and in combination with organic molecules (an iodophor). Iodine combined with polymers such as polyvinylpyrolidone is a particularly effective antiseptic because the iodine is released slowly. Iodine is frequently applied to minor wounds to kill microorganisms on the skin, thereby preventing infection of the wound. Normally it does not seriously harm human tissues, but tincture of iodine stains tissue and may cause local skin irritation and occasional allergic reactions.

Iodophors are less irritating than tincture of iodine and do not stain the skin surface. Antiseptic iodophors are used routinely for preoperative skin cleansing and disinfection. In surgical procedures, the surgical staff often scrubs with soap impregnated with iodine, and the patient's skin in the area of the incision is normally treated with iodine before beginning the surgical procedure (Figure 17.7). Betadine and Isodine are frequently used for this purpose. A standard surgical scrub with a 10 percent solution (1 percent available iodine) decreases the cutaneous bacterial population by 85 percent and is particularly effective against any Gram negative bacteria that may be present.

Phenolics and Related Compounds. Although phenol was perhaps the first antiseptic used to treat wounds, it is no longer used for this purpose because of its toxicity to human tissues. However, several phenolic compounds are widely used antiseptics. Resorcinol (*meta*-dihydroxybenzene) is only about one third as active as phenol, but it is both bactericidal and fungicidal. Resorcinol is used in the treatment of acne, ringworm, eczema, psoriasis, dermatitis, and other cutaneous lesions. It is usually applied as a 10 percent ointment or lotion. Hexylresorcinol is commonly used in mouthwashes and in over-the-counter-drugs used for treating sore throats. Thymol is used in vaginal deodorants at a concentration of 1 percent because of its antibacterial and antifungal activities.

Hexachlorophene is one of the most useful of the phenol derivatives. Combined with a soap it is a highly effective skin disinfectant. Unlike most phenolic compounds, hexachlorophene has no irritating odor and has a high residual action. Hexachlorophene is more effective against Gram positive than against Gram negative bacteria. A 3 percent solution of hexachlorophene will kill *Staphylococcus* within 30 seconds, but up to 24 hours may be required to kill Gram negative bacteria. Because most bacteria found on the skin are Gram positive, hexachlorophene once was commonly used by surgeons, physicians, and other health care workers. It was also used in the 1960s for daily bathing of newborns to prevent fatal *Staphylococcus* infections. However, it was found that extended bathing of these infants could lead to neurological damage and therefore this practice was largely discontinued. Over-the-counter preparations of hexachlorophene were banned in the United States by the Food and Drug Administration, but it is still used for limited purposes in hospitals.

The use of pHisoHex as a surgical scrub has largely been replaced by the use of providone iodine (scrub soap form of betadine) and Hibiclens (chlorhexidine gluconate). These two scrubs are effective against a wide range of microorganisms commonly encountered in hospitals. The antimicrobial action of iodine has already been discussed. Hibiclens is chemically different from other antimicrobials that are commonly used in the United States; chemically it is a biguanide. It maintains a high level of antimicrobial activity in the presence of organic matter, such as blood, and does not irritate or dry the skin.

Peroxides. Hydrogen peroxide (H_2O_2) is an effective and nontoxic antiseptic. The molecule is unstable and degrades into water and oxygen. During the decomposition of peroxides a reactive form of oxygen—the superoxide radical—is formed, which is highly toxic to microorganisms. Anaerobic bacteria are particularly sensitive to peroxides. Hydrogen peroxide concentrations of 0.3 to 6.0 percent are used as antiseptics. A 3 percent solution is often used to cleanse and disinfect wounds. Although its germicidal action is brief, the effectiveness of hydrogen peroxide against anaerobic bacteria is important because several deadly anaerobic bacteria often are associated with soils that may contaminate wounds. Higher concentrations of 6.0 to 25.0 percent can be used in sterilization. Such treatment is useful for surgical implants and contact lenses because it leaves no residual toxicity after a few minutes of exposure.

Dyes. A variety of dyes that are bacteriostatic or bactericidal are widely used as antiseptics. For example, crystal violet, which is also called gentian violet, is a potent bacteriostatic agent for Gram positive bacteria and is bactericidal at concentrations of less than 1:10,000. The mechanism of action of this compound against Gram positive bacteria appears to be very similar to that of penicillin, blocking a final step in the synthesis of cell wall material. Crystal violet has been used for the treatment of vaginal tract infections because both the protozoan *Trichomonas* and the fungus *Candida albicans*, two common etiologic agents of vaginitis, are very sensitive to this dye.

Heavy Metals. Heavy metals are also used in antiseptic formulations. Mercury, zinc, and silver are examples of heavy metals used to kill microorganisms. The inhibitory effect of heavy metals is termed oligodynamic action (Figure 17.8).

Figure 17.8 Photograph showing the oligodynamic action of heavy metals. (From BPS: Richard Humbert, Stanford University.)

Silver nitrate, for example, has classically been applied to the eyes of newborn human infants to kill possible microbial contaminants, in particular to prevent the the transmission of gonococcal infections from mother to newborn. Silver combines with proteins and disrupts bacterial surface structures. Silver nitrate and silver sulfadizine are used to treat severe burns. Preventing infections of tissues exposed by burns is critical to the recovery of the patient.

Organic mercury compounds are effective antiseptics for the treatment of minor wounds and as preservatives in serums and vaccines. The organic mercurials are bacteriostatic and relatively nontoxic. Mercurachrome (merbromin) was the first organic mercurial antiseptic to be introduced. Actually mercurachrome has limited bacteriostatic action and the lowest therapeutic index of the commercial mercurial antiseptics. Metaphen and Merthiolate are more effective.

Salts of zinc are employed as mild antiseptics. Calamine lotion for example contains zinc oxide. Calamine lotion is used in the treatment of ringworm, impetigo, and various other cutaneous diseases. White lotion, which contains zinc sulfate at a concentration of 4 percent, is also used to treat skin diseases and infections. A mixture of a long-chain fatty acid and the zinc salt of the acid is commonly used as an antifungal powder or ointment. It is particularly effective in the treatment of athlete's foot. The zinc salt also acts as an astringent and aids in healing superficial lesions, as does zinc oxide paste, which is commonly recommended for treating diaper rash and its concurrent bacterial or fungal infections.

Quaternary Ammonium Compounds. Several quaternary ammonium cationic detergents are used as antiseptic agents. These compounds are relatively nonirritating to human tissues at concentrations that are inhibitory to microorganisms. However, they act slowly and are inactivated by soaps. They are also adsorbed by cotton and other porous materials, which can severely interfere with their effectiveness as antiseptics. Zephiran, Phemerol, Cepacol and Bradosol are examples of quats that are used as antiseptics. Cepacol and Bradosol, for example, are used in throat lozenges.

Acids. Several acids are used as antiseptics. Acetic acid at a concentration of 5 percent is bactericidal and at lower concentrations is bacteriostatic. It is occasionally used at a concentration of 1 percent in surgical dressings. *Pseudomonas aeruginosa* is particularly susceptible to acetic acid, and this acid may be employed in burn therapy. It is used in vaginal douches and to suppress fungal and protozoan infections of the vaginal tract.

Undecylenic acid is active against a variety of fungi, including the fungi that cause superficial mycoses. It is usually compounded with zinc but may also

Table 17.3 Summary of Chemical Agents Used to Control Microbial Growth

Antimicrobial Agent	Description
Phenolics	Phenol is no longer used as a disinfectant or antiseptic because of its toxicity to tissues; derivatives of phenol such as *o*-phenylphenol, hexylresorcinol, and hexachlorophene are used as disinfectants and antiseptics
Halogens	Chlorination is extensively used to disinfect water—drinking water, swimming pools, and waste treatment plant effluents are disinfected by chlorination; iodine is an effective antiseptic—iodophors are used as disinfectants and antiseptics and the soaps used for surgical scrubs often contain iodophors
Alcohols	Alcohols are bactericidal and fungicidal, but are not effective against endospores and some viruses; ethanol and isopropanol are commonly used as disinfectants and antiseptics; thermometers and other instruments are disinfected with alcohol and swabbing of the skin with alcohol is used before injections
Heavy metals	Heavy metals such as silver, copper, mercury, and zinc have antimicrobial properties and are used in disinfectant and antiseptic formulations; silver nitrate was used to prevent gonococcal eye infections; mercurachrome and merthiolate are applied to skin after minor wounds; zinc is used in antifungal antiseptics; copper sulfate is used as an algicide
Dyes	Several dyes such as gentian violet inhibit microorganisms and are used as antiseptics for treating minor wounds
Surface-active agents	Soaps and detergents are used to remove microbes mechanically from the skin surface; anionic detergents (laundry powders) mechanically remove microbes; cationic detergents, which include quaternary ammonium compounds, have antimicrobial activities—quats are used as disinfectants and antiseptics
Acids and alkalies	Organic acids can control microbial growth and are frequently used as preservatives; sorbic, benzoic acid, lactic, and propionic acids are used to preserve foods and pharmaceuticals; benzoic, salicylic, and undecylenic acids are used to control fungi that cause diseases such as athlete's foot
Aldehydes	Formaldehyde is used as a preservative and disinfectant; glutaraldehyde is used to sterilize some surgical equipment
Ethylene oxide	Ethylene oxide is an excellent sterilizing agent, especially for objects that would be destroyed by heat; ethylene oxide sterilizers are used for the disinfection of plastics and linens
Oxidizing agents	Ozone is a powerful oxidizing agent; ozonation may replace chlorination for the disinfection of drinking water; hydrogen peroxide is a mild antiseptic that is effective against anaerobic bacteria

be used alone. Compound undecylenic acid contains 2–5 percent undecylenic acid and 20 percent zinc undecylenate. This antiseptic agent is very useful for the treatment of ringworm. Undecylenic acid is the active ingredient in Desenex, which is used to treat athelete's foot and other fungus infections of the skin. Benzoic acid and salicylic acid used in combination also inhibit fungal growth. Whitfield's ointment contains benzoic acid and salicylic acid in a ratio of 2:1. This ointment is used to prevent fungal growth on the feet as occurs in athlete's foot.

Summary

A number of chemicals are inhibitory to microorganisms. Chemical inhibitors are used to prevent the spread of disease-causing microorganisms and to preclude the growth of microbes that would cause the spoilage of foods or biodeterioration of industrial products. Some antimicrobial chemicals kill microorganisms (microbicidal agents) and others prevent their growth (microbiostatic agents). These agents are widely used as preservatives, disinfectants, and antiseptics. Antimicrobial agents are classified according to their application and spectrum of action: biocides are agents that kill living organisms; germicides are chemical agents that kill microorganisms; virucides kill viruses, bactericides kill bacteria, algicides kill algae and fungicides kill fungi. Table 17.3 lists the major types of antimicrobial chemicals that are used for these purposes.

Preservatives are added to foods to prevent spoilage of the food due to the action of microbial growth. Disinfectants are used on inanimate surfaces to kill or prevent the growth of nuisance microorganisms. Concentration and contact time are important criteria used in assessing the effectiveness of disinfectants. Antiseptics are used to treat minor wounds and in medical procedures where the skin is punctured to prevent infection. Antiseptics differ from disinfectants in that they are safe for application to the skin surface.

Study Outline

A. Chemical inhibitors are used to prevent the spread of disease-causing microorganisms and to preclude the growth of microbes that would cause the spoilage of foods or biodeterioration of industrial products.

B. Antimicrobial agents are classified according to their application and spectrum of action.
- **1.** Biocides are agents that kill living organisms.
- **2.** Germicides are chemical agents that kill microorganisms.
  - **a.** Virucides kill viruses.
  - **b.** Bactericides kill bacteria.
  - **c.** Algicides kill algae.
  - **d.** Fungicides kill fungi.
- **3.** Microbiostatic agents inhibit the growth of microorganisms but do not kill them.

C. The use of chemical additives is important in the preservation of foods and other products.
- **1.** The addition of salt or sugar to food reduces the amount of available water and alters the osmotic pressure, creating bacteriostatic conditions.
- **2.** Various low molecular weight carboxylic acids are inhibitors of microbial growth.
  - **a.** Propionates are effective against filamentous fungi and are used in milk and bread dough products.
  - **b.** Lactic and acetic acids are effective preservatives that form naturally in such food products as cheeses, pickles, and sauerkraut.
  - **c.** Benzoates are used as preservatives in fruit juices, jams, jellies, soft drinks, salad dressings, tomato catsup, margarine, and pharmaceutical preparations.
  - **d.** Sorbic acid is an effective preservative at pH 4–6 and inhibits fungi and bacteria.
  - **e.** Boric acid is used as a preservative in eyewash and other products.
- **3.** Nitrates and nitrites are added to cured meats to preserve the red meat color and protect against the growth of food spoilage and poisoning microorganisms.

D. Disinfectants are antimicrobial substances that kill or prevent the growth of microorganisms and are used on inanimate objects.

1. Disinfectants should

a. Have high germicidal activity.

b. Rapidly kill a wide range of microorganisms including spores.

c. Be chemically stable and effective in the presence of organic compounds and metals.

d. Be able to penetrate into crevices.

e. Not harm the surface to which it is applied.

f. Be inexpensive and aesthetically acceptable.

2. Concentration and contact time are critical factors that determine the effectiveness of an antimicrobial agent; standardized tests evaluate the effectiveness of disinfectants.

a. The phenol coefficient test compares the activity of a given product with the killing power of phenol under the same test conditions; it does not establish the appropriate concentrations for use of the product.

b. The AOAC use-dilution method is now the standard method for evaluating the effectiveness of disinfectants and also establishes the appropriate dilutions of a germicide for actual conditions; an acceptable use-dilution is one that kills all test organisms at least 95% of the time.

3. Types of disinfectants.

a. Chlorine kills microorganisms by disrupting membranes and inactivating enzymes; chlorination is the standard treatment for disinfecting drinking water and effluents from sewage treatment plants.

b. Phenolics disrupt cytoplasmic membranes, inactivate enzymes, and denature proteins; they are very stable when heated or dried and retain activity in the presence of organic material; they are in common household disinfectants and are routinely used in hospital wards and operating theaters.

c. Detergents, particularly quaternary ammonium compounds, are effective disinfectants used for removing microorganisms from floors and walls; anionic detergents are used as laundry detergents to remove soil and thus lower the numbers of associated microorganisms.

d. Alcohols are the most effective and used agents for sterilization and disinfection; they denature proteins and disrupt membrane function.

e. Aldehydes, such as formaldehyde and glutaraldehyde, are used for disinfecting and sterilizing medical instruments.

f. Ethylene oxide sterilization is used in hospitals to disinfect materials that cannot withstand steam sterilization.

g. Ozone is a strong oxidizing agent that kills microorganisms by oxidizing cellular biochemicals; ozonation can be used to disinfect drinking water.

h. Heavy metals, such as mercury, silver and copper, inhibit microorganisms.

E. Antiseptics kill or prevent the growth of microorganisms and are used for surface applications to biological tissues but are not necessarily safe for consumption.

1. In determining the effectiveness of an antiseptic, the antimicrobial activity of the agent and the lack of toxicity to living tissue must be evaluated.

a. Pieces of filter paper saturated with a test agent are added to agar media seeded with test bacterial cultures; the lack of bacterial growth in the vicinity of the filter paper containing the test agent evaluates the effectiveness of the tested antiseptic.

b. A toxicity index is the ratio of the greatest dilution of the product that can kill chick-heart animal cells in 10 minutes to the dilution that can kill bacterial cells in the same period of time and under identical conditions; antiseptics should have a toxicity index of less than 1.0.

2. Types of antiseptics.
   a. Alcohols, such as isopropanol and ethanol, denature proteins, extract membrane lipids, and act as dehydrating agents.
   b. Iodine is bactericidal and sporocidal and also effective against protozoa; in combination with polymers, it is released slowly; iodophors are used for preoperative skin cleansing and disinfection.
   c. Phenolics include resorcinol and hexachlorophene; resorcinol is both bactericidal and fungicidal and is used in the treatment of acne, ringworm and other skin infections; hexachlorophene combined with soap is a highly effective skin disinfectant and deodorant.
   d. Peroxides, in the form of hydrogen peroxide, are effective nontoxic antiseptics because they are unstable and degrade into a reactive form of oxygen that is toxic to microorganisms; it is useful with surgical implants and contact lenses.
   e. Dyes, such as crystal violet, block a final step in the synthesis of cell wall material.
   f. Heavy metals are inhibitory via oligodynamic action.
   g. Quaternary ammonium cationic detergents are used as antiseptic agents.
   h. Acids used as antiseptic agents include acetic acid and undecylenic acid.

Study Questions

1. Match the following statements with the terms in the list to the right.
   a. Agent applied to skin for treatment of minor wounds.
   b. Agent that prevents microbial growth only as long as it is present.
   c. Agent used to kill microorganisms on the walls of a hospital.
   d. Agent that sterilizes medical instruments.
   e. Suffix meaning able to kill microorganisms.
   f. Suffix meaning able to inhibit growth of microorganisms.
   g. Chemical used to prevent growth of microorganisms in foods.

   (1) disinfectant
   (2) microbiostatic agent
   (3) antiseptic
   (4) germicide
   (5) -static
   (6) -cidal

2. Match the following statements with the terms in the list to the right.
   a. Used as preservative in fruit juices, jams, and jellies.
   b. Mode of preservation based upon alteration of osmotic pressure.
   c. Fermentation product that acts as a natural preservative in cheeses.
   d. Used to prevent fungal growth in bread.
   e. Used to prevent growth of *Clostridium botulinum* in meat.

   (1) Salt
   (2) Lactic acid
   (3) Propionates
   (4) Sorbates
   (5) Benzoates
   (6) Nitrates

3. Match the following statements with the terms in the list to the right.
   a. Alternative to chlorination for disinfection of drinking water.
   b. Classical method for evaluating effectiveness of a disinfectant.
   c. Used to prevent biodeterioration of railroad ties.
   d. Used to prevent algal growth in swimming pools.
   e. Used to sterilize plastics.
   f. Modern AODC method for evaluating disinfectants.
   g. Commonly used today to disinfect floors and walls in hospitals.
   h. Used to disinfect medical instruments.
   i. Used to limit numbers of bacteria in swimming pools.

   (1) Phenol coefficient
   (2) Use-dilution test
   (3) Chlorination
   (4) Ozonation
   (5) Creosol
   (6) Quats
   (7) Alcohols
   (8) Aldehydes
   (9) Copper sulfate
   (10) Ethylene oxide

4. Match the following statements with the terms in the list to the right.
   a. Betaine soap.
   b. Hexachlorophene.
   c. Value less than one often indicates effectiveness.
   d. Crystal violet.
   e. Mercurachrome.
   f. Oligodynamic action.
   g. Zephiran and Cepacol.
   h. Form toxic superoxide radicals.

   (1) Toxicity index
   (2) Phenolics
   (3) Iodophors
   (4) Peroxides
   (5) Dyes
   (6) Heavy metals
   (7) Acids
   (8) Quats
   (9) Zone of inhibition

Some Thought Questions

1. Should there be soap bars in public restrooms?
2. Should chemical preservatives be added to foods?
3. Is food purchased from health food stores always healthy?
4. Can you kill a virus with Lysol?
5. Do alcoholics have fewer bacteria in their guts than teetoatalers?
6. Why is chicken soup good medicine?
7. Why aren't arsenic-containing compounds used to treat syphilis today?
8. Why do health care workers but not the general public still use hexachlorophene soaps?
9. What are the essential characteristics of disinfectants? Does Clorox meet these criteria?
10. Why has the phenol coefficient test been replaced by other methods for assessing the effectiveness of disinfectants?
11. How can the same oral thermometer be used for taking the temperature of numerous patients?

Additional Sources of Information

Block, S. S. 1983. *Disinfection, Sterilization, and Preservation*. Lea & Febiger, Philadelphia. A useful source of information about chemical disinfection and preservation.

Desrosier, N. W. 1977. *The Technology of Food Preservation*. AVI Publishing Co., Westport, Connecticut. This work includes a discussion of the various chemicals that are used to preserve foods.

Gilman, A. G., L. S. Goodman, and A. Gilman (eds.). 1980. *Goodman and Gilman's The Pharmacological Basis of Therapeutics*. Macmillan Publishing Company, New York. This extensive work contains a chapter that discusses the various antiseptics that are of therapeutic use.

Johnson, J. D. (ed.). 1975. *Disinfection of Water and Wastewater*. Ann Arbor Science, Ann Arbor, Michigan. This volume describes the uses of chlorination, ozonation, and other methods for disinfecting drinking water, wastewater, and swimming pool water.

Pratt, W. B. 1977. *Chemotherapy of Infection*. Oxford University Press, New York. This easy-to-read work covers the various chemicals that are used to control infectious microbes.

Russell, A. D., W. B. Hugo, and G. A. J. Ayliffe (eds.). 1982. *Principles and Practice of Disinfection, Preservation and Sterilizaton*. Blackwell Scientific Publications, Boston. A comprehensive work covering many different antimicrobial agents that are used as preservatives, disinfectants, and antiseptics.

18 *Antibiotic Control of Disease-Causing Microorganisms*

KEY TERMS

achievable serum levels
aminoglycosides
antagonism
antibiotic resistance
antibiotics
antimicrobial agents
antimicrobial susceptibility testing
antimicrobic
arsenical
autolysins
β-lactamases
Bauer–Kirby test
broad spectrum antibiotics
cephalosporins
generic drug name
intermediately sensitive (I)
minimum bactericidal concentration (MBC)
minimum inhibitory concentration (MIC)
multiple antibiotic resistance
narrow spectrum antibiotics
penicillinase
penicillins
polyenes
polymyxins
resistant (R)
schizontocidal action
Schlicter test
selective toxicity
sensitive (S)
serum killing power
synergy
tetracyclines
therapeutic value
zone of inhibition

OBJECTIVES

After reading this chapter you should be able to

1. Define the key terms.
2. Discuss why there is concern over excessive use of antibiotics in medical practice.
3. Name and describe the modes of action of the major antibiotics used to treat bacterial infections.
4. Discuss why it is more difficult to find antibiotics for treating fungal and protozoan infections than for treating bacterial infections.
5. Name and describe the modes of action of the major antibiotics used to treat fungal infections.
6. Name and describe the modes of action of the major antibiotics used to treat protozoan infections.
7. Discuss why there are no broad spectrum antiviral drugs.
8. Name and describe the modes of action of the antibiotics that are used to treat viral infections.
9. Describe the test procedures for determining the susceptibility of pathogenic microorganisms to antibiotics.
10. Discuss why the Mininimum Inhibitory Concentration and Minimum Bactericidal Concentration test procedures are replacing other procedures for determining the appropriate antibiotic for treating a disease.

Treating Diseases with Antimicrobial Agents

The control of disease-causing microorganisms is fundamental to the practice of modern medicine. Today there are a large number of antimicrobial agents available for treating diseases caused by microorganisms. Such drugs have enabled physicians to treat many disease conditions that a few years ago were frequently fatal. Many infectious diseases are relatively easily treated with antibiotics. We need only think about the number of times a physician has prescribed an antibiotic for us or our children to appreciate the wide use of antibiotics in medical practice. Many once widespread deadly diseases, such as plague, are rare today because of the effective use of antibiotics. The use of antimicrobial agents for treating infectious diseases has led to greatly increased life expectancies.

The antimicrobial agents used in medical practice are aimed at eliminating infecting microorganisms or at preventing the establishment of an infection. These chemicals should not be confused with the large number of drugs used in medical practice for alleviating the symptoms of disease or for treating diseases not caused by microorganisms. To be of therapeutic use, an antimicrobial agent must exhibit selective toxicity. To have therapeutic value, an antimicrobial agent must inhibit infecting microorganisms and exhibit greater toxicity to the infecting pathogens than to the host organism. A drug that kills the patient is of no use in treating infectious diseases, whether or not it also kills the pathogens! As a rule, antimicrobial agents are of most use in medicine when the mode of action of the antimicrobial chemicals involves biochemical features of the invading pathogens not possessed by normal host cells.

Cure the disease and kill the patient.

—FRANCIS BACON, *Essays of Friendship*

Antibiotics

Antibiotics are biochemicals produced by microorganisms that inhibit the growth of, or kill, other microorganisms (Figure 18.1). The discovery and use of antibiotics have revolutionized medical practice in the twentieth century. By their very nature, antibiotics must exhibit selective toxicity because they are produced by one microorganism and exert varying degrees of toxicity against others. The formal definition of an antibiotic distinguishes biochemicals that are produced by microorganisms from organic chemicals that are synthesized in the laboratory. This distinction is no longer meaningful because organic chemists can synthesize the biochemical structures of many naturally occurring antibiotics. Additionally, many antibiotics in current medical use are chemically modified forms of microbial biosynthetic products. Consequently the term antimicrobic is now widely used in medical fields instead of the term antibiotic. In this chapter both the terms antibiotic and antimicrobic are used to refer to both natural and semisynthetic antimicrobial agents that are available to the physician for treating infectious diseases.

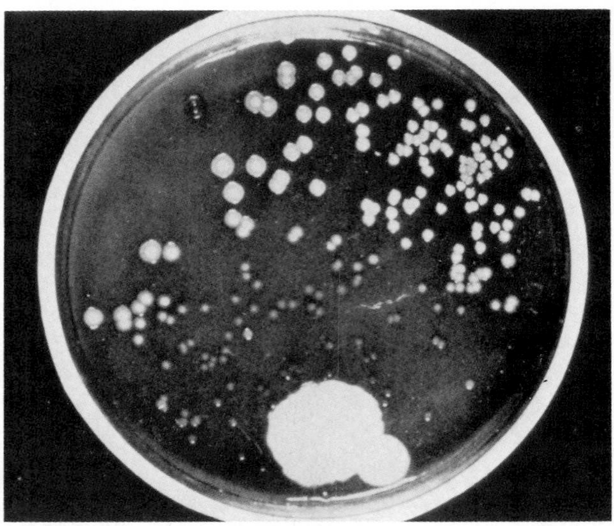

Figure 18.1 This photograph of the original culture plate of Alexander Fleming shows the lack of growth of *Staphylococcus* colonies near a colony of the antibiotic-producing fungus *Penicillium*. Colonies of *Staphylococcus* developing nearest the *Penicillium* show evidence of bacteriolysis. (From E. Chain and H. W. Florey, 1944, *Endeavor*, 3:9.)

Antibiotics Effective Against Bacterial Infections

The biochemical differences in the cell structures of bacterial cells and eukaryotic cells form the basis for the effective use of antibiotics against bacterial infections. The bacterial cell with its unique murein-containing wall and 70S ribosome has two obvious sites against which antibacterial agents may be directed. Most of the common antibiotics used in medicine for treating bacterial infections are inhibitors of cell wall or protein synthesis. Some an-

MICROBIOLOGY HEADLINES

Rush on Penicillin

Penicillin, the wonder drug of 1943, last week made headlines up & down the nation. Newspapers reported a wave of frantic appeals for the drug by blood-poisoning sufferers who suddenly learned that 1) penicillin might save their lives, 2) there was not enough to go around.

The rush began last fortnight when a distraught father phoned the city editor of the New York *Journal-American,* begged him to help get some penicillin for his baby daughter Patricia, who was dying of a staphylococcus blood infection. The city editor made the most of a journalistic opportunity, persuaded an official of WPB, which controls the minuscule U.S. supply of penicillin, to release some for Patricia. Four hours later a *Journal-American* car, with a convoy of screaming police sirens, drove up to the hospital with the drug. Next morning Patricia was much better.

Requests and Rations. It was not the first U.S. use of penicillin for a civilian (among several hundred who have received it were some victims of the Boston Coconut Grove fire), but it was by far the most publicized. Within a few hours after the nation's press had picked up Patricia's story, requests for penicillin began to pour into WPB, the Army, the White House. Only a few applicants, with diseases for which penicillin has been proved effective,* got it (one who did died nonetheless).

Most moving plea was by comely, 19-year-old Marie Barker, of Chicago, who wrote Eleanor Roosevelt: "Won't you ask the Army to send me a little of its precious medicine so that I may have a fighting chance? I am engaged to marry a fine man now serving in the U.S. Army." But Marie got no penicillin: doctors held that it could not save her because her hemolytic staphylococcus infection had affected the heart.

The man who had the unenviable job of deciding who should and who should not get penicillin was Dr. Chester S. Keefer, of Boston's Evans Hospital. He is chairman of the National Research Council's committee on chemical therapy, to which WPB turned over the rationing of the drug. Because it is especially effective in treating battle-wound infections, most of the meager supply (amount: a military secret) goes to the armed forces.

Plants and Processes. Penicillin is scarce because the only way it is being made commercially is by the extremely slow growth of a mold similar to a cheese mold. But plans for expanded manufacturing facilities and experiments on new processes hold out hope for an increase in the supply. In recent months the leading producers have increased their output more than tenfold. Last week Commercial Solvents Corp. and Winthrop Chemical Co. announced that they were building big new penicillin plants.

One hope for increased production is a new and speedier mold process developed by Stanford University's Bacteriologist Charles E. Clifton. Suggested by the method of making vinegar by dripping alcohol through wood shavings inoculated with bacteria, Clifton's laboratory experiments show that penicillin can be made by dribbling a mold-growing solution through shavings inoculated with the mold. In the present commercial process the mold grows in jars without mechanical help. Clifton's process would result in continuous production.

* No cure-all, penicillin has so far been used to treat only a limited group of infections: staphylococcus aureus (causing bone infections, cellulitis, face carbuncles, certain types of pneumonia), hemolytic streptococcus, gonorrhea.

Source: Time, Aug. 30, 1943. Reprinted by permission.

tibiotics are more selective than others with respect to the bacterial species that they can inhibit. A narrow spectrum antibiotic may be targeted at a particular pathogen, for instance at Gram positive cocci, or at a particular bacterial species. In contrast, some antibiotics have a broad spectrum of action, inhibiting a relatively wide range of bacterial species, including both Gram positive and Gram negative types. The choice of a particular antibiotic depends in part on the biochemical properties of the specific infecting bacterial strain.

In most cases, physicians make an educated guess as to which antibiotic is appropriate for treating a particular infection, and the selection of the antibiotic is based on the most likely pathogen causing the given disease symptomology and the antibiotics generally known to be effective against such pathogens. Many times a physician will select a broad spectrum antibiotic in order to ensure effective treatment. Only in special cases, such as when a patient fails to respond to a particular antibiotic and an infection persists, is an attempt usually made to isolate the pathogenic bacterium and to determine the range of specific antibiotic sensitivity of that organism.

However, concern is mounting in the medical field about the overuse of antibiotics because an undesired side effect of such "drug abuse" is the selection for disease-causing antibiotic resistant strains. It is now considered proper medical practice to perform culture and sensitivity studies to determine the proper antibiotic for treating a patient. Only in cases of life threatening infections should antibiotics be used without such testing to avoid the selective pressure for the development of antibiotic resistant pathogens. The importance of the

MICROBIOLOGY HEADLINES

Linking Drugs to the Dinner Table

Antibiotics on the farm may be playing a role in human disease

By ANASTASIA TOUFEXIS

If antibiotics have proved to be wonder drugs for medicine, they have been nothing short of miracle workers in agriculture over the past quarter-century. Today, about 15 million lbs. a year, nearly half of U.S. annual production of antibiotics, are fed to farm animals, primarily cattle, poultry and pigs. Although the drugs help check the spread of bacterial infections among closely penned animals, their use is prompted as much by a happy side effect: for reasons not yet understood, they accelerate animal growth. But the lacing of animal feed with antibiotics is being increasingly challenged by scientists who claim it is a major factor in a fast-growing medical problem: the resistance of disease-causing microbes to antibiotics.

Nearly 25% of *Salmonella* bacteria—organisms that commonly cause food poisoning—are now resistant to many antibiotics. Critics charge that routinely putting antibiotics in feed promotes bacterial resistance by wiping out the less hardy of the vast array of microbes normally present in animals, leaving those that are drug resistant to flourish. If they are transferred to humans through meat and poultry products, these bacteria could then colonize their new hosts or pass on their antibiotic resistance to other bacteria already in residence.

Definitive proof of a link between drug-laced feed and human illness is difficult to obtain. But recently epidemiologists from Minnesota, South Dakota and the federal Centers for Disease Control in Atlanta traced a serious outbreak of gastrointestinal illness that was caused by antibiotic-resistant germs. The source of the bacteria, say the investigators, was hamburger from cattle that had been fed the antibiotic chlortetracycline. Declares Dr. Scott Holmberg of the CDC, who led the disease detectives in the yearlong investigation: "We were able to show for the first time ever how an antibiotic-resistant bacterium can actually make its way from the barnyard to the dinner table."

The elegant piece of sleuthing, reported in the *New England Journal of Medicine,* began on Feb. 19, 1983, with a call to the CDC from Michael Osterholm of the Minnesota health department. In the preceding four weeks, ten people in the Minneapolis–St. Paul area had been struck by gastrointestinal illness. The malady, marked by diarrhea, stomach cramps, high fever and vomiting, was severe enough to hospitalize six victims. The culprit was *Salmonella newport,* a strain of bacteria that normally accounts for only a dozen of the salmonellosis cases in the state in a year. Within days, an investigative team had discovered an upsetting but valuable fact: the bacterium was resistant to the antibiotics ampicillin, carbenicillin and tetracycline.

The most plausible explanation for the outbreak, that there was an unusual food to blame, was quickly rejected. According to Holmberg, "These people had the all-American diet: lots of meat, potatoes and bread." Seven of the ten victims had been taking antibiotics shortly before their illness, raising the possibility that the drugs were tainted. This too was eliminated.

The big break came when Kenneth Senger, the state epidemiologist in South Dakota, reported that there had been four cases of infection with antibiotic-resistant *S. newport* in the state in three months. Interviews established that the victims lived on farms six miles apart and that they got their beef from the same nearby feed lot, which routinely added chlortetracycline to the animals' feed. The CDC traced the path of the meat shipments from the feed lot to eight supermarkets patronized by the ten Minnesota victims. All had reported eating hamburgers within a week of the time they became ill.

The report has added new intensity to the debate about antibiotic additives in livestock feed. Since 1977 the Food and Drug Administration has been proposing a ban on the addition to feed of penicillin and tetracycline, two antibiotics widely used to combat human disease. Farmers would be free to substitute antibiotics not commonly used by people. Great Britain limited the addition of some drugs to livestock feed in 1971, and other European Community countries followed in 1973. But heavy lobbying by livestock breeders and pharmaceutical companies, which supplied antibiotics worth $270.9 million to the feed industry in 1983, has blocked a proposed U.S. ban in Congress. In addition, scientists disagree about whether there is a link between human disease and animal antibiotics.

Most scientists do agree that much of the weakened impact of antibiotics can be blamed on doctors who overprescribe antibiotics, ordering them, for example, for virus-caused colds, and on people who use them indiscriminately. Veterinarian Jerry Brunton of the Animal Health Institute, a lobbying group, finds major flaws in the study: "No meat samples were available to indicate that disease-causing organisms were ever present, nor were such organisms isolated in the meat processing and preparation locations or from the farm where the alleged source animals were raised."

Indeed, the conclusions of the CDC report are "inferential," concedes epidemiologist Reuel Stallones of the University of Texas, who contributed to a 1980 report from the National Academy of Sciences that found the human health hazards of antibiotic feeds "neither proven nor disproven." But, he adds, "this is the best evidence I've seen up to this time that human illness is somehow linked to the use of antibiotics in animals for growth promotion. This study draws the net much tighter around the issue, but it is still a net, not a rope."

problem of excessive use of antibiotics was recently underscored when the American Medical Association called upon physicians to avoid unnecessary use of antibiotics.

The reason for concern about how we use antibiotics is because numerous bacterial strains have acquired the ability to resist the effects of some antibiotics, with some bacterial strains, generally those containing R plasmids, having multiple antibiotic resistance. The basis of resistance in some cases rests with the ability of the particular strain to produce enzymes that degrade the antibiotics, preventing the active form of the antibiotic from reaching the bacterial cells where they could be inhibitory. For example, some bacterial strains produce penicillinase enzymes (β-lactamases) that are able to degrade the antibiotic penicillin, making such strains resistant to penicillin. Resistance may also be due to decreased drug uptake, metabolic transformation of the drug to its nontoxic forms, decreased transformation of the drug to its active form, and/or decreased sensitivity of the microbial structure against which the drug is directed.

The penicillin-susceptible parasites that the penicillin first killed slowly or not so slowly metamorphosed into the penicillin-resistant. They mutated. The seesaw between the penicillin-sensitive and the penicillin-resistant was only a passing example of all seesaws. Already in the nose of the newborn, hovered over by the nose of the nurse, it began, snotty nose in the nurse, then snotty nose in the newborn, snotty noses in both, then snotty noses in neither, and that over and over. "Fie on't! oh fie, fie! 'T is an unweeded garden, that grows to seed."

—GUSTAV ECKSTEIN, *The Body Has a Head*

There is also concern about the cost of antibiotics prescribed by physicians. In many cases the selection of an alternate antibiotic in cases where the pathogen is resistant to the antibiotic of first choice greatly increases the cost of treating that disease. Additionally many antibiotics are produced by several companies under different trade names (Table 18.1). Many municipalities now require pharmacies to fill prescriptions by their generic names to encourage competitive pricing, unless a specific brand is specified. Such laws are controversial because the pharmaceutical manufacturers claim they will lose the incentive to develop new antibiotics and to maintain high and costly quality control when a less stringently regulated and less expensive product will be used to fill a prescription. The public wants cost effective but quality antibiotics.

Table 18.1 Generic and Trade Names for Some Common Antibiotics

Generic Name	Trade Names
ampicillin	Amcill, Omnipin, Penbritin, Polycillin
cephalexin	Keflex
cephalothin	Keflin
chloramphenicol	Chloromycetin, Mychel
chloroquine	Aralen, Avloclor, Resochin
chlorotetracycline	Aureomycin
demeclocycline	Declomycin
doxycycline	Vibramycin
erythromycin	Ilotycin
gentamicin	Garamycin
kanamycin	Kantrex
methacycline	Rondomycin
minocycline	Minocin, Vectrin
nystatin	Mycostatin, Nilstat
oxytetracycline	Teramycin
penicillin G	Crysticillin, Duracillin
tetracycline	Acromycin, Panmycin, Tetracyn, Tetrachel, Rexamycin
trimethoprim sulfamethoxazole	Bactrin, Septra

Modes of Action of Antibacterial Antibiotics

Cell-wall Inhibitors. The penicillins and cephalosporins are two widely used classes of antibiotics that inhibit the formation of bacterial cell-wall structures (Table 18.2). Penicillins are synthesized by strains of the fungus *Penicillium*. The cephalosporins are produced by members of the fungal genus *Cephalosporium*. The penicillins and cephalosporins both contain a β-lactam ring and thus have related biochemical structures (Figure 18.2).

Both the penicillins and cephalosporins inhibit the formation of peptide crosslinkages within the murein backbone of the cell wall. Bacterial cell walls lacking the normal crosslinking peptide chains are subject to attack by autolysins (autolytic enzymes produced by the organism that degrade the cell's own cell-wall structures). The result is that, in the presence of cephalosporins or penicillins, growing bacterial cells are subject to lysis because without functional cell-wall structures, the bacterial cell is not protected against osmotic shock. It should be noted that the penicillin and cephalosporin antibiotics do not themselves remove intact cell walls and thus are ineffective against resting or dormant cells.

Many of the penicillins, such as penicillin G, have a relatively narrow spectrum of activity, being most

Table 18.2 Some Diseases and Their Causative Organisms for Which Penicillins and Cephalosporins Are Recommended

Causative Organism	Disease	Drugs of Choice
Staphylococcus aureus	Abscesses, bacteremia, endocarditis, pneumonia, meningitis, osteomyelitis	Penicillin G; a penicillinase-resistant penicillin
Streptococcus pyogenes	Pharyngitis, scarlet fever, otitis media, pneumonia, bacteremia	Penicillin G
Streptococcus (*viridans* group)	Endocarditis, bacteremia	Penicillin G ± an aminoglycoside;
Streptococcus faecalis (enterococcus)	Endocarditis, urinary tract infection	Ampicillin
Streptococcus pneumoniae (pneumococcus)	Pneumonia, meningitis, endocarditis, otitis	Penicillin G
Neisseria gonorrhoeae (gonococcus)	Genital infections	Ampicillin or amoxicillin or penicillin G
Neisseria meningitidis (meningococcus)	Meningitis	Penicillin G
Clostridium perfringens	Gas gangrene	Penicillin G
Clostridium tetani	Tetanus	Penicillin G
Haemophilus influenzae	Otitis media Sinusitis, bronchitis, epiglottitis	Amoxicillin or Ampicillin
Enterobacter aerogenes	Urinary tract infection	Cephamandole
Klebsiella pneumoniae	Urinary tract infection, pneumonia	Cephalosporin
Pasteurella multocida	Wound infection, abscesses, meningitis	Penicillin G
Bacteroides	Oral disease, brain abscess, lung abscess	Penicillin G
Treponema pallidium	Syphilis	Penicillin G
Treponema pertenue	Yaws	Penicillin G
Leptospira	Weil's disease	Penicillin G

effective against Gram positive cocci, including *Staphylococcus* species. Other penicillin antibiotics, such as ampicillin, have a broader spectrum of activity, inhibiting some Gram negative as well as Gram positive bacteria. Ampicillin, an amino-substituted penicillin, is active against many Gram negative rods, including *Escherichia coli, Haemophilus influenzae, Shigella* sp., and *Proteus* sp.

Penicillin G and various other β-lactam antibiotics are subject to inactivation by penicillinase enzymes (β-lactamases). In fact, none of the broad spectrum penicillins are penicillinase resistant. These antibiotics are ineffective against penicillinase-producing bacterial strains. For example, penicillin G is normally effective against *Neisseria gonorrhoeae*, a Gram negative coccus that causes gonorrhea, but some penicillinase-producing strains of *N. gonorrhoeae* have now been found, requiring the use of antibiotics other than penicillin G in the treatment of cases of gonorrhea caused by these penicillin-resistant strains. There may also be other causes for the penicillin resistance of *N. gonorrhoeae*. About one in 10^9 cells of *N. gonorrhoeae* is resistant to penicillin, and thus sufficiently high antibiotic concentrations must be given for a long enough time to allow the natural body defense mechanisms to eliminate all the infecting bacteria. Structural modifications of penicillin G, such as occur in methicillin, can render the molecule resistant to penicillinases but may also narrow the spectrum of action, limiting the primary use of such antibiotics to the treatment of infections caused by penicillinase-producing *Staphylococcus* species.

In contrast to the penicillins, the cephalosporins generally have a broad spectrum of action, and many of the cephalosporins, such as cefoxitin and cephalothin, are relatively resistant to penicillinase. As such, the cephalosporins are useful in treating

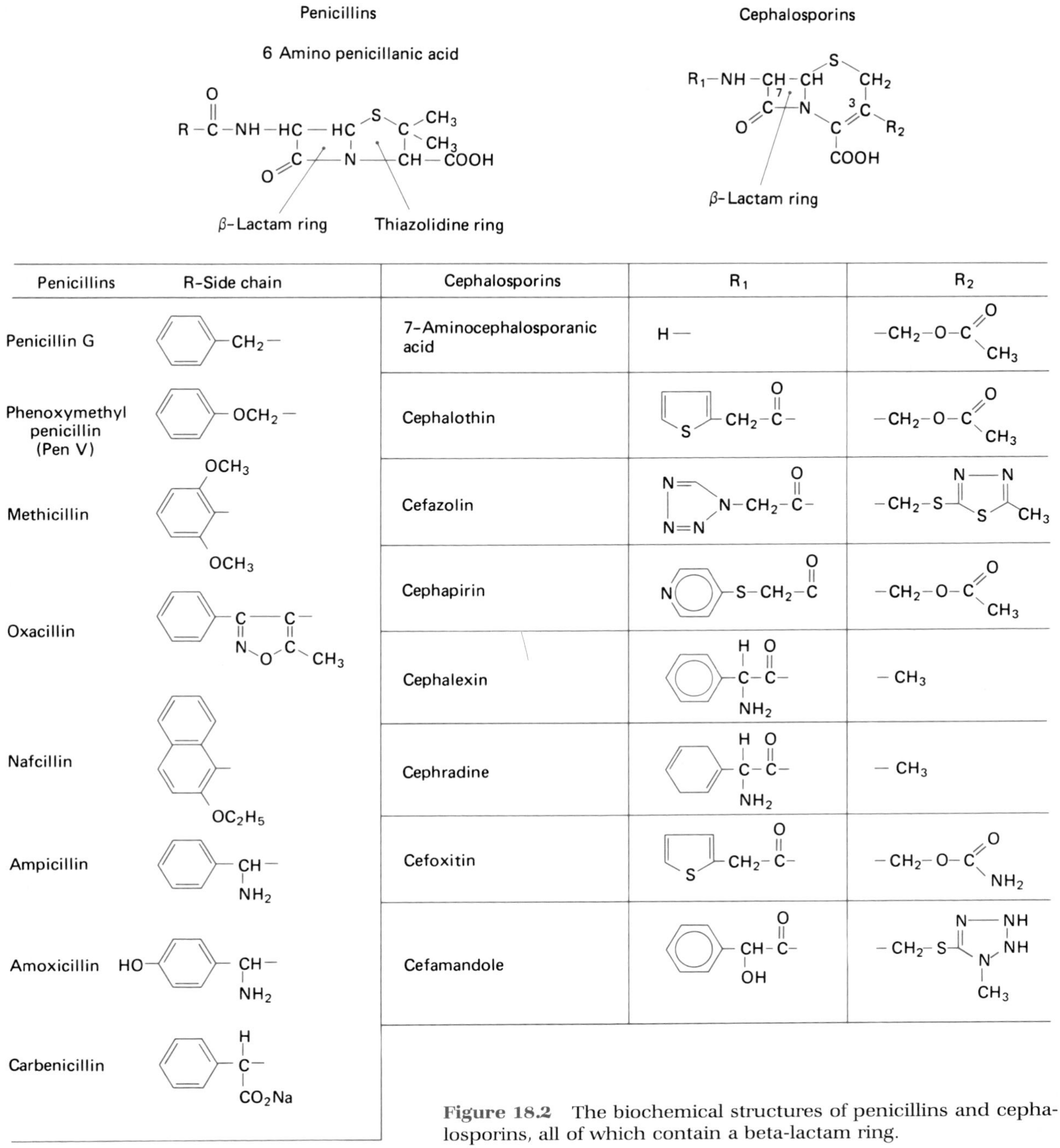

Penicillins

6 Amino penicillanic acid

$R-C(=O)-NH-HC-HC$ (S) $C(CH_3)(CH_3)$; $C(=O)-N-CH-COOH$

β-Lactam ring

Thiazolidine ring

Cephalosporins

$R_1-NH-CH-CH$ (S) CH_2; $C(=O)-N-C(COOH)=C-R_2$ (positions 7 and 3)

β-Lactam ring

Penicillins	R-Side chain
Penicillin G	(benzene ring)–CH_2–
Phenoxymethyl penicillin (Pen V)	(benzene ring)–OCH_2–
Methicillin	(benzene ring with OCH_3, OCH_3)–
Oxacillin	(benzene ring)–C—C–, N, O, C, CH_3 (isoxazole ring)
Nafcillin	(naphthalene ring with OC_2H_5)–
Ampicillin	(benzene ring)–CH(–NH_2)–
Amoxicillin	HO–(benzene ring)–CH(–NH_2)–
Carbenicillin	(benzene ring)–C(H)(–CO_2Na)–

Cephalosporins	R_1	R_2
7-Aminocephalosporanic acid	H—	$-CH_2-O-C(=O)-CH_3$
Cephalothin	(thiophene ring, S)–$CH_2-C(=O)-$	$-CH_2-O-C(=O)-CH_3$
Cefazolin	(tetrazole ring N=, N=N)–N–$CH_2-C(=O)-$	$-CH_2-S-$(thiadiazole ring N—N, S)–CH_3
Cephapirin	N(pyridine ring)–$S-CH_2-C(=O)$	$-CH_2-O-C(=O)-CH_3$
Cephalexin	(benzene ring)–C(H)(NH_2)–C(=O)–	$-CH_3$
Cephradine	(cyclohexadiene ring)–C(H)(NH_2)–C(=O)–	$-CH_3$
Cefoxitin	(thiophene ring, S)–$CH_2-C(=O)-$	$-CH_2-O-C(=O)-NH_2$
Cefamandole	(benzene ring)–CH(OH)–C(=O)–	$-CH_2-S-$(tetrazole ring N—NH, NH, N–CH_3)

Figure 18.2 The biochemical structures of penicillins and cephalosporins, all of which contain a beta-lactam ring.

a variety of infections caused by Gram positive and Gram negative bacteria. Many physicians are now using broad spectrum cephalosporins where the use of narrow range and more specifically directed penicillins would be adequate. Cephalosporins are most prudently used as alternatives to penicillins in cases where the patient is allergic to penicillin and in cases where the pathogen is not penicillin sensitive.

Cephalothin is often the antibiotic of choice for

treating severe staphylococcal infections, such as endocarditis, to avoid complications in cases where the infecting *Staphylococcus* species produces β-lactamases. Cefamandole, another one of the cephalosporins, is widely used in treating pneumonia, as it is active against *Haemophilus influenzae, Staphylococcus aureus*, and *Klebsiella pneumoniae*, which are frequently the causative agents of respiratory tract infections resulting in pneumonia. The cephalosporins may also be used in place of penicillins for the prophylaxis of infection by Gram positive cocci following surgical procedures.

In addition to the penicillins and cephalosporins, several other antibiotics inhibit cell-wall synthesis, including vancomycin, bacitracin, and cycloserine. These antibiotics do not block the enzymes involved in the formation of peptide crosslinkages in the murein component of the wall but rather block other reactions involved in the synthesis of the bacterial cell wall. The use of these antibiotics is limited because of their potentially undesirable side effects. For example, the use of bacitracin is restricted to topical application because this antibiotic causes severe toxicity reactions.

Inhibitors of Protein Synthesis. The antibiotics streptomycin, gentamicin, neomycin, kanamycin, tobramycin, and amikacin are inhibitors of bacterial protein synthesis. This group of antibiotics is known as the aminoglycosides because of related biochemical structures that contain amino sugars linked by glycosidic bonds. The aminoglycosides are used almost exclusively in the treatment of infections caused by Gram negative bacteria; they are relatively ineffective against anaerobic bacteria and facultative anaerobes growing under anaerobic conditions, and their action against Gram positive bacteria is also limited.

The aminoglycoside antibiotics bind to the 30S ribosomal subunit of the 70S prokaryotic ribosome, blocking protein synthesis and decreasing the fidelity of translation of the genetic code. Aminoglycosides disrupt the normal functioning of the ribosomes by interfering with the formation of initiation complexes, the first step of protein synthesis that occurs during translation. Additionally, aminoglycosides induce misreading of the mRNA molecules, leading to the formation of nonfunctional enzymes. The interference of protein synthesis results in the death of the bacterium.

The aminoglycoside antibiotics are useful in treating a variety of diseases (Table 18.3). Streptomycin is used in the treatment of a limited number of bacterial infections. It is, for example, sometimes used in the treatment of brucellosis, tularemia, endocarditis, plague, and tuberculosis. Gentamicin is effective in treating urinary tract infections, pneumonia, and meningitis. Gentamicin is, however, extremely toxic and thus is used only in cases of severe infections that may prove lethal if unchecked, particularly when the infecting bacteria are not sufficiently sensitive to other less toxic antibiotics. Kanamycin, a narrow spectrum antibiotic, is frequently used by pediatricians for infections due to *Klebsiella, Enterobacter, Proteus*, and *E. coli*. Amikacin, which has the broadest spectrum of activity of the aminoglycosides, is the antibiotic of choice for treating serious infections by Gram negative rods acquired in hospitals because such nosocomial infections are often caused by bacterial strains that

Table 18.3 Some Diseases and Their Causative Organisms for Which Aminoglycoside Antibiotics Are Recommended

Causative Organism	Disease	Drug of Choice
Enterobacter aerogenes	Urinary tract, other infections	Gentamicin, tobramycin
Proteus sp.	Urinary tract, other infections	Gentamicin, tobramycin
Pseudomonas aeruginosa	Bacteremia	Gentamicin, tobramycin
Acinetobacter	Various nosocomial infections, bacteremia	Gentamicin
Yersinia pestis	Plague	Streptomycin ± tetracycline
Serratia	Variety of nosocomial and opportunistic infections	Gentamicin
Mycobacterium tuberculosis	Tuberculosis	Streptomycin + other antibiotics

Table 18.4 Some Therapeutic Uses of Tetracyclines, Chloramphenicol, Erythromycin, and Clindamycin

Causative Organism	Disease	Drug of Choice
Salmonella	Typhoid fever, paratyphoid fever, bacteremia	Chloramphenicol
Haemophilus influenzae	Pneumonia, meningitis	Chloramphenicol
Haemophilus ducreyi	Chancroid	A tetracycline
Brucella	Brucellosis	A tetracycline ± streptomycin
Vibrio cholerae	Cholera	A tetracycline
Campylobacter fetus	Enteritis	No treatment or erythromycin
Bacteroides fragilis	Brain abscess	Chloramphenicol
	Lung abscess, intra-abdominal abscess, bacteremia, endocarditis	Clindamycin
Legionella pneumophila	Legionnaire's disease	Erythromycin
Borrelia recurrentis	Relapsing fever	A tetracycline
Mycoplasma pneumoniae	Atypical pneumonia	Erythromycin or a tetracycline
Rickettsia typhi	Typhus fever	Chloramphenicol
Rickettsia prowazekii	Murine typhus	A tetracycline
Rickettsia rickettsii	Rocky Mountain spotted fever	A tetracycline
Chlamydia trachomatis	Trachoma	A sulfonamide + a tetracycline
	Inclusion conjunctivitis	A tetracycline
	Nonspecific urethritis	A tetracycline

are resistant to multiple antibiotics, including other aminoglycosides such as gentamicin.

In addition to the aminoglycoside antibiotics, a number of other antibiotics inhibit bacterial protein synthesis. These antibiotics include the tetracyclines, chloramphenicol, erythromycin, lincomycin, clindamycin, and spectinomycin. Some recommended therapeutic uses of these antibiotics are shown in Table 18.4. Unlike the aminoglycoside antibiotics that are bactericidal, these inhibitors of bacterial protein synthesis are generally bacteriostatic.

The tetracyclines, like the aminoglycosides, bind specifically to the 30S ribosomal subunit, blocking protein synthesis. The tetracyclines are effective against a variety of pathogenic bacteria, including rickettsia and chlamydia species. Tetracylines, for example, are used therapeutically in treating the rickettsial infections of Rocky Mountain spotted fever, typhus fever, and Q fever and the chlamydial diseases of lymphogranuloma venereum, psittacosis, inclusion conjunctivitis, and trachoma. Tetracylines are useful in treating a variety of other bacterial infections, including pneumonia caused by *Mycoplasma pneumoniae*, brucellosis, tularemia, and cholera.

Unlike the antibiotics discussed so far that inhibit bacterial protein synthesis, chloramphenicol acts primarily by binding to the 50S ribosomal subunit. Chloramphenicol is a fairly broad spectrum antibiotic, active against many species of Gram negative bacteria. However, chloramphenicol has a number of toxic effects, limiting its therapeutic uses to those where the benefits of chloramphenicol use outweigh the dangers associated with toxic reactions. Chloramphenicol is the antibiotic of choice for treating typhoid fever as well as various other infections caused by *Salmonella* species; it is also effective against anaerobic pathogens and can be used effectively in treating diseases, such as brain abscesses, normally caused by anaerobic bacteria.

Like chloramphenicol, erythromycin acts by binding to 50S ribosomal subunits, blocking protein synthesis. Erythromycin is most effective against Gram positive cocci, such as *Streptococcus pyogenes*. Erythromycin is not active against most aerobic Gram negative rods but does exhibit antibacterial activity against some Gram negative organisms, such as *Pasteurella multocida, Bordetella pertussis*, and *Legionella pneumophilia*. Therapeutically, erythromycin is recommended for the treatment of Legionnaire's disease and is also effective in treating pneumonia caused by *Mycoplasma pneumoniae*, diphtheria, and whooping cough. Erythromycin may also be used as an alternative to penicillin in treating staphylococcal infections, streptococcal infections, tetanus, syphilis, and gonorrhea. Erythromycin is now routinely applied to the eyes of newborns in the United States to prevent the transmsission of *Neisseria gonorrhoeae* and *Chlamydia trachomatis* from mother to newborn during birth. Erythromycin has replaced silver nitrate because of the rise in sexually transmitted *Chamydia* infections; silver nitrate is effective against *Neisseria* but not *Chlamydia*, whereas erythromycin is effective against both.

Several other antibiotics that inhibit protein synthesis are not generally used for treating bacterial infections because they inhibit mammalian-cell protein synthesis to the same extent as that in bacterial cells. If the mode of action of these antibiotics is not specific toward bacteria, they are not therapeutic antibacterial agents. For example, dactinomycin (actinomycin D) blocks protein synthesis in both bacterial and eukaryotic cells; this antibiotic binds to double-stranded DNA, blocking transcription of the genetic information to form a mRNA molecule. Although not useful in treating bacterial infections, dactinomycin does have a therapeutic role in treating some malignancies where it is desirable to block the rapid division of cancer cells.

> *Diseases desperate grown*
> *By desparate appliance are relieved,*
> *Or not at all.*
>
> —WILLIAM SHAKESPEARE, *Hamlet*

Rifampin, a semisynthetic derivative of rifamycin B, also blocks protein synthesis at the level of transcription. Because this antibiotic is more effective against bacterial RNA polymerase enzymes than mammalian RNA polymerases, it can be used therapeutically in treating some bacterial diseases. Rifampin is used in combination with other antibiotics in the treatment of mycobacterial diseases, such as tuberculosis.

Inhibitors of Membrane Transport

The cytoplasmic membrane is the site of action of some antimicrobial agents. The polymyxins, such as polymyxin B, interact with the cytoplasmic membrane, causing changes in the structure of the bacterial cell membrane and leakage from the cell. Polymyxin B is bactericidal, and its effectiveness is restricted to Gram negative bacteria. The action of polymyxin B is related to the phospholipid content of the cell wall and membrane complex. Sensitive bacteria take up more polymyxin B than resistant strains. The principle use of polymyxin B and colistin (polymyxin E) is in the treatment of infections caused by *Pseudomonas* species and other Gram negative bacteria resistant to penicillins and the aminoglycoside antibiotics. Both polymyxin B and colistin are useful in treating severe urinary tract infections caused by *Pseudomonas aeruginosa* and other Gram negative bacteria, particularly when the infecting bacteria are resistant to other antibiotics.

Inhibitors with Other Modes of Action

There are several other antibiotics that act by different mechanisms, some as antimetabolites and others whose modes of action are unknown. Some of these antibiotics are useful in treating specific infections, such as tuberculosis, and others are particularly useful in treating infections of particular tissues, such as urinary tract infections. For example, the sulfonamides, sulfones, and *p*-aminosalicylic acid are useful antibacterial agents. The sulfones are useful in treating leprosy. The use of sulfonamides and *p*-aminosalicylic acid has declined as the result of the occurrence of resistant strains and the development of more effective antibiotics with lesser toxic side effects. Trimethoprim is a broad spectrum antibacterial agent and is effective in the treatment of many urinary and intestinal tract infections. It is used primarily for the treatment of urinary infections due to *Escherichia coli*, and species of *Proteus, Klebsiella*, and *Enterobacter*. The combined formulation of trimethoprim and sulfamethoxazole greatly enhances the antibacterial activities of these antimetabolites, a phenomenon known as synergy. The trimethoprim–sulfamethoxazole mixture is also effective in treating typhoid fever.

In addition to the use of trimethoprim-sulfamethoxazole, several other compounds are used as antiseptics in treating urinary tract infections. These compounds, which inhibit the growth of many bacterial species, include methenamine, nalidixic acid, oxolinic acid, and nitrofurantoin. The usefulness of these drugs, though, depends on the fact that they are concentrated in the urinary tract tissues and thus can act as antiseptics specifically within that tract. Nalidixic acid acts by blocking DNA synthesis, but transcription and translation (protein synthesis) can still occur. This antimicro-

Table 18.5 Some Therapeutic Uses of Antifungal Agents

Causative Organism	Disease	Drug of Choice
Candida albicans	Skin and superficial mucus membrane lesions	Amphotericin B, nystatin
Cryptococcus neoformans	Meningitis	Amphotericin B + flucytosine
Candida albicans	Pneumonia	Amphotericin B
Aspergillus	Meningitis	Amphotericin B
Mucor	Skin lesions	Amphotericin B
Histoplasma capsulatum	Lung lesions, histoplasmosis	Amphotericin B
Coccidioides immitis	Coccidiomycosis (desert fever)	Amphotericin B
Blastomyces dermatitidis	Blastomycosis	Amphotericin B

bial agent is effective against most Gram negative bacteria that cause urinary tract infections. Nitrofurantoin inhibits several bacterial enzymes, but the specific mode of action of this antimicrobial agent is unknown. Nitrofurantoin is used only in a limited number of cases because it is generally not as effective as other antibiotics, including sulfanomides.

Isoniazid is sometimes used in the treatment of tuberculosis, but it is not particularly effective against *Mycobacterium tuberculosis* when used alone and is generally used in association with other antibiotics. The specific mode of action of isoniazid is not known, but its primary action appears to involve the inhibition of mycolic acid biosynthesis. Mycolic acids are unique components of the cell walls of mycobacteria, and blockage of the biosynthesis of these compounds could specifically inhibit mycobacteria.

Thus, a large number of antibacterial agents with varying modes of action and differing activity spectrums are available for treating infectious diseases caused by bacterial pathogens.

Antifungal Agents

In contrast to the relative ease with which bacterial infections can be treated, the fact that fungi have eukaryotic cells limits the sites against which antimicrobial agents can be selectively directed to control fungal pathogens of human beings. Consequently, there are relatively few antifungal agents of therapeutic value, with particularly few effective for treating systemic infections. Some of the therapeutic uses of antifungal agents are listed in Table 18.5. These agents are directed at the sites that differ between fungal and human cells.

The polyene antibiotics, amphotericin B, and nystatin are used in treating a variety of fungal diseases. The polyene antibiotics act by altering the permeability properties of the cytoplasmic membrane, leading to the death of the affected cells. Interactions of polyenes with the sterols in the cytoplasmic membranes of eukaryotic cells appear to form channels or pores in the membrane, allowing leakage of small molecules through the membrane (Figure 18.3). Differences in the sensitivity of various organisms is determined by the concentrations of sterols in the membrane. Because mammalian cells, like fungi, contain sterols in their cytoplasmic membranes, it is not surprising that polyene antibiotics also cause alterations in the membrane permeability of mammalian cells and toxicity to mammalian tissue as well as the death of fungal pathogens.

Nystatin is primarily used in the treatment of topical infections by members of the fungal genus *Candida*. Vaginitis and thrush are are effectively treated by using nystatin. Amphotericin B has a relatively broad spectrum of activity and is used in the treatment of systemic fungal infections; it is the most effective therapeutic agent for treating systemic infections due to yeast and fungi. The potential toxic side effects of amphotericin B usage, however, such as kidney damage, require careful supervision of its administration. Patients requiring administration of amphotericin B must be hospitalized so that the initial reaction to the therapy can be carefully supervised. Persons who have received amphotericin B almost invariably exhibit some toxic side effects, but without the administration of

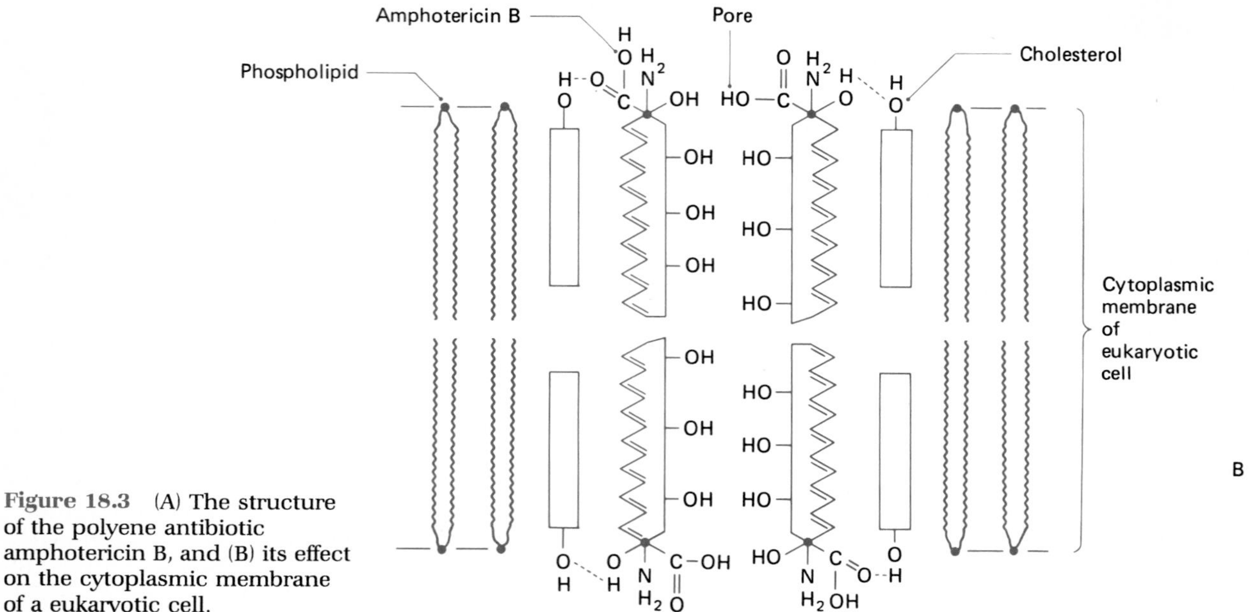

Figure 18.3 (A) The structure of the polyene antibiotic amphotericin B, and (B) its effect on the cytoplasmic membrane of a eukaryotic cell.

this drug, systemic fungal infections are almost invariably fatal. Amphotericin B is used in the treatment of cryptococcosis, histoplasmosis, coccidioidomycosis, blastomycosis, sporotrichosis, and candidiasis.

In addition to the polyene antibiotics, miconazole and clotrimazole have a broad spectrum of antifungal activities and are used in the topical treatment of superficial mycotic infections. These two antimicrobial agents appear to alter membrane permeability, leading to the inhibition and/or death of selected fungal species. Flucytosine is also effective in treating systemic fungal infections; it is less effective, but also less toxic, than amphotericin B and is primarily used in combination with amphotericin B treatment. Within fungal cells flucytosine is converted to an inhibitor of normal nucleic acid synthesis. Mammalian cells do not convert as much flucytosine to this inhibitor as do fungal cells, accounting for the selective toxicity of this antifungal agent.

Griseofulvum is another antibiotic that is effective against some fungal infections. This antibiotic causes a disruption of mitotic spindles, inhibiting fungal mitosis. Griseofulvum is used in the treatment of fungal diseases of the skin, hair, and nails caused by various species of dermatophytic fungi, like *Microsporum*, *Epidermophyton*, and *Trichophyton*. These dermatophytic fungi concentrate griseofulvum by an active uptake process, and their sensitivity is correlated directly with their ability to concentrate the antibiotic.

Antiprotozoan Agents

Treatment of human protozoan diseases with antimicrobial agents presents a special problem because many of the pathogenic protozoa exhibit a complex life cycle, often including stages that develop intracellularly within mammalian cells. Different antimicrobial agents are generally needed for use against different forms of the same pathogenic protozoan, depending on the stage of the life cycle and the involved tissues. For example, the protozoan species of the genus *Plasmodium* that cause malaria exhibit complex life cycles, part of which are carried out in the liver and blood of human beings. The erythrocytic stage of the *Plasmodium* life cycle that occurs within human blood cells is the most sensitive to antimalarial drugs. The life stages that occur within the liver are difficult to treat, and the sporozoites injected into the blood stream by mosquitoes are not affected by antimalarial drugs. The antimalarials effective against the erythrocytic forms of the protozoan include chloroquine and amodiaquine. These drugs are most widely used for suppressing the malarial infection, but neither is effective against the hepatic stages of the *Plasmodium* that occur in the liver. These antimalarial agents appear to interfere with DNA replication. The effect of these drugs is a rapid schizontocidal action, that is, the rapid interruption of schizogony or multiple division that occurs within red blood cells. The sensitivity of malarial protozoa to these drugs depends on the active transport of these compounds into the protozoa and the selective accumulation of the drugs intracellularly.

Chloroguanide is also used in the suppression of malaria. This drug is transformed within the body to a substance that inhibits the proliferation of the malaria protozoa. Chloroguanide binds more strongly to the enzymes of the malarian protozoa than to the comparable mammalian enzymes, accounting for its selective inhibition. Because resistance to the synthetic antimalarial drugs is increasing, quinine, one of the early drugs used for the effective treatment of malaria, is once again being used to treat this disease.

For the radical cure of malaria, that is, the eradication of both the erythrocytic and liver stages of the protozoan, primaquine is normally used. This drug is used in conjunction with chloroquine and chloroguanide. The precise mode of action of primaquine has not been elucidated. Because of the toxic side effects of primaquine, it is primarily used in the treatment of relapsing malarial infections.

Table 18.6 Some Drugs Used in the Treatment of Diseases Caused by Protozoan Pathogens

Infecting Organism and Disease	Drug of Choice
Entamoeba histolytica	
Asymptomatic cyst passer	Diiodohydroxyquin
Mild intestinal disease	Metronidazole
Severe intestinal disease	Metronidazole
Hepatic abscess	Metronidazole
Giardia lamblia	Quinacrine hydrochloride
Balantidium coli	Oxytetracycline
Dientamoeba fragilis	Diiodohydroxyquin
Trichomonas vaginalis	Metronidazole
Pneumocystis carinii	Trimethoprim–sulfamethoxazole
Toxoplasma gondii	Pyrimethamine + trisulfapyrimidines
Malaria *Plasmodium* sp.	
Erythrocytic stage	Chloroquinine, amodiaquine
Hepatic stage	Primaquine
Leishmaniasis	
Leishmania donovani (kala-azar, visceral leishmaniasis)	Sodium stibogluconate
Leishmania tropica (oriental sore, cutaneous leishmaniasis)	Sodium stibogluconate
Leishmania braziliensis (American mucocutaneous leishmaniasis)	Sodium stibogluconate
African trypanosomiasis	
Trypanosoma gambiense	Pentamidine
Trypanosoma rhodesiense	Suramin
Either *T. gambiense* or *T. rhodesiense* in late disease with central nervous system involvement	Melarsoprol
South American trypanosomiasis (Chagas' disease)	
Trypanosoma cruzi	Nifurtimox

Several other drugs are used in the treatment of various other protozoan infections (Table 18.6). As in the case of malaria, the life cycle of the particular protozoan determines which agents will be effective in controlling the infection. Only a few of these antiprotozoan agents will be discussed further here. Metronidazole (Flagyl) is used in the treatment of dysentery caused by the protozoan *Entamoeba histolytica*. Metronidazole specifically inhibits the growth of anaerobic microorganisms, including anaerobic protozoa. Pentamidine and related diamidine compounds are useful in treating infections by members of the protozoan genus *Trypanosoma*. Compounds of this type interfere with DNA metabolism.

Melarsoprol is an arsenical, that is, an arsenic-containing compound useful in treating some stages of human trypanosomiasis, particularly because of its ability to penetrate into cerebrospinal fluid. Arsenicals react with the sufhydryl groups of proteins, inactivating a large number of enzymes. It appears that mammalian cells can metabolize these compounds to nontoxic forms more rapidly than protozoan cells, accounting for the selective toxicity of melarsoprol to trypanosome protozoans. The drug sodium stibogluconate, an antimony-containing compound, is useful in treating diseases caused by members of the protozoan genus *Leishmania*. Other antiprotozoan agents useful in the chemotherapy of protozoan diseases include suramin, a nonmetalic compound that inhibits a wide variety of enzymes, and nifurtimox, which is effective against *Trypanosoma cruzi*, the causative organism of Chagas' disease.

Antiviral Agents

The search for antiviral drugs comparable to the antibiotics used to control bacterial infections has been fruitless for the most part. There are no broad spectrum antiviral agents currently in clinical use, and most viral infections cannot be treated effectively by using antiviral chemicals. The integral role of the mammalian host cell in the process of viral replication complicates the difficulty of finding compounds that specifically inhibit viral replication. Most compounds that prevent the reproduction of viruses also interfere with mammalian cell

MICROBIOLOGY HEADLINES

New Drug is Found to Curb AIDS Virus

Researchers Preparing to Test Ribavirin in Patients With Fatal Immune Disease

A drug investigated as an influenza treatment blocks the growth in laboratory tests of a virus that causes acquired immune deficiency syndrome, or AIDS, a new study has found, and researchers are preparing to test it in people who have AIDS.

Because the drug, called ribavirin, has already been studied in patients with other illnesses, the testing that often delays drug development has already been completed; thus, trials with AIDS patients could begin very soon, researchers said.

The report of ribavirin's effectiveness against AIDS appears in the current issue of *The Lancet*, the British medical journal.

Ribavirin is being studied as a possible flu treatment and is nearing approval by the Food and Drug Administration for use against respiratory syncytial virus infections, which interfere with breathing. The drug is made by ICN Pharmaceuticals of Covina, Calif.

Results Termed Preliminary

Dr. Donald Forthal of the Centers for Disease Control in Atlanta, where the research was done, noted that success in laboratory, or in-vitro, tests did not necessarily translate into success in patients.

"Even the in-vitro results are preliminary, and its clinical usage is completely unknown," said Dr. Forthal, who is drawing up plans for the clinical research in collaboration with Dr. Joseph McCormick, director of the laboratory studies.

Ribavirin is the second drug that has been shown in laboratory tests to prevent reproduction of the virus that causes AIDS. Victims of AIDS gradually succumb to a variety of infectious diseases and unusual cancers. There is now no cure for the disease.

The first drug, suramin, which is used to treat parasitic diseases, was identified as a potential anti-AIDS agent in October at the National Cancer Institute in Bethesda, Md.

A pilot study of suramin's safety in AIDS patients has been completed, and large-scale trials of the drug at medical centers across the country are beginning, said Dr. Samuel Broder, head of the clinical oncology program at the cancer institute.

Other researchers have tried to combat AIDS with interferon and a substance called interleukin-2. While those drugs can delay the inevitable decline of AIDS patients, they do not eliminate the viral infection that is the source of the disease, Dr. Broder said.

Source: The Associated Press, Dec. 21, 1984. Reprinted by permission.

Table 18.7 **Some Antiviral Agents and Their Therapeutic Uses**

Causative Organism	Disease	Drug of Choice
Herpes simplex virus	Keratoconjunctivitis	Vidarabine, trifluridine, acyclovir
	Encephalitis	Vidarabine, acyclovir
	Cold sores	Acyclovir
	Genital herpes	Acyclovir
Influenza virus *A*	Influenza	Amantadine

metabolism, resulting in adverse effects on human cells so as to preclude the therapeutic use of such agents. Very few antiviral agents have been found with clinical applicability, and these generally have a narrow spectrum of antiviral activity.

There are some remedies worse than the disease.

—Publilius Syrus, *Maxim 301*

The best "treatment" for many viral diseases is prevention through the appropriate use of vaccines and by the controlling the vectors that act as carriers for viruses. Generally, recovery from a viral infection is dependent on the natural immune defense response of the body. For most viral infections treatment is aimed at maintaining a physiological state in which an effective immune response can be ensured, by following the sage advice, "rest and drink plenty of fluids." The discovery that interferons play an important role in the natural immune response to viruses holds promise for the future if interferons can be produced commercially and if they can be administered with therapeutic value. Genetic engineering would seem to provide the greatest hope for the commercial production of interferons, making these antiviral substances available for medical treatment of viral infections.

Only a few chemical agents have been developed to date for therapeutic use in treating specific viral infections (Table 18.7). Among these antiviral compounds, idoxuridine, an analogue of thymidine that interferes with DNA metabolism in both viral and mammalian cells, has clinical uses in the treatment of herpes simplex keratitis. Amantadine is recommended for the prophylaxis of high-risk patients in cases of documented influenza A virus epidemics. Amantadine appears to interfere with the absorption and uptake of influenza A viruses. There are several compounds that have antiviral activity against herpes viruses. Trifluridine, a nucleic acid base analogue that inhibits DNA synthesis, is effective against epithelial keratitis of the eye caused by Herpes simplex virus. Vidarabine was originally developed for the treatment of leukemia but has proven to be more effective in treating herpes simplex encephalitis and keratoconjunctivitis. Vidarabine (vira-A) is an adenine arabinoside. The selectivity of vidarabine occurs because this drug inhibits mammalian DNA synthesis to a lesser extent than viral DNA replication.

The most promising antiviral agent that has been discovered recently is acyclovir. This antibiotic is an effective inhibitor of herpes viruses. Acyclovir has been reported to be more effective than idoxuridine and vidarabine for the treatment of herpetic ocular disease. Acyclovir is used for treating herpes-caused encephalitis. It also is effective in treating cold sores and genital herpes caused by Herpes simplex, and also may be of value in treating chicken pox and shingles. Acyclovir, a nucleoside analogue, is converted in vivo to a chemical that inhibits herpes simplex viral DNA polymerase, thus blocking viral DNA replication. The activation of acyclovir is initiated by a viral-directed enzyme. In an uninfected cell there is only very limited conversion of acyclovir to the toxic forms of the chemical. Because an enzyme coded for by the herpes virus is required to activate acyclovir, this compound exhibits selective antiviral activity, making it therapeutically valuable.

Selection of Antibiotics for Treating Infectious Diseases

The selection of a particular antimicrobial agent for treating a given disease depends on several factors, including: (1) the sensitivity of the infecting microorganism to the particular antimicrobial agent; (2) the side effects of the antimicrobial agent, relative to direct toxicity to mammalian cells and to the microbiota normally associated with human tissues; (3) the biotransformations of the particular antimicrobial agent that occur in vivo, relative to whether the antimicrobial agent will remain in its active form for a sufficient period of time to be selectively toxic to the infecting pathogens; and (4)

Table 18.8 Distribution of Antimicrobics to Specific Body Areas

Bone
Penicillins, tetracyclines, cephalosporins, lincomycin, and clindamycin antimicrobics penetrate bone and bone marrow; levels are higher in infected bone than in normal bone.

Central Nervous System
Only lipid-soluble antimicrobics cross the blood–brain barrier and reach brain tissues. Chloramphenicol achieves high brain tissue levels. In the presence of inflammation, such as brain abscess, various penicillins achieve appreciable concentrations in the brain. Levels of most antimicrobics in the cerebrospinal fluid (CSF) are low. Penicillin G and ampicillin can achieve adequate CSF levels in the presence of inflammation; oxacillin, naficillin, and methicillin can be used to treat staphylococcal meningitis. CSF levels of chloramphenicol are adequate to treat *Streptococcus, Neisseria,* and *Haemophilus* but not most Gram negative bacteria. Cefoxamine, moxalactam, and cefoperazone enter CSF in the presence of inflammation in concentrations that are adequate to treat *Streptococcus, Neisseria, Haemophilus, Klebsiella,* and *E. coli.* Amphotericin can be used to treat cryptococcal meningitis.

Ears and Sinuses
Most of the penicillins reach levels in the middle ear fluid in sufficient concentrations for the treatment of otitis media. Concentrations of antimicrobics in sinuses are adequate for ampicillin, amoxicillin, tetracyclines, erythromycin, sulfonamides, and trimethoprim to treat sinus infections.

Eyes
Few antimicrobics penetrate the eye wall. Levels of penicillins and cephalosporins in the aqueous humor are less than 10 percent of the peak serum levels and inhibit only highly sensitive bacteria.

Pleural and Pericardial Fluids
Most of the penicillins, cephalosporins, sulfonamides, macrolides, clindamycin, chloramphenicol, and antituberculosis drugs diffuse into serus cavities.

Pulmonary
Concentrations of most antibiotics within the lung are satisfactory provided there is sufficient blood flow. Penicillins and tetracyclines show variable sputum concentrations. Chloramphenicol reaches high sputum and bronchial secretion concentrations. Antituberculosis agents, such as isoniazid and rifampin, achieve appreciable levels in pulmonary tissue.

Skin
Tetracyclines and clindamycin concentrate in skin tissue and are effective in treatment of acne.

Synovial Fluid
Most antibiotics used in the treatment of joint infections reached inflamed joints in adequate concentrations.

Urinary Tract
Treatment of kidney and other urinary tract infections depends largely on the concentrations in the urine rather than on serum levels. Nalidixic acid and nitrofurantoins are effective in treating urinary tract infections.

the chemical properties of the antimicrobial agent that determine its distribution within the body, relative to whether or not adequate concentrations of the active antimicrobial chemical will be able to reach the site of infection in order to inhibit or kill the pathogenic microorganisms causing the infection. Advances in pharmacology has led to improved selection of antimicrobics that can be used to treat infectious diseases with minimal side effects. In particular, knowing how antimicrobics are distributed to specific body areas (Table 18.8), how long specific antimicrobics remain active in body tissues, the toxicological properties of specific antimicrobics (Table 18.9), and potential drug interactions are all critical in the proper selection of antimicrobial agents for treating disease. For example, although many antibiotics possess antimicrobial activities effective against the pathogenic bacteria that cause urinary tract infections, only a limited number of antibiotics are effective in treating these infections because relatively few antibiotics can reach and be concentrated in the tissues of the urinary tract in their active form. Similarly the direct toxicities of the aminoglycosides limits their use to special cases. Additionally, one antimicrobial agent can influence the effects of another antimicrobial agent. In some cases treatment with two drugs enhances the effectiveness of treatment, and in other cases one drug can interfere with the inhibitory effects of a second antimicrobial agent (Figure 18.4).

Antimicrobial Susceptibility Testing

The determination of the antimicrobial susceptibility of a pathogen is important in aiding the clinician to select the most appropriate agent for treating that disease. It does no good to prescribe an antibiotic that is ineffective against the microorganism causing the disease. Physicians want to avoid indiscriminant administration of antibiotics be-

Table 18.9 **Major Toxicities of Selected Antimicrobics**

Antimicrobic Agent	Mechanism	Signs
Aminoglycosides	Binds hair cells of organ of Corti	Deafness
	Binds vestibular cells	Vertigo
	Competitive neuromuscular blockage	Respiratory paralysis
	Tubular necrosis	Nephrotoxicity
Amphotericin	Distal tubular damage	Nephrotoxicity
	Renal tubular acidosis	Nephrotoxicity
Carbenicillin	Inhibition of platelet aggregation	Bleeding
Cephalosporins	Cortical stimulation	Myoclonic seizures
Cephaloridine	Proximal tubular damage	Nephrotoxicity
Chloramphenicol	Damages stem cell	Aplastic anemia
	Inhibits protein synthesis	Reversible anemia
Clindamycin	Proliferation of *Clostridium difficile*	Diarrhea
Emetine	Permeability changes	Hypotension
Isoniazid	Liver cell damage	Hepatitis
Neomycin	Villous damage	Malabsorption
Penicillins	Cortical stimulation	Myoclonic seizures
Polymyxins	Noncompetitive neuromuscular blockage	Respiratory paralysis
		Nephrotoxicity
	Tubular necrosis	
Rifampin	Liver cell damage	Hepatitis
Sulfonamides	Glucose-6-phosphate deficiency	Hemolytic anemia
	Collecting duct obstruction	Nephrotoxicity
Tetracyclines	Liver cell damage	Hepatitis
	Degradation products	Fanconi syndrome

cause the selective pressures of excessive antibiotic usage can induce problematical antibiotic-resistant strains of pathogenic microorganisms. The clinical microbiology laboratory provides information, through standardized in vitro testing, with regard to the activities of antimicrobial agents against microorganisms that have been isolated and identified as the probable etiologic agents of disease. Antibiotic susceptibility testing, relying on the observation of antibiotics inhibiting the growth and/

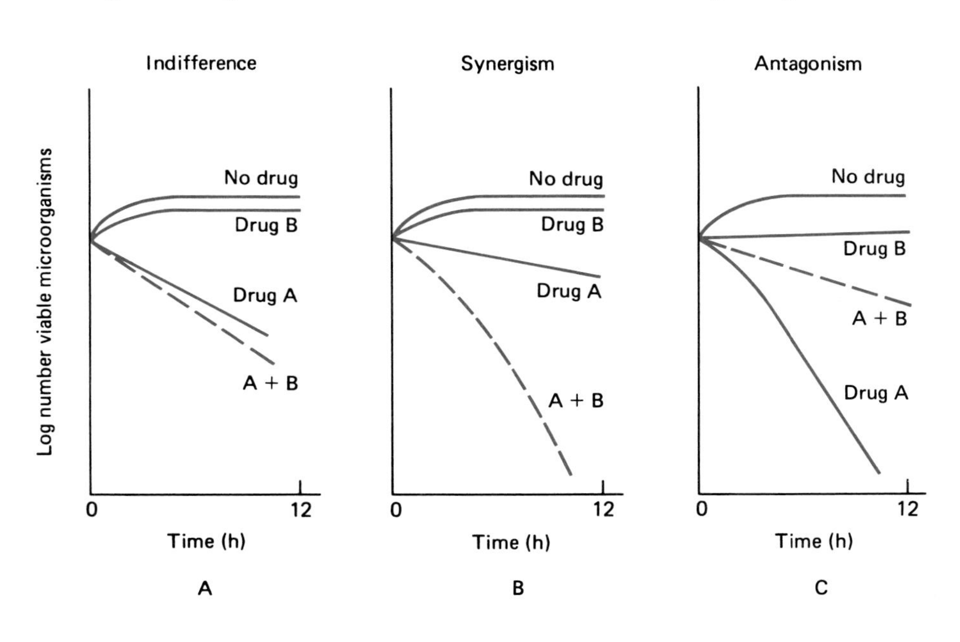

Figure 18.4 This diagram illustrates the possible interactions between antimicrobial agents. (A) Indifference; (B) synergism; and (C) antagonism. Interactions are reflected in the log numbers of microorganisms that are still viable after application of these agents.

or killing cultures of microorganisms in vitro, provides the physician with the information needed to prescribe the proper antibiotics for treating infectious diseases.

> *There are no such things as incurables,*
> *There are only things for which man*
> *has not found a cure.*
>
> —BERNARD BARUCH

Agar Diffusion Methods

The qualitative susceptibility of microorganisms to antimicrobial agents can be determined on agar plates by using filter paper disks impregnated with antimicrobial agents. The Bauer–Kirby test system is a standardized antimicrobial susceptibility procedure in which a culture is inoculated onto the surface of Mueller–Hinton agar, followed by the addition of antibiotic impregnated disks to the agar surface. In agar diffusion methods for determining antibiotic susceptibility, the antibiotics diffuse into the agar, establishing a concentration gradient. Inhibition of microbial growth is indicated by a clear zone (zone of inhibition) around the antibiotic disk (Figure 18.5). The diameter of the zone of inhibition reflects the solubility properties of the particular antibiotic—that is, the concentration gradient established by diffusion of the antibiotic into the agar—and the sensitivity of the given microorganism to the specific antibiotic. Standardized zones for each antibiotic disk have been established to determine whether the microorganism is sensitive (S) intermediately sensitive (I), or resistant (R) to the particular antibiotic (Table 18.10). The results of Bauer–Kirby testing indicate whether a particular antibiotic has the potential for effective control of an infection caused by a particular pathogen.

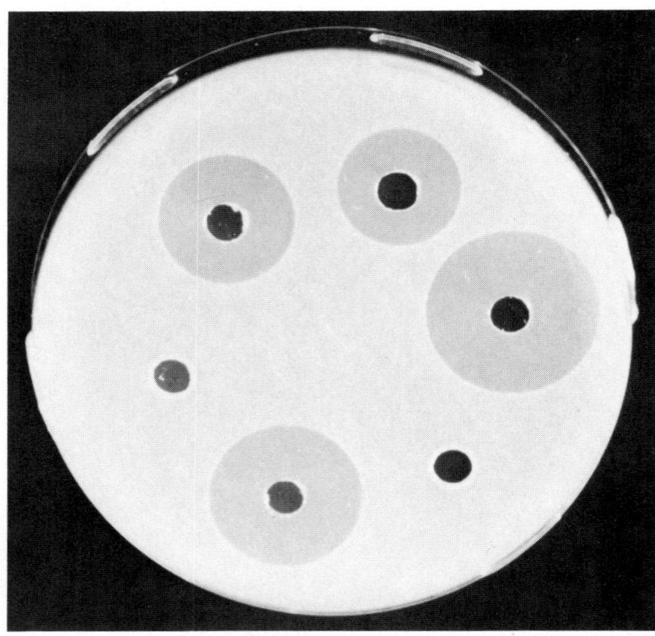

Figure 18.5 Zone of inhibition around an antibiotic disk on an agar plate. This is an antibody-sensitivity testing plate showing sensitivity of a bacterial isolate to various antibiotics. In the Bauer–Kirby procedure the following items are standardized: the amount of each antibiotic impregnated into the disks used in the test procedure; the composition of the test medium; the nature of the inoculum; the incubation time; and the measurement of the size of the zone of inhibition needed to establish antibiotic susceptibility. The Bauer–Kirby antibiotic susceptibility test procedure is the widely accepted standard test employed in most clinical laboratories. (Courtesy Bernard Abbott, Eli Lilly Research Laboratories, Indianapolis.)

Liquid Diffusion Methods

Besides Bauer–Kirby and other agar diffusion testing, many clinical laboratories use the Autobac system for measuring microbial growth, based on light scattering or equivalent automated liquid diffusion methods for antibiotic sensitivity testing. This system uses Eugonic broth and/or low thymidine broth culture and standardized antibiotic disks, but the antibiotics in the disks dissolve into the broth rather than into agar, as in the Bauer–Kirby procedure. The concentrations of the antibiotics and the density and growth phase of the cultures are adjusted so that uniform interpretive guidelines can be used for assessing antibiotic sensitivities. A normalized light scattering index is generated to determine S, I, and R; the light scattering index for resistance is 0.00–0.50, intermediate sensitivity 0.51–0.60, and susceptible 0.60–1.00, except for penicillin G where the index for resistance is 0.00–0.90. Other automated systems are also available for performing this procedure, including the Microscan, BBL sceptre, Vitek AMS, and Abott MSII. These automated systems simplify and enhance the reliability of antimicrobial susceptibility testing.

Minimum Inhibitory Concentration Procedures

Another approach to antimicrobial susceptibility testing is to determine the minimum inhibitory concentration (MIC) using tube dilution procedures. This procedure determines the concentration of an antibiotic that is effective in preventing the growth of the pathogen and gives an indication

Table 18.10 Interpretation of Zones of Inhibition for Bauer–Kirby Antibiotic Susceptibility Testing

Antibiotic	Disc Conc.[a]	Inhibition Zone Diameter (mm)		
		Resistant	Intermediate	Susceptible
Amikacin	0.01 mg	13 or less	12–13	14 or more
Ampicillin	0.01 mg	11 or less	12–13	14 or more
Bacitracin	10 units	8 or less	9–11	13 or more
Cephalothin	0.03 mg	14 or less	15–17	18 or more
Chloramphenicol	0.03 mg	12 or less	13–17	18 or more
Erythromycin	0.015 mg	13 or less	14–17	18 or more
Gentamicin	0.01 mg			13 or more
Kanamycin	0.03 mg	13 or less	14–17	18 or more
Lincomycin	0.002 mg	9 or less	10–14	15 or more
Methicillin	0.005 mg	9 or less	10–13	14 or more
Nalidixic acid	0.03 mg	13 or less	14–18	19 or more
Neomycin	0.03 mg	12 or less	13–16	17 or more
Nitrofurantoin	0.3 mg	14 or less	15–16	17 or more
Penicillin G, (staphylococci)	10 units	20 or less	21–28	29 or more
Penicillin, (other organisms)	10 units	11 or less	12–21	22 or more
Polymyxin	300 units	8 or less	9–11	12 or more
Streptomycin	0.01 mg	11 or less	12–14	15 or more
Sulfonamides	0.3 mg	12 or less	13–16	17 or more
Tetracycline	0.03 mg	14 or less	15–18	19 or more
Vancomycin	0.03 mg	9 or less	10–11	12 or more

[a]mg = milligrams, units = 0.6 micrograms.

of the dosage of the antibiotic that should be effective in controlling the infection in the patient. A standardized microbial inoculum is added to tubes containing serial dilutions of an antibiotic, and the growth of the microorganism is monitored as a change in turbidity. In this way, the break point or minimum inhibitory concentration of the antibiotic that prevents growth of the microorganism in vitro can be determined (Figure 18.6).

The MIC indicates the minimal concentration of the antibiotic that must be achieved at the site of infection to inhibit the growth of the microorganism being tested. By knowing the MIC and the theoretical levels of the antibiotic that may be

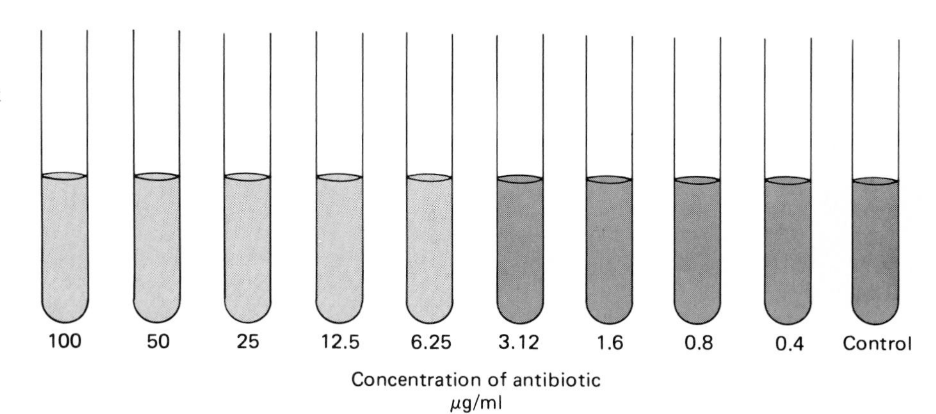

Figure 18.6 Tube dilution procedure for determining break point in MIC (minimum inhibitory concentration) determination. The minimal inhibitory concentration for the test illustrated here is 6.25 micrograms per milliliter. As compared with the Bauer–Kirby procedure, the MIC provides quantitative information that can be used to determine the dose rate and administration schedule for an antibiotic.

Table 18.11 Achievable Levels of Some Common Antibiotics in Various Body Fluids

Antibiotic	Achievable Peak Blood Levels (μg/mL)[a]	Achievable Urine Levels (μg/mL)[a]	Dose[b]/6–12 hours	
Clindamycin	1–4	>20	Oral	150–300 mg
	6–10	>60	IV	300–600 mg
Erythromycin	1–2		Oral	250–500 mg
	10–20		IV	300 mg
Penicillin	2–3	>300	Oral	500 mg
	6–8	>300	IM	1×10^6 Units
	4–7	>300	IV	500 mg
Ampicillin	1–3	>50	Oral	250–500 mg
	2–6	>20	IM	250–500 mg
	10–25	>100	IV	1–1.5 g
Cephalothin	3–18	>300	Oral	250–500 mg
	9–24	>1000	IM	0.5–1 g
	30–85	>1000	IV	1–2 g
Gentamicin	2–10	>20	IM/IV	1–2 mg
Tetracycline	1–2	>200	Oral	250–500 mg
	10–20	>200	IV	500 mg
Chloramphenicol	10–12	>100	Oral	1 g
	20–30	>200	IV	1 g
Nitrofurantoin		>100	Oral	50–100 mg

[a]μg/mL = micrograms per milliliter.
[b]IV = intravenus; IM = intramuscular; mg = milligram; g = gram; units = 0.6 micrograms.

achieved in body fluids, such as blood and urine, the physician can select the appropriate antibiotic, the dosage schedule, and the route of administration (Table 18.11). For systemic infections an achievable serum level of two to four times the MIC is desirable to ensure successful treatment of the disease. For central nervous system infections the MIC should be equal to the achievable level of the antibiotic in cerebrospinal fluid. For urinary tract infection the achievable level of the antibiotic in urine should be 20 times the MIC.

MIC determinations can be used for determining the antibiotic sensitivity of both aerobic and anaerobic microorganisms. The use of microtiter plates and automated inoculation and reading systems makes the determination of MIC's feasible for use in the clinical laboratory (Figure 18.7). MIC's can even be performed on normally sterile body fluids without isolating and identifying the pathogenic microorganisms. For example, blood or cerebrospinal fluid containing an infecting microorganism can be added to tubes containing various dilutions of an antibiotic and a suitable growth medium. An increase in turbidity would indicate the growth of microorganisms and the fact that the antibiotic at that concentration was ineffective in inhibiting microbial growth, whereas a lack of growth would indicate that the pathogenic microorganisms were susceptible to the antibiotic at the given concentration. By determining the minimum inhibitory concentration, the appropriate dosage as well as the right antibiotic can be selected for treating an infectious disease.

Minimal Bactericidal Concentration

MIC determinations are aimed at determining the concentration of an antibiotic that will inhibit growth but are not designed to determine whether the antibiotic is microcidal. The minimal bactericidal concentration (MBC) on the other hand is the lowest concentration of an antibiotic that will kill a defined proportion of viable organisms in a bacterial suspension during a specified exposure period. Typically the minimal bactericidal concentration is about three times higher than the MIC. In general, a 99.9 percent kill of bacteria at an initial concentration of 105–106 organisms per milliliter during an 18–24-hour exposure period is used to define

A

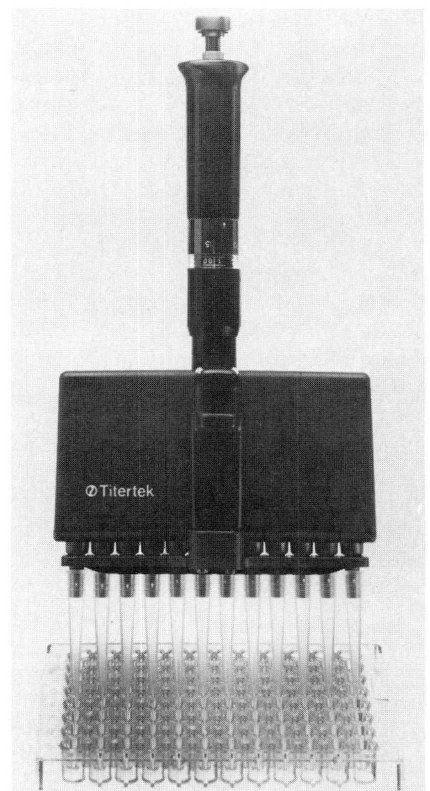

B

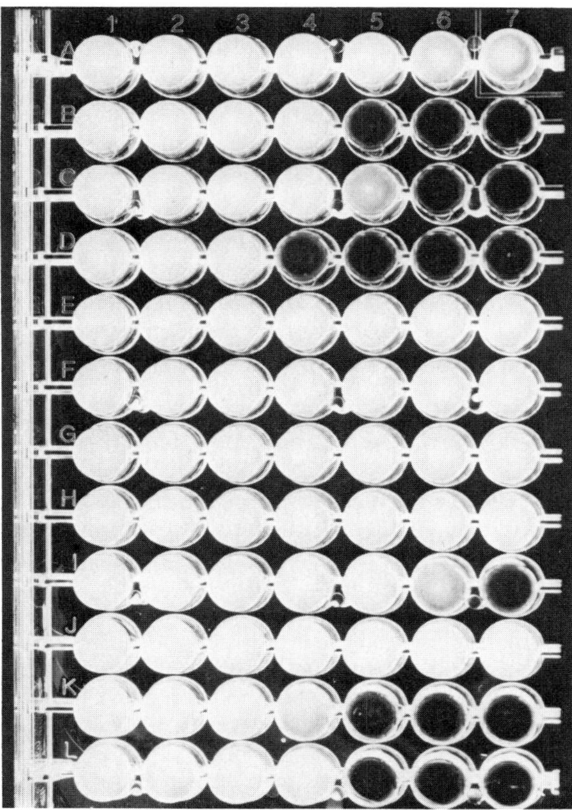

Figure 18.7 Test plates used for automated MIC determination. (A) This Titertek automatically fills the microtiter plates. (Courtesy Flow Laboratories, McLean, Virginia.) (B) In this photograph the results of MIC testing are shown. Lack of growth, that is, inhibition, appears as clear wells. (From BPS: Leon J. LeBeau, University of Illinois Medical Center, Chicago.)

the MBC. In order to determine the minimal bactericidal concentration, it is necessary to plate the tube suspensions showing no growth in tube dilution (MIC) tests to determine whether the bacteria are indeed killed or whether they survive exposure to the antibiotic at the concentration being tested.

Determination of the MBC is essential in cases of heart and bone infections. In these cases the normal body defenses cannot be relied upon to eliminate residual pathogenic microorganisms, and the antibiotic must kill the disease-causing microorganisms. The MBC also is particularly useful in determining the appropriate concentration of an antibiotic for use in treating patients with lowered immune defense responses, such as may occur in patients receiving chemotherapy treatment for cancer.

Serum Killing Power

Another approach to determining the effectiveness of an antibiotic is to measure serum killing power (Schlicter test). Instead of adding dilutions of an antibiotic to suspensions of bacteria in a growth medium, a bacterial suspension is added to dilutions of the patient's serum. The ability of the bacteria to grow in the patient's blood is assessed by measuring changes in turbidity. Assuming that the patient is being treated with an antibiotic, no bacterial growth should occur. The break point in the dilutions where bacterial growth occurs reflects the concentration of the antibiotic in the patient's blood and the in vivo effectiveness of the antibiotic in controlling the infection. Inhibition at dilutions of the patient's serum of greater than or equal to 1:8 is considered an acceptable level.

Summary

When host defense mechanisms fail and exposure to a pathogenic microorganism results in the development of an infection, there are a variety of antimicrobial agents that can be employed for the treatment of disease. The bacterial diseases are perhaps the easiest of the infectious diseases to treat because the differences between prokaryotic and eukaryotic cells provide sites against which chemicals can exert selective inhibition against bacteria without killing or severely inhibiting host mammalian cells. The discovery and use of antibiotics, which are antimicrobial agents produced by microorganisms, have revolutionized medical practice. The choice of a particular antibiotic for treating a disease rests on the nature of the infecting microorganism and the tissues in which the infection occurs. In general, the murein component of the bacterial cell wall and the 70S ribosomes of bacteria provide targets against which many of the therapeutically useful antibiotics are aimed. Prudent medical practice requires that antibiotics be specifically aimed at the infecting microorganisms; that is, wherever possible specific or narrow spectrum antibiotics should be employed, so as to avoid the selective pressure for the development of antibiotic-resistant strains of pathogenic microorganisms. The occurrence of antibiotic-resistant strains is becoming a very serious problem in treating infectious diseases.

The treatment of infections by viruses and eukaryotic microorganisms presents a greater problem than bacterial infections for finding specific inhibitors that are not also directed at mammalian cells. Systemic mycotic and protozoan infections are particularly difficult to treat, often persisting for long periods of time. Many of the drugs used in the treatment of such infections also cause side effects toxic to human tissues.

Microbiologists associated with the pharmaceutical industry are continuously screening for new antibiotics or derivatives of known antimicrobial agents that might be of therapeutic value in treating infectious diseases. Advances continue to be made in our ability to treat diseases caused by microorganisms as new antimicrobial agents are discovered and developed. We have made great advances in our ability to prevent and treat infectious diseases and look forward to achieving a higher standard of human health in the future.

Study Outline

A. The use of antimicrobial agents in modern medical practice is aimed at eliminating microorganisms or preventing the establishment of an infection.

B. To be of therapeutic value an antimicrobial agent must be selective; it must exhibit greater toxicity against the infecting pathogen than against the host organism.

C. The mode of action of most antimicrobial agents involves biochemical features of the pathogen that are not possessed by the normal host cells.

D. Antibiotics are biochemicals produced by microorganisms that inhibit the growth of, or kill, other microorganisms; today they may be produced chemically or may be chemically modified forms of microbial biosynthetic products. The term antimicrobic is often used instead of antibiotic to describe natural and synthetic antimicrobial compounds.

E. Modes of action of antibacterial agents.

- **1.** The most common antibiotics used in treating bacterial infections are effective because they inhibit cell wall or protein synthesis.
- **2.** Cell-wall inhibitors.
  - **a.** Penicillins (from fungus *Penicillium*) and cephalosporins (from fungus *Cephalosporium*) inhibit formation of peptide crosslinkages within the peptidoglycan backbone of the cell wall leaving the bacteria subject to attack by autolysins and rendering them vulnerable to osmotic shock; they do not remove intact cell walls and so are ineffective against resting or dormant cells.
  - **b.** Cycloserine, bacitracin, and vancomycin also inhibit bacterial cell-wall synthesis but side effects limit their therapeutic usefulness.

3. Inhibitors of protein synthesis.
   a. The aminoglycoside antibiotics (produced by actinomycetes) include streptomycin, gentamicin, neomycin, kanamycin, tobramycin and amikacin. They are used against Gram negative bacteria; they must be transported across the cytoplasmic membrane to be effective, and they bind specifically to the 70S ribosomal subunit, blocking protein synthesis and decreasing fidelity to genetic code transmission.
   b. Tetracyclines bind specifically to the 30S ribosomal subunit, preventing the addition of amino acids to the growing peptide chain.
   c. Chloramphenicol acts by binding to the 50S ribosomal subunit; it is a broad spectrum antibiotic, active against many Gram negative species.
   d. Erythromycin binds to 50S ribosomal subunits, blocking protein synthesis and is most effective against Gram positive cocci.
4. The polymyxins inhibit membrane transport; they interact with the cytoplasmic membrane causing changes in the structure of the bacterial cell membrane, resulting in leakage of cellular constituents.
   a. Polymyxin B is bactericidal and effective against Gram negative bacteria.
   b. Polymyxins B and E (colistin) are used in the treatment of infections caused by *Pseudomonas* species and other Gram negative bacteria resistant to penicillin and aminoglycoside antibiotics.
5. Inhibitors with other modes of action.
   a. Sulfonamides, sulfones, and *p*-aminosalicylic acid interrupt normal metabolism.
   b. Trimethoprim blocks bacterial growth by preventing the formation of a coenzyme required for metabolism; it is a broad spectrum antibiotic used primarily for the treatment of urinary tract infections due to *E. coli, Proteus, Klebsiella*, and *Enterobacter*.
   c. Methanamine, nalidixic acid, ocolinic acid and nitrofurantoin are used to treat infections within the urinary tract.
   d. Isoniazid is effective against *Mycobacterium tuberculosis*; its action appears to involve inhibition of biosynthesis of mycolic acid, a component of the cell walls of mycobacteria.

F. Antifungal agents.
1. Polyene antibiotics act by altering the permeability properties of the cytoplasmic membrane.
   a. Nystatin is used in the treatment of topical infections caused by *Candida* species.
   b. Amphotericin B (produced by *Streptomyces nodosus*) is a broad spectrum antibiotic used in the treatment of systemic fungal infections; may cause kidney damage.
2. Miconazole and clotrimazole have a broad spectrum of antifungal activity and are used in the topical treatment of superficial mycotic infections; they act by altering membrane permeability.
3. Flucytosine is used in the treatment of systemic fungal infections because it inhibits normal nucleic acid synthesis.
4. Griseofulvum causes a disruption of the mitotic spindles during fungal mitosis; it is used in the treatment of dermatophytic fungal infections of the skin, hair, and nails.

G. Antiprotozoan agents are difficult to use because many pathogenic protozoa exhibit a complex life style and different antimicrobial agents are needed for different forms of the same pathogen depending on the stage of the life cycle and the tissues involved.
1. Malaria is caused by protozoan species of the genus *Plasmodium*, which exhibit complex life cycles, parts of which are carried out in the liver and blood of humans.
   a. Antimalarial agents include chloroquine and amodiaquine, which appear to interfere with DNA replication and are effective against erythrocyctic forms.
   b. Chloroguanide interferes with essential metabolic reactions.
   c. Primaquine is effective against both erythrocytic and hepatic stages.

2. Metronidazole is used in the treatment of dysentery caused by *Entamoeba histolytica* because it interferes with hydrogen transfer reactions, thus inhibiting the growth of anaerobes.
3. Pentamide and related diamidine compounds are used in the treatment of infections caused by *Trypanosoma*; they interfere with DNA metabolism.
4. Melarsoprol is an arsenic-containing compound used in the treatment of some stages of human trypanosomiasis because of its ability to penetrate into the cerebrospinal fluid, where it inactivates a number of enzymes.
5. Sodium stibogluconate is an antimony-containing compound useful in treating diseases caused by *Leishmania*.
6. Suramin is a nonmetalic compound that inhibits a wide variety of enzymes.
7. Nifurtimox is effective against *Trypanosoma cruzi*, the causative organism of Chagas' disease.

H. Antiviral agents.

1. There are no broad spectrum antiviral agents because the integral role of the mammalian cell in viral replication complicates the difficulty of finding compounds that specifically inhibit viral replication.
2. Indoxuridine is an analogue of thymidine, it interferes with DNA metabolism in both viral and mammalian cells and is used in the treatment of herpes simplex keratitis.
3. Amantadine is used in the prophylactic treatment of high-risk patients during influenza A epidemics; it interferes with adsorption and uptake of influenza A viruses.
4. Trifluridine is a nucleic base analog that inhibits DNA synthesis; it is effective against epithelial keratitis of the eye caused by *Herpes simplex*.
5. Vidarabine inhibits mammalian DNA synthesis to a lesser extent than viral DNA replication; it is effective in treating herpes simplex encephalitis and keratoconjunctivitis.
6. Acyclovir is used in the treatment of herpetic ocular disease, herpes encephalitis, cold sores, and genital herpes; it is a nucleoside analogue, inhibiting herpes simplex viral DNA polymerase, thus blocking viral DNA replication.

I. The selection of an antibiotic depends upon the sensitivity of the infecting microorganism to the particular antimicrobial agent, the side effects, whether the antimicrobial agent will remain in an active form long enough to be selectively toxic to the pathogens, and the chemical properties of the antimicrobial that determine its distribution within the body.

J. Antibiotics are either narrow (targeted at a specific pathogen or species) or broad (capable of inhibiting a wider range of bacterial species) spectrum.

K. There is rising concern about the overuse of antibiotics due to the development of antibiotic-resistant bacterial strains.

L. Culture and sensitivity studies should be done to determine proper antibiotic therapy.

1. The Bauer–Kirby test is an agar diffusion test that is used to determine if a bacterial isolate is sensitive or resistant to particular antibiotics.
2. The autobac system, a liquid diffusion method, is based on light scattering in broth.
3. Minimal inhibitory concentration (MIC) procedures determine the lowest concentration of an antibiotic that will inhibit microbial growth; by comparing the MIC with the achievable levels of the antibiotic that can be achieved at a given body site the appropriate dosage and most effective antibiotic can be selected.
4. The minimal bactericidal concentration (MBC) is the lowest concentration of an antibiotic that will kill a defined proportion of organisms in a bacterial suspension during a specified exposure period; determining the MBC is important in cases of endocarditis and lowered immune defense responses.
5. The Schlicter test determines serum killing power; bacterial suspensions are added to dilutions of the patient's serum to test the effectiveness of the antibiotic in controlling the infection.

Study Questions

1. List and briefly describe the factors involved in selecting an appropriate antibiotic.
2. Match the following statements with terms in the list to the right.
   - **a.** Subject to inactivation by penicillinase enzymes (β-lactamases)
   - **b.** Does not remove intact cell walls, so ineffective against resting or dormant cells.
   - **c.** Therapeutic use limited by toxic reactions involving the central nervous system.
   - **d.** Generally broad spectrum; used as alternatives to penicillin, as for example in individuals who are allergic to penicillin.
   - **e.** Inhibit formation of peptide crosslinkages within the peptidoglycan backbone of the cell wall; bacterial cell walls are then subject to attack by autolysins, rendering the cell vulnerable to osmotic shock.
   - **f.** Contain β-lactam ring; have related biochemical structures.
   - **g.** Has relatively narrow spectrum of activity; most effective against Gram positive cocci.

   (1) Penicillins
   (2) Cephalosporins
   (3) Cycloserine
3. Match the following statements with the terms in the list to the right.
   - **a.** Used almost exclusively against Gram negative bacteria.
   - **b.** Restricted to topical application because of severe toxicity reactions.
   - **c.** Streptomycin, gentamicin, neomycin, kanamycin, tobramycin, amykacin.
   - **d.** Used to treat only serious infections by penicillin-resistant strains of *Staphylococcus aureus* or when patient is allergic to penicillins and cephalosporins.
   - **e.** Bind to 30S ribosomal subunit of 70S ribosome, blocking protein synthesis and decreasing fidelity of genetic code translation.

   (1) Aminoglycoside antibiotics
   (2) Vancomycin
   (3) Bacitracin
4. Match the following statements with the terms in the list to the right.
   - **a.** Bind specifically to 30S ribosomal subunits, apparently blocking the receptor site for attachment of tRNA to mRNA on the ribosome, thereby preventing the addition of amino acids to the growing peptide chain.
   - **b.** Binds to 50S ribosomal subunits, blocking protein synthesis.
   - **c.** Most effective against Gram positive cocci.
   - **d.** Use restricted by side effects.

   (1) Tetracyclines
   (2) Chloramphenicol
   (3) Erythromycin
5. Match the following statements with the terms in the list to the right.
   - **a.** Inhibits several bacterial enzymes; the specific mode of action is unknown.
   - **b.** Bactericidal; effectiveness limited to Gram negative bacteria.
   - **c.** Broad spectrum antimicrobic.
   - **d.** Blocks bacterial growth by preventing the formation of a coenzyme required for metabolism.
   - **e.** Used primarily for the treatment of urinary tract infections due to *Escherichia coli, Proteus, Klebsiella*, and *Enterobacter*.
   - **f.** Acts by blocking DNA synthesis.
   - **g.** Interact with cytoplasmic membrane, causing changes in the structure of the bacterial cell membrane resulting in leakage of cellular constituents.

   (1) Polymyxins
   (2) Polymyxin B
   (3) Sulfonamides, sulfones, *p*-aminosalicylic acid
   (4) Trimethoprim
   (5) Nalidixic acid
6. Match the following statements with the terms in the list to the right.

a. Effective against *Mycobacterium tuberculosis*.
b. Used in the treatment of systemic fungal infections.
c. Primary use in the treatment of topical infections by *Candida* species.
d. Broad spectrum of antifungal activities.
e. Used in topical treatment of superficial mycotic infections.
f. Appears to alter membrane permeability, leading to fungal inhibition and/or death.
g. Polyene antibiotics that act by altering the permeability properties of the cytoplasmic membrane.
h. Inhibits normal nucleic acid synthesis.
i. Causes disruption of mitotic spindles, inhibiting fungal mitosis.
j. Used in the treatment of dermatophytic fungal infections of the skin, hair, and nails.

(1) Isoniazid
(2) Nystatin
(3) Amphotericin B
(4) Miconazole
(5) Clotrimazole
(6) Flucytosine
(7) Griseofulvum

7. Match the following statements with the terms in the list to the right.
   a. Effective against epithelial keratitis of the eye caused by *Herpes simplex*.
   b. Antimony-containing compound.
   c. Used in the treatment of dysentery caused by *Entamoeba histolytica*.
   d. Used in the treatment of infections caused by *Trypanosoma*.
   e. Arsenic-containing compound.
   f. Used in treating some stages of human trypanosomiasis because of its ability to penetrate into the cerebrospinal fluid.
   g. Effective against both erythrocytic and hepatic stages of malaria.
   h. Useful in treating diseases caused by *Leishmania*.
   i. Effective against *Trypanosoma cruzi*, the causative organism of Chagas' disease.
   j. Used in the treatment of genital herpes.
   k. Prophylactic treatment of high-risk patients during influenza A virus epidemics.

   (1) Chloroquine
   (2) Chloroguanide
   (3) Primaquine
   (4) Pyrimethamine
   (5) Metronidazole
   (6) Melarsoprol
   (7) Sodium stibogluconate
   (8) Nufurtimox
   (9) Idoxuridine
   (10) Amantadine
   (11) Trifluridine
8. List four antimicrobial sensitivity tests.
9. For each of the following indicate how many times higher than the MIC should be achieved and in what body fluid.
   a. Endocarditis
   b. Urinary tract infection
   c. Bacterial meningitis
   d. Typhoid fever

Some Thought Questions

1. Should pharmacists be required to dispense generic drugs?
2. Should antibiotics be sold over-the-counter or should they require prescriptions from licensed physicians?
3. In the 1960s hospital walls often were washed with antibiotics. Why? Why has the practice been discontinued?
4. Who should pay for developing new drugs?
5. What role do antibiotics play in nature?
6. Why does it cost so much money to develop an effective antibiotic for human usage?
7. Why are different antibiotics sometimes used for treating infections caused by identical pathogens when different parts of the body are involved?
8. Why is determining the MIC and spectrum of antibiotic sensitivity important?

Additional Sources of Information

Berdy, J. (ed.). 1980–1982. *Handbook of Antibiotic Compounds*. CRC Press, Inc., Boca Raton, Florida. Contains descriptions of the numerous antibiotics.

Bryan, L. E. 1982. *Bacterial Resistance and Susceptibility to Chemotherapeutic Agents*. Cambridge University Press, New York. Discusses of the basis of resistance of microorganisms to antibiotics.

Bryan, L. (ed.). 1984. *Antimicrobial Drug Resistance.* Academic Press, Inc., Orlando, Florida. This text is devoted to the discussion of resistance to antibacterial, antifungal, antiviral, and antimalarial agents, as well as metal ions.

Gilman, A. G., L. S. Goodman, and A. Gilman (eds.). 1980. *Goodman and Gilman's The Pharmacological Basis of Therapeutics*. Macmillan Publishing Company, New York. Extensive descriptions of the antibiotics and other pharmaceuticals used in medical practice.

Lynn, M. and M. Solotorovsky (eds.). 1981. *Chemotherapeutic Agents for Bacterial Infections* (Benchmark Papers in Microbiology). Van Nostrand Reinhold, New York. Discussion of the origin, production, spectrum of activity, and mode of action of antibiotics.

Physician's Desk Reference. published annually. Medical Economics Co., Oradell, New Jersey. A continuously updated compilation of the descriptions of drugs used in medical practice; includes information on the pharmacological properties, available dosages, recommended uses, side-effects, and trade names of antibiotics.

Pratt, W. B. 1977. *Chemotherapy of Infection*. Oxford University Press, New York. An informative text describing the variety of antimicrobial agents that are available for treating infectious diseases.

UNIT SEVEN

Host Defenses Against Disease-Causing Microorganisms

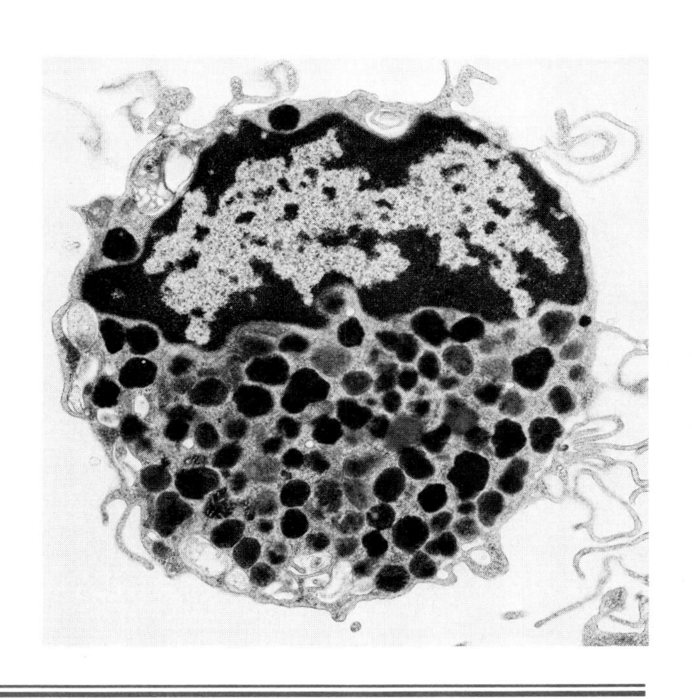

19 *Host Defenses Against Pathogenic Microorganisms*

KEY TERMS

agranulocytes
allelopathic substances
band cells
basophils
bradykinin
complement
degranulation
diapedesis
dust cells
eosinophils
fever
granulocytes
histamine
histiocytes
host-cell specific
immune adherence
immune defense network
immune system
indigenous microbial population
inflammatory exudate
inflammatory response
interferon
keratin
Kupffer cells
lactoferrin
leukocytes
leukocytosis
lymphocytes
lysozyme
macrophages
mast cells
microglia
monocytes
mononuclear phagocyte system
muco-ciliary escalator system
nonspecific defense mechanisms
neutrophilia
neutrophils
normal microbiota
normal microflora
phagocytosis
phagosome
prostaglandins
pus
pyrogens
transferrin
shift to the left
stab cells
vasodilators
wandering cells

OBJECTIVES

After reading this chapter you should be able to

1. Define the key terms.
2. Describe the physical barriers that protect the body against disease-causing microorganisms.
3. Describe the chemical barriers that protect the body against disease-causing microorganisms.
4. Describe what is meant by the term normal human microbiota and how these microorganisms contribute to the healthy state of humans.
5. Describe the different types of phagocytic blood cells and how these cells protect the body against microbial infections.
6. Describe the inflammatory response and how it protects the body against pathogenic microorganisms.

Nonspecific Host Defense Mechanisms

Humans have numerous defense systems that protect against microbial infections. The first lines of defense are nonspecific in that they do not respond to individual species but rather are generally protective against a wide range of potential pathogens. These nonspecific defenses are augmented by a second line of defense, the immune system, which is a complex defense network that recognizes and defends against specific pathogens. Taken together, the nonspecific defenses and the immune system provide an integrated immune defense network with adequate redundancy to protect the body against most potential pathogens.

> Ecological equilibrium with microorganisms is an ideal state but one which is not readily achieved and is frequently disturbed. Microbial diseases are the manifestations of its failures.
>
> —RENÉ DUBOS, *Mirage of Health*

Physical Barriers

Intact body surfaces represent the first line of defense against microorganisms (Figure 19.1). These surface tissues of the body physically block the entry of microbes into the body. The epidermis and linings of various organs act as mechanical barriers to microbial invasion. Most microorganisms are noninvasive, which means they are unable to penetrate the skin and mucous membranes. The outer surface of the skin layer is composed of keratin, a protein not readily enzymatically degraded by microorganisms. Keratin resists the penetration of water and thus proves a formidable external barrier for microorganisms to penetrate.

> By a jungle river an alligator sleeps and sleeps, can risk it because besides the built-in defenses that all of us have it has that outer barricade . . . Man has his outer barricade too, his skin, gentle, soft, yet doing rough work . . . Skin is a defense against assaults of many kinds . . .
>
> —GUSTAV ECKSTEIN, *The Body Has a Head*

The respiratory tract is protected in part against the invasion of pathogenic microorganisms by cilia and secretions of the mucous membranes that line the nose and pharynx of the upper respiratory tract. Microorganisms, dust, pollen, and other particulates tend to become entrapped in mucus and are swept out of the body by ciliated epithilial cells composing the lining of much of the respiratory tract. The ciliated epithelial cells effectively act as traps to prevent potential pathogens from penetrating the surface tissues of the respiratory tract (Figure 19.2). Some of the mucus and microorganisms are swept out of the body through the oral and nasal cavities. This system is referred to as the muco-ciliary escalator system. Sneezing and coughing tend to remove many of these microorganisms from the respiratory tract (Figure 19.3). The gag reflex also helps remove post nasal drip and mucus swept up by the ciliated epithelium of the bronchi, with their associated microorganisms. Additionally, the swallowing reflex removes most particulates from the respiratory tract by moving them

Figure 19.1 The first lines of defense against infection: protection by external body surfaces acting as barriers.

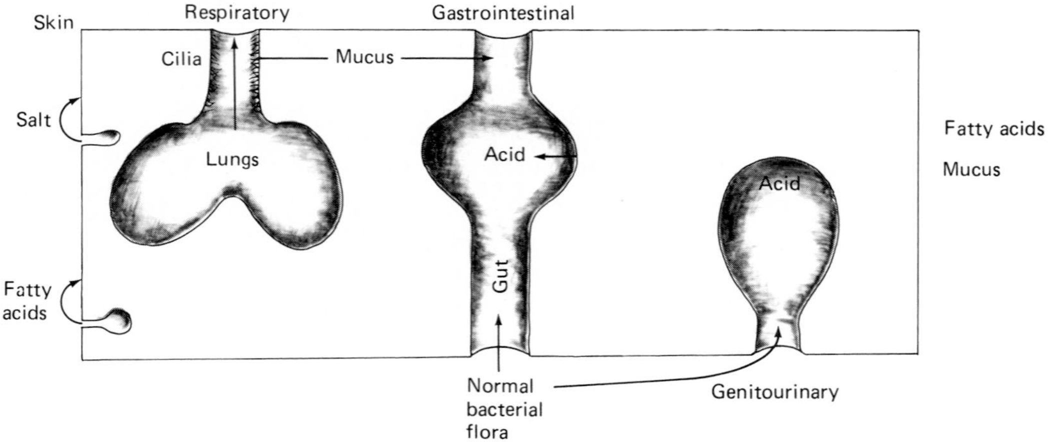

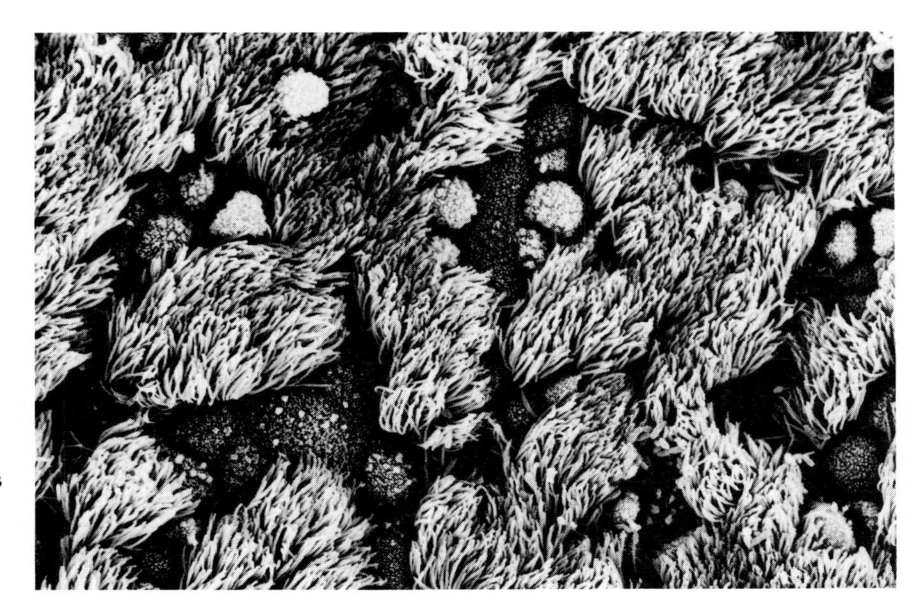

Figure 19.2 Micrograph showing cilia and mucous membrane lining of the trachea that protects the upper respiratory tract against microbial invasion. (From BPS: K. E. Muse, Duke University Medical Center.)

into the digestive tract, including microorganisms that become attached to mucus. The mucous membranes of the intestinal tract make it difficult for pathogenic microorganisms to attach to and penetrate the tract lining.

Air comes in tickly
Through my nose,
Then very quickly —
Out it goes:
Ahhh — CHOO!

With every sneeze
I have to do,
I make a breeze —
Ahh — CHOOH! Ahh — CHOOH!

—MARIE LOUISE ALLEN, *Sneezing*

Various surfaces of the body are protected against accumulations of microorganisms by the movement of fluids across the surface. The flow of fluids across body surfaces tends to wash microorganisms from those surfaces. For example, tears contiunuously remove microorganisms from the eye

Figure 19.3 Coughing and sneezing expel microorganisms from the respiratory tract. Unfortunately this defense mechanism also propels droplets into the air, and thus potentially toward the respiratory tract of another individual, as shown in this British poster that appeared during World War II. Using a handkerchief reduces the spread of droplets. (Reprinted by the kind permission of the Trustees of the Imperial War Museum, London.)

tissues. Urine flushes microorganisms from the surfaces of the urinary tract. Perspiration washes microorganisms from the skin surface.

Chemical Defenses

Some of the fluids that wash body surfaces also contain antimicrobial chemicals. Tears which continously wash the eye, for example, contain lysozyme, an enzyme that protects the tissues of the eye against the undesirable growth of microorganisms. Lysozyme is found in other body fluids, including saliva and mucus. This enzyme degrades the peptidoglycan of the cell walls of bacteria. Body fluids that contain lysozyme therefore possess antimicrobial activity, particularly against Gram positive bacteria with their relatively high cell-wall peptidoglycan content. In a similar way swallowing, coughing, and sneezing expose bacteria to body fluids with antimicrobial activity, thus reducing the number of potential pathogens.

Some chemicals within the body bind iron, thereby withholding this essential growth element from pathogenic microorganisms. By limiting the amount of available iron, these compounds limit the growth of pathogens. Lactoferrin and transferrin are examples of such iron-binding compounds. Lactoferrin is present in tears, semen, breast milk, bile, and nasopharyngeal, bronchial, cervical, and intestinal mucosal secretions. Transferrin is present in serum and the intercellular spaces of various tissues and organs.

The acid of the stomach provides another chemical barrier that acts to prevent microbial invasion of the body. Most microorganisms entering the digestive tract are unable to tolerate the low pH (normally 1–2) of the stomach. Thus, the numbers of viable microorganisms are greatly reduced during passage through the stomach. Bile and digestive enzymes in the intestines further reduce the numbers of surviving microorganisms. As a result, most microorganisms entering the intestinal tract are degraded during passage through the gastrointestinal system.

Other body tissues are also protected against microbial invasion. The genitourinary tract is protected from pathogenic microorganisms by a variety of mechanisms that include the outward flow of mucus, low pH (normally 4.0–4.5), high salt concentrations, and the presence of antimicrobial enzymes. In addition to the physical resistance of skin to microbial penetration, sweat on the skin surface contains high concentrations of salt and fatty acids that inhibit microbial growth.

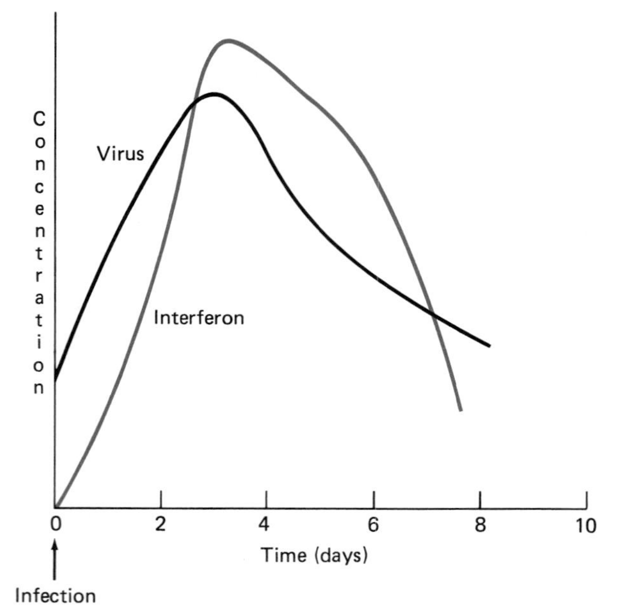

Figure 19.4 Appearance of interferon in relation to numbers of influenza viruses in the course of an infection.

Interferon. Blood contains several chemical factors that defend against microbial infections. Interferon, a glycoprotein found in low concentrations in blood, appears to be the main defense mechanism against viral infections (Figure 19.4). The production of interferon plays a key role in response to viral infections like the common cold. Interferon is produced by infected cells and by cells of the lymphatic system. Once interferon is produced, it limits viral reproduction and in many cases leads to the elimination of the infecting viruses. The production of interferon is considered a nonspecific resistance factor because interferon produced in response to one virus is also effective in preventing the replication of other types of viruses. Interferon, however, is host-cell specific. This means that interferon produced by human cells protects the human body against a variety of viral pathogens but that interferon produced in another animal, such as a mouse, will not protect human cells against viral infections.

Another interesting aspect of interferon is that it does not act directly on viruses nor on virally infected cells. Rather interferon produced by a cell that is infected with a virus migrates to uninfected cells, protecting those cells from viral infections (Figure 19.5). Generally interferon is produced in very limited quantities so that only neighboring cells

MICROBIOLOGY HEADLINES

More Kudos for Interferon

By S. I. BENOWITZ

Interferon burst upon the scientific scene in 1957, billed as a possible cure for everything from cancer to the common cold. But widespread use of this naturally occurring protein was stymied by production difficulties and high costs. By 1980, when mass production of interferon was achieved (SN: 1/26/80, p. 52), its precarious promise gained a toehold on reality.

Following some preliminary successes using the protein against a variety of cancers (SN: 4/3/82, p. 230), scientists have now found it effective against a blood cell cancer called non-Hodgkin's lymphoma. In the Nov. 1 New England Journal of Medicine, scientists from the National Cancer Institute (NCI) in Frederick, Md., and Hoffmann-La Roche in Nutley, N.J., report that "recombinant leukocyte A interferon may be an effective new therapy for some patients with low- and intermediate-grade non-Hodgkin's lymphoma."

Lymphomas are cancers that affect the white blood cells of the immune system. They are characterized by the abnormal growth of lymphocytes, the infection-fighting cells in the lymph nodes, spleen and thymus. Lymphomas are usually classified as either Hodgkin's disease or the less common non-Hodgkin's lymphomas. Approximately 23,000 new cases of non-Hodgkin's lymphomas turn up each year in the United States.

NCI researcher Kenneth Foon and co-workers treated 37 patients who had non-Hodgkin's lymphoma and were no longer benefiting from chemotherapy, the standard treatment for the disease. The patients were given intramuscular injections of leukocyte A (or alpha) interferon, one of the three known human forms of the substance, three times a week for three months or longer. Only one person with high-grade cancer was helped. But of 30 patients with low- or intermediate-grade lymphoma, 10 showed partial responses and five had complete remissions.

The disease returned, however, within a few months after the interferon injections were halted. Foon speculates that long-term maintenance therapy may be needed to hold it in check. "It's hard to say how long the effects will last," Foon says. "We haven't followed patients long enough to be able to tell whether interferon treatments are going to have a major impact on the overall course of the disease. But these results are encouraging."

It's not clear exactly how the interferon works, says NCI immunologist Ronald Herberman. But he notes that interferon's apparent antilymphoma powers may be due to a combination of directly slowing down lymphoma cell growth while boosting the body's immune response.

"It's difficult to generalize about interferon's use in treating cancer," says Herberman. "It's very promising for the lymphomas and leukemias, but so far the results in other types of cancers, like breast and colon, have been disappointing."

Source: Science News, Nov. 10, 1984. Reprinted by permission.

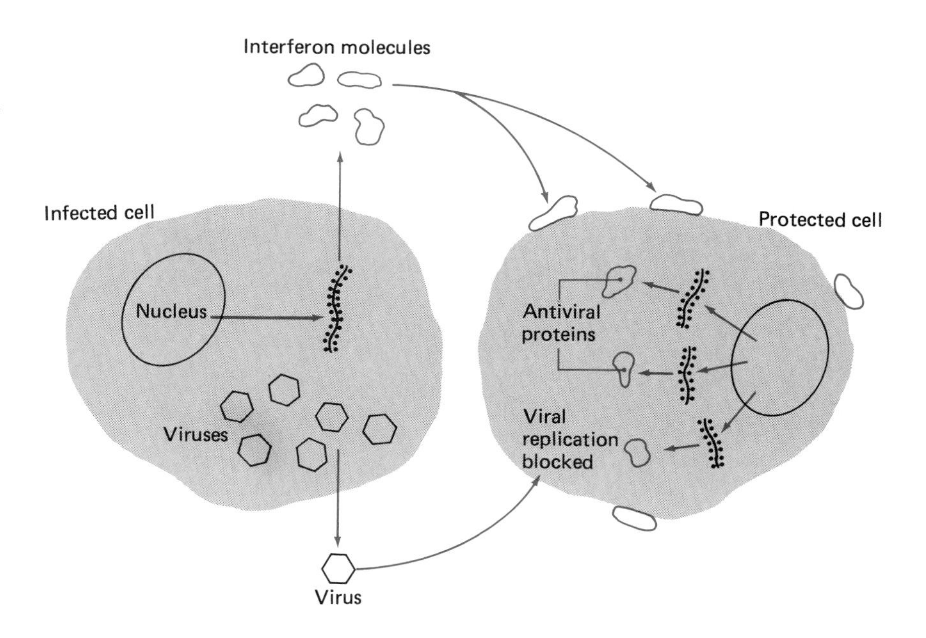

Figure 19.5 The production of interferon activates a system that prevents viral replication within host cells. (A) General sequence of events. (B) Details of interferon action.

are immediately protected. Interferon does not block the entry of the virus into a cell; rather, it prevents the replication of viral pathogens within protected cells by inducing the production of an antiviral protein, which, in turn, blocks the synthesis of the viral genetic macromolecules. Thus, interferon has no direct effect on viruses, and the antiviral action of interferon is mediated by cells in which the antiviral state is induced. For this reason interferon has no effect on cells that are already infected with a virus; for example, interferon cannot eliminate herpes viruses from infected cells.

Besides its role in protecting against viral infections, interferon acts as a regulator of the complex defense network that protects the body against infections and the development of malignant cells. As such, interferon is involved in the control of various phagocytic blood cells that engulf and kill various pathogens (including bacteria) and abnormal or foreign mammalian cells (including cancer cells).

Complement. In addition to interferons, blood contains other glycoproteins called complement that play a role in the removal of invading pathogens. Complement glycoproteins include at least eleven different proteins in the blood serum. As the name implies this substance augments or complements other defenses that protect the body against microbial infections. The complement system normally acts in association with the specific immune response to remove invading microorganisms, but the complement system can act in an alternate pathway as part of the nonspecific defenses of the body (Figure 19.6). A third pathway, the properdin pathway, can produce the same end result. When activated, complement molecules attach to the surface of an invading cell in a cascade fashion, one complement molecule adding after another. The sequential association of complement molecules with the cell surface alters the permeability of the cytoplasmic membrane of that cell, resulting in the

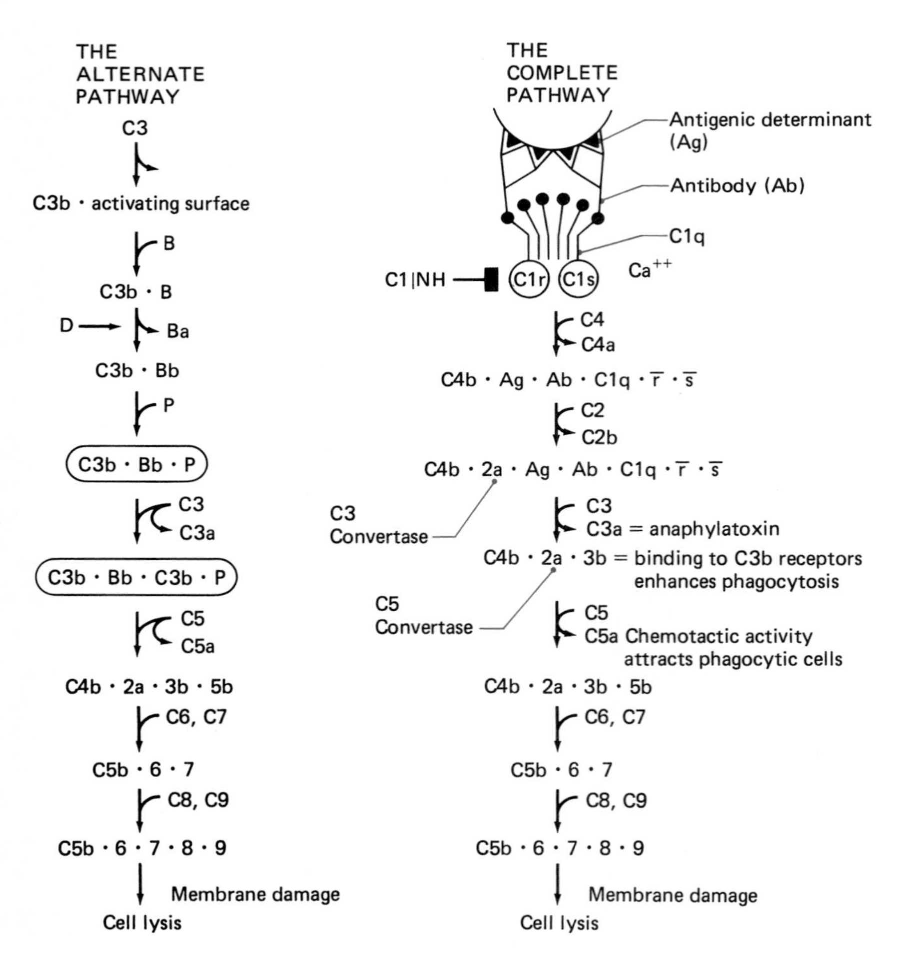

Figure 19.6 Activation pathways and functions of complement molecules.

death of the cell. Biochemicals, such as endotoxin of the Gram negative cell wall, can trigger the nonspecific response of the complement system, and thus, the presence of complement tends to prevent infections of the circulatory system by Gram negative bacteria. Besides cascading onto the surfaces of bacterial cells that results in cell lysis, complement molecules activate phagocytic blood cells. This phenomenon, known as immune adherence, occurs when complement molecules bind to the surface of a microorganism and then interact with receptors on the surface of a phagocytic blood cell to enhance the efficiency by which that blood cell engulfs and destroys the invading microbe. Some of the complement molecules act as chemotactic agents to attract phagocytic cells to the site of infection. This is particularly important in the inflammatory response which will be discussed later in this chapter.

The Normal Human Microbiota

Microorganisms associated with some of the surface tissues of the body also help defend against disease-causing pathogens. Germ-free animals, delivered by caesarean section and reared in the absence of microorganisms, develop gross anatomical abnormalities of the gastrointestinal tract and are also more susceptible to disease than animals with normal associated microbiota. The normal associated microbiota of animals contribute in part to the normal defense mechanisms that protect animals against infection by pathogens. Animals reared in germ-free environments are especially susceptible to disease. Apparently, such animals fail to develop adequate defenses against pathogenic microorganisms. The lymphatic tissues, which play an important role in protecting the body against infections, typically are poorly developed in germ-free animals.

The microbial populations that are associated with particular body tissues are referred to as the indigenous microbial populations, normal microflora, or normal microbiota. Although the term microflora is used extensively, the term microbiota is preferable as it avoids any inference that microorganisms are little plants. These microbes establish a dynamic and mutally beneficial association with body tissues. However, if population imbalances occur, these same microbes can cause disease.

The normal microbiota of the human body inhabits the surface tissues of the skin, oral cavity, respiratory tract, gastrointestinal tract, and genitourinary tract. By colonizing body tissues, the normal microbiota preempt the colonization of those tissues by other microbes, some of which may cause disease. Many of these microorganisms produce antimicrobial substances (also termed allelopathic substances) that act to prevent the establishment of infection on or through the skin by pathogenic microorganisms. For example, lactic acid production by the indigenous *Lactobacillus* species of the vaginal tract lessens the probability of infection with *Neisseria gonorrhoeae*. Other mechanisms of antagonism by the indigenous microbiota that enhance host resistance to disease include alteration of the oxygen tension, production of allelopathic substances, and competition for available nutrients.

When the normal microbiota are adversely affected—for example, by antibiotics that are used to treat a disease condition—an imbalance may occur that leads to the development of disease. For example, the use of antibiotics sometimes disrupts the balance of the microbial community of the gastrointestinal tract, permitting the growth of *Clostridium difficile*, which causes a severe and sometimes fatal gastrointestinal tract infection (antibiotic-associated pseudomembranous enterocolitis). Similarly, women taking anitbiotics sometimes develop vaginitis due to an overgrowth of the fungus *Candida albicans*, which is normally held in check by the indigenous bacteria of the vaginal tract.

Germ-free animals appear essentially normal and are able to reproduce themselves for several consecutive generations in the man-made, germ-free world. But they are highly susceptible to the common ubiquitous microbes, even to those usually unable to cause disease in animals raised under less esoteric conditions. The germ-free animals commonly develop severe infections, of which they die, as soon as they are placed in the open world. It is of interest in this regard that their internal organs are very deficient in certain constituents always present in normal animals, in particular the cells making up the so-called lymphoid tissue, which presumably plays a part in resistance to infection. This tissue appears, therefore, to be produced normally as an adaptive, protective response to the microorganisms which contaminate all ordinary objects and to which all living things are exposed constantly.

—René Dubos, *Mirage of Health*

In addition to their role in preventing certain diseases, the indigenous microbiota contribute to the nutrition of the animal by synthesizing nutrients essential to the welfare of the host. For example,

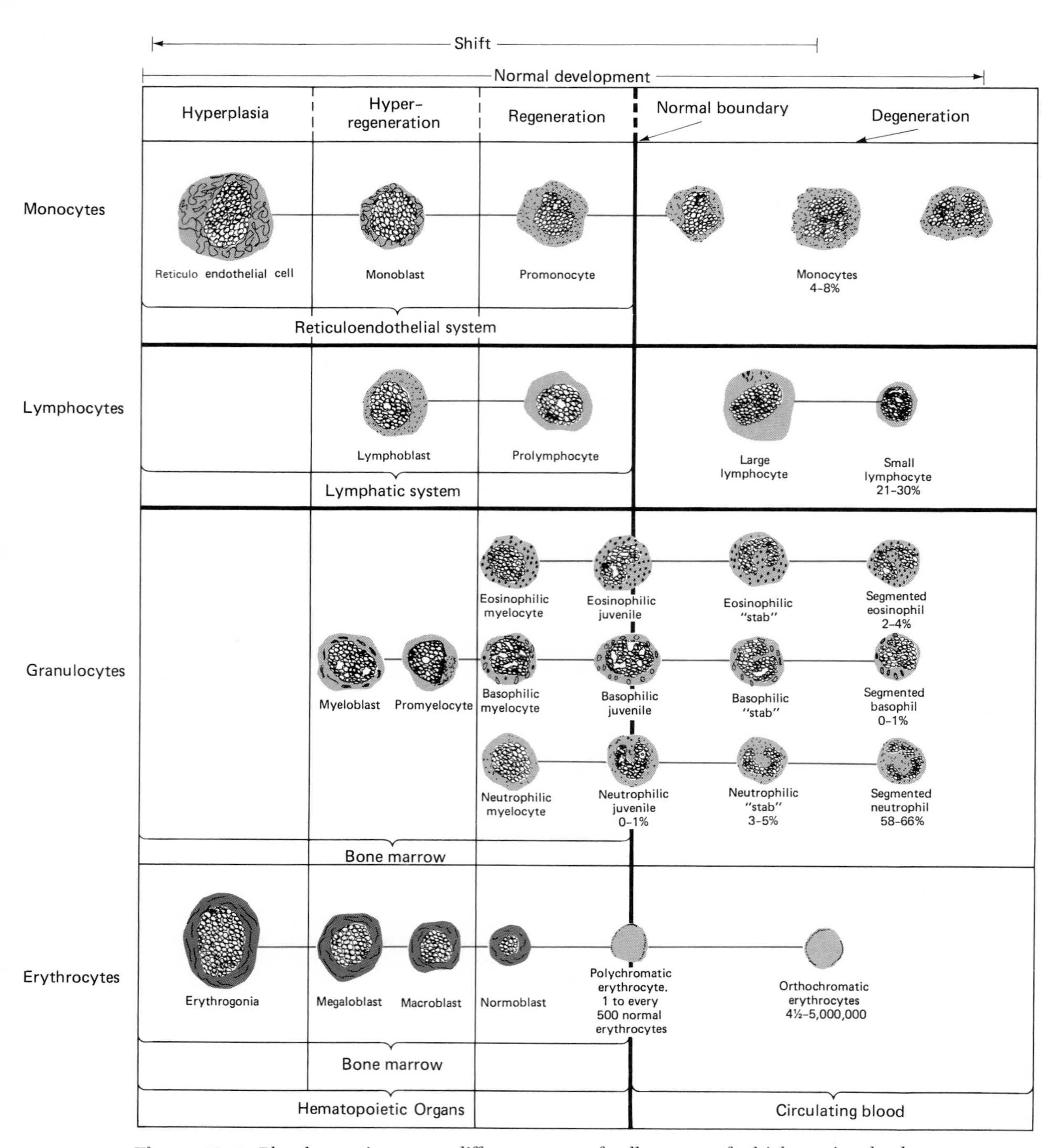

Figure 19.7 Blood contains many different types of cells, many of which are involved in nonspecific protection against invasion by microorganisms. This figure illustrates the different types of blood cells, showing the sequence of normal development and the normal cellular composition of adult human blood.

germ-free animals require dietary vitamin K, which normally is synthesized by the resident microbiota of the gastrointestinal tract. The microbiota of the gastrointestinal tract also synthesize biotin, riboflavin, pantothenate, and pyridoxine, supplying these vitamins to the animal host. Thus, the maintenance of a "healthy" indigenous microbiota is essential to the maintenance of a healthy individual.

Phagocytosis

The normal microbiota normally are noninvasive, whereas most disease-causing microorganisms enter the body tissues. When microorganisms penetrate the outer chemical and physical barriers, they are subject to phagocytosis by a variety of cells of the blood, the mononuclear phagocyte system (formerly known as the reticuloendothelial system) and the lymphatic system. Phagocytosis is a highly efficient host defense mechanism against the invasion of microorganisms. Such defense mechanisms are innate properties of the host organism that are active without prior exposure to microbial pathogens. As a rule, such innate defense mechanisms exhibit relatively low specificity toward particular species of microorganisms.

In blood, several types of white blood cells (leukocytes) are involved in nonspecific resistance against pathogenic microorganisms (Figure 19.7). Some of the leukocytes, called granulocytes, contain cytoplasmic granules. These granules are actually membrane bound lysosomes that have a granular appearance. (Lysosomes contain enzymes that are capable of killing and digesting bacteria.) The leukocytes that occur in blood include: basophils, leukocytes which may be stained with basic dyes; eosinophils, leukocytes that react with acidic dyes becoming red when stained with the dye eosin; and neutrophils (also called polymorphonuclear neutrophils, polymorphs or PMNs) that contain granules that exhibit no preferential staining—that is, they are stained by neutral, acid, and basic dyes. These various types of granulocytes each serve somewhat different functions. Eosinophils, for example, are important in the response to infestations with parasitic worms. The neutrophils are very important in the defense against bacterial infections.

The leukocytes that do not contain granular inclusions (agranulocytes) include the monocytes and the lymphocytes. The monocytes are particularly important in the nonspecific immune response, and lymphocytes are especially important in the specific immune response.

Phagocyte. Microphage. Macrophage. Each is a member of the cellular defense-arm. Leucocyte. Histiocyte. Polymorphonuclear. Large mononuclear. Sinusoidal cell of liver. Cell of marrow. "Fillet of a fenny snake, In the cauldron boil and bake; Eye of newt, and toe of frog . . . " And inside the defending cells were the defending molecules—so it was said, so it was, so it was meant.

—GUSTAV ECKSTEIN, *The Body Has a Head*

Changes in the composition of the blood usually occur as a consequence of microbial infections. Such changes generally are reflected in shifts in the relative quantities and types of white blood cells. An elevated white blood cell count (leukocytosis) is characteristic of many systemic infections. A systemic bacterial infection, for example, is normally characterized by a progressive increase in neutrophil cells (neutrophilia), particularly by an increase of young neutrophil cells known as stab or band cells (Table 19.1). As compared to mature neutrophils, stab cells have a U-shaped nucleus that is slightly indented but not segmented. The increase in stab cells, indicative of neutrophilia, is known as a shift to the left, which refers to a blood cell classification system in which immature blood cells are positioned on the left side of a standard reference chart and mature blood cells are placed on the right. The recovery phase of an infection is characterized by a reduction in fever, a decrease in the total number of leukocytes, and an increase in the number of monocytes. Gradually, the relative numbers of the various white blood cells return to their respective normal ranges.

Neutrophils. Neutrophils, the most abundant phagocytic cells in blood, are produced in the bone marrow and are continuously present in circulating blood, affording protection against the entry of foreign materials. These white blood cells exhibit chemotaxis and are attracted to foreign substances, including invading microorganisms, which they engulf and digest along with particulate matter (Figure 19.8). Neutrophils are short-lived in the body, surviving for only a few days, but are replenished from the bone marrow to the blood in high numbers.

A microorganism engulfed by the pseudopods of a phagocytic blood cell is transported by endocytosis across the membrane of the phagocytic cell where it is contained within a vacuole called a phagosome. The phagosome migrates to and fuses with a lysosome, thus exposing the microorga-

MICROBIOLOGY HEADLINES

Heart Patient Watched for Infection

By LAWRENCE K. ALTMAN

LOUISVILLE, Ky., Dec. 1—As William J. Schroeder enters his second week with an artificial heart, his doctors are hopeful that he will avoid a major hurdle, the risk of infection.

The 52-year-old retired Federal worker is particularly vulnerable to infection because he is a diabetic. But today's progress report from Humana Heart Institute International said his temperature, one indicator of possible infection, had reached normal for the first time since the implant operation last Sunday.

Before this morning's reading of 98.3 degrees, Mr. Schroeder's temperature had been in the range of 100 to 101 degrees. His doctors said there was no evidence of infection.

Mr. Schroeder was reported to be making "slow and steady progress" as he rested in Room 8 of the coronary care unit. He was alert and ate breakfast with his family but continued to be "somewhat tired," Humana officials said.

Nevertheless, Mr. Schroeder's doctors remained alert to the ever-present risk of infection after open-heart operations and to the fact that from 7 to 10 days after surgery is the period when postoperative infections often set in. Mr. Schroeder received his artificial heart last Sunday.

Complications such as infections can show up weeks later, and in Mr. Schroeder's case there are particular concerns about infections developing at any time.

His life depends on the artificial heart.

Any time infection develops in one part of the body it can spread through the blood and seed new sites elsewhere.

If the micro-organisms became lodged on the valves in the device, they could send off showers of bacteria that could end up in the brain, causing strokes. Or micro-organisms could infect the plastic tubes that are the crucial link carrying the air from the compressors near his bed to power the mechanical heart.

Another possibility is that the microorganisms could interfere with the scar tissue that is forming the natural anchor between the remaining portions of the atria, or upper chambers, of Mr. Schroeder's own heart and the Jarvik-7 artificial heart.

And there are other ways infections could cause lethal damage to an artificial heart recipient.

Thus, Mr. Schroeder and his doctors will be on the lookout for infections the rest of his life.

Moreover, control of infection is particularly important in diabetics. A minor skin infection caused by staphylococcal bacteria, for example, could be much more troublesome in a diabetic than a nondiabetic. The reason is that simple infections can set off a vicious cycle in a diabetic.

Diabetes is characterized in part by abnormally large amounts of a sugar called glucose in the blood. High blood glucose levels can impair several functions of the white blood cells, the ones that provide the body's basic defense against bacterial infections. For example, a high blood sugar can impair the ability of white cells to kill bacteria.

Blood Sugar Reported High

Accordingly, doctors pay particular attention to blood sugar test results in treating diabetics.

Friday morning, for instance, Dr. Allan M. Lansing, chief medical spokesman for the Humana artificial heart team, reported Mr. Schroeder's blood sugar at 400 milligrams per deciliter. A normal range is 65 to 120, depending on how soon it is measured after one eats.

Mr. Schroeder's diet is limited to liquids and soft foods, in part because removal of his infected upper teeth one week before the heart implant has made it hard for him to chew. Because of the high blood glucose levels, the doctors have lowered the amounts of carbohydrates in his diet and raised those of proteins and fat.

The doctors also switched the types of insulin they were prescribing, from short-acting regular to a longer-acting form called NPH, to keep the diabetes in better control.

With those measures, Mr. Schroeder's blood sugar dropped to 209 this morning.

Doctors generally treat infections vigorously with antibiotics as soon as they find evidence of infection in diabetics, Mr. Schroeder's doctors stopped prescribing the antibiotics he had been taking in recent days to control infections that had developed in his teeth and gallbladder.

'We Have Nothing to Treat'

Now, Dr. Lansing said, Mr. Schroeder is not taking any antibiotics because "we have nothing to treat" and because unnecessary antibiotic therapy could be dangerous. Indiscriminate use of antibiotics could rid him of many useful micro-organisms and set the stages for infection with bacteria that are resistant to antibiotic therapy.

In their examinations of Mr. Schroeder, doctors inspect all the intravenous and other tubing that can serve as a focus of infection. For that reason, Dr. Lansing said Mr. Schroeder now had only one intravenous needle in place. He also has a catheter to help him drain urine from his bladder, but it could become a source of infection.

An additional sign that Mr. Schroeder does not have an infection is his white blood count. The last two reported were 8,500 and 11,300, which are about normal and compare to a peak of about 17,000 earlier, according to Dr. Lansing.

Meanwhile, other laboratory test results indicate that Mr. Schroeder's kidney and liver function, which had been abnormal, are now returning to normal.

Source: The New York Times, Dec. 2, 1984. Reprinted by permission.

Table 19.1 Some Representative Differential White Blood Cell Counts for Various Infections

Condition	Leukocytes (total)	Basophils	Eosin-ophils	Neutrophils				Lymphocytes	Monocytes
				Myelocytes	Juveniles	Stabs	Segments		
Normal (healthy)	7,500	0–1	2–4	0–1	3–5	58–66	21–30	4–8	
Scarlet fever	16,680	2	0	84	1	15	58	18	7
Appendicitis	13,800	0	0	0	10	59	20	10	0
Staphylococcus septicemia	34,950	0	0	0	12	31	46	8	3
Tularemia	19,550	1	0	0	6	53	23	12	5

nisms to the degradative enzymes within the lysosome and enzymes that catalyze the production of biochemicals such as hydrogen peroxide and superoxide radicals (Figure 19.9). This process is called degranulation. During phagocytosis there is an increase in oxygen consumption by the phagocytic cells associated with elevated rates of metabolic activities. As a consequence, oxygen is converted to the superoxide anion, hydrogen peroxide, singlet oxygen, and hydroxyl radicals, all of which have antimicrobial activity. Catalase-producing microorganisms can decompose hydrogen peroxide and are therefore relatively resistant to phagocytic destruction. However, phagocytic cells also contain digestive enzymes that are involved in killing ingested bacteria. The degraded microorganisms are transported to the cytoplasmic membrane within a vacuole and are removed from the phagocytic cells by exocytosis or are consumed within the phagocytic cell.

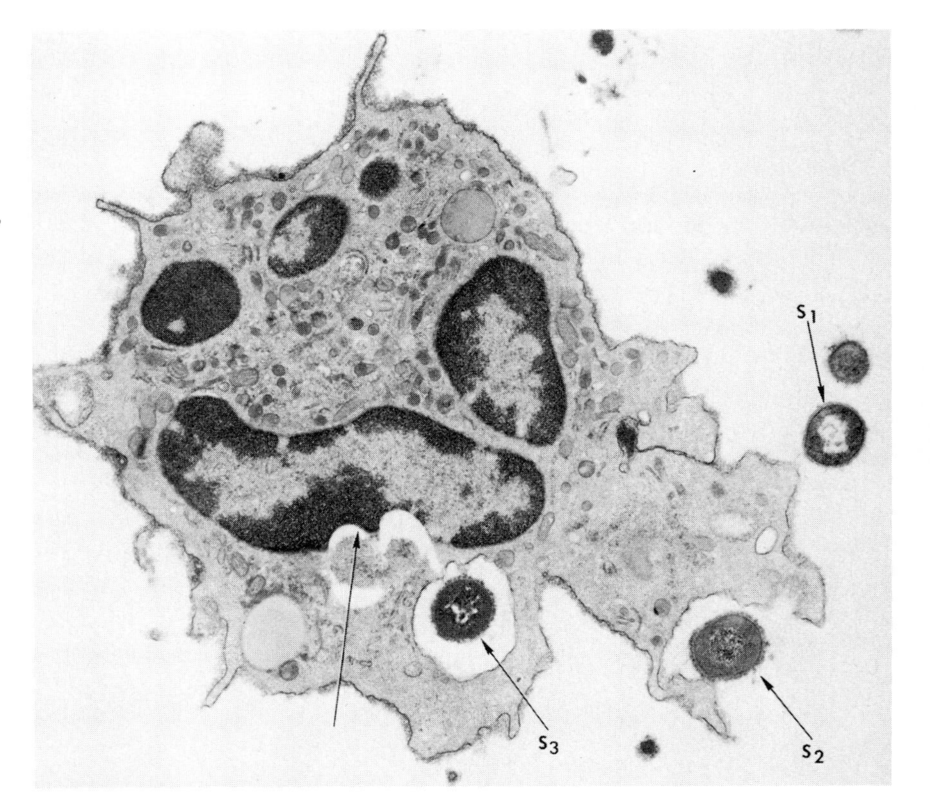

Figure 19.8 Micrograph of a polymorphonuclear leukocyte in the presence of *Streptococcus pyogenes*. S_1 is free, S_2 is being phagocytized, and S_3 has been phagocytized. The arrow shows where the nucleus is being digested. (From BPS: C. L. Sanders, Battelle Pacific Northwest Laboratories.)

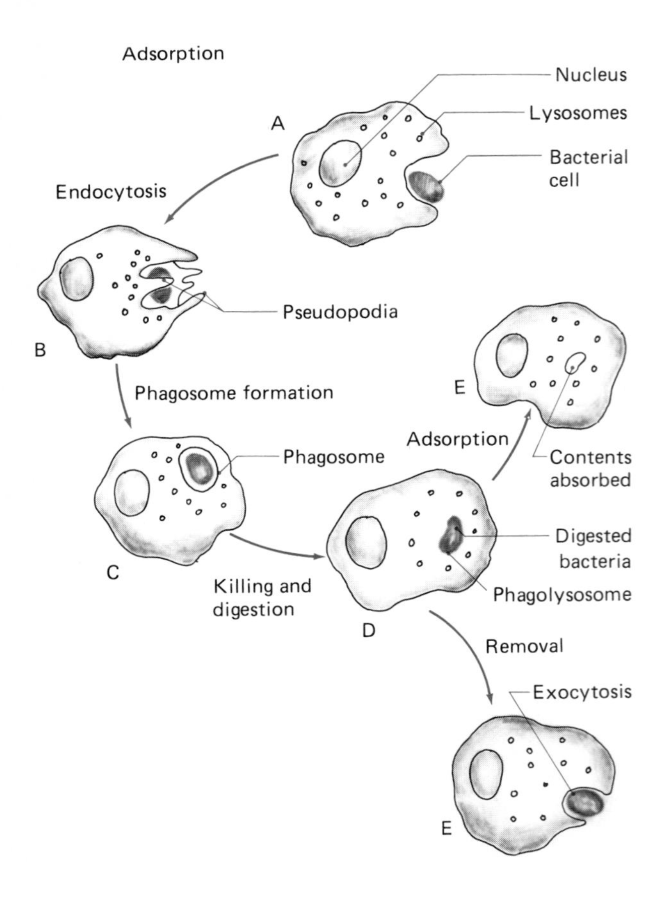

Figure 19.9 The phagocytic killing of pathogenic microorganisms involves the engulfment and digestion of the microorganism by phagocytic blood cells. First, the prey become attached to the surface of the phagocyte membrane; the attachment is assisted by complement and/or antibodies (A). Numerous pseudopodia surround the prey, and the prey enters the cell by phagocytosis (B). The membranes of the pseudopodia fuse to enclose the prey within a phagosome (C). Enzymes from many granules are dumped into the phagosome, and the prey is often digested (D). The fluids and solutes are absorbed by the cell or removed by exocytosis (E).

Macrophages. In addition to neutrophils, monocytes are formed by the stem cells in the bone marrow. These mononuclear cells, which are larger than neutrophils, are the precursors of macrophages and are able to move out of the blood to tissues that are infected with invading microorganisms. Outside the blood monocytes become enlarged, forming phagocytic macrophages. In contrast to neutrophils, macrophages are long-lived, persisting in tissues for weeks or months. Once these differentiated macrophages are formed, they are capable of reproducing additional macrophages, whereas neutrophils are terminal cells that are nonreproductive and must be replenished from the bone marrow. Macrophages, like neutrophils, are able to engulf and digest microorganisms. Taken together, the macrophages and the neutrophils (sometimes called microphages) are the two main types of phagocytic cells in the body.

Although not as numerous as neutrophils, macrophages are distributed throughout the body, including at fixed sites within the mononuclear phagocyte system (Figure 19.10). The mononuclear phagocyte system refers to a systemic network of phagocytic cells distributed through a network of

A

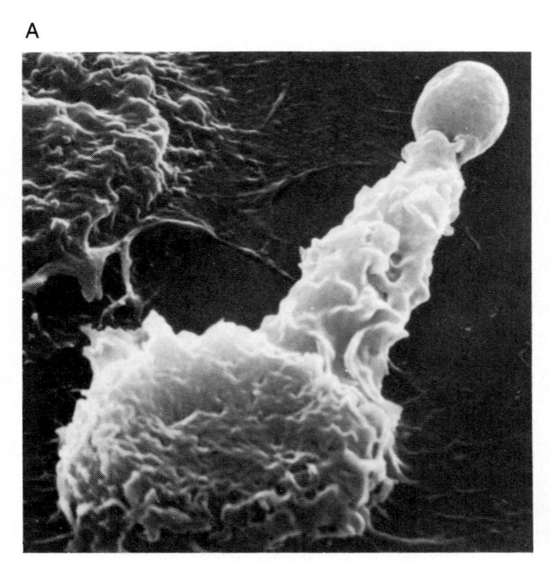

B

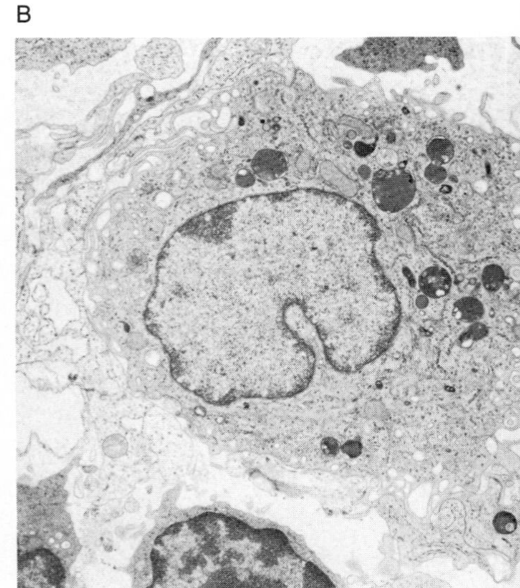

Figure 19.10 (A) This alveolar macrophage is utilizing an extended pseudopod for ingesting a yeast cell (4300 ×). (From BPS: John G. Hadley, Battelle Pacific Northwest Laboratories, Richland, Washington.) (B) This macrophage is from the medulla of a rat lymph node (12,000 ×). (From BPS: R. Rodewald, University of Virginia.)

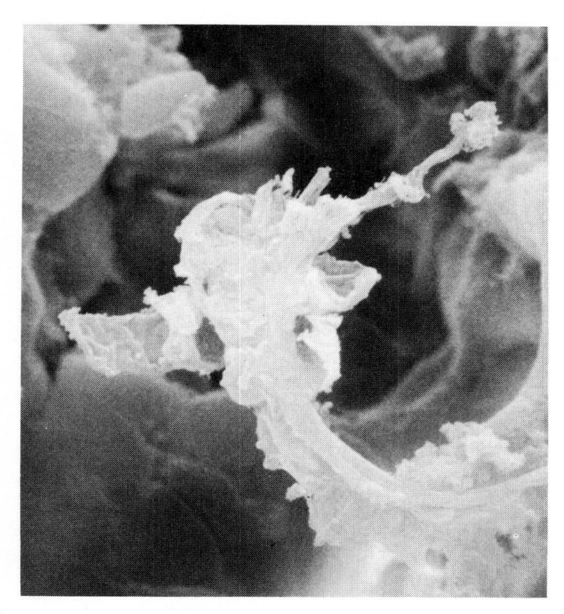

Figure 19.11 Kupffer cells are some of the fixed macrophage cells of the mononuclear phagocyte system. In this scanning electron micrograph of a Kupffer cell (3000×) the pseudopodia of this fixed macrophage of the rat liver are visible. The topmost pseudopod appears to be attached to a foreign body. (Courtesy Robert P. Apkarian, University of Louisville.)

loose connective tissue and the endothelial lining of the capillaries and sinuses of the human body. The phagocytic cells associated with the linings of the blood vessels in bone marrow, liver, spleen, lymph nodes, and sinuses constitute this host defense system. Some of the macrophages in the mononuclear phagocyte system occur at fixed sites and are designated with particular names. For example, microglia refer to macrophages of the central nervous system; Kupffer cells are phagocytic cells that line the blood vessels of the liver; dust cells are macrophages fixed in the alveolar lining of the lungs; and fixed macrophages in connective tissues are referred to as histiocytes (Figure 19.11). Other macrophages of the mononuclear phagocytic system are called wandering cells because they move freely into tissues where foreign substances have entered. Wandering macrophages occur in the peritoneal lining of the abdomen and the alveolar lining of the lung as well as in other tissues. The presence of relatively high numbers of macrophages in the respiratory tract is important in preventing the establishment of both pathogens and a normal indigenous microbiota within the tissues of the lower respiratory tract.

Inflammatory Response

The inflammatory response represents a generalized response to an infection that is designed to localize the invading microorganisms and arrest the spread of the infection. (Inflammation can also be triggered by injury and exposure to various chemicals.) The inflammatory response is characterized by four symptoms: reddening of the localized area, swelling, pain, and elevated temperature (fever). Although an inflammatory response does not necessarily reflect an infectious disease, an elevated body temperature (fever) is often used as presumptive evidence of a microbial infection. The physician observing a patient with a red sore throat and who is also running a fever, assumes that the symptoms are the result of a microbial infection. In many such cases, when the presumptive evidence strongly indicates pharyngitis (infection of the pharynx), treatment normally is administered without rigorous clinical diagnosis and confirmation of the cause. It would be better to perform laboratory studies and culture bacterial pathogens before administering antibiotics, especially since many cases of pharyngitis are caused by viruses, but for practical and economic reasons this often is not done.

The redness associated with inflammation results from capillary dilation that allows more blood to flow. The term dilation is really a misnomer because the capillaries are no more dilated than normal; during "dilation" there are simply fewer constricted capillaries, permitting increased blood circulation through more open capillaries. The elevated temperature also results because the capillary dilation permits increased blood flow through these vessels, carrying with it heat from deep tissues. The dilation of blood vessels is accompanied by increased capillary permeability, causing swelling as fluids accumulate in the interstitial spaces around the affected tissue cells. This condition is called edema. Actually, the swelling is due to increased permeability of the venules, but the term increased capillary permeability is entrenched in the clinical terminology to describe this phenomenon. Pain is due to lysis of white blood cells associated with the release of bradykinin and prostaglandins. Bradykinin decreases the firing threshhold for nerve fibers that mediate pain, and the prostaglandins intensify this effect. Aspirin, which we often use to decrease pain, antagonizes prostaglandin formation but has little or no effect on bradykinin formation. Thus, aspirin can decrease but not eliminate the pain associated with the inflammatory response.

After considering the characteristic symptoms of

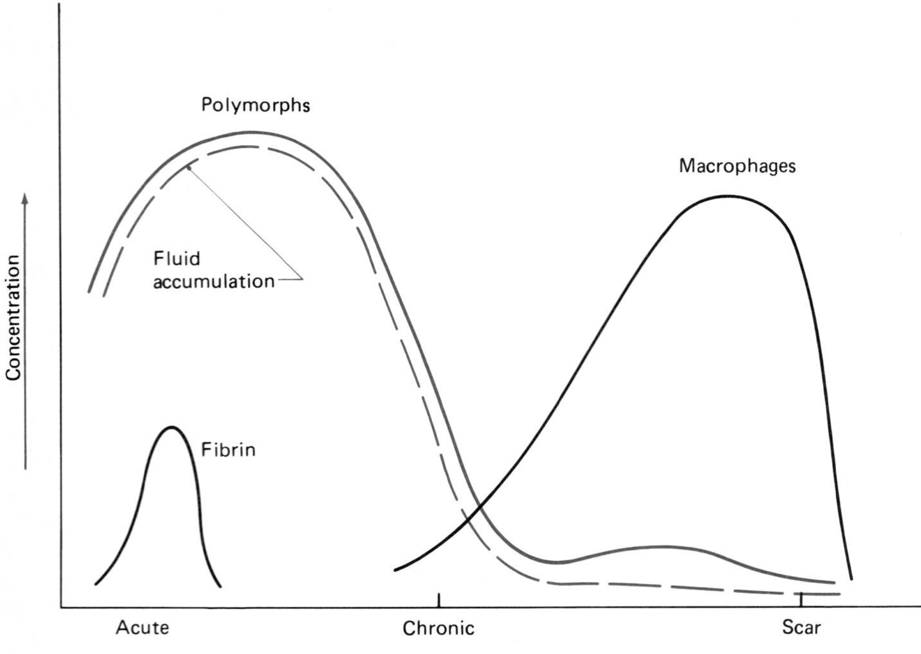

Figure 19.12 During the early stages of the inflammatory response, polymorphs (neutrophils) are particularly important; in the later stages of an inflammatory response the concentration of polymorphs declines and the number of macrophages increases.

the inflammatory response, let us discuss how it helps defend against infections by pathogenic microorganisms. As indicated, the dilation of blood vessels in the area of the inflammation increases blood circulation, allowing increased numbers of phagocytic blood cells to reach the affected area. Neutrophils are initially most abundant, but in the later stages of inflammation, monocytes and macrophages of the mononuclear phagocyte system predominate (Figure 19.12). The phagocytic cells are able to kill many of the ingested microorganisms. Phagocytic blood cells stick to the lining of the blood vessels and migrate to the affected tissues, passing between the endothelial cells of the blood vessel by a process known as diapedesis. Macrophage and neutrophils move in an amoeboid fashion to the site of an infection in a process called migration. They are attracted to the inflammed tissues, which release a chemical factor called leu-

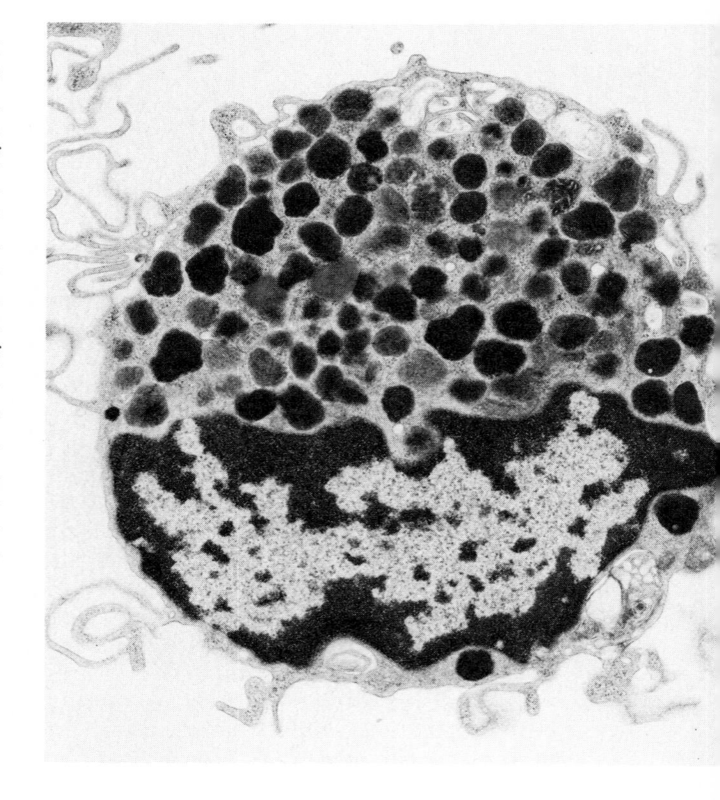

Figure 19.13 A human mast cell from the lung, showing granule inclusions of histamines. The release of histamine from these cells occurs as part of the immune response, leading to altered vascular permeability. (Courtesy Ann Dvorak, Beth Israel Hospital, Boston.)

kocytosis-promoting factor; the release of this factor causes chemotactic migration of phagocytic cells, so that a contunous supply of phagocytic cells are supplied to the site of infection. Many of the neutrophils and macrophage die as a result of the enzymatic activities involved in responding to infecting agents.

The death of phagocytic blood cells involved in combating the infection results in the release of histamine, prostaglandins, and bradykinins which, in addition to their other effects, are **vasodilators** (substances that cause blood vessels to dilate). Additionally, **mast cells**, which are a type of blood cell found throughout the body, react with complement, leading to the release of large amounts of the vasoactive substance **histamine** contained within these blood cells (Figure 19.13). The release of these vasoactive substances causes major changes in the circulation of blood during the inflammatory response (Figure 19.14). Thus, the death of some phagocytic cells and the release of partially degraded microorganisms enhances the inflammatory response.

The area of the inflammation also becomes walled off as result of fibrinous clots; these develop because clotting elements of the blood have been delivered to the injured tissue. The deposition of fibrin isolates the inflamed area, cutting off normal circulation. The fluid that forms in the inflamed area is known as the **inflammatory exudate**, commonly called **pus**. The area of pus accumulation is called an abscess. Boils and pimples are examples of such areas where an abscess has formed. The inflammatory exudate of the abscess contains dead cells and debris in addition to body fluids. After the removal of the exudate, the inflammation can terminate, and the tissues may return to their normal functioning.

The elevation of body temperature (fever) aids in the ability of the inflammatory response to protect against invading microorganisms. Fever is a response to **pyrogens** (substances that cause a rise in body temperature, including those pyrogens released when bacterial endotoxins interact with white blood cells) that enter the blood as a result of the death of microorganisms or that are released by phagocytic blood cells. These pyrogens interact with the hypothalmus, the organ of the body that reg-

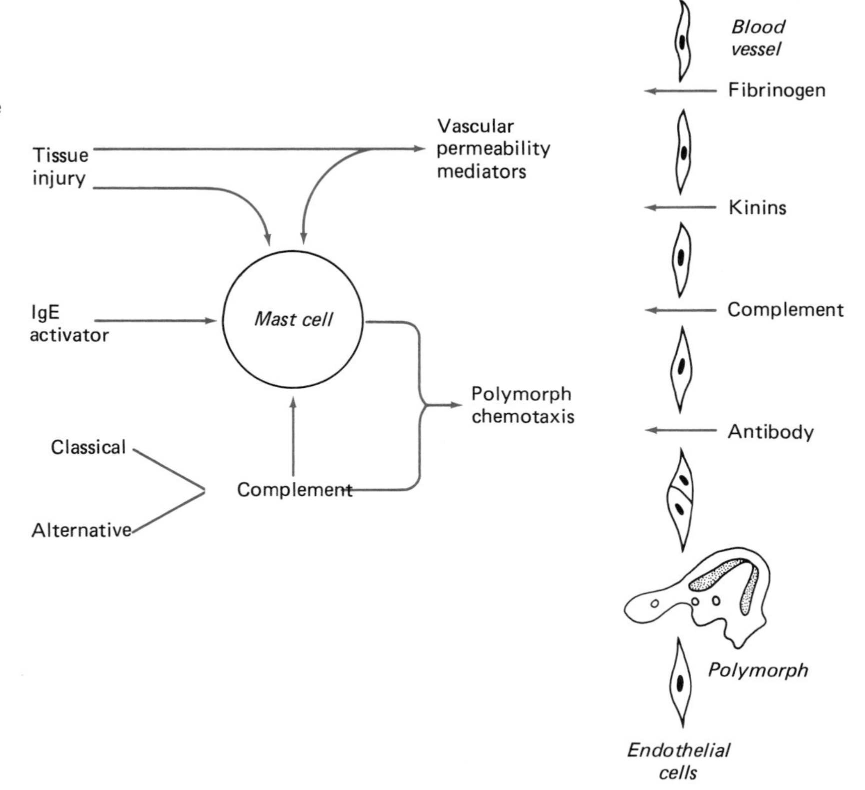

Figure 19.14 The degranulation of mast cells, which may be triggered by various factors, leads to altered vascular permeability and enhanced polymorph chemotaxis; an acute protective inflammatory reaction results.

ulates body temperature, causing a rise in body temperature above the normal 37°C (98.6°F). Elevated body temperatures favor enhanced rates of phagocytic activity and other aspects of the immune response. The elevated temperatures also often reduce the rates of reproduction of pathogenic microorganisms, many of which have optimal temperatures coincident with the normal body temperature. Compared to the elevated temperatures within the body the skin is cooler and thus chills characteristically occur during the rise of temperature. The chills normally disappear when the body reaches a stable albeit elevated temperature. As the infection subsides, heat loss occurs that is reflected in sweating. The phase of an infection is called the crisis and indicates that the body temperature is declining.

The final stage of the inflammatory response involves the repair of the affected tissue. New cells are formed and added to the tissue. In some cases only cells that comprise the functional cells of the tissue (parenchyma cells) are formed, leading to the restoration of normal tissue. In other cases, however, cells of the supportive connecting tissue (fibroblast cells of the stroma) are also produced, producing scar tissue. Thus, the complex reactions of the inflammatory response work in an integrated fashion to contain and eliminate infecting pathogenic microorganisms and to restore normal or near normal tissue.

Summary

The surfaces of body tissues are generally resistant to penetration by microorganisms. In addition to acting as a physical barrier, human surface tissues often contain biochemicals with antimicrobial activity, such as the fatty acids of skin tissues and the enzyme lysozyme in tears. Microorganisms that penetrate the surface barriers are confronted by several nonspecific resistance mechanisms. Several substances in the circulatory system, such as interferon and complement, have antimicrobial activities. Interferons in particular prevent the reproduction of viruses within human cells.

Additionally, various cells that circulate through the body, such as monocytes, neutrophils, and macrophages have phagocytic activity. The ability to engulf and ingest invading microorganisms eliminates most would-be pathogens. The inflammatory response brings together several different mechanisms for preventing the proliferation and spread of infecting pathogenic microorganisms. The inflammatory response tends to limit the spread of an infection, to direct phagocytic cells to the site of infection, and to enhance the effectiveness of phagocytic removal of infecting pathogens. These host defense mechanisms are physiological in nature and as such are subject to individual variation. Numerous factors can act to compromise the host defense mechanisms, rendering an individual susceptible to disease-causing pathogenic microorganisms. For example, a wound breaks the normal surface barrier permitting invasion by microorganisms; emotional and physiological stress (for example, improper nutrition, exposure to extreme temperatures, and tension) lower the physiological potential for successful phagocytic removal of infecting microorganisms; and age may influence the physiological ability of an individual to resist infection, with young children and the elderly being particularly prone to various infectious diseases.

In addition to the mechanisms already described for maintaining a balance between microbial populations and human beings without the occurrence of disease, there is a specific or acquired immune defense mechanism that protects individuals against diseases caused by microorganisms. The preclusion of microbial diseases of humans involves an elaborate and complex integrated system of host resistance mechanisms.

Study Outline

A. Nonspecific host-defense mechansisms do not respond to individual species of invading microorganisms, but rather are generally protective against all potential pathogens.
- **1.** The surface tissues of the body physically block the entry of microbes into the body.
  - **a.** The respiratory tract is protected in part against the invasion of pathogenic microorganisms by secretions of the mucous membranes and cilia that line the upper respiratory tract.
- **2.** Some of the fluids the body produces contain antimicrobial chemicals.
  - **a.** Lysozyme in tears degrades the peptidoglycan of the cell walls of bacteria.

b. Lactoferrin and transferrin bind iron, withholding this essential growth element from pathogens.
c. Stomach acid creates an environment whose pH is intolerable to most microorganisms.
d. The genitourinary tract is protected by the outward flow of mucus, low pH, high salt concentrations, and the presence of antimicrobial enzymes.
e. Interferon, a glycoprotein found in low concentrations in blood, appears to be the main defense mechanism against viral infections. It prevents the replication of viral pathogens within protected cells.
f. The complement system acts in association with the specific immune response to remove invading microorganisms and as an alternate pathway as part of the nonspecific defenses of the body. Complement molecules can attach to the surface of an invading cell in a cascade fashion, altering the permeability of the cytoplasmic membrane of that cell and the cell dies.

3. Microorganisms normally associated with the surface tissues of the body also help defend against disease-causing pathogens. The normal microbiota of the human body inhabits the surface tissues of the skin, oral cavity, respiratory tract, gastrointesinal tract, and genitourinary tract.
   a. Many of these microorganisms produce allelopathic substances that act to prevent the establishment of infection on or through the skin.
   b. The indigenous microbiota contribute to the nutrition of the host by synthesizing essential nutrients.
4. When microorganisms penetrate the outer chemical and physical barriers, they are subject to phagocytosis by a variety of cells of the blood, the mononuclear phagocyte system (formerly known as the reticuloendothelial system) and the lymphatic system. It is an innate property of the host organism that is active without prior exposure to microbial pathogens and it exhibits relatively low specificity toward particular species of microorganisms.
   a. In blood, several types of white blood cells (leukocytes) are involved in nonspecific resistance against pathogenic microorganisms. Some of the leukocytes, called granulocytes, contain cytoplasmic granules. These include basophils, eosinophils, and neutrophils (also called polymorphonuclear neutrophils, polymorphs, or PMNs).
   b. The leukocytes that do not contain granular inclusions (agranulocytes) include the monocytes and the lymphocytes.
   c. Neutrophils are the most abundant phagocytic cells in blood; they are produced in the bone marrow and are continuously present in circulating blood; they exhibit chemotaxis and are attracted to foreign substances, including invading microorganisms, which they engulf and digest along with particulate matter. Neutrophils are short-lived in the body, surviving for only a few days, but are replenished to the blood from the bone marrow in high numbers.
   d. A microorganism engulfed by the pseudopods of a phagocytic blood cell is transported by endocytosis across the membrane of the phagocytic cell where it is contained within a phagosome. The phagosome migrates to and fuses with a lysosome, exposing the microorganisms to the enzymes within the lysosome. Forms of oxygen that have antimicrobial properties are produced as a result of the increased metabolic activity of the hexose monophosphate shunt. Phagocytic cells contain digestive enzymes that also kill ingested bacteria.
   e. Monocytes are formed by the stem cells in the bone marrow. These mononuclear cells are the precursors of macrophages and are able to move out of the blood to infected tissues. Outside the blood monocytes become enlarged, forming phagocytic macrophages. Macrophages are long-lived, persisting for weeks or months, are capable of reproducing to form additional macrophages, whereas neutrophils are nonreproductive terminal cells. Macrophages, like neutrophils, are able to engulf and digest microorganisms.

f. Macrophages are distributed throughout the body, including at fixed sites within the mononuclear phagocyte system, a systemic network of phagocytic cells distributed through a network of loose connective tissue and the endothelial lining of the capillaries and sinuses of the human body. The phagocytic cells associated with the linings of the blood vessels in bone marrow, liver, spleen, lymph nodes, and sinuses constitute this host-defense system.

5. The inflammatory response is a generalized response to infection designed to localize invading microorganisms and arrest the spread of the infection, characterized by four symptoms: reddening of the localized area, swelling, pain, and elevated temperature.
   a. The redness results from capillary dilation, which allows more blood to flow. The elevated temperature is the result of capillary dilation, permitting increased blood flow through these vessels. Swelling is due to increased permeability of the venules.
   b. Pain is due to lysis of blood cells, triggering the production of bradykinin and prostaglandins. Bradykinin decreases the firing threshhold for pain fibers and the prostaglandins, PGE1 and PGE2, intensify this effect. Aspirin antagonizes prostaglandin formation but has little or no effect on bradykinin formation.
   c. The dilation of blood vessels in the area of the inflammation increases blood circulation, allowing increased numbers of phagocytic blood cells to reach the affected area. Neutrophils are initially most abundant, but in the later stages of inflammation, monocytes and macrophages of the mononuclear phagocyte system predominate. Phagocytic blood cells stick to the lining of the blood vessels and migrate to the affected tissues, passing between the endothelial cells of the blood vessel by a process known as diapedesis. The death of phagocytic blood cells involved in combating the infection results in the release of histamine, prostaglandins, and bradykinins, which are vasodilators. Mast cells react with complement leading to the release of large amounts of histamine contained within these blood cells. The release of these vasoactive substances causes major changes in the circulation of blood. Thus, the death of some phagocytic cells and the release of partially degraded microorganisms enhances the inflammatory response.
   d. The area of the inflammation becomes walled off by fibrinous clots that cut off normal circulation. Inflammatory exudate forms in the inflamed area; it contains dead cells and debris in addition to body fluids.
   e. Fever is a response to pyrogens entering the blood as a result of the death of microorganisms or released by phagocytic blood cells. Elevated body temperatures increase rates of phagocytic activity and other aspects of the immune response and reduce the rates of reproduction of pathogenic microorganisms.

Study Questions

1. What is an inflammatory response?
2. How does inflammation act to prevent the spread of pathogens throughout the body?
3. What role does phagocytosis play in defending the body against microbial infections?
4. What are the first lines of defense against disease-causing microbes?
5. Indicate whether the following statements are true or false.
   a. Lack of normal microbiota makes an organism healthier and more resistant to disease.
   b. The human skin is well suited for the growth of most microbes.
   c. Red blood cells are phagocytic.
   d. Neutrophils are part of the human defense system against infection.
   e. Lysozyme is found in tears.
   f. The inflammatory response should be suppressed because it does not help check the spread of disease-causing microbes.
   g. Most microorganisms have enzymes that degrade keratin.

Some Thought Questions

1. Why do sick fish swim to warmer water?
2. Do good bacteria ever do bad things?
3. What function does perspiration serve?
4. Would it be valuable to have thick skin?

5. Why are microbial infections a very serious problem for burn victims?
6. If body tissues are sterile at birth, where did your gut microbiota come from?
7. Is pain good for you?
8. What is meant by the saying "feed a cold and starve a fever?"

Additional Sources of Information

Burke, D. C. 1977. The status of interferon. *Scientific American*, **236**(4):42–62. An interesting article on interferon, its role in protecting the body against infections and its possible role in controlling the growth of malignant cells.

Edelson, R. L., and J. M. Fink. 1985. The immunologic functions of the skin. *Scientific American*, 252(6):46–53. A fascinating discussion of the complex active role of the skin in defending the body against foreign invaders.

Goren, M. 1977. Phagocyte lysosomes: interactions with infectious agents, phagosomes, and experimental perturbations in function. *Annual Review of Microbiology*, **31**:507–533. A detailed review about the way in which phagocytic blood cells eliminate infecting microorganisms.

Immunology: Readings from Scientific American. 1976. W. H. Freeman and Company, Publishers, San Francisco. An excellent collection of articles on all aspects of the defense network that protects the human body against infectious agents.

Mayer, M. M. 1973. The complement system. *Scientific American*, **229**(5):54–66. A detailed article on the functions of complement in defending the body against infectious agents.

Schlesinger, R. B. 1982. Defense mechanisms of the respiratory system. *BioScience*, **32**(1):45–50. An easy-to-read article describing the defense mechanisms that protect the respiratory tract against infectious microorganisms.

Wilkinson, P. C. 1974. *Chemotaxis and Inflammation*. Churchill Livingstone, Edinburgh, Great Britain. A thorough review of the inflammatory response and how various cells involved in this reponse are attracted to the site of inflammation.

Zweifach, B. W., L. Grant, and R. T. McCluskey. 1974. *The Inflammatory Process*. Academic Press, Inc., New York. A detailed review of the inflammatory process detailing the processes involved in this complex integrated protective defense mechanism.

20 The Immune Response: Antibody- and Cell-Mediated Immunity

KEY TERMS

antibodies
antibody-mediated immunity
antigens
antitoxins
autoimmunity
B lymphocytes
B memory cells
blood groups (A, B, AB, O)
cell-mediated immunity
clonal selection theory
cytotoxic T cells
Fab fragment
Fc fragment
flagella antigens
Forssman antigens
H antigens
hapten
hemolytic disease of the newborn
humoral immunity
hybridomas
IgA
IgD
IgE
IgG
IgM
immune interferon
immune response system
immunity
immunogenicity
immunoglobulins
lymphokines
macrophage chemotactic factor
major histocompatibility complex (MHC)
monoclonal antibodies
nonself-antigens
O antigens
opsonization
Rh (Rhesus) antigen
secondary B cells
self-antigens
somatic antigens
T aggressor cells
T helper cells
T killer cells
T lymphocytes
T suppressor cell
thymocytes
toxoids

OBJECTIVES

After reading this chapter you should be able to

1. Define the key terms.
2. Describe the differences between antibody- and cell-mediated immunity.
3. Describe the basic structure of the immunoglobulin molecule.
4. Describe the characteristics and functions of the major classes of immunoglobulins.
5. Describe how the immune system is programmed to differentiate self from nonself antigens.
6. Describe the antigens and antibodies present in individuals with blood types A, B, AB, and O.
7. Describe the functions of lymphokines.
8. Describe the different types of T cells involved in cellular immunity.

The Immune System

The immune response system of human beings and other higher animals provides a mechanism for a specific response to the invasion of particular pathogenic microorganisms and other foreign substances. It is largely this specific physiological response that protects us against disease. In contrast to nonspecific defense mechanisms, the immune response is characterized by specificity, memory, and the acquired ability to detect foreign substances. The human immune response is able to recognize macromolecules that are different in some way from the normal macromolecules of the body. The ability to differentiate "self" from "nonself" at the molecular level is an underlying necessity for the development of the specific immune response.

The specificity of the immune response permits the recognition of even very slight biochemical differences between molecules, and consequently, the macromolecules of one microbial strain can elicit a different response than the macromolecules of even a very closely related strain of the same species. The specific immune response is adaptive or acquired, in that having responded once to a particular macromolecule, called an antigen, a memory system is established that permits a rapid and specific secondary response upon reexposure to that same substance. By being able to recognize rapidly and to respond to pathogenic microorganisms, a state of immunity results that precludes infection with those specific pathogens.

When a population—of plants, animals or men—is exposed to a pathogen with which it has had no past experience, exposure may bring about severe disease in many of its individuals. The generalized epidemic, however, soon calls into play adaptive changes in both the host population and the infective agent which bring about an ecological equilibrium between them. The infective agent may remain widely distributed in the community, but its presence need not be associated with injurious effects.

—René Dubos, *Mirage of Health*

The Basis of Immunity: Antibody- and Cell-mediated Immunity

Examination of the basis of immunity reveals that there are two different forms of the immune response. In one response mode, called antibody-mediated immunity, specific proteins (antibodies) are made when foreign antigens are detected. In antibody-mediated immunity, plasma cells derived from certain white blood cells—called B lymphocytes or simply B cells—synthesize antibodies in response to the detection of a foreign macromolecule with antigenic properties. By definition, an antigen can be any macromolecule that elicits the formation of antibody and that can subsequently react with antibody. In old medical terminology blood and other vital body fluids were considered as "humors," and thus, antibody-mediated immunity is often referred to as humoral immunity because the

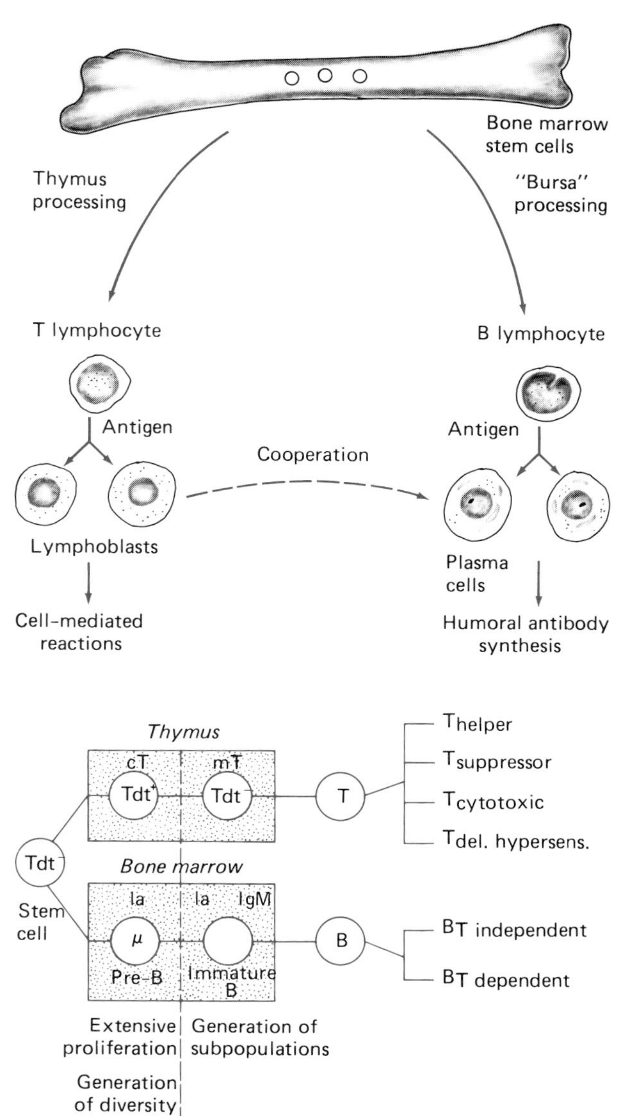

Figure 20.1 The differentiation of T and B cells forms the basis for the two arms of the immune defense network.

MICROBIOLOGY HEADLINES

Gene Related to Immune System Is Purified by 3 Research Groups

By HAROLD M. SCHMECK Jr.

Scientists have detected, purified and reproduced in the laboratory a human gene that is believed to be vital to the body's internal defenses against disease.

The gene is the second of two that are believed to govern the way in which a group of cells interact with the immune system to thwart invasion of the body by viruses and other disease-causing germs. These T cells are probably vital also to the body's natural defenses against cancer.

Both of these genes that are key to the system used by T cells to recognize alien cells have been found for the first time this year.

It has long been known that T cells must have recognition sites, or receptors, on their surface through which they recognize and call attention to the presence of any foreign invader.

Recognition signals are vitally important to immunity. Defense cells in the body must be able to chemically distinguish between self and alien, friend and foe. Failure can lead to infection, cancer or any of several other kinds of serious disease.

Among the most important cells of the immune system are the T-cell types and the B-cell types.

Chemistry of Receptors

The special recognition sites on the B cells have been known for years. Although structurally a part of the cell surface, they are actually immunoglobulin, the substance of which antibodies are made.

The specific chemical nature of the T-cell receptor has long defied discovery. However, analysis of T cells indicated that the important recognition sites on their surfaces probably contained two key substances each made under the direction of a different gene.

For at least 10 years, scientists have been intensively searching for those key genes. One was named the alpha-chain gene, the other the beta-chain gene for two key parts of the receptor they govern.

This year both genes have been found, analyzed and reproduced in the laboratory; first the beta and more recently the alpha. The substances produced by these genes have also been identified and analyzed in detail. Altogether, the accomplishment represents the work of several scientific groups.

Three Groups of Researchers

In addition to being a triumph of basic research, the discoveries are expected, in the long term, to help doctors devise new and better ways to manipulate immunity for management of organ transplantation, resistance to infection and treatment of cancer.

Three groups of scientists, at the National Jewish Hospital and Research Center in Denver, Stanford University and the Massachusetts Institute of Technology, independently discovered the second gene, considered the final piece in a long-standing scientific puzzle.

Dr. Susumu Tonegawa and colleagues of Massachusetts Institute of Technology presented the first formal report on the alpha-chain gene purified from cells of laboratory mice to an international congress on cell biology in Tokyo in August. Last month his group and that of Dr. Mark Davis of Stanford University reported the details of their independent analyses of the alpha-chain gene in the British scientific journal, Nature.

A new report in the Dec. 20 issue of Nature describes analysis and growth in the laboratory of the comparable T-cell receptor gene from cells of a human cancer. Principal investigators in the research described in this report were Dr. Gek Kee Sim and Dr. John W. Kappler of the National Jewish Hospital and Research Center.

Medical Uses of Discovery

In addition to its scientific interest, detailed knowledge of the T-cell receptor, through understanding of its key genes, may eventually have important medical applications, Dr. Kappler said yesterday. He said the knowledge should help doctors devise ways of increasing or decreasing the activity of T cells for the treatment of cancer or other diseases. The advances might also aid efforts to safeguard transplanted immunological defense system, he said. T cells are particularly important in the process through which the body tries to reject transplanted organs such as hearts or kidneys.

Even before the alpha-chain gene was characterized, Dr. Tonegawa's group at Massachusetts Institute of Technology this year had discovered the existence of a different but related gene that seems to play a role in the functions of the T-cell receptor. This has been named the gamma gene, Dr. Tonegawa said yesterday.

The discovery of the gamma gene was unexpected by immunologists because analysis of the T-cell receptor indicated that the products of only two genes would be found to play direct roles in the cells' recognition of foreign molecules.

The function of the gamma gene is still not known, but there are several possibilities in keeping with the known complexity of T cells' functions in the immunological defense system, Dr. Tonegawa said. Research is in progress in his laboratory to find the function of this recently discovered gene.

antibody molecules flow extracellularly through the body fluids. In the second form of the immune response, called cell-mediated immunity, certain cells of the body acquire the ability to destroy other cells that are recognized as foreign or abnormal. In contrast to antibody-mediated immunity, cell-mediated immunity depends on another class of lymphocyte cells, known as T lymphocytes or simply as T cells. Sensitized T lymphocytes interact with "foreign" cells to bring about the destruction of those cells.

Immunity thus depends on B and T lymphocytes, which are cells differentiated in the lymphatic system (Figure 20.1). Both B and T lymphocytes originate from bone stem cells and become differentiated during maturation. The precursors for T cells pass through the liver and spleen before reaching the thymus gland, where they are processed. Within the thymus, antigenic determinants are added to the T cells, which then move through the circulatory system to colonize the lymph nodes, spleen, tonsils, and Peyer's patches of the intestines. Although differentiated in the thymus gland, T cells are inactive until later maturation in other lymphoid tissues. The thymus-dependent differentiation of T cells or thymocytes occurs during the early years of life, and by adolescence the secondary lymphoid organs of the body generally contain their full complement of T cells. In human beings the B lymphocytes appear to develop in the bone marrow. The term B lymphocyte actually refers to Bursa-dependent lymphocytes, so-named because these lymphocytes are differentiated in chickens and other birds in the lymphoid organ known as the Bursa of Fabricius. Even though we do not possess a Bursa of Fabricius, the designation B lymphocyte is applied to lymphocytes that can differentiate into antibody-synthesizing cells in humans. Within lymphatic tissues, B lymphocytes give rise to antibody-secreting plasma cells in response to antigenic stimulation.

Antibody-Mediated Immunity

Antigens

A wide variety of macromolecules can act as antigens, including most proteins, many nucleoproteins and lipoproteins, some polysaccharides, and various small biochemicals if they are attached to proteins or polypeptides. The two essential properties of an antigen are its immunogenicity (ability to stimulate antibody formation) and its specific reactivity with antibody molecules. The antigen molecule consists of one or more reactive portions (haptens) that chemically react with the antibody molecule and a carrier portion, which is necessary for the stimulation of antibody production. A hapten can react with antibody molecules but is unable by itself to elicit the formation of antibody, whereas a complete antigen molecule is both reactive and elicits the production of specific antibodies. Haptens often are small molecules containing antigenic determinants, and if they combine with larger molecules they can become antigenic. For example, penicillin is a hapten that becomes antigenic upon combining with tissue proteins.

Antigenic molecules may be multivalent, having multiple reactive sites, or monovalent, having only one reactive site. Generally, multivalent antigens elicit a stronger immune response than monovalent antigens. In some cases a multivalent antigen can react with antibodies produced in response to a different antigen; such antigens are called heterophile antigens, heterologous antigens, or Forssman antigens.

In many cases, antigens are associated with cell surfaces and are therefore called surface antigens. Human cells have specific surface antigens, and for example, human blood types are determined by the presence or absence of antigens designated A and B on the surfaces of red blood cells. Microorganisms, including viruses and bacteria, also have many surface antigens, some of which may be associated with particular anatomical structures. For example, strains of *Salmonella* have specific antigens associated with their flagella proteins, called flagella or H antigens, and other specific antigens associated with the surface lipopolysaccharides of the cell wall, called somatic or O antigens.

One of the reasons that the immune response is effective in preventing disease is that the toxins contributing to the virulence of pathogenic microorganisms usually have antigenic properties. Antibodies produced in response to toxins are referred to as antitoxins. They bind with the toxin to block its active site, thereby neutralizing the toxin. Most protein toxins are highly antigenic, eliciting the synthesis of high titers (concentrations) of antibody. Denatured proteins often retain their antigenic properties. Of particular importance are denatured toxins called toxoids, which retain their antigenic properties but do not cause the onset of disease symptoms; they are antigenically active macromolecules but are no longer toxic. Proteins can be denatured to produce toxoids by heating or by adding chemicals such as formaldehyde. Toxoids are useful for eliciting antibody-mediated im-

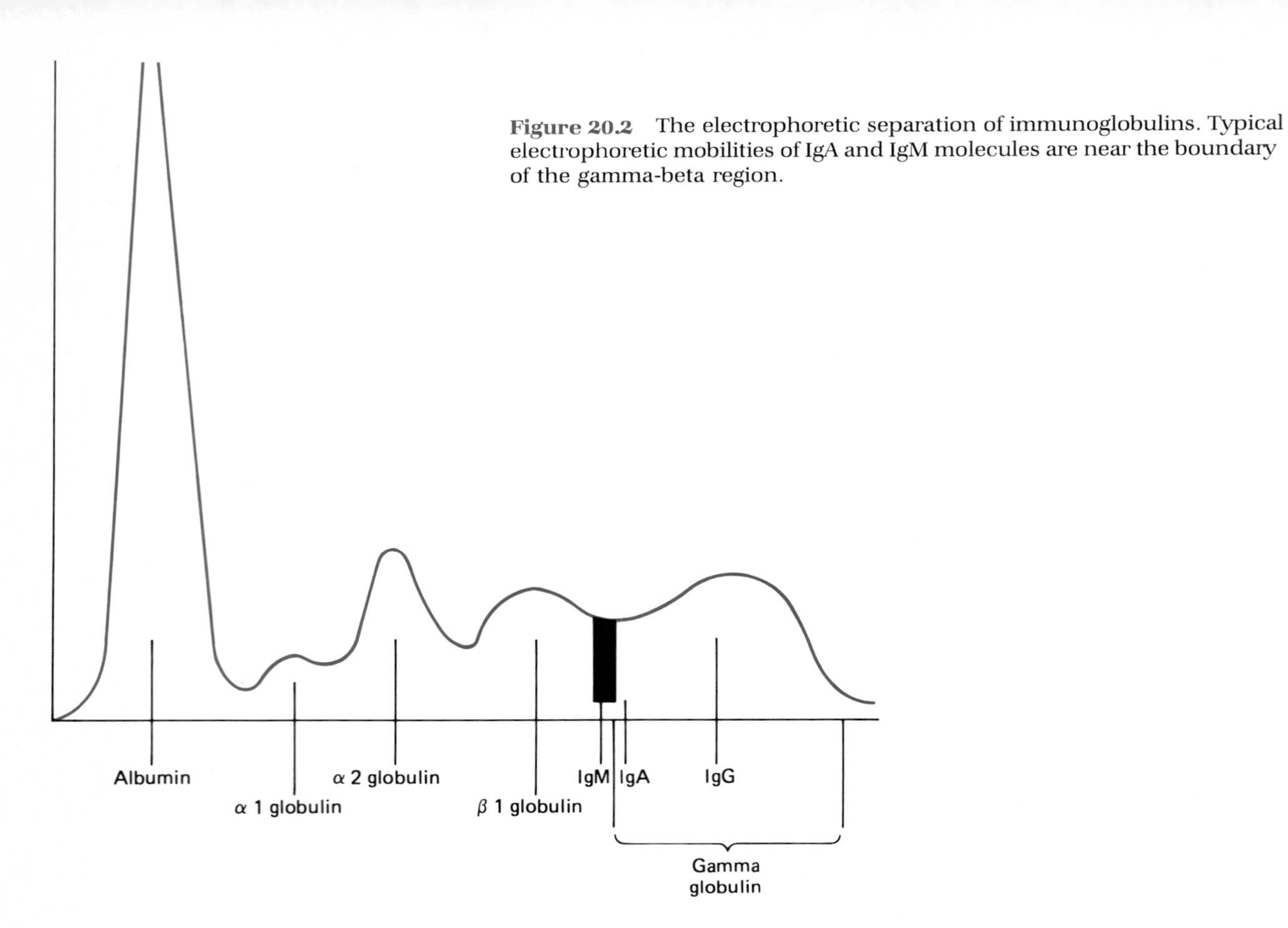

Figure 20.2 The electrophoretic separation of immunoglobulins. Typical electrophoretic mobilities of IgA and IgM molecules are near the boundary of the gamma-beta region.

Table 20.1 Properties of Five Classes of Immunoglobulins

Class	IgG	IgA	IgM	IgD	IgE
Molecular weight	150,000	160,000 and dimer	900,000	185,000	200,000
Number of basic peptide units	1	1, 2	5	1	1
Valency for antigen binding	2	2, 4	5(10)	2	2
Concentration range in normal serum[a]	8–16 mg/mL	1.4–4 mg/mL	0.5–2 mg/mL	0–0.4 mg/mL	17–450 ng/mL
Total immunoglobulin (%)	80	13	6	0–1	0.002
Cross placenta	+	–	–	–	–
Bind to mast cells and basophils	–	–	–	–	+
Bind to macrophages and neutrophils	+	±	–	–	–
Time required for one half the antibody concentration to disappear from serum (half-life in days)	25–35	6–8	5–11	2–3	2–3
Major characteristics	Most abundant Ig of body fluids; combats infecting bacteria and toxins	Major Ig in sero-mucous secretions; protects external body surfaces	Effective agglutinator produced early in immune response	Mostly present on lymphocyte surface	Protects external body surfaces; responsible for atopic allergies

[a]mg/mL = milligrams per milliliter

mune responses without producing the diseases caused by toxins and toxin-producing pathogenic microorganisms. Most commercial toxoid preparations are made by adding 0.2–0.4 percent formalin to the toxin. Toxoids are used as vaccines for protecting individuals against diphtheria, tetanus, and various other diseases caused by pathogenic microorganisms that produce protein toxins.

Immunoglobulin Molecules

Antibody molecules, often referred to as immunoglobulins, are proteins found in the serum fraction of blood, that is, in the liquid portion after blood plasma is clotted. The synthesis of antibody molecules occurs in response to antigenic stimulation. There are five classes of immunoglobulins: IgG, IgA, IgM, IgD, and IgE. These five classes of immunoglobulins can be separated by electrophoresis, a method in which an electric field is used to bring about the differential migration of molecules based upon their size, shape, and electronic charge (Figure 20.2). The characteristics of the five major classes of immunoglobulin molecules, all of which are globular proteins, are summarized in Table 20.1. The various classes of immunoglobulins serve different functions in the immune response.

Immunoglobulin molecules all have the same basic structure, consisting of four peptide chains, two identical heavy chains and two identical light chains, joined by disulfide bridges linking the chains (Figure 20.3). The immunoglobulin molecule can be split by the enzyme papain to form two identical

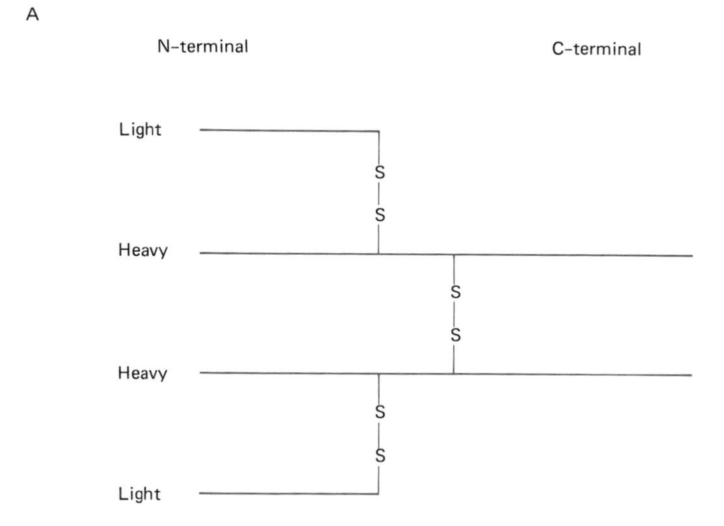

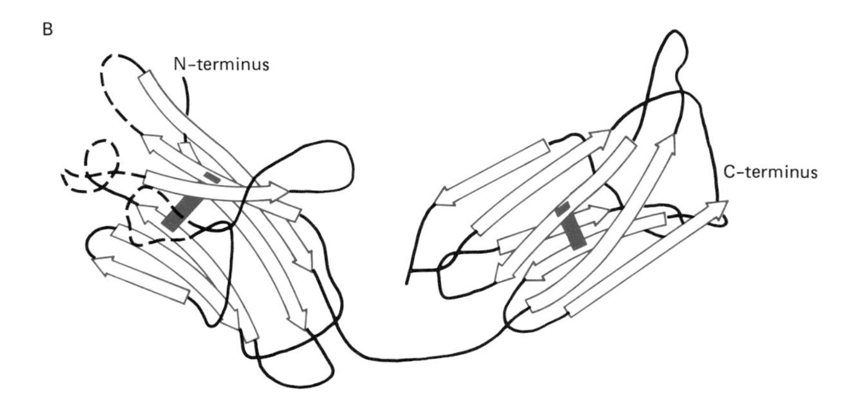

Figure 20.3 (A) An immunoglobulin molecule, showing light and heavy chains. (B) A representation of the folded structure of the immunoglobulin molecule.

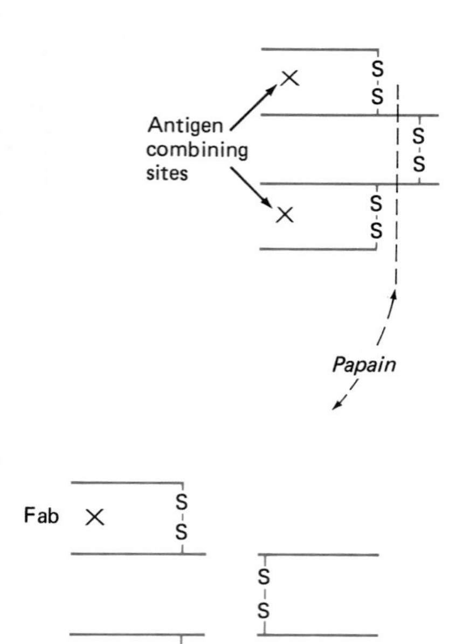

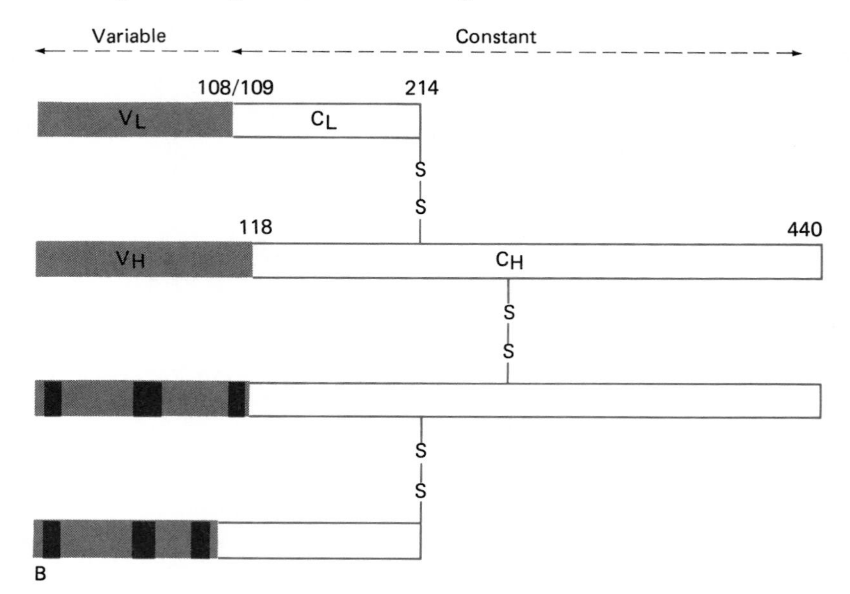

Figure 20.4 (A) Splitting of the immunoglobulin macromolecule by papain produces characteristic fractions with distinct functions. (B) The Fab fragment is characterized by a variable region that accounts for the diversity of immunoglobulin molecules; the Fc fragment is a relatively constant region among all human immunoglobulins.

Fab fragments (antigen binding fragments) and an additional Fc fragment (crystalizable fragment) (Figure 20.4). The two portions of the immunoglobulin macromolecule serve different functions. It is the Fab fragment that actually binds to antigen molecules, whereas the Fc fragment augments the action of the immunoglobulin molecule by binding to complement molecules and/or phagocytic cells.

The flexibility and variability of the Fab ends of immunoglobulin molecules permit the specificity of their reactions with antigenic biochemicals. The three-dimensional structure of the antibody allows the antigen and antibody molecules to fit properly together, something like a lock and key. The ends of the Fab fragments are said to be hypervariable, containing different amino acid sequences that give rise to varying three-dimensional structures for different immunoglobulin molecules. The specific amino acid sequences of the immunoglobulins are determined by the genetics of the antibody-synthesizing cells. The ability of the immune response to differentiate between many different antigens is dependent upon the genetic variability of the cells involved in antibody production; a great deal of genetic variability is necessary to code for the wide diversity of antibody molecules that function as part of the immune defense system.

Types of Immunoglobulins

IgG is generally the largest fraction of the body's immunoglobulin molecules, generally comprising approximately 80 percent. IgG is the predominant antibody that circulates in blood through the body. It readily passes through the walls of small vessels (venules) into extracellular body spaces (interstitial spaces) where it can react with antigen and stimulate the attraction of phagocytic cells to invading microorganisms. Reactions of IgG with surface antigens on bacteria activate the complement system and attract additional neutrophils to the site of the infection. IgG also crosses the placenta and confers immunity upon the fetus through the first months after birth. IgG plays a major role in preventing the systemic spread of infection through the body and in the recovery from many infectious diseases.

IgA occurs in mucus, in semen and in secretions such as saliva, tears, semen, and sweat. IgA is important in the respiratory, gastrointestinal, and genitourinary tracts where it plays a major role in protecting the mucous membrane surface tissues against invasion by pathogenic microorganisms. IgA also is secreted into human breast milk and plays a role in protecting nursing newborns against infectious disease. IgA is the major immunoglobulin molecule in mucus secretions involved in the immune response that protects external body surfaces. In mucus secretion IgA is present as a dimer; that is, IgA in these secretions contains two units of the basic four chain immunoglobulin molecule. The IgA molecules bind with surface antigens of microorganisms, preventing the adherence of such antibody-coated microorganisms to the mucosal cells lining the respiratory, gastrointestinal, and genitourinary tracts. Thus, IgA molecules block the attachment of pathogens to body surfaces.

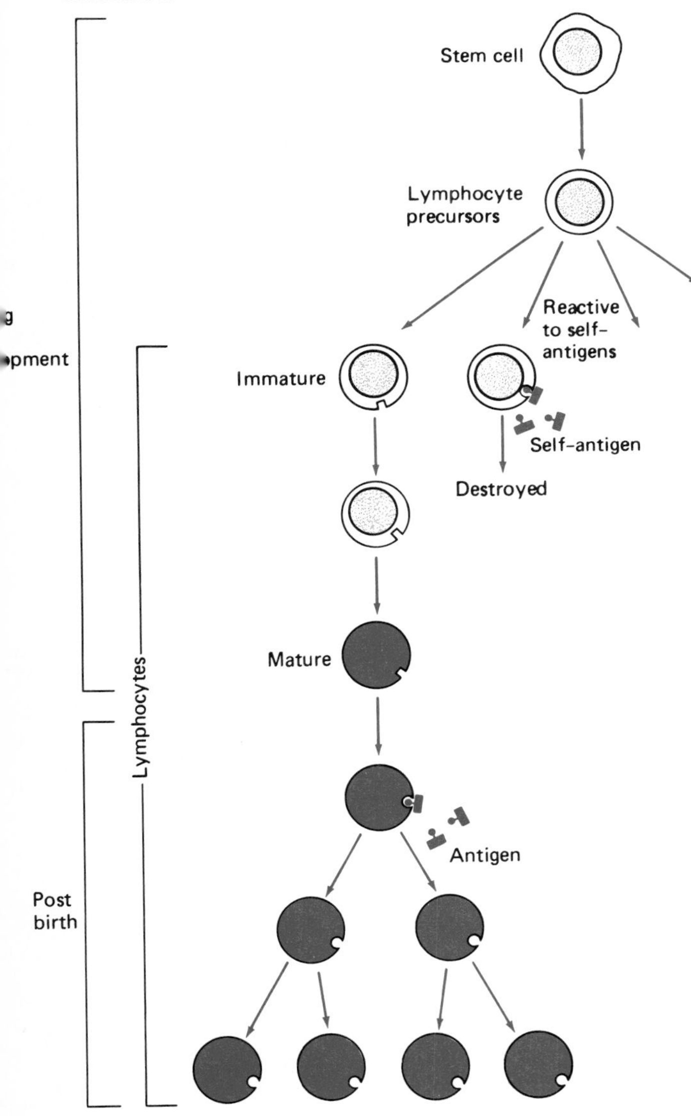

Figure 20.5 Illustration of the clonal selection theory. As shown, there is a critical shift in how lymphocytes react with antigen before and after birth. During fetal development the immune response network loses the ability to react with self-antigens. After birth the detection of foreign antigens leads to the cloning of lymphocytes and differentiation to destroy the foreign substance.

IgM is a high molecular weight immunoglobulin, occurring as a pentamer; that is, IgM contains five monomeric units of the basic four-peptide chain immunoglobulin molecule. IgM molecules are formed prior to IgG molecules in response to exposure to an antigen. Because of its high number of antigen binding sites, the IgM molecule is effective in attaching to multiple cells that have the same surface antigens. As such, it is important in the initial response to a bacterial infection. During the later stages of infection, IgG molecules increase in concentration, thereby becoming more important. IgM molecules occur primarily in the blood serum and, together with IgG molecules, are important in preventing the circulation of infectious microorganisms through the circulatory system.

IgD antibody molecules are present on the surface of some lymphocyte cells together with IgM. Although its precise role remains to be defined, it appears that IgD may play a role in lymphocyte activation and suppression. Within blood plasma, IgD molecules are usually short-lived, being particularly susceptible to enzymatic degradation.

IgE molecules are normally present in the blood serum as a very low proportion of the immunoglobulins. The ratio of IgG:IgE is normally 50,000:1. IgE serum levels, though, are elevated in individuals with allergic reactions, such as hay fever, and in some persons with chronic parasitic infections. The main role of IgE appears to be the protection of external mucosal surfaces by mediating the attraction of phagocytic cells and the initiation of the inflammatory response. IgE molecules are important because they bind to mast cells and assist in mediating hypersensitivity reactions, including allergic responses such as hay fever. The tight binding of IgE to cell membranes in circulating mast cells and leukocytes, based on the characteristics of the Fc portion of the IgE molecule, is an unusual property of IgE.

Synthesis of Antibody

The ability to synthesize the great diversity of antibody molecules that are needed to exhibit the specificity of the immune response can be explained in large part by the clonal selection theory (Figure 20.5). According to the clonal selection theory, B lymphocyte cells are programmed to synthesize a single specific immunoglobulin molecule. The particular immunoglobulin molecule for which the individual lymphocyte is genetically programmed is integrated into the cytoplasmic membrane of that B lymphocyte cell and acts as a specific surface receptor for an antigen molecule. Antibodies formed by B lymphocyte cells are deposited on the surfaces of their membranes, rather than being released extracellularly. The reaction of the antigen molecule with the surface antibody receptor initiates the differentiation and multiplication of that cell to form two different cell populations: plasma cells that are able to synthesize a specific antibody and secondary B cells (Figure 20.6). The secondary B lymphocytes, also called B memory cells, are capable at a later time of initiating the antibody-mediated immune response upon detection of the specific foreign antigen molecule for which they are genetically programmed. Compared to the primary B cells, these secondary B cells circulate more actively from blood to lymph and live longer.

Thus, in addition to leading to the formation of antibody-secreting cells, the reaction of B cells with antigen leads to the establishment of an increased population of memory cells. The presence of a sizable population of secondary B cells reduces the processing time needed to bring about the immune response. Although the B lymphocyte memory cells do not secrete antibody themselves, they are the precursors for plasma cells. The plasma cells derived from the cloning and differentiation of B lymphocytes are responsible for the secretion of extracellular antibodies into the blood serum. Each clone of a plasma cell line secretes a single specific antibody molecule. Thus, according to this mechanism, antibody production is under genetic control coded within the B lymphocyte cells prior to exposure to antigen molecules. The antigen molecules simply trigger the response but do not act as templates for the synthesis of the antibody molecules, nor are the antigen molecules responsible

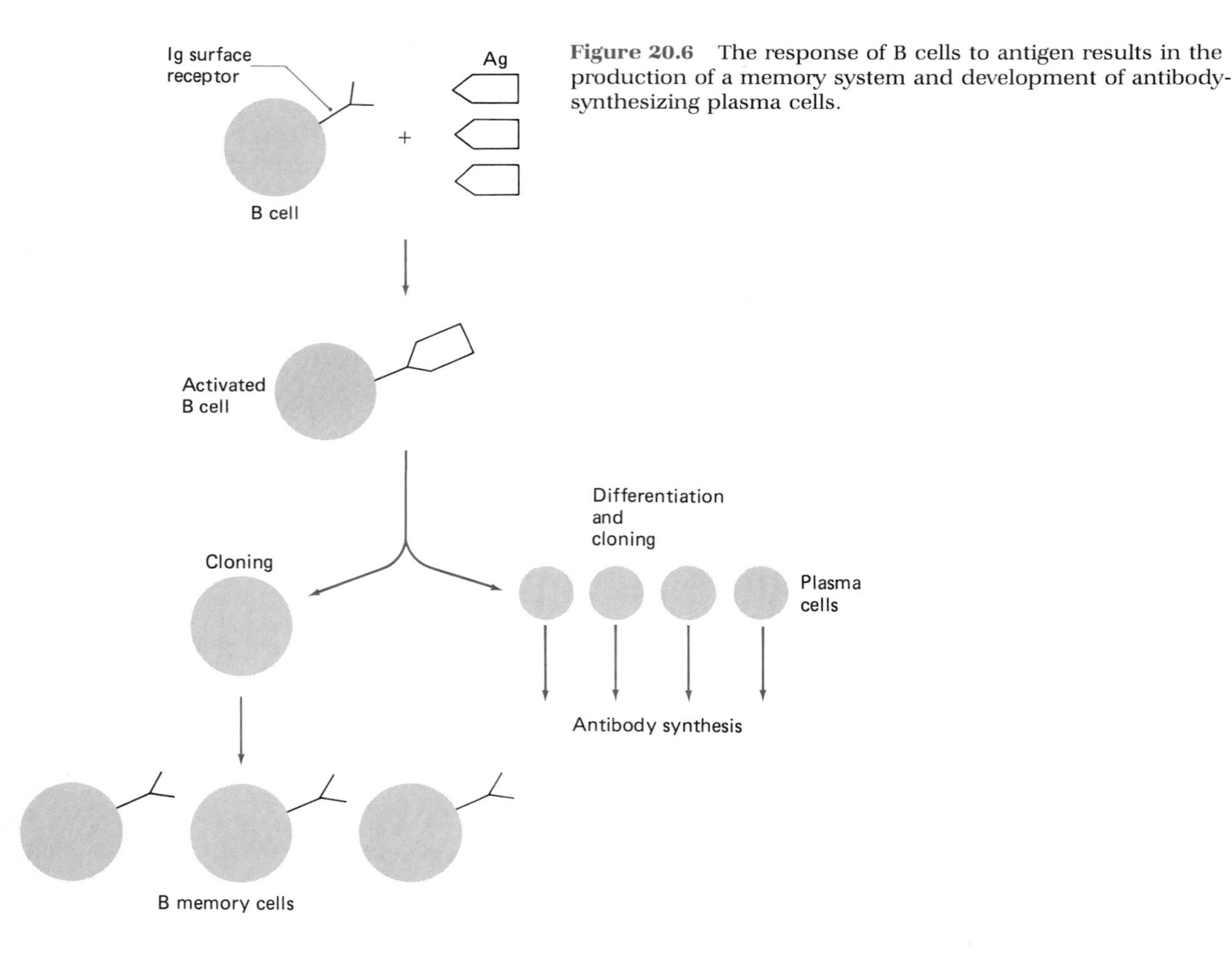

Figure 20.6 The response of B cells to antigen results in the production of a memory system and development of antibody-synthesizing plasma cells.

for modifying and determining the functional properties of the antibody molecules.

Events During Fetal Development. How, though, can specific immunoglobulins be synthesized in response to specific foreign antigens, and how can we avoid synthesizing immunoglobulins that react with our own antigens? To explain this, we hypothesize that interactions between B lymphocytes and antigens during fetal development are quite different from those that occur after birth. The clonal selection theory depends on the development and differentiation of a large population of B lymphocytes during fetal development. It assumes that during the development of the fetus, a complete set of lymphocytes is developed with each individual lymphocyte cell containing the genetic information for initiating an immune response to only a single specific antigen, with the receptors for the the particular antigen for which the cell is programmed located within the cytoplasmic membrane of that differentiated B lymphocyte. According to the clonal selection theory, an individual lymphocyte responds to only one antigen because it has only one type of receptor, and the correspondence between lymphocyte receptor and the antigenic determinant accounts for the specificity of the antibody-mediated immune response.

To explain the development of tolerance to self-antigens, the clonal selection theory says that during fetal development surface receptors of B cells react with antigen but that lymphocyte cells are unable to divide to form a clone of cells. Rather, the reaction of B lymphocytes with antigens during fetal development leads to the destruction of those lymphocytes. As such, there is a negative selection for lymphocytes capable of reacting with the antigens of the human body, and only lymphocytes that fail to react with antigens during the fetal development period survive. Theoretically, any antigen that is detected and reacts with the appropriate fetal lymphocyte is recognized as "self." A state of tolerance to self-antigens develops because all the B lymphocytes that could produce antibody to self-antigens are destroyed during development of the fetus. Thus, at birth the remaining genetically differentiated B lymphocytes should be competent to react only with foreign antigens, that is, with nonself-antigens.

Assuming the clonal hypothesis is correct, one should not exhibit autoimmunity; that is, one should not show an immune response against one's own antigens. Autoimmunity can occur, though, if B cells are not exposed to some human antigens, such as sperm antigens, during fetal development. Autoimmunity can also occur because some antigenic determinants are hidden within proteins but may be exposed at a later time; this may be involved in the aging process. Additionally, autoimmunity will occur if mutations reestablish B cell lines that were eliminated and/or if some B cells programmed for reacting with self-antigens survive fetal development. Indeed, it appears that some B cell lines that are genetically programmed for reacting with self-antigens are present in the body, but that these B cells are normally held in check by the action of T suppressor cells. Thus, the development of self-tolerance does not require the complete elimination of self-reactive precursor lymphocytes, but it is necessary that suppression be dominant.

The clonal theory also indicates that should a fetus develop an infection during development, the lymphocytes programmed for the antigens associated with the pathogen will be destroyed and the individual should never develop immunity to that disease. This indeed appears to occur in the disease congenital rubella. If a mother develops German measles (rubella) during pregnancy, the rubella virus can be transmitted to the fetus, which develops a persistant rubella infection.

Postnatal Events. After completion of fetal development, the reaction of an antigen with the surface-bound receptor of a lymphocyte cell initiates a totally different response than occurs during the fetal development period. The postnatal interaction of an antigen and lymphocyte initiates rapid cell division leading to the cloning of B cells, that is, the positive selection of lymphocyte cells. This hypothesis assumes that a full complement of B cells for binding all possible antigens with which the individual is capable of reacting are formed prior to actual exposure to the antigen.

The clonal theory requires a very large number of genetically different lymphocyte cells. More than 1 million genetically different lymphocyte types may be present at the time of birth. Although there is initially a large diversity of B cells, there are only a few that can react with each given antigen. Because relatively few B cells with the appropriate receptors are present upon the first exposure to a given antigen, the primary response is characteristically slow, producing relatively low yields of antibody. There is a long lag period (normally of the order of days) in the primary response, during which selection, differentiation, and cloning of appropriate B cell lines must occur.

After an antigen has elicited an antibody-mediated immune response, there is an increase in the number of B lymphocytes capable of reacting

Applications of Antibody Discovery Having Broad Impact on Medicine

By LAWRENCE K. ALTMAN

This year's Nobel Prize in medicine honored the discovery of monoclonal antibodies, widely seen as one of the most remarkable research achievements in recent years.

The practical applications of the discovery are already having a broad impact on most areas of medicine. As one of the winners, Dr. Cesar Milstein, has said, they are "a windfall of basic research."

One important part of the windfall is the improved accuracy of diagnosing ailments in fields ranging from obstetrics and pediatrics to geriatrics. The new techniques have opened up fresh avenues of research into infertility, brain disorders and diabetes. Another part affects the research in devising new treatments for such chronic and devastating diseases as cancer and leukemia. The techniques have allowed identification of so-called T cell subsets of lymphocytes, a form of white blood cells. Such subdivisions are important in typing some types of leukemias and lymphomas.

Also, monoclonal antibody techniques, by identifying T-4 lymphocytes as the cells destroyed in acquired immune deficiency syndrome, helped doctors recognize the first cases of AIDS in 1981.

Moreover, monoclonal antibodies have solved the crucial problem of precise reproducibility in the preparation of reagants, the substances used in chemical reactions to measure and produce other substances. As a result, monoclonal antibodies are now used so often in diagnostic tests, therapy and other essentials of medicine that the new techniques will account for an estimated $500 million in sales this year and an estimated $2 billion by 1990.

The windfall has come from the novel technique that Dr. Milstein and one of his co-winners, Dr. George Köhler, developed in 1975. Through newer modifications, scientists can now make unlimited amounts of pure cloned antibody against almost any antigen. Antibodies are the natural protective substances that the body forms to attack antigens, which are foreign substances such as microorganisms.

The body forms a specific antibody against each antigen, and no one knows the limit to the number of antibodies the body can make, "Certainly," Dr. Milstein has said, there are "well over a million."

The monoclonal antibodies are capable of distinguishing molecules with even subtle chemical differences, such as the difference of just a single amino acid in the sequence of hundreds within a substance. Because the monoclonal antibody technique allows scientists to make pure antibodies against any chosen antigen, Dr. Milstein has also said that "it is somewhat like selecting individual dishes out of a very elaborate menu: antibodies a la carte."

Doctors have used antibodies since the end of the last century, at first in the form of antitoxins such as in the prevention of diphtheria. But impurities in the production processes limited or precluded the widespread application of antibodies.

Now, with the new techniques, scientists have come closer to a long-time goal of the "magic bullet" that Paul Erlich sought at the turn of the 20th century. Doctors have been trying to link drugs and toxins to monoclonal antibodies in the hope that they will hit only the cells affected by cancer or other diseases while sparing healthy cells.

Although the development of monoclonal antibodies has revolutionized the field, the antibodies still can be tedious to prepare and there are limitations to the technique because scientists sometimes encounter technical problems in making them.

Ironically, the commercial success of the spin-offs from the basic research has led to increased competition between universities and industry for a researcher's time. The resulting conflicts have become a major issue in academic circles because they have blurred the distinctions between industry and academia.

The following is a partial list of the uses of monoclonal antibodies cited by the Nobel Committee and in medical journals:

Infectious diseases. Tests for hepatitis B and streptococcal infections have been improved and doctors have been guided to more appropriate choices of antibiotics in treating specific conditions as a result of monoclonal antibody techniques. Also, they have led to the recognition of differences in viral strains that were not previously appreciated and the resulting tests have improved the epidemiological study of many infections. Problems of interpretation, reproducibility and standardization, resulting partly from the unavoidable diversity of the antibodies detected by older tests, have always haunted virologists. Monoclonal antibody techniques have helped overcome many of these limitations. Further, the advances leading to the possibiity of a malaria vaccine are dependent on monoclonal antibody techniques that also offer the hope for improvements in vaccines against other diseases.

Cancer. A major limitation of now standard anticancer therapies has been the lack of their specificity for cancer cells. Many anticancer drugs and radiation therapy have a narrow range between their effectiveness and toxicity. Cancer researchers are now trying to harness drugs and toxins to monoclonal antibodies to produce much more potent weapons than are now available. At the same time, cancer researchers have been using these techniques to identify antigens specific to the surface of cancer cells, and thereby to develop tests to detect metastases, the spread of tiny clumps of cells in the body beyond their original source. Such steps would help improve the accuracy of cancer diagnosis and treatment.

Infertility. The cause of many cases of infertility is unknown and doctors are exploring the role of the different antigens detected on the surface of sperm in the process of fertilization. Scientists have found four that are specific for the front and back sides of the head, the entire head, and the tail of sperm.

Endocrinology. Researchers have coupled monoclonal antibody with radioimmunoassay and other techniques to identify hormones and the hormone receptor sites on the surface of cells. Such studies have advanced knowledge about thyroid disorders and have helped doctors understand that juvenile diabetes, also called Type I diabetes, may be a so-called autoimmune disorder.

Autoimmune disorders. Lupus, rheumatoid arthritis and other autoimmune disorders may be caused by the production of antibodies that go awry and attack one's own body. Using monoclonal antibody techniques, anti-antibodies have successfully thwarted experimental autoimmune disease.

Brain. Monoclonal antibody techniques have helped scientists detect the different carbohydrates that are involved in the everyday fuel needs of brain cells and to show their important roles in cell to cell interactions. Such results have led to theories that some inherited brain disorders might also be due to a failure in the maturation of specific brain cells. Also, doctors have detected similar antigens in some brain tumors and cancers that develop in other areas of the body such as the adrenal gland and blood cells. The similarity is presumed to be due to a common origin during early development.

Kidney disease. The cause of most kidney disorders is unknown. Scientists have developed monoclonal antibodies against specific anatomical components of the kidney in the hope that they will improve knowledge of the organ and the diseases that affect it.

Allergy. Antipollen antibodies may produce allergic symptoms when a susceptible individual is exposed to pollen. The production of antipollen antibodies has been prevented in animal experiments by anti-antibodies.

Organ transplantation. Improvement in tissue typing tests may lead to improved results in the transplants of kidneys, hearts and other organs.

Heart disease. Researchers are experimenting with monoclonal antibody techniques to try to preserve cells during a heart attack.

Caution is needed in limiting the expectations of the benefits derived from monoclonal antibody techniques.

The tests may be too specific, recognizing only a particular virus or subtype, when what may be required is an antibody reacting with all viruses of a group. It might be necessary to use a cocktail of several monoclonal antibodies reacting with different subtypes for diagnostic purposes.

Moreover, Dr. Nathaniel B. Brown of the University of California at Los Angeles has noted that it may be difficult to assess the safety and efficacy of monoclonal antibody products when they are coupled with drugs and toxins. Such combinations may be unpredictably dangerous, Dr. Brown has suggested, and thus "the specificity and safety of a monoclonal antibody product needs rigorous evaluation in every application" in medicine, he said.

Source: New York Times, Oct. 16, 1984. Reprinted by permission.

---

with that antigen. Because antigen binds to and selects for cells having receptors of the highest affinity, this process results in an increase in the number of lymphocyte cells with receptors of high affinity for the particular antigenic molecule. The cloning of these cells establishes a bank of memory cells, so that upon subsequent exposure to the same antigen there is a larger population of B cells that can rapidly and efficiently initiate the secondary immune response. The importance of this memory response will be discussed in Chapter 21.

Thus, life in the world of nature, implying as it does endless contact with all kinds of microbes, early brings forth in animals an adaptive response which modifies the internal organs in such a manner as to increase their general resistance to infection.

—René Dubos, *Mirage of Health*

In addition to the activation of B lymphocytes by antigen, a cooperative interaction between B and T cells is necessary effectively to stimulate antibody production (Figure 20.7). A particular type of T cell, called a T helper cell, is required before an activated B cell can initiate cloning and extensive antibody synthesis. T cells are involved in the presentation of antigen so that it can be recognized by B cells. Usually a B cell cannot be triggered to initiate antibody synthesis without interacting with T helper cells. These T helper cells provide a necessary second signal for the initiation of cloning. Macrophages are also involved in the cooperative antibody response; they process antigens to expose determinant sites and to get the fragmented antigens to the proper lymphocytes. Mononuclear monocyte and macrophage cells are involved in presenting the antigen to the T lymphocyte cells in the proper orientation so that the T cells can recognize the antigen.

Culture of Hybridomas—Monoclonal Antibodies

The fact that each specific antibody is synthesized by a different cell line of B lymphocytes and their derived plasma cells makes it possible to culture cells producing monoclonal antibodies. This can be accomplished by fusing myeloma cells—tumor cells of the immune system that can be readily cultured—with lymphocyte cells programmed for the synthesis of a single antibody. The resulting hybrid cell, known as a hybridoma, can be cultured and can synthesize specific antibodies (Figure 20.8). The malignant cells have a defective genetic regulatory mechanism that extends to antibody production. As such, the hybidoma cells produce unusually large amounts of imunnoglobulins. The

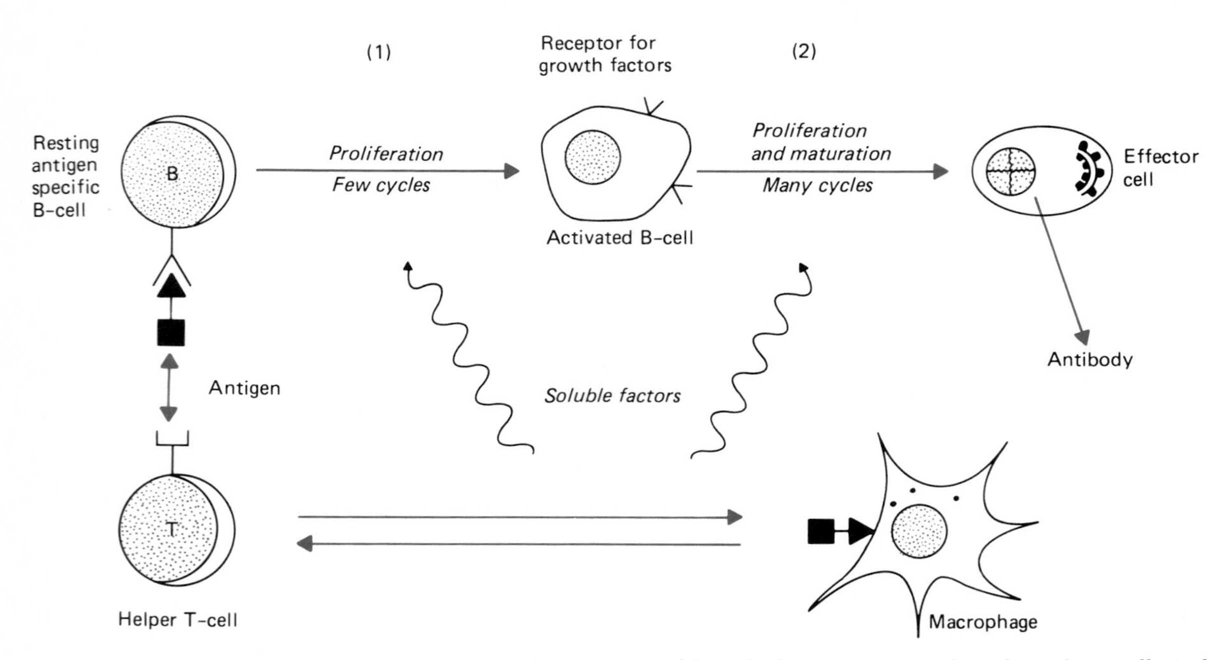

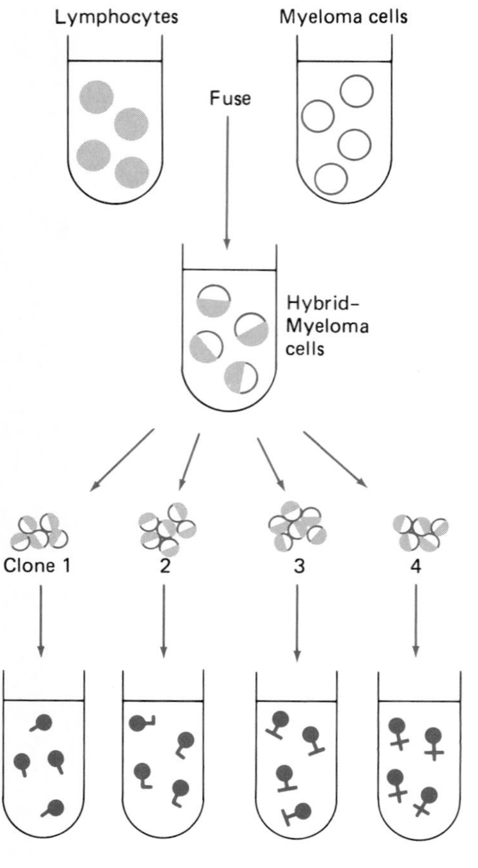

Figure 20.7 Although they were once thought to be totally independent systems, we now recognize that there are extensive cooperative interactions between B and T cells in response to an antigen. Helper T cells are activated by macrophage-processed antigen to stimulate the resting B cell, which has bound antigen through its hapten-specific Ig receptors, to transform after a small number of divisions to an activated B cell with receptors for soluble growth factors. Stage 2 is non-antigen-specific; these soluble growth factors are produced by interactions between the T cell and macrophages, possibly different from the original type initially representing the antigen, and stimulate the activated B cells to repeated division and maturation to become antibody-producing effector cells.

Figure 20.8 Hybridomas are produced by fusing B cells, which are genetically programmed for producing a particular (single) antibody, with a cancer (myeloma) cell, which is capable of poliferating rapidly under laboratory conditions. The resulting hybridomas (hybrid cells) are screened to select those that produce a desired antibody. Grown in culture, the descendants of "clones" of a single hybridoma cell continue to produce only the type of antibody characteristic of the parent B cell, hence the name "monoclonal" antibody. The hybridoma formed is a permanent union; hybridoma cells are uniform, highly specific, and can readily be produced in large quantities. Monoclonal antibodies are already used for diagnosis of allergies and infectious diseases, such as hepatitis, rabies, and some venereal diseases. Early stages of cancer may also be detectable with monoclonal antibodies, because cancer cells have surface antigens that differ from those of normal cells. It is possible that in the future, monoclonal antibodies could be used to treat cancer and infectious disease. The production of hybridomas capable of monoclonal antibody synthesis holds great promise for improved diagnosis and treatment of many diseases.

highly specific monoclonal antibodies produced this way are useful in clinical procedures and may prove useful in the treatment of some diseases. If, for example, monoclonal antibodies could be made into antigens that were unique to particular pathogens, these antibodies could be used to treat specific diseases.

Antigen–Antibody Reactions

The fundamental basis of antibody-mediated immunity depends on the reactions of antigen molecules with antibody molecules. Many of these reactions are used in vitro for the diagnosis of disease. These serological reactions will be discussed in Chapter 30. Here we will only consider the antigen-antibody reactions that occur in vivo as part of the immune defense system. Antibody-mediated immunity plays an important role in preventing and eliminating bacterial infections. The reactions of IgA molecules with bacteria and viruses in the fluids surrounding surface tissues prevent the adsorption of many potential pathogens onto these surface barriers. In this way IgA antibody molecules prevent the establishment of infections. IgG antibody molecules, acting as antitoxins, are able to neutralize toxin molecules by combining with the toxin molecules, blocking their reactions, and preventing the onset of disease symptomology. Even poisonous cobra venom can be neutralized by reaction with appropriate specific antibody molecules. The interactions of antibody with surface antigens of bacterial cells render many pathogenic bacteria susceptible to phagocytosis. In fact, the ingestive phagocytic attack on most bacteria requires an initial antigen–antibody reaction before phagocytic blood cells can engulf the invading bacteria. Antibodies that activate phagocytosis, called opsonizing antibodies, enhance the phagocytic destruction of pathogenic bacteria. The coating of the bacteria by antibodies that enhances destruction of the invading cells is called opsonization.

Opsonin? What the devil is opsonin?
Opsonin is what you butter the disease germs with to make your white blood corpuscles eat them.

—George Bernard Shaw, *The Doctor's Dilemma*

Antigen–antibody reactions are required to overcome infections by bacteria, such as *Haemophilus influenzae*, that are inherently resistant to phagocytosis. These in vivo reactions between antigen and antibody molecules constitute a major line of defense against invading bacteria and other microorganisms.

ABO Blood Groups. As an example of the reactions between antigen and antibody, let us consider the ABO blood groups. When the antigen is located on a cell surface, the reaction of antigen and antibody can produce clumping or agglutination of the cells. Clumping occurs because antigens on cell surfaces are usually multivalent and antibodies are divalent. Thus, the antigen–antibody reaction produces a three-dimensional lattice. Human red blood cells contain antigens on their surfaces. The membranes of human red blood cells may possess either type A or type B polysaccharide antigens, both type A and type B antigens, or neither of these two antigens. Blood types are determined by mixing known antisera (anti-A and anti-B antibodies) with a blood sample (Figure 20.9). An agglutination reaction indicates the presence of the corresponding antigen. Type A blood has type A but not type B surface antigens on the red blood cells. Type B blood has type B but not type A antigens. Type AB blood has both type A and type B red blood cell surface antigens. A person with type O blood has neither A nor B antigens on their red blood cell surfaces.

An individual's blood serum also contains antibodies to any antigens that do not occur in the cytoplasmic membranes of the red blood cells of

Figure 20.9 Illustration of blood typing to determine the major blood groups (A, B, AB, and O) based on agglutination reactions.

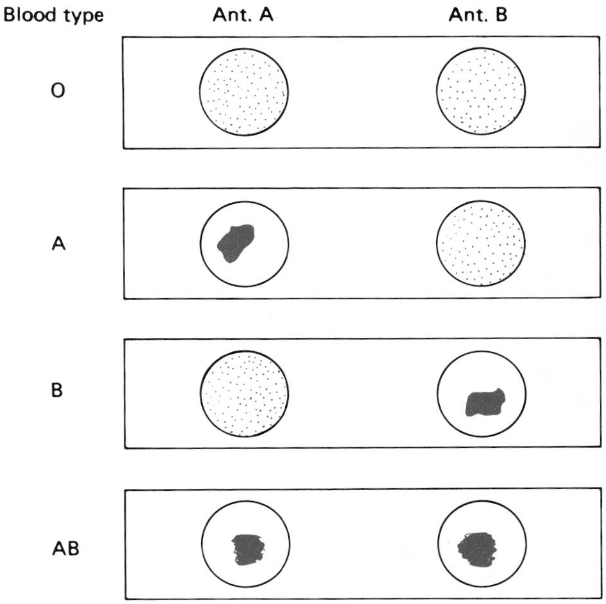

that individual. For example, type O blood contains both anti-A and anti-B antibodies. These antibodies develop after exposure to the nonself blood antigens, generally in food; the exposure to the nonself antigens results in an immune response that produces antibodies.

The antigen–antibody reactions involving these red blood cell surface antigens establish the primary compatibility of blood types; a transfusion with an incompatible blood type can lead to the death of the individual receiving the incompatible blood type. An individual with type O blood is a universal donor but can receive only transfusions of type O blood. If type O blood is given to an individual with type A, B, or AB blood, there may be some side reactions because of the reaction of antibody in the donor serum, but these reactions are normally fairly minor because of the relatively low concentrations of these antibodies when diluted through the circulatory system of the transfusion recipient. Under ideal conditions, assuming the availability of all blood types, type O blood is used only for transfusions of recipients who have type O blood, but in emergencies, type O blood may be used regardless of the blood type of the recipient.

Other Blood Cell Antigens. In addition to the A and B antigens of the red blood cell, various other antigens, including M, N, and Rh, can be identified on the red blood cell surface. Some of these antigens can create incompatibilities between blood types, and thus, before transfusions the blood of donor and recipient is screened for about 810 of the antigens most commonly involved in transfusion reactions. The detailed serotyping of red blood cell antigens can also be used for other purposes, such as determining possible paternity. Because the antigens of the red blood cells are genetically determined, the genetic information is heritable, and the pattern of minor antigens on the red blood cell surface—although not unique—is sufficiently characteristic to reflect genetic linkage.

The Rh (rhesus) antigen of the red blood cell membrane is especially important for determining the compatibility of fetal and maternal antigen–antibody interactions. Rh positive blood indicates the presence of the Rh antigen, whereas Rh negative blood cells lack this antigen. If the red blood cells of the fetus are Rh positive, that is, the fetal cells have the Rh antigen, and the mother produces anti-Rh antibodies, there can be a reaction that causes hemolytic disease of the newborn. This condition produces a cytotoxic hypersensitivity reaction that will be discussed in Chapter 21.

Cell-Mediated Immune Response

Whereas the antibody-mediated immune response system recognizes substances that are outside of host cells, the cell-mediated immune response is effective in eliminating infections by pathogenic microorganisms that develop within host cells and in controlling the proliferation of malignant cells. Cell-mediated immunity is important in controlling infections where the pathogens are able to reproduce within human cells, including infections caused by viruses; some bacteria, such as rickettsias and chlamydias; and some protozoa, such as trypanosomes. The cell-mediated immune response depends on the actions of T lymphocytes. When stimulated by an antigen, T lymphocytes respond by dividing and differentiating into cytotoxic T cells, and various other T cells that release biologically active soluble factors, which mediate the responses of other cells involved in the immune response (Figure 20.10). The soluble factors, collectively termed lymphokines, are effective in mediating the responses of monocytes and macrophages. Like the B lymphocytes, the T lymphocytes have surface receptors that can react with antigen, triggering the cell-mediated response. These surface receptors of T lymphocytes have recently been isolated and identified.

Lymphokines

As part of the cell-mediated immune response, lymphokines are released by T lymphocytes after antigenic stimulation. These soluble factors exhibit different activities (Table 20.2). The lymphokines produced by T cells include: (1) macrophage chemotaxis factor, a lymphokine that causes the attraction and accumulation of macrophages to the site of lymphokine release; (2) migration inhibition factor, which inhibits macrophage cells from migrating farther once they have reached the site of lymphokine attraction; (3) macrophage activating factor that results in an alteration of macrophage cells that increases their lysosomal activities and thus their ability to kill and ingest organisms; (4) skin reactive factor, which enhances capillary permeability and thus movement of monocytes across the vascular spaces; and (5) immune interferon, which activates antiviral proteins, preventing further intracellular multiplication of viruses. In addition to these lymphokines, T helper cells are involved in the stimulation of B cells as part of the antibody-mediated immune response, and T suppressor cells are involved in shutting off the antibody-mediated

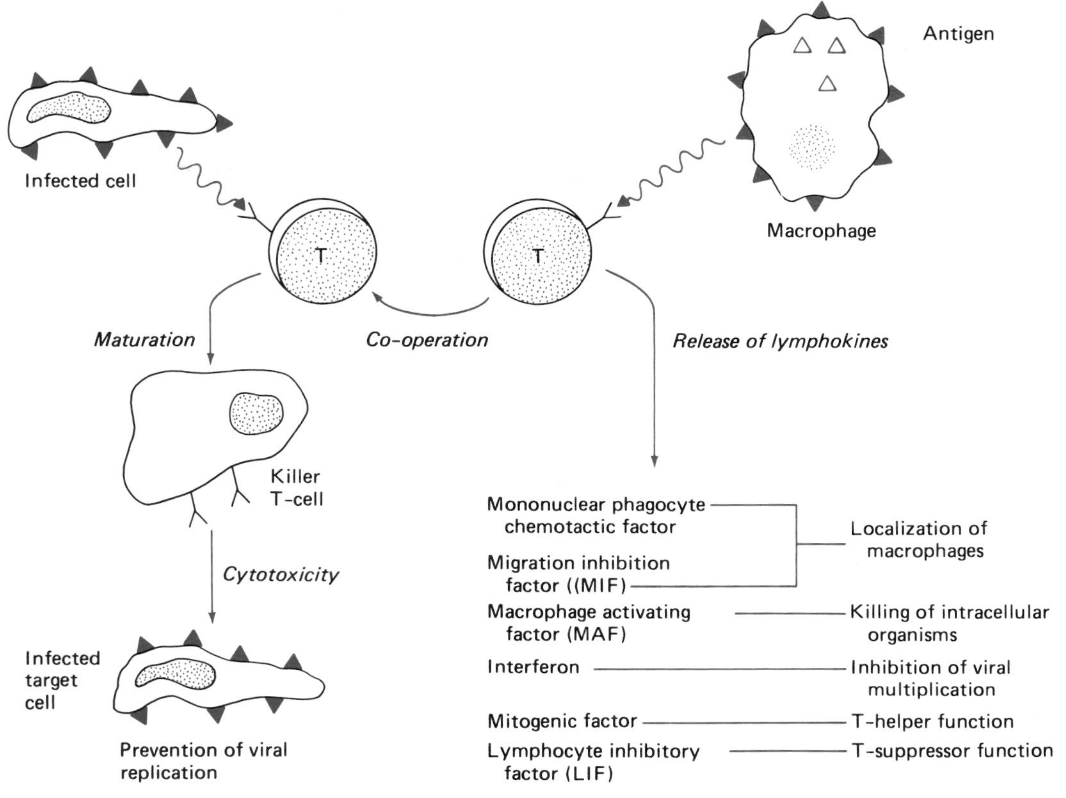

Figure 20.10 The cell-mediated immune response, the response of T cells to antigen, showing lymphokines and cytotoxic cell (killer T cell) formation. This operates through the generation of cytotoxic T cells and the release of lymphokines through the stimulation of two distinct T subpopulations. Different lymphokines may be produced by different lymphocyte subsets.

Table 20.2 Effects of Lymphokines

Cells affected	Lymphokine
Macrophages	Migration inhibitory factor
	Macrophage activating factor
	Chemotactic factor
Neutrophils	Leukocyte inhibitory factor
	Chemotactic factor
Lymphocytes	Mitogenic factor
	T-cell replacing factor
	Suppressor factors
	Chemotactic factor
Eosinophils	Chemotactic factor
	Migration stimulatory factor
Basophils	Chemotactic factor
Other cells	Cytotoxic and cytostatic factors (lymphotoxin)
	Interferons
	Osteoclast activating factor
	Colony stimulating factor

response by suppressing synthesis of antibody. It is thus apparent that lymphokines and T lymphocytes play a major role in regulating the activities of other cells involved in the immune response.

One of the specific lymphokines, immune or gamma interferon is secreted by lymphocytes in response to a specific antigen to which they have been sensitized or stimulated to divide. This immune interferon is different in some of its physiological effects from other interferon molecules and may kill tumor cells. Like other interferons, immune interferon molecules have antiviral activities, stimulating the synthesis of antiviral proteins. The primary function of immune interferon, though, appears be different from that of other interferons. Immune interferon may regulate the proliferation of the lymphoid cells that are stimulated to divide in response to interactions with antigenic biochemicals. Immune interferon may also enhance phagocytosis by macrophages and enhance the cytotoxicity of lymphocytes and the activities of T killer cells.

Cytotoxic T Cells

In addition to the antigen-induced release of lymphokines from T cells, intracellular infections, such as viral infections, can result in the generation of killer T cells that are specifically cytotoxic for intracellularly infected host cells (Figure 20.11). The activities of T killer cells include the production of factors that directly destroy microorganisms and the production of macrophage chemotactic factor (MCF) which attracts macrophages to the site of infection. The killer T cells, also known as cytotoxic T cells and T aggressor cells, appear to direct their activity against the major histocompatibility antigens of the cell surface. These histocompatibility antigens are genetically controlled by the major histocompatibility complex (**MHC**) region of the genome. The MHC antigens are highly variable because allelic genes each code for different antigens. These MHC antigens play an important role in the recognition of "self" and "nonself" tissues and are especially important in the immune response to tumors and tissue transplants.

The ability of cytotoxic T cells to recognize and to react with antigens on cell surfaces depends on the gene products of the major histocompatibility complex. Small variations in the surface antigens that occur in tumor cells and cells infected with viruses can be detected by T cells, leading to the appropriate cytotoxic response by these killer T cells. The ability to recognize antigens on the surfaces of cytoplasmic membranes permits T cells to monitor and regulate the proliferation of other types of cells. This appears to be extremely important not only for preventing the unchecked development of cells harboring intracellular pathogens but also for inhibiting tumor development. As a result of the ability of cytotoxic T cells to recognize transformed human cells, many would-be tumor cells do not lead to malignant growths.

Often our chemical defense-arm might seem to have little to do, like the members of the fire department at the corner, occupies its time with leisurely housework, cleaning up the bacterial and viral world in which all of us are forever immersed as in a dirty ocean or a fetid atmosphere. But let there be a five-alarm fire and every defender can be counted on to toil through day and night . . .

—GUSTAV ECKSTEIN, *The Body has a Head*

In a number of intracellular viral infections, antibodies are ineffective. In such cases, cell-mediated immunity augments the antigen-mediated immune response by controlling infections. Although antibody molecules can neutralize free viruses, antibodies are unable to penetrate host cells and attack viruses multiplying within host cells. It is the cell-mediated immune response that has the capability of eliminating cells infected with viruses. The sensitized cytotoxic T cells are able to recognize virally modified histocompatibility antigens or a complex of histocompatibility antigens with virally associated antigens in order to direct properly the cytotoxic attack of these lymphocytes against only those cells that are infected with a virus. Additionally, the T helper and T suppressor cells interact with the B cells to regulate antibody-mediated immunity.

As the complexity of the specific immune response is unraveled, it becomes increasingly clear that humans have evolved an interactive network

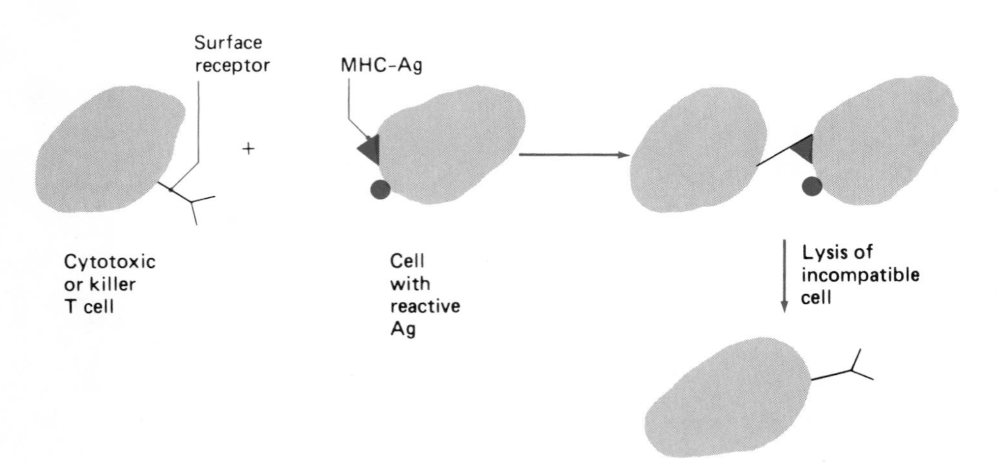

Figure 20.11 The interaction of cytotoxic T cells with cells having foreign HMC antigens. The major histocompatibility complex (MHC) and surface antigens in this region of the genome play an important role in the recognition of incompatible cells.

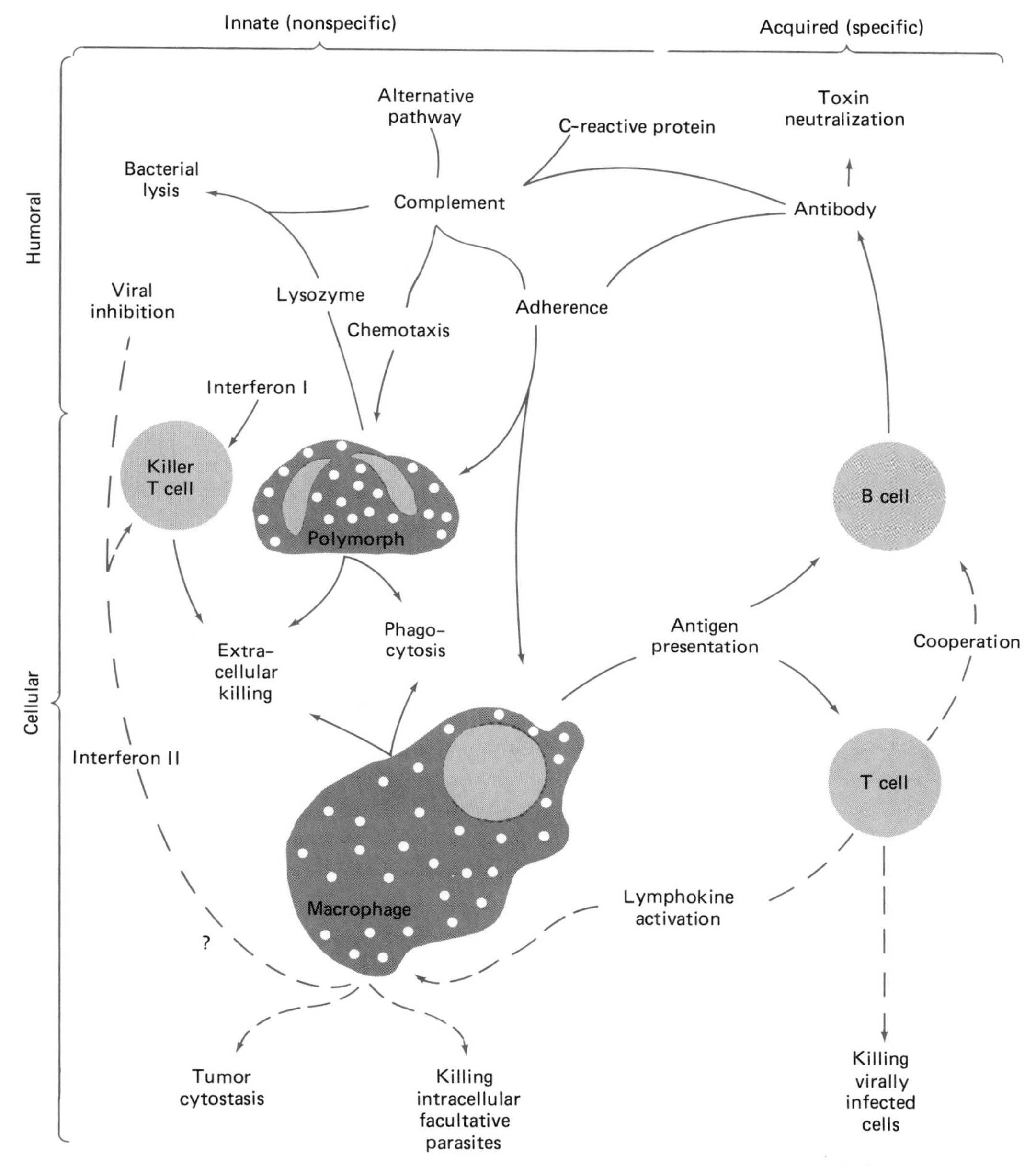

Figure 20.12 Illustration of the complex network of interactions of host-defense mechanisms. Reactions influenced by T cells are indicated by a broken line.

of various cells and macromolecules formed by these cells to regulate the immune response in a functional manner (Figure 20.12). It is also clear that antigens and receptors on cell surfaces permit molecular level communication between cells so that cells can act in an interactive and cooperative fashion.

Summary

The defense mechanisms of a host against pathogenic microorganisms involve a complex network of interactions between various host cells. The specific immune response represents a major line of protection against invading microorganisms and other foreign biochemicals. The immune response is activated by exposure to substances known as antigens. Immunity depends on the ability to detect and respond rapidly to such antigens.

Two types of lymphocyte cells, B lymphocytes and T lymphocytes, are both involved in conferring immunity but differ in their modes of action. B cells are primarily involved in the antibody-mediated immune response. Upon exposure to an appropriate antigen, B cells undergo differentiation, forming plasma cells that are able to secrete antibodies. There is a diversity of B lymphocytes, each containing different surface receptors, and each B cell can establish clones of plasma cells capable of responding to a particular antigenic determinant. The lymphocytes contain the genetic information for producing specific antibodies, and lymphocyte cell lines maintain a memory system for recognizing and responding to foreign antigens. The clonal selection theory explains how they come to possess the lymphocytes that are able to distinguish "self" from "nonself" antigens. Failures of this recognition process can render one susceptible to infectious diseases or may result in autoimmunity.

Five major classes of antibodies are secreted, each serving a somewhat different function in the immune response. IgG antibodies are the major circulating antibodies reacting with antigens, neutralizing toxins, and preventing the spread of pathogenic microorganisms through the circulatory system. IgA antibodies involved in the protection of surface cells from microbial attachment are secreted from mucosal cells. The reaction of antigen and antibody can form a complex that can react further with complement molecules to initiate the classical complement pathway, resulting in the lysis or death of invading microorganisms. The classical complement pathway also enhances phagocytosis and initiates the inflammatory response, both of which contribute to the elimination of invading microorganisms.

Within the human body, antibody-mediated immunity constitutes a major line of defense against toxins and the development of bacterial infections outside host cells. The immunological defense against intracellular pathogens, such as viruses, depends, in large part, on cell-mediated immunity. The cell-mediated immune response involves T cells that may release lymphokines that initiate a variety of defense mechanisms, including increased antibody synthesis, enhanced phagocytosis, and inflammation. Cell-mediated immunity protects the body against a variety of intracellular parasites, most notably viruses. In addition to releasing other lymphokines, activated T cells release interferon, preventing the multiplication of viruses in uninfected cells. T cells may also be differentiated into cell lines with cytotoxic properties that are able to recognize and kill cells harboring viruses. T killer cells eliminate infected cells, and this appears to be the main mechanism by which viral infections are eliminated. The ability of T cells to recognize altered host cells depends on their interaction with surface antigens coded for by the major histocompatibility complex. The MHC antigens are also important in the ability of cytotoxic T cells to recognize tumor cells, preventing the formation of malignancies, and the ability of T cells to recognize foreign tissues.

Study Outline

A. The immune response provides the mechanism for the specific response to the invasion of particular pathogenic microorganisms and other foreign substances. It is characterized by specificity, memory, and the acquired ability to detect foreign substances.
- **1.** In antibody-mediated immunity antibodies are made by plasma cells derived from B lymphocytes when foreign antigens are dectected.
  - **a.** B lymphocytes give rise to antibody-secreting plasma cells in response to antigenic stimulation.
- **2.** In cell-mediated immunity T lymphocytes destroy cells that are recognized as foreign.
  - **a.** B and T lymphocytes are cells differentiated in the lymphatic system; they originate in bone stem cells and become differentiated during maturation.
  - **b.** T cell precursors pass through the liver and spleen before reaching the thymus, where they are processed.

B. Antibody-mediated immunity.
- **1.** Antigens must be immunogenic and specifically reactive; they consist of a reactive portion, the hapten, that chemically reacts with an antibody and a carrier portion that is necessary for the stimulation of antibody production.

a. Antigens can be mono- or multivalent, having one or more reactive sites.
b. Many antigens are associated with cell surfaces.
c. Most toxins are antigenic and elicit the synthesis of antibody.
d. Denatured proteins, toxoids, that are antigentically active but are no longer toxic are used as vaccines because they elicit antibody-mediated immune responses without producing the diseases caused by toxins.

2. Immunoglobulins are proteins found in the serum fraction of the blood and are synthesized in response to antigenic stimulation.
   a. There are five classes of immunogloblubins: IgG, IgA, IgM, IgD, and IgE.
   b. They consist of four peptide chains, two identical heavy chains and two identical light chains, joined by disulfide bridges linking the chains.
   c. They can be split by papain to form two identical Fab fragments and an Fc fragment.
   d. IgG is the largest fraction of immunoglobulins; its reactions with the surface antigens of bacteria activate the complement system and attract additional neutrophils.
   e. IgA occurs in the mucus and other body secretions; it binds with surface antigens of microorganisms and prevents their adherence to the mucosal cells lining the respiratory, gastrointestinal, and genitourinary tracts.
   f. IgM contains five monomeric units of the basic four peptide chain immunoglobulin molecule and so is effective in attaching to multiple cells that have the same surface antigens.
   g. IgD may play a role in lymphocyte activation and suppression.
   h. IgE protects the external mucosal surfaces by mediating the attraction of phagocytic cells and the initiation of the inflammatory response; they bind to mast cells and mediate hypersensitivity reactions.
3. The clonal selection theory explains the ability to synthesize the diversity of antibody molecules that are needed to exhibit the specificity of the immune response.
   a. The reaction of B cells with antigen leads to the establishment of an increased population of memory cells. B lymphocyte memory cells are the precursors for plasma cells which are derived from the cloning and differentiation of B lymphocytes and are responsible for the secretion of extracellular antibodies into the blood serum. Each clone of a plasma cell line secretes a single specific antibody molecule. Antibody production is under genetic control coded within the B lymphocyte cells prior to exposure to antigen molecules.
   b. The clonal selection theory assumes that during fetal development a complete set of B lymphocytes are developed with each individual lymphocyte cell containing the genetic information for initiating an immune response to a single specific antigen, with receptors for each antigen located within the cytoplasmic membranes of the differentiated B lymphocytes. An individual lymphocyte, therefore, responds to only one antigen because it has only one type of receptor.
   c. The clonal selection theory also says that during fetal development only lymphocytes that fail to react with the antigens of the human body survive; all B lymphocyctes that could produce antibody to self-antigens are destroyed.
   d. The postnatal interaction of an antigen and lymphocyte initiates rapid cell division leading to the cloning of B cells, establishing a bank of memory cells, so that upon later exposure to the same antigen there is a larger population of B cells to initiate the secondary immune response. The primary immune response is much slower because there are relatively few B cells with the appropriate receptors.
   e. Helper T cells are required before an activated B cell can initiate cloning and antibody synthesis.
4. Hybridomas, the result of the fusing of myeloma cells and lymphocyte cells programmed for the synthesis of a single antibody, can synthesize large quantities of highly specific monoclonal antibodies.

5. The interactions of antibody with the surface antigens of bacterial cells render many pathogenic bacteria susceptible to phagocytosis. Antigens that activate phagocytosis are called opsonizing antigens and the phagocytic destruction of pathogenic bacteria, opsonization.
6. The membranes of human red blood cells may possess either type A or type B polysaccharide antigens, both of these, or neither, determining our blood types. Blood serum also contains antibodies to any antigens that do not occur in the cytoplasmic membranes of the red blood cells of that person.
7. If the red blood cells of a fetus contain the Rh antigen and the mother's do not, the mother produces anti-Rh antibodies upon the birth of the infant, when bleeding exposes the mother to the antigens and initiates a primary immune response. Subsequent, possibly Rh positive babies must be protected from this anti-Rh IgG by injecting the mother with anti-Rh antibodies after each delivery to clear the Rh positive fetal cells from the mother's blood stream.

C. The cell-mediated immune response depends on the action of T lymphocytes, which, when stimulated by an appropriate antigen, divide and differentiate into cytotoxic T cells and other T cells that release biologically active soluble factors, lymphokines, that mediate the responses of monocytes and macrophages.

1. Lymphokines include:
   - **a.** Macrophage chemotaxis factor, which attracts macrophages to the site of lymphokine release.
   - **b.** Migration inhibition factor, which inhibits macrophages from leaving the site.
   - **c.** Macrophage activating factor, which causes macrophage cells to increase their lysosomal activity.
   - **d.** Skin reactive factor, which enchances capillary permeability and so the movement of monocytes across vascular spaces.
   - **e.** Immune interferon, which activates antiviral proteins. It may kill tumor cells, may regulate the proliferation of lymphoid cells that are stimulated to divide in response to interactions with antigenic biochemicals, and may enhance phagocytosis by macrophages and cytotoxicity of lymphocytes.
2. Cytotoxic, killer, or aggressor T cells are specifically cytotoxic for intracellularly infected host cells and direct their activity against the major histocompatibility antigens of the cell surface. These T cells can recognize even small variations in the MHC antigens and permit them to monitor and regulate the proliferation of other types of cells, including would-be tumors.
   - **a.** Although antibody molecules can neutralize free viruses, antibodies are unable to penetrate host cells and attack the viruses multiplying inside. Sensitized T aggressor cells can recognize virally modified histocompatibility antigens and direct the cytotoxic attack against only those cells that are infected with a virus.

Study Questions

1. Match the following statements with the terms in the list to the right.

   (1) IgA
   (2) IgD
   (3) IgE
   (4) IgG
   (5) IgM

   - **a.** Comprises 80 percent of the immunoglobulins and readily passes through vessel walls and the placenta.
   - **b.** Main role appears to be protection of external mucosal surfaces by mediating the attraction of phagocytic cells and the initiation of the inflammatory response.
   - **c.** Plays a major role in preventing systemic spread of infection.
   - **d.** Occurs in mucus and secretions.
   - **e.** Binds to mast cell membrane and plays a role in allergic reactions.
   - **f.** Present on the surface of some lymphocyte cells.
   - **g.** Important in the initial response to bacterial infection before IgG.
   - **h.** Occurs primarily in the serum.
2. Indicate whether the following statements are true or false.
   - **a.** In the clonal selection theory, B cells are programmed to synthesize a specific immunoglobulin molecule.
   - **b.** Memory cells produce antibodies.

c. Each clone of a plasma cell line secretes a single specific antibody molecule.
d. Specificity of antibody-mediated immune response is due to correspondence between unique lymphocyte receptor and antigenic determinant.
e. Development of tolerance to self-antigens occurs during childhood development.
f. Helper T cells are important in cellular immunity.
g. Lymphokines are the mediators of the cellular immune response.
h. Killer T cells produce antibodies.

Some Thought Questions

1. Why do young and old people contact more infectious diseases than middle-aged individuals?
2. How do we learn to recognize infectious microorganisms?
3. How does cellular immunity differ from antibody-mediated immunity?
4. What controls the integrated immune system?
5. Why is it necessary to match blood types for transfusions?
6. Why are immunologists often called upon to testify in paternity suits?
7. How can we account for the great diversity of antibody molecules having the same basic structure?
8. What infections would someone be particulary susceptible to if they didn't have a functional thymus?
9. Why are there five major classes of antibodies?
10. How has the ability to differentiate clinically the various types of T cells altered the ability to diagnose disease?
11. How can monoclonal antibodies be used to diagnose disease?
12. How can monoclonal antibodies be used to treat cancer?

Additional Sources of Information

Allen, J. C. 1980. *Infection and the Compromised Host*. Williams & Wilkins Company, Baltimore. This discussion of the importance of the immune response emphasizes that diseases may be the result of failures of host defenses.

Benacerraf, B., and E. R. Unanue. 1979. *Textbook of Immunology*. The Williams & Wilkins Company, Baltimore. An excellent advanced text on the immune response.

Bigley, N. J. 1980. *Immunologic Fundamentals*. Year Book Medical Publishers, Inc., Chicago. An easy-to-read overview of the immune response.

Eisen, H. N. 1980. *Immunology*. Harper & Row, Publishers, Hagerstown, Maryland. A thorough, advanced text describing the immune system.

Immunology: Readings from Scientific American. 1976. W. H. Freeman and Company, Publishers, San Francisco. A collection of papers on the immune system.

Kimball, J. W. 1983. *Introduction to Immunology*. Macmillan Publishing Co., Inc., New York. A good introductory text about the immune system.

Milstein, C. 1980. Monoclonal antibodies. *Scientific American*, 243(4):66–74. An interesting article describing the production and potential uses of monoclonal antibodies.

Playfair, J.H.L. 1984. *Immunology at a Glance*. Blackwell Scientific Publications, Oxford, England. This book presents a series of topics in immunology through a series of excellent illustrations and accompanying explanations.

Roitt, I. M. 1980. *Essential Immunology*. Blackwell Scientific Publications, Oxford, England. A superb, thorough yet easy-to-read, text covering all aspects of the immune defense network.

21 The Immune Response: Protection Against Disease and Pathology of the Immune System

KEY TERMS

acquired immune deficiency syndrome (AIDS)
acquired immunity
active immunity
adjuvants
allergens
anamnestic response
anaphylactic hypersensitivity
anaphylactic shock
antibody-dependent cytotoxic hypersensitivities
antihistamines
Arthus reaction
artificially acquired active immunity
artificially acquired passive immunity
atopic allergies
attenuated pathogen
autoimmunity
booster vaccination
Bruton's congenital agammaglobulinemia
cell-mediated hypersensitivities
complex-mediated hypersensitivities
desensitization
DiGeorge syndrome
herd immunity
hemolytic disease of the newborn
hypersensitivity
immune
immunity
immunodeficiencies
immunosuppressant
immunosuppression
intradermal
intramuscular
intravenous
late onset hypogammaglobulinemia
long term immunity
naturally acquired active immunity
passive immunity
primary immune response
secondary immune response
serum sickness
severe combined deficiency
short term immunity
species immunity
subcutaneous
transfusion incompatibility
vaccination
vaccines

OBJECTIVES

After reading this chapter you should be able to

1. Define the key terms.
2. Describe the different types of immunity, indicating the differences between active versus passive and naturally acquired versus artificially acquired.
3. Discuss how vaccination is used to protect against disease; include the advantages and limitations of this approach for the control of disease.
4. Discuss why the immune system must be suppressed in cases of tissue transplants and the side-effects of immunosuppression.
5. Describe the underlying reasons for several immunodeficiency diseases.
6. Discuss the difference between hypersensitivity and autoimmunity.
7. Describe the four types of hypersensitivity reactions.
8. Describe how hemolytic disease of the newborn is treated and prevented.

The Immune Response

The immune response can protect against infectious agents because it can detect and react with foreign substances, including pathogenic microorganisms, in ways that preclude the harmful effects of that foreign agent. Toxins (substances produced by microorganisms that interfere with physiological function) can be neutralized, viruses inactivated, bacteria killed, and so forth. Some aspects of immunity are innate; that is, the individual is born with that state of resistance. For example, humans are resistant to a number of infectious agents that cause diseases in other animals because the human cells and body are incompatible with the requirements of that disease-causing agent. This is called species immunity. There are also racial and individual differences that affect the susceptibility to disease.

Protection against disease also rests with resistance obtained during the life of the individual. In particular, resistance to disease depends upon the ability of the immune system to remember previous exposures to specific antigens associated with pathogenic microorganisms. Having responded once to a particular antigen, a memory system is established that permits a rapid and specific secondary response upon reexposure to that same substance (Figure 21.1). By being able to rapidly recognize and respond to pathogenic microorganisms, a state of immunity results that precludes infection with those same pathogens.

When individuals are immune to a disease, they no longer participate in the chain of disease transmission. Limiting the number of individuals who are susceptible to a disease can protect the entire population by precluding the epidemic spread of infectious agents. This concept is known as herd immunity. The actual proportion of a population that must be immune to prevent an epidemic depends upon the effectiveness with which the pathogen is transmitted from one individual to another and the virulence of the pathogen. Generally, when about 70 percent of the population is immune, the whole population is protected.

Types of Immunity

Acquired immunity refers to the resistance to infection obtained during the life of the individual that results from the production and/or activity of antibodies and T cells against specific microbial antigens. There are four categories of acquired immunity, based upon the source of the antibodies and the means by which an individual acquired them (Table 21.1). In passive immunity there is no long-term protection because no memory response develops. Passively acquired immunity comes about when antibodies produced somewhere else are introduced into the body. Actively acquired immunity, on the other hand, involves production of one's own antibodies. In many cases the acquisition of active immunity also involves the development of memory cells so that the immunity persists for a relatively long time.

> Thus, life in the world of nature, implying as it does endless contact with all kinds of microbes, early brings forth in animals an adaptive response which modifies the internal organs in such a manner as to increase their general resistance to infection.
>
> —RENÉ DUBOS, *Mirage of Health*

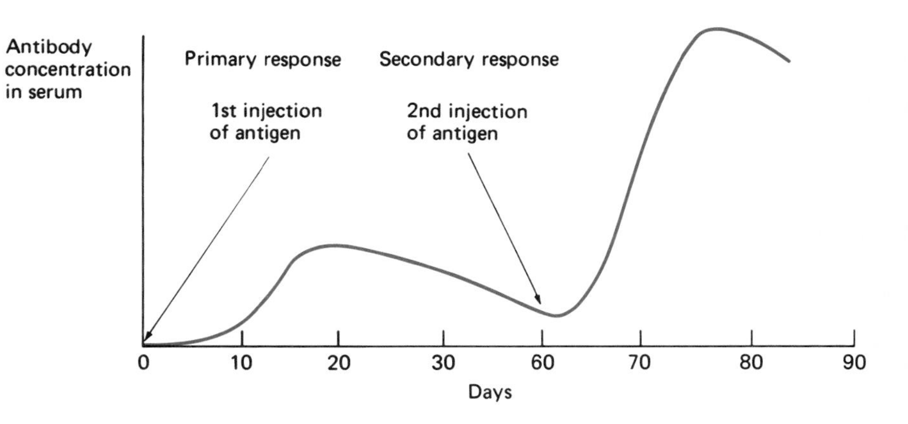

Figure 21.1 The primary and secondary (anamnestic) response to staphylococcal toxin. The primary response is characteristically slow and results in relatively low production of antibody. The secondary response has a characteristically short lag period and high rate of antibody production.

Table 21.1 Characteristics of the Four Types of Acquired Immunity

Naturally acquired passive immunity
- Immunity based upon transfer of antibodies from mother to child
  - Transplacental transfer to fetus
  - Transfer in human milk during nursing
- Offers protection against only diseases to which the mother is immune
- Protects infant during vulnerable period
- Immunity short term—no transfer of memory system

Artificially acquired passive immunity
- Immunity based upon injection of antibodies into an individual
- Immunity short term—no transfer of memory system
- Provides immediate protection against a specific disease agent
- Particularly useful for neutralizing toxins
- Useful for disease prophylaxis when exposure likelihood is high

Naturally acquired active immunity
- Immunity based upon development of secondary immune response after natural exposure to antigens associated with a pathogen
- Immunity long term in many cases—development of memory system

Artificially acquired active immunity
- Immunity based upon development of secondary immune response after artificial exposure to antigens associated with a pathogen
- Immunity long term in many cases—development of memory system
- Many diseases prevented by vaccination

Passive Immunity

Naturally acquired passive immunity is derived from the transfer of antibodies from mother to child (Figure 21.2). Antibodies are transferred to the fetus across the placenta and from mother to newborn in human colostrum and milk during nursing. Naturally acquired passive immunity generally lasts from a few weeks to a few months, protecting the infant against a number of diseases to which the mother is immune. After that time the infant is able to make sufficient immunoglobulin to defend the body against infectious agents.

Artificially acquired passive immunity describes the injection of antibodies into an individual to provide immediate protection against a pathogen or toxin. Artificially acquired passive immunity is immediate but short term immunity—two–three weeks. Administration of antitoxin is critical for averting the life threatening effects of some toxins. For example, a variety of antitoxins can be used to neutralize toxins of microbial or other origin that are responsible for disease symptomology. Antitoxins are used to neutralize the toxins in snake venom, saving the victims of snake bites. They are also used to neutralize bacterial toxins, such as tetanus and botulinum toxins. The toxins in poisonous mushrooms can also be neutralized by administration of appropriate antitoxins. The administration of antitoxins and immunoglobulins to prevent disease generally occurs after exposure to a toxin and/or an infectious microorganism.

Immunoglobulin G (IgG) injections are some-

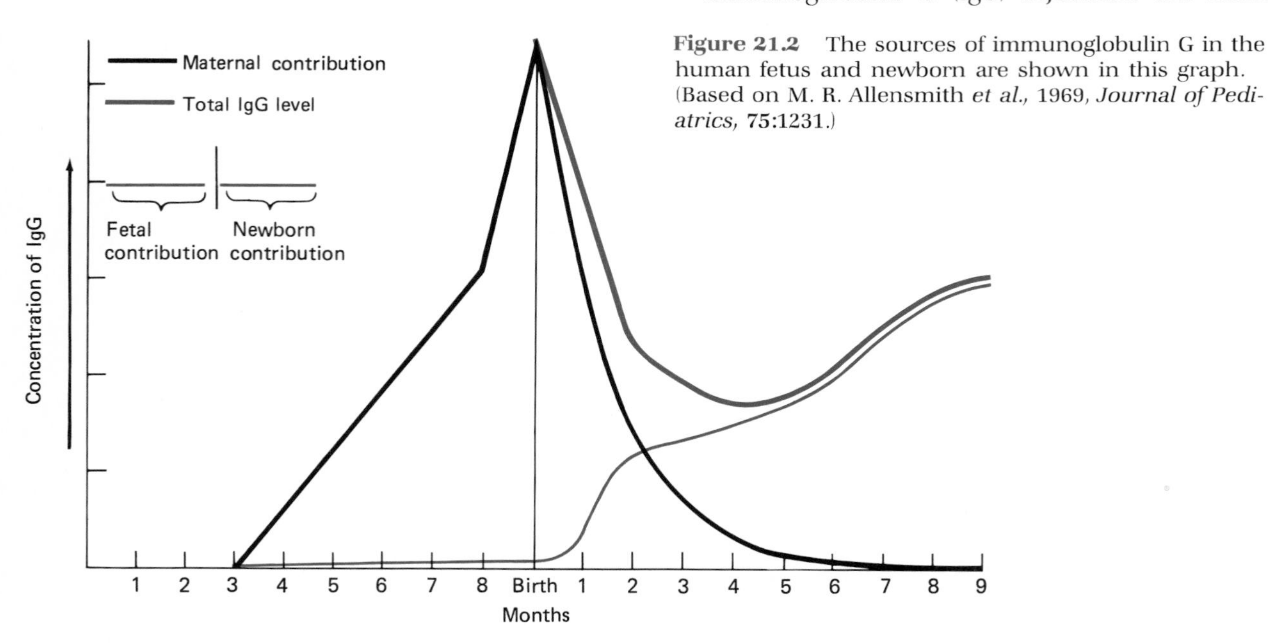

Figure 21.2 The sources of immunoglobulin G in the human fetus and newborn are shown in this graph. (Based on M. R. Allensmith *et al.*, 1969, *Journal of Pediatrics*, **75**:1231.)

times also used for disease prophylaxis. Individuals who are unable to produce adequate levels of immunoglobulins on their own, for example, are sometimes given passive injections of IgG to raise their general level of immunity. Generally pooled human globulin from donated blood that has passed the date of expiration for use in transfusions is used for this purpose. Passive immunization today is used for those exposed to tetanus, hepatitis A, measles, rabies, diphtheria, and also is used to protect against snake venoms, black widow spiders, and to prevent Rh negative mothers from becoming senstized to the Rh antigens of a newborn child. Travellers to rural areas of developing nations with poor sanitary conditions sometimes are given a shot of IgG (gamma globulin or immune serum globulin) to protect against hepatitis and other infectious diseases. Such injections of IgG offer limited protection for a short period of time. Although limited in duration, passive immunity offers immediate protection because the concentration of antibody instantly goes from zero to protective levels.

Active Immunity

In active immunity individuals make their own antibodies. In naturally acquired active immunity, a memory response develops after natural exposure to microbial antigens elicits a primary immune response. For example, in response to an infectious disease, an individual actively produces antibodies against the pathogen and develops clones of memory cells. This occurs even in subclinical cases of the disease in which there are no overt symptoms or signs that the individual has been exposed to a pathogen and its associated antigens. In some cases, long term immunity develops. Therefore we suffer from certain diseases, such as measles, only once. Having recovered from such diseases we need not worry about reexposure to the pathogen. In other cases, the immunity is short term and thereafter the memory system disappears. In some cases protection may last a few years. For example, protection against diphtheria lasts only a few years. In other cases, the secondary immune response lasts only a minimal time. As a result of such limited immunity one can contract certain diseases, such as syphilis, upon each renewed exposure. In yet other cases there are a variety of antigenically different agents that cause a particular disease and immunity to one antigenic variety does not necessarily confer immunity to the others. For this reason we are subject to repeated colds and periodic bouts with the flu.

Fortunately one need not develop the clinical symptoms of many diseases in order to become immune to those diseases; subclinical infections are sufficient to establish active immunity. In artificially acquired active immunity, an antigen is intentionally introduced into the body. For this purpose we use antigens that will not cause disease. Even though they do not cause disease, these antigens stimulate the immune response, eliciting the formation of antibody and memory cells. Immunity can be established for a high proportion of the population by using vaccines to artificially stimulate the immune response system.

Vaccination. Vaccination renders an individual unsusceptible to a particular disease, thereby helping to protect both the individual and the entire population. Vaccines are preparations of antigens designed to simulate the normal primary immune response. Exposure to the antigens in a vaccine results in a proliferation of the memory cells and the ability to exhibit a secondary immune response, also known as a memory response or anamnestic response, upon subsequent exposure to the same antigens. Immunization has become a worldwide practice for preventing various infectious diseases.

All interest in disease and death is only another expression of interest in life.

—THOMAS MANN, *The Magic Mountain*

The antigens within the vaccine need not be associated with active virulent pathogens; thus vaccines can be effective without causing disease. To confer protection, the antigens in the vaccine need only elicit an immune response, with the production of antibodies that can react with the critical antigens associated with the particular pathogen. The usefulness of vaccines rests with the fact that they can confer immunity; that is, they render an individual nonsusceptible to a disease without actually producing the disease, or at least not a serious form of the disease.

It's easier to keep out than get out.

—MARK TWAIN

Some vaccines useful in preventing diseases caused by a variety of microorganisms are listed in Table 21.2. Vaccines may contain antigens prepared by killing or inactivating pathogenic microorganisms, vaccines may use attenuated or weakened strains, or vaccines may contain only portions of microorganisms that are unable to cause the onset of severe disease symptoms. Some vaccines are pre-

Table 21.2 Some Vaccines That Are Useful in Preventing Diseases Caused by Microorganisms

Disease	Vaccine
Smallpox	Attenuated live virus
Yellow fever	Attenuated live virus
Hepatitis B	Attenuated live virus
Measles	Attenuated live virus
Mumps	Attenuated live virus
Rubella	Attenuated live virus
Polio	Attenuated live virus (Sabin)
Polio	Inactivated virus (Salk)
Influenza	Inactivated virus
Rabies	Inactivated virus
Tuberculosis	Attenuated live bacteria
Pertussis	Inactivated bacteria
Cholera	Inactivated bacteria
Diphtheria	Toxoid
Tetanus	Toxoid
Pneumococcal pneumonia	Capsular material

pared by denaturing microbial exotoxins to produce toxoids. Protein exotoxins, such as those involved in the diseases tetanus and diphtheria, are suitable for toxoid preparation, and the vaccines for preventing these diseases employ toxins inactivated by treatment with reagents such as formaldehyde or glutaraldehyde. These toxoids retain the antigenicity of the protein molecules; that is, they elicit the formation of antibody and are reactive with antibody molecules, but because the proteins are denatured, they are unable to initiate the biochemical reactions associated with the active toxins that cause disease conditions.

In some cases whole microorganisms rather than individual protein toxins are used for preparing vaccines. Killing a variety of microorganisms by treatment with chemicals, radiation, or heat, retains the antigenic properties of the pathogen but eliminates the risk that exposure to the vaccine could cause the onset of the disease associated with the virulent live pathogens. The vaccines used for the prevention of whooping cough (pertussis) and influenza are representative preparations containing antigens prepared by inactivating, that is, killing, pathogenic microorganisms.

In yet other vaccines individual components of the cell are extracted and used to elicit an immune response. For example, the capsule of *Streptococcus pneumoniae* is used to make a vaccine against pneumococcus pneumonia. This vaccine is used in high-risk patients, individuals over 50 years old who have chronic systemic diseases. Attempts have been made to make a vaccine against gonorrhea using pili from *Neisseria gonorrhoeae*, the bacterium that causes this disease. This vaccine has not been successful because long lasting immunity against *N. gonorrhoeae* does not develop. It has been used by the military to achieve short term immunity. Yet other vaccines are in development that use ribosomes instead of surface components of the cell.

In contrast to the vaccines just discussed, some vaccine preparations contain living but attenuated strains of microorganisms. Attenuated pathogens are produced by a variety of procedures, including moderate use of heat, chemicals, desiccation, and growth in tissues other than the normal host. The Sabin vaccine for poliomyelitis, for example, uses

Figure 21.3 A child receiving the Sabin oral polio vaccine at a village in Colombia, South America. (From *World Health*, July, 1976. Reprinted by permission of the World Health Organization, Geneva.)

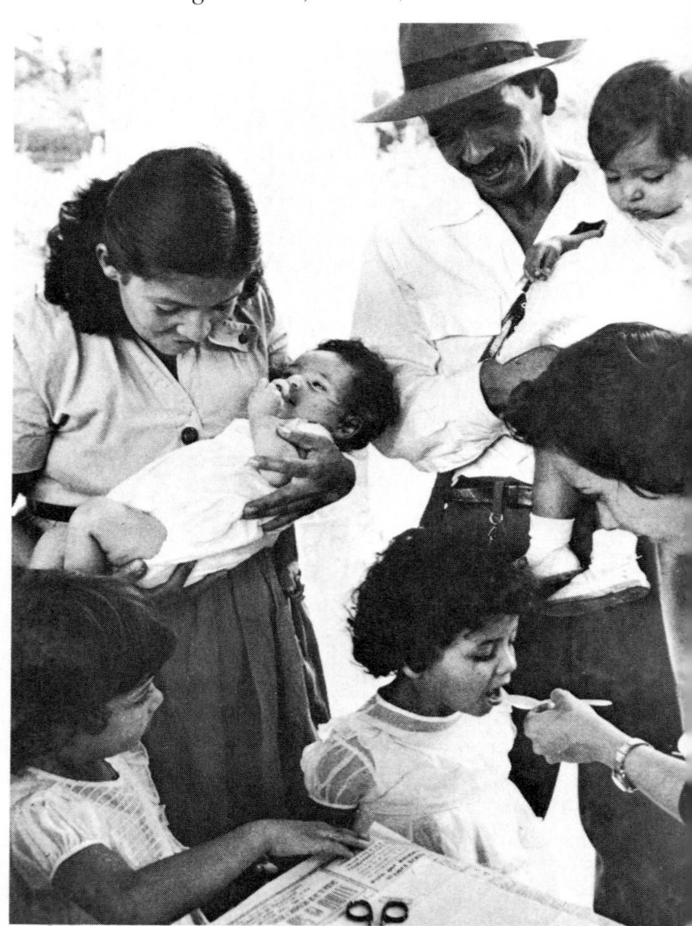

MICROBIOLOGY HEADLINES

Chicken Pox Vaccine Found Effective in Tests

By HAROLD M. SCHMECK Jr.

A vaccine against chicken pox has proved highly effective in protecting healthy children against the potentially dangerous disease, medical scientists reported yesterday.

Chicken pox, known to doctors as varicella, is a highly contagious virus disease with which most people are infected in childhood. While the illness is mild in most of the roughly three million cases that occur every year in the United States, there are cases of severe illness and even death. It is estimated that 100 to 150 Americans a year die of chicken pox.

The disease is particularly dangerous for people whose immune defenses are compromised by illnesses such as leukemia.

A report yesterday in *The New England Journal of Medicine* said a large-scale test of the chicken pox vaccine on children showed it to be safe and 100 percent effective in preventing the disease in the children studied.

Of 914 children in the study, about half received the vaccine while the other half received a harmless, inert substitute, a placebo. A total of 39 of the children later became naturally infected with chicken pox, all of them in the group that received placebo rather than the vaccine, which is not yet on the market.

None of the children who had received the vaccine developed chicken pox, even though their exposure to natural infection was presumably equal to that of the children who received a placebo in the trial. The children ranged in age from 1 to 14 years. The report covered their first 9 months after receiving either the vaccine or a placebo.

The chicken pox vaccine was produced by Merck Sharp & Dohme Research Laboratories, of West Point, Pa., using a virus first isolated more than 10 years ago by scientists in Japan. The efficacy trial of the Oka/Merck vaccine was conducted in collaboration with medical scientists of the department of pediatrics at the University of Pennsylvania.

"The results of this clinical trial provide preliminary evidence that administration of the live attenuated Oka/Merck varicella virus vaccine in healthy children is safe, highly immunogenic and protective against natural varicella, a potentially serious illness," the report in the journal said.

Dr. Edward M. Scolnick, a senior vice president for research at Merck who was an author of the study, said his company hoped to have the vaccine available for widespread use in about 18 months.

The vaccine is also being tested in children suffering from leukemia, a form of cancer that attacks the body's blood-forming system and seriously degrades the patient's ability to withstand infections of all kinds.

Dr. Anne A. Gershon, professor of pediatrics at New York University, who is leading a major study of the vaccine in such patients, said two injections of the vaccine had protected about 95 percent of patients in her study, which totals about 238 children suffering from leukemia and closely related cancers. Even those patients who did catch chicken pox despite the vaccine had mild cases, she said.

In healthy children one dose of the vaccine is considered sufficient to produce immunity.

An editorial in the medical journal described the results of the Pennsylvania study as "excellent and an essential step in the development of this vaccine," but emphasized that longer studies would be needed to prove that the immunity produced by the vaccine was long lasting.

Source: The New York Times, May 31, 1984.

viable polioviruses attenuated by growth in tissue culture (Figure 21.3). These viruses are capable of multiplication within the digestive tract and the salivary glands but are unable to invade the nerve tissues and thus do not produce the symptoms of the disease polio. The vaccines for measles, mumps, rubella, and yellow fever similarly utilize viable but attenuated viral strains. Attenuated strains of rabies virus can be prepared by desiccating the virus after growth in the central nervous system tissues of a rabbit or following growth in a chick or duck embryo. (A newer rabies vaccine uses viruses grown in human diploid tissue culture that are then inactivated.) There is a vaccine against tuberculosis, the BCG (bacillus of Calmette and Guerin vaccine), which is prepared by growing *Mycobacterium bovis* in media containing bile salts. By repeated subculturing of this bacterium, a strain is produced that is antigenically similar to *M. tuberculosis*, the organism that causes human tuberculosis. The immune response elicited by the BCG vaccine protects against tuberculosis and also stimulates the cell-mediated immune response, making it useful for therapy of other diseases, including cancer immunotherapy. The BCG vaccine has been widely used in Europe to protect against tuberculosis for individuals who are exposed to infected individuals for prolonged periods; it has not been extensively used in the United States to protect against tuberculosis but is used in cancer therapy.

Vaccines containing viable attenuated strains require relatively low amounts of the antigens because the microorganism is able to replicate after administration of the vaccine, resulting in a large increase in the amount of antigen available within the host to trigger the immune response mechanism. Quality control is extremely important in preparing vaccines, particularly those using attenuated strains, to prevent disease outbreaks resulting from inadequately attenuated pathogens or from microorganisms that accidentally contaminate the vaccine preparation.

MICROBIOLOGY HEADLINES

Vaccine Gap: 3 Issues Cloud Future

By HAROLD M. SCHMECK Jr.

Public fears, industry reluctance to risk damage suits and a longstanding failure of Government to draft an effective policy have all come together abruptly to contribute to the most serious national shortage of a major vaccine in modern American history.

Experts say they do not recall anything in the past quite comparable with the recommendation earlier this week to delay final booster shots of whooping cough vaccine to conserve supplies in the face of a looming nationwide shortage.

Still, the situation does not constitute a health crisis at this time. Most children have already been immunized against the disease, and failure to receive the last two recommended doses in the normal schedule of five is not expected to reduce immunity in the short term. Specialists expressed concern, however, that some doctors and health agencies might hoard existing vaccine because of publicity concerning the shortage, thus aggravating the existing supply problems.

Much of that risk was removed yesterday by an announcement from Wyeth Laboratories of Radnor, Pa., that it would supply 8.4 million additional doses next year, enough to alleviate the shortage expected for 1985. Altogether, about 19 million doses a year are needed. The drug company had announced earlier this year that it was halting production of the vaccine because of the rising costs of dealing with litigation.

That is clearly a significant step toward preventing a crisis. Nevertheless, the shortage of whooping cough vaccine could still have serious consequences in the long term if policies are not drafted to deal with it and situations like it.

Public health experts say the current vaccine against whooping cough does cause more frequent reactions than any other vaccine in large-scale nationwide use and that there is a real need for an improved product. They emphasize, however, that the discomforts and risks of the present vaccine are far smaller than the risks of the disease whooping cough itself. The disease is particularly hazardous for the very young.

At the Centers for Disease Control in Atlanta, Dr. Walter A. Orenstein, chief of the surveillance and investigation section of the division of immunization, said 15 deaths from whooping cough were reported to the centers in the two previous years. He said those deaths were among 3,159 cases of the illness in 1982 and 1983 that were studied in detail by the centers' staff.

3 Targets of Vaccine

Deaths associated with the vaccine, if any do occur, are so rare that they cannot even be quantified, Dr. Orenstein said. The vaccine is given as the combination called DPT, giving protection against three diseases, diphtheria; pertussis, which is the technical name for whooping cough, and tetanus. It is not always certain which component produces a given adverse effect in a particular patient. Furthermore, just the fact that a patient who took the vaccine experiences a bad reaction does not necessarily prove that the vaccine is responsible.

On the frequency of nonfatal adverse reactions, he cited a study done by scientists at University of California at Los Angeles indicating that 41 percent of children had swelling at the injection sites, 51 percent had pain and 47 percent suffered fever after their immunizations. The estimated risk of severe reactions, such as convulsions, was one in 110,000 doses of vaccine and the risk of permanent injury was about one in 310,000 doses of vaccine.

Improved vaccines against whooping cough are being developed here and abroad, but none is yet ready for marketing in the United States.

Vaccine liability problems have been of major concern to the industry since at least 1970 when a court awarded damages of $100,000 to the family of a child who developed paralytic polio immediately after receiving polio vaccine, even though it was clear that the vaccine was not responsible for the illness. The court said the manufacturer had the responsibility to warn the family of the possibility of harm.

Liability problems came to a head in 1976 when swine flu vaccination was followed by serious illness in several hundred people.

Drug companies have been dropping out of manufacturing vaccine in recent years, asserting that profits are too slim and the risk of liabilities is too great. Effective national policies to deal with this trend have not yet been developed, even though the roots of the present liability problem go back many years.

This week Connaught Laboratories Inc., of Swiftwater, Pa. said it was accepting no new orders for the vaccine, because of the high cost of liability insurance.

"We have had a terrible time getting anybody to cope with the liability problem," said Dr. Frederick C. Robbins, president of the Institute of Medicine of the National Academy of Sciences. He said the institute was preparing a new study of this problem, to be completed early next year.

Representative Henry A. Waxman, Democrat of California, will hold a hearing on the whooping cough vaccine shortage next Wednesday. An aide to the Congressman's Energy and Commerce subcommittee on health and the environment said Mr. Waxman had been working on legislation on the liability problem, but that the hearing next week would focus mainly on the current shortage itself.

Source: The New York Times, Dec. 15, 1984. Reprinted by permission.

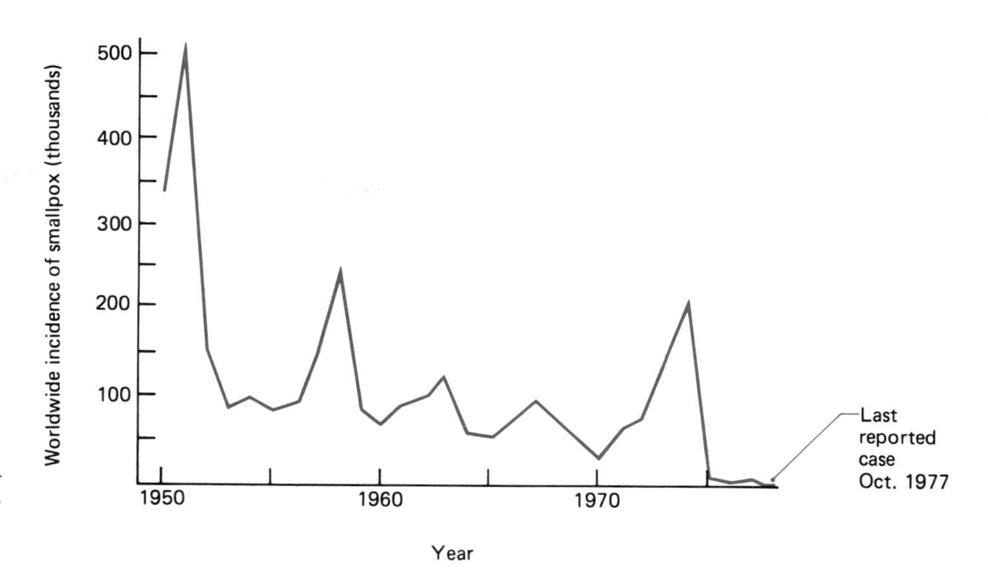

Figure 21.4 This graph shows the decline in the incidence of smallpox as a result of an effective international vaccination program.

The effectiveness of a vaccine depends on a number of factors, including the antigens within the vaccine, the physical form of the antigen, the other chemicals in the vaccine preparation, and the route of administration of the vaccine. Some vaccines must contain multiple antigens to prevent diseases caused by a variety of virulent strains of a given pathogen. For example, the Sabin polio vaccine contains type 1, 2, and 3 antigens associated with the three predominant strains of poliovirus. Three different vaccines, each containing only one of the antigenic types of polio virus, must be administered at different times in order to prevent interference between the three types of antigens with respect to the development of a memory immune response.

Other vaccines, such as those used to prevent influenza, are directed at particular strains of the pathogens, and different vaccines are needed to protect against different forms of the same disease. New vaccines are frequently needed to prevent influenza outbreaks because influenza viruses frequently change genetically, producing different antigens than strains for which previous vaccines have been developed.

The greatest success in preventing disease through the use of vaccines can be seen in the case of smallpox (Figure 21.4). The vaccine used for preventing smallpox contains a live strain of pox (vaccinia) virus. The vaccine most commonly used for preventing smallpox is prepared from scrapings of lesions from cows or sheep. The scrapings are treated with 1 percent formaldehyde to kill bacterial contaminants and 40 percent glycerol to stabilize the viral antigens. The viral antigens are quite labile, which is why live viral preparations are required for successful vaccination to achieve a state of immunity. Various viral strains have been used for the production of commercial vaccines. Although the commercial strains used for vaccine production were presumed to have been derived from cowpox virus, it now appears, based on its antigenic properties, that an attenuated strain of smallpox virus may have been inadvertently substituted for the cowpox virus. Because of the length of time that this virus has been cultivated, it is difficult to positively identify its original source, but certainly the pox virus used for vaccine preparation differs from the cowpox viruses found in nature.

Regardless of the origins of the viral strain used in its vaccines, smallpox, a once dreaded disease, has been completely eliminated through an extensive worldwide immunization program conducted under the auspices of the World Health Organization (WHO). The success of the WHO program depended on the use of lyophilized vaccines to overcome the problem of inactivation of the viral antigens in hot climates. The program was not without risks because the virus used for vaccination was virulent enough to cause a fatality rate of 1 in 1 million vaccinations. By immunizing a sufficient portion of the world's population against smallpox, though, it was possible to interrupt the normal transmission of smallpox virus from infected individuals to susceptible hosts. A consequence of the success of this immunization program is that it is no longer necessary to vaccinate against smallpox. The successful elimination of smallpox through a vaccination program was dependent on the fact that human beings are the only known host for the smallpox virus and that the virus has a relatively short survival time outside human host tissues. Smallpox presumably has been permanently eradicated and as such is the only infectious human disease known to have been eliminated.

Some chemicals, known as adjuvants, greatly enhance the antigenicity of other biochemicals (Figure 21.5). The inclusion of adjuvants in vaccines therefore can greatly increase the effectiveness of the vaccine. An interesting adjuvant is the BCG vaccine. This preparation of attenuated *Mycobacterium bovis* can be used to protect against tuberculosis, but also produces a generalized enhancement of the immune response that makes it more useful as an adjuvant. BCG effects both antibody-mediated and cell-mediated immunity—increasing production of antibodies, enhancing phagocytic activities, and accellerating rejection of foreign cells. This adjuvant has been used to augment the natural immunolgic response to malignant cells and has had some success in the treatment of melanomas.

Another way of enhancing the immune response to a vaccine is to mix the antigens with aluminum compounds. When protein antigens are mixed with aluminum compounds, a precipitate is formed that is more useful for establishing immunity than the proteins alone. Alum-precipitated antigens are released slowly, enhancing the stimulation of the immune response. The use of adjuvants can eliminate the need for repeated booster doses of the antigen. They do so by increasing the intracellular exposure time to the antigens, permiting the use of smaller doses of the antigen in the vaccine.

Figure 21.5 Adjuvants enhance antigenicity, as is illustrated by this graph.

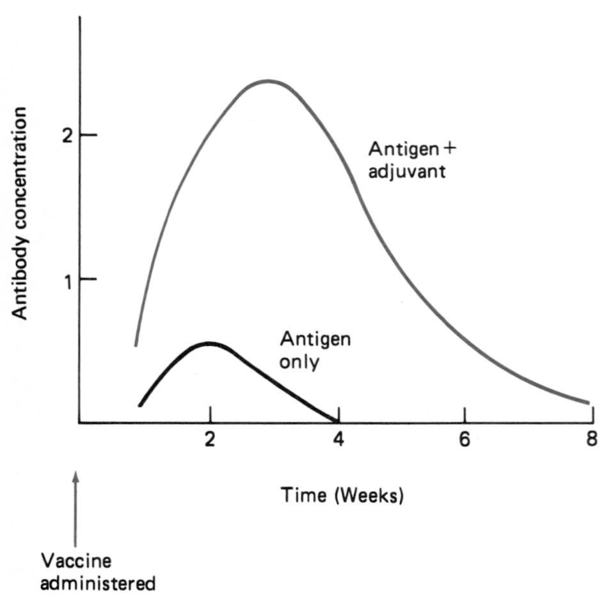

The ability of a vaccine to elicit the desired immune response is also affected by the route of administration. The antigens in the vaccine may be introduced into the body by a number of routes: intradermally (into the skin), subcutaneously (under the skin), intramuscularly (into the muscle), intravenously (into the blood stream), and into the mucosal cells lining the respiratory and gastrointestinal tracts. The effectiveness of a given vaccine depends in part on the normal route of entry for the particular pathogen. For example, polioviruses normally enter via the mucosal cells of the pharynx or gastrointestinal tract and therefore, the Sabin polio vaccine is administered orally, enabling the attenuated viruses to enter the mucosal cells of the gastrointestinal tract. It is likely that vaccines administered in this way stimulate secretory antibodies of the IgA class in addition to other immunoglobulins. Intramuscular administration of vaccines like the Salk polio vaccine, which contains inactivated polioviruses, is more likely to stimulate IgG production; IgG is particularly effective in precluding the spread of pathogenic microorganisms and the toxins produced by such organisms through the circulatory system.

Multiple exposures to antigens are sometimes needed to ensure the establishment and continuance of a memory response. Periodic booster vaccinations are necessary, for example, to maintain immunity against tetanus. No booster vaccinations, though, are needed to establish immunity against measles, mumps, and rubella. For each of these diseases, a single administration of an attenuated virus strain is sufficient to establish lifelong immunity. The recommended schedule for the administration of some vaccines is shown in Table 21.3.

Not all diseases can be prevented by using vaccines. Some antigens confer immunity that lasts only weeks or months. Such short-lived immunity may be effective in preventing disease if there is a known likelihood of exposure to a given pathogen, but it is not feasible to attempt to use vaccines that confer only short term immunity on a wide scale for preventing disease. Travelers to areas with a high incidence of a particular disease and workers likely

Table 21.3 **Recommended Schedule for Vaccine Administration**

Disease	Primary Immunization	Booster Doses
Diphtheria, pertussis, tetanus	Intramuscular DPT injections at 2, 4, 6, and 18 months	One intramuscular booster at 3–6 years (tetanus every 10 years)
Influenza	Seasonally for high-risk elderly and chronically ill	
Measles, mumps, rubella	One subcutaneous injection at 12 months	None
Polio	Sabin vaccine: oral at 2, 4, and 16 months	One oral dose at 5 years
	Salk vaccine: intramuscular injections at 2, 3, 4, and 16 months	Intramuscular injections every few years

to be exposed to a specific pathogen may find vaccination useful, even if the protection is only short term. Some recommended vaccinations for travelers are listed in Table 21.4. These vaccinations are prophylactic and are used when the likelihood of contracting a particular disease is high.

Although vaccines are normally administered prior to exposure to antigens associated with pathogenic microorganisms, some vaccines are administered after suspected exposure to a given infectious microorganism. In these cases the purpose of vaccination is to elicit an immune response before the

MICROBIOLOGY HEADLINES

Vaccination Drive for Adults Begins

Medical Group Warns Doctors Have Lagged in Protecting Against 7 Key Ailments

WASHINGTON, Jan. 16 (UPI)—A national medical society opened a major campaign today to see that adult Americans are as well protected by vaccines as their children are against a variety of diseases.

"Immunization isn't just kid stuff," said Dr. Robert H. Moser, executive vice president of the 60,000-member American College of Physicians.

The doctors said there were safe, effective and largely underused vaccines against seven adult ailments: tetanus, diphtheria, measles, rubella, hepatitis B, influenza and pneumococcal pneumonia.

"We want every physician to get in the habit of routinely considering what vaccines would protect patients against these diseases," Dr. Moser said at a news conference announcing publication of a handbook for doctors on adult immunizations.

Lag in Vaccinating Adults

Although prevention of disease by immunization is routine for children, Dr. Anthony S. Fauci, director of the National Institute of Allergy and Infectious Diseases, said, "The medical community has lagged in applying these same principles to adult diseases."

Dr. Alan R. Hinman, head of immunization for the Federal Centers for Disease Control in Atlanta, said everyone should be immunized against tetanus and diphtheria. These vaccines are combined and Dr. Hinman said booster doses were required every 10 years. He said 40 percent of adults needed such boosters.

Dr. Fauci said two-thirds of tetanus cases in the United States occurred among people over the age of 50, "those least likely to have been immunized as children." From 50 to 100 adults die of tetanus each year in the United States.

Dr. Hinman said adults should also be protected against measles and rubella, or German measles, a particularly serious disease for women of child-bearing age because rubella can cause birth defects.

Health care workers and other people at risk of catching hepatitis B should receive this new vaccine, Dr. Hinman said. Those at risk include homosexuals and intravenous drug abusers.

All people over 65 and other adults who have chronic diseases should be vaccinated against influenza and pneumococcus, the most common cause of bacterial pneumonia. About 25,000 Americans die of pneumococcal pneumonia each year.

Table 21.4 Recommended and Required Vaccinations for International Travelers

Tetanus
A booster dose of tetanus vaccine every 10 years is recommended. This must be the adult formulation and not the DPT triple vaccine.

Diphtheria
A booster dose of tetanus vaccine every 10 years is recommended. This must be the adult formulation and not the DPT triple vaccine.

Polio
For those never vaccinated, a complete set of polio vaccinations is advisable. A booster vaccination is suggested for those who have previously received the polio vaccine.

Measles
For those who have never had measles or been immunized (particularly those over 30 years) a measles vaccine is recommended.

Typhoid
For travelers likely to spend 2 weeks or more in rural areas of developing nations that have poor sanitation, a typhoid vaccine is recommended. Zaire specifically recommends immunization against typhoid. The vaccine does not ensure complete protection but reduces the likelihood of getting typhoid for a 4–6 month period.

Cholera
The cholera vaccine is not very effective and is not generally recommended for American travellers unless required by a specific country. A valid vaccination certificate for cholera requires two initial injections and booster injections every 6 months.

Yellow Fever
Yellow fever vaccine is recommended for travelers to central Africa (south of Sudan and north of Zambia and Mozambique), Panama and most of the northern half of South America (except the most western portions of Columbia, Ecuador, Peru, and Bolivia, and the eastern portions of Brazil). Many countries in South America, Africa, and the Middle East require a valid yellow fever certificate. The yellow fever vaccination certificate is issued only at special centers. A booster every 10 years is necessary for the certificate to be valid.

Bubonic Plague
Recommended only for travelers who expect to encounter wild animals in rural areas of Africa, Asia, and South America where plaque is endemic to the rodent populations.

Rabies
Recommended for children who are likely to come in contact with animals in areas where rabies is endemic.

Smallpox
No longer necessary to travel anywhere in the world.

onset of disease symptomology. For example, rabies vaccine is administered after animal bites that may have introduced rabies virus. The effectiveness of vaccines administered after the introduction of the pathogenic microorganisms depends on the relatively slow development of the infecting pathogen prior to the onset of disease symptoms and the ability of the vaccine to initiate antibody production before the onset of serious disease symptoms.

Suppression of the Immune Response

Sometimes it is necessary to suppress the immune response (immunosuppression). Such is often the case when organs or tissues are transplanted from one individual to another. Tissues contain surface antigens that are coded for by the major histocompatibility complex. Reactions to these antigens generally preclude successful transplantation of tissues unless the transplant tissues come from a compatible donor or unless the normal immune response is suppressed. In a sense, the normal immune defense mechanism is dysfunctional in transplantation and grafting because in these cases it does not serve the desired or useful function. The ability to transplant major organs, such as kidneys and hearts, is dependent on being able to control the normal immune response and prevent rejection of the transplanted tissues.

There are a number of drugs, including cyclosporin and prednisone, that may be used to suppress the normal immune response. The most widely used immunosuppressant for blocking rejection of major organ transplants is cyclosporin. This drug is produced by microorganisms. It blocks rejection of the transplanted tissue by specifically suppressing the cell-mediated immune system. Cyclosporin, however, is extremely expensive and it currently costs $800 per month when this drug is used to prevent rejection of a transplanted heart.

Additionally, suppression of the immune response renders the individual susceptible to a variety of infectious pathogens that are normally excluded by the host immune defense mechanisms. Immunosuppression can also result in graft versus host (GVH) disease. This occurs when the transplanted (grafted) tissue contains T lymphocytes that respond against the antigens of the tissues of the recipient. The depressed immune system of the recipient is unable to control the transplanted tissue and the reaction can be fatal. This commonly is a problem in bone marrow transplants because bone marrow contains large numbers of B and T cells that can initiate an immune response against the immunosuppressed recipient.

Autoimmune Diseases

There are a number of autoimmune diseases that result from the failure of the immune response to recognize properly self-antigens. The inability to differentiate properly between self- and nonself-antigens results in reactions with self-antigens and the killing of some of one's own cells. Autoimmunity can result if an individual is born with lymphocytes that are genetically programmed to react with self antigens. This can occur, according to the clonal selection theory, if lymphocytes fail for any reason to react with antigens during fetal development and therefore survive and remain in the body. Autoimmunity can also occur if genetic mutations occur that reestablish cell lines of lymphocytes that are genetically programmed to react with self-antigens. Regardless of the mechanism, such autoimmunity diseases often result in the progressive degeneration of tissues.

Table 21.5 Some Types of Autoimmunity

Disease	Source of Antigen
Hashimoto's thyroiditis; primary myxoedema	Thyroglobulin
Thyrotoxicosis	Cell surface TSH receptors
Pernicious anemia	Intrinsic factor; parietal cell microsomes
Addison's disease	Cytoplasm adrenal cells
Premature onset of menopause	Cytoplasm steroid-producing cells
Male infertility (some)	Surface of spermatozoa
Juvenile diabetes	Cytoplasm of islet cells; cell surface
Goodpasture's syndrome	Glomerular and lung basement membrane
Myasthenia gravis	Skeletal and heart muscle; acetyl choline receptor
Autoimmune haemolytic anemia	Surface of erythrocytes
Idiopathic thrombocytopenic purpura	Surface of platelets
Ulcerative colitis	Colon "lipopolysaccharide"
Sjögren's syndrome	Ducts, mitochondria, nuclei, thyroid; IgG
Rheumatoid arthritis	IgG; collagen
Systemic lupus erythematosus	DNA; nucleoprotein; cytoplasmic soluble Ag; array of other Ag including elements of blood clotting factors

The characteristics of a number of autoimmune diseases are summarized in Table 21.5. In rheumatoid arthritis, a fairly common inflammatory disease of the joints and connective tissue, the serum contains autoantibodies of the IgM class (rheumatoid factor) that reacts with IgG antibodies in the synovial tissue of joints. The reation of IgM and IgG antibodies initiates an inflammatory response that involves complement molecules and phagocytic neutrophils. Substances released as a result of phagocytic activity cause tissue damage to the affected joints. In systemic lupus erythematosus, numerous autoantibodies are produced that react with self-antigens, including some directed at DNA molecules. In this disease antigen–antibody complexes often circulate and settle in the glomeruli of the kidney, causing kidney failure. In cases of myasthenia gravis, antibodies react with nerve–muscle junctions. In autoimmune hemolytic anemia, antibodies react with red blood cells, causing anemia. Various other disease conditions may reflect the failure of the immune system to recognize self-antigens. These diseases can be treated by using immunosuppressive substances to prevent the "self-destruction" of body tissues by the body's own immune response.

Immunodeficiencies

Because the immune response plays a critical role in the protection against infectious diseases, failures of the immune response (immunodeficiencies) can compromise the ability to resist infection. When the immune system fails it leaves an individual susceptible to a variety of diseases. Indeed, one can consider that many infectious diseases result from failures of the immune response to protect the individual adequately against the invasion of and/or toxicity associated with pathogenic microorganisms. The development of tumor cells can also be viewed as a failure of the immune response, but in this case the failure to recognize and to respond properly to inappropriate cells within the body allows malignant cells to proliferate in an uncontrolled manner.

There are several types of deficiencies that can occur within the immune system. Immunodeficiencies can affect a variety of cells that are part of

Table 21.6 Types of Immunodeficiencies

Deficiency	Example	Immune Response: Humoral	Immune Response: Cellular	Infection	Treatment
Complement	C_3 deficiency	Normal	Normal	Pyogenic bacteria	Antibiotics
B cell	Infantile sex-linked agammaglobulinaemia (Bruton)	Absent	Normal	Pyogenic bacteria; *Pneumocystis carinii*	Gamma-globulin
T cell	Thymic hypoplasia (DiGeorge)	Lower	Absent	Certain viruses; *Candida*	Thymus graft
	Acquired immune deficiency syndrome (AIDS)	Lower	Lower	Numerous opportunistic pathogens	Supportive
Stem cell	Severe combined deficiency (Swiss-type)	Absent	Absent	All the above	Bone marrow graft

MICROBIOLOGY HEADLINES

Boy Who Lacks Immunity Thriving

PLYMOUTH, Mass., Nov. 24 (AP)—Sean Halloran was not much impressed with his first taste of banana, but it was a big event for his parents.

The fact that the 15-month-old victim of severe combined immune deficiency syndrome could eat the once-forbidden fruit is another sign he is beating the rare disease.

"I've been wanting to give it to him for a long time," his mother, June, said in a telephone interview today. "He didn't like it. But that didn't matter."

Sean was born without an immune system, making even the mildest sniffle a potentially fatal illness. His condition was similar to that of a boy in Houston, dubbed the "Bubble Boy," who lived 12 years in a series of sterile, plastic bubbles.

That child, known to the public only as David, received a bone marrow transplant in October 1983 but developed a reaction to the transplant and died 15 days after he was removed from his bubble in February for hospital treatment. Bone marrow is a source of the cells used by the immune system to ward off infection.

Doctors Were Prepared

Doctors were prepared for possible problems with Sean because a brother, Jason, had died from the same problem.

Sean was born by Caesarean section to insure sterile conditions. He was immediately put in a specially sanitized clean room at Children's Hospital in Boston. He was healthy and weighed 7 pounds 6 ounces, but tests disclosed he had the disease.

When Sean was two weeks old he received a bone marrow transplant from his father. Although their genetic profiles were not an exact match, doctors say the transplant appears to be working.

Sean remained in isolation for five months, seeing only nurses and his mother, who had to scrub with antiseptic for five minutes and dress in sterile gown, mask, cap, gloves and boots to see her son.

In January the doctors said Sean could go home, but not to a normal household.

The house had to be washed with a heavy-duty antiseptic and repeatedly vacuumed. Mrs. Halloran, a 31-year-old former nurse, gave up her job to care for Sean.

But after three months the work was cut back to every other day and now it is once a week.

Initially, family members had to wear gowns and masks and there could be no visitors. Sean is still not allowed playmates.

Mrs. Halloran carried a bottle of alcohol everywhere to wipe anything Sean might touch. His hands were constantly washed in the battle to prevent infection.

Regular Hospital Tests

Every three weeks Sean goes back to the hospital for blood tests, and recent results have been encouraging.

He remains on a low-bacteria diet and eats off sterilized dishes with sterilized utensils. He cannot drink or bathe in tap water; it has to be sterilized. He cannot eat in a restaurant.

Fruit once was out because of its bacteria content. But doctors now say he can try such treats as bananas.

Sean, who once wore a surgical mask whenever he went outdoors, can now go shopping with his mother without the mask.

But Mrs. Halloran must still be wary of strangers.

"Little old ladies want to give him a tickle, and I have to move the shopping cart," said Mrs. Halloran. "I try to be polite and say he's been sick and very susceptible."

Source: Associated Press, Nov. 24, 1984. Reprinted by permission.

MICROBIOLOGY HEADLINES

Minister Questions the Ethics of 'Bubble Boy's' Treatment

CHICAGO, Jan. 4 (AP)—The medical profession may have done a splendid technical job with David, the 12-year-old "bubble boy" who died last year after living almost all his life in isolation, but it failed him miserably as a human being, a minister involved in the case contends.

The Rev. Raymond J. Lawrence, a former staff member at Texas Children's Hospital, where David was treated for a rare disorder that left him with no immunity to disease, said the creation of the plastic bubble in which the boy lived suggested the medical world had drifted into a "technocratic imperialism."

Mr. Lawrence's commentary and an editorial response by Dr. Drummond Rennie was published today in *The Journal of the American Medical Association,* of which Dr. Rennie is a senior contributing editor.

Mr. Lawrence said that although David received the most humane treatment possible, doctors in the early stages of the case gave little thought to the human consequences of life in isolation.

'Human' Implications Neglected

Mr. Lawrence, who was director of clinical pastoral education at the hospital, said that from the beginning he was troubled by "what appeared to be a neglect of the potential emotional, psychological and spiritual—that is to say, human—implications of continued life" in the bubble.

"No one has discussed the horror of his existence," Mr. Lawrence said in a telephone interview from Houston.

Mr. Lawrence said in the journal that the medical community's pride in its reliance on scientific knowledge created unfounded optimism about finding a cure for David.

Dr. Rennie, professor of medicine at Rush Medical School in Chicago, said that Mr. Lawrence posed interesting questions but that David's parents, doctors and nurses had "a very human" relationship with the boy.

Ethical Issues Worried Doctors

The medical profession, Dr. Rennie said, was worried about ethical issues raised by this case. He said David's parents had been fully informed of the possible consequences of their son's treatment. He also said that at the time the bubble was made there seemed a good chance it would prove unnecessary or temporary.

Susannah Moore Griffin, a spokesman at the Baylor College of Medicine, where the physicians who treated David were based, declined comment on Mr. Lawrence's remarks and said the physicians felt that Dr. Rennie's comments adequately addressed the issue.

David, whose parents requested that his last name not be disclosed, suffered from severe combined immune deficiency syndrome, a condition that left his body defenseless against disease.

He spent all but the final two weeks of his life in a series of plastic enclosures, and died last Feb. 22 when a type of antibody, probably introduced in a bone marrow transplant in October 1983, unexpectedly proliferated.

Source: Associated Press, Jan. 4, 1985. Reprinted by permission.

---

the immune response system (Table 21.6). Such deficiencies result in the failure of the system to recognize and respond properly to the antigens of pathogenic microorganisms. Individuals with immunodeficiencies are more prone to infection than those who are capable of a complete and active immune response.

Severe Combined Deficiency

The most severe type of immunodeficiency, severe combined deficiency, results from a failure of stem cells to differentiate properly. This is a genetically determined disease and the trait is sex-linked so that it can be acquired only by males. Individuals with severe combined deficiency do not have either B or T lymphocytes. As such they are incapable of any immunological response. Any exposure of such individuals to microorganisms can result in the unchecked growth of the microorganisms within the body, resulting in certain death. Individuals suffering from severe combined deficiency can be kept alive in sterile environments where they are protected from any exposure to microorganisms (Figure 21.6). Bone marrow grafts may be employed to establish normal immune functions, but the grafts must come from siblings with compatible bone marrow.

DiGeorge Syndrome

Less severe immunodeficiencies occur when only B-cell or only T-cell functions are lacking. The DiGeorge syndrome results from a failure of the thymus to develop properly, so that T lymphocytes do not become properly differentiated. Individuals with this condition produce B lymphocytes and therefore have a partially active immune response. Individuals suffering from DiGeorge syndrome do not exhibit cell-mediated immunity and thus are prone to viral and other intracellular infections. Additionally, because T helper cells are involved in enhancing antibody production by B cells, the antigen-mediated or humoral response is de-

Figure 21.6 David was born in 1971 with severe combined immune deficiency. He survived for 13 years by being isolated from the microbe-laden environment. He is shown here just before his ninth birthday playing with a fish tank from inside his isolation bubble where he had lived since birth. He also had a sterile isolator "space suit" that gave him some very limited mobility outside of the bubble that protected him from infections his body could not fight off. David died in 1984 after an attempt to treat his immunodeficiency by injecting bone marrow from a sibling into his body. This treatment has been successfully used to treat other children with immunodeficiencies. The bone marrow from the immunologically compatible sibling provides the missing B and T cell lines that are essential for establishing immunity. Unfortunately, after the bone marrow transplant, there was an uncontrolled proliferation of B cells in David's body that apparently attacked his own tissues, resulting in his tragic death. Much has been learned from David that will help treat future cases of severe combined immune deficiency. (Reprinted with permission of United Press International, Inc.)

pressed in individuals suffering from this disease. The complete absence of the thymus is rare; partial DiGeorge syndrome, in which some T cells are produced although in lower numbers than are formed in individuals with fully functional thymus glands, is more common. This condition can be successfully treated by fetal thymus transplantation.

Bruton's Congenital Agammaglobulinemia

Bruton's congenital agammaglobulinemia results in the failure of B cells to differentiate and produce antibodies. This immunodeficiency disease is sex-limited and affects only males. The cell-mediated response in individuals suffering from this disease is normal. Children with Bruton's agammaglobulinemia are particularly subject to bacterial infections, including those by pyrogenic (fever-inducing) bacteria, such as *Staphylococcus aureus, Streptococcus pyogenes, Streptococcus pneumoniae, Neisseria meningitidis*, and *Haemophilus influenzae*. The treatment of this disease involves the repeated administration of IgG to maintain adequate levels of antibody in the circulatory system.

Late Onset Hypogammaglobulinemia

The most common form of immunodeficiency, known as late onset hypogammaglobulinemia, occurs later in life. In this condition there is a deficiency of circulating B cells and/or B cells with IgG surface receptors. Such individuals are unable to respond adequately to antigen through the normal differentiation of B cells into antibody–secreting plasma cells. Other immunodeficiencies may affect the synthesis of specific classes of antibodies. For example, some individuals exhibit IgA deficiencies, producing depressed levels of IgA antibodies. Such individuals are prone to infections of the respiratory tract and body surfaces normally protected by mucosal cells that secrete IgA.

Acquired Immune Deficiency Syndrome (AIDS)

In recent years a new disease, acquired immune deficiency syndrome (AIDS) has been discovered. This disease is apparently caused by the virus HTLV-III. Years after initial infection with this virus, there is a shift in the balance of T helper and T suppressor cells such that the immune system no longer functions efficiently to eliminate pathogens and to control malignant cells. Individuals with AIDS, therefore, are susceptible to a number of infectious diseases and to an increased incidence of cancer. Unlike the other immunodeficiencies that have been discussed, which are the result of genetic inheritance, AIDS is an immunodeficiency that results from an infection. One of the primary means of transmission of this disease is through sexual contact and the disease will be considered in greater detail in Chapter 27.

Other Immunodeficiencies

Immunodeficiencies may also affect the complement system. Individuals who fail to produce sufficient complement molecules are unable to respond properly to bacterial infections. The lack of an active complement system limits the inflammatory response and lytic killing of pathogenic bacterial cells. Immunodeficiencies may also affect the functioning of monocytes, neutrophils, and macrophages. Phagocytic cells lacking enzymes that produce hydrogen peroxide and other antimicrobial forms of oxygen do not have proper lysosomal functions to kill bacteria, and pathogenic bacteria are able to multiply within such metabolically deficient phagocytic cells. Antibiotics can be used to protect individuals with deficiencies of both the complement system and active phagocytic cells from invading pathogenic bacteria. The use of antibiotics is only marginally successful because immunodeficient individuals are severely compromised.

Hypersensitivity Reactions

An excessive immunological response to an antigen can result in tissue damage and a physiological state known as hypersensitivity. Hypersensitivity reactions occur when an individual is sensitized to an antigen so that further contact with that antigen results in an elevated immune response. The hypersensitivity reaction may be immediate, occurring shortly after exposure to the antigen, or delayed, occurring a day or more after contact with the antigen. There are several types of hypersensitivity reactions, each mediated by different aspects of the immune response.

Anaphylactic Hypersensitivity

Anaphylactic hypersensitivity (type 1 hypersensitivity)—a systemic, life-threatening condition—occurs when an antigen reacts with antibody bound to mast or basophil blood cells (Figure 21.7), leading to disruption of these cells with the release of vasoactive mediators, such as histamine (Figure 21.8). This condition is also known as immediate hypersensitivity because it occurs shortly after exposure to the antigen that triggers this response. In anaphylactic hypersensitivity mast cells or basophils generally are coated with antibody whose Fc region binds to specific sites on the mast cell surface. IgE antibodies are generally involved in such binding to mast and basophil cells. Some individuals inherit the genetic trait for producing high levels of IgE

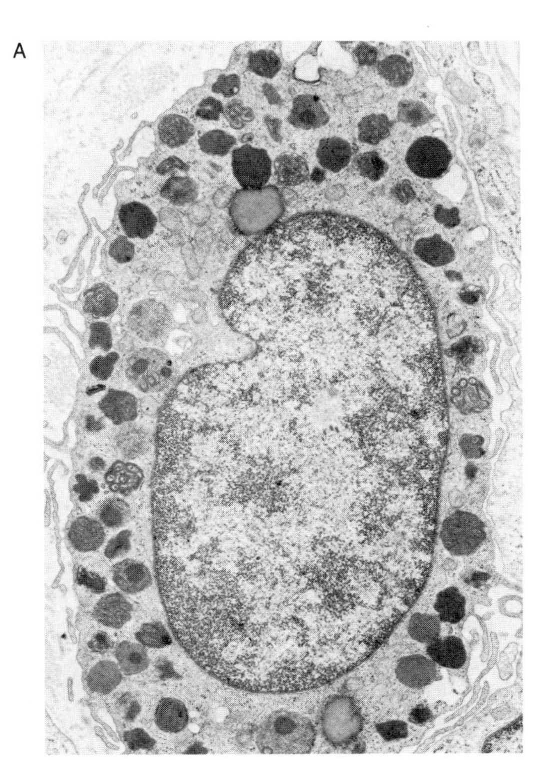

Figure 21.7 (A) Electron micrograph of a typical mast cell, showing extensive granular inclusions of the potent vasoactive agent, histamine (14,000×). (B) Higher magnification of granular inclusions of a mast cell, showing their characteristic scrolls, which consist of coils of a crystalline or fibrillar structure (150,000×). (Courtesy D. Zucker-Franklin, New York University Medical Center. Reprinted by permission of Grune and Stratton, Inc., and the author, from D. Zucker-Franklin, 1980, "Ultrastructural evidence for the common origin of human mast cells and blastophils," *Blood,* **56**:536.)

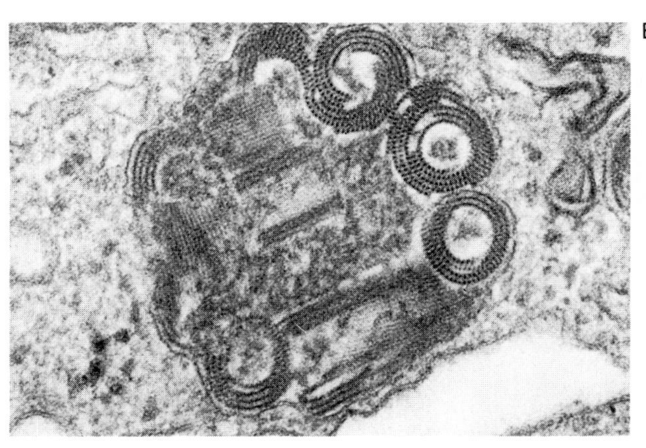

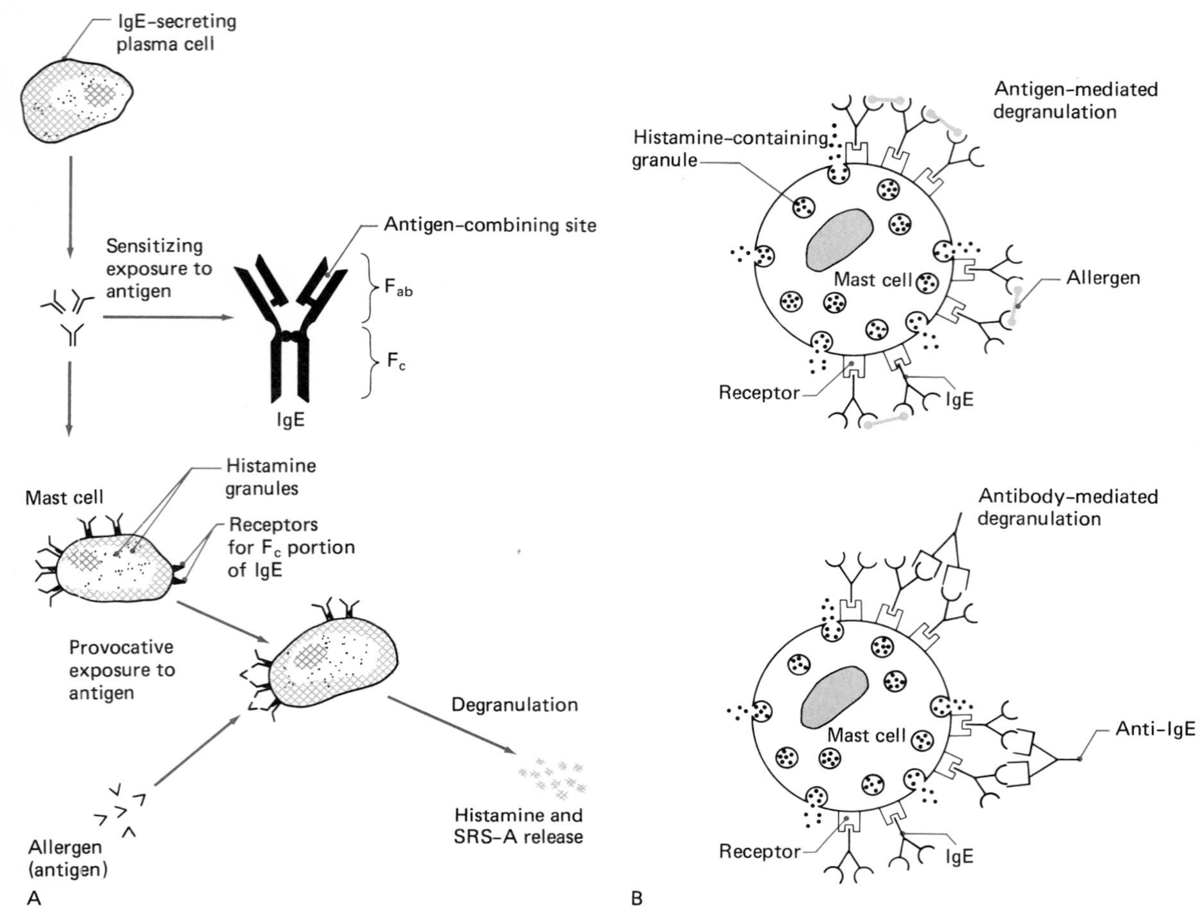

Figure 21.8 (A) Illustration of type 1 hypersensitivity reaction. An immune system response to an allergen. Allergens come into contact with specific IgE antibody molecules on the surface of a basophil or a mast cell. A subsequent membrane response causes the release of chemical mediators from granules in the basophil or mast cell. (B) The mast cell or basophil will degranulate when a crosslinkage forms between receptors on the surface of the cell. This may occur upon exposure of IgE-sensitized basophils or mast cells to an allergen with which the IgE will react (**antigen-mediated degranulation**), or when IgE sensitized cells react with anti-IgE antibodies (**antibody-mediated degranulation**).

and these individuals are prone to develop type 1 hypersensitivities.

The reaction of IgE and an antigen can lead to the formation of a chemical bridge between two mast or basophils cells. When this occurs, the basophils or mast cells lyse. The release of the contents of basophil or mast cells, after the reaction of antigen with antibody bound to the cell surface, establishes the basis for a severe physiological response. The sudden release of a large amount of histamine (a potent vasodilator) and other pharmacologically active compounds—such as heparin, platelet-activating factors (PSFs), eosinophil chemotactic factor of anaphylaxis (ECF-A), slow-reacting substance of anaphylaxis (SRS-A), and serotonin—into the bloodstream can produce anaphylactic shock, causing respiratory or cardiac failure. Bee stings and the administration of antibiotics, such as penicillin, can trigger such anaphylactic hypersensitivity reactions (Figure 21.9). Symptoms may include hives, abdominal cramps, diarrhea, nausea, vomiting, respiratory difficulties, and rapid death.

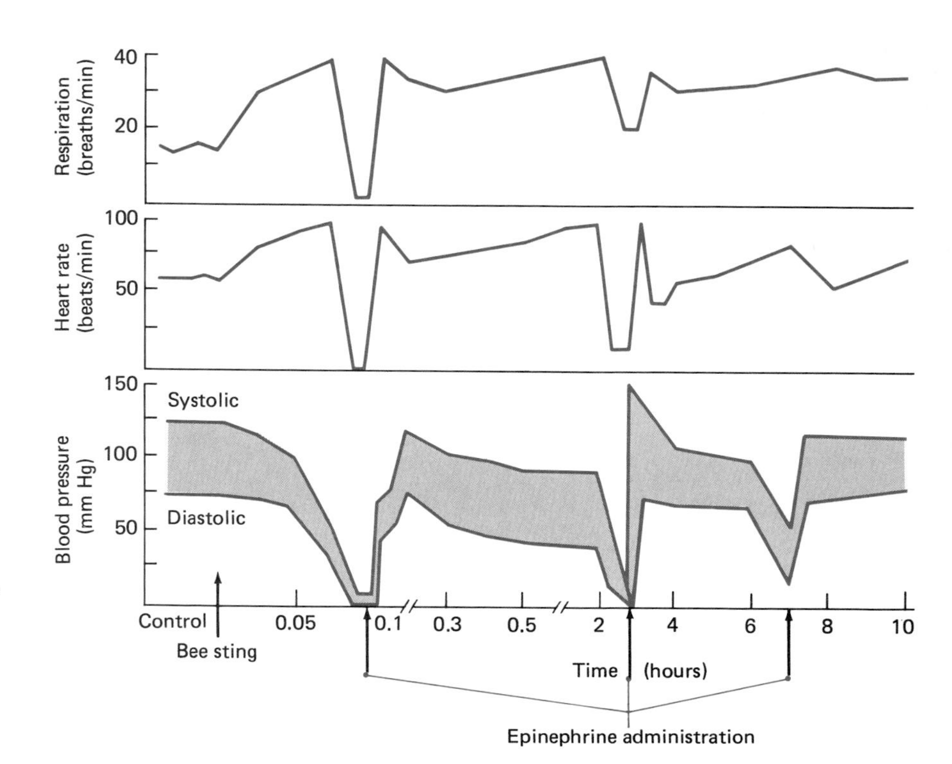

Figure 21.9 Systemic anaphylaxis triggered by the sting of a single honeybee. (Based upon J. Vick and L. M. Lichtenstein, 1977, "Allergic responses to airborne allergens and insect venoms," *Federation of American Societies for Experimental Biology Proceedings*, 36:1727.)

Prompt administration of adrenalin (epinephrine) is used to counter anaphylactic hypersensitivity reactions.

Atopic allergies result from a localized expression of type 1 hypersensitivity reactions. The interaction of antigens (allergens) with cell-bound IgE on the mucosal membranes of the upper respiratory tract and conjunctival tissues initiates a localized type 1 hypersensitivity reaction. Hay fever and allergies to certain foods are examples of such atopic allergies. When the allergen interacts with sensitized cells of the upper respiratory tract, the symptoms often include coughing, sneezing, congestion, tearing eyes, and respiratory difficulties. In cases where the allergen enters the body through the gastrointestinal tract, the symptoms often include vomitting, diarrhea, or hives. Antihistamines are useful in treating many such allergic reactions because the main mediator of the physiological response, the antihistamine, blocks the vasoactive action of the histamine released from sensitized mast and basophil cells.

In some cases the allergic reaction primarily effects the lower respiratory tract, producing a condition known as asthma. Asthma is characterized by shortness of breath and wheezing. These symptoms occur because the allergic reaction causes a constriction of the bronchial tubes and produces spasms. The primary mediator of asthma is not histamine but rather slow-reacting substance of anaphylaxis (SRS-A). This substance is released by activated basophils and mast cells; it causes increased blood capillary permeability and also smooth muscle contractions of the bronchial tubes. Because histamine is not the primary mediatory of asthma, antihistamines are not of therapeutic value in treating this condition. The treatment of asthma generally involves administration of epinephrine or aminophylline.

Atopic allergies can be diagnosed by using skin testing (Figure 21.10). The subcutaneous or intradermal injection of antigens results in a localized inflammation reaction if the individual is allergic to that antigen, that is, if that individual exhibits a type 1 hypersensitivity reaction. In this way allergens for a particular individual are identified. The symptoms of atopic allergies can be controlled, at least in part, by the use of antihistamines. The antihistamines combine with the histamines that are released from mast cells as part of the allergic response, thereby blocking the action of a principal chemical mediator of the allergic reaction.

I heard them speak of allergy,
I asked them to explain,
Which when they did, I asked them
To please explain again.

I found the pith of allergy
In Bromides tried and true;
For instance, you like lobster,
But lobster don't like you.

Does aspirin cause your eyes to cross?
Do rose-leaves make you nervy?
Do old canaries give you boils?
Do kittens give you scurvy?

Whatever turns your skin to scum,
Or turns your blood to glue,
Why, that's the what, the special what,
That you're allergic to.

O allergy, sweet allergy,
Thou lovely word to me!
Swift as an heiress Reno-bound
I called on my M.D.

This doctor was obliged to me
For reasons I must edit.
(I knew he had two extra wives,
And neither did him credit.)
I spoke to him of allergy;
Perhaps I clenched my fist;
But when I left his domicile
I had a little list.

I can't attend the opera now,
Or sleep within a tent;
I cannot ride in rumble seats;
My allergies prevent.

Oh, garden parties speed my pulse
And pound my frame to bits;
I'd mind the child on Thursdays,
But children give me fits.

When Duty sounds her battlecry,
Say never that I shirk;
It isn't laziness at all,
But an allergy to work.

—OGDEN NASH, *Allergy Met a Bear*

In addition to treating the immediate symptoms of an allergic reaction, attempts can be made to desensitize the individual. Desensitization usually is achieved by identifying and then by administering repeated doses of the allergen. The procedure generally is time consuming and costly. Over a period of time, however, the intensity of the allergic response can be reduced or eliminated. The mechanism by which desensitization works may be to direct immunoglobulin production in the direction

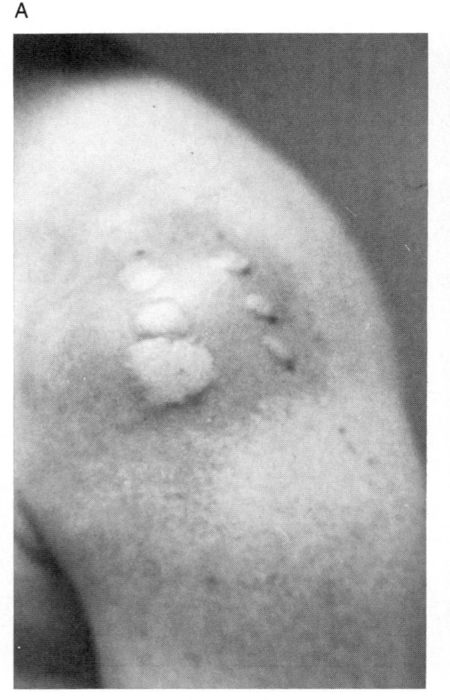

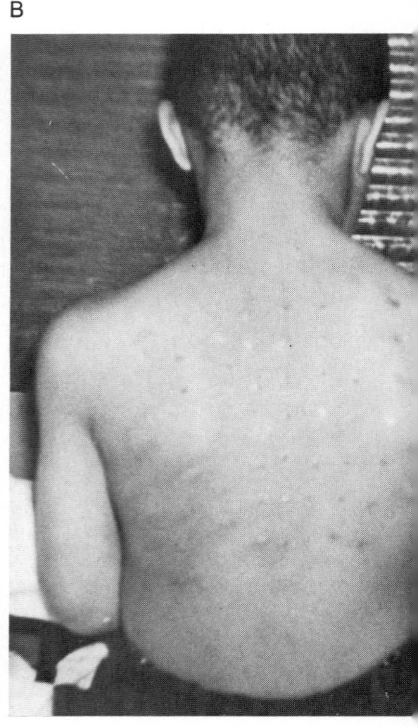

Figure 21.10 Photographs showing skin testing for allergies; the positive reactions show inflammation and development of pustules; results are rated on a scale of 1–4+. (A) Results of intradermal inoculation of several allergens into the arm; each test shows erythema and a weal; the pseudopods on the bottom right indicate a clear 4+ reaction to intradermal antigens. (B) The scratch test is performed on the back or forearm; the scratches should be 1 centimeter long and 2.5 centimeters apart and the allergens then applied to the torn skin surface. As shown in this photograph, a large number of tests against different allergens, about 50, can be performed simultaneously on the back of the patient; in the series of tests shown, positive scratch tests—with erythema, weal, and pseudopods—are evident in row nine, and negative tests are exemplified in the first row and column. (By permission of Dr. John Karibo, Department of Allergy and Immunology, University of Louisville.)

of IgG, which is not involved in the allergic response, and away from IgE, which is a critical mediator of atopic allergies. Once made, the IgG molecules can react with the allergen molecules before they reach IgE, preventing degranulation of mast cells and basophils and blocking the allergic reaction.

Antibody-dependent Cytotoxic Hypersensitivity

Antibody-dependent cytotoxic hypersensitivity reactions (type 2 hypersensitivity) occur by a different mechanism. In type 2 hypersensitivity reactions an antigen present on the surface of the cell combines with an antibody, resulting in the death of that cell by stimulating phagocytic attack or by initiating the sequence of the complement pathway that results in cell lysis. This type of antibody-dependent cytotoxic response occurs in transfusion incompatibility, that is, as a result of transfusions with incompatible blood types. If, for example, a person with type A blood (antigen A on blood cell surface and antibody B circulating) were given a transfusion with type B blood (antigen B on blood cell surface and antibody A in serum) the circulating antibodies in the recipient would react with the surface antigens of the donor cells, initiating the addition of complement molecules and the lysis of the donated cells. Symptoms of such incompatible transfusions include fever, chills, chest pain, nausea, vomiting, jaundice, and sometimes death. It is therefore essential that blood transfusions be made wtih compatible blood types.

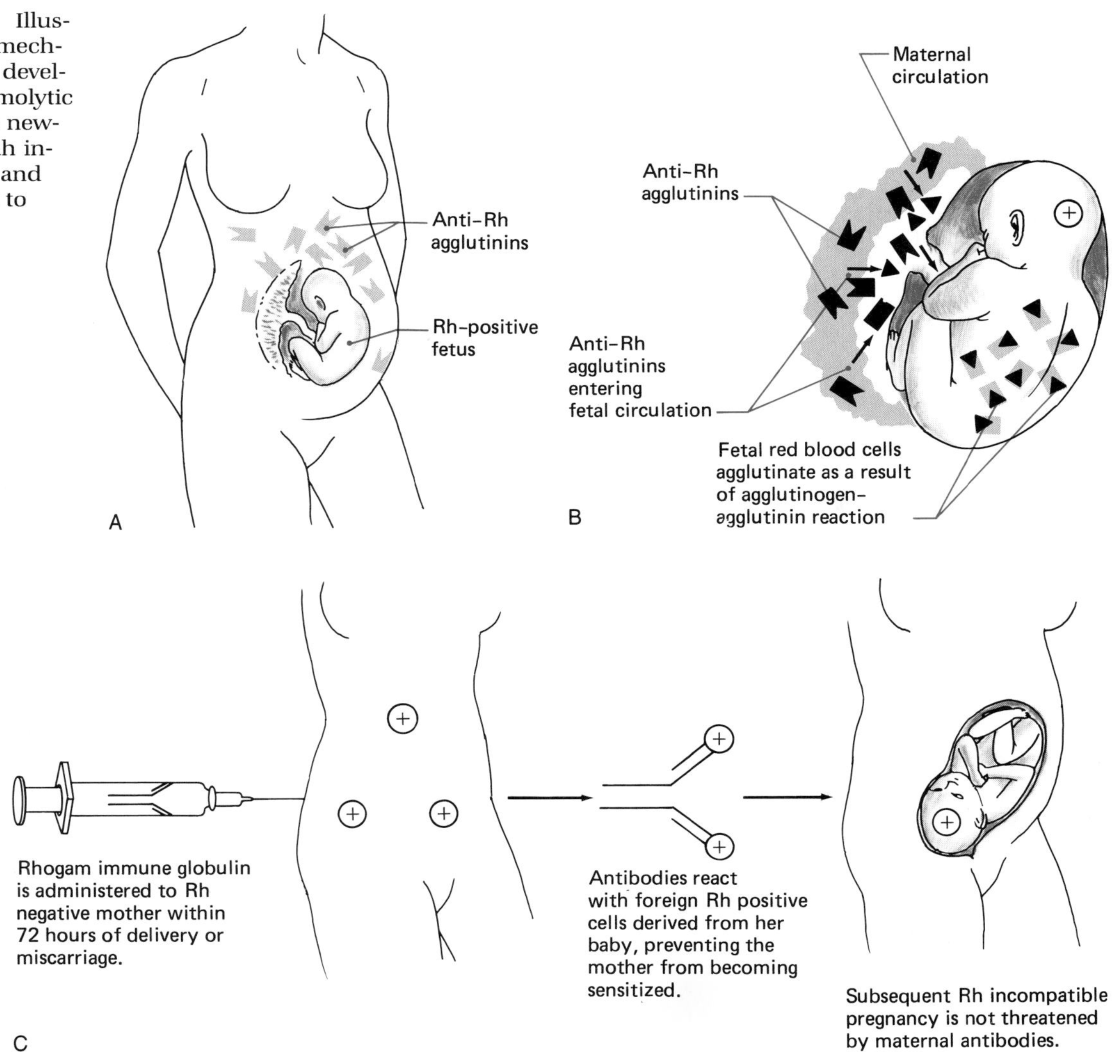

Figure 21.11 Illustration of the mechanism for the development of hemolytic disease of the newborn due to Rh incompatibility and the treatment to prevent it.

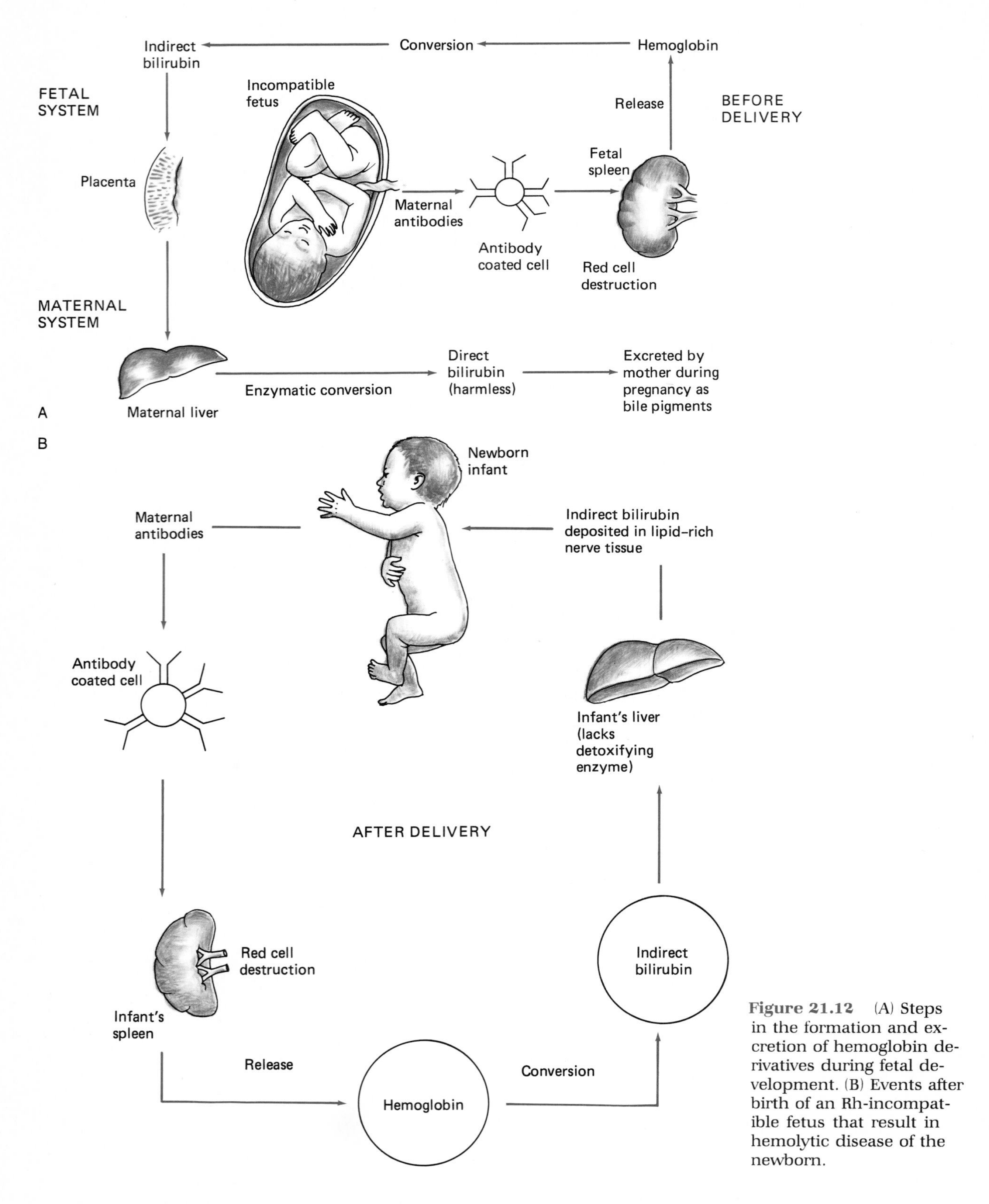

Figure 21.12 (A) Steps in the formation and excretion of hemoglobin derivatives during fetal development. (B) Events after birth of an Rh-incompatible fetus that result in hemolytic disease of the newborn.

With respect to compatible blood types, persons with type O blood are sometimes called universal donors and individuals with type AB blood are sometimes called universal recipients. The reason for this is that type O blood cells lack both A and B antigens on their surfaces, and therefore lack the antigens generally associated with transfusion incompatibility; regardless of the circulating antibodies in the recipient, the donated blood cells do not have the antigens to react and the anti-A and anti-B antibodies in the donated blood are rapidly diluted when introduced into the larger volume of blood in the recipient. Similarly persons with type AB blood do not have circulating antibodies against either A or B antigens, and therefore lack antibodies that would react with the A and B antigens on blood cells that are introduced regardless of the cell type. However, this concept of universal donor and recipient only refers to the major A and B antigens. There are various other antigens on blood cell surfaces, including the Rh antigen, that can cause incompatibility reactions. Therefore, except in emergencies, transfusions are given only after adequate analysis of cell antigens and are given only with matching blood types.

Rh incompatibility between mother and fetus is another example of type 2 hypersensitivity (Figure 21.11). Rh incompatibility occurs when the father is Rh positive, the mother is Rh negative, and the fetus is Rh positive. In this case the mother develops Rh antibodies in response to exposure to the Rh antigens of the fetus. Generally the mother is exposed to the fetal Rh antigens at the time of birth so that she does not develop an immune response until after the birth of the first child. In subsequent pregancies however, the anti-Rh antibodies (IgG) circulating through the mother's body can cross the placenta and attack the cells of the fetus. Anemia develops as a result of this condition. During development of the fetus, the fetal blood is purified by the mother's liver (Figure 21.12). At birth, the fetal blood is no longer purified by the maternal circulatory system and the infant also develops jaundice. This disease, hemolytic disease of the newborn (previously called erythroblastosis fetalis), can be treated by removal of the fetal Rh-positive blood and replacement by transfusion with Rh-negative blood. These cells are later replaced by Rh-positive cells produced by the infant. Cell destruction does not occur at this time because the maternal antibodies have been eliminated from the infant's circulatory system.

To prevent hemolytic disease of the newborn, passive immunization of the Rh-negative mother with Rhogam (anti-Rh antibodies) is used at the time of birth. The anti-Rh antibodies react with the fetal Rh-positive cells that enter the mother at the time of birth through the traumatized tissue. The reaction of anti-Rh antibodies with the Rh-positive cells limits the development of an anamnestic immune response in the mother by binding to the Rh antigens that have been introduced, thereby preventing their recognition by the immune system of the mother. Thus, artificial passive immunization is used to prevent the development active natural immunity. This treatment is repeated at each birth when the baby is Rh positive and the mother is Rh negative. As a result of this treatment we are able to prevent a serious antibody-dependent cytotoxic hypersensitivity reaction.

Complex-mediated Hypersensitivity

Complex-mediated hypersensitivity (type 3 hypersensitivity) reactions, involve antigens, antibodies, and complement, which initiate an inflammatory response. These reactions occur when the formation of antibody–antigen complexes trigger the onset of an inflammatory response (Figure 21.13). Such an inflammatory response is part of the normal immune response, but if there are large excesses of antigen, the antigen–antibody–complement complexes may circulate and become deposited in various tissues. Inflammatory reactions from such deposition of immune complexes can cause physiological damage to kidneys, joints, and skin. Various examples of type 3 hypersensitivity or

Figure 21.13 Pathways to inflammation triggered by antigen–antibody complexes in type 3 hypersensitivity reactions. (After J. W. Kimball, 1983, *Introduction to Immunology*, Macmillan, New York.)

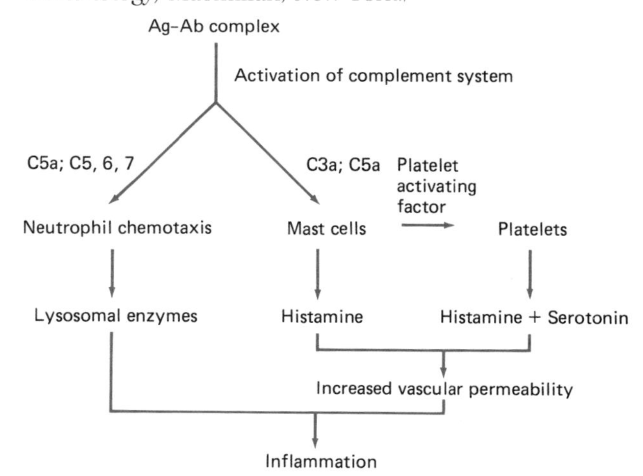

Table 21.7 Examples of Type 3 Hypersensitivity Reactions

Disease Condition	Cause
Farmer's lung	Actinomycete spores
Cheese washer's disease	*Penicillium casei* spores
Furrier's lung	Fox fur protein
Maple bark stripper's disease	*Cryptostroma* spores
Pigeon fancier's disease	Pigeon antigens
Serum sickness	Foreign blood serum

Arthus immune-complex reactions are listed in Table 21.7.

In the Arthus reaction, the site becomes infiltrated with neutrophils, and there is extensive injury to the walls of the local blood vessels. This sometimes occurs in the lungs because of repeated exposure to antigens on the surfaces of inhaled particulate matter, causing extrinsic allergic alveolities. When an antigen–antibody complex initiates such a reaction in the alveoli, the symptoms generally include cough, fever, and difficulty in breathing. These symptoms typically develop over a period of 4–6 hours and the attck usually subsides within a few days after the removal of the source of antigen. Various occupations have a high risk of developing this condition. For example, farmers often develop this reaction because of repeated exposure to the airborne spores of actinomycetes growing on hay. Sugarcane workers, mushroom growers, cheesemakers, and pigeon fanciers are also prone to this condition because of exposure to airborne antigens associated with their activities.

Serum sickness is another type of immune complex disorder. This disease results when patients are given large doses of foreign serums, such as

Figure 21.14 Sequence of events in the development of contact dermatitis due to exposure to poison ivy.

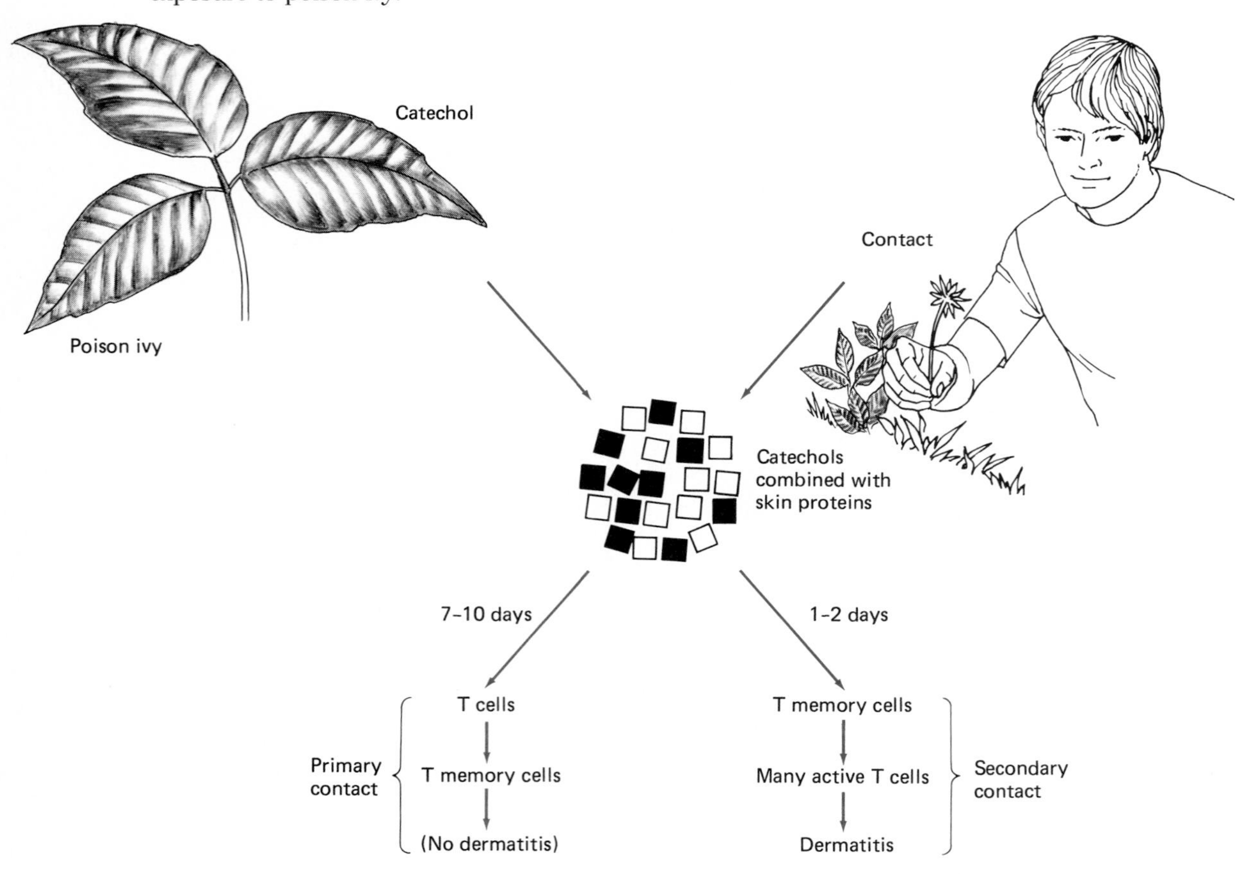

horse serum antitoxins to protect against tetanus and diphtheria—a once widely used practice—and antilymphocyte serum for immunosuppression to protect against rejection of transplanted tissues. The antigens in these foreign serums stimulate an immune response. Because large infusions of serum are given, these antigens have not been degraded and cleared from the body by the time that circulating antibodies appear. Immune complexes form between the residual antigens and the circulating antibodies. These antigen–antibody complexes are deposited at various body sites, including the joints, kidneys, and blood vessel walls. Symptoms of serum sickness generally appear 7–10 days after injection of the foreign serum and include fever, nausea, vomiting, malaise, hives, and pain in muscles and joints. In many cases of serum sickness the immune complexes are carried to the kidneys and cause nephritis (inflammatory disease of the kidneys).

This condition of glomerulonephritis can also be brought about by persistent infections that result in the formation of antigen–antibody complexes that are deposited within the glomeruli of the kidneys. Immune complexes—formed by antibody reactions with antigens produced by *Streptococcus pyogenes* (causative agent of strep throat, which produces protein toxins that may circulate through the body and cause other diseases), hepatitis B virus (cause of serum hepatitis), *Plasmodium* species (protozoa that cause malaria), and *Shistosoma* (helminthic worms that cause shistosomiasis)—may lead to this condition. The persistence of these infections provides a continuing supply of antigen to react with circulating antibodies produced by the infected individual. The immune complex that forms accumulates in the kidneys, eventually causing nephritis due to complex-mediated hypersensitivity.

Cell-mediated (Delayed) Hypersensitivity

Cell-mediated or delayed hypersensitivity (type 4 hypersensitivity) reactions involve T lymphocytes. As the name implies, these reactions occur only after a prolonged delay following exposure to the antigen; such reactions often reach maximal intensity 24–72 hours after initial exposure. Delayed hypersensitivity reactions occur as allergies to various microorganisms and chemicals. Contact dermatitis, resulting from exposure of the skin to various chemicals, is a typical delayed hypersensitivity reaction. Poison ivy is one of the best-known examples of such contact dermatitis (Figure 21.14). Contact with catechols in the leaves of the poison ivy plant leads to the development of a characteristic rash with itching, swelling, and blistering. The catechols, which lipids in the skin help to retain, appear to act as haptens and react with skin proteins to form an active antigen. Combined with skin proteins, the catechols bring about a cell-mediated response inolving T cells. Upon primary exposure to poison ivy, no dermatitis (skin rash) occurs; however, subsequent exposure of sensitized individuals to the oils of the poison ivy plant result in dermatitis after an initial delay of several days. Various other agents—including metals, soaps, cosmetics, and biological materials—can also cause contact dermatitis. The treatment of contact dermatitis often involves the administration of corticosteroids, which depress cell-mediated immune reactions.

A wicked witch
Is Mizzable Scratch,
And it's TROUBLE she grows
In her garden patch.
And her garden patch
Lies all around,
For she grows Poison Ivy!
By ditch and fence
She leaves her trail,
As she sows her seed
Over hill and dale,
And her crop of mischief
Can never fail,
For she grows Poison Ivy!
So listen, my children,
Take heed, be good,
And if ever you roam
Through a tangled wood,
Or follow a road,
Some lovely day,
Over the hills and far away,
Please keep out
Of the garden patch
Sown and grown by Mizzable Scratch,
Or it's TROUBLE you will surely catch,
For she grows Poison Ivy!

—KATHARINE GALLAGHER, *Poison Ivy*

The delayed hypersensitivity reaction involves the release of lymphokines from activated T cells (Figure 21.15). These lymphokines initiate a typical series of cell-mediated immune responses that include the attraction of phagocytic cells and the initiation of the inflammatory response. The delayed type of hypersensitivity reaction is a normal reaction to the intracellular invasion of host cells by a variety of

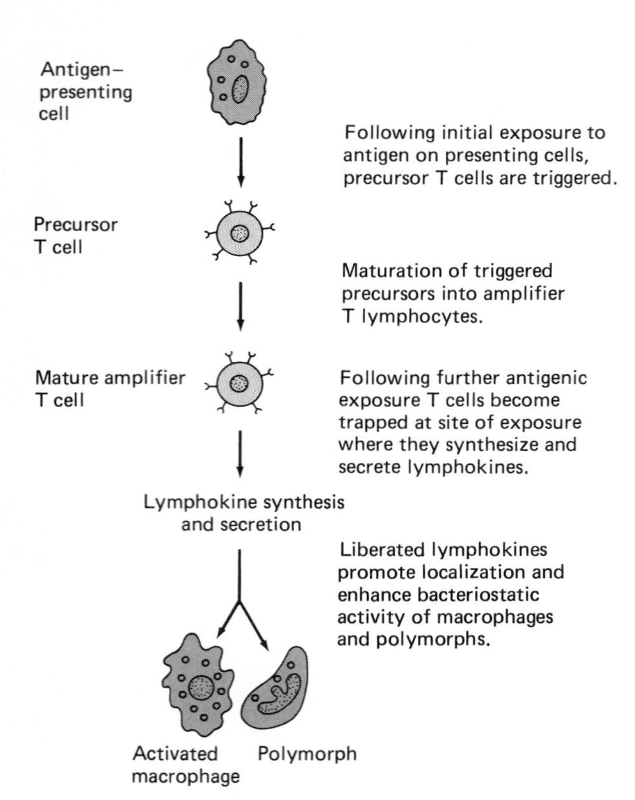

Figure 21.15 Illustration of the mechanism of delayed-type hypersensitivity response.

pathogens. This reaction can also be used for diagnostic testing (Figure 21.16). For example, in the tuberculin test, the delayed onset of an inflammatory response after the subcutaneous injection of an appropriate antigen is indicative of a positive delayed hypersensitivity reaction. A positive tuberculin test reveals previous infection with *Mycobacterium* but does not establish the presence of an active case of tuberculosis.

Summary

The immune system protects the body against various disease-causing microorganisms. A state of immunity exists when the body is protected against an infectious agent. There are several types of immunity (Figure 21.17). Some forms of immunity are acquired and others are innate. Some states of acquired immunity are active (immunoglobulins made by the individuals themselves) and others are passive (immunoglobulins acquired from external sources). Additionally, some states of immunity are natural (acquired by infection or placental transfer) and others are artificial (acquired by injection of foreign immunoglobulins or intentional exposure to antigens in vaccines).

The use of vaccines to render the host unsusceptible to specific pathogenic microorganisms is

A

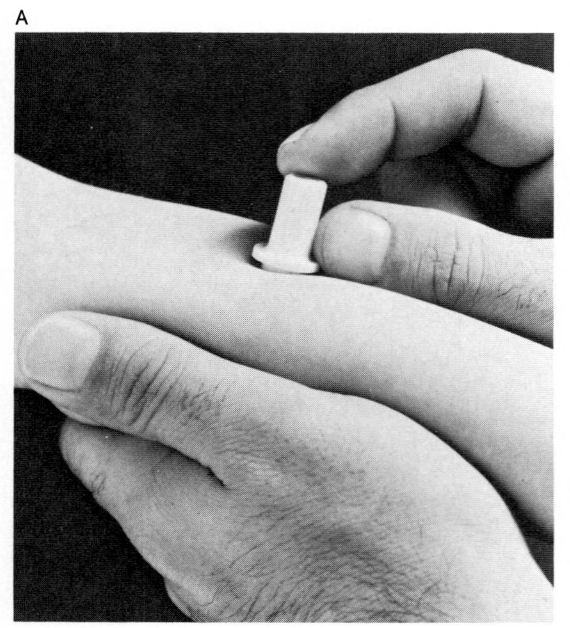

B

TUBERCULIN TINE TEST® RECORD

Accurate for use only with Lederle TINE TEST® Tuberculin (Old)

Date________

Name________

Address________

Telephone________

Instructions: Feel the skin where the test was given between 48 and 72 hours (2 to 3 days) after testing. ________ DAY AND TIME

Mark an X in the circle of the box below in which the raised bumps feel most like those on the skin. *Mark only 1 circle.*

(If you cannot feel any bumps on the skin, mark this box.)

Printed in U.S.A. January 1978 7764-1 A528

Figure 21.16 The results of a skin test, showing a positive tuberculin reaction. Development of inflammation after a delay is indicative of a positive reaction. (A) Application of the antigen to the skin in the Tine test. (B) Record card for the Tine test showing development of positive reactions. (Courtesy Lederle Laboratories, Wayne, New Jersey.)

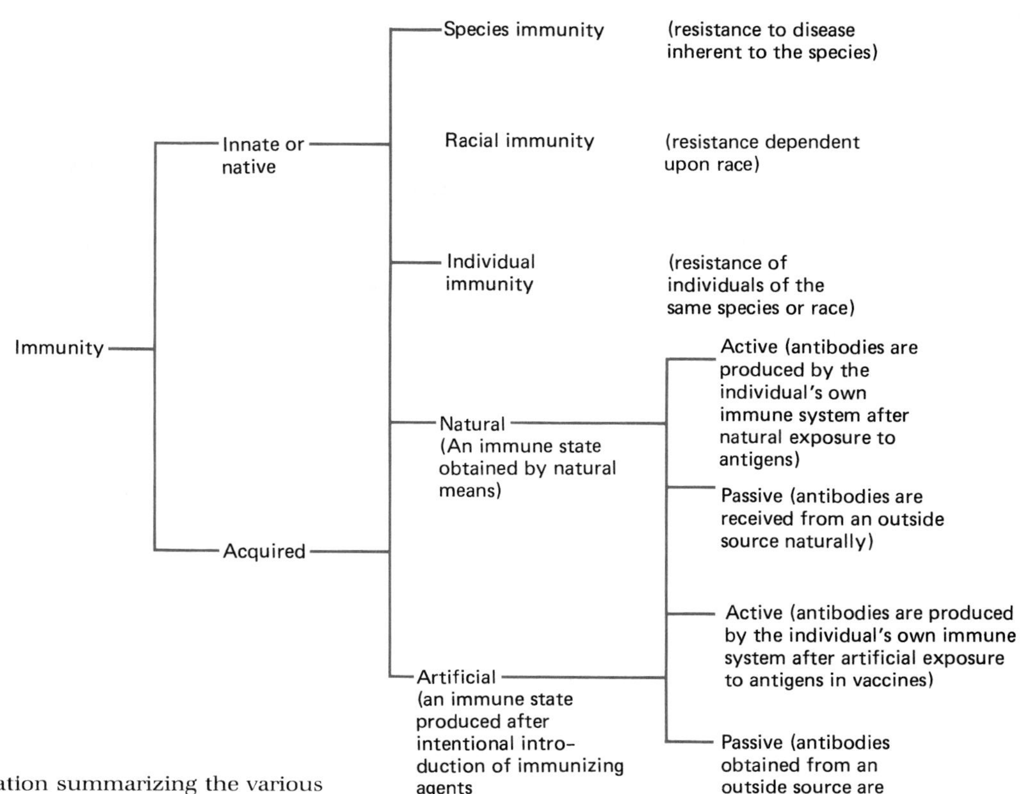

Figure 21.17 Illustration summarizing the various types of immunity.

of great use in controlling some microbial diseases. Smallpox, for example, has been eliminated as a human disease as a consequence of the extensive use of vaccination. Polio, measles, mumps, rubella, yellow fever, tetanus, and rabies are all diseases that can be prevented by the appropriate use of vaccines. The effectiveness of vaccines depends on the normal memory response of the human immune system and on establishing active immunity through the exposure to antigens that elicit the desired immune response without causing disease. Both inactivated and attenuated microorganisms are employed in different vaccines used for immunization purposes.

The regulation and expression of the immune response is an extremely complex system that is developed only in higher vertebrate animals, such as human beings. The primary function of this highly developed and integrated system seems to be the protection of "self" from invasion by disease-causing foreign agents. Failures of the immune response to produce the necessary differentiated tissue lines, such as macrophages, B lymphocytes, and T lymphocytes, result in immunodeficiency diseases in which an individual is unable to respond properly and thus defend against invasion by microorganisms. Many human infections occur when the host defense mechanisms are compromised and the immune response fails to respond optimally to potential pathogens.

Study Outline

A. Acquired immunity refers to the resistance to infection obtained from the production and/or activity of antibodies and T cells against specific microbial antigens.

 1. Passively acquired immunity comes about when antibodies produced in another organism are introduced into the body.

a. Naturally acquired passive immunity is derived from the transfer of antibodies from mother to child across the placenta and in the colostrum; it lasts from a few weeks to a few months.
b. Artificially acquired passive immunity describes the injection of antibodies into an individual to provide immediate protection against a pathogen or toxin; it lasts 2–3 weeks.

2. Active immunity can be naturally or artificially acquired.
   a. Naturally acquired active immunity is a memory response that develops after exposure to microbial antigens when the individual actively produces antibodies against the pathogen and develops clones of memory cells; the immunity may be long term or short term.
   b. In artificially acquired immunity an antigen is intentionally introduced into the body; this antigen will not cause disease but will stimulate the immune response eliciting the formation of antibody and memory cells.
3. Vaccines are preparations of antigens designed to stimulate the normal primary immune response. Exposure to the antigens in a vaccine results in a proliferation of memory cells and the ability to exhibit a secondary memory or anamnestic response upon subsequent exposure to the same antigens.
   a. Vaccines may contain antigens prepared by killing or inactivating pathogenic microorganisms or may use attenuated strains that are unable to cause the onset of severe disease symptoms.
   b. Some vaccines are prepared by denaturing microbial exotoxins to produce toxoids; these toxoids can elicit the formation of antibody and are reactive with antibody molecules but are unable to initiate the biochemical reactions associated with the active toxins that cause disease conditions.
   c. Microorganisms used for preparing inactivated vaccines are killed by treatment with chemicals, radiation, or heat; they retain their antigenic properties but cannot cause the onset of the disease caused by the live virulent pathogen.
   d. Living, but attenuated, strains of microorganisms used in vaccines are weakened by moderate use of heat, chemicals, desiccation, and growth in tissues other than the normal host. Such vaccines can be administered in relatively low doses because the microorganism is still able to replicate and increase the amount of antigen available to trigger the immune response mechanism.
   e. Components of microorganisms, such as capsules and pili can be used to make vaccines.
   f. Smallpox has been completely eliminated through an extensive worldwide immunization program conducted by the World Health Organization using a lyopholized vaccine containing a live strain of pox virus. Smallpox was permanently eliminated because human beings are the only known hosts for the virus and because the virus has a short survival time outside human host tissues.
   g. Adjuvants are chemicals that enhance the antigenicity of other biochemicals; they are used in vaccines and can eliminate the need for repeated booster doses and permit the use of smaller doses of the antigen.
   h. The antigens in a vaccine may be introduced into the body intradermally, subcutaneously, intramuscularly, intravenously, or into the mucosal cells lining the respiratory and gastrointestinal tracts.
   i. Some vaccines are administered after suspected exposure to a given infectious microorganism in order to elicit an immune response before the onset of disease symptomology. In these cases the development of the infecting pathogen must be slow, and the vaccine must be able to initiate antibody production quickly before active toxins are produced and released.

B. It is necessary to suppress the immune response when organs or tissues are transplanted from one individual to another. Immunosuppressant drugs block the rejection of transplanted tissue by suppressing the cell-mediated immune response that is reacting to the presence of foreign surface antigens.

C. Autoimmune diseases result from the failure of the immune response to recognize properly self-antigens and can lead to the progressive degeneration of tissues.

D. Immunodeficiencies can occur within the immune response, can affect a variety of cells, and can result in the failure of the system to recognize and respond properly to the antigens of pathogenic microorganisms.

1. Severe combined deficiency results from the failure of stem cells to differentiate properly; neither B nor T lymphocytes are produced and individuals so afflicted are incapable of any immunological response.
2. DiGeorge syndrome results from the failure of the thymus to develop properly so that T lymphocytes do not become properly differentiated. Individuals with this condition do produce B lymphocytes but do not exhibit cell-mediated immunity and thus are prone to viral and other intracellular infections. Since T helper cells enhance antibody production by B cells, the antigen-mediated response is also depressed.
3. Bruton's congenital agammaglobulinemia results from the failure of B cells to differentiate and produce antibodies. It occurs only in males who are then particularly subject to bacterial infections. Treatment of this disease involves repeated administration of IgG to maintain adequate levels of antibody in the circulatory system.
4. Late onset hypogammaglobulinemia is a deficiency of circulating B cells and/or B cells with IgG surface receptors.
5. Individuals who do not produce sufficient complement molecules are unable to respond properly to bacterial infections; the inflammatory response and the lytic killing of pathogenic bacterial cells is then limited.

E. Hypersensitivity is a physiological state resulting from an excessive immunological response to an antigen. Such reactions occur when an individual is sensitized to an antigen so that further contact with that antigen results in an elevated immune response.

1. Anaphylactic hypersensitivity occurs when an antigen reacts with antibody bound to mast or basophil blood cells, leading to disruption of these cells with the release of vasoactive mediators. The sudden release of a large amount of these substances into the bloodstream can cause anaphylactic shock, resulting in respiratory or cardiac failure. This condition can be treated by injection of epinephrine.
   - **a.** Atopic allergies, such as hay fever and food allergies, result from a localized expression of these type 1 hypersensitivity reactions. Such allergies are diagnosed using skin testing and can be controlled by the use of antihistamines.
   - **b.** Allergies can be eliminated by desensitization in which repeated exposure to the allergen is used to reduce the intensity of the allergic reaction.
2. Antibody-dependent cytotoxic hypersensitivity reactions occur when an antigen present on the surface of the cell combines with an antibody, resulting in the death of that cell by stimulating phagocytic attack or by initiating the sequence of the complement pathway.
   - **a.** Transfusion with incompatible blood can cause antibody-dependent hypersensitivity reactions.
   - **b.** Rh incompatibility between mother and fetus occurs when the father is Rh positive, the mother is Rh negative, and the fetus is Rh positive. The mother will develop Rh antibodies upon exposure to the Rh antigens of the infant at birth.
   - **c.** To prevent these anti-Rh antibodies from attacking a possibly Rh positive fetus through the placenta in a subsequent pregnancy, the mother is passively immunized with Rhogam after each birth. The reaction of anti-Rh antibodies with the Rh-positive cells limits the development of an anamnestic immune response in the mother.
   - **d.** If the response is unchecked the new fetus will be anemic, at birth the fetal blood is no longer purified by the maternal circulatory system and the infant develops jaundice. This disease is called hemolytic disease of the newborn.

3. Complex-mediated hypersensitivity reactions involve antigens, antibodies, and complement, which initiate an inflammatory response. If there are large excesses of antigen, the complexes of antigen and antibody may circulate and deposit in various tissues. This type of sensitivity causes serum sickness and nephritis due to persistent infections.
4. Cell-mediated (delayed) hypersensitivity reactions involve T lymphocytes and the release of lymphokines from activated T cells.
   a. They occur after a prolonged delay following exposure to the antigen and appear as allergies to microorganisms and chemicals.
   b. The lymphokines initiate a typical series of cell-mediated immune responses that include the attraction of phagocytic cells and the initiation of the inflammatory response.

Study Questions

1. Indicate whether the following statements are true or false.
   a. Smallpox is a prevalent disease today.
   b. There is a vaccine to prevent the common cold.
   c. Measles can be prevented by vaccination.
   d. Chicken pox can be prevented by vaccination.
   e. Influenza can be prevented by vaccination.
   f. All travelers should receive a cholera vaccine.
   g. Vaccines cause acute cases of diseases.
   h. Rh incompatibility occurs when both mother and father are Rh positive.
   i. Allergies are autoimmune responses.
   j. Adjuvants enhance immunogenicity of antigens.
   k. Passive immunity is long term.
2. How does vaccination prevent disease?
3. Why is vaccination against syphilis not useful?
4. Name and briefly describe four types of hypersensitivity reactions.
5. Identify and briefly describe three types of immune deficiency syndromes.

Some Thought Questions

1. Should immunization be required?
2. What should you do if you're allergic to penicillin?
3. Why do some people move to the southwestern United States?
4. Why is "drink plenty of fluids and rest" sage advice?
5. Should "David," the bubble boy with severe combined immune deficiency disease, have been given a bone marrow transplant?
6. What can you do to alleviate allergies?
7. How has cyclosporin changed medical practice?
8. Should Baby Fae have been given a baboon heart?
9. Why are bone marrow transplants used to treat certain types of immunodeficiencies?
10. How does immunity relate to infertility?
11. Why is the immune network sometimes dysfunctional?
12. How does the tetanus booster shot differ from the DPT vaccine?
13. Why is IgG and not IgM routinely used for passive immunization?
14. Why is the new rabies vaccine better than the older vaccine?
15. Why do some people with no family history of allergies develop allergies to penicillin?
16. How are anti-Rh antibodies used to prevent hemolytic disease of the newborn?
17. How are anti-snake venoms made?

Additional Sources of Information

Benenson, A. S. (ed.). 1975. *Control of Communicable Diseases in Man*. American Public Health Association, Washington, D. C. This series of articles describes the use of vaccines to control infectious diseases.

Donelson, J. E., and M. J. Turner. 1985. How the trypanosome changes its coat. *Scientific American*, 252(2):44–51. This article considers recent advances in our understanding of the immune response to parasites and how some parasites change their outer antigenic structure to evade the host's immune defense system.

Kimball, J. W. 1983. *Introduction to Immunology*. Macmillan Publishing Co., Inc., New York. An introductory textbook covering the immune response.

Roitt, I. M. 1980. *Essential Immunology*. Blackwell Scientific Publications, Oxford, England. An excellent introduction to the immune response and its role in protection against disease.

Rose, N. R. 1981. Autoimmune diseases. *Scientific American*, **244**(2):80–103. An interesting article describing diseases that result from failures of the immune response to properly recognize self antigens.

Sabin, A. B. 1981. Evaluation of some currently available and prospective vaccines. *Journal of the American Medical Association*, **246**:236–241. This article describes the vaccines that are currently available for protecting against bacterial and viral diseases and the prospects for developing new vaccines.

UNIT EIGHT
Microorganisms and Human Diseases

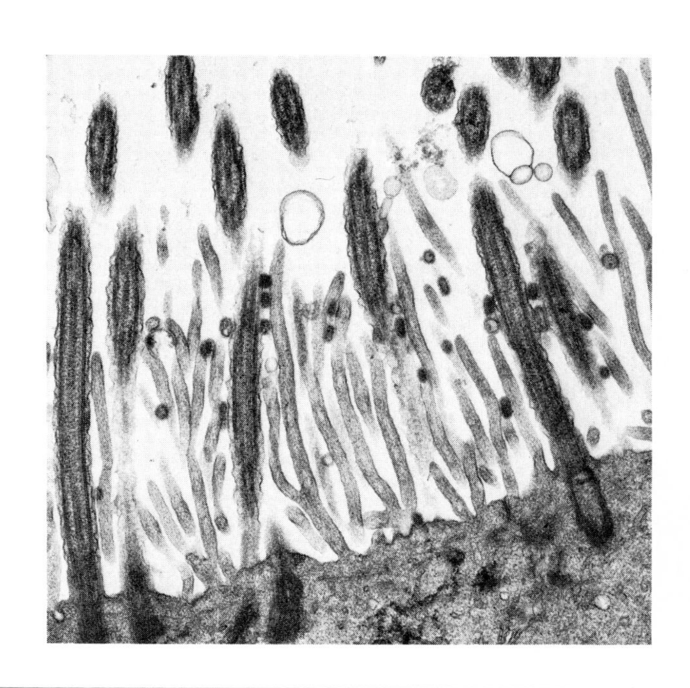

A

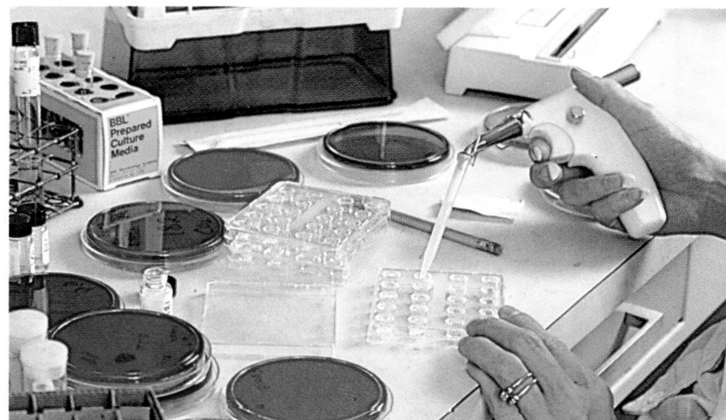

B

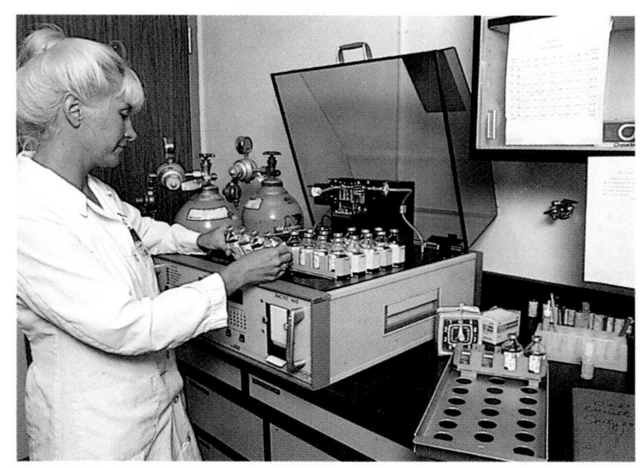

C

D

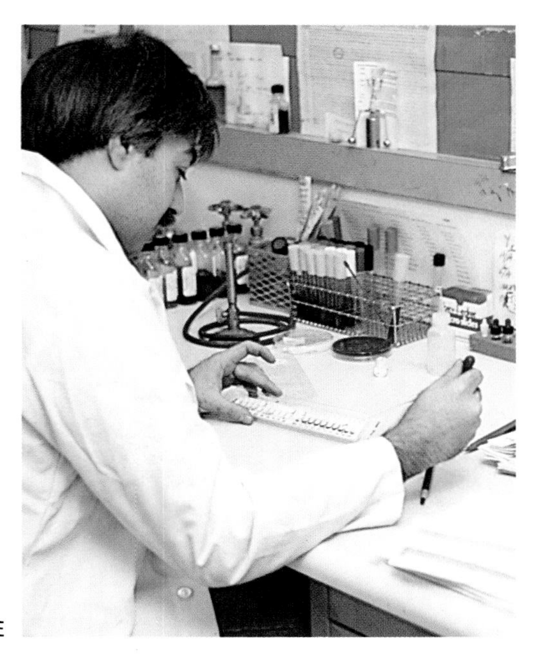

E

Plate 1 The clinical microbiology laboratory aids the physician in properly diagnosing diseases and determining the appropriate treatment. In the clinical laboratory, microorganisms are (1) isolated using various media; (2) identified using microscopical observations, biochemical tests, and serological reactions; (3) enumerated, when this is critical to establishing the presence of an infection; and (4) their antibiotic sensitivity profiles are determined to help establish the course of treatment. These photographs show some of the activities of the clinical microbiology laboratory staff. (A) Isolating bacteria in the clinical microbiology laboratory is important for properly diagnosing various human diseases. (From Medichrome/The Stock Shop: David York.) (B) Running a variety of biochemical tests is a necessary step in the identification of most bacterial pathogens. (From Medichrome/The Stock Shop: David York.) (C) To detect microbial infections of blood, samples are incubated with radiolabeled glucose and the production of labeled carbon dioxide is detected using an automated Bactec system; this method detects infections rapidly. (From Medichrome/The Stock Shop: David York.) (D) Counting bacteria with a colony counter and recording data is performed for quantitative assays; such assays are particularly important in the diagnosis of urinary tract infections. (From Camera MD Studios: Carroll H. Weiss.) (E) Various commercial test systems are used to identify isolates in the clinical microbiology laboratory; here an API test system is being inoculated. (From Taurus Photos: Martin Rotker.)

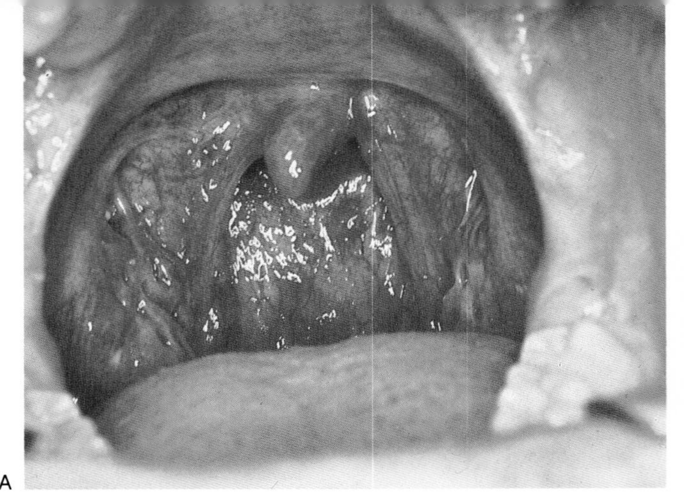

A

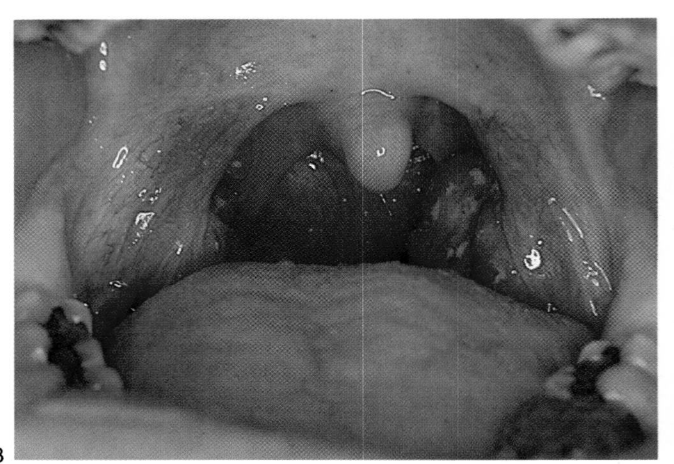

B

Plate 2 The respiratory tract is continuously exposed to numerous microorganisms. Despite the elaborate defense systems of the respiratory tract, some pathogenic microorganisms are able to establish respiratory tract infections. *Streptococcus* species are often involved in bacterial infections of the respiratory tract. (A) Pharyngitis, inflammation of the pharynx, is shown in this view of a "red throat." (From Camera MD Studios: Carroll H. Weiss.) (B) Tonsillitis due to streptococcal infection, showing characteristic white lesions on tonsils. (From Camera MD Studios: Carroll H. Weiss.) (C) Growth of bacteria from throat culture showing presence of β-hemolytic streptococci, which is diagnostic of a strep throat. *Streptococcus pyogenes*, which is β-hemolytic, causes pharyngitis or strep throat, tonsillitis, and various other human diseases. On blood agar *S. pyogenes* forms small pinpoint colonies and also exhibits β-hemolysis. (From BPS: Leon J. LeBeau, University of Illinois Medical Center.) (D) The Streptex system uses passive agglutination for identifying *Streptococcus* groups. (Courtesy Wellcome Diagnostic, Research Triangle, North Carolina.) (E) A Gram stain of a sputum sample from a patient with pneumonia reveals the presence of *Streptococcus pneumoniae*. (From BPS: Leon J. LeBeau, University of Illinois Medical Center.) (F) On blood agar *Streptococcus pneumoniae* exhibits partial clearing (greening) indicative of α-hemolysis.

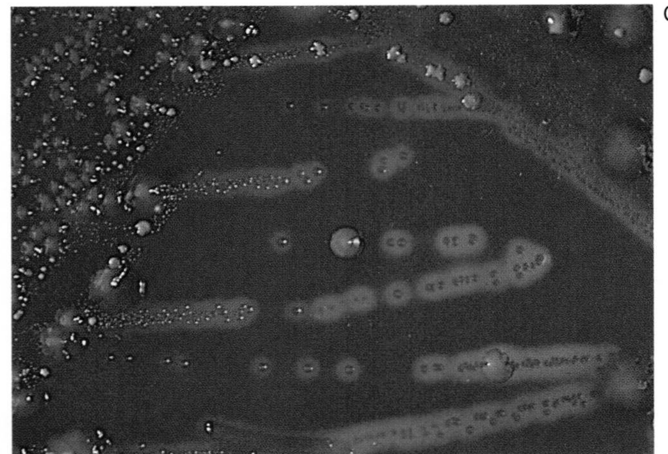

C

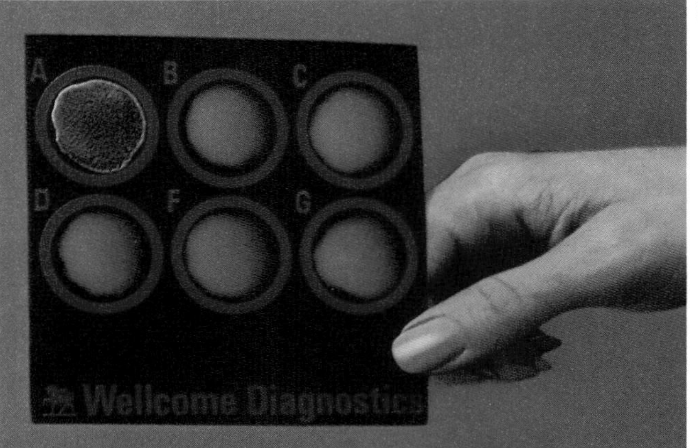

D

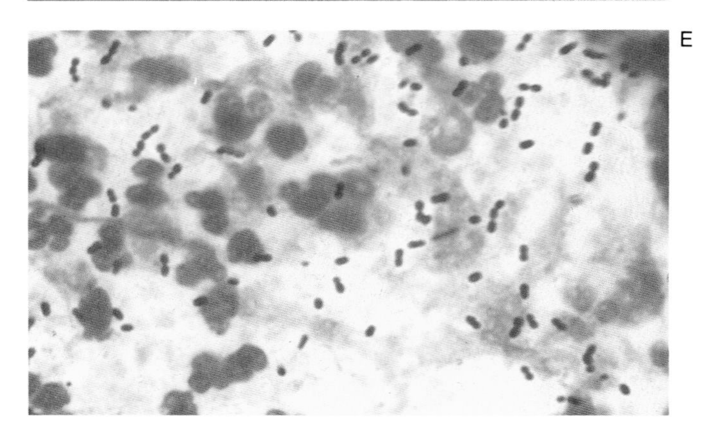

E

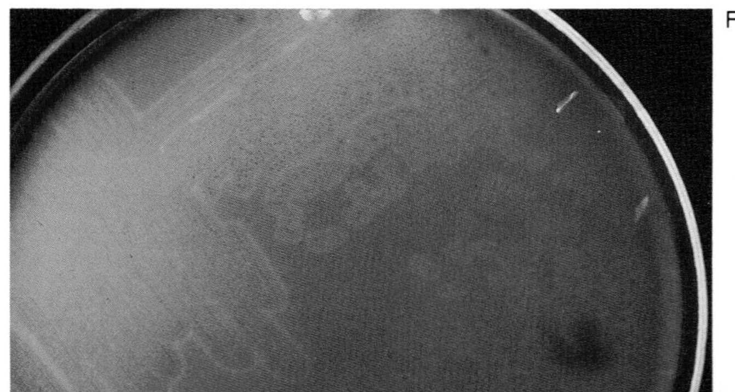

F

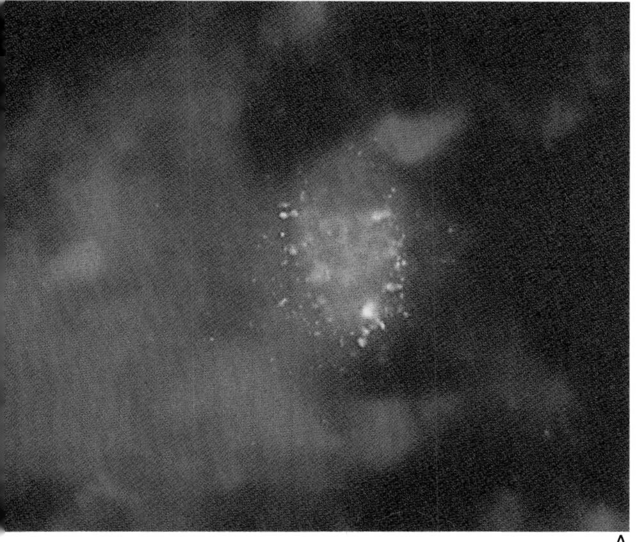

A

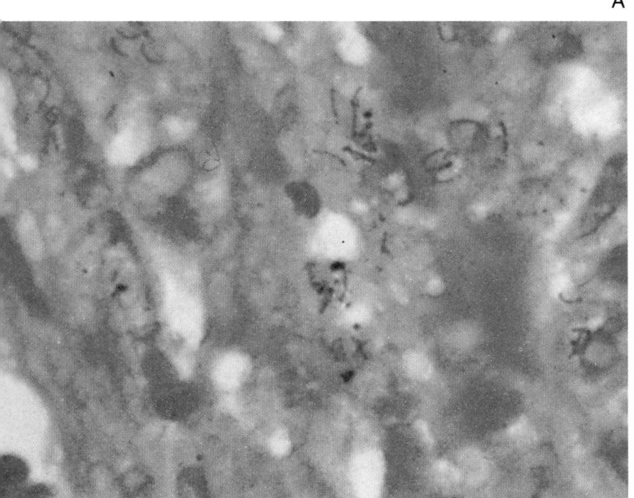

B

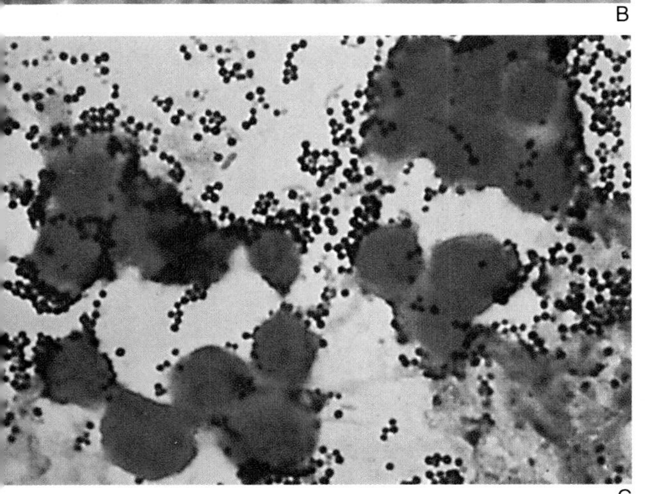

C

Plate 3 (A) At first the etiology of Legionnaire's disease was unknown; later, it was discovered that this disease is caused by *Legionella pneumophila,* shown here in lung tissue as revealed by direct fluorescent antibody staining. (Courtesy James Snyder, Norton Kosair Children's Hospitals, Louisville, Kentucky.) (B) The bacteria that cause tuberculosis are able to avoid the normal host defense systems and establish a progressive infection. In this photomicrograph, the presence of *Mycobacterium tuberculosis* (red rods) in lung tissue is revealed by the acid fast staining procedure. (Courtesy James Snyder, Norton Kosair Children's Hospitals, Louisville, Kentucky.) (C) This micrograph of a Gram stain preparation shows staphylococci in the stool from a case of staphylococcal food poisoning. (Courtesy James Snyder, Norton Kosair Children's Hospitals, Louisville, Kentucky.) (D) *Klebsiella pneumoniae,* a common cause of bacterial pneumonia and other diseases, exhibits characteristic pink colonies when growing on MacConkey agar. (E) The beef tapeworm (*Taenia saginata*) is one of several worms that cause parasitic human infections. (From Medichrome/The Stock Shop: Martin M. Rotker.)

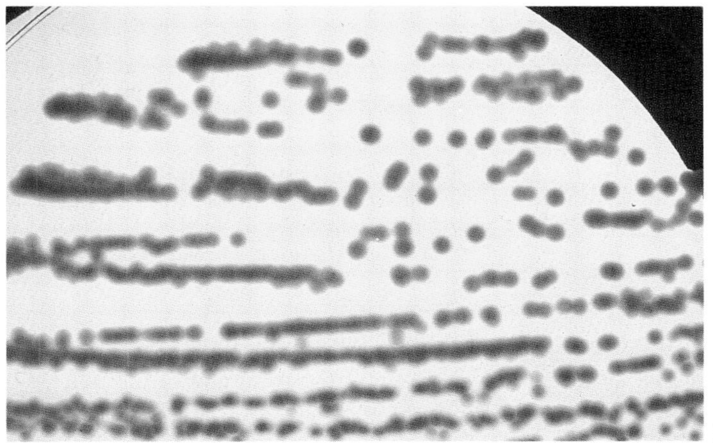

D

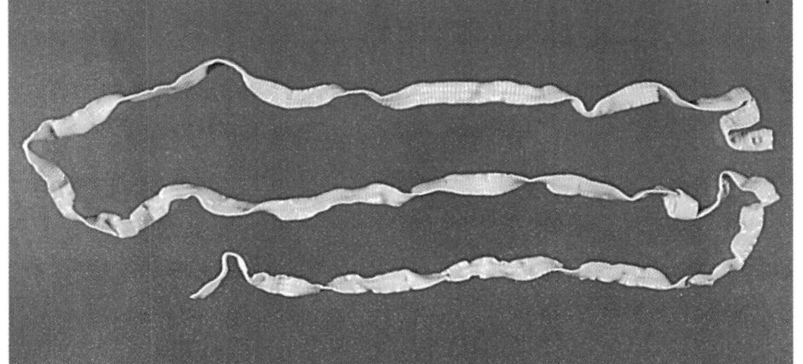

E

Plate 4 The identification of enteric bacteria, several of which cause gastrointestinal tract infections, employs growth on differential and selective media. These media permit suppression of the large normal microbiota of the gastrointestinal tract and the detection of specific diagnostic features of the isolates. (A) *Salmonella oranienburg* on xylose lysine desoxycholate agar (XLD) showing black growth due to production of hydrogen sulfide and yellow due to acid production. (From Camera MD Studios: Carroll H. Weiss.) (B) *Salmonella typhimurium* on xyline desoxycholate agar (XLD) showing black growth due to production of hydrogen sulfide. (C) *Vibrio cholerae* on TCBS agar showing yellow colonies due to fermentation of sucrose. (D) *Salmonella enteritidis* on Hektoen agar showing black colonies due to production of hydrogen sulfide.

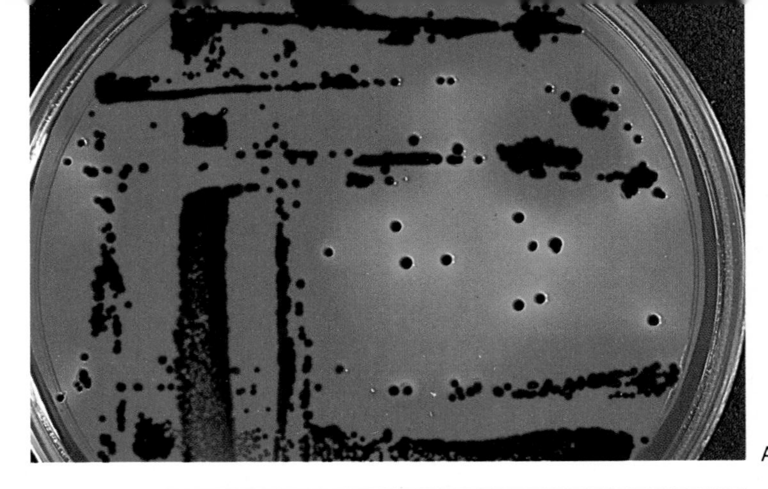

A

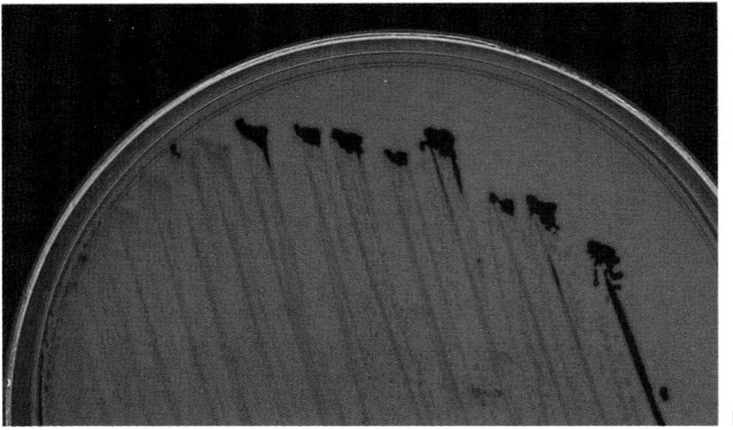

B

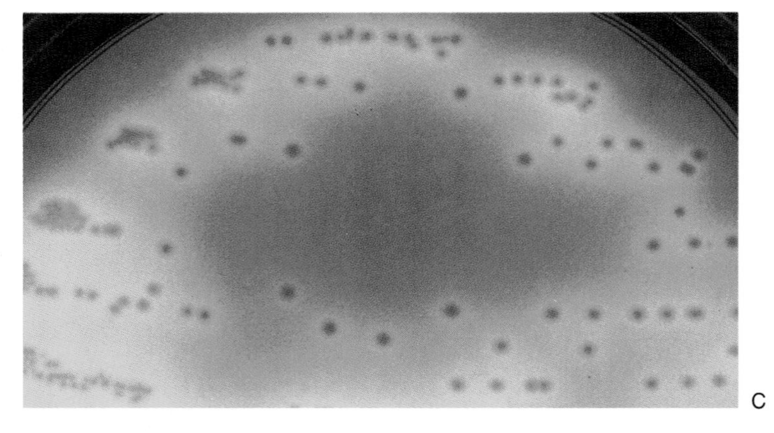

C

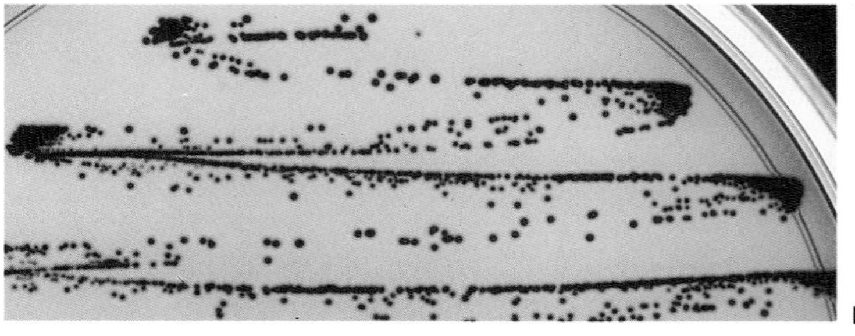

D

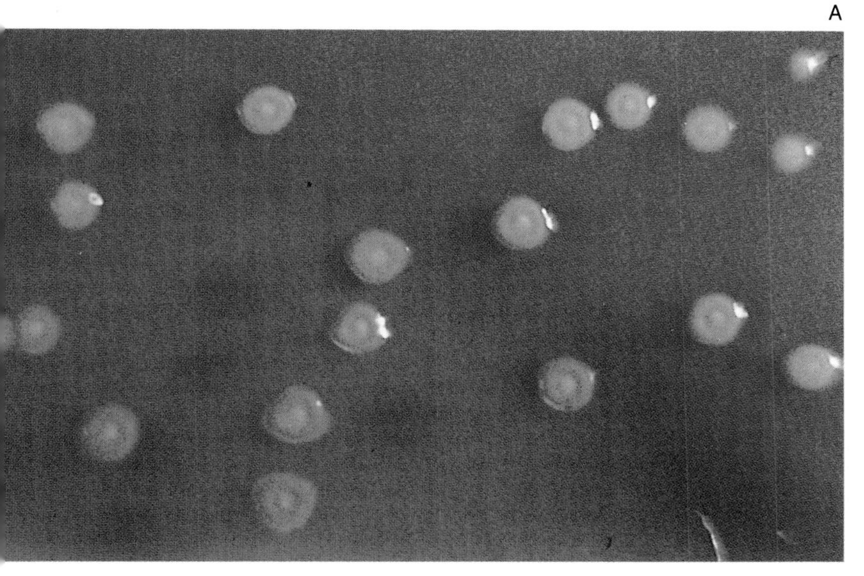

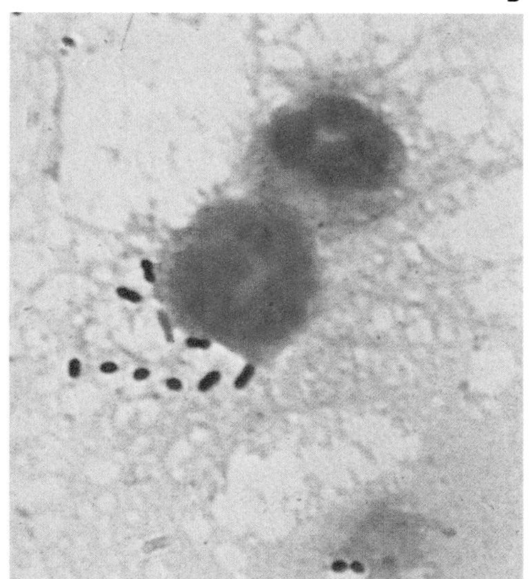

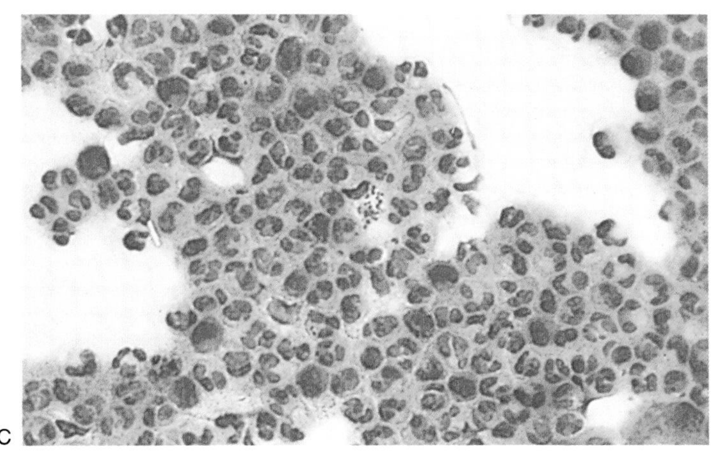

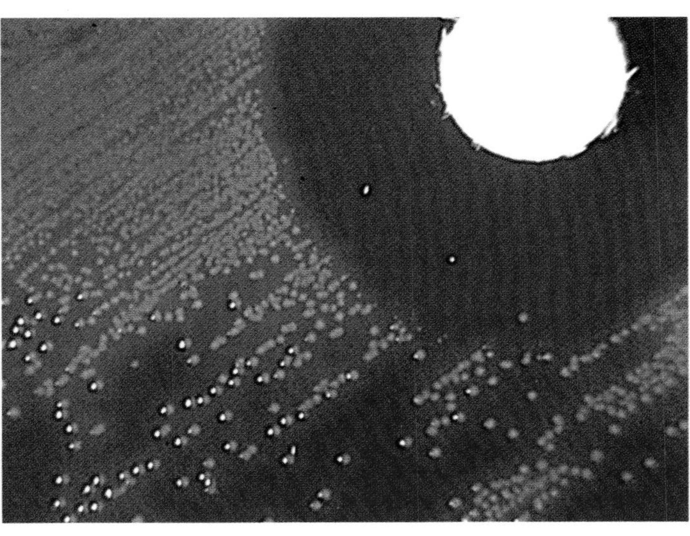

Plate 5 When infectious of the central nervous system occur, rapid and accurate diagnosis is often critical. (A) Colonies of *Escherichia coli* on MacConkey agar, showing characteristic coloration. *E. coli* is the most common pathogen causing bacterial meningitis in children. (Courtesy James Snyder, Norton Kosair Children's Hospitals, Louisville, Kentucky.) (B) Gram stain showing the presence of *Neisseria meningitidis* (meningococcus) in cerebrospinal fluid (CSF) in a case of bacterial meningitis. (Courtesy James Snyder, Norton Kosair Children's Hospitals, Louisville, Kentucky.) (C) *Streptococcus pneumoniae* and white blood cells in cerebrospinal fluid (CSF) in a case of bacterial meningitis. (Courtesy James Snyder, Norton Kosair Children's Hospitals, Louisville, Kentucky.) (D) *Streptococcus pneumoniae* isolated from cerebrospinal fluid (CSF) in a case of bacterial meningitis growing on blood agar. The photograph shows α-hemolysis (greening), which is characteristic of this species, and inhibition by optochin, which is a characteristic used in the clinical identification of this organism. (Courtesy James Snyder, Norton Kosair Children's Hospitals, Louisville, Kentucky.)

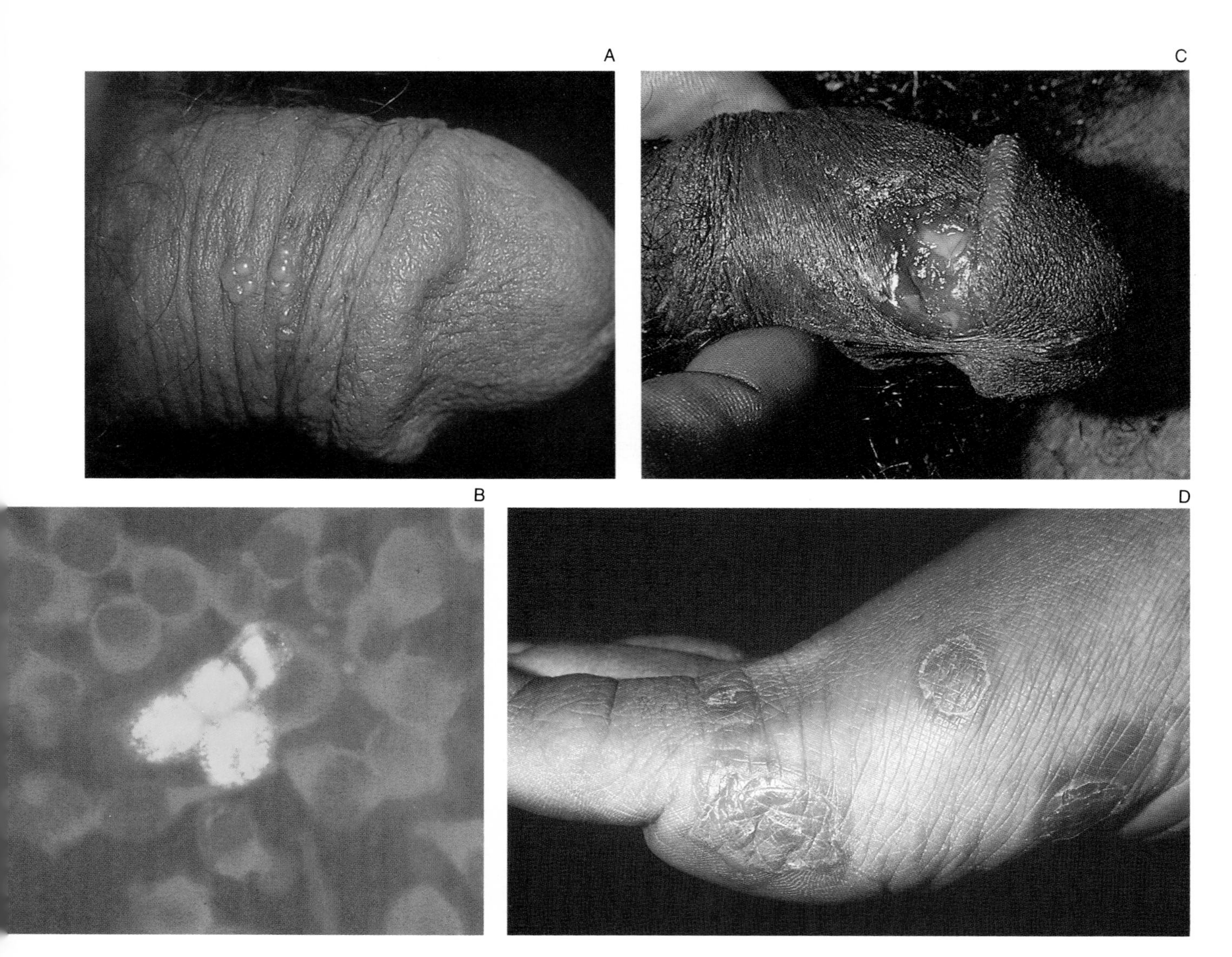

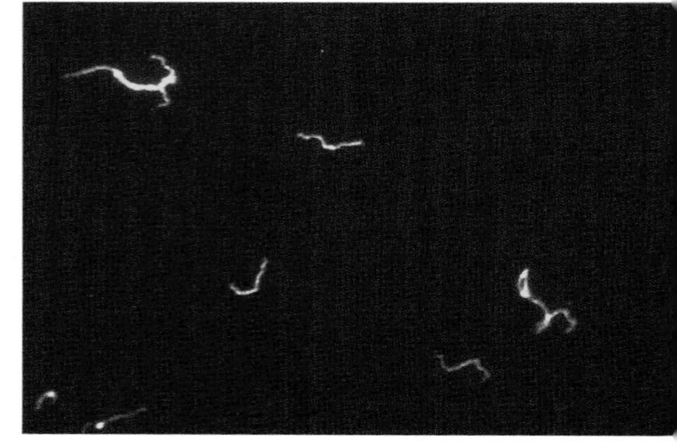

Plate 6 Several pathogens cause sexually transmitted diseases (STDs). These pathogens often have specific physiological requirements that restrict their growth to special conditions. The incidence of some of the sexually transmitted diseases has increased in epidemic fashion in recent years. (A) Genital herpes lesions on a penis; outbreaks of such lesions periodically occur on the genitals of individuals infected with this disease. (From Camera MD Studios: Carroll H. Weiss.) (B) Herpes simplex type 2 virus revealed by indirect immunofluorescent staining. This test is applicable to direct clinical smears and cell culture isolates. The infected cells fluoresce bright green, indicating a positive test, whereas uninfected cells appear red or orange. (Courtesy Electro-Nucleonics, Inc., Columbia, Maryland.) (C) Primary chancre on the penis of an individual in a case of syphilis. (From Camera MD Studios: Carroll H. Weiss.) (D) Secondary syphilis lesions of a foot. (From Camera MD Studios: Carroll H. Weiss.) (E) *Treponema pallidum,* the bacterium that causes syphilis, viewed after fluorescent antibody staining. (From Centers for Disease Control, Atlanta.)

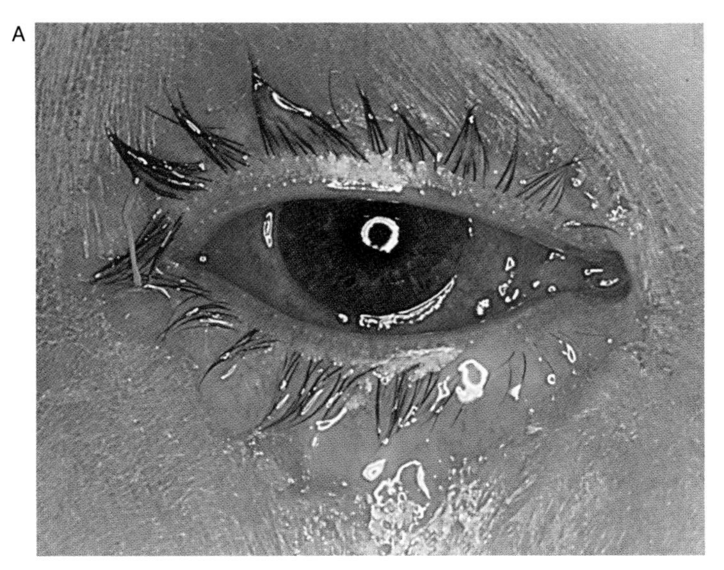

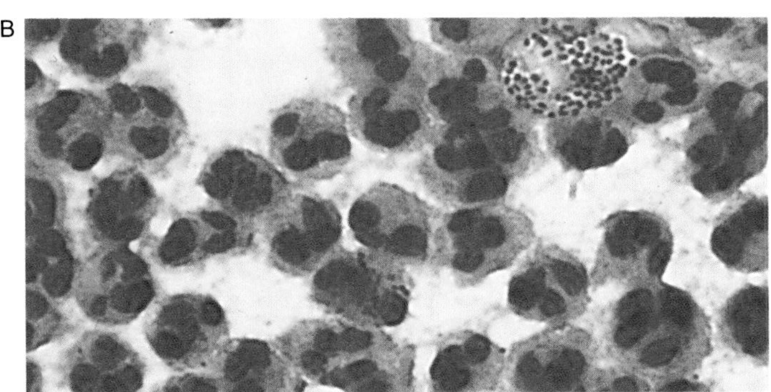

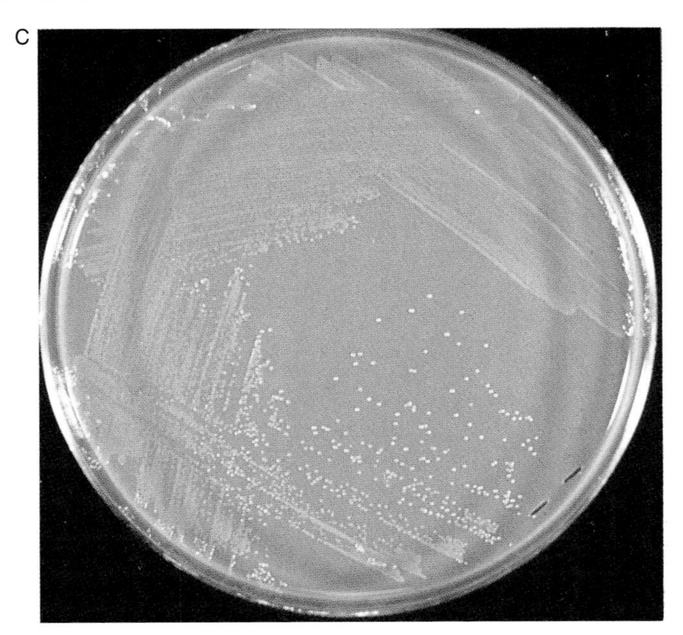

Plate 7 Some of the sexually transmitted pathogens can also be transmitted to eye tissues. Because the presence of these pathogens is not always apparent in a woman, it is routine practice to wash the eyes of newborns with erythromycin sulfate to prevent possible transmission. (A) Appearance of eye infected by *Neisseria gonorrhoeae. N. gonorrhoeae* can be transmitted sexually and to eye tissues. (From BPS: Leon J. LeBeau, University of Illinois Medical Center.) (B) *Neisseria gonorrhoeae* and white blood cells in pus from a case of gonorrhea. The bacteria can be seen as diplococci within one of the white blood cells. (Courtesy James Snyder, Norton Kosair Children's Hospitals, Louisville, Kentucky.) (C) Culture of *Neisseria gonorrhoeae* growing on a plate of chocolate agar. (Courtesy James Snyder, Norton Kosair Children's Hospitals, Louisville, Kentucky.) (D) *Chlamydia trachomatis* revealed by indirect immunofluorescent staining. The presence of a region of brilliant apple-green fluorescence in the cytoplasm of infected cells indicates a positive reaction. Uninfected cells appear red with a reddish-black nucleus. (Courtesy Electro-Nucleonics, Inc., Columbia, Maryland.) (E) Inclusion body of *Chlamydia trachomatis* from a conjunctival scraping. (Courtesy James Snyder, Norton Kosair Children's Hospitals, Louisville, Kentucky.)

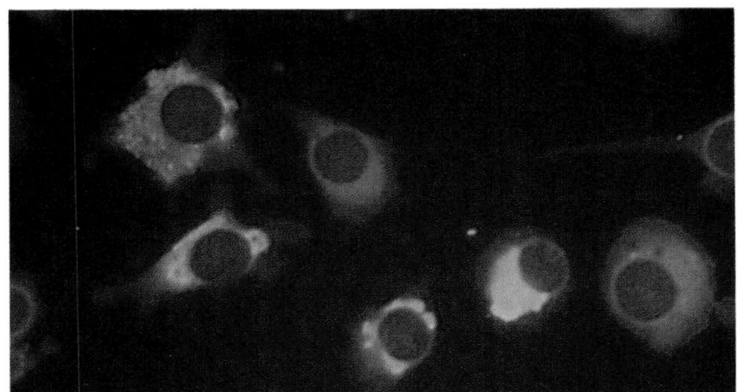

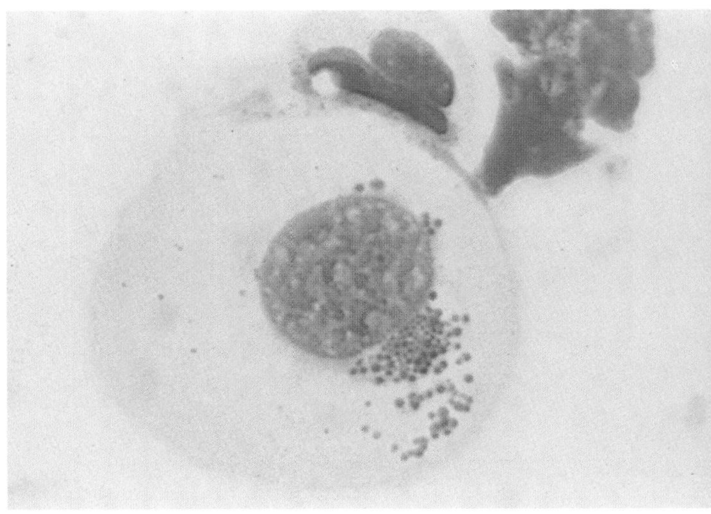

A

Plate 8 (A) AIDS (acquired immune deficiency syndrome), caused by HTLV-III virus, is characterized by a loss of immune system function so that the body is subject to numerous opportunistic infections. Also there is a high incidence of a form of the malignancy Kaposi's sarcoma, shown here in the oral cavity, in patients with AIDS. (Courtesy James Snyder, Norton Kosair Children's Hospitals, Louisville, Kentucky.) (B) The herpes zoster virus causes chicken pox and shingles. There is no vaccine as yet to prevent these diseases. In this photograph the typical chicken pox rash is seen covering the face of a two-year-old child. (From Camera MD Studios: Carroll H. Weiss.) (C) Various bacteria and viruses cause infections of the eye. This photograph shows the characteristic reddening of the eye in a case of infectious conjunctivitis. (From Camera MD Studios: Carroll H. Weiss.) (D) Rubella (German measles) can be very serious in pregnant women because the rubella virus can cross the placenta and cause congenital rubella in the newborn. Vaccination is now used to protect against this disease. Here the characteristic rubella rash is seen covering the chest of an adult woman. (From Camera MD Studios: Carroll H. Weiss.)

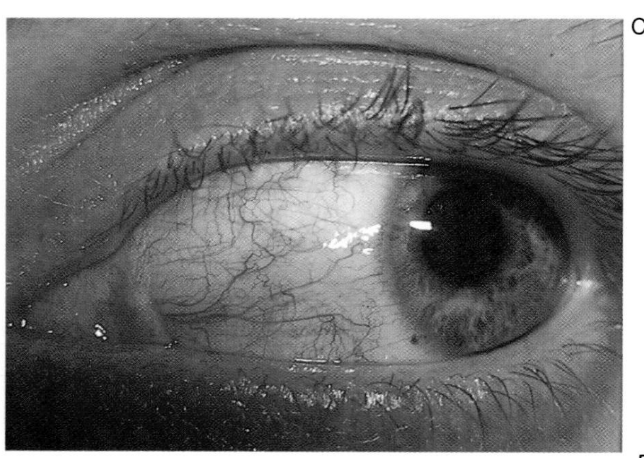

C

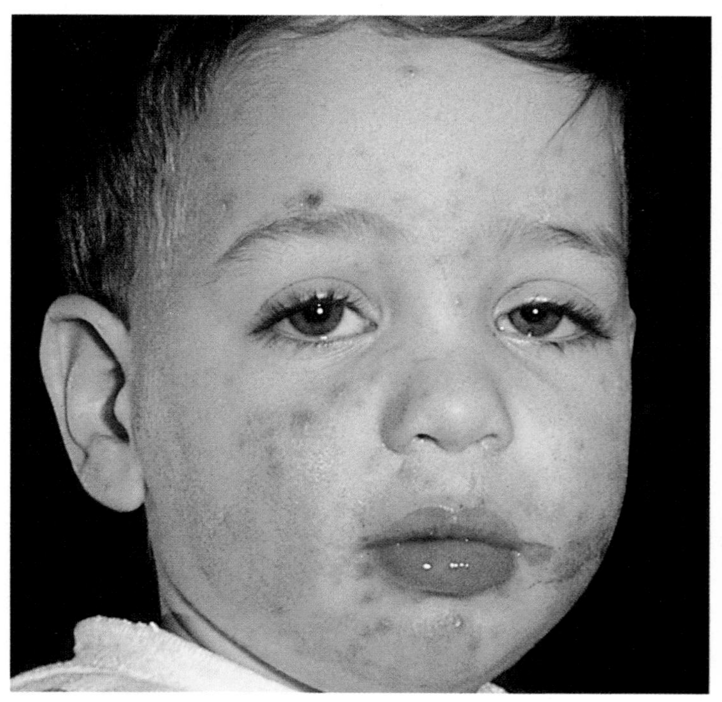

B

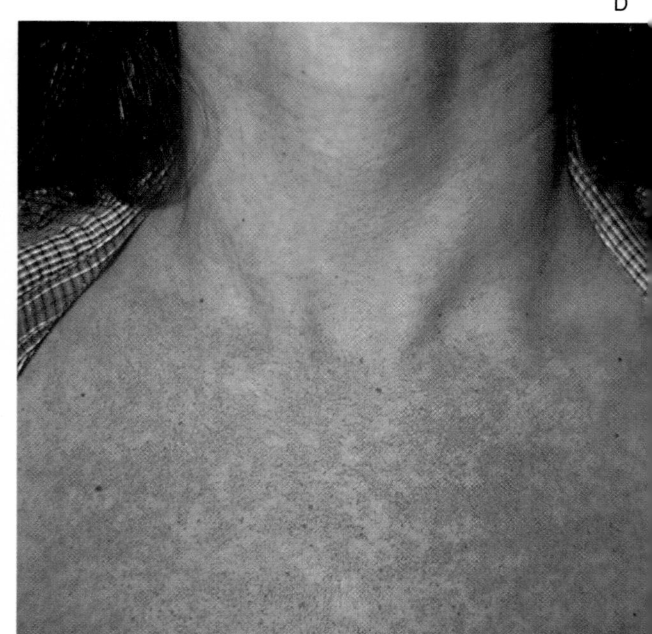

D

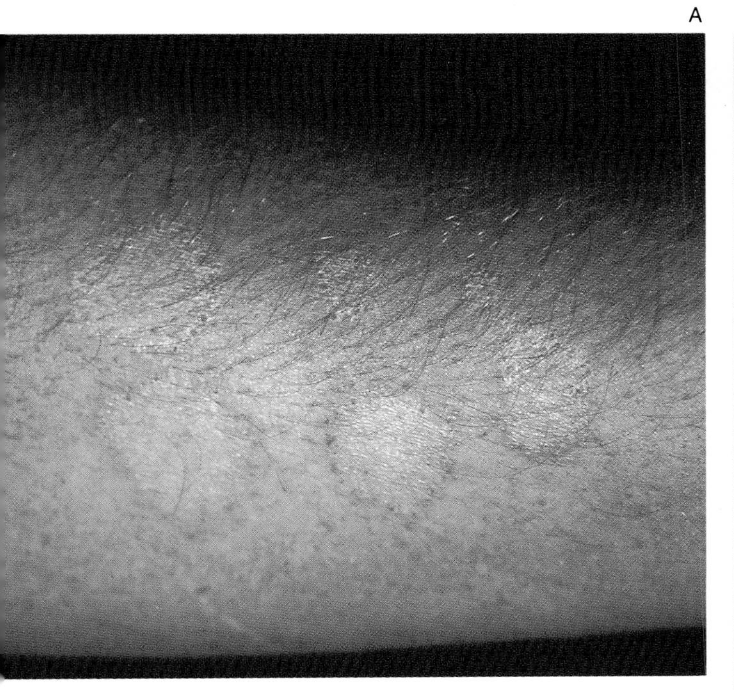

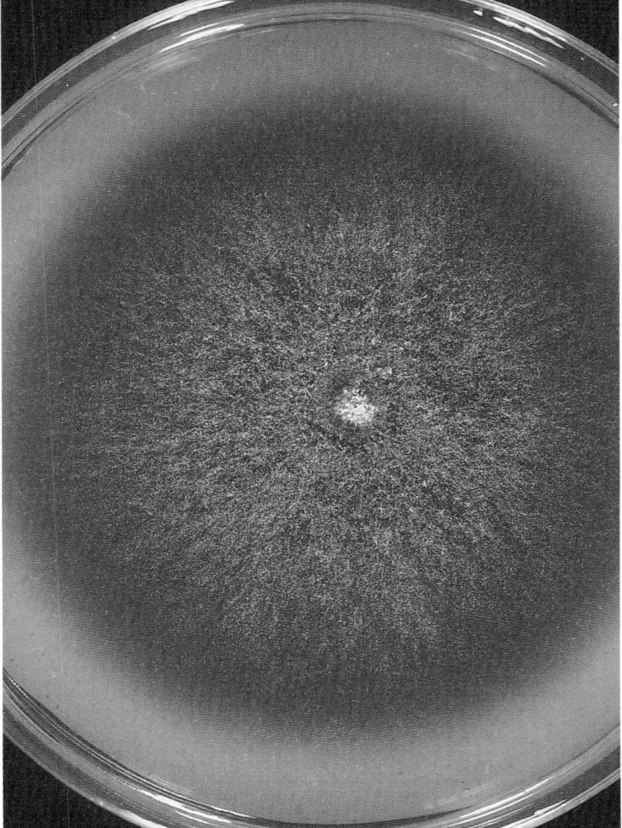

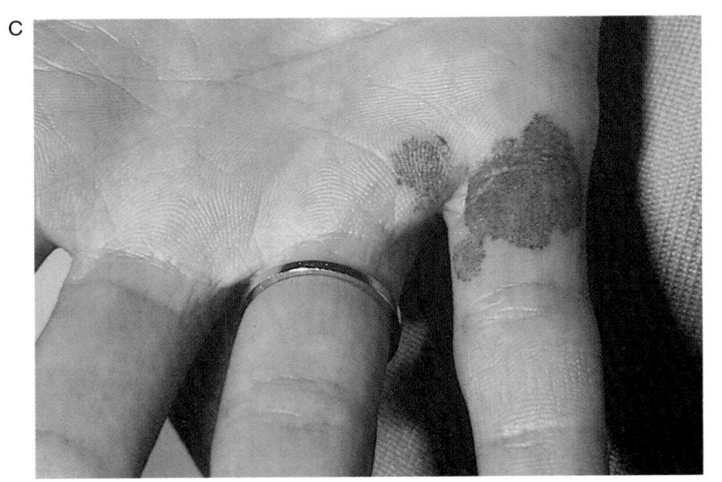

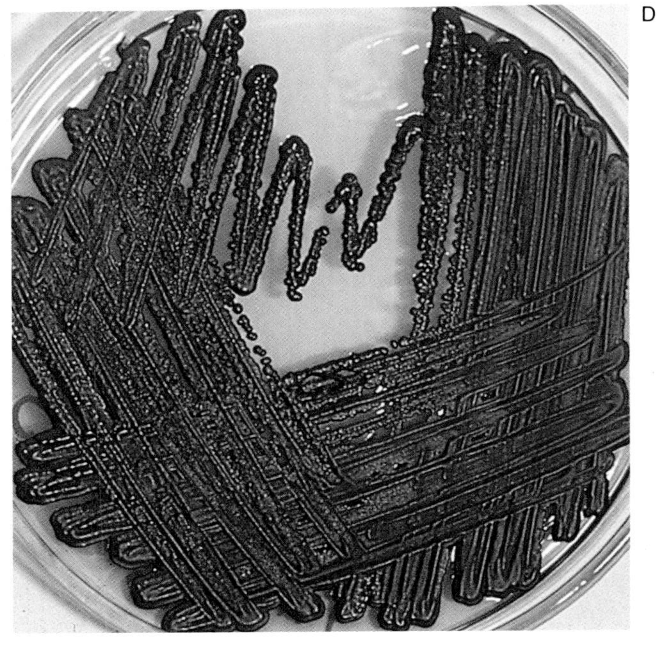

Plate 9 Several fungi, called dermatophytes, are able to infect cutaneous tissues. These fungi cause tinea or ringworm. (A) The forearm of a patient with tinea corporis (ringworm of the body). (From Camera MD Studios: Carroll H. Weiss.) (B) *Trichosporon rubrum,* a fungus that causes various forms of tinea, growing on cornmeal glucose agar. (From Camera MD Studios: Carroll H. Weiss.) (C) Tinea nigra of a South American woman. This disease is caused by *Exophiala werneckii* and is common in the tropics where it has been called Caribbean beach stigmata. (From Camera MD Studios: Carroll H. Weiss.) (D) *Exophiala werneckii* culture plate. The fungus has a melanin pigment in its cell wall, which causes a brownish discoloration of infected areas. (From Camera MD Studios: Carroll H. Weiss.)

A

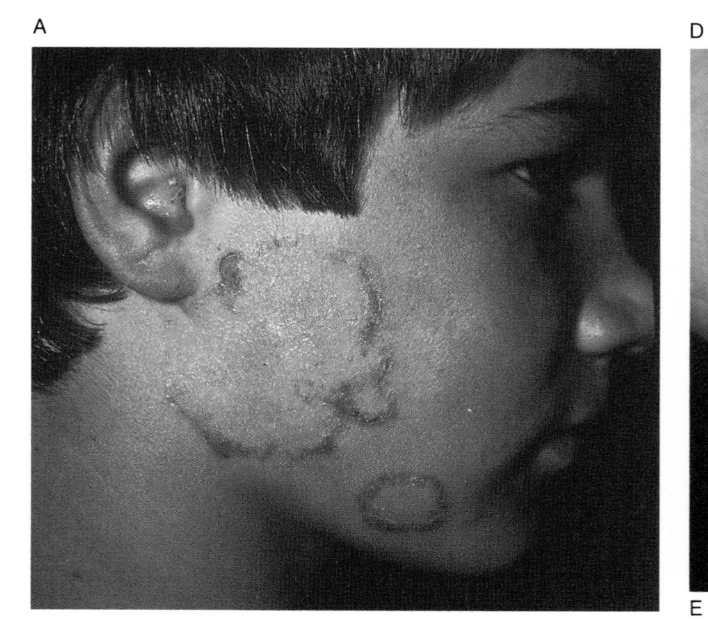

B

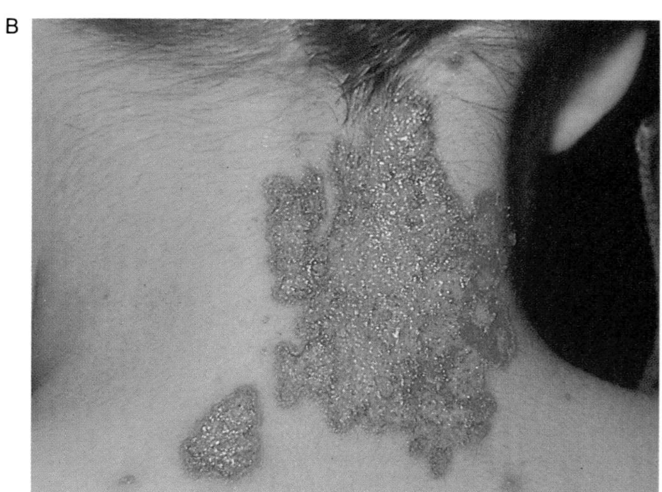

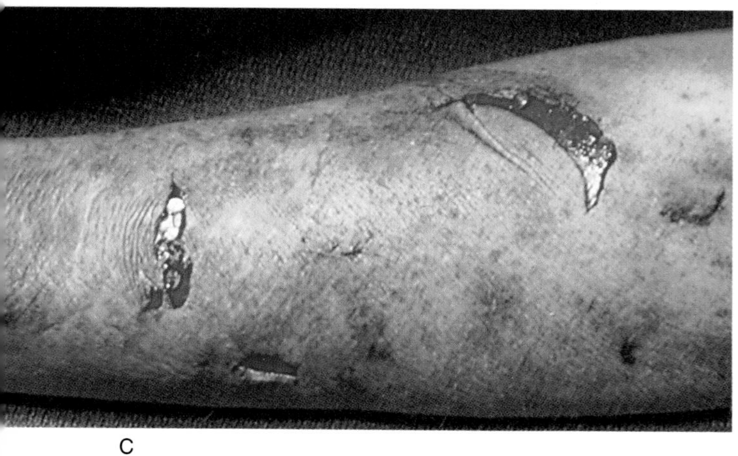

C

D

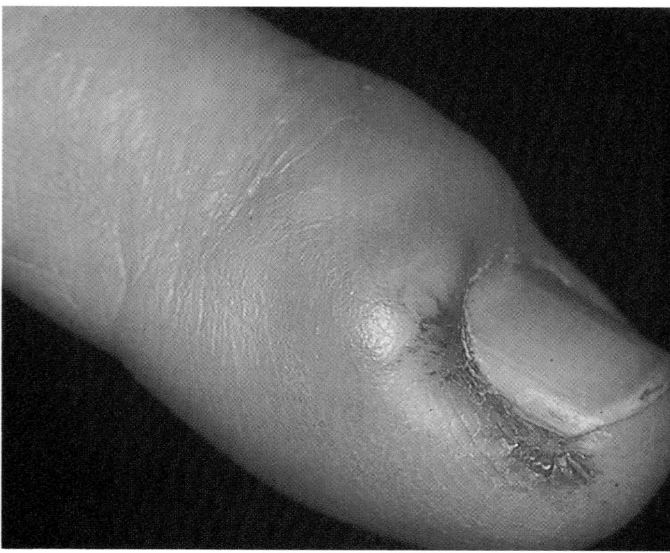

E

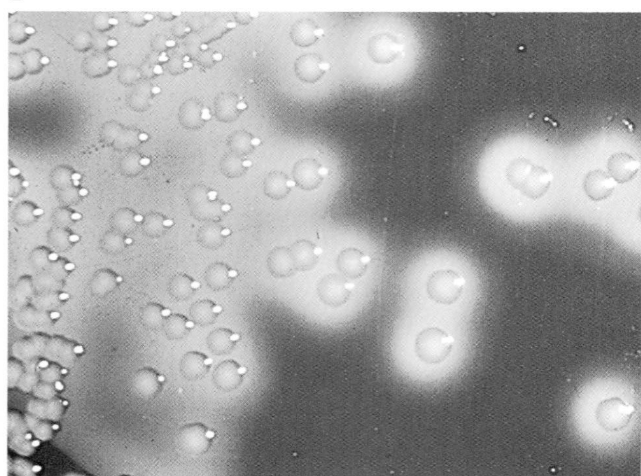

Plate 10 Staphylococci are indigenous to the normal microbiota inhabiting the skin surface. In some cases β-hemolytic streptococci cause skin infections. (A) Characteristic appearance of impetigo on the cheek caused by infection with *Staphylococcus aureus.* (From Camera MD Studios: Carroll H. Weiss.) (B) Impetigo of neck caused by staphylococcal infection showing characteristic yellowing around lesion. (From Camera MD Studios: Carroll H. Weiss.) (C) Appearance of forearm in a case of staphylococcal scalded skin syndrome (SSSS). (Courtesy James Snyder, Norton Kosair Children's Hospitals, Louisville, Kentucky.) (D) Staphylococcal infection (paronychia) of the index finger. (From Camera MD Studios: Carroll H. Weiss.) (E) Culture of *Staphylococcus aureus* on blood agar plate showing characteristic colony morphology and β-hemolysis (clearing around colonies). (Courtesy James Snyder, Norton Kosair Children's Hospitals, Louisville, Kentucky.)

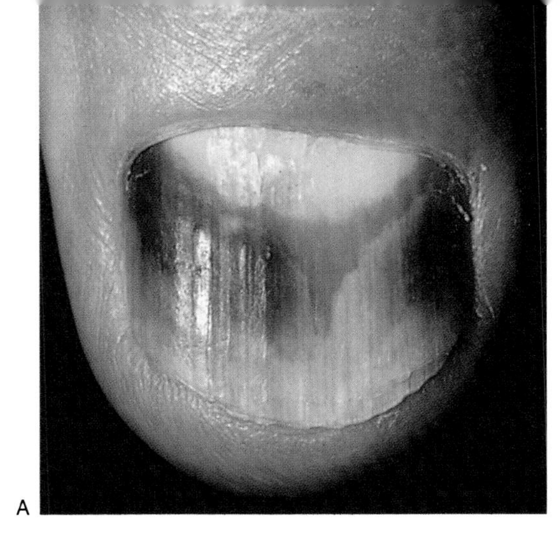
A

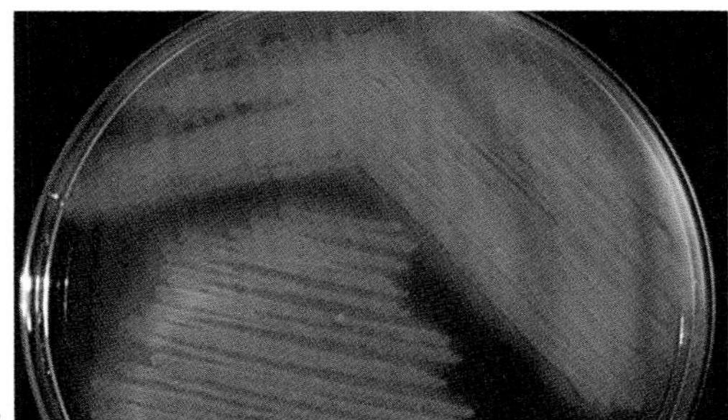
B

Plate 11 (A) Fingernail infected with *Pseudomonas*, a condition called "green nail." (From Camera MD Studios: Carroll H. Weiss.) (B) *Pseudomonas aeruginosa* showing fluorescent growth. (From BPS: Leon J. LeBeau, University of Illinois Medical Center.) (C) Gas gangrene is a very serious disease that often requires radical surgery to prevent the spread of the pathogenic bacteria. If unchecked, gangrene causes death. This photograph shows the characteristic appearance of gangrene of toes. (From Medichrome/The Stock Shop: Dianora Niccolini.) (D) Cells of *Clostridium* causing gangrene. (From Medichrome/The Stock Shop: Martin M. Rotker.) (E) *Candida* species cause various infections of the body. Often these are topical infections, such as of the vaginal tract. Here we see a *Candida* infection of the oral cavity before and after treatment with clotrimazole (Mycelex Troche). (Courtesy Miles Laboratories, West Haven, Connecticut.)

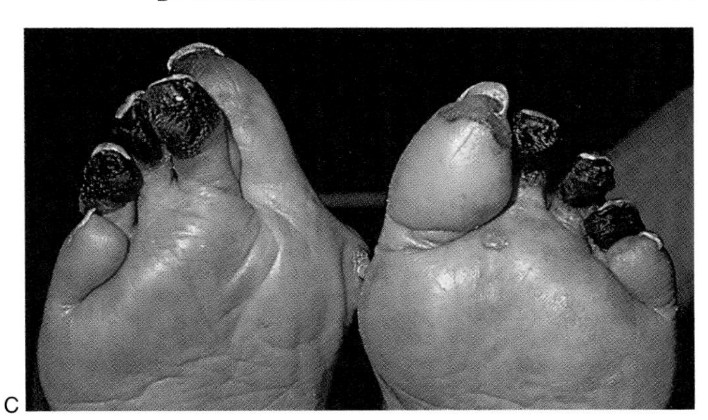
C

D
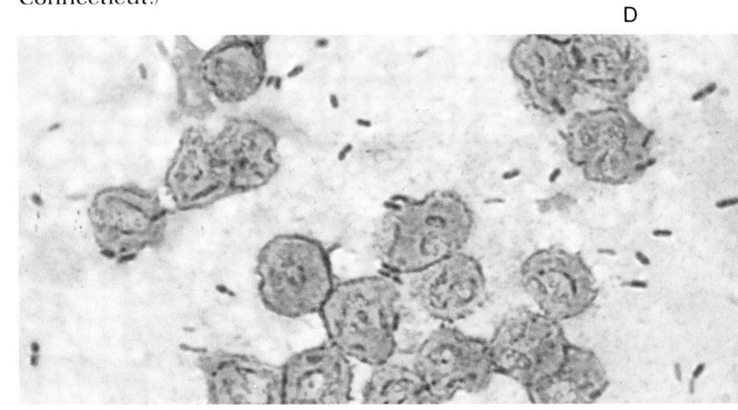

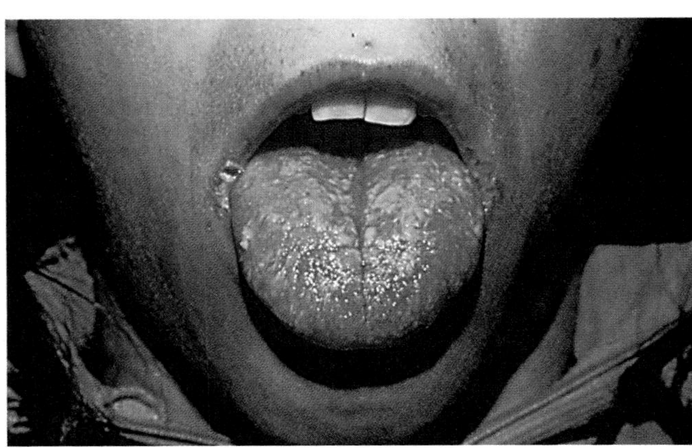
E

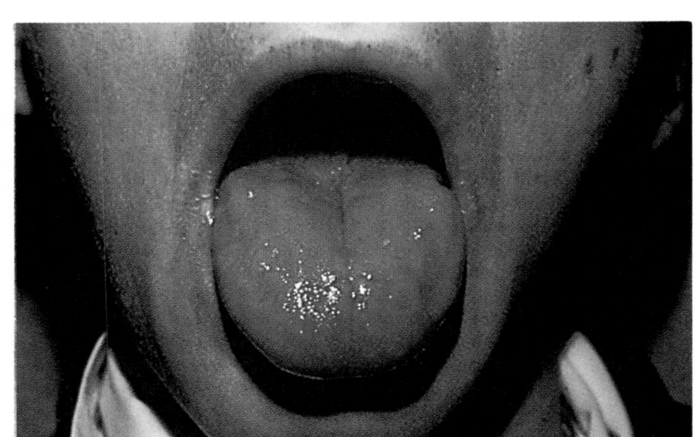

Plate 12 Infections of the oral cavity are the most common human diseases. The excessive growth of microorganisms causes inflammation of the gingiva and eventual loss of bone. (A) Appearance of normal gingiva. (B) Appearance of inflamed gingiva in a case of gingivitis. (C) Appearance of gingiva in a case of periodontitis. (D) Appearance of gingiva in a case of advanced periodontitis. (A–D Copyright American Dental Association. Reprinted by permission.) (E) Appearance of oral cavity in a case of gingivitis of Vincent's disease, which is also known as trench-mouth and necrotizing ulcerative gingivitis (NUG). This disease is characterized by necrotic projections and inflamed and swollen gingiva. (From Camera MD Studios: Carroll H. Weiss.)

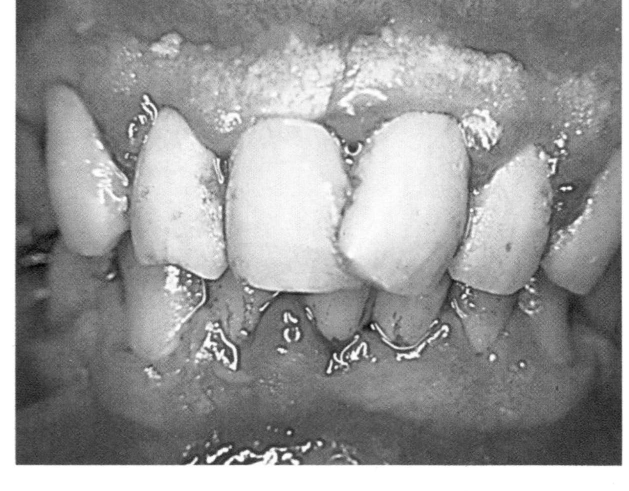

C

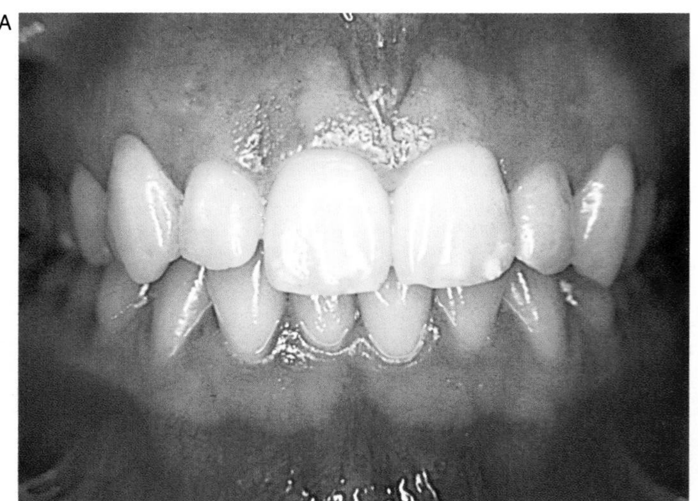

A

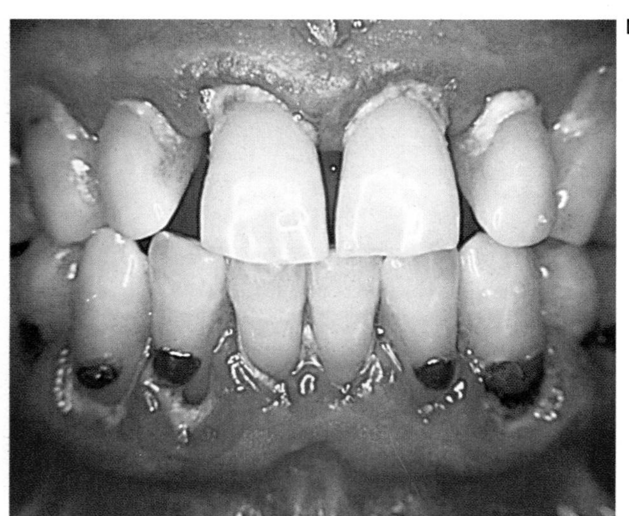

D

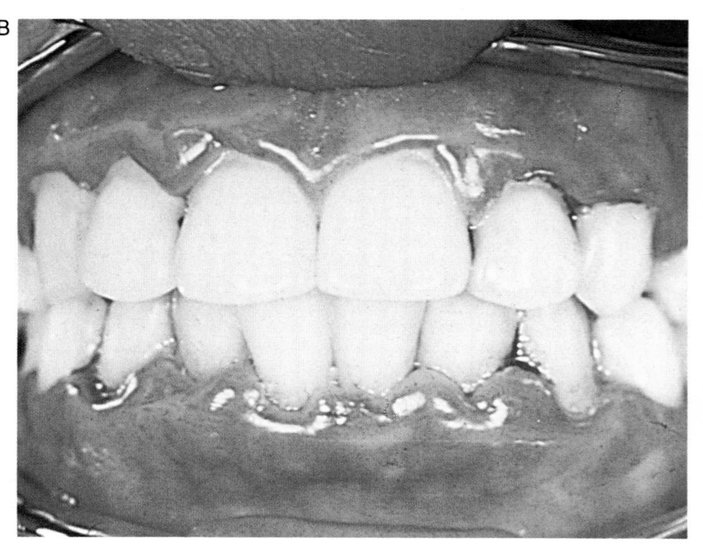

B

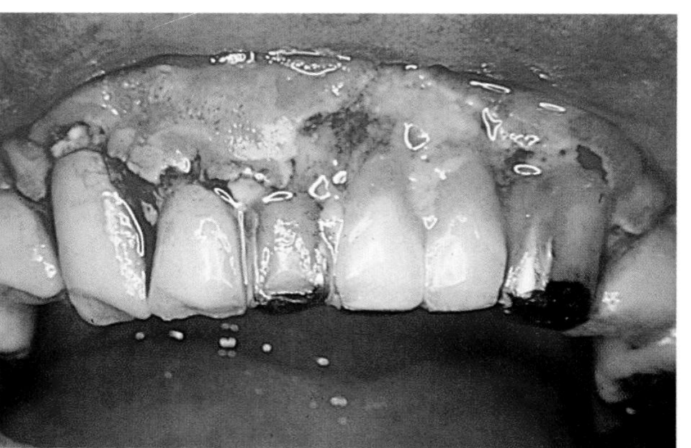

E

22 *Microorganisms and Human Disease*

KEY TERMS

acute disease
acute stage
adhesins
aerosols
alpha hemolysis
avirulent
beta hemolysis
carriers
Center for Disease Control
chronic disease
coagulase
collagenase
common source outbreak
contagious disease
convalescence
cytopathic effects
cytotoxins
disease syndrome
endemic
endotoxins
enterotoxins
epidemics
epidemiology
etiology
exotoxins
fibrinolysis
fomites
hemolysins
hyaluronidase
incubation period
infections
infectious disease
infectious dose
invasiveness
lipopolysaccharide toxins
medical asepsis
morbidity rate
mortality rate
neurotoxins
noninvasive
nosocomial infection
parenteral route
pathogens
period of illness
person-to-person epidemic
person-to-person transmission
portals of entry
prodromal stage
propagated transmission
protein toxins
reservoirs
signs
spreading factor
surgical asepsis
symptoms
toxigenicity
vectors
virulence factors
zoonoses

OBJECTIVES

After reading this chapter you should be able to

1. Define the key terms.
2. List Koch's postulates and describe how they are used to establish the etiology of infectious diseases.
3. Describe the stages of an infectious disease.
4. Describe the portals of entry for pathogenic microorganisms and the routes of disease transmission.
5. Describe the properties of microorganisms associated with virulence.
6. Describe some factors that enhance invasiveness of pathogens.
7. Describe the differences between endotoxins and exotoxins.
8. Describe the effects of several different toxins and discuss how the actions of the toxin are related to human diseases.

Causes of Disease

The interactions of microorganisms and humans represent a continuum from those that are needed to maintain good health to those that cause human disease and even death. Some of the interactions between human beings and microbial populations are essential for the well-being of the individual, whereas others result in diseases. The nature of the interaction between a specific microorganism and a specific human host depends on the physiological properties of both the microorganism and the host. There is a balance between the properties of microorganisms that cause disease and the defenses of the human body that protect against disease. Disease results when humans are exposed to microorganisms that possess particular virulence factors and/or when the defense systems of the body fail.

> All living things, from men to the smallest microbe, live in association with other living things . . . an equilibrium is established which permits the different components of biological systems to live at peace together, indeed often to help one another. Whenever the equilibrium is disturbed by any means whatever, either internal or external, one of the components of the system is favored at the expense of the other . . . and then comes about the process of disease.
>
> —RENÉ DUBOS, *The Germ Theory Revisited*

There are many causes of human disease, only some of which are associated with microorganisms. Infections occur when certain microorganisms invade or colonize body tissues. Changes in the state of a person's health (disease) sometimes occur when microorganisms infect the body. Infectious diseases, which are caused by microorganisms, probably account for less than half of all human diseases. In Chapters 23–29 we will consider a number of human diseases caused by microorganisms. Most of these diseases are associated with specific pathogenic (disease-causing) microorganisms that cause infections when they reproduce within the human body.

The requirements for scientifically proving that a specific disease is caused by a specific microorganism were introduced in the late nineteenth century by Robert Koch. Today, we refer to this process for establishing the etiology (cause) of infectious diseases as Koch's postulates. Koch successfully used this process to show that anthrax is casued by *Bacillus anthracis* and that tuberculosis is caused by *Mycobacterium tuberculosis*. More recently the same basic procedure was used to elucidate the etiology of Legionnaire's disease and AIDS.

In order to satisfy Koch's postulates

1. The organism should be present in all animals suffering from the disease and absent from all healthy animals.
2. The organism must be grown in pure culture outside the diseased animal host.
3. When such a culture is inoculated into a healthy susceptible host, the animal must develop the symptoms of disease.
4. The organism must be reisolated from the experimentally infected animal and shown to be identical with the original isolate.

It is not always possible to rigorously follow Koch's postulates. In some cases microorganisms are opportunistic pathogens, only causing disease under specific conditions. Many such opportunistic pathogens are normally associated with healthy individuals in contradiction to Koch's first postulate. Sometimes the pathogen cannot be grown outside the host organisms, making it impossible to satisfy Koch's second postulate. For example, *Treponema pallidum*, the causative agent of syphilis, has yet to be grown on artificial media. In yet other cases we consider it unethical to inoculate human subjects with suspected pathogens and have been unable to find suitable animal model systems. For example, we have determined that the virus HTLV-3 is the probable cause of acquired immune deficiency syndrome (AIDS), but have yet to rigorously demonstrate this fact because we have yet to inject an animal with this virus and observe the development of AIDS.

Virulence Factors of Pathogenic Microorganisms

What is it about some microorganisms that allows them to establish the infections that cause human disease? To answer this question we need to examine the intrinsic properties of pathogenic microorganisms that allow them to overcome host defenses and initiate the physiological changes inherent in the disease process. An infection of the human body by a pathogen results in disease when the potential of the microorganism to disrupt normal bodily functions is fully expressed. In many cases, though, infections with potentially pathogenic microorganisms are limited so that the overt clinical manifestation of the disease does not occur. Many healthy individuals are carriers of potentially pathogenic microorganisms. Although we understand today that pathogenicity is not a property of the microorganism alone, it is still useful to examine the intrinsic properties of pathogenic microorganisms that contribute to their potential for causing human disease.

Some disease-causing microorganisms possess properties, referred to as virulence factors, that enhance their pathogenicity and allow them to invade human tissues and disrupt normal bodily functions. The virulence of pathogenic microorganisms, that is, their ability to induce human disease, depends in large part on two properties of the microorganisms, invasiveness and toxigenicity. Toxigenicity refers to the microbial production of agents that interfere with normal physiological functions of the infected individual. Invasiveness refers to the ability of the microorganism to invade human tissues, attach to cells, and multiply within the cells or tissues of the human body. Most of the microorganisms of the normal microbiota of humans are noninvasive, that is, they do not invade the tissues on whose surfaces they grow. Microorganisms that possess invasive properties are able to establish infections within host cells and tissues.

Imagine an assault, an infection, established via a packet of bacteria in a kiss. No, via a scratch. No, that lawn mower! Sixty minutes later so-and-so-many million had been multiplied by such-and-such a factor. That's many. Unseen cohorts, trained to Marine deportment, had come to the defense of the host, us. Enemy cohorts were escalating—enemies always escalate. They multiplied, multiplied, multiplied, microscopic bacteria. Combat was on. It had exploded over the whole terrain, was producing effects beyoud the terrain, host molecules versus parasite molecules, jungle warfare.

—GUSTAV ECKSTEIN, *The Body Has a Head*

Invasiveness

Enzymes that enhance invasiveness. Several microbial enzymes can destroy body tissues (Table 22.1). For example, hyaluronidase breaks down hyaluronic acid, the substance that holds together the cells of connective tissues. Pathogens that produce hyaluronidases spread through body tissues, and therefore, hyaluronidase is referred to as the "spreading factor." Various species of *Staphylococcus*, *Streptococcus*, and *Clostridium* produce hyaluronidase enzymes. Some *Clostridium* species also produce collagenase, an enzyme that breaks down the proteins of collagen tissues. The κ toxin of *C. perfringens*, which can cause gas gangrene, for example, is a collagenase that contributes to the spread of this organism through the human body. The breakdown of fibrous tissues enhances the invasiveness of pathogenic microorganisms. Thus, the actions of some microbial enzymes contribute to the virulence of pathogens by enhancing the ability of the microorganisms to proliferate within body tissues and by interfering with the normal defense mechanisms of the host organism.

Some *Staphylococcus* and *Streptococcus* species produce fibrinolysin (kinase). The fibrinolytic enzymes staphylokinase and streptokinase catalyze the lysis of fibrin clots. The action of these two fibrinolytic enzymes may enhance the invasiveness of pathogenic strains of *Staphylococcus* and *Streptococcus* by preventing fibrin in the host from walling off the area of bacterial infection. Without the action of fibrin, the pathogens are free to spread to surrounding areas.

In a somewhat different way the production of coagulase enhances the virulence of some *Staphylococcus* species. The enzyme coagulase converts fibrinogen to fibrin. Some *Staphylococcus* species, such as *Staphylococcus aureus*, produce this enzyme, and the deposition of fibrin around the

Table 22.1 Some Extracellular Enzymes Involved in Microbial Virulence

Enzyme	Action	Examples of Bacteria Producing Enzyme
Hyaluronidase	Breaks down hyaluronic acid (spreading factor)	*Streptococcus pyogenes*
Coagulase	Blood clots; coagulation of plasma	*Staphylococcus aureus*
Phospholipase	Lyses blood cells	*Staphylococcus aureus*
Lecithinase	Destroys red blood cells and other tissue cells	*Clostridium perfringens*
Collagenase	Breaks down collagen (connective tissue fiber)	*Clostridium perfringens*
Fibrinolysin (kinase)	Dissolves blood clots	*Streptococcus pyogenes*

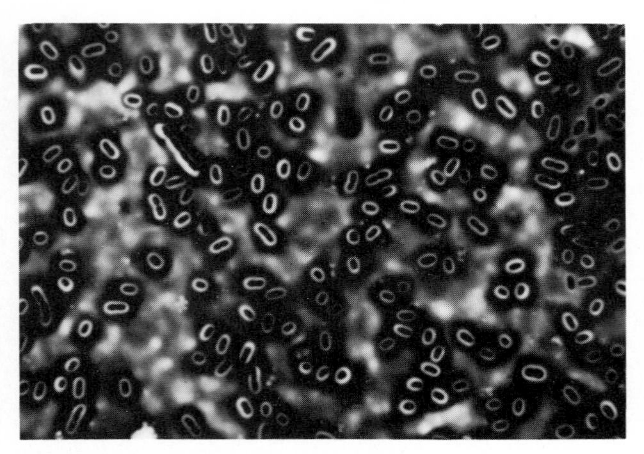

Figure 22.1 Micrographs showing the encapsulated bacterium *Klebsiella pneumoniae.* The presence of a capsule increases the virulence of microbial pathogens because they are more resistant to host defense mechanisms. (From BPS: Leon J. LeBeau, University of Illinois Medical Center, Chicago.)

staphylococcal cells presumably protects the cells against the circulatory defense mechanisms of the host. Phagocytic cells apparently fail to recognize the fibrin-coated *Staphylococcus* cells as foreign agents. Coagulase negative strains of *Staphylococcus aureus*, however, still have been found to be virulent pathogens. It is thus difficult to associate virulence with the activity of a single enzyme, even though these enzymes appear to play a role in the virulence of a variety of pathogenic microorganisms.

Factors that interfere with phagocytosis. Surface layers of a pathogen contribute to its virulence. The surface layers surrounding a pathogenic microorganism can interfere with the ability of phagocytic blood cells to engulf and destroy that pathogen when it invades the human body. Capsules surrounding the cells of strains of *Streptococcus pneumoniae*, for example, are antiphagocytic and permit these bacteria to evade the normal defense mechanisms of the host, allowing them to reproduce and cause the disease symptomology of pneumonia. Nonencapsulated strains of *S. pneumoniae* are avirulent (do not cause disease) whereas capsule-producing strains can cause pneumonia and death. The virulence of other bacteria, including *Haemophilus influenzae* and *Klebsiella pneumoniae*, is also enhanced by capsule production (Figure 22.1). Here again the capsule appears to protect the pathogen against the phagocytic defenses of the body, giving the bacterium an edge that allows it to successfully invade the body.

Adhesion factors. In addition to their role in avoiding phagocytosis, capsules and slime layers contribute to the ability of bacteria to attach or adhere to particular host cells or tissues. Attachment is extremely important in many disease processes. Many pathogenic bacteria must adhere to mucous membranes in order to establish an infection (Figure 22.2). Specific factors that enhance the ability of a microorganism to attach to the surfaces of mammalian cells are termed adhesins, and the production of such substances is another important factor that determines the virulence of particular pathogens. The pili of several pathogenic bacteria and their associated adhesins appear to play a key role in permitting the bacteria to adhere to host cells and establish infections. For example, enteropathogenic strains of *Escherichia coli* have particular adhesins associated with their pili that permit

Figure 22.2 Micrograph of an enteropathogenic strain of *E. coli* attached to the microvilli of the intestine by the glycocalyx surrounding the bacterial cells. (Courtesy William Costerton, University of Calgary.)

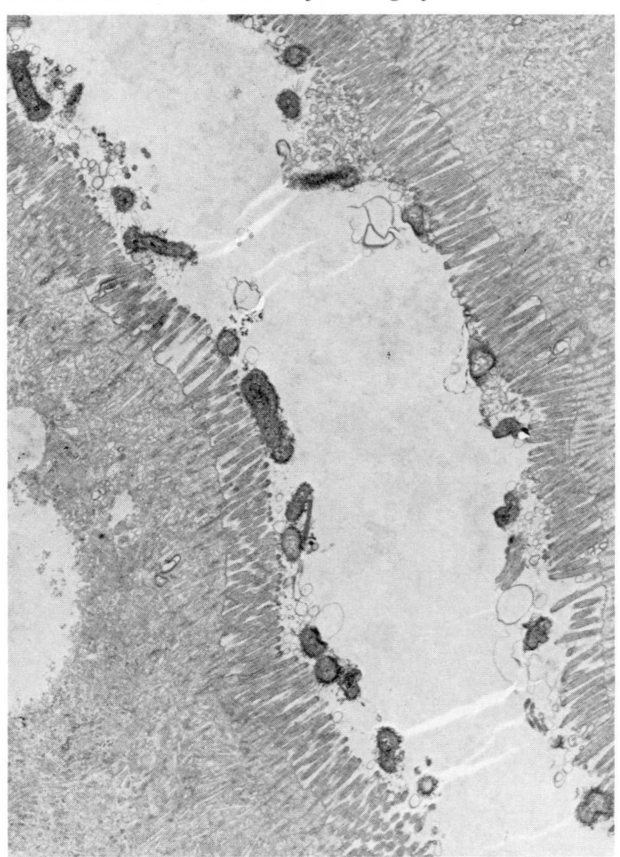

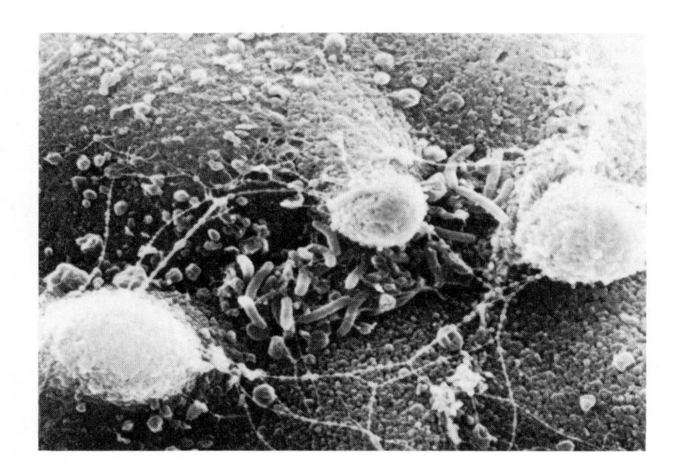

Figure 22.3 Micrograph showing *Vibrio cholerae* attached to the lining of the gut of a mouse at the tip of the villus. The ability to attach to the intestine is an important characteristic of many pathogens that enter the body via the gastrointestinal tract. (From BPS: Garry T. Cole, University of Texas, Austin.)

them to bind to the mucosal lining of the intestine. In a similar manner, *Vibrio cholerae* is able to adhere to the mucosal cells lining the intestine, allowing the establishment of an infection (Figure 22.3).

Likewise, the adsorption of certain viruses onto specific receptor sites of human cells establishes the necessary prerequisite for the uptake of the viruses by those cells, leading to the reproduction of the viruses, disruption of normal host cell function, and the production of disease symptoms by the invading viral pathogens. Some viruses, such as adenoviruses, have external spikes that aid in their attachment to host cells. Similarly, the spikes of orthomyxoviruses and paramyxoviruses attach to receptors of N-acetylneuraminic acid on the surfaces of human red blood cells. The ability of pathogenic microorganisms, including viruses, to attach to and invade particular cells and tissues establishes specific tissue affinities for pathogenic microorganisms.

Toxigenicity

Toxigenicity refers to the ability of a microorganism to produce biochemicals, known as toxins,

Table 22.2 **Some Toxins Produced by Microorganisms Causing Disease in Humans**

Microorganism	Toxin	Disease	Action
Clostridium botulinum	Several neurotoxins	Botulism	Paralysis, blocks neural transmission
Clostridium perfringens	α toxin	Gas gangrene	Lecithinase
	κ toxin		Collagenase
	θ toxin		Hemolysin
Clostridium tetani	Neurotoxin (tetanospasm)	Tetanus	Spastic paralysis, interferes with motor neurons
	Tetanolysin		Hemolytic cardiotoxin
Corynebacterium diphtheriae	Diphtheria toxin	Diphtheria	Blocks protein synthesis at level of translation
Streptococcus pyogenes	Streptolysin O	Scarlet fever	Hemolysin
	Streptolysin S		Hemolysin
	Erythrogenic		Causes rash of scarlet fever
Shigella dysenteriae	Neurotoxin	Bacterial dysentery	Hemorrhagic, paralytic
Staphylococcus aureus	Enterotoxin	Food poisoning	Intestinal inflammation
Aspergillus flavus	Aflatoxin B_1	Aflatoxicosis	Blocks protein synthesis at level of transcription
Amanita phalloides	α-Amanitin	Mushroom food poisoning	Blocks protein synthesis at level of transcription

that disrupt the normal functions of cells or are generally destructive to human cells and tissues (Table 22.2). Toxins cause discernable damage to human systems and in some cases cause death. Some toxin-producing microorganisms need not establish an infection within the human body to cause disease; they can grow outside of the host and still cause disease symptoms if the toxins enter human tissues. In the past we have classified toxins produced by microorganisms as endotoxins if they comprise part of the microbial cell, and as exotoxins if they are secreted by the cell, but we now know that some exotoxins are not released until the cell is disrupted, and substances classified as endotoxins are sometimes released from the cell without lysis. Therefore, a better classification system for toxins is based on the biochemical nature of the toxin.

Lipopolysaccharide toxins (endotoxins). The term bacterial endotoxin refers to the lipopolysaccharide (LPS) portion of the cell wall of Gram negative bacteria. LPS can be detected by the *Limulus* amebocyte test. When Gram negative bacteria die, their cell walls disintegrate, releasing the lipopolysaccharide toxin. Some growing Gram negative bacteria also release lipopolysaccharide toxin. The physiological effects of lipopolysaccharide toxins include fever, circulatory changes that produce shock, and other general symptoms, such as weakness and nonlocalized aches. The injury to the circulatory system by the lipopolysaccharide of the Gram negative cell is basic to the action of this toxin, but the mechanism of its action is not yet understood. The lipopolysaccharide toxins of all Gram negative pathogens play a part in the infectious process but in some cases the role of endotoxins is greater than in others. The effects of lipopolysaccharide toxins are not specific to the particular species of Gram negative bacteria, and thus, there is no specific characteristic disease symptomology associated with the endotoxin of a particular bacterial species. Lipopolysaccharide toxins of *Salmonella* and *Shigella* species are responsible in part for diseases, such as gastroenteritis, caused by these pathogens, but these pathogens also produce protein toxins that are more important in the pathogenicity of these organisms.

Protein toxins (exotoxins). In contrast to lipopolysaccharide toxins, the effects of protein toxins (exotoxins) are specific to the microorganism producing the toxin. Exotoxins are produced by both Gram negative and Gram positive bacteria. Protein toxins are more readily inactivated by heat than are lipopolysaccharide toxins. A protein toxin can normally be inactivated by exposure to boiling water for 30 minutes, whereas lipopolysaccharide toxins can withstand autoclaving (Table 22.3). Typically, protein toxins are excreted into the surrounding medium. For example, *Clostridium botulinum*, the causative organism for botulism, can secrete a potent exotoxin into canned food products; the ingestion of even minute amounts of this toxin is lethal. Protein toxins are generally more potent than lipopolysaccharide toxins, with far less protein toxin needed to produce serious disease symptoms than is required for disease symptoms due to lipopolysaccharides. As examples of the potency of protein toxins, about 30 grams of diphtheria toxin could kill 10 million people and 1 gram of botulinum toxin could kill everyone in the United States (over 225 million people).

Protein toxins cause distinctive clinical symp-

Table 22.3 Comparison of Selected Characteristics of Bacterial Lipopolysaccharide Toxins (Endoxins) and Protein Toxins (Exotoxins)

Characteristic	Endotoxin	Exotoxin
Chemical composition	Lipopolysaccharide–protein complex	Protein
Source	Cell walls of Gram negative bacteria; released upon death and autolysis of the bacteria	Gram negative and Gram positive bacteria; excretion products of growing cells, or in some cases, substances released upon autolysis and death of the bacteria
Effects on host	Nonspecific	Generally affects specific tissues
Thermostability	Relatively heat stable (may resist 120°C for 1 hr)	Heat labile; most are inactivated at 60–80°C

toms. Often, protein toxins are referred to by the disease they cause, such as diphtheria toxin or botulinum toxin. They may also be categorized according to the symptoms that they cause. As examples, neurotoxins affect the nervous system, enterotoxins cause excessive secretions of fluid and electrolytes from the lining of the gastrointestinal tract, and cytotoxins interfere with cellular functions, generally resulting in cell death.

Neurotoxins. The neurotoxins produced by *Clostridium botulinum*, which are responsible for the symptomology of botulism, bind to nerve synapses. These toxins block the release of acetylcholine from nerve cells of the central nervous system, causing the loss of motor function (Figure 22.4). The toxin blocks conduction of the nerve impulse at or near the end of the point of final branching between the nerve filaments and the presynaptic part of the end plate complex of the muscle cell, preventing the proper transmission of the nerve impulse to the muscle. The inability to transmit impulses through motor neurons can cause respiratory or cardiac failure and death.

The neurotoxin tetanospasmin produced by *Clostridium tetani*, the causative agent for the disease tetanus, interferes with the peripheral nerves of the spinal cord (Figure 22.5). Tetanospasmin blocks the ability of these nerve cells properly to transmit signals to the muscle cells, causing the symptomatic spasmatic paralysis of tetanus. Like the neurotoxin produced by *C. botulinum*, the neurotoxin of *C. tetani* paralyzes motor neurons, but unlike botulinum toxin, tetanospasmin acts only on the nerves of the cerebrospinal axis. It is postulated that tetanus toxin inhibits the release of glycine from the inhibitory neurons (interneurons) in the anterior horn of the spinal cord. Because glycine is the inhibitory neurotransmitter in these interneurons, the result is convulsions similar to those produced by strychnine, which is known to compete with glycine for receptor sites.

Enterotoxins. Enterotoxins cause an inflammation of the gastrointestinal tract. Generally, enterotoxins cause gastrointestinal upset. In some cases enterotoxins are responsible for severe human diseases. For example, cholera is caused by the enterotoxin cholaragen produced by *Vibrio cholerae* (Figure 22.6). Cholaragen blocks the action of an enzyme involved in the conversion of cyclic AMP to ATP. This results in elevated concentrations of cyclic AMP, which causes the release of inorganic ions, including chloride and bicarbonate ions, from the mucosal cells that line the intestine into the intestinal lumen. The change in the ionic balance resulting from the action of this toxin causes the movement of large amounts of water into the lumen, leading to severe dehydration that sometimes results in the death of infected individuals.

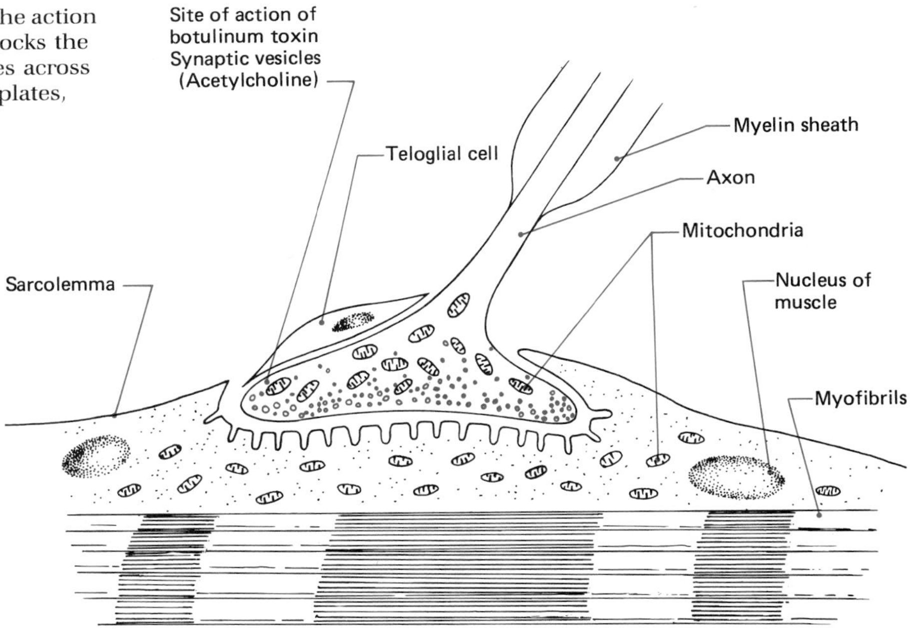

Figure 22.4 Diagram showing the action of botulinum toxin. The toxin blocks the transmission of nervous impulses across the synapses of the motor end plates, producing flaccid paralysis.

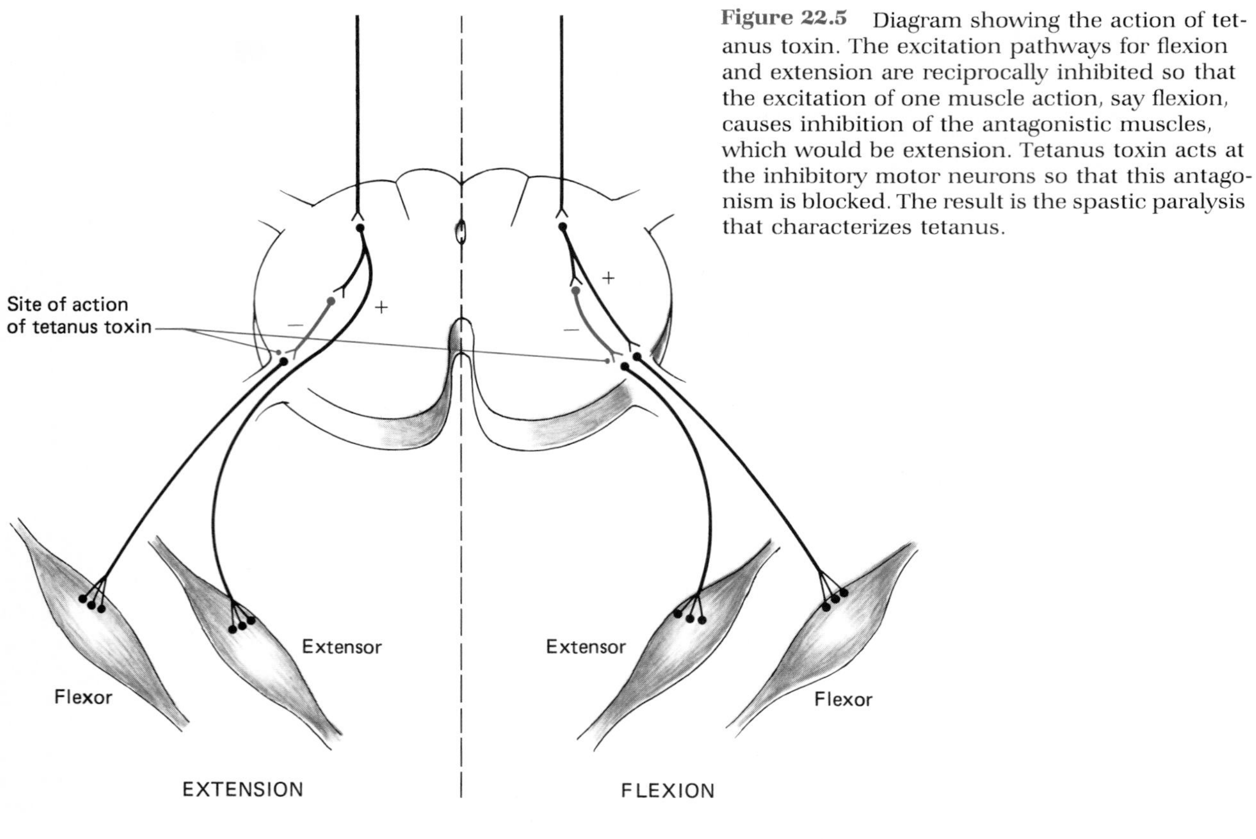

Figure 22.5 Diagram showing the action of tetanus toxin. The excitation pathways for flexion and extension are reciprocally inhibited so that the excitation of one muscle action, say flexion, causes inhibition of the antagonistic muscles, which would be extension. Tetanus toxin acts at the inhibitory motor neurons so that this antagonism is blocked. The result is the spastic paralysis that characterizes tetanus.

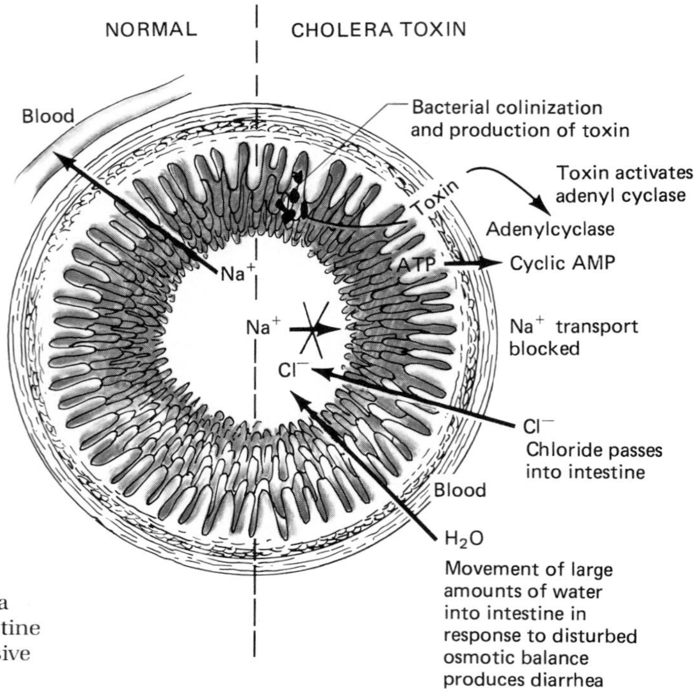

Figure 22.6 Diagram showing the action of cholera toxin. The action of the toxin at the lining of the intestine causes an osmotic imbalance and the flow of excessive amounts of water out of the body.

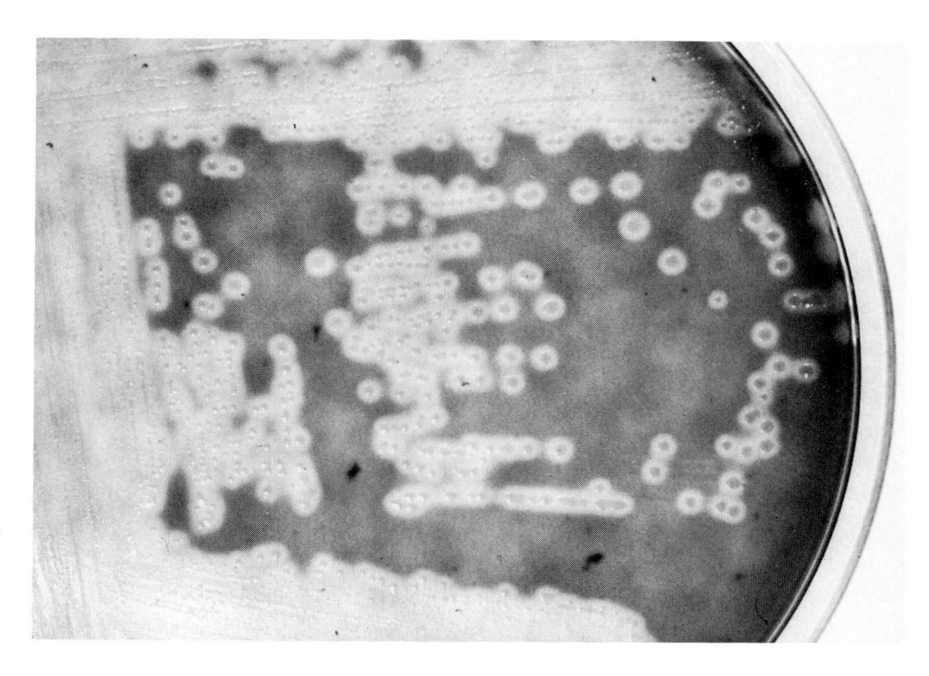

Figure 22.7 Beta-hemolytic growth of *Streptococcus pyogenes* on blood agar plates, showing zones of clearing where hemolysis has occurred around the colonies. (From BPS: L. M. Pope and D. R. Grote, University of Texas, Austin.)

Cytotoxins. Cytotoxins cause cell death, often by lysis and/or interference with protein synthesis. For example, diphtheria toxin, produced by *Corynebacterium diphtheriae,* inhibits protein synthesis in mammalian cells. This toxin blocks transferase reactions during the translation of mRNA, preventing the addition of amino acids and thus the elongation of the peptide chain. The production of diphtheria toxin is particularly interesting because only lysogenized cells of *C. diphtheriae* produce diphtheria toxin proteins. The protein toxin is coded for by the phage genome. Thus, only when *C. diphtheriae* is infected with a virus does a human infection with *C. diphtheriae* result in disease.

Some proteins produced by pathogenic microorganisms contribute to their virulence by causing the lysis of red blood cells. Cytotoxins that cause the lysis of human erythrocytes are termed hemolysins because their action results in the release of hemoglobin from these red blood cells. For example, *Streptococcus* species produce various hemolysins, including streptolysin O, an oxygen-labile and heat stable protein, and streptolysin S, an acid-sensitive and heat-labile protein. The hemolytic action of these cytotoxins produces zones of clearing when these bacteria are grown on blood agar plates. A complete zone of clearing around a bacterial colony growing on a blood agar plate is referred to as beta hemolysis), and the partial clearing of the blood agar plate around the bacterial colony is referred to as alpha hemolysis) (Figure 22.7). Alpha hemolysis involves the conversion of hemoglobin to methemoglobin, with the production of a zone of greening (partial clearing) around the colony.

Hemolytic activity is associated not only with *Streptococcus* species but also with various other bacterial genera, including *Staphylococcus* and *Clostridium* as well as other bacterial species. In addition to red blood cells, white blood cells are killed by some microbial cytotoxins. For example, leukocidin produced by *Staphylococcus aureus* causes lysis of leukocytes, contributing to the pathogenicity of this organism. The alpha toxin of *C. perfringens* is a lecithinase (also known as phospholipase C or phosphatidylcholine phosphohydrolase). This enzyme hydrolyzes lecithin, which is a lipid component of eukaryotic membranes. Lecithinase activity, associated with *C. perfringens,* is in part responsible for the ability of this organism to cause gas gangrene. The hydrolysis of lecithin by the alpha toxin of *C. perfringens* destroys many cells and is the primary cause of the extensive tissue damage in this disease. Lecithinase also acts as a hemolysin, causing lysis of red blood cells in addition to destroying cells of various other tissues.

Cytopathic Effects of Viruses

When viruses reproduce within host cells they also can produce substances that can destroy or interfere with the normal functioning of cells. The observable changes in the appearance of cells in-

A

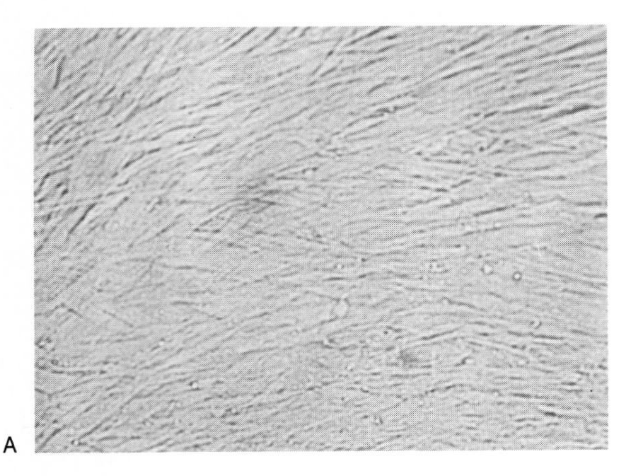

B

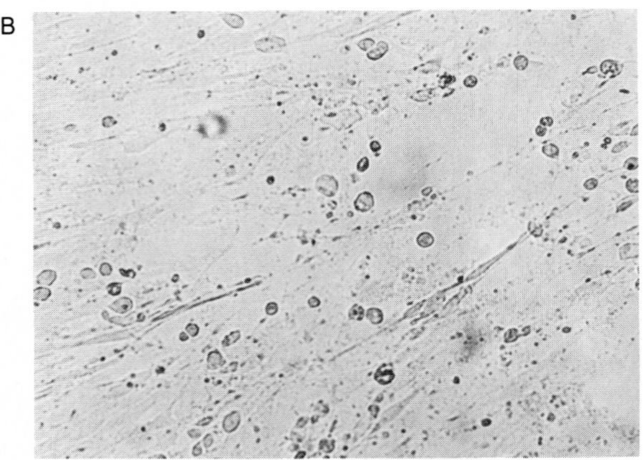

Figure 22.8 The characteristic cytopathic effect (CPE) of Rhinoviruses in tissue culture is indicated by cell shrinkage, nuclear pyknosis, and loss of adsorption to glass. (A) Micrograph of normal uninfected cell line (60×). (B) Micrograph showing CPE in cells after infection with Rhinovirus. (Courtesy Jack H. Schieble, Viral and Rickettsial Disease Laboratory, California Department of Health Services, Berkeley, California.)

fected with viruses are collectively known as cytopathic effects (CPE) (Figure 22.8). In some cases human cells infected with viruses die. For example, polio viruses kill the human cells they infect. In other cases the infected cells develop abnormalities. Inclusions sometimes occur within the nucleus or cytoplasm of infected cells. These inclusions may be stained with basic or acid dyes. For example, cells infected with measles virus develop acidophilic inclusions in the nucleus and cytoplasm; cells infected with rabies virus develop acidophilic inclusions only within the cytoplasm; and cells infected with adenovirus develop basophilic inclusion within the nucleus. Some viruses cause infected cells to fuse, forming multinuclear giant cells. Measles virus causes the formation of such giant cells.

Pattern of Disease

In many cases the reproduction of pathogenic microorganisms within the body produces specific diseases that are associated with characteristic signs and symptoms. Signs are objective changes, such as a rash or fever that a physician can observe. Symptoms are subjective changes in body function, such as pain or loss of appetite, that are experienced by the patient. A characteristic group of signs and symptoms constitutes a disease syndrome. Often the physician is able to diagnose a disease based exclusively on the symptoms reported by the patient and the signs that they observe. In other cases, more elaborate laboratory tests are necessary to identify the cause of the disease.

"I cannot go to school today,"
Said little Peggy Ann McKay.
"I have the measles and the mumps,
A gash, a rash and purple bumps.
My mouth is wet, my throat is dry,
I'm going blind in my right eye.
My tonsils are as big as rocks,
I've counted sixteen chicken pox
And there's one more—that's seventeen,
And don't you think my face looks green?
My leg is cut, my eyes are blue—
It might be instamatic flu.
I cough and sneeze and gasp and choke,
I'm sure that my left leg is broke—
My hip hurts when I move my chin,
My belly button's caving in,
My back is wrenched, my ankle's sprained,
My 'pendix pains each time it rains,
My nose is cold, my toes are numb,
I have a sliver in my thumb.
My neck is stiff, my spine is weak,
I hardly whisper when I speak.
My tongue is filling up my mouth,
I think my hair is falling out.
My elbow's bent, my spine ain't straight,
My temperature is one-o-eight.
My brain is shrunk, I cannot hear,
There is a hole inside my ear.
I have a hangnail, and my heart is—what?
What's that? What's that you say?
You say today is . . . Saturday?
G'bye, I'm going out to play!"

—SHEL SILVERSTEIN, *Sick*

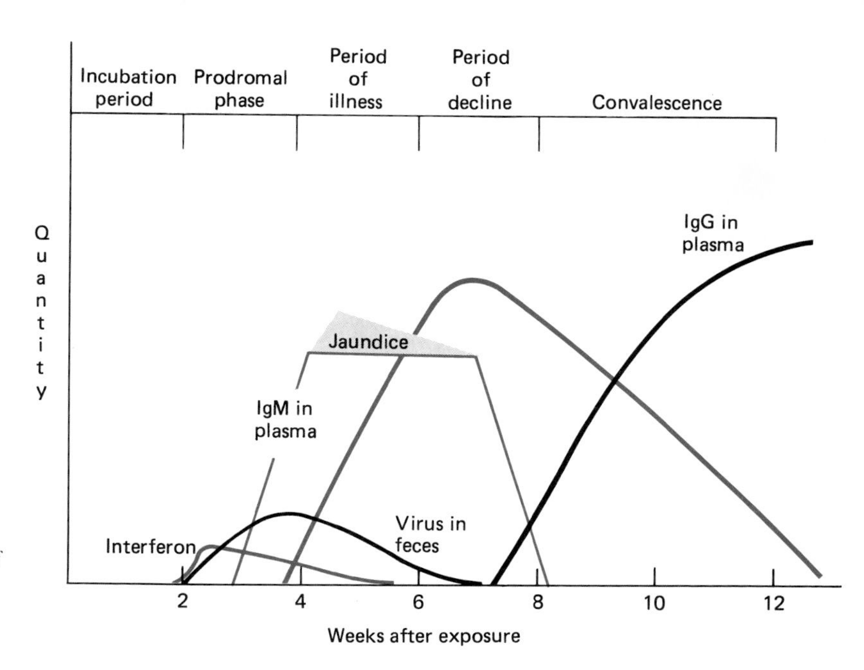

Figure 22.9 Time course of a hepatitis type A infection, showing stages of the disease, changes in concentrations of the pathogenic virus, and changes in host defense responses.

In **acute diseases** the symptoms and signs develop rapidly, reaching a height of intensity, and end fairly quickly. Measles, cholera, and influenza are all examples of acute diseases. In **chronic diseases** the symptoms persist, for a prolonged period of time. The persistent cough associated with chronic bronchitis is typical of the long-term signs associated with a chronic disease.

Whether a disease is acute or chronic, when it is caused by an infectious agent, there is a characteristic pattern to the disease. The progress of any infectious disease in a given patient can be divided into several stages: incubation, prodromal, period of illness (also called onset of acute phase), period of decline (also called progression of acute phase), and convalescence (Figure 22.9).

In the normal course of disease there will be an **incubation period** after the pathogen enters the body and before any signs or symptoms appear. The incubation period varies for different diseases (Table

Table 22.4 Incubation Periods of Representative Diseases

Disease	Incubation period (days)	Disease	Incubation period (days)
Chicken pox (varicella)	14–16	Meningitis (bacterial)	1–7
Cholera	1–3	Mumps	14–21
Coccidiodomycosis	7–21	Plague	2–6
Diphtheria	2–5	Poliomyelitis	4–7
Dysentery (shigellosis)	1–2	Rabies	10–365
Gas gangrene	1–5	Rocky Mountain spotted fever	3–12
German measles (rubella)	14–21	Salmonellosis	less than 1
Gonorrhea	3–8	Serum hepatitis (hepatitis type B)	40–180
Histoplasmosis	5–18	Syphilis	10–90
Infectious hepatitis	15–50	Tetanus	3–21
Infectious mononucleosis	14–40	Tuberculosis	28–42
Influenza	1–2	Typhoid fever	5–14
Leptospirosis	2–20	Undulant fever (brucellosis)	6–14
Leprosy (Hansen's disease)	300–1800	Whooping cough (pertussis)	12–20
Malaria	14	Yellow fever	3–6
Measles (rubeola)	11–14		

22.4). During the incubation period the microorganism has invaded the host, is migrating to various tissues, but has not yet begun to increase to sufficient numbers to cause discomfort. Generally numbers of pathogens are insufficient during the incubation period for the individual to be infective. The onset of symptoms marks the end of the incubation period and the start of the prodromal stage. The patient has become aware of discomfort but does not yet have adequately precise symptoms to permit the clinician to make a diagnosis. However, sufficient replication of the pathogen has occurred to render the patient contagious to others. Moreover, the nonspecific inflammatory defenses have become operative.

The period of illness occurs next, during which time the disease is most severe. Various signs and symptoms that characterize the particular disease occur in this period. During the period of illness (acute stage), the patient often is sufficiently ill to alter his or her normal work or school activities. Clones of B or T cells are being selected to initiate immune defense. This phase of the disease progresses either toward death or convalescence. Recovery depends upon whether the immune systems or medical treatments are adequate. Assuming the disease is not fatal or chronic, the signs and symptoms begin to disappear during the period of decline. Convalescence progresses either to a carrier stage or to freedom from the pathogen. In some cases, the immmune memory system may protect the person from reoccurrence of the infection for several months, several years, or life. Full recovery marks the end of the disease syndrome.

Transmission of Disease—Epidemiology

The likelihood of disease transmission depends on the concentration and virulence of the pathogen, the distribution of susceptible individuals, and the potential sources of exposure to the pathogenic microorganisms. The examination of disease transmission is part of the field of epidemiology. The epidemiologist considers the etiology (cause) of the disease and the factors involved in disease transmission, especially as these factors relate to populations. With this information the epidemiologist evaluates how a disease outbreak in a population can be effectively controlled.

Epidemiology is based on the statistical proba-

MICROBIOLOGY HEADLINES

Disease Risk Found Greater for Early Ages in Day Care

CHICAGO, Nov. 9 (AP)—Children 5 years old and younger who attend day-care centers are almost twice as likely as those who stay home to contract a bacterial disease that is the leading cause of meningitis and epiglottitis in children, according to a new study.

The study in today's issue of *The Journal of the American Medical Association* found that the younger the child the greater the risk was but that day-care participants older than 5 were no more likely to contract the disease than their stay-at-home counterparts.

The disease, hemophilus influenza type B, has a mortality rate of about 10 percent and leaves neurological problems in about one-third of its victims.

It can cause meningitis, an inflammation of the nerve lining around the brain and spinal cord that could prove fatal if untreated, and epiglottitis, a respiratory infection that impairs the ability to swallow.

The study by two doctors in Rochester, N.Y., Stephen Redmond of the Monroe County Department of Health and Michael Pichichero of University of Rochester School of Medicine, examined the records of all cases of the bacterial disease in Monroe County in 1982 and 1983.

The researchers said 9,933 children, or 22 percent of the 44,289 children 5 and younger in the county, attended licensed day-care facilities.

The study found that among children younger than a year the risk of getting the bacterial disease was 12.3 times greater for those in day care than for those kept at home, but the researchers said the risk declined quickly, to 3.8 times greater for day-care enrollees 2 years to 3 years old.

The researchers suggested the leveling off resulted from a gradual immunization to one another's infections among the children attending day care.

"This corresponds to the anecdotal information gleaned from parents of day-care center attendees that once their children enter school, they do not seem to be as sick as other children" who stayed at home, the researchers wrote.

Source: Associated Press, Nov. 9, 1984. Reprinted by permission.

bility that a susceptible individual will be exposed to a particular pathogen that would result in disease transmission. In many cases a disease outbreak can be traced to a single source of exposure. The epidemiologist often acts as a detective in order to locate the origin of a disease outbreak, in some cases, searching for a source of tainted food, in others, direct contact with infected individuals and so forth. The epidemiologist is aided by the knowledge that disease transmission normally occurs via the air, the ingestion of contaminated food or water, or by direct contact with infected individuals or contaminated inanimate objects (fomites). By determining where and what people have eaten, where they have been, and with whom they have been in contact, the epidemiologist can establish a pattern of disease transmission.

The number of cases reported each day and the locations of disease occurrences enables the epidemiologist to distinguish between a common source outbreak, which is characterized by a sharp rise and rapid decline, and a person-to-person epidemic, which is characterized by a relatively slow and prolonged rise and decline (Figure 22.10). In the United States local and state health departments and the Centers for Disease Control (CDC) in Atlanta, Georgia, compile the statistics necessary for such determinations. CDC publishes *Morbidity and Mortality Weekly Report (MMWR)*, which contains data on the incidence of certain diseases (morbidity rates) and the deaths that result from these diseases (mortality rates) (Table 22.5). The incidence of a disease is the fraction of a population that contracts that disease during a specified time period.

Figure 22.10 The shape of the curve of reported cases of a disease is indicative of whether the epidemic originates from a common source or is propagated from person to person.

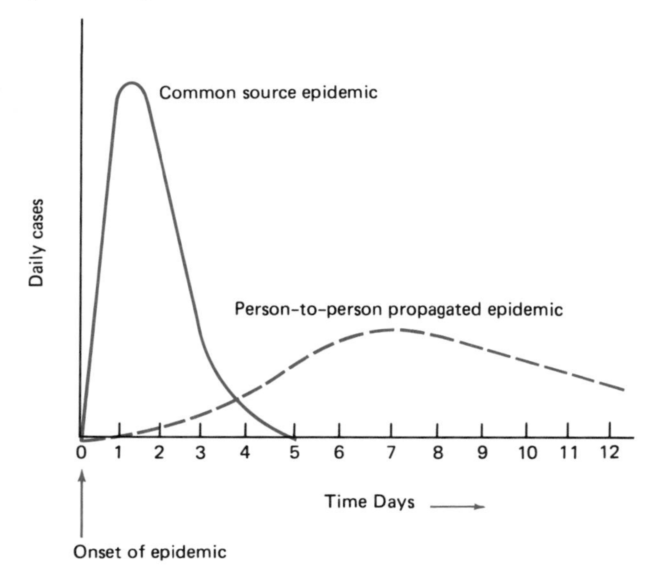

When common sources (reservoirs) or vectors (carriers involved in disease transmission—often arthropods) of pathogens are identified, action can be taken to break the chain of disease transmission, for instance by recalling potentially contaminated foods from the marketplace. When person-to-person (propagated) transmission is responsible for disease outbreaks in a population, steps can be taken to reduce the numbers of susceptible individuals, for example, by immunization, thereby breaking the chain of transmission. Indeed, it is the aim of the epidemiologist to identify the sources of disease outbreaks and to advise public health officials on the steps that should be taken to prevent such disease outbreaks.

No easy thing seeing that there are above 1500 diseases to which Man is subjected

—Sidney Smith, *Letter to Lord Grey*

Some diseases are endemic; these diseases are constantly present, normally at relatively low rates within certain geographic areas. For example, plague is endemic to the southwestern United States. Epidemics are outbreaks of disease in which unusually high numbers of individuals within a population contract the disease. There are annual epidemic outbreaks of influenza as well as other diseases around the world. During outbreaks of some infectious diseases, the pathogen is transmitted from infected to susceptible individuals (Figure 22.11). As individuals recover from the disease and develop immunity to that pathogen, the chain of transmission is broken. The number of individuals who must be immune to prevent an epidemic is a function of the infectivity of the disease (I), the duration of the disease (D), and the proportion of susceptible individuals within the population (s). High values for s, I, and D favor epidemic outbreaks.

Some people are so sensitive they feel snubbed if an epidemic overlooks them.

—Frank Hubbart, *Abe Martin's Broadcast*

Occasionally, the worldwide outbreak of a disease is so high that it is said to be pandemic. Fortunately pandemics occur infrequently. During the 1919 pandemic outbreak of influenza, over half a million

Table 22.5 Typical Summaries Reported in MMWR for Cases of Notifiable Diseases in the United States

Summary—cases of specified notifiable diseases, United States

Disease	51st Week Ending			Cumulative, 51st Week Ending		
	Dec. 22, 1984	Dec. 24, 1983	Median 1979–1983	Dec. 22, 1984	Dec. 24, 1983	Median 1979–1983
Acquired Immunodeficiency Syndrome (AIDS)*	188	63	N	4,386	2,075	N
Aseptic meningitis	77	132	121	7,943	12,299	9,358
Encephalitis: Primary (arthropod-borne & unspec.)	31	18	20	1,164	1,796	1,500
Post-infectious	—	4	1	82	95	95
Gonorrhea: Civilian	15,649	14,741	15,859	822,089	883,944	978,492
Military	388	380	380	19,999	23,534	26,230
Hepatitis: Type A	385	393	453	21,070	20,915	25,034
Type B	512	524	419	25,592	23,654	20,466
Non A, Non B	72	63	N	3,664	3,355	N
Unspecified	82	151	179	5,324	7,102	10,345
Legionellosis	17	13	N	645	747	N
Leprosy	8	3	3	234	236	230
Malaria	9	5	9	955	774	1,019
Measles: Total**	5	8	21	2,531	1,450	2,957
Indigenous	5	7	N	2,239	1,146	N
Imported	—	1	N	292	305	N
Meningococcal infections: Total	44	41	60	2,609	2,641	2,660
Civilian	44	41	60	2,604	2,625	2,641
Military	—	—	—	5	16	16
Mumps	48	58	140	2,857	3,295	5,208
Pertussis	18	58	44	2,164	2,328	1,636
Rubella (German measles)	13	9	37	742	947	2,281
Syphilis (Primary & Secondary): Civilian	456	579	489	26,974	31,579	30,435
Military	2	7	3	277	382	360
Toxic Shock syndrome	11	5	N	457	421	N
Tuberculosis	445	432	503	21,197	23,095	26,660
Tularemia	3	4	4	287	293	255
Typhoid fever	7	5	8	365	451	495
Typhus fever, tick-borne (RMSF)	2	5	5	848	1,100	1,100
Rabies, animal	30	69	69	5,124	5,788	6,058

Notifiable diseases of low frequency, United States

	Cum. 1964		Cum. 1984
Anthrax	1	Plague	31
Botulism: Foodborne (Calif. 1)	20	Poliomyelitis: Total	4
Infant (Calif. 2)	92	Paralytic	4
Other	6	Psittacosis (N.C. 1, Tex. 1)	89
Brucellosis (Fla. 1)	120	Rabies, human	3
Cholera	1	Tetanus	64
Congenital rubella syndrome	4	Trichinosis	61
Diphtheria	1	Typhus fever, flea-borne (endemic, murine) (Tex. 1)	37
Leptospirosis (Mass. 1)	31		

*The 1983 reports which appear in this table were collected before AIDS became a notifiable condition.

**There were no cases of internationally imported measles reported for this week.

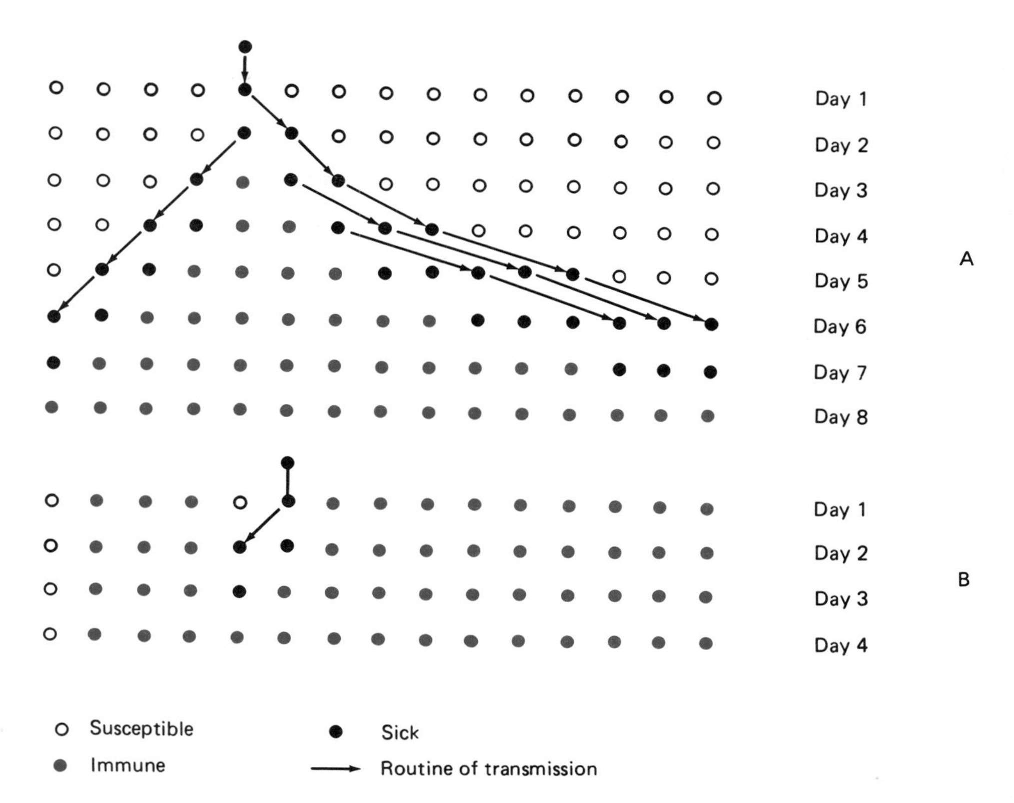

Figure 22.11 The kinetics of the spread of an infectious disease and the effect of increasing the number of immune individuals in the population in limiting epidemic outbreaks of disease. (A) The individuals in this population are all susceptible. The introduction of a sick individual initiates an epidemic outbreak of the disease. At the height of the epidemic 50 percent of the population is ill. Eventually, all individuals develop the disease and are subsequently immune. (B) If eighty percent of the population is immune to the disease when a sick individual enters the population, no epidemic occurs and even a susceptible individual fails to contact and contract the disease.

people died in the United States and 20 million people died worldwide.

Epidemics have often been more influential than statesmen and soldiers in shaping the course of political history and diseases may also color the moods of civilizations.

—René and Jean Dubos, *The White Plague*

Portals of Entry

Pathogenic microorganisms gain access to the body through a limited number of routes known as portals of entry. The routes of entry are the respiratory tract, gastointestinal tract, genitourinary tract, skin, wounds, and animal bites (Figure 22.12). Most pathogenic microbes will only cause disease if they enter the body via a specific route. For example, depositing *Clostridium tetani* on the intact skin surface does not result in disease, but deposition of *C. tetani* into deep wounds results in the deadly disease tetanus. The reasons that pathogens are restricted with respect to how they enter the body rests with the nonspecific and immune defenses associated with different body tissues and the inherent properties of the microorganism.

What geography is to history such is anatomy to medicine—it describes the theater.

—J. Fernel

The restrictive nature of the portals of entry also means that a sufficient number of microorganisms is necessary to initiate an infective process. The number of pathogens needed to establish a disease is known as the infectious dose (ID). For some pathogens the infectious dose may be as low as one organism, but for others hundreds of thousands of microbes may be necessary to overwhelm the host defenses and to allow the invading microorganisms to reproduce within the body. Various factors influence the infectious dose required to initiate a disease, including the nature of the pathogen, the portal of entry, and the state of the host defenses. In many cases diminished host defenses permit relatively low numbers of potential pathogens to establish an infection. Malnutrition, for example, results in lessened amounts of antimicrobial body fluids and inadequate host defenses to protect against infective microbes.

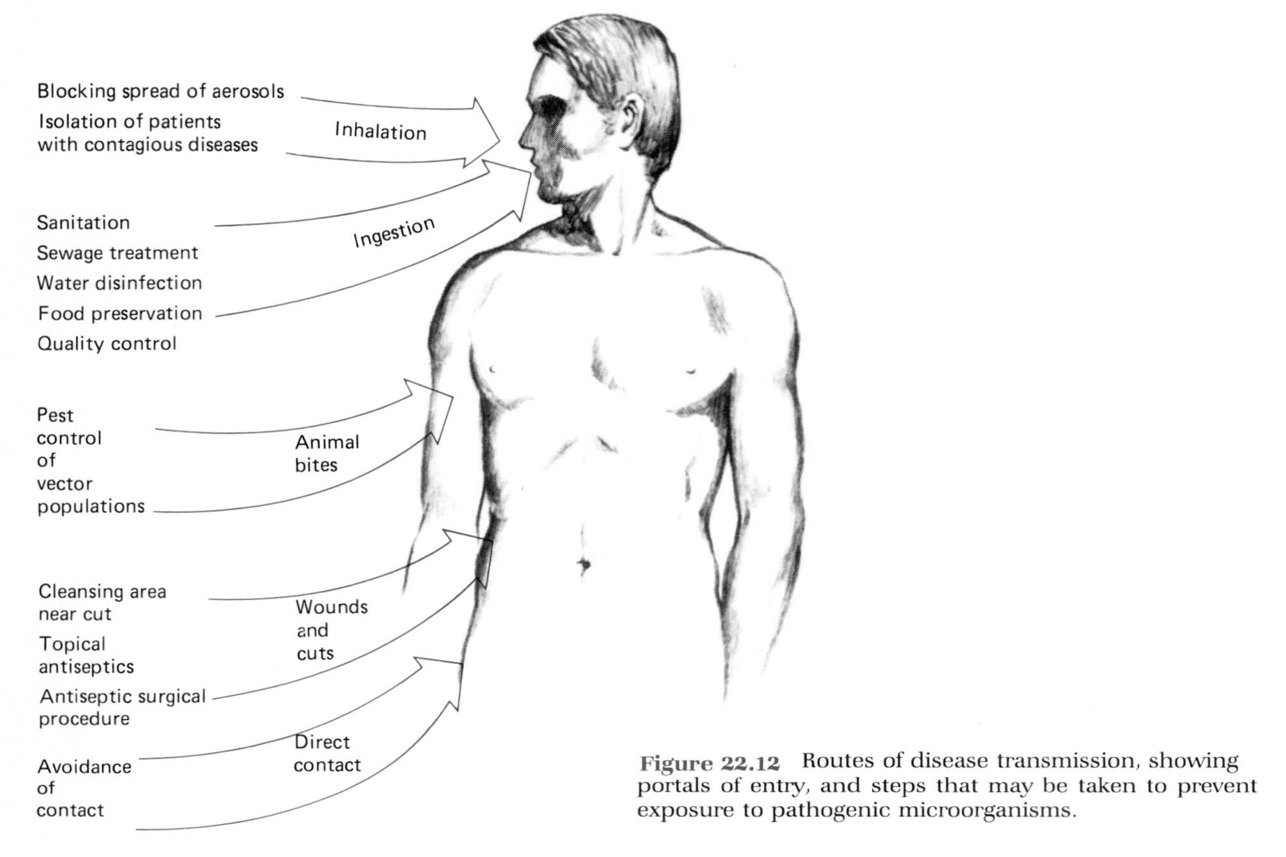

Figure 22.12 Routes of disease transmission, showing portals of entry, and steps that may be taken to prevent exposure to pathogenic microorganisms.

MICROBIOLOGY HEADLINES

Big Drop Foreseen in Infant Deaths

Unicef Says Third World Drive Starting to Yield Results

By JO THOMAS

LONDON, Dec. 19—Unicef reported today that its effort to revolutionize children's health care, begun two years ago, was starting to save lives in dramatic numbers.

It said there was now hope that over the next 10 to 15 years, infant death rates in many third world countries could fall by as much as 5 percent or more a year.

The report, which is issued annually, was made public in London today. It comes at a time when 15 million children in Africa, Asia, and South America are thought to die each year—about 40,000 a day—from malnutrition, measles, tetanus and other vaccine-preventable diseases because of a lack of basic health services.

'The Silent Emergency'

"The loud emergency—the Ethiopia famine—hits the news, but the 'silent emergency' takes the great majority of these 15 million small children's lives each year," James P. Grant, executive director of Unicef, the United Nations Children's Fund, said at a news conference today.

The worldwide cost of putting into effect the immunization and health education techniques for what Unicef calls a "child survival revolution" is $1 billion, Mr. Grant said. "Basically," he said, "what we're talking about is a billion dollar bargain."

The report, titled "The State of the World's Children 1985," noted that Unicef began a campaign in 1982 aimed at enabling parents themselves to cut the child death rate in half by using four simple, inexpensive techniques.

These include breast-feeding for at least the first six months of life, a simple 10-cent growth chart that enables a mother to keep track of a child's weight and detect malnutrition and a full course of immunization that would, according to Unicef, cost $5 a child.

Only 20 percent of the developing world's children are now immunized, the report said. Preventable diseases kill five million each year and leave five million others with serious disabilities, it said.

The fourth technique, called Oral Rehydration Therapy, uses a solution of salt, sugar and water to replace vital fluids lost through diarrhea. The solution can be made cheaply at home, using ordinary household ingredients. The packet of salts used in the preparation costs 10 cents.

The report estimated that this simple treatment for dehydration caused by diarrhea—the biggest single killer of children—has saved the lives of 500,000 children this year.

An Optimistic Forecast

"The child survival revolution is no longer a theory," the report said. "Many thousands of children's lives are being saved. And there is now a realistic basis for hope that, over the next 10 to 15 years, infant death rates will fall by as much as 5 percent or more a year in countries such as Tanzania, Nigeria, Algeria, Turkey, India, Pakistan, Indonesia, Haiti, Nicaragua, Brazil and Colombia."

The Unicef report said dramatic efforts to inaugurate these four health measures had been seen this year in some of the countries that need them most.

In India, which has more children than all the nations of Africa together and where one child in seven dies before reaching the age of 5, successful local child immunization campaigns have led several state governments to move to immunize all children, it said.

Unicef's regional office in New Delhi, the report said, estimates that if all state governments adopt such a policy, India can achieve its goal of vaccinating 85 percent of all infants by 1990.

In Pakistan, where 500,000 children die each year from diarrheal dehydration and preventable diseases, a new accelerated health program has lifted the immunization rate from 5 percent to almost 50 percent, produced 30 million packets of oral rehydration salts and trained more than 12,000 traditional birth attendants in these low-cost techniques.

In Colombia, more than 800,000 young children were immunized on each of three National Vaccination Days this year, the report said, while in Baguio, the Philippines, a campaign centering on the recommended techniques has helped reduce infant and child death rates by 50 percent in five years.

In Haiti, where half of all child deaths are from diarrhea, the number of mothers using oral rehydration rose from 2 percent to 34 percent after a six-month information campaign, much of it by radio.

Deaths from "Controlled" Diseases

Estimated number of children who died in 1983 from some diseases for which immunization was available.

	Neonatal Tetanus	Measles	Whooping Cough	TOTAL
India	298,000	782,000	189,000	1,269,000
Pakistan	132,000	163,000	66,000	361,000
Bangladesh	119,000	173,000	69,000	361,000
Indonesia	71,000	218,000	63,000	352,000
Nigeria	64,000	171,000	68,000	303,000
Mexico	31,000	57,000	19,000	107,000
Ethiopia	16,000	60,000	25,000	101,000
Zaire	21,000	45,000	19,000	85,000
Philippines	12,000	59,000	12,000	83,000
Brazil	28,000	34,000	18,000	80,000
Burma	20,000	43,000	16,000	79,000
Thailand	10,000	57,000	11,000	78,000
Vietnam	12,000	46,000	19,000	77,000
Kenya	9,000	37,000	15,000	61,000
Egypt	16,000	32,000	13,000	61,000
South Africa	11,000	35,000	14,000	60,000
Sudan	8,000	36,000	15,000	59,000
Afghanistan	11,000	27,000	11,000	49,000
Iran	17,000	19,000	9,000	45,000
Algeria	10,000	25,000	8,000	43,000
Morocco	10,000	21,000	5,000	36,000
Turkey	8,000	16,000	5,000	29,000
Colombia	9,000	14,000	6,000	29,000
Tanzania	6,000	7,000	6,000	19,000
Korea	5,000	10,000	2,000	17,000
Other developing countries	181,000	411,000	139,000	731,000
TOTAL	**1,135,000**	**2,598,000**	**842,000**	**4,575,000**

Source: World Health Organization

In Brazil, a program using more than 400,000 volunteers this year achieved immunization of two million infants and small children against measles, a million and a half against diphtheria, whooping cough and tetanus and almost all the nation's children against polio.

In Nigeria, where 800,000 children die each year, a national vaccination campaign has been begun in an effort to repeat the success of an effort in one locality that used improved refrigeration techniques to preserve vaccines and drew on community leaders, teachers, the churches and sound trucks that went from village to village. The percentage of children immunized rose from 9 percent last year to more than 80 percent.

In Turkey, where the low-cost techniques reduced infant deaths in Van Province by 65 percent in four years, a five-year nationwide program will begin in 1985.

Routes of Disease Transmission

The transmission of disease involves the movement of pathogens from a source to the appropriate portal of entry (Figure 22.13). The source of an infectious agent is known as the reservoir. In some cases, the reservoirs of human pathogens are nonliving sources such as soil and water. For example, tetanus is generally acquired when spores of *Clostridium tetani*, which are widely distributed in soil, contaminate a wound. Often diseases acquired from such sources are noncommunicable; that is, they are singular events and are not normally transmitted from one infected individual to the next. For example, health care workers treating a patient with tetanus are not at any greater risk of contracting this disease than is the rest of the poulation.

Although water and soil are sometimes the reservoirs of human pathogens, they are more frequently only indirectly involved in the transmission of pathogens from an infected individual to the next suceptible host. Humans are the principal reservoirs for microorganisms that cause human diseases. People infected with a pathogen act as a source of cantagion for others. The term contagious disease (infectious disease) indicates that a pathogen will move with ease from one infected individual to the next. People who come in contact with someone suffering from a contagious disease are at risk of contracting that disease unless they are immune.

Figure 22.13 Routes of disease transmission.

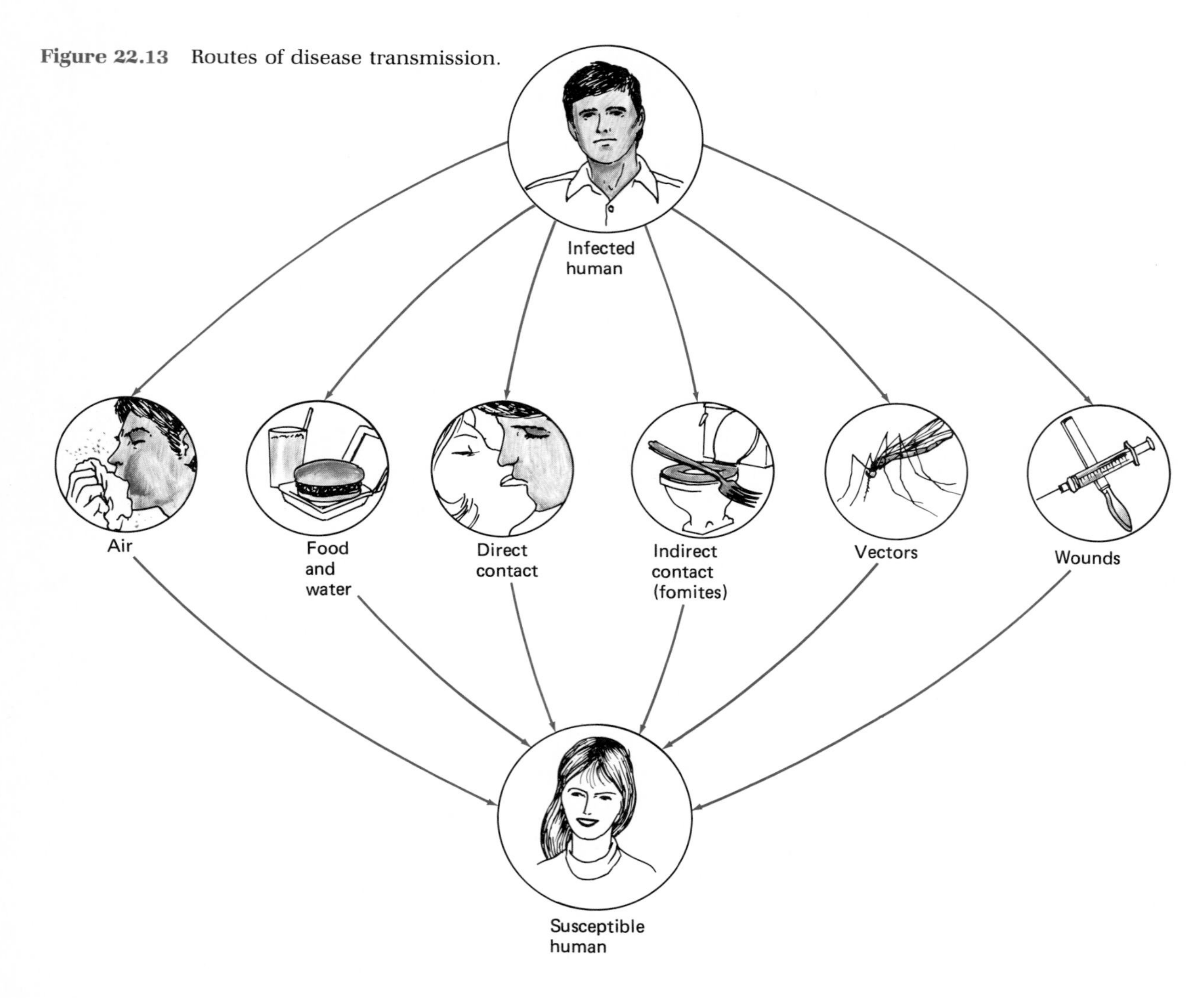

Contagion has this illness widely spread;
And, I feel sure, will further spread it yet;
As in the fields one scabby sheep the flock
Destroys, and one infected pig the stock.

—JUVENAL, *Satires*

In some cases infected individuals do not develop disease symptomology. Such individuals are called aymptomatic carriers or are simpley referred to as carriers. Although they do not become sick themselves, carriers are an important reservoir of infectious agents. The classic case of disease transmission by such a carrier occurred in the early 1900s when a cook, Mary Mallon, spread typhoid fever from one community to another. In 10 years Mary worked in seven different households in which 28 cases of typhoid fever occurred. She would quit and disappear each time there was a case of typhoid in the household of her employer. When she was finally tracked down in 1907, she was diagnosed as a carrier of *Salmonella typhi*. She refused to undergo the surgical removal of infected tissues and was imprisoned. She was released in 1910 due to public outcry protesting her confinement, only to continue spreading typhoid, including to a sanitorium in New Jersey and a hospital in New York. In 1915, she was recaptured and forcibly hospitalized for the rest of her life. Typhoid Mary, as she became known, died in 1938 at the age of 70.

Humans, both carriers and those showing overt signs of disease, are not the only sources of infectious agents. In some cases nonhuman animal populations are the source. Diseases that primarily affect wild and domestic animals are known as zoonoses. Some zoonoses can be transmitted to humans by direct contact with infected animals, by ingesting contaminated meat, or more frequently by vectors. Vectors are carriers of disease agents. The vector need not develop disease; it need only transmit it from a reservoir to a susceptible individual. Arthropods, such as mosquitoes, are frequently the vectors of human disease.

Because the transmission of pathogens occurs via restricted routes, it is possible to control our interactions with microbial populations in ways that reduce the probability of contracting infectious diseases. The methods employed for preventing exposure to specific disease-causing microorganisms vary depending on the particular route of transmission (see Figure 22.12). Many modern sanitary practices are aimed at reducing the incidence of diseases by preventing the spread of pathogenic microorganisms or by reducing their numbers to concentrations that are insufficient to cause disease. Mosquito control, rodent control, sanitary waste disposal, sewage treatment, chlorination of water supplies, pasteurization, and various other methods are used to restrict the spread of pathogens. The greatly diminished incidence of many diseases caused by microorganisms is the consequence of an adequate understanding of the modes of transmission of pathogenic microorganisms and preventive measures that reduce exposure to disease-causing microorganisms.

Respiratory Tract and Airborne Transmission

Many microorganisms are carried through the air. Some survive for long periods of time, even crossing oceans; others survive in the air for more limited time periods. Many pathogens are transmitted from one host to the next through aerosols (Table 22.6). When an individual with a respiratory tract infection coughs and sneezes, they generally propel saliva droplets into the air containing large numbers of infectious microorganisms. Rapidly evaporating water bodies, such as air conditioner cooling towers, also can carry large numbers of potential pathogens into the air as aerosols. Inhalation of aerosols containing pathogens is a major route of disease transmission when the respiratory tract is the portal of entry.

Several steps can be taken to reduce the likelihood of exposure to airborne pathogenic microorganisms. Covering one's nose and mouth with a

Table 22.6 Causative Agents of Representative Airborne Diseases

Disease	Causative Agent
Diphtheria	*Corynebacterium diphtheriae*
Bacterial meningitis	*Neisseria meningitidis*
Pneumococcal pneumonia	*Streptococcus pneumoniae*
Tuberculosis	*Mycobacterium tuberculosis*
Whooping cough (pertussis)	*Bordetella pertussis*
Influenza	Myxovirus
Measles (rubeola)	Paramyxovirus
German measles (rubella)	Togavirus
Chicken pox (varicella)	Zoster (herpesvirus)
Smallpox (variola)	Poxvirus
Coccidiodomycosis (primary infection)	*Coccidiodes immitis* (fungus)
Histoplasmosis	*Histoplasma capsulatum* (fungus)

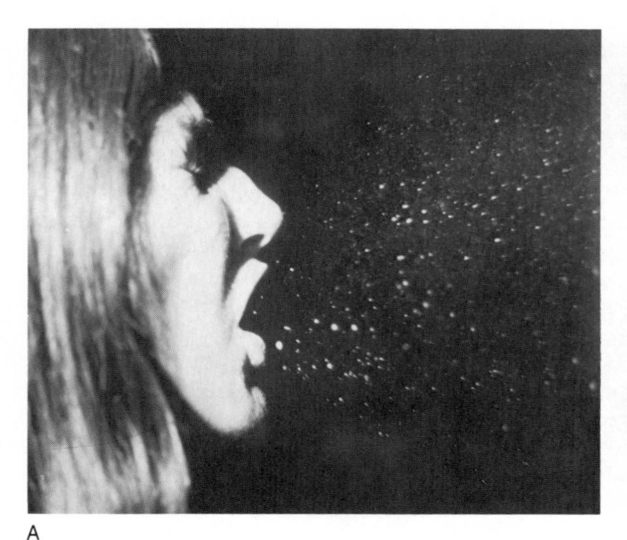

A

B

Figure 22.14 These photographs show droplet spread by sneezing and the effect of covering one's nose and mouth with a handkerchief when sneezing. (A) An unstifled sneeze; (B) the use of a handkerchief blocks the spread of droplets.

handkerchief when sneezing and coughing reduces the number of potential pathogens that may become airborne in aerosols and that therefore may be transmitted from an infected individual to a susceptible host (Figure 22.14). Recently handkerchiefs have been developed that contain antimicrobial agents; the use of such treated handkerchiefs can reduce the airborne spread of pathogens.

The use of surgical masks when visiting individuals who are particularly susceptible to airborne pathogenic microorganisms, such as newborn infants, is an important precaution for preventing the spread of infectious diseases. The use of ultraviolet lights and air filtration devices, including electrostatic filters, is employed in some facilities to limit the spread of airborne pathogens. Preventing the exposure of individuals whose immunological defense mechanisms are compromised as a result of a variety of conditions—such as treatment for cancer or a recent organ transplant—to airborne pathogenic microorganisms is an important aspect of patient management practice. Similarly, masks should be worn in the presence of patients with tuberculosis or other pulmonary diseases or else other precautions, such as maintaining a safe distance from the infected individual, should be taken to minimize exposure to the pathogenic microorganisms. Isolation of individuals with contagious microbial diseases, in which the infectious agent is airborne, is often practiced. For example, children with measles, chicken pox, or mumps are often kept away (isolated) from other children who are not immune to these diseases. Such practices decrease the probability of exposure to pathogenic organisms and prevent, or at least reduce, the transmission of disease.

Gastrointestinal Tract and Water and Foodborne Transmission

Contaminated food or water are most often associated with disease transmission when the portal of entry is the gastrointestinal tract (Table 22.7). Often the source of contamination can be traced to fecal contamination. High concentrations of pathogenic microorganisms are shed in feces from individuals infected with some diseases, such as cholera. If such pathogen-containing matter contaminates food or water supplies, the disease can easily be transmitted to other susceptible individuals. This is often referred to as the fecal–oral route of transmission.

Proper sewage treatment and drinking water disinfection programs reduce the likelihood of contracting a disease through contaminated water. Recognition of the fact that many serious diseases,

Table 22.7 Causative Agents of Representative Water- and Food-borne Diseases

Bacillary dysentery (shigellosis)	*Shigella* species
Brucellosis (undulant fever)	*Brucella*
Cholera	*Vibrio cholerae*
Salmonellosis	*Salmonella* species
Typhoid fever	*Salmonella typhi*
Infectious hepatitis	Hepatitis A virus
Poliomyelitis	Polio virus
Trichinosis	*Trichinella spiralis* (helminth)

such as typhoid, are transmitted through water contaminated with fecal material is the basis for strict water quality control. Chlorination of municipal water supplies is widely used to prevent exposure to the pathogenic microorganisms that occur in water supplies and thus to ensure the safety of drinking water.

Quality control measures are also applied throughout the food industry to prevent the transmission of disease-causing microorganisms through food products. A number of methods are used to preserve food products, preventing the growth of microorganisms that spoil food and may also cause human disease. Pasteurization of milk is a good example of a process designed to reduce exposure to pathogenic microorganisms that occur and proliferate in untreated milk. The aim of washing one's hands before eating is to avoid the accidental contamination of one's food with soil or other substances that may harbor populations of disease-causing microorganisms.

A dirty cook gives diarrhea quicker than rhubarb.

—TUNG-SU PA

Failure to maintain quality control of water and food supplies often results in outbreaks of disease; for example, cholera outbreaks often occur when sewage contaminates drinking water supplies, such as frequently occurs in the Far East when monsoon rains cause flooding. Outbreaks of botulism are associated with improperly canned food products, when food products have not been heated long enough to kill contaminating spores of *Clostridium botulinum*; bacterial growth occurs in the anaerobic conditions within the canned food, resulting in the exposure of the individuals who ingest it to the lethal toxins produced by this bacterium. Extensive quality control testing is required in most countries to prevent the outbreaks of diseases associated with contaminated water and food supplies.

Genitourinary Tract and Sexual Transmission

The genitourinary tract is the portal of entry for only a limited number of pathogenic microorganisms. Most pathogens that enter the body by this route are transmitted only by direct sexual contact. Avoiding sexual contact with infected individuals and exercising appropriate prophylactic measures is essential to control the transmission of these diseases. Major public health programs in various countries are aimed at raising public awareness about these diseases and the means that can be used to prevent their transmission. Avoiding sexual contact, though, with individuals suffering from venereal diseases, such as syphilis and gonorrhea, interrupts the transmission of the pathogens that cause these diseases; such avoidance of contact with infected individuals is essential and will undoubtedly remain the main method for controlling the spread of sexually transmitted diseases.

Skin and Direct Contact

The skin is normally an effective barrier to most microorganisms. Some organisms, such as the parasitic worms that cause shistosomiasis, however, can penetrate through the skin. This is why entering a stream containing worm-infested snails is so dangerous. Similarly, some bacteria and fungi can establish infections when they are deposited directly on the skin. In cases where the pathogen is transmitted in this manner, it is important to avoid direct contact with infected individuals to prevent the spread of disease. Historically, the isolation of leprosy patients in remote colonies is an example of the extreme steps that have been taken to prevent contact of sick individuals with the general population. Today, this extreme practice is not needed to prevent the spread of leprosy because of the use of antimicrobial agents. Much concern has surfaced about avoiding contact with children who have congenital herpes because of possible transmission of herpes viruses by direct contact. In some communities children with congenital herpes are not permitted in the same classroom with other children; in other school districts children with congenital herpes are checked daily for the presence of lesions that could act as a source of infection and are only permitted in the classroom during periods of remission of the disease.

Wounds, Bites, and the Parenteral Route

When the intact skin is broken, pathogens can enter deep body tissues and the circulatory system. This route of entry is called the parenteral route. Breaks in the skin can occur as a results of cuts, wounds, punctures, surgical incisions, injections, and animal bites. A variety of procedures are employed to reduce the probability of exposure to pathogenic microorganisms that enter the body through breaks in the skin surface. For example, wounds are cleansed to prevent the introduction of foreign material that may harbor potential pathogens, and antiseptics are applied to skin surfaces to minimize the entry of pathogenic microorganisms into tissues normally protected by an intact skin covering.

Practices are normally employed to control in-

Table 22.8 Representative Diseases of Humans Transmitted by Arthropod Bites

Disease	Etiologic Agent	Biological Vector	Reservoir
Yellow fever	Yellow fever virus	Mosquito (*Aedes aegypti, Haemagogus* spp.)	Humans, monkeys
Dengue fever	Dengue fever virus	Mosquito (*Aedes* spp., *Armigeres obturbans*)	Humans
Eastern equine encephalitis	Encephalitis viruses	Mosquito (*Aedes* spp., *Culex* spp., *Mansonia titillans*)	Humans, horses, birds
Colorado tick fever	Colorado tick fever virus	Wood ticks (*Dermacentor andersoni*)	Golden mantle ground squirrel
Plague	*Yersinia pestis*	Rodent fleas (*Xenopsylla cheopis*); human fleas (*Pulex irritans*)	Rodents (rats)
Tularemia	*Francisella tularensis*	Ticks (*Dermacentor* spp., *Amblyomma* spp.); deerflies (*Chrysops discalis*)	Rodents, ticks
Rocky Mountain spotted fever	*Rickettsia rickettsii*	Ticks (*Dermacentor* spp., *Amblyomma* spp., *Ornithodoros* spp., etc.)	Rodents
Endemic typhus fever	*Rickettsia typhi*	Fleas (*Xenopsylla cheopis* and others)	Humans
Relapsing fever	*Borrelia recurrentis* and other species	Body louse (*Pediculus humanus*)	Humans, ticks
Chagas' disease	*Trypanosoma cruzi*	Cone-nosed bugs (*Triatoma* spp., *Panstronglyus* spp., *Rhodnius* spp.)	Dogs, cats, opossums, rats, armadillos
African trypanosomiasis (sleeping sickness)	*Trypanosoma gambiense; T. rhodesiense*	Tsetse flies (*Glossina* spp.)	Humans, wild mammals
Malaria	*Plasmodium vivax, P. malariae, P. falciparum, P. ovale*	Mosquito (*Anopholes* spp.)	Humans
Leishmaniasis	*Leishmania donovani; L. tropica; L. braziliensis*	Sandflies (*Phlebotomus* spp.)	Dogs, foxes, rats, mice, two-toed sloth, gerbils, man

sect and other animal populations that act as vectors (carriers) for the transmission of diseases caused by pathogenic microorganisms. The most notable vectors of pathogenic microorganisms are mosquitoes, lice, ticks, and fleas (Table 22.8). Some public health measures, such as mosquito control programs, are aimed at reducing the sizes of these vector populations and thus lowering the probability of exposure to the pathogenic microorganisms capable of causing diseases such as plague, typhus fever, yellow fever, malaria, and a variety of other diseases transmitted by animal vectors.

Hospital-Acquired (Nosocomial) Infections

The hospital represents a unique environment that requires special epidemiological considerations. Some infections are acquired in hospitals. These are called nosocomial infections. Factors that

MICROBIOLOGY HEADLINES

Four Ways to Fight Hospital Infection

By DAN SPERLING

NEW YORK—Almost 2 million Americans a year catch hospital-caused infections—costing us $2.5 billion—but the number can be drastically reduced if hospitals follow a few simple procedures, according to a new study.

The infections—known as nosocomial infections—are caused by surgery, intravenous feeding, catheters and certain drugs, says Dr. Robert Haley, associate professor at the University of Texas Southwestern Medical School in Dallas. They cause an estimated 20,000 deaths a year and are linked to 60,000 more.

Haley's 10-year study found that about 5 percent of all patients got such infections.

But those hospitals taking infection-control measures decreased this rate by an average of 32 percent.

The four keys to cutting hospital-caused infection:

- Surveillance programs to detect infections and report them to hospital staff.
- Changing patient care to prevent infections.
- Full-time nurses trained in infection control assigned for every 250 beds.
- A doctor trained in infection control in the program.

"Over the last 20 years, hospital medicine has become increasingly technological, with more invasive devices like catheters and respirators and more drugs that impair the body's ability to fight off infection, and also we're keeping people alive longer. So a hospital today has an increasingly large number of people who are more susceptible to infection," says Haley, who presented his findings in New York City this week at a symposium on infectious diseases sponsored by the National Institute of Allergies and Infectious Diseases and Hoffman-LaRoche.

Source: USA Today, Dec. 13, 1984. Reprinted by permission.

influence the spread of disease within a hospital include the lowered resistances of patients, the presence of concentrated reservoirs of infectious agents, and the movement of personnel and materials among areas acting as reservoirs and areas where individuals are susceptible to infection (Figure 22.15).

> In one day: Seven thousand hospital patients become infected with an illness other than the one that sent them to the hospital.
>
> —TOM PARKER, *In One Day*

Hospital health care personnel are important sources of infectious agents because their bodies and clothing may transport pathogens as they move about the hospital. Medical asepsis is employed in an effort to prevent the transmission of infection. Medical asepis refers to those techniques used to reduce transmission of pathogenic microorganisms by reducing their number and by hindering their transfer from one place to another.

Correct and thorough hand washing is one of the most important measures in controlling the spread of infection in a hospital. Isolation gowns are worn by those members of the hospital staff likely to come into contact with contaminated material. Some hospitals require the use of masks when dealing with infectious patients. It is important that the masks be dry and fit well, as a wet mask or one that fits loosely will not block the passage of microorganisms.

Also, the equipment and materials used in the course of their professional activities may be involved in the transmission of nosocomial diseases. Floors, walls, and bedding must frequently be disinfected. The employment of disposable equipment and supplies, followed by the correct mode of discarding used materials is critical in order to control of the spread of hospital infections. All contaminated material should be removed by the double-bagging technique, whereby the material is placed in a paper or plastic bag which is then passed into another clean bag held outside the contaminated room. Terminal disinfection, sterilizing or disinfecting all possibly contaminated material and equipment, is performed when a patient has recovered, been transferred, or died. For equipment that is reused, chemical disinfection or sterilization is essential. For example, before the introduction of disposable electronic thermometer probes, it was critical to disinfect mercury thermometers with alcohol between use by different patients or to use the thermometer for only one patient.

Additionally it is important that patients who are particularly susceptible to infection not be exposed to pathogenic microorganism. Special precautions are often taken to protect these patients with compromised immune systems by increasing the control over the microbial content of their environment. This category of patients can include premature infants, organ transplant patients, patients undergoing radiation therapy, and severe burn

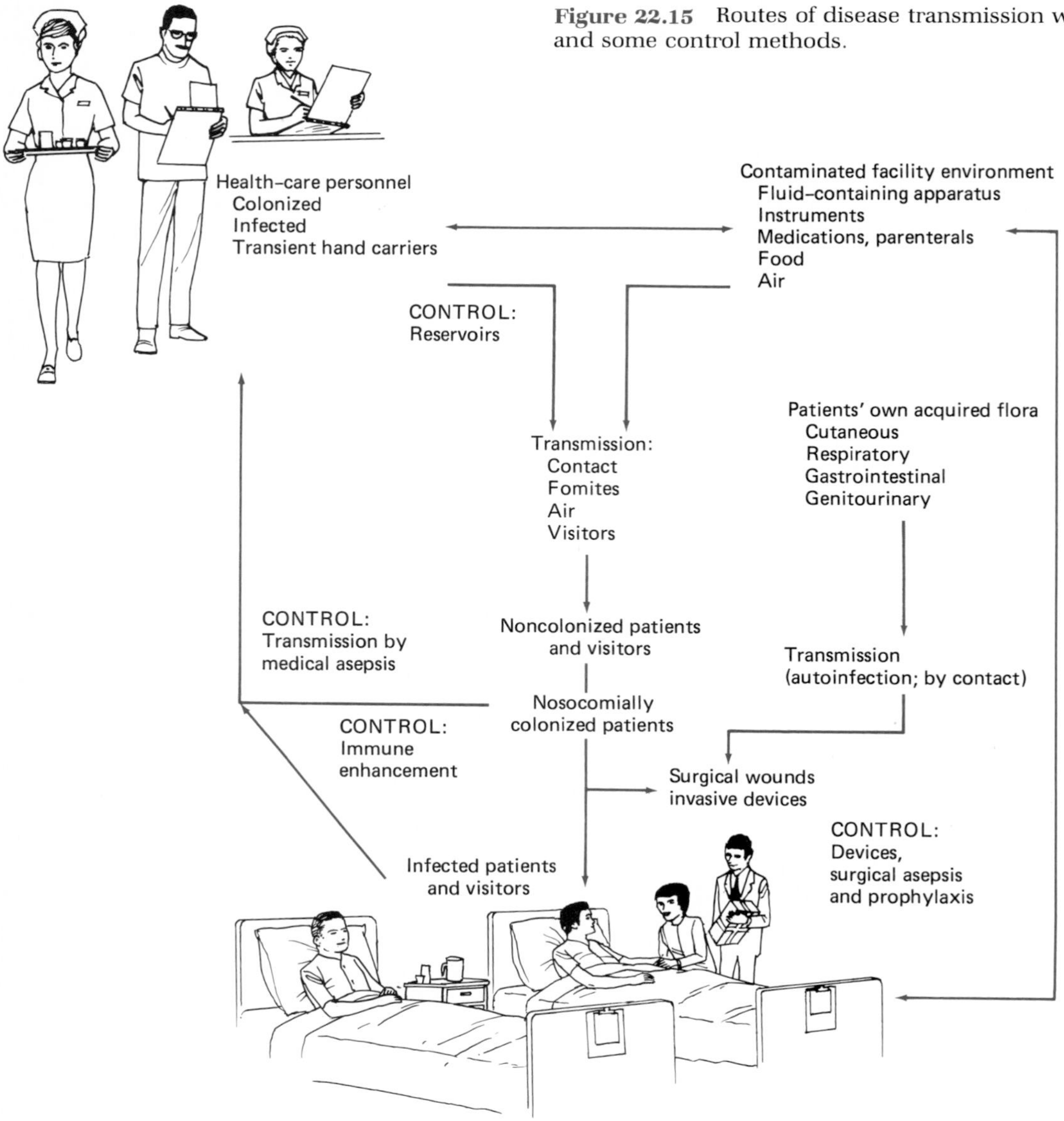

Figure 22.15 Routes of disease transmission within hospitals and some control methods.

victims. In some cases this requires great restrictions on visitors, the required use of masks and gowns, special air filters, and even complete isolation. Few visitors, for example, are permitted to come in contact with newborns. Isolation of unusually susceptible patients protects them from exposure to disease agents.

Great care also is taken in burn wards and during surgical procedures to prevent contamination by accidental introduction of microorganisms into the exposed tissues. Sterile operating rooms, instruments, garments, gloves, and masks are all used by a hospital's surgical staff (Figure 22.16). Surgical asepsis refers to those practices that make and keep objects and areas that come into contact with susceptible patients sterile.

When patients have infectious diseases, it is necessary to prevent their contact with other patients and also to limit the number of staff involved in their treatment. Isolation of an infectious patient prevents the spread of a communicable disease to others. This is particularly important when treating

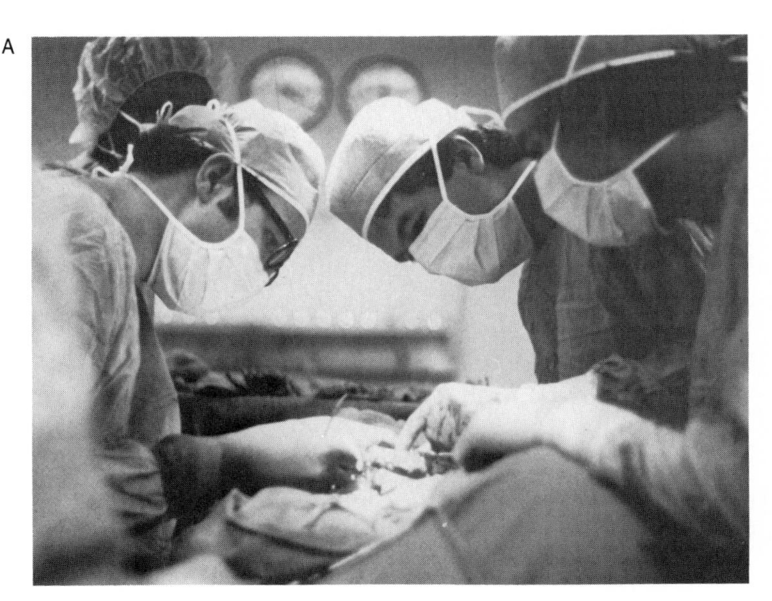

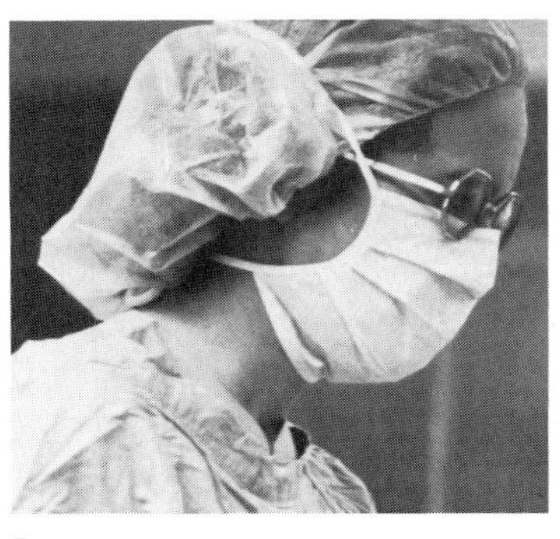

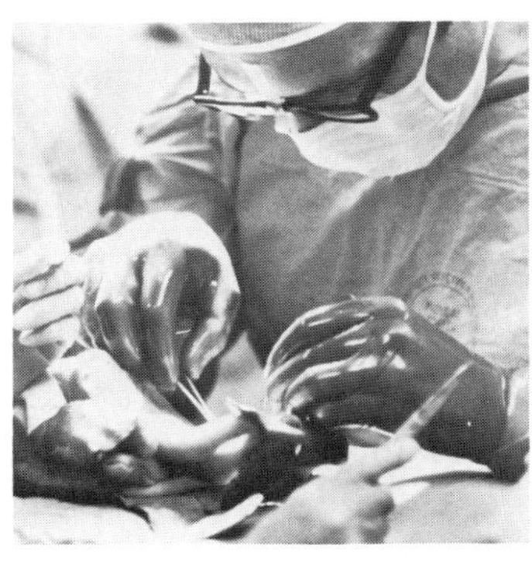

Figure 22.16 Operating room surgical staff take many precautions to preclude microbial contamination of tissue exposed during the surgical procedure. (A) Everyone is masked, gloved, capped, and gowned. (B) Note how much of the face is covered by the mask. (C) Note that the parts of the patient's body not involved in the operation are also covered. (Courtesy Michelle Ising, University of Louisville.)

a patient with a serious disease when the pathogens may be transmitted through the air or by direct contact.

Another epidemiological problem for hospitals is the widespread use of antibiotics. The use of antibiotics is essential for the control of the diseases that are treated in hospitals but extensive use of antibiotics leads to the evolution of antibiotic resistant microorganisms within hospitals. Excessive use of antibiotics is a serious problem because noscomial infections caused by antibiotic resistant microorganisms are particularly difficult to treat, especially if the patient is in a debilitated state owing to another disease condition.

As a result of these special considerations hospitals employ resident epidemiologists and use standardized procedures to help control the spread of disease.

Summary

The interactions between microorganisms and human beings represent a delicate balance between biological populations. The normal microbiota of humans play a role in maintaining a "healthy" condition, but various microorganisms can act as pathogens causing disease. The normal microbiota of human beings are usually indigenous to the surfaces of body tissue, most notably the oral cavity, gastrointestinal tract, vaginal tract, and skin surface. These resident microorganisms are generally noninvasive. The environmental conditions of particular body tissues and the physiological properties of particular microbial populations determine the species composition of the resident microbial community of a given body tissue. In contrast to the normal human microbiota, pathogenic microorganisms that cause human disease generally are either invasive, toxigenic, or both. There are a variety of factors that confer virulence on microbial populations, including the ability to produce toxins, which disrupt the normal functioning of mammalian cells; invasiveness factors, such as enzymes that permit the microorganisms to spread through body tissues; and adaptations, such as capsules, that permit the pathogenic microorganisms to evade the normal host defense mechanisms.

Study Outline

A. Infectious diseases, those which are caused by microorganisms, probably account for less than half of all human illness.

  1. The process for establishing the etiology of infectious diseases is called Koch's postulates. Koch's four postulates are

    a. The organism should be present in all animals suffering from the disease and absent from all healthy animals.

    b. The organisms must be grown in pure culture outside the diseased animal host.

    c. When such a culture is inoculated into a healthy susceptible host, the animal must develop the symptoms of disease.

    d. The organism must be reisolated from the experimentally infected animal and shown to be identical with the original isolate.

B. Virulence factors are properties, possessed by some disease-causing microorganisms, that enhance their pathogenicity and allow them to invade human tissues and disrupt normal bodily functions.

  1. Invasiveness is the ability of a microorganism to invade human tissues, attach to cells, and multiply within cells or tissues.

    a. Various pathogens produce enzymes that destroy body tissues, enhance the pathogen's ability to proliferate within body tissue, and interfere with the host's normal defense mechanisms. Hyaluronidase breaks down the substance that holds together the cells of connective tissues, and since pathogens that produce it spread through body tissues, it is called the "spreading factor;" fibrinolytic enzymes prevent fibrin in the host from walling off the area of bacterial infection, allowing the pathogens to spread.

    b. The surface layers, such as capsules, surrounding a pathogenic microorganism interfere with the activity of phagocytic cells.

    c. Specific factors that enhance the ability of a microorganism to attach to the surfaces of mammalian cells are called adhesins; examples are pili and viral spikes.

  2. Toxigenicity is the ability of a microorganism to produce biochemicals, toxins, that disrupt the normal functions of cells. They can destroy human cells and tissues and can cause death.

    a. Lipopolysaccharide toxins (exotoxins) come from the lipopolysaccharide portion of the cell wall of dead and disintegrating Gram negative bacteria; they produce fever, circulatory changes, and other general symptoms.

    b. Protein toxins (exotoxins) are specific to the microorganism producing the toxin. They are produced by both Gram positive and negative bacteria, can be inactivated by heat, and cause distinctive clinical symptoms.

    c. Neurotoxins bind to nerve synapses causing the loss of motor functions.

    d. Enterotoxins cause an inflammation of the gastrointestinal tract.

    e. Cytotoxins cause lysis of blood cells and/or interfere with protein synthesis. Hemolysins cause the lysis of human erythrocytes; their action results in the release of hemoglobin. Hemolytic action produces zones of clearing around a bacterial colony growing on blood agar; complete clearing is called beta hemolysis and partial clearing is alpha hemolysis.

  3. The observable changes in the appearance of cells infected with viruses are called cytopathic effects.

C. Pathogenic microorganisms produce specific diseases that are associated with characteristic signs, objective changes, and symptoms, subjective changes, in body functions. A disease syndrome is a characteristic group of signs and symptoms.

  1. In acute diseases the symptoms and signs develop rapidly, reach a height of intensity, and end quickly. In chronic diseases the symptoms persist for a long time.

  2. The course of an infectious disease is divided into stages:

    a. The incubation period occurs after the pathogen has entered the body and before any signs or symptoms appear.

    b. The prodromal period begins with the onset of symptoms, the patient is now contagious and the nonspecific inflammatory defense system becomes operative.

    c. The period of illness, the acute phase, occurs next; the characteristic signs and symptoms occur, and the immune defense system is operative. The acute phase progresses toward death or convalescence.

    d. Recovery depends upon whether the immune systems and/or medical treatments are adequate. Signs and symptoms begin to disappear during the period of decline and convalesence progresses to either a carrier stage or freedom from the disease.

D. The likelihood of disease transmission depends upon the concentration and virulence of the pathogen, the distribution of susceptible individuals, and the potential sources of exposure to the pathogenic microorganisms. These factors are studied by the epidemiologist in order to evaluate how a disease outbreak in a population can be effectively controlled.

  1. Epidemiology is based on the statistical probability that exposure of a susceptible individual to a particular pathogen will result in disease transmission. By determining where and what people have eaten, where they have been, and with whom they have been in contact, the epidemiologist can establish a pattern of disease transmission.

    a. A common source outbreak is characterized by a sharp rise and rapid decline. A person-to-person epidemic is characterized by a relatively slow rise and decline.

    b. Morbidity rate is the incidence of a disease. Mortality rate is the number of deaths resulting from that disease.

    c. Endemic diseases occur at relatively low rates within certain geographic areas. Epidemics are outbreaks of disease in which unusually high numbers of individuals within a population contract the disease.

E. Pathogenic microorganisms gain access to the body through a limited number of routes known as portals of entry. Most pathogens will cause disease only if they enter the body via a specific route because of the nonspecific and immune defenses associated with different body tissues and the inherent properties of the microorganisms.

  1. The number of pathogens needed to establish a disease is the infectious dose, which is determined by the nature of the pathogen, the portal of entry, and the state of the host defenses.

F. The transmission of disease involves the movement of pathogens from a source, the reservoir, to the appropriate portal of entry.

  1. A contagious disease is one in which the pathogen will move with ease from one host to another.

  2. Humans are the principal reservoirs for microorganisms that cause human disease.

    a. Carriers are infected individuals who do not themselves develop disease symptomology.

  3. Zoonoses are diseases that primarily affect wild and domestic animals. They can be transmitted to humans by direct contact with infected animals, by ingesting contaminated meat, or by vectors.

    a. Vectors are carriers of disease agents.

  4. Because the transmission of pathogens occurs via restricted routes, it is possible to control our interactions with microbial populations in ways that reduce the probability of contracting infectious diseases.

5. Airborne microorganisms can survive for a long time and can travel long distances. They enter human hosts through the respiratory tract via aerosols and droplets spread by coughs and sneezes.
   a. To reduce the likelihood of exposure to airborne pathogenic microorganisms, we use handkerchiefs, masks, and isolation.
6. Contaminated food or water are most often associated with disease transmission when the portal of entry is the gastrointestinal tract.
   a. Often the source is fecal contamination, and proper sewage treatment and drinking water disinfection programs reduce the likelihood of disease.
   b. Quality control measures in the food industry prevent the transmission of disease-causing microorganisms through food products, for example, pasteurization.
7. Most pathogens that enter the body via the genitourinary tract are transmitted by direct sexual contact.
   a. Avoiding sexual contact with infected individuals and exercising appropriate prophylactic measures controls the transmission of these diseases.
8. Some microbes can establish infections when they are deposited directly on the skin.
   a. Avoiding direct contact with infected individuals prevents the spread of these diseases.
9. The parenteral route is used by pathogens infecting deep body tissues and the circulatory system when the intact skin is broken as a result of cuts, wounds, animal bites, punctures, surgical incisions, and injections.
   a. To prevent these types of infections surgical rooms, instruments, garments, gloves, and masks are kept sterile and wounds are cleansed and have antiseptics applied to them.
   b. Public health measures are taken to control insect and other animal populations that act as vectors for the transmission of disease caused by pathogenic microorganisms.

G. Hospital-acquired (nosocomial) infections.
   1. Hospitals represent unique epidemiological situations because of the potential reservoirs of disease and the susceptibility of patients to infection.
   2. Personnel are important in the transmission of microorganisms within the hospital and therefore practice medical asepsis to limit the spread of pathogens.
   3. Surgical asepsis involves major precautions for preventing the introduction of microorganisms into exposed wound tissues.
   4. In some cases patients are isolated to prevent transmission of pathogens; this may be done if the patient has a contagious disease or if the patient is particularly susceptible because of compromised immunity.

Study Questions

1. Match the following statements with the terms in the list to the right. Choose the most specific term that is correct.
   a. Causes spasms followed by convulsion.
   b. This toxin occurs in Gram negative but not Gram positive bacteria.
   c. Secreted into the surrounding medium.
   d. Inactivated by boiling for 30 minutes.
   e. Produced by *Clostridium* species.
   f. Prevents the release of glycine and inhibits neurotransmission.
   g. Causes distinct clinical symptoms.
   h. Blocks ATP–cyclic AMP conversion; leads to severe loss of water.

   (1) Endotoxin
   (2) Exotoxin
   (3) Neurotoxin
   (4) Botulinum toxin
   (5) Tetanus toxin
   (6) Cholera toxin

2. Indicate whether the following statements are true or false.
   a. Pathogenicity is a property of a microorganism alone.
   b. Exposure to pathogens always causes disease.
   c. Virulence is the ability to induce disease.
   d. The lipopolysaccharides of the bacterial cell wall are exotoxins.
   e. Endotoxins can withstand autoclaving.

f. Capsules protect some bacteria from phagocytosis.
g. Pili are involved in attachment to body tissues.
h. Enterotoxins affect the nervous system.
i. Hyaluronidase enhances invasiveness.

3. Describe the phases of an acute infection. For each stage indicate what is happening with respect to numbers of pathogens, components of the immune response, and symptoms of the patient.
4. Describe the role of the epidemiologist in controlling the spread of infectious diseases.
5. List the portals of entry for infectious pathogens and describe the routes of disease transmission associated with each.
6. List Koch's postulates. Describe how these postulates are applied for determining the etiology of a disease.

Some Thought Questions

1. How can we tell if it is going to be a good or bad year for the flu?
2. How can microorganisms get from New York to London?
3. Should microbiologists carry out research on germ warfare?
4. Why isn't it beneficial for parasites to kill their hosts?
5. Should the budget for Centers for Disease Control be cut? Why?
6. What should you do when your child gets sick?
7. Is there an Andromeda strain?
8. Which disease is the most serious human health threat?
9. Why can going to the hospital be dangerous?

Additional Sources of Information

Benenson, A. L. (ed.). 1981. *Control of Communicable Diseases in Man*. American Public Health Association, Washington, D. C. A useful reference to the public health measures available to control the spread of infectious diseases.

Bernheimer, A. W. 1976. *Mechanisms in Bacterial Toxicology*. John Wiley & Sons, Inc., New York. A detailed treatise on the toxins produced by bacteria.

Chu, F. S. 1978. Mode of action of mycotoxins and related compounds. *Advances in Applied Microbiology*, 22:83–143. A detailed article describing a variety of toxins produced by fungi.

Cliff, A., and P. Haggert. 1984. Epidemics. *Scientific American*, 250(5):138–147. This discussion of the spread of measles in Iceland serves as a model for understanding the epidemic spread of disease.

Costerton, J. W., R. T. Irvin, and K. -J. Cheng. 1981. The bacterial glycocalyx in nature and disease. *Annual Reviews of Microbiology*, 35:299–324. A well-illustrated review of the role of the glycocalyx in the attachment of bacteria, including its importance in disease processes.

Cuatrecasas, P. (ed.). 1977. *The Specificity and Action of Animal, Bacterial and Plant Toxins*. Chapman and Hall, London. A series of articles on the actions of toxins, including those produced by microorganisms

Easmon, C., and J. Jeljaszewicz (eds.). 1982—. *Medical Microbiology*. Academic Press, Inc., Orlando, Florida. This significant new series of review articles brings together a summary of specific areas of subjects within medical microbiology.

Finegold, S. 1977. *Anaerobic Bacteria in Human Disease*. Academic Press, Inc., Orlando, Florida. Infections of the central nervous system, abdomen and perineal areas, respiratory tract and cardiovascular system are covered extensively.

Isenberg, H. D., and A. Balows. 1981. Bacterial pathogenicity in man and animals. In *The Prokaryotes* (M.P. Starr, H. Stolp, H.G. Truper, A. Balows, and H.G. Schlegel, eds.), Vol. 2, pp. 83–122. Springer-Verlag, Berlin. An interesting discussion of the pathogenicity of bacteria.

Kass, E. H., and S. M. Woolf. 1973. *Bacterial Lipopolysaccharides: The Chemistry, Biology and Clinical Significance of Endotoxins*. The University of Chicago Press, Chicago. An advanced discussion of bacterial endotoxins and their role in producing human disease.

Mausner, J. S., and A. K. Bahn. 1974. *Epidemiology: An Introductory Text*. W. B. Saunders Company, Philadelphia. An excellent introductory textbook on the epidemiology of infectious diseases.

Robbins, S. L., and R. S. Cotran. 1974. *The Pathological Basis of Disease*. W. B. Saunders Company, Philadelphia. An advanced medical review on the pathological effects of infectious microorganisms.

Smith, H. 1968. Biochemical challenge of microbial pathogenicity. *Bacteriological Reviews*, 32:164–184. A detailed review of the biochemical basis of microbial virulence factors.

Smith, H., and J. H. Pearce (eds.). 1972. *Microbial Pathogenicity in Man and Animals*. Society for General Microbiology, Cambridge, England. An excellent series of articles on the mechanisms by which microorganisms cause disease.

23 *Respiratory Tract Diseases*

KEY TERMS

acute epiglottitis
blastomycosis
bronchitis
coccidioidomycosis
common cold
diphtheria
dust cells
Ghon complex
H spikes
histoplasmosis
influenza
Legionnaire's disease
macrophages
miliary tuberculosis
muco-ciliary escalator
mucus
N spikes
nosocomial infection
pandemics
phagocytosis
pharyngitis
pneumonia
primary atypical pneumonia
psittacosis
Q fever
sore throat
strep throat
tonsillitis
tubercules
tuberculosis
wandering macrophage
whooping cough

OBJECTIVES

After reading this chapter you should be able to

1. Define the key terms.
2. Describe the structure of the respiratory tract.
3. Describe the resident microbiota of the respiratory tract.
4. Describe the defense systems of the respiratory tract that protect against infectious microorganisms.
5. Discuss the causative agents, routes of transmission, key symptoms, and treatments for the following diseases: blastomycosis, bronchitis, coccidioidomycosis, common cold, diphtheria, histoplasmosis, influenza, Legionnaire's disease, pneumonia, primary atypical pneumonia, psittacosis, Q fever, sore throat, tuberculosis, and whooping cough.

Structure of the Respiratory Tract

The respiratory tract (Figure 23.1) is designed to permit the exchange of gases between cells, the blood, and the environment. Humans require oxygen for cellular metabolism and must dispose of the carbon dioxide produced during respiratory metabolism. Air enters the body through the nose and nasal cavity. Hairs at the entrance of the nose and the cilia on its epithelial lining entrap dust and other foreign particulates. The nose is designed so that incoming air is swirled, giving the maximal exposure to the moist and sticky mucus lining of the nasal chambers and the hairs in the nose. As a result, the air reaching the lungs has greatly reduced numbers of microorganisms. When you inhale through the mouth this process is not accomplished. The numerous blood vessels embedded in the lining of the nasal cavity also warm the incoming air and mucous secretions moisten it before it flows into the lungs. From the nasal cavity air moves through the pharynx and down through cartilage-reinforced tubes, the larynx and the trachea.

The trachea, or windpipe, branches into two bronchi, which lead to the paired lungs. Lungs are elastic sponge-like sacs that move freely within the protection of the chest cavity; they are attached only at the region where air tubes and major pulmonary blood vessels connect them. In the lungs the bronchi branch into increasingly smaller tubes that divide the lungs into compartments, the smallest of which are called bronchioles. These bronchioles end in thin-walled pouches called alveoli. Human lungs contain more than 300 million alveoli. Gas exchange takes place across the alveolar epithelium, passing through the interstitial fluid of a thin layer of connective tissue and across the epithelium of the capillaries into the blood.

Air entering the alveoli has a high concentration of oxygen and a low concentration of carbon dioxide. The blood within adjacent lung capillaries is low in oxygen and rich in carbon dioxide. These two dissolved gases exchange by diffusion in response to these concentration gradients. As oxygen moves into the blood plasma and then into red blood cells, it forms weak, reversible bonds with hemoglobin. When oxygen-rich blood reaches a systemic tissue capillary bed, oxygen diffuses outward and carbon dioxide moves from tissues into the capillaries. Once blood returns to lung capillaries, the lower carbon dioxide concentration in the alveoli reverses the direction of the exchange reactions. Hemoglobin has greater affinity for oxygen than for carbon dioxide and makes this exchange possible. The released carbon dioxide is removed by exhalation.

Defenses of the Respiratory Tract Against Microbial Invasion

The continuous movement of air into the body, 10,000–20,000 liters per day, exposes the tissues of the respiratory tract to numerous microorganisms. Multiple defense mechanisms act to protect the respiratory tract from infection by disease-causing microorganisms (Figure 23.2). The respiratory tract is protected in part against the invasion of pathogenic microorganisms by secretions of mucus. Mucus is secreted from goblet cells and subepithelial glands. Lysozyme, which is found in mucus, degrades the cell walls of bacteria, conferring antimicrobial activity on these body fluids.

The tracheal epithelium is lined with cilia and mucus-secreting goblet cells are embedded within it. This protective system is called the muco-ciliary escalator because trapped particles are swept upward and out of the respiratory tract. The muco-ciliary escalator effectively acts as a filtration system to prevent potential pathogens from penetrating the surface tissues of the respiratory tract (see Figure 19.2). As inhaled particles become stuck in the mucus of the tracheal membrane, the upward-beating motion of the cilia sweeps this debris-laden mucus back to the mouth or the nasal cavity where it can be expelled.

Some of the mucus and microorganisms are swept out of the body through the oral and nasal cavities, and other entrapped microorganisms are swept to the back of the throat where they are swallowed and enter the digestive tract. Sneezing and coughing tend to remove many of these microorganisms from the respiratory tract (Figure 23.3). Coughing and sneezing also expose bacteria to mucus with its antimicrobial activity, thereby reducing the number of potential pathogens. Mucus also contains plasma cells that produce antibodies, which enhance the phagocytosis of microorganisms attempting to invade the body. Additionally, the swallowing reflex moves most particulates, including microorganisms that become attached to mucus, into the digestive, rather than into the respiratory tract.

Microorganisms that do manage to reach the lower respiratory tract are subject to further phagocytosis. Phagocytosis is a highly efficient host defense mechanism against the invasion of microorga-

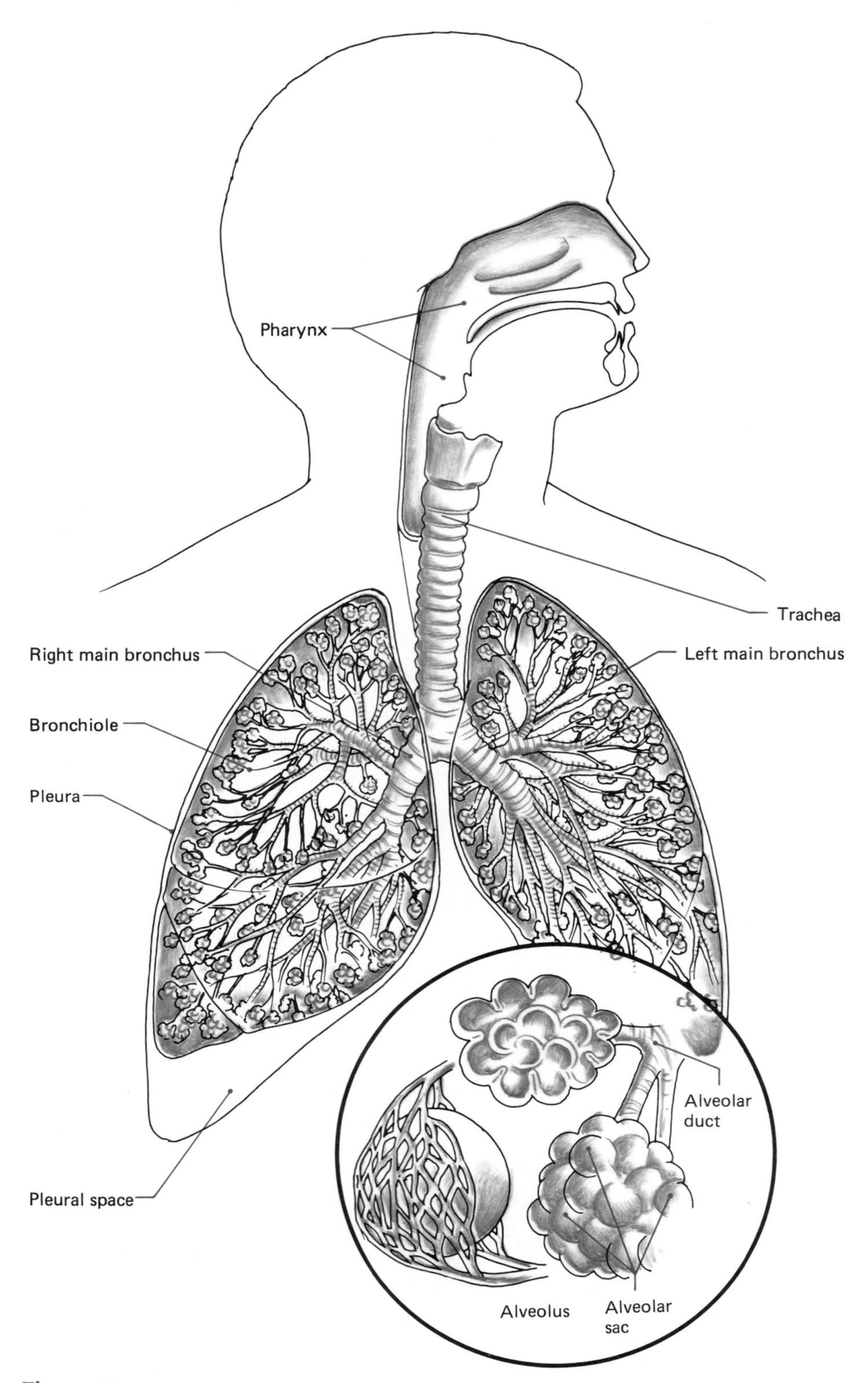

Figure 23.1 Structure of the respiratory tract.

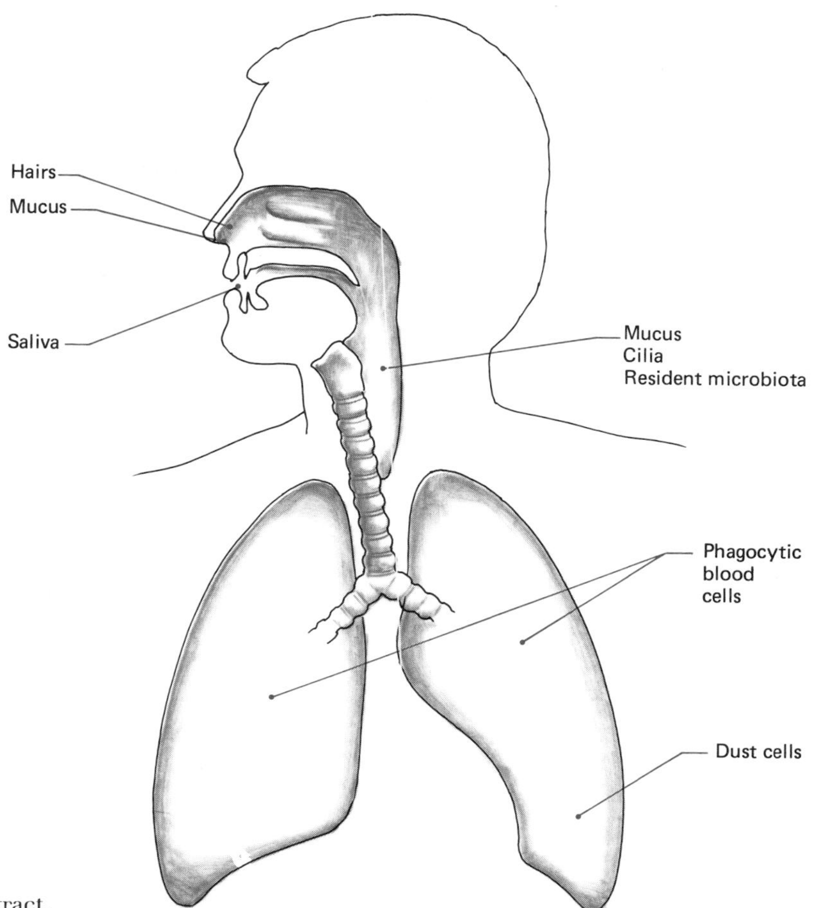

Figure 23.2 Defenses of the respiratory tract.

Figure 23.3 A violent cough expels microorganisms from the respiratory tract. Unfortunately this defense mechanism propels droplets into the air, and thus potentially toward the respiratory tract of another individual. (A) Note the distant spread of droplets as this young boy coughs. (B) Note how the spread of droplets is inhibited by this little girl's placing her hand over her mouth.

A

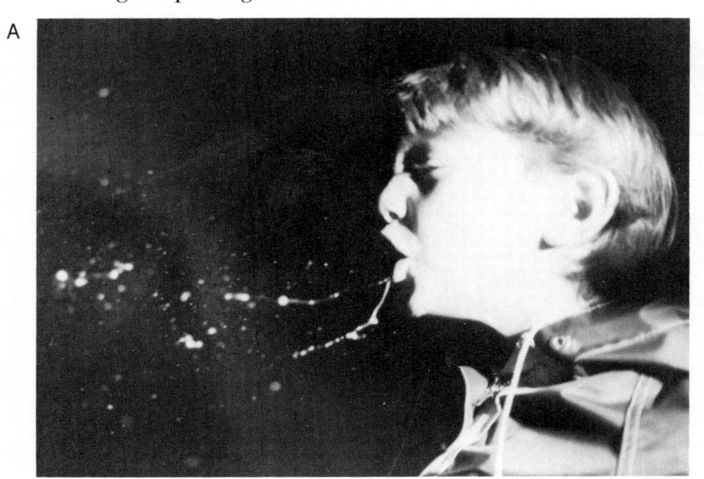

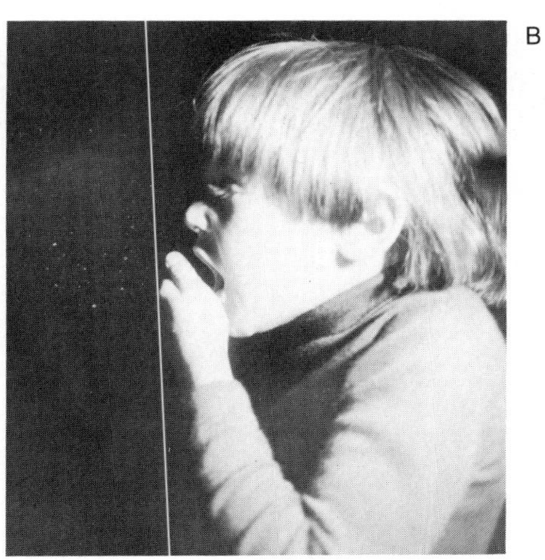

B

nisms. In phagocytosis the microorganism is engulfed by the phagocytic cell and digested. Several types of blood cells that circulate through the numerous blood vessels of the lungs are phagocytic. Some blood cells (monocytes) migrate out of the circulatory system and become enlarged, forming phagocytic cells called macrophages. Macrophages are long-lived, persisting in tissues for weeks or months. Once differentiated macrophages are formed, they are capable of reproducing to form additional macrophages. The lungs contain macrophages called dust cells that are fixed in the alveolar lining. Wandering macrophages also occur in the alveolar lining of the lung. Dust cells and wandering macrophages engulf and digest most microorganisms that enter the lower respiratory tract. The presence of relatively high numbers of these macrophages in the respiratory tract is important in limiting infections of the lower respiratory tract.

Normal Microbiota of the Respiratory Tract

Even though the respiratory tract has various defense mechanisms for preventing microbial infections, the upper respiratory tract, including the nasal cavity and nasopharynx, is normally inhabited by various bacterial species. These resident microorganisms help prevent colonization of the upper respiratory tract by pathogens. The resident microbiota of the upper respiratory tract typically include the genera *Streptococcus, Staphylococcus, Branhamella, Neisseria, Haemophilus, Bacteroides*, and *Fusobacterium* as well as members of the spirochete and coryneform groups. The lower respiratory tract may be colonized temporarily by microorganisms inhaled on dust particles or water droplets and microorganisms that normally colonize the upper respiratory tract. However, because of the abundance of phagocytic cells, the lower respiratory tract lacks a resident microbiota in healthy individuals.

Diseases that Primarily Affect the Respiratory Tract

Because potential pathogens freely enter the respiratory tract through the normal inhalation of air and when substances are placed in the mouth, the respiratory tract provides a portal of entry for many human pathogens. Various viruses, bacteria, and

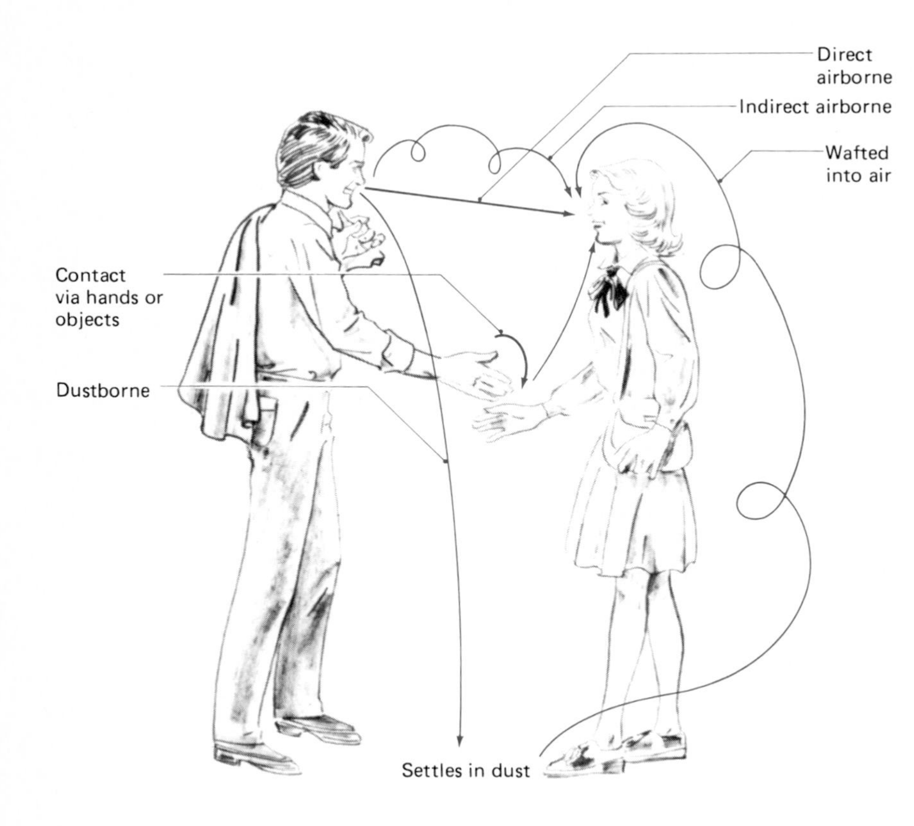

Figure 23.4 Routes of transmission of pathogens to the respiratory tract. Close contact with an individual emitting pathogens from the respiratory tract increases the probability of successful transmission.

MICROBIOLOGY HEADLINES

Scientists Say Common Cold Spread Mainly by Hand Contact

By CRISTINE RUSSELL

WASHINGTON—After more than a decade of work, an increasing body of scientific evidence suggests that the hundreds of viruses that cause the common cold are spread chiefly by hand contamination rather than coughing or sneezing, researchers said yesterday.

Teams from the University of Virginia and the University of Wisconsin reported on studies showing that a new chemically treated version of facial tissue is highly effective in stopping the spread of the most common illness by killing cold viruses before they get on the hands.

But the Virginia research also shows that a less costly, old-fashioned approach—wiping one's nose frequently with regular facial tissue and keeping one's hands clean—may be about as effective.

A cold sufferer may contaminate his hands while blowing his nose or sneezing and then transfer the organisms by touching others or by touching household objects where the viruses may live for hours or days.

The strongest proponent of a new "viricidal" or virus-killing tissue is its pioneer, Dr. Elliot C. Dick, a University of Wisconsin researcher who said yesterday that the tissue has proved 100 percent effective in stopping the spread of cold viruses under experimental conditions.

In the tests, in which student volunteers purposely infected with colds played poker with healthy men for 12-hour stretches, the new tissue far surpassed the performance of cotton handkerchiefs.

Dick, who began testing the idea in Antarctica in the 1970s, said that in two recent tests with the virus-killing tissue, none of the 24 healthy volunteers got sick. But in the cotton handkerchiefs trial, 58 percent—14 out of 24—of the people who were exposed to a cold caught one.

He said a single square inch of the specially treated three-ply tissue can destroy 100,000 virus particles in one minute—or about 80 percent of the viruses present.

The tissue is treated with three compounds—citric and malic acids, found in fruits, and sodium lauryl sulfate, used in toothpaste—that are considered non-toxic to humans but deadly to "rhinoviruses," the most common group of cold viruses.

In a separate experiment at the University of Virginia, Drs. J. Owen Hendley and Jack Gwaltney Jr. also found that the chemically treated tissue was highly effective, with none of 24 exposed research subjects developing a cold.

But Hendley found to his surprise that the "control" tissue, regular tissue without any special treatment, was also quite effective.

Only three of the 25 persons who were exposed to cold sufferers who frequently used regular tissue became infected. But if cold sufferers used no tissue at all, about half of the people they exposed to the viruses got sick.

Hendley noted that it has been difficult to prove exactly how colds spread, but he and his colleagues have concluded that the most likely route is through the hands of a cold sufferer touching infected nostrils where the viruses are concentrated.

While other Virginia experiments supported this hand-transmission route, Hendley was surprised to find that it was difficult to catch a cold by breathing in cold viruses spread into the air by a cough or sneeze.

Both Hendley and Dick presented their latest findings at the 24th Annual Interscience Conference on Antimicrobial Agents and Chemotherapy, sponsored in Washington by the American Society for Microbiology.

Their work with the new virus-killing tissue has been conducted under the auspices of Kimberly-Clark Corp., makers of Kleenex, which is beginning test marketing of the virus-killing tissues in New York.

Called Avert, it is expected to be marketed in a blue-and-white carton of 60 tissues at a cost of about $1.30, about twice the price of regular tissues, says company representative Russell Carpenter. He says it may not be available nationally for "several years."

Source: The Washington Post, Oct. 10, 1984. Reprinted by permission.

fungi are able to multiply within the tissues of the respiratory tract, sometimes causing localized infections and sometimes entering the circulatory system through the numerous blood vessels associated with the respiratory tract and spreading through the bloodstream to other sites in the body. In order to establish an infection of the respiratory tract, a pathogen must overcome the natural defense mechanisms that are particularly extensive in the lower respiratory tract.

Transmission through the air is undoubtedly the main route of transmission of pathogens that infect the respiratory tract. Airborne transmission often occurs when droplets containing pathogenic microorganisms are transferred from an infected to a susceptible individual (Figure 23.4). Droplets regularly become airborne during normal breathing. Coughing and sneezing associated with respiratory tract infections are especially responsible for the spread of pathogens in aerosols, and hence the airborne transmission of disease. The incidence of these diseases can be reduced by covering one's nose and mouth while coughing and sneezing, and avoiding contact with contagious individuals, a practice we are taught to follow at an early age.

Diseases Caused by Viral Pathogens

The common cold. The common cold is one of the most frequent infectious human diseases, and it is safe to assume that we all have had a cold at sometime. More than 200 million work and school days are lost each year in the United States alone because of the common cold.

The common cold actually refers to a cluster of diseases primarily caused by viruses, characterized by a localized inflammation of the upper respiratory tract, the release of mucus secretions, and generally accompanied by sneezing and sometimes coughing. The etiologic agents responsible for the majority of cases of the common cold have yet to be identified, but we do know that in adults approximately 25 percent of all colds are caused by rhinoviruses, whereas only about 10 percent of colds in children are caused by rhinoviruses. There are least 90 immunologically distinct types of rhinoviruses capable of causing the common cold; hence, it is not surprising that immunity does not offer continuous protection against all of the distinct strains of viruses capable of causing the disease.

In one day: 137,000 Americans stay at home from work because they have a cold. 164,000 children stay home from school with the same excuse.

—TOM PARKER, *In One Day*

Viruses causing the common cold infect the cells lining the nasal passages and pharynx, producing an inflammatory response with associated tissue damage in the infected region. The initial viral infection can be followed by a secondary bacterial infection as the normal microbiota of the upper respiratory tract invade the damaged tissues. Symptoms during the course of the common cold include nasal stuffiness, sneezing, coughing, headache, malaise (a vague feeling of discomfort), sore throat, and sometimes a slight fever. There is no specific treatment for the common cold, which is a self-limiting clinical syndrome. Recovery usually occurs within 1 week without complications as a result of the natural immune defense response.

If you think that you have caught a cold, call in a good doctor.
Call in three good doctors and play bridge.

—ROBERT BENCHLEY, *From Bed to Worse*

As with many other respiratory diseases, colds occur primarily during the winter months, in part because of the physiological stress posed by exposure to cold temperatures and the excessive drying of mucous membranes in low humidity heated buildings, and in part because of increased contact between individuals during indoor winter activities. Recently investigations have produced evidence that suggests that the respiratory tract is not the initial portal of entry for rhinoviruses. Few healthy individuals contracted colds when they were intentionally exposed to the airborne droplets from

Go hang yourself, you old M.D.
You shall no longer sneer at me.
Pick up your hat and stethoscope,
Go wash your mouth with laundry soap;
I contemplate a joy exquisite
In never paying you for your visit.
I did not call you to be told
My malady is a common cold.

By pounding brow and swollen lip;
By fever's hot and scaly grip;
By these two red redundant eyes
That weep like woeful April skies;
By racking snuffle, snort and sniff;
By handkerchief after handkerchief;
This cold you wave away as naught
Is the damnedest cold man ever caught.

Give ear, you scientific fossil!
Here is the genuine Cold Colossal;
The Cold of which researchers dream,
The Perfect Cold, the Cold Supreme.
This honored system humbly holds
The Super-cold to end all colds;
The Cold Crusading for Democracy;
The Fuehrer of the Streptococcracy.

Bacilli swarm within my portals
Such as were ne'er conceived by mortals,
But bred by scientists wise and hoary
In some Olympian laboratory;
Bacteria as large as mice,
With feet of fire and heads of ice
Who never interrupt for slumber
Their stamping elephantine rumba.

A common cold, gadzooks, forsooth!
Ah, yes. And Lincoln was jostled by Booth;
Don Juan was a budding gallant,
And Shakespeare's plays show signs of talent,
The Arctic winter is rather coolish,
And your diagnosis is fairly foolish.
Oh what derision history holds
For the man who belittled the Cold of Colds!

—OGDEN NASH, *The Common Cold*

sneezing and coughing people with colds. However, colds were transmitted when healthy individuals shook hands with people with colds and then rubbed their eyes. This series of experiments suggests that the eyes are the portal of entry for rhinoviruses and that the viruses migrate to the tissues of the respiratory tract. Disinfectant companies have been quick to respond to this possibility, advertising the advantages of spraying disinfectant on bathroom counter tops to prevent the spread of cold-causing rhinoviruses that you may pick up and rub into your eyes. It remains possible, however, that colds are transmitted via droplets entering the respiratory tract when the muco-ciliary escalator protection system fails.

Influenza. Compared to the common cold, influenza often is a more serious and debilitating disease. Influenza is caused by the virus *Myxovirus influenzae*, commonly referred to as the "flu virus" (Figure 23.5). Influenza is transmitted by inhalation of droplets containing flu viruses, released into the air as droplets originating from the respiratory tracts of infected individuals. The genome of the influenza virus is segmented, consisting of eight fragments of RNA. This genome easily recombines to form new genetic combinations. Minor antigenic changes occur because of mutations, but the major antigenic shifts probably result from gene reassortment between animal and human strains of flu viruses.

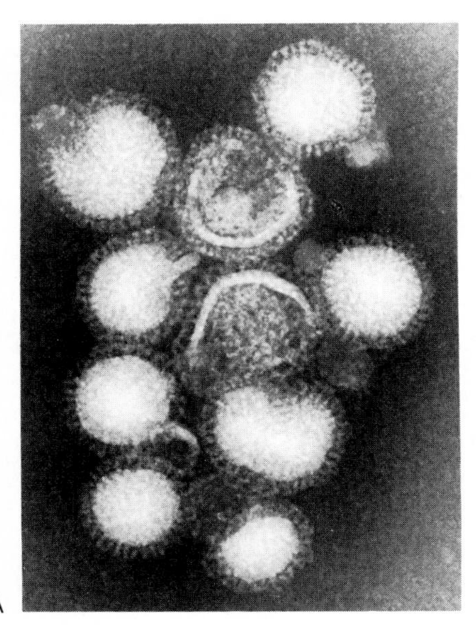

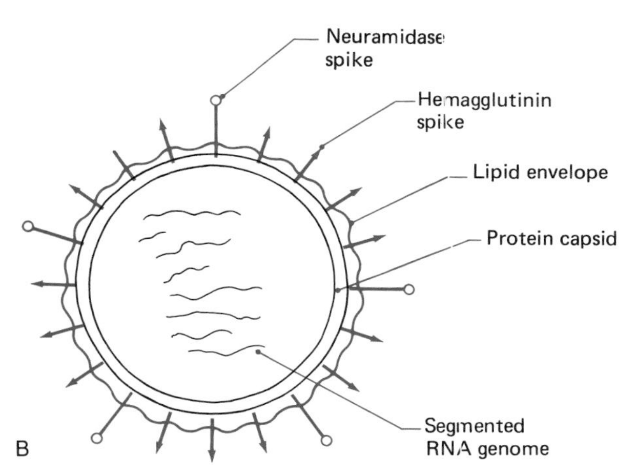

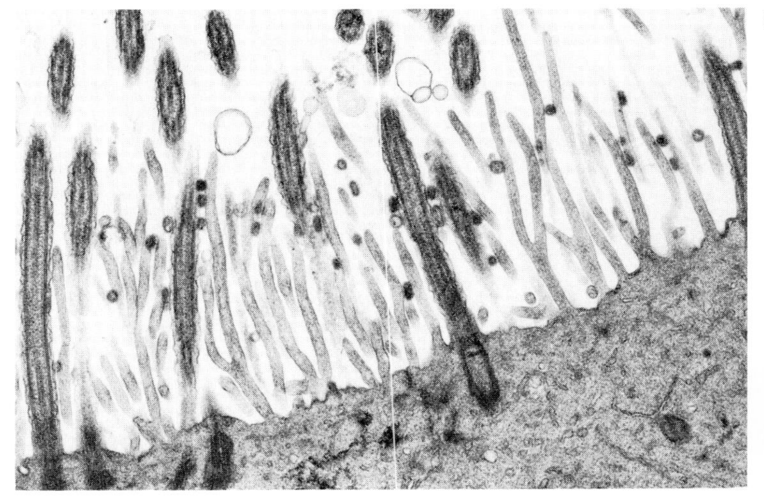

Figure 23.5 (A) Electron micrograph of the influenza A Hong Kong virus. (From BPS: F. A. Murphy, Centers for Disease Control, Atlanta.) (B) Diagram of the structure of an influenza virus. (C) Electron micrograph showing filtering of influenza viruses by the cilia lining the respiratory tract. The influenza viruses appear as small black particles. Many microorganisms are trapped on the surface mucus of the projecting cilia and fail to penetrate the respiratory tract, but in the case of influenza viruses, some virions may penetrate, leading to the onset of influenza. (Courtesy Dr. R. Dourmashkin, Laboratory for Paediatric Gastroenterology, St. Bartholomew's Hospital, London, England.)

The outer envelope of an influenza virus has numerous protruding spikes that affect both the pathogenicity and antigenicity of the particular viral strain. Changes in the combinations of genes that cause antigenic shifts and the production of new strains of influenza viruses generally are associated with changes in the biochemistry of these spikes. There are two types of spikes designated H (hemagglutinin) and N (neuramidase) spikes. The **H spikes** cause clumping (agglutination) of red blood cells; they presumably are important in the ability of the influenza virus to attach to cells in the body and are also a valuable aid in the serological identification of the particular strain of influenza virus. Antibodies against the H spikes are very important in the resistance against infection by that particular strain of influenza virus. The **N spikes** are less important in resistance to influenza infections and appear to be involved in the release of viruses from infected cells following viral replication.

Major groups of influenza viruses are designated according to the antigens associated with their capsids. There are three major groups of influenza viruses, designated types A, B, and C. Specific strains of influenza virus are further designated by variations in the protein composition of their H and N spikes. Additionally strains of influenza viruses are often described by the location where outbreaks of

Figure 23.6 Influenza outbreaks show regular cyclic fluctuations because of antigenic changes in the virus and associated variations in the susceptibility of the population. The highest incidence of influenza occurs during the winter, as shown in this graph of reported deaths in 121 United States cities compared with influenza isolates reported to WHO (World Health Organization) Collaborating Centers in the United States from September 1972 to August 1979. Outbreaks of influenza occur each year. In some years the severity and incidence of influenza are greater than others. By detecting the primary outbreak and identifying the particular strain of influenza virus responsible for that outbreak, steps can be taken to minimize the extent of the influenza outbreak. For example, the Asian flu epidemic of 1957 was the most severe outbreak of influenza since the great pandemic of 1918–1919, but far fewer people died in 1957 because the 1957 outbreak did not come as a surprise. A new and potent strain of type A influenza virus appeared in southern China early in 1957. It reached Amsterdam, Sydney, and San Francisco within weeks, but even before it spread from the Far East, a worldwide network of scientists was mobilized. Researchers worked in about 30 laboratories to identify the new strain and to trace the pattern of influenza outbreaks. The early detection of this outbreak of a new form of influenza, Asian flu, allowed time for the preparation of an effective vaccine. Within 10 months of detecting the first cases of Asian flu, pharmaceutical manufacturers had produced 10 million doses of vaccine. The use of the vaccine to immunize large numbers of individuals, who otherwise would have been susceptible to influenza, and the availability of antibiotics to combat secondary bacterial infections minimized the mortality due to this serious health threat. The Surgeon General of the United States announced that "for the first time in history the medical community was ahead of an impending epidemic."

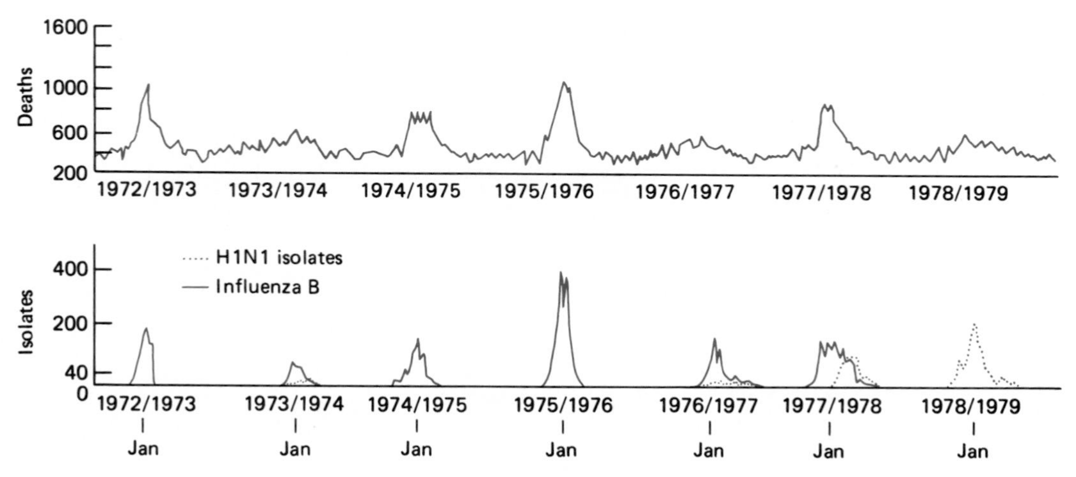

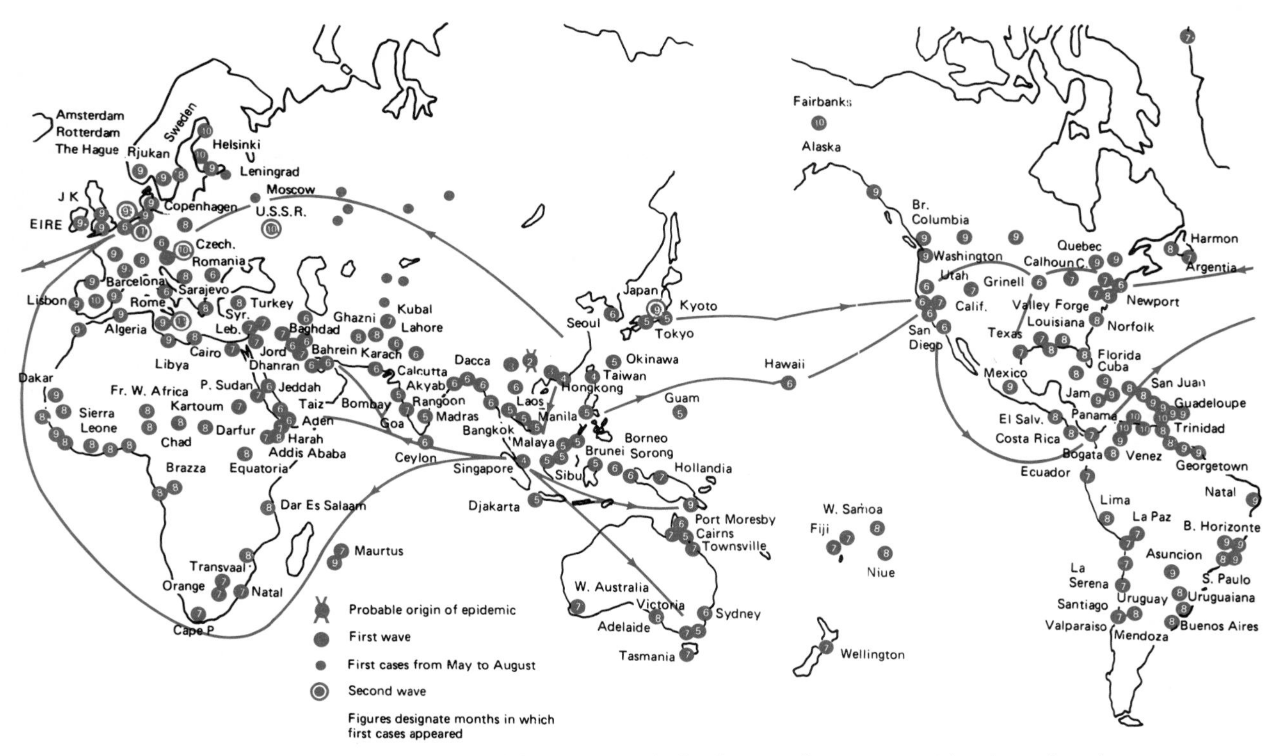

Figure 23.7 Map showing worldwide spread of influenza during an epidemic outbreak. This epidemic originated in southeast Asia and spread in waves to all other population centers of the world.

the disease associated with that particular antigenic variety of the flu virus was first detected. For example, the Hong Kong strain of influenza virus, first seen in the Orient in 1968, is a type A influenza virus designated as H_3N_2. The antigenic designation is important because it indicates to epidemiologists whether there has been a substantial change in the antigenic properties of the virus and thus whether a sufficient proportion of the population will be susceptible to that strain so that an epidemic outbreak is likely.

There are regular periodic outbreaks of influenza when a sufficient proportion of the population lacks resistance to the particular antigenic strain. Outbreaks of influenza are cyclical, with major outbreaks caused by type A virus occurring every 2–4 years, those caused by type B virus every 4–6 years, and outbreaks caused by type C virus occurring only rarely. Within each of the major types of influenza viruses there are various antigenic subtypes responsible for different outbreaks of influenza (Figure 23.6). The continued genetic drift with major antigenic shifts in the virus resulting from genetic recombination permits these periodic epidemic outbreaks. The introduction of a virus into a population most of whose members are susceptible to infection provides the conditions necessary for the establishment of an epidemic.

Influenza outbreaks spread worldwide via airborne transmission from the site of an initial outbreak with a new strain, and it is possible to watch the disease spread from one area to another (Figure 23.7). Each year epidemiologists make prognostications about the severity of influenza outbreaks, and public health officials take the necessary steps to immunize high risk individuals and warn the general public about the dangers of this disease. Even in a nonepidemic year, influenza causes a significant number of deaths; for example, the death rate in the United States due to influenza in 1980—a nonepidemic year—was 0.3 deaths per 100,000 population.

There have been several major, or pandemic outbreaks of influenza during the nineteenth and twentieth centuries. During 1918–1919, outbreaks of influenza resulted in the deaths of over 20 million people. Many of these deaths may have been the result of secondary infections with *Streptococcus pneumoniae*, rather than from the primary influenza infection. Because the 1919–1920 influenza pandemic was associated with a type C influenza virus, the detection of two deaths at Fort Dix, New Jersey in 1976 from a type C flu virus caused sufficient fear to initiate an ill-conceived and inadequately controlled immunization program. Some of the batches of the swine flu vaccine used in this program were not adequately treated and caused active infections of the type they were designed to prevent.

The disease influenza is characterized by the sudden onset of a fever, with temperatures abruptly reaching 102–104°F approximately 1–3 days after actual exposure and onset of infection. The development of the disease is further characterized by malaise, headache, and muscle ache. In uncomplicated cases of influenza, the viral infection is self-limiting and recovery occurs within a week. However, infection with influenza virus can lead to complications, such as bacterial pneumonia and several neurological syndromes such as Guillain Barre's and Reye's syndromes (see discussion of these neurological diseases in Chapter 26). Complications associated with influenza infections are prevalent among the elderly and individuals with compromised host defense responses. Such individuals should be immunized against the prevalent strains of influenza virus prior to the outbreak of influenza epidemics because complications can result in death.

The clinical diagnosis of influenza depends on the isolation of an influenza virus by using tissue culture, with a cell line such as monkey kidney cells, and the serological detection of increased titers of antibody in the patient's serum that are reactive with flu viral antigens. As with most other viral diseases, treatment of uncomplicated cases of influenza centers on treating the symptoms, with recovery from the disease dependent upon the immune response of the infected individual. The antiviral drug amantadine is effective against influenza A viruses, but only during the incubation period. Once the symptoms of the disease have appeared, amantadine is ineffective against influenza viruses, and thus, this drug is of limited clinical use. Primary control of influenza is achieved by vaccinating individuals who are prone to the complications resulting from this disease, leaving others unprotected to suffer periodically from influenza.

Diseases Caused by Bacterial Pathogens

Pharyngitis (sore throat). Several viral and bacterial species can cause an inflammation of the upper respiratory tract known as pharyngitis or more commonly as sore throat. The common symptoms of this disease are pain in the throat, redness of the lining of the throat, and fever. In some cases the tonsils are also inflamed, in which case the disease is called tonsillitis. The tonsils are sometimes surgically removed from individuals suffering from repeated cases of tonsillitis. The removal of tonsils used to be more prevalent before the widespread use of antibiotics to control bacterial infections of the upper respiratory tract.

My sore-throats, you know, are always worse than anybodys

—JANE AUSTEN, *Persuasion*

The most commonly found bacterial causative agents of pharyngitis and tonsillitis are *Staphylococcus aureus* (Gram positive cocci that form grape-like clusters and are coagulase positive) and *Streptococcus pyogenes* (Gram positive cocci that tend to form chains and produce toxins including hemolysins that cause beta hemolysis of red blood cells). Although these are undoubtedly the most frequent bacterial causes of sore throats in children, recent reports indicate that up to 40% of all cases of sore throats in adults may be the result of *Chlamydia* infections. (These reports are based upon serological diagnostic studies in Europe of adult patients showing signs of pharyngitis.) When the causative agent is *Streptococcus pyogenes* the disease is commonly called strep throat. The diagnosis and treatment of strep throat are important because the toxins produced by *S. pyogenes* can spread to other parts of the body and cause more serious diseases, notably scarlet fever and rheumatic fever.

He prayeth best who loveth best
All things both great and small.
The *Streptococcus* is the test—
I love it least of all.

—WALLACE WILSON, *Letters to Dr. E. P. Scarlett*

It is now routine practice in cases of suspected strep throat to streak a throat swab onto blood agar. The development of small (pinpoint) white colonies surrounded by a zone of clearing is diagnostic of beta-hemolytic streptococci, such as *S. pyogenes*. The zone of clearing is due to the lysis of the red blood cells in the medium. The test may be supplemented by the inclusion of a disc impregnated with bacitracin because group A strains of *S. pyogenes* are particularly sensitive to this antibiotic and will not grow in the vicinity of the disc. The use of bacitracin discs are not as definitive as serological tests and therefore are used when serological test procedures cannot be readily performed. Recently developed serological procedures permit identification of *Streptococcus* infections without culturing, permitting rapid diagnosis within minutes that can aid the physician in selecting the appropriate method of treatment. These tests are likely to replace culturing on blood agar because all the physician need do is to collect a throat swab from the inflamed area of the pharynx, mix with the reagents, and read the results. When strep throat infections are detected, they are readily treated with penicillin or alternative drugs such as erythromycin.

Pneumonia. Whereas pharyngitis is a disease of the upper respiratory tract, pneumonia is the result of an infection of the lower respiratory tract. Pneumonia is an inflammation of the lungs involving the alveoli and can be caused by a number of bacteria and other microorganisms. This disease often occurs as a complication to another disease condition, particularly when the host defense mechanisms are compromised. Viral infections of the respiratory tract often precede occurrences of bacterial pneumonia. Frequently, pneumonia is a nosocomial infection; that is, an infection acquired while a patient is hospitalized. When patients are "run down" and their physiologically impaired state depresses the effectiveness of the immune response system, such as happens after surgery or during the course of treatment for another disease, they are prone to secondary infections. The lack of movement and deep breathing in postsurgical patients reduce the efficiency of the normal defense mechanisms to clear the lungs of mucus and bacteria, and the accumulation of fluids favors the establishment of a microbial infection of the lower respiratory tract. To prevent the development of pneumonia in postsurgical and other patients, inhalation devices—which may simply be a tube with a ball that moves when air is blown into the tube—are frequently used by the patient to ensure that they are performing adequate deep breathing to clear their lungs of fluids.

There is a high rate of mortality in cases of pneumonia (Figure 23.8). Bacteria that cause pneumonia most frequently enter the lungs in air, although transport of pathogens to the lungs through the blood stream can also result in pneumonia. More than half of all pneumonia cases are caused by bacteria, and this disease ranks among the top causes of death from infectious diseases. The most frequent etiologic agent of bacterial pneumonia is *Streptococcus pneumoniae* (pneumococcus), a Gram

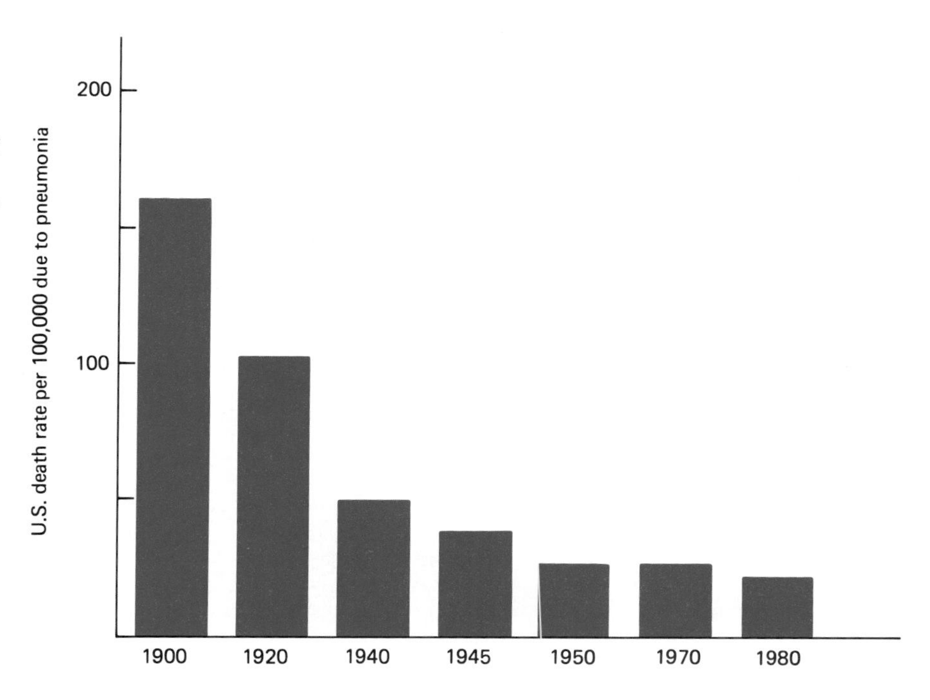

Figure 23.8 Death rate due to pneumonia in the United States in the twentieth century. Note that the dramatic decreases in the death rate in the first half of this century, which occurred after the use of antibiotics began, have now leveled off.

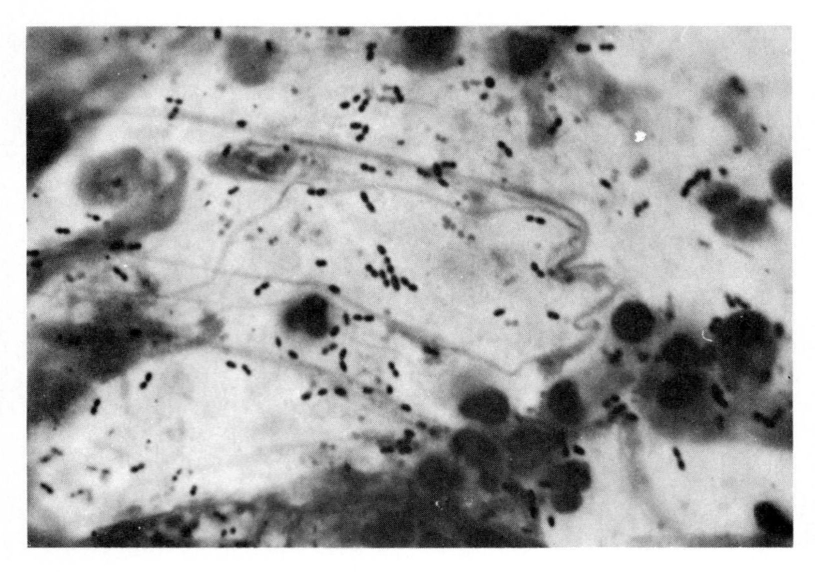

Figure 23.9 Photomicrograph of *Streptococcus pneumoniae* in a sputum sample. (Courtesy Ramon Kranwinkel, Danbury Hospital, Danbury, Connecticut.)

positive, capsule-forming coccus (Figure 23.9; Table 23.1). Often, pneumococcal pneumonia is an endogenous disease; that is, the infection originates from the individual's own normal throat microbiota. Several other bacteria, including *Staphylococcus aureus, Haemophilus influenzae*, and *Klebsiella pneumoniae*, are also responsible for a significant number of cases of pneumonia. These bacteria are normally acquired by an endogenous route when bacteria are carried along with secretions from the upper tract to the lower respiratory tract.

A vaccine (Pneumovax) has been developed from the capsular material of 14 common strains of *S. pneumoniae*. These 14 strains are responsible for 80 percent of the cases of pneumococcal pneumonia in the United States. The vaccine is effective in preventing pneumonia and is currently used as a preventative measure for susceptible groups, such as the elderly and debilitated individuals with impaired immune defense systems.

The symptoms of pneumococcal pneumonia, which is most prevalent in adult males, include the sudden onset of a high fever, production of rust-colored purulent sputum, and congestion. In most patients this is a secondary infection, and an upper respiratory tract infection with the characteristic symptom of a sore throat precedes the development of pneumococcal pneumonia.

During the development of pneumonia, bacteria reproduce in the lung tissue, forming a lesion. The phagocytic portion of this inflammatory response results in the accumulation of fluid. A chest X ray is used to make a preliminary diagnosis based upon characteristic shadows that result from the accumulation of fluids within the lungs. Coughing causes movement of fluid containing bacteria to the ad-

Table 23.1 Frequency of Major Types of Bacterial Pneumonia

Clinical Entity	Etiologic Agent	Percent of Cases	Indicated Antimicrobial Drug
Pneumococcal lobar pneumonia	*Streptococcus pneumoniae*	Over 90	Penicillin
Klebsiella (Friedlander's) pneumonia	*Klebsiella pneumoniae*	1–5	Gentamicin
"Flu" pneumonia	*Haemophilus influenzae* type b	1–5	Ampicillin

jacent alveoli, spreading the infection. Bacteria spread through the alveoli and into the circulatory system and pleural cavity surrounding the lungs. The exudate that develops during pneumonia also interferes with gas exchange in the lungs. Without treatment the death rate from pneumococcal pneumonia is about 30 percent.

Antibiotic treatment is effective for curing bacterial pnuemonia, with penicillin the antibiotic of choice for treating pneumonia caused by *Streptococcus pneumoniae*. The specific antibiotic treatment for pneumonias caused by bacteria other than *S. pneumoniae* varies in accord with their specific antibiotic sensitivities. Several bacteria, however, cause atypical pneumonias requiring special treatments. Identification of the pathogen and determination of its antibiotic susceptibility are essential for selecting the best antimicrobial agents in cases of atypical pneumonia.

Primary atypical pneumonia. *Mycoplasma pneumoniae* causes an atypical self-limiting pneumonia called primary atypical pneumonia. It is called atypical because conventional cultural procedures produce negative results and because the disease does not respond to penicillin treatment. *M. pneumoniae* lacks a cell-wall structure and is not sensitive to penicillin. This organism, however, is sensitive to tetracycline and erythromycin, which can be used effectively in treatment. Unlike other mycoplasmas, *M. pneumoniae* can attach to the epithelial surface of the respiratory tract. This bacterium does not penetrate the epithelial cells nor does it produce a protein toxin, but the hydrogen peroxide released by the bacterium causes cell damage, including loss of the cilia lining the respiratory tract and death of surface endothelial cells.

Legionnaire's disease (Legionnellosis). Another atypical form of pneumonia, Legionnaire's disease or Legionnellosis, is caused by *Legionella pneumophila* and related species in this genus. *L. pneumophila* is a Gram negative, fastidious (having complex nutritional and physiological requirements for growth), rod-shaped organism, whose nutritional requirements for growth complicated early isolation attempts, initially confounding epidemiologists who were trying to discover the etiologic agent of Legionnaire's disease. In addition to the typical symptoms of pneumonia, Legionnaire's disease is often characterized by kidney and liver involvement and by an unusually high incidence of associated gastrointestinal symptoms. The fever associated with this disease starts low but then typically reaches 104–105°F. If untreated the fatality rate is about one in six. *L. pneumophila* produces β-lactamases (enzymes that attack penicillins and other β-lactam antibiotics) and therefore is not sensitive to most penicillins and cephalosporins. It is sensitive to other antibiotics, such as erythromycin and tetracycline. Erythromycin is the antibiotic of choice when Legionnaire's disease is diagnosed.

Legionnaire's disease received its name because the first detected outbreak of this disease occurred during a convention of the American Legion in Philadelphia during July 1976. In the investigation of this outbreak of the disease, the first 90,000 man-hours of investigation, costing over $2 million and employing virtually all conventional isolation procedures, failed to reveal the causative agent of this disease. Finally the breakthrough, revealing that this disease is of bacterial etiology, came by using indirect immunofluorescent staining with antibodies from the sera of affected individuals. Later, it was discovered that the bacterium, subsequently named *Legionella pneumophila*, could be grown on a chocolate agar medium if iron and cysteine were included as growth factors.

It was also later found by examining stored blood sera that a 1968 outbreak of a disease in Pontiac, Michigan, the etiology of which had not been identified, had been caused by a different strain of *Legionella*. Various other outbreaks of this disease have since been identified, and it is now clear that there are several different clinical manifestations of respiratory tract infections caused by *Legionella*. Mild infections include Pittsburgh fever, whereas the more severe manifestations occur as Legionnaire's disease.

Species of *Legionella* appear to be natural inhabitants of bodies of water. During periods of rapid evaporation, such as occur during summer, the bacteria can become airborne in aerosols, and inhalation of contaminated aerosols can lead to the onset of the illness. In some cases, outbreaks of Legionnaire's disease have been traced to air-conditioning cooling systems. These bacteria multiply in the cooling system waters, which are rapidly evaporated to provide cooling, and inadvertently are permitted to become airborne and circulate through the air-conditioning system. Addition of biocides to air-conditioning systems is used to prevent spread of this disease.

Psittacosis. Another type of atypical pneumonia, psittacosis, is caused by an obligate intracellular bacterium, *Chlamydia psittaci*. Psittacosis is also known as ornithosis or parrot fever because birds act as a reservoir for *C. psittaci*. Parakeets, canaries, other pet birds, and domestic fowl are frequently the sources of human infection. Psittacosis is contracted through inhalation of *C. psittaci*. The primary route of transmission of psittacosis is from birds to human beings via aerosol dispersal of droplets and contaminated dust particles. *C. psittaci* multiplies in the cells of the mononuclear phagocyte system prior to systemic dissemination through the bloodstream. The symptoms of psittacosis include fever, headache, malaise, and coughing. This disease is generally mild and in uncomplicated cases recovery normally occurs within 1 week, aided by the use of tetracyclines in the treatment of this disease.

Bronchitis. Bronchitis, an inflammatory disease involving the bronchial tree that does not extend into the pulmonary alveoli, can be caused by several microorganisms, including *Streptococcus* sp., *Staphylococcus* sp., *Haemophilus influenzae, Mycoplasma pneumoniae*, and various other types of viruses. *M. pneumoniae* and various viruses appear to be the most frequent causative organisms of bronchitis; however, it is difficult to define the specific etiologic agent for this disease because bronchitis almost always occurs as a complication of another disease condition, such as pharyngitis (sore throat). Bronchitis is often a secondary infection that involves the endogenous movement of bacteria from the upper to the lower respiratory tract. The symptoms of bronchitis are normally preceded by those associated with a normal upper respiratory tract infection, such as malaise, headache, and sore throat. The onset of bronchitis is marked by the development of a cough that eventually yields mucopurulent sputum, reflective of the development of bronchial congestion.

Acute bronchitis can be effectively treated with antibiotics, such as penicillin and tetracycline. The development of chronic bronchitis is not because of microbial infection alone, but rather the etiology of this disease appears to depend on irritation of the bronchii by repeated microbial infections and/or the inhalation of irritants, such as cigarette smoke. These irritations compromise the normal secretory and ciliary function of the bronchial mucosa, and the resulting excessive mucus secretion in the bronchii favors bacterial growth and the establishment of infection.

Acute epiglottitis. *Haemophilus influenzae* not only can cause bronchitis and pneumonia, it can also cause acute epiglottitis. This often is a severe disease that can result in death within 24 hours. The disease occurs when *H. influenzae* infects the epiglottis. The epiglottis is the structure that covers the the larynx (voice box) and prevents the flow of fluids, intended to move into the gastrointestinal tract, from entering the lower respiratory tract. When the epiglottis is infected with *H. influenzae* it becomes severely inflamed and no longer properly opens and closes the entrance to the lower respiratory tract. Blockage of the entrance to the larynx can result in asphyxiation. Often the onset of acute epiglottitis is sudden and a tracheotomy must be performed immediately to maintain an airway and prevent death.

Diphtheria. Diphtheria is caused by the bacterium *Corynebacterium diphtheriae*. *C. diphtheriae* is a club-shaped bacterium that stains irregularly and forms pallisade layers because of the incomplete separation of cells after division. Actually only strains of *C. diphtheriae* harboring a lysogenic phage produce the protein toxin that causes diphtheria. The process by which a microorganism is induced to produce toxin by a phage is called lysogenic or phage conversion.

Just as men imitate each other, germs imitate each other. There is the genuine diphtheria bacillus discoverd by Loeffler; and there is the pseudo-bacillus, exactly like it, that you could find, as you say in my own throat . . . if the bacillus is the genuine Loeffler, you have diphtheria; and if it's the pseudo-bacillus, you're quite well. Nothing simpler. Science is always simple and always profound. It is only the half-truths that are dangerous.

—GEORGE BERNARD SHAW, *The Doctor's Dilemma*

Diphtheria toxin is a potent protein exhibiting toxicity against almost all mammalian cells. *C. diphtheriae* is normally transmitted via droplets from an infected individual to a susceptible host, establishing a localized infection on the surface of the mucosal lining of the upper respiratory tract. There is generally a localized inflammatory response, pharyngitis, in the vicinity of bacterial multiplication in the upper respiratory tract. The bacteria generally do not invade the tissues of the respiratory tract. Rather the toxin, which inhibits protein synthesis, produced by the bacteria on the surface of the respiratory tract disseminates through the body, causing the severe symptoms of this disease. The toxin causes swelling of the throat. In severe infections with *C. diphtheriae*, symptoms include low grade fever, cough, sore throat, difficulty in swallowing, swelling of the lymph glands, and the formation of grey-white pseudomembranes on the tonsils and the back of the throat. The development of pseudomembranes is indicative of diphtheria. Complications from diphtheria can block respiratory gas exchange and result in death due to suf-

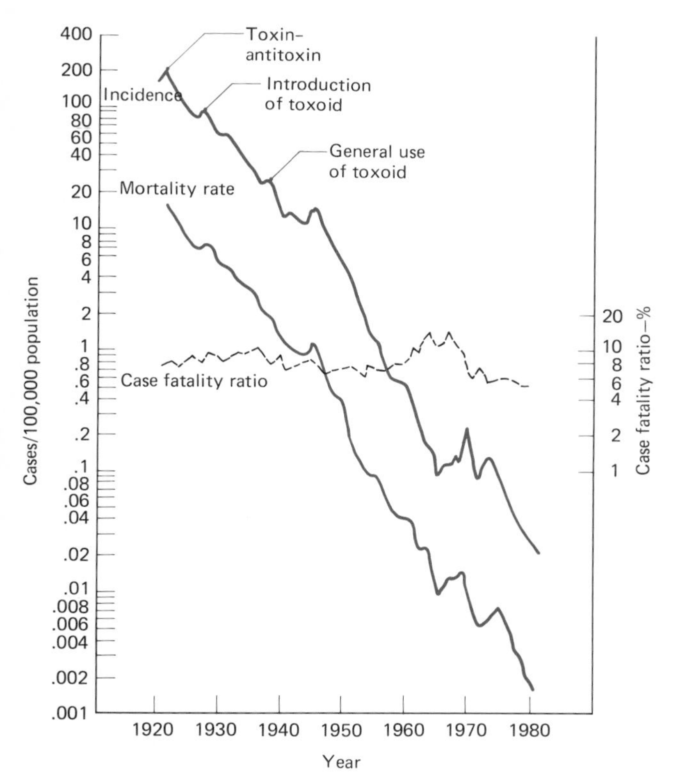

Figure 23.10 Incidence, mortality rates, and case fatality ratio of diphtheria in the United States from 1920 to 1980. The rates of this disease have declined dramatically since effective vaccination programs were begun. However, the case fatality ratio has remained relatively constant, underscoring the importance of prevention of this disease and the difficulty in its treatment.

focation; toxemia due to the diphtheria toxin can also cause death. A tracheotomy can be performed to restore breathing and to prevent death due to suffocation.

The extensive use of vaccines to prevent diphtheria has greatly reduced the incidence of this disease but has not altered the case fatality ratio (Figure 23.10). In immunized individuals, infection with toxicogenic strains of *C. diphtheriae* are generally restricted to a localized pharyngitis with no serious complications. Diphtheria, however, remains a serious problem in socioeconomically depressed regions of the world where extensive immunization is not practiced. Treatment of diphtheria involves the use of antitoxin to block the cytopathic effects of diphtheria toxin; the use of antitoxin prevents the occurrence of serious symptoms associated with diphtheria toxin. This immunological treatment is augmented by the use of antibiotics, like erythromycin, to eliminate the bacterial infection.

Whooping cough (Pertussis). Whooping cough or pertussis derives its name from the distinctive symptomatic cough associated with this disease. Other symptoms of pertussis resemble those of the common cold, although vomiting often occurs after severe coughing episodes. Strangulation on aspirated vomitus is often the cause of death for the few fatalities associated with this disease. Whooping cough is caused by *Bordetella pertussis*, a Gram negative cocco-bacillus, which exhibits fastidious nutritional requirements. *B. pertussis* is capable of reproducing within the respiratory tract, and high numbers of *B. pertussis* are found on the surface tissues of the bronchi and trachea in cases of pertussis (Figure 23.11). *B. pertussis* produces several toxins that establish the pathogenicity of this organism.

Erythromycin and tetracyclines are effective in eliminating the infecting bacteria, although treatment of whooping cough primarily involves maintenance of an adequate oxygen supply. The administration of pertussis vaccine has greatly reduced the occurrence of whooping cough, and prevention of the disease is accomplished by routine immunization of infants. Recently several manufacturers have stopped making this vaccine because of liability over occasional severe side effects that are associated with the administration of pertussis vaccine. This has resulted in uncertainty and concern over the ability to meet demand for routine immunization.

Little Willie from his mirror
Sucked the mercury all off,
Thinking in his childish error,
It would cure his whooping-cough.
At the funeral Willie's mother
Smartly said to Mrs. Brown:
"Twas a chilly day for William
When the mercury went down."

—ANONYMOUS

Tuberculosis. The frequency of tuberculosis has been greatly reduced from the time when this disease affected so many Americans that most cities were forced to construct separate hospitals for its treatment. Today separate tuberculosis treatment facilities have, for the most part, been eliminated. Nevertheless, tuberculosis remains a common dis-

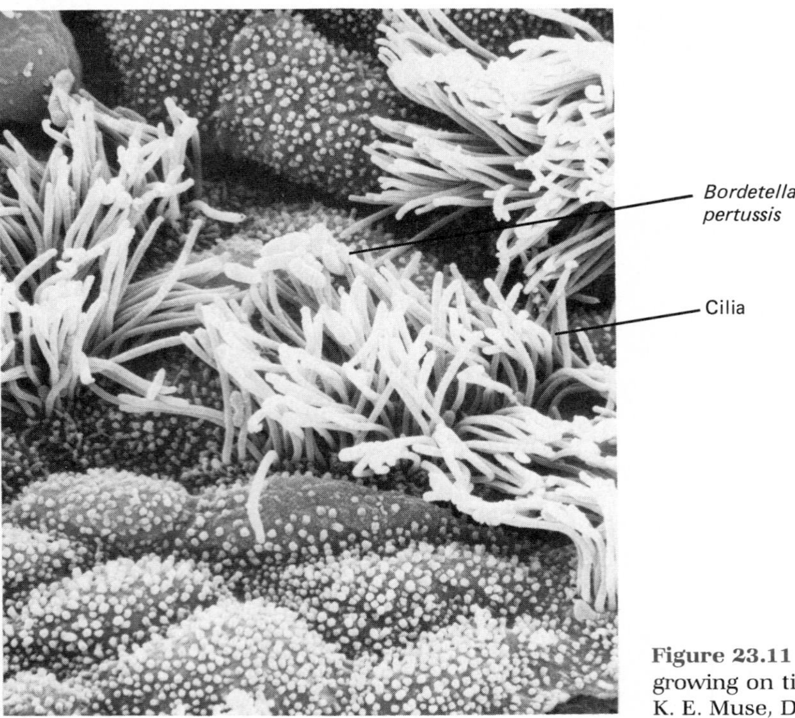

Figure 23.11 Photomicrograph of *Bordetella pertussis* growing on tissues of the respiratory tract. (From BPS: K. E. Muse, Duke University Medical Center.)

ease with about 25,000 cases occurring annually in the United States and a worldwide incidence of about 5 million new cases per year.

One sniffed at what I spat, another tapped at the place whence I spat it, and the third poked and listened while I spat . . . The first said that I would die, the second that I was dying, and the third that I was already dead.

FREDERIC CHOPIN, *December 3, 1838, letter to his musical colleague Julius Fontana*

Tuberculosis, which is caused by *Mycobacterium tuberculosis* and related mycobacterial species, is primarily transmitted via droplets from an infected to a susceptible individual. Mycobacteria are slender acid-fast rods. Tuberculosis can also be transmitted through the ingestion of contaminated food, and before the extensive use of pasteurization, milk contaminated with *M. tuberculosis* or *M. bovis* was associated with outbreaks of this disease. The principal portal of entry for *M. tuberculosis*, however, is through the respiratory tract. Much lower numbers of bacteria are required to establish an infection via the respiratory tract than are needed to initiate tuberculosis through the gastrointestinal system. The global occurrence of tuberculosis is estimated to be 1500 million persons infected with tubercule bacilli, 20 million sputum-positive individuals capable of disseminating the disease, 3–5 million new cases each year, and 600,000 deaths per year.

The common form of tuberculosis involves an infection of the pulmonary system, with multiplication of *M. tuberculosis* occurring in the lower respiratory tract. *M. tuberculosis* escapes the phagocytic activity of macrophage that protect the lower respiratory tract from infection by most potential bacterial pathogens. The outer lipid-rich layer of mycobacteria protect them against normal host defenses. *M. tuberculosis* does not elicit a strong humoral response, and opsonizing antibodies do not react with the mycobacteria to bring about their effective destruction by phagocytes.

The pulmonary form of tuberculosis involves inflammation and lesions of lung tissue, which can be detected by chest X rays (Figure 23.12). When lung tissue is infected with *M. tuberculosis*, exudative lesions initially form. These lesions contain the mycobacteria, phagocytic leukocytes, and exudate fluid. This reaction appears as an area of nonspecific inflammation that resembles pneumonia. Infection with *M. tuberculosis* elicits a cellular immune response because bacteria are able

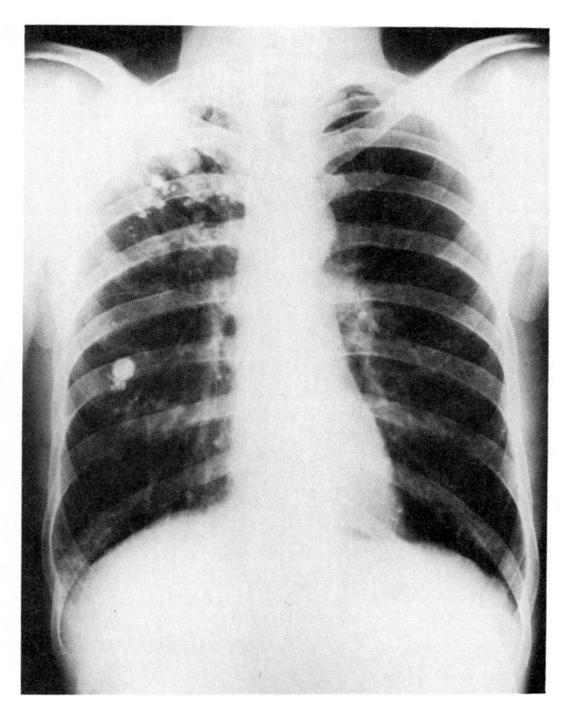

Figure 23.12 X ray showing development of tuberculosis in lungs of an adult. The left lung is normal, but the right lung shows a large calcified tubercle, smaller tubercles, and infiltration in the apex of the right lung. (From BPS: R. B. Morrison, M.D., Austin, Texas.)

to reproduce within phagocytic cells, and a delayed hypersensitivity reaction is typical of infection with *M. tuberculosis*. This leads to a second type of reaction, which leads in turn to the development of granulomas as a result of a cell-mediated immune (hypersensitive) reaction to tuberculoprotein. The macrophages responding to the infection undergo a major modification when this occurs, becoming concentrically arranged in the form of elongated cells to create the tubercules characteristic of this disease. This granulomatous inflammation occurs 2–4 weeks after the formation of an initial lesion and only after a cell-mediated immune response occurs. Outside the concentric layers of elongated cells, lymphocytes and fibroblasts proliferate, causing extensive fibrosis. Dormant mycobacteria can remain viable for indefinite periods within the tubercules, and the infectious process can be reactivated at a later time, with various physiological factors probably contributing to the reactivation of the disease.

In the normal course of tuberculosis, caseation necrosis occurs in which the centers of the tubercules remain semisolid (cheese-like) rather than softening to form pus. Caseous lesions heal by fibrosis and calcification, which can produce extensive scar tissue. The healed calcified regions of the tubercules, called Ghon complexes, can be seen in chest X rays for the remainder of the individual's lifetime. The formation of tubercules and caseous lesions appears to be an attempt by the body to wall off or contain the infecting mycobacteria.

In some instances this isolation mechanisms fails and the caseous lesions rupture, allowing the mycobacteria to spread from the lesions through lymph and blood to other parts of the body. This condition is called miliary tuberculosis, so named because of the numerous tubercules formed that are the size of millet seeds. Miliary tuberculosis leads to a progressive disease. If a caseous lesion opens into the airway when it ruptures, as opposed to a blood or lymph vessel, mycobacteria spread into other parts of the body and are released from the individual in aerosols by coughing.

It was the fashion to suffer from the lungs; everybody was consumptive, poets especially; it was good form to spit blood after each emotion that was at all sensational, and to die before reaching the age of thirty.

—Alexandre Dumas, *Memoirs*

The course of tuberculosis varies greatly between infected individuals. Host resistance is related to genetic factors and the extent of other infections. Malnutrition and stress are important factors relating to the susceptibility to tuberculosis and the course of the disease. In some cases the infection is restricted to the area of primary lesions, but in others it spreads into various other tissues. Tuberculosis can occur in bone and central nervous system tissues. Disease symptoms, including fatigue, weight loss, and fever, generally do not appear until there are extensive lesions in the lung tissues. As a result of the slow growth rate of *M. tuberculosis* and the ineffectiveness of phagocytic cells in killing this bacterial species, tuberculosis is generally a persistent and progressive infection. Without treatment, the disease is often fatal.

I should like to die of a consumption because the ladies would all say: "Look at that poor Byron, how interesting he looks in dying!"

—Lord George Gordon Byron

Diagnosis of tuberculosis is based on chest X rays and skin testing to establish potential tuberculosis

MICROBIOLOGY HEADLINES

Eleven Died of Airborne Fungus at Hospital Amid a Delay Over New Air Filter

By ROBERT HANLEY

BUFFALO—After a second patient had become infected by an airborne fungus and had died in a new bone marrow transplant unit at a state cancer hospital here, hospital officials asked the state to install special air filtration systems in each of the unit's six rooms.

But the request became stalled in the state's Office of General Services. In the next 8 months, 9 more patients out of 29 in the special unit also became infected with the fungus and died, according to interviews with hospital officials and an examination of documents at the hospital and the State Department of Health.

After the 11th death last April 2, officials of the hospital, Roswell Park Memorial Institute, closed the unit—a wing of the building—until the new air filtration systems could be installed. The wing remains closed.

In the year that the wing was used exclusively for bone marrow transplant patients, there was a marked increase in the incidence of the fungal infection, called aspergillosis, according to the State Health Department. Before then—between September 1977, when Roswell Park started to do marrow transplants, and April 1982, when the special unit opened—the transplant patients were mingled with other cancer patients, and only 2 of 80 contracted the infection, the Health Department said.

The Office of General Services, which designs and oversees most of the state's capital projects, said Roswell Park's request for the $23,000 filtration devices had arrived at a time when the agency was devoting nearly all its efforts to easing overcrowding in prisons.

Thomas Cooper, a spokesman for the office, said: "At no time did they indicate there was an urgency here. If they had, we would have treated this with more dispatch. If Roswell Park had said, 'We need these things in a hurry,' they would have had these new designs in less than six months."

The first six deaths occurred between June 27 and Nov. 29, 1982.

While transplant patients are prone to infections, the percentage of deaths at Roswell Park is considerably higher than at other bone marrow transplant facilities that have installed the latest air-purification systems, according to experts.

Last week the State Health Department said it had started an investigation to determine if the infectious fungi spores at Roswell Park had been spread by an old air ventilation system that served the six-room transplant wing. The department said it would also investigate infection-control procedures at the hospital.

Roswell Park officials said that since the late 1970's they had considered the wing's air ventilation system outmoded. But the state never provided money to upgrade it. In mid-1981 the hospital asked that the system be replaced. The money was allocated and the state was ready to proceed, but then the transplant unit opened before the work was done, according to Mr. Cooper.

No testing for spores was conducted before the wing began receiving transplant patients, Roswell Park officials said. Before the first death, the officials said, no thought had been given to installing the special air filters. Dr. C. William Aungst, the hospital's assistant director for clinical affairs, said, "We didn't think there was a problem or would be a problem."

Documents at the State Health Department, which approved both the request for repairs in 1981 and the July 1982 application for the modern air filtration system, show that Roswell Park's main complaint with the system was that it provided air that was too hot in summer and too cold in winter.

Fungus Harmless to Healthy People

The disease at issue, aspergillosis is caused by the aspergillus fungus, of which there are different varieties. Aspergillus spores are found everywhere in the environment. Ordinarily they are harmless to healthy people. But to leukemia victims and bone marrow transplant patients with severely weakened immunological systems, the fungus is almost always fatal, experts say. Lungs are the most vulnerable to the infection, according to doctors.

By the time the third death had occurred at Roswell Park on July 31, 1982, the hospital had found aspergilli spores on and near the bone marrow transplant wing, according to Dr. Aungst.

"There were very low levels of aspergillus," Dr. Aungst said. "The levels were very low on all of them." The interior spore counts, he said, were the same as those found in the air outside Roswell Park. He said that hospital officials had decided to keep the transplant wing open after the tests.

Over the next eight months eight more transplant patients, all of whom had leukemia, were stricken with the fungal disease and died.

While Health Department and Roswell officials have refused to identify the patients who died, citing patient confidentiality, they said the victims included a 15-year-old girl and a 16-year-old boy; four women, aged 27, 41, 44 and 45, and five men, aged 24, 26, 32, 44, and 45.

Since the last death the ceilings in the six rooms of the marrow transplant wing have been torn out to await installation of the new air filters, known as hepa-filters. A number of other cancer facilities around the country have installed such air systems in the last few years to shield bone marrow transplant patients from infectious airborne contaminants.

The Federal Centers for Disease Control in Atlanta and air purification specialists say the hepa-filters are 95 to 99.9 percent effective in screening out germs. The old air-handling system in the Roswell Park transplant rooms was only 30 to 35 percent effective, according to Gerald K. Schofield, the hospital's deputy director for administration.

Wide Investigation Planned by State

The Health Department's investigation, announced last week, will include reviews of the medical records of all bone marrow transplant patients at Roswell Park since 1977, as well as scrutiny of the wing's air handling system, its infection control procedures and what was described as the hospital's failure to notify the Health Department of the increasing infection cases, as required by the State Hospital Code.

The state's Health Commissioner, Dr. David Axelrod, also said he would ask the Federal Centers for Disease Control to "insure that appropriate environmental evaluations are conducted."

In a statement, Dr. Axelrod said: "While we recognize that patients undergoing bone marrow transplant therapy are highly susceptible to a myriad of common infectious agents, our primary concern is to ascertain whether there are any environmental issues or patient care practices that need to be addressed at Roswell Park to more effectively safeguard high-risk patients."

State regulations do not require installation of the hepa-filter systems in the rooms of patients who are highly prone to airborne infection.

Spore Mystery Remains

After the Roswell Park wing was closed, a strain of spores called *Aspergillus fumigatus* was found in the ducts of the ventilation system, according to Dr. Aungst, the assistant director for clinical affairs. A different strain, called *Aspergillus flavus,* was found during autopsies of the victims, he says.

Because the strains in the ducts differed from those in the patients, Dr. Aungst said, the hospital is not convinced the ventilation system was responsible for the exposure.

"I don't know if it's something that can be proven or disproven," he said.

A number of cancer hospitals that have had aspergillosis cases or have learned of such cases at other institutions have usually installed hepa-filter purification systems, according to experts in the field. These filters have been used for years by pharmaceutical companies, the manufacturers of computer microchips, medical research laboratories and the nation's space program to guarantee sterile air.

Robert J. Weeks, an environmental microbiologist with the Centers for Disease Control, said the filters could block the entry of airborne fungal spores down to 1 micron in size. A micron is equivalent to the diameter of the point of a common pin.

"Aspergilli-particle spores are about 3 to 4 microns," Dr. Weeks said. "There are very few fungi smaller than 2 microns."

He said about 95 percent of the spores would be snagged by a hepa-filter before they got into a room. The high volume of filtered air, he went on, creates a positive pressure force in the room. Any time a door to the room is opened, the pressurized air inside the room rushes out into the hallway, carrying with it any remaining spores and preventing any from getting into the room.

Choice of Filters: Simple or Complex

Since the mid-1970's there have been two hepa-filter systems on the market. The simpler one uses a forced-air device with a hepa-filter mounted at its opening. It is attached to a ceiling duct. As the air is forced vertically through the filter, fungi, bacteria and viruses are trapped on the filter, a fiberglass matrix shaped like an accordion.

The more complex system is called a laminar airflow room. In such rooms the entire wall behind a patient's bed is equipped with a series of hepa-filters, and the room is constantly bathed with horizontal jets of purified air. Experts say this system makes rooms more sterile than those with just one hepa-filter.

Roswell Park is planning to install the simpler filter system.

Some experts say there are two main drawbacks to a laminar airflow room. It is more expensive than a single ceiling unit. And, doctors say, a patient in a laminar airflow room remains in isolation and is not supposed to be touched by relatives or medical personnel.

Dr. Dean Buckner, assistant director of the 40-bed bone marrow transplant unit at the Fred Hutchinson Cancer Center in Seattle, said that in its seven years of operation, that center's transplant facility had never had a case of aspergillosis. Fourteen of the 40 rooms have laminar airflow systems. Some of the others have single hepa-filters, and some have no special air filtration systems.

Twenty additional rooms, to be built soon, will have laminar airflow systems, he said. "We're aware this problem could happen to us, and we want to be prepared for it." Dr. Buckner said.

The Memorial Sloan-Kettering Cancer Center in Manhattan, the only other hospital in New York State that performs bone marrow transplants, has had laminar airflow equipment in its seven transplant rooms since 1978, five years after it began the transplants. Between 1973 and 1978 it had simpler air purification systems.

Susan Rauffenbart, a spokesman at Sloan-Kettering, said two of the center's 400 transplant patients had died of aspergillosis between 1973 and 1978 and two others had died since 1978.

At the University of Minnesota's 12-bed bone marrow transplant unit, the ceiling hepa-filters were installed in May 1981 after a flurry of fatal aspergillosis cases. Of 66 patients who had received transplants there in 1979, a total of 12, or about 18 percent, died of aspergillosis. In the year after the hepa-filters went in, 3 of 64 transplant patients, or 4.6 percent, died of the fungus, according to Andrew Streifel, a hospital environmental specialist.

A Problem Not Expected

Roswell Park had 12 fatal cases of aspergillosis in 1969, according to Dr. Edward S. Henderson, chief of the section that treats leukemia patients. In subsequent years the number of infections declined. There were no cases in 1980 and "a few" in 1981, Dr. Henderson said.

In June 1982, before the new unit was two months old, Dr. Harvey D. Preisler, a member of Dr. Henderson's staff, wrote to Michael Murphy, head of Roswell Park's Bio-Hazards Control Department, to express concern about aspergillosis and to ask for a sampling of air at the unit, officials said.

Mr. Schofield said that the sampling had found low levels of the fungus.

On June 27, 1982 the first patient died, and on July 1 the second died. On July 12 Mr. Schofield asked the state for the hepa-filter system.

Last August, almost five months after the unit had closed, Dr. Gerald P. Murphy, the director of Roswell Park, wrote to John C. Egan, Commissioner of General Services, asking for a speed-up of work on the filter project. A month later he made a similar request to Dr. Axelrod, the State Commissioner of Health.

The designs for installation of the filters arrived at Roswell Park on Oct. 4, about 15 months after they had been requested.

Source: The New York Times, Nov. 3, 1983. Reprinted by permission.

infections, and direct microscopy and cultural examination of sputum samples for the presence of acid-fast rods. Confirmation of a positive, active case of tuberculosis is difficult and time consuming. Culturing of mycobacteria from sputum samples takes weeks because of the slow growth of this bacterium.

The skin test is particulary important for identifying individuals who may have contracted this disease. Skin testing often is performed when there is reason to suspect that an individual, for example, a family member or health care worker, may have been exposed to someone with an active case of tuberculosis. The skin test involves the introduction of tuberculin proteins into the skin and observation for a delayed hypersensitivity reaction.

The Tine and Mantoux tests both are used for this purpose. In the Mantoux test a standard amount of purified protein derivative (PPD) from *M. tuberculosis* is introduced intradermally in the forearm. The test is examined 48 hours later and, if negative or doubtful, is reread at 72 hours. The test is read by measuring the extent of induration (hardening) and reddening around the site of the injection (0–5 millimeters diameter = negative test; 6–9 millimeters = doubtful test; greater than 10 millimeters = positive test). A positive test indicates previous exposure to tuberculosis-causing mycobacteria but does not indicate an active infection. A false positive test can occur, for example, if an individual has been immunized against tuberculosis with the BCG vaccine or has been treated and recovered from a case of tuberculosis. Additionally there are many false negative tests that occur because of improper adminstration of PPD, inactive PPD preparations, erroneous reading of results, and depressed cellular immune responses in the individual.

When tuberculosis is diagnosed, it can be successfully treated with antibiotics. Once antibiotic therapy is begun the individuals are not considered to be contagious and therefore are not isolated. Effective treatment of tuberculosis is generally prolonged and is achieved by using multiple antibiotics such as streptomycin, rifampin, and isoniazid. Multiple antibiotics are needed because during the prolonged treatment of this persistent infection, the mycobacteria can mutate and become antibiotic resistant. The disease would then progress despite the administration of antibiotic therapy. Statistically, however, it is highly unlikely that the mycobacteria will undergo multiple mutations during the treatment period that would make them resistant to two or more different antibiotics. Using multiple antibiotics therefore provides the necessary margin of safety in treating tuberculosis in order to ensure that the mycobacteria will remain susceptible to the antibiotics and that the treatment will efectively eliminate the infection.

Antibiotics are also used for chemoprophylaxis when an individual has been exposed to tuberculosis and when individuals exhibit positive skin test but negative cultural tests for active infections. In these cases 12 months of isoniazid alone is beneficial. Chemoprophylaxis is used in the United States when individuals are known to have recently converted from tuberculosis negative to tuberculosis positive skin tests and when there is a high risk of contracting this disease because of exposure to an active case. Additionally, people who are likely to be exposed can be immunized by a single intradermal injection of BCG (bacillus of Calmette and Guerin) vaccine. This vaccine has been widely used in Europe and many developing nations for mass immunizations of infants. Widespread use of the BCG vaccine is not used in countries such as the United States where the incidence of tuberculosis is relatively low and where occurrences of the disease can be effectively treated with antibiotics. The reason for the limited use of the vaccine in the United States is owing to questions about its effectiveness for conferring long-term immunity on large numbers of individuals.

Q fever. Q fever, caused by the rickettsia *Coxiella burnetii*, is unique because this rickettsial disease does not manifest itself as a rash and is the only one normally transmitted via the respiratory tract. The disease can also be transmitted through direct contact of the bacteria with the eyes and via the gastrointestinal tract through the ingestion of contaminated food such as unpasteurized milk. *C. burnetii* is normally maintained within nonhuman animal populations, such as cattle and sheep, where it is transmitted via tick vectors. The bacteria can become airborne on fomites (inanimate objects), such as hair and dust particles, and establish human infections by invading the lower respiratory tract, leading to a systemic infection. The symptoms often include fever, headache, chest pain, nausea, and vomiting. Tetracyclines and cloramphenicol are normally used in treating this disease.

Diseases Caused by Fungal Pathogens

Histoplasmosis. Histoplasmosis, one of several fungal diseases of humans, is caused by the fungus *Histoplasma capsulatum*. *H. capsulatum* is a dimorphic fungus that grows in a filamentous form in the environment but changes and grows as a yeast-like fungus when it infects the human body.

A

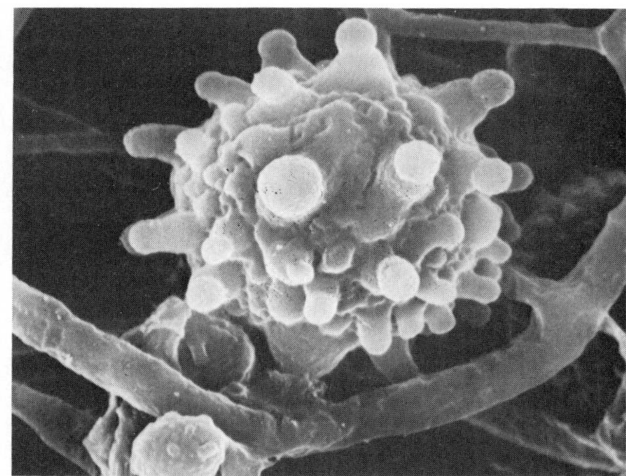

Figure 23.13 (A) Scanning electron micrograph of *Histoplasma capsulatum,* the dimorphic fungus that causes histoplasmosis. (From BPS: Garry T. Cole, University of Texas, Austin.) (B) This map, based on the incidence of skin reactivity to histoplasmin among naval recruits between 1958 and 1965, shows that histoplasmosis in the United States occurs primarily in the Mississippi and Ohio valleys. In southern Kentucky, middle Tennessee, and southern Missouri, the incidence of skin reactivity is as high as 90–95 percent. (After L. B. Edwards et al., 1969, "An atlas of sensitivity of tuberculin, PPD-B, and histoplasmosis in the United States," *American Review of Respiratory Disease,* 99:1.)

B

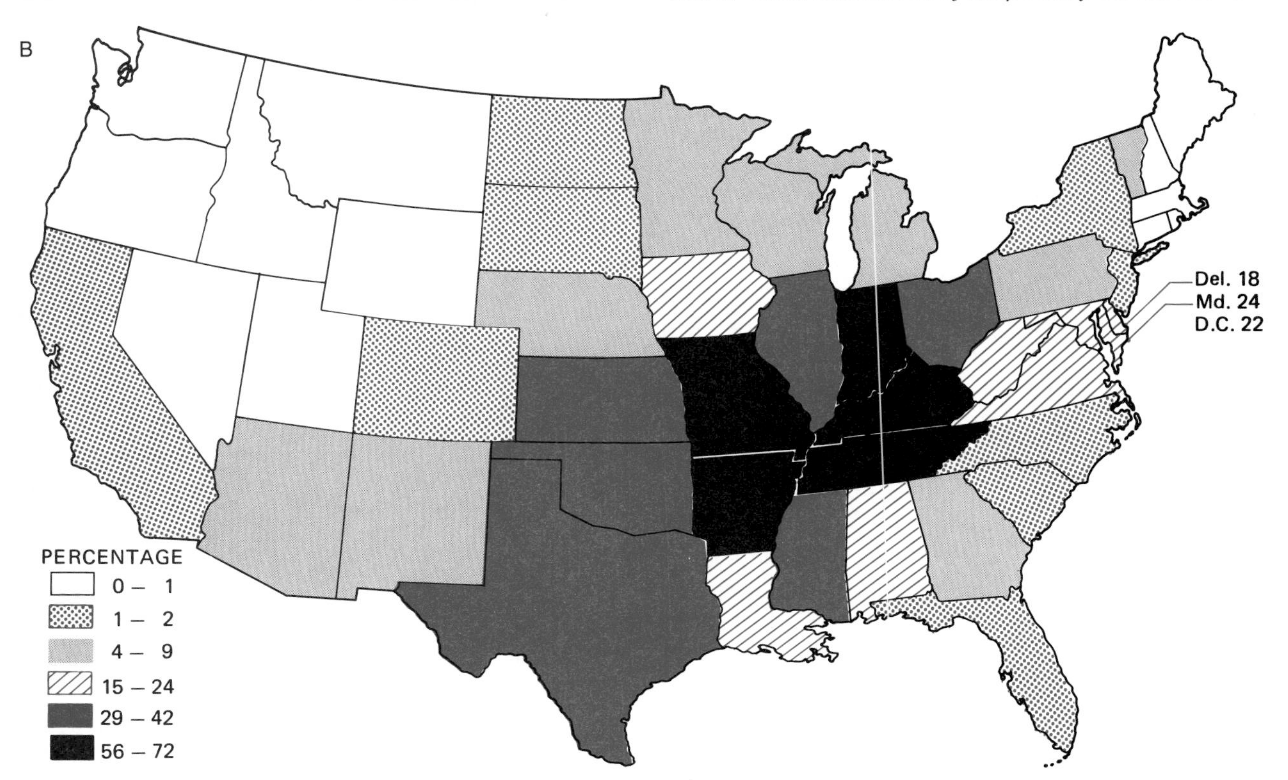

Histoplasma enters the respiratory tract through the inhalation of spores, which are subsequently deposited within the lungs. Histoplasmosis is endemic to certain regions of the world, such as the Ohio and Mississippi River Valleys of the United States (Figure 23.13). The fungus is found in soils contaminated with bird droppings, and dust particles released from abandoned bird roosts appear to be involved in some outbreaks of this disease. The apparent association of bird roosts with histoplasmosis has been used as the justification for large-scale kills of blackbirds, but the usefulness of this procedure has not been conclusively demonstrated.

Normally, histoplasmosis is a self-limiting disease in which the symptoms may be absent or resemble a mild cold. In some cases, however, the systemic distribution of the fungus to different organs of the body may prove fatal. The pathology of disseminated cases of histoplasmosis resembles that of tuberculosis. There is granuloma formation as a result of the cell-mediated immune response. The fungus can grow within phagocytic cells and is able to persist within the lungs and the tissues and cells of the mononuclear phagocyte system. Amphotericin B is used in the treatment of cases of progressive disseminated histoplasmosis, but mild cases are not treated because of the potential side effects of this antibotic.

Coccidioidomycosis. Coccidioidomycosis, caused by the fungus *Coccidioides immitis*, is also referred to as Valley or San Joaquin Fever because of the localized geographic distribution of this disease. Because of the association of the spores of *C. immitis* with the arid soils of the southwestern United States, soils from this region must be disinfected before shipping to other areas. Many individuals living in the southwestern part of the United States show a positive skin test, indicating that they have been infected at some time with this fungus. Visitors to the southwestern states of Nevada, California, Utah, Arizona, and New Mexico often develop symptoms of a mild cold because of infection with *C. immitis*.

Figure 23.14 Micrograph showing a section of lung infected with *Coccidioides immitis*. The fungus occurs as endospores within a body known as a spherule. Such fungal infections initiate a major inflammatory response. (Reprinted by permission from J. W. Jones, N. G. Miller, H. W. McFadden *et al.*, 1967, *The Atlas of Medical Mycology*, American Society of Clinical Pathologists, Chicago.)

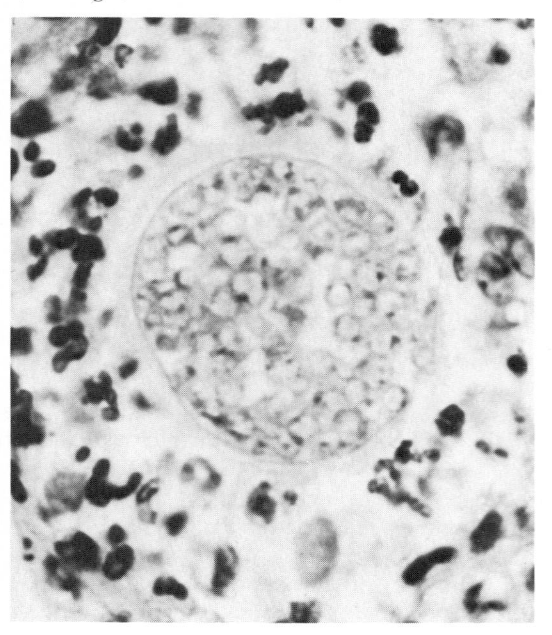

Normally, *C. immitis* occurs in soil, and transmission of coccidioidomycosis involves inhalation of dust particles containing conidia of this fungus. When deposited in the bronchi or alveoli, the conidia of *C. immitis* elicit an inflammatory response (Figure 23.14). Within host tissues, *C. immitis* appears as spherules containing multiple spores. In some cases, *C. immitis* remains localized in the area of the primary lesion, but the organism can be distributed to other parts of the body. Symptoms of coccidioidomycosis include chest pain, fever, malaise, and a dry cough. In most cases, no special treatment is required for the cure of localized coccidioidomycosis, and upon recovery the individual is immune to this disease. However, when there is evidence of systemic dissemination of *C. immitis*, amphotericin B is normally used in effectively treating the disease.

Blastomycosis. North American blastomycosis is a systemic disease caused by *Blastomyces dermatidis*. The primary site of infection is the lungs, from which the fungus can be disseminated to many other body tissues. *B. dermatidis* is a dimorphic fungus that normally inhabits soils. Blastomycosis occurs most frequently in the southeastern United States. The symptoms of this disease, which are generally mild and self-limiting, include cough, fever, and general discomfort. In severe cases, the cell-mediated immune response to the fungal infection produces granulomas. Skin lesions (cutaneous ulcers) are common, and there can be extensive abscess formation and tissue destruction in cases of blastomycosis. The demonstration of nonencapsulated, thick-walled, multinucleate yeast cells in pus, sputum, or tissue sections and their isolation by culture on Sabouraud's agar establish the diagnosis of blastomycosis. As with other systemic mycoses, blastomycosis can be treated with amphotericin B but hydroxystilbamidine is sometimes preferred because of its lower toxicity.

Diseases Caused by Protozoa

Pneumocystis pneumonia. One type of pneumonia, *Pneumocystis* pneumonia, is caused by the protozoan *Pneumocystis carinii*. This is a moderately rare form of pneumonia that has received prominance because it occurs with relatively high frequency in patients with aquired immune deficiency syndrome (AIDS). The occurrence of *Pneumocystis* pneumonia appears to be associated with failures of the immune system, as occurs in cases of AIDS. Patients receiving drugs that suppress the immune system are prone to *Pneumocystis* pneumonia. Additionally, hospitalized patients whose immune systems are stressed owing to disease conditions such as leukemia are susceptible to this protozoan infection. The reservoir for *P. carinii* within hospitals has not been identified; in most hospitals patients with this infection are isolated. If untreated *Pneumocystis* pneumonia generally is fatal. During the course of this disease, the alveoli characteristically fill with fluid, thereby preventing gas exchange. The disease can be treated with antiprotozoan drugs such as pentamidine.

Summary

The respiratory tract is constantly exposed to microorganisms in the air. The upper respiratory tract is colonized by a resident microbiota, which contributes to the defense against pathogens, but

Table 23.2 Summary of Diseases Associated with the Respiratory System

Disease	Causative Agent	Mode of Transmission	Treatment
Viral			
Common cold	Rhinoviruses	Aerosols, direct contact with eye tissues	None
Influenza	Influenza viruses	Aerosols containing respiratory secretions	Amantadine in some cases
Bacterial			
Sore throat	*Streptococcus pyogenes* and other bacteria	Aerosols containing respiratory secretions	Penicillin, erythromycin
Tonsillitis	*Streptococcus pyogenes* and other bacteria	Aerosols containing respiratory secretions	Penicillin, erythromycin
Diphtheria	*Corynebacterium diphtheriae*	Aerosols containing respiratory secretions	Antitoxin and antibiotics: penicillin, tetracyclines, erythromycin
Whooping cough	*Bordetella pertussis*	Aerosols containing respiratory secretions	Erythromycin, tetracyclines, chloramphenicol
Tuberculosis	*Mycobacterium tuberculosis*	Aerosols containing respiratory secretions; food, e.g., milk	Isoniazid, rifampin
Pneumonia (pneumococcal)	*Streptococcus pneumoniae*	Indigenous bacteria after another infection or stress to defense systems	Penicillin
Primary atypical pneumonia	*Mycoplasma pneumoniae*	Aerosols containing respiratory secretions	Tetracycline, erythromycin
Legionnaires' disease	*Legionella pneumophila*	Aerosols from evaporating water, e.g., from air-conditioning cooling tower water	Erythromycin
Psittacosis	*Chlamydia psittaci*	Aerosols of bird droppings	Tetracyclines
Q fever	*Coxiella burnetii*	Aerosols with dust contaminated by infected animals	Tetracyclines
Fungal			
Histoplasmosis	*Histoplasma capsulatum*	Aerosols of bird droppings	Amphotericin B
Coccidioidomycosis	*Coccidioides immitis*	Aerosols of desert soil	Amphotericin B
Blastomycosis	*Blastomyces dermatitidis*	Aerosols of soil	Amphotercin B
Protozoan			
Pneumocystis pneumonia	*Pneumocystis carinii*	Aerosols or direct contact in cases of compromised defense systems	Pentamidine

the defense systems of the lower respiratory tract lack indigenous microorganisms. Several different defense systems protect the respiratory tract from microbial infections. Notable among these are the muco-ciliary escalator system, which traps and expels microorganisms entering the upper respiratory tract and the phagocytic cells that engulf and digest microorganisms reaching the lower respiratory tract.

Despite these elaborate defense systems, some pathogenic microorganisms are able to establish infections in the respiratory tract. Many infections of the respiratory tract occur as a result of airborne transmission, particularly owing to the spread of aerosols containing pathogenic microorganisms from infected to susceptible individuals. Crowded conditions favor airborne transmission of pathogens that enter the body this way. As a result, outbreaks of diseases caused by such pathogens often affect individuals who work or live together, such as children attending the same school. Respiratory tract infections tend to be more prevalent during the winter months when people are in closer contact and when the defense systems of the respiratory system are stressed.

Table 23.2 presents a summary of microbial infections that primarily involve the respiratory tract.

Study Outline

A. The respiratory tract is designed to permit the exchange of gases between cells, the blood, and the environment.
  1. Air entering the body through the nose and nasal cavity is filtered and warmed as it moves through the pharynx and down the larynx and trachea to the lungs.
  2. In the lungs are the alveoli where gas exchange takes place by diffusion across a concentration gradient established because air entering the alveoli has a high concentration of oxygen and a low concentration of carbon dioxide, and the blood within adjacent lung capillaries is low in oxygen and rich in carbon dioxide.
  3. Oxygen, moving into the blood plasma and then into red blood cells forms weak reversible bonds with hemoglobin, and when this oxygen-rich blood reaches a systemic tissue capillary bed, the oxygen diffuses outward and carbon dioxide moves from tissues into the capillaries.

B. The respiratory tract is protected against the invasion of pathogenic microorganisms by secretions of mucus from goblet cells and subepithelial glands, which contain lysozyme that degrades the cells walls of bacteria; sneezing and coughing remove many microorganisms caught in the respiratory tract; microorganisms that do manage to reach the lower respiratory tract are subject to phagocytosis.
  1. In phagocytosis the microorganism is engulfed by the phagocytic cell and digested.
    a. Macrophages are phagocytic cells that begin as blood cells that migrate out of the circulatory system and become enlarged; in the lungs they are called dust cells and are fixed in the alveolar lining of the lung. There are also wandering macrophages in the lungs.

C. The resident microbiota of the upper respiratory tract typically include the genera *Streptococcus, Staphylococcus, Branhamella, Neisseria, Haemophilus, Bacteriodes*, and *Fusobacterium*, as well as members of the spirochete and coryneform groups, whereas the lower respiratory tract lacks a resident microbiota.

D. Passage through the air is the main route of transmission of pathogens that infect the respiratory tract; potential pathogens enter through the normal inhalation of air and when substances are placed in the mouth.

E. Diseases of the respiratory tract caused by viral pathogens.
  1. The common cold.
    a. Causative organism(s): 25% of adult colds are caused by rhinoviruses of which there are at least 90 immunologically distinct types.
    b. Mode of transmission: via droplets entering the respiratory tract when the muco-ciliary escalator protection system fails; recent evidence suggests that the eyes

are the portal of entry for rhinoviruses and that the viruses migrate to the tissues of the respiratory tract.

c. Symptoms: nasal stuffiness, sneezing, coughing, headache, malaise, sore throat, slight fever.

d. Treatment: none, the common cold is a self-limiting clinical syndrome with recovery usually occurring within a week without complications as a result of the natural immune defense system.

2. Influenza.

a. Causative organism(s): *Myxovirus influenzae* (flu virus) types A, B, and C with different antigenic subtypes that are responsible for different outbreaks of influenza; these antigenic shifts probably result from gene reassortment between animal and human strains of flu viruses and produce new strains of influenza viruses.

b. Mode of transmission: via inhalation of droplets containing flu viruses released into the air as droplets originating from the respiratory tracts of infected individuals.

c. Symptoms: sudden onset of fever, malaise, headache, muscle ache.

d. Treatment: in uncomplicated cases the viral infection is self-limiting and recovery occurs within a week, but complications can lead to bacterial pneumonia and several neurological syndromes; control is achieved by vaccinating individuals prone to the complications resulting from this disease.

F. Diseases caused by bacterial pathogens.

1. Sore throat (pharyngitis)

a. Causative organism(s): most commonly *Staphylococcus aureus* (Gram positive cocci, grape-like clusters, coagulase positive) and *Streptococcus pyogenes* (Gram positive cocci, chains, beta hemolysis on blood agar) in children and *Chlamydia* (obligate intracellular inhabitants detected by serology) in adults.

b. Symptoms: pain in the throat, redness of the lining of the throat, fever, sometimes inflammation of the tonsils (tonsillitis).

c. Treatment: throat swabs from infections suspected of being caused by *S. pyogenes* are taken and cultured because its toxins can spread to other parts of the body and cause more serious diseases, such as scarlet and rheumatic fevers; these infections can be treated with penicillin and erythromycin.

2. Pneumonia.

a. Causative organism(s): most frequently *Streptococcus pneumoniae* (Gram positive cocci, chains of cells, alpha hemolysis) but also *Staphylococcus aureus, Haemophilus influenzae* (Gram negative rods), and *Klebsiella pneumoniae* (Gram negative encapsulated rods).

b. Mode of transmission: often occurs as a complication of another disease condition, especially when the host defense mechanisms are compromised; bacteria that cause pneumonia most frequently enter the lungs in air and also through the blood stream; often acquired by the endogenous route from bacteria in the upper respiratory tract.

c. Symptoms: sudden onset of high fever, production of colored, purulent sputum, and congestion; in most cases a sore throat preceeds the development of pneumococcal pneumonia.

d. Treatment: antibiotics, with penicillin the drug of choice for treating pneumonia caused by *S. pneumoniae*. Vaccine available for pneumococcal pneumonia.

3. Primary atypical pneumonia.

a. Causative organism(s): *Mycoplasma pneumoniae* (lack cell wall and form colonies with a fried egg appearance on agar).

b. Treatment: tetracycline and erythromycin.

4. Legionnaire's disease (Legionnellosis).

a. Causative organism(s): *Legionella pneumophila* (Gram negative rods, culturable on complex media and detectable by fluorescent antibody staining).

b. Mode of transmission: bacteria become airborne in aerosols, the inhalation of which can lead to onset of illness; adequate treatment of rapidly evaporating water bodies can be used to control spread of this disease.
c. Symptoms: typical pneumonia symptoms, may also be characterized by kidney and liver involvement and associated gastrointestinal symptoms.
d. Treatment: erythromycin.

5. Psittacosis (ornithosis or parrot fever).
   a. Causative organism(s): *Chlamydia psittaci* (obligate intracellular inhabitants that can be demonstrated in tissue culture).
   b. Mode of transmission: from birds to human beings via aerosol dispersal of droplets and contaminated dust particles.
   c. Symptoms: fever, headache, malaise, and coughing.
   d. Treatment: tetracyclines.
6. Bronchitis.
   a. Causative organism(s): most frequently, *Mycoplasma pneumoniae* and various types of viruses; can also be caused by *Streptococcus* sp., *Staphylococcus* sp., *Haemophilus influenzae*.
   b. Mode of transmission: almost always occurs as a complication of another disease condition.
   c. Symptoms: onset is marked by development of a cough that eventually yields mucopurulent sputum.
   d. Treatment: antibiotics, such as penicillin and tetracycline.
7. Diphtheria.
   a. Causative organism(s): *Corynebacterium diphtheriae* strains that harbor a lysogenic phage that produce a protein toxin (Gram positive rods that exhibit snapping division and Chinese letter formation).
   b. Mode of transmission: via droplets from an infected individual to a susceptible host.
   c. Symptoms: pharyngitis, low grade fever, cough, difficulty in swallowing, and swelling of the lymph glands.
   d. Treatment: antitoxin to block the cytopathic effects of diphtheria toxin, antibodies to prevent the occurrence of serious symptoms, and antibiotics to eliminate the bacterial infection. Vaccine available.
8. Whooping cough (pertussis).
   a. Causative organism(s): *Bordetella pertussis* (Gram negative rods).
   b. Symptoms: resemble those of the common cold but include the distinctive characteristic cough.
   c. Treatment: erythromycin and tetracyclines; maintain adequate oxygen supply. Vaccine available.
9. Tuberculosis.
   a. Causative organism(s): *Mycobacterium tuberculosis* (acid-fast rods).
   b. Mode of transmission: via droplets from an infected to a susceptible individual; also through the ingestion of contaminated food.
   c. Symptoms: fatigue, weight loss, fever, lesions in the lung tissues.
   d. Treatment: multiple antibiotics, such as streptomycin, refampin, and isoniazid. Vaccine available.
10. Q fever.
   a. Causative organism(s): *Coxiella burnetii*.
   b. Mode of transmission: direct contact of the bacteria with the eyes and via the gastrointestinal tract; *C. burnetii* is maintained within cattle and sheep populations where it is transmitted via tick vectors; bacteria also become airborne on fomites.
   c. Symptoms: fever, headache, chest pain, nausea, and vomiting.
   d. Treatment: tetracyclines and cloramphenicol.

G. Diseases caused by fungal pathogens.

1. Histoplasmosis.

a. Causative organism(s): *Histoplasma capsulatum*.

b. Mode of transmission: via inhalation of spores that are deposited in the lungs.

c. Symptoms: resemble those of a mild cold.

d. Treatment: amphotericin B.

2. Coccidioidomycosis.

a. Causative organism(s): *Coccidioides immitis*.

b. Mode of transmission: inhalation of dust particles containing conidia of this fungus.

c. Symptoms: chest pain, fever, malaise, and a dry cough.

d. Treatment: amphotericin B when there is evidence of systemic dissemination.

3. Blastomycosis.

a. Causative organism(s): *Blastomyces dermatidis*

b. Symptoms: skin lesions.

c. Treatment: amphotericin B.

H. Diseases caused by protozoa.

1. *Pneumocystis* pneumonia.

a. Causative organism(s): *Pneumocystis carinii*.

b. Mode of transmission: associated with failures of the immune system as occurs in cases of AIDS and other patients whose immune defense systems are impaired.

c. Treatment: antiprotozoan drugs such as pentamidine.

Study Questions

1. For each of the following diseases, what are the causative organisms and characteristic symptoms?

a. Influenza
b. Tuberculosis
c. Histoplasmosis
d. Legionnaire's disease
e. Whooping cough

2. Name five diseases that are routinely prevented today through the use of prophylactic immunization.

3. Match the following statements with the terms in the list to the right.

a. Causative agents are three antigenic types of *Myxovirus*, designated A, B, C.

b. Characterized by localized inflammation of upper respiratory tract and release of mucus secretions, generally accompanied by sneezing and sometimes coughing.

c. Most frequent infectious human disease.

d. Caused by *Chlamydia psittaci*.

e. Caused by *Mycoplasma pneumoniae*.

f. Inflammation of lungs involving alveoli.

g. Most frequent etiologic agent is *Streptococcus pneumoniae*.

h. Caused by airborne transmission of *Legionella pneumophila*.

i. Results from protein toxin produced by strains of *Corynebacterium diphtheriae* harboring a lysogenic phage.

j. In addition to usual symptoms of pneumonia, it is often characterized by kidney and liver involvement and unusually high incidence of associated gastrointestinal symptoms.

k. Multiple antibiotic treatment is generally prolonged; streptomycin, rifampin, and isoniazid are used.

l. Caused by *Mycobacterium tuberculosis*.

m. Symptoms include distinctive cough and others resembling those of the common cold.

(1) Common cold
(2) Influenza
(3) Pneumonia
(4) Primary atypical pneumonia
(5) Legionnaire's disease
(6) Diphtheria
(7) Tuberculosis
(8) Whooping cough
(9) Histoplasmosis
(10) Coccidioidomycosis
(11) Blastomycosis

n. Caused by *Bordetella pertussis*.

o. Systemic disease caused by *Blastomyces dermatidis*.

p. Caused by *Coccidioides immitis*.

q. Normally occurs in southwestern United States; transmission involves inhalation of dust particles containing conidia of *C. immitis*.

r. Caused by *Histoplasma capsulatum*.

Some Thought Questions

1. If your child has symptoms of the flu, should you give him or her aspirin?
2. Should the western custom of shaking hands with strangers be replaced with the Japanese custom of bowing or the French custom of kissing on the cheeks?
3. Why was Legionnaire's disease not detected until 1976?
4. Why do influenza outbreaks recur?
5. Why are respiratory tract infections more common in the winter months?
6. Can we find a cure for the common cold?
7. Why are pigeons a public health nuisance?
8. Should we kill blackbirds and starlings to protect human health?
9. Why is pneumonia caused by *Mycoplasma pneumoniae* called atypical pneumonia?

Additional Sources of Information

Boyd, R. F., and B. G. Hoerl. 1981. *Basic Medical Microbiology*. Little, Brown and Company, Boston. A basic text emphasizing microbes and disease that includes several chapters on respiratory tract diseases.

Braude, A. I. (ed.). 1980. *Medical Microbiology and Infectious Disease*. W.B. Saunders Company, Philadelphia. An advanced work with individual chapters dealing with specific diseases of the respiratory tract; chapters dealing with infectious diseases are organized according to the taxonomy of the causative agent.

Davis, B. D., R. Dulbecco, H. N. Eisen, H. S. Ginsberg, and W. B. Wood. 1981. *Microbiology*. Harper & Row, Publishers, Hagerstown, Maryland. An excellent text used in many graduate and medical school microbiology courses dealing with all aspects of microbiology; chapters dealing with infectious diseases are organized according to the taxonomy of the causative agent.

Eron, C. 1981. *The Virus that Ate Cannibals*. Macmillan Publishing Co. Inc., New York. This popular work includes a chapter on the anatomy of a cold that describes experiments suggesting transmission of the common cold occurs via the eyes.

Fraser, D. W., and J. E. McDade. 1979. Legionellosis. *Scientific American*, **241**(4):82–99. This article discusses the fascinating investigations that led to the discovery of the cause of Legionnaire's disease.

Hoeprich, P. D. (ed.). 1977. *Infectious Diseases: A Modern Treatise of Infection Processes*. Harper & Row, Publishers, Hagerstown, Maryland. This classic treatment of infectious diseases makes an excellent reference volume.

Jawetz, E., J. L. Melnick, and E. A. Adelberg. 1980. *A Review of Medical Microbiology*. Lange Medical Publications, Los Altos, California. A very good review of infectious diseases.

Kaplan, M. M., and R. G. Webster. 1977. The epidemiology of influenza. *Scientific American*, **237**(6):88–106. This article discusses the antigenic changes in influenza viruses that lead to major outbreaks of influenza.

Joklik, W. K., and H. P. Willett (eds.). 1980. *Zinsser Microbiology*. Appleton Century Crofts, New York. A classic text used in many medical school programs that includes chapters on infectious diseases.

Rose, N. L., and A. L. Rose (eds.). 1983. *Microbiology: Basic Principles and Clinical Applications*. Macmillan Publishing Co., Inc., New York. An advanced work on medical microbiology; chapters are organized for the most part according to taxonomic groups but one chapter is devoted to the diagnosis of diseases affecting the respiratory tract.

Schlesinger, R. B. 1982. Defense mechanisms of the respiratory system. *BioScience*, **32**(1):45–50. An interesting article on the defenses that protect the respiratory tract against microbial infections.

Stuart-Harris, C. 1981. The epidemiology and prevention of influenza. *American Scientist*, **69**:166–172. This well-written article describes the epidemiology of influenza and how the spread of this disease can be prevented.

Youmans, G. P., P. Y. Paterson, and H. M. Sommers. 1980. *The Biological and Clinical Basis of Infectious Disease*. W. B. Saunders Company, Philadelphia. This is an excellent text covering most human infectious diseases.

24 *Gastrointestinal Tract Diseases*

KEY TERMS

amebiasis
amebic dysentery
appendicitis
bacterial dysentery
balantidiasis
cestodes
cholera
cytomegalovirus inclusion disease
cytomegaly
cysticercosis
definitive hosts
enterotoxin
flukes
food intoxication
food poisoning
gastroenteritis
giardiasis
hepatitis
hookworms
hydatid cysts
intermediate hosts
metacercariae
mumps
nematodes
orchitis
perfringens food poisoning
pinworms
proglottid
roundworm disease
roundworms
salmonellosis
scolex
shigellosis
tapeworms
trematodes
trichinosis
trophozoites
whipworms
yersiniosis

OBJECTIVES

After reading this chapter you should be able to

1. Define the key terms.
2. Describe the structure of the gastrointestinal tract.
3. Describe the resident microbiota of the gastrointesinal tract.
4. Describe the defense systems of the gastrointestinal tract that protect against infectious microorganisms.
5. Discuss the causative agents, routes of transmission, key symptoms, and treatments for the following diseases: mumps, cytomegalovirus inclusion disease, viral gastroenteritis, infectious hepatitis, perfringens food poisoning, staphylococcal food poisoning, salmonellosis, shigellosis, yersiniosis, juvenile dysentery, traveler's diarrhea, appendicitis, cholera, amebic dysentery, giardiasis, balantidiasis.

Structure of the Gastrointestinal Tract

The gastrointestinal tract consists of a series of interconnected organs that are involved in food reception, digestion, nutrient absorption, and waste elimination (Figure 24.1). The alimentary tract effectively is a tube that runs through the body carrying food. This digestive tube consists of the mouth, oral cavity, pharynx, esophagus, stomach, small and large intestine, rectum, and anus. These structures have an inner lining of mucosa backed by connective tissue. The accessory organs of the digestive system develop from the wall of the digestive tube. These accessory organs, which aid in the process of digestion, include the teeth, tongue, salivary glands, liver, and pancreas. Although it is a part of the digestive system, we will consider the oral cavity separately in a later chapter.

The process of digestion begins in the mouth where the food is macerated by the teeth and mixed with saliva. The salivary glands secrete saliva, which contains amylase, to hydrolyze starch and a glycoprotein that lubricates the chewed and moistened food, easing its passage to the stomach. The tongue is used to force the moistened ball of food into the pharynx and subsequently into the esophagus. The epiglottis blocks the passage of food into the respiratory tract, ensuring that the food passes from the pharynx into the esophagus. Food moves through the esophagus by peristalsis, the alternating progression of contracting and relaxing muscle movements along the length of the tube that pushes the food mass into the stomach.

The stomach is a muscular, elastic sac where the initial digestion of proteins begins. Glands in the stomach mucosa secrete mucus, hydrochloric acid, and enzymes. The mucus protects the lining of the stomach against the digestive action of the acid and enzymes. The hydrochloric acid denatures the incoming proteins, and the enzyme pepsin begins the digestion of proteins in the stomach. Although broken down in the stomach, proteins are not absorbed into the body across the lining of the stomach. Rather, the partially digested food leaves the stomach at a carefully controlled rate, entering the small intestine where enzymatic digestion continues.

Most of the digestive processes occurs within the small intestine. Here a series of enzymes, many of which come from accesory organs, mix with the food. Enzymes from the pancreas that function within the small intestine include: trypsin, chymotrypsin, and peptidases for protein digestion; lipases for fat digestion; and amylase and disaccharidases for carbohydrate digestion. Bile produced by the gall bladder enters the small intestine via the liver and emulsifies fats, rendering these water-insoluble materials more susceptible to enzymatic attack. The nutrients that are released by the digestion of foods are absorbed across the lining of the small intestine. Absorption of most nutrients occurs through the villi, which are dense outfoldings of the mucosa of the small intestine. Villi are covered with millions of thread-like projections, called microvilli, which greatly increase the membrane surface area available for absorption of nutrients from the digested food. As nutrients move across the epithelial lining of the villus, amino acids and sugars enter the capillaries and lipids enter the lymph vessels that are closely associated with each villus.

Within 2–5 hours after eating, undigested residues enter the large intestine or colon. The colon functions in absorbing water and minerals. These residues are now in a highly liquid state, having been mixed with ingested water and gastric and intestinal juices. The colon secretes mucus, which helps lubricate the remaining residues as they move to the rectum. From the rectum the residues move out through the anus, the terminal opening of the digestive tube.

Defenses of the Gastrointestinal Tract

Although we regularly ingest viable microorganisms with our food, relatively few survive passage through the gastrointestinal tract. The very low pH of the stomach creates a barrier for pathogens entering the digestive tract because they are unable to tolerate the acidity. Exposure to the gastric juices secreted into the stomach kills most microorganisms. Bile and digestive enzymes in the intestines further reduce the numbers of surviving microorganisms. Thus, the numbers of viable microorganisms are greatly reduced during passage through the stomach and upper portion of the small intestine.

Some microorganisms, however, are protected within food particles from the acidity and digestive enzymes of the stomach and upper portions of the small intestine. Spores may also survive passage through the stomach. Some of these microorganisms reach the regions of the gastrointesinal tract that are more permissive to microbial growth. The

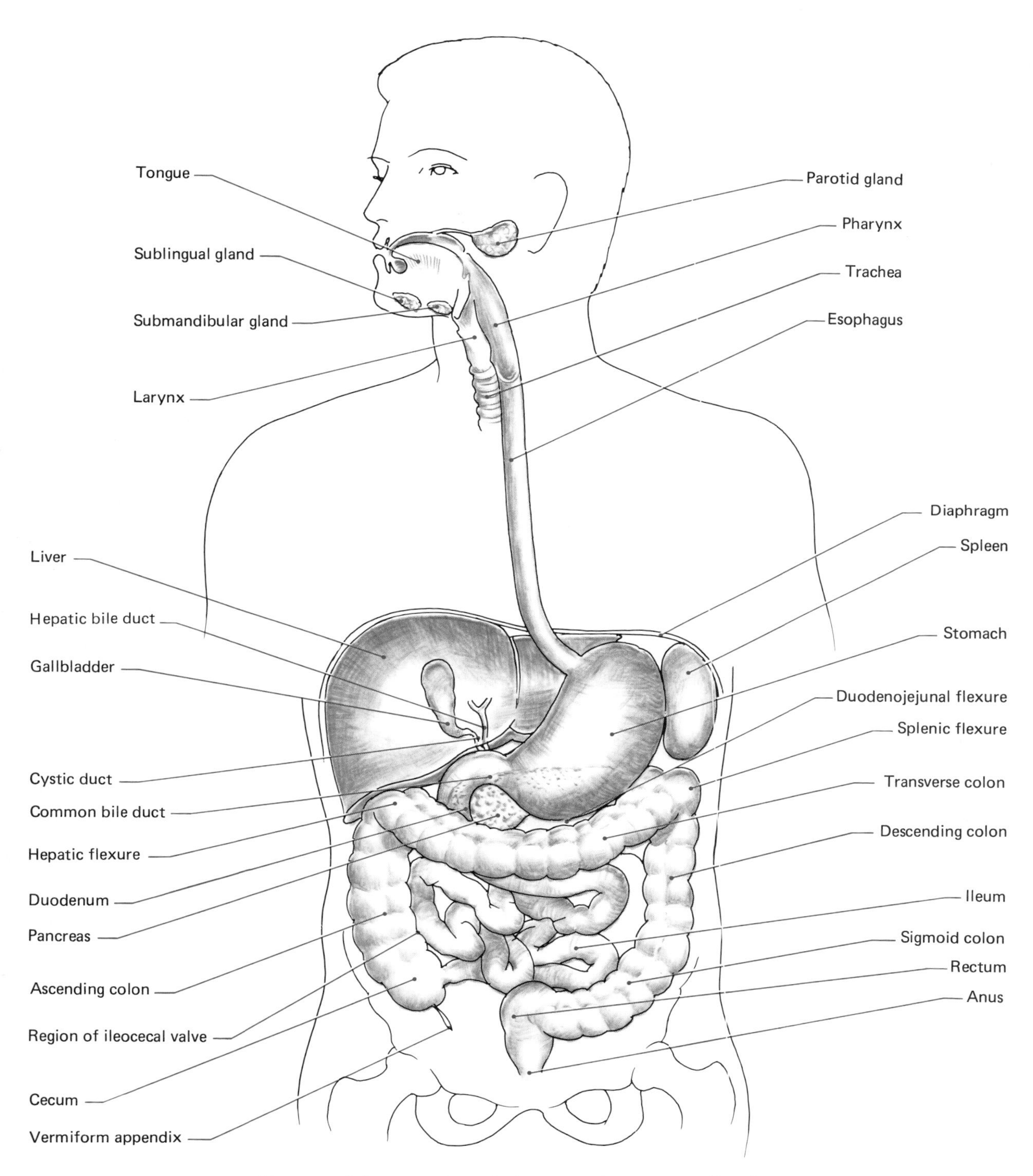

Figure 24.1 Diagram of the gastrointestinal tract and accessory digestive organs.

mucous membranes that line the intestinal tract make it difficult for pathogenic microorganisms to attach to and penetrate the tract lining, thereby limiting infections of the gastrointesinal tract. Additionally, the presence of large populations of indigenous microbiota in the lower intestinal tract protect the host against invasion by pathogens; the natural microbiota of the gastrointestinal tract enter into antagonistic relationships with nonindigenous microorganisms. There are also plasma cells in the mucus of the digestive tract that produce antibodies which help keep the pathogens in check. As a result, most nonindigenous microorganisms entering the intestinal tract are degraded during passage through the gastrointestinal system or are removed along with large numbers of indigenous microorganisms in the passage of fecal material from the body.

Resident Microbiota of the Gastrointestinal Tract

A large resident microbiota develops in the human intestinal tract after birth. These resident microorganisms are important for the maintenance of good health and are usually not involved in disease processes. The resident microbiota of the gastrointestinal tract are normally noninvasive and are associated with the surface tissues and ingested food material. The initial residents of the intestinal tract in breast-fed infants are members of the genus *Bifidobacterium*, whereas in bottle-fed infants *Lactobacillus* species initially colonize the intestinal tract.

These initial colonizers of the intestinal tract are later displaced by other bacterial species, which include both obligate and facultative anaerobes and members of the genera *Lactobacillus, Streptococcus, Clostridium, Veillonella, Bacteroides, Fusobacterium*, and coliform bacteria (Gram negative rods that ferment lactose with acid and gas formation, for example, *Escherichia coli*). The enteric bacteria, including members of the genera *Escherichia, Proteus, Klebsiella*, and *Enterobacter*, are Gram negative facultative anaerobes. Although these facultative anaerobes were once thought to comprise the majority of the microorganisms inhabiting the intestinal tract, methodological improvements for the isolation and culture of obligate anaerobes have revealed that up to 99 percent of the intestinal mi-

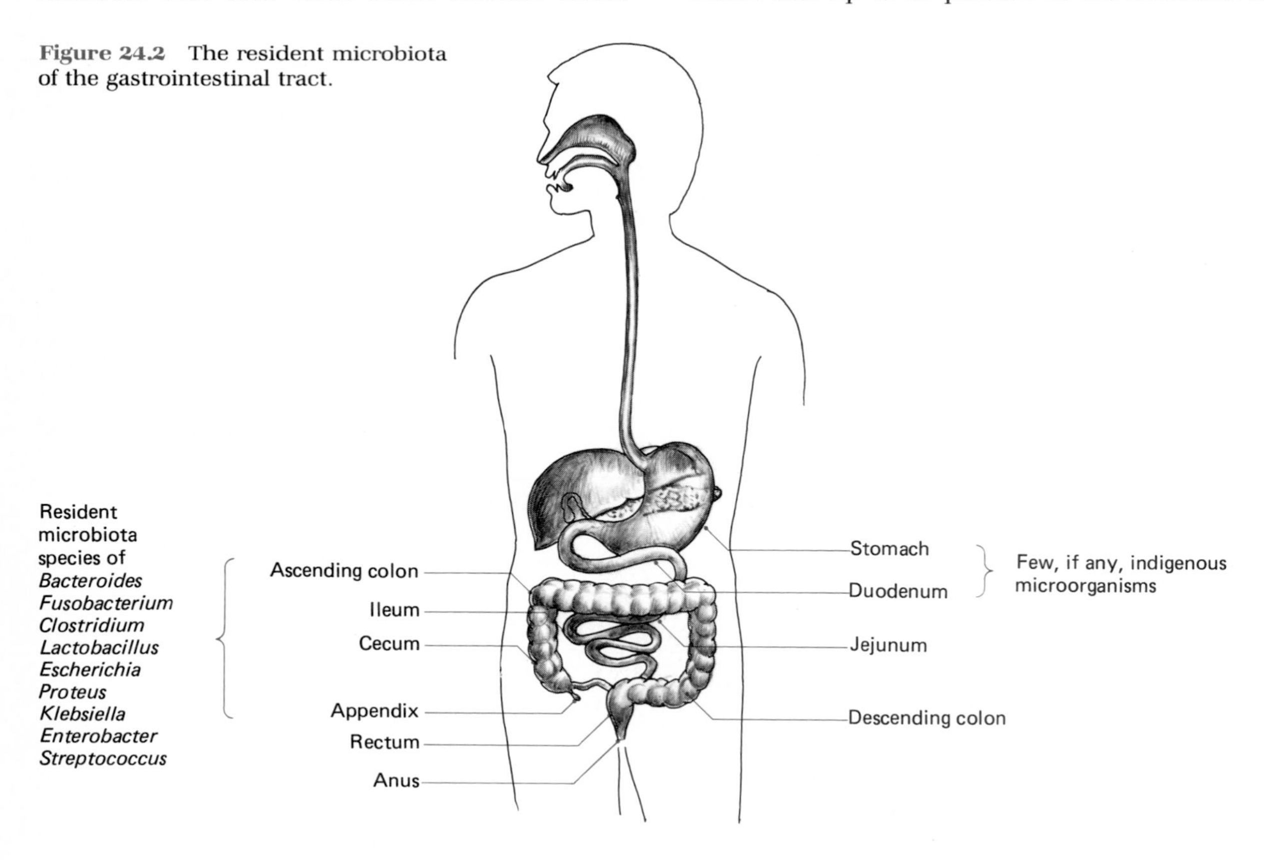

Figure 24.2 The resident microbiota of the gastrointestinal tract.

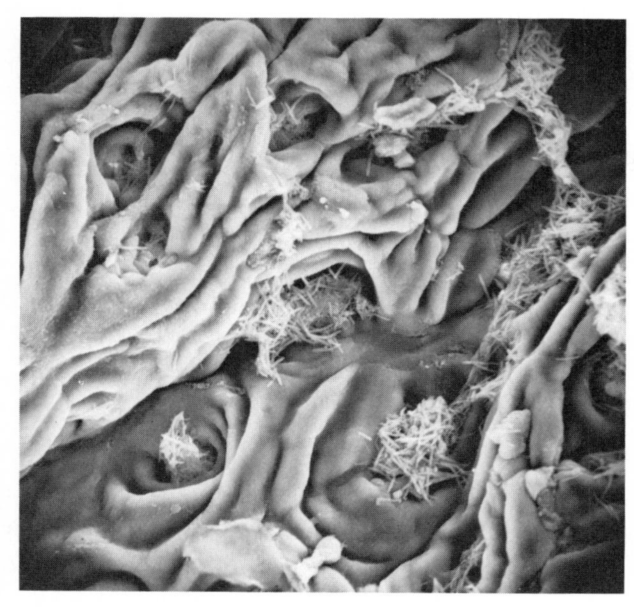

Figure 24.3 The microbiota of the intestine are shown in the mucous membrane of the large intestine. The bacteria (400 ×) are in crypts. (From BPS: D. C. Savage, University of Illinois, Champaign-Urbana.)

crobiota may be obligate anaerobes of the genera *Bacteroides* and *Fusobacterium*. The actual proportions of the individual bacterial populations within the indigenous microbiota of the intestinal tract vary, depending in part on the diet of the host.

Additionally the indigenous microbiota are not evenly distributed within the gastrointesinal tract. The low pH of the stomach precludes the existence of an indigenous microbiota associated with the lining of that organ in individuals with normal gastric acidity. Similarly the low pH and abundant enzymes of the upper reaches of the small intestine do not favor the development of a resident microbiota. In cases of pernicious anemia, where the pH of the gastrointestinal tract is higher than normal, a resident microbiota does develop in the stomach. In normal individuals an abundant indigenous microbiota is associated with the lower regions of the small intestine and the colon or large intestine where the pH is higher than within and near the stomach (Figure 24.2). The constant temperature, 37°C, and availability of water and nutrients makes the large intestine a particularly favorable habitat for the growth of a variety of microbial populations. Food, partially digested by the body's enzymes, is continuously supplied to the microorganisms inhabiting the intestinal tract, supporting the growth of a large resident microbiota. Consequently, the highest numbers of resident microbiota are associated with the large intestine (Figure 24.3).

Diseases of the Gastrointestinal Tract

Despite the relatively effective defenses of the gastrointestinal tract, some pathogenic microorganisms possess toxigenic or invasive properties that permit them to cause gastrointestinal tract diseases. Most diseases of the gastrointestinal tract originate with tainted food or water although other routes of disease transmission are associated with some diseases of the digestive system.

There are two distinct processes that generally initiate the onset of diseases of the gastrointestinal tract. In one case microorganisms growing in food or water can produce toxins, and the ingestion of these microbial toxins initiates a disease process. Such diseases are classified as food poisoning or food intoxication because the etiologic agents of the disease need not grow within the body. In cases of food poisoning there is no true infectious process. Toxins absorbed through the gastrointestinal tract can cause neural damage and death in some cases, as well as localized inflammation and gastrointestinal upset in others. In the second type of disease-causing process, invasive pathogens establish an initial infection through the gastrointestinal tract and cause localized gastrointestinal upset or systemic disease symptomology. Generally, the establishment of an infection of the gastrointestinal tract requires a relatively large infectious dose. A high infectious dose containing a relatively large number of pathogenic microorganisms can overcome the inherent defense mechanisms of the gastrointestinal tract. Quite different measures are required to prevent and treat infectious gastrointestinal diseases and those specific microorganisms responsible for food poisoning.

Diseases Caused by Viral Pathogens

Mumps. Mumps is characterized by the enlargement of one or more of the salivary glands. The enlargement of the salivary glands results from the replication of an enveloped RNA virus (paramyxovirus) within those glands. Swelling on both sides (bilateral parotitis) occurs in about 75 percent of patients, with swelling in one gland usually preceding swelling on the other side by about 5 days. This disease primarily occurs during childhood.

I had a feeling in my neck,
And on the sides were two big bumps;
I couldn't swallow anything
At all because I had the mumps . . .

—ELIZABETH MADOX ROBERTS, *Mumps*

The mumps virus is transmitted via contaminated droplets of saliva. The initial infection appears to occur in the upper respiratory tract, with subsequent dissemination to the salivary glands and other organs. The average incubation period for mumps is 18 days, and the swollen salivary glands generally persist for less than 2 weeks. The mumps virus may spread to various body sites, and although the effects of the disease are normally not long lasting, there can be several complications; for example, mumps is a major cause of deafness in childhood. In males past puberty, the mumps virus can cause orchitis (inflammation of the testes), but old wive's tales to the contrary, mumps rarely results in male sterility. There is no specific treatment for mumps, but the disease can be prevented by vaccination with MMR (measles-mumps-rubella) vaccine.

Cytomegalovirus Inclusion Disease. Cytomegalovirus (CMV) is the cause of the severe and often fatal disease of newborns known as cytomegalovirus inclusion disease. This disease can be transmitted across the placenta from mother to fetus or during the process of birth. The congenital transmission of this virus can cause various abnormalities. CMV is a herpesvirus that establishes a continuous latent infection within the body, probably within lymphocytes. The virus may remain inactive within the body or may multiply within a variety of organs to produce this serious disease. In newborns, CMV generally affects the salivary glands and liver as well as the kidneys and lungs. The reproduction of CMV causes abnormalities of the infected human cells characterized by an increase in size (cytomegaly) and the development of inclusion bodies within the nucleus. Between 10 and 20 percent of stillborn infants show abnormal cells characteristic of cytomegalovirus inclusion disease. Up to 80 percent of adults over 35 years show antibodies that indicate prior or active infection with CMV. In most cases the virus is inactive, producing a latent infection. The viral infection can be activated when the immune system is suppressed, such as when immunosuppresive drugs are used after transplants. Pregnancy and blood transfusions also activate CMV infections.

Viral Gastroenteritis. Gastroenteritis involves an inflammation of the lining of the gastrointestinal tract. In most cases viral gastroenteritis is a self-limiting disease, often referred to as the 24-hour or intestinal flu. Viral gastroenteritis is not caused by an influenza virus and is not related to true cases of flu, but rather this disease can be caused by several different viruses including adenoviruses, coxsackieviruses, polioviruses, and members of the ECHO virus group. Viruses causing gastroenteritis normally replicate within cells lining the gastrointestinal tract, and large numbers of viruses are released in fecal matter. Contamination of food with fecal matter is an important route of transmission of viral gastroenteritis as well as many other diseases caused by microorganisms that enter via the gastrointestinal tract.

The Norwalk agent, a small DNA virus identified as being responsible for an outbreak of "winter vomiting disease" that occurred in Norwalk, Ohio in 1968, appears to be an important etiologic agent of various viral gastroenteritis outbreaks. Rotavirus, a large RNA virus, also appears to be a very common etiologic agent for diarrhea in infants, particularly in socioeconomically depressed regions of the world.

The characteristic symptoms of a viral gastroenteritis include sudden onset of gastrointestinal pain, vomiting, and/or diarrhea (Figure 24.4). Recovery normally occurs within 12–24 hours of the onset of disease symptomology. As a result of the vomiting and diarrhea, there can be a severe loss of body fluids and dehydration. The loss of water and resultant imbalances in electrolytes can have serious consequences, particularly in infants, where viral gastroenteritis is sometimes fatal. Treatment generally is supportive and may involve replenishment of body fluids and maintenance of proper electrolyte balance.

Hepatitis Type A. Hepatitis is a systemic viral infection that primarily affects the liver. There are several types of hepatitis viruses designated types A, B, and nonA–nonB. Type A hepatitis virus normally enters the body via the gastrointestinal tract and causes type A hepatitis (formerly called infectious hepatitis). There are nearly 25,000 cases of hepatitis A reported each year in the United States. Hepatitis type A virus is usually transmitted by the fecal–oral route and is prevalent in areas with inadequate sewage treatment. The virus is shed in the feces of infected individuals. Several outbreaks of viral hepatitis have been associated with contaminated shellfish that have concentrated viruses from sewage effluents. Outbreaks have also occurred when untreated rural water sources become contaminated with fecal matter. Even in areas with chlori-

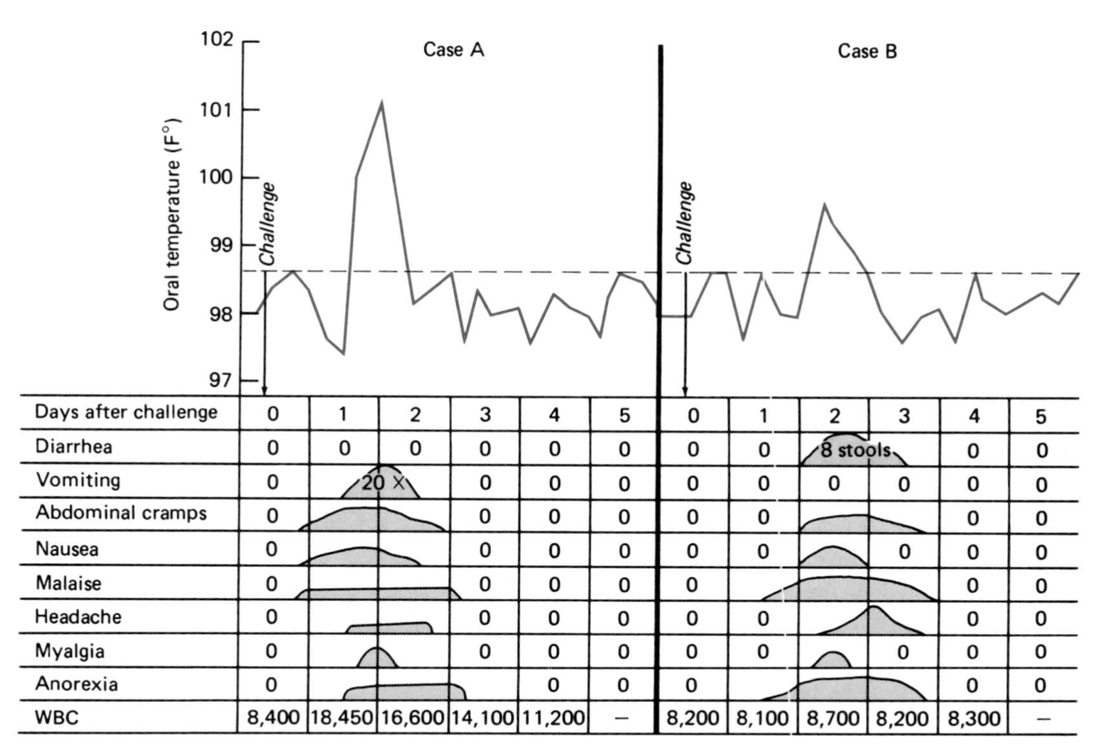

Figure 24.4 The time course of symptoms in viral gastroenteritis is shown in this illustration of the response of two volunteers to oral administration of stool filtrate derived from a volunteer who received an original Norwalk rectal-swab specimen. The height of the shaded curve is roughly proportional to the severity of the sign or symptom. Essentially, this sort of experiment both confirms the cause of disease by Koch's postulates and demonstrates the variability of symptomology; in viral diseases such as this, human subjects rather than experimental animals are used because of the specificity of the virus–host relationship. (Reprinted by permission of the University of Chicago Press from R. Dolin *et al.*, 1971, "Transmission of acute infectious nonbacterial gastroenteritis to volunteers by oral administration of stool filtrates," *Journal of Infectious Diseases,* 123:307.)

nated water supplies outbreaks of hepatitis A can occur because of the relative resistance of the virus to chlorine.

The incubation period for hepatitis A virus is several weeks before disease symptoms appear. The initial symptoms of type A hepatitis include fever, abdominal pain, and nausea, followed by jaundice, the yellowing of the skin indicative of liver impairment caused by the virus (Figure 24.5). The time course of the body's response to this disease was shown in Figure 22.9. Damage to liver cells also results in increased serum levels of enzymes, such as transaminases, normally active in liver cells. The detection of increased serum levels of these enzymes is used in diagnosing this disease. Type A hepatitis tends to be less serious than types B and nonA–nonB hepatitis. In most cases of type A hepatitis, the infection is self-limiting and recovery occurs within 4 months. There is no specific treatment for type A hepatitis. The incidence of type A hepatitis, however, can be reduced by using passive immunization with immunoglobulin G. Passive immunization is used when there is a high probability of exposure to the hepatitis A virus.

Diseases Caused by Bacterial Pathogens

Staphylococcal Food Poisoning. Strains of *Staphylococcus aureus*, which cause food poisoning, produce an enterotoxin, that is, a toxin that causes an inflammation of the lining of the gastrointestinal tract. *S. aureus* are Gram positive cocci that form grape-like clusters. Staphylococcal enterotoxins do

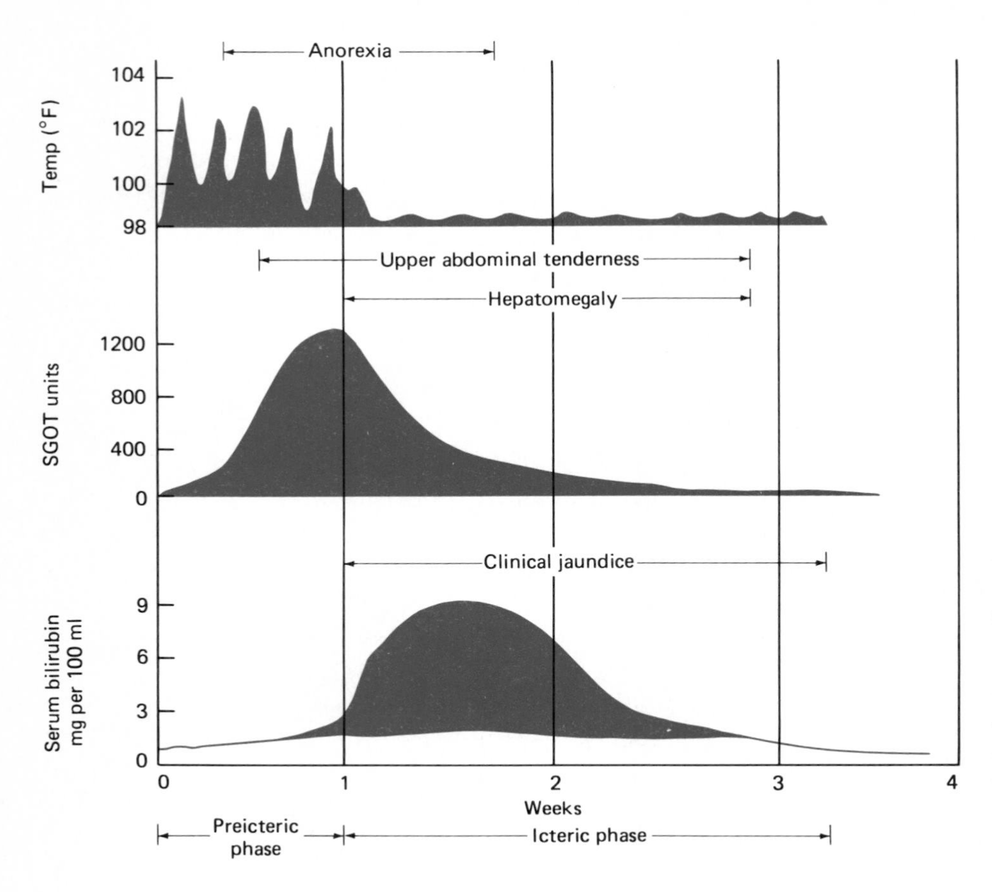

Figure 24.5 The time course of symptoms in infectious hepatitis shows that jaundice occurs beginning 1 week after the initial fever. (After W. P. Havens, Jr., and R. Paul, in *Viral and Rickettsial Infections of Man*, F. L. Horsfall Jr. and I. Tamm, eds., J. B. Lippincott Company, Philadelphia.)

not exert a direct local effect but are absorbed through the blood stream and circulate back to the digestive tract to initiate a staphylococcal food poisoning syndrome. The symptoms of staphylococcal food poisoning occur relatively rapidly after ingestion of toxin-containing food, usually within 1–6 hours. The greater the dose of toxin, the sooner the onset of disease symptoms and generally the more severe the disease. These symptoms generally include nausea, vomiting, abdominal pain, and diarrhea. Fever does not occur. Symptoms generally subside within 8 hours of their onset, and complete recovery usually occurs within a day or two.

In one day: Spoiled or improperly handled food makes 27,000 Americans sick.

—TOM PARKER, *In One Day*

Enterotoxin-producing strains of *S. aureus* often enter foods from the skin surfaces of people handling food. The prevention of staphylococcal food poisoning depends on proper handling and preservation of food products to prevent contamination and subsequent growth of enterotoxin-producing strains of staphylococci. *S. aureus* is able to reproduce within many different types of food products. Custard-filled bakery goods, dairy products, processed meats, potato salad, and various canned foods are frequently found to be the source of the toxin in cases of staphylococcal food poisoning. Salads prepared for a summer picnic can easily be contaminated (inoculated) with *S. aureus*, and when left in the sun in a wicker basket (incubated), the bacteria can multiply, producing a sufficient amount of enterotoxin to provide an unexpected nighttime encore to the day's fun.

Perfringens Food Poisoning. *Clostridium perfringens*, a Gram positive endospore-forming anaerobic rod–shaped bacterium, accounts for over 10 percent of the outbreaks of food-borne disease in the United States. The ingestion of food containing toxin produced by *C. perfringens* and the adsorption of the toxin into the cells lining the gastrointestinal tract initiate the disease known as perfringens food poisoning. Toxin type A of *C. perfringens* is asso-

Publicity About Airline Incident Leads to Crucial Diagnosis

By LAWRENCE K. ALTMAN, M.D.

SAN FRANCISCO—One day after Kim Wah Chan flew here on a visit from London last March, he developed severe abdominal pain, nausea, vomiting and diarrhea. Doctors at San Francisco General Hospital rushed him to the operating room because tests indicated his life was being threatened by a condition called small bowel obstruction.

The diagnosis was reasonable because Mr. Chan, a British citizen who is 42 years old, had had surgery for that problem 10 years ago in England. It was due to a long-term complication of an earlier automobile accident, and now the San Francisco surgeons thought it had recurred.

However, to their surprise this time there was no blockage. What followed was an extraordinary medical story.

After the exploratory operation, Mr. Chan's condition got worse. His blood pressure fell drastically, he bled severely and he could not breathe without the aid of a mechanical respirator. The surgeons immediately operated again to determine if they had done something wrong during the first operation. Again they found nothing to explain his medical problem. His kidneys failed, making his condition even worse.

Mr. Chan's life hung in the balance as the doctors added dialysis treatment to a regimen that included several drugs.

Then about three days later the hospital's microbiologists identified a bacterium called *Salmonella enteritidis* in several specimens taken from Mr. Chan.

But the meaning of the report was unclear. *Salmonella* bacteria often cause a common condition called salmonellosis; it usually produces much milder symptoms. Also, some people are carriers of salmonella, harboring the bacteria in their bowel without becoming ill.

For those and other reasons, Mr. Chan's doctors were puzzled. Was the detection of salmonella merely an incidental finding unrelated to his illness? Or were all of Mr. Chan's symptoms due to salmonellosis? If so, why was it so severe? Where did he get the infection?

Two days later while Dr. Steven H. Fugaro, one of Mr. Chan's physicians, was sipping coffee in the cafeteria he read a newspaper article about an outbreak of food poisoning aboard British Airways jets.

"How horrible to fly somewhere and to get sick on arrival," Dr. Fugaro recalls saying to himself. A moment later he read that the outbreak was due to *Salmonella enteritidis*.

Dr. Fugaro raced upstairs to the intensive care unit and asked Mr. Chan, "Did you fly British Airways."

Mr. Chan could not speak because he had a tube in his throat and "he must have thought I was crazy because I yelled so," Dr. Fugaro said in an interview.

Mr. Chan nodded, and Dr. Fugaro became excited. For the first time the diagnosis was clear. Mr. Chan had an extraordinarily severe case of salmonellosis.

As it turned out, he was one of at least 766 passengers and crew members who suffered salmonellosis in what probably was the worst airline food poisoning epidemic in history. The worst previous outbreak involved almost 300 passengers who flew aboard several flights of a French airline in 1976.

Most people shrug off food poisoning as a common nuisance because the symptoms, although often distressing, usually disappear in a few hours or days without any serious consequences. But Mr. Chan's case is one of the extraordinary ones that illustrate how at times food poisoning can be severe, even deadly, and how its diverse symptoms can fool even the best doctors into thinking they are due to something else. Mr. Chan's doctors were faculty members of the University of California at San Francisco, a leading medical school.

Today, Mr. Chan continues to fight for his life. He has had at least a dozen more operations because of complications of salmonellosis. At one point his abdomen and bowel swelled to the point where Mr. Chan's young son said: "It's like a pumpkin about to burst."

Mr. Chan has overcome pneumonia, the bleeding problem and kidney failure. But because his muscles have wasted so much he cannot breathe normally; a breathing tube in his throat is connected to a mechanical respirator. Also, there is a large hole in his abdomen that will not heal.

As a result, Mr. Chan has been in the intensive care unit for more than three months since he became sick from eating the contaminated food.

His doctors estimate the cost of his care to date to be at least $300,000. That figure does not include expenses involved in bringing his wife and two young children here to stay with him during his illness.

Moreover, Mr. Chan's case illustrates one of the little known facts of medicine: that doctors occasionally make diagnoses through news accounts, particularly in public health emergencies.

Dr. Robert V. Tauxe of the Centers for Disease Control in Atlanta said that "many times epidemiologists have no other mechanism besides the press to alert people rapidly to public health hazards and we are absolutely dependent on it."

Dr. Tauxe, who heads the Atlanta team that is investigating the British Airways outbreak, said that immediately after the disease detectives learned of the outbreak—British Airways had notified the Atlanta centers of the contamination—they sent messages by telex to all state health departments, which in turn notified doctors and hospitals.

Nevertheless, of the 186 cases in the United States that have been reported to the Atlanta diseases center, the diagnoses of many were made only as a result of news accounts, Dr. Tauxe said.

The global epidemiological investigation is still incomplete and no one yet knows how many, if any, people died.

So far, Dr. Tauxe said, the outbreak seems to have been caused by contamination of several food items served aboard British Airways flights from March 12 to March 14.

For reasons that are not yet clear, the illnesses also seem to have been unusually severe, causing about 15 percent of the ill passengers to be admitted to hospitals.

For many of these patients, the news accounts came in time to spare them from misdiagnoses. Had the San Francisco doctors known about the nature of the airline food-poisoning outbreak earlier, Dr. Fugaro said, Mr. Chan might possibly have been spared the emergency operation.

Source: The New York Times, June 26, 1984. Reprinted by permission.

ciated with most cases of clostridial food poisoning, particularly with cooked meats if a gravy is prepared with the meat. The spores of *C. perfringens* type A can survive the temperatures used in cooking many meats, and if incubated in a warm gravy, there is sufficient time for the spores to germinate and the growing bacteria to elaborate sufficient toxin to cause this disease.

The symptoms of food poisoning associated with *C. perfringens* generally appear within 10–24 hours after ingestion of food containing toxin. These symptoms usually include abdominal pain and diarrhea, but vomiting, headache, and fever normally do not occur. In cases of food poisoning caused by *C. perfringens*, recovery generally occurs within 24 hours.

Salmonellosis. Gastroenteritis caused by members of the genus *Salmonella* (Gram negative, nonlactose-fermenting, rods) is called salmonellosis Many different *Salmonella* species, especially the numerous serotypes of *Salmonella enteritidis*, are commonly the etiologic agents of this disease. *Salmonella* species causing gastroenteritis are normally transmitted by ingestion of contaminated food. Birds and domestic fowl, especially ducks, turkeys, and chickens, including their eggs, are commonly identified as the sources of *Salmonella* infections. Inadequate cooking of large turkeys and the ingestion of raw eggs contribute to a significant number of cases of salmonellosis. Some cases of salmonellosis are associated with the handling of pet turtles, which are often contaminated with high numbers of *Salmonella*, followed by injestion of the bacteria from contaminated hands.

Like many other bacteria that cause gastrointestinal tract infections, *Salmonella* species have pili that enable them to adhere to the lining of the gastrointestinal tract (Figure 24.6). *Salmonella* species are able to reproduce within the intestines, causing inflammation. Most *Salmonella* species do not penetrate the mucosal lining of the gastrointestinal tract. Some *Salmonella* species, however, such as *S. paratyphi* and *S. typhimurium* which cause paratyphoid fever, move into the circulatory system where they continue to reproduce, causing bacteremia. Paratyphoid fever is characterized by gastroenteritis and a relatively high rate of bacteremia (Figure 24.7). Typhoid fever, a systemic infection caused by *S. typhi*, will be discussed in Chapter 27.

The less serious forms of *Salmonella* infections are normally characterized by abdominal pain, fever, and diarrhea that lasts for 3–5 days. The onset of disease symptoms normally occurs 8–30 hours after ingestion of contaminated food. Nausea and vomiting may be initial symptoms but usually do not persist once pain and diarrhea begin. The feces may contain mucus and blood. During acute salmonellosis the feces may contain 1 billion *Salmonella* cells per gram. Fecal contamination of water and food supplies can contribute to the transmission of this disease. Salmonellosis normally does not require the use of antibiotics because the prognosis is normally for relatively rapid recovery. Generally, the disease is self-limiting, with recovery occurring within 1 week, although a carrier state sometimes is established.

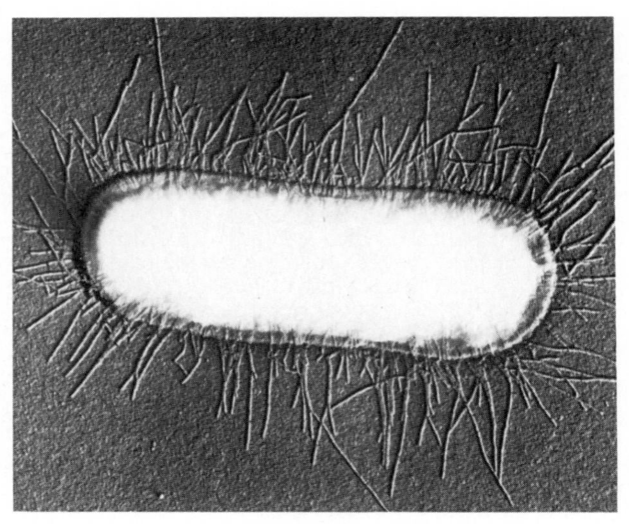

Figure 24.6 Micrograph of *Salmonella anatum* surrounded by pili. The pili or fimbriae, as they are often called, permit the bacteria to adhere to the lining of the gastrointestinal tract. (Courtesy J. P. Duguid. Reprinted by permission of the *Journal of Pathology* and J. P. Duguid from J. P. Duguid, *et al.*, 1966, *Journal of Pathology and Bacteriology*, **92**:107.)

Shigellosis. Shigellosis, or bacterial dysentery, is an acute inflammation of the intestinal tract caused by species of *Shigella* (Gram negative nonmotile rods). Several species of *Shigella* cause bacterial dysentery, including *S. flexneri, S. sonnei*, and *S. dysenteriae*. The transmission of *Shigella* species normally occurs by the direct fecal–oral route, although water and food supplies are involved in some outbreaks of bacterial dysentery. *Shigella* species are able to penetrate the mucosal cells of the large intestine and multiply in the submucosa. Even in fatal cases, however, the bacteria do not leave the gastrointestinal tract. Areas of intense inflammation develop around the multiplying bacteria and

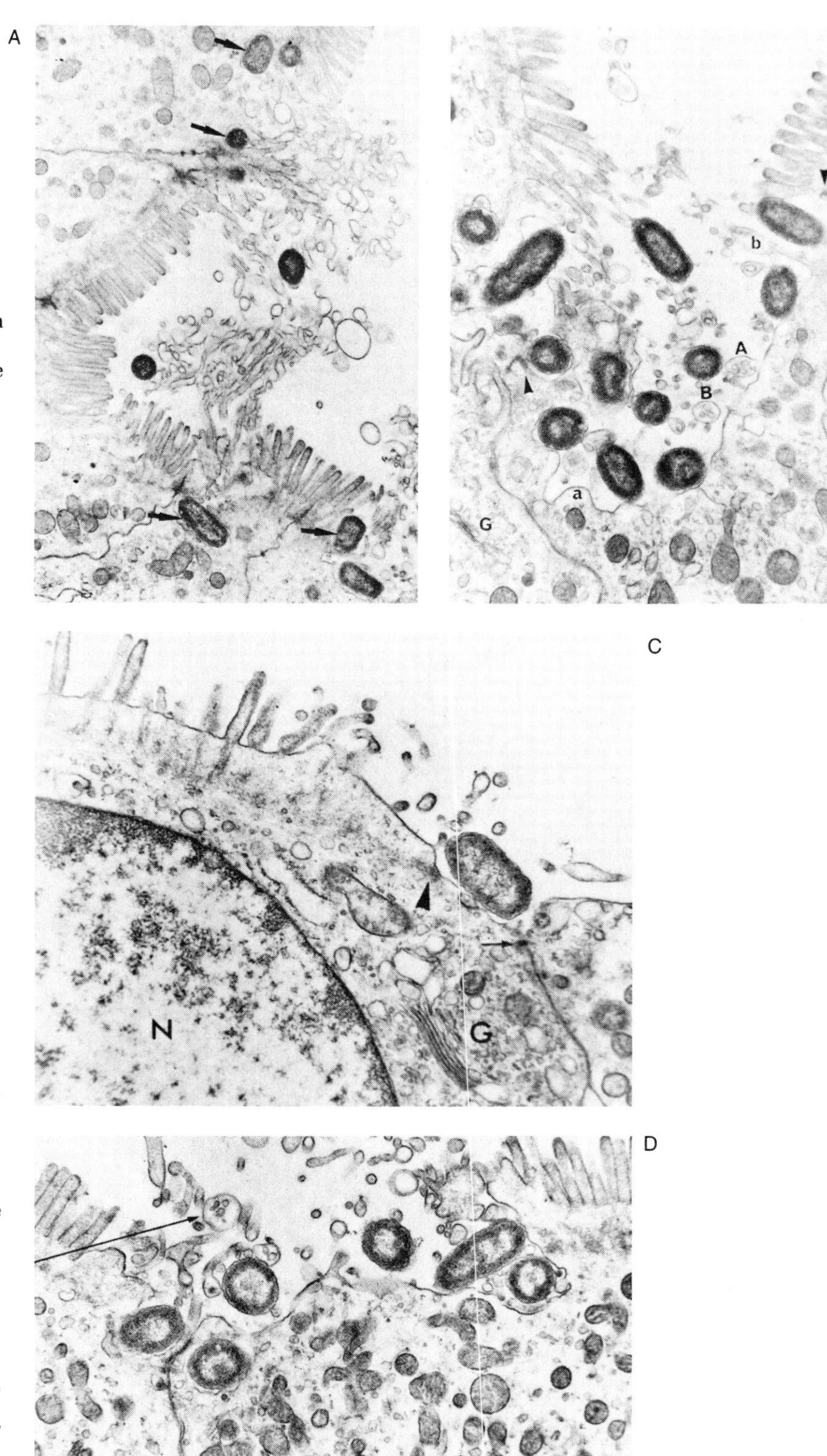

Figure 24.7 Micrographs showing invasion of the intestine by *Salmonella.* (A) Note extensive degeneration of the microvilli terminal well and apical cytoplasm localized at the site of bacterial penetration (arrows). Other cytoplasmic components and adjacent cells are unaltered (11,000 ×). (B) Note the projections (referred to as blebs) with (A) or without (a) vesicles that arise from the host cell cytoplasm and are pinched off (B, C and b, c) into a cavity that also contains degenerating microvilli. There are increasing numbers of small vesicles around the Golgi (G). An intercellular junctional complex is laterally displaced (27,000 ×). (C) A *Salmonella* cell lies close to a junctional complex (thin arrow). The terminal web (thick arrow) of the cell at the left ends abruptly at the lumen surface. The corresponding terminal web of the cell at the right is absent. The degeneration of the microvilli and apical cytoplasm is severe near the bacterium (15,000 ×). G = Golgi; N = nucleus. (D) The microvilli, terminal web, and apical cytoplasm are replaced by a shallow cavity in which degenerated microvilli, blebs, and vesicles are present. The adjacent intercellular tight junctions are laterally displaced. The arrow indicates a striking similarity between a multivesicular body (MVB) and a bleb containing small vesicles. The remaining cytoplasmic organelles are intact (12,000 ×). (A: Reprinted by permission of the American Society for Microbiology, Washington, D.C., from A. Takeuchi, 1975, in *Microbiology—1975,* p. 176; B–D: reprinted by permission of Hoeber Medical Division of Harper & Row, Publishers, Hagerstown, Maryland, from A. Takeuchi, 1967, *Americam Journal of Pathology,* 50:125, 127, 123.)

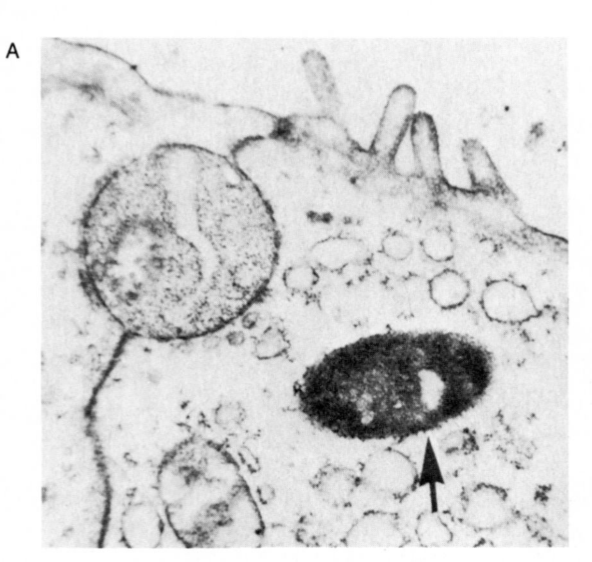

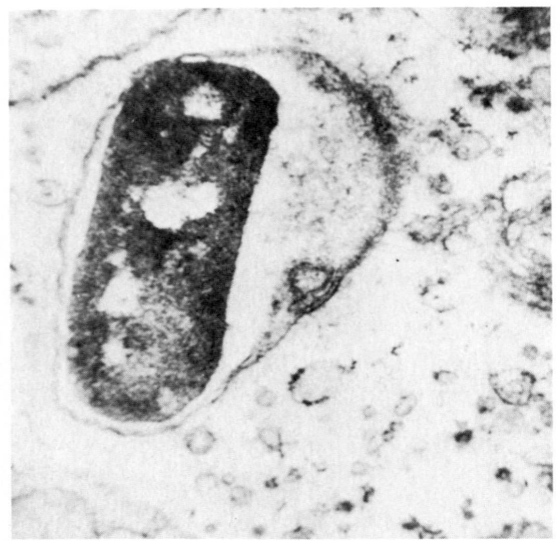

Figure 24.8 Micrographs showing intestine infected with *Shigella.* (A) Apical portion of intestinal epithelium 24 hours after Shigella infection. A *Shigella* organism (arrow) lies free in the cytoplasm near the lumen surface. The microvilli and terminal web have undergone regressive changes. The endoplasmic reticulum is swollen. Membrane-bound intracytoplasmic inclusions contain osmophilic granular material resembling ribosomes and a mitochondrion. A similar structure is present in the intracellular space (8000 ×). (B) A composite of intraepithelial, membrane-enclosed *Shigella* organisms (10,000 ×). The membranes frequently enclose cytoplasmic components. G = Golgi apparatus. (Reprinted by permission of Hoeber Medical Division of Harper & Row, Publishers, Hagerstown, Maryland, from A. Takeuchi *et al.*, 1965, *American Journal of Pathology,* **47**:1037, 1041.)

microabscesses form and spread, leading to bleeding ulceration (Figure 24.8).

The symptoms of shigellosis include abdominal pain, fever, and diarrhea, with mucus and blood in the excretion. Bacterial dysentery normally is a self-limiting disease, with recovery occurring 2–7 days after onset. The severe dehydration associated with this disease can cause shock and lead to death in children, among whom the incidence of bacterial dysentery is highest. In cases of diagnosed shigellosis in children, antibiotics are used in treating the disease. However, many *Shigella* strains now contain plasmids coding for antibiotic resistance, and the spectrum of antibiotics needed to effectively combat shigellosis has increased. As a result, combined therapy with tetracycline, ampicillin, and nalidixic acid is often now employed for treating bacterial dysentery.

Campylobacter Infections. *Campylobacter fetus* var. *jejuni* has been found to be the causative agent of many cases of gastroenteritis in infants. In fact *C. fetus* may be more important in juvenile gastroenteritis than *Salmonella* or other bacterial species. The transmission of *C. fetus* appears to be via contaminated food or water. *C. fetus* is a Gram negative, motile, curved bacterium, formerly known as *Vibrio fetus,* that also causes fetal abortion in cattle and sheep. The disease symptoms range from mild to severe and include abdominal pain, diarrhea, fever, and bloody stools. Treatment may include administration of antibiotics, such as erythromycin and tetracyclines and maintenance of proper fluid and electrolyte balance. In mild cases the disease is self-limiting and no treatment is necessary.

Traveler's Diarrhea. Enterotoxin-producing strains of *Escherichia coli* (Gram negative, lactose-fermenting rods) are also capable of causing both mild and severe forms of gastroenteritis. In most cases, enterotoxin-producing strains of *E. coli* do not invade the body through the gastrointestinal tract; rather, the toxin released by cells growing on the surface lining of the gastrointestinal tract causes diarrhea. Many cases of severe diarrhea in children are caused by noninvasive, enterotoxin-producing strains of *E. coli*. Aside from diarrhea, abdominal

U.S. Panel Issues Guideline on Travelers' Diarrhea

By PHILIP M. BOFFEY

WASHINGTON, Jan. 30—A panel of experts today issued the Federal Government's first authoritative instructions on how to cope with a problem that affects a large portion of the more than eight million Americans who travel to developing countries each year: diarrhea.

The panel urged travelers not to take antibiotics, Pepto-Bismol and other popular medicines in an effort to prevent the onset of diarrhea, as many now do, because most travelers would not come down with the ailment and all of the medicines carry some small risk of adverse side effects.

Instead, the panel said, travelers should wait for the first signs of diarrhea, perhaps two or more episodes of loose or watery stool, and then take the medicines, which are generally powerful enough to end most cases of diarrhea in less than 30 hours.

Left untreated, such diarrhea typically lasts three to five days before complete recovery occurs naturally. Thus, treatment should be considered optional, depending on how keenly the traveler wants to get rid of the diarrhea, cramps or nausea that are causing discomfort, the panel said.

Recommendations from Panel

The recommendations were issued today by a "travelers' diarrhea consensus development panel" convened by the National Institutes of Health, the Government's chief biomedical research agency.

Public relations officials at the agency said the group's recommendations were the first high-level Federal advice on how best to treat and prevent the ailment that afflicts millions of people each year but is generally mild and thus attracts little official attention.

The 13-member panel was headed by Sheldon L. Gorbach, professor of medicine and microbiology at Tufts University School of Medicine and chief of the infectious diseases division at New England Medical Center in Boston.

The panel noted that more than eight million residents of the United States were expected to travel to developing countries this year and that about one-third of them would come down with diarrhea.

The panel defined "travelers' diarrhea" as an ailment resulting in four to five loose, watery stools per day and typically lasting only three to four days, although a tiny percentage of cases may last for weeks or months. Commonly associated symptoms, it said, include abdominal cramps, nausea, bloating, urgency and malaise.

The episodes of diarrhea generally begin abruptly, occur in travel or soon afterward, and generally end of their own accord. The disease is rarely life-threatening, the panel said, citing an extensive survey of several hundred thousand Swiss travelers, which found no deaths attributable to diarrhea.

The panel said it had found no data to support the notion, held by many individuals and even some doctors, that travelers' diarrhea was caused by such factors as jet lag, altitude, fatigue or a change in the normal diet.

The disease is caused by infectious microorganisms, it said, chiefly exotic forms of the common bacterium, Escherichia coli, which adheres to the small intestine and produces a poison that causes the intestine to shed copious amounts of fluid, or "weep." Other microorganisms implicated in some cases of travelers' diarrhea include various other bacteria and parasites, but viruses do not appear to be a frequent cause of the ailment, the panel said.

The panelists noted at a news conference that dysentery was a particularly severe form of diarrhea, often caused by shigellae or salmonellal bacteria, in which the stools contain blood and mucusa. Such cases, the panelists said, require medical attention and should not simply be treated by the traveler himself.

Dr. Gorbach said that the "most controversial" recommendation was the advice against using drugs as a preventive measure, a practice currently followed by countless travelers who ingest daily doses of medicine to ward off any attacks. The panel acknowledged that bismuth subsalicylate, taken in liquid form as the active ingredient of Pepto-Bismol in two-ounce doses four times a day, had decreased the incidence of diarrhea by 60 percent in one study.

Warning on Side Effects

It also noted that two antibiotics, trimethoprim/sulfamethoxazole, sold under the brand names Bactrim and Septra, and doxycycline, brand name Vibramycin, had consistently reduced the incidence of travelers' diarrhea in various parts of the developing world by 50 to 86 percent. But the panel warned that all these drugs had some risk of side effects. Large doses of bismuth can cause neurological effects, and subsalicylate can cause ringing in the ears, the panel said.

Moreover, the various antibiotics can cause allergic reactions and other side effects, including skin rashes, sensitivity to sunburn, blood disorders, vomiting and staining of the teeth in children. Although these side effects appear to be relatively uncommon, the panelists said, they make it impossible to recommend that eight million Americans take the drugs preventively.

However, some experts who participated in the conference contended that the risks were slight and that individuals who feel it is very important to avoid even a few days of diarrhea should be advised to consult with a doctor and make their own judgments as to whether or not to take the risk. Such individuals might include heads of state going to an important conference, businessmen negotiating a crucial deal, and newlyweds on a honeymoon, one expert suggested.

Source: The New York Times, Jan. 31, 1985. Reprinted by permission.

cramping is normally the only other clinical symptom of this disease. Travelers from the United States to Mexico often suffer severe diarrhea as a result of ingestion of strains of *E. coli* foreign to their own microbiota and therefore generally avoid drinking the water. The resulting disease often is descriptively called Montezuma's Revenge, Aztec Two Step, and New Delhi Belly.

In some cases, pathogenic strains of *E. coli* invade the body through the mucosa of the large intestine and cause a serious form of dysentery (frequent watery stools). Invasive strains of *E. coli* are primarily associated with contaminated food and water in Southeast Asia and South America. The ability to invade the mucosa of the large intestine depends on the presence of a specific K antigen in enteropathogenic serotypes of *E. coli*. The enterotoxins produced by *E. coli* cause a loss of fluids from intestinal tissues. With proper replacement of body fluids and maintenance of the essential electrolyte balance, infections with enterotoxic *E. coli* normally are not fatal.

Vibrio Parahaemolyticus *Gastroenteritis.* *Vibrio parahaemolyticus* is responsible for a large number of cases of gastroenteritis in Japan and perhaps in the United States. *V. parahaemolyticus* occurs in marine environments, and the ingestion of contaminated seafood, particularly the eating of raw fish, is the main route of transmission. Gastroenteritis caused by *V. parahaemolyticus* requires establishment of an infection within the gastrointestinal tract, rather than simple ingestion of an enterotoxin. The symptoms of gastroenteritis generally appear 12 hours after ingestion of contaminated food and include abdominal pain, diarrhea, nausea, and vomiting. Recovery from this form of gastroenteritis normally occurs in 2–5 days, and the mortality rate is very low.

Yersiniosis. *Yersinia enterocolitica* (Gram negative short rods) and related species produce a severe form of gastroenteritis. Outbreaks of yersiniosis are most common in western Europe but have also been confirmed in the United States. *Y. enterocolitica* is widely distributed and has been found in water, milk, fruits, vegetables, and seafoods. This organism is psychrotrophic and thus is able to reproduce within refrigerated foods, where it can multiply and reach an infectious dose. In fact, *Y. enterocolitica* grows better at 25°C than at 37°C.

The symptoms of an infection with *Y. enterocolitica* resemble appendicitis and include abdominal pain, fever, diarrhea, vomiting, and leukocytosis. Often an appendectomy is performed before this disease is properly diagnosed as yersiniosis. In an outbreak of yersiniosis in New York involving over 200 school children, the infection was traced to a common source of contaminated chocolate milk. Ten children underwent unnecessary appendectomies before the true etiology of the disease was established.

Cholera. Cholera is caused by the Gram negative, curved rod, *Vibrio cholerae*, serotypes *cholerae* and *eltor*. This disease is primarily transmitted via water supplies contaminated with fecal matter from infected individuals. Cholera is a particular problem in socioeconomically depressed countries, where there is poor sanitation and inadequate sewage treatment and where medical facilities have a limited capacity to deal with outbreaks of this disease (Figure 24.9). *V. cholerae* is also a natural inhabitant of estuaries, and some cases of cholera along the Gulf coast of the United States have been traced to contaminated shellfish.

We typically associate cholera with Asia, sometimes referring to the disease as Asiatic cholera. This disease is endemic in the Ganges delta, and there are annual epidemic outbreaks of cholera in India and Bangladesh associated with the monsoon floods, which wash fecal matter into drinking water supplies. In endemic areas of Asia the death rate is normally 5–15 percent of those affected, but during the seasonal epidemic outbreaks, the mortality rate may reach 75 percent.

V. cholerae is able to multiply within the small intestine and produces the enterotoxin responsible for the symptoms of cholera, which occur suddenly and include nausea, vomiting, abdominal pain, diarrhea with large amounts of mucus in stools (rice water stools), and severe dehydration, followed by collapse, shock, and in many cases death. *V. cholerae* itself does not invade the body and is not disseminated to other tissues. Rather, the enterotoxin produced by *V. cholerae* binds to the epithelial cells of the small intestine, causing changes in the membrane permeability of the mucosal cells, which initiates secretion of water and electrolytes into the lumen of the small intestine. The rapid loss of fluid from the cells of the gastrointestinal tract associated with this disease often produces shock, and if untreated there is a high mortality rate. The initial diarrhea that results from infection with *V. cholerae* can cause the loss of several liters of fluid within a few hours. The treatment of cholera centers on replacing fluids and maintaining the electrolyte balance, that is, on combating shock. Treatment with

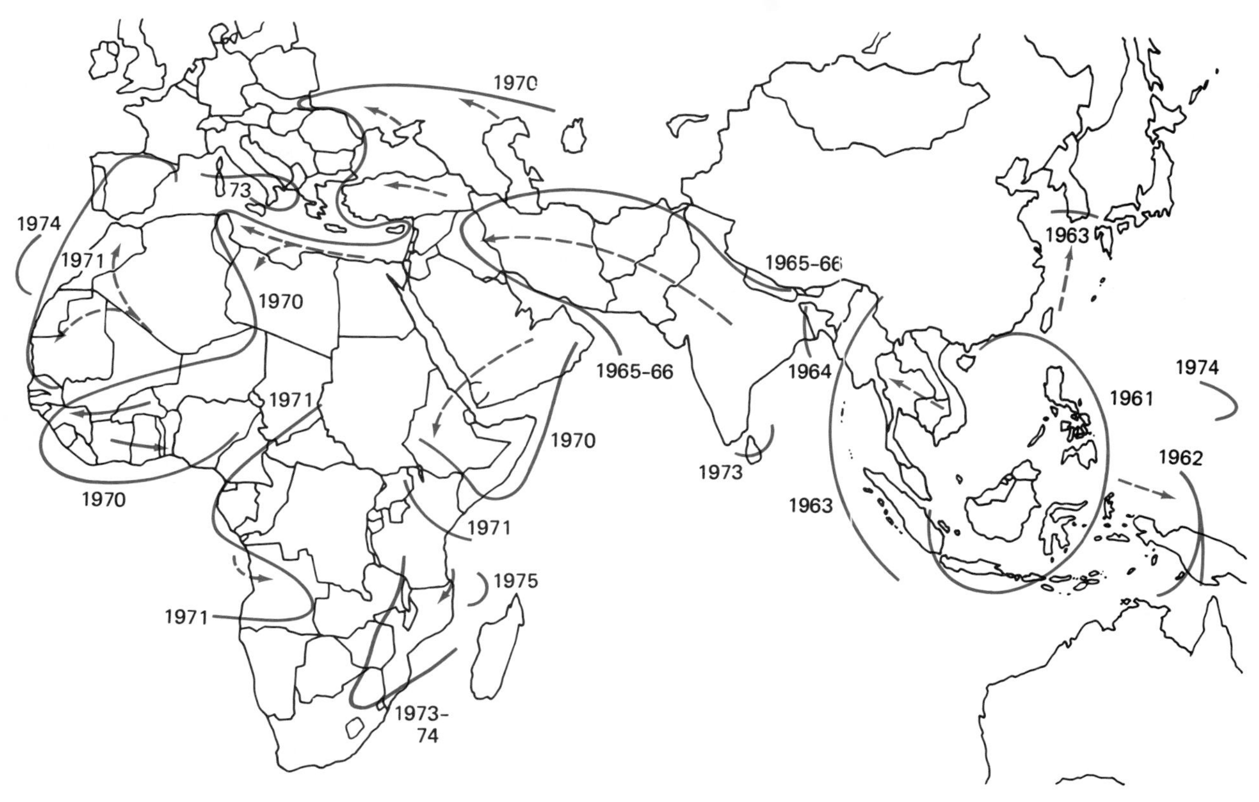

Figure 24.9 Map showing worldwide incidence of cholera from 1961 to 1975.

tetracycline generally reduces the duration of the disease. There is a vaccine against cholera but it is not very effective.

Appendicitis. There are more than 200,000 cases per year in the United States of appendicitis, an inflammation of the appendix occurring when there is an obstruction of the lumen of the appendix. Appendicitis is caused by a mixture of bacterial populations that constitute the normal microbiota of the intestinal tract, and virtually all members of the normal microbiota of the gastrointestinal tract can contribute to appendicitis. The symptoms of appendicitis normally include abdominal pain, localized tenderness, fever, nausea, vomiting, and leukocytosis. Treatment of appendicitis usually involves the surgical removal of the appendix. Serious complications from appendicitis arise if the infection is permitted to progress and the appendix ruptures, releasing bacteria into the abdominal cavity. In such cases, peritonitis and systemic bacteremia occur, and there is a high rate of mortality. Antibiotics are not effective in treating uncomplicated acute appendicitis and are unable to stop the development of infection before rupture of the appendix can occur. In cases where appendicitis is complicated by rupture of the appendix, antibiotic treatment is, however, essential in controlling the spread of infection to other tissues.

Diseases Caused by Protozoan Pathogens

Amebic Dysentery. Amebic dysentery or amebiasis is caused by the protozoan *Entamoeba histolytica*. Amebic dysentery occurs as a result of inadequate sewage treatment and contamination of water with *E. histolytica*, whose cysts are not killed by the normal chlorination methods used to treat municipal drinking water. Infection is acquired by ingesting contaminated food or water containing cysts of *E. histolytica*. Infections with *E. histolytica* may be asymptomatic or may involve mild or severe diarrhea and abdominal pain.

Infestation occurs in the small intestine without causing any disease symptoms. The trophozoites (motile flagellate forms) penetrate and multiply within the colon where they appear to ingest red blood cells, yeasts, and bacteria (Figure 24.10). When

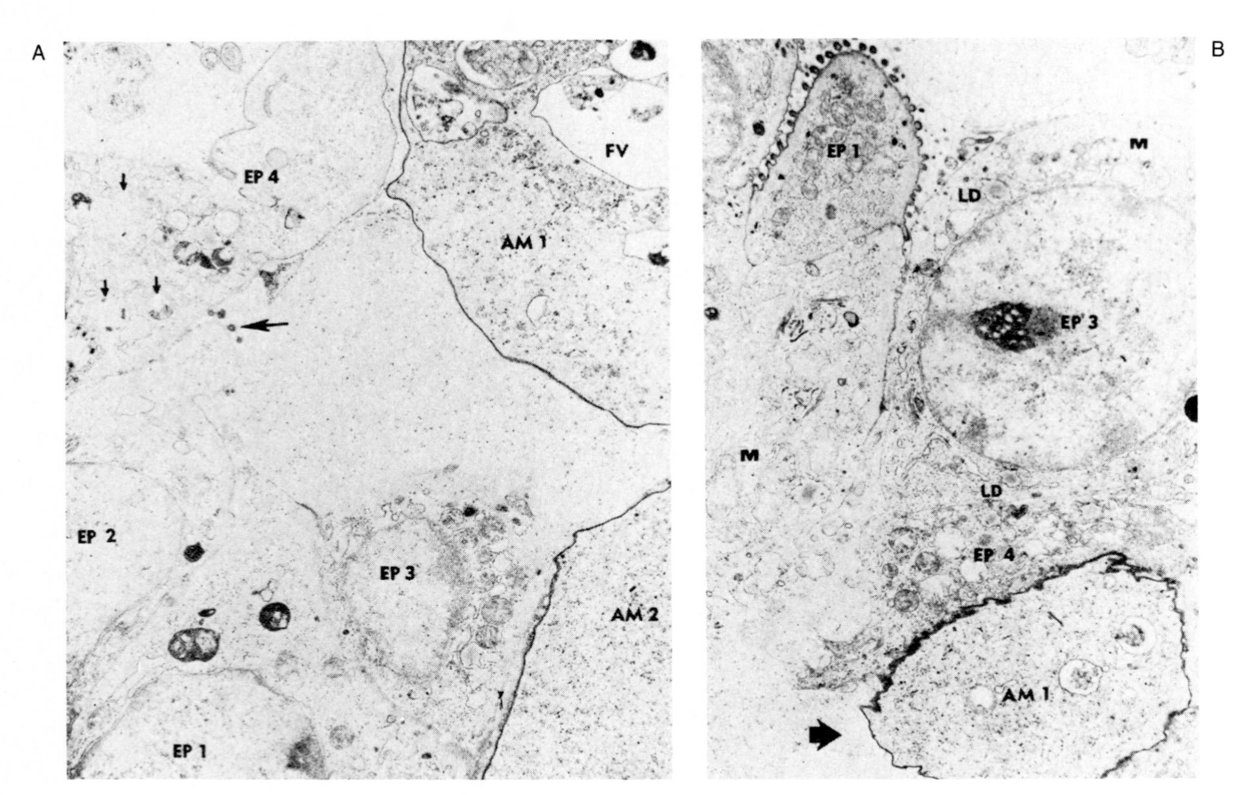

Figure 24.10 Micrograph of intestinal epithelium of a guinea pig infected with *Entamoeba histolytica.* (A) Amoeba (AM) in contact with the luminal surface of interglandular epithelial cells (EP) (10,200×). The varying degrees of cytoplasmic change range from nearly normal in one cell (EP1) to severe alterations (EP4). The microvilli and terminal web have disappeared for the most part, with a remnant remaining (large arrow) in a host cell (EP2). The cell membrane of the amoeba is thicker and denser than the host cytoplasmic membrane. AM1 has adhered to the projected apical cytoplasm of host cell EP3. Note spectrum of mitochondrial changes (small arrows) within a single cell (EP4). (B) The basal cytoplasm of a host cell lacks cellular organelles and shows cytoplasmic projection (arrow) toward the pseudopodium (AM1). Note close contact between membrane of host cell (EP4) and amoeba (AM1). Increasing degeneration process ranging in degree from EP1 to EP4 is evident. A third amoeba (AM3) is in the lumen (9,100×). M = altered mitochondria; LD = liquid droplet. (Reprinted by permission of the American Society for Tropical Medicine and Hygiene, from A. Takeuchi and B. P. Phillips, 1975, *American Journal of Tropical Medicine and Hygiene,* **24**;39, 41. ©1975, The Williams and Wilkins Co., Baltimore.)

immune surface defense mechanisms are lowered, the trophozoites are able to invade the epithelial cells of the colon. The invasion of the colon lining results in the formation of ulcers. The trophozoites spread through the submucosa, cutting off the blood supply through the mucosal lining, which leads to sloughing off of mucosal cells and enlargement of ulcers. Mucus and blood characteristically appear in the feces from these lesions. Cysts are shed with the fecal matter and can enter and contaminate water supplies. In some cases *E. histolytica* can spread to the liver, lung, or skin, causing ulcer or abscess formation in these tissues. Several antiprotozoan drugs, such as metronidazole, are effective in treating amebiasis, and the choice of which one to use depends on whether the infection is restricted to the intestinal tract or whether other organs, such as the liver, are involved.

Giardiasis. *Giardia lamblia*, a flagellated protozoan, is responsible for most cases of giardiasis, a diarrhea infection caused by protozoa. *G. lamblia* forms both motile trophozoites and nonmotile cysts, and the cysts are the infective form. The cysts of *G. lamblia* can enter the gastointestinal tract through contaminated water. A high incidence of giardiasis occurred among groups touring Leningrad during the 1970s as a result of contaminated water supplies. In 1973 a major outbreak of giardiasis occurred in upstate New York, with an estimated 4800 individuals developing symptoms of the disease. The following year the disease was the most common waterborne disease in the United States. *G. lamblia* can live saprophytically within the small intestine without causing any symptoms of giardiasis, and in the United States almost 4 percent of the population appears to be infected by this organism. Excessive growth of the organism, however, can cause disease symptoms that include diarrhea, dehydration, mucus secretion, and flatulence. Metronidazole is generally used in the treatment of this disease.

Balantidiasis. *Balantidium coli*, a ciliated protozoan, causes the disease balantidiasis or balantidial dysentery. This disease is transmitted via food and water contaminated with feces containing cysts of *B. coli*. The typical symptoms of balantidial dysentery are abdominal pain, nausea, vomiting, diarrhea, and loss of weight. In some cases *B. coli* causes ulcerations of the colon that can be fatal. In most cases the trophozoites of this protozoan reproduce within the large intestine causing a mild form of the disease.

Diseases Caused by Multicellular Parasites

In addition to microorganisms, some multicellular parasites cause diseases of the gastrointestinal tract. The clinical microbiologist sometimes finds evidence for such parasitic diseases in the samples that are sent to the laboratory for examination. Eggs and cysts of multicellular parasites can easily be confused with protozoa and other microorganisms. Therefore it is important for the microbiologist to have some knowledge of diseases caused by intestinal worms and other multicellular parasites.

Tapeworms. Tapeworms or cestodes are flatworms (platyhelminths) that infect the intestinal tracts of animals (Figure 24.11). Each tapeworm has a definitive host in which the larval stages can develop into an adult worm. For example, the larval stages of the both the beef tapeworm (*Taenia saginata*) and the pork tapeworm (*Taenia solium*) only develop into adult forms within the human intestinal tract, and therefore humans are the only de-

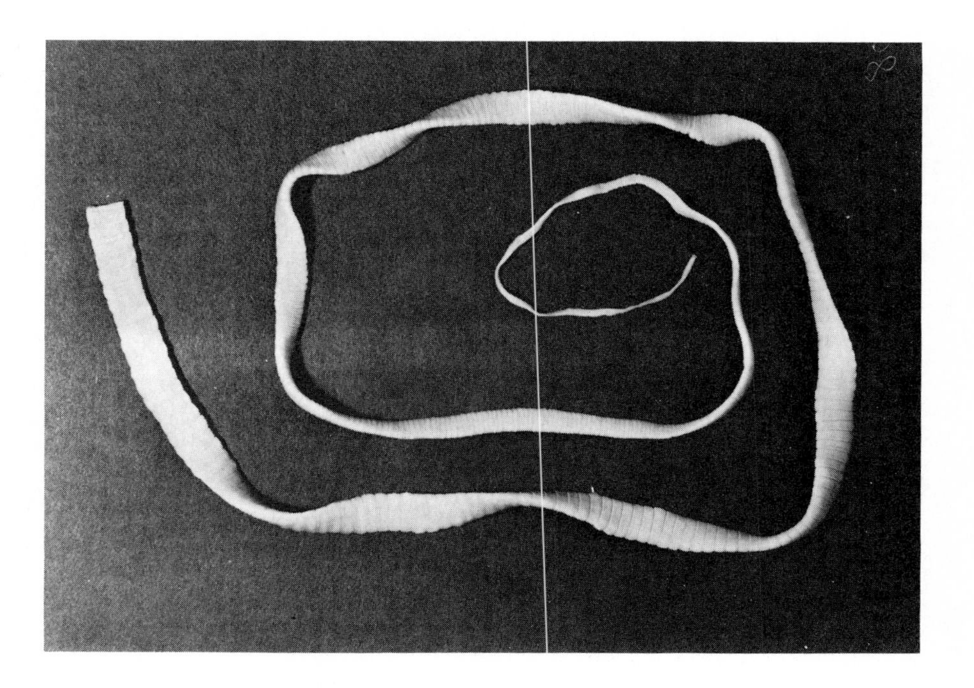

Figure 24.11 Photograph of the beef tapeworm (*Taenia segmata*). (From United States Armed Forces Institute of Pathology, neg. no. 69-7765-2.)

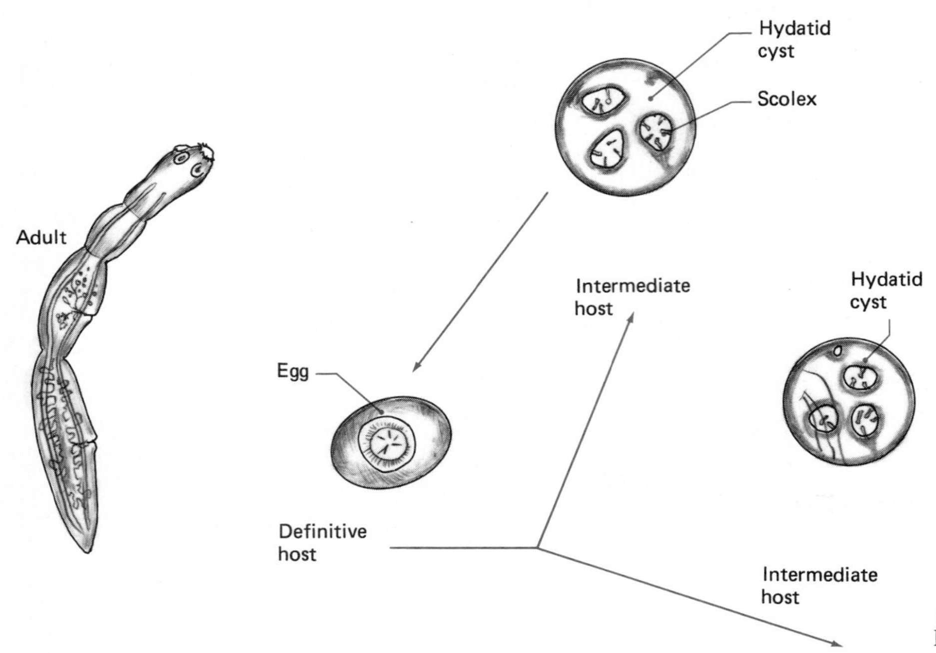

Figure 24.12 Diagram of the life cycle of the beef tapeworm.

finitive hosts for these parasites (Figure 24.12). Eating raw or inadequately cooked beef containing the larval stage of *T. saginata* is responsible for transmission of beef tapeworm to humans. Similarly, eating inadequately cooked pork containing the larval forms of *T. solium* is the source of transmission of pork tapeworm to humans.

Within the human intestinal tract the scolex or head of the tapeworm emerges from the larval form and attaches to the lining of the intestine. The scolex has several suckers that allow it to attach to the mucosal lining of the small intestine. Deriving its nutrition from its human host, the tapeworm matures and grows to a length of several meters. The name tapeworm is based on the resemblence of these parasites to a tapemeasure. Tapeworms are segmented and the posterior segments (proglottids) become filled with eggs. These egg-bearing segments break off and are shed with human feces. Surprisingly, most infected individuals show few if any symptoms. When symptoms occur they generally include loss of weight, anemia, and abdominal pain. Niclosamide is used to treat intestinal infestations with these tapeworms. Observation of proglottids and eggs in human fecal matter is diagnostic of tapeworm infestations.

The contaminated fecal matter is also the source of transmission of these tapeworms to their intermediate hosts, within which the eggs develop into larval stages. Once inside the intermediate host, the eggs hatch and develop into larvae. The larvae become disseminated throughout the body of the intermediate host. The larvae of the beef tapeworm can remain viable in the musculature of the cow for only about 9 months. The larvae of the pork tapeworm, however, can remain viable for years. The passage between humans and the intermediate hosts for the worms perpetuates the life cycles of these tapeworms.

The King of Parasites, delicate, white and blind,
Ruling his world of fable even as they,
Dreams out his greedy and imperious dream
Immortal in the bellies of mankind

In a rich bath of pre-digested soup
Warm in the pulsing bowel, safely shut
From the bright ambient horror of sun and air,
His slender segments ripening loop by loop,
Broods the voluptuous monarch of the gut,
The Tapeworm, the prodigious Solitaire . . .

Herds are his nurseries till the mouths of men
At public feasts, or the domestic hearth,
Or by the hands of children at their play,
Transmit his line to human flesh again . . .

No eagle with a serpent in his claw . . .
But the great, greedy, parasitic worm,
Sucking the life of nations from within,
Blind and degenerate, snug in excrement.

—A. D. HOPE, *The Kings*

The development of tapeworms within humans is not always restricted to the intestines. For example, ingestion of the larvae of the pork tapeworm results in intestinal infestation, but ingestion of the eggs of this tapeworm leads to an infection in which larvae can develop throughout the body, much in the same way the tapeworm develops within pigs. This condition is called cysticercosis. The liver, muscles, heart, and brain are commonly the sites of larval development. In cases where the larvae develop within the heart or central nervous system, the infestation is usually fatal. There is no satisfactory treatment for cysticercosis.

The tapeworm *Echinococcus granulosus* also causes a human infestation that is not restricted to the intestines. The larval stages of this tapeworm form within a fluid-filled bladder called a hydatid cyst. This tapeworm develops in a number of animals, including dogs. In dogs *E. granulosus* grows within the small intestine with no apparent ill effects. Humans become accidental intermediate hosts for this tapeworm by close association with dogs. In humans, the cysts most frequently develop in the liver because that is where most ingested eggs become trapped. Bone, brain, lungs, and kidneys are other sites where the hydatid cysts may develop. The development of the hydatid cyst causes an inflammatory reaction that tends to isolate the cysts within a walled-off region. Development of the cyst can cause liver impairment. If the cyst ruptures, the release of fluids from the cyst can cause a hypersensitivity reaction and death from anaphylactic shock. Diagnosis of hydatidosis prior to rupture of the cysts is important. The cysts are normally detected by X rays. Surgery then is performed to remove the intact, walled-off, hydatid cysts.

Flukes. Flukes or trematodes are much smaller than the tapeworms. The largest human parasite trematode is 8 centimeters in length. Humans are alternate hosts for these parasitic worms, most of which have complex life cycles that involve alternation of hosts. Trematodes can cause several human diseases and these diseases can be widespread. For example, in some villages in China, there is a 100 percent infection rate of the local population with the intestinal fluke *Fasciolopsis buski*. This high infection rate occurs because night soil (human feces) is used as a fertilizer and because sewage drains directly into the adjacent water supplies. Ingestion of contaminated plants leads to dissemination of this parasite. In humans, *F. buski* matures within the small intestine. Once the infec-

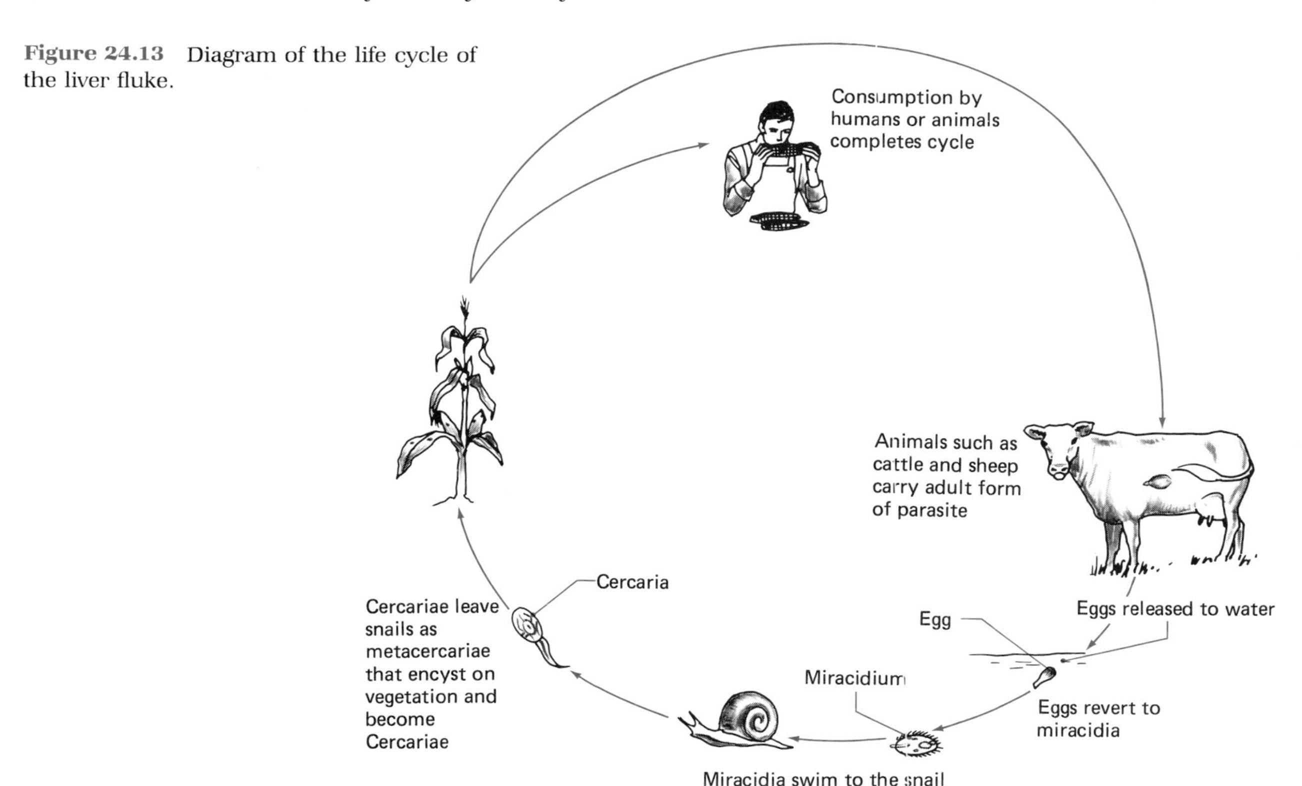

Figure 24.13 Diagram of the life cycle of the liver fluke.

tion progresses, diarrhea and abdominal pain occur; loss of body fluids can be fatal, particularly in children. This parasitic disease can be treated with hexylresorcinol.

The liver fluke (*Fasciola hepatica*), which is normally associated with sheep, can infest the human liver. Eggs from the fluke can be deposited in water bodies. If snails are present, *Fasciola* will pass through several life stages, eventually being deposited as encysted metacercariae (a resting stage of trematodes) on vegetation. Human infections with *F. hepatica* most commonly occur in Latin America and Mediterranean countries (Figure 24.13). Liver flukes normally enter the body on contaminated vegetation, such as watercress, and migrate from the intestine to the liver. These flukes develop within the liver, producing digestive disturbances. Fever, pain, and liver malfunction (hepatitis) result as the infestation progresses. Extensive liver damage eventually produces cirrhosis, which can be fatal. Identification of eggs in feces is important to the proper diagnosis of this disease. Several drugs, such as dehydroemetine, are used in the treatment of this disease.

Roundworms. Several roundworms cause human diseases. The pinworm, *Enterobius vermicularis*, is responsible for one of the more common worm infections of humans (Figure 24.14). The adult worm lives in the cecum and colon, where they attach to the mucosal lining. Usually, gravid female worms migrate toward the anus and emerge nocturnally. Large numbers of worms can be seen and many eggs are extruded around the anus each night. The eggs hatch and within 6 hours the larvae are infective for the same or other individuals. Larvae may reenter the digestive tract through the anus. Contamination of the hands with infective eggs also leads to introduction of these worms through the mouth. Hand-to-mouth contact and ingestion of contaminated food and drink are responsible for

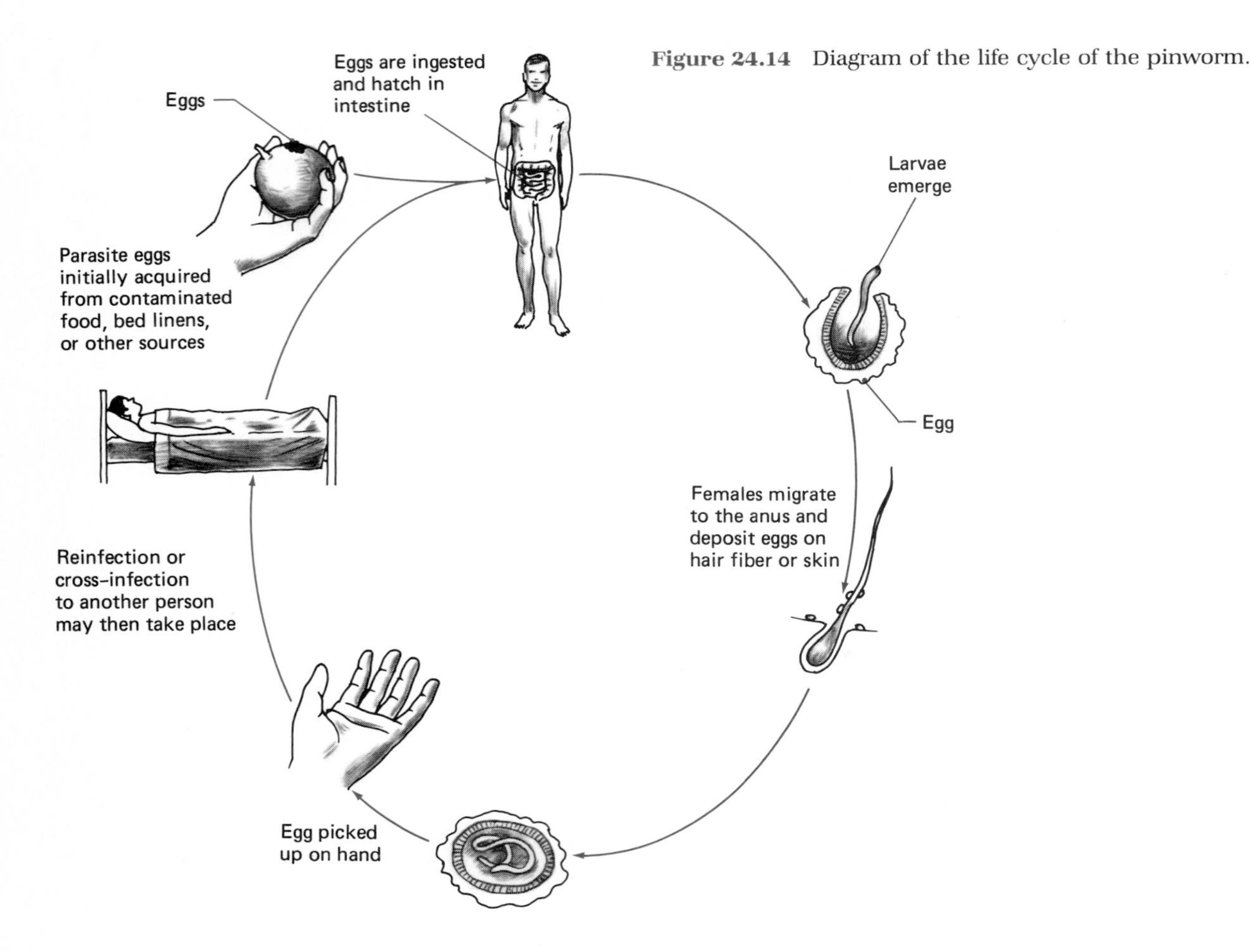

Figure 24.14 Diagram of the life cycle of the pinworm.

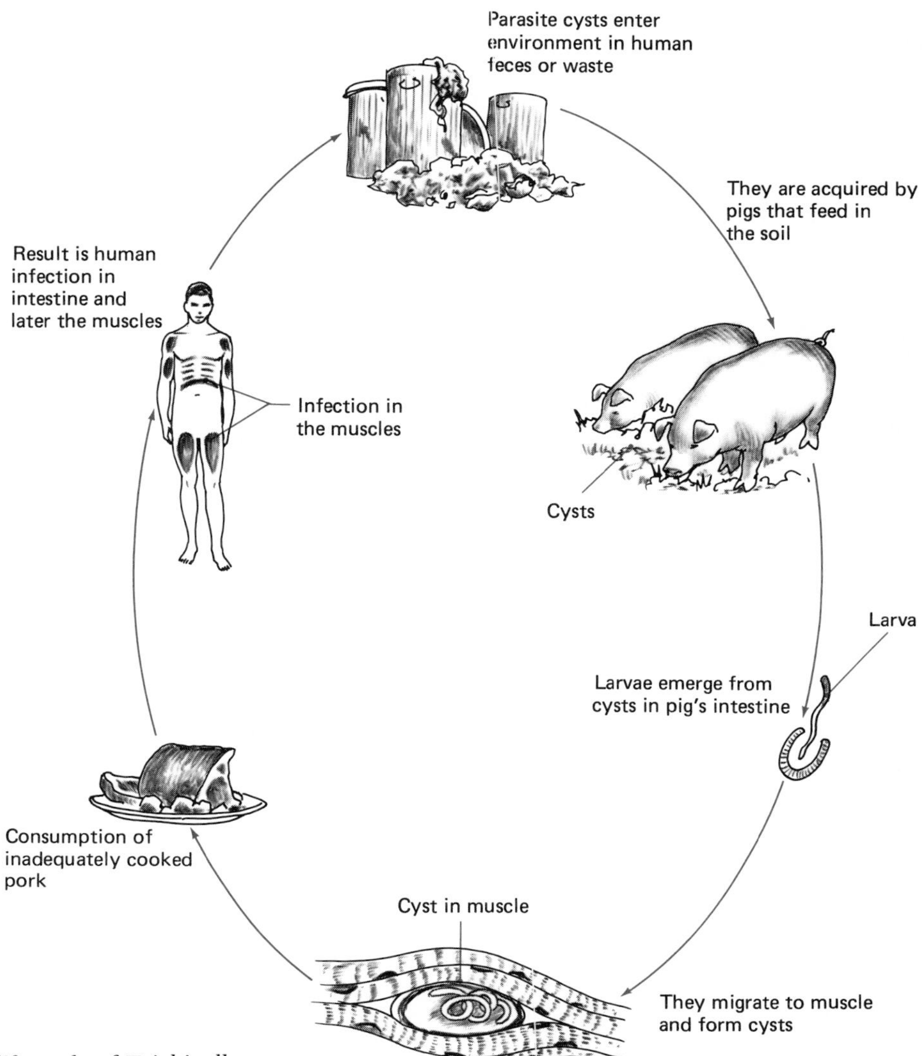

Figure 24.15 Diagram of the life cycle of *Trichinella.*

the maintenance of a high rate of infection. About one third of pinworm-infected persons suffer no symptoms. When symptoms occur, they normally include nausea, vomiting, and diarrhea. Poor sleep resulting from the irritation associated with the nocturnal migration of gravid females often leads to nervous restlessness in children infected with this worm. Pyrantel pamoate (Antiminth) is the drug of choice used to treat cases of pinworm. Good hygienic practices are also needed to prevent reinfection. Often it is necessary to treat an entire family to eliminate pinworm infestations.

Whipworm is another disease caused by a nematode. In the United States this disease is prevalent in the south. The causative agent, *Tricursis trichiura*, is shaped like a whip. This adult worms live primarily within the large intestine. Hand-to-mouth contact and ingestion of food contaminated with fecal matter is important in the transmission of this disease. Light infections often are asymptomatic. Symptoms may resemble appendicitis and result in emaciation. The broad spectrum antihelminthic mebendazole (Vermox) is considered the drug of choice for treating this disease.

Trichinosis is another human disease caused by roundworms. Human infections with the nematode *Trichinella spiralis*, the causative agent of trichinosis, generally is associated with the ingestion of undercooked pork (Figure 24.15). Occasionally outbreaks of trichinosis are also associated with eating raw bear and walrus meat. Adequate cooking or freezing meat long enough to kill these worms will prevent trichinosis. The infective larval stage of *T. spiralis* is found in an encysted form in the muscle tissues of infected pigs. When ingested, the larvae within the cyst are released and become attached to the lining of the small intestine. The worms reproduce within the small intestine, and larvae spread to other parts of the body. Symptoms of this disease vary from one person to another. Early symptoms usually involve the digestive tract, for example, diarrhea, but as the worms are disseminated, other symptoms, such as muscle pain and difficulty in breathing, become prevalent. Various manifestations of trichinosis can be fatal. Treatment depends upon the specific particular tissues that are involved.

Yet another roundworm disease occurs when *Ascaris lumbricoides* infests the intestines. This infestation, which is called roundworm disease of man, is prevalent throughout the tropics and occurs in the southeastern United States. The eggs of this parasite can remain viable in the moist soils of these regions. Hundreds of millions of individuals are infested with this nematode on a worldwide basis. *A. lumbricoides* is a very large nematode that can measure 30 centimeters. Large masses of worms in the intestines can obstruct the gastrointestinal tract. Surgical removal may be necessary if the infested individual becomes too emaciated. Transmission of this disease is often associated with fecal contamination of foods and waters. In regions where human feces (night soil) are used as fertilizer, the worms may be transmitted to crops. Ingestion of contaminated vegetables, drinking contamined water, or oral contact with contaminated fingers are all po-

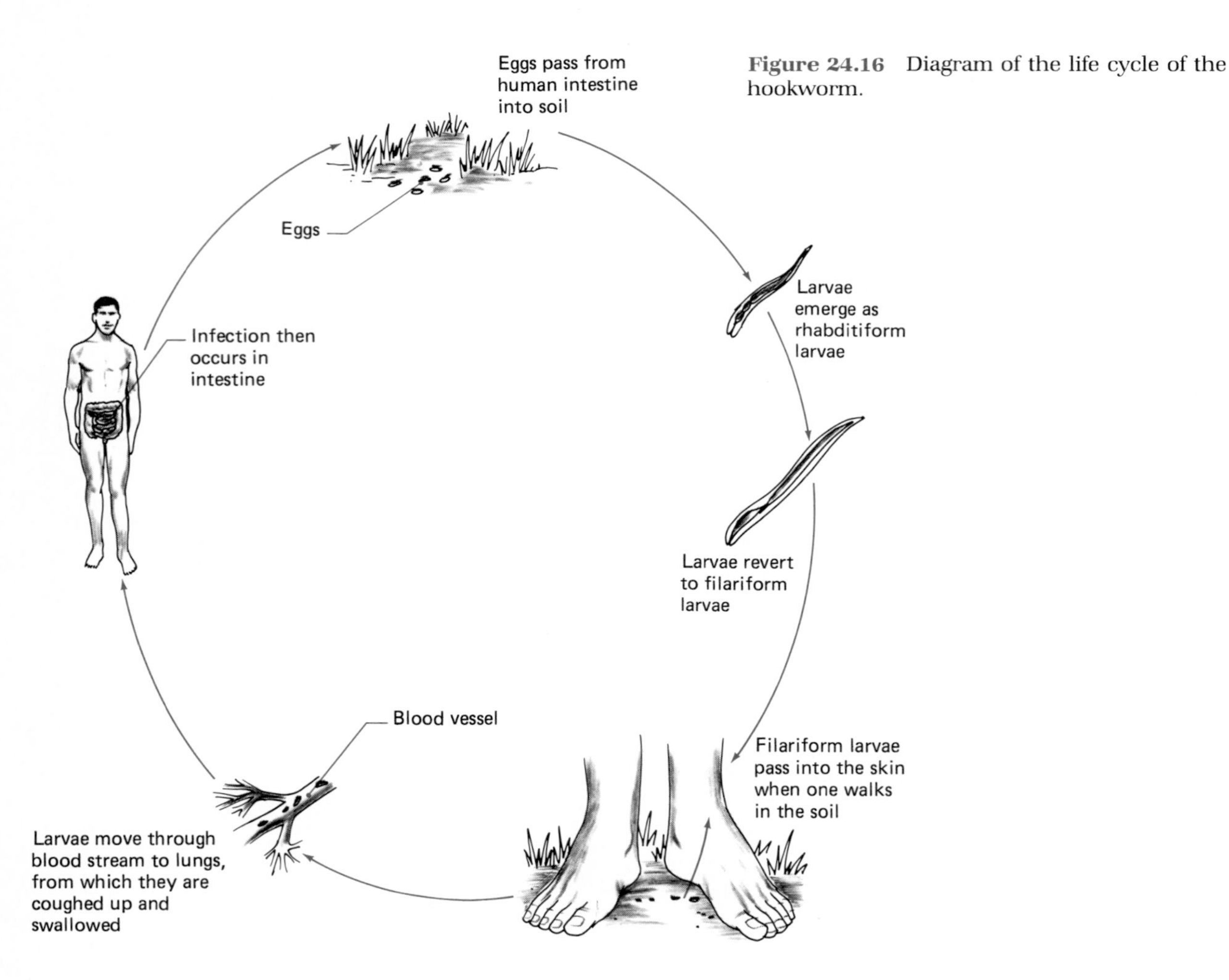

Figure 24.16 Diagram of the life cycle of the hookworm.

tential routes of transmission of this parasite. Children playing or working in fields fertilized with night soil often develop this disease. Once ingested, the worms infest the intestines. In some cases other parts of the body, such as the lungs, may also be sites of infestation. Treatment of intestinal infestations often involves use of piperazine citrate (syrup of Antepar), which paralyzes the worm. The live worm is then excreted in feces.

Unlike the previously discussed nematodes, hookworms (*Necator americanus* and *Ancylostoma duodenale*) enter the body through the skin and then migrate to the intestinal tract (Figure 24.16). Hookworm infections are most common in the tropics. Transmission most frequently occurs by walking barefoot on soil contaminated with these worms.

> The almost universal infestation of hookworm in many lands of the South has had consequences greater than those of other factors more frequently discussed by historians and economists. The blood loss in human beings infested with hookworms long wasted the strength of these countries. Men lack physical vigor and initiative when they are made anemic through loss of blood or malnutrition.
>
> —RENÉ DUBOS, *Mirage of Health*

Hookworms are able to burrow through the skin and migrate through the pulmonary system in order to reach the intestines. The hookworms live in the small intestine, firmly attached to the intestinal lining. Eggs are passed with the feces, and larvae develop in moist soil. The larvae sit on soil particles, surviving for up to several weeks. If they come in contact with a warm-blooded animal, they penetrate the skin and are carried through the circulatory system to the lungs. They are too large to pass through the capillaries of the lungs and therefore break out of the alveolar space. They are carried to the pharynx and are swallowed.

Symptoms of hookworm infestations often resemble those of a duodenal ulcer. A major pathologic condition of hookworm infections is anemia, caused because the worms suck blood while they are attached to the intestinal mucosa. Treatment with iron will correct the anemia. Pyrantel pamoate (Antiminth) is the drug of choice for eliminating hookworm infection.

Summary

Many microorganisms enter the digestive tract along with ingested food and water. Acid in the stomach acts as a first line of defense against microbial infections of the gastrointestinal tract. The mucus lining of the alimentary canal and the digestive enzymes involved in this system also limit possible microbial infections of the gastrointestinal system. An entensive resident microbiota, however, develops in the intestines. This microbiota is noninvasive and contributes to the defenses against infections by pathogenic microorganisms.

Some pathogenic microorganisms, however, are able to cause gastrointestinal tract diseases. Many of these pathogens are associated with contaminated food and water. Fecal contamination of food and water is an important route of transmission of serious diseases of the gastrointestinal tract, necessitating proper sanitation methods. In some cases the growth of microorganisms within food results in the accumulation of toxin; the ingestion of toxin causes food poisoning. In other cases, pathogens enter the body with the food and initiate an infection of the gastrointestinal tract.

Table 24.1 presents a summary of microbial infections that primarily involve the gastrointestinal tract.

Table 24.1 Summary of Diseases Associated with the Gastrointestinal System

Disease	Causative Agent	Mode of Transmission	Treatment
Mumps	Mumps virus	Saliva and respiratory secretions	None; vaccine for prevention
Viral gastroenteritis "24 hour flu"	Various viruses; Norwalk agent, rotavirus	Ingestion of contaminated food and water	None; maintain electrolyte balance
Cytomegalovirus inclusion disease	*Cytomegalovirus* (CMV)	Mother to fetus transfer	None
Infectious hepatitis	Hepatitis A virus (HAV)	Ingestion of contaminated food and water	None
Perfringens food poisoning	*Clostridium perfringens*	Ingestion of contaminated food and water	None; maintain electrolyte balance
Staphylococcal food poisoning	*Staphylococcus aureus*	Ingestion of food containing *Staphylococcus* toxin	None; maintain electrolyte balance
Traveler's diarrhea	*E. coli*	Ingestion of contaminated water	None; doxycyclin in severe cases
Salmonellosis	*Salmonella* species	Ingestion of contaminated food and water	None; maintain electrolyte balance
Shigellosis (bacillary dysentery)	*Shigella* species	Ingestion of contaminated food and water	Antibiotics; maintain electrolyte balance
Juvenile gastroenteritis	*Campylobacter fetus*	Ingestion of contaminated food and water	Antibiotics
Yersiniosis	*Yersinia enterocolitica*	Ingestion of contaminated food	Antibiotics
Vibrio parahaemolyticus gastroenteritis	*Vibrio parahaemolyticus*	Ingestion of contaminated shellfish	Antibiotics in severe cases
Cholera	*Vibrio cholerae*	Ingestion of contaminated food and water	Maintain electrolyte balance
Appendicitis	Various indigenous bacteria	Blockage of appendix opening	Surgical removal
Amebic dysentery (amoebiasis)	*Entamoeba histolytica*	Ingestion of contaminated food and water	Metronidazole
Giardiasis	*Giardia lamblia*	Ingestion of contaminated food and water	Metronidazole
Balantidiasis (balantidial dysentery)	*Balantidium coli*	Ingestion of contaminated food and water	Chlortetracycline, oxytetracycline
Tapeworms	*Taenia saginata, Taenia solium*	Ingestion of contaminated food and water	Niclosamide
Cysticercosis	*Taenia solium*	Ingestion of pork containing tapeworm eggs	None
Hydatid cysts	*Echinococcus granulosus*	Contact with infested dogs	Surgical removal of cyst
Intestinal flukes	*Fasciolopsis buski*	Ingestion of contaminated food	Hexylresorcinol
Liver flukes	*Faciola hepatica*	Ingestion of contaminated food	Dehydroemetine
Pinworms	*Enterobius vermicularis*	Ingestion of contaminated food and drink, hand-to-mouth contact	Pyrantel pamoate
Whipworms	*Trichuris trichuira*	Ingestion of contaminated food and drink, hand-to-mouth contact	Mebendazole
Trichinosis	*Trichinella spiralis*	Ingestion of contaminated pork	Varied
Roundworm disease	*Ascaris lumbricoides*	Ingestion of contaminated food or water, fecal–oral transmission via contaminated hands	Piperazine citrate
Hookworms	*Necator americanus, Ancyclostoma duodenale*	Through skin from contaminated soil	Pyrantel pamoate

Study Outline

A. The gastrointestinal tract consists of a series of interconnected organs that are involved in food reception, digestion, nutrient absorption, and waste elimination.

  1. The alimentary tract consists of the mouth, oral cavity, pharynx, esophagus, stomach, small and large intestine, rectum, and anus.

  2. Digestion begins in the mouth where food is macerated by the teeth and mixed with saliva, then the tongue forces it through the pharynx into the esophagus to the stomach.

  3. Digestion of proteins begins in the stomach where glands secrete degradative mucus, hydrochloric acid, and enzymes; food leaves the stomach only partially digested and goes to the small intestine.

  4. Most of the digestive processes occur in the small intestine where a series of enzymes mix with the food. Nutrients that are released by the digestion of foods are absorbed across the lining of the small intestine through the villi and microvilli.

  5. Undigested residues enter the large intestine or colon, which secretes mucus that helps lubricate the remaining residues as they move to the rectum, from which they move out through the anus.

B. The large resident microbiota of the gastrointestinal tract is necessary for the maintenance of good health. Resident microbiota include both obligate and facultative anaerobes and members of the genera *Lactobacillus, Streptococcus, Clostridium, Veillonella, Bacteroides*, and *Fusobacterium*; enteric bacteria, including members of the genera *Escherichia, Proteus, Klebsiella*, and *Enterobacter*; and coliform bacteria.

C. The defenses of the gastrointestinal tract include: the low pH of the stomach, which establishes an acidity most pathogens cannot tolerate; bile and digestive enzymes reduce numbers of surviving microorganisms; and the natural microbiota are antagonistic to the nonindigenous microorganisms.

D. Most diseases of the gastrointestinal tract originate with tainted food or water, although other routes of disease transmission are associated with some diseases of the digestive system.

  1. Microorganisms growing in food and water can produce toxins, and ingestion of these microbial toxins initiates a disease process called food poisoning or intoxication, which does not represent a true infectious process.

  2. Invasive pathogens can also establish an initial infection through the gastrointestinal tract and cause localized upset or systemic disease symptomology; this requires a large infectious dose.

E. Diseases of the gastrointestinal tract caused by viral pathogens.

  1. Mumps.
    a. Causative organism(s): mumps virus.
    b. Mode of transmission: via contaminated droplets of saliva.
    c. Symptoms: swelling of the salivary glands.
    d. Treatment: none; vaccine available.

  2. Cytomegalovirus inclusion disease.
    a. Causative organism(s): cytomegalovirus (CMV).
    b. Mode of transmission: across the placenta from mother to fetus or during the birth process; blood transfusions.
    c. Symptoms: affects the salivary glands and liver of the digestive system as well as the kidneys and lungs causing abnormalities of the infected human cells characterized by an increase in size and the development of inclusion bodies within the nucleus.
    d. Treatment: none.

  3. Gastroenteritis (24-hour or intestinal flu).
    a. Causative organism(s): Norwalk agent, rotavirus.
    b. Mode of transmission: ingestion of food contaminated with fecal matter.

c. Symptoms: sudden onset of gastrointestinal pain, vomiting, and/or diarrhea.
d. Treatment: supportive.

4. Type A hepatitis (infectious hepatitis).
   a. Causative organism(s): hepatitis virus type A.
   b. Mode of transmission: ingestion of food contaminated with fecal matter.
   c. Symptoms: fever, abdominal pain, and nausea, followed by jaundice.
   d. Treatment: none but can be prevented by prophylactic use of gamma globulin.

F. Diseases caused by bacterial pathogens.
1. Perfringens food poisoning.
   a. Causative organism(s): *Clostridium perfringens*.
   b. Mode of transmission: ingestion of food containing toxin produced by *C. perfringens*.
   c. Symptoms: abdominal pain and diarrhea.
   d. Treatment: none.
2. Staphylococccal food poisoning.
   a. Causative organism(s): *Staphylococcus aureus*.
   b. Mode of transmission: ingestion of food containing enterotoxin produced by *S. aureus*.
   c. Symptoms: nausea, vomiting, abdominal pain, and diarrhea.
   d. Treatment: none.
3. Traveler's diarrhea.
   a. Causative organism(s): enterotoxin-producing strains of *Escherichia coli*.
   b. Mode of transmission: ingestion of water containing strains of *E. coli* foreign to one's own microbiota.
   c. Symptoms: abdominal cramping and diarrhea.
   d. Treatment: none.
4. Salmonellosis.
   a. Causative organism(s): *Salmonella enteritidis* and other *Salmonella* species.
   b. Mode of transmission: ingestion of contaminated food, usually birds and domestic fowl or their eggs; also fecal contamination of water and food supplies.
   c. Symptoms: abdominal pain, fever, and diarrhea.
   d. Treatment: none.
5. Shigellosis (bacterial dysentery).
   a. Causative organism(s): several species of *Shigella*, including *S. flexneri, S. sonnei*, and *S. dysenteriae*.
   b. Mode of transmission: direct anal–oral route, although contaminated food and water supplies are involved in some outbreaks.
   c. Symptoms: abdominal pain, fever, and diarrhea with mucus and blood in the excretion.
   d. Treatment: none for adults; antibiotics for children.
6. *Campylobacter* infections (infant gastroenteritis).
   a. Causative organism(s): *Campylobacter fetus* var *jejuni*.
   b. Mode of transmission: via contaminated food or water.
7. *Vibrio parahaemolyticus* gastroenteritis.
   a. Causative organism(s): *Vibrio parahaemolyticus*.
   b. Mode of transmission: ingestion of contaminated seafood.
   c. Symptoms: abdominal pain, diarrhea, nausea, and vomiting.
   d. Treatment: none.
8. Yersiniosis.
   a. Causative organism(s): *Yersinia enterocolitica*.
   b. Mode of transmission: ingestion of contaminated water, milk, fruits, vegetables, or seafood.
   c. Symptoms: abdominal pain, fever, diarrhea, vomiting, and leukocytosis.

9. Cholera.
   a. Causative organism(s): *Vibrio cholerae* serotypes *cholerae* and *eltor*.
   b. Mode of transmission: via water supplies contaminated with fecal matter from infected individuals.
   c. Symptoms: abdominal pain, diarrhea with "rice water stools," nausea, vomiting, and severe dehydration followed by collapse, shock, and death.
   d. Treatment: replace fluids and maintain the electrolyte balance; tetracycline.
10. Appendicitis.
   a. Causative organism(s): a mixture of bacterial populations from the normal microbiota of the intestinal tract.
   b. Symptoms: abdominal pain, localized tenderness, fever, nausea, vomiting, and leukocytosis.
   c. Treatment: surgical removal of appendix.

G. Diseases caused by protozoan pathogens.
1. Amebic dysentery.
   a. Causative organism(s): *Entamoeba histolytica*.
   b. Mode of transmission: ingestion of contaminated food or water containing cysts of *E. histolytica*.
   c. Symptoms: mild or severe diarrhea and abdominal pain.
   d. Treatment: antiprotozoan drugs, such as metronidazole.
2. Giardiasis.
   a. Causative organism(s): *Giardia lamblia*.
   b. Mode of transmission: ingestion of contaminated food or water containing cysts of *G. lamblia*.
   c. Symptoms: diarrhea, dehydration, mucus secretion, and flatulence.
   d. Treatment: metronidazole.
3. Balantidiasis.
   a. Causative organism(s): *Balantidium coli*.
   b. Mode of transmission: via food or water contaminated with feces containing cysts of *B. coli*.
   c. Symptoms: abdominal pain, nausea, vomiting, diarrhea, and loss of weight.

H. Diseases caused by multicellular parasites.
1. Tapeworm.
   a. Causative organism(s): *Taenia saginata* (beef tapeworm), *Taenia solium* (pork tapeworm), other tapeworms.
   b. Mode of transmission: ingestion of food containing larvae.
   c. Symptoms: asymptomatic, mild abdominal pain, weight loss.
   d. Treatment: niclosamide.
2. Cysticercosis.
   a. Causative organism(s): *Taenia solium* (pork tapeworm).
   b. Mode of transmission: ingestion of food containing worm eggs.
   c. Symptoms: varied depending upon sites of infestation.
   d. Treatment: no satisfactory treatment.
3. Hydatid cysts.
   a. Causative organism(s): *Echinococcus granulosus*.
   b. Mode of transmission: contact with infested dogs.
   c. Symptoms: varied depending upon sites of infestation.
   d. Treatment: surgical removal of intact cyst.
4. Intestinal fluke.
   a. Causative organism(s): *Fasciolopsis buski*.
   b. Mode of transmission: ingestion of contaminated vegetation.
   c. Symptoms: diarrhea and abdominal pain.
   d. Treatment: hexylresorcinol.

5. Liver fluke.
   a. Causative organism(s): *Faciola hepatica*.
   b. Mode of transmission: ingestion of contaminated vegetation.
   c. Symptoms: cirrhosis.
   d. Treatment: dehydroemetine.
6. Pinworm.
   a. Causative organism(s): *Enterobius vermicularis*.
   b. Mode of transmission: ingestion of contaminated food and drink, hand-to-mouth contact.
   c. Symptoms: nausea, vomiting, diarrhea.
   d. Treatment: pyrantel pamoate.
7. Whipworm.
   a. Causative organism(s): *Trichuris trichiura*.
   b. Mode of transmission: ingestion of contaminated food and drink, hand-to-mouth contact.
   c. Symptoms: nausea, vomiting, constipation, emaciation.
   d. Treatment: mebendazole.
8. Trichinosis.
   a. Causative organism(s): *Trichinella spiralis*.
   b. Mode of transmission: ingestion of undercooked pork.
   c. Symptoms: varied.
   d. Treatment: varied.
9. Roundworm disease.
   a. Causative organism(s): *Ascaris lumbricoides*.
   b. Mode of transmission: ingestion of contaminated food or water, fecal–oral transmission via contaminated hands.
   c. Symptoms: malnutrition, fever, cough, bloody sputum.
   d. Treatment: piperazine citrate.
10. Hookworm.
   a. Causative organism(s): *Necator americanus, Ancylostoma duodenale*.
   b. Mode of transmission: transmission through skin from contaminated soil.
   c. Symptoms: gastrointestinal upset, anemia.
   d. Treatment: pyrantel pamoate.

Study Questions

1. What is the difference between a food-borne infection and food poisoning (intoxication)? How is this difference reflected in how we control different types of food-borne diseases?
2. For each of the following diseases, what are the causative organisms, characteristic symptoms, and prescribed treatments?
   a. Appendicitis
   b. Yersiniosis
   c. Cholera
3. Match the following statements with the terms in the list to the right.
   a. Characterized by enlargement of salivary glands (parotitis).
   b. Enterotoxins released by the bacterial cells growing on the surface lining of gastrointestinal tract cause diarrhea and abdominal cramps.
   c. Systemic viral infection primarily of the liver.
   d. Caused by adenoviruses, coxsackieviruses, polioviruses, ECHO viruses, Norwalk agent, rotavirus.

   (1) Mumps
   (2) Viral gastroenteritis
   (3) Staphylococcal food poisoning
   (4) Perfringens food poisoning
   (5) Hepatitis
   (6) Cholera
   (7) Giardiasis
   (8) Yersiniosis
   (9) Appendicitis
   (10) Amoebic dysentery
   (11) *Salmonella* spp.
   (12) *Shigella* spp.
   (13) *Vibrio parahaemolyticus*

e. Caused by ingestion of food containing *Clostridium perfringens* toxin, and absorption of toxin into cells lining gastrointestinal tract.
f. Caused by *Staphylococcus aureus* toxin.
g. Resembles appendicitis, prompting unnecessary appendectomies.
h. Caused by *Vibrio cholerae*.
i. Caused by *Giardia lamblia*.
j. Caused by *Entamoeba histolytica*.
k. Symptoms include nausea, vomiting, abdominal pain, "rice water" diarrhea, and severe dehydration followed by collapse, shock, and, in some cases, death.
l. Caused by members of normal intestinal microbiota.
m. Causes juvenile gastroenteritis.
n. Causes bacterial dysentery.
o. Responsible for a large number of gastroenteritis cases associated with eating raw fish.

(14) *Campylobacter fetus*
(15) *Yersinia enterocolytica*
(16) *Vibrio cholerae*

Some Thought Questions

1. Why are most cases of botulism associated with home canned food?
2. Would you eat a turkey that had been left in the trunk of a car for 2 weeks before Thanksgiving? Why?
3. When microbiologists go out to dinner they often have difficulty deciding what to eat. Why?
4. Why don't we wait to culture *Yersinia* before performing appendectomies?
5. Why do Jewish dietary laws prohibit the eating of pork?
6. Why do health statistics severely underestimate the incidence of gastroenteritis?
7. Should restaurant workers be required to wear plastic gloves?

Additional Sources of Information

Beck, J. W., and J. E. Davies. 1976. *Medical Parisitology*. The C. V. Mosby Company, Saint Louis. A concise text covering the major parasitic diseases of humans.

Boyd, R. F., and B. G. Hoerl. 1981. *Basic Medical Microbiology*. Little, Brown and Company, Boston. A basic text emphasizing microbes and disease that includes sections on gastrointesinal tract diseases.

Braude, A. I. (ed.). 1980. *Medical Microbiology and Infectious Disease*. W. B. Saunders Company, Philadelphia. An advanced work with individual chapters dealing with specific diseases of the gastrointestinal tract; chapters dealing with infectious diseases are organized according to the taxonomy of the causative agent.

Davis, B. D., R. Dulbecco, H. N. Eisen, H. S. Ginsberg, and W. B. Wood. 1981. *Microbiology*. Harper & Row, Publishers, Hagerstown, Maryland. An excellent text used in many graduate and medical school microbiology courses dealing with all aspects of microbiology; chapters dealing with infectious diseases are organized according to the taxonomy of the causative agent.

Feachem, R. G., D. J. Bradley, H. Garelick, and D. D. Mara. 1983. *Health Aspects of Excreta and Wastewater Management*. John Wiley, Chichester. This thorough reference volume, which was prepared for the World Bank, covers the major pathogenic and parasitic diseases that are transmitted through contaminated water and foods.

Hoeprich, P. D. (ed.). 1977. *Infectious Diseases: A Modern Treatise of Infection Processes*. Harper & Row, Publishers, Hagerstown, Maryland. This classic treatment of infectious diseases makes an excellent reference volume.

Jawetz, E., J. L. Melnick, and E. A. Adelberg. 1980. *A Review of Medical Microbiology*. Lange Medical Publications, Los Altos, California. This is a very good review of infectious diseases.

Joklik, W. K., and H. P. Willett (eds.). 1980. *Zinsser Microbiology*. Appleton-Century-Crofts, New York. A classic text used in many medical school programs that includes chapters on infectious diseases.

Riemann, H., and F. L. Bryan (eds.). 1979. *Food-Borne Infections and Intoxications*. Academic Press, Inc., New York. This book presents detailed reviews of those infectious diseases that are transmitted through food and water.

Rose, N. L., and A. L. Rose (eds.). 1983. *Microbiology: Basic Principles and Clinical Applications*. Macmillan Publishing Co., Inc., New York. An advanced work on medical microbiology; chapters are organized for the most part according to taxonomic groups.

Savage, D. C. 1977. Microbial ecology of the gastrointestinal tract. *Annual Review of Microbiology*, **31**:107–133. This extensive review examines the resident microbiota of the gastrointesinal tract.

Youmans, G. P., P. Y. Paterson, and H. M. Sommers. 1980. *The Biological and Clinical Basis of Infectious Disease*. W. B. Saunders Company, Philadelphia. This is an excellent text covering most human infectious diseases.

25 *Genitourinary Tract and Sexually Transmitted Diseases*

KEY TERMS

chancroid
cystitis
genital herpes
genital warts
glomerulonephritis
gonococcal urethritis
gonorrhea
granuloma inguinale
leptospirosis
lymphogranuloma venerum
nephron
nongonococcal urethritis (NGU)
nonspecific urethritis (NSU)
pelvic inflammatory disease
prostatitis
pyelonephritis
salpingitis
sexually transmitted disease (STD)
syphilis
urethritis
vaginitis
venereal disease
vulvovaginitis
Weil's disease

OBJECTIVES

After reading this chapter you should be able to

1. Define the key terms.
2. Describe the structure of the genital and urinary tracts.
3. Describe the resident microbiota of the genitourinary tract.
4. Describe the defense systems of the genitourinary tract that protect against infectious microorganisms.
5. Discuss the causative agents, routes of transmission, key symptoms, and treatments for the following diseases: chancroid, cystitis, genital herpes, glomerulonephritis, gonorrhea, granuloma inguinale, leptospirosis, lymphogranuloma venerum, nongonococcal urethritis (NGU), prostatitis, pyelonephritis, syphilis, and vaginitis

Structure of the Genitourinary Tract

Urinary System

The urinary system consists of two glandular kidneys, which remove substances from the blood and form urine; two tubular ureters, which transport urine away from the kidneys; a sac-like urinary bladder, which serves as a urine reservoir; and a tubular urethra, which conveys urine to the outside of the body (Figure 25.1). The primary function of the urinary system, particulary the kidneys, is to regulate the concentrations of a variety of substances circulating in the blood. Blood is filtered within the kidneys and the excess materials are removed via the urinary system. The kidneys are reddish brown, bean-shaped, smooth-surfaced organs that lie on either side of the vertebral column in a depression high on the back wall of the abdomen.

The functional unit of the kidney is the nephron

Figure 25.1 Diagram of the urinary tract showing detailed structure of kidneys.

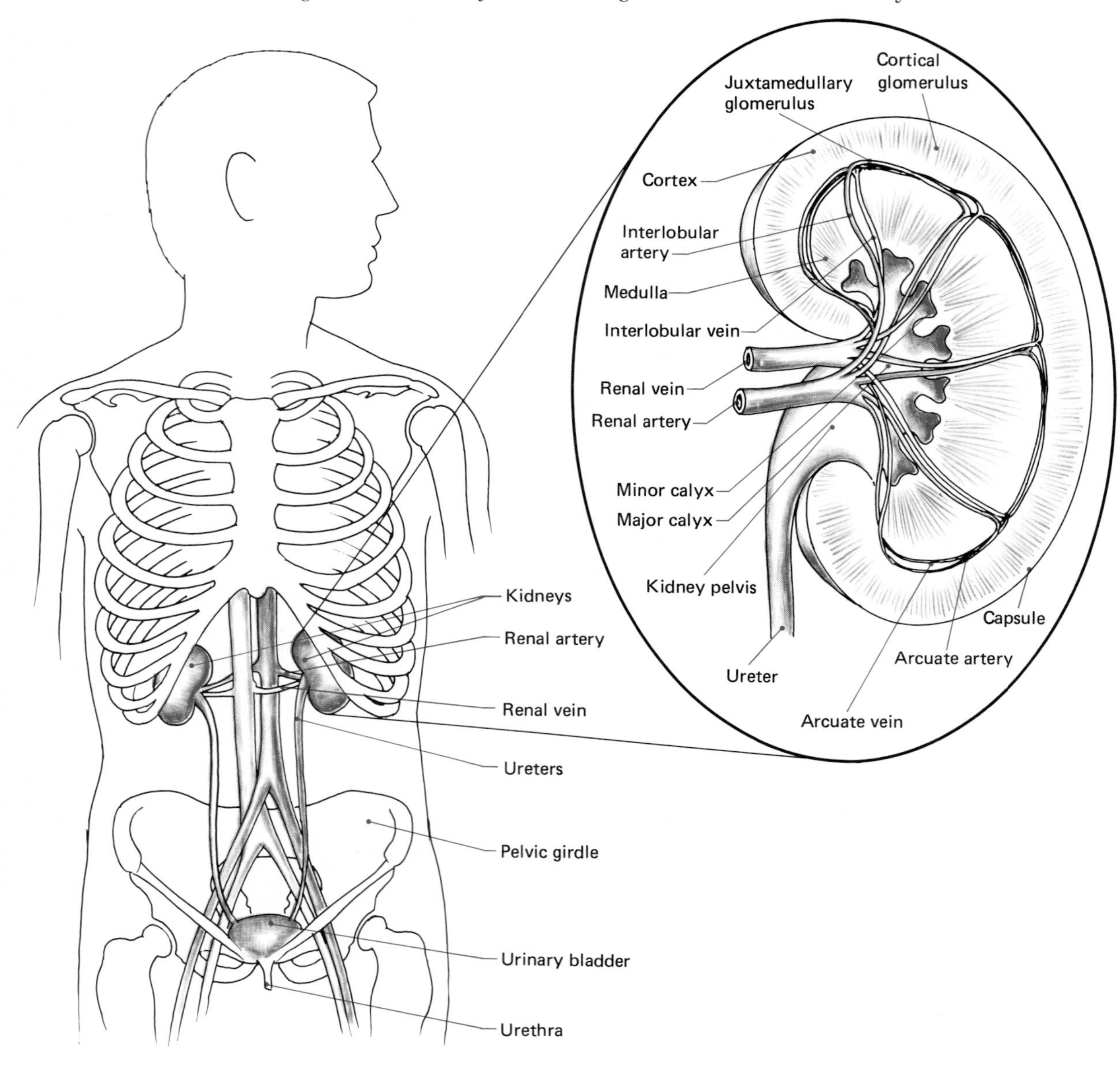

Each kidney contains about 1 million nephrons, each consisting of a renal corpuscle and a renal tubule. A renal corpuscle is composed of a tangled cluster of blood capillaries, called a glomerulus, surrounded by a thin-walled, sac-like structure called Bowman's capsule. The primary function of the nephrons include the removal of waste substances from the blood and regulation of water and electrolyte concentrations within body fluids. The end product of these functions is urine, which carries wastes, excess water and electrolytes. Urine formation involves filtration into renal tubules of various substances from the plasma within glomerular capillaries, readsorption of some of these substances into the plasma, and secretion into the renal tubules of other substances from the plasma.

Urine is conveyed by a ureter from the kidneys to the urinary bladder. The ureter is a 25 centimeter long tube whose muscular wall helps move urine. The urinary bladder is a hollow, distentible muscular organ located within the pelvic cavity. Urine accumulates within the urinary bladder until it is excreted from the body. The urethra connects the urinary bladder to the outside of the body, and urine is excreted to the outside of the body through the urethra.

Genital system

In addition to conveying urine from the bladder to the exterior, the urethra in males also is part of the reproductive system, serving to carry semen. The male reproductive system consists of three parts (Figure 25.2). The paired testes are the essential organs of the system, responsible for the production of male sex cells, sperm, and the male sex hormone, testosterone. The testes are contained in the scrotum. A system of ducts (vas deferens and ejaculatory duct) conveys the sperm and secretions contributed by glands to the exterior. The system also includes accessory glands, the paired seminal vesicles (add secretions rich in prostaglandins) and the prostate gland (secretes alkaline fluid into the vas deferens). The mixture of sperm and secretions is called semen. The penis serves the reproductive system as the copulatory organ to introduce sperm into the vagina of the female.

The female reproductive system consists of the ovaries, paired structures that produce eggs and sex hormones; the uterine tubes, which convey eggs to the uterus; the uterus, which houses the fetus; and the vagina, which receives the penis and serves as the birth canal (Figure 25.3).

Figure 25.2 Diagram of the male genital system.

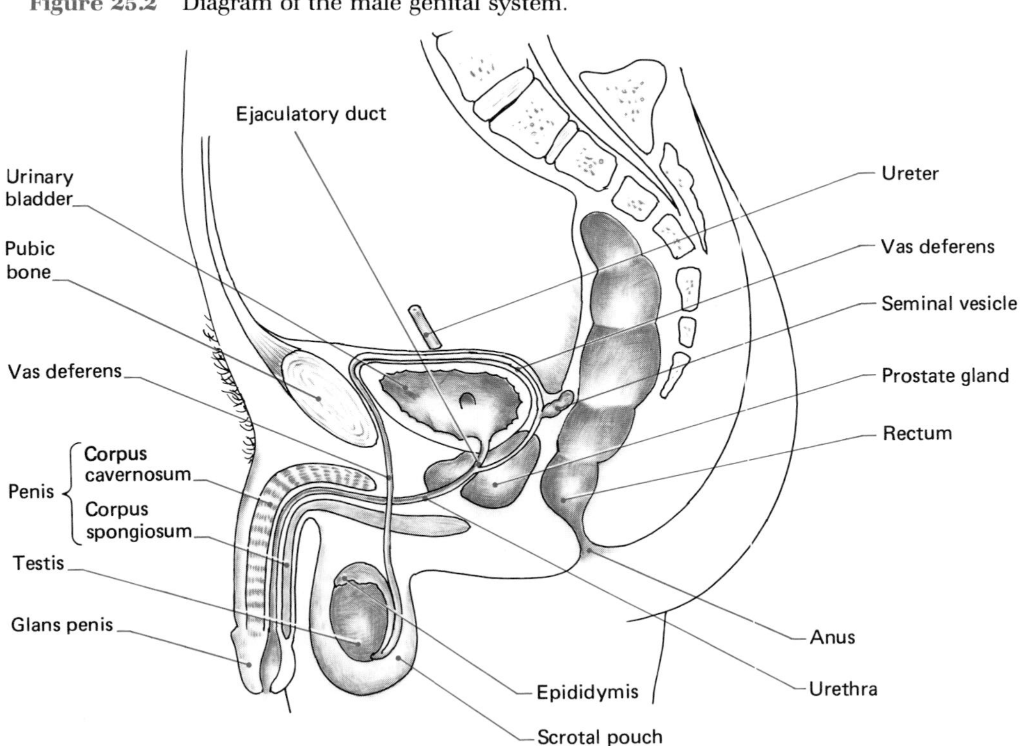

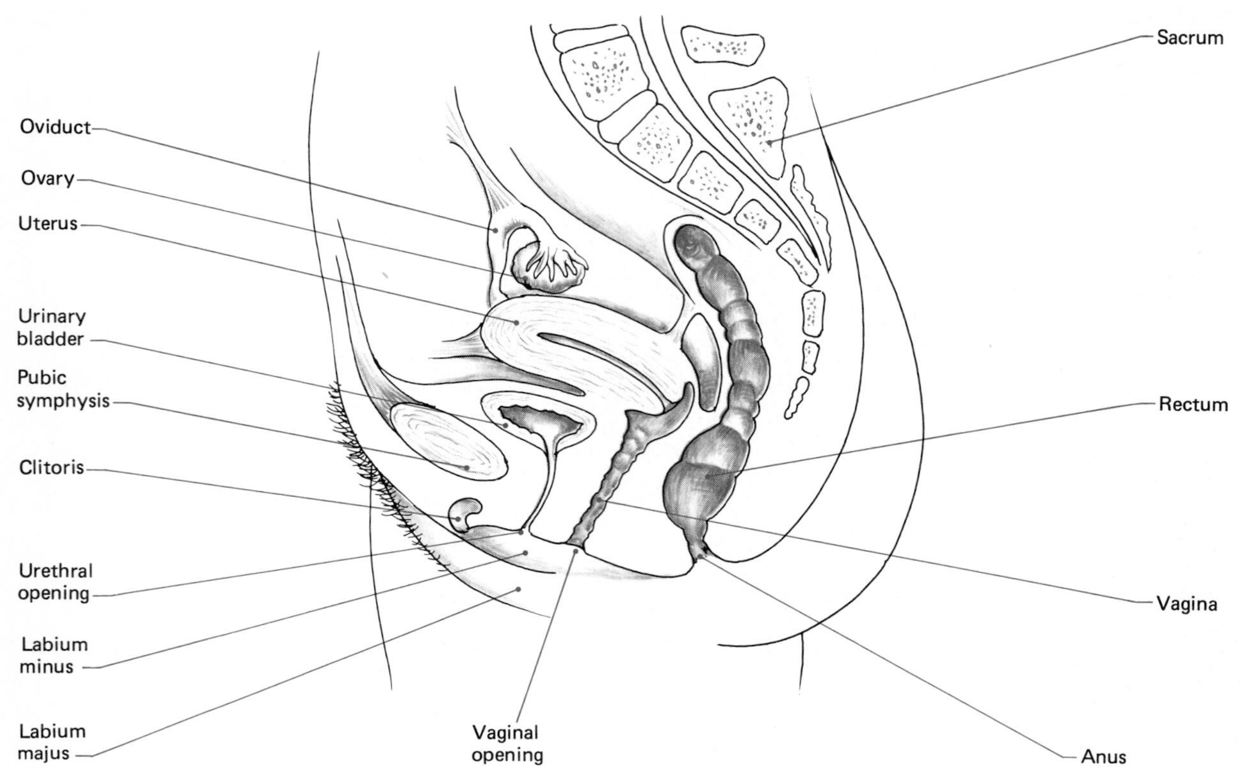

Figure 25.3 Diagram of the female genital system.

Resident Microorganisms of the Genitourinary Tract

Although most of the genitourinary tract, including the kidneys and urinary bladder, is normally free of microorganisms, the external genital regions of both sexes contain various indigenous microbial populations. For example, the terminal areas of the urethra in both males and females are colonized with bacteria. Because of its relatively large surface area and the nature of mucoidal secretions, the vaginal tract harbors an especially large variety of resident microbiota. The normal microbiota of the vaginal tract include bacteria such as *Streptococcus, Lactobacillus, Bacteroides*, and *Clostridium*; coliforms; spirochetes; yeasts, including members of the genus *Candida*; and flagellated protozoans of the genus *Trichomonas* (Figure 25.4). The low pH of the vaginal tract, 4.4–4.6, limits the species of microorganisms that can reproduce there, and the metabolic activities of the lactic acid producing resident microbiota contribute to the maintenance of the low pH. Thus, the microorganisms that multiply within the vaginal tract are normally acid tolerant. The resident microbiota of the vaginal tract is a carefully balanced complex microbial community in which large fluctuations occur during the normal menstrual cycle. Occasionally, population imbalances occur, for instance, the proliferation of the yeast *Candida albicans* or the protozoan *Trichomonas vaginalis*, leading to inflammation or vaginitis. Such population imbalances often occur when there are major shifts in hormonal levels, such as during puberty and pregancy and after menopause.

Defenses of the Genitourinary Tract

The genitourinary tract is protected from pathogenic microorganisms by a variety of mechanisms. These defenses include the urinary sphincters, outward flow of mucus, presence of antibodies in the mucus, low pH, high salt concentrations, the presence of antimicrobial enzymes, and antagonistic re-

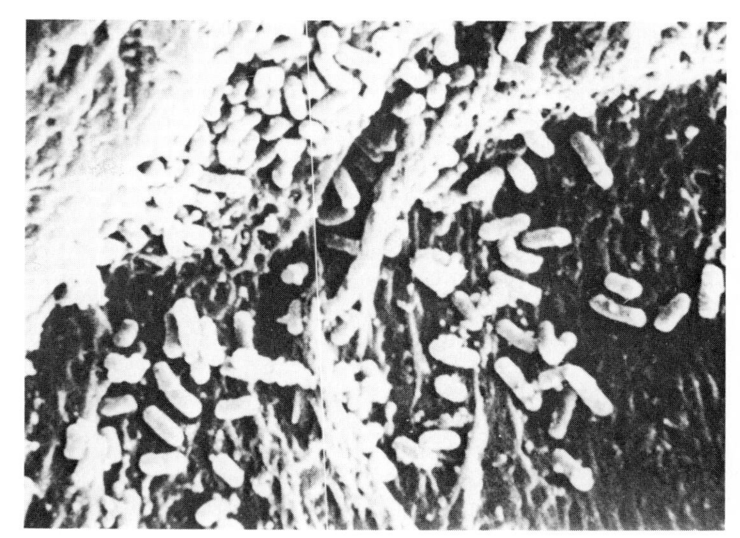

Figure 25.4 The normal microbiota of the vaginal tract are represented by this micrograph of a microcolony (9500 ×). (Reprinted by permission of the American Society for Microbiology, Washington, D.C., from B. Larsen, A. J. Markovetz, and R. P. Galask, 1977, *Applied and Environmental Microbiology*, 34:85.)

lationships with the natural microbiota of the lower genitourinary tract. The inner tissues of the genitourinary tract, including the bladder and kidneys, are normally sterile. The outward flow of urine, with its low pH and toxic components, protects these inner organs for the most part from microbial infections. However, these tissues are subject to infection with opportunistic pathogens if they become contaminated by microorganisms from the indigenous microbiota of the extremities of the genitourinary tract or the gastrointestinal tract. Additionally, the genitourinary tract provides the portal of entry for pathogens that are directly transmitted during sexual intercourse. Infections with such pathogens are known as venereal, or sexually transmitted, diseases (STDs).

Urinary Tract Infections

Urethritis and Cystitis

Although many microorganisms can cause urinary tract infections, the most common etiologic agents are Gram negative bacteria, particularly those normally occurring in the gastrointestinal tract. Accidental contamination of the urinary tract with fecal matter appears to be one of the most important means of transmission of such infections. *Escherichia coli* probably is the most common cause of urinary tract infections, but *Klebsiella, Enterobacter, Serratia, Proteus,* and *Pseudomonas* are also isolated relatively frequently from urinary tract infections (Table 25.1). Nosocomial infections with *Serratia* species and *Pseudomonas aeruginosa* most

Table 25.1 Representative Relative Proportions of Bacterial Isolates from Hospital-acquired and Community-acquired Urinary Tract Infections

Community-acquired Organisms	Percent	Hospital-acquired Organisms	Percent
Escherichia coli	57	*Escherichia coli*	39
Klebsiella pneumoniae	13	*Proteus mirabilis*	18
Proteus mirabilis	8	*Pseudomonas aeruginosa*	17
Streptococcus sp.	8	*Streptococcus* sp.	11
Pseudomonas aeruginosa	7	*Klebsiella pneumoniae*	10
Proteus rettgeri	5	*Serratia marcescens*	8
Others	2		

often occur after catheterization procedures because the insertion of a catheter can carry microorganisms from the extremities of the genitourinary tract, contaminating the inner tissues and resulting in urethritis and cystitis.

The urethra and urinary bladder are most frequently the sites of infections within the urinary tract, with the resulting infections referred to as urethritis and cystitis, respectively. The symptoms of urethritis normally include pain and a burning sensation during urination. Cystitis is characterized by suprapubic pain and the urge to urinate frequently.

Various antibiotics, such as sulfonamides, nitrofurantoin, and nalidixic acid, are used in treating infections of the lower urinary tract. Depending on whether there is an obstruction or another underlying cause involved in the establishment of an infection, the prognosis is normally good.

Pyelonephritis

Infections of the urethra may spread to the kidney, causing serious life-threatening disease. Pyelonephritis can cause kidney failure and death, but as with other urinary tract infections, the use of antibiotics is generally effective in curing this disease. *Escherichia coli* is the most frequent etiologic agent of pyelonephritis in patients who develop infections outside of hospital settings. *Proteus mirabilis, Enterobacter aerogenes, Klebsiella pneumoniae, Pseudomonas aeruginosa, Streptococcus* sp., and *Staphylococcus* sp. are additional causative agents of pyelonephritis. Infections with *Proteus* sp. are especially significant because they have the enzymatic capacity to decompose urea in the urine to ammonia and carbon dioxide. Urgent and frequent urination is normally symptomatic of pyelonephritis. Pyelonephritis and other urinary tract infections are more frequent in adult women than in males or young females, in part because of the size of the anatomical opening to the structure of the urinary tract of females, and also because of the physiological changes that occur during menstruation and pregnancy that favor the establishment of microbial infections.

Leptospirosis

Leptospirosis, or Weil's disease, is principally a disease of nonhuman animals caused by *Leptospira interrogans* (Figure 25.5). Infected animals excrete leptospires in their urine, and people generally contract the disease through contact with contaminated material harboring viable *Leptospira* species. This spirochete can enter the human body through the mucous membranes of the eyes and nasopharynx or through abrasions of the skin.

During the acute phase of this illness, the symptoms normally include a high spiking fever, chills, headache, muscle ache, malaise, abdominal pain, nausea, and vomiting. Various body organs can be involved, and most fatal cases result from infection of the kidney and subsequent renal failure. In most cases, leptospirosis is subclinical, and the prognosis for complete recovery is excellent. When severe symptoms develop, however, the death rate from Weil's disease is about 10–40 percent. Penicillin may provide a useful treatment and may shorten the duration and severity of the illness. However, in most cases antibiotics are not effective in curing this disease. Avoidance of potentially con-

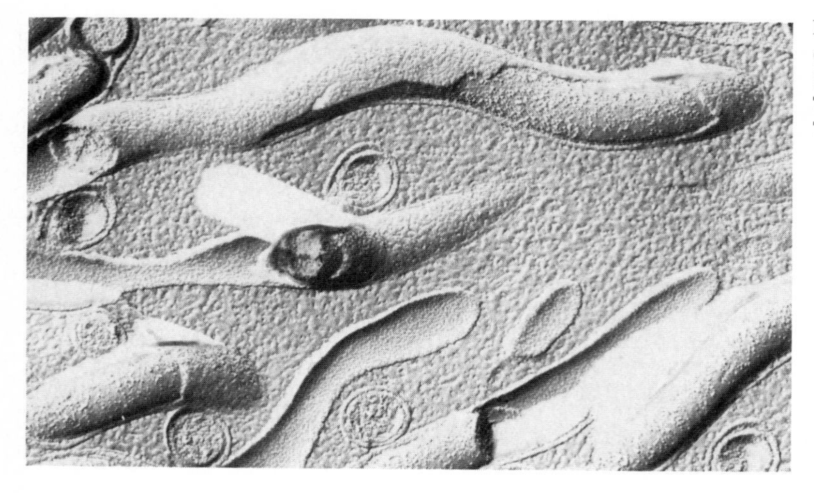

Figure 25.5 A freeze-etched micrograph of the spirochete *Leptospira interrogans*. (From BPS: Stanley C. Holt, University of Texas Health Science Center, San Antonio, Texas.)

taminated environments, such as sewers and contaminated streams, is effective in preventing outbreaks of this disease.

Glomerulonephritis

Infections with *Streptococcus pyogenes* can result in glomerulonephritis, which impairs kidney function. In this disease there is an inflammation of the glomeruli, which are the blood capillaries involved in filtering blood as it passes through the kidneys. The best evidence now suggests that glomerulonephritis is an immune-complex disease that follows a streptococcal infection and is not owing to an active kidney infection. The symptoms of glomerulonephritis include fever, elevated blood pressure, and the presence of protein and red blood cells in the urine. The blood and protein in the urine are a result of the altered permeability of the inflamed glomeruli in the kidney. Antibiotics, such as chloramphenicol and penicllin, are useful in treating the initial infection. In most cases there is no chronic damage to the kidneys and recovery is complete.

Genital Tract Infections

Vaginal Tract Infections

The female genital tract is subject to infection by a variety of microorganisms. The trauma associated with sexual intercourse, menstrual bleeding, normal changes during pregnancy, use of contraceptives (such as oral contraceptives and intrauterine devices), and trauma due to childbirth contribute to occurrences of infections of the female genital tract. Infections of the vulva and vagina are common and generally not serious. Vaginitis is an inflammation of the vagina; vulvovaginitis is an inflammation of both the vulva and vagina and can be caused by viruses, bacteria, fungi, and protozoa. In about 20 percent of the cases of vulvovaginitis, the flagellate protozoan *Trichomonas vaginalis* is the etiologic agent. The fungus *Candida albicans* is also frequently implicated. Bacterial and fungal overgrowth of the vaginal tract is common during pregnancy when the pH of the vaginal tract increases. The use of birth control pills also tends to result in higher pH values in the vagina, and the administration of antibacterial antibiotics, such as tetracyclines, which can adversely affect the indigenous bacterial populations, also favors the development of *Candida* infections of the vaginal tract.

The excessive growth of *Trichomonas vaginalis, Candida albicans*, or various bacteria causes changes in the mucosal cells lining the vaginal tract. The symptoms of vulvovaginitis include increased vaginal discharge and burning. In cases where *T. vaginalis* is the causative agent, there is normally a profuse, greenish, odorous discharge, and when vulvovaginitis is caused by *Candida* infections, the discharge is normally thicker and cheesier and is released in lesser quantity. Treatment with antimicrobics is effective in controlling vulvovaginitis, with the specific drug choice depending on the identification of the specific etiologic agent. Metronidazole can be used in cases of protozoan infections, and nystatin generally is used for *Candida* infections.

Infections of the lower parts of the female genital tract may also spread to higher structures, including the cervix and fallopian tubes. One result of such infections is pelvic inflammatory disease (PID). Infections of the upper regions of the genital tract are particularly serious during pregnancy and can lead to septic abortion. Some infections of the gestating female genital tract can also be transferred to the fetus, resulting in congenital abnormalities. Microbial infection with herpes viruses, for example, can occur during childbirth. In some cases of female genital tract infections, it is necessary to perform a Caesarean section delivery in order to protect the infant from contamination.

Prostatitis

In males the prostate gland is subject to infection by various microorganisms. Microbial infections of the prostate cause acute inflammation. In middle aged males, bacterial infection of the prostate frequently causes chronic prostatitis. Often several opportunisitic bacteria are involved in infections of the prostate. *E. coli, Pseuodomonas aeuriginosa, Staphylococcus aeureus*, and *Streptococcus* species have frequently been identified as the causative agents of prostate infections. Besides these infections, the prostate is often involved in gonococcal infections.

Sexually Transmitted Diseases

Sexually transmitted diseases (STDs) are contracted by direct sexual contact with an infected individual, generally during sexual intercourse. These diseases previously were called venereal diseases, which means diseases of love. These diseases have little to do with love and therefore the name has been changed to reflect the involvement of sexual contact in their transmission. The physiological properties of the pathogens causing STDs restrict their transmission, for the most part, to direct physical contact because the etiologic agents of the sexually transmitted diseases have very limited natural survival times outside infected tissues.

Don't give a dose to the one you love most.
Give her some marmalade; give her some toast.
You can give her the willies or give her the blues,
But the dose that you give her will get back to you.

I once had a lady as sweet as a song.
She was my darlin' and she was my dear.
But she had a dose and she passed it along.
Now she's gone, but the dose is still here.

So don't give a dose to the one you love most.
Give her some marmalade; give her some toast.
You can give her a partridge up in a pear tree,
But the dose that you give her might get back to me.

So if you've got an itchin'—if you've got a drippin'
Don't sit there wishin' for it to go away, cause it won't.
If there's a thing on the tip of your thing or your lip,
Run down to the clinic today and say:

I won't give a dose to the one I love most . . .

—SHEL SILVERSTEIN, *Don't Give a Dose*

The rate of incidence of sexually transmitted diseases reflects contemporary sexual behavioral patterns but may also in part reflect changes in reporting and recording cases of these diseases. At present, outbreaks of some sexually transmitted diseases are considered to be reaching epidemic proportions. The social implications inherent in the transmission of these diseases often overshadow the fact that they are infectious diseases. STDs must be treated as medical problems, with emphasis on curing the patient and reducing the incidence of disease by preventing spread of the infectious agents. The overall control of sexually transmitted diseases rests with breaking the network of transmission, which necessitates public health practices that seek to identify and treat all sexual partners of anyone diagnosed as having one of the sexually transmitted diseases.

Have been in London several weeks without ever enjoying the delightful sex . . . Many fold are the reasons for this my present wonderful continence . . . I have suffered severely from the loathsome distemper, and therefore shudder at the thoughts of running any risk of having it again. Besides, the surgeons' fees in this city come very high.

—SAMUEL PEPYS, *December 14, 1762 diary entry*

It should be noted that health care workers run the risk of contracting pathogens that cause sexually transmitted diseases by direct, but nonsexual, contact. A dentist or dental hygienist, for example, may contact a lesion in the oral cavity containing infective pathogens that cause one of the sexually transmitted diseases. Rubbing ones eyes with hands that had been in contact with lesions of syphilis or other sexually transmitted diseases can also lead to nonsexual transmission. Fortunately this mode of transmission is not too common and health care workers run only a slightly higher risk of nonsexual contraction of one of the sexually transmitted diseases than the general public.

Bacterial Diseases

Gonorrhea. Gonorrhea is a sexually transmitted disease caused by the Gram negative, kidney bean-shaped, diplococcus, *Neisseria gonorrhoeae*, often referred to as gonococcus. *N. gonorrhoeae* is a fastidious organism readily killed by drying and exposure to metals. The sensitivity of *N. gonorrhoeae* to desiccation makes the chances of transmission of gonorrhea through inanimate objects, such as toilet seats in public restrooms, negligible. The adherence of *N. gonorrhoeae* to the lining of the genitourinary tract during sexual transmission and the ensuing spread of the infection depends on the pili of this bacterial species. *N. gonorrhoeae* is able to penetrate to the subepithelial connective tissue during spread of the infection (Figure 25.6). *N. gonorrhoeae* infects the mucosal cells lining the epithelium. This bacterium is able to infect the urethra, cervix, anal canal, pharynx, and conjunctivae. Transmission can occur at any point in sexual contact, and while transmission to the genitals is most frequent, pharyngeal and anal gonorrhea are not uncommon.

In one day: 2,700 Americans discover that they have gonorrhea. Two hundred discover that they have syphilis.

—TOM PARKER, *In One Day*

There are about 1 million cases of gonorrhea reported in the United States each year. Many more cases undoubtedly occur that are never reported to public health authorities. There has been an alarming increase in the number of cases of gonorrhea in the United States since 1960 (Figure 25.7).

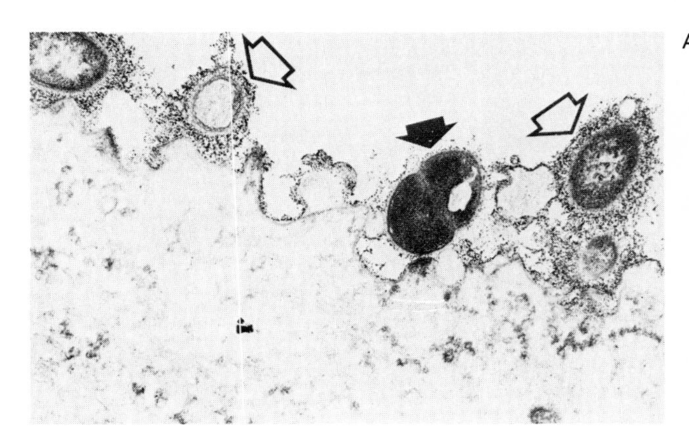

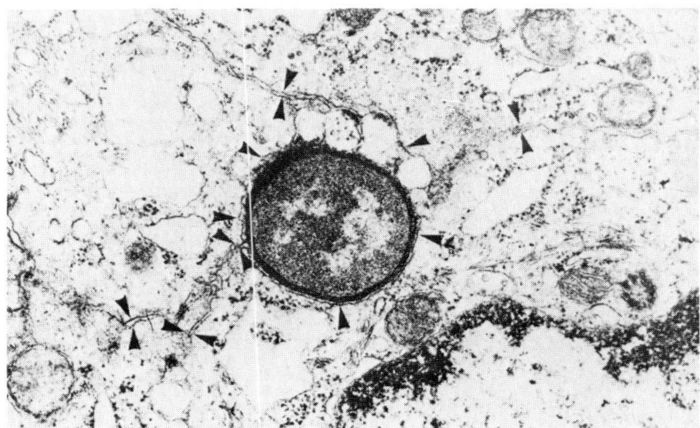

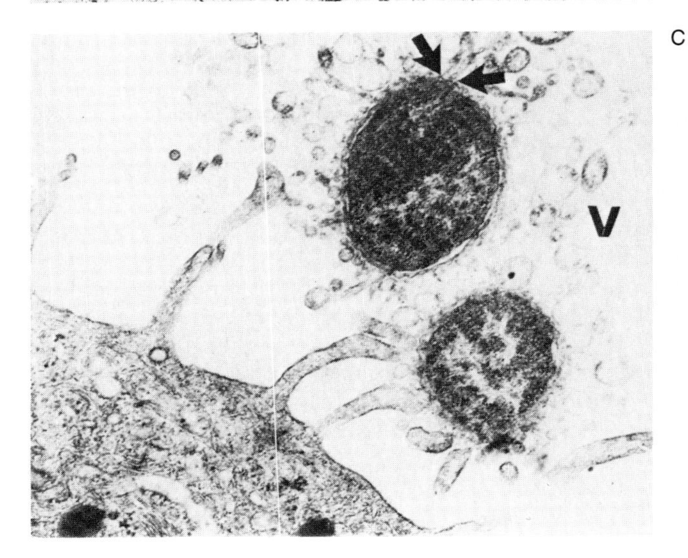

Figure 25.6 Electron micrographs of gonococcus in the genital tracts of patients with gonorrhea. (A) Gonococcus (solid arrow) and commensal bacteria lying on the cervical epithelium of a female (25,650 ×). Colloidal thorium treatment reveals well-defined acidic capsules on the commensal bacteria (hollow arrows). Glycoproteins of the epithelial cell coat and in cervical mucus have also been stained. By contrast, there is no evidence of a capsule in the gonococcal surface. Light staining with colloidal thorium can be attributed to contaminating host cell material. (B) Gonococcus lying between epithelial cells in a human fallopian tube organ culture 8 hours after challenge (21,900 ×). The gonococcus is tightly enclosed by the surrounding host cell surfaces (arrows) that are linked in places by desmosomes. (C) Gonococci approaching the surfaces of a urethral epithelial cell from a male with early symptomatic gonorrhea (29,200 ×). Microvilli from the host cell have made contact with the bacteria. The vesicles (V) are characteristically associated with host-grown gonococci and appear at some points (arrows) to be budded off from the bacterial surface. (Reprinted by permission of the American Society for Microbiology, Washington, D.C., from M. E. Ward *et al.*, 1975, in *Microbiology—1975*, D. Schlessinger, ed., pp. 188–199.)

This increase coincides with the widespread introduction of birth control pills, the hormones in which cause pH values in the vaginal tract to increase, removing the normal protection afforded against infections by acid-sensitive bacteria, including *N. gonorrhoeae*. Gonorrhea is normally contracted from someone who is asymptomatic or who has symptoms but does not seek treatment. The rate of gonorrhea acquisition for males is about 35 percent after a single exposure to an infected partner and

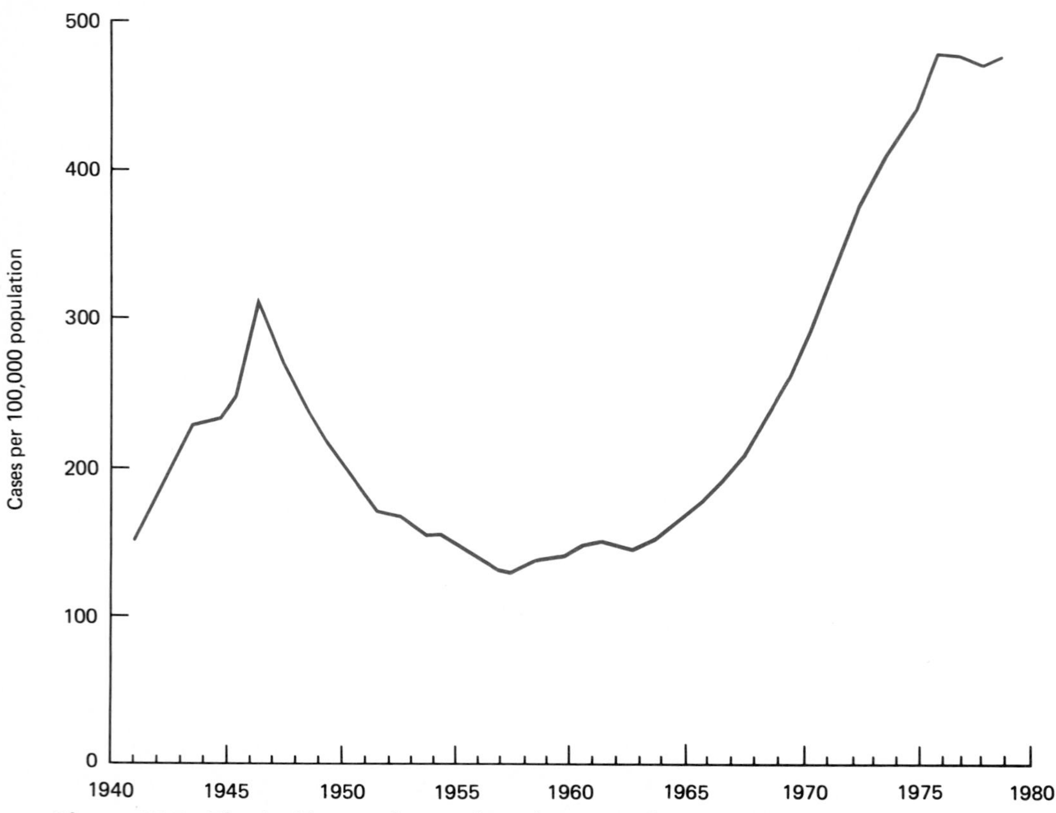

Figure 25.7 The incidence of gonorrhea shows an alarming, epidemic increase in recent years, as shown by these reported civilian cases in the United States from 1941 to 1978. Note the sharp decline from 1946 to 1956, followed by the dramatic increase since then. The precise cause for this rise is unknown; some point to increased sexual activity in a permissive society; it seems likely that the use of oral contraceptives, which lead to altered vaginal pH, has played a role in the increased ability of *Neisseria gonorrhoeae* to survive, resulting in increased transmission rates for this disease.

rises to 75 percent after multiple sexual contacts with the same individual. It is more difficult to assess the rate of transmission of gonorrhea to females because some become asymptomatic carriers.

In most cases gonorrhea is a self-limiting disease, but in both sexes the infection may spread to contiguous parts of the genitourinary tract, and *N. gonorrhoeae* may be disseminated to other parts of the body. For example infections with *N. gonorrhoeae* can spread to the joints, causing gonorrheal arthritis; to the heart, causing gonorrheal endocarditis; and the central nervous system, causing gonorrheal meningitis.

Often, in women the early stages of gonorrhea are not associated with any overt symptoms, and many women with gonorrhoea in fact remain asymptomatic carriers. The cervix often is the site of gonococcal infection. Various other tissues may be involved if the infection spreads. If the infection spreads to the uterus, it causes a chronic infection of the uterine tubes called salpingitis. This condition typically is characterized by abdominal pain. Salpingitis can cause infertility and can also cause fertilization to occur outside the uterus, a life-threatening situation called ectopic pregnancy. A gonococcal infection may spread to the urethra, causing an inflammation called gonococcal urethritis. Gonorrhea also can lead to pelvic inflammatory disease (PID), which results from a generalized bacterial infection of the uterus, pelvic organs, uterine tubes, and ovaries (Figure 25.8). PID may occur without overt symptoms of gonorrhea but nevertheless may cause infertility. Occlusions of the fallopian tubes due to scarring produce sterility in some cases of gonorrhea.

Female infections with *N. gonorrhoeae* exhibit a

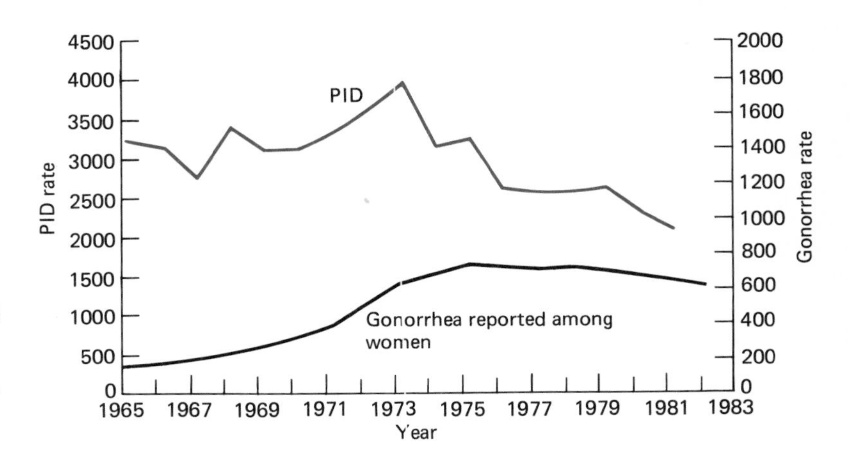

Figure 25.8 Total number of consultations for pelvic inflammatory disease (PID) and reported cases of gonorrhea among women per 100,000 population. (Based upon *Morbidity and Mortality Weekly Reports,* Centers for Disease Control, Atlanta.)

wide variety of symptoms. Most commonly, symptomatic cases exhibit inflammation with some pain and swelling, abnormal vaginal discharge, and abnormal menstrual bleeding. Because many women with gonorrhea are asymptomatic, the eyes of all infants are routinely washed immediately after birth with erythromycin sulfate to prevent infections of the eye, which could result from the transmission of *N. gonorrhoeae* from mother to infant during passage through the vaginal tract. *N. gonorrhoeae* can also be transmitted to the eyes of adults by rubbing with hands contaminated with *N. gonorrhoeae*.

In men gonorrhea results in a characteristic painful, purulent urethral discharge. The pus results from the migration of phagocytic leukocytes to the site of infection. Symptoms of gonorrhea in males are usually apparent less than 1 week after infection. If the disease is untreated, occlusions of the vas differens due to scarring may produce sterility in some males.

Gonorrhea is readily treated with antibiotics, with penicillin being the antibiotic of choice. Other antibiotics, such as tetracycline, are also effective. In recent years there has been an increase in the tolerance of *N. gonorrhoeae* to antibiotics, creating a major concern in the treatment of gonorrhea (Figure 25.9). The recent identification of beta-lactamase-producing strains of *N. gonorrhoeae,* which are resistant to most penicillins, may in some cases lead to a preference for antimicrobics other than penicillin for treating gonorrhea. Resistant strains are called PPNGs (penicillinase-producing *N. gonorrhoeae*). In cases of penicillin-resistant strains of *N. gonorrhoeae,* spectinomycin is the antibiotic of choice.

Culturing for laboratory diagnosis is accomplished using chocolate agar (blood agar that has been heated) and a 10 percent carbon dioxide atmosphere. The observation of Gram negative diplococci in pus also is diagnostic for this disease.

Diagnosis and treatment of gonorrhea is not necessarily the end of an individual's involvement with gonorrhea. No long-term immunity to the disease develops and so one can become reinfected by sexual contact with an infected individual shortly after

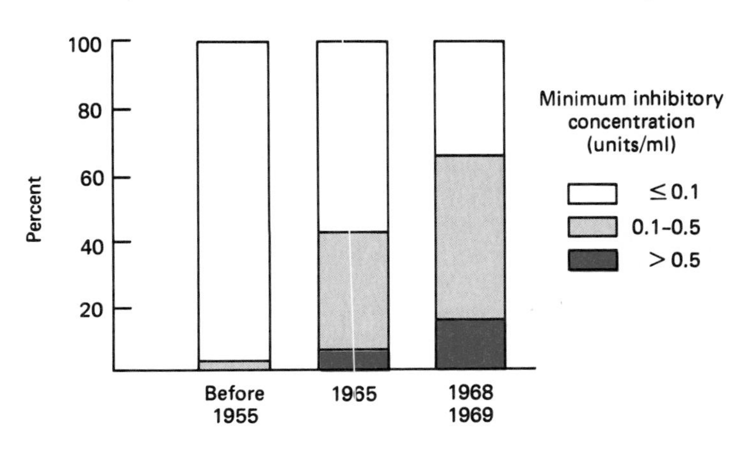

Figure 25.9 Histogram showing increased resistance of *Neisseria gonorrhoeae* to penicillin from 1950 to 1969. The increasing occurrence of resistant strains coincides with increased use of antibiotics and reflects selective pressure.

A

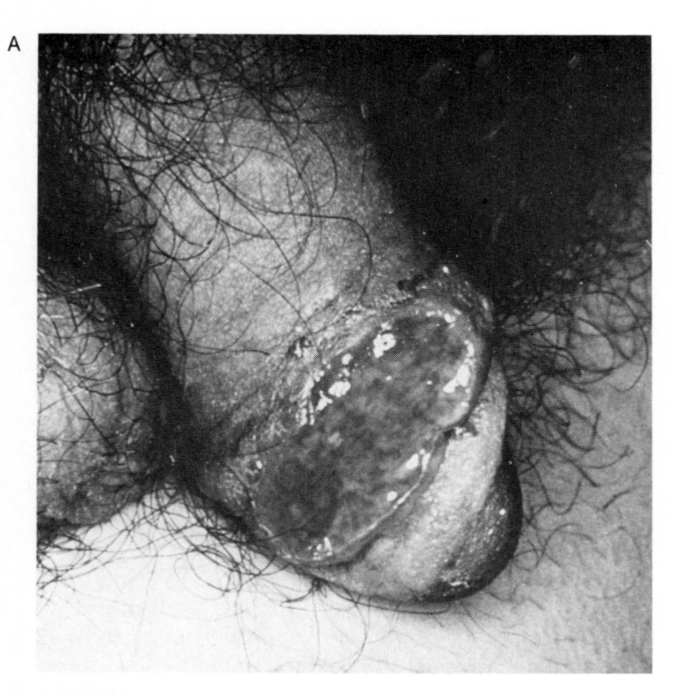

B

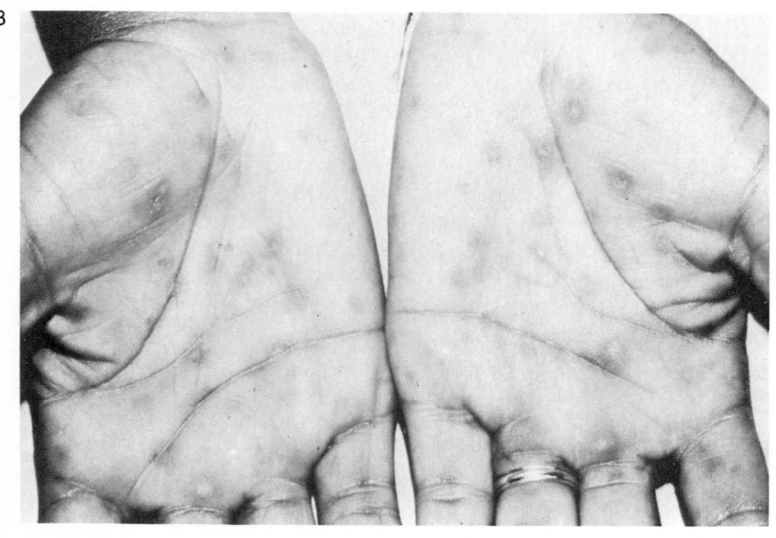

Figure 25.10 (A) Primary syphilis lesion on the genitals. (Centers for Disease Control, Atlanta.) (B) Secondary syphilis lesions on the palms of hands resulting from syphilis infection. (From BPS: Leonard Winograd, Stanford University.)

being cured. Tracking the chain of transmission and treating infected individuals is an important activity of public health workers. Education is important if the public is to help eliminate gonorrhea. In Scandinavia large billboards portray the virtues of using prophylactic condoms for blocking the transmission of gonorrhea. Routine screening of women during gynecological examinations also is useful in identifying asymptomatic carriers and treating this disease.

Syphilis. Syphilis is a sexually transmitted disease caused by *Treponema pallidum*. Nearly 30,000 cases of syphilis are reported in the United States each year. The causative organism, *T. pallidum*, is a bacterial spirochete, fastidious in its growth requirements and readily killed by drying, heat, and disinfectants such as soap, arsenicals, and mercurial compounds. Historically immersion in hot bath spas and arsenic- and mercury-containing compounds have been used in the treatment of syphilis. The inability of *T. pallidum* to survive long outside the body makes transmission through inanimate objects (fomites) virtually nonexistent. Transmission depends on direct contact with the infective syphilitic lesions containing *T. pallidum*. *Treponema pallidum* enters the body via abrasions of the epithelium and by penetrating into mucous membranes. The bacteria migrate to the lymphatic system shortly after penetrating the dermal layers.

Syphilis manifests itself in three distinct stages. During the primary stage of syphilis, a chancre develops at the site of *Treponema* inoculation. Primary lesions generally occur on the genitalia but in 10 percent of the cases the primary lesions occur in the oral cavity. The average incubation time for the manifestation of primary syphilis is 21 days after infection. The primary lesions typically heal within 3–6 weeks. The secondary stage of syphilis normally begins 6–8 weeks after the appearance of the primary chancre. During this stage there are cutaneous lesions and lesions of the mucus membranes that contain infective *Treponema pallidum*. Lesions may appear on the lips, tongue, throat, penis, vagina, and numerous other body surfaces (Figure 25.10). There may be additional symptoms of systemic disease during this stage, such as headache, low grade fever, enlargement of the lymph nodes, and so on. The secondary stage is the most infectious because of the numerous spirochetes in the multiple lesions. After the secondary stage, syphilis enters a characteristic latent period during which there is an absence of any clinical symptoms of the disease. The latent phase can be detected by serological means because the VDRL (Venereal Disease Research Laboratory) and similar tests are still positive for antibodies even in the latent period. The latent phase marks the end of the infectious period of syphilis.

The tertiary phase of syphilis, also known as late syphilis, usually does not occur until years after the initial infection. This stage is generally noninfectious. During tertiary syphilis damage can occur to any organ of the body. In many cases of tertiary syphilis lesions develop that are similar to those of tuberculosis. These lesions, called gumma granulomas, have a rubbery consistency. They can occur anywhere on the body, sometimes causing extensive tissue damage. In about 10 percent of the cases of untreated syphilis, the tertiary phase involves the aorta, and damage to this major blood vessel can result in death. In approximately 8 percent of the cases of untreated syphilis, there is central nervous system involvement with a variety of neurological manifestations, including personality changes, paralysis, deafness, blindness, and loss of speech. Blindness, insanity, and bone deformations are characteristic of advanced stages of syphilis.

There was a young man from Back Bay
Who thought syphilis just went away.
He believed that a chancre
Was only a canker
That healed in a week and a day.
But now he has "acne vulgaris"—
(Or whatever they call it in Paris);
On his skin it has spread
From his feet to his head,
And his friends want to know where his hair is.
There's more to his terrible plight:
His pupils won't close in the light
His heart is cavorting,
His wife is aborting,
And he squints through his gun-barrel sight.
Arthralgia cuts into his slumber;
His aorta's in need of a plumber;
But now he has tabes,
And saber-shinned babies,
While of gummas he has quite a number.
He's been treated in every known way,
But his spirochetes grow day by day;
He's developed paresis,
Has long talks with Jesus,
And think's he's the Queen of the May.

—ANONYMOUS, *The Strange Story of Venereal Disease*

The risks of debilitating symptoms and death make syphilis a very serious form of venereal disease. Fortunately, syphilis can be treated with penicillin and other antibiotics, particularly during the early stages. Unlike gonorrhea, the number of cases of syphilis in the United States has been relatively constant (Figure 25.11). However, there has been a large increase in reported cases of syphilis among homosexual men; for example, over 60 percent of the cases of syphilis reported in the state of Washington during 1971 involved homosexual transmission. No long-term immunity develops after infections with *Treponema pallidum*, and individuals who are cured by treatment with antibiotics remain susceptible to contracting syphilis again. As with other sexually transmitted diseases, the overall control of syphilis, however, rests with finding and treating all sexual contacts who may have contracted this disease and may be involved in the further transmission of it.

In addition to sexually transmitted syphilis, *Treponema pallidum* can be transmitted across the placenta of pregnant women with syphilis, infect-

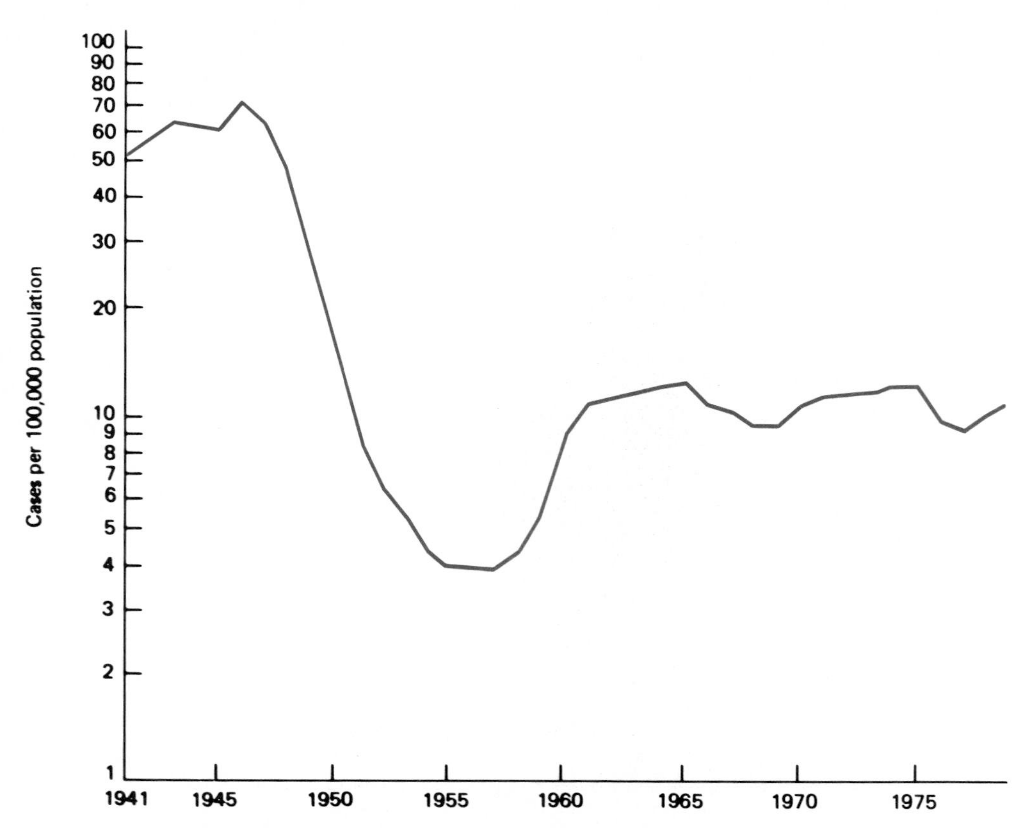

Figure 25.11 Incidence of primary and secondary syphilis among the civilian population in the United States from 1941 to 1979. The incidence of this sexually transmitted disease has not shown the same dramatic increase as has occurred for gonorrhea.

ing the fetus and causing stillbirth or congenital syphilis in the newborn. Stillbirth is likely if pregnancy occurs during the primary or secondary stages of syphilis. Congenital syphilis is most likely during the latent period of the disease. Congenital syphilis has very serious consequences, usually resulting in mental retardation and neurological abnormalities in the infant; the probability of such infants surviving depends on the specific nature of the neurological impairment.

The secret awareness of syphilis, and the utter secret terror and horror of it, has had an enormous and incalculable effect on the English consciousness and on the American. Even when the fear has never been formulated, there it has lain, potent and overmastering . . . I am convinced that some of Shakespeare's horror and despair, in his tragedies, arose from the shock of his consciousness of syphilis. Some of Shakespeare's father-murder complex, some of Hamlet's horror of his mother, of his uncle, of all old men came from the feeling that fathers may transmit syphilis, or syphilis-consequences, to children . . .

—D. H. LAWRENCE

Serological tests are generally employed for the diagnosis of syphilis because *Treponema pallidum* cannot be cultured using cell-free media. The organism can be observed in exudates from lesions by using darkfield microscopy. Its characteristic shape can aid in an early presumptive diagnosis. *T. pallidum* does not stain well with conventional dyes; therefore bright field microscopy is not used in diagnostic procedures. Fluorescent antibody stains, however, are useful in viewing *T. pallidum* and positively diagnosing cases of syphilis. Several additional serological procedures for the diagnosis of syphilis will be discussed in Chapter 30.

Nongonococcal Urethritis (NGU). Nongonococcal urethritis is the term used to describe sexually transmitted diseases that result in inflammation of the urethra caused by bacteria other than *Neisseria* *gonorrhoeae*. This disease is also called nonspecific urethritis (NSU). It is estimated that between 4 and 9 million people in the United States have nongonococcal urethritis. Most cases are mild. Females often are asymptomatic and males usually notice some pain and discharge during urination. In serious cases the inflammation associated with this condition can cause infertility. In males the epididymis may become inflamed, and in females the cervix or fallopian tubes may become blocked. Most cases of NGU appear to be caused by *Chlamydia trachomatis*, a small, obligately intracellular parasitic bacteria. *Mycoplasma hominis* and *Ureaplasma urealyticum* also have been reported to cause NGU frequently. These two bacterial species lack cell walls. It has been difficult to diagnose the specific causes of NGU because of problems in culturing and identifying these bacteria. All of these bacteria are inhibited by antibiotics such as tetracyclines but not by penicillin. Compared to gonococcal urethritis, NGU has a longer incubation period. Individuals treated with penicillin for cases diagnosed as gonorrhea may later develop NGU if multiple bacteria were associated with the sexually transmitted disease.

Chlamydia trachomatis, like *Neisseria gonorrhoeae*, can be transmitted during birth from mother to the eyes of the newborn. *Chlamydia* infections of the eye can be serious. Because many females are asymptomatic, it is difficult to take selective measures to prevent this occurrence. Therefore erythromycin is applied to the eyes of newborns shortly after birth to protect them against both *N. gonorrhoeae* and *C. trachomatis*. Before the prevalence of *C. trachomatis* was recognized, silver nitrate was used to treat the eyes of newborns. Silver nitrate inhibits *N. gonorrhoeae* but not *C. trachomatis*, whereas erythromycin is effective against both these infectious agents.

Lymphogranuloma Venereum and Other Chlamydial Infections. Lymphogranuloma venereum is a more serious form of sexually transmitted disease caused by *Chlamydia trachomatis*. This disease is more common in the tropics than in the United States, where only a few hundred cases a year occur. In lymphogranuloma venereum, *C. trachomatis* enters the body through small abrasions in the genitourinary tract. The chlamydial cells are phagocytized and carried to lymph nodes. The lymph nodes are the primary sites affected in this disease, with swelling and tenderness usually occurring 5–21 days after healing of the primary lesions in the area of initial infection. In about 25 percent of the cases, there is a genitoanorectal syndrome, characterized by a bloody, mucopurulent rectal discharge. Lymphogranuloma venereum can cause serious consequences, but this rarely occurs. Although treatment with tetracycline and other antibiotics is usually effective in preventing the onset of late and serious symptoms, it does not seem to shorten the usual time of 4–6 weeks required for enlarged lymph nodes to heal.

Chancroid. Chancroid is a relatively rare sexually transmitted disease caused by *Haemophilus ducreyi*, a Gram negative rod-shaped bacterium. Chancroid occurs most frequently in the underdeveloped nations of Africa, the Caribbean, and Southeast Asia, but the incidence of this disease has been increasing in the United States. Soft chancres develop 3–5 days after sexual exposure, and untreated lesions may persist for months. Chancroidal ulcers heal quickly but often leave deep scars. Sulfanilamide and tetracycline are used in the treatment of this disease.

Granuloma Inguinale. Granuloma inguinale, also known as granuloma venereum, is caused by a Gram negative rod-shaped bacterium, *Calymmatobacterium granulomatis*. Granuloma inguinale most frequently occurs in India, the west coast of Africa, islands of the South Pacific, and some South American countries. The disease is relatively rare in the United States. It appears to be sexually transmitted, as evidenced by the fact that initial lesions occur on the genitalia and the first lesions appear 9–50 days after sexual intercourse with an infected individual. Scrapings from the lesions reveal the presence of *C. granulomatis* within large mononuclear cells called Donovan bodies. There appears to be a very low rate of infection for this sexually transmitted disease, and in many cases sexual partners do not contract this disease. The genitalia develop characteristic ulcers, older portions of which exhibit loss of pigmentation. Chloramphenicol, erythromycin, and tetracycline, as well as other antibiotics, are used in the treatment of this disease.

Viral Diseases

Genital herpes. It is estimated that 20 million Americans now have genital herpes and that there will be at least a half a million new cases per year unless effective means are found for controlling this disease. Genital herpes, caused by herpes simplex type 2 virus, is most frequently transmitted by sexual contact and causes infection of the genitalia (Figure 25.12). (Herpes simplex type 1 viral infections generally are not sexually transmitted and most frequently occur above the waist in contrast to type 2 genital infections.) Unlike the fastidious sexually

MICROBIOLOGY HEADLINES

Drug Proves Effective Against Genital Herpes

By VICTOR COHN

WASHINGTON—Medical authorities are reporting the first long-term, "meaningful" treatments for genital herpes, a disease that now affects 5 million to 20 million Americans.

Doctors from five medical centers have used an oral form of acyclovir, a drug first used on genital herpes in ointment form two years ago, to ease the recurring attacks most victims suffer.

Other doctors reportedly are finding that daily use of the oral drug sometimes can prevent recurrences for up to eight months.

Researchers have also discovered that the use of acyclovir shortens the period in which victims shed the herpes virus, which can infect others.

"From a public-health standpoint (this) may have great impact" in reducing the spread of the disease, two scientists wrote in an editorial in a recent issue of the *Journal of the American Medical Association,* because surveys show that a third of the sex partners of patients are exposed to the virus during recurrences.

The report on acyclovir, which appeared in the same journal, describes the use of the drug for five days at a time to lessen the severity of herpes attacks and to help heal the sores such attacks typically cause.

Dr. Steven Straus of the National Institute of Allergy and Infectious Diseases, part of the National Institutes of Health in Bethesda, Md., said during a scientific meeting last year that the daily use of the drug for four months prevented recurrences in 13 of 17 patients.

Those studies have been extended and confirmed at other centers, according to an official of Burroughs Wellcome, the company that manufactures acyclovir.

Straus said he cannot give further details until they have been published. However, Drs. William Whittington and Willard Cates of the disease centers have written that oral acyclovir "offers a genuine ray of hope," both as a treatment and as a preventive.

Research findings, they say, suggest that the drug may be useful "to avoid recurrences," because "several preliminary reports have documented significantly fewer recurrences in a majority of patients" who took daily doses of acyclovir.

All the doctors say several cautions should be kept in mind:

- Acyclovir is publicly available now mainly as a prescription ointment, although an injectable form is used in hospitals.

The ointment is highly effective in reducing the severity of initial attacks, but not so effective in easing later ones.

- The more powerful oral form of the drug is available only at research centers such as those whose research was the subject of the AMA report: the universities of Vermont, Washington and Alberta, Canada, the San Diego Veterans' Hospital and the Sir Montimer Davis-Jewish General Hospital of Montreal.
- The Food and Drug Administration could approve the oral form for general use sometime this year, but long-time daily use may prove expensive.

A two- to three-week supply of ointment costs $18 to $24.

- The drug's long-term safety hasn't been established, and it isn't known whether resistant forms of the virus might spring up.

No safety problems have appeared so far, but "caution is warranted," according to the disease-control centers, whose experts say acyclovir's transition from brief use "by a few hundred patients in (careful) studies to prolonged use by perhaps 1 million persons" must be carefully monitored.

Source: The Washington Post, May 5, 1984. Reprinted by permission.

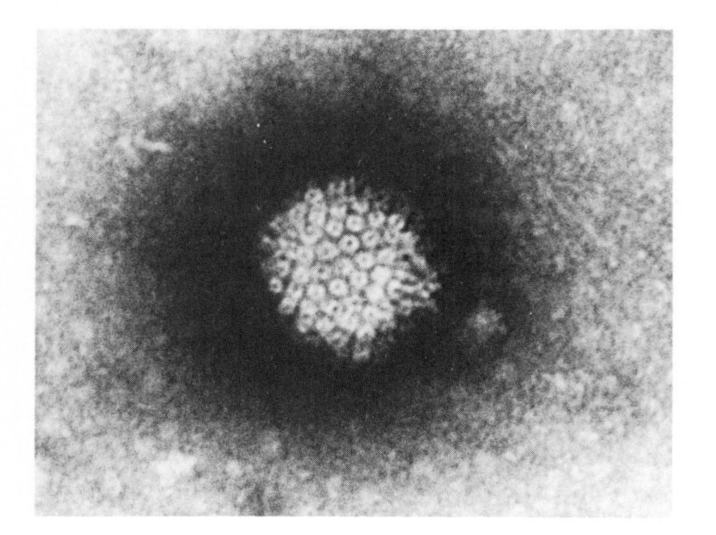

Figure 25.12 Electron micrograph of herpes virus. (From BPS: B. Roizman, University of Chicago.)

transmissible bacteria, herpes viruses can survive for some period of time in moist environments. Some herpes outbreaks have been traced to hot spas where the virus contaminates the water and surrounding wood decks.

In women the primary site of herpes infection is the cervix but may also involve the vulva and vagina. In men the herpes simplex virus frequently infects the penis. The primary infection exhibits symptoms of genital soreness and ulcers in the infected areas. The virus and manifestations of infection may be transmitted to other areas of the body, most notably the mouth and anus. Genital herpes may have particularly serious repercussions in pregnant women because the virus can be transmitted to the infant during vaginal delivery and in rare cases may cross the placenta, causing damage to the infant's central nervous system and/or eyes (Figure 25.13). Herpes is lethal to up to 60 percent of infected newborns, and for surviving babies there is a 50 percent risk of blindness or neurological damage. Caesarean deliveries are often performed during active outbreaks of herpes to prevent transmission from mother to newborn.

The ulcers produced from herpes simplex type 2 infections generally heal spontaneously in 10–14 days. However, because of the budding mode of reproduction of herpes viruses, the infection is not eliminated when the ulcers heal; rather, a reservoir of infected cells remains in the body within nerve cells. At subsequent times multiplication of the viruses can produce new secondary ulcers, even in the absence of additional sexual activity. It is not known exactly what initiates subsequent attacks of herpes, but such recurrences may be triggered by sunlight, sexual activity, menstruation, and stress. Generally the secondary ulcers heal more rapidly than the primary lesion. The disease remains transmissible, which interferes with establishing stable sexual relationships; there are many adverse psychological effects associated with genital herpes. Genital herpes disrupts marital relationships, and the epidemic outbreak of genital herpes may reverse the sexual revolution.

Who'd go to bed with me if they thought they'd get herpes? Look, I don't want to give it to anyone else, but I don't want to ruin my whole sex life either.

—KAREN B

I've had every damn thing you can get . . .
I've had syph twice, gonorrhea maybe four times, yeast and urinary tract infections a hundred times. But the real bitch is herpes. You can't get rid of it, ever . . . My doctor told me there's nothing else of this God's earth that's more contagious than herpes during an infection. The rest of the time it's not infectious—it lays quiet in the nerves, they say.

—JOANNE

—FRANK FREUDBERG, *Herpes: A Complete Guide to Relief and Reassurance*

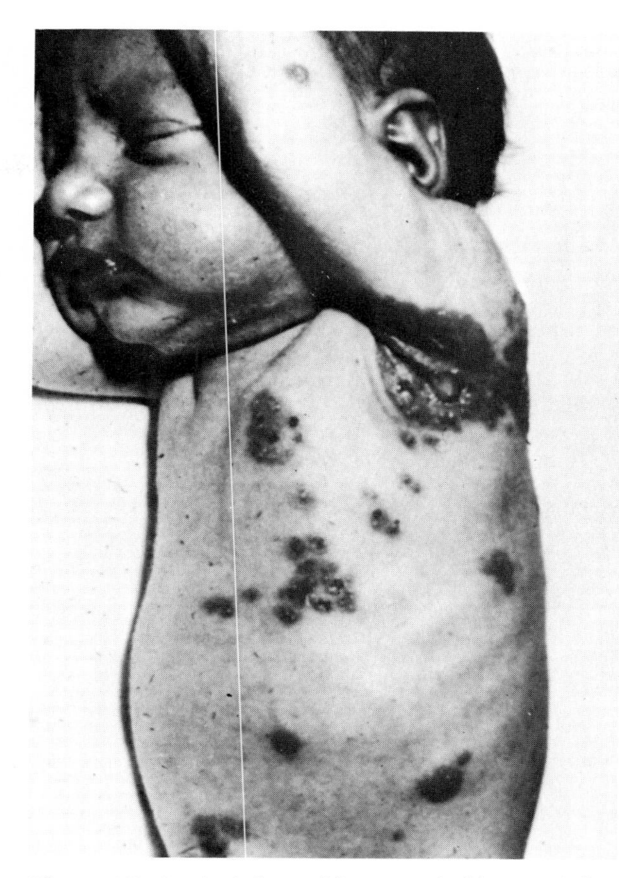

Figure 25.13 An infant with congenital herpes infection acquired during birth from its sexually infected mother. (From Centers for Disease Control, Atlanta.)

Although a direct relationship between genital herpes viruses and cancer has not been established, adolescent females who have had extensive sexual contacts and have developed genital herpes infections have an elevated rate of development of cervical cancer. Frequent pap smears are suggested for women diagnosed with genital herpes. Several of the newly developed antiviral drugs should be useful in the treatment of herpes viral infections. In particular, acyclovir and interferon show promise of reducing the severity of the symptoms; neither of these drugs, though, promises a cure for the disease.

Genital Warts. Warts including genital warts, are benign tumors caused by infections with papilloma viruses. These viruses are transmitted by direct contact, normally infecting the skin and mucous membranes. Warts can develop on the genitals and direct sexual contact usually is the source of the infecting papilloma viruses when this occurs. Warts generally do not appear for several weeks after infection. Chemical (for example, acid) and physical (for example, freezing) methods can be used for the

Table 25.2 Summary of Diseases Associated with the Genitourinary System

Disease	Causative Agent	Mode of Transmission	Treatment
Urethritis	*E. coli, Proteus, Enterobacter, Klebsiella*	Opportunistic infections	Nitrofurantoin or other antibiotics
Cystitis	*E. coli, Proteus, Enterobacter, Klebsiella, Pseudomonas*	Opportunistic infections	Nitrofurantoin or other antibiotics
Pyelonephritis	*E. coli, Enterobacter, Klebsiella, Proteus, Pseudomonas, Streptococcus, Staphylococcus*	Systemic infection	Various antibiotics
Glomerulonephritis	*Streptococcus pyogenes*	Infection by *S. pyogenes* elsewhere in the body leading to later immune complex hypersensitivity reaction	Chloramphenicol, erythromycin, and other antibiotics to eliminate initial infection and prevent development of hypersensitivity reaction
Leptospirosis	*Leptospira interrogans*	Direct contact with infected animals, contaminated water	Penicillin
Vaginitis	*Candida albicans*	Opportunistic pathogen	Nystatin
	Trichomonas vaginalis	Opportunistic pathogen	Metronidazole
Prostatitis	*E. coli, Pseudomonas, Streptococcus, Staphylococcus*	Opportunistic pathogen	Various antibiotics
Gonorrhea	*Neisseria gonorrhoeae*	Sexual contact	Penicillin, spectinomycin
Syphillis	*Treponema pallidum*	Sexual contact	Penicillin, erythromycin tetracycline
Nongonococcal urethritis	*Chlamydia trachomatis, Mycoplasma hominis, Ureaplasma urealyticum*	Sexual contact	Tetracyclines and other antibiotics
Lymphogranuloma venereum	*Chlamydia trachomatis*	Sexual contact	Tetracycline and other antibiotics
Chancroid	*Haemophilus ducreyi*	Sexual contact	Tetracycline and other antibiotics
Granuloma inguinale	*Calymmato-bacterium granulomatis*	Sexual contact	Tetracycline and other antibiotics
Genital herpes	Herpes simplex	Sexual contact	Acyclovir
Genital warts	Papilloma virus	Sexual contact	None or chemical or physical removal

removal of warts, but these benign tumors also will disappear without treatment.

Summary

In the case of pathogens that enter the body through the genitourinary tract, contamination with microorganisms comprising the normal microbiota of the gastrointestinal tract and sexually transmitted pathogens are principally involved in such infections. The treatment of sexually transmitted diseases requires public health measures to identify the source of infection in order to interrupt the chain of transmission. Recognition that sexually transmitted diseases represent a medical, rather than a social, problem is important for effectively treating and reducing the incidence of these diseases.

Table 25.2 presents a summary of diseases that primarily effect the genitourinary tract and those that are sexually transmitted diseases.

Study Outline

A. The urinary system consists of two kidneys, which remove substances from the blood and form urine; two ureters, which transport urine away from the kidneys; the urinary bladder, which serves as a urine reservoir; and a tubular urethra, which conveys urine to the outside of the body. The primary function of the urinary system is to regulate the concentrations of a variety of substances circulating in the blood.

B. The genital system.
- **1.** Male: Male testes are responsible for the production of sperm and the hormone testosterone. They are contained in the scrotum. The penis introduces the sperm into the vagina of the female.
- **2.** Female: The female reproductive system consists of the ovaries, which produce eggs and sex hormones; the uterine tubes, which convey eggs to the uterus; the uterus, which houses the fetus; and the vagina, which receives the penis and serves as the birth canal.

C. Although most of the genitourinary system is normally free of microorganisms, the external genital regions of both sexes contain various indigenous microbial populations. The normal microbiota of the vaginal tract include bacteria, coliforms, spriochetes, yeasts, and flagellated protozoans.

D. The genitourinary system is protected from pathogenic microorganisms by mucus, low pH, high salt concentrations, the presence of antimicrobial enzymes, and the antagonism of the resident microbiota.

E. Urinary tract infections.
- **1.** Urethritis and cystitis.
  - **a.** Causative organism(s): most commonly *Escherichia coli*, but also *Klebsiella, Enterobacter, Serratia, Proteus*, and *Pseudomonas*.
  - **b.** Symptoms: urethritis = pain and a burning sensation during urination; cystitis = suprapubic pain and the urge to urinate frequently.
  - **c.** Treatment: various antibiotics, such as nitrofurantoin and nalidixic acid.
- **2.** Pyelonephritis.
  - **a.** Causative organism(s): most commonly *Escherichia coli*, but also *Klebsiella pneumoniae, Proteus mirabilis, Enterobacter aerogenes, Pseudomonas aeruginosa, Streptococcus* sp., and *Staphylococcus* sp.
  - **b.** Symptoms: urgent and frequent urination.
  - **c.** Treatment: antibiotics.
- **3.** Leptospirosis (Weil's disease).
  - **a.** Causative organism(s): *Leptospira interrogans*.
  - **b.** Mode of transmission: can enter through the skin, especially if there are small abrasions.
  - **c.** Symptoms: high spiking fever, chills, headache, muscle ache, malaise, abdominal pain, nausea, and vomiting.
  - **d.** Treatment: penicillin.

4. Glomerulonephritis.
   a. Causative organism(s): *Streptococcus pyogenes*.
   b. Mode of transmission: this is an immune complex disease that follows a streptococcal infection.
   c. Symptoms: high spiking fever, chills, headache, muscle ache, malaise, abdominal pain, nausea, and vomiting.
   d. Treatment: antibiotics, such as chloramphenicol and penicillin are used to treat *S. pyogenes* infections and thereby prevent the development of the hypersensitivity reaction that causes this disease.

F. Genital tract infections.
1. Vaginal tract infections (vulvovaginitis).
   a. Causative organism(s): *Trichomonas vaginalis* (a flagellate protozoan), *Candida albicans* (fungus).
   b. Mode of transmission: increased pH of the vaginal tract caused by pregnancy, the use of birth control pills, and the administration of antibacterial antibiotics can lead to bacterial and fungal overgrowth.
   c. Symptoms: increased vaginal discharge and burning.
   d. Treatment: antibiotics, metronidazole for protozoan infections and nystatin for *Candida* infections.
2. Prostatitis (infections of the prostate).
   a. Causative organism(s): *E. coli, Pseuodomonas aeuriginosa, Staphylococcus aeureus*, and *Streptococcus* sp. have frequently been identified as the causative agents of prostate infections.

G. Sexually transmitted diseases are contracted by direct sexual contact with an infected individual.
1. Gonorrhea.
   a. Causative organism(s): *Neisseria gonorrhoeae*.
   b. Mode of transmission: direct sexual contact.
   c. Symptoms: male = characteristic, painful purulent urethral discharge; female = inflammation with some pain and swelling of the cervix, abnormal vaginal discharge, and abnormal menstrual bleeding. Many women with gonorrhea are asymptomatic so all newborn's eyes are washed with erythromycin sulfate to prevent eye infections.
   d. Treatment: antibiotics, usually penicillin.
2. Syphilis.
   a. Causative organism(s): *Treponema pallidum*.
   b. Mode of transmission: direct contact with infective syphilitic lesions containing *T. pallidum*, which enters the body via abrasions of the epithelium and by penetrating into mucous membranes. Syphilis can also be transmitted across the placenta from an infected mother to her fetus causing mental retardation and neurological abnormalities in the infant.
   c. Symptoms: Primary stage: chancre develops at site of inoculation, generally on the genitalia; secondary stage: cutaneous lesions and lesions of the mucus membranes form on the lips, tongue, throat, penis, vagina, and other body surfaces, also headache, low grade fever, and enlargement of the lymph nodes; latent phase: no symptoms appear; tertiary stage or late syphilis: damage can occur to any organ.
   d. Treatment: antibiotics, usually penicillin.
3. Nongonococcal urethritis (NGU).
   a. Causative organism(s): *Chlamydia trachomatis* and also *Mycoplasma hominis* and *Ureaplasma urealyticum*.
   b. Mode of transmission: sexually transmitted, also during birth from mother to the eyes of the newborn.

   c. Symptoms: female = often asymptomatic; male = pain and discharge during urination.
   d. Treatment: antibiotics, such as tetracyclines.
4. Lymphogranuloma venereum.
   a. Causative organism(s): *Chlamydia trachomatis* carried to lymph nodes.
   b. Mode of transmission: sexually transmitted.
   c. Symptoms: swelling and tenderness of the lymph nodes, bloody, mucopurulent rectal discharge.
   d. Treatment: antibiotics, such as tetracyclines.
5. Chancroid.
   a. Causative organism(s): *Haemophilus ducreyi*.
   b. Mode of transmission: sexually transmitted.
   c. Symptoms: soft chancres.
   d. Treatment: tetracycline and sulfanilamide.
6. Granuloma inguinale.
   a. Causative organism(s): *Calymmatobacterium granulomatis*.
   b. Mode of transmission: sexually transmitted.
   c. Symptoms: genitalia develop characteristic ulcers, older portions of which lose their pigmentation.
   d. Treatment: tetracycline, chloramphenicol, and erythromycin.
7. Genital herpes.
   a. Causative organism(s): herpes simplex type 2 virus.
   b. Mode of transmission: sexually transmitted.
   c. Symptoms: genital soreness and ulcers in the infected areas.
   d. Treatment: acyclovir and interferon may reduce the severity of the symptoms.
8. Genital warts.
   a. Causative organism(s): papilloma virus.
   b. Mode of transmission: sexually transmitted.
   c. Symptoms: development of benign tumors (warts).
   d. Treatment: none or chemical or physical removal.

Study Questions

1. What organisms often cause urinary tract infections?
2. What organisms normally cause vaginitis?
3. What is a sexually transmitted disease? What are some examples of these diseases and what are their causative organisms? Why is it difficult to control the spread of these diseases, considering that many are readily treated with antibiotics?
4. Match the following statements with the terms in the list to the right.
   a. Caused by fastidious *Neisseria gonorrhoeae*.
   b. Caused by *Trichomonas vaginalis* and *Candida albicans*.
   c. Most common etiologic agent is *Escherichia coli* from accidental contamination of urinary tract with fecal material.
   d. Caused by *Treponema pallidum*.
   e. Disease that involves three distinct stages.
   f. Caused by *Chlamydia trachomatis*.
   g. Caused by herpes simplex type 2 virus.
   h. Also called Weil's disease.
   i. Caused by the spirochete *Leptospira interrogans*.

   (1) Gonorrhea
   (2) Syphilis
   (3) Genital herpes
   (4) Leptospirosis
   (5) Lymphogranuloma venereum

Some Thought Questions

1. What would you do if you thought you had syphilis?
2. Why are prophylactics used?
3. Why is herpes called the gift that keeps on giving?
4. Did Henry VIII have syphilis?

5. Based upon epidemiological data, did the pill lead to a sexual revolution?
6. Why should you after defecation wipe your rectum with a backward rather than with a forward movement of your hand?
7. Will the movement to have babies at home instead of in a hospital lead to increased incidence of postpartum sepsis?

Additional Sources of Information

Boyd, R. F., and B. G. Hoerl. 1981. *Basic Medical Microbiology*. Little, Brown and Company, Boston. A basic text emphasizing microbes and disease that includes several chapters on genitourinary tract diseases.

Braude, A. I. (ed.). 1980. *Medical Microbiology and Infectious Disease*. W. B. Saunders Company, Philadelphia. An advanced work with individual chapters dealing with specific diseases of the genitourinary tract; chapters dealing with infectious diseases are organized according to the taxonomy of the causative agent.

Catterall, R. D., and C. S. Nicol (eds.). 1976. *Sexually Transmitted Diseases*. Academic Press, Inc., New York. A series of papers on the control of sexually transmitted diseases.

Davis, B. D., R. Dulbecco, H. N. Eisen, H. S. Ginsberg, and W. B. Wood. 1981. *Microbiology*. Harper & Row, Publishers, Hagerstown, Maryland. An excellent text used in many graduate and medical school microbiology courses dealing with all aspects of microbiology; chapters dealing with infectious diseases are organized according to the taxonomy of the causative agent.

Hoeprich, P. D. (ed.). 1977. *Infectious Diseases: A Modern Treatise of Infection Processes*. Harper & Row, Publishers, Hagerstown, Maryland. This classic treatment of infectious diseases makes an excellent reference volume.

Jawetz, E., J. L. Melnick, and E. A. Adelberg. 1980. *A Review of Medical Microbiology*. Lange Medical Publications, Los Altos, California. This book provides a very good review of infectious diseases.

Joklik, W. K., and H. P. Willett (eds.). 1980. *Zinsser Microbiology*. Appleton-Century-Crofts, New York. A classic text used in many medical school programs that includes chapters on infectious diseases.

Kunin, C. M. 1974. *Detection, Prevention, and Management of Urinary Tract Infections*. Lea & Febiger. A thorough reference work on the causes and treatment of urinary tract infections.

Rose, N. L., and A. L. Rose (eds.). 1983. *Microbiology: Basic Principles and Clinical Applications*. Macmillan Publishing Co., Inc., New York. An advanced work on medical microbiology; chapters are organized for the most part according to taxonomic groups.

Rosebury, T. 1971. *Microbes and Morals*. Viking Press, New York. A discussion of various aspects of sexually transmitted diseases.

Schachter, J. 1981. Chlamydiae. *Annual Review of Microbiology*, 34:285–309. An advanced review of the chlamydiae and their involvement in human disease.

Youmans, G. P., P. Y. Paterson, and H. M. Sommers. 1980. *The Biological and Clinical Basis of Infectious Disease*. W. B. Saunders Company, Philadelphia. This is an excellent text covering most human infectious diseases.

26 *Diseases of the Central Nervous System*

KEY TERMS

African sleeping sickness
aseptic meningitis
botulism
brain abscess
central nervous system
cerebrospinal fluid
Creutzfeldt–Jakob disease
cryptococcosis
dementia
encephalitis
ergotism
Guillain–Barré syndrome
herpes encephalitis
hydrophobia
Kuru
lockjaw
lumbar puncture
meninges
meningitis (bacterial)
mushroom poisoning
paralytic shellfish poisoning
peripheral nervous system
poliomyelitis
rabies
Reye's syndrome
tetanus

OBJECTIVES

After reading this chapter you should be able to

1. Define the key terms.

2. Describe the structure of the central nervous system.

3. Describe the defense systems of the central nervous system that protect against infectious microorganisms.

4. Discuss the causative agents, routes of transmission, key symptoms, and treatments for the following diseases: rabies, poliomyelitis, aseptic meningitis, encephalitis, herpes encephalitis, Reye's syndrome, Guillain–Barré syndrome, Creutzfeldt–Jakob disease, meningitis (bacterial), brain abscess, tetanus, botulism, mushroom poisoning, ergotism, paralytic shellfish poisoning, and African sleeping sickness.

Structure of the Central Nervous System

The nervous system is composed of the brain, the spinal cord, and numerous peripheral nerves (Figure 26.1). The peripheral nerves connect the central nervous system to the other parts of the body and provide the means of detection of changes in internal and external environments of the body. All of the nervous tissue lying outside the bony structures of the skull and spinal column constitutes the peripheral nervous system. The peripheral nervous system transmits information to the central nervous system for action and then delivers the messages to muscles and glands for response.

The brain and the spinal cord make up the central nervous system, which controls the integration, interpretation, thought, and transmission of messages to and from the periphery. Both the brain and the spinal cord are protected by two coverings. The outer cover consists of bone; cranial bones encase the brain and vertebrae encase the spinal cord. The inner cover consists of membranes called meninges. The meninges are composed of three distinct layers; the dura mater is the outermost layer and the arachnoid membrane lies between the dura mater and the pia mater, which is the innnermost layer. The pia mater adheres to the outer surface of the brain and cord and contains blood vessels. Between the dura mater and the arachnoid membrane is a small space called the subdural space, and between the arachnoid and the pia mater is the subarachnoid space, which contains the cerebrospinal fluid. Cerebrospinal fluid also protects the brain and spinal cord by providing a cushion of fluid both around and within them. The subarachnoid spaces around the brain and the cord, the ventricles and aqueduct inside the brain, and the

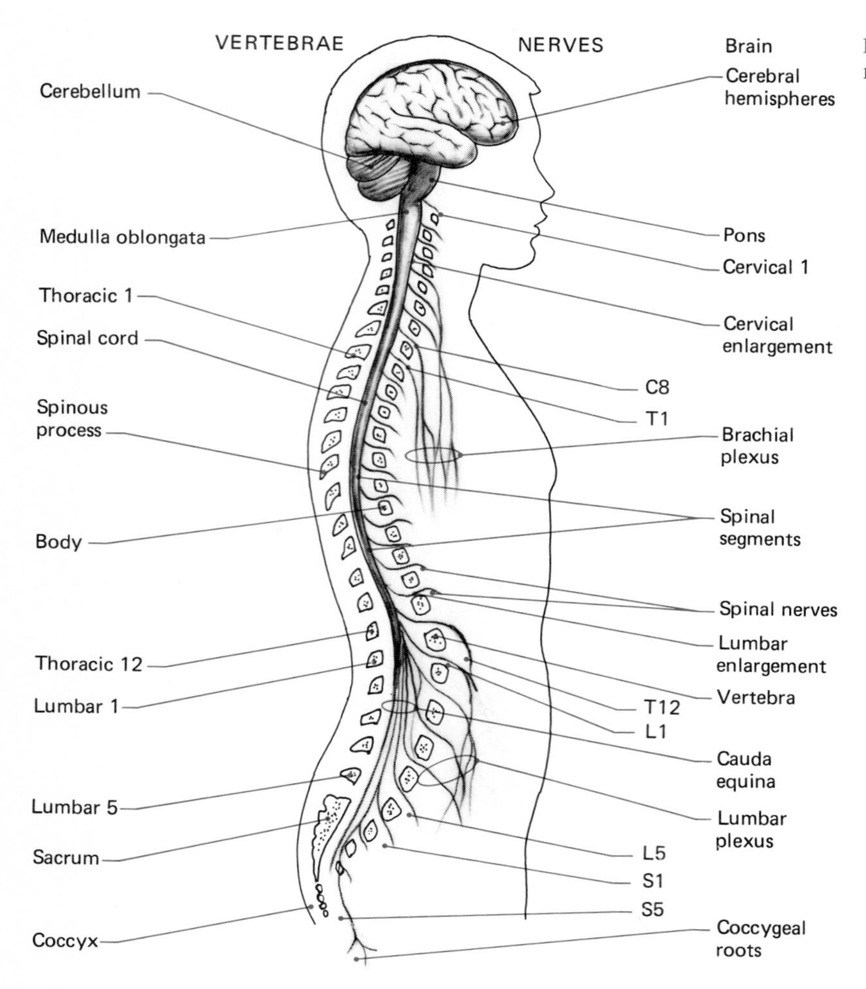

Figure 26.1 Diagram of the central nervous system.

central canal inside the cord all contain cerebrospinal fluid.

Cerebrosinal fluid can be removed for diagnosing diseases of the central nervous system by performing a lumbar puncture. This procedure is also known as a spinal tap. In this procedure the patient lies on his or her side so that the back is arched and the vertebrae are separated. A needle is then inserted just above or just below the fourth lumbar vertebra, and a sample of cerebrospinal fluid is withdrawn. Analysis of the cerebrospinal fluid can reveal whether there is a microbial infection of the central nervous system and the nature of the infecting microorganism. The presence of relatively high numbers of white blood cells and increased protein and decreased sugar levels, for example, would be indicative of a bacterial infection.

Defenses of the Central Nervous System

The central nervous system is well protected against microbial invasion by various physical barriers; and it does not have an associated indigenous microbiota. Infections of the central nervous system are associated with failures of the physical defense systems to keep out microorganisms. Some infections of the central nervous occur as a result of extensions of other infections, such as from the middle ear to mastoid to brain. Other diseases occur when pathogens, such as rabies virus, migrate from peripheral nerves to the central nervous system. Trauma, such as a fracture of the skull or failure of the blood–brain barrier, is often involved in infections of the central nervous system. The blood–brain barrier refers to the restricted movement of substances from the blood to the brain. Unlike other capillaries of the body, the capillaries supplying blood to the brain have limited permeability. Some substances, such as oxygen, pass from the blood to the brain. Other substances, including microorganisms and most antibiotics, however, cannot move from the blood to the central nervous system.

Diseases of the Central Nervous System

Diseases Caused by Viral Pathogens

Rabies. Rabies is principally a disease of warm-blooded animals other than humans. The rabies virus, a bullet-shaped, single-stranded RNA virus, can be transmitted to people through the bite of an infected animal (Figure 26.2). In lower animals rabies appears in two forms. Animals with the "furious" type of rabies snap and bite; animals with the "dumb" form of rabies exhibit paralysis. In urban settings, dogs and cats are most frequently involved in the transmission of rabies to humans. Many municipalities require that pet dogs be immunized against rabies and require a certificate of rabies vac-

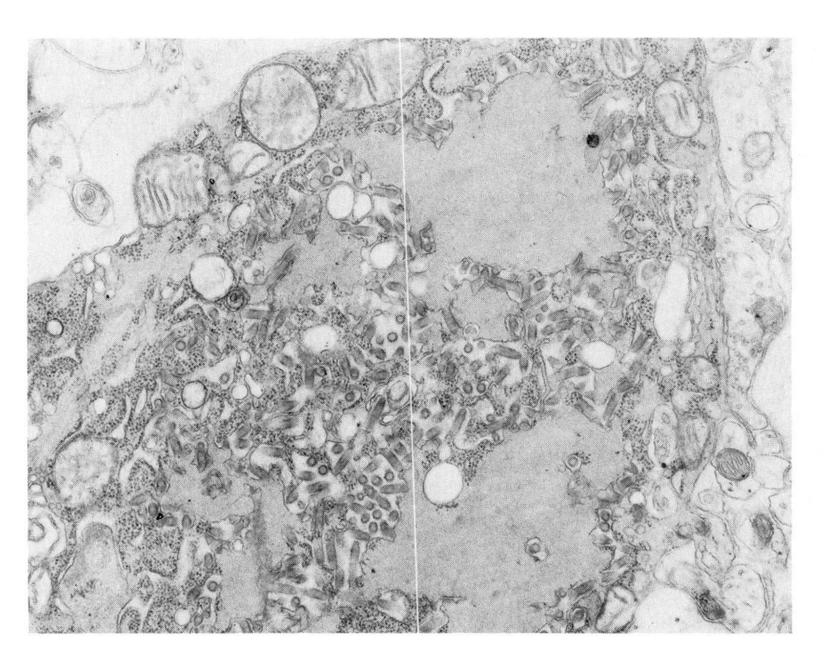

Figure 26.2 Micrograph of rabies virus from the neuron of an inoculated mouse (30,500 ×). Note how the cell is crowded with virus particles. (From BPS: A. K. Harrison, Centers for Disease Control, Atlanta.)

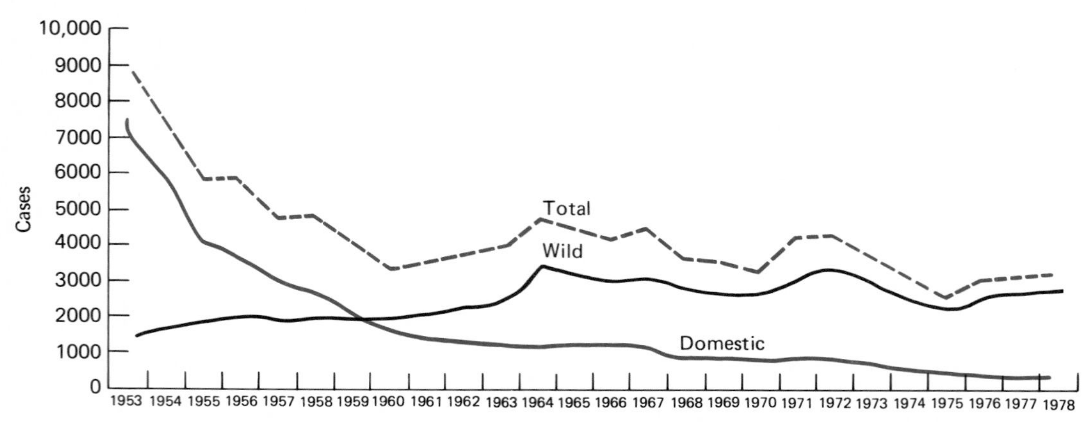

Figure 26.3 Incidence of rabies in wild and domestic animals in the United States from 1953 to 1978.

cination to obtain a pet license; some also require vaccination of cats. Other wild animals involved in the transmission of rabies include foxes, skunks, jackals, mongooses, squirrels, raccoons, coyotes, badgers, and bats (Figure 26.3). Cats, racoons, and other wild animals are increasingly involved in the transmission of rabies to humans. Rabies viruses multiply within the salivary glands of infected animals, and normally enter humans in the animal's saliva through the portal of entry established by the animal's bite. The rabies virus is not able to penetrate the skin by itself, and deposition of infected saliva on intact skin does not necessarily result in transmission of the disease. Transmission from bats can also occur via the respiratory tract involving transmission via aerosols formed in the atmosphere around dense populations of infected bats. Additionally, at least one case of rabies transmission occured as the result of a corneal transplant from an infected individual. Veterinarians and animal handlers need be aware of possible transmission from animals they are treating, and vaccination is sometimes recommended to avoid the risk of contracting rabies for those likely to be exposed to the virus.

When rabies viruses enter the human body through an animal bite, they are normally deposited within muscle tissues. The rabies viruses reach peripheral nerve endings and migrate to the central nervous system. Cytoplasmic inclusion bodies, known as Negri bodies, develop within the neurons of the brain. Multiplication of rabies viruses within the nervous system causes a number of abnormalities, manifested as the symptoms of this disease. The initial symptoms of rabies, which do not occur until the viruses reach the brain, include anxiety, irritability, depression, and sensitivity to light and sound. These symptoms are followed by difficulty in swallowing and development of hydrophobia (fear of water). This fear of water really is a fear of swallowing because of the occurrence of a convulsive reflex that can result in choking. As the infection progresses, there is paralysis, coma, and death.

Once the clinical symptoms of rabies begin, the disease is considered to be almost invariably fatal, and therefore, treatment of rabies requires vaccination and the use of prophylactic antibodies before the symptoms become manifest. There have been a few individuals who have survived symptomatic cases of rabies, but such cases are extraordinarily unusual. The symtomatic disease is very rare in the United States, with less than 10 deaths per year, because of the use of immune treatment to protect individuals who may have been infected with the rabies virus.

Rabies in humans has been regarded as uniformly fatal when associated with hydrophobia symptom. A 25-year-old woman, Candida de Sousa Barbosa of Rio de Janeiro, Brazil was believed to be the first to survive the disease in Nov 1968, though some sources give priority to Matthew Winkler, 6, on Oct 10, 1970, who was bitten by a rabid bat.

—1985 GUINNESS BOOK OF WORLD RECORDS

The rabies vaccine stimulates antibody synthesis that prevents proliferation of the virus before it can cause irreversible damage to the central nervous system. It is critical to diagnose the disease in the animal that has bitten a human to determine if vaccination is necessary. Frequently, the public is asked to help identify and locate a dog that has bitten a child in order to determine whether it is necessary to carry out the vaccination procedure. Only if the animal is conclusively shown not to be infected with rabies can the immunization procedure be safely omitted for the bitten individual.

This dog and man at first were
friends;
But when a pique began,
The dog, to gain his private ends
Went mad and bit the man . . .

But soon a wonder came to light
That show'd the rogues they lied,
The man recovered of the bite,
The dog it was that died.

—OLIVER GOLDSMITH, *Elegy on the Death of a Mad Dog*

If someone is bitten by an animal with rabies, antibodies are applied to the wound area to inactivate the virus and vaccination is initiated to establish active immunity against rabies. For many years the vaccine used was prepared from rabies virus propagated in embryonated duck eggs and inactivated by β-propiolactone. The treatment involved 21 daily injections followed by booster inoculations 10 and 20 days later. These injections were administered through the abdomen and were very painful. Recent improvements in the vaccine have reduced the number of administrations that are required to establish immunity. The new rabies vaccine, prepared by growing the virus in human diploid fibroblast tissue culture, has a higher concentration of the necessary antigens for eliciting an immune response and generally only three injections over a 7-day period are required to establish immunity. These injections are administered intramuscularly through the thigh.

Poliomyelitis. Polioviruses may enter the body either through the gastrointestinal or respiratory tracts and are able to multiply within the tissues of the oropharynx and intestines (Figure 26.4). The fecal–oral route via contaminated water supplies is probably the most common route of transmission. Viruses entering the body through the gastrointesinal tract move into the blood stream and are disseminated to other tissues, including lymphatic tissues where additional viral replication occurs. Polioviruses have the ability to cross the blood–brain barrier, where they continue to multiply within neural tissues and cause varying degrees of damage to the nervous system.

The initial symptoms of poliomyelitis (commonly referred to as polio) include headache, vomiting, constipation, and sore throat. In many cases, these early symptoms are followed by obvious neural involvement, including paralysis due to the injury of motor neurons. Although the paralysis can affect any motor function, in over half of the cases of paralytic poliomyelitis, the arms and/or legs are involved. Fortunately, the paralytic symptoms occur in only 1–2 percent of polio cases—1000 times less frequently than nonparalytic infections; in many cases, poliovirus infections fail to show any evidence of clinical symptomology.

The question of why Roosevelt caught infantile paralysis is of less practical importance than its effect on him . . . Before he was paralyzed, Roosevelt was a brash young man who had been insulated from reality by wealth, protection, and facile success. After it, he was a wiser, much older man, who had seen life closely, although briefly, in the disguise of death.

—NOEL F. BUSCH, *What Manner of Man?*

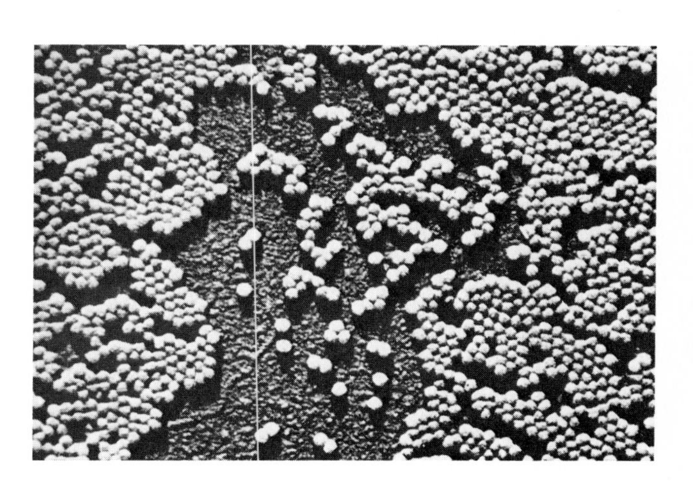

Figure 26.4 Electron micrograph of polioviruses. (From Photoresearchers Inc.)

Figure 26.5 Child with paralytic polio. (From *World Health*, July, 1976. Reprinted from *World Health* by permission of the World Health Organization, Geneva.)

Once I spent two years lying in bed, trying to move my big toe. That was the hardest job I ever had to do. After that, anything else seems easy.

—FRANKLIN DELANO ROOSEVELT

During outbreaks of poliomyelitis the incidence is particularly high among children, and because of this the disease is also called infantile paralysis (Figure 26.5). The disease also strikes adults, and in fact, the fatality rate in adults is much higher than in children (Table 26.1). The use of the Salk and Sabin polio vaccines has dramatically reduced the incidence of this disease (Figure 26.6). It is important that preschool children be immunized because major outbreaks of poliomyelitis have traditionally been associated with transmission among children in close contact in a schoolroom. Despite the ability to prevent this serious disease, many children are not immunized voluntarily, even in economically affluent countries such as the United States. Many school systems now require evidence of polio vaccination before a child may be enrolled. This is essential because the lowered rate of occurrence of poliomyelitis means that there are fewer paralyzed individuals to serve as visible reminders of the seriousness of this disease. Constant efforts to reinforce parental awareness of the importance and success of vaccination against poliomyelitis are worthwhile.

Table 26.1 Age-specific Case Fatality Rates of Paralytic Poliomyelitis in the United States, 1960–1968

Age Group	Cases	Deaths	Case Fatality Rate (%)
0–4	1899	92	4.8
5–9	959	65	6.8
10–14	379	33	8.7
15–19	205	18	8.8
20–29	462	71	15.4
30–39	275	65	23.6
40 and over	136	41	30.1

Aseptic Meningitis. Invasion of the central nervous system by various viruses can cause an inflammation of the meninges, a condition known as aseptic meningitis. There are about 10,000 cases of this meningitis in the United States each year. Frequently outbreaks of aseptic meningitis affect several members of a community or family. This aids public health officials in identifying the source of the viruses causing this disease.

The term aseptic meningitis indicates that no microorganisms are observed in smears and no bacteria or fungi are recovered in cultures. The spinal fluid does not have depressed levels of glucose in cases of this disease. Examination of spinal fluid using serological tests can confirm the viral origin

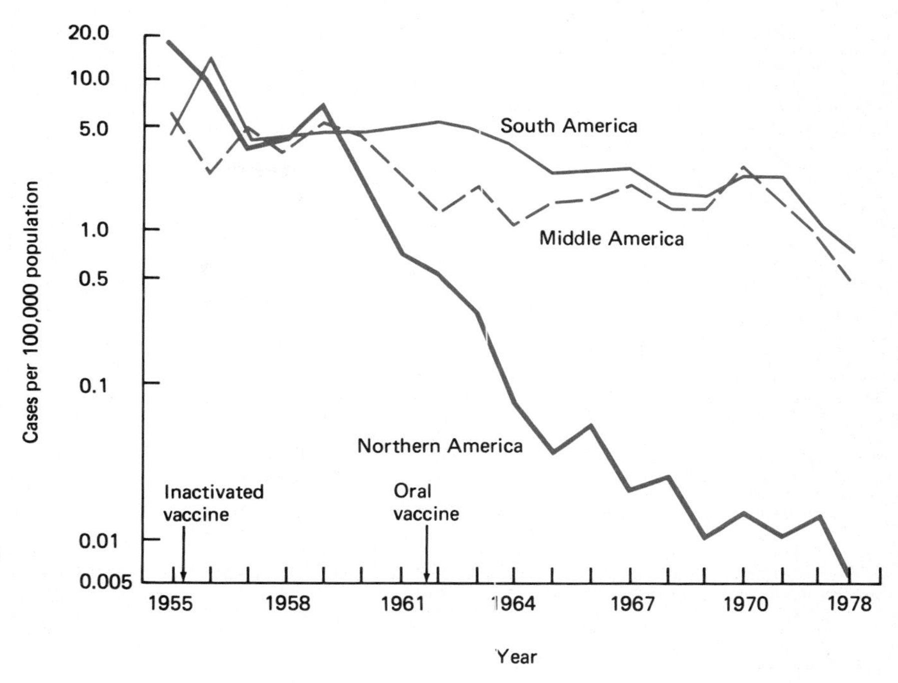

Figure 26.6 Incidence of poliomyelitis per 100,000 in the Americas, 1955 to 1973. Note that whereas the rates of incidence were similar in 1965, the occurrence of this disease in North America dramatically declined after the introduction of vaccines; no similar decline has occurred in Central and South America, where extensive vaccination has not been carried out.

of this disease. These tests reveal the presence of antigens associated with particular viruses known to cause aseptic meningitis. Often echoviruses and coxsackieviruses are the causative agents.

The symptoms of aseptic (nonbacterial) meningitis include fever, stiffness of the neck, fatigue, and irritability. Treatment involves maintenance of essential body functions and reliance upon natural body defenses for recovery. The infection tends to be of short duration, with most patients recovering without complication in 3–14 days.

Viral Encephalitis. Encephalitis, a disease de-

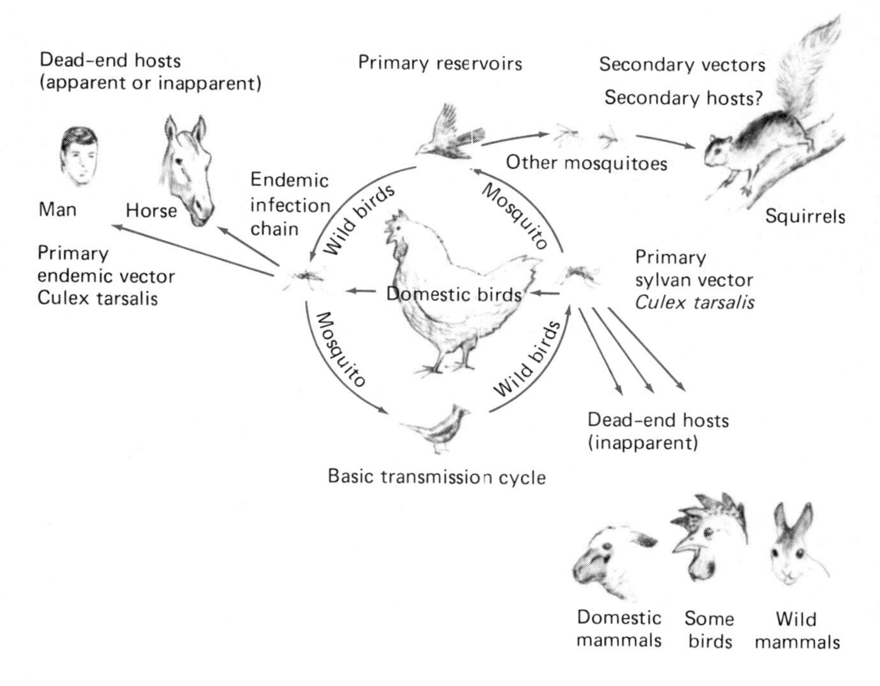

Figure 26.7 Epidemiological transmission of equine encephalitis. The chains for rural St. Louis encephalitis are similar, except that horses are inapparent, rather than apparent, hosts. Eastern equine encephalitis infections also have a similar summer infection chain, but a few significant differences exist: the identity of the vector infecting man is unknown; domestic birds do not appear to be a significant link in that chain; and it has a bird-to-bird secondary cycle in pheasants, whose role is unclear.

fined by an inflammation of the brain, can be caused by various viruses, many of which are arthropod-borne. Postinfection encephalitis sometimes occurs after measles, rubella, influenza, and other viral infections. Many of the viruses capable of causing encephalitis in humans are maintained in populations of various vertebrates, particularly birds and rodents, as well as populations of arthropods. Transmission to human beings, via an arthropod vector in which the virus has multiplied, generally represents a dead end in the transmission cycle (Figure 26.7). Viral multiplication within the arthropod initially occurs in the gut and is followed by dissemination through the hemolymph and multiplication within the salivary glands. Viruses acccumulate in the saliva of the arthropod, and this facilitates transfer to a person bitten by the arthropod. Outbreaks of viral encephalitis exhibit seasonal cycles, with increased numbers of cases occurring during summer when the vector populations are at their peak (Figure 26.8).

The different forms of viral encephalitis include eastern equine, western equine, Venezuelan equine, St. Louis, Japanese B, Murray valley, California, and tick-borne encephalitis (Table 26.2). The specific viral etiologic agent, arthropod vector, and geographic distribution are different for each of these forms of encephalitis. Infections with encephalitis-causing viruses begin with viremia, followed by localization of the viral infection within the central nervous system, where lesions develop. The locations of the lesions within the brain are characteristic for each type. With the exception of St. Louis encephalitis, where kidney damage also occurs, the pathologic changes in cases of encephalitis are normally restricted to the central nervous system.

Encephalitis symptoms are often subclinical. When the disease is symptomatic, the manifestation of encephalitis begins with fever, headache, and vomiting, followed by stiffness and then paralysis, convulsions, psychoses, and coma. Different forms of viral encephalitis have different outcomes. For example, in symptomatic cases of eastern equine encephalitis the mortality rate is approximately 80 percent, whereas in western equine encephalitis, the fatality rate is under 15 percent. Individuals who recover from symptomatic encephalitis may have permanent neurological damage.

No antiviral drug has been developed yet for the treatment of encephalitis caused by arthropod-borne viruses. Vaccines have been developed, though, that are effective in establishing immunity against the

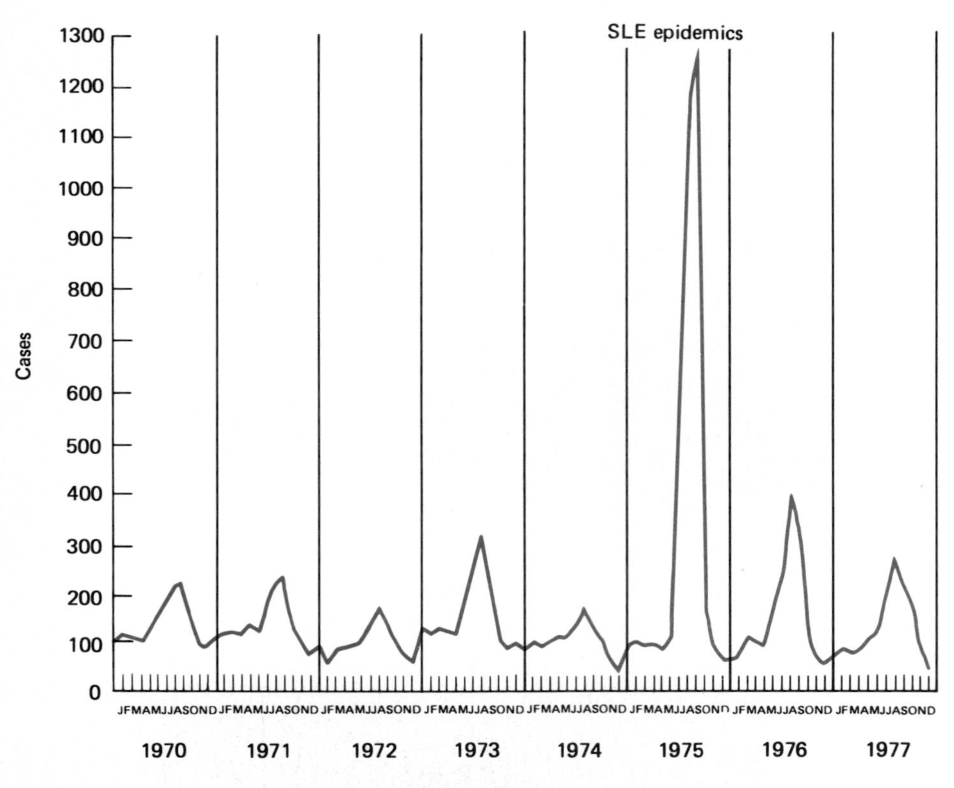

Figure 26.8 Incidence of encephalitis in the United States from 1970 to 1977. The major outbreak of encephalitis in 1975 was due to the St. Louis B virus and attests to the difficulty in controlling this disease.

specific viruses that cause encephalitis, and there is an effective vaccine against Japanese B virus. As a rule, control of arthropod-borne viral encephalitis depends on control of vector populations. Insecticides have been widely used in public health programs to reduce the population levels of the mosquitoes and ticks that are the vectors of viral encephalitis. Mosquito eradication programs are often intensified when positive diagnoses of encephalitis raise the possibility of widespread outbreaks.

Not all forms of encephalitis are arthropod-borne. For example, Herpes simplex virus, which is not normally transmitted through vectors, can cause both meningitis and encephalitis (herpes encephalitis). Acyclovir has been demonstrated to be effective in treating Herpes simplex virus encephalitis and approved for use in treating this disease, but this drug has not been proven effective against other encephalitis-causing viruses.

Reye's Syndrome. One serious complication associated with outbreaks of influenza and chicken pox is the development of Reye's syndrome, an acute pathological condition affecting the central nervous system. Reye's syndrome also occurs after infections with other viruses, and the specific relationship to influenza and chicken pox viruses is not clear. Occurrences of Reye's syndrome are highly, but inexplicably, correlated with outbreaks of influenza B virus and of chicken pox. Reye's syndrome principally is associated with children. It was thought to be largely restricted to those under 16 years of age but is now known to affect those up to 19 years old. Fever and severe vomiting often occur at the onset of Reye's syndrome; this is followed by the accumulation of fluids in the brain (cerebral edema), coma, and death due to intracranial pressure. For reasons that have yet to be elucidated, there is a high incidence of Reye's syndrome when aspirin is used for treating the symptoms of a viral infection. Consequently, some pediatricians have warned against the use of aspirin for children with influenza and other viral infections of the respiratory tract. Because a cause and effect relationship between the use of aspirin during viral infections has not been established, the issuance of this warning, which may soon be required on bottles of aspirin, is very controversial. Nevertheless, the most recent studies indicate that young victims of chicken pox or flu who are given aspirin are 25 times more likely to develop Reye's syndrome than those young victims of the diseases who are not treated with aspirin.

Guillain–Barré Syndrome. Another condition involving the central nervous system associated with influenza infections is Guillain–Barré syndrome. This disease normally is rare, but in 1976 there was an increased incidence of the syndrome associated with an active immunization program against a predicted outbreak of swine flu, which did not occur. Swine flu is caused by an influenza virus that normally occurs in pigs but can be transmitted to humans with fatal consequences. The outbreak of Guillain–Barré syndrome appears to have resulted from contamination of some batches of vaccine with influenza viruses that had not been adequately inactivated. This disease involves a rapidly developing inflammation of several nerves that results in spreading muscular weakness of the extremeties and possible paralysis.

Table 26.2 Vectors and Distribution of Human Viral Encephalitis

Viral Disease	Vector	Geographic Distribution
Eastern equine encephalitis (EEE)	Mosquito	Eastern U.S., Canada, Brazil, Cuba, Panama, Dominican Republic, Trinidad, Philippines
Venezuelan equine encephalitis (VEE)	Mosquito	Brazil, Colombia, Ecuador, Trinidad, Venezuela, Mexico, Florida, Texas
Western equine encephalitis (WEE)	Mosquito	Western U.S., Argentina, Canada, Mexico, Guyana, Brazil
St. Louis encephalitis	Mosquito	U.S., Trinidad, Panama
Japanese B encephalitis	Mosquito	Japan, Guam, Eastern Asian mainland, India, Malaya
Murray Valley encephalitis	Mosquito	Australia, New Guinea
Ilheus	Mosquito	Brazil, Guatemala, Honduras, Trinidad
Russian spring–summer encephalitis	Tick	USSR, Canada, Malaya, U.S., Central Europe, Finland, Japan, India, Great Britain
Rio Bravo	Unknown	California, Texas

Diseases Caused by Prions

Creutzfeldt–Jakob and Related Diseases. Creutzfeldt–Jakob disease is a rare disease that results in the loss of intellectual capacity (dementia). The

MICROBIOLOGY HEADLINES

Makers of Aspirin Agree to Labels Warning of Deadly Child Ailment

WASHINGTON, Jan. 11 (UPI)—Aspirin makers agreed Friday to develop labels warning of a possible link between the painkiller and the deadly Reye's syndrome in children, Joseph White, president of the Aspirin Foundation, said. The foundation represents major aspirin manufacturers.

After a meeting with officials of the Food and Drug Administration, Mr. White said he hoped the industry and Government could reach agreement on the voluntary labeling by the end of next week.

Margaret M. Heckler, Secretary of Health and Human Services, asked for the warning labels Wednesday, citing a study by the Centers for Disease Control that she said showed a possible link between aspirin and development of Reye's Syndrome in children treated for chicken pox or flu.

U.S. Had 190 Cases Last Year

Reye's Syndrome is a viral ailment, characterized by vomiting and fever, that occurs primarily in children and teen-agers, usually after they have had influenza or chicken pox. There were 190 cases in the United States last year. About a quarter of the victims died.

Mr. White said the aspirin manufacturers had told the Federal officials "that our organization was interested in working with them to develop labeling for aspirin products in reply to the request of the Secretary of Health and Human Services."

The aspirin industry challenged earlier reports that aspirin was linked to Reye's Syndrome.

Mr. White said the Government officials had agreed to provide more data from the study by the Centers for Disease Control but added, "We agreed we would go forward with our part of it without waiting for the data to be made available."

Mrs. Heckler said she was "delighted with the companies' public-spirited response." Among the companies the foundation represents are Sterling Drug Inc., the Bristol-Myers Company, Whitehall Laboratories, Miles Laboratories Inc., the Burroughs Wellcome Company, the Squibb Corporation and Procter & Gamble.

Issues Must Be Worked Out

Mr. White said some issues must still be worked out, including the wording of the warning, its placement and arrangements to discontinue it should it be found unnecessary.

Mr. White said officials from the Government and the aspirin industry planned to meet again next week.

In Memphis, Plough Inc., maker of St. Joseph children's aspirin, said it would "cooperate with the Secretary's request for label changes pending further studies."

Asked whether that meant the company would change its labels, Lewis Nolan, a spokesman said, "We're really not elaborating beyond that statement." Plough is not a member of the Aspirin Foundation.

Source: United Press International, Jan. 11, 1985. Reprinted by permission.

worldwide incidence of Creutzfeldt–Jacob disease is about one in a million. Interest in this disease centers on the finding that it is caused by a prion. Prions are infectious proteins that do not appear to contain any nucleic acids. This sets prions apart from all other infectious agents. It is not understood how they reproduce, but it is clear that when prions infect mammalian cells, new prions are produced and abnormalities develop in the mammalian system. Diseases caused by prions develop slowly and therefore have been termed slow viral diseases. These slow infections are characterized by prolonged incubation periods of months, years, or possibly decades, during which there are no symptoms. Once the illness begins, however, it progresses steadily and generally leads to death.

Creutzfeldt–Jacob disease is one of several apparently related diseases that are characterized by similar clinical and pathological symptoms. Scrapie, a disease of sheep and goats, has similar symptoms and is known to be caused by prions. Alzheimer's disease, which is the commonest form of human senile dementia and fourth leading cause of death in the United States, has yet to be proven to be caused by prions. Other apparently related human diseases are Gerstmann Sträussler syndrome and Kuru. The occurence of Kuru is very restricted because this disease is transmitted through the ingestion of human brain tissue during ritual cannibalism in New Guinea. Such tribal customs are no longer practiced and the incidence of Kuru has been declining. The initial symptoms of scrapie, Kuru, and Gerstmann Sträussler syndrome are difficulty in walking and loss of coordination. These symptoms indicate the involvement of the cerebellum. Late symptoms of Kuru include dementia. Creutzfeldt–Jacob disease sometimes shows early symptoms that resemble Kuru but usually begins with loss of memory. Pathological changes in these diseases are restricted to the central nervous system. These pathogical indicators include proliferation of astrocytes (supporting cells in the brain), depletion of dendritic spines within neurons (structures involved in nerve impulse transmission), vacuolization of brain tissue, and amyloid plaque development. The vacuolization gives the brain tissue a spongy appearance. The amyloid plaques, which characterize diseases such as Alzheimer's disease, may actually be aggregates of prions in an almost crystalline array. Additional studies are needed to positively identify the cause of Alzheimer's disease and to find ways of curing or preventing degenerative diseases of the nervous system.

Diseases Caused by Bacterial Pathogens

Bacterial Meningitis. Meningitis, an inflammation of the meninges (membrane surrounding the brain and spinal cord) can be caused by various bacteria. The typical transmission of bacteria causing meningitis is via droplet spread, with the initial infection occurring in the respiratory tract followed by transmission via the bloodstream to the meninges. Usually high numbers of bacteria in the bloodstream for a prolonged period is necessary before transmission to the meninges will occur. Injuries that expose the central nervous system to bacterial contaminants provide an alternate portal of entry. *Neisseria meningitidis, Haemophilus influenzae, Streptococcus pneumoniae*, and *Escherichia coli* are the most common etiologic agents of bacterial meningitis. There is a high degree of correlation between the age of the patient and the specific etiologic agent (Table 26.3). For example, *E. coli* often causes meningitis in infants but not in adults. *Haemophilus influenzae* is the leading cause of bacterial meningitis in children ages 2–5. A vaccine, prepared from the capsular polysaccharide of *H. influenzae* type b, has recently been developed and is effective in preventing meningitis caused by this bacterium in children older than 18 months.

N. meningitidis (Gram negative diplococci), also known as the meningococcus, on the other hand, is often the causative agent of bacterial meningitis in patients between 5 and 40 years of age but is rarely found in cases of meningitis in younger children. The meningococcus is the main cause of bacterial meningitis in adults. There are about 2500 cases of meningococcal infections in the United States each year. This organism frequently causes infections in those in close contact with an infectious individual. At particular risk are family members, schoolmates, and medical personnel with intimate contact (for example, mouth to mouth resuscitation). The lipoplysaccharides of the cell wall of *N. meningitidis* are responsible for the disease symptomology. The virulence of the organism is largely determined by the presence of a capsule that protects it against phagocytosis.

Bacterial meningitis is characterized by sudden fever, severe headache, painful rigidity of the neck, nausea, vomiting, and frequently by convulsions, delirium, and coma. If untreated, bacterial meningitis is usually fatal. An infection with *N. meningitidis* can progress so rapidly that death occurs within a few hours of the onset of symptoms. Because death may occur within hours of recognition of the onset of neurological symptoms, accurate and rapid diagnosis and speedy initiation of treatment is essential. The diagnosis of the particular etiologic agent of meningitis depends on the isolation and identification of the pathogens from the cerebrospinal fluid. Meningitis can be fatal within hours or days and delays in instituting antibiotic therapy risk brain

Table 26.3 Correlation of Age with the Etiologic Agents of Meningitis

	Percent of Isolates			
Etiologic Agent	**under 2 months**	**2–60 months**	**5–40 years**	**over 40 years**
Neisseria meningitidis	—	20	45	10
Haemophilus influenzae	—	60	5	2
Escherichia coli and other Enterobacteriaceae	55	—	—	10
Pseudomonas aeruginosa	2	—	—	—
Streptococcus pneumoniae and other *Streptococcus* sp.	28	12	25	55
Staphylococcus sp.	5	—	10	13
Other	10	8	10	10

MICROBIOLOGY HEADLINES

The Mystery of Balanchine's Death Is Solved

By LAWRENCE K. ALTMAN, M.D.

In a laboratory at Columbia-Presbyterian Medical Center a few months ago, a pathologist leaned over a microscope and peered at an illuminated slice of brain tissue about 10 microns thick. In the center of the minute specimen of brain cells was a pink circle, known as a kuru plaque, one sign of a strange family of diseases known as slow viruses.

Those brain cells were George Balanchine's, and therein lay the solution to the mystery of his death last year at 79 years of age. The once-athletic choreographer died after a period of several months that had been awash in pathos. He could hardly move, let alone dance, and he could hardly think, let alone choreograph.

Certainly, he was suffering from some degenerative neurological disorder. But what was it?

In the weeks following his death, pathologists determined that Mr. Balanchine had suffered from one of the world's most unusual diseases—Creutzfeldt–Jakob disease. It is categorized in the group of so-called slow virus diseases because of its extremely long incubation period.

The story of Mr. Balanchine's rare affliction is being told publicly now with the permission of Barbara Horgan, the personal assistant to Mr. Balanchine and executor of his estate.

The other day the doctors who ministered to him during the last months of life and the pathologists who examined him after he died gathered to discuss what they finally knew about the great choreographer.

In a conference room on the 15th floor of Columbia's College of Physicians and Surgeons, the whole story of Mr. Balanchine's life's end and the post-mortem diagnosis emerged. In attendance were more than a dozen doctors. Slides were projected in front of them, as Dr. Philip E. Duffy, the medical center's director of neuropathology, went over each of the clues that were seen only after Mr. Balanchine's death.

Dr. Duffy spoke without interruption for about a half hour, and then the other doctors joined in conversation and speculation on what they now knew about the rare disease that struck Mr. Balanchine, and what it told them about the one-in-a-million others who are similarly affected.

This gathering was held on the anniversary of Mr. Balanchine's death April 30, 1983. Except for the fame of the patient, it resembled the clinical-pathological conferences in which pathologists pass along to physicians many facts that would have been impossible to determine in life.

Also, it was a chance for Dr. Edith J. Langner, a Manhattan internist who had been Mr. Balanchine's physician since 1978, to resolve the mystery of what killed him and why. Two questions uppermost in Dr. Langner's mind were: Did her team of many prominent specialists fail to recognize a treatable condition? If so, could his life been saved?

It is now clear that Mr. Balanchine noticed the first signs of Creutzfeldt–Jakob disease about 1978; he was slightly unsteady while standing, something that limited his ability to pirouette. A neurologist who examined Mr. Balanchine could detect nothing abnormal, but Mr. Balanchine, whose fame was made by mastering the subtleties of motion, was aware that something was wrong.

However, the overriding medical problems at that time were related to a recent heart attack and the need for tighter control of his high blood pressure. He took large doses of drugs for angina, but still he suffered the crushing chest pains and they severely restricted his physical activity.

He balked at coronary bypass surgery until the spring of 1980, when he was no longer able to turn over in bed because of the angina pains. But his recovery from the operation was slow. When he was back to full activity, he still had difficulty in keeping his balance. He was particularly self-conscious about it when he walked alone. Mr. Balanchine no longer walked through the steps with his dancers but talked them through their routines, Dr. Langner said.

All the medical tests were done and repeated, and nothing wrong was found. Yet his sense of balance deteriorated further, and a few people around him began to notice.

Dr. Langner was one. She observed that there were more red marks on the wall left by the red elbow patch on his jacket as he stumbled his way along the hall to her office.

Clearly there was something wrong with Mr. Balanchine's cerebellum, the portion of the brain that controls balance. The doctors called it cerebellar degenerative disease. The cause of it was unknown. Because of the suspicions that it might be the effects of arteriosclerosis, the doctors treated Mr. Balanchine with aspirin to reduce the chances of a stroke.

His eyesight and hearing began to fail. Music no longer sounded the same to him. Cataracts distorted his appreciation of blue coloring, which was distressing because he designed the lighting for his productions. He had two eye operations.

One specialist was convinced that Mr. Balanchine had a tumor called an acoustic neuroma and wanted to operate, but Dr. Langner and other specialists stopped him because they seriously doubted that possibility. If he did have it, they reasoned, it would have accounted for only one or two symptoms, not for the generalized nature of his malady.

In September 1982 Mr. Balanchine suffered "the flu" on a trip to Washington, D.C., and received an unexpected extra set of opinions from doctors at George Washington University Hospital. They came up with the same nonspecific diagnosis: cerebellar degenerative disease.

Doctors in two cities now had done every esoteric test they could think of except for a brain biopsy, and Mr. Balanchine rejected that possibility.

Meanwhile, Mr. Balanchine became increasingly confused and he fell often. He broke several ribs despite the attention of companions round the clock. In November 1982, when it was no longer possible to care for him at home, he entered Roosevelt Hospital where his course varied day to day and hour by hour.

One day, the nurses brought him to the cafeteria because they thought that Mr. Balanchine would benefit from some diversion, and might feel more comfortable eating his meals there. But he refused to eat anything because he did not want anyone to see him being so clumsy.

"There was a lot of pressure on both of us," Dr. Langner said. "People from the ballet would call up and ask if he had seen doctor so and so."

Dr. Langner called each one of the doctors. Some experts came. One declined to come at all.

Soon Mr. Balanchine could not recall events that happened a few minutes before. He could not walk and soon he could not use his hands. In the end he had great difficulty swallowing, and this led to the complication that killed him, pneumonia.

In those final weeks, as Dr. Sidney E. Bender, one of Mr. Balanchine's neurologists, said: "We stood at the foot of his bed and shook our heads a lot. We thought he was dying of his own disease—one he invented."

An autopsy was done at Roosevelt Hospital and Mr. Balanchine's brain was put in a jar of formalin and sent to Dr. Duffy's team of experts at a sister institution, Columbia-Presbyterian Medical Center.

There, after 10 days, the brain was removed from the jar and sliced across in layers. From those slices, small blocks of tissue were prepared so that the brain could be studied under the microscope. Chemical stains were added to some to help detect the pattern of appearance of certain brain cells and abnormalities particularly the kuru plaques.

As Dr. Duffy clicked the projector at the conference to show slide after slide of Mr. Balanchine's brain, he commented on the findings. The appearance of the intact brain to the eye "appears normal and there is very little arteriosclerosis," he said.

Then he switched to pictures taken through the microscope and said: "These are very abnormal. Notice that nerve cells are visible but there are regions where the number of nerve cells is dramatically reduced."

Inflammation, so characteristic of most infections, was absent in Mr. Balanchine's brain. It was a subtle but valuable clue because slow virus diseases are characterized by the absence of inflammation.

Next he pointed to a significant feature and a key lead to the diagnosis: small spaces in the astrocytes and neuron cells of the brain. "You see them everywhere in the gray matter of the brain," Dr. Duffy said.

It was a condition called "the spongy state," which is common in a number of so-called slow virus diseases such as Creutzfeldt–Jakob disease, kuru and scrapie. The damage was most striking in the cerebellum.

In addition, star-shaped astrocytes were increased in size and number, evidence of brain injury. Astrocytes help the brain heal and modulate nerve function by "picking-up" certain chemicals called neuro-transmitters, among other functions.

Now the audience was staring at several pink circles. On closer examination there were little threads extending radially from the center. These were kuru plaques. "These plaques were first described in kuru, but they occur in about 10 percent of cases of Creutzfeldt–Jakob disease," Dr. Duffy said.

Dr. Duffy then moved to the blackboard to discuss the history of slow virus diseases. It began with the recognition 200 years ago of scrapie in sheep by Icelandic shepherds who noted that the diseased animals would become irritable, stagger and scrape themselves against trees and rocks before they died.

In recent years, scientists have shown that scrapie can be transmitted to animals after a long incubation period. The agent is in the range of the size of small viruses. Some researchers are exploring the possibility that a slow virus might cause the common disorder known as Alzheimer's disease, which destroys the mind.

The scrapie agent resists radiation, formalin and autoclaving—the standard methods of sterilizing medical equipment. However, the agent seems to be killed by more stringent methods such as using the chemical sodium hypochloride, and by autoclaving for longer periods of time and under higher pressure or temperature.

The neurologists turned their attention to Mr. Balanchine's case. He showed none of the usual features that give a physician the clues to the diagnosis of Creutzfeldt–Jakob. He developed cerebellar and motor problems before his mind began to deteriorate, rather than afterwards as is the usual case.

Mr. Balanchine's electroencephalogram, or brain wave test, did not show any abnormal pattern that can be a clue to this disease. "Nor did he have what is known as the exaggerated startle response, in which a loud noise causes the muscles to suddenly jerk and which is one of the real tipoffs to neurologists about Creutzfeldt–Jakob disease," Dr. Bender said.

The talk turned to other cases of the disease. Dr. Duffy mentioned a prominent neurosurgeon who had developed it, possibly from contact with a patient. He also recalled how in 1974 he had studied one patient who developed Creutzfeldt–Jakob disease after receiving a corneal transplant. On further investigation, his team found that the donor had an undetected case of the same disease. It was the first documentation of person-to-person transmission of the disease.

Someone asked if Mr. Balanchine could have acquired his disease from a contaminated medical instrument. The doctors went through the details of his medical history as they knew it. But because they recognized that they did not know every last detail, Dr. Duffy summarized the group's feeling by saying, "The question of transmission from medical instruments is not excluded, but there is no evidence for it, either."

The conference ended when Edward Bigelow, a dancer and long-time friend of Mr. Balanchine's, said: "Even if you had known this diagnosis before Mr. Balanchine died, you couldn't have done anything because there was no treatment. Correct?"

"Yes," Dr. Duffy replied.

Source: The New York Times, May 8, 1984. Reprinted by permission.

Figure 26.9 This painting of a soldier dying of tetanus, by Charles Bell, shows the characteristic position of the jaw and neck associated with this disease. (By the kind permission of the President and Council of the Royal College of Surgeons of Edinburgh.)

damage, retardation, and death. Meningococcal infections are reported to public health officials who then try to identify people who have been in direct contact with the infected indiviual. Prophylactic use of antibiotics can protect individuals who have been exposed to someone with a *N. meningitidis* infection against developing meningococcal meningitis.

A number of antibiotics are used in the treatment of meningitis, and the specific antibiotic of choice is determined by the antibiotic susceptibility of the causative agent of the disease and its ability to cross the blood–brain barrier. In many cases penicillin is the antibiotic of choice for treating bacterial meningitis. Culture and sensitivity tests traditionally, however, take at least a day so that antibiotic therapy generally must be instituted before a final diagnosis is made. Knowing the most likely etiologic agents for a particular age group aids the physician in selecting an appropriate antibiotic. Also, recent serological methods can provide the necessary identification of the etiologic agent within minutes or hours, permitting rapid positive diagnosis and institution of proper therapy.

Brain abscess. Under some conditions bacteria are able to invade the brain and cause an abscess. The development of a brain abscess usually occurs as a complication of an infection at another site in the body, most frequently from bone and ear infections. The bacteria usually found in cases of brain abscesses include species of *Streptococcus, Staphylococcus, Haemophilus, Nocardia, Bacteroides, Peptostreptococcus, Peptococcus, Fusobacterium,* and *Actinomyces*. Most brain abscesses are caused by anaerobes. When these bacteria invade the brain, they initiate an inflammatory response and the formation of pus, that is, a characteristic abscess. Brain abscesses are obviously serious and potentially life threatening. Fever, headache, nausea, and various neurological symptoms, such as blurred vision, are characteristic of this disease. Diagnosis of the causative agent and initiation of treatment with appropriate antibiotics is critical. Penicillin, chloramphenicol, and metronidazole are frequently used for treating brain abscesses.

Tetanus. Tetanus is caused by *Clostridium tetani*. This organism is able to produce a neurotoxin that can produce severe muscle spasms. Tetanus is sometimes referred to as lockjaw because the muscles of the jaw and neck contract convulsively so that the mouth remains locked closed, making swallowing difficult (Figure 26.9). *C. tetani* is widely distributed in soil. Transmission to humans normally occurs as a result of a puncture wound that inoculates the body with spores of *C. tetani*. If anaerobic conditions develop at the site of the wound, the endospores of *C. tetani* germinate and the multiplying bacteria can produce neurotoxin. *C. tetani* is noninvasive and multiplies only at the site of inoculation. The neurotoxin produced by *C tetani*, however, spreads systemically, causing the symptoms of this disease.

Tetanus—The most terrifying and unnecessary complicaton of any wound

—W. FIROR, 1938

Virtually any type of wound into which foreign material is introduced may carry spores of *C. tetani* and lead to the development of tetanus. Tetanus may occur months after contamination of a wound due to the outgrowth of subcutaneous spores of *C. tetani*. Tales of the association of rusty nails with this disease probably originated because farmers often developed tetanus after stepping on nails that were contaminated with soil and the endospores of *C. tetani*, but clearly rusty nails are not the cause of this disease. If untreated, tetanus is frequently fatal, but if recovery does occur, there are no lasting effects. Many newborns in developing countries die of tetanus neonatorum, which results from postnatal infection of the degenerating remnant of the severed umbilical cord. Tetanus can be treated by the administration of tetanus antitoxin to block the action of the neurotoxin. The disease can be prevented by immunization with tetanus toxoid to preclude the development of infection with *C. tetani*, and tetanus booster vaccinations are frequently given after wound injuries to ensure immunity against this disease. Public awareness and the effective use of vaccines have reduced the incidence of tetanus in the United States to fewer than 100 cases per year.

Botulism. Botulism, the most serious form of bacterial food poisoning, is caused by neurotoxins produced by *Clostridium botulinum*. The toxins are absorbed from the intestinal tract and transported via the circulatory system to motor nerve synapses where their action blocks normal neural transmissions. Various strains of *C. botulinum* elaborate different toxins. Types A, B, and E toxins cause food poisoning of humans. Type E toxins are associated with the growth of *C. botulinum* in fish or fish products, and most outbreaks of botulism in Japan are caused by type E toxins because large amounts of fish are consumed there. Type A is the predominant toxin in cases of botulism in the United States, and type B toxin is most prevalent in Europe.

Fortunately cases of botulism occur relatively infrequently (Figure 26.10). Over 90 percent of the cases of botulism involve improperly home-canned food.

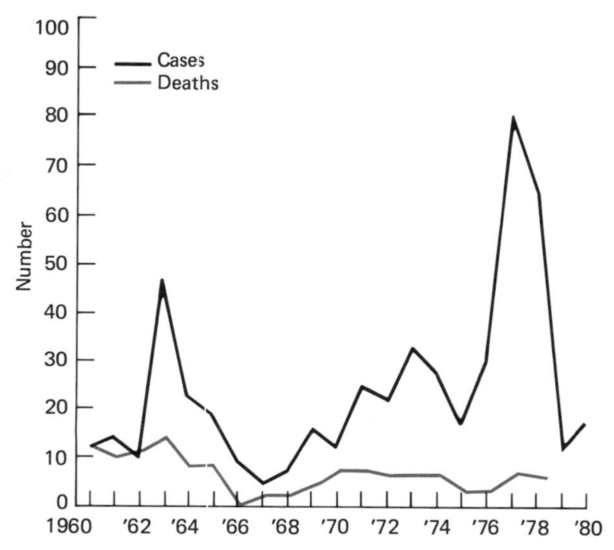

Figure 26.10 Reported cases and deaths due to foodborne botulism in the United States between 1960 and 1980. (From *Morbidity and Mortality Weekly Reports*, Centers for Disease Control, Atlanta.)

Of 236 outbreaks of botulism in the United States between 1899 and 1974, 57 percent were caused by contaminated vegetables, 15 percent by contaminated fish, and 12 percent by fruit containing *C. botulinum*. The endospores of *C. botulinum* are heat resistant and can survive prolonged exposure at 100°C. Contaminated canned foods provide an optimal anaerobic environment for the growth of *C. botulinum*. This bacterium, though, cannot grow and produce toxin at low pH and thus is not a problem in acidic food products. Symptoms of botulism can appear 8–48 hours after ingestion of the toxin, and an early onset of disease symptomology normally indicates that the disease will be severe. Type A toxin botulism is generally more severe than the disease caused by other types of toxin. In severe cases of botulism, there is paralysis of the respiratory muscles, and despite improved medical treatment, the mortality rate is still about 25 percent. The use of trivalent ABE antibodies is useful in treating this disease, but it is of paramount importance in the treatment to ensure continued respiratory functioning.

Infant Botulism. *Clostridium botulinum* is normally not capable of establishing an infection in adults because of the low pH of the stomach and upper end of the small intestine. However, in infants, prior to the colonization of the intestinal tract by *Lactobacillus* species, *C. botulinum* can reproduce and elaborate neurotoxin from the gastrointestinal tract tissues. There is evidence that a few cases of sudden infant death syndrome, or crib death, can be attributed to *C. botulinum*. Accordingly, additional concern is being given to the food products, such as honey, that infants may consume with respect to the possible ingestion of *C. botulinum* endospores.

Diseases Caused by Algae

Paralytic Shellfish Poisoning. Algae are rarely considered as the etiologic agents of disease, but paralytic shellfish poisoning is caused by toxins produced by the dinoflagellate *Gonyaulax*. Blooms of *Gonyaulax* cause red tides in coastal marine environments. During such algal blooms, the algae and the toxins they produce can be concentrated in bivalve shellfish, such as clams and oysters. The ingestion of shellfish containing algal toxins can lead to symptoms that resemble botulism. Shellfishing is banned in areas of *Gonyaulax* blooms to prevent this form of food poisoning.

Diseases Caused by Fungi

Mushroom Poisoning. Toxins produced by some fungi are responsible for some serious cases of food poisoning (mushroom poisoning). Many of these mycotoxins are potent neurotoxins that can be absorbed through the gastrointestinal tract. The ingestion of poisonous mushrooms, such as *Amanita phalloides*, is normally fatal. The toxins produced by *A. phalloides*, and other species of *Amanita*, produce symptoms of food poisoning 8–24 hours after their ingestion. Initial symptoms include vomiting and diarrhea; later, degenerative changes occur in liver and kidney cells, and the patient often becomes comatose. Death may occur within a few days of ingesting as little as 5–10 mg of toxin.

Ergotism. Ergotism (St. Anthony's Fire), another disease caused by fungi, results from ingesting grain containing ergot alkaloids produced by *Claviceps purpurea*. The fungus can grow in grain that becomes moist because of improper storage. Fungal growth results in the accumulation of ergot alkaloids in the grain. The toxins of *C. purpurea* cause degeneration of the capillary blood vessels, resulting in neurological impairment due to inadequate circulation to the nervous system. This type of fungal food poisoning has a relatively high mortality rate. Symptoms of ergotism may include vomiting, diarrhea, thirst, hallucinations, convulsions, and lesions at the body extremities. The convulsions and hallucinations are common symptoms reflecting the effects of the ergot alkaloids on the central nervous system. Various outbreaks of mass hallucinations have been traced to contamination of food with ergot alkaloids, and there even are theories that the Salem Witch Hunts were related to grain contamination and widespread ergotism.

Cryptococcal Meningitis. *Cryptococcus neoformans* is a yeast-like fungus that can cause meningitis. This fungus is associated with soil contaminated by pigeon droppings. Transmission of *C. neoformans* to humans appears to be the result of inhalation of dust associated with pigeon droppings. Infection with *C. neoformans*, called cryptococcosis, initially effects the lungs. In most cases cryptococcosis is restricted to the respiratory tract, but in some cases the fungus spreads systemically. *C. neoformans* can invade the brain and meninges, especially in individuals receiving immunosuppressive therapy for other disease conditions. If untreated the meningitis that results is usually fatal. The disease can be treated with amphotericin B.

Diseases Caused by Protozoan Pathogens

African Sleeping Sickness. African trypanosomiasis, also known as African sleeping sickness, is caused by infections with *Trypanosoma gambiense* and *T. rhodesiense*. The etiologic agents of African trypanosomiasis are transmitted to people through the tsetse fly vector. Tsetse flies acquire *Trypanosoma* species from various vertebrate animals, such as cows, which act as reservoirs of the pathogenic protozoa. In humans, infections with *T. gambiense* or *T. rhodesiense* are disseminated through the mononuclear phagocyte system, and there is evidence of localization within regional lymph nodes. Multiplication of the protozoa can cause damage to heart and nerve tissues. Progression through the central nervous system takes from months to years. If untreated the initially mild symptoms, which include headaches, increase in severity and lead to fatal meningoencephalitis. Diagnosis is by clinical symptoms and by detection of trypanosomes in Giemsa-stained blood smears, cerebrospinal fluid, or lymph node aspirate. If the disease is diagnosed before there is central nervous system involvement, it can be successfully treated with antiprotozoan agents, such as suramin. If there is central nervous system involvement, melarsoprol, an arsenical, is used for treating the disease. Prevention of African trypanosomiasis involves controlling population levels of the tsetse fly, accomplished by clearing vegetation to destroy the natural habitats of the tsetse fly.

Summary

The central nervous system is not directly exposed to microorganisms. Diseases of the central nervous system usually develop when microorganisms or their toxins become systemically disseminated through the body and when there is injury to the protective layers of the central nervous system. Some central nervous system diseases, such as rabies and encephalitis, are associated with animal bites; some diseases such as botulism, mushroom poisoning, and paralytic shellfish poisoning result from the ingestion of toxin-containing foods; other diseases such as bacterial and fungal meningitis often originate as infections of the respiratory system; yet other diseases of the central nervous system, such as tetanus, are transmitted via wounds. Most of these central nervous system diseases are very serious and if not properly treated they are life threatening.

Table 26.4 presents a summary of the causes and treatments of various diseases of the central nervous system that are caused by microorganisms.

Table 26.4 Summary of Diseases Associated with the Central Nervous System

Disease	Causative Agent	Mode of Transmission	Treatment
Rabies	Rabies virus	Bite by infected animal	Rabies vaccine and human rabies immune serum
Poliomyelitis	Poliovirus	Ingestion or inhalation of virus	Prevention by vaccination
Aseptic meningitis	Virus	Complication of other disease	Supportive
Encephalitis	Various encephalitis viruses	Arthropod bites	Supportive
Herpes encephalitis	Herpes simplex virus	Unknown	Acyclovir
Reye's syndrome	Influenza virus, other viruses	Aerosols	Supportive
Guillain–Barré syndrome	Virus?	Unknown	Supportive
Creutzfeldt–Jakob disease	Prion	Unknown	Supportive
Meningitis (bacterial)	*Neisseria meningitidis, E. coli, Haemophilus influenzae, Streptococcus pneumoniae*	Complication of other disease	Antibiotics
Brain abscess	*Streptococcus, Staphylococcus, Haemophilus, Nocardia, Bacteroides, Peptostreptococcus, Peptococcus, Fusobacterium*	Complication of other disease	Penicillin, chloramphenicol, metronidazole
Tetanus	*Clostridium tetani*	Wound	Antitoxin
Botulism	*Clostridium botulinum*	Ingestion of food containing toxin	Antitoxin
Infant botulism	*Clostridium botulinum*	Ingestion of spores	Antitoxin
Paralytic shellfish poisoning	*Gonyalaux* and other dinoflagellates	Ingestion of contaminated shellfish	Antitoxin
Mushroom poisoning	*Amanita* and other poisonous mushrooms	Ingestion of toxin-producing mushrooms	Antitoxin
Ergotism	*Claviceps purperea*	Ingestion of contaminated grain and other foods	Supportive
Meningitis (fungal)	*Cryptococcus neoformans*	Inhalation of dust contaminated by pigeon droppings	Amphotericin B
African sleeping sickness	*Trypanosoma gambiense, T. rhodesiense*	Tsetse fly bite	Suramin

Study Outline

A. The nervous system is composed of the brain, the spinal cord, and numerous peripheral nerves that connect the central nervous system to the other parts of the body and provide the means of detection of changes in internal and external environments of the body.

  1. The peripheral nervous system consists of all the nervous tissue lying outside of the skull and spinal column; it transmits information to the central nervous system and delivers messages to muscles and glands for response.

  2. The brain and the spinal cord make up the central nervous system, which controls the integration, interpretation, thought, and transmission of messages to and from the periphery.

  3. Cerebrospinal fluid may be removed for diagnosing diseases of the central nervous system by performing a lumbar puncture or spinal tap.

B. The central nervous system is protected against microbial invasion by physical barriers. It does not have an indigenous microbiota and so infections of the central nervous system are associated with the failure of the physical defense system.

C. Diseases of the nervous system caused by viral pathogens.

  1. Rabies.
    a. Causative organism(s): rabies virus.
    b. Mode of transmission: via the depositon of saliva through the bite of an infected animal; also via the respiratory tract through aerosols in the atmosphere around dense populations of infected bats.
    c. Symptoms: anxiety, irritability, depression, sensitivity to light and sound, hydrophobia, difficulty in swallowing, paralysis, and coma.
    d. Treatment: vaccination.

  2. Poliomyelitis.
    a. Causative organism(s): poliovirus.
    b. Symptoms: headache, vomiting, constipation, and sore throat, followed by neural involvement including paralysis.
    c. Treatment: vaccine available.

  3. Aseptic meningitis.
    a. Causative organism(s): various viruses that inflame the meninges, often echoviruses and coxsackieviruses.
    b. Symptoms: fever, stiff neck, fatigue, and irritability.
    c. Treatment: supportive.

  4. Encephalitis.
    a. Causative organism(s): various viruses.
    b. Mode of transmission: via the bite of an infected arthropod vector.
    c. Symptoms: often subclinical; fever, headache, and vomiting, followed by stiffness, paralysis, convulsions, psychoses, and coma.
    d. Treatment: supportive depending upon symptoms; vaccines available for prevention and control by vector eradication programs.

  5. Reye's syndrome.
    a. Causative organism(s): associated with outbreaks of influenza and other viral infections.
    b. Mode of transmission: following viral infections especially influenza and chicken pox.
    c. Symptoms: sudden fever, nausea, vomiting, delirium, and coma.
    d. Treatment: symptomatic; prevention by limiting use of aspirin for treating symptoms of children with viral infections.

  6. Guillain–Barré syndrome.
    a. Associated with influenza infections.
    b. 1976 outbreak appears to have resulted from contamination of vaccines with viable

influenzae viruses prepared for an expected outbreak of swine flu, which did not materialize.

c. Treatment: supportive.

D. Diseases caused by prions.

1. Creutzfeldt–Jakob and related diseases (Gerstmann Straussler syndrome and Kuru)
   a. Causative organism(s): prions, infectious proteins that do not appear to contain any nucleic acids.
   b. Mode of transmission: characterized by very long periods of incubation, from months to possibly decades.
   c. Symptoms: none during incubation period, then difficulty in walking and general loss of coordination, loss of memory, dementia, and death.
   d. Treatment: supportive.

E. Diseases caused by bacterial pathogens.

1. Meningitis.
   a. Causative organism(s): various bacteria, most commonly *Neisseria meningitidis, Haemophilus influenzae, Streptococcus pneumoniae*, and *Escherichia coli*.
   b. Mode of transmission: via droplet spread, with the initial infection occurring in the respiratory tract followed by transmission via the bloodstream to the meninges.
   c. Symptoms: sudden fever, severe headache, painful rigidity of the neck, nausea, vomiting, and frequently convulsions, delirium, and coma.
   d. Treatment: antibiotics, in many cases penicillin.
2. Brain abscess.
   a. Causative organism(s): bacteria, including species of *Streptococcus, Staphylococcus, Haemophilus, Nocardia, Bacteroides, Peptostreptococcus, Peptococcus, Fusobacterium*, and *Actinomyces*.
   b. Mode of transmission: develops as a complication of an infection at another body site.
   c. Symptoms: fever, headache, nausea, and various neurological symptoms.
   d. Treatment: antibiotics, frequently penicillin, chloramphenicol, and metronidazole.
3. Tetanus.
   a. Causative organism(s): *Clostridium tetani*.
   b. Mode of transmission: via puncture wound that introduces spores from the soil.
   c. Symptoms: severe muscle spasms, difficulty swallowing.
   d. Treatment: tetanus antitoxin; vaccine available.
4. Botulism
   a. Causative organism(s): *Clostridium botulinum*.
   b. Mode of transmission: ingestion of food, particularly home-canned items, containing neurotoxins produced by *C. botulinum*.
   c. Symptoms: paralysis of the respiratory muscles.
   d. Treatment: trivalent ABE antibodies.

F. Diseases caused by algae.

1. Paralytic shellfish poisoning.
   a. Causative organism(s): dinoflagellate *Gonyaulax* toxins.
   b. Mode of transmission: ingestion of shellfish containing algal toxins.
   c. Treatment: antitoxins; prevention by avoiding ingestion of shellfish collected during times of red tide blooms.

G. Diseases caused by fungi.

1. Mushroom poisoning.
   a. Toxins produced by some fungi are responsible for serious cases of food poisoning.
   b. Symptoms: vomiting and diarrhea, followed by degenerative changes in liver and kidney cells, coma.
   c. Treatment: antitoxins.

2. Ergotism.
   a. Causative organism(s): *Claviceps purpurea*.
   b. Mode of transmission: ingestion of grain or grain products contaminated with ergot alkaloids produced by *Claviceps purpurea*.
   c. Symptoms: vomiting, diarrhea, thirst, hallucinations, convulsions, and lesions at the body extremities.
   d. Treatment: supportive and antitoxins.
3. Cryptococcal meningitis.
   a. Causative organism(s): *Cryptococcus neoformans*.
   b. Mode of transmission: via inhalation of dust associated with pigeon droppings.
   c. Symptoms: initially affects the lungs, may spread systemically.
   d. Treatment: amphotericin B.

H. Diseases caused by protozoan pathogens.
1. African sleeping sickness (African trypanosomiasis).
   a. Causative organism(s): *Trypanosoma gambiese* and *T. rhodesiense*.
   b. Mode of transmission: via tsetse fly vector.
   c. Symptoms: initially headaches, increasing in severity, leading to fatal meningoencephalitis.
   d. Treatment: if diagnosed before central nervous system involvement, antiprotozoan agents, such as suramin; if there is central nervous system involvement, melarsoprol, an arsenical.

Study Questions

1. For each of the following diseases, what is the causative organism, the reservoir for that etiologic agent, and the mode of disease transmission?
   a. African sleeping sickness
   b. Botulism
   c. Tetanus
   d. Paralytic shellfish poisoning
   e. Rabies
   f. Ergotism
2. What are the differences between the toxins produced by *Clostridium botulinum* and *C. tetani?*
3. Match the following statements with the terms in the list to the right.
   a. Caused by neurotoxins produced by *Clostridium botulinum*.
   b. Disease caused by virus that can cross blood–brain barrier, multiplying within neural tissues and causing varying degrees of nervous system damage.
   c. Can be caused by various viruses and bacteria; most common bacterial etiological agents are *Neisseria meningitidis, Haemophilus influenzae, Streptococcus pneumoniae*, and *Escherichia coli*.
   d. Over 90 percent of the cases involve improperly prepared home-canned food.
   e. Caused by toxins produced by dinoflagellate *Gonyaulax*, the organism that causes red tides in coastal marine environments.
   f. Caused by *Clostridium tetani*.
   g. Initial symptoms include anxiety, irritability, depression, and sensitivity to light and sound; followed by hydrophobia (fear of water) and difficulty swallowing; then paralysis, coma, and death.
   h. Cytoplasmic inclusion bodies (Negri bodies) develop within neurons of the brain.
   i. Virus transmitted in animal's saliva to humans through bite of infected animal.
   j. Viral infection of brain transmitted by mosquitoes and associated with horses.

   (1) Meningitis
   (2) Paralytic shellfish poisoning
   (3) Tetanus
   (4) Rabies
   (5) Polio
   (6) Botulism
   (7) Encephalitis

Some Thought Questions

1. Why aren't there as many urgent requests today as there were a few years ago to locate dogs that have bitten people?
2. How did the tale about rusty nails causing tetanus originate?
3. How can we prove that fungi were at the roots of the Salem witch hunts?
4. Why is encephalitis more prevalent in the summer than in the winter?
5. Why is it important for the clinician to know the age of the patient when trying to help diagnose a case of meningitis?
6. Why do people still eat honey when we know that it often contains *Clostridium botulinum*?
7. Why are tetanus booster injections administered whenever someone has a serious wound?
8. Why are Rotary Clubs International rather than the World Health Organization attempting to eliminate polio from tropical areas?

Additional Sources of Information

Boyd, R. F., and B. G. Hoerl. 1981. *Basic Medical Microbiology*. Little, Brown and Company, Boston. A basic text emphasizing microbes and disease with sections covering central nervous system infecions.

Braude, A. I. (ed.). 1980. *Medical Microbiology and Infectious Disease*. W. B. Saunders Company, Philadelphia. An advanced work with individual chapters dealing with specific diseases of the central nervous system; chapters dealing with infectious diseases are organized according to the taxonomy of the causative agent.

Davis, B. D., R. Dulbecco, H. N. Eisen, H. S. Ginsberg, and W. B. Wood. 1981. *Microbiology*. Harper & Row, Publishers, Hagerstown, Maryland. An excellent text used in many graduate and medical school microbiology courses dealing with all aspects of microbiology; chapters dealing with infectious diseases are organized according to the taxonomy of the causative agent.

Eron, C. 1981. *The Virus that Ate Cannibals*. Macmillan Publishing Co. Inc., New York. This popular work includes a chapter on the discovery that Kuru was an infectious disease.

Hoeprich, P. D. (ed.). 1977. *Infectious Diseases: A Modern Treatise of Infection Processes*. Harper & Row, Publishers, Hagerstown, Maryland. This classic treatment of infectious diseases makes an excellent reference volume.

Jawetz, E., J. L. Melnick, and E. A. Adelberg. 1980. *A Review of Medical Microbiology*. Lange Medical Publications, Los Altos, California. This book provides a very good review of infectious diseases.

Joklik, W. K., and H. P. Willett (eds.). 1980. *Zinsser Microbiology*. Appleton-Century-Crofts, New York. A classic text used in many medical school programs that includes chapters on infectious diseases.

Rose, N. L., and A. L. Rose (eds.). 1983. *Microbiology: Basic Principles and Clinical Applications*. Macmillan Publishing Co., Inc., New York. An advanced work on medical microbiology; chapters are organized for the most part according to taxonomic groups.

Youmans, G. P., P. Y. Paterson, and H. M. Sommers. 1980. *The Biological and Clinical Basis of Infectious Disease*. W. B. Saunders Company, Philadelphia. This is an excellent text covering most human infectious diseases.

27 Diseases of the Cardiovascular and Lymphatic Systems

KEY TERMS

acquired immune deficiency syndrome (AIDS)
amastigote
babesiosis
bartonellosis
brucellosis
buboes
Burkitt's lymphoma
cancer
cardiovascular system
carditis
Chagas' disease
childbed fever
dengue fever
elephantiasis
endemic (marine) typhus
endocarditis
epidemic typhus
infectious mononucleosis
kala-azar disease
leishmaniasis
leukemia
Lyme disease
lymph vascular system
lymphoma
malaria
malignant cells
merozoites
myocarditis
oncogenes
oncogenic viruses
plague
postpartum sepsis
promastigote
puerperal fever
relapsing fever
retrovirus
rheumatic fever
rickettsialpox
Rocky Mountain spotted fever
schistosomiasis
schizogamy
scrub typhus
septicemia
serum hepatitis
sporozoites
systemic infection
toxic shock syndrome
toxoplasmosis
transformation
trypanosomiasis
tularemia
typhoid fever
undulant fever
yellow fever

OBJECTIVES

After reading this chapter you should be able to

1. Define the key terms.

2. Describe the structure of the cardiovascular and lymphatic systems.

3. Describe the resident microbiota of the cardiovascular and lymphatic systems.

4. Describe the defense systems of the cardiovascular and lymphatic systems that protect against infectious microorganisms.

5. Discuss the causative agents, routes of transmission, key symptoms, and treatments for the following diseases: leukemia, Burkitt's lymphoma, infectious mononucleosis, acquired immune deficiency syndrome (AIDS), serum hepatitis, dengue fever, yellow fever, rheumatic fever, carditis, typhoid fever, toxic shock syndrome, puerperal fever, plague, relapsing fever, bartonellosis, Rocky Mountain spotted fever, epidemic typhus, endemic typhus, scrub typhus, rickettsialpox, tularemia, brucellosis, Lyme disease, malaria, babesiosis, leishmaniasis, toxoplasmosis, trypanosomiasis, schistosomiasis, and elephantiasis.

Structure of the Cardiovascular and Lymphatic Systems

The human circulatory system consists of the cardiovascular (blood vascular) and lymphatic (lymph vascular) systems (Figure 27.1). In both systems a fluid (blood or lymph) circulates through a series of vessels (arteries, veins, capillaries, lymphatic vessels, and so on) and organs (heart, liver, spleen, and so on). The blood vascular system includes a pump, the heart, which provides the force needed to circulate the blood. The system carries essential nutrients to all body organs, removes their waste products, and carries products of cellular activity to other cells or to appropriate organs of excretion. The blood, the heart, and blood vessels together are the cardiovascular system

Blood is a tissue that contains various types of blood cells and a liquid portion called plasma. The adult human body contains 5–6 liters of blood, the major functions of which are transporting materials through the body and regulating homeostasis (the maintenance of an optimal internal environment). The plasma also contains chemicals that regulate body pH and temperature. Red blood cells (erythrocytes) carry oxygen from the lungs to cells and some of the carbon dioxide produced in metabolism to the lungs. Leukocytes are white blood cells that function primarily in providing protection against microorganisms. These white blood cells include neutrophils, eosinophils, basophils, lymphocytes, and monocytes (See Chapter 19). Platelets in the blood contain several factors important in blood clotting.

The heart is basically a hollow muscular pump that creates a pressure on the blood, causing it to circulate through the body's blood vessels. The heart is actually two pumps in one; the right side of the heart receives blood from the body in general and sends it to the lungs for gas exchange, the left side receives the blood from the lungs and sends it through the blood vessels to the body. The blood vessels include the arteries, arterioles, blood capillaries, venules, and veins. An artery carries blood away from the heart. It has thick, impermeable walls containing smooth muscle and elastic connective tissue. These components allow the artery to expand under the surges of fluid pressure from blood leaving the heart and recoil elastically, forcing the blood into smaller tubes called arterioles. The arterioles in turn feed blood to the capillaries. A capillary is a very narrow blood vessel, so narrow that red blood cells must flow through it single file. Its walls are thin and permeable to many materials, enabling capillaries to serve as the exchange point between blood and surrounding tissue. The direction in which a substance moves depends on its concentration gradient. Capillaries come together in venules and veins, which are the blood vessels leading back to the heart. Veins are distensible elastic tubes with impermeable walls.

The blood circulation system is supplemented by the vessels comprising the lymph vascular system. Pressure generated by the heart forces slightly more water out of capillaries than osmotic pressure drives in. Given the tremendous number of capillaries in the human body, considerable fluid would accumulate in interstitial regions if it were not for lymph vessels functioning as drainage tubes taking care of the overflow. The fluid in these tubes is called lymph. Lymph is formed from the tissue fluid. It has no red cells, few platelets and little protein, but may contain from 500 to 50,000 nongranular leukocyctes per milliliter. The vessels themselves, called lymphatics, converge into larger vessels and ducts, which eventually return the fluid to the blood circulation. The lymphatic vessels serve to return some 10 percent of the fluid filtered from the blood vessels to the blood vascular system, carry large lipid molecules absorbed from the intestines to the blood circulation, return protein from tissue spaces, and provide a route for lymphocyctes and monocytes produced in lymph organs to enter the blood circulation.

The lymph organs include the lymph nodes, tonsils, spleen, and thymus. Lymph nodes are located throughout the course of collecting lymphatic vessels and any particulate matter in the lymph flow is removed by the lymphocyctes and macrophages within the node. Thus nodes provide an important means of defense against organisms that may enter through epithelial wounds or be absorbed through organ walls. There are three lymphoid organs in the body, all of which are involved in the production of specialized cells. The spleen, the largest lymph organ, is situated in the upper abdomen. It is involved in producing nongranular white blood cells and other functions related to the blood. The thymus plays an important role in the differentiation of T lymphocytes, which are critical to the body's immune defense system. The lymph nodes are aggregates of lymphoid tissue along the course of the lymphatic vessels. They are involved in the storage, diffentiation, and distribution of lymphoctes throughout the body.

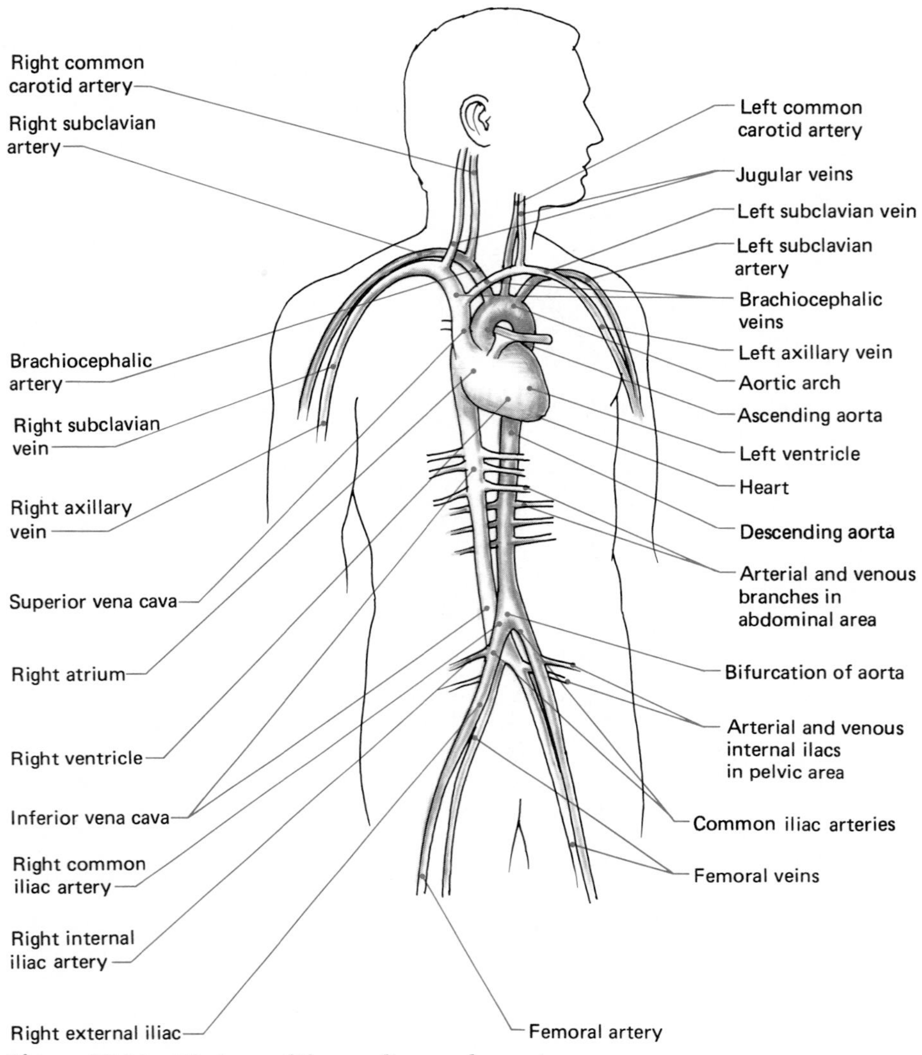

Figure 27.1A Diagram of the cardiovascular system.

Defenses of the Cardiovascular and Lymphatic Systems

As discussed in Chapters 19 and 20, blood and lymph contain numerous types of cells that effectively protect the body against the systemic spread of microorganisms. These same defenses act to prevent microbial infections of the cardiovascular and lymphatic systems. The various phagocytic white blood cells (leukocytes) and macrophages engulf and eliminate many microbes that enter the circulatory system. Circulating antibodies and the lymphocytes of the immune system also prevent the reproduction of most microbes within the circulatory system.

Blood is considered a sterile body fluid because the circulatory system does not possess a resident microbiota. In reality, various microorganisms frequently enter the bloodstream; normally, however, they do not establish growing populations within the circulatory system. For example, a segment of the circulatory system associated with the liver,

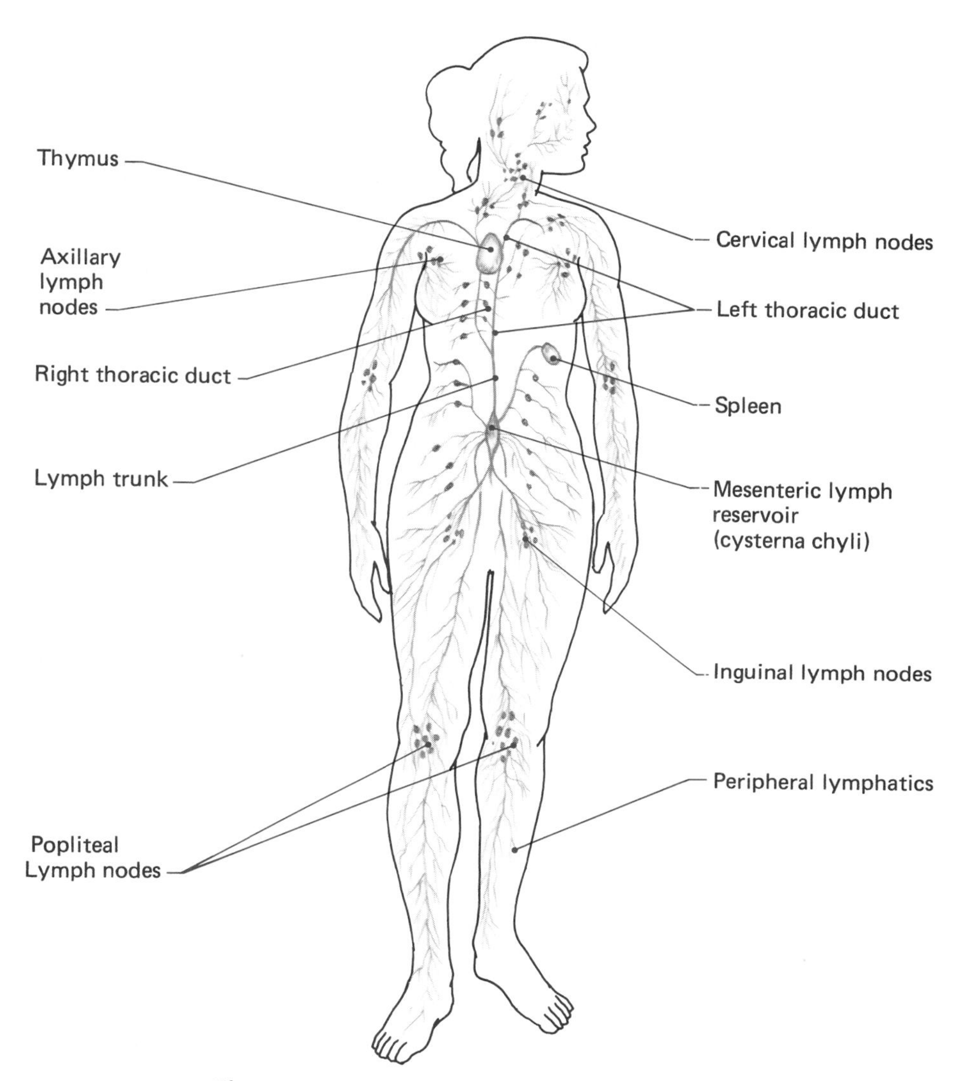

Figure 27.1B Diagram of the lymphatic system.

known as the hepatic portal system, normally contains low numbers of bacteria that pass through the intestinal wall as a result of abrasions to the lining of the intestinal tract caused by food particles. These bacteria are routinely eliminated from the circulatory system by macrophages in the liver. Such defense cells and other antimicrobial factors in blood prevent the establishment of a resident microbiota within the circulatory system. Transient bacteremia (bacteria in the bloodstream), however, can occur throughout the circulatory system, even in healthy individuals. For example, following some dental procedures, such as oral prophalaxis (cleaning of teeth) and tooth extractions, bleeding of the gums results in a transient systemic bacteremia lasting about 24 hours. Neutrophils within the blood are primarily responsible for the rapid removal of bacteria that enter the circulatory system in this way. Additionally bacteria may enter the circulatory system from breaks in the skin (for example, wounds and medical procedures) and from infected areas. Systemic infections occur when these defenses fail and pathogens circulate throughout the body.

Diseases of the Cardiovascular and Lymphatic Systems

Diseases Caused by Viral Pathogens

Leukemia and Other Forms of Cancer. Although most forms of human cancer have not been shown to be associated with viruses, a few types of human cancer are apparently caused by viruses. Cancer is the result of uncontrolled reproduction of malignant cells. When the malignancy occurs in the white blood cells the disease is called leukemia. Leukemia is characterized by the excessive proliferation of white blood cells. As with other forms of cancer, what causes transformation of blood cells to become malignant is not fully understood. Various factors such as exposure to carcinogenic chemicals can transform normal cells into malignant cells.

Several viruses have been found that can transform normal human white blood cells into malignant cells. Viruses that cause malignancies are called oncogenic viruses. The particular viruses that have been demonstrated to cause human leukemias are designated human T-cell leukemia viruses (HTLV-I and HTLV-II). The condition produced by these viruses is a T-cell lymphoma in which an excessive number of lymphocytes is produced. These human T-cell leukemia viruses are RNA viruses that reproduce by initially reversing the transcription process to form a DNA molecule that then acts as a template for the production of new viruses. Viruses that employ a reverse transcriptase for their reproduction are called retroviruses. In several cases the segments of the DNA that are produced by the reverse transcriptase have been found to be virtually identical to oncogenes; that is, they are identical to naturally occurring animal cell genes that code for the transformation of the cell into a cancer cell line. Retroviruses are known to produce various types of cancer in lower animals, including mammary tumors in mice, leukemia in cats and chickens, and sarcomas in birds, monkeys, and rodents.

In addition to the retroviruses, some DNA viruses cause malignancies in humans and lower animals. Here again there appears to be an identity between viral genes and animal cell oncogenes. The herpes viruses are among those DNA viruses that have been associated with human cancers. There is a high statistical correlation between genital herpes in adolescents caused by herpes simplex II virus and the later occurrence of cervical cancer. The Epstein–Barr virus, which is a herpes virus known to cause infectious mononucleosis, also appears to be associated with the occurrence of nasopharyngeal cancer in southeast China and of Burkitt's lymphoma. In this disease the excessive production of lymphocytes becomes localized in tissues of the lymphatic system. Most commonly Burkitt's lymphoma develops into a tumor of the lower jaw; other organs, including the thyroid, liver, and kidneys, may also be the site of tumor development. Burkitt's lymphoma is prevalent in a restricted region of Africa. Children in Uganda, as well as Caucasians working there, exibit an abnormally high incidence of Burkitt's lymphoma. The Epstein–Barr virus has been observed in tumor tissues and has been demonstrated to induce tumors in monkeys. Thus, both DNA and RNA viruses appear to be capable of transforming animal cells. Clearly there is a relationship between oncogenic viruses and oncogenes of animal cells. Several factors may induce the development of malignancies, and the mechanisms by which transformation occurs remain to be elucidated.

Infectious Mononucleosis. In addition to its association with Burkitt's lymphoma, the Epstein–Barr (EB) virus causes infectious mononucleosis. Burkitt's lymphoma and infectious mononucleosis appear to be very different manifestations of infections with Epstein–Barr virus, and there does not appear to be any increase in the incidence of malignancy in individuals who contract infectious mononucleosis. The EB virus occurs in oropharyngeal secretions of infected individuals. The virus appears to be transmitted primarily by direct and indirect exchange of oropharyngeal secretions containing the EB virus. Sharing a drinking cup or eating utensil can be involved in transmission of this disease. Infectious mononucleosis most commonly occurs in young adults 15–25 years of age, a fact that is explained by the exchange of saliva during kissing, a prevalent activity often involving more partners for this age group than others.

A little time for laughter
A little time to sing
A little time to kiss and cling,
And no more kissing after.

—PHILIP BOURKE MARSTON, *After*

In the course of infectious mononucleosis, mononuclear white blood cells are affected, leading to the characteristic changes in the white blood cells used in diagnosing the disease (Figure 27.2). The symptoms of infectious mononucleosis include a sore throat, low grade fever that generally peaks in the early evening, enlarged and tender lymph nodes, general tiredness, and weakness. The liver and

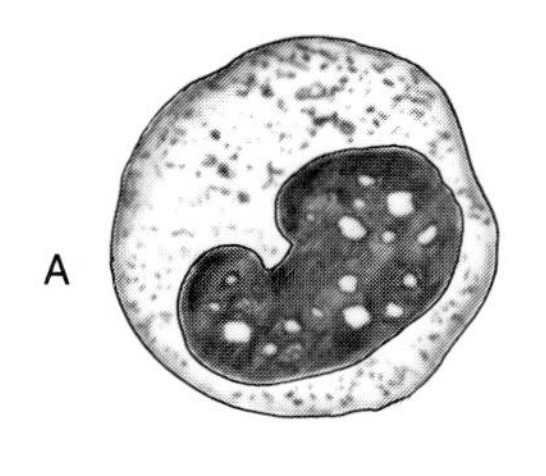

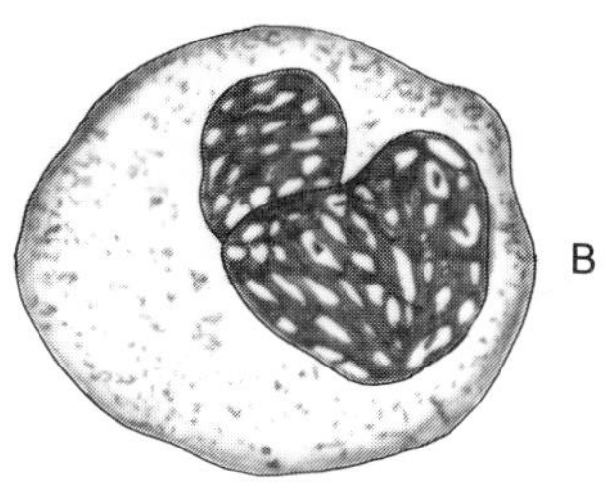

Figure 27.2 (A) Drawing of a normal monocyte. (B) Drawing of an atypical monocyte, indicative of infectious mononucleosis; the monocyte is enlarged, pleomorphic, and vacuolated.

spleen may also be affected by this condition. In most cases of infectious mononucleosis the symptoms are relatively mild and the acute stage of the illness lasts less than 3 weeks. In young children, for example, infectious mononucleosis often is very mild or asymptomatic.

Once infected the virus nomally remains within the body indefinitely. Up to 90 percent of people in the United States have antibodies indicating prior infection with Epstein–Barr virus. Recently it has been reported that chronic active Epstein–Barr virus infection can occur, producing flu-like symptoms of fatigue, headache, depression, sore throat, fever, aches, and pains. Acyclovir may be useful in treating this condition.

Acquired Immune Deficiency Syndrome (AIDS). Acquired immune deficiency syndrome or AIDS was first detected in the United States in 1979. The disease was detected because of an unusually high incidence of relatively rare diseases, such as Karposi's sarcoma and *Pneumocystis* pneumonia among homosexual men in major cities of the United States. It was found that these individuals had a deficiency in their cell-mediated immune systems. Specifically, individuals with AIDS had more T suppressor cells than T helper cells, effectively shutting off the immune response network.

AIDS has been shown to be caused by a human T-cell leukemia virus, designated HTLV-3 (Figure 27.3). Presumably the virus infects T helper cells, leading to the death and depletion of these cells that form a necessary part of the healthy immune defense network. However, the virus has been found to be associated with a large number of individuals who never develop AIDS. The reason why some people develop this disease and others do not is not known at this time.

The incidence of AIDS has been increasing in epidemic fashion in the last few years (Figure 27.4). There are certain groups that have exhibited a particularly high incidence of AIDS (Table 27.1). The high-risk groups include homosexual men, drug abusers, hemophiliacs, and Haitians. It is not clear why Haitians exhibit a high incidence of AIDS. Intravenous drug users are at risk because the virus can be transmitted through blood. The disease can also be transmitted by sexual contact, which is the main means of transmission among homosexuals within infected communities. Heterosexual transmission also occurs, and if AIDS develops during pregnancy it can be transmitted through the placenta to the fetus. Unlike other sexually transmitted diseases, it appears that prolonged or repeated ex-

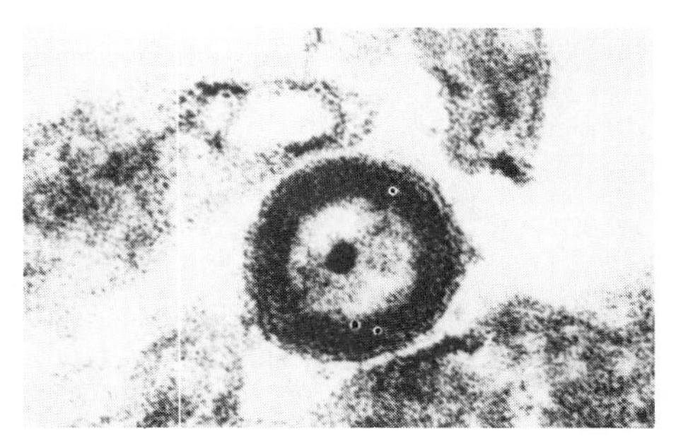

Figure 27.3 Electron micrograph of HTLV-3. (From Photoresearchers Inc.: R. Gallo, National Cancer Institute.)

Figure 27.4 Graph showing incidence of AIDS.

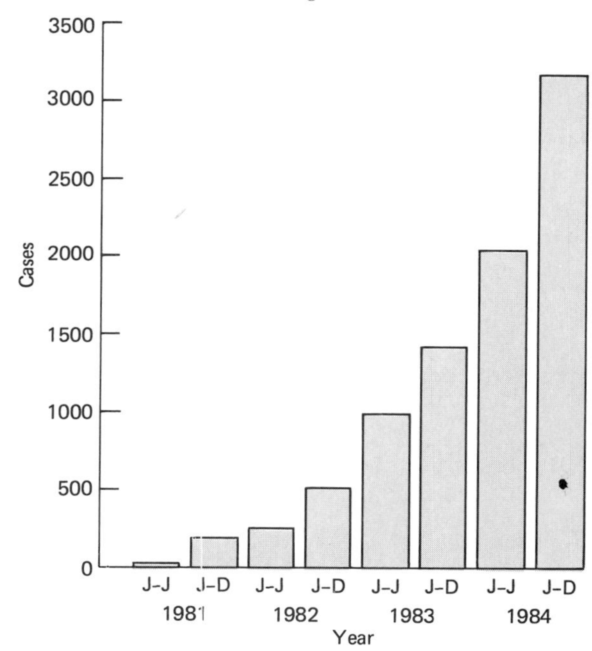

The Age Old Fear of Contagion Arises When Treating AIDS Patient

By PERRI KLASS

The patient has AIDS. Therefore, in obedience to the signs posted on his door, we are putting on gloves and masks. Finally we troop into his room and stand around his bed, our faces obscured by surgical masks, our hands encased in plastic skins. The resident looks down at the sick man in the bed, and suddenly sounds a little bit abashed. "Um, this is hospital policy," he says. "We're all wearing these masks because that's the policy when anyone has a serious infection." So the patient is questioned through masks and responds to the faceless doctors gathered around him, and then he is touched through thin plastic. Then we troop out of the room, strip off our masks and gloves and, one after another, solemnly wash our hands.

Recent evidence suggests that in addition to the known routes of sexual contact and blood transfusion, AIDS may be transmitted through saliva, so it makes sense to take precautions when you do oral or rectal exams, when you draw blood, when you handle secretions. But no one thinks you can get AIDS from listening to a patient's lungs through a stethoscope or taking his pulse. In particular, you can't get AIDS by being the medical student standing in the back of the room while the resident and the interns ask the patient questions and listen to his lungs. Indeed, there is no evidence that any medical person has contracted the disease from taking care of an AIDS patient.

• • •

Why, then, are we all wearing masks and gloves? Sometimes we're told it's a reverse precaution—we're wearing the masks to protect the patients, who have severely compromised immune responses, from catching diseases from us. On the other hand, the patient's family and friends never wear masks, not at home and not when they visit the hospital. The gloves make no sense at all in this context. So in the end we have to face the fact that we are going through these little rituals of sanitary precaution partially because we are all terrified of this disease and are not willing to listen to anything our own dear medical profession may tell us about how it actually is or is not transmitted.

Medical people are sometimes unwilling to accept that deadly infections persist. It does not seem right to us. Throughout history there have been deadly infectious diseases, epidemics and plagues, microbes that disrupted whole civilizations, germs that killed children. Nowadays, in most of the world, this is still true. It was true when my parents were growing up in New York City; polio epidemics left empty seats in the classroom. When I was little and my parents talked about those times, they seemed very remote to me, at least as remote as Beth's death in "Little Women." I asked my parents, "Are there still sicknesses I can catch which could kill me?" No, I was assured, because now medicines can cure all those sicknesses, and vaccines can keep you from catching them.

In recent years, for at least a certain privileged segment of the world's population, epidemics of deadly, incurable infectious diseases have not been major threats. Many diseases are dangerous to already debilitated people. Exceptionally severe infections can carry off children—or indeed anyone. But we don't live in fear of a plague.

Watching myself and the other medical people in the hospital, I wonder if we are able to accept what has always been a central fact of the medical profession: doctors are exposed to infections. When those infections are deadly, when they are beyond the reach of vaccines and antibiotics, we are still exposed to them. When you are talking about colds and sore throats you are talking about a mild occupational hazard, an annoyance, and when you are talking about incurable diseases, then you are talking about what you might fairly call the heroism of the profession. But either way, the concept of medical professionalism does not allow for a refusal to take care of someone because you are afraid of catching a disease.

It's legitimate to feel slightly nervous drawing blood from someone with AIDS or hepatitis B or any other disease that could, theoretically, be transmitted by an accident with a needle. It's legitimate to put on a mask before examining someone with infectious tuberculosis. But when you come to the AIDS patient's door, meaning to ask him a question or two, and find a box of gloves and a box of masks and even a box of disposable surgical gowns sitting there waiting for you, and most especially when you find yourself willingly complying with the whole rigamarole and washing your hands afterward, then you have to acknowledge that something less legitimate is going on.

• • •

On the simplest level, doctors are getting spooked, responding not to a new disease but to the ancient fear of contagion, and responding with an atavistic terror that, paradoxically, uses as charms all the most modern disposable sterile safeguards. I talked with medical students who worked in other hospitals, and we all agreed that the precautions are highly erratic, varying from hospital to hospital and even within a given institution, from floor to floor, from patient to patient—local rituals that bring comfort and security to the staff.

What do they do to the patient? Masks and gloves tend to remove what little human contact is provided by the attentions of the doctor; as an intern said, "It's like he's in surgery, except he's awake."

I talked with a medical student who watched a good friend die of AIDS this year. The student, whose perspective lay somewhere between nondoctor and doctor, viewed the precautions partly in terms of the delicate politics of the hospital. When the policy was masks he wore a mask, specifically not to antagonize the people taking care of his friend. He also felt strongly that the precautions were the doctors' way of creating distance between themselves and the patient: "It's a way of saying, you have AIDS and I don't, you're gay and I'm not, you're going to die and I'm not, and I'm not gonna get attached to you."

This, of course, is a perennial issue for doctors, this creation of distance, and there are many methods, tangible and intangible. Every dying patient is by definition a reminder of mortality. When that patient is dying because of an infectious agent, and the mortality is theoretically communicable, the need for distance may transcend anything that can be established with emotional dead space. Physical barriers are needed, not because doctors think they will actually protect, but as comfortingly palpable extensions of those mental mechanisms. The fear that an epidemic like AIDS engenders may drive doctors beyond the bounds of consistency and rationalism, perhaps because they seem to come out of a more frightening past, when medicine had less power over microbes.

Source: The New York Times, Oct. 25, 1984. Reprinted by permission.

posure is necessary to cause AIDS. Although the virus has been detected in the saliva of infected individuals, direct transmission of the disease by kissing, airborne droplets, and via eating utensils has not been demonstrated. It is likely that the apparent necessity for extended exposure precludes such means of transmission by such casual contacts.

Individuals with AIDS are subject to infection by a wide variety of disease-causing microorganisms and to development of a form of cancer called Kaposi's sarcoma. Generally one opportunistic infectious disease follows another in victims of AIDS until death occurs. The onset of the disease is sudden; early symptoms include low-grade fever, swollen glands, and general malaise. Early treatment with antibiotics can offer protection against infection, and interferon is being tested for control of Kaposi's sarcoma. AIDS itself is incurable at this time, and at best patients with this disease can be treated with passive immunization and antibiotics to protect them against life-threatening diseases.

Clearly AIDS is a serious disease. Our limited understanding of the routes of transmission of HTLV-3 viruses and the factors that lead to development of AIDS has led to excessive public reactions. In some hospitals staff members have refused to treat patients with AIDS. Some hospital administrators have responded by intentionally concealing occurrences of AIDS in their facilities. Schools have refused admission to children with this disease. Establishments in San Francisco catering to the homosexual community have been closed on the grounds that they constitute a public health hazard. These reactions are reminiscent of those in our dark past when people labeled lepers were driven from society. Until we fully understand how to treat and prevent the spread of AIDS, unfounded public reactions are likely to continue. Once we understand the disease, we will undoubtedly find a method for treating AIDS like other infectious diseases.

Serum Hepatitis (hepatitis B and hepatitis nonA–nonB). Hepatitis caused by hepatitis B virus and by hepatitis nonA–nonB virus is typically transmitted via serum. The clinical manifestations of serum hepatitis may include the development of jaundice and generally are very similar to those described for hepatitis A infections. The time course of the disease, however, is much more prolonged. Approximately 25,000 cases of hepatitis B and hepatitis nonA–nonB viral infections are reported each year in the United States. Many cases of hepatitis, however, are unreported and the United States Centers for Disease Control estimate that 100,000 new cases of heptitis B and nonA–nonB hepatitis may actually occur each year in the United States alone. The incidence is much higher in Africa and Asia.

The normal route of transmission of hepatitis B and hepatitis nonA–nonB is by any transfer of virus-carrying blood from one individual to another. These hepatitis viruses may also be transmitted via secretions such as saliva, sweat, breast milk, and semen. Sexual, oral, and mosquito vector transmission also occur. The incubation period is about 60–180 days, which makes tracing the origin of an outbreak difficult. Many cases of hepatitis are unspecified in terms of the type of hepatitis and source of transmission.

Blood transfusions and contaminated equipment such as syringes are common modes of transmission. There is a high rate of transmission of serum hepatitis among drug addicts, who frequently use

Table 27.1 Adult AIDS Patients in the United States Through November 1984 (Based on data from Morbidity and Mortality Weekly Reports (MMWR), Centers for Disease Control)

Patient Group	Total Number of Cases	Percent of Total Cases
Homosexual/bisexual	5038	72.8
IV drug user	1190	17.2
Haitian	249	3.6
Hemophilia patient	46	0.6
Heterosexual contacts	54	0.8
Transfusion recipients	81	1.2
Other	263	3.8

Mysterious Form of Hepatitis Seen as Widespread Threat

By LAWRENCE K. ALTMAN

Scientists have discovered a deadly and mysterious form of hepatitis caused by two viruses that act in concert in a manner never previously detected.

The disease, called delta hepatitis, poses "an ominous specter for much of the world," according to a team of American and Venezuelan researchers who reported a devastating outbreak of the disease among Venezuelan Indians.

It is estimated that hundreds of thousands of people the world over have the newly diagnosed form of hepatitis and that perhaps 200 million more are at high risk either because they have hepatitis B or are latent, but otherwise healthy, carriers of it. However, precise figures are not known.

Delta virus has been detected in all areas of the United States, according to surveys by the Red Cross and Abbott Laboratories.

As many as 800,000 people in the United States are carriers of hepatitis B virus and thus vulnerable to delta infection, according to Dr. Stephen C. Hadler, an epidemiologist at the Centers for Disease Control in Atlanta, who participated in the investigation of the Venezuelan outbreak.

Doctors have diagnosed several delta hepatitis cases in Los Angeles. Since last September seven cases—three of them fatal—have been reported among drug users in Worcester, Mass. But no cases have been reported to the New York City health department.

The discovery that delta hepatitis results from the interaction of two viruses raises the prospect that similar viral combinations, called piggybacking, might be at the root of other diseases for which infectious causes have long been sought.

Delta hepatitis results from a virus that cannot cause infection by itself. But when it piggybacks with another virus—the one that causes hepatitis B—the two viruses can cause an illness more severe than that caused by hepatitis B virus alone. And hepatitis B infection can be severe. Further, delta hepatitis can often become chronic and lead to cirrhosis, another incurable liver disease.

Severe outbreaks can occur anywhere hepatitis B is common, the American and Venezuelan doctors warn in the *Annals of Internal Medicine,* reporting on the team's investigation of an epidemic that killed 34 of 149, or almost 25 percent, of infected Yucpa Indians near Maracaibo. The epidemic left 22 Indians with chronic hepatitis.

The World Health Organization in Geneva expressed concern earlier this month about a need to learn more about which population groups are at highest risk of contracting delta hepatitis, more about how it is spread and what the long-term consequences are.

Researchers have determined that delta hepatitis can be spread by contaminated blood leading to infections among drug addicts and hemophiliacs. But it is not clear whether the disease is spread in other ways and, if so, how.

Hepatitis, a word that doctors use to describe an inflammation of the liver, can be caused by many infectious agents. The hepatitis A, hepatitis B and hepatitis non-A-non-B viruses are the chief ones in the United States, though the ailment can be caused by such other microorganisms as the Epstein-Barr, cytomegalovirus and yellow fever viruses and the leptospirosis bacterium.

Alcohol is one of many chemicals that can cause hepatitis.

The discovery of the delta virus does not mean a new treatment is imminent for delta hepatitis. Doctors are at present powerless to prevent or treat delta infections.

Moreover, little can be done now to prevent delta hepatitis among those who have already been infected with hepatitis B, or recovered from its symptoms but then became carriers of the virus. Nor can much be done for those who have become silent carriers of the hepatitis B virus without ever becoming clinically sick.

Accordingly, efforts are under way to identify components of the delta virus that induce immunity in the body and then to develop a vaccine against delta. Researchers do not know how soon one can be produced.

However, experts do believe that delta hepatitis could be avoided by widespread use of a marketed vaccine that prevents hepatitis B, one of the most common infections in the world. But the vaccine works only for people who have not developed hepatitis B. Also, although newer, synthetic vaccines are on the horizon, the existing one cannot be afforded in many of the countries where hepatitis B is most prevalent.

The delta virus itself, now called a defective virus because it cannot act alone, was discovered in 1977 by Dr. Mario Rizzetto in Turin, Italy. It is one of the smallest viruses known, not much larger than the viroids that cause disease in plants.

The discovery came through astute observations he made in a study of the livers of individuals whose blood tests showed they had hepatitis B infection. Hepatitis B virus contains antigens on its surface and in its core, and at the time doctors were trying to determine more precisely how they correlated with active infection.

Improving the Diagnosis

Dr. Rizzetto said in an interview that he had used a fluorescent test to identify the antigens present in the liver and a different test to detect antigens in the blood. By correlating the results he hoped to help improve the accuracy of the diagnosis of hepatitis.

Instead, he noted discrepancies in the matches and concluded that he had found a new antigen system. The discovery did not attract immediate attention.

"We were confused because we didn't know how to interpret the finding" except to say that the antigen was somehow linked with hepatitis B, Dr. Rizzetto said. "We could go no further in human studies."

In 1978 Dr. Rizzetto moved to the United States. He brought along his samples to pursue further research with Dr. John L. Gerin at Georgetown University in Washington and with Dr. Robert H. Purcell at the National Institutes of Health in Bethesda, Md.

Those collaborations led to ways to isolate the delta antigen, an important development in itself but one that proved crucial to the next steps in the research. Until then, all Dr. Rizzetto's team could do was use the fluorescent test on biopsies of the liver. Now, using the delta antigen, they could develop diagnostic blood tests and use them as well as liver biopsies to follow what happened to chimps injected with specimens obtained from a patient with delta infection.

'The Crucial Experiment'

From these animal studies, the researchers reported in 1980 that the delta antigen was actually a virus that would produce disease in animals only under specific conditions. Specifically, the delta virus caused illness only in the chimpanzees infected with hepatitis B. "This was the crucial experiment, and everything else then followed," Dr. Rizzetto said.

The researchers hypothesized that the delta antigen was a defective virus, one that could only replicate in the presence of hepatitis B virus. The researchers continued collaborating when Dr. Rizzetto returned to Italy. Dr. Gerin and Dr. Purcell, working with other members of Dr. Rizzetto's team, pursued research characterizing the molecular biology of the virus and its epidemiological and clinical features based on samples sent by colleagues around the world.

Soon the researchers noted an unusually high incidence of delta infection in southern Italy, a finding that, though puzzling, did not spark much immediate interest among public health officials. As Dr. James E. Maynard, chief hepatitis epidemiologist at the Centers for Disease Control, put it: "Until recently we didn't know quite what to make of it—whether it was just a bizarre disease in southern Italy or what it meant from a global perspective."

Clues From Other Nations

Clues to the significance of delta infection have emerged from several studies throughout the world. A dramatic one came from a team headed by Dr. Bengt G. Hansson in Malmo, Sweden. His team documented the introduction of the delta agent into the Swedish drug addict community in 1973 and then showed how it steadily infected more addicts to the point where it now has reached 72 percent of the members tested.

However, it was the epidemic among Venezuelan Indians that has drawn the attention of public health experts, because it was so devastating and was the first outbreak in a general population.

"The delta virus causes a lot of disease, but how much disease we don't know yet," said Dr. Gerin, the Georgetown virologist who is a leading expert in the subject.

It has become clear that delta infection occurs throughout the world. Infections have been reported from Colombia, the Amazon basin in Brazil, and other areas in northern South America. Evidence of it also has been found in Eastern Europe, the Soviet Union, the Middle East, sub-Saharan Africa, and most recently in Western Samoa and Nauru and Nuie islands in the South Pacific.

Scientists who have been trying to fill in the delta map have noted "some quirks," said Dr. Purcell, the National Institutes of Health researcher. One quirk is that delta appears relatively rare in Asia, where hepatitis B is so prevalent.

"How did delta get to Pacific islands and escape other Asian areas where there also is a lot of hepatitis B?" Dr. Hadler of the Federal Centers for Disease Control asked.

An Old Disease Newly Detected

All available evidence indicates that delta hepatitis is a newly recognized disease that has been around for a long time. It was present in the United States during World War II, according to recent tests performed by Dr. Antonio Ponzetto at Georgetown on samples of gamma globulin saved for about 40 years. Gamma globulin, which is used to provide temporary protection against some infections, is derived from blood donated by thousands of people.

Severe hepatitis has long been diagnosed in the Amazon basin of Brazil and in the Santa Marta region of Colombia. Much of that hepatitis is now believed to be caused by the delta virus, according to research done by Dr. Karin Ljunggren and other researchers at Georgetown and the National Institutes of Health.

Discovery of the delta virus has also provided doctors with a new perspective on a condition usually called chronic active hepatitis and one they have long diagnosed. Many people with this condition suffer what appear to be sudden relapses, a phenomenom that has puzzled doctors.

Now, it appears that some such relapses are due to infection with the delta agent, and studies are under way to gain more precise statistics.

A Problem of Blood Tests

Undoubtedly, many cases of delta hepatitis have gone undiagnosed in the United States and elsewhere because a blood test has not been generally available beyond the ones devised by a few researchers for their own use. Abbott Laboratories now distributes a delta test but only to researchers doing epidemiologic and clinical studies. However, Cheryl Staruck, an official of the drug company in North Chicago, Ill., said Abbot expects to market a diagnostic delta test that would be available to all physicians by the fall of 1985.

Although delta and hepatitis B viruses must link to produce delta infection, researchers said they are not certain that each of the two infections is spread in exactly the same way. One possible reason, Dr. Purcell said, is that the delta agent may be more fragile than the hepatitis B virus.

Another mystery is the scarcity of delta infection among homosexuals, a group in which hepatitis B is very common, Dr. Cladd E. Stevens of the New York Blood Center said Dr. Ponzetto at Georgetown has tested samples from the New York homosexual community for the delta agent but has not found any cases.

Scientists are now looking to the woodchuck to solve some of the mysteries of delta hepatitis, since they found that they could experimentally infect woodchucks that were infected with a virus closely related to human hepatitis B virus.

Further experiments with those animals will be part of the research needed in the future to learn more about the molecular virology of the delta agent, to know how it depends on the helper virus and the mechanism by which it works.

Source: The New York Times, Aug. 28, 1984. Reprinted by permission.

contaminated syringes. Perhaps 10 percent of those infected with hepatitis B become chronic carriers. The carrier rate of hepatitis B virus in blood donors in the United States appears to be between 1/2 and 1 percent, but in other countries the carrier rate may be as high as 5 percent. The surface envelope antigen of hepatitis B virus, the Australian antigen, can be detected in the blood serum of an infected individual. The blood of an infected individual may remain infective for months or years.

The major type of hepatitis associated with transfusions in the United States is not hepatitis B but rather nonA–nonB hepatitis. NonA–nonB hepatitis indicates that the virus does not have the major surface antigens of either type A or type B hepatitis viruses. Proper screening of blood for hepatitis viral antigens has reduced the transmission rate of hepatitis B but not of hepatitis nonA–nonB. This is because detection methods for screening blood for hepatitis B antigens are available but screening procedures for antigens associated with hepatitis viruses that are not type A and not type B have yet to be developed.

Health care workers who contact blood from patients have a considerably higher incidence of serum hepatitis than does the population at large. Doctors, for example, have five times, and dentists two or three times, the normal rate of infection. The mortality rate from hepatitis B normally is low, usually less than 1 percent for the population at large, but it is higher for certain groups, such as the elderly, and during certain outbreaks. Recently it has been shown that a severe outbreak of hepatitis B in Massachusetts actually involved a simultaneous infection with both hepatitis B virus and a delta hepatitis virus.

Special precautions are used by health care workers to limit the transmission of serum hepatitis. Whenever possible, disposable equipment should be used for routine hospital and clinical procedures and these items should be properly sterilized before being discarded. Nondisposable syringes, needles, tubing, and other types of equipment used for obtaining blood specimens or for the administration of therapeutic agents should be adequately sterilized before reuse. Most common means of chemical sterilization are not reliable for destroying hepatitis-causing viruses. Boiling in water for at least 20 minutes, heating in a drying oven at a temperature of 180°C for 1 hour, and autoclaving are effective in destroying hepatitis viruses. Additionally individuals with a history of jaundice, one of the clinical symptoms of hepatitis, are generally rejected as blood donors. Blood is screened for known hepatitis antigens. Unfortunately, no method used today is effective in rendering whole blood containing hepatitis virus totally free of the agent. Plasma can be treated with the viricide betapropiolactone to inactivate any hepatitis viruses that may be present.

When a health care worker or other individual is believed to have been exposed to blood containing hepatitis viruses, immune prophylaxis is used as a precautionary measure. Both IgG and hepatitis B immune globulin are used for the prevention of hepatitis B infection. Hepatitis B immune globulin has a high level of immune globulins and often is recommended for health care workers who accidently stick themselves with a needle or whose mucous membranes are exposed to blood containing hepatitis viruses.

Additionally health care workers as well as other high-risk individuals today can be vaccinated against hepatitis B. The vaccine for hepatitis B is produced from viral particles isolated from the blood of human carriers of the disease. This vaccine, made available in 1982, was the first completely new viral vaccine in a decade and the first ever licensed in the United States made directly from human blood. The production of this vaccine involves a 65-week cycle of purification and safety testing to preclude the inclusion of viruses or other undesired factors from the blood of hepatitis B carriers. The cost is about $100 for the three doses needed to establish immunity. Several recent advances in genetic engineering, however, indicate that a second-generation hepatitis B vaccine may be available soon at a considerably lower price. At present the approved hepatitis B vaccine is intended for high-risk individuals, including health care workers and drug addicts.

Dengue Fever. Dengue fever is caused by an RNA virus transmitted by the mosquito *Aedes aegypti*. Outbreaks of dengue fever occur in the Caribbean, South Pacific, and Southeast Asia (Figure 27.5). There are several different serotypes of dengue virus and three distinct disease syndromes: classic, hemorrhagic, and shock. A characteristic rash and fever develop during all forms of this disease. The classic form of dengue is a self-limiting disease, but in the hemorrhagic and shock syndromes the fatality rate can reach 8 percent in children.

The dengue virus replicates within the circulatory system, causing viremia (viral infection of the bloodstream) that persists for 1–3 days during the febrile period. Multiplication of the dengue virus within the circulatory system causes vascular damage. It has been postulated that immune reactions

Figure 27.5 Incidence of dengue fever in Puerto Rico in 1977 and 1978.

contribute to the formation of complexes that initiate intravascular coagulation or hemorrhagic lesions. Previous exposure to dengue virus and the presence of crossreacting antibody seem to be important in determining the severity of the disease symptomology. Damage to the circulatory vessels appears to occur when antigen–antibody complexes activate the complement system with the release of vasoactive compounds.

Yellow Fever. Yellow fever also is caused by a small RNA virus transmitted by mosquito vectors, predominantly by *Aedes aegypti*. There are two epidemiologic patterns of transmission. Urban transmission involves vector transfer by *Aedes aegypti* from an infected to a susceptible individual, and jungle yellow fever normally involves transmission by mosquito vectors among monkeys, with transfer via mosquito vectors to human beings representing an occasional deviation from the normal transmission cycle (Figure 27.6). Outbreaks of yellow fever were a major problem in the construction of the Panama Canal, leading to Walter Reed's instrumental work in establishing the relationship between yellow fever and mosquito vectors in 1901.

He gave to man control over that dreadful scourge, yellow fever

—Epitaph on Walter Reed's tomb at Arlington National Cemetery

Today, yellow fever occurs primarily in remote tropical regions, and current outbreaks of yellow fever occur primarily in Central America, South America, the Carribean, and Africa (Figure 27.7). The onset of yellow fever is marked by anorexia (loss of appetite), nausea, vomiting, and fever. The multiplication of the virus results in liver damage, causing the jaundice from which the disease derives its name. The symptoms generally last for less than 1 week, after which either recovery begins or death occurs. The fatality rate for yellow fever is about 5 percent but may be as high as 40 percent in some epidemics. There is no effective antiviral drug at present for treating yellow fever, but the disease can be prevented by vaccination. There are two strains of yellow fever virus that are used for vaccination. The 17D vaccine uses an attenuated strain of yellow fever virus produced by culture in chick embryos and is administered subcutaneously. The

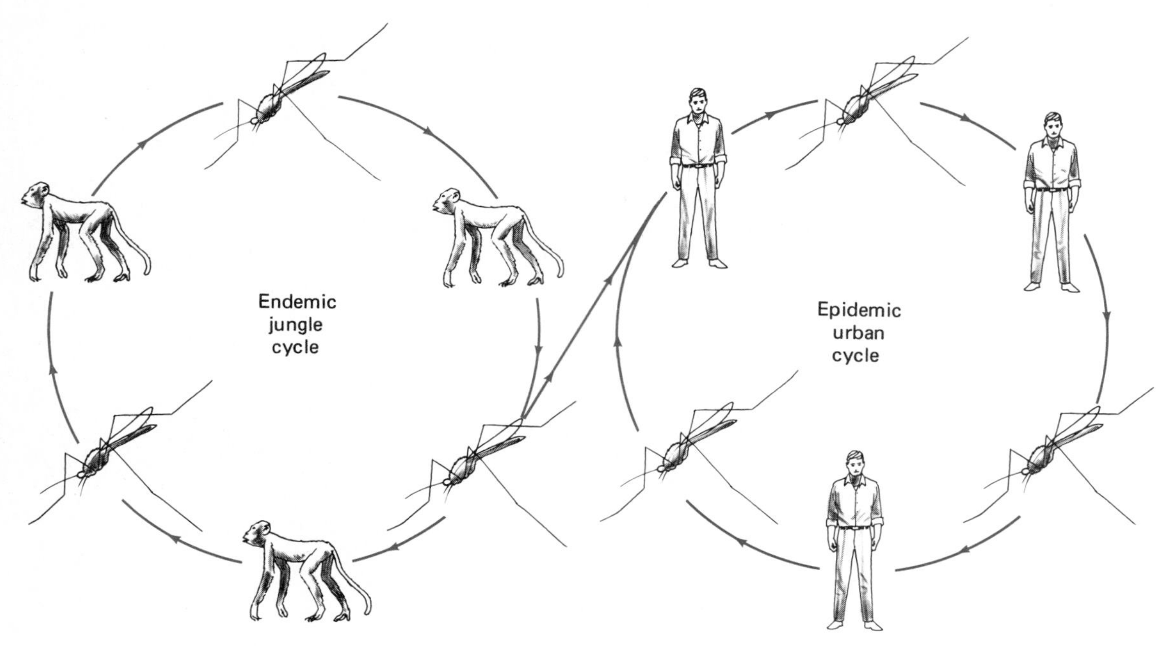

Figure 27.6 Transmission cycle patterns for yellow fever, showing the relationship between endemic and epidemic transmission cycles of the disease.

Dakar strain is grown in mouse brain tissue and is introduced by scratching into the skin. The urban form of transmission has been largely controlled by effective mosquito eradication programs, but the jungle form of transmission cannot easily be interrupted because of the large natural reservoir of yellow fever viruses maintained in monkey populations.

Figure 27.7 The geographic distribution of yellow fever, showing regions where it is epidemic in South America and Africa.

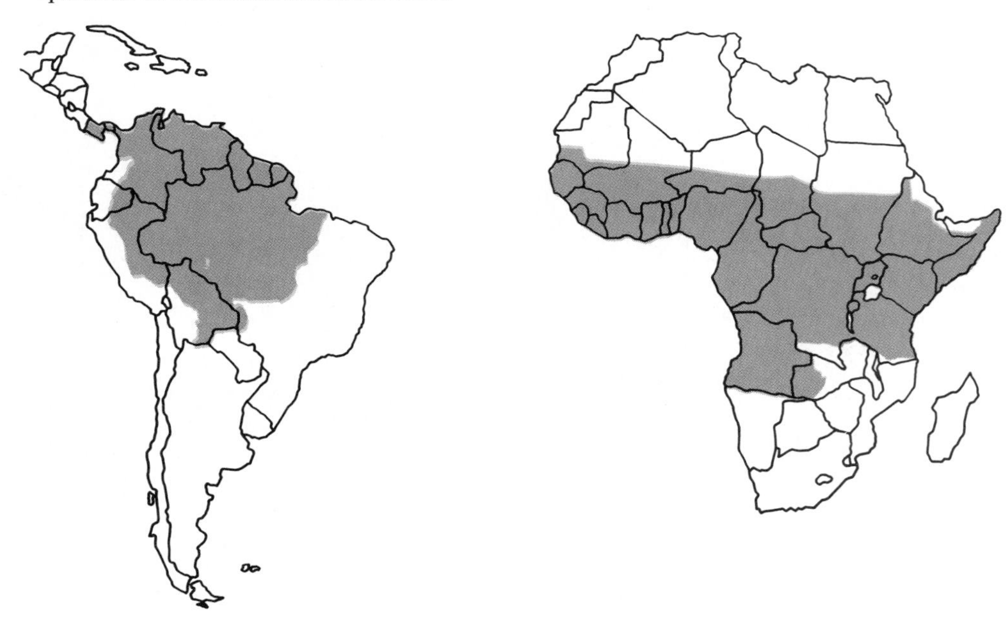

Diseases Caused by Bacterial Pathogens

Septicemia. When microorganisms overcome the defenses of the body and reproduce within the cardiovascular and lymphatic system, they can produce a condition called septicemia or blood poisoning. Most cases of septicemia are associated with Gram negative bacterial infections of the blood. *Escherichia coli, Enterobacter aerogenes, Pseudomonas aeruginosa, Serratia marcescens*, and *Proteus* sp. are frequently identified as the causative agents. These are opportunistic bacterial infections that result when the immune defense systems fail.

The symptoms of septicemia are often associated with the release of lipopolysaccharides (endotoxin) from the cell walls of these Gram negative bacteria when they lyse in the circulatory system. The release of large amounts of endotoxin within the circulatory system damages blood vessels. This causes inflammation of the cardiovascular and lymphatic systems. Characteristically red streaks appear under the skin along the arms and legs due to the inflammation of the lymph vessels (lymphangitis). There is also a high fever and a drop in blood pressure. The resulting low pressure results in shock and may be fatal. Treatment of septicemia must be supportive, maintaining body functions and limiting the effects of shock. Antibiotics are used to eliminate the infecting bacteria; in some cases, however, this treatment may initially exacerbate the disease by causing the lysis of more bacteria, with the release of additional endotoxin into the circulatory system. Therefore, the use of bacteriostatic antibiotics is sometimes used, as these allow phagocytic killing and reduce the amount of endotoxin released from the cells into the circulatory system. Often treatment is prolonged and may last 4–6 weeks.

Rheumatic Fever. Rheumatic fever is a serious disease that can result from *Streptococcus pyogenes* infections. The symptoms of rheumatic fever normally begin to occur a little over 2 weeks after a characteristic sore throat, associated with an upper respiratory tract infection with *S. pyogenes* that causes pharyngitis. In rheumatic fever the systemic spread of *S. pyogenes* toxins affects multiple body sites. The symptoms of this disease vary, but characteristically there is a high fever, painful swelling of various body joints, and cardiac involvement, including subsequent development of heart murmurs due to permanent damage to heart valves in affected children. Because of the serious manifestations of rheumatic fever, it is important to diagnose the etiologic agents of sore throats in children. Penicillin is effective in treating group A streptococcal infections as well as other streptococcal infections, and its use in treating streptococcal pharyngitis can prevent the occurrence of rheumatic fever. Children sometimes are given penicillin as a prophylactic measure to prevent occurrences of rheumatic fever.

The specific causal relationship between *S. pyogenes* and the symptoms of rheumatic fever has not been established. It is possible that antibodies produced in response to a group A streptococcal infection are crossreactive with cardiac antigens and that it is an "autoimmune" or immune-complex response that actually results in cardiac damage. The treatment of rheumatic fever, therefore, includes the use of anti-inflammatory drugs to reduce tissue damage as well as the use of antibiotics to remove the infecting streptococci if there is still evidence of an active streptococcal infection.

Carditis. In addition to cardiac involvement as one manifestation of rheumatic fever, various microorganisms can infect the heart, causing an inflammation of the heart called carditis. Infections of the heart are divided into endocarditis, myocarditis, and pericarditis depending upon which specific portions of the heart are involved. Each of these conditions is different in terms of etiology and pathology.

Infective endocarditis is an infection of the endocardium, a specialized membrane of epithelial and connective tissue that lines the cardiac chambers and forms much of the heart valve structure. Acute endocarditis occurs when *Staphylococcus aureus* invades the heart. Generally death occurs in days to weeks as a result of this type of infection.

Subacute endocarditis most frequently occurs when bacteria that are normally indigenous to body surfaces enter the circulatory system and attach to cardiac tissues. Microbial growth on the heart tissues causes serious abnormalities, and if untreated is often fatal (Figure 27.8). Subacute endocarditis generally has a time course of several months to years. Over 50 percent of the reported cases of endocarditis in the United States during the 1960s were attributed to streptococcal infections, although rickettsia and fungi also cause endocarditis. Viridans streptococci (a heterogeneous group of alpha hemolytic *Streptococcus* species that cannot be placed into other groups based upon specific antigens, negative on bile-esculin test, resistant to optichin disc) are the most frequently identified etiologic agents of endocarditis.

Subacute endocarditis is most common among individuals with underlying heart disease, such as rheumatic fever. Individuals with prosthetic heart

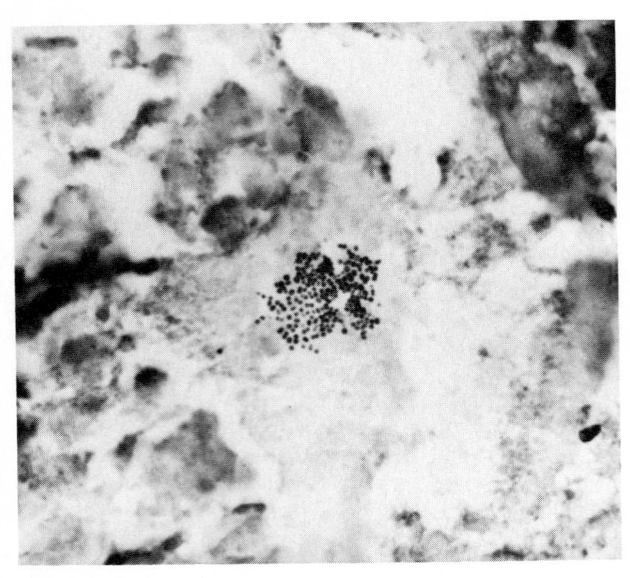

Figure 27.8 This micrograph shows a section of heart tissue infected with viridans streptococci. (Courtesy John J. Buchino, Norton's Kosair Children's Hospital, Louisville, Kentucky.)

valves also have an increased risk of developing subacute endocarditis. Because viridans streptococci are present in high numbers in the oral cavity, and dental treatments are often implicated in initiating entry of these bacteria into the blood stream, individuals with prosthetic heart valves are given prophylactic doses of antibiotics—generally penicillin—just before dental treatments.

Infective endocarditis normally begins with a gradual onset of fever and general tiredness over a period of days or weeks. High fever occurs in acute myocarditis and low grade fever is characteristic of subacute endocarditis. A heart murmur is detectable in 95 percent of the cases of endocarditis. Proper treatment depends upon rapid diagnosis and identification of the etiolgic agent. Treatment of bacterial infections of the heart tissues usually employs broad spectrum antibiotics that are not inactivated by β-lactamases and often requires prolonged administration of high doses of antibiotics to ensure removal of the infecting bacteria.

Myocarditis (inflammation of the myocardium, which is principally cardiac muscle) can also result from microbial infections. Most cases of myocarditis are caused by enteroviruses. The group B coxsackieviruses are most often implicated, although other viruses and bacteria can also cause this disease. Enteroviruses initially infect the gastrointestinal tract. They can enter the circulatory system and be transported to the heart where they cause this infective process. The disease is seasonal, with the highest incidence in the late summer and early fall. Clinical manifestions of myocarditis include chest pain, fever, joint pain, and muscle pain. Neonatal cases of myocarditis have a high mortality rate, but older children and adults normally recover. Treatment for viral myocarditis is supportive and relies upon the individuals' own immune defenses.

Typhoid Fever. Outbreaks of typhoid fever, a systemic infection caused by *Salmonella typhi*, are associated with contaminated water supplies and the handling of food products by individuals infected with *S. typhi*. A relatively low infectious dose is required for *S. typhi* to establish an infection. Although the portal of entry for *S. typhi* normally is the gastrointestinal tract, infections with this organism do not initially cause gastroenteritis; rather, the bacteria simply enter the body via this route and cause infections at other sites. The infecting bacteria rapidly enter the lymphatic system and are disseminated through the circulatory system. Phagocytosis by neutrophil cells does not kill *S. typhi*, and the bacteria continue to multiply within phagocytic blood cells. The surface Vi antigen of *S. typhi* apparently interferes with phagocytosis, and elimination of infecting cells depends on the antibody-mediated immune response.

After invasion of the mononuclear phagocyte system, the infection with *S. typhi* becomes localized in lymphatic tissues, particularly in Peyer's patches of the intestine, where ulcers can develop. Localized infections always develop and cause damage to the liver and gall bladder and sometimes also to the kidneys, spleen, and lungs. The symptoms of typhoid fever include fever (104°F), headache, apathy, weakness, abdominal pain, and a rash with rose-colored spots. The symptoms develop in a stepwise fashion over a 3 week period. If no complications occur, the fever begins to decline at the end of the third week. There is a high carrier rate because the bacteria can remain and multiply within infected phagocytic blood cells. The mortality rate averages 10 percent in untreated cases of typhoid fever. Chloramphenicol is effective in the treatment of this disease. Although there is reluctance to use chloramphenicol because it may cause aplastic anemia, its use and the use of other antibiotics has reduced the death rate due to typhoid to approximately 1 percent.

MICROBIOLOGY HEADLINES

Toxic Shock: A Close Brush With Death

By MARGARET ENGEL

When Olivia Massey had to crawl to her parents' room for help, she realized that her symptoms of nausea, dehydration, raging fever and chills were more than just a severe cold.

By 4 a.m. on Sept. 17, the 25-year-old Lanham market researcher was in the emergency room of Walter Reed Army Medical Center, where doctors worked around the clock to keep her alive.

Massey is the area's latest victim of Toxic Shock Syndrome, the rare, life-threatening disease that strikes at least several hundred people each year. Although publicity about the disease has faded, the infection has not.

It has afflicted 2,492 women and 133 men since the federal Centers for Disease Control in Atlanta began recording it as a unique infection in 1979. Of those stricken, 114 have died.

Olivia Massey was lucky. After 16 days in the hospital, she is at home, still weak, but recovering.

"I heard a nurse say, 'She has no blood pressure. Her heart has stopped beating,' " said Massey, whose skin is peeling as a reaction to the toxins. She has scars from the six tubes inserted in her chest, thigh, arms and nose.

"Fortunately, she responded to the agents we gave her to stabilize her blood pressure," said Dr. Robert Decker, the medical resident whose experience with three other toxic shock cases helped in diagnosing Massey's condition in the emergency room. "Hers was the classic case with vomiting, muscle aches and her heart beating fast and her blood pressure low, signifying shock."

The disease made headlines in 1980 when researchers found that it most often struck young women during or just after their menstrual periods. Tampons, particularly those of high absorbency and ones that were not changed frequently, were implicated. One national brand, Rely, was removed voluntarily from the market by its manufacturer, Procter & Gamble.

In the years that followed, scientists have found the culprit is a blood toxin produced by a bacterium called *Staphylococcus aureus*. The patient profile expanded as several older women and some men contracted Toxic Shock Syndrome, many from infections of surgical wounds.

But most of its victims are young women who use tampons. Recently, according to federal researchers, contraceptive sponges have been implicated in cases of Toxic Shock Syndrome.

A debate also has continued on the labeling of tampons, as the CDC and others have found highly absorbent tampons create more of a risk, perhaps by encouraging the production of the blood toxin.

A two-year effort by the FDA to persuade manufacturers to create a voluntary plan failed. The agency now is developing rules to force the nation's five tampon manufacturers to tell consumers the absorbency of their products. The rules are expected to be final next year.

In the meantime, the CDC has found that when the publicity surrounding the disease diminished, so did the number of doctors who reported it.

Margaret Oxtoby, an epidemiologist in the CDC division of bacterial diseases, said that only Wisconsin, Minnesota and Utah "have maintained strong interest in an active surveillance." As a result, the CDC determines the nation's incidence rate from those three states.

Decker said that an infection control committee at Walter Reed reports its cases of Toxic Shock Syndrome to federal authorities. But Oxtoby said the government's records show that only one case of Toxic Shock Syndrome has been reported to the CDC from the District of Columbia since 1979.

Virginia doctors and hospitals have reported 39 cases in those same years, and Maryland has noted 24 cases.

"We had much better case-finding in the past," said Oxtoby. "But it certainly is a continuing problem."

The problem remains rare, with an estimated 5 to 15 cases each year per 100,000 menstruating women, slightly fewer than the 6 to 17 cases per 100,000 estimated in previous years.

But many experts think that under-reporting is a serious problem. Jill Wolhandler of Boston, a participant in the tampon task force of the American Society for Testing and Materials, which is working with the manufacturers on absorbency ratings, said exact incidence cannot be determined "when we think only 10 percent of the cases are sent on to the federal government."

Whatever the numbers, for those who are stricken it is one case too many.

"This has ravaged me from head to toe," said Massey, a former overseas Army communications specialist who plans to return to Germany in January after she and her fiance are married. "I hadn't heard anything about toxic shock for a while and felt that if I've been using tampons all this time, they won't hurt me."

Massey was using Playtex Super Plus tampons when she became ill. They are the most absorbent brand on the market, according to a chart prepared by the manufacturers for the American Society for Testing and Materials.

Because the disease recurs in 30 percent of the cases, Massey has been instructed never to use tampons.

Her mother, Betty Massey, was so upset about her daughter's experience that she stood up in her church the following Sunday and urged the women in the congregation to give up tampons.

The experience of surviving such a serious threat has given Massey a new respect for emergency medicine.

"I will never forget those four days in intensive care," Massey said.

"I am so grateful to the doctors for saving my life. I was born at Walter Reed and they kept me alive, too."

Source: The Washington Post, Oct. 21, 1984. Reprinted by permission.

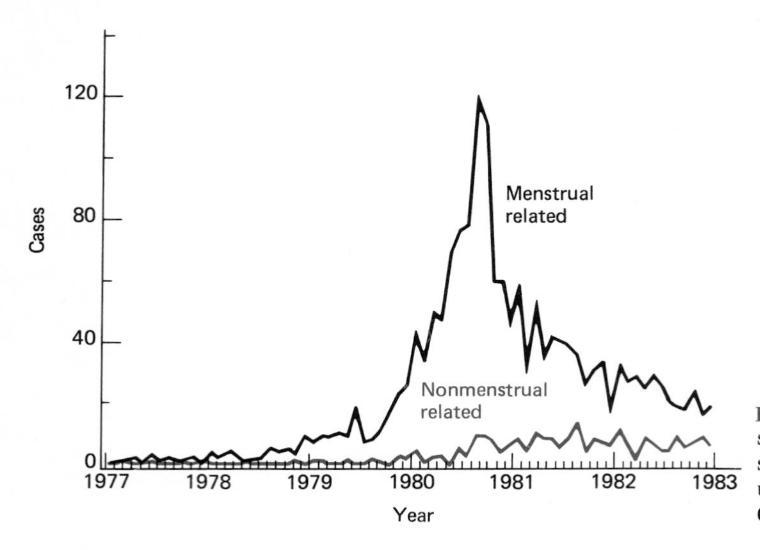

Figure 27.9 Reported cases of toxic shock syndrome showing the high incidence of menstrual related cases of this disease in 1980. (Based upon *Morbidity and Mortality Weekly Reports*, Centers for Disease Control, Atlanta.)

Toxic Shock Syndrome. Toxic shock syndrome is caused by strains of *Staphylococcus aureus*. The occurrence of toxic shock syndrome is especially correlated with the use of tampons during menstruation, particularly if these devices are left in place for a long period of time (Figure 27.9). The association of this disease with the use of tampons received a great deal of publicity in the early 1980s, forcing one major manufacturer to remove its product from the market. The fibres of some tampons absorb magnesium, permitting the proliferation of *Staphylococcus* with the production of high amounts of toxin. This disease is not restricted to women and can occur after introduction of *S. aureus* via other portals of entry, including surgical wounds. Toxins produced by strains of *S. aureus* cause high fever, nausea, vomiting, and in many cases death.

Puerperal Fever. Puerperal or childbed fever is a systemic bacterial infection that may be acquired via the genital tract during childbirth or abortion. The most frequent etiologic agents of postpartum sepsis are β-hemolytic group A and B *Streptococcus* species. *Staphylococcus, Pseudomonas, Bacteroides, Peptococcus, Peptostreptococcus*, and *Clostridium* species as well as other bacteria can also cause this disease. The source of infection normally comes from the obstetrician, obstetrical instruments, or bedding. The bacteria causing puerperal fever are not normally part of the resident microbiota of the vaginal tract. Prior to the introduction of aseptic procedures, puerperal fever was often a fatal complication after childbirth and remains an important complication following childbirth and abortion procedures. It was the leading cause of maternal death in Massachusetts in the mid-1960s. Penicillin is usually effective in treating postpartum sepsis. The use of proper obstetric procedures generally prevents this disease.

Plague. Plague is caused by *Yersinia pestis*, a Gram negative, nonmotile, pleomorphic rod that exhibits bipolar staining. *Y. pestis* is normally maintained within populations of wild rodents and is transferred from infected to susceptible rodents by fleas. *Y. pestis* is able to multiply within the gut of the flea, which blocks normal digestion, causing the flea to increase the frequency of feeding attempts and so to bite more animals and increase the probability of disease transmission. Plague is endemic to many rodent populations; for example, *Y. pestis* is permanently established in rodent populations from the Rocky Mountains to the west coast of the United States (Figure 27.10).

The transmission of plague was extremely widespread during the middle ages because of poor sanitary conditions and the abundance of infected rat populations in areas of dense human habitation (Figure 27.11).

> *This now commencing, is the fourth year,*
> *Since first, the greatest plague that raged here,*
> *Within our time, was sent for our correction,*
> *To scourge us, with a pestilent infection.*
>
> —GEORGE WITHER, *History of the Pestilence, 1625*

The development of rat control programs and improved sanitation methods in urban areas has greatly reduced the incidence of this disease. It is not possible, however, to completely eliminate plague in human beings because of the large num-

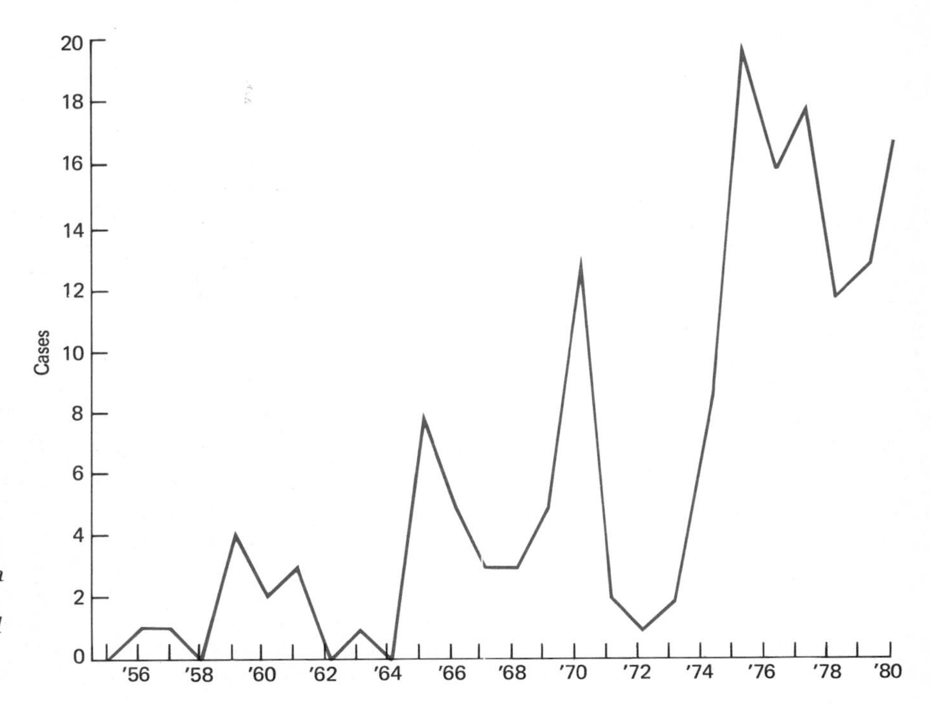

Figure 27.10 The incidence of plague in the United States has been increasing in recent years, particularly in the southwestern part of the country, where *Yersinia pestis* is endemic to wild animal populations. (From *Morbidity and Mortality Weekly Reports*, Centers for Disease Control, Atlanta.)

ber of alternate hosts in which *Y. pestis* is maintained. In rural environments, for example, *Y. pestis* is found in ground squirrels, prairie dogs, chipmunks, rabbits, mice, rats, and other animals.

The introduction of *Y. pestis* into humans through flea bites initiates a progressive infection that can involve any organ or tissue of the body. Phagocytosis is effective in killing many of the invading bacteria, but some cells of *Y. pestis* are resistant to phagocytosis and continue to multiply and spread through the circulatory system. In bubonic plague, *Y. pestis* becomes localized and causes inflammation in the regional lymph nodes, especially about the armpits, neck, groin, and upper legs. The en-

Figure 27.11 Engraving by C. Andran after P. Mignard depicting the devastating effects of plague during the Middle Ages. (From the collections of the Library of Congress.)

larged lymph nodes are called buboes, from whence the name of the disease is derived. The symptoms of bubonic plague include malaise, fever, and severe pain in the areas of the infected regional lymph nodes. Severe tissue necrosis can occur in various areas of the body, and it appears as blackened skin. It was this symptom that gave the name "black death" to the disease in the middle ages. As the infection progresses, the symptoms become quite severe, and without treatment the fatality rate is 60–100 percent.

Plague can also progress in humans to involve the pulmonary system, leading to pneumonic plague. Transfer of *Y. pestis* then can occur through droplet spread, establishing outbreaks of primary pneumonic plague. The pneumonic form of plague has been very important in plague epidemics. If the bacteria invade the lungs, as in primary pneumonic plague, the disease often progresses rapidly and is manifest by severe prostration, respiratory difficulties, and death within a few hours of onset. Plague can be effectively treated with antibiotics, and streptomycin generally is the drug of choice against *Y. pestis*, although other antibiotics, such as chloramphenicol and tetracycline, can also be used.

Relapsing Fever. Relapsing fever is caused by various species of *Borrelia*, a Gram negative spirochete (Figure 27.12). These pathogenic bacteria are

Figure 27.12 Micrograph of *Borrelia hermsi* from a smear of mouse blood stained by the Giemsa technique, followed by counterstaining with crystal violet (2200×). The characteristic morphology of the bacteria is shown in this preparation. (Courtesy Richard T. Kelly, Baptist Memorial Hospital, Memphis, Tennessee.)

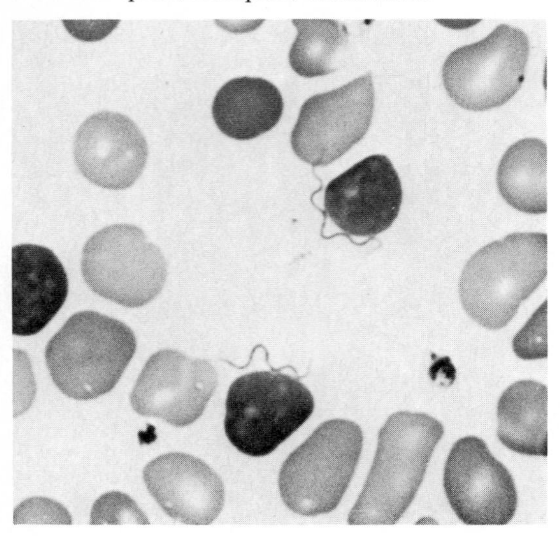

normally transmitted to people through a bite by the body louse, *Pediculus humanus*. Epidemics of louse-borne relapsing fever occur under conditions that favor the proliferation of body lice, such as crowded conditions with relatively poor sanitation. The lice act strictly as vectors and do not develop the disease symptomology when they acquire *Borrelia* from an infected human. Crushing an infected louse can release *Borrelia* into the break in the skin caused by the louse bite. *Borrelia* species can also be transmitted through tick vectors, accounting for periodic outbreaks of relapsing fever in the western United States.

The clinical manifestations of relapsing fever occur intermittently. There is normally a sudden onset of fever approaching 105°F that lasts for 3–6 days. The fever then falls rapidly and remains normal for 5–10 days, followed by a second onset of fever that generally lasts 2–3 days. Additional relapses frequently occur in tick-borne disease but do not normally occur when the disease is louse-borne. Relapses appear to be associated with the appearance of new antigenic types of the bacterium.

During the course of relapsing fever, *Borrelia* sp. are able to multiply in the blood and various other body tissues. The natural removal of *Borrelia* from the body depends on an antibody-mediated immune response rather than on phagocytosis. Anatomical abnormalities develop in the spleen, and lesions may also occur in various other body organs. Penicillin, tetracycline, and chloramphenicol are effective against *Borrelia* and are used in treating this disease. With proper treatment over 95 percent of patients with relapsing fever recover. Control of this disease depends on avoiding contact with vectors carrying *Borrelia*, particularly body lice and ticks. Pesticides are useful in many areas of the world in maintaining low populations of rodents and other animals that act as nonhuman vertebrate hosts of vector ticks.

Bartonellosis. Bartonellosis is an infection caused by obligately intracellular, Gram negative bacterium, *Bartonella bacilliformis*. *B. bacilliformis* is unusual in its restricted geographic distribution and its affinity for red blood cells. Bartonellosis is transmitted to people through the bites of infected sand flies that occur only in the valleys of the Andes Mountains. Thus, this disease occurs only in geographically restricted regions of Peru, Ecuador, and Columbia. *B. bacilliformis* is able to proliferate within the endothelial cells of blood vessels after inoculation by the bite of the sand fly vector. The bacteria reenter the circulatory system and infect erythrocytes, generally causing severe anemia accom-

Table 27.2 **Summary of Certain Important Epidemiologic and Clinical Characteristics of Rickettsial Diseases**

Disease (etiologic agent)	Epidemiologic Features		
	Usual Mode of Transmission to Man	Reservoir	Geographic Occurrence
Spotted fever group			
Rocky Mountain spotted fever (*Rickettsia rickettsii*)	Tick bite	Ticks; rodents	Western Hemisphere
Tick typhus (rickettsias related to *R. rickettsii*)	Tick bite	Ticks; rodents	Mediterranean littoral, Africa, Asia
Rickettsialpox (*R. akari*)	House mouse mite bite	Mites; mice	USA, Russia, Korea
Typhus group			
Epidemic typhus (*R. prowazekii*)	Infected louse feces rubbed into broken skin or as aerosol to mucous membranes	Man	Worldwide
Brill–Zinsser disease	Occurs months or years after primary attack of louse-borne typhus	—	Worldwide
Murine (Endemic) typhus (*R. typhi*)	Infected flea feces rubbed into broken skin or as aerosols to mucous membranes	Rodents	Worldwide (scattered pockets)
Scrub typhus	Mite bite	Mites; rodents	SE Asia, W and SW Pacific, Japan

panied by fever, headache, and delirium.

Bartonellosis can be manifest in two stages: Oroya fever, which is an acute febrile stage, and a later cutaneous stage, known as Verruga peruana. The mortality rate from bartonellosis can be as high as 40 percent during the Oroya fever stage. Chloramphenicol is effective in treating Oroya fever and greatly reduces the fatality rate. If the patient survives the Oroya fever stage, there is a subclinical period that may be followed by the development of reddish lesions at various sites on the body, particularly the extremities. Prevention of Bartonellosis depends on avoiding contact with infected sand fly vectors. The sand flies carrying *Bartonella* are nocturnal, and the use of insect repellents and protective clothing after dark are useful for people in the Andes.

Rocky Mountain Spotted Fever. Rocky Mountain spotted fever is caused by a rickettsia and is the first of several diseases to be discussed in this section that are caused by rickettsias transmitted to humans via biting arthropod vectors (Table 27.2). In the case of Rocky Mountain fever, *Rickettsia rickettsii* is transmitted to human beings through the bite of a tick. *R. rickettsii* is normally maintained in various tick populations, such as the wood tick and dog tick (Figure 27.13). The bacteria multiply within the midgut of the tick and are passed congenitally from one generation of ticks to the next. Humans are accidental, not necessary, hosts of *R. rickettsii* as a result of occasional bites of infected ticks.

Rocky Mountain spotted fever occurs in areas of North and South America, most commonly in spring and summer when ticks and people are most likely to come in contact, and outbreaks normally occur in well-defined localized regions. During the mid-twentieth century many cases of Rocky Mountain spotted fever occurred in the United States in the region of the Rocky Mountains, but relatively few cases have been reported there in recent years. On the other hand, outbreaks of Rocky Mountain spotted fever have risen dramatically in the eastern United States, where most cases now occur (Figure 27.14). There are over 1000 cases of Rocky Mountain

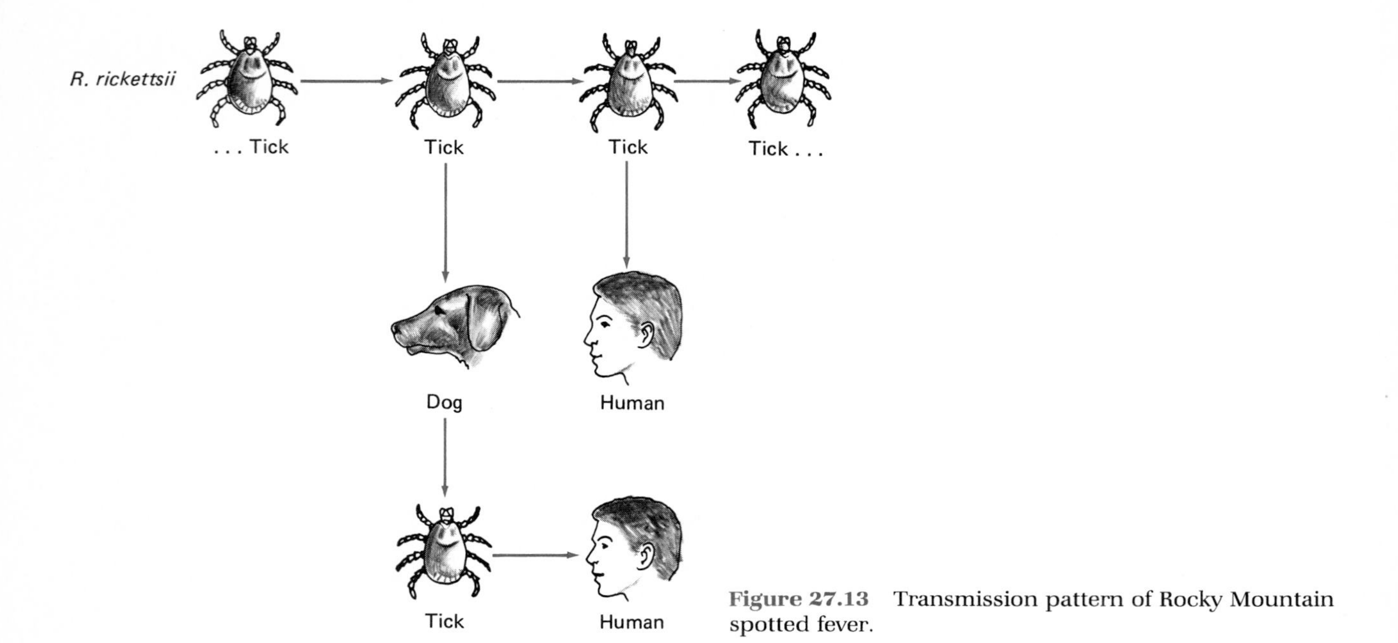

Figure 27.13 Transmission pattern of Rocky Mountain spotted fever.

Figure 27.14 Map showing incidence of Rocky Mountain spotted fever in the continental United States in 1977. Note that the highest rate of incidence is in the eastern half of the country.

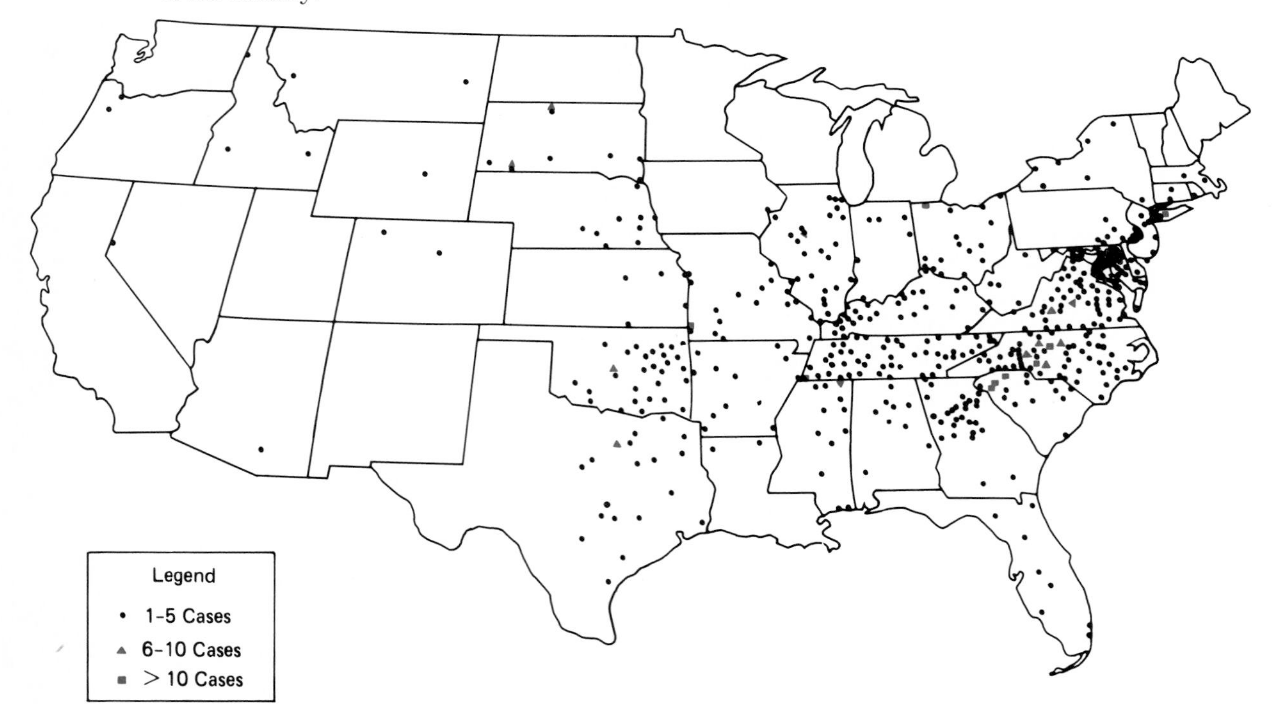

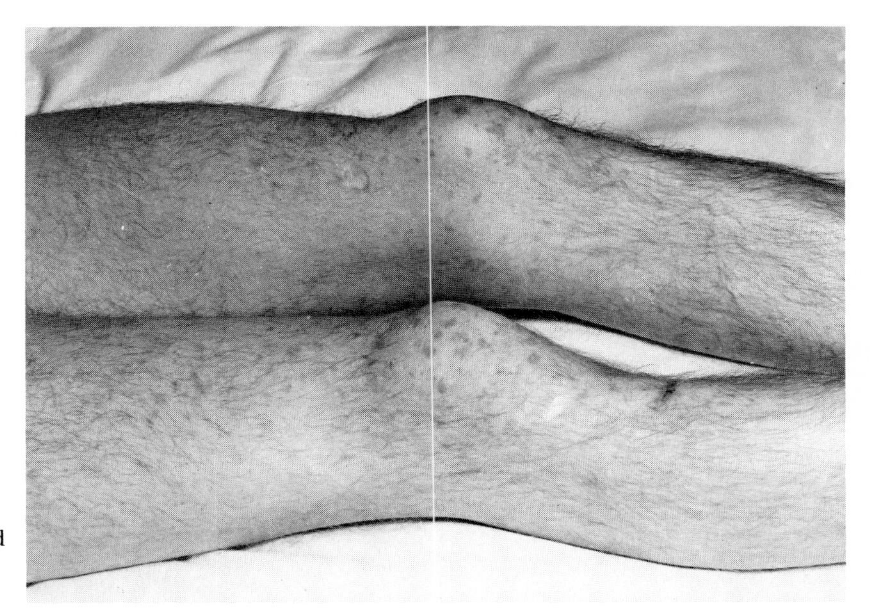

Figure 27.15 Rocky Mountain spotted fever rash on the leg. (From United States Armed Forces Institute of Pathology, neg. no. 531141-2.)

spotted fever in the eastern part of the United States each year.

When injected into human beings, *R. rickettsii* multiply within the endothelial cells lining the blood vessels. Vascular lesions occur and account for the production of the characteristic skin rash associated with this disease (Figure 27.15). The rash is most prevalent at the extremities, particularly on the palms of the hands and the soles of the feet. Lesions probably also occur in the meninges, causing severe headaches and a state of mental confusion. If treated with antibiotics, such as chloramphenicol and tetracycline, the disease is rarely fatal; if untreated, the overall mortality rate is probably greater than 20 percent. Prevention of Rocky Mountain spotted fever primarily involves control of population levels of infected ticks and the avoidance of tick bites. However, control is difficult to achieve, and in 1975 there were almost 900 cases of Rocky Mountain spotted fever in the United States.

Typhus Fever. There are several types of typhus fever, all of which are caused by rickettsias transmitted to humans via biting arthropod vectors (Figure 27.16). Epidemic typhus fever (also called infectious or classical typhus fever) is caused by *Rickettsia prowazekii* and is transmitted to humans via the body louse. The chain of transmission is restricted to humans and lice. The lice contract the disease from infected humans and in turn pass the rickettsia on to susceptible human hosts. *R. prowazekii* multiplies within the epithelium of the midgut of the louse. When an infected body louse bites another human it defecates at the same time, depositing feces containing *R. prowazekii*, which enter through the wound created by the bite; rubbing the wound due to the irritation introduces louse feces and associated *R. prowazekii* into the body.

The name epidemic typhus is derived from the fact that this form of typhus is transmitted from person to person only by the body louse. Under crowded conditions with lice infestation, the disease can spread easily and cause large numbers of cases. For example, millions of cases occurred in World War I and in the concentration camps during World War II. The onset of epidemic typhus involves fever, headache, and rash. The heart and kidneys are frequently the site of vascular lesions. If untreated, the fatality rate in persons 10–30 years old is approximately 50 percent. The Weil–Felix test is useful in diagnosing classic typhus fever. Chloramphenicol, tetracycline, and doxycycline are effective in treating epidemic typhus.

Relapses of louse-borne typhus can occur years after the primary attack. Such recurrences are referred to as the Brill–Zinsser disease. *R. prowazekii* apparently can survive in some cells of the body for a prolonged period of time and give rise to this later infective phase. The factors that initiate recurrence of typhus are unknown, and there is no known way of preventing these recurrences. Prevention of louse-borne typhus fever involves controlling louse infestation, and delousing infected patients effectively interrupts the chain of transmission.

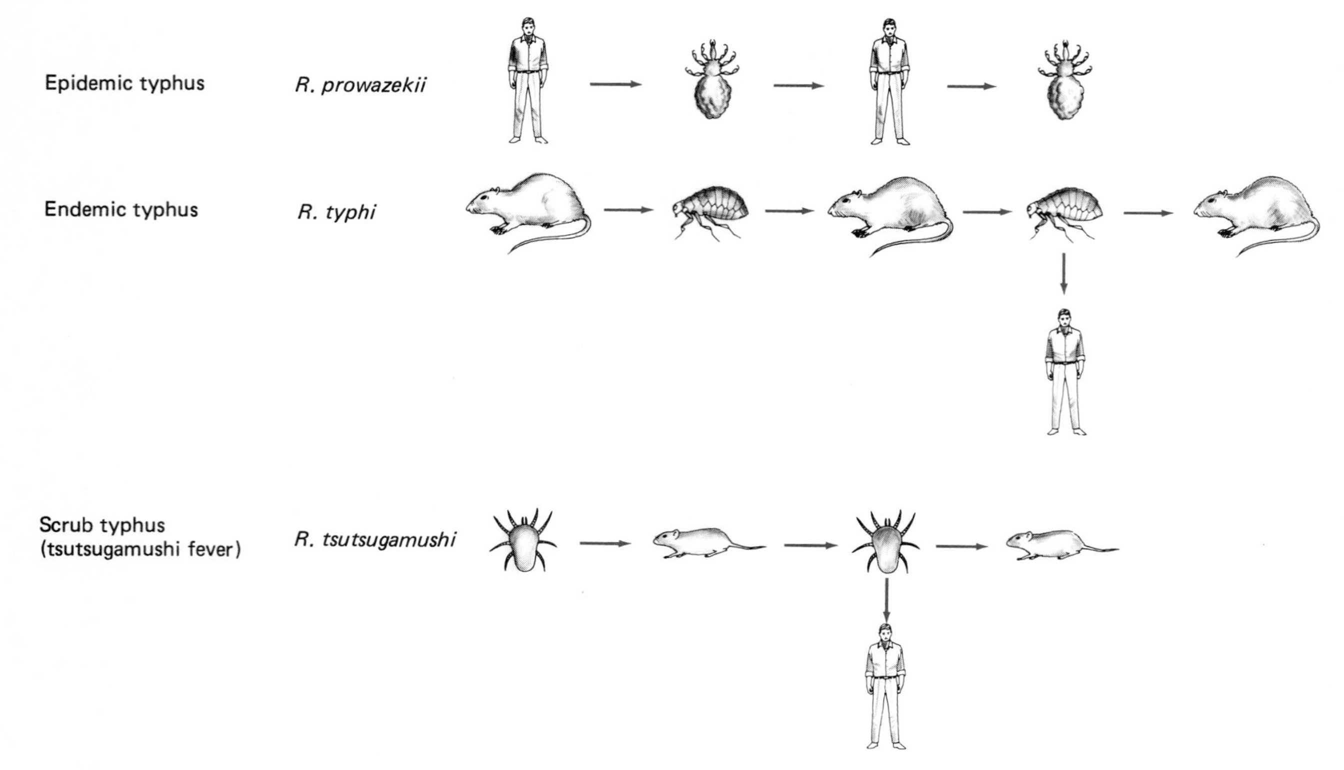

Figure 27.16 Transmission patterns for different types of typhus fever.

Murine or endemic typhus fever is caused by *Rickettsia typhi* and is transmitted to humans by rat fleas. Murine typhus is normally maintained endemically in rat populations through transmission by rat fleas. Occasionally, rat fleas will attack people, and if they are infected with *R. typhi* the disease can be transmitted. As with louse-borne typhus, the flea deposits pathogenic bacteria in the fecal matter, which is rubbed into the flea bite by the host because of the local irritation caused by the bite. The symptoms of murine typhus are similar to those of classical typhus fever but are generally milder. Chloramphenicol and tetracycline are effective in treating this disease, and there is a relatively low mortality rate. Prevention of murine typhus depends on limiting rat populations, which also limits the size of the vector rat flea population.

The vermin only tease and pinch
Their foes superior by an inch.
So, naturalists observe, a flea
Hath smaller fleas that on him prey,
And these have smaller fleas to bite 'em,
A so proceed ad infinitum

—J. SWIFT, *On Poetry: A Rhapsody*

Scrub typhus, caused by *Rickettsia tsutsugamushi*, is almost exclusively restricted to Japan, Southeast Asia, and the western Pacific Islands. The disease is transmitted to people through the bite of mite vectors. *R. tsutsugamushi* is normally transmitted congenitally in mite populations and is only accidently introduced into the human population. In humans multiplication of *R. tsutsugamushi* occurs at the site of inoculation and is later distributed through the circulatory system; eventually the infection becomes localized in the lymph nodes. In scrub typhus there is a characteristic lesion of the skin, known as the eschar, that occurs at the site of initial infection. The eschar may be difficult to find because mite bites may occur anywhere on the body. The symptoms of scrub typhus include fever, severe headache, and rash. Chloramphenicol and tetracycline are effective in treating this form of typhus. Controlling vector populations and minimizing the chance of being bitten by an infected mite are means by which this disease can be prevented.

Rickettsialpox. Rickettsialpox, caused by *Rickettsia akari*, is transmitted to humans by the mouse mite (Figure 27.17). *R. akari* is congenitally maintained in populations of the mouse mite, and in

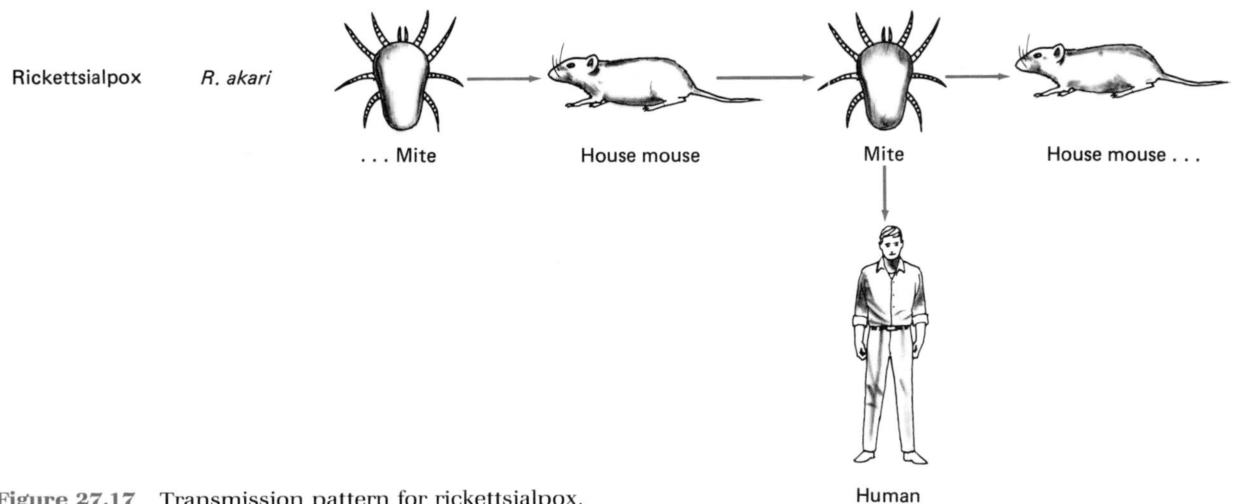

Figure 27.17 Transmission pattern for rickettsialpox.

addition to serving as a reservoir for *R. akari*, the mouse mite acts as a vector. Normally, mouse mites do not attack people, but under certain conditions humans can be accidental recipients of *R. akari* from an infected mouse mite, leading to the onset of the disease rickettsialpox. Inoculation occurs through the oral secretions of the mite. Rickettsialpox occurs almost exclusively in urban environments of the United States and the Soviet Union, where there are not enough mice for the mites to feed on, and it is under these conditions that infected mites will bite people.

The infection of *R. akari* leads to the formation of a primary lesion at the site of the bite. As with other rickettsial diseases, the bacterial infection can spread along the vessels of the circulatory system, causing vascular lesions. Systemic spread of the infection is followed by the onset of the illness, which is manifest by fever, headache, and secondary lesions. The rash associated with this disease may cover any part of the body, but unlike Rocky Mountain spotted fever, the rash rarely develops on the palms or soles. Rickettsialpox is a benign disease, and symptoms disappear without treatment beginning approximately 10 days after the onset of disease. The use of tetracyclines and other antibiotics are effective in ending the manifestation of disease symptomology within 1 day of beginning antimicrobial therapy. Prevention of rickettsialpox depends on effective rodent control programs; the elimination of mice from urban settings can effectively end this disease.

Tularemia. Tularemia is caused by *Francisella tularensis*, a Gram negative, fastidious coccobacillus. *F. tularensis* may gain entry to the body directly through the skin, particularly through minute openings, such as at hair follicles and near the fingernails. Transmission to humans can occur through ingestion of contaminated meat and handling of infected animals. Tularemia is a particular occupational problem for individuals who handle potentially contaminated animals, such as hunters, butchers, cooks, agricultural workers, and so on.

Often, ulcers of the fingers are symptomatic of tularemia that is contracted through the handling of infected animals. The localized ulcers result from the multiplication of *F. tularensis*, as does a systemic infection with localization within the regional lymph nodes. In most cases of tularemia, there is an elevated fever, development of skin ulceration, and enlargement of the lymph nodes. The death rate in untreated cases of tularemia acquired through the skin is approximately 5 percent, but higher mortality rates approaching 30 percent occur when the disease is contracted through inhalation. Streptomycin, tetracycline, and chloramphenicol are effective in controlling this disease. Prevention of tularemia can be achieved by avoiding contaminated animals, and because rabbits have an especially high rate of infection with tularemia, special precautions should be taken in their handling.

Brucellosis. Brucellosis is caused by several *Brucella* species, all of which are Gram negative, small, nonmotile, aerobic rods. Brucellosis is an infectious disease of nonhuman animals that can be transmitted to humans through ingestion of contaminated milk and the handling of infected animals. The use of pasteurization has greatly reduced the transmission of brucellosis via the gastrointestinal

MICROBIOLOGY HEADLINES

Stinging a Global Killer

Malaria is the world's leading infectious disease: it chronically afflicts at least 210 million people in Asia, Africa and Latin America, and public-health officials can't even count the number it kills. What's more, malaria is fast becoming a greater menace than ever before. The anopheles mosquitoes that carry the disease have grown resistant to pesticides, and the tiny parasites that cause it have grown resistant to drugs. But last week there were finally significant signs of progress against this persistent plague. Three teams of U.S. researchers reported in the journal *Science* that they had taken the first critical steps toward a vaccine against the most vicious type of malaria parasite, *Plasmodium falciparum*.

What has thwarted development of a vaccine in the past is the complicated life cycle malaria parasites undergo in their insect and human hosts. When an anopheles carrier bites a human, it injects one-celled organisms called sporozoites into the bloodstream. On reaching the liver, the sporozoites enter a new stage of development—becoming merozoites that then attack and destroy red blood cells, eventually producing the fever, chills and weakness that are the hallmarks of malaria. The new vaccine would help the body produce antibodies to attack and destroy sporozoites, just as current vaccines built immunity to viruses and bacteria.

Protein Coat. Using the new techniques of genetic engineering, Drs. Victor and Ruth Nussenzweig and Vincenzo Enea of New York University Medical Center isolated the gene, or DNA, that produces a specific protein on the sporozoite's surface. Subsequently they, along with teams at the National Institutes of Health and Walter Reed Army Institute of Research, determined the exact chemical structure of this protective protein coat. The next step will be to synthesize the protein in quantity. Injected into humans, it would act as an antigen, stimulating the production of antibodies to kill sporozoites as soon as the anopheles injects them. A spokesman for the Agency for International Development, which has spent $35 million on the research, predicted that a vaccine will be ready for testing within 18 months.

Source: Newsweek, Aug. 13, 1984.

tract, and most human infections today result from direct contact with infected animals. Brucellosis is an occupational hazard in farming, veterinary practice, and meat packing plants, where *Brucella* species can enter through the skin, particularly in areas of minor abrasions.

Once *Brucella* species enter the body, the bacteria are able to spread rapidly through the mononuclear phagocyte system and can multiply within phagocytic cells. Infecting *Brucella* species normally become localized in the regional lymph nodes. There may be enlargement of the spleen and liver, and other symptoms include weakness, chills, malaise, headache, backache, and fever. The fever may rise and fall in some cases, and thus, this disease is often referred to as undulant fever. The death rate in untreated cases of brucellosis is about 3 percent. The use of antibiotics is effective in treating this disease and reduces the death rate to near 0 percent. Antibiotic treatment, however, requires prolonged adminstration of combinations of drugs, such as tetracyclines and streptomycin, because *Brucella* can live intracellularly. The prevention of brucellosis involves eliminating the disease in animals such as cattle, sheep, and goats, the reservoirs of *Brucella*. It is important to segregate and treat infected animals to prevent the spread of this disease through animal herds. Vaccines are effective in limiting the spread of brucellosis through some animal populations, such as cattle, but there are no effective vaccines for other animals, such as hogs.

Lyme Disease. Lyme disease, named for the small Connecticut community in which the diease was first recognized in 1975, is an inflammatory disorder that characteristically begins with a distinctive skin lesion. The circularly expanding annular skin lesions are indurated with wide borders and central clearing. Lesions may reach diameters of 12 inches or more. Weeks to months later, the patient often develops chronic arthritis in the knees; neurologic or cardiac abnormalities may also occur.

Recent epidemiologic and laboratory investigations indicate that Lyme disease may be a different manifestation of erythema chronicum migrans, a syndrome long recognized in Europe. Both diseases are caused by a spirochete transmitted to humans by the bites of *Ixodes* ticks. Similar spirochetes have been recovered from the blood of Lyme disease patients and cultured from *Ixodes* ticks, and patients with Lyme disease have antibodies to the cultured spirochetes. When this disease is diagnosed, penicillin is an effective treatment.

Diseases Caused by Protozoan Pathogens

Malaria. On a worldwide basis malaria is one of the most common human infectious diseases (Fig-

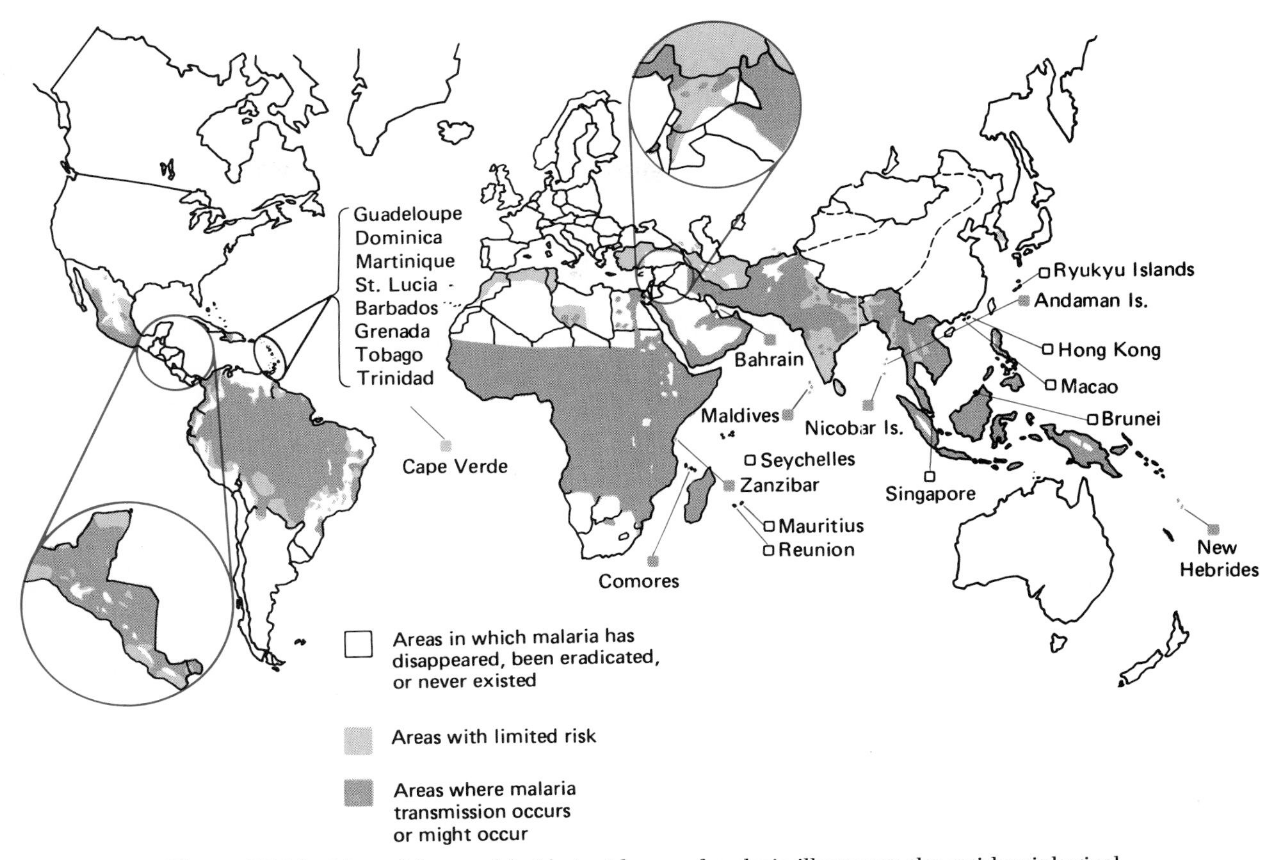

Figure 27.18 Map of the worldwide incidence of malaria illustrates the epidemiological assessment of the status of the disease in December 1976. Malaria remains one of the most serious infectious diseases in the world today.

ure 27.18). The annual incidence of malaria is about 150,000,000 cases. Malaria has been largely eliminated from North America and Europe but remains the most serious problem of infectious disease in tropical and subtropical regions of the world (Figure 27.19). This disease is caused by four species of *Plasmodium* (Table 27.3). *P. vivax* and *P. falciparum* are most frequently involved in human infections. The *Anopheles* mosquito is the vector responsible for transmitting malaria to human beings. After inoculation into the body, the sporozoites (cells produced by the division of the zygote) of *Plasmodium* begin to reproduce within liver cells (Figure 27.20). Multiplication of the *Plasmodium* sporozoites occurs by schizogony (a form of asexual division in which many new protozoa are produced through multiple fission) by which a single sporozoite can produce as many as 40,000 merozoites (feeding stage in the life cycle of this organism). The merozoites are capable of invading blood erythrocyte cells. Invasion of erythrocytes by hepatic merozoites begins the erythrocytic phase of malaria. The *Plasmodium* merozoites are able to reproduce within the red blood cells, killing large numbers of red blood cells and thereby causing anemia (Figure 27.21).

In one day: Four people in this country discover that they have malaria.

—TOM PARKER, *In One Day*

The symptoms of malaria begin approximately 2 weeks after the infection established by the mosquito bite. The symptoms include chills, fever, headache, and muscle ache. The symptoms of malaria are periodic, with symptomatic periods generally lasting less than 6 hours. Schizogony occurs every 48 hours for *P. vivax* and *P. ovale*, and every 72 hours for *P. malariae*, resulting in a synchronous rupture of infected erythrocytes. The symptoms of

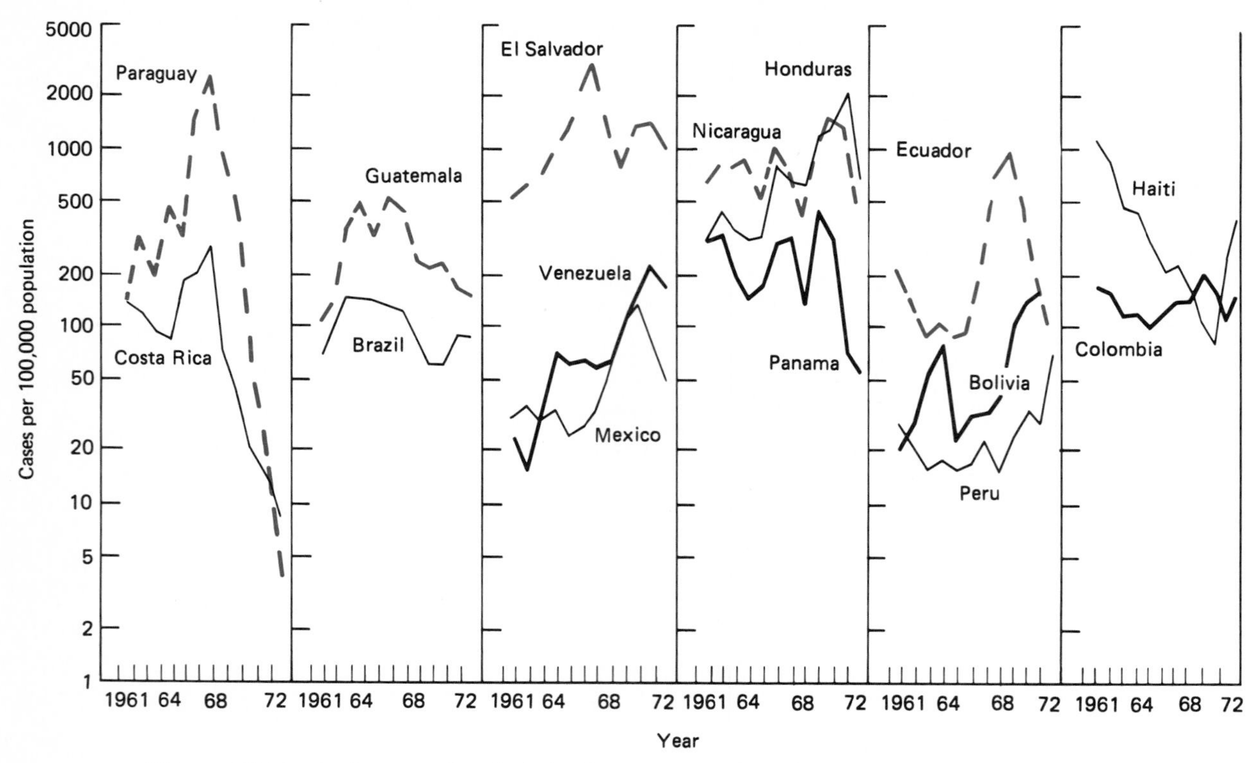

Figure 27.19 Incidence of malaria in the Americas, based on reported cases in 15 countries.

the disease are caused by the asexual erythrocytic cycle, but the periodic onset of disease symptoms coincide with the rupture of infected erythrocytes.

In the walled world of darkness and his fever the
chairs were beasts, the lamp-shade was a
lowered head
The sheet was an expanse of ice enfolding
the clenched fright and the weakness in the
bed . . .

As the apes cried beneath bright candelabra
he sailed far on the incandescent ship, alone,
and reeled with vertigo, and followed downward
the firmament of blood about the bone.

Four other nights rose and were populated
by rivers, priests, a conflagration and a ghost,
but afterward the circular warm tide
and the apes' voices he remembered most.

—CLARK MILLS, *Child with Malaria*

Table 27.3 Summary of Important Characteristics of Human Malarias

	P. falciparum	*P. vivax*	*P. ovale*	*P. malariae*
Incidence	Common	Common	Uncommon	Uncommon
Primary hepatic schizogony	1–40,000 in 5.5–7 days	1–10,000 in 6–8 days	1–15,000 in 9 days	1–2000 in 13–16 days
Secondary hepatic schizogony	1–8 to 24 (av. 16) in 48 hr	1–12 to 24 (av. 16) in 48 hr	1–6 to 16 in 48 hr	1–6 to 12 (av. 8) in 72 hr
Incubation period	8–27 days (av. 12)	8–27 days (av. 14) (rarely months)	9–17 days (av. 15)	15–30 days
Mortality	High in nonimmunes	Uncommon	Rarely fatal	Rarely fatal

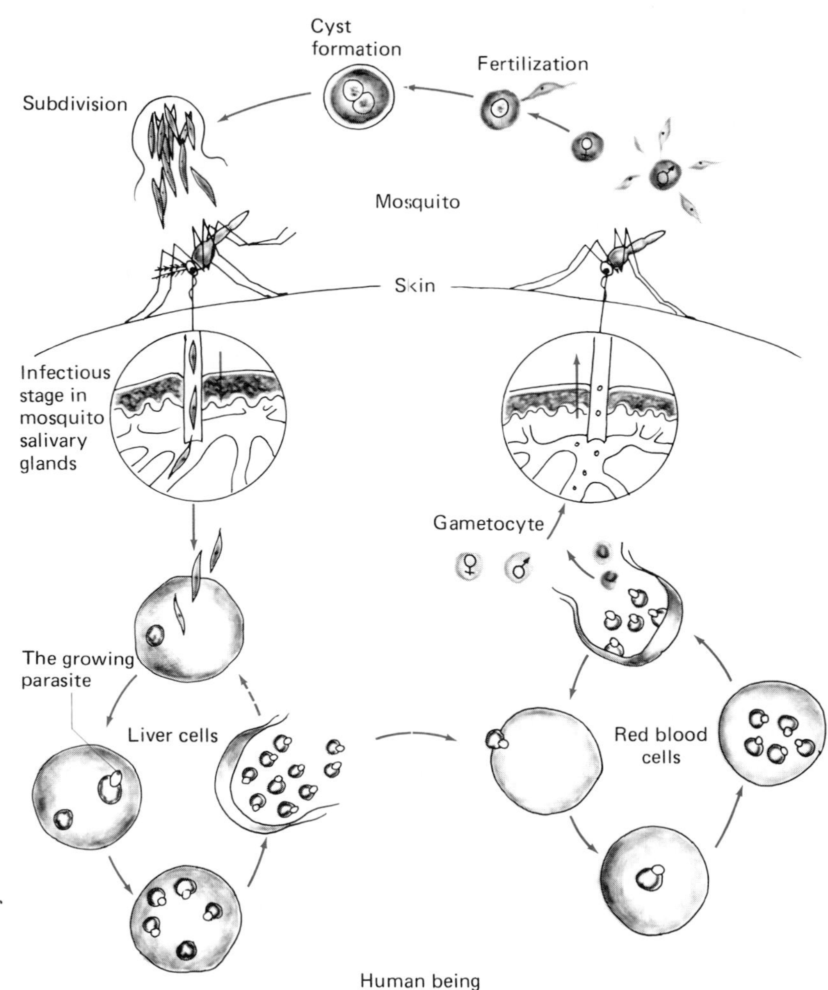

Figure 27.20 Reproduction of a malaria parasite within the mosquito vector and the human body. The protozoan exhibits two distinct reproductive phases.

During World War I all the eastern armies were affected, but it was especially during World War II with reference to the Pacific theater that malaria played an important role in the plans of statesmen and military commanders. By seizing the Dutch Indies early in the war and thus gaining control of the supplies of quinine, the Japanese thought they would paralyze Allied operations in the malarious regions of the South Pacific and Asia. Indeed, this strategic move might have met with success had not the synthetic drug atabrine become available in unlimited quantities to provide an effective antimalarial substitute for quinine. At one time in Burma the casualties from malaria were some ten times those caused by Japanese weapons of war, making the functions of the Director of Medical Services more important for the conduct of the campaign than those of the General Staff. It is estimated that about half a million American soldiers were at some time hospitalized with malaria during that war. Still more recently, malaria was common among the combatants in Korea and Indochina.

—RENÉ DUBOIS, *Mirage of Health*

Malarial infections persist for long periods of time and are rarely fatal, except when the disease is caused by *P. falciparum*. There is no vaccine for malaria, but the disease can be prevented by drug prophylaxis. Individuals traveling to areas with high rates of malaria, such as southeast Asia and Africa, often use antimalarial drugs, such as chloroquine, to avoid contracting this disease. However, many strains of *Plasmodium* are relatively resistant to an-

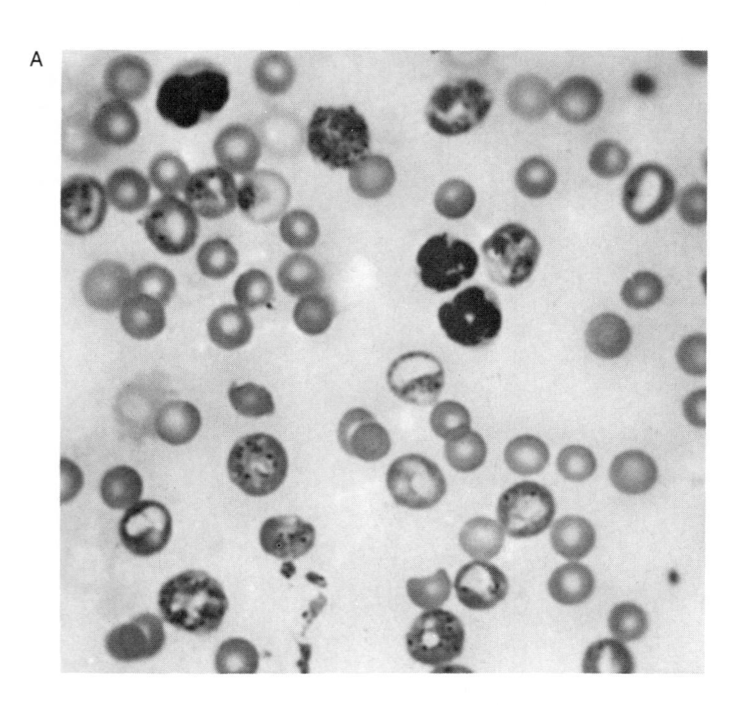

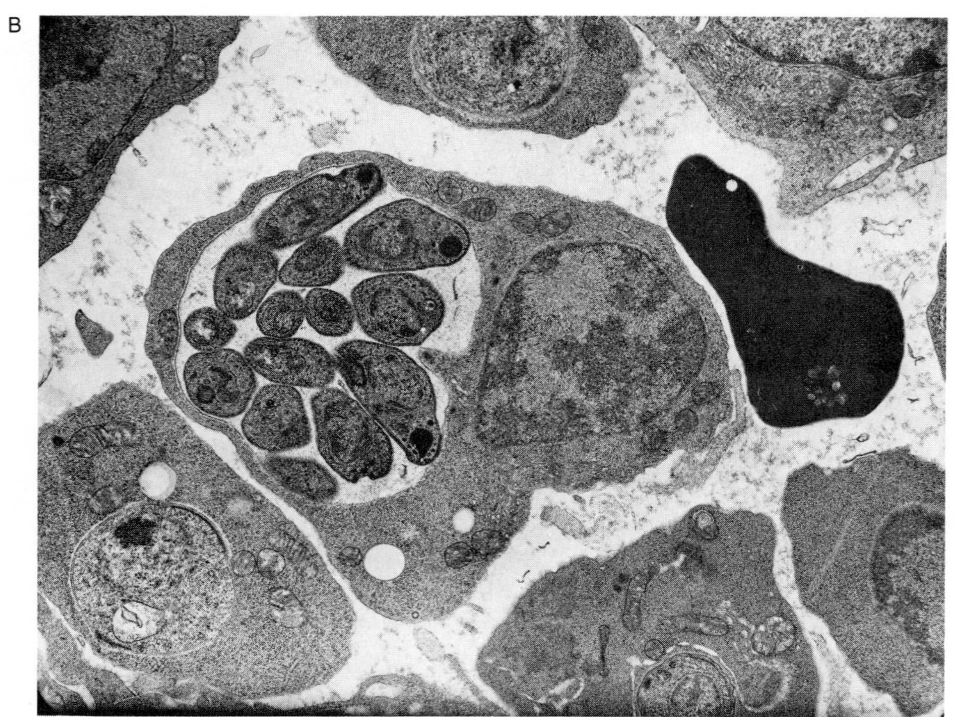

Figure 27.21 Malaria parasites infecting blood cells. (A) Light micrograph; (B) electron micrograph. (From Photoresearchers Inc.: Omikrom.)

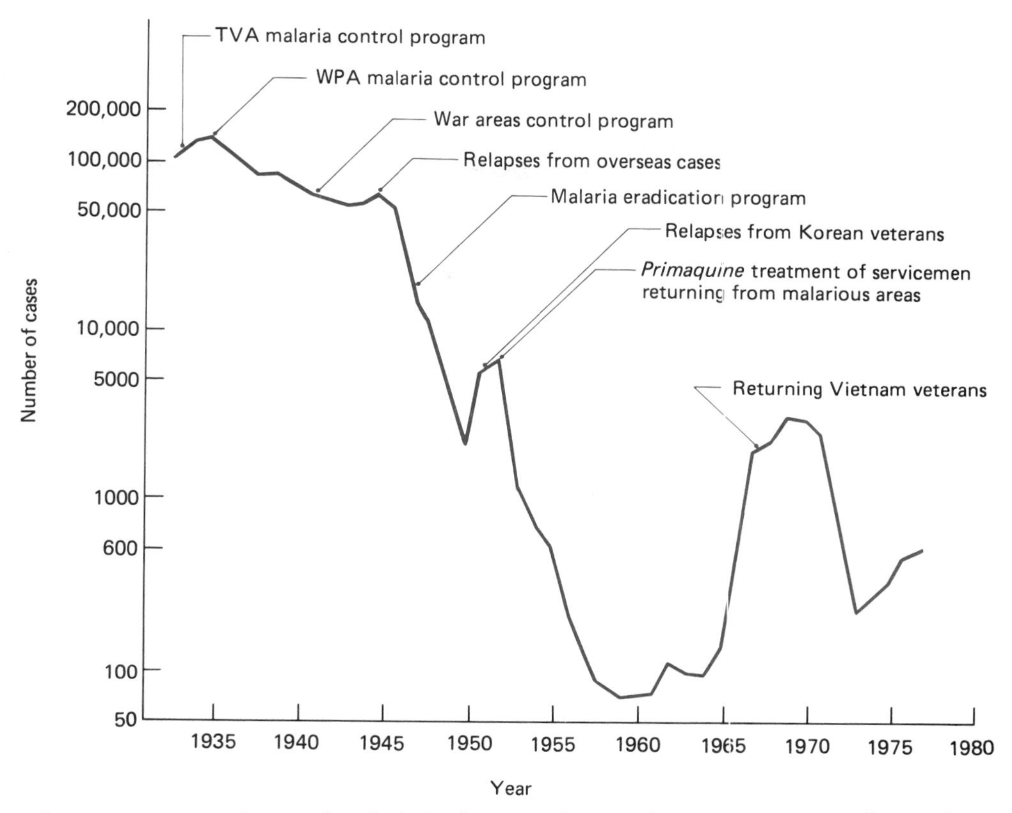

Figure 27.22 Incidence of malaria in the United States from 1933 to 1977. The total number of cases has fluctuated with the application of control measures and the return of military personnel from areas of the world where malaria is common.

timalarial drugs, so drug prophylaxis is not always effective. The use of insect netting and other measures to prevent being bitten by an infected mosquito are extremely important in avoiding contracting malaria. In the United States control measures have been effective, but periodic morbidity increases have occurred after overseas military ventures (Figure 27.22).

Babesiosis. Babesiosis is a protozoan disease that occurs on the offshore islands near Cape Cod, Massachussetts. It is caused by *Babesia microti*, a parasite that normally lives within the red blood cells of rodents. *Babesia* can be transmitted to humans by tick vectors. When this occurs, the infection causes a relatively mild, self-limiting febrile disease. The parasites in the red blood cell take on several forms, the most common of which are two to four small, pear-shaped figures. These may easily be mistaken for *Plasmodium falciparum*. Individuals who have had their spleens removed and elderly individuals who are infected may develop more severe symptoms that resemble malaria. These symptoms include fever, sweating, joint and muscle pain, nausea, and vomiting. Chloroquine has been used to treat babesiosis.

Leishmaniasis. Leishmaniasis, caused by infections with members of the protozoan genus *Leishmania*, is transmitted to human beings by sand fly vectors. *Leishmania* species reproduce in people and other animals intracellularly as a nonmotile form, the amastigote. In sand flies the protozoa exists in a flagellated form, the promastigote. Four species of *Leishmania* cause leishmaniasis in humans (Table 27.4). *L. mexicana* and *L. tropica*, sometimes referred to as Old World cutaneous leishmaniasis or Oriental Sore, cause infections limited to the skin. *L. braziliensis* causes infections of skin and mucocutaneous junctions. The most serious form of leishmaniasis is caused by *L. donovani*, which multiplies throughout the mononuclear phagocyte system and causes kala-azar disease

Leishmaniasis is geographically restricted to re-

Table 27.4 Epidemiology of Leishmaniasis

Etiologic Agent	Disease	Geographic Distribution
Leishmania donovani	Kala-azar, visceral leishmaniasis	Mediterranean, southern Europe, central Asia, China, India, Sudan
Leishmania tropica	Oriental sore, old world cutaneous leishmaniasis	Mediterranean basin, central Asia, India
Leishmania mexicana	New world cutaneous leishmaniasis	Mexico, Guatemala, Central America, Amazonia
Leishmania braziliensis	New world cutaneous leishmaniasis	Central America, South America

gions where sand flies can reproduce and acquire *Leishmania* species from infected canines and rodents. The protozoa are able to reproduce within the sand flies and are present in the saliva. The lesions caused by *Leishmania* infections may be minor or extensive. The formation of extensive ulcers can produce permanent scars. The cutaneous leishmaniasis syndromes are self-limiting. The kala-azar syndrome, however, can be fatal. Untreated cases of kala-azar disease normally last several years and in the terminal stages can involve liver and heart damage and hemorrhages. The cutaneous forms of leishmaniasis can be treated with amphotericin B. The kala-azar syndrome can be treated with sodium stibogluconate, an antimony-containing drug, or with amphotericin B. The prevention of leishmaniasis involves controlling the vector and reservoir populations. The use of DDT has been effective in some regions in eliminating the sand fly vector. In other cases infected rodent populations have been controlled, greatly reducing the incidence of leishmaniasis.

Figure 27.23 Life cycle of *Toxoplasma gondii.* Cats play an important role in transmission of this disease, appearing to be the definitive host.

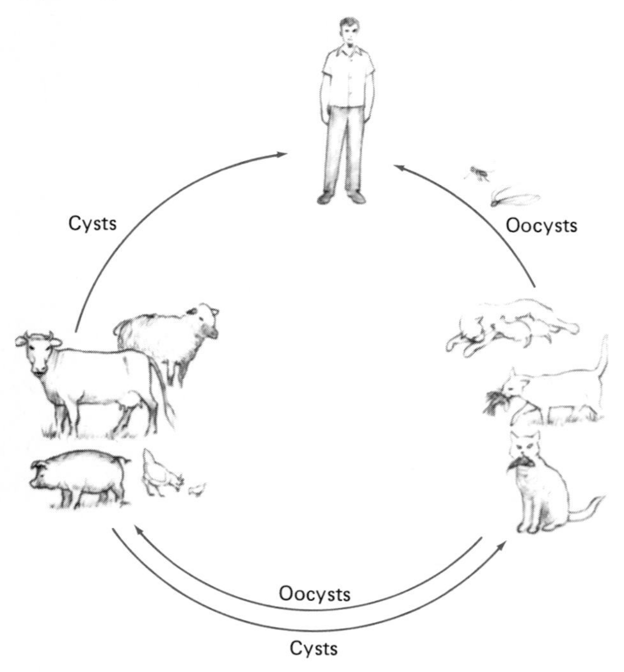

Toxoplasmosis. Toxoplasmosis is caused by the protozoan *Toxoplasma gondii,* a member of the sporozoa. Meat containing cysts of *T. gondii* is frequently involved in the transmission of this disease, although *T. gondii* can also be transmitted congenitally (Figure 27.23). Cats are the definitive hosts for *T. gondii*. Oocysts are passed in the feces of infected cats. Contact with cats is involved in most cases of toxoplasmosis. Pregnant women are advised not to handle cats or clean their litter boxes and not to eat undercooked meat because of possible contraction of toxoplasmosis, which can be transmitted congenitally to the fetus.

The protozoa normally multiply at the site of entry and then are disseminated via the bloodstream to other organs and tissues. Growth of the trophozoites of *T. gondii* cause the death of infected cells. Cell-mediated immunity is involved in containing infections of *T. gondii*. The symptoms and prognosis of toxoplasmosis depend on the virulence of the infecting strain of *T. gondii* and on the immune state of the infected individual. In most cases toxoplasmosis is asymptomatic. Recent immunological surveys suggest that 50 percent of adults in the United States have been infected. When multiple organs are involved in the infection, the consequences are serious and mortality can occur. This is most likely to occur in immunocompromised individuals. Some infections with *T. gondii* have been successfully controlled with antimalarial drugs, such as pyrimethamine. Serious consequences of toxo-

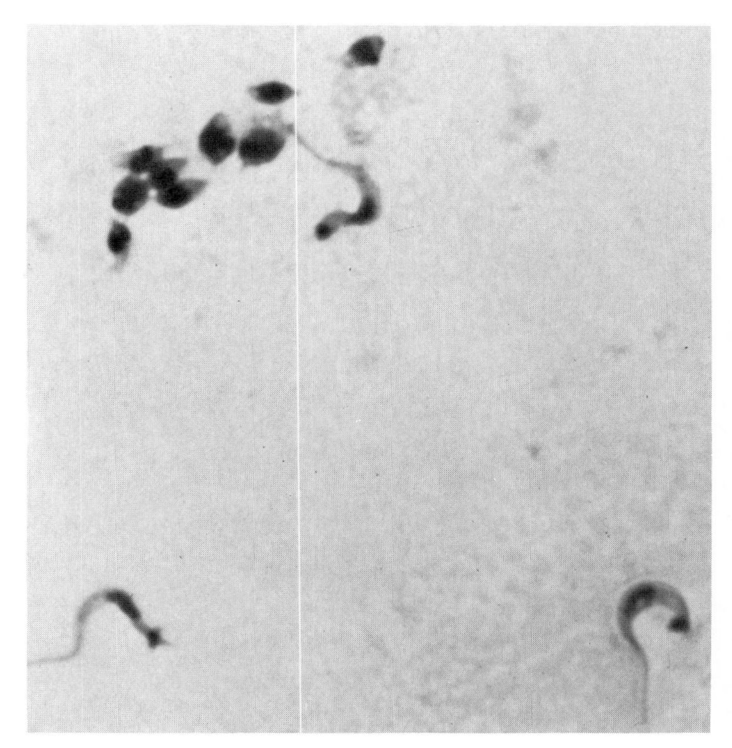

Figure 27.24 Micrograph showing flagellate (trypomastigote or promastigote) and nonflagellate (spheromastigote or amastigote) forms of *Trypanosoma cruzi*. The trypomastigote, formerly known as the trypanosomal or promastigote stage, occurs within the blood and within the insect vector. The spheromastigote stage, formerly known as the amastigote and prior to that as the Leishmanial stage, occurs within tissues, such as heart muscle and brain. Each spheromastigote has one kinetoplast and one nucleus, which appear as the two dark bodies within each cell. (Courtesy Centers for Disease Control, Atlanta.)

plasmosis occur when there is central nervous system involvement, which is particularly prevalent in the congenital transmission of this disease.

Trypanosomiasis. Trypanosomiasis is caused by infections with species of the protozoan genus *Trypanosoma*. American trypanosomiasis, or Chagas' disease, occurs in Latin America and is caused by *T. cruzi*, usually transmitted to humans by infected tryatomine (cone-nosed) bugs. *T. cruzi* is a flagellate protozoan, but in vertebrate hosts it forms a nonflagellate form, the amastigote (Figure 27.24). Dogs and cats are reservoirs of *T. cruzi*. The vectors of Chagas' disease normally live in the mud and wood houses of South America, and construction of better housing eliminates the habitat for vector populations that brings them into close contact with humans. Cone-nosed bugs feed at night, biting sleeping people on the lips, face, and forearms.

When *T. cruzi* infects human hosts, the protozoa initially multiply within the mononuclear phagocyte system. Later, the myocardium and nervous systems are invaded. Organisms localize in the heart and interfere with the impulse conduction system of the atrioventricular node, causing the myocardial fibers to contract less efficiently and sometimes to undergo atrophy. Damage to the heart tissue occurs as a result of this infection. In 90 percent of the cases, there is spontaneous remission of the disease, but 10 percent of the hospitalized patients die during the acute phase because of myocardial failure. Death due to heart disease as a result of Chagas' disease may also occur well after recovery from the acute phase. Chagas' disease is the leading cause of cardiovascular death in South America, and the incidence of this disease in Brazil is extraordinarily high. Several antiprotozoan drugs, such as aminoquinoline, are effective in treating Chagas' disease if the symptoms are recognized early, but once the progressive stages have begun, treatment is supportive rather than aimed at eliminating the infecting agent.

Diseases Caused by Multicellular Parasites

Schistosomiasis. Three species of blood flukes (*Schistosoma mansoni, S. japonicum, S. haematobium*) are responsible for infestations of the human cardiovascular system. These blood flukes cause shistosomiasis. The World Health Organization reports that shistosomiasis is second only to malaria as the cause of disease in the tropics.

The life cycle of the shistosomes involves specific species of snails and this restricts the distribution of this disease (Figure 27.25). Eggs deposited with urine or fecal matter hatch in contaminated waters.

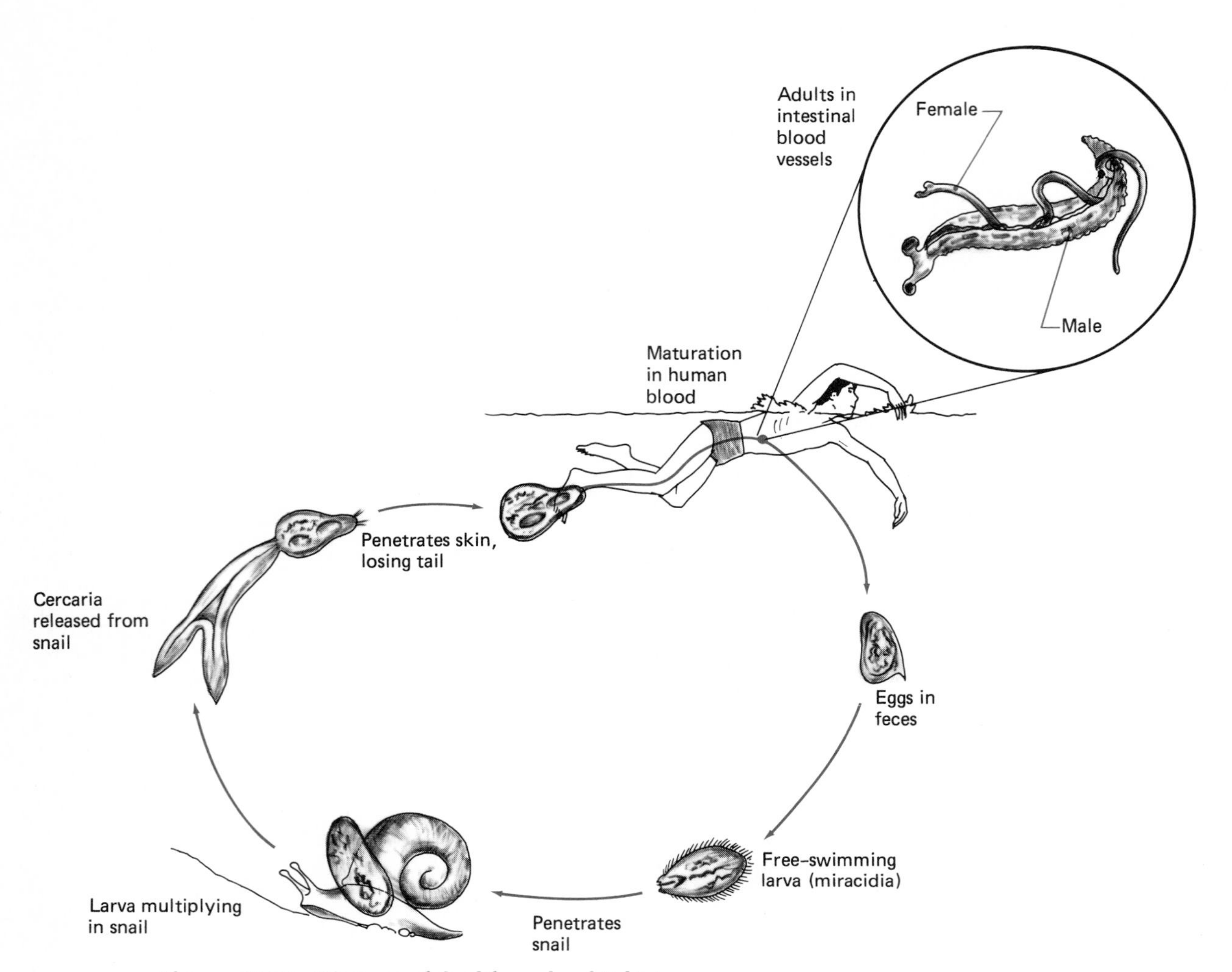

Figure 27.25 Diagram of the life cycle of *Schistosoma.*

The immature larvae enter a suitable snail host where they develop to form larvae called cercariae. The cercariae escape from the snails and swim freely and if they contact humans they are able to penetrate the skin. After penetrating the skin, the cercariae migrate through the circulatory system, moving through the heart and lungs. The cercariae settle in the hepatic portal of the liver where maturation to the adult form occurs. When they approach or reach adulthood, they migrate to the mesenteric venules of the intestines. Fertilized eggs laid in the venules move to the gastrointestinal or urinary tracts and are passed in urine or feces.

The symptoms of shistosomiasis vary with the stage of infection and the schistosome species. The penetration phase is usually asymptomatic. Most of the symptoms occur during the acute phase when the worms mature and start to lay eggs. Symptoms usually include diarrhea, fever, and discomfort. In the chronic stage there is a tissue reaction to the eggs that are deposited within the body. The eggs become walled off, forming granulomas. Cirrhosis occurs because this late stage involves the hepatic system of the liver. Diagnosis of schistosomiasis is based upon the detection of eggs in the urine or feces of individuals who have been exposed to the cercariae in infested areas of the world. Various antimony-containing drugs are used for treatment of this disease.

Elephantiasis. Elephantiasis is a grotesque disease caused by infestation of the circulatory system by the nematode *Wuchereria bancrofti*. This worm is transmitted to humans by mosquito vectors. The distribution of the disease is restricted to regions where there are carrier mosquitoes. Symptoms of this disease are associated with adult forms of the

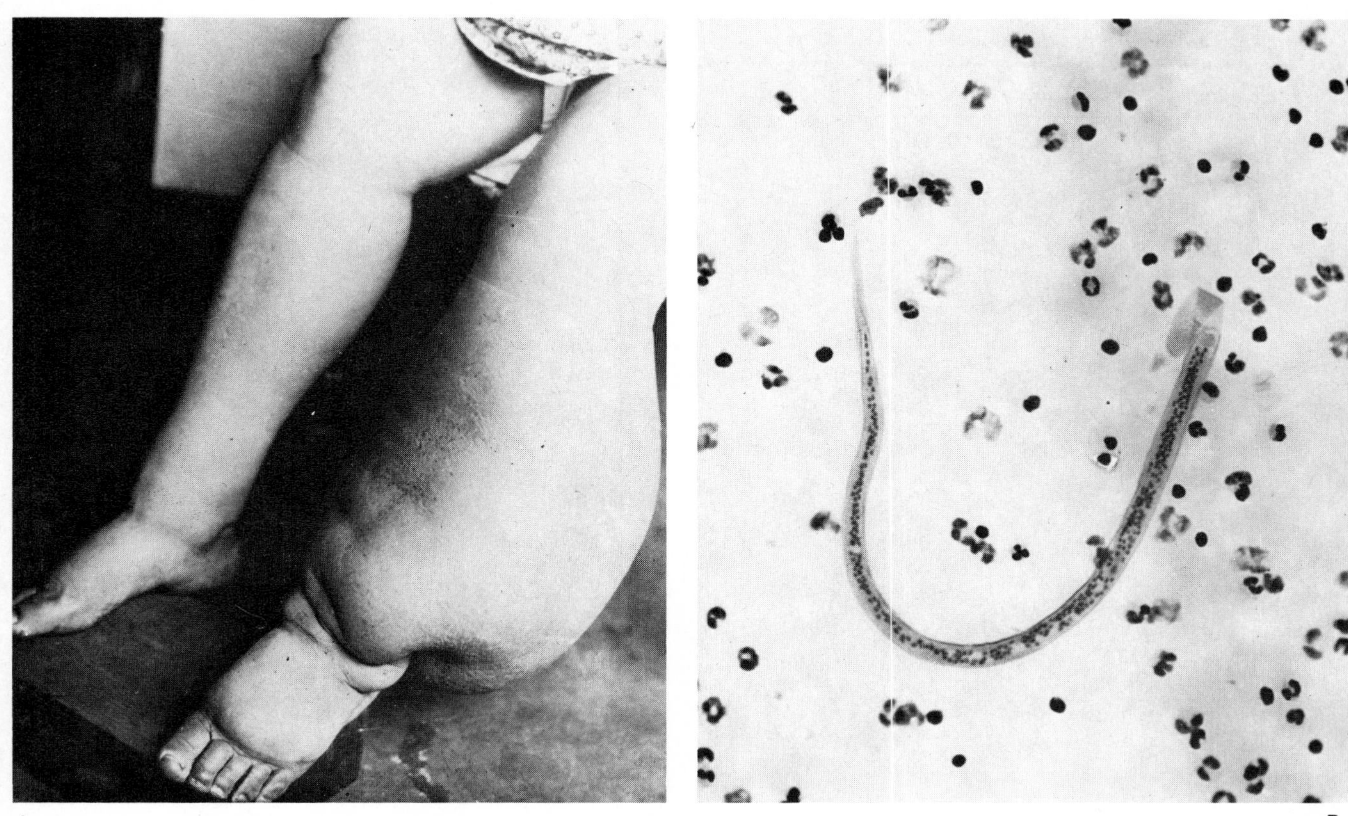
A B

Figure 27.26 (A) Photograph of an enlarged limb caused by elephantiasis. (B) Micrograph showing parasites within tissue. (From United States Armed Forces Institute of Pathology, neg. nos. 78873 and 70-10516, respectively.)

nematode. Repeated bites by infected mosquitoes result in the buildup of numerous adult worms. A hypersensitivity reaction appears to be involved in the progression of the disease. Vessels break down and lymph accumulates in tissue resulting in severe swelling (Figure 27.26). There is no suitable chemotherapy for the disease because the adult worms are resistant to drugs. The development of massive limbs and other body parts can cause severe mental problems, and psychotherapy is an important part of the supportive treatment of this disease.

Summary

The cardiovascular and lymphatic system circulate blood and lymph throughout the body. Because the circulatory system contacts virtually all parts of the body, microorganisms frequently enter the cardiovascular and lymphatic systems. Even a simple dental cleaning allows large numbers of microorganisms to enter the circulatory system. Blood and lymph are protected against microbial infections by a variety of phagocytic white blood cells and by the presence of antibodies and other components of the immune defense system. Nevertheless some microorganisms escape these defenses, causing systemic infections and travelling through lymph and blood to various parts of the body. In some of these systemic diseases the organs, vessels, and tissues of the cardiovascular and lymphatic systems are affected.

Table 27.5 presents a summary of systemic microbial infections that involve the cardiovascular and lymphatic systems.

Table 27.5 Summary of Diseases Associated with the Cardiovascular and Lymphatic Systems

Disease	Causative Agent	Mode of Transmission	Treatment
Leukemia	Human leukemia virus	Unknown	Methotrexate, vincristine, L-aparaginase, other anticancer drugs
Infectious mononucleosis	Epstein–Barr virus	Oropharyngeal secretions	Supportive
Acquired immune deficiency syndrome (AIDS)	HTLV-3 virus	Sexual contact, serum	Supportive
Serum hepatitis	Hepatitis B virus	Transfusions with contaminated blood and contaminated needles	Supportive
Dengue fever	Dengue virus	Bites of *Aedes aegypti*	Supportive
Yellow fever	Yellow fever virus	Bites of *Aedes aegypti*	Supportive
Septicemia	*E. coli, Enterobacter aerogenes, Pseudomonas aeruginosa, Proteus* spp., *Serratia marcescens*	Opportunistic infections	Supportive, antibiotics
Rheumatic fever	*Streptococcus pyogenes*	Complication of respiratory tract infection	Penicillin
Carditis	*Streptococcus* sp.	Complication of respiratory tract infection; abrasions of the oral cavity	Broad spectrum antibiotics
Typhoid fever	*Salmonella typhi*	Ingestion of contaminated food or water	Chloramphenicol
Toxic shock syndrome	*Staphylococcus aureus*	Wound or abrasion	Antitoxin
Puerperal fever	*Streptococcus, Staphylococcus,* other bacteria	Trauma of genital tract during birth	Penicillin
Plague	*Yersinia pestis*	Flea bites	Streptomycin and other antibiotics
Relapsing fever	*Borrelia*	Louse bites	Tetracycline, chloramphenicol, pesticides to eliminate lice
Bartonellosis	*Bartonella bacilliformis*	Sand fly bites	Chloramphenicol
Rocky Mountain spotted fever	*Rickettsia rickettsii*	Tick bites	Chloramphenicol, tetracycline
Epidemic typhus	*Rickettsia prowazekii*	Louse bites	Chloramphenicol, tetracycline, doxycycline
Endemic typhus	*Rickettsia typhi*	Flea bites	Chloramphenicol, tetracycline
Scrub typhus	*Rickettsia tsutusgamushi*	Mite bites	Chloramphenicol, tetracycline
Rickettsial pox	*Rickettsia akari*	Mite bites	Tetracyclines
Tularemia	*Francisella tularensis*	Skin abrasions; contact with infected animals	Streptomycin, chloramphenicol, tetracycline
Brucellosis	*Brucella*	Skin abrasions; contact with infected animals; ingestion of contaminated food	Tetracyclines
Lyme disease	Spirochete	Tick bites	Penicillin
Malaria	*Plasmodium* sp.	Mosquito bites	Chloroquinine and other antimalarials
Babesiosis	*Babesia microti*	Tick bites	Chloroquinine
Leishmaniasis	*Leishmania* sp.	Sand fly bites	Amphotericin B, sodium stibogluconate

Table 27.5 (*cont.*)

Disease	Causative Agent	Mode of Transmission	Treatment
Toxoplasmosis	*Toxoplasma gondii*	Ingestion of contaminated meat	Pyrimethamine and other antimalarials
Trypanosomiasis (Chaga's disease)	*Trypanosoma cruzi*	Cone-nosed bug bites	Aminoquinoline and other antiprotozoan drugs
Schistosomiasis	*Schistosoma mansoni, S. japonicum, S. haematobium*	Penetration of skin in waters inhabited by infected snails	Antimony-containing drugs
Elephantiasis	*Wuchereria bancrofti*	Mosquito bites	Supportive

Study Outline

A. The human circulatory system consists of the cardiovascular and lymphatic systems. Both have a fluid, blood or lymph, which circulates through blood or lymphatic vessels and lymph organs. The blood vascular system includes the heart, which provides the force needed to circulate the blood. The system carries essential nutrients to all body organs, removes their waste products, and carries products of cellular activity to other cells or to appropriate organs of excretion. The blood, the heart, and blood vessels together are the cardiovascular system.

  1. Blood is a tissue; its major functions are transportation of materials through the body and regulation of homeostasis.

    a. Erythrocyctes, red blood cells, carry oxygen from the lungs to cells and some carbon dioxide to the lungs.

    b. Leukocytes, white blood cells, function to protect against microorganisms.

  2. The heart is a pump that creates pressure on the blood, causing it to circulate through the body's blood vessels, which include the arteries, arterioles, blood capillaries, venules, and veins.

    a. Arteries carry blood away from the heart.

    b. Capillaries serve as the exchange point between blood and surrounding tissue; the direction of exchange is determined by the concentration gradient.

    c. Veins are the blood vessels leading back to the heart.

  3. The lymph vascular system functions as drainage tubes taking care of the overflow of water forced out of the capillaries. This lymph water also contains nongranular leukocyctes.

    a. The lymphatic vessels return about 10 percent of the fluid filtered from the blood vessels to the blood vascular system, carry large lipid molecules absorbed from the intestines to the blood circulation, return protein from tissue spaces, and provide a route for lymphocytes and monocytes produced in lymph organs to enter the blood circulation.

    b. The lymph organs include the lymph nodes, tonsils, spleen, and thymus. Lymph nodes are located along the course of collecting lymphatic vessels and any particulate matter in the lymph flow is removed by the lymphocyctes and macrophages within the node, providing an important means of defense.

B. In the blood several types of leukocytes are involved in the defense against pathogens.

  1. Basophils contain cytoplasmic granules, as do eosinophils and neutrophils, which are the most abundant phagocytic cells in the blood.

  2. Monocytes are important in the nonspecific immune response and are precursors of macrophages.

3. Macrophages, like neutrophils, are able to engulf and digest microorganisms. Macrophages are distributed throughout the body, including at fixed sites within the mononuclear phagocyte system.
4. Lymphocytes are involved in the specific immune defense system. T and B lymphocytes recognize and eliminate specific infectious microbes. They produce antibodies and interferon.

C. The circulatory system does not possess a resident microbiota; the blood is considered sterile.

1. The hepatic portal system contains low numbers of bacteria that pass through the intestinal wall, but these are eliminated by macrophages in the liver.
2. Transient bacteremia can occur throughout the circulatory system.

D. Diseases of the cardiovascular and lymphatic systems caused by viral pathogens.

1. Leukemia and other forms of cancer.
   a. Cancer is the result of the uncontrolled reproduction of malignant cells; when this occurs in the white blood cells it is called leukemia.
   b. Viruses that cause malignancies are called oncogenic viruses. Human T cell leukemia viruses, RNA viruses that reproduce by initially reversing the transcription process to form a DNA molecule which then acts as the template for the production of new viruses, cause leukemia. Such viruses are called retroviruses.
2. Infectious mononucleosis.
   a. Causative organism(s): Epstein–Barr (EB) virus.
   b. Mode of transmission: via aerosols and by direct exchange of oropharyngeal secretions containing the virus.
   c. Symptoms: sore throat, low grade fever, enlarged and tender lymph nodes, fatigue, and weakness.
3. Acquired immune deficiency syndrome (AIDS).
   a. Causative organism(s): HTLV-3 virus.
   b. Mode of transmission: sexually transmitted, also through blood and across placenta.
   c. Symptoms: suppression of immune response, malaise, swollen lymph nodes, various secondary infections such as *Pneumocystis* pneumonia, Kaposi sarcoma.
   d. Treatment: supportive.
4. Serum hepatitis (hepatitis B and nonA–nonB hepatitis)
   a. Causative organism(s): hepatitis B virus; hepatitis nonA–nonB virus.
   b. Mode of transmission: transfusions with contaminated blood; use of contaminated syringe needles.
   c. Symptoms: jaundice, similar to those of hepatitis A.
   d. Treatment: supportive; immune prophylaxis and vaccine used for prevention.
5. Dengue fever.
   a. Causative organism(s): RNA virus.
   b. Mode of transmission: via mosquito *Aedes aegypti* vector.
   c. Symptoms: rash and fever.
6. Yellow fever
   a. Causative organism(s): RNA virus.
   b. Mode of transmission: via mosquito vectors.
   c. Symptoms: anorexia, nausea, vomiting, fever, jaundice.
   d. Treatment: vaccine available.

E. Diseases caused by bacterial pathogens.

1. Septicemia (blood poisoning).
   a. Causative organism(s): frequently *Escherichia coli, Enterobacter aerogenes, Pseudomonas aeruginosa, Serratia marcescens*, and *Proteus* sp.
   b. Symptoms: red streaks under the skin along the arms and legs, high fever, low blood pressure, shock.
   c. Treatment: maintain body functions, limit effects of shock; antibiotics.

2. Rheumatic fever.
   a. Causative organism(s): *Streptococcus pyogenes*.
   b. Symptoms: high fever, painful swelling of various body joints, cardiac involvement.
   c. Treatment: penicillin, anti-inflammatory drugs.
3. Carditis (inflammation of the heart muscle).
   a. Causative organism(s): most frequently *Streptococcus viridans*.
   b. Mode of transmission: occurs when normally nonpathogenic members of the microbiota associated with body surfaces enter the circulatory system and attach to cardiac tissues.
   c. Treatment: broad spectrum antibiotics.
4. Typhoid fever.
   a. Causative organism(s): *Salmonella typhi*.
   b. Mode of transmission: via ingestion of contaminated water and food handled by infected individuals.
   c. Symptoms: fever, headache, apathy, weakness, abdominal pain, and a rash.
   d. Treatment: chloramphenicol.
5. Toxic shock syndrome.
   a. Causative organism(s): *Staphylococcus aureus*.
   b. Mode of transmission: the use of tampons produces lesions in the vaginal wall providing a portal of entry; surgical wounds.
   c. Symptoms: high fever, nausea, vomiting.
   d. Treatment: supportive.
6. Puerperal fever (childbed fever).
   a. Causative organism(s): most frequently β-hemolytic group A and B *Streptococcus* species.
   b. Mode of transmission: via the obstetrician, obstetrical instruments, or bedding.
   c. Treatment: penicillin.
7. Plague.
   a. Causative organism(s): *Yersinia pestis*.
   b. Mode of transmission: via transfer within rodent populations by fleas who then bite people.
   c. Symptoms: inflammation and enlargement of lymph nodes, malaise, fever, pain, tissue necrosis.
   d. Treatment: antibiotics, generally streptomycin.
8. Relapsing fever.
   a. Causative organism(s): various species of *Borrelia*.
   b. Mode of transmission: via bite of the body louse, also tick vectors.
   c. Symptoms: occur intermittently; a sudden onset of fever approaching 105°F lasts for 3–6 days, then falls rapidly and remains normal for 5–10 days, followed by a second onset of fever that generally lasts 2–3 days.
   d. Treatment: tetracycline and chloramphenicol.
9. Bartonellosis.
   a. Causative organism(s): *Bartonella bacilliformis*.
   b. Mode of transmission: through bites of infected sand flies.
   c. Symptoms: severe anemia, fever, headache, delirium.
   d. Treatment: chloramphenicol.
10. Rocky Mountain spotted fever.
    a. Causative organism(s): *Rickettsia rickettsii*.
    b. Mode of transmission: via tick bite.
    c. Symptoms: skin rash, severe headaches, mental confusion.
    d. Treatment: antibiotics, such as chloramphenicol.
11. Typhus fever.
    a. Causative organism(s): *Rickettsia prowazekii*.

b. Mode of transmission: via the body louse.
c. Symptoms: fever, headache, rash, heart and kidney lesions.
d. Treatment: chloramphenicol, doxycycline, and tetracycline.
e. Other forms: Brill–Zinsser disease—relapse of louse-borne typhus; murine or endemic typhus—caused by *R. typhi*, transmitted by rat fleas; scrub typhus—caused by *R. tsutsugamushi*, transmitted by the bite of mite vectors, producing a characterisitic skin lesion called an eschar.

12. Rickettsialpox.
- **a.** Causative organism(s): *Rickettsia akari*.
- **b.** Mode of transmission: mouse mite vector.
- **c.** Symptoms: fever, headache, primary lesion at the bite site and secondary lesions, rash.
- **d.** Treatment: tetracyclines for symptoms.

13. Tularemia.
- **a.** Causative organism(s): *Francisella tularensis*.
- **b.** Mode of transmission: ingestion of contaminated meat and handling of infected animals.
- **c.** Symptoms: ulcers on fingers and skin, elevated fever, enlarged lymph nodes.
- **d.** Treatment: streptomycin, tetracycline, and chloramphenicol.

14. Brucellosis (undulent fever).
- **a.** Causative organism(s): several *Brucella* species.
- **b.** Mode of transmission: via ingestion of contaminated milk and handling infected animals.
- **c.** Symptoms: enlargement of spleen and liver, chills, weakness, headache, back-ache, and fever, which may rise and fall.
- **d.** Treatment: tetracyclines; vaccines available to limit spread through some animal populations.

15. Lyme disease.
- **a.** Causative organism(s): spirochete.
- **b.** Mode of transmission: tick bite.
- **c.** Symptoms: rash followed by arthritis of the knee and possible neurologic and cardiac abnormalities.
- **d.** Treatment: penicillin.

F. Diseases caused by protozoan pathogens.

1. Malaria.
- **a.** Causative organism(s): most frequently *Plasmodium vivax* and *P. falciparum*.
- **b.** Mode of transmission: via *Anopheles* mosquito vector.
- **c.** Symptoms: periodic chills, fever, headache, muscle ache.
- **d.** Treatment: antimalarial drugs, such as chloroquine.

2. Babesiosis.
- **a.** Causative organism(s): *Babesia microti falciparum*.
- **b.** Mode of transmission: via tick vector.
- **c.** Symptoms: periodic chills, fever, headache, muscle ache.
- **d.** Treatment: chloroquine.

3. Leishmaniasis.
- **a.** Causative organism(s): *Leishmania* sp.
- **b.** Mode of transmission: via sand fly vectors.
- **c.** Symptoms: *L. mexicana* and *L. tropica* cause skin infections (Old World cutaneous leishmaniasis or Oriental Sore); *L. donovani* multiplies throughout the mononuclear phagocyte system (kala-azar disease).
- **d.** Treatment: skin infections treated with amphotericin B; kala-azar treated that or with sodium stibogluconate.

4. Toxoplasmosis.
- **a.** Causative organism(s): *Toxoplasma gondii*.

b. Mode of transmission: ingestion of meat containing cysts of *T. gondii*; also congenitally.
c. Symptoms: often asymptomatic; can lead to central nervous system involvement.
d. Treatment: antimalarial drugs, such as pyrimethamine.

5. Trypanosomiasis (Chagas' disease).
   a. Causative organism(s): *Trypanosoma cruzi*.
   b. Mode of transmission: via infected tryatomine (cone-nosed) bugs.
   c. Symptoms: affects mononuclear phagocyte system, later the myocardium and nervous systems.
   d. Treatment: antiprotozoan drugs, such as aminoquinoline.

G. Diseases caused by multicellular parasites.
1. Schistosomiasis.
   a. Causative organism(s): blood flukes.
   b. Mode of transmission: able to penetrate human skin.
   c. Symptoms: penetration stage is asymptomatic, acute state occurs when the organisms start to lay eggs and include diarrhea, fever, and discomfort, late stage includes cirrhosis.
   d. Treatment: antimony-containing drugs.
2. Elephantiasis.
   a. Causative organism(s): nematode *Wuchereria bancrofti*.
   b. Mode of transmission: via repeated bites of infected mosquitoes.
   c. Symptoms: accumulation of lymph results in severe swelling of limbs and other body parts.
   d. Treatment: Supportive.

Study Questions

1. For each of the following diseases, what are the causative organisms and characteristic symptoms?
   a. Rocky Mountain spotted fever
   b. Endemic typhus
   c. Rickettsialpox
   d. Epidemic typhus
   e. Malaria
   f. AIDS
2. Match the following statements with the terms in the list to the right.
   a. Caused by Epstein–Barr (EB) virus.
   b. Usually caused by attachment to cardiac tissues of normally nonpathogenic members of body surface microbiota that have entered the circulatory system.
   c. Generally most serious consequence of *Streptococcus pyogenes* infections.
   d. Characteristic symptoms include high fever, painful swelling of various body joints, and cardiac involvement; normally begin to occur two weeks after characteristic sore throat associated with upper respiratory tract infection.
   e. Viridans streptococci most frequently identified etiologic agent.
   f. Systemic infection caused by *Salmonella typhi*.
   g. Associated with contaminated water supplies and the handling of food products by contaminated individuals.
   h. Correlated with use of tampons during menstruation.
   i. Caused by strains of *Staphylococcus aureus* that contain lysogenic phage.
   j. In the bubonic form, the infection becomes localized, causing inflammation in re-

   (1) Infectious mononucleosis
   (2) Relapsing fever
   (3) Rocky Mountain spotted fever
   (4) Rheumatic fever
   (5) Toxic shock syndrome
   (6) Plague
   (7) Serum hepatitis
   (8) Typhoid fever
   (9) Malaria
   (10) Endocarditis
   (11) Endemic typhus fever
   (12) Epidemic typhus fever
   (13) Rickettsialpox
   (14) Scrub typhus
   (15) Leishmaniasis
   (16) Trypanosomiasis

gional lymph nodes; symptoms include malaise, fever, and pain in areas of infected lymph nodes; severe tissue necrosis may appear as blackened skin.

k. Flea vector.
l. Caused by *Yersinia pestis*.
m. Caused by various species of *Borrelia*.
n. Permanently established in rodent populations from Rocky Mountains to the West Coast of the United States.
o. Caused by *Rickettsia prowazekii*.
p. Caused by *Rickettsia rickettsii*.
q. Caused by *Rickettsia typhi*.
r. Sand fly vector (occurs only in valleys of Andes Mountains).
s. Body louse vector.
t. Rat reservoir; rat flea vector.
u. Leading cause of cardiovascular death in South America.
v. After inoculation into the body, sporozoites begin to reproduce within liver cells (single sporozoite can produce up to 40,000 merozoites); merozoites can invade erythrocytes, causing anemia.
w. Caused by *Leishmania* species.
x. Caused by four species of *Plasmodium*; *P. vivax* and *P. falciparum* most often involved in human infections.
y. One of most common infectious diseases worldwide.
z. Chills and fever are periodic, generally lasting less than 6 hours; coincide with rupture of infected erythrocytes.

Some Thought Questions

1. Why are only drug abusers, homosexuals, and health care workers being given the hepatitis B vaccine?
2. Why were we able to eliminate smallpox but not plague?
3. What is the basis for the legend of the Pied Piper of Hamlin?
4. Should cultures of smallpox virus be maintained?
5. Why is malaria the most prevalent human disease?
6. Why are cases of malaria in the United States associated with military actions?
7. Would you volunteer to help identify the etiologic agent of AIDS?
8. Should health care workers be required to treat patients with AIDS?
9. Why should women discontinue use of tampons and call a physician if they develop a high fever or other symptoms of disease during their menstruation period?

Additional Sources of Information

Boyd, R. F., and B. G. Hoerl. 1981. *Basic Medical Microbiology*. Little, Brown and Company, Boston. A basic text emphasizing microbes and disease that includes chapters on systemic infections.

Braude, A. I. (ed.). 1980. *Medical Microbiology and Infectious Disease*. W. B. Saunders Company, Philadelphia. An advanced work with individual chapters dealing specific diseases of the cardiovascular and lymphatic systems; chapters dealing with infectious diseases are organized according to the taxonomy of the causative agent.

Burgdorfer, W., and R. Anacker. 1982. *Rickettsiae and Rickettsial Diseases*. Academic Press, Inc., New York. An advanced treatise on diseases caused by bacteria that are obligate intracellular parasites.

Davis, B. D., R. Dulbecco, H. N. Eisen, H. S. Ginsberg, and W. B. Wood. 1981. *Microbiology*. Harper & Row, Publishers, Hagerstown, Maryland. An excellent text used in many graduate and medical school microbiology courses dealing with all aspects of microbiology; chapters dealing with infectious diseases are organized according to the taxonomy of the causative agent.

Faust, E. C., P. C. Beaver, and R. C. Jung. 1975. *Animal Agents and Vectors of Human Disease*. Lea & Febiger, Philadelphia. This work examines the role of insects and other vectors in the transmission of human diseases, many of which involve the cardiovascular and lymphatic systems.

Hoeprich, P. D. (ed.). 1977. *Infectious Diseases: A Modern Treatise of Infection Processes*. Harper & Row, Publishers, Hagerstown, Maryland. This classic treatment of infectious diseases makes an excellent reference volume.

Jawetz, E., J. L. Melnick, and E. A. Adelberg. 1980. *A Review of Medical Microbiology*. Lange Medical Publications, Los Altos, California. This text presents a very good review of infectious diseases.

Joklik, W. K., and H. P. Willett (eds.). 1980. *Zinsser Microbiology*. Appleton Century Crofts, New York. A classic text used in many medical school programs that includes chapters on infectious diseases.

Rose, N. L., and A. L. Rose (eds.). 1983. *Microbiology: Basic Principles and Clinical Applications*. Macmillan Publishing Co., Inc., New York. An advanced work on medical microbiology; chapters are organized for the most part according to taxonomic groups.

Youmans, G. P., P. Y. Paterson, and H. M. Sommers. 1980. *The Biological and Clinical Basis of Infectious Disease*. W. B. Saunders Company, Philadelphia. This is an excellent text covering most human infectious diseases.

28 *Skin and Wound Infections*

KEY TERMS

acne
anthrax
chicken pox
dermatophytic fungus
dermis
epidermis
folliculitis
gas gangrene
German measles
Hansen's disease
impetigo
leprosy
measles
mycetoma
provirus
pyoderma
ringworm
scarlet fever
shingles
skin
smallpox
sporotrichosis
tinea
warts
wound botulism

OBJECTIVES

After reading this chapter you should be able to

1. Define the key terms.
2. Describe the structure of the skin.
3. Describe the resident microbiota of the skin.
4. Describe the defense systems of the skin that protect against infectious microorganisms.
5. Discuss the causative agents, routes of transmission, key symptoms, and treatments for the following diseases: measles, German measles, chicken pox, shingles, smallpox, warts, leprosy, anthrax, pyoderma, impetigo, scarlet fever, acne, mycetoma, ringworm, sporotrichosis, gas gangrene, and wound botulism.

Structure of the Skin

The average adult human has a skin surface area of nearly 2 square meters. The thickest skin, 5 to 6 millimeters, occurs on the palms of the hands and the soles of the feet. The skin covering the rest of the body is much thinner, less than 3 millimeters thick.

The outer layer of the skin is the epidermis, which is composed largely of epithelial cells. The outermost layer of the epidermis consists of dead cells that contain the protein keratin. Keratin, which also occurs in hair and nails, is quite hydrophobic, making it water repellant. Most microorganisms are unable to metabolize keratin, which makes the intact skin an effective barrier to microorganisms. As the external covering of the body, the skin surface serves as a physical barrier against microbial invasion. It also guards against the loss of vital internal body constituents, such as water, salts, and organic molecules. Breaks in the continuity of the skin, such as result from a variety of types of wounds, allow fluids to leave the body and provide an opening through which many disease-causing microorganisms may enter.

The inner, thicker layer of the skin is called the dermis. This layer is composed of a thick layer of dense fibrous connective tissue. Two layers, the papillary layer and the reticular layer, subdivide the dermis. The papillary layer gets its name from the papillae, small nipple-shaped projections, that are arranged in ridges on its surface. Because the epidermis conforms to the surface of the dermis underlying it, it too has ridges which appear as the well-defined, unique fingerprint patterns. The reticular layer is the innermost layer of the dermis and consists of a reticulum or network of fibers, most of which are collagenous fibers that give the skin its toughness. The dermis contains an abundant supply of blood vessels, lymph vessels, and nerve endings. It also contains hair follicles (small tubes from which hair grows) and sweat glands. Hairs penetrate out from the skin surface. Sweat glands have ducts that end at the surface of the epidermis. Sweat from these glands is secreted as perspiration onto the surface of the skin. Excretion of sodium chloride and urea takes place via sweat glands. The evaporation of sweat on the skin surface carries heat with it, cooling the body when the body's temperature is too high. Both hairs and sweat ducts provide possible portals of entry by which microorganisms can penetrate into the skin tissues and enter the body.

Factors Influencing Growth of Microorganisms on the Skin

Human skin surfaces are not especially favorable habitats for the growth of microorganisms (Figure 28.1). Even though the skin is continuously exposed to microorganisms, most of these microorganisms are unable to reproduce there. A major factor that determines the distribution of microbial populations on the skin surface is the microenvironment. Environmental factors, such as temperature, water activity, pH, and salinity, represent a severe environmental stress that can prevent many microbial populations from colonizing the skin surface. The lack of available water is a major limiting factor controlling the extent of such microbial growth. Although sweat glands secrete fluids containing water onto the skin surface, sweat contains high concentrations of salt and organic wastes that inhibit most microbial growth. The microorganisms that become established within the resident microbiota of the skin must tolerate the osmotic stress caused by the high salt concentrations there.

Additionally, the presence of antimicrobial agents, including free fatty acids produced by the host animal and antimicrobial compounds produced by those microorganisms that do successfully establish themselves as the normal microbiota of the skin, act to prevent foreign microorganisms from growing on the skin surface. Lipids on the skin surface are derived mainly from the secretion of sebum from the sebaceous glands. Sebum is secreted into hair follicles, and although some bacteria are able to colonize these depressions in the skin surface, many of the unsaturated free fatty acids in sebum have antimicrobial activity. Those microorganisms inhabiting the skin surface must not only tolerate the natural antimicrobial biochemical secretions but also metabolize these compounds as their source of nutrients. Furthermore, many microorganisms that successfully inhabit the skin surface produce antimicrobial substances; these compounds, some of which are low molecular weight fatty acids, act to prevent the invasion of the skin surface by other microbial populations. Another factor that limits microbial growth on the skin is desquamation; as the microorganisms become attached to the skin some of it sloughs off, physically removing the microbes.

Proper cleanliness habits and good hygienic practices tend to prevent the establishment of nonindigenous microorganisms among the natural skin

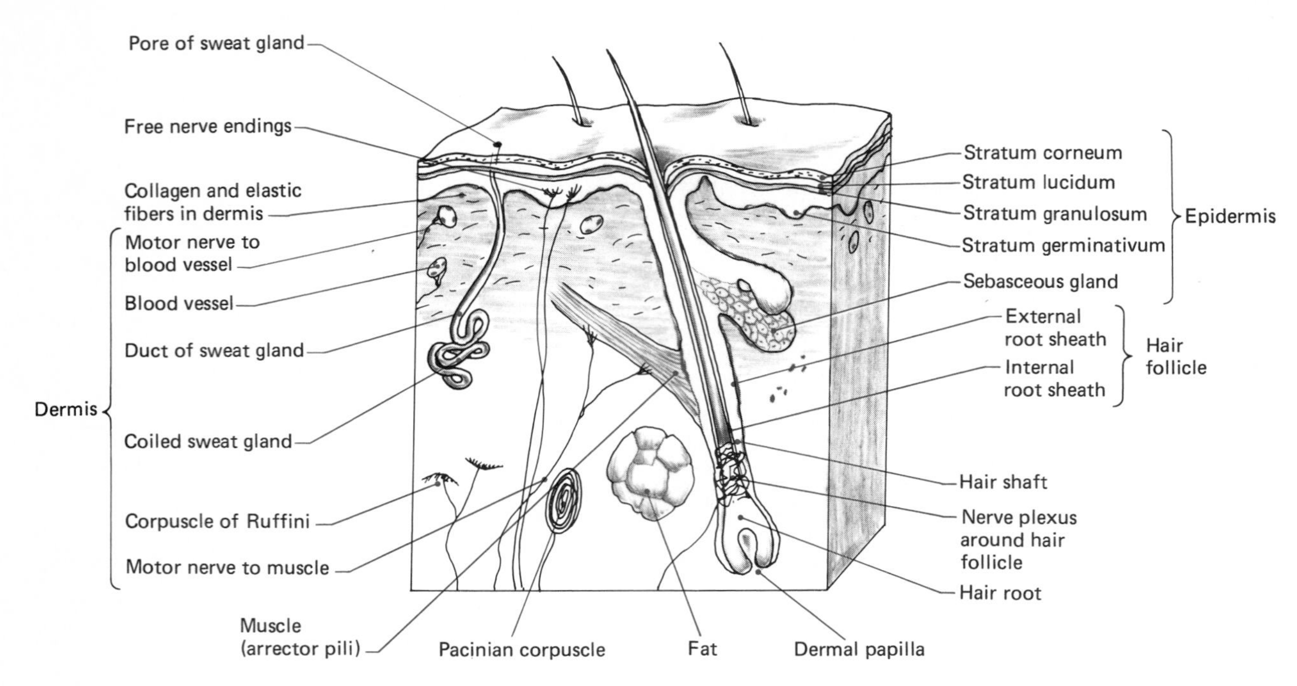

Figure 28.1 Drawing of the surface of the skin, indicating some of the indigenous microbiota found on the human skin surface.

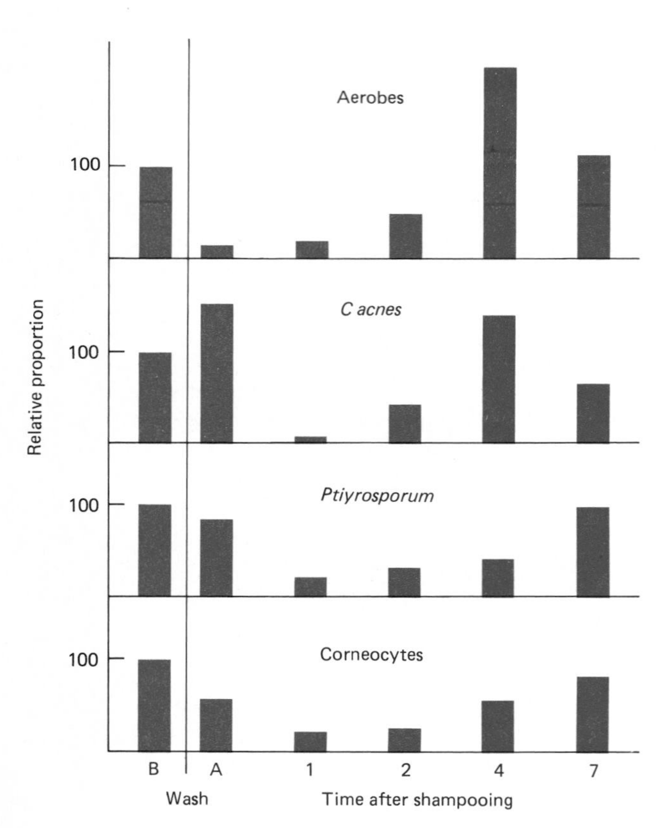

Figure 28.2 The changes in microbial populations on the human scalp after shampooing. (Reprinted by permission of the Society for Applied Bacteriology from R. P. Marples, 1974, "Effects of germicides on skin flora," in *The Normal Microbial Flora of Man,* F. A. Skinner and J. G. Carr, eds., © 1974, Society for Applied Bacteriology.)

microbiota by preventing the buildup of excessive concentrations of organic matter. The presence of high concentrations of organic matter can make the skin surface a more favorable environment for microorganisms, permitting the growth of many microorganisms that otherwise could not grow there. Washing the skin surface, however, removes excess organic matter and temporarily reduces the numbers of both resident and transient microorganisms (Figure 28.2). Shortly after the skin surface is washed, the populations of indigenous microbiota on the skin surface begin to return to their original numbers.

Resident Microbiota of the Skin

The dominant microbial populations on the skin surface are Gram positive bacteria, which are normally relatively resistant to desiccation compared to Gram negative bacteria. Members of the genera *Staphylococcus* (cocci occurring in clusters) and *Micrococcus* (cocci occurring singly or in small clusters) are frequently the most abundant microorganisms on the skin surface (Figure 28.3) because they are generally salt tolerant and can utilize the lipids present on the skin surface. Other Gram positive, rod-shaped bacteria usually found as part of the normal microbiota of the skin surface include *Corynebacterium*, *Brevibacterium*, and *Propionibacterium* species. Gram negative bacteria generally occur primarily in the moister regions of the skin surface, such as in the arm pits and between the toes. Relatively few fungi are included in the normal microbiota of the skin surface. Two yeasts, *Pityrosporum ovale* and *P. orbiculare*, though, are able to metabolize the lipids found on the skin surface and normally occur on scalp tissues. Additionally, some dermatophytic fungi occasionally grow on the skin surface, producing diseases such as athlete's foot and ringworm.

The most abundant microbial growth on the skin occurs in moist regions, such as under the arms, between the toes, and on the scalp. The excessive growth of microbes on the organic compounds carried in sweat in these regions can result in the production of odoriferous compounds, which are major contributors to body odor. Smelly feet, for example, occur when bacteria grow on moist sweaty skin surface and produce volatile fatty acids that have strong odors.

Diseases that Affect the Skin

Diseases Caused by Viral Pathogens

Measles (rubeola). Measles or rubeola is a highly contagious disease occurring on a worldwide basis almost exclusively in children. The measles virus is readily transmitted from an infected child to a susceptible host, as illustrated by the fact that there is greater than a 90 percent incidence of an acute infection after exposure to measles virus by sus-

Figure 28.3 Scanning electron micrographs of bacteria on the surface of human skin. (A) Skin, showing coccal-shaped bacteria (5000 ×). (B) Skin showing coccal-shaped bacteria and crystalline deposits from sweat (5000 ×). (Courtesy Robert P. Apkarian, University of Louisville.)

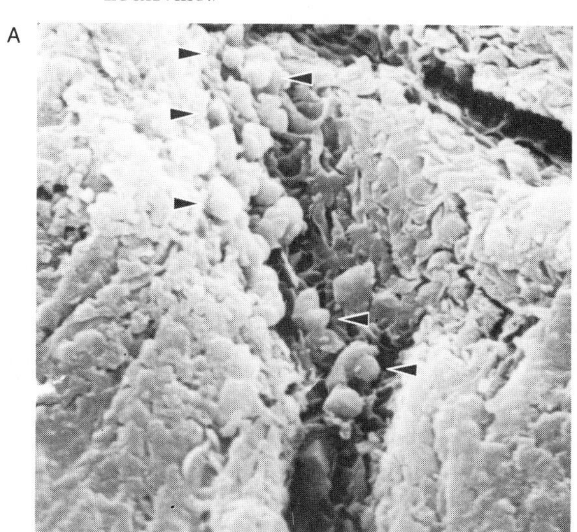

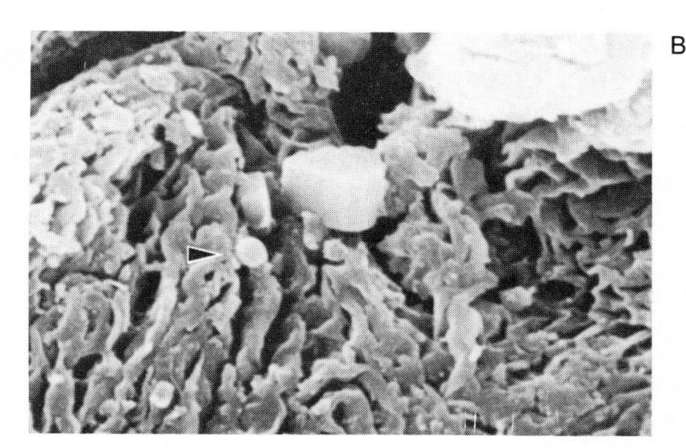

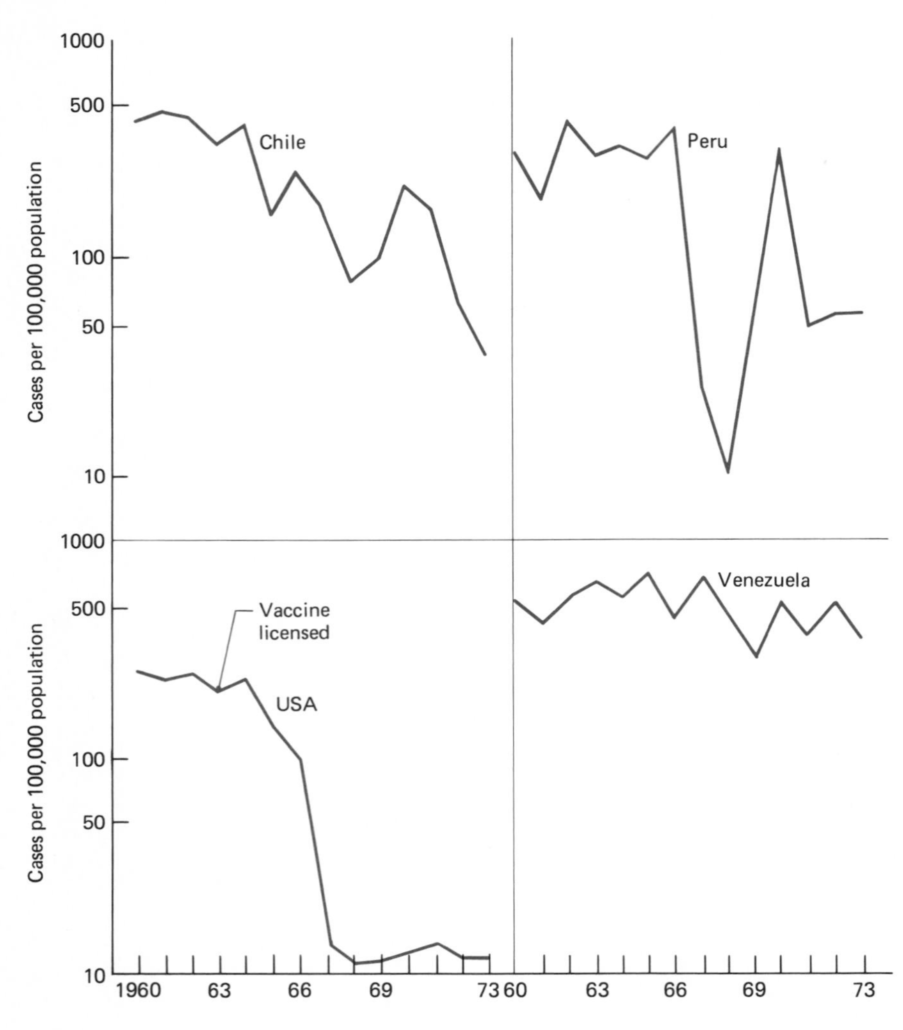

Figure 28.4 Reported cases of measles per 100,000 population in four countries from 1960 to 1973. Although the incidence of this disease has been greatly reduced in the United States through immunization practices, it remains a major problem in other countries.

ceptible children. Measles can be prevented by childhood immunization, and the rate of measles infections in the United States, at least, has been declining regularly in recent years, although in other countries measles has not declined in the same way (Figure 28.4). After initial viral multiplication in the mucosal lining of the upper respiratory tract, measle viruses appear to be disseminated to lymphoid tissues where further multiplication occurs. Before the onset of symptoms, large numbers of measles viruses are shed in secretions of the respiratory tract, the eyes, and in urine, promoting the rapid epidemic spread of this disease. Infection with measles viruses can involve a number of organs, and there is a high rate of mortality associated with measles in regions of the world where malnutrition and limited medical treatment facilities predominate. When measles is fatal, the virus generally invades the central nervous system causing encephalitis.

Measles is characterized by the eruption of a maculopapular skin rash approximately 14 days after exposure to the measles virus (Figure 28.5). The rash generally appears initially behind the ears, spreading rapidly to other areas of the body during the next 3 days. Disease symptomology often begins a few days before the onset of the characteristic measles rash. These initial symptoms include high fever, coughing, sensitivity to light, and the appearance of Koplik's spots (red spots with a white dot in the center that occur in the oral cavity generally first appearing on the inner lip). Treatment is normally supportive, including rest and the intake of sufficient fluids. In uncomplicated cases, the fever disappears within 2 days, and the individual returns to normal activities a few days later. If the

MICROBIOLOGY HEADLINES

56 Shun Measles Vaccine, Stay Quarantined at Camp

ORR, Minn.—Fifty-six teenagers isolated at a church camp by a measles outbreak two weeks ago will stay in "voluntary quarantine" for two more weeks because they have refused to be vaccinated on grounds it would violate their religious beliefs, officials said Saturday.

Another 280 campers, from the United States, Europe, South Africa and Australia, already have accepted measles immunizations and left the Youth Opportunities United camp, said Kevin Dean, who runs the camp for the California-based Worldwide Church of God.

The remaining 56 teens could leave if they agreed to be vaccinated, but as long as healthy teens remain isolated in camp with ill ones, the disease could spread, said Lon Anderson, communicable disease specialist for the St. Louis County Health Department.

Anderson said the camp, about 100 miles north of Duluth, had reported a second measles outbreak, bringing to 20 the number of cases since Aug. 3.

Those who refused vaccination were advised to stay under "voluntary quarantine," he said. "They would have been infectious to their communities."

The teens have refused immunization for "religious or personal reasons," said Dean.

"It's the church's philosophy that God heals," and some church members may interpret that to mean they cannot receive vaccinations, but the church does not have such a policy, he said.

There were 336 campers when the outbreak began, and many left at that time, Dean said. But others "didn't want to go home . . . they wanted a longer vacation."

Dean said the church, a fundamentalist Christian sect with about 150,000 members worldwide, has run the camp since 1972 at an annual cost of $900,000.

Source: Associated Press, Aug. 19, 1984. Reprinted by permission.

---

fever persists for more than 2 days after the eruption of the rash, it is likely that a complication, such as bronchitis or pneumonia, has developed. In these cases additional treatment is needed to cure secondary infections. Prolonged or latent measles infections may also cause immune complex reactions. Such immune complex hypersensitivity reactions may be responsible for neurological complications of measles infections, including the condition subacute sclerosing panencephalitis (SSPE).

The night it was horribly dark
The measles broke out in the Ark;
Little Japheth, and Shem, and all the young Hams,
Were screaming at once for potatoes and clams.
And 'What shall I do,' said poor Mrs. Noah,
'All alone by myself in this terrible shower?
I know what I'll do: I'll step down in the hold,
And wake up a lioness grim and old,
And tie her close to the children's door,
And give her a ginger-cake to roar
At the top of her voice for an hour or more;
And I'll tell the children to cease their din,
Or I'll let that grim old party in,
To stop their squeazles and likewise their measles.'
She practiced this with the greatest success:
She was everyone's grandmother, I guess.

—SUSAN COOLIDGE, *Measles in the Ark*

German Measles. Like measles, transmission of rubella virus, the causative agent of German measles (rubella), appears to be via droplet spread, with the initial infection occurring in the upper respiratory tract. In contrast to the measles virus, however, the rubella virus exhibits a relatively low rate of infectivity, and thus, relatively prolonged exposure appears to be needed for the establishment of an infection. After multiplication in the mucosal cells of the upper respiratory tract, rubella viruses appear to be disseminated systemically through the blood. Approximately 18 days after initiation of the infection by the rubella virus, a characteristic rash,

Figure 28.5 Photograph showing the characteristic papular measles rash. (From Centers for Disease Control, Atlanta.)

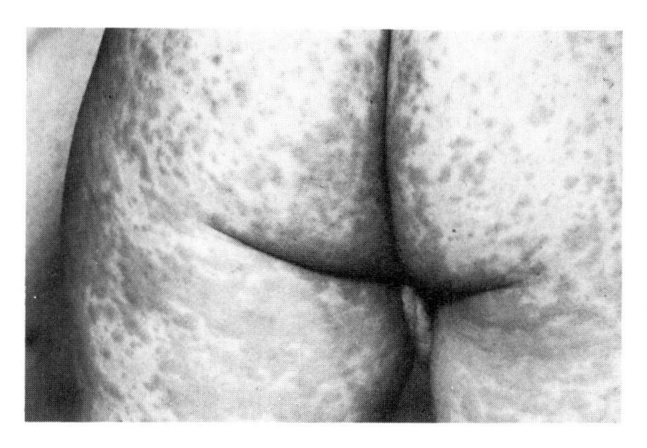

appearing as flat pink spots, occurs on the face and subsequently spreads to other parts of the body. Enlarged and tender lymph nodes and a low-grade fever characteristically precede the occurrence of the German measles rash. In children and adolescents, rubella is usually a mild disease. If it is acquired during pregnancy, however, the fetus can become infected with the rubella virus, resulting in congenital rubella syndrome, characterized by the development of multiple abnormalities in the infant, such as mental retardation and deafness. Generally the earlier in pregancy the German measles occurs, the more serious are the effects on the fetus. There is a very high rate of mortality, exceeding 25 percent, in cases of congenital rubella syndrome. Vaccination has greatly reduced the incidence of rubella (Figure 28.6) in children and is also used to confer immunity on women of childbearing age who had not contracted the disease at an earlier age.

Chicken Pox and Shingles. Chicken pox (varicella) is caused by the varicella-zoster virus, a member of the herpesvirus group, which also causes shingles (Herpes zoster) (Figure 28.7). Ninety percent of all cases of chicken pox occur in children under 9 years of age. In children, chicken pox is generally a relatively mild disease, but when this disease occurs in adults the symptoms are more frequently severe. Chicken pox is a highly contagious disease and probably is transmitted via con-

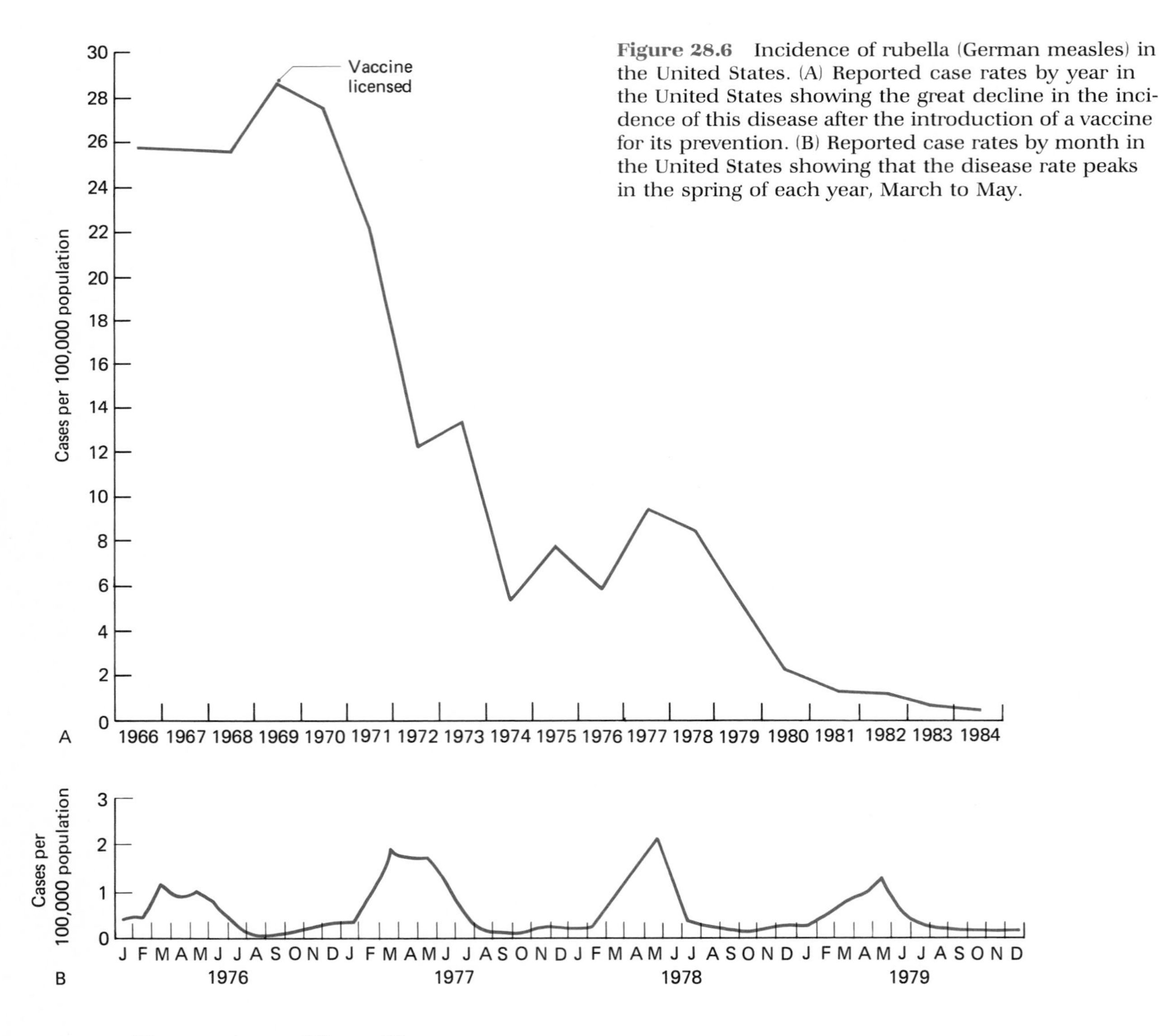

Figure 28.6 Incidence of rubella (German measles) in the United States. (A) Reported case rates by year in the United States showing the great decline in the incidence of this disease after the introduction of a vaccine for its prevention. (B) Reported case rates by month in the United States showing that the disease rate peaks in the spring of each year, March to May.

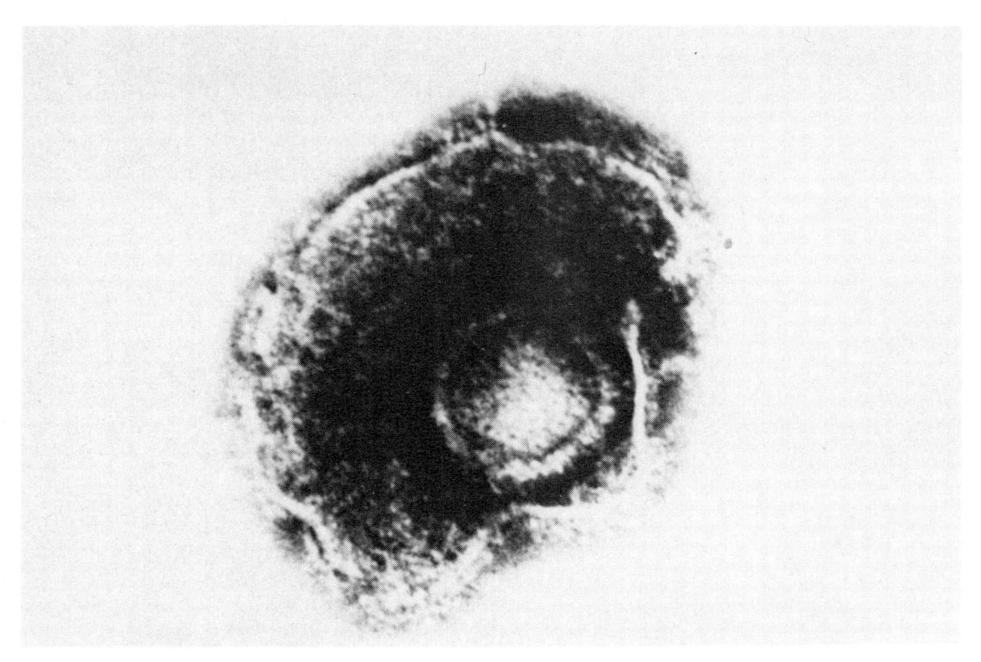

Figure 28.7 Electron micrograph of varicella (chicken pox) virus. (From Photoresearchers Inc.)

taminated droplets and direct contact with vesicle fluid containing varicella-zoster viruses. The initial site of viral replication has not been positively established but appears to be in the upper respiratory tract. Local vesicular lesions occur in the skin after dissemination of the virus through the body (Figure 28.8). These skin lesions become encrusted, and the crusts fall off in about 1 week. Vesicles also occur on mucous membranes, especially in the mouth.

In some cases, the varicella-zoster virus spreads to the lower respiratory tract, resulting in pneumonia; in this way several other tissues, including the central nervous system, can also be involved in complicated cases of chicken pox. Approximately 100 deaths occur each year in the United States due to encephalitis caused by the spread of chicken pox viruses to the central nervous system. Complications from a case of chicken pox are a particular danger in immunocompromised individuals and gamma-2-immune globulin is used as a prophylactic measure when such individuals develop a case of chicken pox. Additionally, there is an elevated incidence of Reye's syndrome associated with occurrences of chicken pox, especially if aspirin is used during the chicken pox. Unlike the other childhood viral diseases, vaccination practices against chicken pox have not yet been introduced, and outbreaks of chicken pox continue to show regular seasonal cycles of the same magnitude (Figure 28.9). Clinical trials of an experimental vaccine have produced promising results, and this disease may also soon be controlled by vaccination.

In adults, shingles is the principal disease resulting from reactivation of infections with the varicella-zoster virus. The varicella-zoster virus ac-

Figure 28.8 Photograph showing the characteristic skin eruptions of chicken pox. (From United States Armed Forces Institute of Pathology, neg. no. AMH-10529-E.)

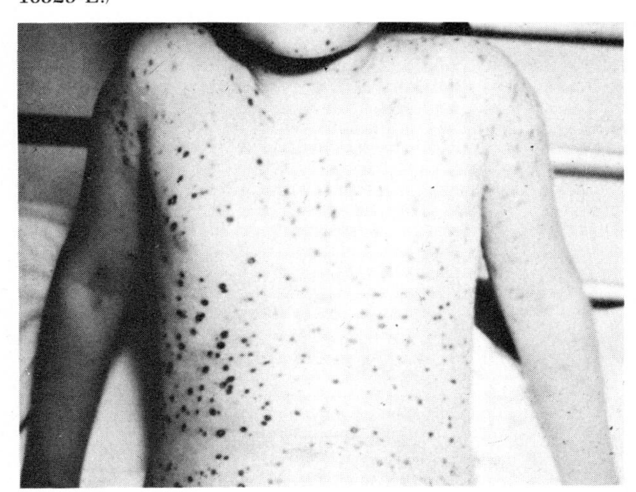

quired in childhood can remain as a provirus within the body; that is, the viral DNA can be incorporated into human chromosomes in a state of lysogeny. The presence of such proviruses can later become manifest as a new disease. In cases of shingles the virus reaches the sensory ganglia of the spinal or cranial nerves, producing an inflammation. There is usually an acute onset of pain and tenderness along the affected sensory nerves. A rash develops along the skin overlying affected nerves usually lasting for 2–4 weeks, but the pain may last for weeks or months. The skin lesions contain chicken pox viruses, and susceptible individuals who come in contact with these lesions can acquire chicken pox. Usually the symptoms of shingles are restricted to one side of the body at a time because the disease follows the branches of the cutaneous sensory nerves.

Warts. Warts are benign tumors of the skin caused by papillomaviruses, which are small icosahedral DNA viruses. Transmission of wart viruses appears to occur primarily by direct contact of the skin with wart viruses from an infected individual, although indirect transfer also may occur through fomites. The human papillomaviruses appear to infect only humans and no other animals, with children developing warts more frequently than adults. The development of warts can occur on any of the body surfaces and the appearance of the warts varies, depending on its location (Figure 28.10). At present there is no effective antiviral treatment for human warts, and therapy often involves destruction of in-

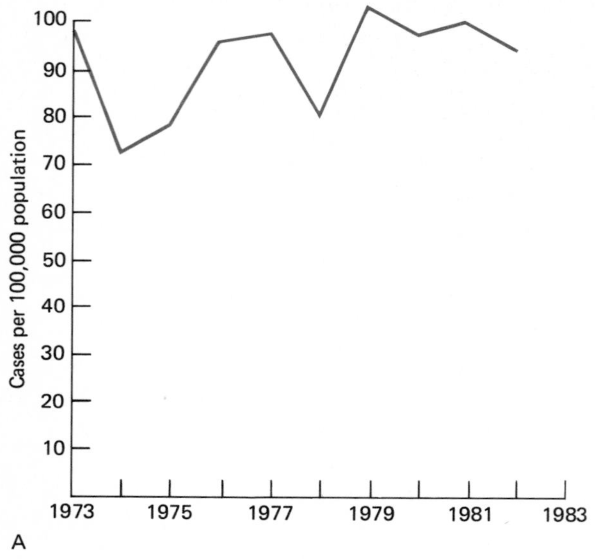

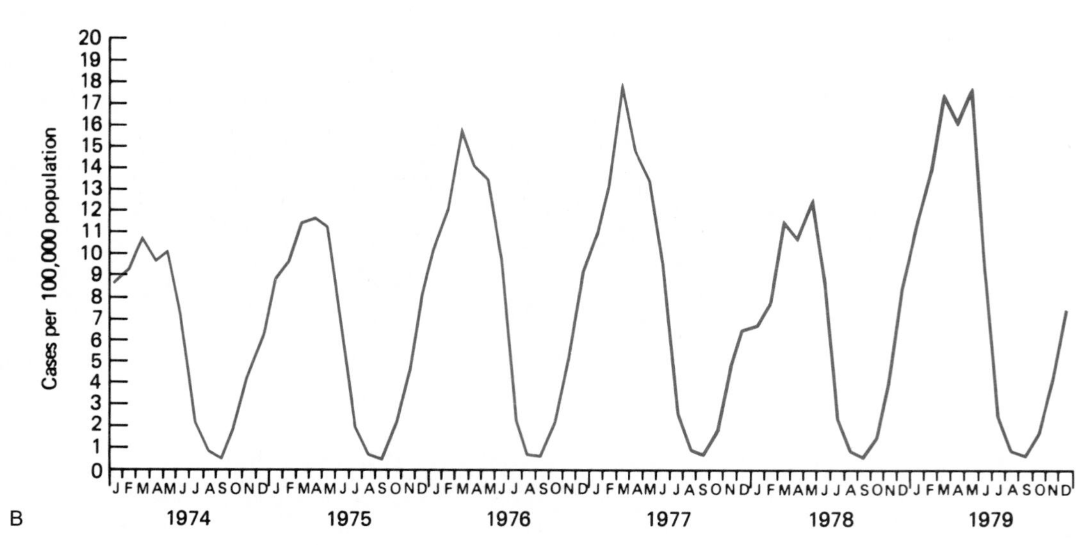

Figure 28.9 Reported case rates of chicken pox by month in the United States. At present there is no licensed vaccine for preventing this disease. (A) The annual incidence of chicken pox has not significantly declined. (B) The seasonal pattern has remained fairly constant with the peak incidence occurring between March and May.

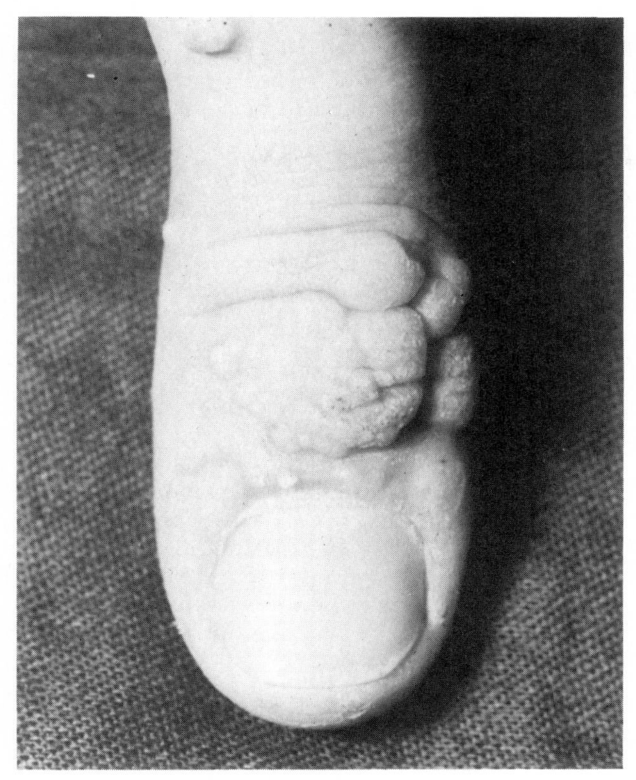

Figure 28.10 Appearance of warts on the thumb. This infection commonly occurs on the hands and face. (From United States Armed Forces Institute of Pathology, neg. no. AMH 10737-2.)

fected tissues by applying acid or freezing. In general, human warts is a self-limiting disease and recovery can be expected without treatment within 2 years.

Smallpox. Smallpox is the only human disease that has been totally eliminated. This is a remarkable event considering that in the Middle Ages there was an 80 percent chance of contracting smallpox. The last naturally occurring case of smallpox occurred in Somalia, Africa, in 1977. One later case in 1978 was the result of a laboratory accident in England. The successful elimination of smallpox was the result of an extensive worldwide vaccination program, coordinated by the World Health Organization, the fact that humans are the only known host for the smallpox (variola) virus, and there are no asymptomatic carriers of the disease. As a result of the elimination of succeptible hosts the virus and the dread disease was eliminated. Consequently, it is no longer necessary to vaccinate individuals against this disease. (The only exceptions are certain workers at laboratories such as the Centers for Disease Control, where smallpox viruses are still maintained for research purposes.)

When it existed, smallpox was transmitted via aerosols and entered the body through the respiratory tract. The virus moved through the circulatory system, reaching the skin, where it caused the noticeable symptoms of this disease. The growth of the virus in the epidermal layers of the skin caused the development of a rash that eventually formed pustules (raised blisters containing pus). These lesions developed over the entire body (Figure 28.11). Lesions also occur internally, where the virus infects mucous membranes. In addition to the obvious lesions over the entire body, the symptoms of smallpox included fever, malaise, and pain in the head, back, and stomach. In one form of this disease, variola major, the mortality rate was greater than 20 percent, but in another form, variola minor, less than 1 percent of the individuals infected with

Figure 28.11 A smallpox victim in Africa—this disease no longer occurs. (From *World Health,* courtesy of the World Health Organization, Geneva.)

the poxvirus died. Individuals who recovered from smallpox were immune from this disease.

Diseases Caused by Bacterial Pathogens

Leprosy (Hansen's Disease). In 1980 there were about 3 million cases of leprosy, caused by *Mycobacterium leprae*, in the world (Figure 28.12). Today this disease is more commonly today called Hansen's disease to avoid the fearful reaction to the name "leprosy." There are two forms of leprosy. The lepromatous, or cutaneous form, is characterized by disfiguring nodules in the skin all over the body. This form of the disease is progressive and can produce deformations of various parts of the body such as the hands and nose. The tubercular, or neural form, of leprosy produces lesions around the peripheral nerves. There is a loss of sensation in the regions of the nodules. Bacteria present in the lesions are infective and can be transmitted through the skin or mucous membranes to susceptible individuals. There may be 1 billion viable cells of *M. leprae* per gram of skin in advanced cases of lepromatous leprosy, and direct skin contact appears to be very important in the transmission of this disease. In tubercular leprosy there are fewer lesions and lower numbers of bacteria and therefore lower infectivity.

> The wrath of the Lord was against the city with a very great destruction, and He smote the men of the city, both small and great and they had emerods in their secret parts and the hand of God was very heavy there, and the men that died not were smitten with the emerods
>
> —SAMUEL 1:5

Leprosy may also be spread through droplets, which can have as many as 107 cells per milliliter of *M. leprae*, from infected individuals to susceptible hosts where the primary infection occurs within the lung tissues. There is an extremely long incubation period for leprosy, usually 3–5 years, before the onset of disease symptomology. The symptoms of leprosy vary, but the earliest detectable symptoms generally involve skin lesions or loss of sen-

Figure 28.12 Map showing the worldwide incidence of leprosy. Today, this disease is concentrated in subequatorial zones.

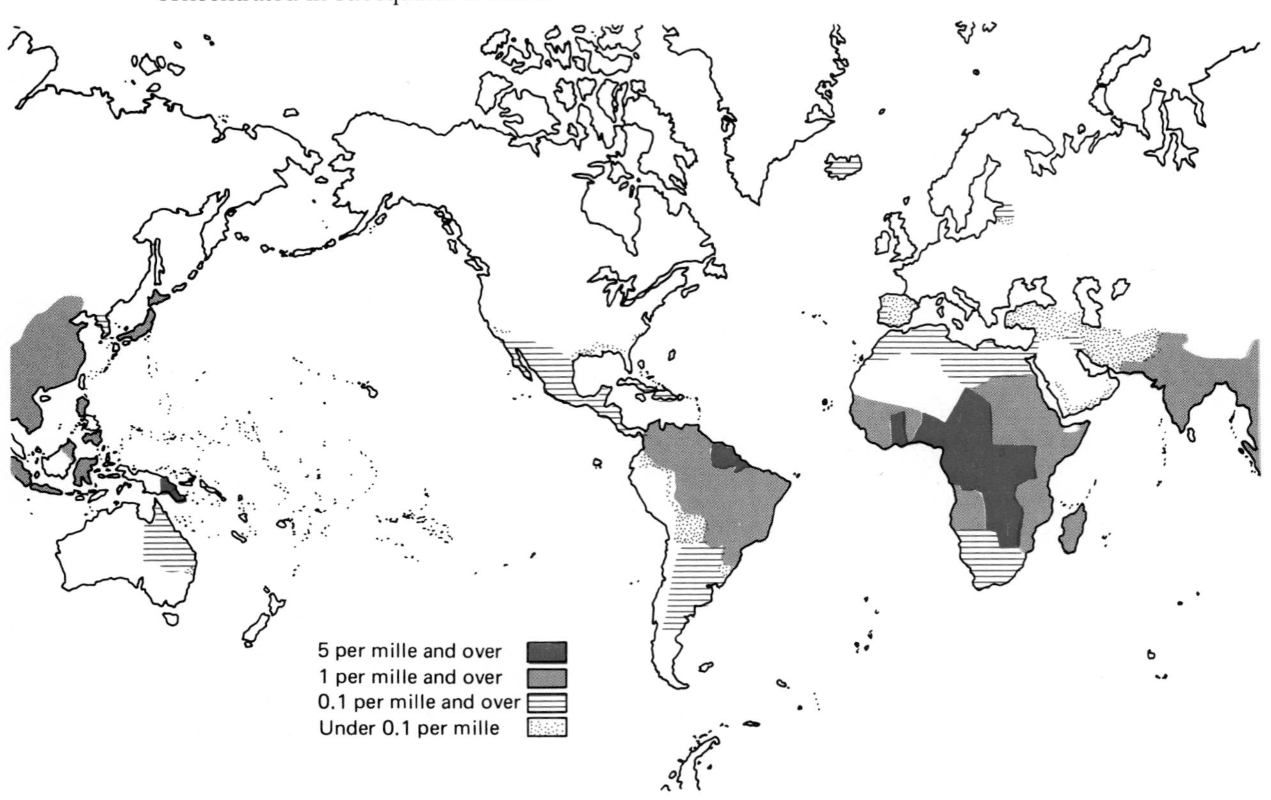

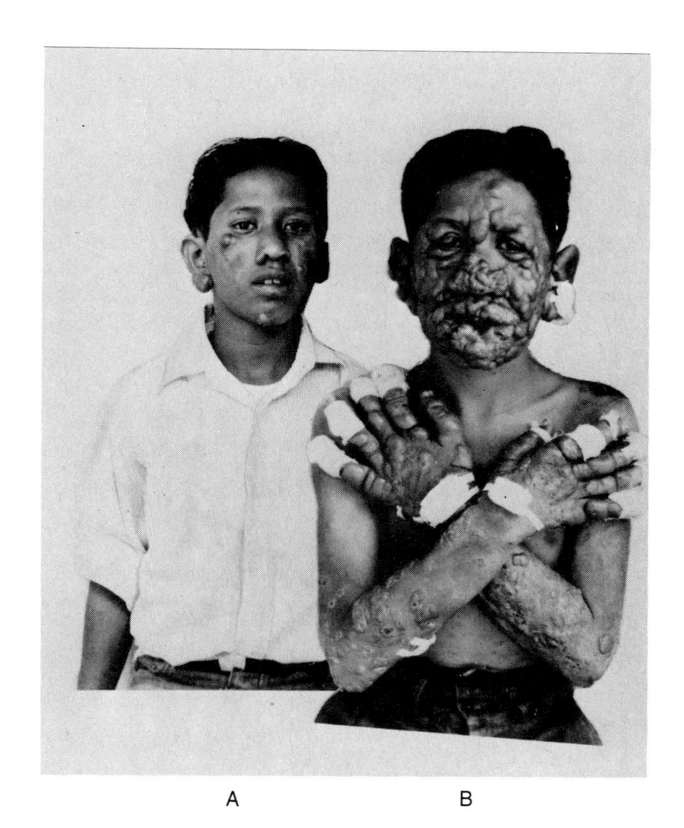

A B

Figure 28.13 Throughout history leprosy has been treated as a dread disease, and infected individuals have been outcast from societies. (A) A child with early stage of leprosy before the discovery of antibiotics. (B) This same boy showing the progression of the disease. It is no wonder that leprosy was feared. (A and B from United States Armed Forces Institute of Pathology, neg. no. 75-2479-2.) Today this disease can be cured with antibiotics. The recovery of a child with leprosy treated with antibiotics is illustrated in C and D. (C) A young boy with borderline leptomatous leprosy. (D) The same child 2 years later after treatment with antibiotics. (C and D courtesy American Leprosy Missions, Elmwood Park, New Jersey.)

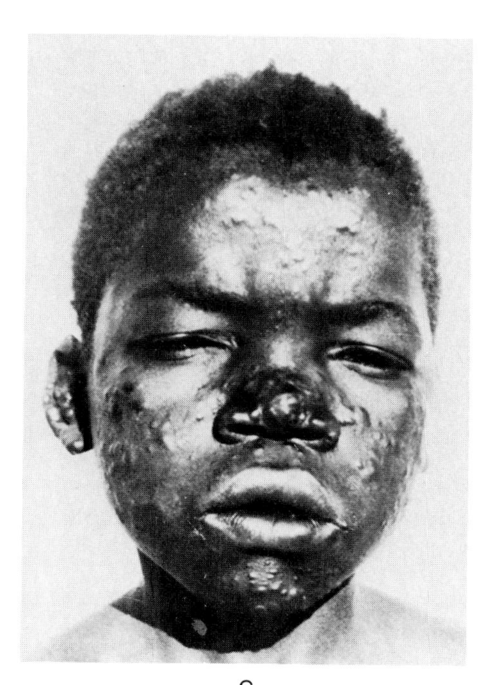

C

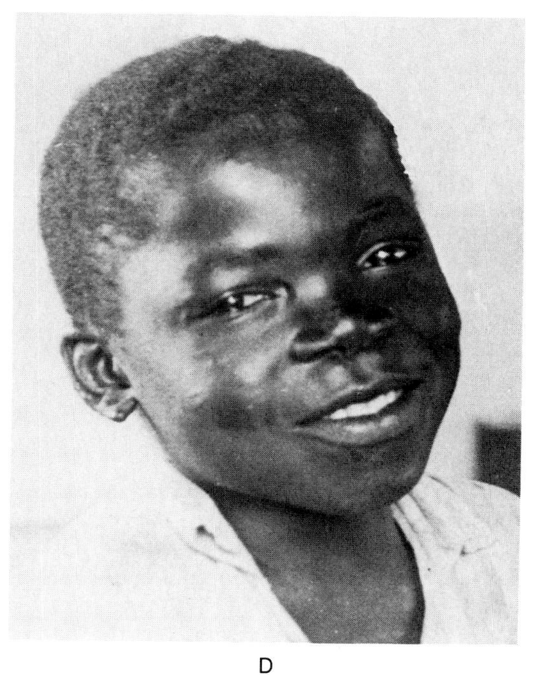

D

sation in a localized region of the skin. Unlike other mycobacterial species, *M. leprae* is able to reproduce within nerve nerve cells (called Schwann cells), so that damage to the nervous system occurs.

And if he see the leprosy in his skin,
and the hair turned white, and
the place where the leprosy appears lower than the skin and
the rest of the flesh it is the stroke of leprosy,
and upon his judgment he shall be separated.

—LEVITICUS 13:3

During the course of this disease, many organs and tissues of the body may be infected in addition to the infection of nerve cells characteristic of all forms of leprosy. In the tubercular form, there are relatively few nerves and skin areas involved; but in lepromatous leprosy, multiplication of *M. leprae* is not contained by the immune defense mechanisms, and the bacteria are disseminated through many tissues. Treatment of leprosy can be achieved by using dapsone, which is bacteriostatic, or rifampin, which is bactericidal. Prolonged treatment with antimicrobial agents is needed to maintain control of leprosy infections. Before the use of antibiotics, leprosy was a dread and deadly disease; today, fortunately, leprosy is rarely fatal, and complete recovery occurs after treatment in many cases (Figure 28.13).

Anthrax. Anthrax is primarily a disease of animals other than humans, but this disease can occasionally be transmitted to people. The disease is caused by *Bacillus anthracis*, a Gram positive, aerobic, endospore-forming rod. Transmission to humans can occur by direct contact of a skin abrasion with the endospores of *B. anthracis*, via the respiratory tract through inhalation of spores, and via the gastrointestinal tract through the ingestion of spores. The cutaneous route of transmission accounts for 95 percent of the cases of anthrax in the United States. Contact with animal hair, wool, and hides containing spores of *B. anthracis* is often implicated in transmission of anthrax, and therefore, it is also known as woolsorter's disease. Deposition of spores of *B. anthracis* under the epidermis permits germination with subsequent production of toxin by the growing bacteria.

The localized accumulation of toxin causes necrosis of the tissue in the formation of a blackened lesion called an eschar. The development of cutaneous anthrax can initiate a systemic infection, and untreated cutaneous anthrax has a fatality rate of 10–20 percent. Cutaneous anthrax can be treated with penicillin and other antibiotics, reducing the death rate to under 1 percent. Avoiding contact with infected animals and preventing the development of anthrax in farm animals through the use of anthrax vaccine have effectively reduced the incidence of this disease.

Pyoderma and Impetigo. Pyoderma is an infection of the skin caused primarily by *Staphylococcus* and *Streptococcus* species. Impetigo, which occurs almost exclusively in children during warm weather, is a type of pyoderma caused by infections with *Staphylococcus aureus* and/or *Streptococcus pyogenes*. Direct contact with infected material appears to be important in the transmission of impetigo. Strains of *S. pyogenes* deposited on the normal skin surface can remain viable for some time, and minor trauma to the skin permits invasion by the streptococci. *S. pyogenes* produces hyaluronidase, which permits spread of the bacteria. The infection results in the formation of a lesion that can be invaded in some cases by *S. aureus* from the skin surface, establishing a secondary infection. Typical manifestations of impetigo are localized lesions that progress to form pustules and eventually to form amber crusts (Figure 28.14). Pyoderma caused by *Streptococcus* and/or *Staphylococcus* species tends to be benign, although the pustules may persist and spread, forming satellite infections. The disease normally begins on the extremities and intense itching is associated with the lesions that develop. Scratching facilitates spread to other parts of the body. In some cases of impetigo, layers of skin peel off in sheets. This condition is known as scalded skin syndrome and apparently is a reaction to toxins produced by *Staphylococcus*.

Scarlet Fever. Scarlet fever is the result of a β-hemolytic *Streptococcus* infection in which the toxin spreads to the skin. Generally scarlet fever occurs as a secondary complication to pharyngitis caused by *Streptococcus pyogenes*. Scarlet fever is characterized by the development of a characteristic rash over the body. The generalized rash is due to the toxins produced by certain *Streptococcus* strains that contain lysogenic phage. Mild cases of this disease generally are called scarlatina although scarlet fever and scarlatina are really different names for the same disease. As with other β-hemolytic streptococcal infections, penicllin is the drug of choice and most frequently used for the treatment of scarlet fever. The antibiotic eliminates the streptococcal infection and thus the spread of toxin through the body to the skin.

Acne. Acne is a common problem during adolescence. More than 65 percent of teenagers develop this problem to some extent. Acne occurs

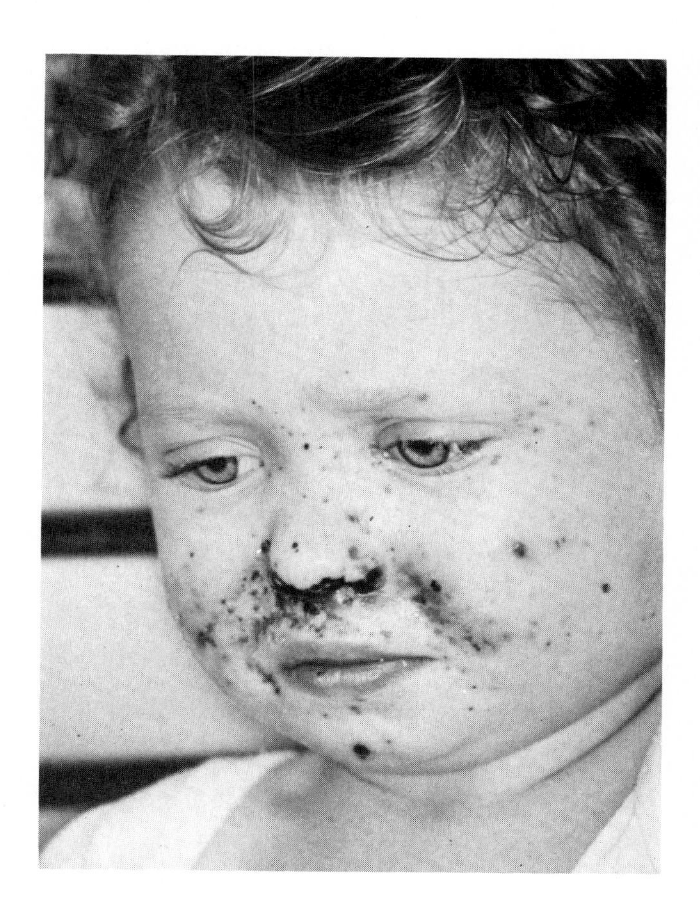

Figure 28.14 Photograph of skin in a case of impetigo. (From BPS: Leon J. LeBeau, University of Illinois Medical Center, Chicago.)

when microbial invasion at the base of hair follicles is associated with excessive secretions by the sebaceous glands. Various bacteria, including *Propionibacterium acnes* (*Corynebacterium acnes*) and *Staphylococcus epidermidis*, have been implicated as the etiologic agents of acne. *P. acnes* is able to metabolize sebum, producing fatty acids that elicit an inflammatory response. The inflammation can produce scarring if the lesions are scratched, picked, or even just pressed too hard. Scarring from acne can be permanent. Diet appears to have little effect on the development of this disease. Makeup, however, can aggrevate the condition.

Outbreaks of acne most commonly occur on the face, upper back, and chest. The disease is characterized by inflammatory papules, pustules, and cysts. Acne most frequently occurs during puberty because of the hormonal changes that occur during that period. Once adaptation to mature levels of sex hormones occurs, adults are normally not susceptible to this inflammatory disease. Several drugs can be used in the treatment of acne. The United States Food and Drug Administration has reported that sulfur, benzoyl peroxide, and salicylic acid, alone or in combination with resorcinol, are safe and effective in the treatment of acne. Many nonprescription remedies for acne contain these ingredients, and under a proposed FDA rule all acne preparations would be required to contain at least one of these effective ingredients.

Folliculitis. Folliculitis is a bacterial infection of the hair follicle that causes the formation of a pustule. The appearance of pustules is diagnostic of this disease. Folliculitis may also lead to the development of furuncles (boils) or carbuncles. *Staphylococcus aureus* is the most common cause of folliculitis. Staphylococcal folliculitis most commonly occurs when there is poor personal hygiene, exposure to chemicals such as cutting oils, or abrasions to the skin. Some cases of folliculitis are caused by *Pseudomonas aeruginosa*. Outbreaks of pseudomonad folliculitis often are associated with bathing in hot spas and swimming pools. *Pseudomonas* is relatively resistant to chlorine and can reach high numbers in heated pools, leading to the development of skin infections in individuals bathing in such contamined facilities. Topical antibiotics, such as bacitracin and polymyxin B, are used in the treatment of staphylococcal folliculitis. Topical application of polymyxin B is also used to treat pseudomonad folliculitis. Systemic antibiotics, such as gentamicin, are also used to eliminate persistent cases of folliculitis.

Mycetoma. Mycetoma, an infectious disease of the skin, can progress to involve subcutaneous tissues and bones. This disease is caused by various species of actinomycetes, including *Nocardia madurae*, and by various fungal species. The feet are the most common sites of involvement in cases of mycetoma because the actinomycetes and fungi causing the disease apparently frequently enter the skin through abrasions acquired while walking barefoot on soil; however, the infectious agents may penetrate the skin in other ways. The multiplication of the bacteria or fungi within the infected area results in swelling, which is frequently grotesque. One form of mycetoma, known as madura foot, is important in the Madura Province of India.

It is necessary to identify the specific etiologic agent of mycetoma in order to determine the appropriate treatment. When the causative organism is a fungus, amphotericin B can be used; a *Nocardia* species or other actinomycete is responsible, sulfonamides are normally effective in the treatment. The remaining worldwide incidence of this disease is largely a function of socioeconomic conditions because the occurrence of mycetoma can be greatly reduced by wearing shoes.

Diseases Caused by Fungal Pathogens

As just discussed, fungi as well as bacteria can cause mycetoma. Additionally, various fungal species are responsible for a number of superficial infections of the skin. Many of the fungi that cause superficial skin infections are dermatophytes; that is, they only infect the skin and its appendages, such as hair and nails. Even though the growth of dermatophytic fungi is restricted to the skin, the host–parasite interaction can cause serious manifestations. Contact of the skin with a spore of a

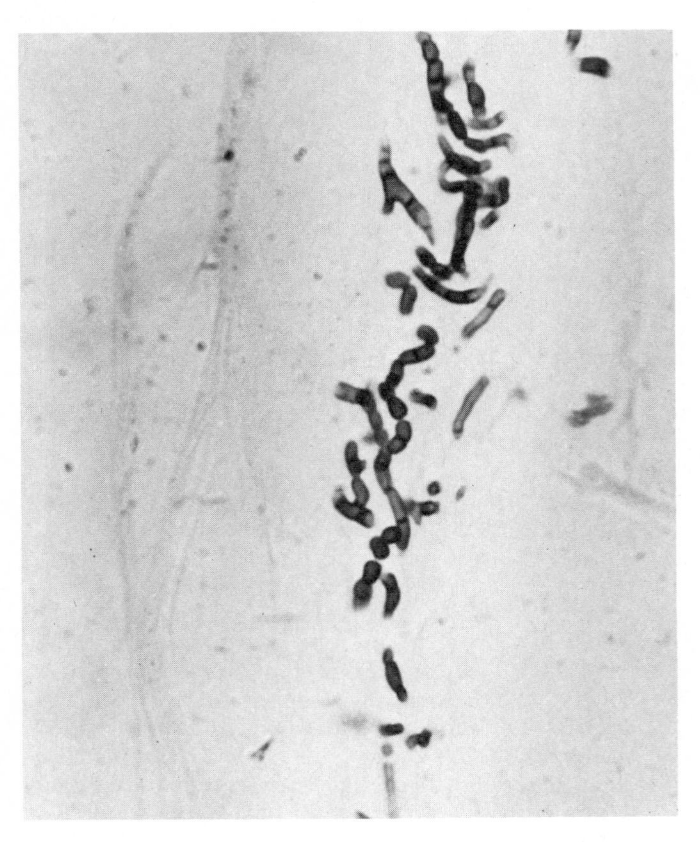

Figure 28.15 Fungal hyphae in a case of tinea. (From United States Armed Forces Institute of Pathology, neg. no. 751175.)

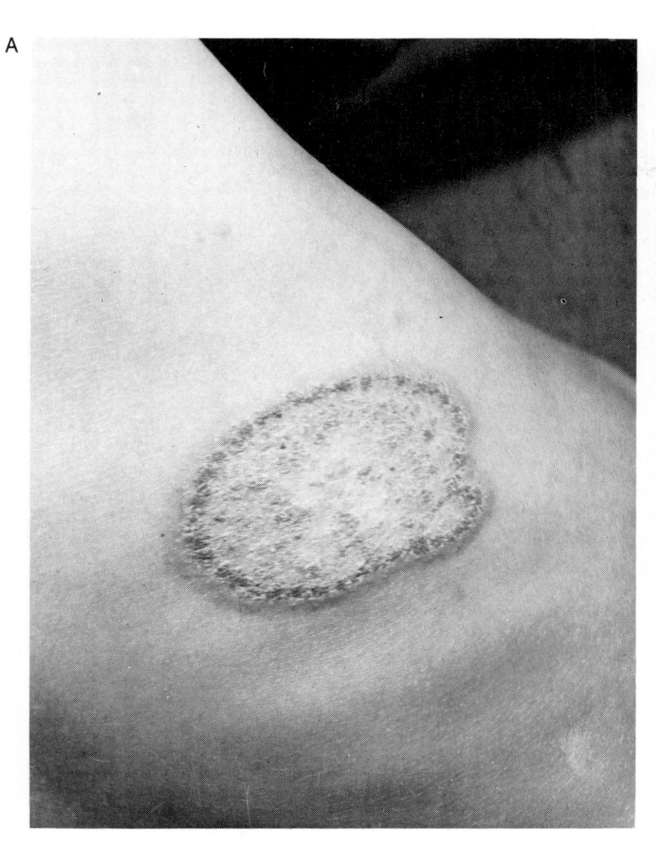
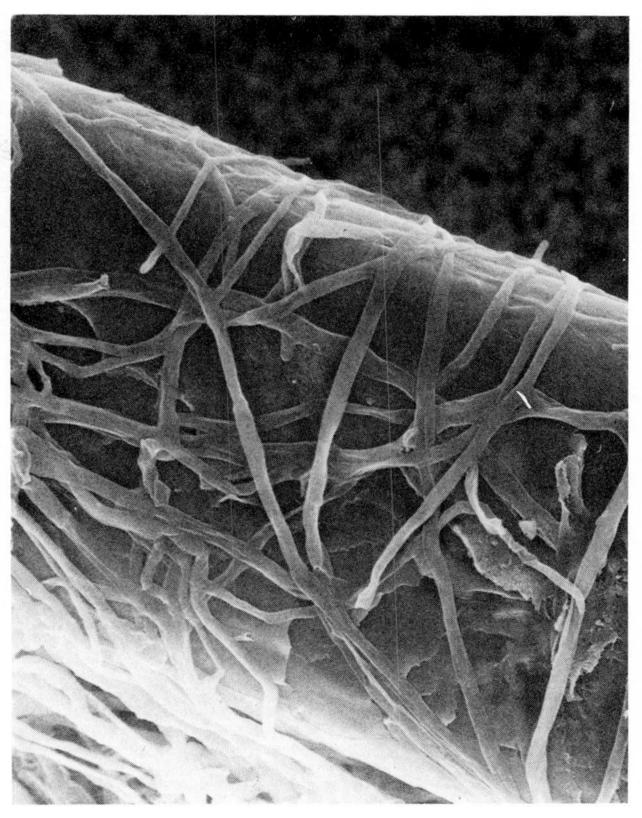

Figure 28.16 (A) Photograph of a case of tinea. (From United States Armed Forces Institute of Pathology, neg. no. 55-19008.) (B) Micrograph of *Trichophyton* growing on a human hair in a case of tinea barbae or ringworm of the beard. (From BPS: Garry T. Cole, University of Texas, Austin.)

dermatophyte can initiate infection, with the dermatophytic fungi growing filamentously within the dead keratin-containing layers of the skin (Figure 28.15). The colonization of the skin by dermatophytic fungi initiates a cell-mediated immune response that generally occurs 10–35 days after infection. The development of the cell-mediated immune response causes inflammatory damage to the skin tissue; however, it also prevents further lateral spread of the fungus.

Tinea (ringworm). Diseases caused by dermatophytic fungi are manifest as tinea or ringworm (Figure 28.16). These diseases are normally well localized and never fatal. The specific diseases are distinguished based on which regions of the body are infected (Table 28.1). Most dermatophytic fungi are members of the genera *Microsporum, Epidermophyton,* and *Trichophyton*. These fungi can grow on skin, nails, and hair.

Transmission of dermatophytic fungi is enhanced by conditions of high moisture and sweating, and retention of moisture increases the probability of contracting superficial infections of the skin. The transmission of athlete's foot, for example, is often associated with the high moisture levels and bare feet of athletes in a locker room, although it is now known that this disease is not acquired unless the individual has skin abrasions that the fungus can infect. Drying feet and using antifungal agents, however, can reduce the spread of athlete's foot. It is virtually impossible to protect all body areas against potential infection with superficial dermatophytic fungi, resulting in a high incidence of dermatomycoses. Most cases of tinea can be treated by topical application of antifungal agents, such as miconazole and amphotericin B. In some cases, such as when the nails are involved, prolonged treatment with griseofulvin is employed to eliminate the infection.

Sporotrichosis. In contrast to the superficial fun-

Table 28.1 Epidemiology of Dermatomycoses

Disease	Causative Agent	Transmission	Examples of Sources
Tinea capitis (ringworm of the scalp)	*Microsporum* sp., *Trichophyton* sp.	Direct or indirect contacts	Lesions, combs, toilet articles, headrests
Tinea corporis (ringworm of the body)	*Epidermophyton* sp., *Microsporum* sp., *Trichophyton* sp.	Direct or indirect contacts	Lesions, floors, shower stalls, clothing
Tinea pedis (ringworm of the feet—Athlete's foot)	*Epidermophyton* sp., *Trichophyton* sp.	Direct or indirect contacts	Lesions, floors, shoes and socks, shower stalls
Tinea unguium (ringworm of the nails)	*Trichophyton* sp.	Direct contact	Lesions
Tinea cruris (ringworm of the groin—jock itch)	*Trichophyton* sp., *Epidermophyton*	Direct or indirect contacts	Lesions, athletic supports

gal diseases just discussed, sporotrichosis is a subcutaneous mycosis caused by the fungus *Sporothrix schenckii*. Inoculation of *S. schenckii* into the skin as a result of a minor injury initiates this infection. Sporotrichosis occurs worldwide and is especially widely distributed in South Africa, France, and Mexico. The infection begins with the formation of a subcutaneous nodule, with secondary nodules developing later as the infection spreads through the lymphatic system. Lesions due to sporotrichosis normally occur on the extremities and spread to other parts of the body. The spread of the disease occurs slowly, permitting time for diagnosis and therapy. As a result, the death rate due to sporotrichosis is very low. The cutaneous form of sporotrichosis can be treated with a solution of iodine, but if the fungus is widely disseminated, administration of amphotericin B is necessary.

Infections Associated with Wounds and Burns

Infections after Wounds

Wounds disrupt the protective barrier of the skin and provide a portal of entry through which microorganisms can enter the circulatory system and deep body tissues. Microorganisms on the skin surface can readily pass through the opening of a wound; as a result, many infections associated with wounds are caused by opportunistic pathogens derived from the normal microbiota of the skin. To avoid entry of bacteria into wounds, the area is usually covered with gauze to protect against contamination.

Staphylococcus aureus is the most frequent etiologic agent of wound infections, and *Streptococcus pyogenes* is also often associated with infections of wounds. In cases of severe wounds, where the integrity of the gastrointestinal tract is disrupted, enteric bacteria are frequently the causative agents of wound infections (Table 28.2). Wound infections are often localized at the site of the wound, but infections established in this manner can spread systemically and may involve many body tissues and organs. For example, infections with *Staphylococcus aureus* established through skin wounds can spread and form abscesses in bone marrow (osteomyelitis) and other body tissues, including the spine and brain. Superficial wounds can generally be treated with topical antiseptics or antibiotics to prevent the establishment of infections. Serious deep wounds, however, may require the prophylactic use of systemic antibiotics to prevent the onset of serious infections. Infections of deep wounds may involve anaerobic bacteria of the genera *Clostridium, Bacteroides*, and *Fusobacterium*, as well as *Staphylococcus* and *Streptococcus* species.

Gas Gangrene. Deep wounds not only provide a portal of entry for microorganisms, but the tissue damage often interrupts circulation to the area, creating conditions that permit the growth of obligately anaerobic bacteria. Gas gangrene is a serious infection that may result from the growth of *Clostridium perfringens* and other *Clostridium* species. It is often a mixed infection of organisms capable of producing a very large number of enzymes that greatly enhance their invasive capabilities. The development of gas gangrene is dependent on the deposition of endospores of *Clostridium* in the wound tissue and the development of anaerobic conditions due to necrosis of local tissues that permit the germination and multiplication of these obligately anaerobic bacteria.

> Let a single clot of blood or a single fragment of dead flesh lodge in a corner of the wound, sheltered from the oxygen of the air, where it remains surrounded by carbonic acid gas, even though it might be over a small area, and beginning at once the septic germs will give rise in less than twenty-four hours to an infinite number of microbes, multiplying by binary fission and capable of causing in a very short time a mortal septisemia.
>
> —Louis Pasteur

The *Clostridium* species that cause gas gangrene produce toxins, the diffusion of which extends the area of dead and anaerobic tissues. The exotoxins produced by these *Clostridium* species are tissue necrosins and hemolysins that kill the infected tissues and account in part for the rapid spread of infection (Figure 28.17). The growing *Clostridium* species produce carbon dioxide and hydrogen gases and the formation of odoriferous low molecular weight metabolic products. The gas that accumulates is primarily hydrogen because it is less soluble than carbon dioxide. The buildup of gas pockets in the tissues is called crepitation. In most cases, the onset of gas gangrene occurs within 72 hours of the occurrence of the wound, and if untreated the disease is fatal. Even with antimicrobial treatment, there is a high rate of mortality, and therefore, radical surgery—amputation—is often employed to prevent the spread of infection. If treated rapidly enough, localized areas of necrotic tissue can be excised—a process called debridgement—and high doses of penicillin administered to block the spread of the infection. The prevention of gas gangrene depends on ensuring that wounds are not suitable for the growth of the anaerobic *Clostridium* species. This requires that wounds have adequate drainage to prevent establishment of anaerobic conditions and that foreign material and dead tissue be removed. Hyperbaric oxygen chambers are sometimes used to treat gas gangrene in an effort to kill the oxygen-sensitive *Clostridium* species before they spread to the point where radical surgery is needed to prevent the death of the patient.

Table 28.2 Bacterial Causes of Wound Infections Recorded at an Urban Hospital

Staphylococcus aureus	48%
Enteric and other bacteria associated with the gastrointestinal tract	49%
Escherichia coli	
Proteus sp.	
Klebsiella sp.	
Enterobacter sp.	
Bacteroides sp.	
Streptococcus faecalis	
Streptococcus pyrogenes, group A	3%

Wound Botulism. Wound botulism is caused when *Clostridum botulinum* is inoculated into a wound. Spores of *C. botulinum* are abundantly distributed in soils, and if inoculated into a deep wound, this bacterium can initiate an infection. *C. botulinum* is noninvasive but can produce a potent neurotoxin that spreads throughout the body. The toxin blocks neural transmissions, causing symptoms that often

Figure 28.17 Photograph of a foot in a case of gas gangrene showing the black necrotic tissue due to the infection with *Clostridium perfringens.* (From United States Armed Forces Institute of Pathology, neg. no. 79-18280-2.)

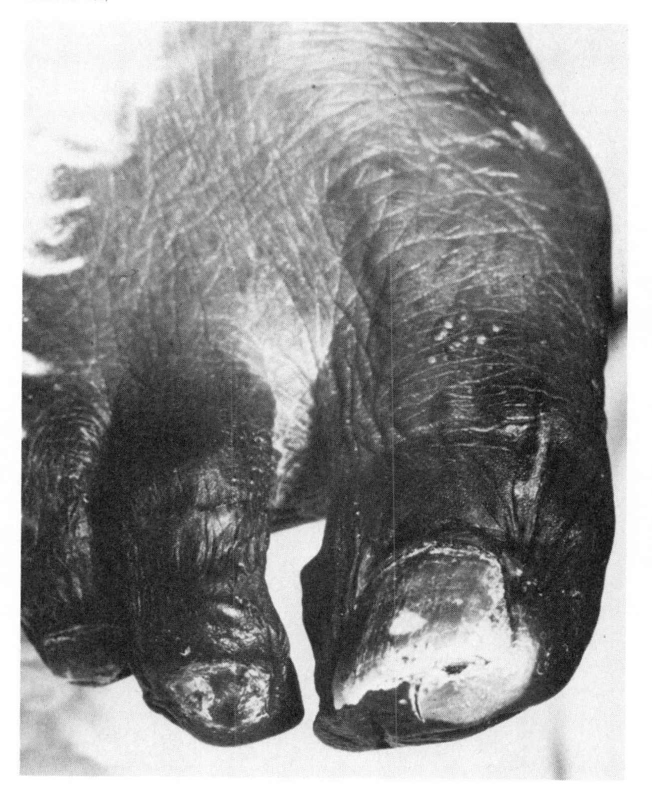

include respiratory failure and death. This is a relatively rare disease, but there has been a recent increase in the incidence of wound botulism. If the disease is diagnosed in time, antitoxins can be used to neutralize the toxins.

Infections after Burns

Burns remove the protective skin layer, exposing the body to numerous potential pathogens. Microbial infection after extensive burns, where a large portion of the skin is damaged, is a very serious complication that often results in the death of the patient. *Staphylococcus aureus, Streptococcus pyogenes, Pseudomonas aeruginosa, Clostridium tetani,* and various fungi often cause infections in burn victims. These organisms are widely disseminated and exposure is difficult to control. They produce toxins that can spread systemically and cause death. It is important to avoid contamination of the burn area, which can introduce opportunistic pathogens into exposed tissue (Figure 28.18). To insure that infections are detected in time to permit treatment with antimicrobial agents, microbiological tests are frequently performed; typically two times a week a moist swab is run over a 1 square centimeter area and cultured for quantitation of bacteria and fungi; a count of greater than 105 is indicative of colonization and pending invasion. For prophylaxis, silver-containing antimicrobials, such as silvidene, are frequently used. When infections are detected, specific topical agents are selected based on the sensitivity of the infecting agent. If infections do develop, the prognosis depends on the size of the burn, the extent of infection, and the physiological state of the patient.

Infections after Surgical Procedures

Surgical procedures often expose deep body tissues to potentially pathogenic microorganisms. A surgical incision circumvents the normal body defense mechanisms. Great care is therefore taken in modern surgical practices to minimize microbial contamination of exposed tissues (Figure 28.19). These practices include the use of clean operating

Figure 28.18 Burn victims are especially vulnerable to infection because of the destruction of the surface skin barrier. The open air procedure is currently the method of choice used at many burn units; in this procedure the burn wound for the most part is left open to the air to speed healing. As a result extreme precautions are taken to minimize exposure to microorganisms, including keeping burn victims in isolation rooms and ensuring that individuals in contact with such patients are properly attired to prevent the transmission of potentially pathogenic microorganisms. (A) The vulnerability of burn tissues to infection is apparent in this photograph showing the leg of a man with severe burns over most of the body. (B) When applying dressings all staff must be masked, gloved, and gowned. (Courtesy Norton Kosair Children's Hospitals, Louisville, Kentucky.)

A

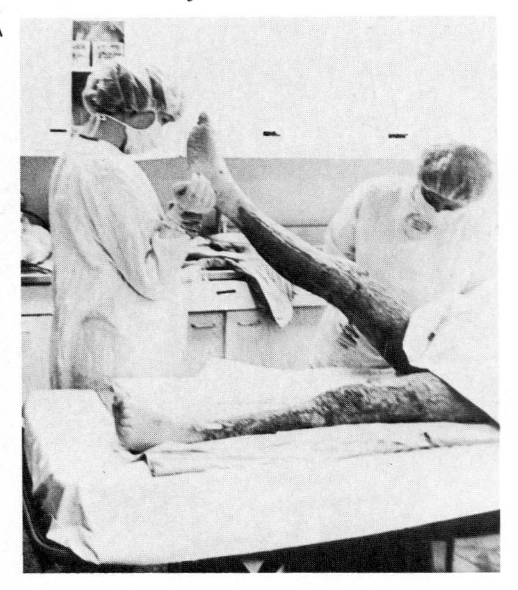

B

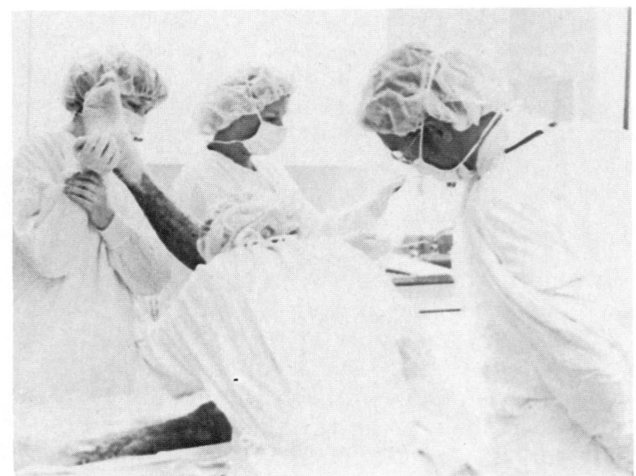

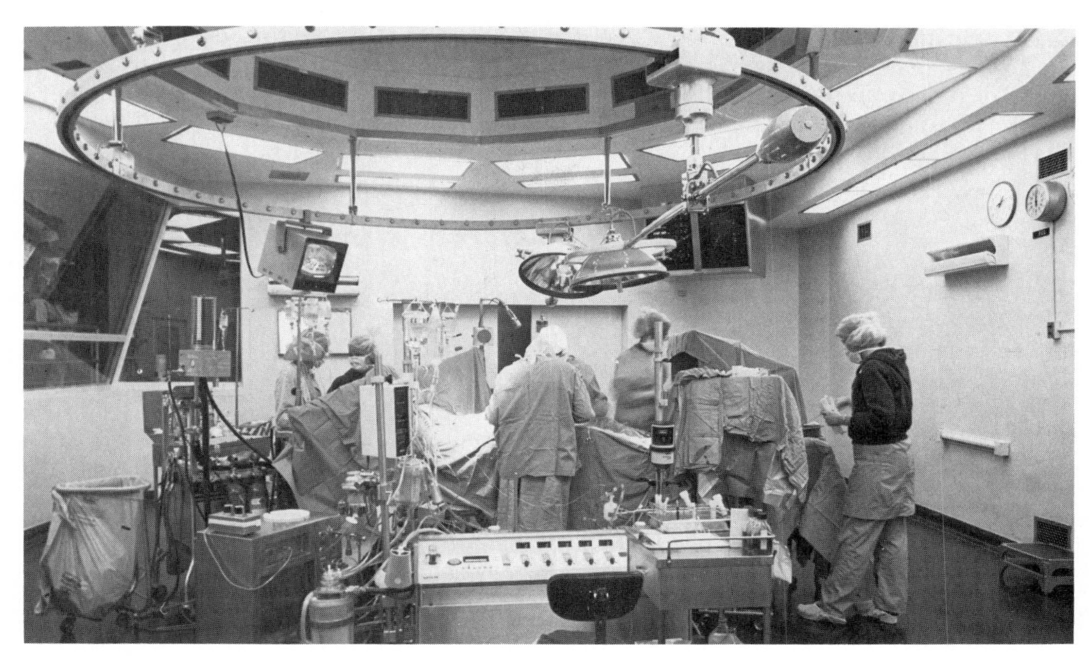

Figure 28.19 In operating theaters, walls are disinfected, instruments sterilized, and staff appropriately garbed to prevent infection of exposed tissues. Note that the surgical staff are all attired so as to minimize bringing any microorganisms in from the outside. Surgical staff wear hats, masks, gloves, gowns, and even shoe covers and extensively scrub (wash) their skin before entering the operating room. (Courtesy National Institutes of Health, Washington, D.C.)

rooms with minimal numbers of airborne microorganisms, sterile instruments, masks, and gowns, all of which prevent the spread of microorganisms from the surgical staff to the patient, and the application of topical antiseptics prior to making the incision in order to prevent accidental contamination of the wound with the indigenous skin microbiota of the patient. After many surgical procedures antibiotics are given for several days as a prophylactic measure.

Despite all of these precautions, infections still sometimes occur after surgery (Table 28.3). Infections after surgery can be quite serious, as the patient is already in a debilitated state. The onset of such infections is generally marked by a rise in fever. A purulent lesion may develop around the wound. Particularly, serious complications may follow open heart surgery if the patient develops endocarditis, caused by *Staphylococcus* or *Streptococcus* species. In surgical procedures involving cutting the intestines, the normal gut microbiota may contaminate other body tissues, causing peritonitis and infections of other tissues caused by enteric bacteria unless great care is taken to minimize such contamination; antibiotics are used to prevent microbial growth. The specific microorga-

Table 28.3 Some Factors in the Development of Surgical Wound Infections

Situation	*Staphylococcus aureus* (%)	Enteric bacteria (%)	*Streptococcus pyogenes* (%)
Emergency operation	10	21	1
In wound of second operation	11	12	0
First operation of day	17	16	0
Other	10	0	2
Total:	48	49	3

nisms causing infections of surgical wounds and the specific tissues that may be involved depend on the nature of the surgery and the tissues that are exposed to potential contamination with opportunistic pathogens.

Summary

The skin presents an effective physical barrier that blocks the entry of most microorganisms into the body. Some microorganisms colonize the skin but these generally are noninvasive and do not cause infections of the intact skin. In some cases the systemic spread of microorganisms that enter the body through the respiratory tract or other portals of entry results in diseases that involve the skin. Often such diseases are evidenced by the development of a rash or swelling of the skin. For example, measles and many of the so-called childhood diseases are systemic infections that characteristically involve the skin. In addition to these infective processes, some pathogenic microorganisms can directly colonize the skin, causing dermatophytic infections. For example, ringworm is a disease caused by dermatophytic fungi that can affect various parts of the body.

Wounds and animal bites that disrupt the protective skin layers provide the main portals of entry by which pathogenic microorganisms penetrate the skin and gain entry to the body. Minimizing contamination of wounds with foreign matter and associated microorganisms is important in controlling infections that may occur when the skin barrier is broken. Surgical practices utilize elaborate aseptic procedures to minimize potential infection, but nevertheless, infections sometimes occur after surgery, attesting to the vulnerability of the body to microbial infection when the skin barrier is disrupted and host defense mechanisms are impaired.

Table 28.4 presents a summary of microbial infections that primarily involve the skin and those that are associated with wounds.

Table 28.4 Summary of Diseases of the Skin and Diseases Associated with Wounds

Disease	Causative Agent	Mode of Transmission	Treatment
Measles	Measles virus	Aerosols	Supportive
German measles (rubella)	Rubella virus	Aerosols	Supportive
Chicken pox	Varicella-zoster virus	Aerosols	Supportive
Shingles	Varicella-zoster virus	Reactivation of chicken pox	Supportive
Smallpox	Variola virus	Aerosols	No longer needed; disease eliminated by extensive vaccination program
Warts	Papiloma virus	Direct contact	Destruction of infected tissue
Leprosy	*Mycobacterium leprae*	Direct contact, aerosols	Dapsone, Rifampin
Anthrax	*Bacillus anthracis*	Direct contact, aerosols, ingestion	Penicillin
Pyoderma	*Staphylococcus aureus, Streptococcus pyogenes*	Direct contact	Penicillin and other antibiotics
Impetigo	*Staphylococcus aureus, Streptococcus pyogenes*	Direct contact	Penicillin and other antibiotics
Scarlet fever	*Streptococcus pyogenes*	Aerosols	Penicillin
Acne	*Propionibacterium acnes, Streptococcus epidermidis*	Opportunistic infections	Antibiotics
Mycetoma	Actinomycetes	Direct contact	Sulfonamides and other antibiotics
Ringworm	Dermatophytic fungi, *Microsporum, Trichophytum*	Direct contact	Nystatin
Sporotrichosis	*Sporothrix schenckii*	Direct contact	Amphotericin B
Gas gangrene	*Clostridium perfringens*, other anaerobes	Wound	Penicillin, radical surgery
Wound botulism	*Clostridium botulinum*	Wound	Antitoxin

Study Outline

A. The outer layer of the skin, the epidermis, is composed largely of epithelial cells and its outermost layer consists of dead cells that contain the protein keratin.

- **1.** Keratin is quite water repellant and most microorganisms are unable to metabolize keratin making the intact skin, the external covering of the body, an effective barrier to microorganisms.
- **2.** The skin also guards against the loss of vital internal body constituents, such as water, salts, and organic molecules.
- **3.** Breaks in the continuity of the skin, such as result from a variety of types of wounds, allow fluids to leave the body and provide an opening through which many disease-causing microorganisms may enter.
- **4.** The inner layer of the skin is called the dermis, and it is composed of a thick layer of dense fibrous connective tissue.
  - **a.** Two layers, the papillary layer and the reticular layer, subdivide the dermis.
  - **b.** The dermis contains blood vessels, lymph vessels, and nerve endings. It also contains hair follicles and sweat glands. Sweat-gland ducts end at the surface of the epidermis. Hairs penetrate out from the skin surface. Sweat is secreted as perspiration onto the surface of the skin. Both hairs and sweat ducts provide possible portals of entry by which microorganisms can penetrate into the skin tissues and enter the body.

B. Human skin surfaces are not favorable habitats for the growth of microorganisms, even though the skin is continuously exposed to microorganisms.

- **1.** Environmental factors, such as temperature, water activity, pH, and salinity, create a stress that prevents many microbial populations from colonizing the skin surface.
  - **a.** The lack of available water is a major limiting factor. Sweat glands secrete fluids containing water, but this water contains high concentrations of salt and organic wastes that reduce the water activity sufficiently to inhibit most microbial growth.
  - **b.** Skin microbiota must tolerate the osmotic stress caused by the high salt concentrations there.
  - **c.** Antimicrobial agents are produced by the host animal and by those microorganisms that do successfully establish themselves as the normal microbiota of the skin and act to prevent the growth of foreign microorganisms on the skin surface.
- **2.** Proper cleanliness habits and good hygienic practices tend to prevent the establishment of nonindigenous microorganisms among the natural skin microbiota by preventing the buildup of excessive concentrations of organic matter.
  - **a.** Washing the skin surface removes excess organic matter and temporarily reduces the numbers of both resident and transient microorganisms.

C. The dominant microbial populations on the skin surface are Gram positive bacteria, with members of the genera *Staphylococcus* and *Micrococcus* frequently being the most abundant because they are generally salt tolerant and can utilize the lipids present on the skin surface.

- **1.** Other bacteria usually found on the skin surface include *Corynebacterium, Brevibacterium*, and *Propionibacterium* species.
- **2.** Relatively few fungi are included in the normal microbiota of the skin surface. Two yeasts, *Pityrosporum ovale* and *P. orbiculare* normally occur on scalp tissues. Some dermatophytic fungi occasionally grow on the skin surface, producing diseases such as athlete's foot and ringworm.

D. Diseases that affect the skin caused by viral pathogens.

- **1.** Measles.
  - **a.** Causative organism(s): measles virus.
  - **b.** Mode of transmission: from infected individual to susceptible host.
  - **c.** Symptoms: skin rash, high fever, coughing, sensitivity to light, Koplik's spots.
  - **d.** Treatment: supportive, including rest and fluids; vaccine available.

2. German measles.
   a. Causative organism(s): rubella virus.
   b. Mode of transmission: via droplet spread.
   c. Symptoms: characteristic spots, enlarged and tender lymph nodes, low grade fever.
   d. Treatment: vaccine available.
3. Chicken pox (generally a disease of childhood) and shingles (generally a disease of adults).
   a. Causative organism(s): varicella-zoster virus.
   b. Mode of transmission: via contaminated droplets and direct contact with vesicle fluid containing the virus.
   c. Symptoms: skin lesions in chicken pox; in shingles inflammation of the ganglia of the spinal or cranial nerves, pain and tenderness and also a rash along these nerves.
4. Warts.
   a. Causative organism(s): papillomaviruses.
   b. Mode of transmission: direct contact.
   c. Treatment: destruction of infected tissues.
5. Smallpox.
   a. Causative organism(s): smallpox (variola) virus.
   b. Mode of transmission: via aerosols.
   c. Symptoms: rash, fever, malaise, pain in head, back, and stomach.
   d. Treatment: disease eradiacated through international vaccination program.

E. Diseases caused by bacterial pathogens.
1. Leprosy.
   a. Causative organism(s): *Mycobacterium leprae*.
   b. Mode of transmission: through the skin or mucous membranes via direct contact, also through droplets.
   c. Symptoms: skin lesions, damage to nerve tissue.
   d. Treatment: dapsone and rifampin.
2. Anthrax.
   a. Causative organism(s): *Bacillus anthracis*.
   b. Mode of transmission: direct contact of skin with endospores of *B. anthracis*, via the respiratory tract through inhalation of spores, and via the gastrointestinal tract through ingestion of spores.
   c. Symptoms: tissue necrosis.
   d. Treatment: penicillin.
3. Pyoderma and impetigo.
   a. Causative organism(s): pyoderma by *Staphylococcus* and *Streptococcus* species; impetigo by *Streptococcus pyogenes* or *Staphylocccus aureus*.
   b. Mode of transmission: direct contact.
   c. Symptoms: localized lesions that progress to form pustules, itching.
4. Scarlet fever.
   a. Causative organism(s): *Streptococcus pyogenes*.
   b. Mode of transmission: secondary complication to pharyngitis caused by *S. pyogenes*.
   c. Symptoms: characteristic rash.
   d. Treatment: penicillin.
5. Acne.
   a. Causative organism(s): *Propionibacterium acnes* and *Staphylococcus epidermidis*.
   b. Mode of transmission: microbial invasion at the base of hair follicles initiates excessive secretions by sebaceous glands.
   c. Symptoms: infammatory papules, pustules, and cysts on the face, upper back, and chest.

6. Folliculitis.
    a. Causative organism(s): *Staphylococcus aureus, Pseudomonas aeruginosa*, other bacteria.
    b. Mode of transmission: invasion of follicules.
    c. Symptoms: furnucles or carbuncles.
    d. Treatment: bacitracin, polymyxin B.
7. Mycetoma.
    a. Causative organism(s): various species of actinomycytes, including *Norcardia madurae* and various fungal species.
    b. Mode of transmission: through abrasions in the skin, especially the feet.
    c. Symptoms: swelling.
    d. Treatment: fungus by amophotericin B; acintomycyte by sulfonamides.

F. Diseases caused by fungal pathogens.

1. Tinea (ringworm).
    a. Causative organism(s): *Microsporum* and *Trichophyton* species.
    b. Mode of transmission: enhanced by conditions of high moisture and sweating.
    c. Treatment: antifungal agents.
2. Sporotrichosis.
    a. Causative organism(s): *Sporothrix schenkii.*
    b. Mode of transmission: through a skin abrasion.
    c. Symptoms: lesions, nodules.
    d. Treatment: iodine and amphotericin B.

G. Diseases associated with wounds. *Staphylococcus aureus* is the most frequent etiologic agent of wound infections; *Streptococcus pyogenes* is also often involved. In a severe wound, when the gastrointestinal tract is disrupted, enteric bacteria frequently cause infections. These infections are often localized at the site of the wound, but can spread systemically. Superficial wounds can generally be treated with topical antiseptics or antibiotics to prevent the establishment of infections. Serious deep wounds may required the prophylactic use of systemic antibiotics to prevent serious infections. Infections of deep wounds may involve anaerobic bacteria of the genera *Clostridium, Bacteroides*, and *Fusobacterium*, as well as *Staphylococcus* and *Streptococcus* species.

1. Gas gangrene.
    a. Causative organism(s): *Clostridium perfringens* and other *Clostridium* species.
    b. Mode of transmission: deposit of *Clostridium* endospores in wound tissue and development of anaerobic conditions.
    c. Treatment: excision of necrotic tissue, amputation, penicillin.
2. Wound botulism.
    a. Causative organism(s): *Clostridium botulinum*.
    b. Mode of transmission: spores of *C. botulinum* from soil inoculated into a deep wound.
    c. Symptoms: neurotoxin blocks neural transmissions.
    d. Treatment: administration of antitoxin.

H. Infections after burns. Burns remove the protective skin cover and can lead to serious infections, most commonly by *Staphylococcus aureus, Streptococcus pyogenes, Pseudomonas aeruginosa, Clostridium tetani*, and various fungi. Silver-containing antimicrobials are often used to control infections of burn tissues.

I. Infections after surgical procedures. The onset of such infections is generally marked by a rise in fever; purulent lesions may also develop. Following open heart surgery, endocarditis may be caused by *Staphylococcus* or *Streptococcus* species. In surgical procedures involving cutting the intestines, the normal gut microbiota may contaminate other body tissues, causing such infections as peritonitis.

Study Questions

1. For each of the following diseases what are the causative organisms and characteristic symptoms?
   a. Ringworm of the body
   b. Measles
   c. Chicken pox
   d. Gas gangrene
   e. Wound botulism

2. Match the following statements with the terms in the list to the right.

   a. Symptoms begin a few days before appearance of a rash; high fever, coughing, sensitivity to light, appearance of Koplik's spots (red spots with a white dot in the center that occur in the oral cavity).
   b. Vaccination is not yet available for this childhood disease.
   c. Childhood disease with relatively low rate of infectivity; prolonged exposure required.
   d. Characteristic rash (flat pink spots) appears on face and spreads to other parts of body.
   e. If acquired during pregnancy, multiple abnormalities may develop in the fetus.
   f. Caused by varicella-zoster virus.
   g. Symptoms include local skin lesions and vesicles on mucous membranes, especially in the mouth.
   h. Principal disease resulting from varicella-zoster viral infections in adults who have had chicken pox as a child.
   i. Rubella.
   j. Caused by *Mycobacterium leprae*.
   k. Caused by papillomaviruses.
   l. Benign skin tumors.
   m. Caused by *Propionibacterium (Corynebacterium) acnes*.
   n. Exotoxins produced are tissue necrosins and hemolysins that extend the area of dead and anaerobic tissues; carbon dioxide and hydrogen gases also formed.
   o. Superficial fungal diseases caused by dermatophytes.
   p. Caused by *Clostridium perfringens* and other *Clostridium* species.

   (1) Acne
   (2) Measles
   (3) Gas gangrene
   (4) German measles
   (5) Chicken pox
   (6) Shingles
   (7) Herpes simplex 1
   (8) Ringworm
   (9) Leprosy
   (10) Warts

Some Thought Questions

1. Why were there serious outbreaks of measles on college campuses in the United States during 1983?
2. Should the Nobel prize be awarded to the person who discovers the cure for jock itch or athlete's foot?
3. Why is tinea called ringworm?
4. Why is measles still prevalent in South America?
5. Why should women be immunized against rubella before reaching childbearing age?
6. Why do some people develop acne and not others?

Additional Sources of Information

Boyd, R. F., and B. G. Hoerl. 1981. *Basic Medical Microbiology*. Little, Brown and Co., Boston. A basic text emphasizing microbes and disease that includes sections on diseases of the skin and diseases associated with wounds.

Braude, A. I. (ed.). 1980. *Medical Microbiology and Infectious Disease*. W. B. Saunders Company, Philadelphia. An advanced work with chapters dealing with infectious diseases organized according to the taxonomy of the causative agent.

Davis, B. D., R. Dulbecco, H. N. Eisen, H. S. Ginsberg, and W. B. Wood. 1981. *Microbiology*. Harper & Row, Publishers, Hagerstown, Maryland. An excellent text used in many graduate and medical school microbiology courses dealing with all aspects of microbiology; chapters dealing with infectious diseases are organized according to the taxonomy of the causative agent.

Hoeprich, P. D. (ed.). 1977. *Infectious Diseases: A Modern Treatise of Infection Processes*. Harper & Row, Publishers, Hagerstown, Maryland. This classic treatment of infectious diseases makes an excellent reference volume.

Jawetz, E., J. L. Melnick, and E. A. Adelberg. 1980. *A Review of Medical Microbiology*. Lange Medical Publications, Los Altos, California. This text presents a very good review of infectious diseases.

Joklik, W. K., and H. P. Willett (eds.). 1980. *Zinsser Microbiology*. Appleton-Century-Crofts, New York. A classic text used in many medical school programs that includes chapters on infectious diseases.

Marples, M. J. 1969. Life on the human skin. *Scientific American*, **220**(1):108–115. An interesting article describing the resident microbiota of the human skin.

Rose, N. L., and A. L. Rose (eds.). 1983. *Microbiology: Basic Principles and Clinical Applications*. Macmillan Publishing Co., Inc., New York. An advanced work on medical microbiology; chapters are organized for the most part according to taxonomic groups.

Youmans, G. P., P. Y. Paterson, and H. M. Sommers. 1980. *The Biological and Clinical Basis of Infectious Disease*. W. B. Saunders Company, Philadelphia. This is an excellent text covering most human infectious diseases.

29 Diseases of the Eye, Ear, and Oral Cavity

KEY TERMS

cold sore
conjunctivitis
dental caries
dental plaque
fever blister
gingivitis
herpes corneales
inclusion conjunctivitis
juvenile periodontitis
keratitis
keratoconjunctivitis
ophthalmia neonatorum
oral cavity
otitis externa
otitis media
periodontal disease
periodontal pocket
periodontitis
pink eye
shipyard eye
swimmer's ear
thrush
trachoma
trench mouth

OBJECTIVES

After reading this chapter you should be able to

1. Define the key terms.
2. Describe the structure of the eye, ear, and oral cavity.
3. Describe the resident microbiota of the oral cavity.
4. Describe the defense systems of the eye, ear, and oral cavity that protect against infectious microorganisms.
5. Discuss the causative agents, routes of transmission, key symptoms, and treatments for the following diseases: pink eye, inclusion conjunctivitis, trachoma, shipyard eye, epidemic hemorrhagic conjunctivitis, swimmer's ear, otitis media, thrush, dental caries, periodontitis, and trench mouth.

Structure of the Eye

The eye is the sensory organ of vision (Figure 29.1). Most of the eyeball lies recessed in the orbit, protected by this bony socket, leaving only the small anterior surface of the eyeball exposed. The eyeball is composed of three layers of tissues or coats, the sclera, the choroid, and the retina. The outermost layer, the sclera, is composed of tough white fibrous tissue. The anterior portion of the sclera, called the cornea, lies over the colored part of the eye, the iris. The cornea is transparent, whereas the rest of the sclera is white and opaque. The sclera and cornea are covered by a thin mucous membrane, the conjunctiva, that lines the eyelid. The middle or choroid coat of the eye contains many blood vessels and a large amount of pigment. Its anterior portion is modified into three separate structures, the ciliary body, the suspensory ligament, and the iris. The colored part of the eye, the iris, consists of circular and radial smooth muscle fibers arranged in a doughnut shape. The hole in the middle is the pupil.

The retina is the innermost coat of the eyeball. Three layers of neurons make up the major portion of the retina; in the order in which they conduct impulses, they are the photoreceptor neurons, the bipolar neurons, and the ganglion neurons. The distal ends of the photoreceptor neurons are shaped like rods and cones and constitute our visual receptors. They are highly specialized structures for converting light images into neurological impulses. Neurons extend back to a small circular area in the posterior part of the eyeball known as the optical disk. This part of the sclera contains perforations through which the fibers emerge from the eyeball as the optic nerve. The optic nerves pass from each eye and proceed to the optic chiasma, where the nerve fibers from the basal half of each retina cross to the opposite side. The fibers form the optic tracts that convey the visual impulses to the visual cortex region of the brain.

Defenses of the Eye

The eye is continuously bathed in fluids that have antimicrobial activities. Tears contain lysozyme and IgA that protect the eyes from microbial infections. The continuous washing of the exposed eye surface is effective in preventing microorganisms from colonizing or invading the eye tissues. Blinking blocks the entry of foreign matter and helps wash the eye surface. Only rarely are microorganisms able to overcome these defenses and cause eye infections.

Infections of the Eye

Bacterial Infections

Conjunctivitis. Conjunctivitis is an inflammation of the conjunctiva (mucous membrane) covering the

Figure 29.1 Diagram of the structure of the eye.

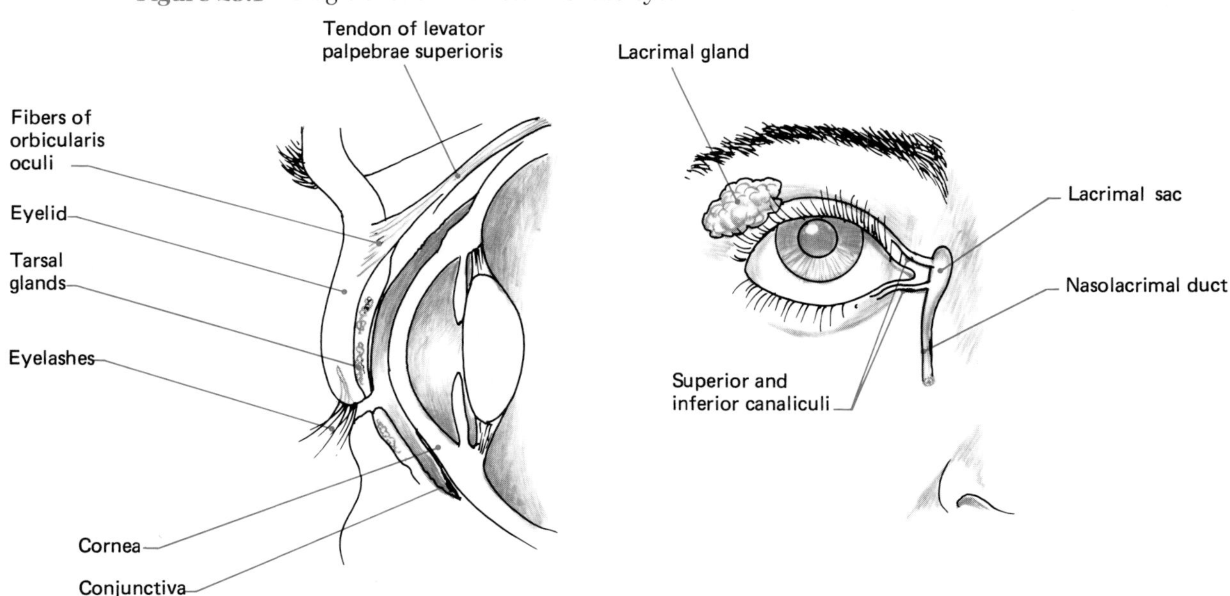

cornea and under the eyelids that results from bacterial infections of the eye. The establishment of an infection of the eye requires not only inoculation with opportunistic pathogens but also that the bacteria overcome the natural defenses of the eye. Contamination of the eye with foreign material and injuries of the eye tissue favor the establishment of eye infections. Bacteria causing conjunctivitis must resist washing out by tears and inactivation by lysozyme and antibodies. Bacterial conjunctivitis can be caused by a variety of bacteria, including *Haemophilus* sp., *Moraxella* sp., *Staphylococcus aureus, Streptococcus pneumoniae, Pseudomonas aeruginosa, Corynebacterium* sp., *Chlamydia* sp., and *Neiserria gonorrhoeae*.

The manifestations of conjunctivitis and the seriousness of this disease depend on the particular pathogenic microorganisms. The symptoms of conjunctivitis generally include swelling and reddening of the eyelids and the formation of purulent discharges. The eye becomes red and itchy because of extensive dilation of the capillaries. There may also be some photophobia and blurring of vision. Many cases of bacterial conjunctivitis are self-limiting, and the disease symptoms disappear within 1 week after onset. In cases of bacterial conjunctivitis, the use of antimicrobial agents are useful in limiting the severity of the disease. Sulfacetamide, neomycin, polymyxin B, tobramycin, and gentamicin are frequently used in treating infections of the eye.

Pink Eye. *Haemophilus aegyptius*, a fastidious, Gram negative, rod-shaped bacterium known as the Koch–Weeks bacillus, is the common etiologic agent of epidemic conjunctivitis or pink eye among school children. This bacterium is highly infectious and pink eye can rapidly spread among the members of a school. Symptoms of pink eye can include extreme swelling of the eyelids, extensive discharge from the eye, and bleeding within the conjunctiva, as well as the redness and itching associated with conjunctivitis inflammations. Pink eye can be treated with sulfonamides and tetracyclines.

Ophthalmia Neonatorum. Some forms of conjunctivitis are caused by sexually transmitted pathogens. Adults, for example, can acquire gonococcal opthalmia by rubbing their eyes with contaminated fingers. Newborns are at particular risk of acquiring sexually transmitted pathogens from lesions within the vaginal tracts of their mothers during normal vaginal tract deliveries. *Neisseria gonorrhoeae*, the causative agent of ophthalmia neonatorum (conjunctivitis of the newborn) can be transmitted in this manner (Figure 29.2). Often gonorrhea is asymptomatic in women, making it difficult to assess the risk of transmission of *N. gonorrhoeae* to the eye of the newborn. The prophylactic use of silver nitrate and, more recently, erythromycin sulfate to treat the eyes of newborns is a common measure to prevent the transmission of *N. gonorrhoeae*. When ophthalmia neonatorum occurs, it almost always involves both eyes. The symptoms of this disease include discharge of blood and pus and swelling of the eyelids; if untreated, the disease can cause blindness. Erythromycin, penicillin, and tetracyclines are effective in treating cases of ophthalmia neonatorum.

Inclusion Conjunctivitis. In addition to *Neisseria*

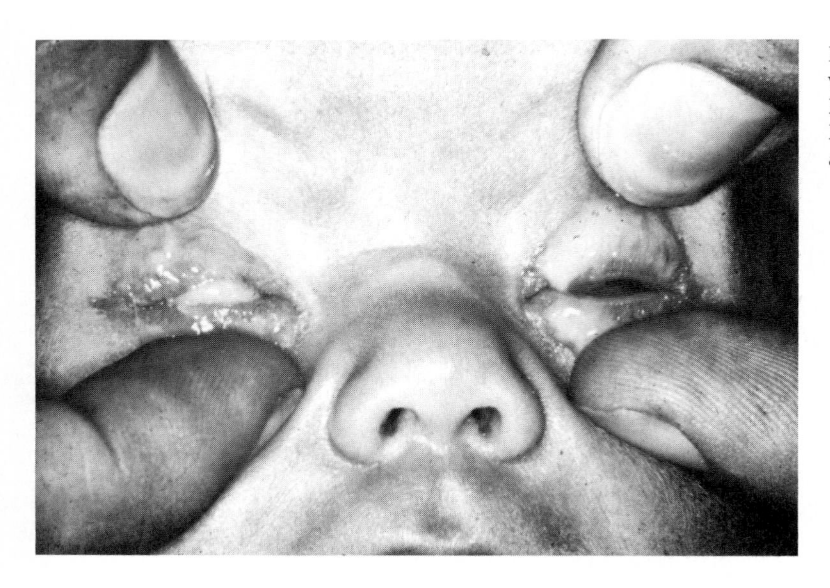

Figure 29.2 Photograph of a newborn with ophthalmia neonatorum. (From H. Hunter Handsfield, Department of Medicine, U.S. Public Health Hospital, Seattle.)

gonorrhoeae, Chlamydia trachomatis can be transmitted from infected mother to the eyes of the newborn during birth. *Chlamydia* only grows within cells and is an obligate intracellular inhabitant. Like *N. gonorrhoeae, C. trachomatis* is a sexually transmitted pathogen that can be asymptomatic in women. Although silver nitrate treatment of the eyes of newborns can prevent the transmission of *N. gonorrhoeae*, this treatment does not block transmission of *C. trachomatis*. Erythromycin prevents both *Neisseria* and *Chlamydia* infections, and therefore washing the eyes of newborns with erythromycin sulfate is now recommended.

When *C. trachomatis* infects eye tissues it causes inclusion conjunctivitis. The symptoms of this disease include reddening of the eyelid and a pussy discharge. In newborns these symptoms usually appear within 36 hours of birth. Inclusion conjunctivitis caused by *C. trachomatis* can also be acquired by adults and children by direct contact with infected individuals and from swimming in nonchlorinated waters contaminated with this bacterial species from genital sources. The disease can be treated with tetracyclines and sulfonamides.

Trachoma. *Chlamydia trachomatis* also is the causative agent for trachoma (Figure 29.3). This disease is a type of keratoconjunctivitis, so named because the conjunctivae becomes inflamed and the cornea become covered with dead keratin protein (keratinized). Trachoma occurs worldwide but is most prevalent in dry regions of the world, including the southwest United States, where this disease is a particular problem on Indian reservations. The worldwide incidence of trachoma may be as high as 500 million cases.

Transmission of the disease is through direct or indirect contact of the eye with infectious material. The development of the chlamydial infection in cases of trachoma initially involves a localized infection of the conjunctivae, but this is followed by spread to other areas of the eye, including the cornea. In *C. trachomatis* infections, vascular papillae and lymphoid follicles are formed, eye tissues can become deformed and scarred, and a chronic infection can lead to partial or total loss of vision. Scarring of the eye often is due to scratching the cornea with lesions that develop on the eyelid. Several antibiotics, including sulfonamides, tetracyclines, and erythromicin, are useful in treating trachoma.

Keratitis. In some cases, infections of the eye cause inflammation of the cornea, generally called keratitis. These infections are particularly serious because they can produce scarring of the cornea with permanent vision impairment. Symptoms of infections involving the cornea often include fever, headache, swelling of the tissues around the eye, and pussy discharge. Occasionally, as when the cornea is infected with *Pseudomonas aeruginosa* (Gram negative rods), perforation may result within 24 hours after onset of infection. *Streptococcus pneumoniae* (Gram negative, α-hemolytic cocci) can cause the formation of cataracts (clouding of the lens of the eye obstructing the passage of light). Other bacterial species, including *Moraxella lacunata* (Gram negative coccobacilli), *Neisseria meningitidis* (Gram negative diplococci), *Staphylococcus aureus* (Gram positive cocci in clusters), and *Streptococcus* species (Gram positive cocci in chains), may also cause damage of the cornea. Treatment of keratitis can employ a variety of antibiotics depending on the specific etiologic agent.

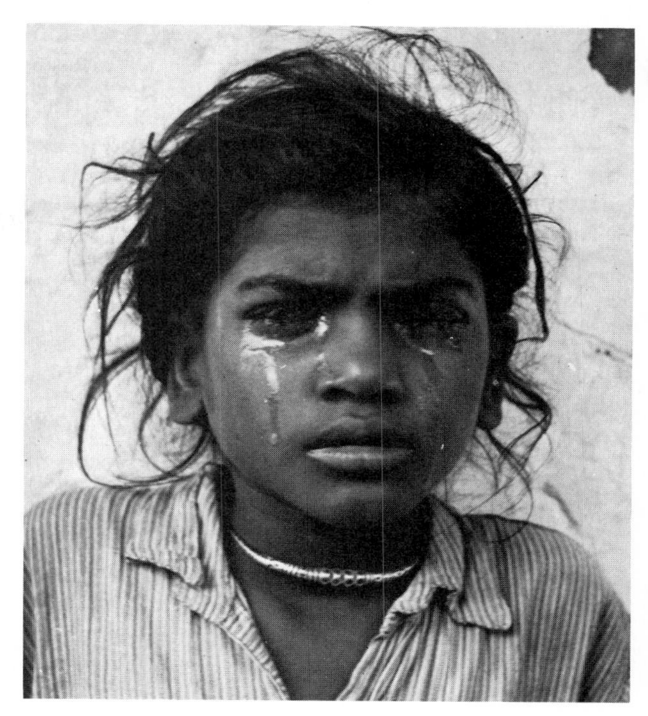

Figure 29.3 Photograph of a girl with trachoma. (From *World Health*, courtesy of the World Health Organization, Geneva.)

Viral Diseases

Epidemic Keratoconjunctivitis (shipyard eye). In addition to bacterial conjunctivitis, several viruses cause acute conjunctivitis. Adenoviruses are the etiologic agents of epidemic keratoconjunctivitis or shipyard eye. The disease derives its name from the fact that outbreaks have occurred in shipyards and industrial plants where trauma to the eye may occur as an occupational hazard. Adenoviruses also may occur in swimming pools, and transmission of the disease can be waterborne. Shipyard eye can result in the keratinization of the cornea as well as the inflammation associated with the conjunctivae.

Epidemic Hemorrhagic Conjunctivitis. Epidemic hemorrhagic conjunctivitis is caused by an enterovirus. This disease occurs in epidemic outbreaks in Africa, Asia, and Europe and is transmitted by direct or indirect contact with discharges from infected eyes and by droplets containing the infective enteroviruses. This form of conjunctivitis is characterized by subconjunctival hemorrages. The disease is self-limiting, and antimicrobial agents are not useful in treating this form of conjunctivitis.

Herpetic Keratoconjunctivitis. Herpes simplex virus type 1 can cause an ocular infection called herpes corneales. This disease, which occurs in both children and adults, affects the cornea. Ulcerations of the cornea caused by infection with herpes simplex virus can cause blindness in some cases. The use of antiviral drugs, such as iododeoxyuridine and acyclovir, is effective in controlling herpes simplex viral infections of the eye. However, herpes infections persist within the body and recurrences can occur despite treatment with antiviral drugs.

Structure of the Ear

The ear is divided into three portions: the outer or external ear; the middle ear or tympanic cavity; and the inner or internal ear (Figure 29.4). The outer ear consists of the auricle (the part of the ear that you see) and the auditory canal. The auditory canal penetrates the temporal bone and ends at the eardrum (tympanic membrane), which separates the outer and middle ear. The eardrum contains a membrane that vibrates as sound waves strike it. Disturbances in the air (sound waves) are funneled into the auditory canal where they cause the eardrum to vibrate. The outer surface of the tympanic membrane, which is part of the outer ear, is covered by a thin layer of skin.

The middle ear consists of the tympanic cavity, the ear bones, the eustachian tube, the tympanic

Figure 29.4 Diagram of the structure of the ear.

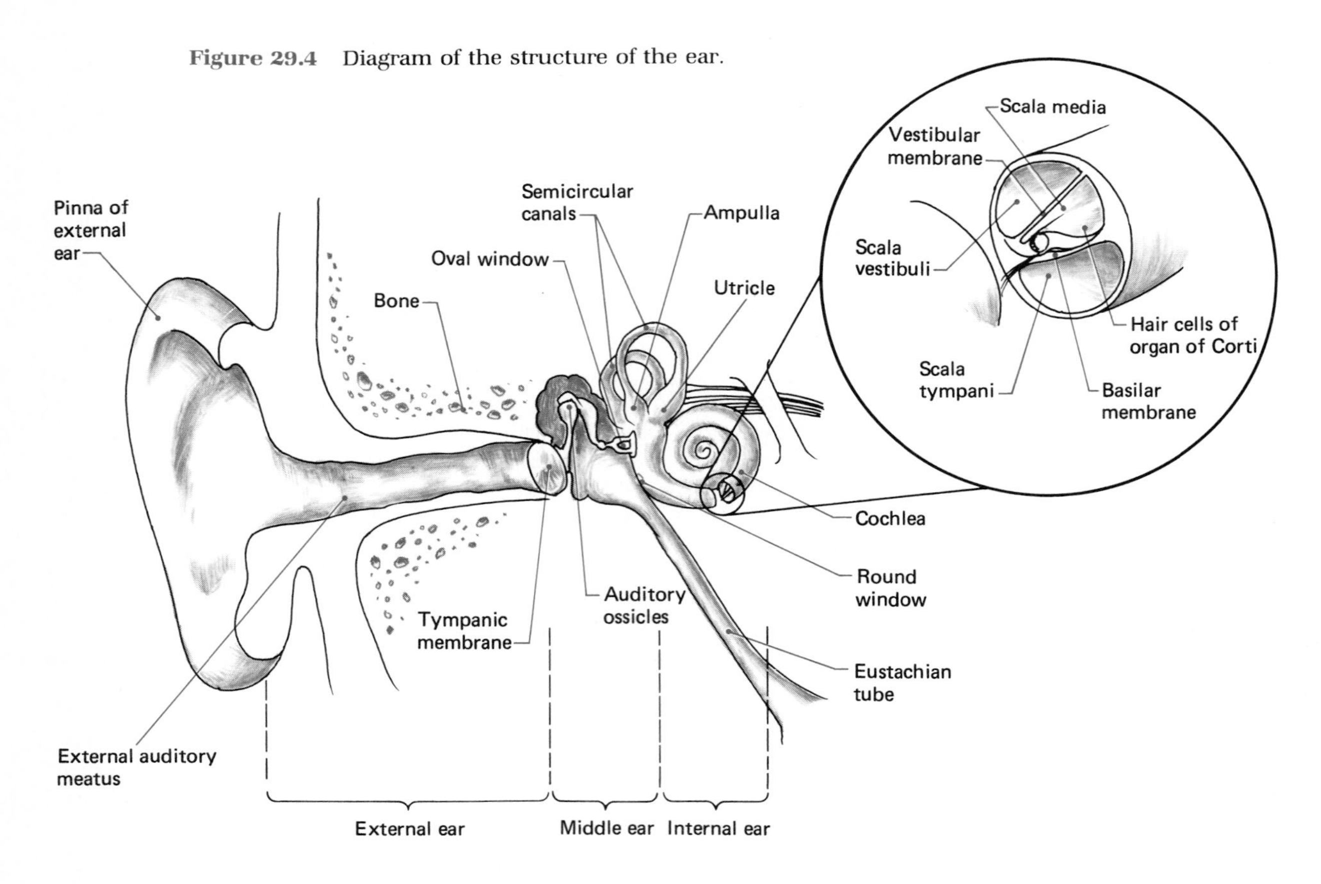

membrane (eardrum), and the tendons of small muscles that attach to two of the ear bones. The inner side of the tympanic membrane, which is part of the middle ear, is covered by a mucous membrane. The tympanic cavity is an irregular air-filled space in the temporal bone. The cavity is connected to the back part of the pharynx by way of the eustachian tube, which allows for the equalization of air pressure on both sides of the eardrum. Three of the small bones of the middle ear, the hammer, anvil, and stirrup, form a system of levers that transfer vibrations of the eardrum to the organ of hearing.

The inner ear consists of interconnected fluid-filled membranous structures that float in three fluid-filled cavities in the temporal bone: the vestibule, the semicircular canals, and the cochlea. The membranous structures in the vestibule and semicircular canals house sensory receptors associated with balance. The structures and receptors associated with hearing are in the cochlea. Vibrations originating with the tiny bones of the middle ear establish wave motions in the fluid-filled ducts on either side of the cochlear duct. These waves induce the hair-like projections on the hearing receptors in the floor of the cochlear duct to be stretched, initiating nerve impulses that are transmitted by the auditory nerve to the brain.

Defenses of the Ear

The three distinct regions of the ear are protected from microbial infections by different mechanisms. The inner portion is essentially a sealed unit. Microorganisms would have to migrate through bone or nerve tissues to invade the inner ear. This physical barrier generally is effective in protecting the inner ear. The outer ear, which is in contact with the air, is covered with skin. This skin covering acts as a physical barrier and even protects the tympanic membrane. Additionally, the brown waxy substance called cerumen or earwax is secreted into the outer ear from a type of sweat gland. This substance physically limits the movement of bacteria toward the eardrum and, like other sebaceous secretions, has antimicrobial activities. Microorganisms cannot move from the outer ear to the middle ear because the path is blocked by the tympanic membrane. However, microorganisms can enter the middle ear from the respiratory tract by passage through the eustachian tubes. Many microorganisms indeed do enter the middle ear, but most are unable to colonize or penetrate the tissues of the middle ear which are covered by mucous membranes. The secretion of mucus and the presence of cilia retard the movement of microorganisms to the surfaces of the tissues of the middle ear. Also, the mucous membranes of the middle ear contain IgA, which limits microbial infections.

Infections of the Ear

Otitis Externa

Several microorganisms can cause inflammations of the outer ear, or otitis externa. In some cases these infections involve the tympanic membrane. *Escherichia coli, Proteus* sp., *Streptococcus pyogenes*, and *Staphylococcus aureus* are opportunistic pathogens of the outer ear. In many cases outer ear inflammation is the result of a mixed infection involving these organisms. Ampicillin, penicillin, chloramphenicol, erythromycin, and various other antibiotics are used to treat such infections.

One type of outer ear infection, known as swimmer's ear, is caused by *Pseudomonas aeruginosa*. This disease is prevalent among swimmers, which accounts for its name. Bathing in *Pseudomonas*-contaminated waters is responsible for the transmission of this disease. *P. aeruginosa* is highly resistant to chlorine and hence swimming pools and hot spas are easily contaminated with this bacterium. In cases of swimmer's ear there is an inflammation of the lining of the external ear. Generally a broad spectrum antibiotic is used to treat this condition.

Otitis Media

Otitis media is an inflammatory disease of the mucosal lining of the middle ear. Bacterial infections of the middle ear normally originate from an upper respiratory infection or from the normal microbiota of the upper respiratory tract, with the bacteria entering the ear through the auditory (eustachian) tube, the principal portal of entry to the ear. Certain individuals are more prone to middle ear infections than others. In particular children are predisposed to middle ear infections if they have adenoids (enlarged masses of lymphoid tissue) in the nasopharynx. The middle ear infections in these childen is an extension of the tendency in such children to experience repeated upper respiratory tract infections. Today, plastic tubes are placed into the ears of some children to keep the auditory tubes open and limit the occurences of ear infections.

Streptococcus pneumoniae is the etiologic agent in over 50 percent of the cases of otitis media, and ampicillin is effective in treating such infections. *Streptococcus pyogenes* and *Haemophilus influenzae* are also frequently the causative agents of middle ear infections. *Haemophilus* is more frequently a problem in children than adults. A variety of antibiotics, including tetracyclines, chloramphenicol, and penicillin, are effective in treating middle ear infections caused by these organisms. Manifestations of middle ear infections normally include severe pain and fever and sometimes loss of balance due to accumulation of fluids. In cases caused by *Streptococcus* species, the tympanic membrane is usually fiery red. There is generally a loss of hearing. Otitis media can become a chronic condition and the pressure on the tympanic membrane can lead to thickening, scarring, or even rupture. Such disruptions of the tympanic membrane cause significant losses of hearing.

Structure of the Oral Cavity

The mouth is delimited by the lips, the cheeks, the hard palate, the tongue, and the soft palate (Figure 29.5). The oral cavity is that part of the mouth enclosed by the teeth. Humans have four types of teeth: incisors, canines, premolars, and molars. Teeth are living tissues. The teeth grind and cut food that enters the oral cavity, beginning the process of digestion. The crown of each tooth projects into the mouth cavity and is covered with enamel. The root of the tooth lies within the jawbone (lower teeth) or skull (upper teeth) and is covered with a bone-like material called cementum. Between the crown and the root is the cervix or neck of the tooth, a narrow region where enamel and cementum are thinnest. This is an area particularly liable to attack by decay-producing substances. The bulk of the tooth is formed of dentin, a yellowish soft material surrounding the pulp cavity of the tooth. This cavity is connected to the outside via the root canal and is filled with dental pulp. Pulp consists of a loose type of connective tissue containing blood vessels, nerves, and lymphatic vessels that nourish the tooth. On the walls of the pulp cavity are odontoblasts, cells that produce dentin. The tooth is held firmly, but not rigidly in its socket by a periodontal membrane that also acts as the source of cementum and as a periosteum (source of bone) for the jawbone.

Thy teeth are like a flock of sheep that are even shorn, which came up from the washing; whereof every one bears twins, and none is barren among them.

—Song of Solomon

The gums are composed of dense fibrous tissue, continuous with the periodontal membrane, covered with the oral mucus membrane. They also aid in holding the teeth in place and prevent food from lodging between the teeth.

The tongue is the main organ of the sense of taste, is important in word formation, and aids in guiding food between the teeth for chewing and swallowing. Three pairs of salivary glands empty their secretions into the mouth. Although saliva production is continuous, the greatest amount is secreted when food is in the mouth. Little absorption of digestion products occurs in the mouth.

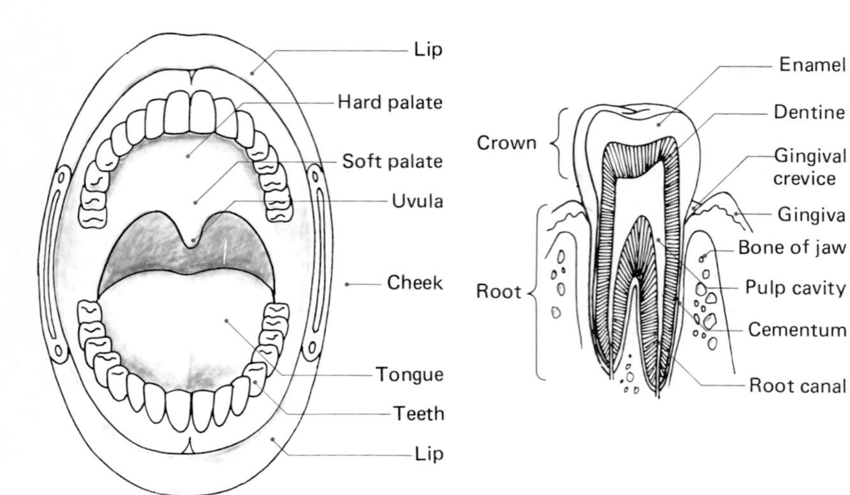

Figure 29.5 Structures within the human oral cavity.

Resident Microbiota of the Oral Cavity

An abundance of microorganisms develops within the oral cavity. The oral cavity contains a great variety of surfaces, with differing environmental conditions providing the varied habitats for colonization by diverse microbial populations. Within the oral cavity, microorganisms can grow on various surfaces, including the gums (gingiva) and teeth (Table 29.1). The availability of water and nutrients, including growth factors, provides an environment favorable for the proliferation of microorganisms.

Streptococcus species normally constitute a high proportion of the normal microbiota of the oral cavity. Various *Streptococcus* species, including *S. mutans*, produce slime layers and adherence factors that allow them to stick onto tooth surfaces. Other *Streptococcus* species, such as *S. sanguis* colonize saliva. Some of these lactic acid bacteria form dental plaque on the surfaces of teeth and are implicated in the formation of dental caries. Although *Lactobacillus* species are normally present in low numbers, they also are frequently found in association with dental caries. Besides *Streptococcus* species, obligate anaerobic bacteria are found in high numbers in the oral cavity, including Gram negative coccoid members of the genus *Veillonella* and Gram negative rod-shaped species of *Bacteroides* and *Fusobacterium*. Additional bacteria that are normally indigenous within the oral cavity are species of *Actinomyces, Neisseria, Staphylococcus, Micrococcus, Vibrio, Leptotrichia*, and *Rothia*; fungi, such as the yeast *Candida albicans*; and protozoa, such as species of *Entamoeba* and *Trichomonas*. The somewhat surprisingly high incidence of obligate anaerobes within the resident microbiota of the oral cavity results from the high rates of metabolism by the facultatively anaerobic members of this microbial community that scavenge the free oxygen, producing conditions that favor the growth of obligate anaerobes.

Table 29.1 Bacterial Populations Found in Different Parts of the Oral Cavity

	Percent Cultivatable Bacteria		
	Tooth	Tongue	Gingival Crevice
Streptococcus	25	45	25
Bacteroides	7	5	17
Fusobacterium	5	1	4
Corynebacterium	0	4	17
Veillonella	13	15	11
Actinomyces	33	2	9
Peptostreptococcus	11	5	9

Effects of Brushing

Although we routinely attempt to cleanse tooth surfaces of food particles and dental plaque, daily brushing represents only a temporary disturbance to the normal microbiota of the oral cavity. Within minutes of even the complete removal of microorganisms from the tooth surface, populations of *Streptococcus sanguis, S. salivarius*, and *Actinomyces viscosus*, which colonize the tongue and saliva, once again cover the tooth surface. Within a week or two after a thorough dental cleaning, the microbial community colonizing the tooth surface includes species of *Bacteroides, Veillonella, Neisseria, Fusobacterium*, and *Rothia*, in addition to *Streptococcus* and *Actinomyces* species. Although cleaning teeth does not eliminate these resident microorganisms, brushing does remove substrates that permit the overgrowth of microorganisms, the acidic products of microbial metabolism, and dental plaque, which are responsible for dental caries and periodontal disease.

Defenses of the Oral Cavity

As discussed in the previous section, the oral cavity is heavily colonized by microorganisms. Most of these organisms are noninvasive and except for diseases caused by excessive growth of the indigenous microorganisms, diseases of the oral cavity are relatively rare. The presence of enzymes and immunoglobulins in saliva is one of the major reasons that microorganisms do not frequently establish disease-causing infections of the oral cavity. Saliva also washes microorganisms out of the mouth. The gums, tongue, and other soft tissues of the oral cavity are protected by a mucus covering that limits attachment of potential pathogens. The enamel of the tooth surface is resistant to microbial invasion. Additionally, many of the resident microbiota that are abundant in the mouth produce lactic acid and other compounds that are antagonistic toward potential pathogens. Therefore, most microorganisms entering the oral cavity simply pass through into the gastrointestinal or respiratory tracts.

Infections of the Oral Cavity

Herpes Simplex Infections

Herpes simplex 1 virus is most frequently involved in nongenital herpes infections. Infections with herpes simplex 1 virus most commonly involve the lips, mouth, and skin above the waist. The virus appears to be transmitted through direct contact of the surface epithelial tissues with the virus. A focal infection develops around the site of inoculation, and dissemination of the virus occurs from the primary focal lesion. In most cases the lesions are limited to the epidermis and surface mucous membranes; however, in some cases the herpes simplex virus can spread systemically, causing central nervous system infections.

The development of herpes viruses in the mouth and on the lips is seen as the development of lesions known as cold sores or fever blisters (Figure 29.6). Similar lesions also develop in infected regions of the skin. The lesions generally heal within several days of their appearance, but new lesions develop periodically, indicative of the persistence of herpes simplex infections. Active lesions contain infectious herpes simplex viruses and direct contact should be avoided. Dental and other health care workers must be aware of the possible spread of infection to their own fingers. Several of the newly developed antiviral drugs, such as acyclovir, are effective against herpes viruses and may prove useful in treating cutaneous herpes infections.

Thrush

Candida albicans is a yeast-like fungus that can grow on the oral mucosa, causing thrush. The growth of *Candida* in the oral cavity can cover the tissues, producing a white frothy ("cheesey") covering of the tongue and filling the oral cavity (Figure 29.7). This disease occurs mostly in newborns. It is prevalent in some developing countries. Transmission to newborns is common because *Candida* is a normal inhabitant of the vaginal tract and often grows excessively during pregnancy. Transfer to newborns during birth results in development of superficial infections. Nystatin and other antifungal drugs are effective in controlling this disease.

Figure 29.6 Photograph showing a fever blister on the lip. (From N. H. Rickles, Department of Pathology, University of Oregon Dental School.)

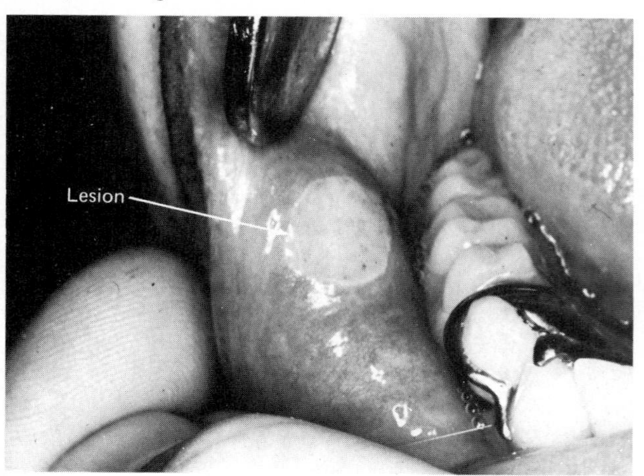

Dental Caries

The surfaces of the oral cavity are heavily colonized by microorganisms. Excessive growth of microorganisms in the mouth can cause diseases of the tissues of the oral cavity.

The Crocodile
Went to the dentist
And sat down in the chair,
And the dentist said, "Now tell me, sir,
Why does it hurt and where?"
And the Crocodile said, "I'll tell you the truth,
I have a terrible ache in my tooth,"
And he opened his jaws so wide, so wide,
That the dentist, he climbed right inside,
And the dentist laughed, "Oh isn't this fun?"
As he pulled the teeth out, one by one.
And the crocodile cried, "You're hurting me so!
Please put down your pliers and let me go."
But the dentist just laughed with a Ho Ho Ho,
And he said, "I still have twelve to go—
Oops, that's the wrong one, I confess,
But what's one crocodile's tooth, more or less?"
Then suddenly, the jaws went SNAP,
And the dentist was gone, right off the map,
And where he went one could only guess . . .
To North or South or East or West . . .
He left no forwarding address.
But what's one dentist, more or less?

—SHEL SILVERSTEIN, *The Crocodile's Toothache*

One of the most common human diseases caused by microorganisms is dental caries (Figure 29.8). Caries is initiated at the tooth surface as a result of the growth of *Streptococcus* species that are normal ingenous inhabitants of the oral cavity. These streptococci can initiate caries because they have the following essential properties: (1) they can adhere to the tooth surface; (2) they produce lactic acid as a result of their fermentative metabolism, thereby dissolving the dental enamel surface of the tooth; and (3) they produce a polymeric substance,

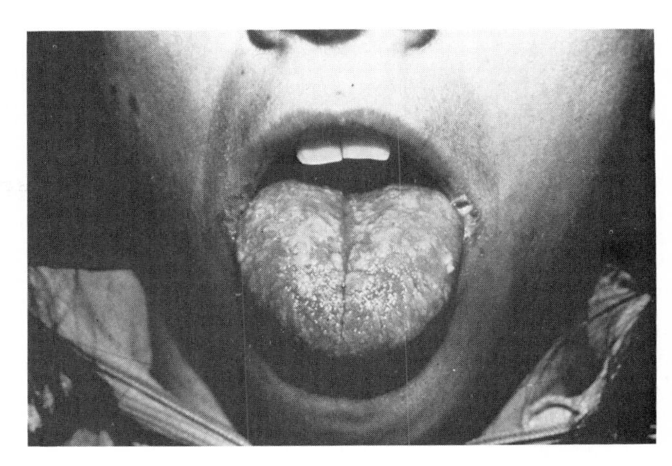

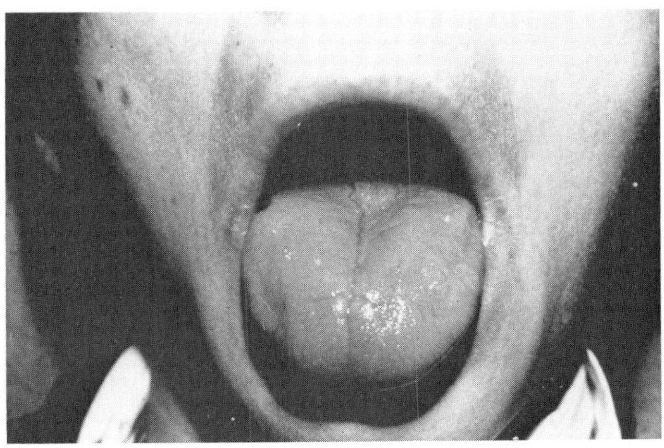

Figure 29.7 Photograph of the oral cavity in a case of thrush before and after treatment with the antimicrobic clotrimazole. (Courtesy Miles Laboratories, West Haven, Connecticut.)

which causes them to remain in contact with the tooth surface. *S. mutans, S. sanguis,* and *S. salivarius* are implicated as the causative agents of dental caries. These bacteria produce a dextran sucrose enzyme that catalyzes the formation of extracellular glucans from sucrose, responsible for the accumulation of dental plaque (Figure 29.9). Dental plaque is an accumulation of a mixed bacterial community in a dextran matrix. The accumulation of bacteria in plaque may be 500 cells thick. There is a high degree of structure within plaque, indicative of sequential colonization by different bacterial populations and the different positions of each population within this complex bacterial community.

Figure 29.8 Photograph of dental cavities. (From N. H. Rickles, Department of Pathology, University of Oregon Dental School.)

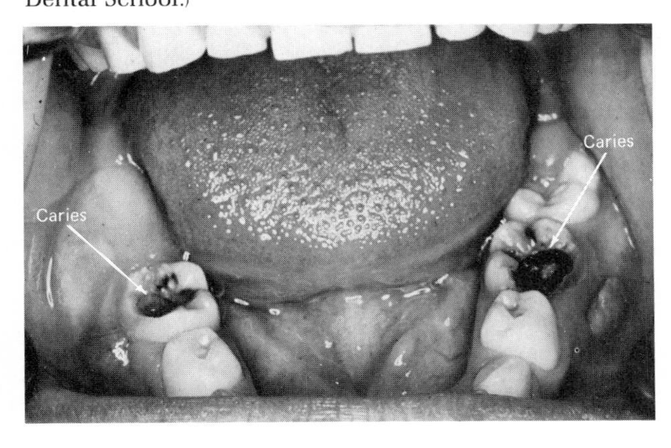

Evidence of dental diseases has been found in only a very few fossil animals, but studies on human neolithic skulls in France have revealed an incidence of some three to four per cent of caries. Similar findings made at the prehistoric site of Tepe Hissar in Iran in a population dating from 4000 B.C. to 2000 B.C. make clear that, despite common belief, caries afflicted man long before the advent of soft food and candy.

—René Dubos, *Mirage of Health*

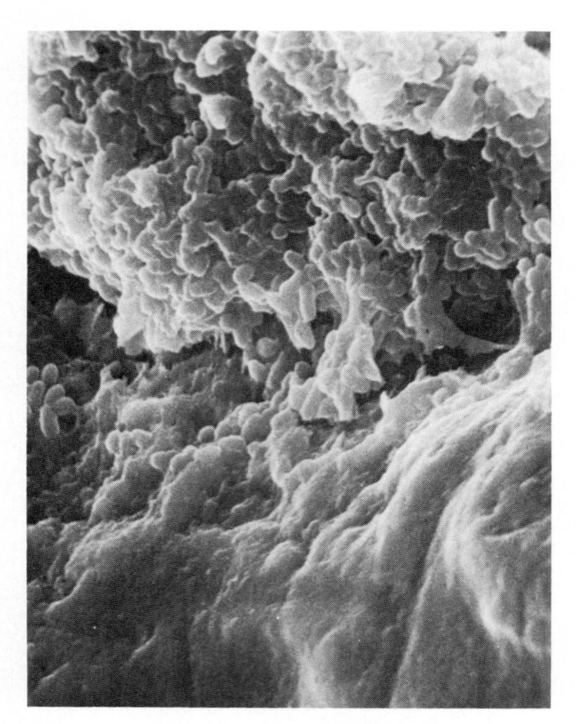

Figure 29.9 *Streptococcus mutans* in dental plaque. The bacterial accumulation on the tooth, is seen in this carious lesion of tooth enamel (3600×). These bacteria produce polymers and an enzyme, dextran sucrase, that create a sticky film on the tooth surface, providing a favorable environment for bacteria to grow and attack tooth enamel; it also restricts the movement of their metabolic product, lactic acid, away from the tooth surface, resulting in tooth decay. Of particular importance is sucrose in the diet, which is necessary for formation of the plaque polymer. Reducing intake of sucrose and cleaning the tooth surface to remove plaque buildup reduces dental caries. (From BPS: Z. Skobe, Forsythe Dental Center.)

Dental caries can be prevented by limiting dietary sugar substrates from which the bacteria produce acid and plaque and by removing accumulated food particles and dental plaque by periodic brushing and flossing of the teeth. Such hygienic practices are particularly effective if performed after eating meals and snacks. Additionally, tooth surfaces can be rendered more resistant to microbial attack by including calcium in the diet, as by drinking milk, and by fluoride treatments of the tooth surface. The administration of fluorides in the diet, such as by consumption of fluoridated water, during the period of tooth formation can reduce dental caries by as much as 50 percent.

Periodontal Disease

In addition to causing dental caries, microorganisms growing in the oral cavity can cause diseases involving the supporting tissues of teeth (periodontal disease). Periodontal disease can be manifest as gingivitis (inflammation of the gingiva), acute necrotizing ulcerative gingivitis (trench mouth), periodontitis (inflammation of the periodontium, also known as pyorrhea), and juvenile periodontitis (noninflammatory degeneration of the periodontium leading to bone regression) (Figure 29.10). Periodontal disease develops when dental plaque accumulates between the tooth and the surrounding tissues. Bacteria associated with plaque clearly play a role in the development of periodontal disease. Antibiotics that suppress microbial populations in the oral cavity limit the the advance of periotonal disease, indicating the role of bacteria in the etiology of gingivitis and periodontitis. Although it is unlikely that Koch's postulates can ever be satisfied to identify specific etiological agents of periodontal disease, it appears that different defined bacterial populations are involved in different stages of development of periodontal disease.

The commonest non-contagious disease is periodontal disease, such as gingivitis, which afflicts some 80% of the US population. In Great Britain 13% of the people have lost all their teeth before reaching 21. During their lifetime few completely escape its effects.

—1985 GUINNESS BOOK OF WORLD RECORDS

Healthy periodontal tissues in humans appear to be associated with relatively few indigenous microorganisms. Most of these organisms are Gram positive cocci that are located supragingivally on the tooth surface. Microorganisms commonly associated with these tissues include *Streptococcus mitis, Streptococcus sanguis, Staphylococcus epidermidis, Rothia dentocariosa, Actinomyces viscosus, Actinomyces naeslundii*, and occasionally species of *Neisseria* and *Veillonella* among others. The onset of gingivitis is marked by characteristic changes in the bacterial populations associated with gingival plaque. Initially there appears to be an overgrowth of the normal supragingival plaque accompanied by a large increase in the proportion of *Actinomyces* species, such as *A. vicosus*. These *Actinomyces* species may be important in the attachment of other bacterial populations. As the inflammation of the gingiva progresses, anaerobic Gram negative bacteria, such as species of *Veillonella, Campylobacter*, and *Fusobacterium*, become prevalent. These

MICROBIOLOGY HEADLINES

Tests with Humans Could Lead to Vaccine Against Tooth Decay

By SALLY SQUIRES

Thirty college students in Boston are adding an unusual test to their curriculum this month through March.

They are to travel to the Forsyth Dental Clinic to drink a liquid protein produced by the bacteria that cause tooth decay. Then they will have their blood and saliva tested for the presence of special antibodies that ultimately fight tooth decay.

If the results are positive, a vaccine against cavities may be available within the next decade for routine use.

Tooth decay afflicts just about everyone, costs billions of dollars to treat and is the leading cause of tooth loss in adults 35 and younger.

Two advances—fluoridation and dental sealants—are capable of temporary protection against tooth decay. But the real breakthrough, a vaccine that could immunize people against cavities, has remained elusive.

There are two promising leads.

Some researchers, such as immunologist Daniel Smith of Forsyth Dental Clinic, are concentrating on the protein produced by the bacteria that cause tooth decay.

Others, including biologist Roy Curtiss III of Washington University, are attempting to control the bacterium itself.

Known as *Streptococcus mutans*—or *S. mutans*—the bacterium gradually colonizes the mouth during tooth development. Infants "don't seem to have the bacteria in their mouths until they develop teeth," says Curtiss. "Studies suggest very strongly that infants acquire the bacteria from other family members."

Once the bacteria take hold they reproduce slowly but persistently. "Only 50 percent of 3-year-olds have detectable levels of *S. mutans* in their mouths," says Smith. But "*S. mutans* grows in the mouths of some 95 percent of teen-agers.

"If we could devise a scheme where we could immunize kids before they have *S. mutans* and then keep antibodies high against the bacteria during the cavity-prone years, we think we could have a more complete system for protection from tooth decay."

S. mutans is one of the more benign members of the large bacterial family that causes strep throat and rheumatic fever. This bacterium erodes teeth in a three-step process, explains Curtiss.

First, it attaches to a thin layer of saliva that bathes the teeth with a filmy protein.

Next, *S. mutans* takes advantage of sugar in the diet. The bacteria break down sugar for their own energy use and also to produce a gluey substance that enables them to stick to the tooth and to other bacteria. This buildup is known as dental plaque and can be removed only by professional dental cleaning, Curtiss says.

Finally, cavities develop as *S. mutans* produces lactic acid, which gradually eats away tooth enamel.

Curtiss and his colleagues are attempting to produce a vaccine that will exploit the bacteria's use of a protein called Spa-A. This protein gives the bacterial surface a fuzzy coat and seems to be a crucial part of all three stages of tooth decay.

Curtiss is working to prompt the immune system to make antibodies to Spa-A, which would—he hopes—interfere with plaque buildup and beat tooth decay.

Despite some promising advances for a vaccine, Curtiss believes that routine immunization for tooth decay is still "three to 10 years away."

One problem that has hampered researchers is an unexpected complication that arose in animal testing and has forced investigators to concentrate on more difficult methods of vaccine production.

Injection of *S. mutans* into animals—to prompt the production of antibodies—occasionally has caused the animal's immune systems to mistake heart cells for the bacteria. In this case, the antibodies could attack the heart.

So instead of promoting a reaction to the bacteria, researchers are trying to prompt an attack on substances necessary for the bacteria's survival—such as Spa-A. The problems have been worked out sufficiently to gain federal Food and Drug Administration approval for the human testing in Boston.

Even if a vaccine against tooth decay were available today, experts caution that it still would be only "one of the weapons" in the dental arsenal, Smith says.

Preventive Measures Include Fluoride Use

Until dental vaccines are available, tooth decay can still be battled successfully with a combination of dental weapons.

Yet surveys suggest that "the general public simply does not know the best way to protect against tooth decay," says Alice Horowitz of the National Institute of Dental Research.

Among the steps recommended for prevention are:

- Fluoride, which Horowitz calls "the first line of defense against tooth decay." Fluoride is available in topical solutions, but the best way to get adequate fluoride protection, she says, is by drinking fluoridated water.

Fluoride protects teeth by making them tougher and more resistant to erosion by the acid that bacteria produce in the mouth.

- Dental sealants, described by Horowitz as "second only to fluoride" in offering protection. Unlike fluoride, sealants must be applied to the teeth by a dentist or a dental hygienist.

Sealants, a fairly recent technique, are generally used to seal pits and fissures in the upper tooth surface.

- Diet. The most potent cavity promoters are foods high in sugar, especially sweet, sticky foods, like raisins and chewy candy, that cling to the teeth.

Dentists recommend limiting consumption of these foods to meals and encourage brushing immediately afterward.

- Toothpaste. The American Dental Association evaluates toothpaste on request of the manufacturers. Based on tests of safety and effectiveness, the ADA allows manufacturers to put a seal of approval on the label.

Five fluoride-containing toothpastes—Aim, Aquafresh, Colgate, Crest and Macleans—carry the ADA seal, as do three toothpastes for sensitive teeth: Denquel, Protect and Sensodyne.

Source: The Washington Post, Jan. 3, 1985.

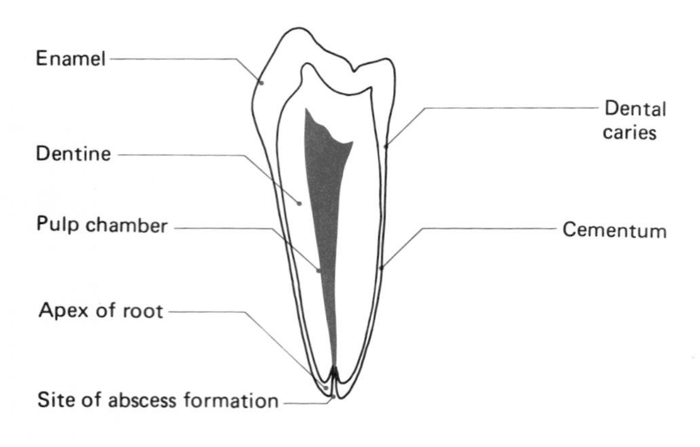

Figure 29.10 (A) The anatomy of a tooth, showing sites of microbial attack. (B) Diagram showing how the development of periodontal disease involves the tissues surrounding the tooth.

Figure 29.11 Photograph showing characteristic symptoms of trench mouth, showing the "punched out" inflamed gingiva. (From F. Howell, Department of Pathology, University of Oregon Dental School.)

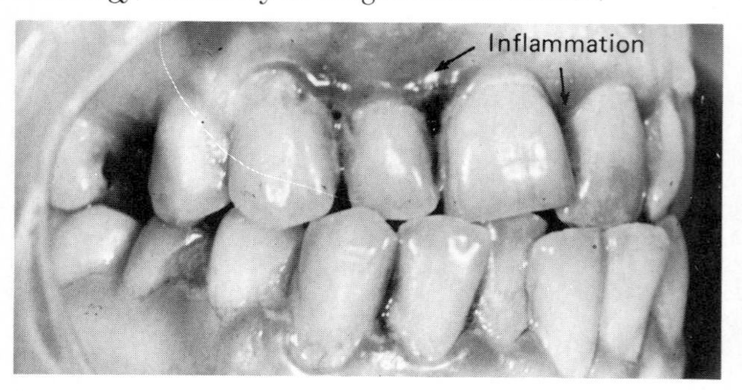

Gram negative bacteria appear to be located primarily on the surface of the plaque in subgingival sites.

The development of acute necrotizing ulcerative gingivitis (also known as trench mouth or Vincent's infection) is the only form of periodontal disease in which invasion of microorganisms into the tissues occurs. Clinical manifestations of trench mouth include fever, swelling, ulceration, and bleeding of the gums with tissue necrosis, causing ulcers between the teeth (Figure 29.11). Ultrastructural studies have shown that an unknown spirochete infiltrates the tissues and is the likely cause of this disease. This infection occurs mostly in adolescents and young adults. The name trench mouth is derived from the prevalence of this condition in the trenches of World War I. Fatigue, poor diet, poor oral hygiene, and anxiety are important predisposing factors that play a role in the establishment of this disease. Trench mouth can be treated with penicillin and metronidazole.

The development of advanced forms of periodontal disease, involving tissues beyond the gingiva, is serious and can lead to the loss of teeth. Microbial infection of the subgingiva can establish abscesses and pockets between the tooth and supporting tissues in which microorganisms can proliferate (peridontal pocket) (Figure 29.12). Microorganisms growing in such protected pockets erode the alveolar bone and destroy periodontal membranes, causing loosening of the teeth that can lead to tooth loss. The control of periodontal disease can be achieved by frequent removal of accumulations of dental plaque, by daily flossing, and pe-

Figure 29.12 Photograph showing abscess in a case of advanced periodontal disease. (From N. H. Rickles, Department of Pathology, University of Oregon Dental School.)

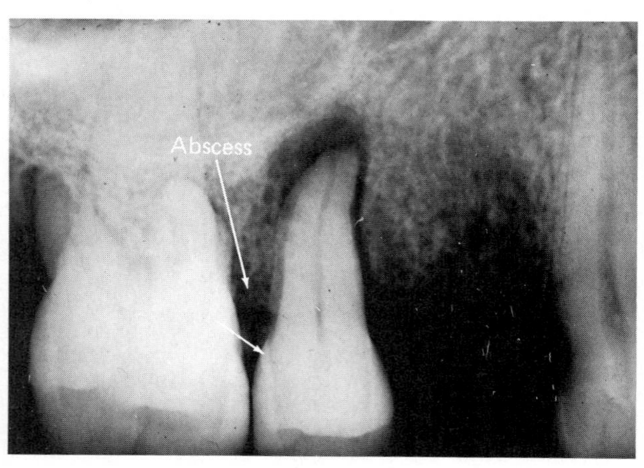

riodic professional dental cleanings. Flossing removes plaque which plays an important role in the development periodontal disease; prophylactic dental treatment also removes calcified plaque (calculus) that may play a role in the advance of periodontal disease.

In some cases there are large accumulations of plaque for years without destruction of the supporting structures. In contrast, some individuals develop dramatic loss of the periodontium with little evidence of plaque accumulation or gingival inflammation. It may be that local factors or the development of specific bacterial populations within the plaque determine the course of periodontal disease. The progression of inflammation into the periodontal membrane involves the development of a mixed community that is dominated by Gram negative bacteria.

The specific bacterial populations vary in different syndromes of periodontitis. For example, up to 75 percent of the isolates in cases of advanced periodontitis have been found to be Gram negative rods, predominantly *Bacteroides melaninogenicus* subspecies *asaccharolyticus* and *Fusobacterium nucleatum*. *Eikenella corrodens*, spirochetes and various unidentifed motile bacteria also occur in relatively high numbers in cases of periodontitis.

The development of juvenile periodontitis (periodontosis) is a distinct disease associated with different Gram negative bacterial populations. Juvenile periodontitis involves a rapidly progressing periodontal breakdown that affects young patients. The periodontal lesions are localized to permanent first molars and incisors or are generalized. The deep pocket bacteria in cases of juvenile periodontitis are comprised of approximately 65 percent Gram negative bacteria. *Capnocytophaga* species (formerly *Bacteroides* species) and *Actinobacillus actinomycetemcomitans* are the predominant isolates, suggesting that these baceria play an important role in the etiology of this disease.

Summary

The eyes are protected from infections by constant washing with tears and the IgA-containing mucous membranes. Some microorganisms are able to overcome these defenses and cause inflammations of the eye tissues. Eye infections are serious because if they progress to the cornea they can cause blindness.

As with eye tissues, the ears are relatively well protected against infections. Most infections of the ear originate from the respiratory tract and involve the inner ear. Outer ear infections sometimes occur by direct contact with contaminated water.

In the case of diseases of the oral cavity caused by microorganisms indigenous to the mouth, it is not possible to avoid contact with the pathogens. Consequently, the prevention of dental caries and periodontal disease depends on the effective removal of dental plaque with its associated microbial populations. Essentially, the prevention of these diseases depends on the elimination of microbial metabolic products before they can accumulate in concentrations sufficient to cause diseases involving the tissues of the oral cavity. Inclusion of elements in the diet, such as fluoride, that render the host tissues more resistant to microbial attack, and the reduced intake of foods containing high carbohydrate levels, which microorganisms can readily utilize as substrates to proliferate and form dental plaque, can greatly reduce the incidence of dental caries and periodontal disease.

Table 29.2 presents a summary of the diseases caused by microorganisms that involve the eye, ear, and oral cavity.

Table 29.2 Summary of Diseases of the Eye, Ear, and Oral Cavity

Disease	Causative Agent	Mode of Transmission	Treatment
Conjunctivitis	*Haemophilus* sp., *Moraxella* sp., *Staphylococcus* sp., *Streptococcus* sp., *Pseudomonas* sp., *Corynebacterium* sp., *Neisseria* sp., *Chlamydia* sp.	Direct contamination of eye tissues, aerosols	Various antibiotics
Pink eye	*Haemophilus aegypticus*	Direct contamination with infected discharge	Sulfonamides, tetracyclines
Ophthalmia neonatorum	*Neisseria gonorrhoeae*	Direct contamination of newborn with vaginal secretions	Erythromycin, penicillin
Inclusion conjunctivitis	*Chlamydia trachomatis*	Direct contamination with genital secretions	Sulfonamides, tetracyclines, erythromycin
Trachoma	*Chlamydia trachomatis*	Direct contact with infectious material	Sulfonamides, tetracyclines, erythromycin
Keratitis	*Moraxella lacunata*, *Neisseria meningitidis*, *Staphylococcus aureus*, *Streptococcus* sp., *Pseudomonas aeruginosa*	Direct contact with infectious material	Various antibiotics
Shipyard eye	Adenoviruses	Contaminated fomites	Supportive
Epidemic hemorrhagic conjunctivitis	Enteroviruses	Direct contact with aerosols with infectious material	Supportive
Herpes conjunctivitis	Herpes simplex 1	Direct contact and aerosols with infectious material	Acyclovir, iododeoxyuridine
Otitis externa	*E. coli*, *Proteus* sp., *Staphylococcus aureus*, *Streptococcus pyogenes*	Opportunistic	Various antibiotics
Swimmer's ear	*Pseudomonas aeruginosa*	Contaminated water	Broad spectrum antibiotic
Otitis media	*Streptococcus pneumoniae*, *Haemophilus influenzae*	Opportunistic, respiratory tract	Various antibiotics
Herpes fever blisters	Herpes simplex 1	Direct contact	Acyclovir
Thrush	*Candida albicans*	Mother to newborn	Nystatin
Dental caries	*Streptococcus*	Opportunistic infection	Restoration of damaged region
Gingivitis	*Actinomyces* sp., *Veillonella* sp., *Campylobacter* sp., *Fusobacterium* sp., *Bacteroides* sp., *Eikenella* sp.	Opportunistic infections	Various antibiotics
Trench mouth	*Borrelia vincentii*, *Fusobacterium fusiforme*, spriochetes	Opportunistic infections	Penicillin and other antibiotics
Periodontitis	*Actinomyces* sp., *Veillonella* sp., *Campylobacter* sp., *Fusobacterium* sp., *Bacteroides* sp., *Eikenella* sp.	Opportunistic infections	Various antibiotics

Study Outline

A. The eye is the organ of vision; its eyeball is composed of three coats, the sclera, the choroid, and the retina. The outermost, the sclera, is composed of tough white fibrous tissue. The anterior portion of the sclera, called the cornea, lies over the colored part of the eye, the iris. The cornea is transparent. The middle or choroid coat contains blood vessels and a large amount of pigment. Its anterior portion is modified into three separate structures, the ciliary body, the suspensory ligament, and the iris.
  1. The colored part of the eye, the iris, consists of muscle fibers arranged in a doughnut shape. The hole in the middle is the pupil.
  2. The retina is the innermost coat of the eyeball and mostly is made up of three layers of neurons. The distal ends of the photoreceptor neurons are shaped like rods and cones that convert light images into neurological impulses. Neurons extend back to the optical disk, which contains perforations through which the fibers emerge from the eyeball as the optic nerve.
  3. The eye is bathed in fluids that contain antimicrobial substances. Tears contain lysozyme and IgA.

B. Bacterial infections of the eye.
  1. Conjunctivitis.
    a. Causative organism(s): variety of bacteria, including *Haemophilus* sp., *Moraxella* sp., *Staphylococcus aureus, Streptococcus pneumoniae, Pseudomonas aeruginosa, Corynebacterium* sp., *Chlamydia* sp., and *Neiserria gonorrhoeae*.
    b. Symptoms: swelling and reddening of the eyelids, formation of purulent discharges, photophobia, blurring.
    c. Treatment: antimicrobial agents, including sulfacetamide, neomycin, and gentamicin.
  2. Pink eye.
    a. Causative organism(s): *Haemophilus aegyptius*.
    b. Symptoms: swelling of the eyelids, discharge from the eye, bleeding within the conjunctiva, redness and itching.
    c. Treatment: sulfonamides and tetracyclines.
  3. Ophthalmia neonatorum.
    a. Causative organism(s): *Neisseria gonorrhoeae*.
    b. Mode of transmission: from infected mother to infant during childbirth.
    c. Symptoms: discharge of blood and pus and swelling of the eyelids.
    d. Treatment: erythromycin, penicillin, and tetracyclines.
  4. Inclusion conjunctivitis.
    a. Causative organism(s): *Chlamydia trachomatis*.
    b. Mode of transmission: from infected mother to the eyes of the newborn during birth.
    c. Symptoms: reddening of the eyelid and a pussy discharge.
    d. Treatment: tetracyclines and sulfonamides.
  5. Trachoma.
    a. Causative organism(s): *Chlamydia trachomatis*.
    b. Mode of transmission: direct contact of the eye with infectious material.
    c. Symptoms: inflammed conjunctivae and covering of the cornea with dead keratin protein.
    d. Treatment: antibiotics, including sulfonamides, tetracyclines, and erythromycin.
  6. Keratitis (inflammation of the cornea).
    a. Causative organism(s): *Pseudomonas aeruginosa, Streptococcus pneumoniae, Moraxella lacunata, Neisseria meningitidis, Staphylococccus aureus*, and *Streptococcus* sp.

b. Symptoms: fever, headache, swelling of the tissues around the eye and pussy discharge, perforation of the cornea.
c. Treatment: choice of antibiotic dependent on specific etiologic agent.

C. Viral diseases of the eye.
1. Shipyard eye (keratoconjunctivitis).
a. Causative organism(s): adenoviruses.
b. Mode of transmission: associated with eye trauma.
c. Symptoms: keratinization of the cornea and inflammation of the conjunctivae.
2. Epidemic hemorrhagic conjunctivitis.
a. Causative organism(s): enterovirus.
b. Mode of transmission: direct or indirect contact with discharges from infected eyes and droplets containing infective enteroviruses.
c. Symptoms: subconjunctival hemorrages.
d. Treatment: none.
3. Herpes corneales.
a. Causative organism(s): herpes simplex virus type 1.
b. Symptoms: ulceration of the cornea.
c. Treatment: antiviral drugs, such as iododeoxyuridine and acyclovir.

D. The ear is divided into the outer or external ear; the middle ear or tympanic cavity; and the inner or internal ear.
1. The outer ear consists of the auricle and the auditory canal. The auditory canal penetrates the temporal bone and ends at the eardrum (tympanic membrane), which separates the outer and middle ear. The eardrum contains a membrane that vibrates as sound waves strike it.
2. The middle ear consists of the tympanic cavity, the ear bones, the eustachian tube, the tympanic membrane (eardrum), and the tendons of small muscles that attach to two of the ear bones. The tympanic cavity is an irregular air-filled space in the temporal bone. Three small bones in the middle ear, the hammer, anvil, and stirrup, form a system of levers that transfer vibrations of the eardrum.
3. The inner ear consists of interconnected fluid-filled membranous structures that float in three fluid-filled cavities in the temporal bone. The structures and receptors associated with hearing are in the cochlea. Vibrations originating with the tiny bones of the middle ear establish wave motions in the fluid-filled ducts on either side of the cochlear duct. These waves induce the hair-like projections on the hearing receptors in the floor of the cochlear duct to be stretched, initiating nerve impulses that are transmitted by the auditory nerve to the brain.

E. The inner portion of the ear is protected against microbial invasion because it is essentially sealed off. The outer ear is covered with skin, and cerumen secreted into the outer ear physically limits the movement of bacteria. Cerumen also has antimicrobial properties.
1. Microorganisms can enter the middle ear from the respiratory tract via the eustachian tubes, but secretions of mucus and the presence of cilia retard their movement.

F. Infections of the ear.
1. Otitis externa
a. Causative organism(s): *Escherichia coli, Proteus* sp., *Streptococcus pyogenes*, and *Staphylococcus aureus*.
b. Mode of transmission: swimmer's ear by swimming in *Pseudomonas*-infected water.
c. Treatment: ampicillin, penicillin, chloramphenicol, erythromycin, or other antibiotics.
2. Otitis media.
a. Causative organism(s): usually *Streptococcus pneumoniae*, also *S. pyogenes* and *Haemophilus influenzae*.
b. Mode of transmission: bacteria from an upper respiratory infection entering the ear through the eustachian tube.

c. Symptoms: severe pain and fever.
d. Treatment: antibiotics, including tetracyclines or chloramphenicol, and penicillin.

G. The oral cavity is that part of the mouth enclosed by the teeth.
1. The teeth grind and cut food. The crown of each tooth projects into the mouth cavity and is covered with enamel.
a. The root of the tooth lies within the jawbone and is covered with cementum. Between the crown and the root is the cervix or neck of the tooth, a narrow region where enamel and cementum are thinnest and where attack by decay-producing substances is most likely.
b. The bulk of the tooth is formed of dentin, a yellowish soft material surrounding the pulp cavity of the tooth. This cavity is connected to the outside via the root canal and is filled with dental pulp. Pulp consists of a loose type of connective tissue containing blood vessels, nerves, and lymphatic vessels that nourish the tooth.
c. The tooth is held in its socket by a periodontal membrane.
2. The gums are composed of dense fibrous tissue, continuous with the periodontal membrane, covered with the oral mucus membrane.
3. The tongue is the main organ of the sense of taste, is important in word formation, and aids in guiding food between the teeth for chewing and swallowing. Three pairs of salivary glands empty their secretions into the mouth.

H. The oral cavity contains a great variety of surfaces, with differing environmental conditions providing the varied habitats for colonization by diverse microbial populations.
1. *Streptococcus* species normally constitute a high proportion of this microbiota. Various *Streptococcus* species, including *S. mutans*, produce slime layers and adherence factors that allow them to stick onto tooth surfaces. Other *Streptococcus* species, such as *S. sanguis* colonize saliva. Some of these lactic acid bacteria form dental plaque on the surfaces of teeth and are implicated in the formation of dental caries.
2. Although *Lactobacillus* species are normally present in low numbers, they are frequently found in association with dental caries.
3. Obligate anaerobic bacteria are found in high numbers in the oral cavity, including Gram negative coccoid members of the genus *Veillonella* and Gram positive species of *Bacteroides* and *Fusobacterium*.
4. Additional members of the normal microbiota of the oral cavity include species of the bacteria *Actinomyces, Neisseria, Staphylococcus, Micrococcus, Vibrio, Leptotrichia*, and *Rothia*; fungi, such as the yeast *Candida albicans*; and protozoa, such as species of *Entamoeba* and *Trichomonas*.
5. Daily brushing represents only a temporary disturbance to the normal microbiota of the oral cavity. Within minutes, populations of *Streptococcus sanguis, S. salivarius*, and *Actinomyces viscosus* that colonize the tongue and saliva once again cover the tooth surface. Brushing does, though, remove substrates that permit the overgrowth of microorganisms, the acidic products of microbial metabolism, and dental plaque that are responsible for dental caries and periodontal disease.

I. The presence of enzymes and immunoglobulins in saliva is one of the major reasons microorganisms do not frequently establish disease-causing infections in the oral cavity.
1. Most of the natural microbiota are noninvasive and produce lactic acid and other compounds that are antagonistic toward potential pathogens.
2. The gums, tongue, and other soft tissue are protected by a mucus covering.
3. Tooth enamel is resistant to microbial invasion.

J. Infections of the oral cavity.
1. Herpes simplex infections.
a. Causative organism(s): herpes simplex 1 virus.
b. Mode of transmission: direct contact of the surface epithelial tissues with the virus.
c. Symptoms: cold sores or fever blisters.
d. Treatment: acyclovir.

2. Thrush.
   a. Causative organism(s): *Candida albicans*.
   b. Mode of transmission: transfer from the vaginal tract to newborns during delivery.
   c. Symptoms: white frothy covering of the tongue.
   d. Treatment: nystatin.
3. Dental caries.
   a. Causative organism(s): *Streptococcus* species., *S. mutans, S. sanguis*, and *S. salivarius*.
   b. Treatment: prevention by limiting sugars, brushing and flossing, including calcium in diet, and using fluoride treatments.
4. Periodontal disease.
   a. Causative organism(s): *Actinomyces* species may initiate gingivitis, more and different microbes involved as disease progresses.
   b. Mode of transmission: disease develops when food particles and dental plaque are allowed to accumulate between the tooth and surrounding tissues.
   c. Symptoms: gingival inflammmation.
5. Trench mouth.
   a. Causative organism(s): various invasive bacteria.
   b. Mode of transmission: associated with fatigue and anxiety in adolescents.
   c. Symptoms: fever, swelling, ulceration, and gum bleeding.
   d. Treatment: penicillin and metronidazole.

Study Questions

1. For each of the following diseases what are the causative organisms and characteristic symptoms?
   a. Thrush
   b. Trachoma
   c. Shipyard eye
   d. Pink eye
   e. Swimmer's ear
2. Match the following statements with the terms in the list to the right.
   a. Caused by *Pseudomonas aeruginosa*.
   b. Causes epidemic hemorrhagic conjunctivitis.
   c. Symptoms include severe pain, fever, fiery red tympanic membrane.
   d. Caused by *Haemophilus aegyptius*.
   e. Inflammatory disease of middle ear mucosal lining.
   f. Cause of epidemic keratoconjunctivitis.
   g. Organism that causes eye infections of newborn that can be excluded by using silver nitrate or erythromycin.
   h. Symptoms include fever, swelling, ulceration, and bleeding of gums with tissue necrosis, causing ulcers between the teeth.
   i. Caused by *Chlamydia trachomatis*.
   j. Transmission from mother to eyes of newborn can be prevented by erythromycin but not by silver nitrate.
   k. Also called Vincent's infection.
   l. Develops when food particles and plaque are allowed to accumulate between tooth and surrounding tissues.
   m. Initiated at tooth surface as result of growth of *Streptococcus* species (*S. mutans, S. sanguis, S. salivarius*).

   (1) Otitis media
   (2) Pink eye
   (3) Enterovirus
   (4) Adenovirus
   (5) *Neisseria gonorrhoeae*
   (6) Trachoma
   (7) Dental caries
   (8) Trench mouth
   (9) Periodontal disease
   (10) Swimmer's ear
   (11) *Chlamydia trachomatis*

Some Thought Questions

1. Are contact lens wearers at higher risk than others for contracting disease?
2. Can we find a vaccine for tooth decay?
3. Should fluoride be added to municipal water supplies?
4. Why are most ear infections associated with upper respiratory tract infections?
5. Why are certain eye diseases associated with particular occupations?
6. Why do some people have more cavities than others?

Additonal Sources of Information

Boyd, R. F., and B. G. Hoerl. 1981. *Basic Medical Microbiology*. Little, Brown and Company, Boston. A basic text emphasizing microbes and disease that includes sections on diseases of the eye, ear, and oral cavity.

Braude, A. I. (ed.). 1980. *Medical Microbiology and Infectious Disease*. W. B. Saunders Company, Philadelphia. An advanced work with chapters dealing with infectious diseases organized according to the taxonomy of the causative agent.

Burnett, G. W., H. W. Scherp, and G. S. Schuster. 1976. *Oral Microbiology and Infectious Disease*. Wiliams & Wilkins Company, Baltimore. A text dealing with infections of the oral cavity.

Davis, B. D., R. Dulbecco, H. N. Eisen, H. S. Ginsberg, and W. B. Wood. 1981. *Microbiology*. Harper & Row, Publishers, Hagerstown, Maryland. An excellent text used in many graduate and medical school microbiology courses dealing with all aspects of microbiology; chapters dealing with infectious diseases are organized according to the taxonomy of the causative agent.

Hoeprich, P. D. (ed.). 1977. *Infectious Diseases: A Modern Treatise of Infection Processes*. Harper & Row, Publishers, Hagerstown, Maryland. This classic treatment of infectious diseases makes an excellent reference volume.

Jawetz, E., J. L. Melnick, and E. A. Adelberg. 1980. *A Review of Medical Microbiology*. Lange Medical Publications, Los Altos, California. This text presents a very good review of infectious diseases.

Joklik, W. K., and H. P. Willett (eds.). 1980. *Zinsser Microbiology*. Appleton-Century-Crofts, New York. A classic text used in many medical school programs that includes chapters on infectious diseases.

Nolte, W. A. (ed.). 1980. *Oral Microbiology*. The C. V. Mosby Company, St. Louis. A work covering diseases of the oral cavity.

Rose, N. L., and A. L. Rose (eds.). 1983. *Microbiology: Basic Principles and Clinical Applications*. Macmillan Publishing Co., Inc., New York. An advanced work on medical microbiology; chapters are organized for the most part according to taxonomic groups.

Youmans, G. P., P. Y. Paterson, and H. M. Sommers. 1980. *The Biological and Clinical Basis of Infectious Disease*. W. B. Saunders Company, Philadelphia. This is an excellent text covering most human infectious diseases.

UNIT NINE

Applied Microbiology

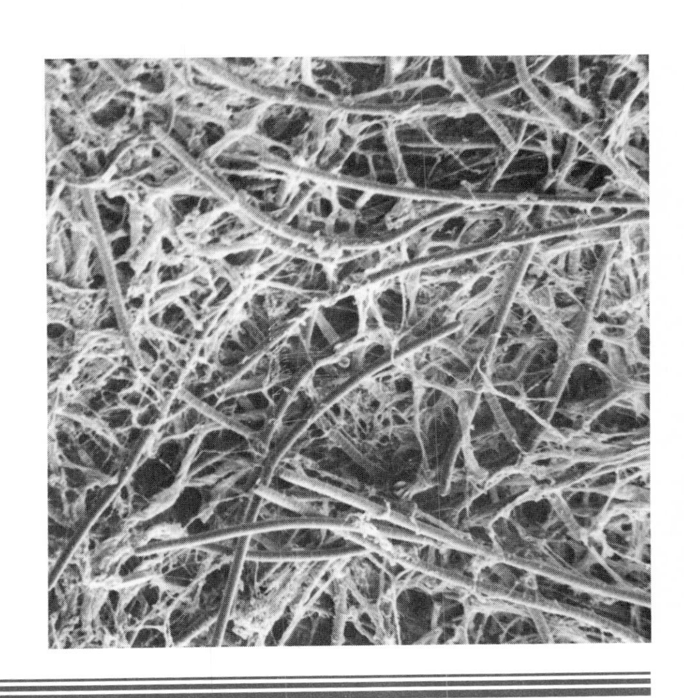

30 *Clinical Identification of Pathogenic Microorganisms*

KEY TERMS

agglutination
alpha (α) hemolysis
beta (β) hemolysis
biotype
cerebrospinal fluid
coagglutination
coagulase
complement fixation
counter immunoelectrophoresis (CIE)
defined media
differential media
direct fluorescent antibody test
enrichment media
enzyme-linked immunosorbant assay (ELISA)
etiologic agent
fluorescent antibodies
fluorescent antibody staining (FAB)
gamma (γ) hemolysis
hemagglutination
hemagglutination inhibition
heterophile antibodies
immunoelectrophoresis
immunofluorescent microscopy
indirect fluorescent antibody
lumbar puncture
miniaturized commercial identification systems
passive agglutination
precipitin band
radioimmunoassay (RIA)
selective media
septicemia
sputum
transtracheal aspiration
venous puncture
Wassermann test
Weil–Felix test
Widal test

OBJECTIVES

After reading this chapter you should be able to

1. Define the key terms.
2. Describe the initial culture methods used for detecting pathogens from the upper respiratory tract.
3. Describe the initial culture methods used for detecting pathogens from the lower respiratory tract; include a description of how to determine if a sputum sample is suitable.
4. Describe the initial culture methods used for detecting pathogens from the gastrointestinal tract.
5. Describe the initial culture methods used for detecting pathogens from the urinary tract.
6. Describe the initial culture methods used for detecting pathogens from the genital tract.
7. Describe the approaches that are used for the identification of pathogens based upon culturing microorganisms.
8. Compare the classical approach to microbial identification with computer-assisted commercial identification systems.
9. Describe the various serological procedures that are used for diagnosing the etiology of disease.
10. Discuss why serological procedures are replacing cultural procedures for determining the causes of infectious diseases.

Clinical microbiology and immunology laboratories are involved in determining the causes of disease and assisting the physician in selecting appropriate treatment methods (Figure 30.1). A major task of a clinical laboratory is to determine whether a given disease condition is caused by a microbial infection, with the objective of documenting the presence or absence of infectious agents in samples collected from an ailing individual. If a disease is of microbial origin, the clinical laboratory has the

Figure 30.1 Flow chart showing the diagnostic cycle between the physician and the clinical diagnostic laboratory.

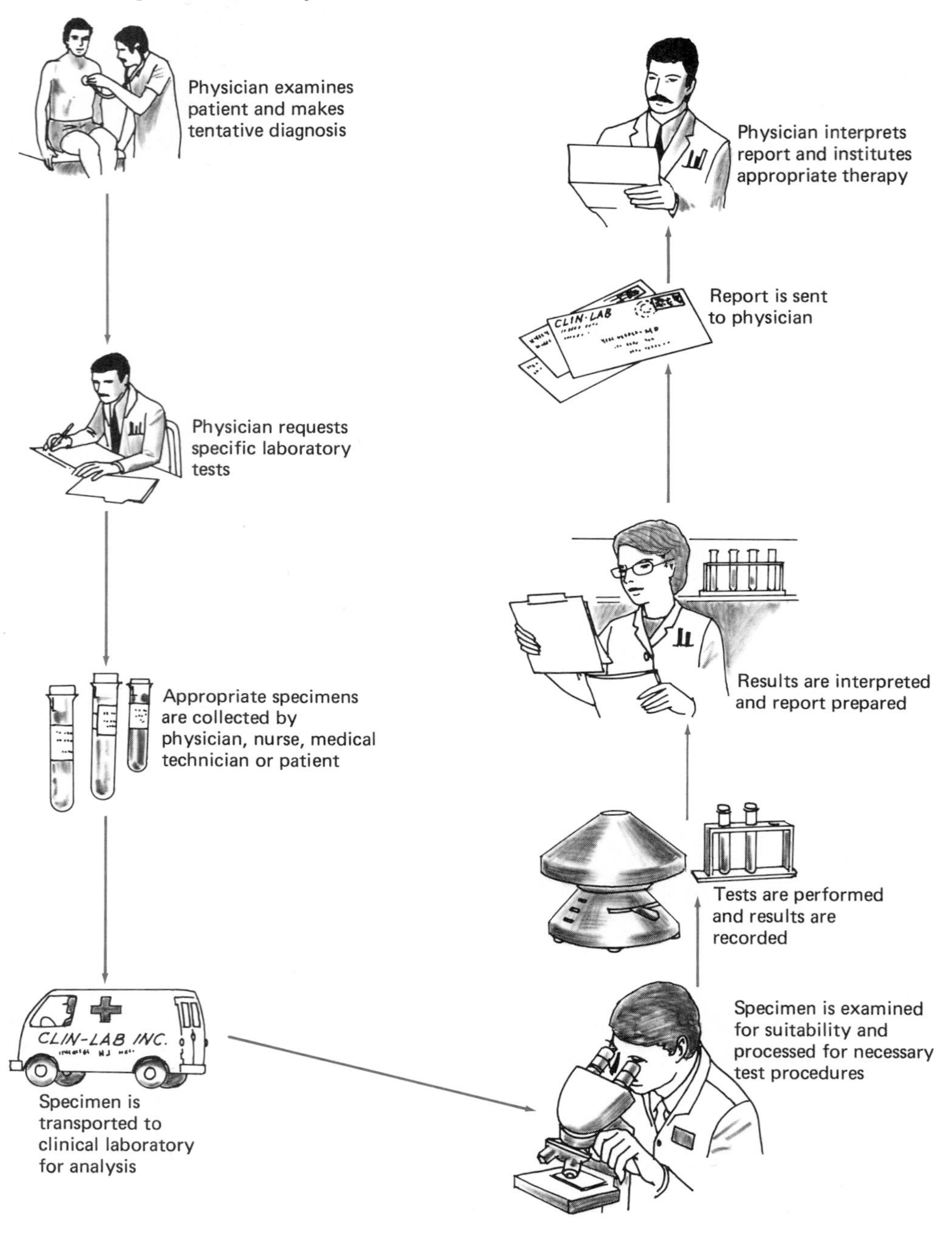

responsibility of identifying the causative microorganism (etiologic agent).

Medici causa morbi inventa curationen esse inbentam putant
Physicians consider that when the cause of a disease is discovered, the cure is discovered

—CICERO, *Tusculanarium Desputationum*

The clinical microbiology laboratory applies the full scope of our basic understanding of microorganisms, the immune response, and the relationships between host and pathogen with regard to the disease process. Diagnostic methods include various types of microscopic observation, isolation, and culture methods for the identification of pathogenic microorganisms. Clinical identification schemes employ a variety of morphological, physiological, and biochemical features to provide accurate and timely identification of a variety of human pathogens. Serological (in vitro) antigen–antibody reactions are also employed for the detection and identification of various etiologic agents.

The beginning of health is to know the disease

—MIGUEL DE CERVANTES, *Don Quixote*

Screening and Isolation Procedures

The proper collection of a specimen for analysis is an extremely important step in the identification of the etiology of a disease condition. Failure to collect and handle specimens correctly can lead to failure in correct diagnosis of a disease and the institution of inappropriate therapy. In some cases specimens are collected by patients themselves but in most cases trained medical staff collect and transport the specimen to the clinical laboratory. It is critical that the specimen material be from the site of infection and not contaminated by material from adjacent tissues or secretions. Analyzing upper respiratory tract secretions, for example, does not aid in the diagnosis of lower respiratory tract infections. The sample must be collected at the right time in order to detect the causative organism or the appropriate physiological indicator of a disease condition. For example, early morning sputum and urine samples generally contain the highest numbers of infecting microorgansims because of the overnight inactivity of the patient. Whenever possible specimens should be collected before antibiotic therapy is instituted, as the presence of antibiotics may preclude the recovery of the pathogen.

A sufficient quantity of specimen must be collected for all analyses that must be performed. The clinical laboratory cannot fulfill requests for multiple analyses when there is insufficient sample. Appropriate collection devices and specimen containers must be employed to ensure lack of contamination and survival of microorganisms from the site of infection. In many cases, where cultures are needed, a transport medium is necessary to protect the collected microorganisms while they are transferred to the laboratory for analysis. It is important to prevent desiccation and to protect anaerobes against exposure to oxygen. Specimens must be transported quickly to the laboratory to avoid changes that will occur on extended storage. Finally, the specimen must be properly labelled. The minimal information that must be included is the name of the patient, name of the physician, type of specimen, time of collection, and identification number. This information is essential if the laboratory is to accurately provide the physician with the complete information requested and needed to make a diagnosis and to institute appropriate therapy.

The positive diagnosis of an infectious disease requires the isolation and identification of the pathogenic microorganism or the identification of antigens specifically associated with a given microbial pathogen. In this abbreviated way the clinical microbiologist meets the requirements of Koch's postulates for establishing the etiology of the disease. A variety of procedures are employed for the isolation of pathogenic microorganisms from different tissues. Very different procedures are required for the isolation of different types of microorganisms. For example, the procedures used for the isolation of pathogenic bacteria are not applicable for the isolation of viruses, and the procedures designed for the recovery of aerobic microorganisms preclude the isolation of obligate anaerobes.

Most clinical procedures have been designed to screen and facilitate the recovery of those etiologic agents of disease that predominate within specific tissues. When the symptomology suggests that the disease may be caused by a rare pathogen and/or routine screening fails to detect a probable causative microorganism, additional specialized isolation procedures may be required. In the pages following, we will examine some of the routine procedures employed for the isolation of pathogens from various sites of the human body so that the etiologic agents of infectious diseases can be identified.

Upper Respiratory Tract Cultures

For the isolation of pathogens from the upper respiratory tract, throat and nasopharyngeal cultures are collected using sterile swabs made of polyester or other synthetic material. The swabs are placed in sterile transport media to prevent desiccation (drying out) during transit to the laboratory. These cultures are streaked onto blood agar plates and incubated, sometimes in an atmosphere of 5–10 percent carbon dioxide, for isolation of microorganisms. Sheep rather then human red blood cells are used in the preparation of blood agar because the natural human antibodies inhibit the recovery of bacteria, particularly *Streptococcus* species. Blood agar plates permit the detection of α-hemolysis (greening of the blood around the colony due to partial hemolysis), β-hemolysis (zones of clearing around the colony due to complete hemolysis), and γ-hemolysis (no clearing and no hemolysis). *Streptococcus pyogenes*, which forms relatively small colonies and demonstrates β-hemolysis on blood agar due to the production of hemolysin (an important virulence factor in the pathogenicity of this bacterial species), is often detected by using throat swabs and blood agar plates. The detection of β-hemolytic streptococci is important because these organisms can cause rheumatic fever. β-Hemolytic *Staphylococcus aureus* cultures as well as any other hemolytic bacteria may also be detected by using this procedure.

Haemophilus influenzae, Neisseria meningitidis, and *N. gonorrhoeae* can also be detected by using throat swabs and plating on various media. These organisms grow better on chocolate agar (a medium prepared by heating blood agar until it turns a characteristic brown color) than on plain blood agar. These species require heme but are incapable of hemolysis of red blood cells, which is why the blood cells must be prehemolyzed by heating to supply the growth factors required for culturing these pathogens. Thayer–Martin medium is preferred for the isolation of *N. gonorrhoeae* because the growth of normal throat microbiota is inhibited on this medium. This medium contains three antibiotics—vancomycin, colistin, and nystatin—that inhibit most Gram positive bacteria, Gram negative rods, and fungi that occur in the normal microbiota of the upper respiratory tract. Early diagnosis is important because an infection of the upper respiratory tract with these pathogens can lead to serious diseases, such as meningitis, if the infection spreads. Also, *Haemophilus influenzae* may cause acute epiglottitis, the rapid diagnosis of which is important, because death can result within 24 hours.

Figure 30.2 The typical Chinese-letter formation of *Corynebacterium diphtheriae* is shown in this micrograph. (From Centers for Disease Control, Atlanta.)

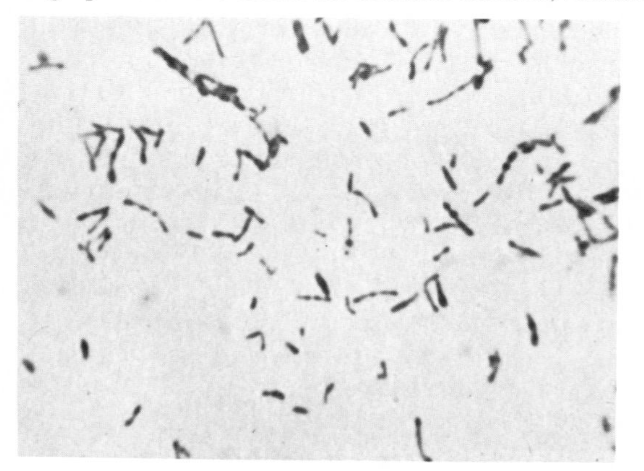

Nasopharyngeal swabs may also be used for determining the presence of *Bordetella pertussis*, the causative organism of whooping cough. The isolation of *B. pertussis* requires plating on a special medium, such as Bordet–Gengou potato–glycerol–blood agar. *B. pertussis* will grow on this medium, forming colonies that are similar in appearance to a drop of mercury (smooth with a pearly sheen). Cultures of *B. pertussis* also may be obtained by having the patient cough directly onto a plate of Bordet–Gengou medium. Cough plates are prefered to swabs for identification purposes because they eliminate most of the normal upper respiratory tract microorganisms, but in many cases the organism that causes whooping cough is not detectable by the the time the characteristic cough appears and the cultures are therefore negative.

When diphtheria is suspected, additional special procedures must be carried out. The presence of bacteria demonstrating typical snapping division (Chinese letter formation) indicates the possible presence of *Corynebacterium diphtheriae* (Figure 30.2). However, this is not a positive diagnosis because other bacteria with similar morphology may be present. Usually, several media are employed in the culture of *C. diphtheriae*. It can be cultured on Loeffler's medium and can be identified on tellurite agar where it forms black colonies due to the reduction of tellurite. Because individuals may carry nontoxigenic strains of *C. diphtheriae* in their throat or nasopharynx, the definitive laboratory identification of pathogenic strains must include tests for toxigenicity. These tests can be run in vivo using guinea pigs or rabbits. Comparison of the effects of inoculation with a culture into animals with and without the injection of diphtheria antitoxin can demonstrate whether the strain is producing toxin. For example, if death occurs when the culture is inoculated into a test guinea pig not treated with antitoxin and not when antitoxin is given then the the interpretation is that the strain being tested is pathogenic. Tests for diphtheria toxin can also be run in vitro by streaking plates with the isolate and adding paper strips impregnated with diphtheria antitoxin. A white line of precipitation that extends out from the line of bacterial growth indicates the formation of an antibody–antigen complex between the antitoxin and the toxin produced by the isolate; formation of a precipitin line indicates pathogenicity, that is, toxin production, by the isolated strain of *C. diphtheriae*.

In addition to culturing for bacterial pathogens, throat swabs can be used to collect viral pathogens. The laboratory growth of viruses employs tissue cultures rather than bacteriological media. Primary rhesus monkey kidney cells are used most frequently for viral tissue culture. Antibiotics are added to the tissue culture to prevent bacterial and fungal growth. Viruses are washed from throat swabs by using appropriate synthetic medium such as Hanks' or Earls' basal salt solutions, and the solution is then added to tissue culture tubes or plates. The viral infection of the tissue culture cells normally produces cytopathic effects that can be observed readily by microscopic observation.

Lower Respiratory Tract Cultures

Isolating microbial pathogens from the lower respiratory tract is a more difficult task than culturing organisms from the upper respiratory tract. Sputum, an exudate containing material from the lower respiratory tract, is frequently used for the culture of lower respiratory tract pathogens. Unfortunately, sputum samples vary greatly in quality and should therefore be examined microscopically before screening is carried out in order to determine whether they are suitable for culturing lower respiratory tract organisms. Acceptable sputum samples should have a high number of neutrophils, should show the presence of mucus, and should have a low number of squamous epithelial cells (Table 30.1). A large number of epithelial cells generally indicates contamination with oropharyngeal secretions, and such samples are not suitable because they are likely to contain bacteria that are part of the indigenous microbiota of the upper respiratory tract.

Table 30.1 Rating System for Determining Quality of Sputum Samples

Number of Cells per Low-power Field		
Epithelial	Leukocytes	Score
>25	<10	−3
>25	10–25	−2
>25	>25	−1
10–25	>25	+1
<10	>25	+2[a]

[a]High quality specimen that is acceptable for complete identification of microbiota.

In order to ensure the quality of lower respiratory tract specimens—free of normal upper respiratory tract microorganisms—transtracheal aspiration may be employed. In this technique a needle is passed through the neck, a catheter is extended into the trachea, and samples are then collected by aspiration through the catheter tube. Collection of sputum using the transtracheal procedure avoids contact with oropharyngeal microorganisms so that the clinician is certain of the source of any isolates. It is important to know where the isolated strains originate; some microorganisms are not likely to be associated with disease when they occur among the normal microbiota of the upper respiratory tract, but if these same organisms are found in the lower respiratory tract, they are prime candidates for the etiologic agents of disease.

The routine examination of sputum involves plating on appropriate selective, differential, and enriched media, for example, on blood agar, chocolate agar, and MacConkey's agar (a medium used for the isolation of Gram negative enteric bacteria). The bacterial pathogens normally detected by using this technique include *Streptococcus pneumoniae, Staphylococcus aureus, Streptococcus pyogenes, Klebsiella pneumoniae*, and *Haemophilus influenzae*, among the other pathogens that cause lower respiratory tract infections.

In cases of suspected tuberculosis and diseases caused by fungi, additional procedures are necessary. Examination of stained smears of the sputum are useful in detecting such infections. Fungal cells can be recognized in such smears. Also plating on Sabouraud's agar is used for the recovery of fungi. The acid-fast stain procedure can reveal the presence of *M. tuberculosis* in sputum samples. Members of the genus *Mycobacterium* are acid-fast (they do not decolorize in this staining procedure as do most other bacteria) and appear red when stained by this procedure. When fungi or mycobacteria are detected in the stained sputum sample, special culture media, such as Lowenstein–Jensen medium, should be used to isolate and positively identify the suspected pathogen.

Cerebrospinal Fluid Cultures

In cases of suspected infection of the central nervous system, cerebrospinal fluid (CSF) can be obtained by performing a lumbar puncture. It is important to determine rapidly whether there is a microbial infection of the cerebrospinal fluid because such infections (meningitis) can be fatal if not rapidly and properly treated. Several chemical tests can be performed immediately to determine whether a CSF infection is of probable bacterial, fungal, or viral origin. Most bacterial and fungal infections of the CSF greatly reduce the level of glucose, whereas viral infections do not, and this difference can be determined rapidly by measuring glucose and lactic acid concentrations in the CSF. Also the observation of an elevated white blood cell count is suggestive of a bacterial infection.

The CSF can be screened further for possible fungal infection by observing india ink stained smears for the presence of encapsulated yeasts. The observation of such yeast cells indicates infection with *Cryptococcus neoformans*, a fungus that causes meningitis. Bacterial infections can be detected by observing Gram stained slides and by culture techniques. The growth of bacteria in a liquid enrichment medium can easily and rapidly be detected as an increase in turbidity (Figure 30.3). Cultures can also be obtained from CSF by plating on blood and chocolate agar. In order to provide a sufficient innoculum, the CSF is routinely centrifuged to concentrate the bacteria, and the sediment is used for inoculation. The organisms are also usually cultured using 5–10 percent carbon dioxide atmospheres as *Neisseria* and *Haemophilus* require high carbon dioxide concentrations for optimal growth. Bacteria commonly associated with cases of bacterial meningitis include *Streptococcus pneumoniae, Neisseria meningitidis, Haemophilus influenzae, Streptococcus pyogenes, Staphylococcus aureus, Escherichia coli, Klebsiella pneumoniae*, and *Pseudomonas aeruginosa*. Bacterial pathogens can also be identified by serological examination of CSF using procedures described later in this chapter; viral meningitis can also be diagnosed by serological examination of CSF.

Blood Cultures

The detection of microorganisms in blood, likewise, is important in diagnosing a number of diseases, including septicemia (systemic bacterial infection of the blood stream, that is, blood poisoning). Some pathogens, such as the protozoa that cause malaria, can be observed directly by microscopic examination of blood samples. Also, differential changes in the numbers of various types of white blood cells can indicate bacterial and parasitic infections. For culturing bacteria, blood should be collected by venous puncture, using aseptic technique. The numbers of infecting bacteria in the blood are often low and may vary with time, and therefore, it is necessary to collect and examine blood samples at various time intervals. For example, in respiratory infections bacteria, may follow a 45-minute to 1-hour cycle of entry into the blood, removal, and reentry; whereas in endocarditis, the

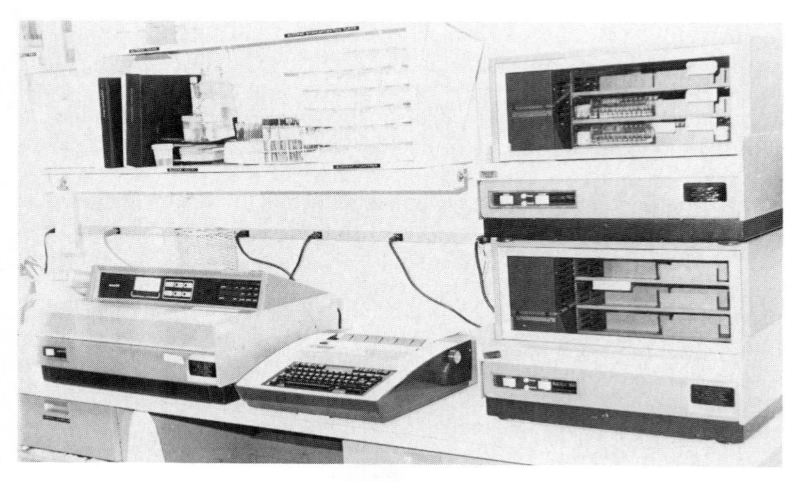

Figure 30.3 An autobac system for rapid screening of body fluids for bacterial contamination. The use of such automated systems has decreased the time needed by the clinical laboratory to make a diagnosis and recommend to the physician an appropriate course of treatment.

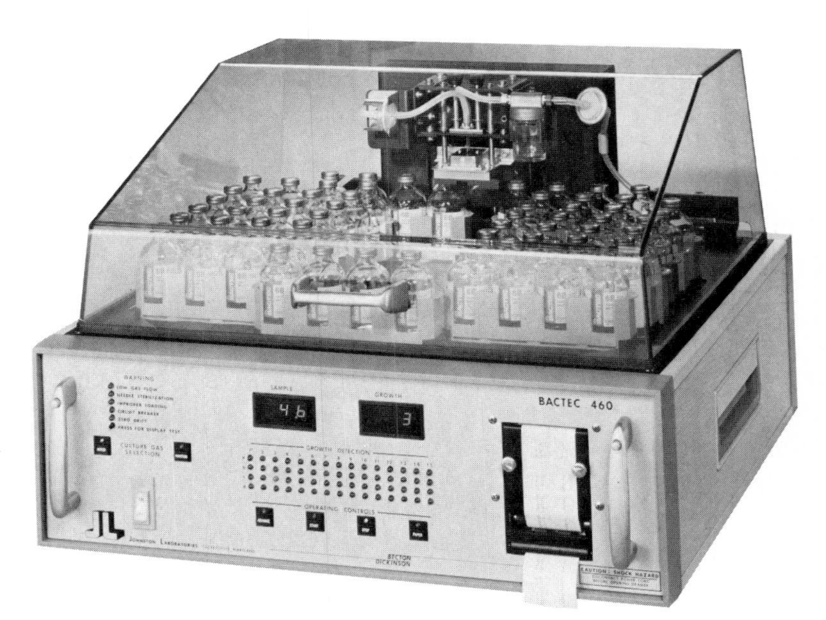

Figure 30.4 The Bactec system for detecting microbial growth uses a radiolabeled substrate. Growth of microorganisms on the labeled substrate releases radiolabeled gaseous products that are detected in this instrument. The Bactec system is advantageous because of its sensitivity and speed. (Courtesy Johnston Laboratories, Cockeysville, Maryland.)

numbers of bacteria in the blood generally remain relatively constant. Both aerobic and anaerobic culture techniques are needed to ensure the growth of any bacteria present in the blood. Often the blood sample is inoculated into trypticase soy broth medium containing 0.1 percent agar; the agar flocculates in the bottom of the flask creating anaerobic conditions so that aerobes can grow in the top of the broth and anaerobes at the bottom. One widely used, rapid-screening procedure for the presence of bacteria in blood employs a medium containing radiolabeled glucose. The conversion of glucose to radiolabeled carbon dioxide is rapidly determined with great sensitivity by using automated instrumentation (Figure 30.4).

Urine Cultures

In order to detect urinary tract infections, a midstream catch of voided urine is usually employed to minimize contamination with the normal microbiota of the genitourinary tract. A sample collected with a catheter would be preferable because it would eliminate possible contamination by microorganisms that occur at the outer extremeties of the urinary tract. However, because of the trauma to the patient of this procedure, voided urine samples generally are used. Precautions are normally taken to avoid contamination with exogenous bacteria during voiding of the urine sample, for instance, by washing the area around the opening of the urinary tract. Urine, though, normally becomes contaminated with bacteria during discharge through the urethra, particularly in females. Therefore, culture of the urine should be performed both qualitatively and quantitatively.

High numbers of a given microorganism are indicative of infection rather than contamination of the urine during discharge. In general, greater than 105 bacterial per milliliter indicates a urinary tract infection. Plating should be performed on a general medium, like blood agar, and on selective and differential media, such as cysteine lactose electrolyte deficient agar (CLED), MacConkey agar, or eosine-methylene blue (EMB) agar. Selective media contain agents that inhibit the growth of some microorganisms, permitting the selective growth of others. Differential media contain agents that permit the identification of particular microorganisms. EMB agar, for example, contains methylene blue to inhibit the growth of Gram positive bacteria and an indicator to permit the detection of lactose fermenters such as *Escherichia coli*. Bacteria of clinical significance that may be found in the urine include *E. coli, Klebsiella, Proteus, Pseudomonas, Salmonella, Serratia, Streptococcus*, and *Staphylococcus* species. Gram negative enteric bacteria are found most frequently to be the etiologic agents of urinary tract infections.

Urethral and Vaginal Exudate Cultures

The examination of urethral and vaginal exudates centers on the detection of microorganisms that cause sexually transmitted diseases, most notably *Neisseria gonorrhoeae, Chlamydia* sp., and

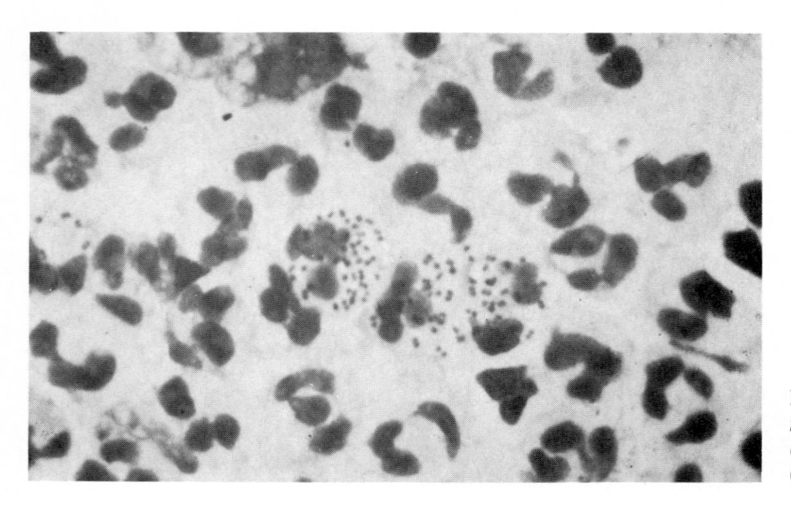

Figure 30.5 Micrograph of *Neisseria gonorrhoeae* in pus; the presence of bean-shaped diplococci is diagnostic of gonorrhea. (From Centers for Disease Control, Atlanta.)

Treponema pallidum. In males, one symptom of gonorrhea is the painful release of a urethral exudate. Gram stained slides of the exudate are made, and if Gram negative, kidney-bean-shaped diplococci are present within cells (intracellular diplococci), this suggests a diagnosis of gonorrhea (Figure 30.5). In females, gonorrhea is more difficult to detect because of the high numbers and variety of normal microbiota associated with the vaginal tract. Culture techniques using inoculation with swabs collected from the cervix, vagina, and anal canal can be employed for detecting *N. gonorrhoeae* in females. Thayer–Martin medium and chocolate agar, incubated under an atmosphere of 5–10 percent carbon dioxide, are employed for the culture of *N. gonorrhoeae*.

Figure 30.6 Micrograph of *Treponema pallidum* using darkfield microscopy, showing its characteristic corkscrew shape. (From Centers for Disease Control, Atlanta.)

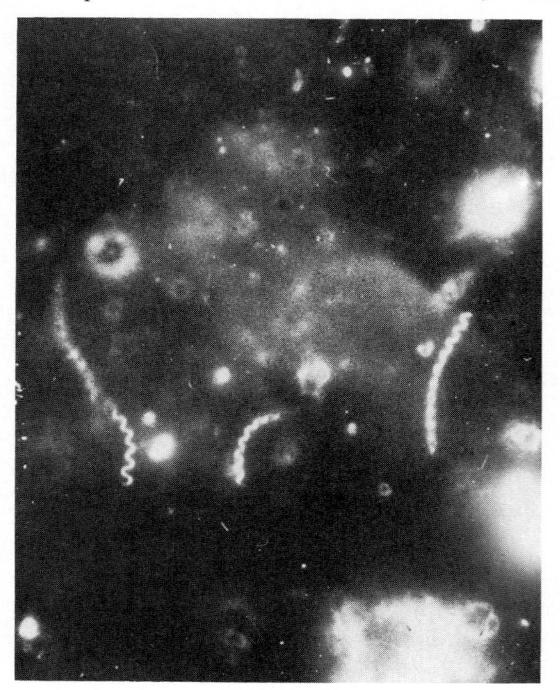

T. pallidum, the causative organism of syphilis, cannot be cultured on laboratory media, but exudates from primary or secondary lesions can be examined by using dark-field microscopy and fluorescent antibody staining for the detection of this organism (Figure 30.6). The characteristic morphology and movement of spirochetes can be readily detected by dark-field microscopic observation, but caution should be employed in the diagnosis because nonpathogenic spirochetes may also be present.

Fecal Cultures

Stool specimens are normally used for the isolation of microorganisms that cause intestinal tract infections. The common enteric bacterial pathogens are *Salmonella* species, *Shigella* species, enteropathogenic *E. coli, Vibrio cholerae, Campylobacter* species, and *Yersinia enterocolytica.*

Because fecal matter contains a large number of nonpathogenic microorganisms, it is necessary to employ selective and differential media for the isolation of intestinal tract pathogens. Usually, selective media contain some toxic factor that selectively inhibits some microorganisms, often preventing the overgrowth of pathogenic microorganisms that are present in low numbers by other more numerous

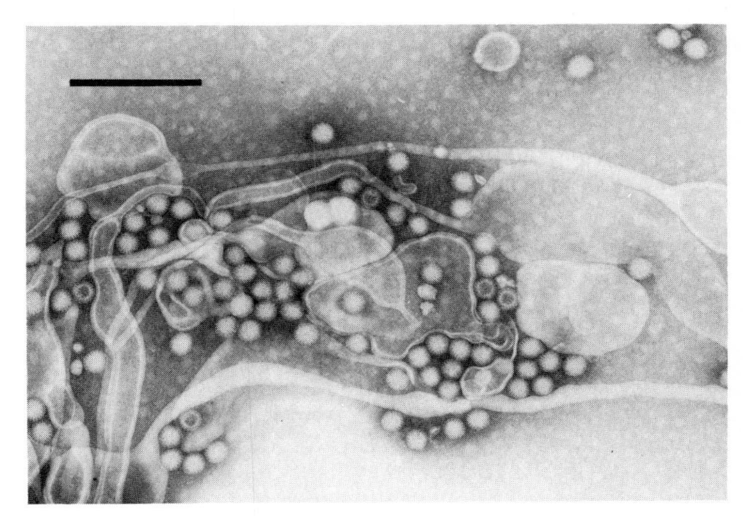

Figure 30.7 Virions of Rotavirus in the stool of an infant with acute gastroenteritis are shown in this negatively stained slide. Bar = 400 nanometers. (Courtesy F. P. Williams, Jr., U.S. Environmental Protection Agency, Cincinnati.)

microbial populations. Selective media contain components that select for the growth of particular microorganisms. Differential media contain indicators that permit the recognition of microorganisms with particular metabolic activities. For example, pH indicators are often incorporated into media for the detection of acidic metabolic products. Common selective and differential media that are employed for the isolation of intestinal tract pathogens include *Salmonella–Shigella* (SS), Hektoen enteric (HE), xylose–lysine–desoxycholate (XLD), brilliant green, eosin methylene blue (EMB), Endo, and MacConkey agars. It is often necessary to culture organisms in an enrichment medium before *Salmonella* and *Shigella* species can be isolated by using differential or selective solid media. For example, in cases of suspected typhoid fever it may be necessary to carry out an enrichment in appropriate selective medium, such as GN (Gram negative) broth or selenite F broth, before isolation of *Salmonella typhi* can be achieved.

In cases of suspected viral infections, tissue cultures can be inoculated with fecal matter for the culture of enteric viral pathogens. Characteristic cytopathic effects and serological procedures can then be employed in the identification of viral pathogens. Additionally, electron microscopy can be used for the direct detection of viruses, such as Rotavirus, that cause viral gastroenteritis (Figure 30.7).

Eye and Ear Cultures

Fluids collected from eye and ear tissues can be inoculated onto blood agar, chocolate agar, and MacConkey's agar or other defined media to culture bacterial pathogens commonly found in these tissues. Additionally, a Gram stain slide can be prepared and observed to detect the presence of bacteria. The microscopic observation of stained slides can give an indication as to whether the infection is due to bacterial pathogens. In the case of eye infections, it is particularly important to differentiate between bacterial and viral infections in order to determine the appropriate treatment.

Skin Lesion Cultures

Material from skin lesions, including wounds and boils, can be collected for culture purposes with swabs, aspirates, or washings. Such material, though, often is contaminated with endogenous bacteria. A variety of bacteria can infect wounds and cause localized skin infections. Both aerobic and anerobic culture techniques are required for the screening of wounds for potential pathogens. Particular concern must be given to the possible presence of *Clostridium tetani* and *Clostridium perfringens* because these anaerobes cause serious diseases.

Various fungi may also be involved in skin infections, and appropriate fungal culture media are required for the isolation of these organisms. Dermatophytic fungi and actinomycetes that cause skin infections can also be detected by direct microscopic examination of skin tissues. Because skin is thick it is first emulsified in a drop of 10 percent potassium hydroxide to permit visualization of microorganisms. The characteristic morphological appearance of filamentous fungi and bacteria often permits rapid, presumptive diagnosis of the disease by this direct observation method.

Identification of Cultures of Pathogenic Microorganisms

Determining that a disease is of microbial etiology and isolating microorganisms are only part of the job of diagnosing a specific disease. It is also necessary to identify the specific causative organism. A wide range of biochemical and serological procedures are available for the definitive identification of microbial isolates of clinical significance. Accuracy, reliability, and speed are important factors governing the selection of clinical identification protocols. The selection of the specific procedures to be employed for the identification of pathogenic isolates is guided by the presumptive identification of the organism at the genus or family level, based on the observation of colonial morphology and other growth characteristics on the primary isolation medium and on the microscopic observation of stained specimens. Some examples of the protocols and criteria used for the identification of various clinical isolates will be discussed in the following pages.

Conventional Tests for Diagnosing Pathogenic Microorganisms

Despite the development of rapid miniaturized commercial diagnostic systems, some clinical laboratories still rely on conventional testing procedures, particularly for the identification of Gram positive bacteria. The identification of *Staphylococcus* species, for example, is based on the observation of typical grape-like clusters, catalase activity, coagulase production, and mannitol fermentation. Members of the genus *Staphylococcus* are strongly catalase positive. Catalase activity can be readily observed by adding a drop of hydrogen peroxide to a loopful of bacteria from a colony that is suspended in saline and observing the production of gas bubbles. Pathogenic strains of *Staphylococcus aureus* are generally coagulase positive and produce acid from mannitol, in contrast to strains of *Staphylococcus epidermidis*, which are negative for both coagulase production and mannitol fermentation. Coagulase activity, the ability to clump or clot blood plasma (Figure 30.8), is associated with virulent strains of *S. aureus* and other staphylococci.

Figure 30.8 The tube test for coagulase production is used for the detection of pathogenic *Staphylococcus aureus*. Both control and positive reactions are shown in this photograph. (From BPS: Leon J. LeBeau, University of Illinois Medical Center, Chicago.)

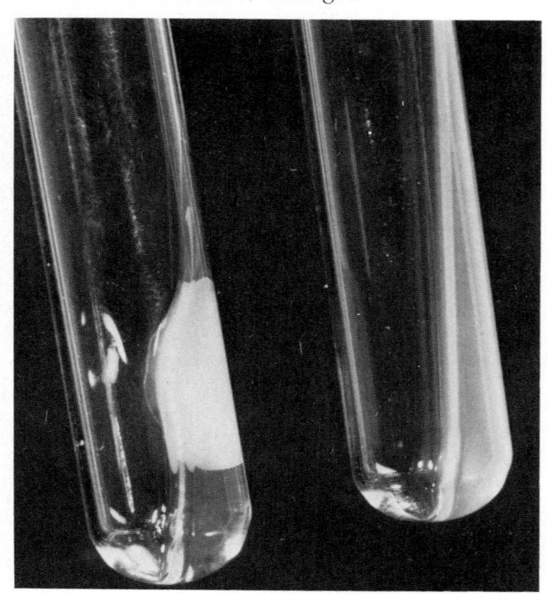

Identification of Anaerobes

A novel approach to the identification of obligate anaerobes used in clinical laboratories involves the gas–liquid chromatographic (GLC) detection of metabolic products. The anaerobes are grown in a suitable medium, and the short chain, volatile fatty acids produced are extracted in ether. Fatty acids detected in this procedure include acetic, propionic, isobutyric, butyric, isovaleric, valeric, isocaproic, and caproic acids (Figure 30.9). The pattern of fatty acid production can be used to differentiate and identify various anaerobes. When coupled with observations of colony and cell morphology and a limited number of biochemical tests, the common anaerobes isolated from clinical specimens can be identified (Table 30.2).

Miniaturized Commercial Identification Systems

Several commercial systems have been developed for the identification of members of the family Enterobacteriaceae. These systems are widely used in clinical microbiology laboratories because of the high frequency of isolation of Gram negative rods indistinguishable except for characteristics determined by detailed biochemical and/or serological testing. Systems commonly used in clinical laboratories include the Enterotube, API 20-E, Minitek, Micro-ID, Enteric Tek, and r/b enteric systems (Figure 30.10). Many of the tests in these systems involve the color change of an indicator that can easily be observed visually or automatically by using a spectrophotometer. Some of the indicators detect

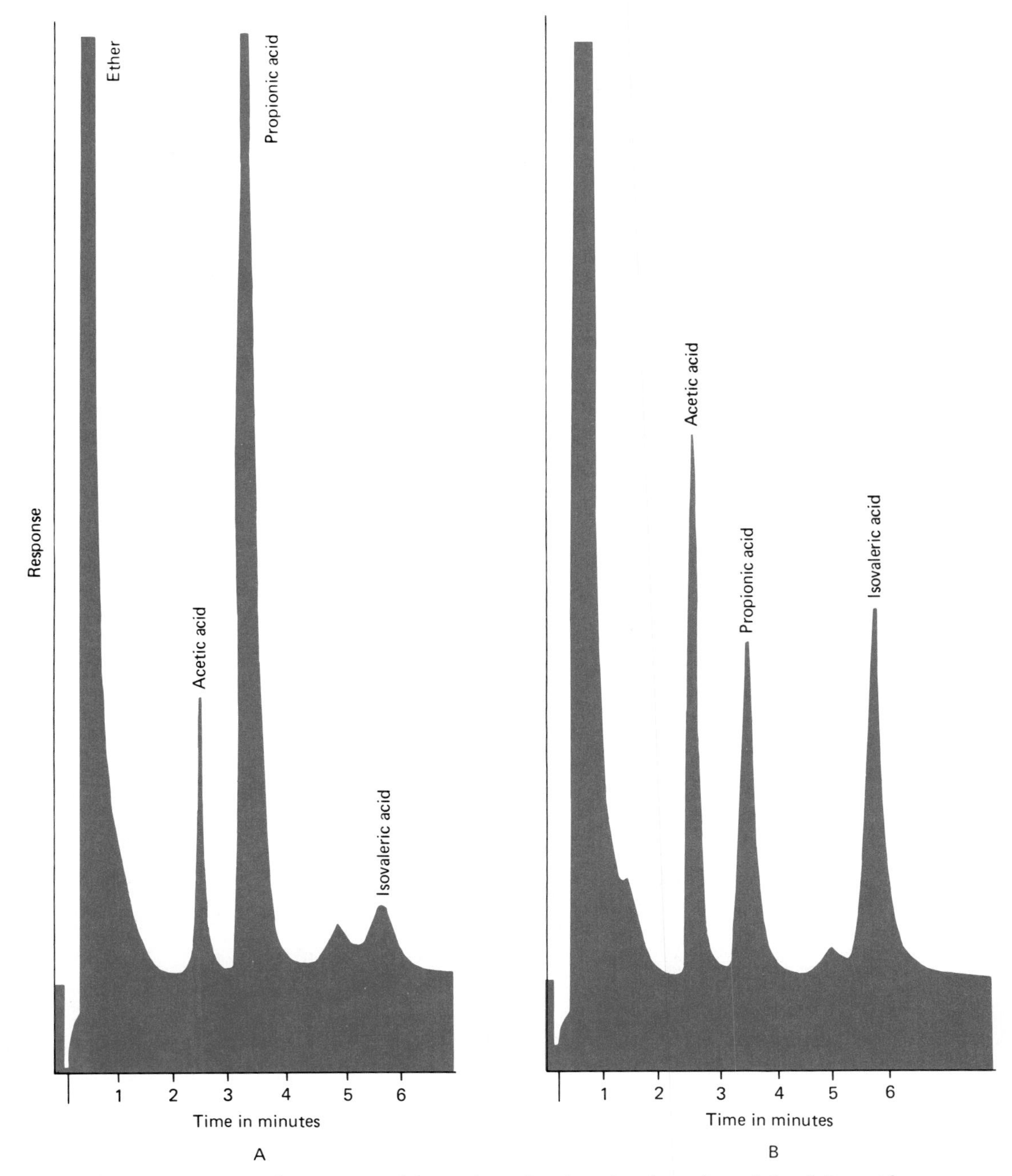

Figure 30.9 Gas chromatographic tracing, showing the detection of the fatty acids produced by anaerobes: (A) *Proprionibacterium* sp.; (B) *Bacteroides fragilis*.

pH changes and others detect specific metabolic products.

The pattern of test results obtained in these systems is converted to a numerical code that can be used to calculate the identity of the isolate. The numerical code describing the test results obtained for a clinical isolate is compared with results in a data bank describing test reactions of known organisms. Some of the commercial systems list a series of possible identifications, indicating the statistical

Table 30.2 Diagnostic Features of Some Anaerobic Bacteria

Organism	Indole Produced	Esculin Hydrolyzed	Catalase Produced	Glucose Fermented	Fatty Acids Produced[a,b]
Actinomyces israelii	−	+	−	+	A, L, S
Propionibacterium acnes	+	−	+	+	A, P
Peptostreptococcus anaerobius	−	−	−	−	A, B, IB, IV, IC
Streptococcus intermedius	−	+	−	−	(A), L
Veillonella alcalescens	−	−	+	−	A, P
Clostridium botulinum	−	±	−	+	A, (P), (IB), B, IV, (V), (IC)
C. perfringens	−	±	−	+	A, B, (P)
C. tetani	±	−	−	−	A, B, (P)

[a]A = acetic acid; B = butyric acid; P = propionic acid; IB = isobutyric acid; IC = isocaproic acid; IV = isovaleric acid; L = lactic acid; S = succinic acid.

[b]() = variable.

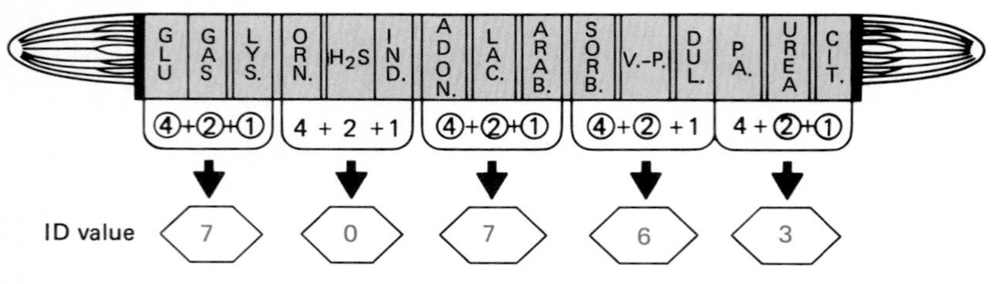

Figure 30.10 Conversion of Enterotube test results to a numerical code. The speed and ease of identification makes these systems very popular in clinical laboratories.

Features Examined in the API 20-E System

Characteristics	Visual reactions: Positive	Visual reactions: Negative
ONPG (β-galactosidase)	Yellow	Colorless
Arginine dihydrolase	Red-orange	Yellow
Lysine decarboxylase	Red-orange	Yellow
Ornithine decarboxylase	Red-orange	Yellow
Citrate	Dark blue	Light green
Hydrogen sulfide	Blackening	Colorless
Urease	Cherry red	Yellow
Tryptophan deaminase (add 10% $FeCl_3$)	Red-brown	Yellow
Indole	Red ring	Yellow
Voges-Proskauer (add KOH plus α-naphthol)	Red	Colorless
Gelatin	Pigment diffusion	No pigment diffusion
Glucose	Yellow	Blue-green
Mannitol	Yellow	Blue-green
Inositol	Yellow	Blue-green
Sorbitol	Yellow	Blue-green
Rhamnose	Yellow	Blue-green
Sucrose	Yellow	Blue-green
Melibiose	Yellow	Blue-green
Amygdalin	Yellow	Blue-green
Arabinose	Yellow	Blue-green

probability that a given organism (biotype) could yield the observed test results. All of the commercial systems employ miniaturized reaction vessels, and some are designed for automated reading and computerized processing of test results. The systems differ in how many and which specific biochemical tests are included. They also differ in whether they are restricted to identifying members of the family Enterobacteriaceae or whether they can be utilized for identifying other Gram negative rods. The test results obtained with all these systems show excellent correlation with conventional test procedures, and these package systems yield reliable identifications as long as the isolate is one of the organisms the system is designed to identify.

Serological Testing in the Identification of Pathogenic Microorganisms

In addition to using growth and biochemical characteristics for the identification of pathogenic microorganisms, a variety of serological test procedures are employed for identifying disease-causing microorganisms. Serology refers to immunological reactions in vitro. Serological tests are particularly useful in identifying pathogens that are difficult or impossible to isolate on conventional media and for identifying pathogenic strains of microorganisms of which there are many varieties not easily distinguished by biochemical testing. For example, over 2000 serotypes in the genus *Salmonella* are defined by the O (somatic cell) and H antigens (flagella), with each serotype defined by a constellation of O and H antigens. The identification of pathogenic viruses and nonculturable bacteria, such as *Treponema pallidum*, generally depend on serological testing. In addition to their specificity, serological tests can be performed rapidly, permitting identification of pathogens within minutes instead of hours or days as with cultural procedures.

Agglutination

When the antigen is located on a cell surface, the reaction of antigen and antibody can produce clumping or agglutination of the cells. This occurs when the antigens on the cell surface react with the multivalent antibody to form a lattice network that holds the cells together (Figure 30.11). The agglutination reaction is probably best known for its use in blood typing (see Figure 20.9). Agglutination reactions can also be used for the rapid identification of microorganisms. If bacteria have been cultured, serotyping can be used to identify the species. In this procedure a suspension of the culture is mixed with known antibodies; agglutination indicates the identity of the culture with the

Figure 30.11 Illustration of agglutination reaction showing how this procedure can be used in the identification of bacteria.

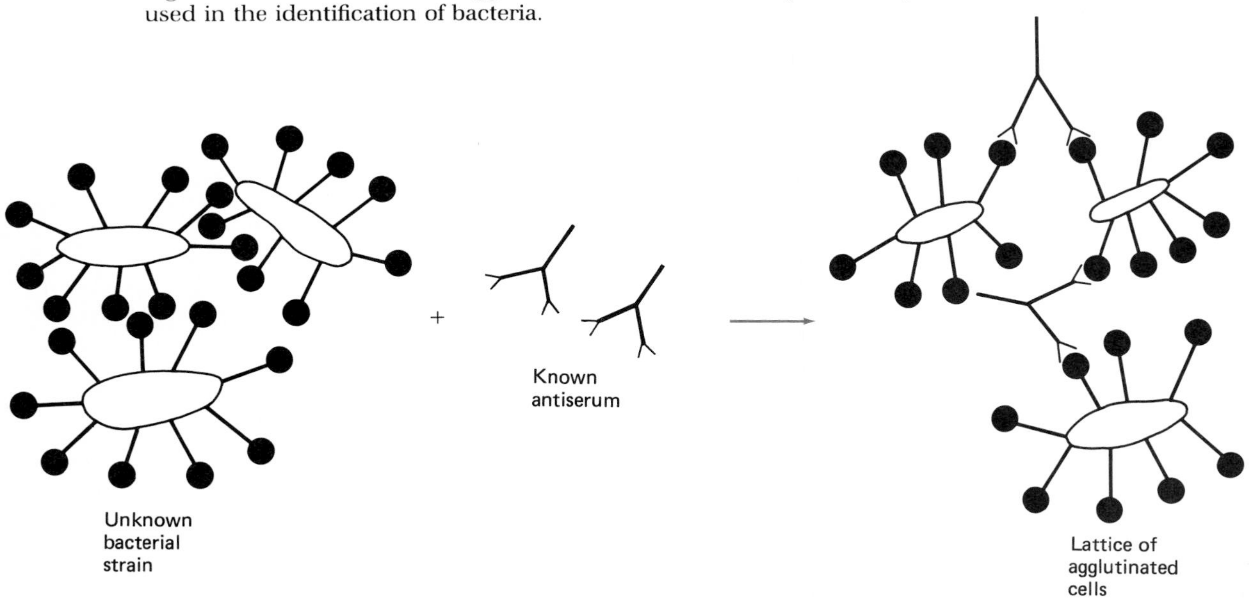

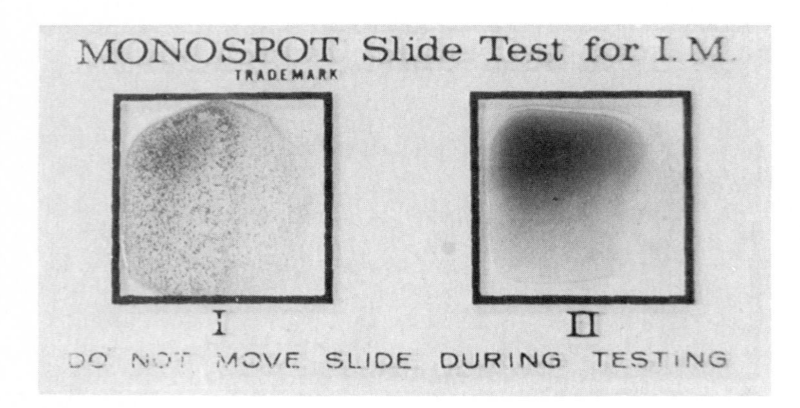

Figure 30.12 An agglutination slide test for the diagnosis of infectious mononucleosis provides a simple, rapid, and accurate method for detecting the presence of antibodies formed in response to infection with Epstein–Barr virus. In this photograph I is a positive reaction and II is a negative reaction.

antibody used. Species of *Salmonella*—including *S. typhi*, the causative organism of typhoid fever, and *Neisseria gonorrhoeae*, the causative organism of gonorrhea, as well as many other pathogenic bacteria—can be identified rapidly by using such agglutination reactions.

Heterophile Antibodies. A more widely used approach is to determine whether the patient has circulating antibodies to a particular pathogen. This eliminates the need to culture the pathogen from that patient. A number of diseases can be rapidly diagnosed by employing agglutination tests in which the patient's serum is mixed with a known antigen to determine whether it contains antibodies that will agglutinate the known antigens. For example, infectious mononucleosis, caused by the Epstein–Barr virus, is routinely diagnosed using serological agglutination tests. The blood serum of individuals infected with this virus contains antibodies that will agglutinate sheep red blood cells. Such antibodies are called heterophile antibodies because they crossreact with antigens other than the ones that elicited their formation. Heterophile antibodies may be present in normal sera, but in individuals with infectious mononucleosis, the concentration of such antibodies is greatly elevated.

By testing the ability of serial dilutions of blood serum to cause agglutination of sheep red blood cells, the titer (concentration) of heterophile antibodies can be determined. In performing these tests the patient's serum is heated to 56°C for 30 minutes to destroy complement that could result in lysis rather than agglutination. A titer of 1:224 or greater is considered as presumptive evidence of infectious mononucleosis.

Additional differential tests for infectious mononucleosis can be carried out to confirm the diagnosis. In these tests the patient's serum is mixed with guinea pig kidney tissue in one tube and beef erythrocyte antigen in another. The guinea pig tissue does not absorb heterophile antibodies produced in response to the Epstein–Barr virus and does not reduce the ability of the serum to agglutinate sheep cells, whereas erythrocyte antigens do absorb these heterophile antibodies, lowering the ability of the serum to agglutinate red blood cells (Figure 30.12).

The detection of heterophile antibodies, that is, crossreactive antibodies, is also useful in diagnosing several other diseases. For example, the Weil–Felix test is used to diagnose some diseases caused by *Rickettsiae* species. In these tests antibodies produced in response to a particular rickettsial infection agglutinate bacterial strains of *Proteus*, designated OX-19, OX-2, and OX-K. The results of the Weil–Felix test are considered to give presumptive diagnoses of rickettsial diseases (Table 30.3). Febrile agglutinins, which are antibodies produced in response to various fever-producing bacteria, can similarly be detected by agglutination tests. The Widal test uses antigens from *Salmonella* species for performing these tests, permitting the diagnosis of typhoid and paratyphoid fevers.

Coagglutination. *Streptococcus* species can also be serotyped by using coagglutination procedures. This approach is used in Streptex and Phadabac systems. In coagglutination, dead *Staphylococcus* cells are coated with antibodies against streptococci and mixed with the streptococci and antistreptococcal antibodies. The reaction forms a lattice matrix of agglutinated cells. The larger clumps of cells produced by coagglutination provide greater test sensitivity than simple agglutination tests.

Hemagglutination Inhibition. Some viruses cause agglutination of red blood cells (hemagglutination). This fact is the basis of hemagglutination inhibition (HI) tests for determining whether a patient has been exposed to a specific virus. In the

Table 30.3 Some Typical Weil–Felix Reactions

Causative Organism	Agglutination with Protein Strain			Diseases
	OX-19	OX-2	OX-K	
Rickettsia prowazekii	+ + + +	+	−	Epidemic typhus
R. mooseri	+ + + +	+	−	Murine typhus
R. rickettsii	+/+ + + +	+/+ + + +	−	Rocky Mountain spotted fever
R. tsutsugamushi	−	−	+ +	Scrub typhus
R. akari	−	−	−	Rickettsialpox
Rochalimaea quintana	−	−	−	Trench fever
Coxiella burnetii	−	−	−	Q fever

HI test the patient's serum is incubated with viral antigens, and the mixture is then added to red blood cells (Figure 30.13). If the patient's serum contains antibody for the specific virus, the viral antigens are neutralized or inactivated, and agglutination of the red blood cells is reduced or absent; that is, the viral antigens are bound by the antibody molecules, which are no longer free to form the lattice that would cause the red blood cells to clump. Agglutination occurs if the patient does not have antibodies to the antigens of the particular virus. The degree of hemagglutination can thus be employed to diagnose viral diseases. Dilutions of the antigen-treated sera are used to determine the end point of hemagglutination inhibition (Figure 30.14), a useful procedure for diagnosing diseases such as rubella and influenza.

Passive Agglutination. The agglutination test procedures discussed so far depend on the occurrence of antigens on cell surfaces. In cases where the antigen to be detected is soluble and not associated with a cell or other particle, passive agglutination tests can be employed. In such tests the antigen is attached to a particle surface before run-

Figure 30.13 Diagram illustrating the hemagglutination inhibition test procedure. (A) Some viruses normally cause the agglutination of red blood cells. (B) However, if a patient's serum contains antibodies against that virus, incubation of the virus with the serum neutralizes the antigens on the viral surface so that hemagglutination does not occur when red blood cells are mixed with the virus. This process of hemagglutination inhibition can be assayed quantitatively.

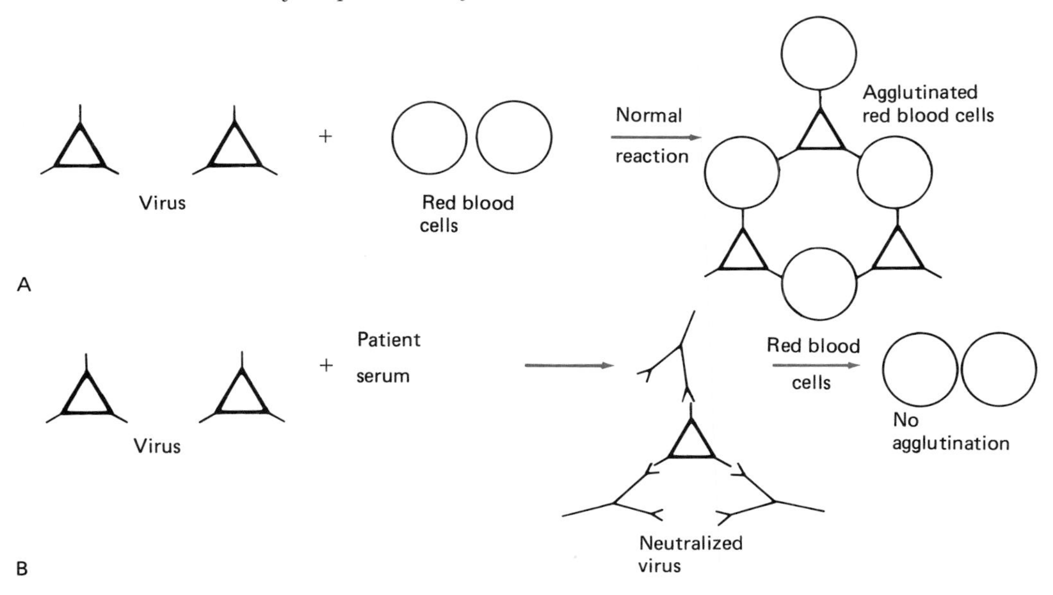

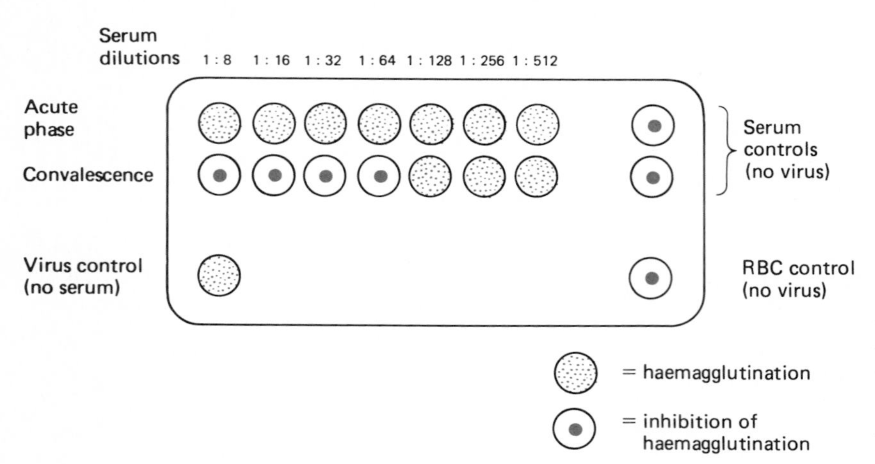

Figure 30.14 This is a diagram of a hemagglutination test for virus in a microtiter plate. Titer of virus hemagglutination is 128. This quantitative test can detect acute and convalescent cases of viral diseases such as hepatitis.

ning an agglutination test (Figure 30.15). The particle may be a cell or, more commonly, a latex bead. Very small quantities of antigen-coated beads are required for these tests.

Passive agglutination using latex beads is rapidly replacing various older procedures and is now employed for detecting *Haemophilus influenzae, Streptococcus pneumoniae, Neisseria meningitidis, Staphylococcus aureus*, and rubella virus (Figure 30.16). Latex beads may also be coated with soluble antibodies and used in agglutination tests. Pneumonia caused by *Streptococcus pneumoniae* can be detected in this manner. Blood serum from patients with pneumococcal pneumonia possesses a protein called C reactive protein (CRP). The CRP reacts with the C polysaccharide of the capsular material of *S. pneumoniae*. Anti-CRP attached to latex beads will react with C reactive protein, causing readily observed agglutination. A positive agglutination test of the patient's serum with anti-CRP coated latex beads gives presumptive evidence of pneumococcal pneumonia. The inflammatory response is monitored by measuring the CRP and is considered more reliable than other methods.

Passive agglutination can also be used to detect *Treponema pallidum* in the diagnosis of syphilis. In the MHA-TP (microhemagglutination *Treponema pallidum*) test, the antigens from *T. pallidum* can be conjugated with red blood cells in passive agglutination tests run to detect the presence of anti-

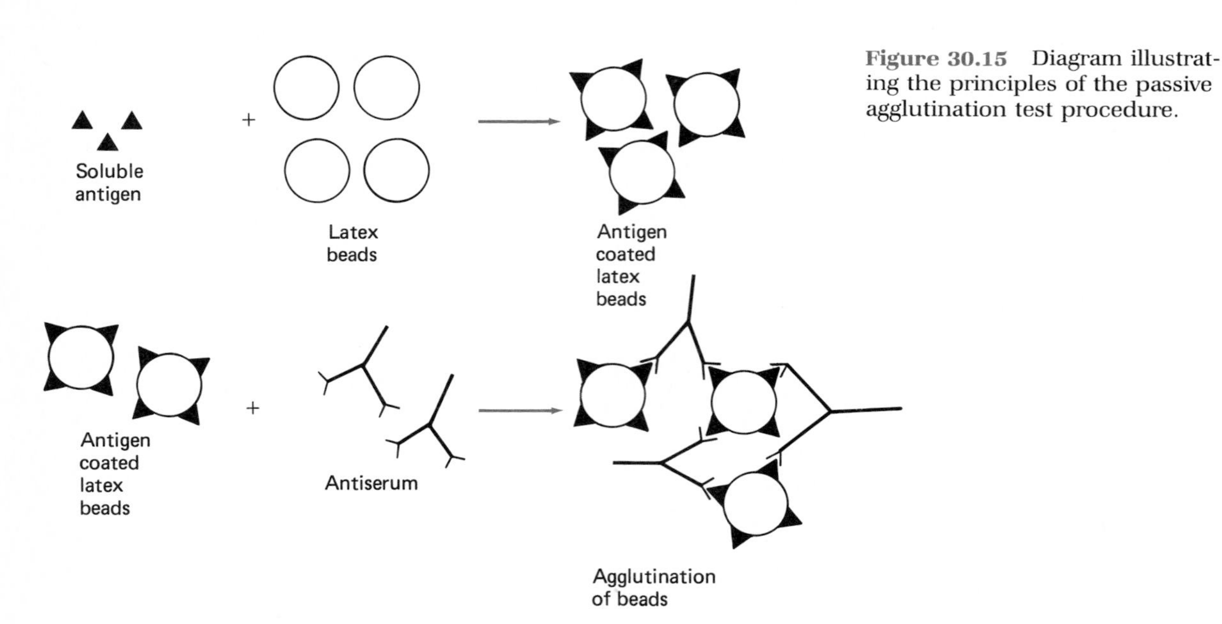

Figure 30.15 Diagram illustrating the principles of the passive agglutination test procedure.

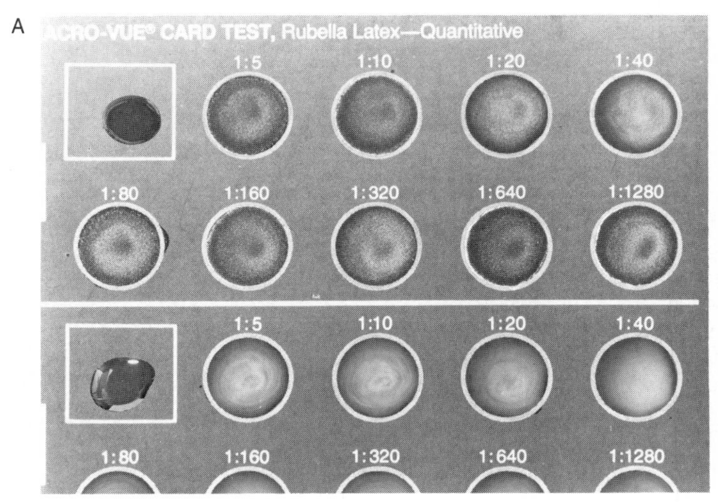

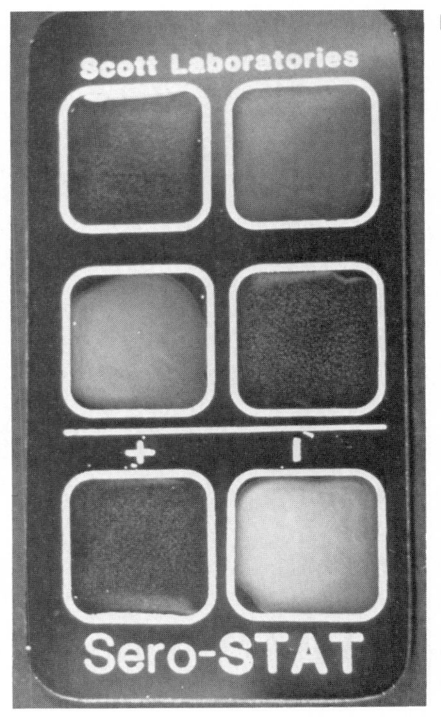

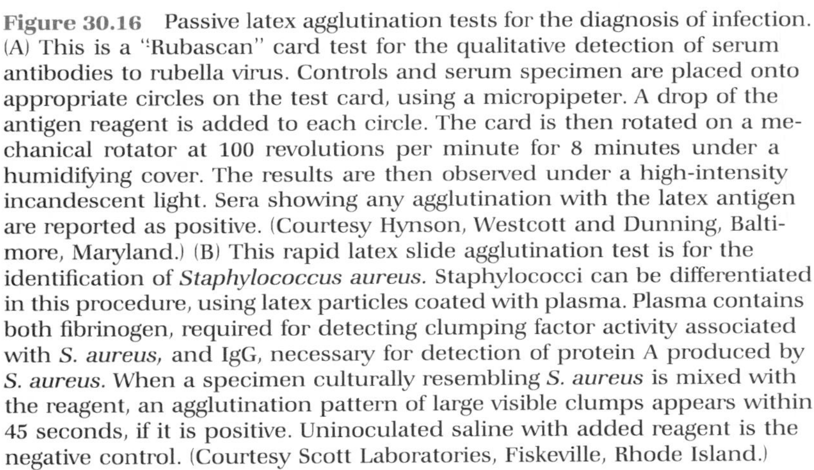

Figure 30.16 Passive latex agglutination tests for the diagnosis of infection. (A) This is a "Rubascan" card test for the qualitative detection of serum antibodies to rubella virus. Controls and serum specimen are placed onto appropriate circles on the test card, using a micropipeter. A drop of the antigen reagent is added to each circle. The card is then rotated on a mechanical rotator at 100 revolutions per minute for 8 minutes under a humidifying cover. The results are then observed under a high-intensity incandescent light. Sera showing any agglutination with the latex antigen are reported as positive. (Courtesy Hynson, Westcott and Dunning, Baltimore, Maryland.) (B) This rapid latex slide agglutination test is for the identification of *Staphylococcus aureus.* Staphylococci can be differentiated in this procedure, using latex particles coated with plasma. Plasma contains both fibrinogen, required for detecting clumping factor activity associated with *S. aureus,* and IgG, necessary for detection of protein A produced by *S. aureus.* When a specimen culturally resembling *S. aureus* is mixed with the reagent, an agglutination pattern of large visible clumps appears within 45 seconds, if it is positive. Uninoculated saline with added reagent is the negative control. (Courtesy Scott Laboratories, Fiskeville, Rhode Island.)

bodies against *T. pallidum* in a patient's serum. Another test used for the diagnosis of syphilis is designed to detect the presence of an IgM antibody produced in individuals infected with *T. pallidum.* This IgM antibody, known as reagin, reacts with phospholipids and can be detected by using cardiolipin, an alcoholic extract of beef hearts. The reaction of reagin with cardiolipin results in an agglutination or flocculation, which can be visualized by using low power microscopy (Figure 30.17). The VDRL (Venereal Disease Research Laboratory) and the RPR (rapid plasma reagin) tests, which are conveniently performed through using a prepared kit, are widely used for the detection of syphilitic reagin. There are some differences in sensitivities of the various serological methods at different stages of syphilis infections (Table 30.4).

Figure 30.17 Degrees of flocculation reaction for syphilis diagnosis as microscopically observed in the cardiolipin microflocculation test.

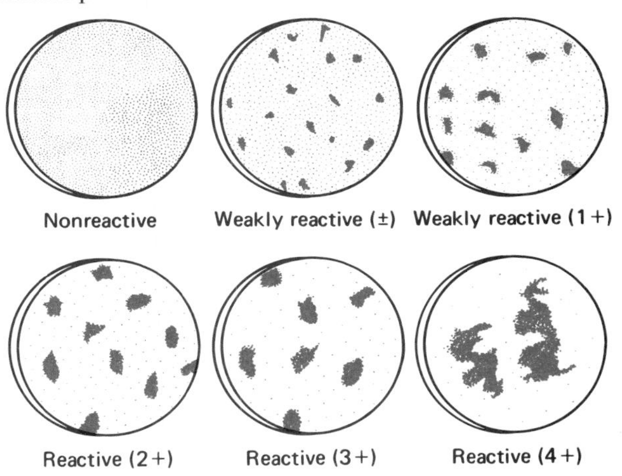

Table 30.4 Percentage of People Showing Positive Reactions at Different Stages of Syphilis with Various Serological Tests

Serologic Test	Primary Stage	Secondary Stage	Latent Stage	Late Stage
VDRL (floculation test based upon reaction of antigen with reaginic antibody)	76	100	73	70
MHA-TP (passive agglutination procedure using erythrocytes coated with *Treponema pallidum* antigens)	64	96	96	98
FTA-ABS (indirect fluorescence antibody absorption test using killed *Treponema pallidum* as the antigen)	86	99	96	100

Precipitin Reactions

When antigen and antibody molecules react, they sometimes form large polymeric macromolecules whose solubility properties result in their precipitation. The precipitation of the antigen–antibody complex can be readily visualized by allowing the antigen and antibody molecules to mix in an agar gel (Figure 30.18). The formation of insoluble complexes that aggregate and precipitate is dependent on the relative concentrations of the antigen and antibody molecules. In general, precipitation occurs when there is an excess of antibody molecules, whereas complexes of antigen–antibody tend to be soluble when fromed in the presence of excess antigen molecules. Thus, to achieve maximal precipitation of the antigen–antibody complex, the reacting antigen and antibody molecules must be present in the proper relative concentrations. To achieve these optimal concentrations, the antigen and antibody molecules are generally allowed to diffuse together, establishing a concentration gradient. A characteristic precipitation line (precipitin band) forms where the optimal relative concentrations of antigen and antibody occur.

Precipitin tests have been used in serotyping various bacteria and in diagnosing the causative agents of some diseases. The clinically significant *Streptococcus* species, for example, can be separated into groups—called Lancefield antigenic groups—based upon precipitin test reactions. In this procedure cells of an unknown β-hemolytic *Streptococcus* species are collected by centrifugation and autoclaved to release their antigens. The cellular debris is removed by centrifugation and the supernatant fluid is added to a capillary tube containing antibodies (antiserum) for a known antigenic *Streptococcus* group. A precipitation reaction indicates that the unknown *Streptococcus* belongs to the indicated *Streptococcus* antiserum group.

Counter Immunoelectrophoresis

Antigen–antibody reactions that result in precipitation reactions also can be used in conjunction with electrophoresis. The method makes use of the ability to visualize the formation of an antigen–antibody complex by the development of a precipitin band in an agar gel together with the

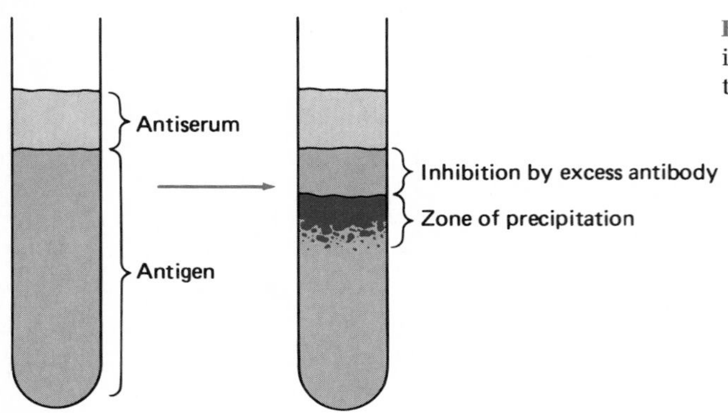

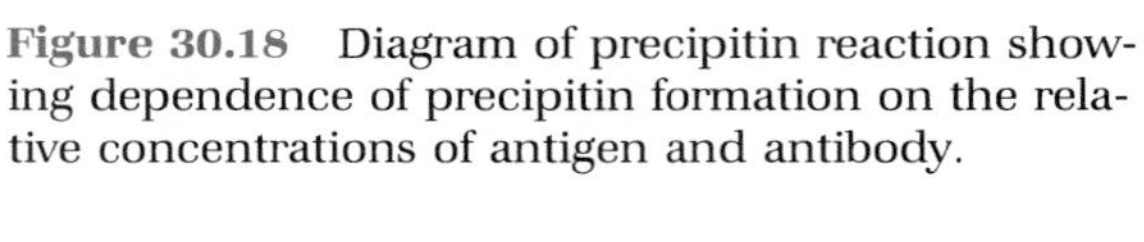
Figure 30.18 Diagram of precipitin reaction showing dependence of precipitin formation on the relative concentrations of antigen and antibody.

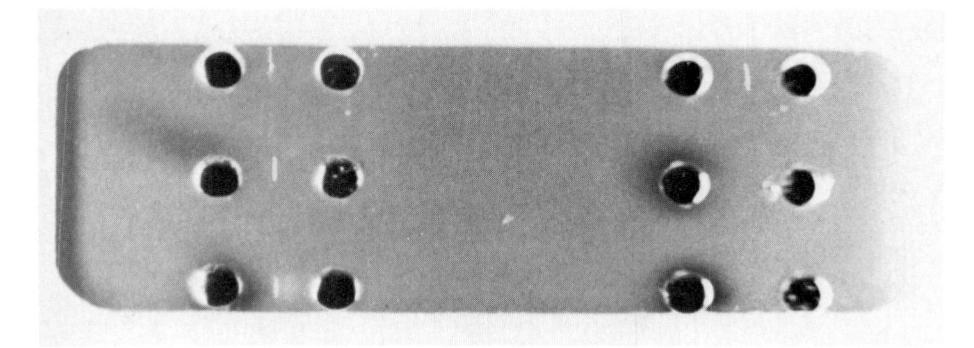

Figure 30.19 Counter immunoelectrophoresis (CIE) testing is used for the rapid diagnosis of various infections. For example, CIE is used to screen spinal fluid for bacteria, as shown in this photograph of actual test results. At left are positive controls, showing precipitation bands between wells filled with antibodies and antigens: Top, *Haemophilus influenzae;* middle, *Neisseria meningitidis;* and bottom, *Streptococcus pneumoniae.* At right are reactions with patient's serum, showing a positive test for the presence of *H. influenzae* and negative results for the other two antigens.

ability of electrophoresis to separate complex mixtures of macromolecules. Electrophoretic separation depends on the differing rates of migration of molecules in an electric field determined by the size (molecular weight and shape) and electronic charge of the molecule. Immunoelectrophoresis combines the efficiency of electrophoretic separation with the selectivity and sensitivity of detection of the immunological reaction between antigen and antibody molecules.

Counter immunoelectrophoresis (CIE) is used in a number of diagnostic procedures. CIE can be useful in detecting the presence of various microorganisms in body fluids, such as cerebrospinal fluid, urine, and blood. This procedure depends on a precipitin reaction to identify homology between antigen and antibody, relying on immunodiffusion with electrophoresis driving the antigen and antibody toward each other. CIE is used to detect both antigens and antibodies in body fluids and antigens from microbial cultures. *Steptococcus pneumoniae, Haemophilus influenzae, Neissera meningitidis, Pseudomonas aeurginosa, Klebsiella pneumoniae, Escherichia coli, Staphylococcus aureus, Mycoplasma,* and *Legionella pneumophila* all can be detected clinically by using counter immunoelectrophoresis (Figure 30.19).

Complement Fixation

Because complement molecules are fixed (bound by antigen–antibody complexes) as a consequence of their reaction in the immune response, it is possible to use a measure of the degree of complement fixation for diagnostic procedures (Figure 30.20). Complement fixation tests are useful in cases where the reaction of antigen and antibody does not result in a reaction, such as agglutination or precipitation, that can be easily visualized. By adding complement, however, the antigen–antibody reaction can be detected. Complement fixation tests were once widely used for the detection of syphilis and are still used for some diagnostic procedures for the detection of viral, fungal, and protozoan pathogens.

Although it has been replaced with more reliable methods for diagnosing syphilis, the Wassermann test still represents a classic example of a diagnostic complement fixation (CF) test. The complement fixation test employs an indicator system, normally consisting of sheep red blood cells, and homologous antibodies for the red blood cell antigens. The test system is designed to detect the formation of an antigen–antibody–complement complex that leads to cell lysis. The patient's serum is heated to 56°C for 30 minutes to inactivate any available free complement. Eliminating the patient's complement permits the quantitative addition of a specified amount of complement and hence the ability to quantify the amount of antibody. The serum is then mixed with a known antigen and free complement, the mixture is added to the indicator system, and the system is observed to determine if complement molecules have added to an antigen–antibody complex on the cell surface, causing lysis of the cells.

Lysis of the sheep red blood cells (hemolysis) in-

MICROBIOLOGY HEADLINES

Premarital Blood Tests: Mass Screening for the Wrong Population

For millions of Americans a blood test has been part of the legal preliminaries to marriage. Mandatory premarital syphilis serologies were legislated, beginning in Connecticut in 1935, because of a general enthusiasm for the presumed public health benefits of mass screening and a conviction that venereal disease could be eliminated by the detection of current cases.

The Wasserman test had been developed in 1906, and the earliest laws that followed required either a physician's certificate or an affidavit from the marriage partners that they were disease-free. Some of the first laws required this information only from men, reflecting the Victorian assumption that only they risked the dread disease. Because of their promiscuous bachelor escapades, it was feared that men might infect their unsuspecting brides.

Economic arguments in favor of mass screening were invoked during the 1930s on grounds that it was cheaper to screen than to provide long-term institutional care for patients suffering the effects of advanced syphilis. By the end of World War II all states had mandatory screening requirements. Now, however, the utility of this screening and the assumptions on which it is based are being questioned.

Yehudi M. Felman, a physician with New York City's Bureau of Venereal Disease Control, argues (*JAMA,* August 4, 1978, pp. 459–60) that premarital syphilis screening is an inefficient and costly way to detect the disease. In New York City, for example, approximately 116,000 such serologies were performed in 1976, but only 39 cases of syphilis were detected—at a cost of $60,000 per case. (New York's laws do not prevent people with positive test results from getting married and do not mandate treatment.) From 4 million such tests nationally in the same year, only 456 cases were detected.

Such a low cost-benefit ratio is tolerated by public health officials, says Felman, because the cost is borne entirely by patients. On the average, counting both laboratory and physician fees, each serology costs $20, or $80 million annually—money that Felman believes is "wasted."

The number of cases detected by the premarital test is so small because these individuals are not at high risk for syphilis. Rather it is the male homosexual population that accounts for half the active syphilis cases. Because an openly homosexual life style is now more tolerated by society, such men are no longer marrying as frequently as they once did, and hence are not being routinely screened. Felman believes that more money should be spent on screening and treating them.

Furthermore, since most states mandate first-trimester prenatal serologies, pregnant women and their babies can be protected from the effects of congenital syphilis. Felman believes that instead of being tested before marriage, women should be tested twice when pregnant—once in the first trimester, and once in the third, since infection in middle and late pregnancy is potentially as damaging to the fetus as untreated syphilis existing at the beginning of pregnancy.

In fact, a few states have already repealed their premarital screening laws. But some public health officials are resisting the change, because the money does not come from their budgets, and because they believe that the test acts as a surveillance against inadequate case-reporting from other sources. Perhaps more important, they fear that the public might interpret repeal of the requirement as a sign that syphilis is under control.

Source: Hastings Center Report, December 1978. Reprinted by permission.

dicates that there is free complement, that an antigen–antibody–complement complex did not form when the patient's serum was added to the known antigen and complement. This constitutes a negative test for the diagnosis of the disease caused by the organism from which the known antigen was derived. If on the other hand, lysis of the sheep red blood cells does not occur, it indicates that complement was fixed in the reaction, and thus, there is presumptive evidence that the patient had the given infection.

In the case of the Wassermann test, it was found that the IgM antibody, reagin, was responsible for fixing complement. Reagin is released from damaged liver cells during syphilis infection, and the test was therefore diagnosing a symptom of the dis-

How bald that microscope was,
eye without lashes
except for my student gaze
fringed at the peephole
where pale, swimming commas arrived
on the stage,
trapped on a slide
as naked as a mirror.

A woman had been surprised,
the smear alive
with the richness of a recent lover.
Positive Wasserman; positive guilt . . .

—ADRIEN STOUTENBURG, *V D Clinic*

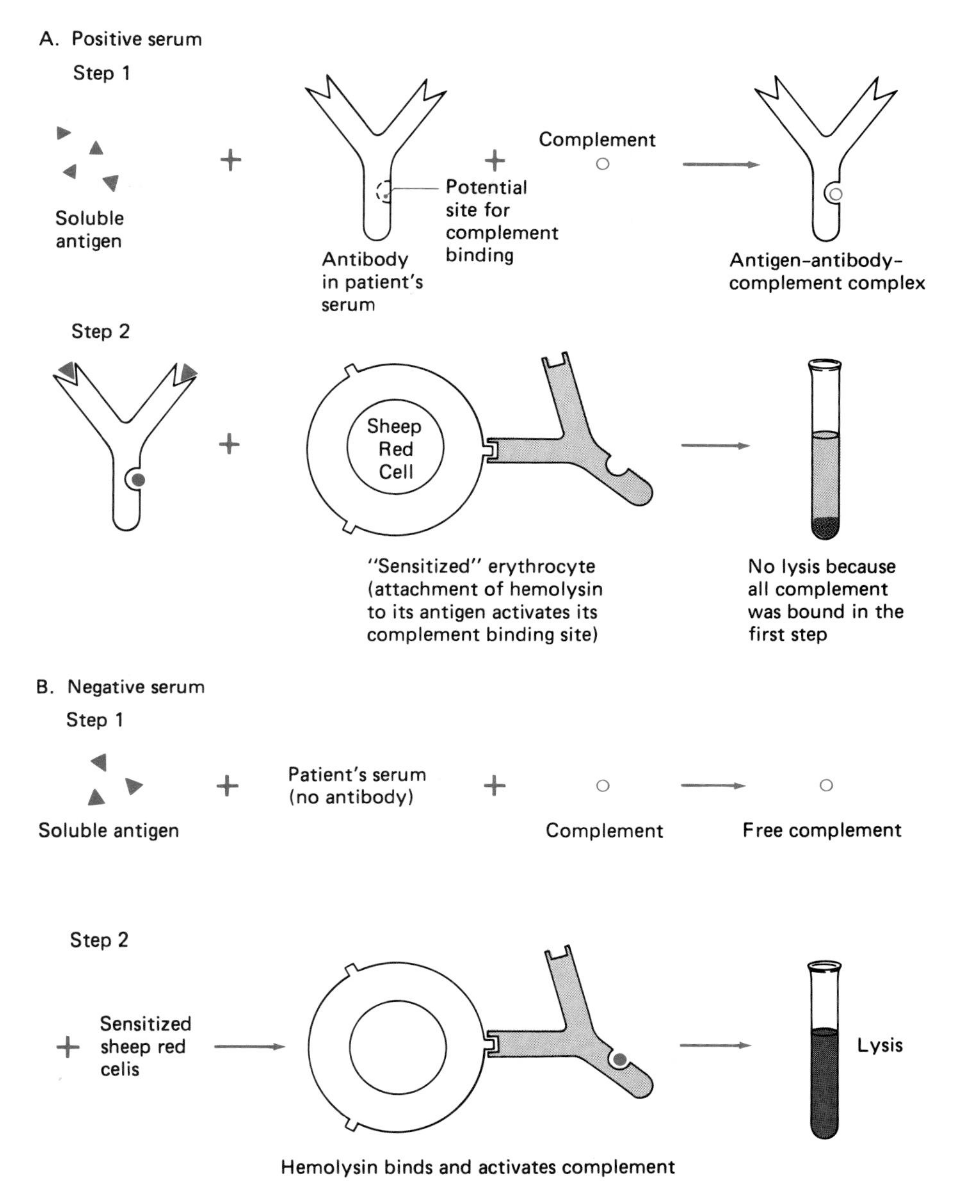

Figure 30.20 Illustration of complement fixation test procedure. (A) Typical positive serum, containing antibody; (B) typical negative serum, antibody absent.

ease rather than specific antibodies that are produced for *Treponema pallidum*. Although no longer used for the diagnosis of syphilis, complement fixation tests are used clinically for the diagnosis of various diseases, including a number of fungal diseases such as histoplasmosis and coccidioidomycosis. Complement fixation tests can also be used for the detection of viral and rickettsial antigens and antibodies.

Enzyme-linked Immunosorbant Assay

Still another approach to serological diagnosis of disease-causing microorganisms is the enzyme-linked immunosorbant assay (ELISA). This method permits the detection of antibody by combining antigen–antibody with an enzyme conjugant that gives a color-producing reaction that can be read spectrophotometrically (Figure 30.21). In the ELISA the antigen is bound to an enzyme, such as peroxidase

A

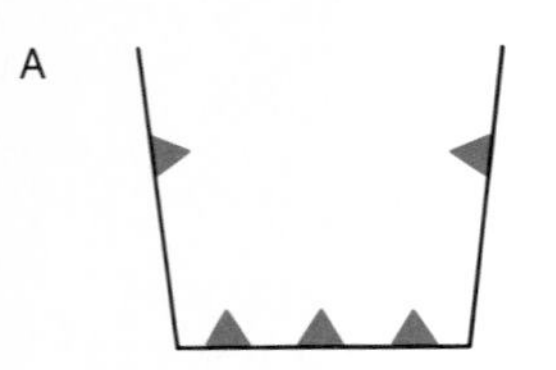

Antigen ▲ is adsorbed to solid phase.

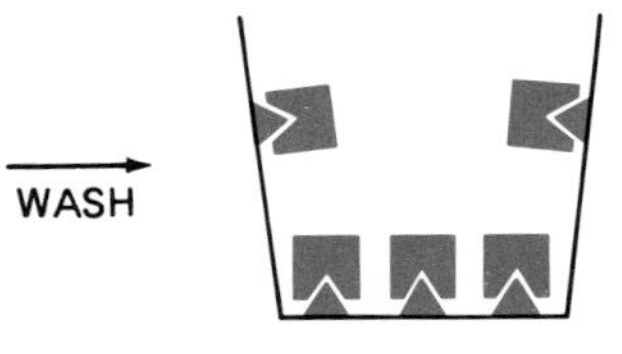

Serum containing specific antibody ■ is added.

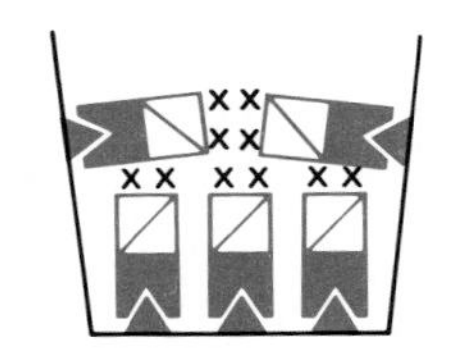

Anti-species antibody–enzyme conjugate ☒ is added.

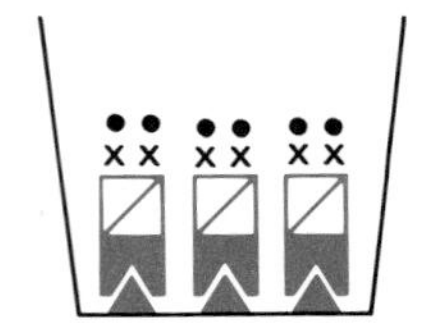

Substrate •• is added to the sandwich and following incubation a color reaction occurs.

B

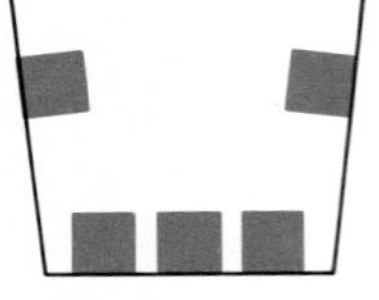

Specific antibody ■ is bound to the solid phase.

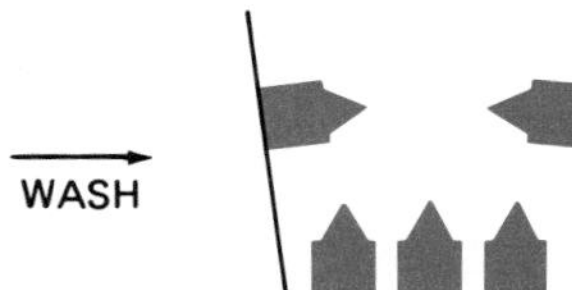

Antigen-containing specimen is added ▲.

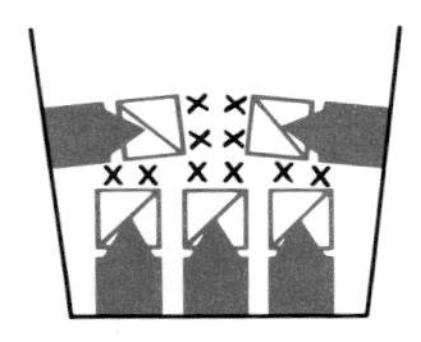

Enzyme conjugated specific antibody ☒ is added.

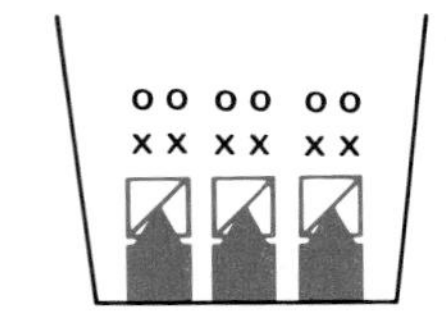

Enzyme substrate ∞ is added and following incubation a color reaction occurs.

C

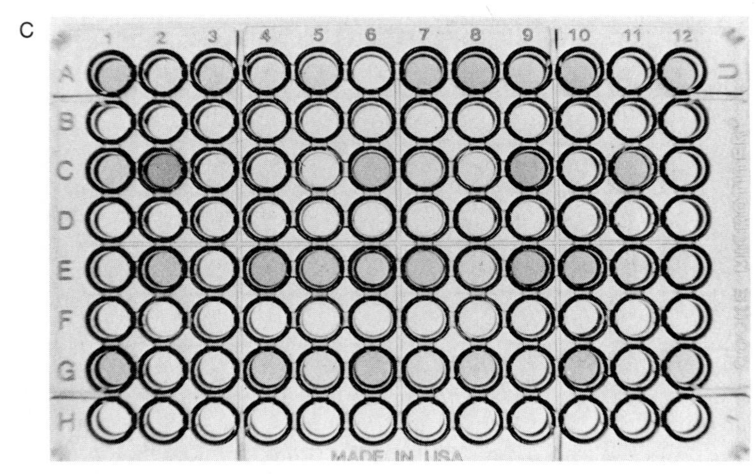

Figure 30.21 ELISA (enzyme-linked immunosorbant assay) procedures are used for the detection of various pathogens. (A) Antibody detection by ELISA. (B) Double antibody sandwich technique for antigen detection by enzyme-linked immunosorbant assay. (C) Results of ELISA procedure for the detection of *Rubella* virus. *Rubella* antigen is attached to the surfaces of the wells of a plastic microtiter plate. The patient's serum is added to the well. Specific antibody, if present in the patient's serum, binds to the attached antigen. Enzyme-conjugated antihuman IgG is added next. It binds to the antibody–antigen complex. Then the enzyme substrate is added and is hydrolyzed by the bound enzyme conjugate. After a specified time the enzyme substrate reaction is stopped with sodium hydroxide. The absorbance of the hydrolyzed substrate is measured at 405 nanometers with a spectrophotometer and is directly proportional to the amount of antibody in the patient's serum. In this photograph the darker the well the greater the amount of antibody in the patient's serum. (Courtesy Karen Cost, Norton Kosair Children's Hospitals, Louisville, Kentucky.)

or phosphatase, which can produce colored reaction products from an appropriate substrate. Today, many different labeled immunoassay procedures are run routinely in clinical laboratories for the diagnosis of the underlying causes of various human disease conditions. The ELISA assays can be used for the detection of antibodies, such as those produced in response to infections with *Salmonella, Yersinia, Brucella, Rickettsia, Treponema, Legionella, Mycobacterium*, and *Steptococcus* species. The ELISA assay can also be used to detect the toxins (antigens) produced by *Vibrio cholerae, Escherichia coli*, and *Staphylococcus aureus*.

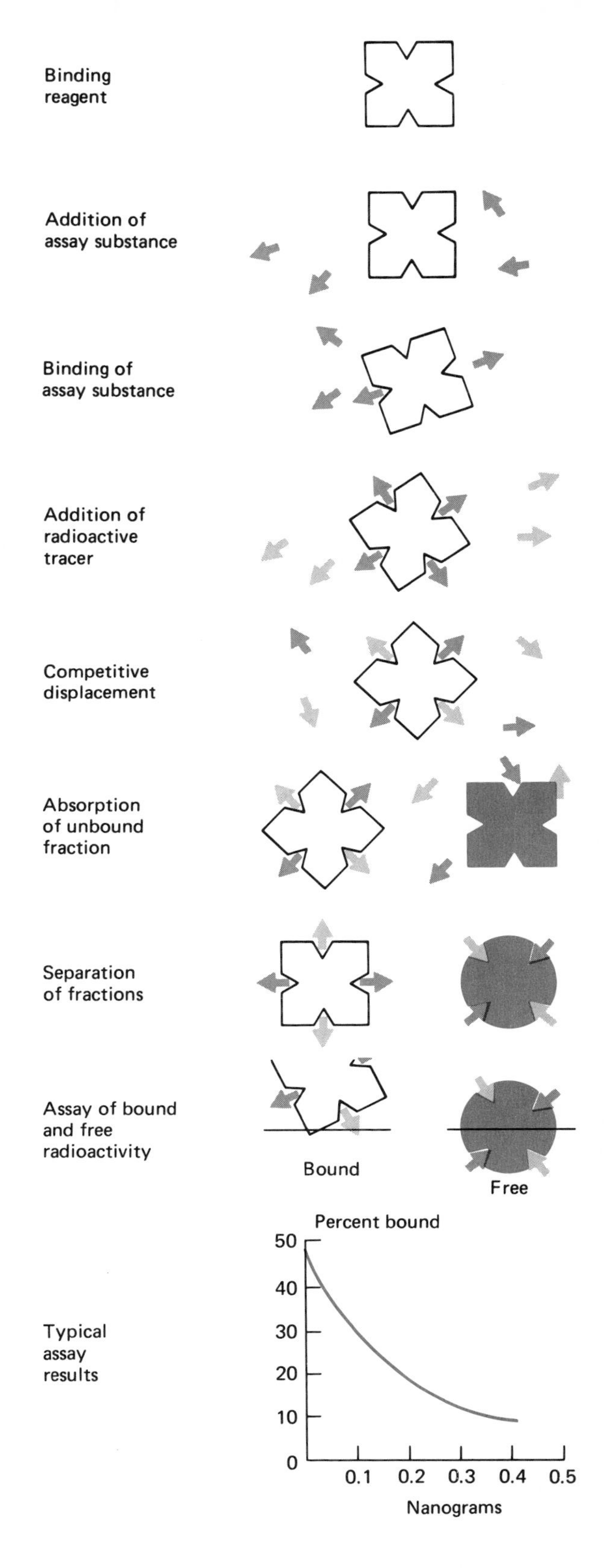

Radioimmunoassays

Radioimmunoassay (RIA) techniques permit the sensitive quantitation of antigen molecules (Figure 30.22). The radioimmunoassay method was initially developed by using radioactive iodine-125 for the detection of very low levels of hormones, such as insulin. The steps of the radioimmunoassay are (1) reaction of specific antibody with a sample containing an unknown quantity of antigen; (2) the addition of radiolabeled antigen, which combines with any unreacted antibody; (3) the separation of radiolabeled antigen–antibody complex from uncombined radiolabeled antigen; (4) determination of the amount of radiolabeled antigen–antibody complex formed; and (5) the calculation of the concentration of the unknown antigen. Radioimmunoassays are applicable to the detection of pathogenic bacteria. In this approach a radioisotopic label, rather than a color reaction, is used to assay the extent of antigen-antibody reaction. RIA tests are extremely sensitive. RIA assays can be carried out for the detection of *Staphylococcus* enterotoxins, *Clostridium botulinum* toxin, *Neiserria meningitidis, Haemophilus influenzae, Pseudomonas aeuruginosa*, and *Escherichia coli*.

Immunofluorescence

Fluorescent dyes, such as fluorescein and rhodamine, can be coupled to antibody molecules to form conjugated fluorescent dyes. Such dyes can be used specifically to stain the antigenic reactive sites of microorganisms. The stained preparations

Figure 30.22 Radioimmunoassay (RIA) has become an important technique for quantitating polypeptides, steroids, thyroid hormones, and vitamin B_{12}. The higher the bound radioactivity, the lower the amount of unlabeled substance; the lower the bound radioactivity fraction, the higher the competitive displacement and the greater the concentration of the substance being assayed in the sample.

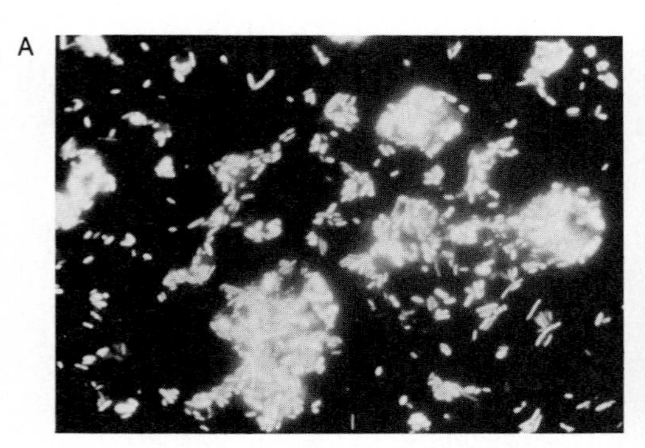

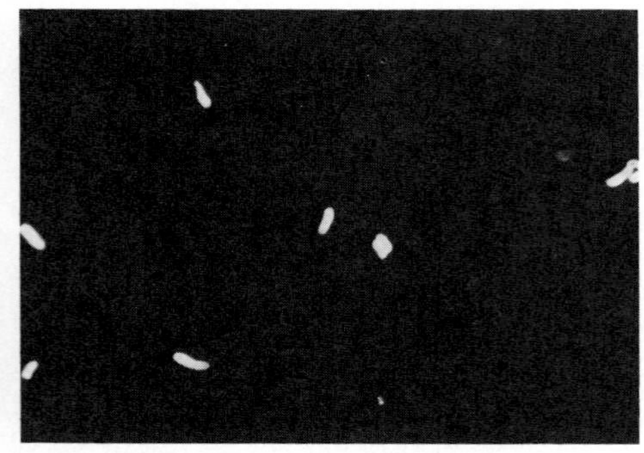

Figure 30.23 Fluorescent antibody staining allows the detection of specific microorganisms. As such this method is useful in clinical identification of pathogenic microorganisms. (A) Photograph showing conjugated-antibody fluorescent staining of *Salmonella typhi* in a human fecal sample; (B) the same human fecal sample stained with acridine orange to show the total microbial content. Comparing A and B, the specificity of immunofluorescence is obvious. (Courtesy B. Bohlool, reprinted by permission of Plenum Press from B. B. Bohlool and E. L. Schmidt, 1980, "The immunofluorescence approach in microbial ecology," *Advances in Microbial Ecology,* 4:203–241.)

are visualized by using a fluorescence microscope with an excitation wavelength appropriate for the dye moiety of the conjugated fluorescent antibody stain. The conjugated fluorescent antibody dye molecules retain the specificity of the antigen binding site of the antibody molecule. Thus, these dyes can be used to react specifically with microorganisms possessing particular antigens, providing a powerful tool for the visualization and identification of specific microorganisms (Figure 30.23).

This technique is particularly useful in identifying specific strains of microorganisms within a mixed microbial community; it is also invaluable in identifying pathogenic microorganisms that are difficult or impossible to culture in the laboratory. Immunofluorescent microscopy permits the visualization and specific identification of *Treponema pallidum,* the causative organism of syphilis, in tissues of infected individuals. This type of microscopy also permits the detection of pathogens such as *Legionella pneumophilia,* the causative organism of Legionnaire's disease, in both infected individuals and environmental samples.

Direct fluorescent antibody staining (FAB) uses a defined antibody conjugated with a fluorescent dye to stain the unknown microorganisms (Figure 30.24). In this procedure microorganisms are stained only if the antibody reacts with the surface antigens of the microorganisms on the slide. Viewing with a fluorescence microscope permits the specific detection of microorganisms that react with the fluorescent antibody stain. In this procedure specific pathogenic microorganisms can be identified even in the presence of other microorganisms. The fluorescence antibody staining method is useful for both the identification of clinical isolates and for detecting pathogens, such as *Treponema pallidum* (Figure 30.25) or *Legionella pneumophila,* within exudates from infected tissues.

An important new development in direct flu-

Figure 30.24 Diagram illustrating the direct fluoresecent antibody method. An antibody that has been conjugated with a fluorescent dye reacts with antigens located on the surfaces of the microorganisms so that the cells become stained with the fluorescent dye.

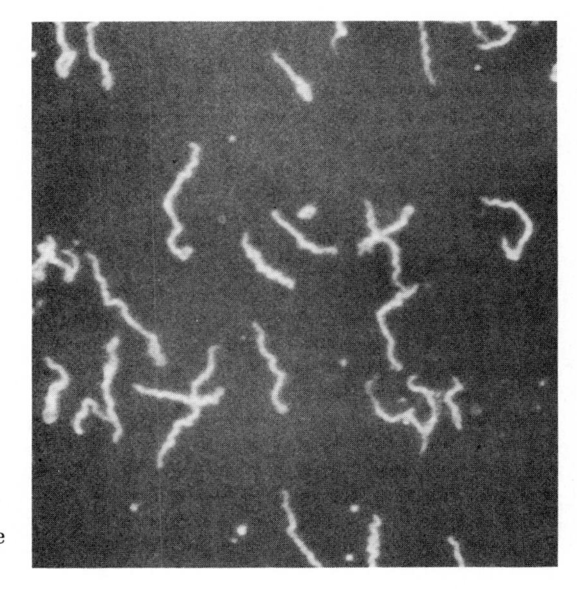

Figure 30.25 Micrograph of *Treponema pallidum* stained with conjugated-fluorescent-antibody stain. (From Centers for Disease Control, Atlanta.)

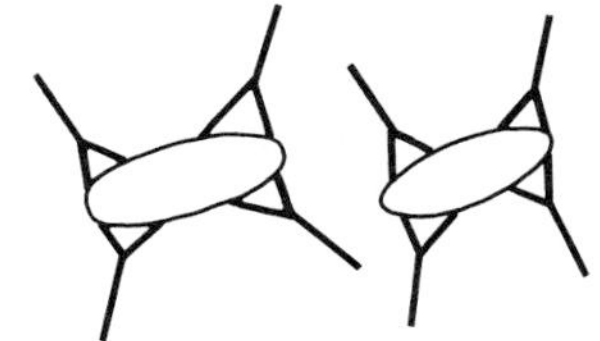

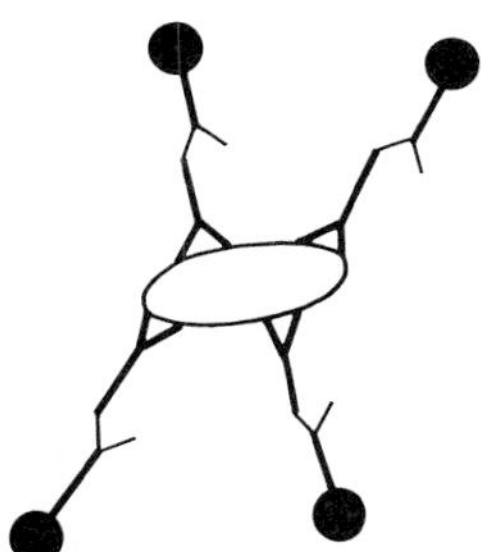

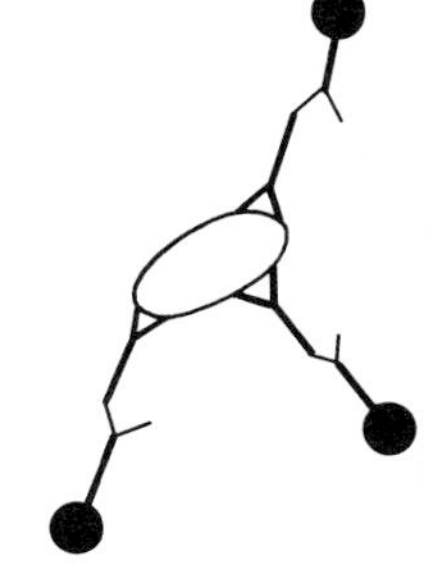

Figure 30.26 Diagram illustrating the indirect fluorescent antibody method. A human immunoglobulin first reacts with antigens on the surfaces of the microorganisms. Then an antibody that has been conjugated with a fluorescent dye reacts with human immunoglobulins that are now affixed to the surfaces of the microorganisms.

Chlamydia: The Silent Epidemic

CLAUDIA WALLIS

Which of the following is the most common venereal disease in the U.S.: 1) syphilis, 2) gonorrhea, 3) herpes? Answer: none of the above. In fact, the most prevalent of all sexually transmitted diseases is one that few people have ever heard of: chlamydia. This disease, named for the tiny bacterium *(Chlamydia trachomatis)* that causes it, strikes between 3 million and 10 million Americans each year. The bug is also a hidden agent in as many as one out of every two cases of pelvic inflammatory disease (PID), a painful, sometimes sterilizing infection that affects about 1 million American women each year. Chlamydia, like herpes, is rapidly becoming the bane of the middle class; up to 10% of all college students are afflicted with it. Says Dr. Mary Guinan of the Centers for Disease Control (CDC) in Atlanta: "Chlamydia is the disease of the '80s."

Then why are most Americans unfamiliar with it? For one thing, the bacterium that causes chlamydia could not be easily isolated and studied until 1965. "The bugs couldn't be grown in the laboratory, and people don't want to work with things that are difficult," says Julius Schachter, an epidemiologist at the University of California, San Francisco. Then, too, the presence of chlamydial infections until recently could be detected only by a complicated laboratory test that took up to seven days to complete and was offered at few medical centers in the U.S. As a result, chlamydia was rarely diagnosed. Says Dr. Ward Cates Jr., head of the CDC's division of sexually transmitted diseases: "Most physicians went to medical school when it was not recognized, and would have trouble spelling it, much less treating it."

In addition, chlamydia's signs are sometimes subtle and easily misinterpreted. Men with chlamydia can experience a burning sensation during urination and a mucoid discharge, but their illness is often diagnosed as gonorrhea. In women, chlamydia may also mimic gonorrhea, causing a vaginal discharge, or result in the frequent and sometimes painful urination associated with a urinary-tract infection.

When doctors confronted with chlamydial symptoms mistakenly prescribe penicillin, the usual treatment for gonorrhea, or order the drugs most commonly used for urinary-tract infections, the chlamydial infection rages on. Repeated doses of the antibiotics tetracycline or erythromycin over a period of one week are required to knock out the bug.

If the wrong drugs are used for chlamydia or if it is left untreated, the infection can spread throughout the reproductive tract. In men, it generally leaves no lasting effects, though many continue to harbor the bacteria and can infect their sexual partners. In women, the bacteria may travel through the uterus into the fallopian tubes, which become inflamed and eventually scarred. While the infection in some cases causes severe lower abdominal pain, thus sending a clear danger signal, the symptoms in other women are barely noticeable. Many of these women remain unaware of their infection. Only after trying unsuccessfully to become pregnant do they discover that their fallopian tubes have been blocked by scar tissue.

Even when the scarring is not severe enough to cause sterility, it can prove troublesome. It can cause a fertilized egg, normally implanted in the uterus, to become embedded and begin dividing in a fallopian tube, leading to a potentially life-threatening condition known as tubal or ectopic pregnancy. Since 1967 the incidence of ectopic pregnancy in the U.S. has tripled. According to Dr. King Holmes, chief of medicine at Harborview Medical Center in Seattle, chlamydia may be a factor in at least one-quarter of the cases.

Women who contract chlamydia during a normal pregnancy face yet another serious problem: transmitting the disease to their babies, who are infected while passing through the birth canal. In infants, chlamydia manifests itself in the form of conjunctivitis, an inflammation of the eye, or as pneumonia. There is also some evidence, Holmes says, that chlamydial infection during pregnancy increases the risks of premature and stillborn births.

Because chlamydia is easy to treat once it is diagnosed, the key to preventing its spread—and its sometimes serious consequences—is better detection. That could be provided by two recently developed diagnostic tests that are both inexpensive and rapid. One, called Micro-Trak, has been marketed for a year by the Syva Co. of Palo Alto, Calif., and provides a diagnosis in less than half an hour. The other, devised by Abbott Laboratories of North Chicago, Ill., takes from 3½ to four hours but is better suited for testing large numbers of people. It became available nationwide last week.

Chlamydia is so widespread that some doctors have begun administering the appropriate antibiotics to suspect patients even before the results of diagnostic tests are in. For example, because 40% of women and 20% of men with gonorrhea also have chlamydia, the CDC's Cates recommends that anyone with a confirmed case of gonorrhea be treated for chlamydia as well. Schachter suggests the same policy for women with PID. "You can't just sit around and wait for a lab diagnosis," he says. "The patient could wind up sterile."

Alarmed by the spread of the disease, Washington State's King County, which includes Seattle, last December allocated $120,000 for a community-wide chlamydia-screening program. Says the county health department's Dr. Hunter Handsfield, who heads the effort: "The problem nationwide is that chlamydia is an out-of-sight, out-of-mind phenomenon." Only when more public health officials recognize and face up to it, he says, will the silent epidemic be brought under control.

Source: Time, Feb. 4, 1985. Reprinted by permission.

orescent antibody methods is the use of monoclonal antibodies conjugated with fluorescent dyes. This method provides the specificity necessary for identifying pathogens in the presence of human tissues and other microbes. One such method has been developed for the detection of *Chlamydia trachomatis*. This method uses fluorescein-labelled monoclonal antibodies formulated to enhance the visibility of *C. trachomatis* for the direct examination of cervical and urethral smears. A smear is placed on a slide, incubated with the fluorescent-monoclonal antibody reagent, and the slide is then examined by fluorescence microscopy. The monoclonal antibodies are able to recognize all varieties of *C. trachomatis* that are human pathogens with no recognition of other organisms, virtually eliminating the nonspecific staining that is a serious problem in other microscopic procedures for the detection of *C. trachomatis*. The one-step procedure takes less than an hour as opposed to days required to culure *Chlamydia* in McCoy cells for identification. Consequently many more patients are being examined for possible *Chlamydia* infections. This has already drastically altered our recognition of the importance of *Chlamydia* as a sexually transmitted pathogen.

Indirect immunofluorescence tests are also employed in identifying bacteria, such as *T. pallidum*. The advantage of this approach is that one labelled antibody can be used as the diagnostic reagent for a number of different test procedures rather than having to label each specific antibody. In an indirect immunofluorescence test a specific unlabelled antibody is first added to a slide with affixed bacterial cells (Figure 30.26). If this antibody reacts with the antigens associated with the bacterial cells, it will react to form an antigen–antibody complex around the surfaces of the cells. A fluorescent-labelled antibody is then used to detect the presence of the unlabelled antibody that was added. For example, in the indirect fluorescent test for treponemal antibodies (FTA), dead *T. pallidum* cells are fixed to a slide, and a patient's serum is added. If the patient has syphilis, the serum will contain human immunoglobulin that will coat the bacterial cells. The slide is washed and a fluorescent antihuman immunoglobulin is added. If the antibodies in the patient's serum reacted with the *T. pallidum* on the slide, the bacteria will be stained (positive test); whereas if the patient's serum did not contain antitreponemal antibodies, no reaction occurs and the fluorescent antibody is washed off the slide so that no fluorescing bacteria are visible (negative test).

Antibiotic Sensitivity Testing

In addition to identifying disease-causing microorganisms, the clinical microbiology laboratory performs tests to aid the physician in the selection of suitable drugs to be used in the treatment of infectious diseases. These tests include determining the sensitivities of isolated microorgansisms to various antibiotics. The procedures used to determine antibiotics potentially of therapeutic value were discussed in Chapter 18. These procedures, which include the Kirby–Bauer disc-diffusion assay and the minimum bactericidal concentration determination, are routinely performed in clinical microbiology. Many of these procedures have been automated and results of some antibiotic sensitivity tests are now available to the physician within only a few hours of sample collection.

Summary

It has not been possible in this chapter to consider all the body tissues and fluids that may be examined in the clinical laboratory for evidence of microbial infection or to consider the variety of techniques that may be employed for the isolation and identification of many specific microbial pathogens. The general procedures for diagnosing microbial infectious diseases have been discussed, including various methods for the isolation and identification of microbial pathogens. The clinical identification of the etiologic agents of disease relies on both biochemical and serological testing. The specific culture methods and identification systems that are employed depend on the preliminary diagnosis of the disease and the microorganisms that are presumed to be the causative agents of the disease based on preliminary screening procedures.

The procedures used in clinical microbiology and immunology laboratories are subject to change as new and better screening procedures aimed at the isolation of pathogenic microorganisms are developed. Speed and accuracy in determining the nature of the disease, the disease etiology, and the appropriate antimicrobial agents that may be effective in controlling the infection are of the essence because an error or delay in diagnosing a disease can prove fatal. Automation has greatly improved speed and reliability in the clinical laboratory, and pathogenic microorganisms can now be isolated and identified within hours. The development of various serological procedures has greatly improved the accuracy and speed of many diagnoses. Similarly, antibiotic susceptibility testing can be accomplished rapidly. At present, hospital clinical laboratories attempt to provide a complete diagnostic report, identifying the pathogen and its antibiotic sensitivities, within 24 hours. In the future, serological testing may permit diagnosis of many diseases at bedside, and improved antibiotic susceptibility testing may permit in vivo monitoring of the effectiveness of antibiotic treatment. Indeed, it has been predicted that by the year 2000 virtually all diagnoses will be made at bedside by using serological methods, providing the physician with immediate information needed to initiate proper therapy.

Study Outline

A. The positive diagnosis of an infectious disease requires the isolation and identification of the pathogenic microorganism or the identification of the antigens specifically associated with a given microbial pathogen.

B. Upper respiratory tract pathogens are isolated by the following procedure:

1. Throat and nasopharyngeal cultures are collected using sterile swabs.
2. Swabs are placed in sterile transport media to prevent desiccation during transit to the laboratory.
3. Cultures are streaked onto blood agar plates that are prepared using defibrinated sheep red blood cells.
4. Plates are incubated in an atmosphere of 5–10 percent carbon dioxide.
5. α-Hemolysis is greening of the blood around the colony.
6. β-Hemolysis is zones of clearing around the colony. It is indicative of *Streptococcus pyogenes*.
7. γ-Hemolysis occurs when there is no clearing.
8. Other media are often used.
   - **a.** *Haemophilus influenzae, Neisseria meningtidis*, and *N. gonorrhoeae* grow better on chocolate agar.
   - **b.** Thayer–Martin medium is preferred for isolation of *N. gonorrhoeae* because growth of normal throat microbiota is inhibited by this medium.
   - **c.** Isolation of *Bordetella pertussis*, causative organism of whooping cough, requires plating on special medium, such as Bordet–Gengou potato medium.
9. Throat swabs can also be used to collect viral pathogens.
   - **a.** Laboratory growth of viruses employs tissue cultures, usually from primary rhesus monkey kidney cells; antibiotics are added to prevent bacterial and fungal growth.
   - **b.** Viruses are washed from throat swabs by synthetic medium and the solution is then added to tissue culture tubes or plates.
   - **c.** Viral infection of tissue culture cells produces morphological changes (cytopathic effects or CPE) which can be readily observed by microscopic examination and can be so characteristic that they can be used in viral identification.

C. Lower respiratory tract cultures are more difficult to obtain than organisms from upper tract.

1. Sputum (exudate containing material from the lower respiratory tract) samples vary greatly in quality and should be screened microscopically to determine suitability. Acceptable samples should contain high numbers of neutrophils, mucus, and low numbers of squamous epithelial cells.
2. Transtracheal aspiration ensures a better quality specimen. A needle is passed through the neck, a catheter is extended into the trachea, and samples are collected through the catheter tube.
3. Routine examination of sputum involves plating on appropriate selective, differential, and enriched media.
4. Stained smears of sputum are also examined for the detection of fungal pathogens.

D. Cerebrospinal fluid can be obtained by performing lumbar puncture in cases of suspected infection of central nervous system.
  - **1.** To determine if the infection is of bacterial or viral origin the glucose level is tested. Most bacterial CSF infections greatly reduce glucose levels; viral infections do not alter glucose levels.
  - **2.** Growth of bacteria in liquid enrichment medium causes an increase in turbidity.

E. Blood cultures are collected by venous puncture using aseptic technique, at various time intervals.
  - **1.** Both aerobic and anaerobic culture techniques are required.
  - **2.** Rapid screening procedure for presence of bacteria employs medium containing radiolabeled glucose; conversion of glucose to radiolabeled carbon dioxide is rapidly determined with great sensitivity using automated instrumentation.

F. Urine cultures are collected by a midstream catch of voided urine to minimize contamination with normal genitourinary microbiota.
  - **1.** High numbers (greater than 105 bacteria per milliliter) of a given microorganism indicates an infection.
  - **2.** Plating should be performed on both general and selective media.
  - **3.** Gram negative enteric bacteria are the most frequent etiologic agents of urinary tract infections.

G. Urethral and vaginal exudate cultures are collected to detect microorganisms that cause sexually transmitted diseases (*Neisseria gonorrhoeae, Chlamydia* sp., *Treponema pallidum*).
  - **1.** Thayer–Martin medium and chocolate agar, incubated under an atmosphere of 5–10 percent carbon dioxide, are employed for culture of *N. gonorrhoeae*.
  - **2.** *T. pallidum* cannot be cultured on laboratory media; dark-field microscopy and fluorescent antibody staining are used.

H. Fecal cultures are collected to isolate microorganisms that cause intestinal tract infections.
  - **1.** Selective and differential media are employed since fecal matter contains a large number of nonpathogenic microorganisms.
    - **a.** Selective media contain components that select for the growth of a particular microorganism.
    - **b.** Differential media contain indicators that permit recognition of microorganisms with particular metabolic activities.
  - **2.** It is often necessary to carry out an enrichment culture before *Salmonella* and *Shigella* species can be isolated.
  - **3.** In cases of suspected viral infections tissue cultures can be inoculated with fecal matter for culture of enteric viral pathogens; characteristic cytopathic effects and serological procedures can then be employed. Electron microscopy can be used for direct detection.

I. Eye and ear fluids can be cultured on defined media.
  - **1.** Gram stain slides are identified.
  - **2.** Microscopic observation of stained slides indicates whether infection is due to bacterial pathogens and appropriate treatment of eye infections is determined.

J. Skin lesion culture material can be collected with swabs, aspirates, or washings, but it is often contaminated with endogenous bacteria.
  - **1.** Both aerobic and anaerobic culture techniques are required for screening for potential pathogens.
    - **a.** Anaerobes *Clostridium tetani* and *C. perfringens* cause serious diseases.
  - **2.** Appropriate fungal culture techniques are required for the isolation of pathogenic fungi.
    - **a.** Dermatophytic fungi and actinomycetes can be detected by direct microscopic examination of skin tissues.

K. Identification of pathogenic microorganisms from the collected cultures is guided by the presumptive identification of the organism at the genus or family level, based on the observation of colonial morphology and other growth characteristics on the primary isolation medium, and on the microscopic observation of stained specimens.

1. Miniaturized commercial identification systems produce a pattern of test results that is converted to a numerical code, which is compared with results in a data bank describing the test reactions of known organisms.

a. All systems employ miniaturized reaction vessels; some are designed for automated reading and computerized processing of test results.

b. Systems differ in how many and which specific biochemical tests are included; whether they are restricted to identifying members of the family Enterobacteriaceae or can be utilized for identifying other Gram negative rods.

2. Conventional tests for the identification of Gram positive bacteria:

a. *Staphylococcus* species by observation of typical grape-like clusters, strongly positive catalase activity (formation of gas bubbles when a drop of hydrogen peroxide is added to the colony), coagulase production and mannitol fermentation.

b. Pathogenic strains of *S. aureus* are generally coagulase positive (clump or clot blood plasma) and produce acid from mannitol.

c. *S. epidermidis* is negative for both coagulase production and mannitol fermentation.

3. Identification of obligate anaerobes involves gas–liquid chromatographic (GLC) detection of metabolic products (fatty acids).

4. Serological test procedures observe immunological reactions in vitro.

a. Agglutination is the clumping reaction of surface antigen and antibody.

b. The presence of heterophile antibodies is determined by agglutination tests in which the patient's serum is mixed with a known antigen; they crossreact with antigens other than the ones that elicited their formation. Such tests are used in the confirmation of infectious mononucleosis, rickettsial diseases, and typhoid and paratyphoid fevers.

c. Coagglutination is used to serotype *Streptococcus* species; dead *Staphylococcus* cells are coated with IgG and mixed with streptococci and antistreptococcal antibodies. The reaction forms a lattice matrix of agglutinated cells; the larger clumps of cells produced by coagglutination provide greater test sensitivity than simple agglutination tests.

d. Hemagglutination is the agglutination of red blood cells caused by viruses. Patient's serum is incubated with viral antigens and the mixture is added to red blood cells. If antibody is present, viral antigens are neutralized or inactivated and hemagglutination is reduced or absent. Agglutination occurs if antibodies are absent.

e. Passive agglutination tests are used when the antigen to be detected is soluble and not associated with a cell or other particle. The antigen is attached to a particle surface (cell or latex bead). Latex beads may also be coated with soluble antibodies and used in agglutination tests. This test can be used to detect *Treponema pallidum* in diagnosis of syphilis.

5. Counter immunoelectrophoresis (CIE) is useful in detecting the presence of various microorganisms in body fluids; it is based on precipitin reaction to identify homology between antigen and antibody and relies on immunodiffusion with electrophoresis driving the antigen and antibody toward each other.

6. Complement fixation can be used as a diagnostic measure because complement molecules are fixed as a consequence of their reaction in the immune response.

a. These tests employ an indicator system, consisting of sheep red blood cells and homologous antibodies, for red blood cell antigens. Serum is heated to inactivate available free complement and then mixed with a known antigen and free complement and added to the indicator system.

b. If lysis of the sheep red blood cells does not occur, it indicates that complement was fixed and provides evidence that the patient had the infection.

7. Enzyme-linked immunosorbant assay (ELISA) permits detection of antibody by combining antigen–antibody with an enzyme conjugant that gives a color-producing reaction, which can be read spectrophotometrically.

a. ELISA can also be used to detect some toxins (antigens).

8. Radioimmunoassays (RIA) use radioisotope labels to assay the extent of antigen–antibody reactions. They are extremely sensitive.

9. Immunofluorescence.

a. Direct fluorescent antibody staining (FAB) uses defined antibody conjugated with fluorescent dye to stain an unknown organism. Microorganisms are stained only if antibody reacts with the organism's surface antigens. The slides are viewed with a fluorescence microscope.

b. Indirect immunofluorescence tests are used for identifying *Treponema pallidum*. Dead *T. pallidum* cells are fixed to a slide to which the patient's serum and fluorescent anti-immunoglobulin are added. If the antibodies in the patient's serum react with the *T. pallidum* on the slide, the bacteria will be stained, constituting a positive test.

Study Questions

1. Match the following statements with the terms in the list to the right.

a. Separated into serological groups based on antigenic differences

b. *Streptococcus epidermidis*.

c. Pathogenic strains of *Staphylococcus aureus*.

d. Used to identify *Treponema pallidum*; involves anti-immunoglobulin.

e. Useful in detecting presence of various microorganisms in body fluids.

f. Can be used to detect *Treponema pallidum* in diagnosis of syphilis.

g. Used in serotyping various bacteria and in diagnosing causative agents of some diseases.

h. Uses *Salmonella* antigens for diagnosis of typhoid and paratyphoid fevers.

i. Employs indicator system, normally consisting of sheep red blood cells and homologous antibodies for the red blood cell antigens.

j. An older test for syphilis; a classic example of diagnostic CF test; replaced by more reliable methods.

k. Diagnosis of some rickettsial diseases.

(1) Coagulase positive
(2) Coagulase negative
(3) Catalase negative
(4) Agglutination
(5) Weil–Felix test
(6) Widal test
(7) Coagglutination
(8) Passive agglutination
(9) Precipitin test
(10) CIE
(11) Complement fixation
(12) Wassermann test
(13) Indirect immunofluorescence

Some Thought Questions

1. How will diseases be diagnosed in the year 2010?

2. Are clinical microbiologists becoming obsolete?

3. Should cultures from patients be sent through the mail to diagnostic laboratories?

4. Why have many different serological methods been developed for the diagnosis of syphilis?

5. Why is it critical that the nursing staff understand microbiological principles if diseases are to be properly diagnosed?

Additional Sources of Information

Bailey, W. R., and E. G. Scott. 1975. *Diagnostic Microbiology*. The C. V. Mosby Company, St. Louis. An excellent text on the classical approaches to the identification of pathogenic microorganisms used in clinical microbiology laboratories.

Fundenberg, H. H., D. P. Sites, J. L. Caldwell, and J. V. Wells (eds.). 1980. *Basic and Clinical Immunology*. Lange Medical Publications, Los Altos, California. A review of the methods used in clinical immunology laboratories, including those used to identify pathogens.

Koneman, E. W., S. D. Allen, V. R. Dowell, Jr. and H. M. Sommers (eds.). 1979. *Color Atlas and Textbook of Diagnostic Microbiology*. J. B. Lippincott Company, Philadelphia. A superb, well-illustrated text on the approaches to the identification of pathogenic microorganisms used in clinical microbiology laboratories.

Kunz, J. C. L., and M. Ferraro (eds.). 1983. *The Direct Detection of Microorganisms in Clinical Samples*. Academic Press, Inc., Orlando, Florida. A critically oriented look at the various methods used to detect microorganisms in clinical samples.

Lennette, E. H., A. Balows, W. J. Hausler, Jr., and J. P. Truant (eds.). 1980. *Manual of Clinical Microbiology*. American Society for Microbiology, Washington, D. C. A thorough manual describing the procedures used by clinical microbiology laboratories for the identification of disease-causing microorganisms.

McMichael, A., and J. Fabre (eds.). 1982. *Monoclonal Antibodies in Clinical Medicine*. Academic Press, Inc., Orlando, Florida. This book brings together reviews on all the major areas of current clinical applications and speculates on future clinical cures.

Rose, N. R., and H. Friedman (eds.). 1980. *Manual of Clinical Immunology*. American Society for Microbiology, Washington, D. C. A thorough manual describing the procedures used by clinical immunology laboratories for the identification of disease-causing microorganisms.

Stansfield, W. D. 1981. *Serology and Immunology*. Macmillan Publishing Co., Inc., New York. An easy-to-read text describing the serological reactions that form the basis for diagnostic procedures.

31 *Water Quality and Biodegradation of Wastes and Pollutants*

KEY TERMS

activated sludge
aerated piles
anaerobic digesters
biodisc
biological oxygen demand (BOD)
chlorination
completed tests
composting
confirmed tests
continuous feed reactors
disinfection
effluent
eutrophication
filtration
indicator organisms
most probable number
potable
presumptive tests
primary sewage treatment
sanitary landfills
secondary sewage treatment
sedimentation
septic tank
sewage fungus
sewage treatment
tertiary sewage treatment
trickling filter
windrows
xenobiotics

OBJECTIVES

After reading this chapter you should be able to

1. Define the key terms.
2. Describe how the bacteriological safety of potable water supplies is assessed.
3. Discuss how potable water supplies are treated to ensure their bacteriological safety.
4. Compare primary, secondary, and tertiary sewage treatment.
5. Discuss the importance of BOD and how sewage treatment processes alter the BOD.
6. Describe three types of secondary sewage treatment processes.
7. Describe two methods for the disposal of solid wastes that exploit the biodegradative capacities of microorganisms.
8. Describe the importance of microorganisms in preventing adverse effects of pollutants.

Treatment and Safety of Water Supplies

The release of human waste materials into the environment sometimes causes serious health problems. Each year more than 500 million people are afflicted with water-borne diseases, and more than 10 million of these individuals die. Most of these disease outbreaks stem from fecal contamination of drinking water supplies. Fecal contamination of potable (suitable for drinking) water supplies from untreated or inadequately treated sewage effluents entering lakes, rivers, or groundwaters that serve as municipal water supplies creates conditions for rapid dissemination of pathogens.

Filth of all hues and odours seem to tell
What street they sail'd from, by their sight and smell . . .
Sweepings from butcher's stalls, dung, guts, and blood,
Drown'd puppies, shaking sprats, all drenched in mud,
Dead cats, and turnip tops, come tumbling down the flood.

—THE TATLER, OCTOBER 17, 1710, DESCRIBING THE THAMES

Disinfecting potable water supplies in order to kill contaminating pathogenic microorganisms can greatly improve the safety and quality of water supplies and the status of human health. Sanitation practices have led to the virtual elimination of water-borne infections in the developed countries, but such infections continue to be major causes of sickness and death in underdeveloped regions. Such sanitation practices are essential to maintain water quality and to permit the multiple uses of water. Rivers flowing past the sewage outfalls of one city continue downstream and serve as the source of drinking water for other communities.

. . . its waters, returning
Back to the springs, like the rain,
shall fill them full of refreshment;
That which the fountain sends forth
returns agains to the fountain.

—HENRY WADSWORTH LONGFELLOW, *Evangeline*

Unfortunately, much of the world's drinking water supplies are contaminated with enteric pathogens owing to inadequate sewage treatment and water purification facilities. Even in developed countries over 85 percent of the population in 1970 did not have reasonable access to uncontaminated water supplies. In many regions of the world, there are no sewage treatment nor water purification facilities and human fecal contamination of water supplies is a routine matter. In developed countries, such as the United States, there are established standards of quality based on indicator organisms to ensure that the water is free of potential disease-causing organisms. These water quality standards are subject to review as better standards are revealed to assess more efficiently the safety of water supplies for different purposes. The standards need to be extended to other developing nations to improve worldwide water quality. Major worldwide efforts are needed to reduce significantly the problem of water pollution.

Disinfection of Potable Water Supplies

Most potable water supplies come from rivers and underground wells and springs. Water from underground sources is partially purified because water is filtered as it passes through the soil column, removing particulate matter and microorganisms. This does not, however, preclude the possibility of bacterial or viral contamination of the water supply, particularly if the source of the water is near a sewage effluent. In some rural areas water is boiled or treated with antimicrobial chemicals to ensure its safety.

In densely populated areas municipal water treatment facilities are designed to ensure the safety of the drinking water supply. The principal processes of a water treatment facility are sedimentation, filtration, and disinfection (Figure 31.1). Sedimentation is carried out in large reservoirs, where the water is held for a sufficient period of time to permit large particulate matter to settle out. Sedimentation rates can be increased by the addition of aluminum sulfate, alum, which forms a floc that precipitates and carries with it to the bottom of a settling basin, microorganisms and suspended organic matter. Filtration, by passing the water through sand filter beds, removes up to 99 percent of the bacteria. The water may also be filtered through activated charcoal to remove potentially toxic organic compounds and organic compounds that impart undesirable colors and/or tastes to the water.

Disinfection is accomplished by treating the water with a disinfecting agent to ensure that it does not contain any pathogens that could be a source of water-borne disease. Chlorination is the usual method employed for disinfecting municipal water supplies. This treatment method is relatively inexpensive, and the free residual chlorine content of the treated water represents a built-in safety factor against pathogens surviving the actual treatment period and causing recontamination.

Bacterial Indicators of Water Safety

The importance to public health of clean drinking water requires objective test methods to establish high standards of water safety and to evaluate the effectiveness of treatment procedures. To monitor water routinely for the detection of actual enteropathogens, such as *Salmonella* and *Shigella*, would be a difficult and uncertain undertaking because of the number of different species and varieties of these organisms and because they may be present in relatively low numbers and still represent a human health threat. Instead, bacteriological tests of drinking water establish the degree of fecal contamination of a water sample by demonstrating the presence of **indicator organisms**. The ideal indicator organism should (1) be present whenever the pathogens concerned are present, (2) be present only when there is a real danger of pathogens being present, (3) occur in greater numbers than the pathogens to provide a safety margin, (4) survive in the environment as long as the potential pathogens, and (5) be easy to detect with a high degree of reliability of correctly identifying the indicator organism, regardless of what other organisms are present in the sample.

The most frequently used indicator organism is the coliform bacterium *Escherichia coli*. Positive tests for *E. coli* do not prove the presence of enteropathogenic organisms but establish the possibility of the presence of such pathogens. Because *E. coli* is more numerous and easier to grow than the enteropathogens, the test has a built-in safety factor for detecting potentially dangerous fecal contamination. *E. coli* meets many of the criteria for an ideal indicator organism, but there are limitations to its use as such, and various other species have been proposed as additional or replacement indicators of water safety.

The conventional test for the detection of fecal contamination involves a three-stage test procedure (Figure 31.2). In the first stage, lactose broth tubes are inoculated with undiluted or appropriately diluted water samples. The tubes showing gas formation are recorded as positive and are used to calculate the **most probable number** (MPN) of coliform bacteria in the sample. Gas formation, detected in small inverted test tubes called Durham tubes, gives positive presumptive evidence of contamination by fecal coliforms (called a **presumptive test**). In the conventional presumptive test five tubes

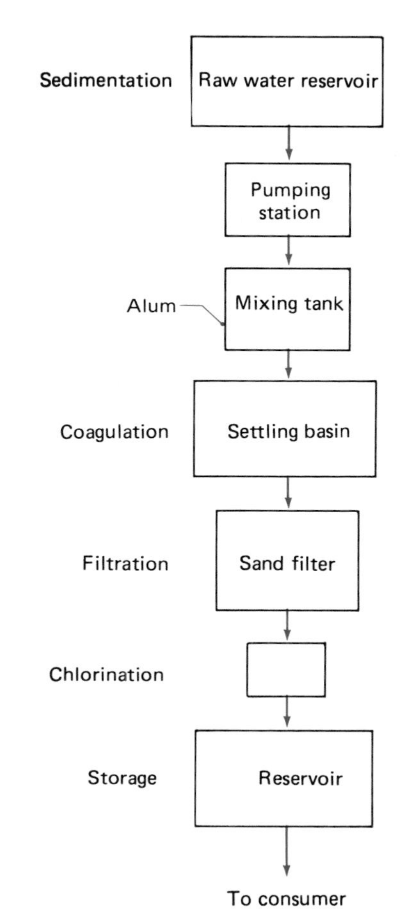

Figure 31.1 A flow diagram of the water purification process used by most municipalities.

of lactose broth are inoculated with water samples of 10 milliliter (mL) volume, five with 1 mL, and five with 0.1 mL. The legal standard for drinking water quality in the United States requires that most samples have less than 5 coliforms per 100 mL which corresponds to no acid and gas production in any of the 15 tubes used in this test protocol. The precise rigor of the test and the required frequency of testing depends upon the size of the particular water system.

Because gas formation in lactose broth at 37°C is not only characteristic of fecal *Salmonella, Shigella*, and *E. coli* strains but also of the nonfecal coliform *Enterobacter aerogenes* and some *Klebsiella* species, a second stage of the test procedure is carried out to confirm the presence of coliforms. Therefore, in the second test stage of this procedure, the presence of enteric bacteria is confirmed by streaking

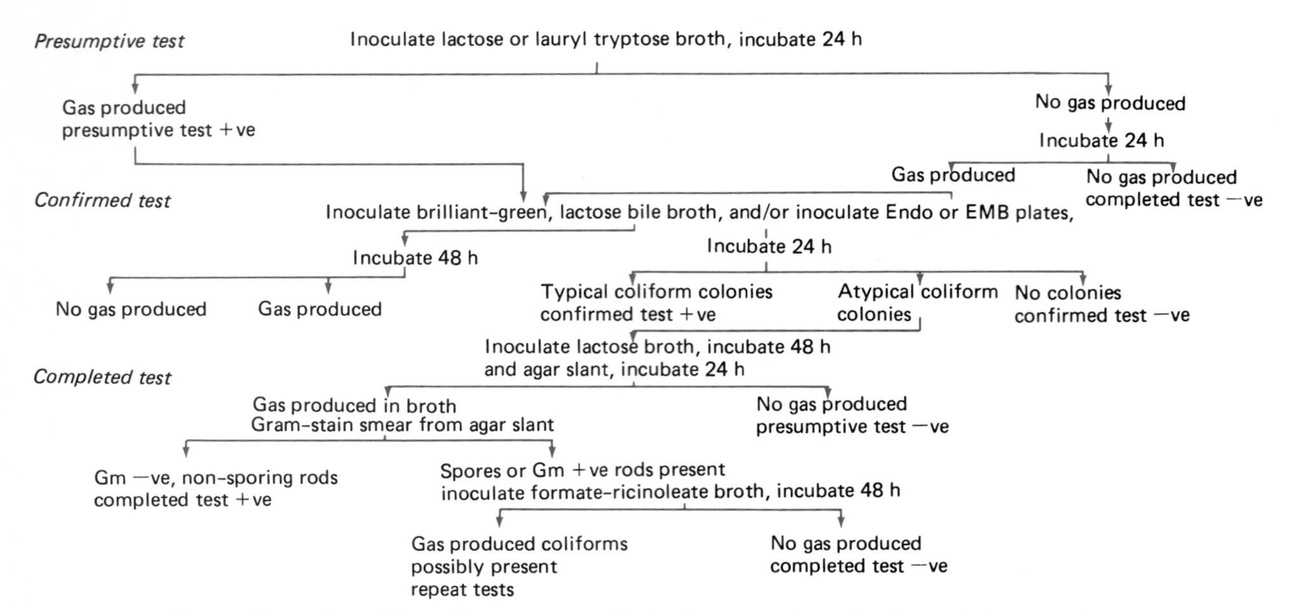

Figure 31.2 An outline of the water quality testing procedures for determining numbers of coliforms by the most probable number (MPN) procedure. Such tests are used for determining the safety of drinking and recreational waters.

samples from the positive lactose broth cultures onto a medium, such as eosin methylene blue agar (EMB). Fecal coliform colonies on this medium acquire a characteristically greenish metallic sheen, *Enterobacter* species form colonies with dark (reddish) centers, and nonlactose fermenters form colorless colonies, respectively (called a confirmed test). Alternatively, the confirmed test can be accomplished by using brilliant green lactose bile broth (BGLB); growth in these tubes is indicative of coliforms. If BGLB is used it is then subcultured onto EMB. Subculturing colonies showing a green metallic sheen on EMB into lactose broth incubated at 35°C should produce gas formation, completing a positive test for fecal coliforms (called a completed test).

In addition to the MPN procedure, numbers of coliform bacteria can be determined by filtering a water sample through a bacteriological filter with a pore size of 0.45 micrometers and incubating the filters on a suitable medium (Figure 31.3). The bacteria in the water sample can be concentrated on the filter in this way, permitting detection of low concentrations of coliforms. Typically a medium such as EMB would be used. The filter would be floated on the medium, which would diffuse into the filter and support the growth of bacteria on the filter surface. Colonies that developed with a characterisitic green-metallic sheen would be counted as coliforms. The test could then be completed by picking colonies and incubating in lactose broth. The membrane filter method is widely used in water quality testing because of its ease and minimal space requirements.

The standards for tolerable limits of fecal contamination vary with the intended type of water use and are somewhat arbitrary with large built-in safety margins, the usefulness of which has been borne out by long experience. Some general standards are listed in Table 31.1. The most stringent standards are imposed on the municipal water supplies to be used by many people. Somewhat higher coliform counts are sometimes tolerated in private wells, used by only one family, because such wells would not become a source of a widespread epidemic. Maintenance of a high drinking water standard does not absolutely exclude the possibility of ingesting enteropathogens with the water but helps keep this possibility to a statistically tolerable minimum. The built-in safety factors are twofold, in that enteropathogens are very likely to be present in much lower numbers than fecal coliforms, and in addition, a few infective bacteria are unlikely to be able to overcome natural body defenses. A minimum infectious dose of several hundred to several thousand bacteria is usually necessary for an actual infection to be established. Drinking water supplies meeting the less than 5/100 mL coliform standard have never been demonstrated to be the source of a water-borne bacterial infection.

Fecal coliform counts are also used to establish the safety of water in shellfish harvesting and recreational areas. Because shellfish tend to concentrate bacteria and other particles acquired through their filter-feeding activity and are sometimes eaten raw, they can become a source of infection by waterborne pathogens. Therefore, there are relatively stringent standards for waters used for shellfishing. Clinical evidence for infection by enteropathogenic coliforms through recreational use of waters for bathing, wading, and swimming is unconvincing, but as a precaution, beaches are usually closed when fecal coliform counts exceed the recreational standard of 1000/100 mL. Some regional standards require that disinfected sewage discharges not exceed the same limit.

Water quality standards based on fecal coliforms do not account for the possible transmission of all microorganisms associated with fecal matter through municipal water supplies. Some viruses, protozoa, and bacteria survive in water supplies despite chlorination or other treatments aimed at assuring the safety of the water. Many destructive epidemics caused by enteroviruses are associated with drinking untreated water contaminated with fecal matter in various underdeveloped countries. Enteroviruses are somewhat more resistant to disinfection by chlorine or ozone than bacteria and, occasionally, active virus particles are recovered from treated water that meets fecal coliform standards. Thus, the possibility exists that water that meets accepted quality standards may still occasionally be a source of a viral infection. Indeed hepatis A outbreaks sometimes occur in this manner.

Additionally, the protozoa *Entamoeba* and *Giardia*, both of which cause gastrointestinal tract infections, can survive in water supplies even when coliforms are not detectable. Diseases associated with these protozoa have been increasing in the United States and the U. S. Environmental Protection Agency (EPA) is considering establishing separate test criteria and statutory limitations for levels of these protozoa in water supplies.

The EPA also is debating how to ensure that water supplies are not the source of *Legionella pneumophila*. This bacterium survives relatively high levels of chlorination, and although not a threat if ingested, inhalation of aerosols containing this pathogenic bacterium can cause Legionnaire's disease. Shower heads are designed to form aerosols, and *Legionella* growing within a water system can thus be inhaled in aerosols while showering. The EPA

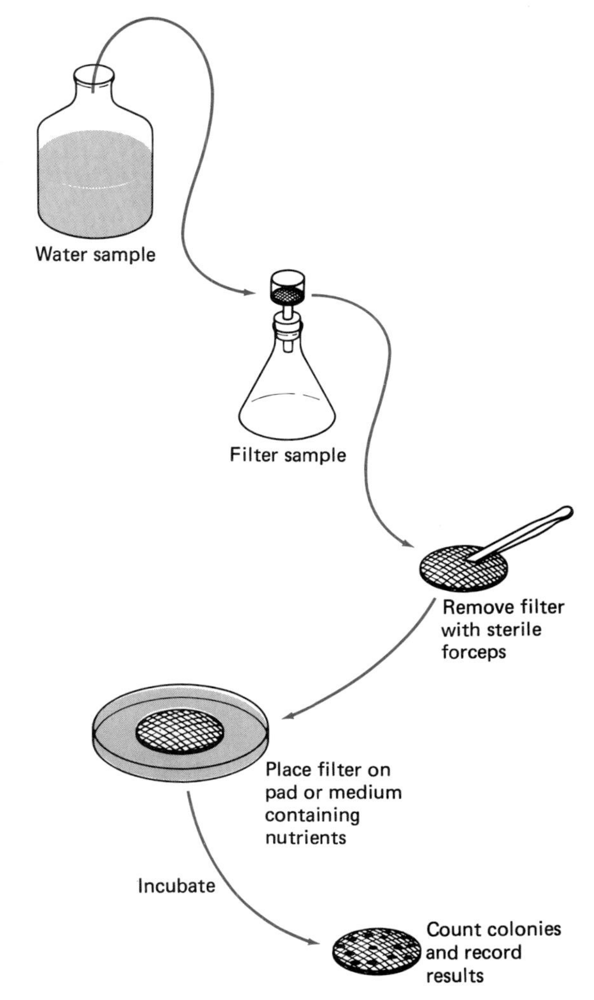

Figure 31.3 A diagram illustrating the membrane filter method for determining numbers of coliforms in water samples.

Table 31.1 United States Water Standards for Coliform Contamination

Water Use	Maximal Permissible Coliform Count (number/100 mL)
Municipal drinking water	<5
Waters used for shellfishing	70
Recreational waters (primary contact)	200
Recreational waters (general contact)	1000

has not yet established safety standards that would protect against such risks to human health, but consideration is being given to extending the drinking water regulations to protect against such occurrences.

Treatment of Liquid Wastes

Liquid waste discharges, including domestic sewage, flow through natural drainage patterns or sewers. These wastes eventually enter natural bodies of water, such as groundwater, rivers, lakes, and oceans. In theory the liquid wastes disappear when they are flushed into such water bodies, according to the adage "the solution to pollution is dilution." Bodies of water into which sewage flows must also serve local communities as the source of water for drinking, household use, industry, irrigation, fish and shellfish production, swimming, boating, and other recreational purposes, making the maintenance of the acceptable high quality of these natural waters essential. Fortunately, self-purification is an inherent capability of natural waters, based on the biogeochemical cycling activities and interpopulation relationships of the indigenous microbial populations.

Low amounts of raw sewage can be accepted by natural waters without any significant decline in

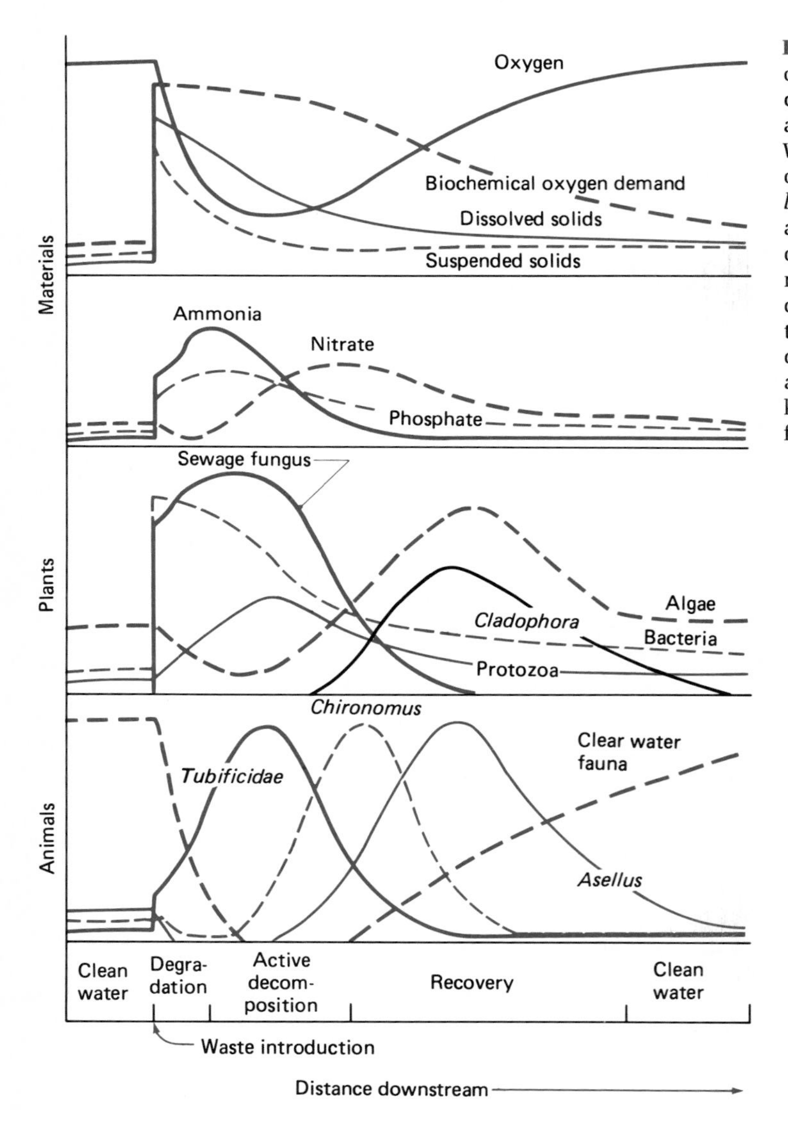

Figure 31.4 The effects of organic effluent on a stream and the changes that occur downstream of the outfall on materials, plants, and animals are represented in this graph. Water quality studies have found that near the outfall there is a high abundance of *Sphaerotilus natans* (the sewage fungus), high ammonium and nitrate concentrations, very low oxygen concentration, and a specialized macrofauna not found in clean water. Depending on the rate of sewage discharge, water flow rate, temperature, and other environmental factors, the water quality may return to close to its original state at a typical distance from a few to several dozen kilometers downstream from the sewage outfall.

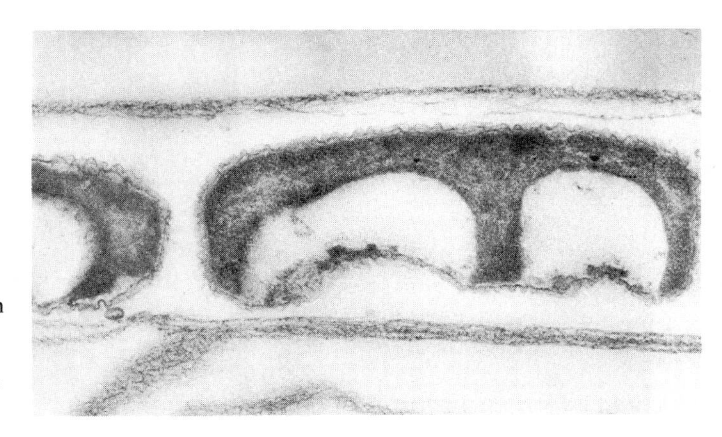

Figure 31.5 Electron micrograph of a thin section of *Sphaerotilus natans* showing the sheath surrounding the cells. Cell width is approximately 500 nanometers. (From BPS: T. J. Beveridge, University of Guelph.)

the level of water quality. Human demographic patterns of densely populated areas, large-scale agricultural operations, and major industrial activities result in the production of liquid wastes on a scale that routinely overwhelms the self-purification capacity of aquatic ecosystems, causing an unacceptable deterioration of water quality. The relative changes in some environmental parameters and populations in a river receiving sewage are illustrated in Figure 31.4. A prominent feature of river water receiving sewage effluents is the presence of the filamentous aerobic bacterium *Sphaerotilus natans*, known as the "sewage fungus" (Figure 31.5).

Exhaustion of the dissolved oxygen content is the principal impact of a sewage overload on natural waters. When the oxygen supply is depleted, the self-purification processes are drastically slowed down. Several measures of water quality are used that aid in the management of aquatic ecosystems. One widely used measure of water quality, the biological oxygen demand (BOD), is a measure of the amount of oxygen required for the microbial decomposition of the organic matter in the water. The polluting power of different sources of wastes is reflected in the BOD of the material (Table 31.2). A high BOD generally indicates the presence of excessive amounts of organic carbon. The dissolved oxygen in natural waters seldom exceeds 8 milligrams per liter (mg/L) because of its low solubility, and it is often considerably lower because of heterotrophic microbial activity, making oxygen depletion a likely consequence of adding wastes with high BOD values to aquatic ecosystems.

Modern methods of sewage treatment are aimed at reducing the BOD and the numbers of potential human pathogens associated with the waste before it is discharged into a water body in order to maintain water quality. There are several different approaches to reducing the BOD and the numbers of human pathogens, employing combinations of physical, chemical, and microbiological methods. Most communities in developed countries have facilities for treating sewage, which is the used water supply containing domestic waste, together with human excrement and washwater; industrial waste, including acids, greases, and oils; animal matter; vegetable matter; and stormwaters. The use of household garbage disposal units also increases the organic content of domestic sewage.

The treatment of liquid wastes is aimed at removing organic matter, human pathogens, and toxic chemicals. The treatment of domestic sewage reduces the biological oxygen demand (BOD) caused by suspended or dissolved organics and the numbers of enteric pathogens so that the discharged sewage effluent will not cause an unacceptable deterioration of environmental quality.

Sewage is subjected to different treatments, depending on the quality of the effluent deemed necessary to be achieved to permit the maintenance of acceptable water quality (Figure 31.6). Primary treatments rely on physical separation procedures to lower the BOD; secondary treatments rely on microbial biodegradation to further reduce the concentration of organic compounds in the ef-

Table 31.2 BOD Values for Different Types of Wastes

Type of Waste	BOD (mg/L)
Domestic sewage	200–600
Slaughter house wastes	1000–4000
Piggery effluents	25000
Cattle shed effluents	20000
Vegetable processing	200–5000

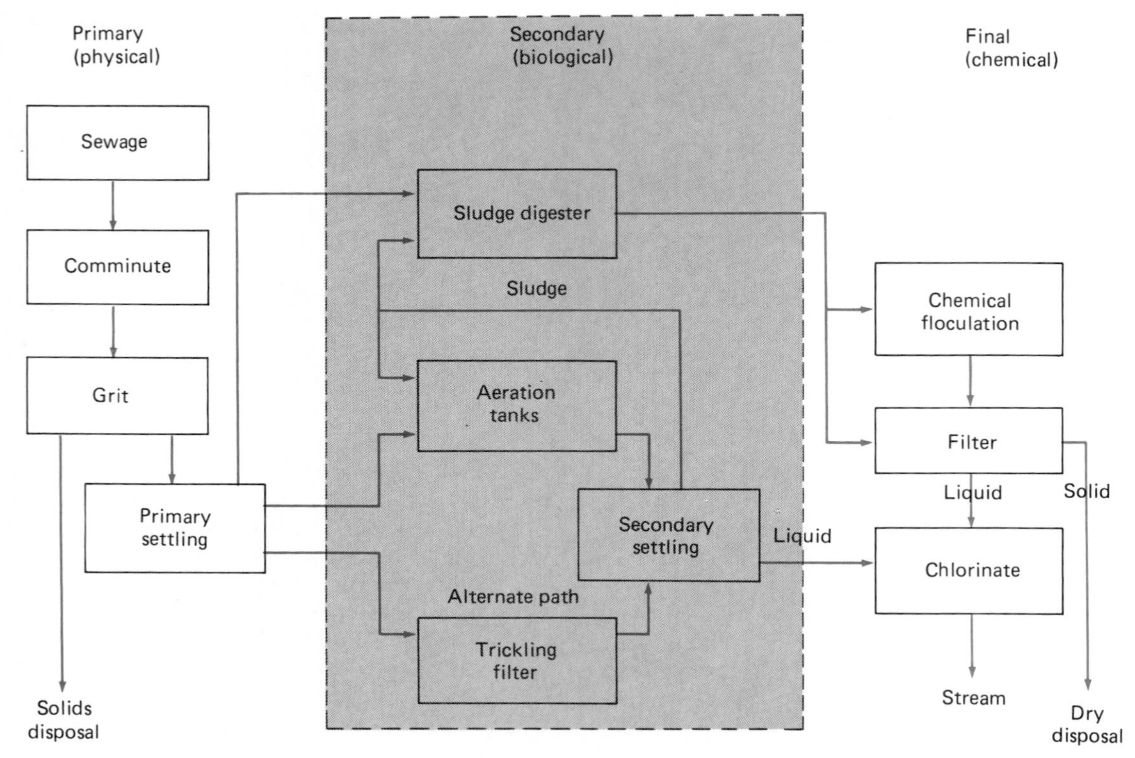

Figure 31.6 Flow chart of the stages of sewage treatment. Primary treatment is principally physical, secondary treatment principally biological, and tertiary treatment principally chemical.

fluent; and tertiary treatments use chemical methods to remove inorganic compounds and pathogenic microorganisms. Municipal sewage treatment facilities are designed to handle organic wastes but are normally incapable of dealing with industrial wastes containing toxic chemicals, such as heavy metals. Industrial facilities frequently must operate their own treatment plants for dealing with waste materials.

Primary Treatment

Primary sewage treatment removes suspended solids by retaining the liquid waste in settling tanks or basins for a period of time (Figure 31.7) or by passing the sewage through a wire mesh screen. For typical domestic sewage (Table 31.3) primary treatment normally removes only 30–40 percent of the BOD. Many microorganisms are adsorbed onto the solid matter and are removed by this physical

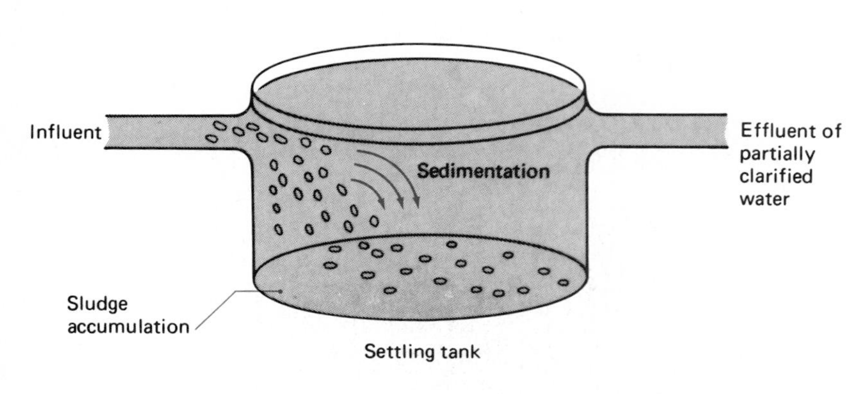

Figure 31.7 Diagram of a settling tank for primary sewage treatment.

Table 31.3. Characteristics of Typical Municipal Waste Water

Component	Concentration (mg/L)
Solids	
Total	700
Dissolved	500
fixed	300
volatile	200
Suspended	200
fixed	50
volatile	150
BOD (biochemical oxygen demand)	300
TOC (total organic carbon)	200
COD (chemical oxygen demand)	400
Nitrogen (as N)	
Total	40
Organic	15
Ammonia	25
Nitrate	0
Nitrite	0
Phosphorus (as P)	
Total	10
Organic	3
Inorganic	7
Grease	100

purification process. The solids are drawn off from the bottom of the tank and may be subjected to anaerobic digestion and/or composting prior to final deposition in landfills or as soil conditioner. In primary treatment waste is removed by settling, and as a result the disposal problem is merely "displaced" to the solid waste area rather than being solved. Nevertheless, this displacement is essential because of the detrimental effects of discharging effluents with high BOD into aquatic ecosystems with naturally low dissolved oxygen contents.

Secondary Treatment

To achieve an acceptable reduction in the BOD, secondary treatment by a variety of means is necessary (Table 31.4). In secondary sewage treatment a small portion of the dissolved organic matter is mineralized (converted to the inorganic compounds carbon dioxide and water), and the larger portion is converted to removable solids. The combination of primary and secondary treatment reduces the original sewage BOD by 80–90 percent. The secondary sewage treatment step relies on microbial activity, may be aerobic or anaerobic, and is conducted in a large variety of devices. A well-designed and efficiently operated secondary treatment unit should produce effluents with BOD and/or suspended solids less that 20 mg/L. The number of human pathogens should also be significantly reduced. Reduction of pathogens occurs as a result of competition with microorganisms involved in the decomposition of the wastes. The decomposer microbial populations win out in the competition and displace (eliminate) the human pathogens that are better adapted for growth within the human body than in the environment of a waste treatment facility.

Septic Tank. The septic tank is one of the simplest treatment systems (Figure 31.8). This system combines primary settling with anaerobic secondary treatment. Many rural and suburban single-family dwellings use septic tanks. A septic tank acts largely as a settling tank, within which the organic components of the waste water undergo limited anaerobic digestion. The accumulated sludge is maintained under anaerobic conditions and is degraded by anaerobic microorganisms to organic acids and hydrogen sulfide. Residual solids settle to the bottom of the septic tank, and the clarified effluent is allowed to percolate into the soil where the dissolved organic compounds in the effluent undergo biodegradation. These products are distributed into the soil along with the clarified sewage effluent. The soil must have suitable hydraulic properties that permit the needed flow of the effluent through the soil column. Septic tank treatment does not reliably destroy intestinal pathogens, and it is important that the soils receiving the

Table 31.4. Efficiency of Various Types of Sewage Treatment

Treatment	BOD (% reduced)	Suspended Solids (% removed)	Bacteria (% reduced)
Sedimentation	30–75	40–95	40–75
Septic tank	25–65	40–75	40–75
Trickling filter	60–90	0–80	70–85
Activated sludge	70–96	70–97	95–99

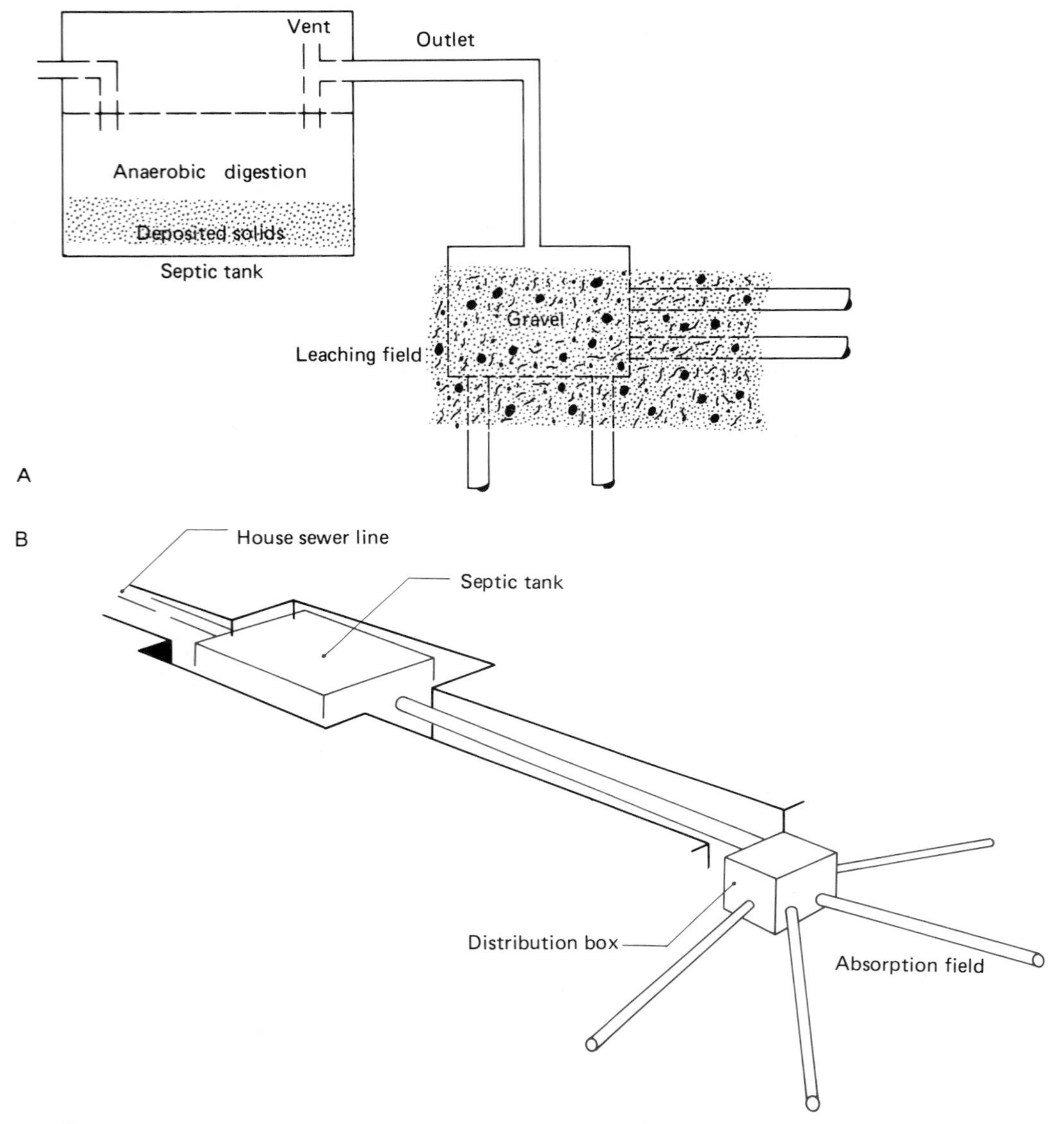

Figure 31.8 (A) Diagram of a typical septic tank used for the disposal of sewage in a rural home. The sewage is degraded by anaerobic digestion. The treated effluent is disposed into a leaching field of permeable soils, with the effluent pipes usually set in gravel. (B) Diagram of the installation of a septic tank for sewage disposal from a home. The sewage effluent is distributed into the surrounding soil.

clarified effluents not be in close proximity to drinking wells to prevent contamination of drinking water with enteric pathogens.

Trickling Filter. The trickling filter system is a simple and relatively inexpensive film-flow type of aerobic sewage treatment method (Figure 31.9). The sewage is distributed by a revolving sprinkler suspended over a bed of porous material. The sewage slowly percolates through this porous bed, and the effluent is collected at the bottom. The porous ma-

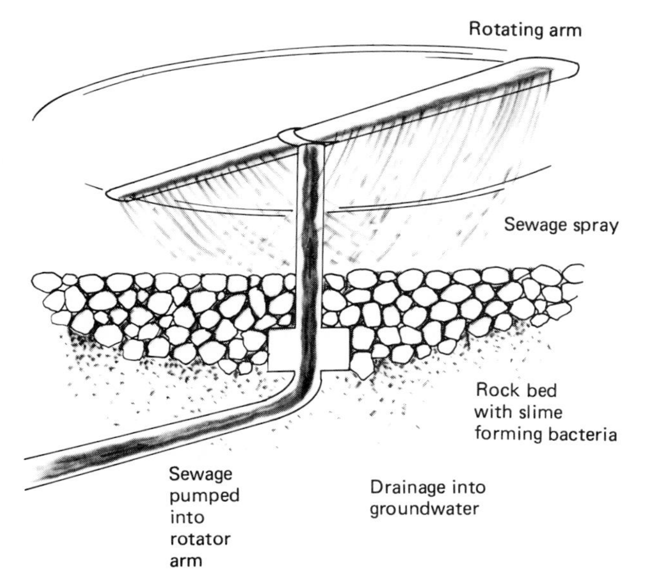

Figure 31.9 In a trickling filter the clarified sewage is sprayed over rocks. The bacterial film on the rocks aerobially decomposes the dissolved organic matter.

terial of the filter bed becomes coated by a dense, slimy bacterial growth, principally composed of *Zooglea ramigera* and similar slime forming bacteria (Figure 31.10). The slime matrix thus generated accommodates a heterogeneous microbial community, including bacteria, fungi, protozoa, nematodes, and rotifers. This microbial community absorbs and mineralizes dissolved organic nutrients in the sewage, reducing the BOD of the effluent (Figure 31.11).

Biodisc System. The rotating biological contactor or biodisc system is a more advanced type of aerobic film-flow treatment system. This system requires less space than trickling filters and is more efficient and stable in operation but needs a higher initial financial investment. The biodisc system is used in some communities for the treatment of both domestic and industrial sewage effluents. In the biodisc system, closely spaced discs—usually made of plastic—are rotated in a trough containing the sewage effluent (Figure 31.12). The discs are partially submerged and become coated by a microbial slime, similar to that which develops in trickling filters. Continuous rotation of the discs keeps the slime well aerated and in contact with the sewage. The thickness of the microbial slime layer in all film-flow processes is governed by the diffusion of

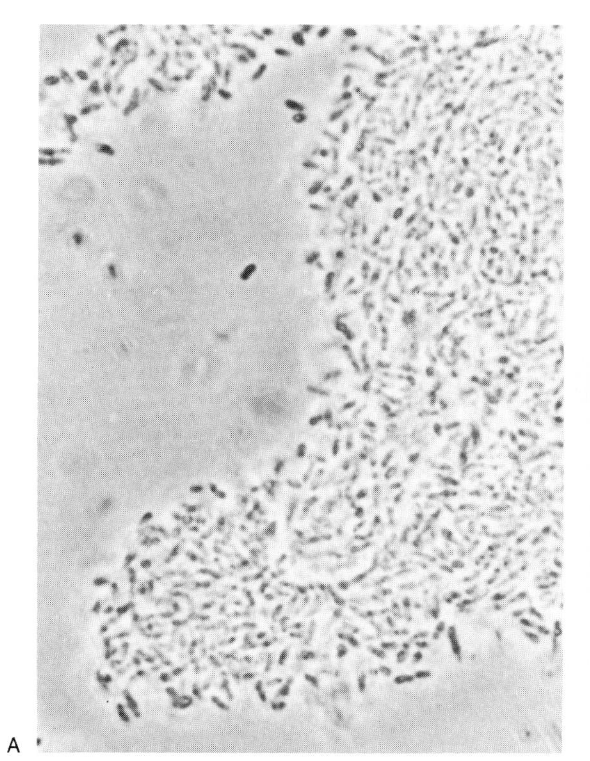

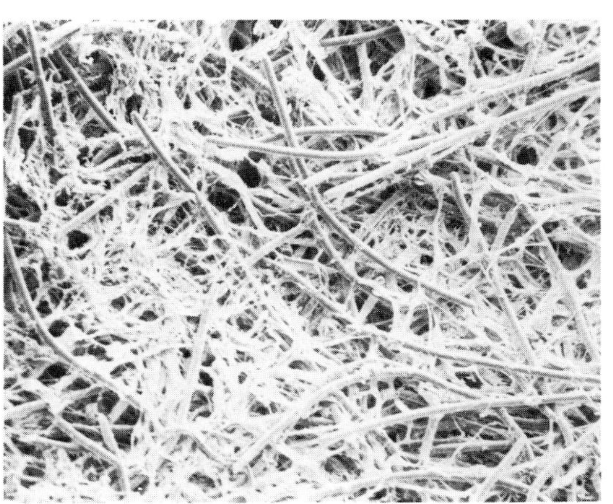

Figure 31.10 Photomicrograph of slime-forming bacteria on a trickling filter. (A) *Zooglea ramigera* surrounded by extensive slime layer. (Courtesy Patrick Dugan, Ohio State University) (B) Scanning electron micrograph of microbial community on the surface of a rock below a trickling filter; note the abundance of filamentous algae and fungi that develop on the surface film (500×). (Courtesy Robert P. Apkarian, University of Louisville.)

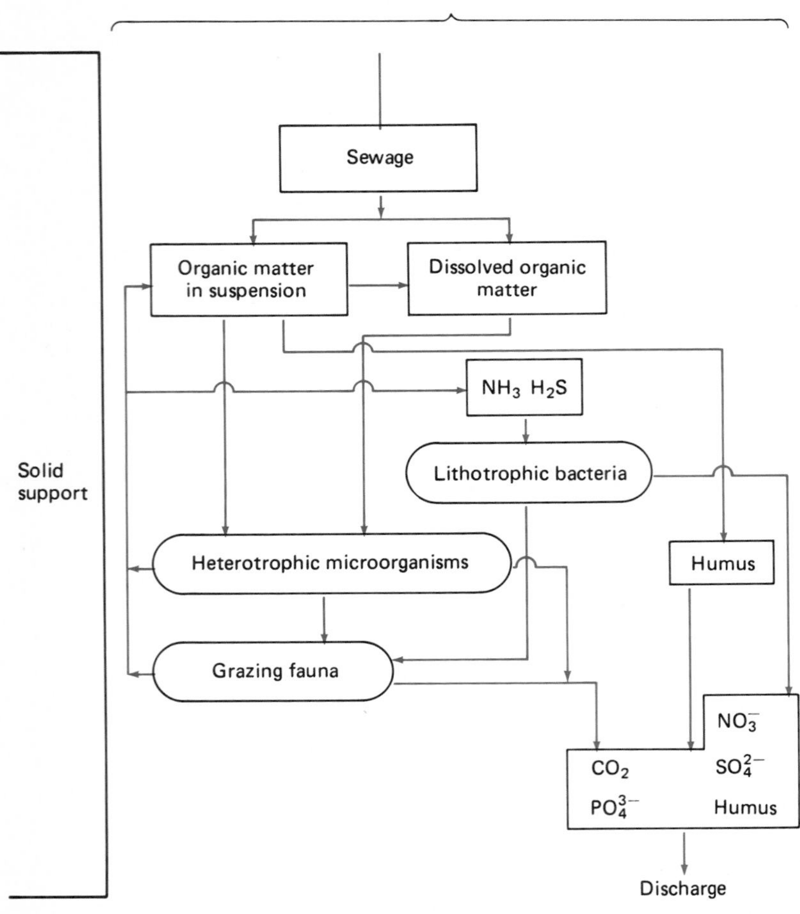

Figure 31.11 The mineral cycling of organic matter by the microbial film in a trickling filter system.

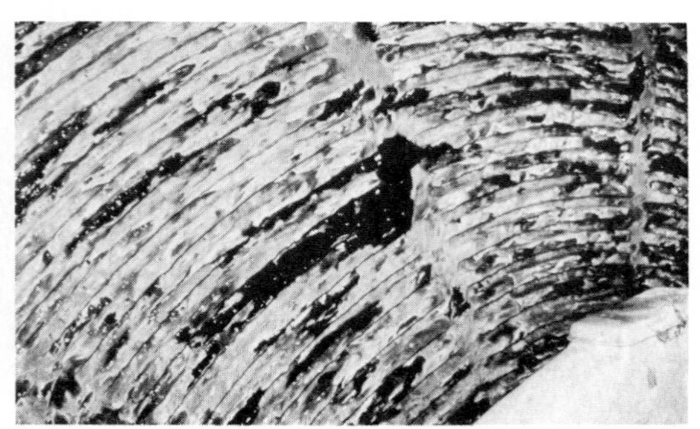

Figure 31.12 (A) A biodisc assembly; the discs rotate through the sewage and microbes growing on the surfaces of the polyethylene cups degrade the organic matter in the sewage. (B) Close-up view of the normal appearance of the surface of the biodisc with a film of aerobic microorganisms. (C) When the system is overloaded, a heavy film of anaerobic microorganisms develops. Note the black area of anaerobic biomass and the occurrence of gelatinous, filamentous sulfur-oxidizing growths on the outer layer. (Courtesy Autotrol Corp., Milwaukee, Wisconsin.)

Figure 31.13 An activated sludge tank in a sewage treatment plant. Note the bubbling due to forced aeration. (From BPS: Carl May, Moss Beach, California.)

nutrients through the film. Microbial growth on the surface of the discs is sloughed off gradually and is removed by subsequent settling.

Activated Sludge. The activated sludge process is a very widely used aerobic suspension type of liquid waste treatment system (Figure 31.13). After primary settling, the sewage, containing dissolved organic compounds, is introduced into an aeration tank. Air injection and/or mechanical stirring provides the aeration. The rapid development of microorganisms is also stimulated by reintroduction of most of the settled sludge from a previous run (Figure 31.14), and the process derives its name from this inoculation with "activated sludge."

As a consequence of extensive microbial metabolism of the organic compounds in sewage, a significant proportion of the dissolved organic substrates is mineralized, and another portion is converted to microbial biomass. In the advanced stage of aeration, most of the microbial biomass becomes associated with flocs that can be removed from suspension by settling. Combined with primary settling, the activated sludge process tends to reduce the BOD of the effluent to 10–15 percent of that of the raw sewage. The treatment also drastically reduces the numbers of intestinal pathogens in the sewage. Numbers of *Salmonella, Shigella,* and *Escherichia coli* typically are 90–99 percent lower in the effluent of the activated sludge treatment process than in the incoming raw sewage (Table 31.5). Enteroviruses are removed to a similar degree, and the main removal mechanism appears to be adsorption of the virus particles onto the settling sewage sludge floc.

Anaerobic Digesters. Large-scale anaerobic digesters are used for further processing the sewage sludge produced by primary and secondary treatments (Figure 31.15). Although anaerobic digesters could be used for direct treatment of sewage, economic considerations favor aerobic processes for relatively dilute wastes, and the use of anaerobic digesters is restricted to treating concentrated organic wastes. Therefore, in practice large-scale anaerobic digesters are used only for processing settled sewage sludge and the treatment of very high BOD industrial effluents. Anaerobic digesters are large fermentation tanks designed for continuous operation under anaerobic conditions.

A properly operating anaerobic digester yields a greatly reduced volume of sludge compared to the starting material. The product obtained by using an anerobic sludge digester, however, still causes odor and water pollution problems and can be disposed of at only a few restricted landfill sites. Aero-

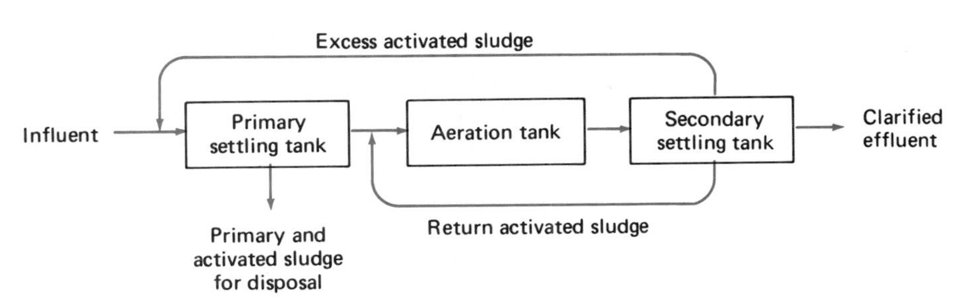

Figure 31.14 The flow of materials through an activated-sludge secondary sewage treatment system. A portion of the sludge is recycled as inoculum for the incoming sewage.

Table 31.5. Percentage Reductions in the Numbers of Indicator Organisms in Different Types of Sewage Treatment Processes

Treatment	*Escherichia coli*	Coliforms	Fecal Streptococci	Viruses
Sedimentation	3–72	13–86	44–60	—
Activated sludge	61–100	13–83	84–93	79–100
Trickling filter	73–97	15–100	64–97	40–82
Lagoons	80–100	86–100	85–99	95

bic composting can be used to further consolidate the sludge, rendering it suitable now for disposal in any landfill site or for use as a soil conditioner. Several gases are produced as a result of the anaerobic biodegradation of sludge, primarily methane and carbon dioxide. The methane can be used within the treatment plant to drive the pumps and/or to provide heat for maintaining the temperature of the digester, or after purification it may be sold through natural gas distribution systems. Thus, in addition to its primary function in removing wastes, anaerobic digesters can produce needed fuel resources.

Tertiary Treatment

The aerobic and anaerobic biological liquid waste treatment processes are designed to reduce the BOD of biodegradable organic substrates and oxidizable inorganic compounds and all represent secondary treatment processes. Tertiary treatment (any practice beyond a secondary one) is designed to remove nonbiodegradable organic pollutants and mineral nutrients, especially nitrogen and phosphorus salts. Tertiary treatment prevents the release of sewage effluents containing phosphates and fixed forms of nitrogen that can cause serious eutrophication in aquatic ecosystems.

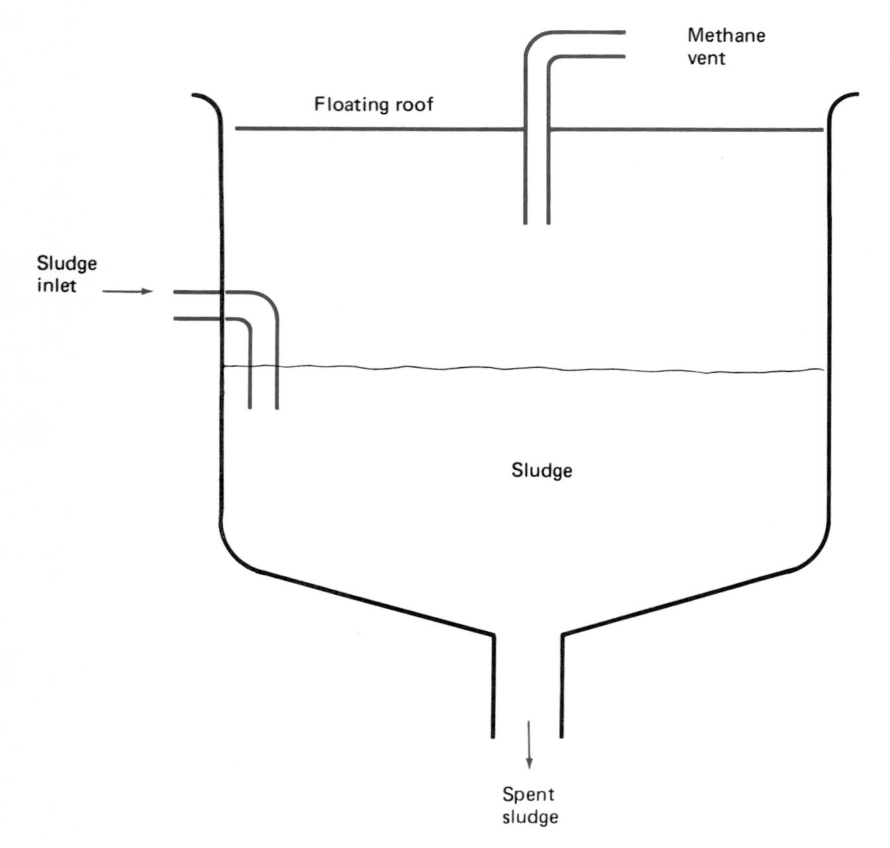

Figure 31.15 Diagram of an anaerobic sludge digester.

Eutrophication, the input of excessive levels of nutrients, triggers explosive algal blooms. Because of a variety of causes—some unknown but including mutual shading, exhaustion of micronutrients, presence of toxic products and/or antagonistic populations—the algal population usually "crashes," and the subsequent decomposition of the dead algal biomass by heterotrophic microorganisms exhausts the dissolved oxygen in the water, precipitating extensive fish kills and septic conditions. Even if the eutrophication does not proceed to this extreme, algal mats, turbidity, discoloration, and shifts in the fish population from valuable species to more tolerant but less valued forms represent undesirable changes due to eutrophication. To prevent eutrophication, phosphate is commonly removed from sewage by precipitation as calcium or iron phosphate. Some local communities have also banned the use of phosphate detergents to minimize problems of eutrophication.

Disinfection

The final step in the sewage treatment process is disinfection, designed to kill enteropathogenic bacteria and viruses that were not eliminated during the previous stages of sewage treatment. Disinfection is commonly accomplished by chlorination, using either chlorine gas (Cl_2) or hypochlorite ($CaOCl_2$ or $NaOCl$). Chlorine gas reacts with water to yield hypochlorous and hydrochloric acids, the actual disinfectants. Hypochlorite is a strong oxidant, which is the basis of its antibacterial action. As an oxidant, it also reacts with residual dissolved or suspended organic matter, ammonia, reduced iron, manganese, and sulfur compounds. The oxidation of these compounds competes for available hypochlorite, reducing its disinfecting power. Amounts of hypochlorite sufficient to satisfy these reactions and to allow excess-free residual chlorine to remain in solution for disinfection would result in high salt concentrations in the effluent. Therefore, it is desirable to remove nitrogen and other contaminants by alternate means and use chlorination for disinfection only. A disadvantage of disinfection by chlorination is that the more resistant organic molecules, such as some lipids and hydrocarbons, are not completely oxidized but instead become partially chlorinated. Chlorinated hydrocarbons tend to be toxic and difficult to mineralize. Because alternative means of disinfection, such as ozonation, are more expensive, chlorination remains the principal means of sewage disinfection; some municipalities, however, are requiring the use of alternative methods of disinfection.

Solid Waste Disposal

Methods for the disposal of solid wastes differ from those used to treat liquid wastes. Urban solid waste production in the United States amounts to roughly 150 million tons per year. Much of this material is inert, composed of glass, metal, and plastic, but the rest is decomposable organic waste, such as kitchen scraps, paper, and other household and industrial garbage. Sewage sludge derived from treatment of liquid wastes and animal waste from cattle feed lots and large-scale poultry and swine farms are also major sources of solid organic waste. In traditional small, family farm operations, most organic solid waste is composted and recycled into the land as fertilizer. In highly populated urban centers and areas of large-scale agricultural production, the disposal of massive amounts of organic waste becomes a difficult and expensive problem.

It costs more to dispose of the *New York Sunday Times* than it does for a subscriber to buy it.

—MELVIN W. FIRST, *Urban Solid Waste Management*

There are several options for dealing with solid waste problems. Today, many of the inert components of solid waste, such as aluminum and glass, are recycled. Even paper, which is relatively resistant to microbial degradation, can be recovered from solid waste, and many books and newspapers are printed on recycled paper. The remaining bulk of the solid waste may be incinerated, creating potential air pollution problems, or the organic components can be subjected to microbial biodegradation in aquatic or terrestrial ecosystems. In many cases the solid waste is dumped at sea or discarded on land, allowing biodegradation to occur naturally without any special treatment, but excessive dumping of organic wastes into terrestrial and marine ecosystems can cause untoward problems unless the operation is carefully managed and monitored.

Sanitary Landfills

The simplest and least expensive way to dispose of solid waste is to place the material in landfills and to allow it to decompose. In landfill procedures both organic and inorganic solid wastes are deposited together in low-lying land that has minimal real estate value. Because exposed waste can cause aesthetic and odor problems, attract insects and rodents, and pose a fire hazard, each day's waste deposit is covered over with a layer of soil, creating a sanitary landfill (Figure 31.16). When the

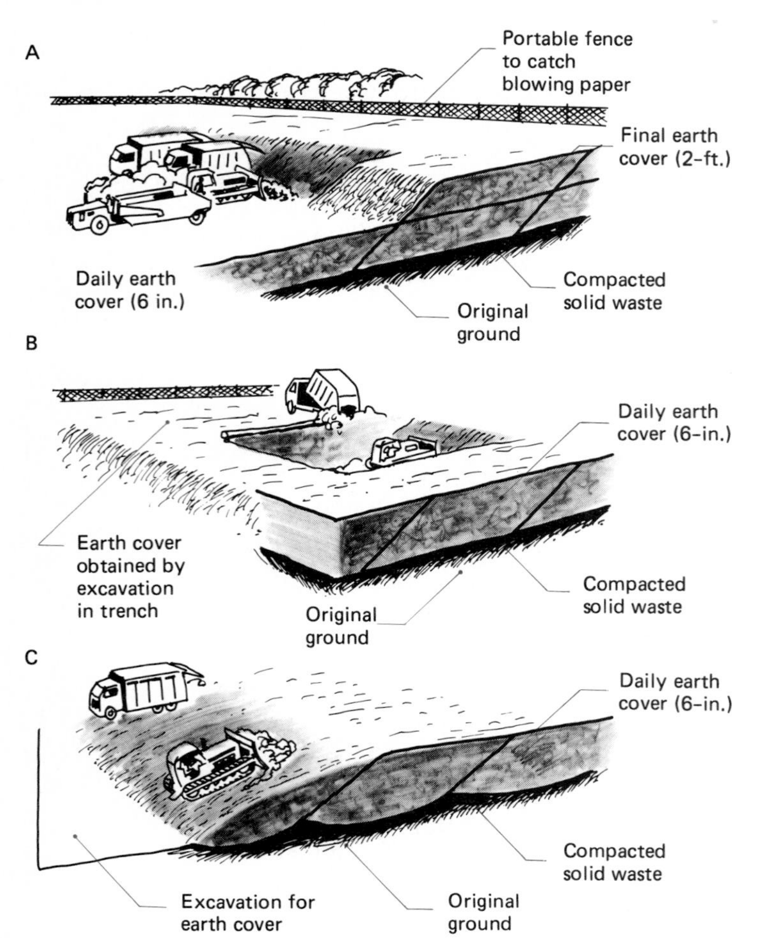

Figure 31.16 Sanitary landfills are used for the inexpensive disposal of solid waste. Several different methods, as shown in these drawings, can be employed. (A) The area method—a bulldozer spreads and compacts solid wastes. Cover material is hauled in and spread at the end of the day's operations. Note that a fence to catch any blowing debris is present in all landfill methods. (B) The trench method—the waste collection truck deposits its load into a trench where it is spread and compacted. At the end of the day, soil is excavated from a future trench and used as the daily cover material. (C) Ramp variation—solid wastes are spread and compacted on a slope. The daily deposit is covered with earth from the base of the ramp. This variation is used with either the area or trench method.

landfill is full, the site can be used for recreational purposes and may eventually provide a foundation for construction.

For 30—50 years after the establishment of a landfill, the organic content of the solid waste undergoes slow, anaerobic microbial decomposition. The products of anaerobic microbial metabolism include carbon dioxide, water, methane, various low molecular weight alcohols, and acids, which diffuse into the surrounding water and air, causing the landfill to settle slowly. Extensive amounts of methane are produced during this decomposition process, potentially providing a source of needed natural gas. Eventually, the decomposition slows greatly, signaling completion of the biodegradation of the solid waste, subsidence ceases, and the land is stabilized and suitable as a site for construction.

Although the use of sanitary landfills is simple and inexpensive, there are several problems associated with this waste disposal method. Premature construction on a still biologically active landfill site may result in structural damage to the buildings because of movement of the land base, and an explosion hazard may exist due to methane seeping into basements and other below ground structures. Above ground plantings may also be damaged because of methane seepage. The number of suitable disposal sites available in urban areas is very limited, often necessitating long and expensive hauling of the solid waste to outlying sites. The possible seepage of anaerobic decomposition products, heavy metals, and a variety of nonbiodegradable hazardous pollutants from the landfill site into underground aquifers, which are used in many urban areas as water sources, has caused many municipalities to place severe restrictions on the location and operation of landfills and to seek alternatives to the landfill technique for disposing of solid waste.

Composting

Composting is an alternative to sanitary landfills. In this process, solid heterogeneous organic matter is degraded by aerobic microorganisms. Various mesophilic and thermophilic bacteria and fungi are involved in the decomposition process. Like incineration, composting requires sorting the solid waste into its organic and inorganic components. This can be accomplished either at the source, by the separate collections of garbage (organic waste) and trash (inorganic waste), or at the receiving facility, by using magnetic separators to remove ferrous metals and mechanical separators to remove glass, aluminum, and plastic materials. The remaining largely organic waste is ground up, mixed with sewage sludge and/or bulking agents, such as shredded newspaper or wood chips, and then composted.

Composting is a microbial process that converts noxious organic waste materials into a stable, sanitary, humus-like product. Reduced in bulk, it can be used for soil improvement. The different composting methods are differentiated by the physical arrangement of the the solid waste; composting can be accomplished in windrows (long rows of waste), aerated piles (discrete heaps of waste), or continuous feed reactors (continuous addition of waste to the reaction) (Figure 31.17). The windrow method is a simple but relatively slow process, typically requiring several months to achieve biodegradation of the metabolizable components and stabilization of the waste material. Odor and insect problems are controlled in this process by covering the windrows with a layer of soil or finished compost. Unless the decomposing material is turned several times during the process, the quality of the finished compost product is uneven. Because the process is so slow, large amounts of land must be used, causing the same problems as sanitary landfills in densely populated urban areas. Aerated piles, as many homeowners use for the disposal of leaves, are also slow. Continuous feed reactors are more sophisticated and optimize conditions to accelerate the decompostion process.

Figure 31.17 These photographs show compost heaps used for decomposition of solid wastes. (A) Steam is rising from this compost heap because of the heat buildup that results from the rapid decomposition of organic matter by thermophilic microorganisms. (B) In this heap a straight row is formed by a machine, which is also used periodically to turn the heap to maintain good aeration. (Courtesy Atal E. Eralp, U.S. Environmental Protection Agency, Cincinnati.) (C) This is a commercial compost facility owned by Paygro Inc. in South Charlestown, Ohio. The T–shaped poles have thermocouples attached to them to monitor continuously the temperature at various depths within the pile. Air is forced through the pile from underneath by fans to provide aeration. (Courtesy J. R. Vestal and V. L. McKinley, University of Cincinnati.)

A

B

C

Biodegradation of Environmental Pollutants

In the twentieth century humans have introduced into the environment many compounds that microorganisms normally do not encounter and thus are not prepared to biodegrade. These compounds include petroleum hydrocarbons, pesticides, plastics, and various other novel-synthetic (xenobiotic) compounds. Many of these compounds are toxic to living systems, and their presence in aquatic and terrestrial habitats often has serious ecological consequences, including major kills of indigenous organisms. The disposal and accidental spillage of these compounds has created serious modern environmental pollution problems, particularly when microbial biodegradation activities fail to remove these pollutants quickly enough to prevent environmental damage. Sewage treatment and water purification systems are usually incapable of removing these substances if they enter municipal water supplies, where they pose a potential human health hazard.

We generally require today that new pesticides and various other novel compounds be biodegradable. Careful testing is performed to ensure that unsafe levels of compounds that are intentionally released into the environment do not accumulate. When xenobiotics are recalcitrant (not subject to microbial degradation) and microorganisms fail in their role as "biological incinerators," environmental pollutants accumulate. The proper chemical design of synthetic compounds, making them susceptible to microbial attack, is essential for maintaining environmental qualtity.

When chemicals are released directly into the environment as a result of careless or accidental spillages, we must generally rely on the metabolic activities of microorganisms to remove the polluting substances. It is possible to create microorganisms, using recombinant DNA technology, capable of degrading complex mixtures of toxic environmental pollutants, and the development of such microorganisms may prove useful in treating the myriad of chemicals in industrial wastes. The human race depends on microorganisms to make the world a better place in which to live and, in the end, we remain dependent on microbial biodegradation for recycling our wastes and maintaining the environmental quality of the biosphere.

Messiers, c'est les microbes qui auront le dernier mot—Gentlemen, it is the microbe that will have the last word.

—Louis Pasteur

Summary

We have a limited freshwater supply, and it is essential that we use that water supply effectively for multiple purposes. It is particularly critical that drinking water supplies be free of human pathogens and that our use of natural water bodies for waste disposals not jeopardize human health and environmental quality. The adequate treatment of sewage to remove pathogens, reduce nutrient levels, and eliminate toxic compounds and the disinfection of potable water supplies have become very important for lowering rates of water-borne infectious diseases.

Sewage treatment facilities permit aerobic and/or anaerobic biodegradation of organic wastes by a complex assemblage of microorganisms within a controlled environment, where maximal rates of biodegradation can be achieved and environmental damage can be minimized. Sewage is treated using primary (physical), secondary (biological), and tertiary (chemical) methods. The biological treatment of sewage involves the microbial decomposition of organic matter, which reduces the biological oxygen demand (BOD). The various methods differ in their efficiencies of reducing the BOD. The choice of a particular system is dictated by cost and the system receiving the waste effluents. The installation of advanced sewage treatment facilities permits the reduction of the BOD of sewage effluents and is essential in the abatement of eutrophication problems in aquatic ecosystems as well as the reduction in the dissemination of human pathogens.

In addition to liquid wastes, microorganisms can be used for the disposal of solid wastes. Sanitary landfills and composting are widely used for this purpose.

Pollution results when substances enter environments and undesirable reactions occur, such as the depletion of oxygen in the water column of an aquatic ecosystem receiving organic wastes. We rely on microorganisms to biodegrade our waste materials and have come to assume that anything thrown out into the environment will disappear and to an incredibly large extent, this is true. Environmental problems occur when microbes fail to degrade a particular compound. Microorganisms have a vast capacity for rapidly degrading organic materials and thus can be relied upon for the most part to act as biological incinerators.

Study Outline

A. Water from underground sources is partially purified because it is filtered as it passes through the soil, removing particulate matter and microorganisms.

- **1.** Water treatment facilities carry out
  - **a.** Sedimentation in large reservoirs, where the water is held for a period of time sufficient to permit large particulate matter to settle out.
  - **b.** Filtration in sand filter beds that remove up to 99 percent of the bacteria. Activated charcoal removes potentially toxic compounds and organic compounds that impart undesirable color and/or taste to the water.
  - **c.** Disinfection which eliminates any pathogens that could be sources of water-borne disease. Chlorination is the usual method. Ozone treatment reliably kills pathogens, but does not have any residual antimicrobial activity.

B. Objective test methods are necessary to establish standards of water safety and to evaluate the effectiveness of treatment procedures.

- **1.** Indicator organisms are used to establish the degree of fecal contamination of a water sample.
  - **a.** Characteristics of ideal indicator organism: they are present when pathogens are present; they are present only when there is a real danger of pathogens being present; they occur in greater numbers than pathogens to provide a safety margin; they survive in the environment as long as the potential pathogens; they are easy to detect with a high degree of reliability.
  - **b.** *Escherichia coli* is the most frequently used indicator organism. Positive tests for *E. coli* only establish the possibility of the presence of enteropathogenic organisms.
- **2.** Detection of fecal contamination involves a three-stage procedure.
  - **a.** Presumptive test: gas formation when lactose broth is inoculated with a water sample.
  - **b.** Confirmed test: samples from positive lactose broth cultures are streaked onto eosin methylene blue (EMB) agar; fecal coliform colonies acquire a greenish metallic sheen.
  - **c.** Completed test: colonies showing green metallic sheen on EMB agar, when cultured into lactose broth, produce gas.
- **3.** Standards for tolerable levels of fecal contamination vary with intended type of water use.
  - **a.** The most stringent standards are applied to municipal water supplies.
  - **b.** Fecal coliform counts are also used to establish safety of water in shellfish-harvesting and recreational areas.
  - **c.** Water quality standards based on fecal coliforms do not account for possible transmission of enteroviruses, which are more resistant to disinfection by chlorine or ozone than bacteria.

C. Liquid wastes eventually enter natural bodies of water, an inherent capability of which is self-purification, but this capacity is overwhelmed by liquid waste production of densely populated areas, large-scale agricultural operations, and major industrial activities.

- **1.** Sewage treatment is aimed at maintaining water quality by reducing the BOD of waste before discharge into a water body.
  - **a.** BOD is a measure of water quality that represents the amount of oxygen required for the microbial decomposition of the organic matter in that water.
- **2.** Sewage treatment employs combinations of physical, chemical, and microbiological methods to remove organic matter, human pathogens, and toxic chemicals.
- **3.** Primary sewage treatment relies on physical separation procedures to lower the BOD by removing suspended solids in settling tanks and basins.
  - **a.** Solids may then be subjected to anaerobic digestion and/or composting.

b. The liquid portion, containing dissolved organic matter, can be subjected to further treatment.

4. Secondary sewage treatment relies on microbial degradation to further reduce the concentration of organic compounds in the effluent. A small portion of the dissolved organic matter is mineralized; a larger portion is converted to removable solids.
   a. The septic tank, a simple anaerobic treatment system, acts largely as a settling tank; accumulated sludge is degraded anaerobically, and the clarified effluent percolates into the soil, where the dissolved organic compounds undergo biodegradation.
   b. In the trickling-filter system, dissolved organic nutrients in the sewage are adsorbed and mineralized by the bacterial growth coating the porous filter bed through which the sewage percolates.
   c. The rotating biological contactor or biodisc system utilizes bacteria-coated plastic disks, which are rotated in a trough containing sewage effluent, with results similar to a trickling-filter system.
   d. In the activated sludge process, sludge from previously treated sewage is introduced into sewage in an aeration tank; extensive microbial metabolism of the organic compounds in the sewage occurs so that a significant proportion is mineralized, and another portion is converted to microbial biomass.
5. Tertiary treatment uses chemical methods to remove inorganic compounds and pathogenic microorganisms.
   a. Activated carbon filters remove toxic, nonbiodegradable organic pollutants.
   b. Phosphates are removed by precipitation to prevent eutrophication.
6. Disinfection is designed to kill enteropathogenic bacteria and viruses not previously eliminated in sewage treatment. It is most commonly accomplished by chlorination.

D. Solid waste, approximately 150 million tons per year of which is produced in the United States, is inert material, composed of glass, metal, and plastic, and other decomposable organic garbage.
   1. Today we recycle of many of the inert components; others are subjected to incineration, while others undergo microbial degradation in aquatic or terrestrial ecosystems.
   2. Sanitary landfills are the simplest and cheapest means of solid waste disposal.
      a. Organic and inorganic solid wastes are buried in low-lying land for 30–50 years, during which organic wastes undergo slow, anaerobic microbial decomposition. Following decomposition, the land is stabilized and can be used.
      b. Disadvantages to this system include structural damage to buildings constructed on a biologically active landfill due to movement of the land base, explosion hazard from methane produced by the anaerobic decomposition, expense of hauling waste to disposal sites, and seepage of anaerobic decomposition products and recalcitrant hazardous pollutants into underground aquifers.
   3. Composting is a process by which solid heterogenous organic matter is degraded by aerobic, mesophilic, and thermophilic microorganisms into a stable, sanitary, humus-like product.
      a. In windrows organic waste is covered with a layer of soil or finished compost to achieve biodegradation. The process is slow; large amounts of land must be used.

E. The use of fossil fuels and synthetic compounds (xenobiotics) have introduced into the environment many compounds that microorganisms do not encounter and are not prepared to biodegrade. The disposal and accidental spillage of these compounds, many of them toxic, have created many environmental pollution problems.

Study Questions

1. What are the differences between composting and sanitary landfill operations?
2. Discuss how we treat sewage. What is primary, secondary, and tertiary sewage treatment?
3. What is an indicator organism? Why are coliform counts used to assess the safety of potable water supplies?

4. What is disinfection of a water supply? How is the disinfection of municipal water supplies normally achieved?
5. What is BOD? Why is it important to reduce the BOD of liquid wastes before they are discharged into rivers or lakes?
6. Compare activated sludge and anaerobic digestors for treating sewage. What roles do each play in an integrated liquid waste removal system?
7. Match the following statements with the terms in the list to the right.

a. Sewage percolates through a porous filter bed.	**(1)** Primary treatment
b. Relies upon microbial degradation to reduce the concentration of organic compounds.	**(2)** Secondary treatment
c. Uses heterotrophic bacteria to degrade sewage sludge.	**(3)** Tertiary treatment
d. Lowers the BOD by removing suspended solids.	**(4)** Trickling-filter
e. Plastic disks are rotated in a trough containing sewage effluent.	**(5)** Rotating biological contactor
f. Uses chemical methods to remove ammonia and phosphates.	**(6)** Activated sludge
	(7) Septic tank

Some Thought Questions

1. Is wellwater on a farm better than city tapwater?
2. Why don't we drink sewage-contaminated water?
3. How can we know if the water is safe to drink?
4. Why is water-borne disease more prevalent in India than in the United States?
5. What can developing nations do to decrease the incidence of water-borne disease?
6. What standards should be set and what methods should be used for determining water safety standards?
7. How can the same water body that receives sewage effluents serve as the source of potable water?

Additional Sources of Information

Bonde, G. J. 1977. Bacterial indicators of water pollution. *Advances in Aquatic Microbiology*, **1**: 273–364. A detailed and thorough discussion of the use of bacteria as indicators of water pollution.

Finstein, M. S. 1975. Microbiology of municipal solid waste composting. *Advances in Applied Microbiology*, 19:113–151. A detailed and thorough discussion of the use of microorganisms to degrade solid wastes.

Gaudy, A., and E. Gaudy. 1980. *Microbiology for Environment Science Engineers*. McGraw-Hill Book Company, New York. An excellent text describing wastewater treatment and other microbial processes used in waste treatment as seen from the perspective of a sanitary or chemical engineer.

Greenberg, A. (ed.). 1980. *Standard Methods for the Examination of Water and Wastewater*. American Public Health Association, Washington, D. C. This volume is continuously updated to reflect the current methods used to examine water and wastewater; it contains protocols used to assess water safety.

Mitchell, R. (ed.). 1972. *Water Pollution Microbiology*. John Wiley & Sons, Inc., New York. An interesting series of articles concerning the microbial degradation of environmental pollutants.

Mitchell, R. 1974. *Introduction to Environmental Microbiology*. Prentice-Hall, Inc., Englewood Cliffs, New Jersey. A simple, introductory text that describes the role of microorganisms in degrading wastes.

APPENDIX I

The Metric System and Some Useful Conversion Factors

Metric System Prefixes

pico (p) = 10^{-12}
nano (n) = 10^{-9}
micro (μ) = 10^{-6}
milli (m) = 10^{-3}
centi (c) = 10^{-2}
deci (d) = 10^{-1}
deka (da) = 10
hecto (h) = 10^{2}
kilo (k) = 10^{3}
mega (M) = 10^{6}

Length

1 kilometer (km) = 0.62 mile (mi)
1 mi = 1.609 km
1 meter (m) = 3.28 feet (ft) = 1.09 yard (yd)
1 ft = 0.305 m
1 yd = 0.914 m
1 centimeter (cm) = 0.394 inches (in)
1 in = 2.54 cm
1 ft = 30.5 cm
1 millimeter (mm) = 0.039 in
1 angstrom (Å) = 10^{-10} m

Area

1 hectare (ha) = 10,000 m^2 = 2.471 acres
1 acre = 0.4047 ha
1 sq. km (km^2) = 100 ha = 0.3861 sq. mi (mi^2)
1 mi^2 = 2.590 km^2
1 sq. m (m^2) = 10,000 cm^2 = 1.1960 sq. yd (yd^2) = 10.764 sq. ft (ft^2)
1 yd^2 = 0.8361 m^2
1 ft^2 = 0.0929 m^2
1 cm^2 = 100 sq. mm (mm^2) = 0.155 sq. in (in^2)
1 in^2 = 6.4516 cm^2

Mass

1 metric ton (t) = 1000 kilograms (kg) = 1.10 short tons
1 short ton = 0.91 t
1 kg = 1000 grams (g) = 2.205 pounds (lb)
1 lb = 453.60 g
1 g = 1000 milligrams (mg) = 0.0353 ounce (oz)
1 oz = 28.35 g
1 mg = 10^{-3} g = 0.02 grains
1 microgram (μg) = 10^{-6} g
1 nanogram (ng) = 10^{-9} g
1 picogram (pg) = 10^{-12} g

Volume (solids)

1 cubic m (m^3) = 1,000,000 cu. cm (cm^3) = 35.315 cu. ft (ft^3) = 1.3080 cu. yd (yd^3)
1 ft^3 = 0.0283 m^3
1 yd^3 = 0.7646 m^3
1 cm^3 = 10^3 cu. millimeter (mm^3) = 0.0610 cu. in (in^3)
1 in^3 = 16.387 cm^3

Volume (liquids)

1 liter (L) = 10^3 milliliters (mL) = 1.06 quarts (qt)
1 gal = 3.785 L
1 kiloliter (kL) = 1000 = 264.17 gal
1 qt = 0.94 L
1 mL = 10^{-3} L = 0.034 fluid oz.
1 fl. oz = 29.57 mL
1 microliter (μL) = 10^{-6} L

Temperature

degrees Fahrenheit (°F) = 9/5°C + 32
degrees centigrade (°C) = 5/9(°F − 32)
0°C = 32°F (freezing point of water)
100°C = 212°F (boiling point of water)

APPENDIX II

Chemical and Biochemical Principles

Atoms

Atoms are the building blocks of all matter. Each of the known chemical elements is composed of a single kind of atom and each has a distinct weight as well as a unique structure that confers special properties on that element. There are now over 106 known elements, 92 of which occur naturally; of these, only a few are incorporated into living organisms (Table A.1). Elements are designated by a symbol that is an abbreviation of its English or Latin name; for example, oxygen O, nitrogen N, carbon C, hydrogen H, and sodium Na (Latin *natrium*).

Atomic Number and Mass

Atoms have a central nucleus made up of neutrons (neutrally charged particles) and protons (positively charged particles), both of which have approximately one atomic mass unit (amu). The atomic number of an element is the number of protons in the nucleus. Thus, ${}_{8}O$ has 8 protons and ${}_{7}N$ has 7 protons. Protons and neutrons have almost identical weights, and the mass of an atom is approximately the sum of the number of protons and neutrons in the nucleus of the atom. The atomic weight of the atom is determined by the number of protons and neutons. The atomic weight of the carbon atom with 6 protons and 6 neutrons is used as a standard reference and assigned a molecular weight of 12.000. Similarly ${}^{14}N$ (nitrogen with an atomic weight of 14) has 7 protons and 7 neutrons.

Elements having the same atomic number can have different numbers of neutrons; such forms of the same element are known as isotopes. For example, in addition to ${}^{14}N$ we also find heavy nitrogen, ${}^{15}N$, which has 7 protons and 8 neutrons. The number of neutrons in the nucleus does not alter the chemical properties of the element and thus hydrogen (${}^{1}H$), deuterium (${}^{2}H$), and tritium (${}^{3}H$) have essentially identical chemical properties. However, the physical properties of isotopes vary, and physical differences between isotopes can be used as labels or markers to tag and to follow a particular element. In some cases isotopes of an element such as ${}^{3}H$ and ${}^{14}C$ are radioactive, emitting particles and energy. Biologists make extensive use of radioactive

Table A.1 The Periodic Table of the Elements

${}^{1}_{1}\mathrm{H}$																	${}^{4}_{2}\mathrm{He}$
${}^{7}_{3}\mathrm{Li}$	${}^{9}_{4}\mathrm{Be}$											${}^{11}_{5}\mathrm{B}$	${}^{12}_{6}\mathrm{C}$	${}^{14}_{7}\mathrm{N}$	${}^{16}_{8}\mathrm{O}$	${}^{19}_{9}\mathrm{F}$	${}^{20}_{10}\mathrm{Ne}$
${}^{23}_{11}\mathrm{Na}$	${}^{24}_{12}\mathrm{Mg}$											${}^{27}_{13}\mathrm{Al}$	${}^{28}_{14}\mathrm{Si}$	${}^{31}_{15}\mathrm{P}$	${}^{32}_{16}\mathrm{S}$	${}^{35}_{17}\mathrm{Cl}$	${}^{40}_{18}\mathrm{Ar}$
${}^{39}_{19}\mathrm{K}$	${}^{40}_{20}\mathrm{Ca}$	${}^{45}_{21}\mathrm{Se}$	${}^{48}_{22}\mathrm{Ti}$	${}^{51}_{23}\mathrm{V}$	${}^{52}_{24}\mathrm{Cr}$	${}^{55}_{25}\mathrm{Mn}$	${}^{56}_{26}\mathrm{Fe}$	${}^{59}_{27}\mathrm{Co}$	${}^{59}_{28}\mathrm{Ni}$	${}^{64}_{29}\mathrm{Cu}$	${}^{65}_{30}\mathrm{Zn}$	${}^{70}_{31}\mathrm{Ga}$	${}^{73}_{32}\mathrm{Ge}$	${}^{75}_{33}\mathrm{As}$	${}^{79}_{34}\mathrm{Se}$	${}^{80}_{35}\mathrm{Br}$	${}^{84}_{36}\mathrm{Kr}$
${}^{85}_{37}\mathrm{Rb}$	${}^{88}_{38}\mathrm{Sr}$	${}^{89}_{39}\mathrm{Y}$	${}^{91}_{40}\mathrm{Zr}$	${}^{93}_{41}\mathrm{Nb}$	${}^{96}_{42}\mathrm{Mo}$	${}^{99}_{43}\mathrm{Te}$	${}^{101}_{44}\mathrm{Ru}$	${}^{103}_{45}\mathrm{Rh}$	${}^{106}_{46}\mathrm{Pd}$	${}^{108}_{47}\mathrm{Ag}$	${}^{112}_{48}\mathrm{Cd}$	${}^{115}_{49}\mathrm{In}$	${}^{119}_{50}\mathrm{Sn}$	${}^{122}_{51}\mathrm{Sb}$	${}^{128}_{52}\mathrm{Te}$	${}^{127}_{53}\mathrm{I}$	${}^{131}_{54}\mathrm{Xe}$
${}^{133}_{55}\mathrm{Cs}$	${}^{137}_{56}\mathrm{Ba}$	${}^{139}_{57}\mathrm{La}$	${}^{178}_{72}\mathrm{Hf}$	${}^{181}_{73}\mathrm{Ta}$	${}^{184}_{74}\mathrm{W}$	${}^{186}_{75}\mathrm{Re}$	${}^{190}_{76}\mathrm{Os}$	${}^{192}_{77}\mathrm{Ir}$	${}^{195}_{78}\mathrm{Pt}$	${}^{197}_{79}\mathrm{Au}$	${}^{201}_{80}\mathrm{Hg}$	${}^{204}_{81}\mathrm{Ti}$	${}^{207}_{82}\mathrm{Pb}$	${}^{209}_{83}\mathrm{Bi}$	${}^{210}_{84}\mathrm{Po}$	${}^{210}_{85}\mathrm{At}$	${}^{222}_{86}\mathrm{Rn}$
${}^{223}_{87}\mathrm{Fr}$	${}^{226}_{88}\mathrm{Ra}$	${}^{227}_{89}\mathrm{Ae}$	${}^{257}_{104}\mathrm{Ku}$	${}^{260}_{105}\mathrm{Ha}$													
				${}^{140}_{58}\mathrm{Ce}$	${}^{141}_{59}\mathrm{Pr}$	${}^{144}_{60}\mathrm{Nd}$	${}^{145}_{61}\mathrm{Pm}$	${}^{150}_{62}\mathrm{Sm}$	${}^{152}_{63}\mathrm{Eu}$	${}^{157}_{64}\mathrm{Gd}$	${}^{159}_{65}\mathrm{Tb}$	${}^{163}_{66}\mathrm{Dy}$	${}^{165}_{67}\mathrm{Ho}$	${}^{167}_{68}\mathrm{Er}$	${}^{169}_{69}\mathrm{Tm}$	${}^{173}_{70}\mathrm{Yb}$	${}^{175}_{71}\mathrm{Lu}$
				${}^{232}_{90}\mathrm{Th}$	${}^{231}_{91}\mathrm{Pa}$	${}^{238}_{92}\mathrm{U}$	${}^{237}_{93}\mathrm{Np}$	${}^{242}_{94}\mathrm{Pu}$	${}^{243}_{95}\mathrm{Am}$	${}^{247}_{96}\mathrm{Cm}$	${}^{247}_{97}\mathrm{Bk}$	${}^{251}_{98}\mathrm{Cf}$	${}^{254}_{99}\mathrm{Es}$	${}^{253}_{100}\mathrm{Fm}$	${}^{258}_{101}\mathrm{Md}$	${}^{255}_{102}\mathrm{No}$	${}^{256}_{103}\mathrm{Lr}$

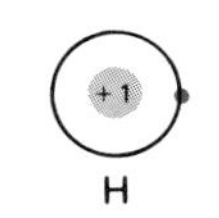

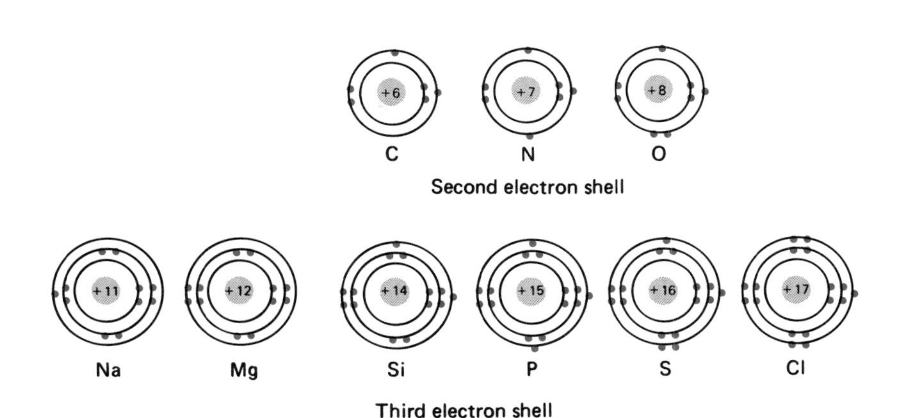

Figure A.1 The electrons of an atom are arranged in orbitals. Each orbital represents a different energy level.

isotopic tracers in following the movement of elements during biochemical reactions.

Electron Orbitals

Electrons determine the chemical reactivity of the atom. A neutrally charged atom has an equal number of protons and electrons. Thus, $_6C$ has 6 electrons in addition to its 6 protons. The electrons of an atom are not stationary, but rather move around (orbit) the nucleus. The electron shell, or orbital of electrons, is defined as the volume in which the electron can be found 90 percent of the time. The electrons in an atom are specifically arranged in a defined order, with electrons located at specified energy levels or shells (Figure A.1).

Each orbital has a maximum number of electrons that it can accommodate. The maximum number of electrons that can occupy a particular energy level increases with distance from the nucleus; for example, the first orbital holds a maximum of 2 electrons, the second orbital holds a maximum of 8 electrons, the third orbital holds up to 18 electrons, the maximum number of electrons in the fourth orbital is 32, and so forth. Orbitals are filled successively with the inner orbitals filling first; the outer orbitals are not always filled, but when orbitals are completely filled the elements are particularly stable. Chemical reactions and the formation of stable molecules is largely based on trying to achieve completely filled orbitals.

Molecules

Molecules are specific combinations of chemical elements, formed when atoms establish chemical bonds by sharing or exchanging electrons in their outer orbitals. The molecular weight of a molecule is simply the sum of the atomic weights of the atoms that compose that molecule. Alternatively the molecular weight may be given in daltons; 1 dalton is equal to the molecular weight of a single hydrogen atom which is 1.008.

The chemical bonds that hold the molecule together are formed by the electrons in the outermost orbitals (valence shells) of the atoms that form the molecule. Stable chemical bonds normally occur when atoms completely fill their valence shells. The number of electrons that an element must gain or lose to achieve a filled or stable outer orbital establishes the valency of that element. Elements with either 1, 2, or 3 electrons in their outer shells tend to donate electrons; those with 5, 6, or 7 electrons in the outer shell tend to accept electrons. Elements that normally gain electrons to fill their outer shells have negative valences; for example, oxygen has a valence of -2 because it needs 2 electrons to fill its outer shell. Elements that normally lose electrons to achieve a complete outer orbital have positive valences; for example, sodium has a valence of $+1$ because by losing 1 electron the remaining shells are complete (see Figure A.1).

The number of bonds that an element can form to produce a molecule depends on the number of electrons required by that atom to fill or deplete its outer electron shell. Carbon, the central atom in organic molecules, for example, can establish four bonds because it has 4 electrons in its outer electron shell, which can hold a maximum of 8 electrons. To fill their outer electron shells, nitrogen requires 3 electrons, oxygen requires 2 electrons, and hydrogen requires 1 electron; therefore, nitrogen can establish three chemical bonds, oxygen two bonds, and hydrogen one bond (see Figure A.1).

Types of Chemical Bonds

There are several types of chemical bonds that hold molecules together: ionic bonds, covalent bonds, and weak bonds. These bonds differ in their relative strengths (Table A.2) and serve different functions in biological systems. The ionic bond provides a mechanism whereby elements can easily dissociate from each other and move from place to place in aqueous solutions. The covalent bond provides the strength needed to support long-lived structural biochemicals—covalent bonds provide permanency to molecules. Weak bonds permit short-lived interactions between intact chemicals—they provide a mechanism for dynamic interactions between molecules.

Ionic Bonds

Ionic bonds are based on charge differences between chemical elements that develop when atoms donate or acquire electrons to fill their outer electron shells. For example, the salt sodium chloride represents a molecule formed by an ionic bond between the elements sodium and chlorine (Figure A.2). The sodium atom donates an electron forming a positively charged sodium ion (cation); the chlorine accepts the electron, forming a negatively charged chloride ion (anion); the positively charged sodium ions and negatively charged chloride ions are held together by electrostatic forces. Ionic bonds are relatively weak, and in aqueous solution the ionic bond is readily broken, resulting in the dissociation of charged ions. Consequently, ionic bonds are not strong enough to hold together the macromolecules of living systems.

Table A.2 Relative Strengths of Some Chemical Bonds

Bond	Relative Strength
Covalent (energy of interaction 30–100 kcal/mole)[a]	
H—H	1.0
C—C	0.8
C—H	1.0
C—O	0.8
C—N	0.7
Ionic (energy of interaction 10–20 kcal/mole)[a]	
Na · · · Cl	0.3
Hydrogen (energy of interaction 2–10 kcal/mole)[a]	
H— · · · O	0.1
H—N · · · N	0.1

kcal/mole = kilocalories per mole.

Covalent Bonds

In contrast to the ionic bond, where an electron is completely transferred from one element to another, covalent bonds form when elements share electrons (Figure A.3). The number of covalent bonds an element is capable of forming depends on its electronic structure. Each covalent bond involves the sharing of a pair of electrons, with each atom contributing one electron. In some cases atoms share two pairs of electrons, giving rise to a double bond. Once covalent bonds are formed, a relatively high amount of energy is required to break them; they provide the stability needed to establish the macromolecules required for the existence of all living organisms.

Covalent bonds exhibit differences in relative strength, depending on which elements are sharing electrons (see Table A.2). There also are differences in the distribution of electrons between atoms forming covalent bonds. If two atoms of the same element share electrons to form a molecule, the electrons are evenly distributed between the two atoms. This occurs in a molecule such as hydrogen (H_2). However, when a covalent bond forms between the atoms of two different elements, the electrons are shared unevenly; usually the electrons exhibit a greater affinity for one of the atoms and are drawn closer to that atom (Figure A.4). This gives rise to polarity, in which one of the atoms has a greater positive charge and the other atom a greater negative charge. The molecule is said to have a dipole resulting from the separation of charge, with the polarity of the covalent bond depending on the specific elemental atoms involved in establishing the bond. The polarity of covalent bonds establishes the basis for charge interactions and the formation of additional weak bonds between organic molecules.

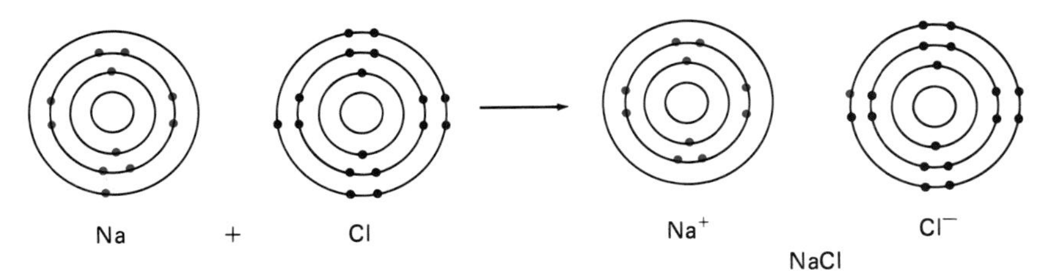

Figure A.2 The formation of ionic bonds involves the loss and gain of electrons. In the formation of NaCl, the chlorine atom gains an electron to fill its outer electron shell and the sodium atom loses an electron, so that all the remaining electron shells are filled. After the formation of the ionic bond, the sodium ion has a positive charge and the chloride ion a negative charge.

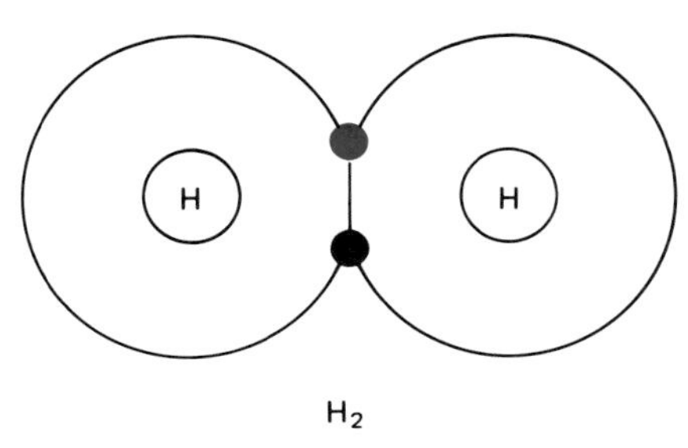

Figure A.3 The formation of the covalent bond involves the sharing of electrons. In methane each hydrogen atom shares an electron with a carbon atom, completing the orbitals of the carbon and hydrogen atoms.

Figure A.4 The covalent bonds of methylene chloride show the formation of dipole moment caused by the uneven sharing of electrons. The shared electron draws closer to the chlorine (Cl) than to the carbon (C), thus establishing a charge separation.

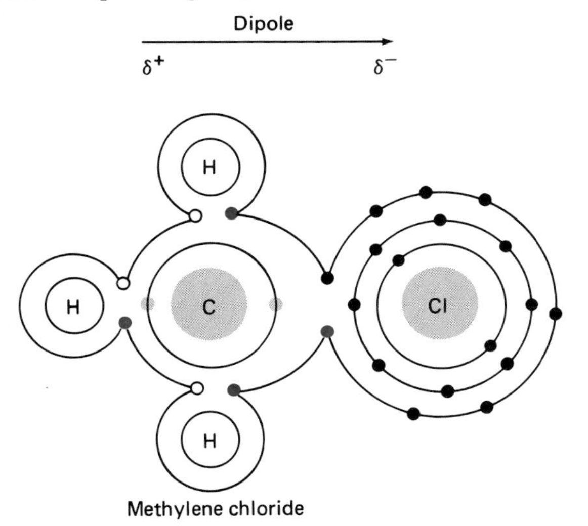

Weak Bonds

Molecules form a variety of weak bonds that result from charge interactions. The properties of water, for example, are determined by the dipole moment of the water molecule and its ability to form weak bonds. Water is an excellent polar solvent because it has a large dipole moment; that is, it has a large charge separation due to the unequal sharing of electrons between the oxygen and hydrogen atoms (Figure A.5). The internal separation of charge in the water molecule permits intermolecular associations with other polar molecules. The positive regions of solute molecules associate with the negative oxygen atom of water, and the negative regions of the solute associate with the positive hydrogen atoms of water. These electrostatic interactions are responsible for separating polar solute molecules, enhancing their dissolution in aqueous solution. Molecules that do not separate into polar groups or do not have charged polar regions do not dissolve in water. The solvent properties of water are essential for the occurrence of biochemical reactions in living systems.

Hydrogen Bonds

Weak intermolecular bonds may occur between any neighboring molecules. A hydrogen atom that is covalently bonded to oxygen or nitrogen within a molecule may be simultaneously attracted to a nitrogen or oxygen atom of a neighboring molecule, forming a weak link between the two molecules; this is known as a hydrogen bond (Figure A.6). The strength of the hydrogen bond is only about one tenth that of the covalent bond, but such weak bonds have very important functions in living systems. The properties of water and many of the macromolecules that are essential for life are determined by hydrogen bonding. Hydrogen bonds are necessary for maintaining the structure of DNA and proteins. The fact that weak hydrogen bonds can be readily

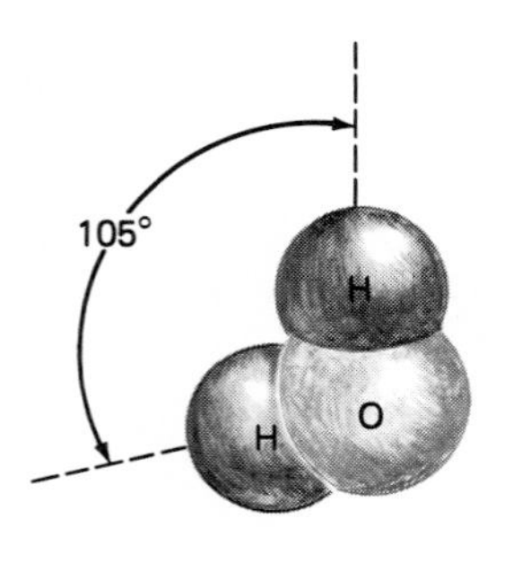

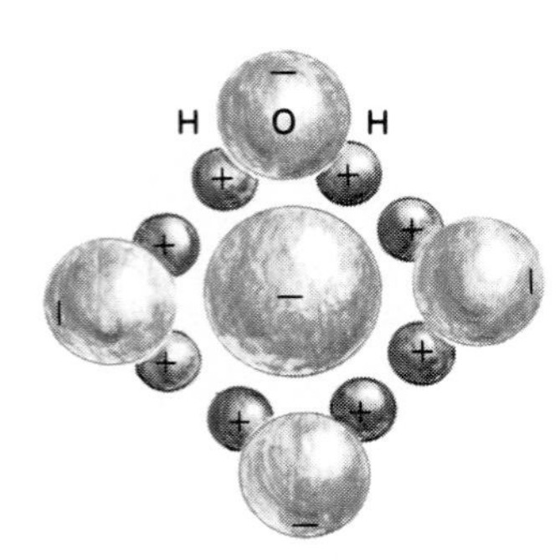

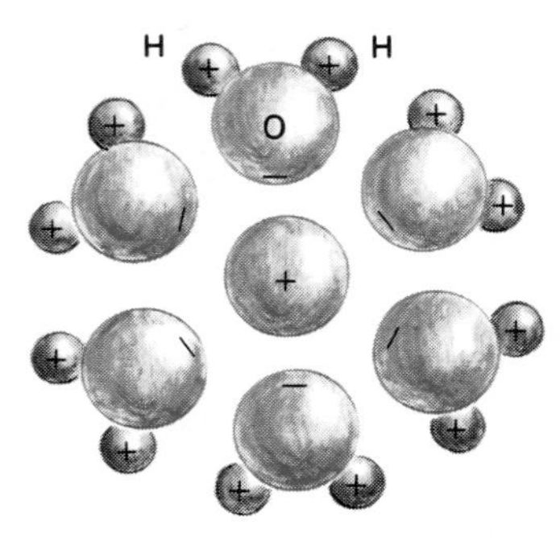

Figure A.5 The spatial arrangement of atoms in a water molecule results in a dipole moment because of unequal distribution of electrons between hydrogen and water. As a result, water is a good polar solvent, because water can surround both positively and negatively charged ions.

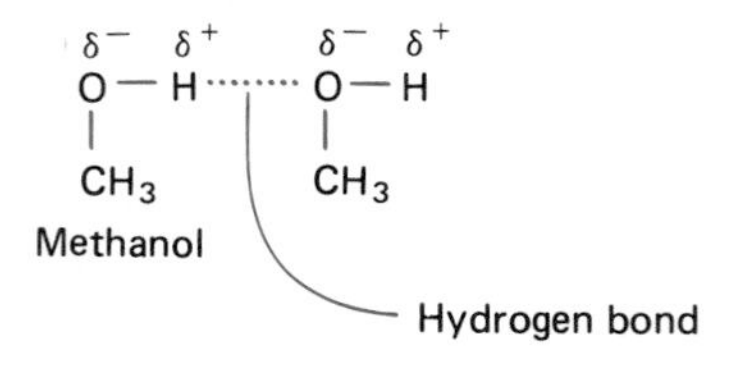

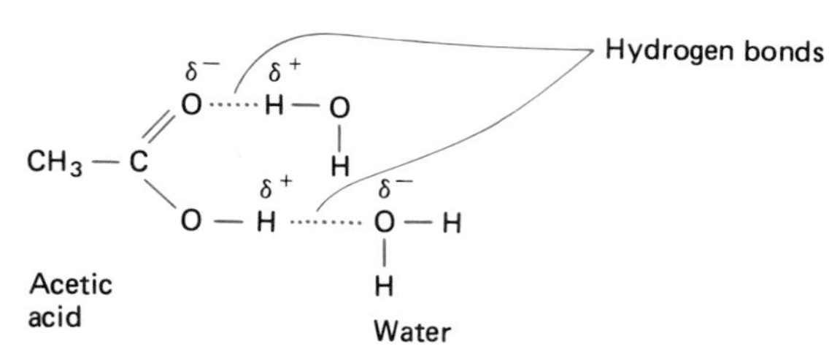

Figure A.6 Hydrogen bonding occurs between molecules as a result of charge interactions stemming from the dipole moments of the molecules.

formed and broken allows the flexibility needed for the various biochemical functions of DNA and protein macromolecules.

Hydrophobic Interactions

Hydrophobic interaction is a type of weak bonding that occurs because the interactions between polar groups exclude nonpolar molecules, or nonpolar portions of large molecules, creating hydrophobic interactions between nonpolar molecules. Hydrophobic interactions are very weak. Although they are weak bonds, hydrophobic interactions are very important for determining the structural orientations of macromolecules. For example, the separation of the polar or hydrophilic ("water-loving") and nonpolar or hydrophobic ("water-fearing") ends of phospholipids establishes the basis for the spontaneous formation of biological membranes and contributes to the selective permeability of biological membranes.

Organic Chemicals

Organic chemicals, by definition, contain carbon, which forms the backbone structure of the molecule; all carbon-containing molecules are organic except for those, such as carbon dioxide and carbon monoxide, that do not also contain hydrogen bonded to the carbon. The element carbon is uniquely suited for establishing the macromolecules needed by living systems. Carbon atoms are able to establish strong covalent bonds with hydrogen, oxygen, nitrogen, and other carbon atoms—these four elements are the most abundant elements found in all organisms. Additionally carbon–carbon bonds are strong enough to hold together the very large macromolecules of living systems. Carbon can form double bonds with oxygen; double bonds are critical in molecules such as carbon dioxide, which is transformed by living systems into the organic molecules that are necessary for the support of life.

Table A.3 **Common Functional Groups of Organic Molecules and Some Representative Classes of Molecules in Which They Occur**

Group	Name	Class of Molecules
R—OH	Hydroxyl group	Alcohols, carbohydrates
R=O (with OH)	Carboxyl group	Carboxylic acids, fatty acids, amino acids
R=O (with R)	Keto group	Ketones, sugars
R—C=O (with H)	Aldehyde group	Aldehydes, sugars
R—C—H (with H, H)	Methyl group	Hydrocarbons, lipids
R—N—H (with H)	Amino group	Amines, amino acids
R—P=O (with OH, OH)	Phosphate group	Organic phosphates, phospholipids, nucleotides
R—SH	Sulfhydryl group	Mercaptans, a few amino acids

Functional Groups of Organic Molecules

The organic molecules that enter into biochemical reactions generally possess functional groups, and the major classes of organic molecules comprising living systems are based on the presence of particular functional groups (Table A.3). Functional groups are distinct groups of atoms other than carbon and hydrogen that occur in organic compounds. Functional groups are attached to a hydrocarbon residue designated R. The functional group confers specific properties on an organic molecule. An important property of a functional group is its ability to react with a different functional group of another molecule. The interactions between functional groups can involve the formation of weak bonds. The nature of these weak bonds ensures that only complementary functional groups can react to form a bond, establishing a specificity of chemical reactions.

Acids, Bases, and pH. Among the most important functional groups of organic molecules are the groups that dissociate in water solution to produce hydrogen ions (H^+)—acids—and those that dissociate in water solution to produce hydroxyl ions (OH^-)—bases; acid–base reactions form the basis for the formation of many of the biochemicals of living systems (Figure A.7). In organic compounds the carboxyl group is the common acidic group and the hydroxyl group the common basic functional unit. The concentration of hydrogen ions (written H^+) determines the **pH** of a solution, which is defined as $-\log H^+$. Pure water contains 10^{-7} moles of hydrogen ions per liter and thus has a pH of 7.0. At pH 7.0 the concentrations of H^+ and OH^- are equal; water, therefore, has a neutral pH—pH values below 7 are acidic, and those above 7 are basic. The pH affects the degree of dissociation of acids and bases, which has a marked influence on the properties of biochemical molecules. It is because of this fact that the pH of the solutions surrounding organisms is critical to their survival.

Structure of Organic Molecules

Organic molecules have three-dimensional shapes, and these shapes, together with the functional groups, are extremely important in determining the function or use of a molecule. Because the shapes and functional groups are so important and slight changes in either of these two parameters can significantly alter the properties of a molecule, several

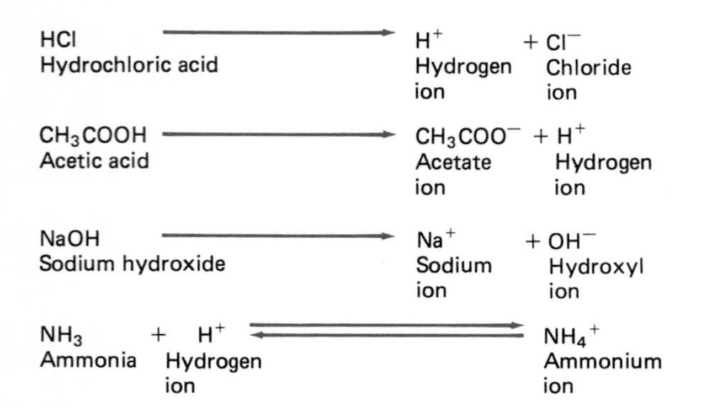

Figure A.7 Diagram of dissociation of acids and bases.

conventional systems have been designed to specify the position and location of functional groups for the various classes of biologically important organic molecules. Although the specific numbering systems for organic molecules vary for different classes of molecules, they all permit the unambiguous specification of all the atoms in the molecule. An example of the numbering of a simple hydrocarbon and a carboxylic acid is shown in Figure A.8.

In some molecules, in addition to the formal numbering system, the term **α-carbon** is used to describe the carbon atom to which the functional group that typifies that class of compounds is directly attached, and the succeeding carbons are also designated by continuing the Greek alphabet. By establishing formal conventions, the exact arrangement of atoms within the biochemical can be specified, permitting discussion of the properties and function of that specific molecule in the biochemistry of living systems.

Isomerism. Molecules with identical molecular formulas but different spatial arrangements are called **isomers**. There are several types of isomers that are important in biological molecules. The carbon atoms of isomers can be bonded at different positions, producing structures with different spatial arrangements (**positional isomers**). For example, one-glucose phosphate and 6-glucose phosphate both have the same molecular formula, but have different properties that cause them to function differently in biological systems; in one case the phosphate is attached to the number 1 carbon atom and in the other case to the number 6 carbon atom of the glucose molecule (Figure A.9).

Figure A.8 An example of the numbering of a carbon chain of a hydrocarbon and a carboxylic acid and the use of Greek letters to designate position relative to functional group.

$$\overset{6}{H_3C}-\overset{5}{CH_2}-\overset{4}{CH_2}-\overset{3}{CH_2}-\overset{2}{CH_2}-\overset{1}{CH_3}$$

Hexane

$$\underset{\epsilon}{\overset{6}{H_3C}}-\underset{\delta}{\overset{5}{CH_2}}-\underset{\gamma}{\overset{4}{CH_2}}-\underset{\beta}{\overset{3}{CH_2}}-\underset{\alpha}{\overset{2}{CH_2}}-\overset{1}{COOH}$$

Hexanoic acid

Isomers with markedly different properties can occur, even when identical substituent groups are bonded at identical positions. This can occur because organic molecules are three-dimensional, allowing compounds that have the same structural formulas still to have different spatial arrangements.

The spatial arrangements of the atoms in the molecule are defined by the configuration and conformation of the molecule. **Configuration** describes the relative arrangement in space of the atoms that make up the molecule, and **conformation** describes the different spatial arrangements of the molecule that result from rotation about single covalent bonds (Figure A.10).

To understand the concept of differing spatial arrangements consider that the image we see in a mirror is reversed—it is a "mirror image." In a related sense, organic compounds with the same molecular formula and the same skeletal structure that have different three-dimensional arrangements of the atoms in space are termed **stereoisomers**. Organic molecules can have two different forms (configurations) about a carbon atom with different groups attached (**asymmetric carbon atom**) that represent mirror images of the molecule; the mirror images of the molecule are not superimposable and represent truly different configurations (Figure A.11).

The different configurations of the stereoisomers can also have differing **optical activities**, and when a plane of polarized light shines through the two different isomers, it will be rotated in opposite directions. The direction of rotation of the light by

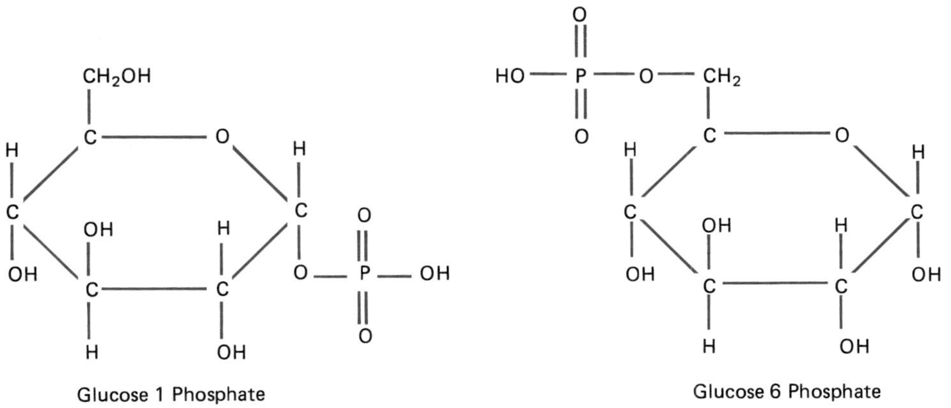

Glucose 1 Phosphate

Glucose 6 Phosphate

Figure A.9 Positional isomers have identical molecular formulas, but the substituents are located at different positions. In this example of glucose phosphate, the only difference is the position of the phosphate group. Nevertheless, these geometric isomers have different physicochemical properties.

the optical isomers is termed dextrorotatory (+) if it is bent to the right, or levorotatory (−) if it is bent to the left. The letters D and L were assigned to describe the absolute configurations of the two optically active isomers of glyceraldehyde. The configurations of the stereoisomers of carbohydrates and amino acids can be related to the arbitrarily chosen reference molecule glyceraldehyde, the smallest sugar that has an asymmetric carbon atom (Figure A.12). All stereoisomers related to the ab-

Figure A.10 Organic molecules can vary in their configuration and conformation, as shown in this illustration. (Note that the carbon molecules within the ring structures in this diagram are not specifically shown but rather are understood to occur at each of the apices within the ring—this is a standard chemical abbreviation.) The stabilities of the differing configurations and conformations vary, and in each case there is a preferred (most stable) form. (A) The chair and boat conformations of cyclohexane. (B) The two possible configurations of the chair conformation of D-glucose.

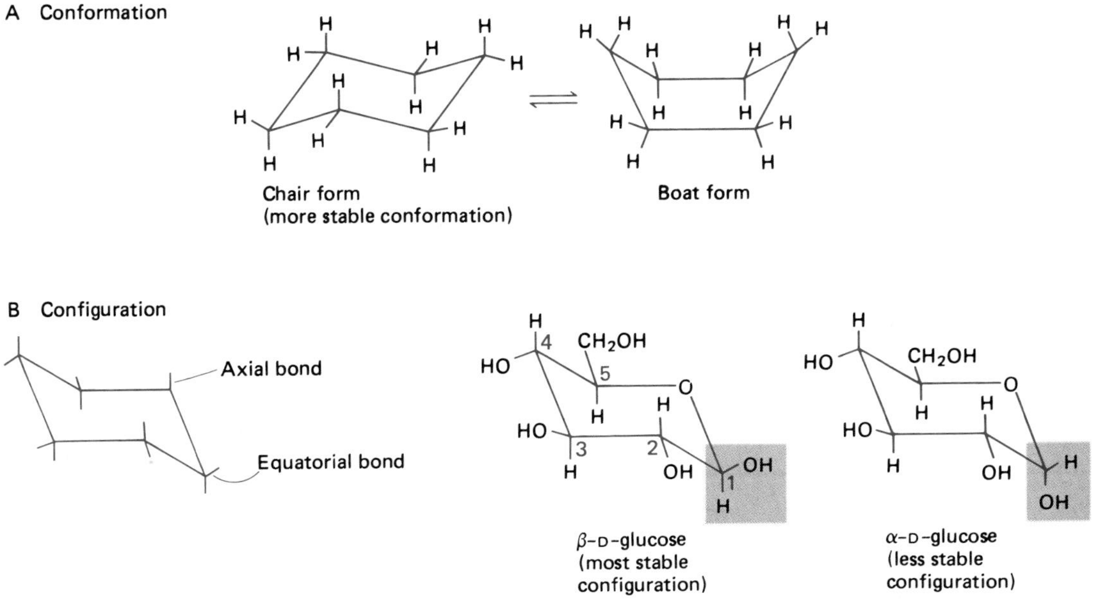

A Conformation

Chair form
(more stable conformation)

Boat form

B Configuration

Axial bond

Equatorial bond

β-D-glucose
(most stable
configuration)

α-D-glucose
(less stable
configuration)

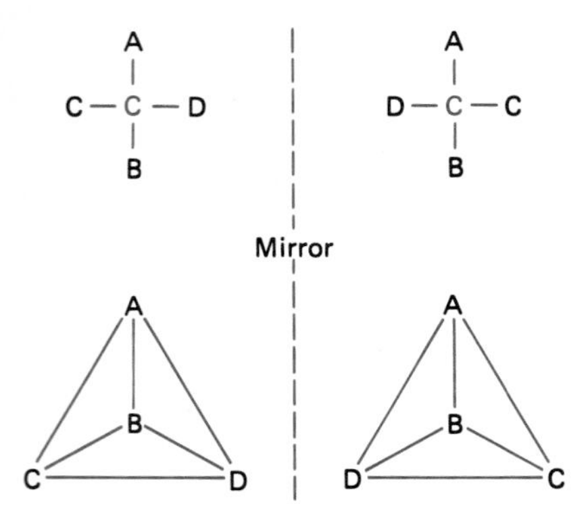

Figure A.11 An asymmetric carbon allows molecules to exist in two different forms that are mirror images.

Figure A.12 The D and L forms of a glyceraldehyde molecule and its relation to L-alanine. In determining the absolute configuration of a carbohydrate, the D designation is used to indicate that the groups H, CHO, and OH, in that order, are situated in a clockwise fashion about the asymmetric carbon atom, when the CH_2OH group is directed away from the viewer. The designation L is used if the order is counterclockwise. In determining the configurations of amino acids, we still use glyceraldehyde as the reference, with the NH_2 group subsituting for OH and the CO_2H group for CHO. We should note that according to this system some D forms are levorotory, designated D-(−), and some L forms are dextrorotatry, designated L-(+).

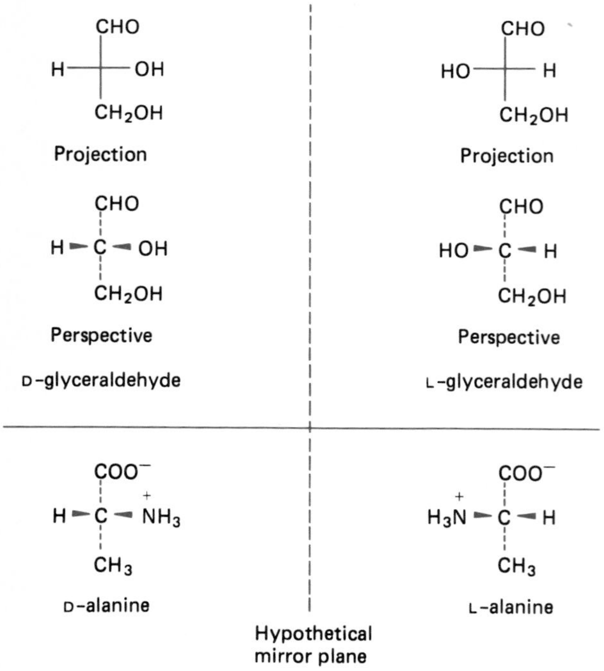

solute configuration of L-glyceraldehyde are designated L, and those related to D-glyceraldehyde designated D, regardless of the direction of rotation of polarized light caused by the isomer.

The reason that designating the configuration of a molecule is important is that organisms differentiate between stereoisomers. Many biochemicals found in living systems contain either D or L forms of a molecule but not both; for example, proteins contain L-amino acids exclusively. Different stereoisomers of a molecule exhibit different reactivities in chemical reactions; generally, only one of the optical isomers is capable of participating in a given biochemical reaction. Thus, organisms select among the many different possible forms of molecules; they organize molecules in specific arrangements that constitute the structures of living systems.

Macromolecules

In the biochemical molecules of living systems, elements are linked together by covalent bonds to form relatively high molecular weight macromolecules with differing functional groups and spatial arrangements. Large polymeric molecules are composed of multiple (poly) units of individual molecules called monomers. There are four major classes of macromolecules that make up the structural and functional units of all living systems: carbohydrates, lipids, proteins, and nucleic acids (Table A.4). Each of these classes of macromolecules are composed of subunits that are covalently bonded together.

Carbohydrates

Carbohydrates (carbon plus water) have the basic chemical formula $(CH_2O)_n$; they include the simple sugars (monosaccharides), such as glucose, and the macromolecules formed from such simple monosaccharides. By definition, carbohydrates are either polyhydroxy aldehydes or polyhydroxy ketones or molecules that yield polyhydroxy aldehydes or polyhydroxy ketones upon hydrolysis (decomposition owing to the addition of a water molecule); carbohydrates must have at least three carbon atoms.

Those carbohydrates that cannot be hydrolyzed are called monosaccharides (Figure A.13). As the number of carbons increases, the prefix of the simple sugar name changes to reflect the number of carbons (tetrose, four carbon; pentose, five carbon;

Table A.4 The Four Major Classes of Macromolecules

Carbohydrates	A polyhydroxy aldehyde, a polyhydroxy ketone, a derivative of these, or a substance that yields them upon hydrolysis. Used to store and to transport energy in living systems.
Lipids	Various substances that have in common their insolubility in water, including fats and oils. Major components of biological membranes; also used for storage of energy.
Proteins	High molecular weight macromolecules composed of amino acids linked together by peptide bonds. Important structural components of biological systems and catalysts of biochemical reactions.
Nucleic acids	Composed of purine and pyrimidine bases, pentose sugars, and phosphoric acid. The molecule ATP is the central carrier of energy within the cell. DNA and RNA are large macromolecules composed of nucleic acid bases linked by phosphate-ester bonds; used to store and to transmit genetic information within biological systems.

Figure A.13 Structural formulas of common monosaccharides.

```
   1 CHO
     |
H — 2C — OH
     |
   3 CH2OH
D-glyceraldehyde
```

```
   1 CHO
     |
H — 2C — OH
     |
H — 3C — OH
     |
   4 CH2OH
D-erythrose
```

```
    1 CHO           1 CHO            1 CHO            1 CHO
      |               |                |                |
 H — 2C — OH     H — 2C — H      HO — 2C — H      H — 2C — OH
      |               |                |                |
 H — 3C — OH     H — 3C — OH      H — 3C — OH     HO — 3C — H
      |               |                |                |
 H — 4C — OH     H — 4C — OH      H — 4C — OH      H — 4C — OH
      |               |                |                |
    5 CH2OH         5 CH2OH          5 CH2OH          5 CH2OH
 D-ribose        D-deoxyribose    D-arabinose      D-xylose
```

```
    1 CHO           1 CH2OH          1 CHO            1 CHO
      |               |                |                |
 H — 2C — OH        2 C=O        HO — 2C — H      H — 2C — OH
      |               |                |                |
HO — 3C — H     HO — 3C — H      HO — 3C — H     HO — 3C — H
      |               |                |                |
 H — 4C — OH     H — 4C — OH      H — 4C — OH     HO — 4C — H
      |               |                |                |
 H — 5C — OH     H — 5C — OH      H — 5C — OH      H — 5C — OH
      |               |                |                |
    6 CH2OH         6 CH2OH          6 CH2OH          6 CH2OH
 D-glucose       D-fructose       D-mannose        D-galactose
```

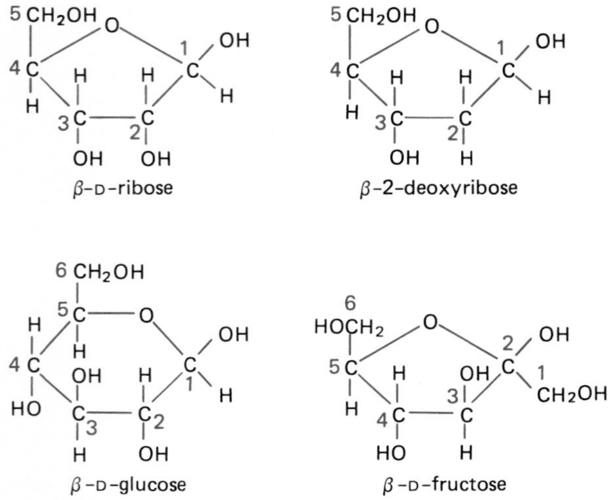

β-D-ribose
β-2-deoxyribose
β-D-glucose
β-D-fructose

Figure A.14 The ring structure of some representative carbohydrates.

hexose, six carbon; and so on). In aqueous solution the pentoses and hexoses (the five- and six-membered monosaccharides) are actually found as a cyclic or ring structure more of the time than as a straight chain (Figure A.14). As a result of ring formation, the functional group (either an aldehyde or a ketone) becomes an extra asymmetric carbon, and therefore, in addition to being designated as D or L, these compounds are designated α or β.

The simple sugars or monosaccharides may be linked to form larger polymeric carbohydrate molecules—a disaccharide contains two monosaccharide units, an oligosaccharide contains from two to ten monosaccharide units, and a polysaccharide contains more than ten monosaccharide units. The bond formed between monosaccharides to form disaccharides is called a glycosidic bond (Figure A.15). The glycosidic bond is specified by the position numbers of the carbons that are linked by the bond and by the three-dimensional orientation of the bond. When a glycosidic bond forms, the functional aldehyde or ketone group of one of the monosaccharide units is generally left free and therefore remains reactive; the disaccharides sucrose and trehalose are exceptions in which the

Figure A.15 The formation of a glycosidic bond showing the α and β orientations.

Glucose
Glucose
α-1,4 Glycosidic bond
α Maltose
Glucose
Glucose
β-1,4 Glycosidic bond
β Cellobiose
Glucose
Fructose
α-Glycosidic bond
Sucrose
Galactose
Glucose
β-Glycosidic bond
Lactose

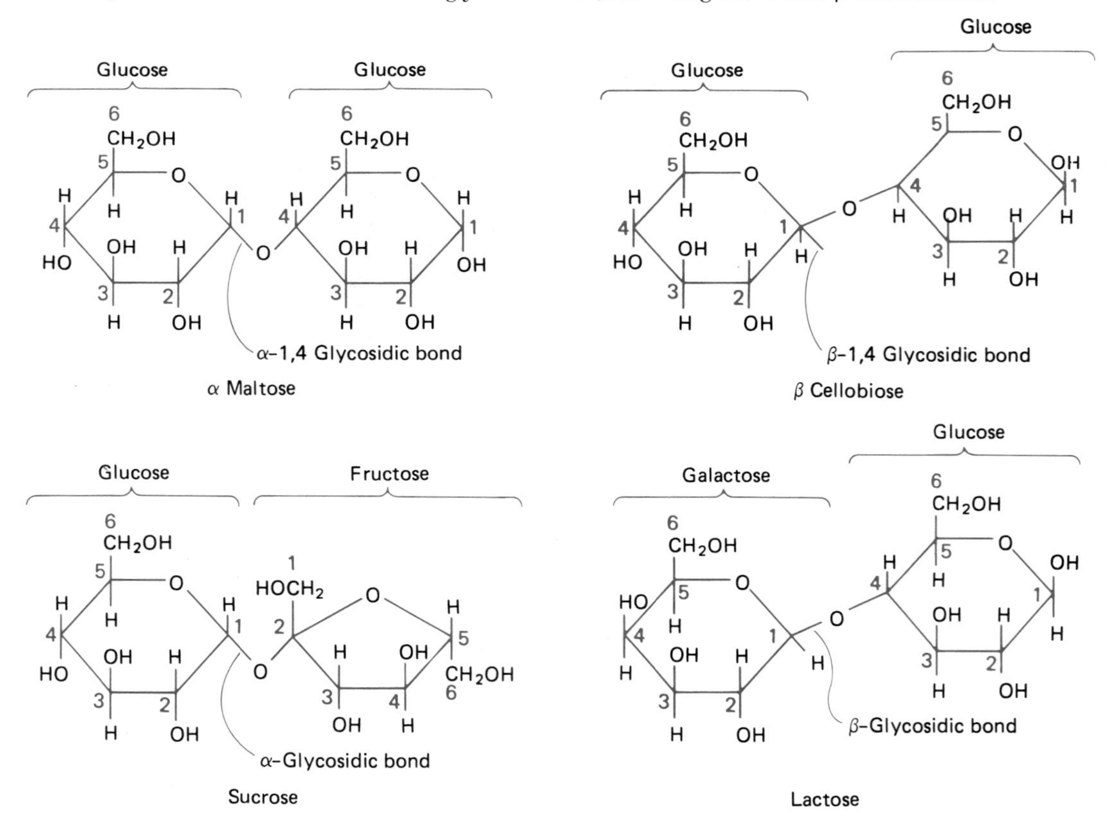

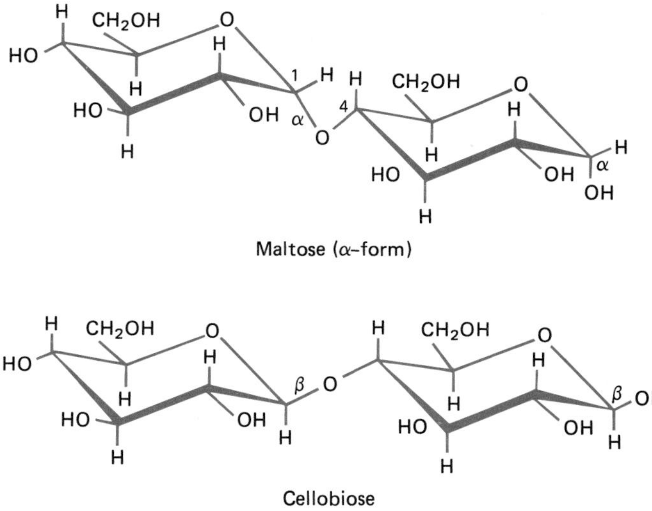
Maltose (α-form)
Cellobiose

Figure A.16 Diagram of structures of maltose and cellobiose.

Figure A.17 Diagram of several polymers of glucose.

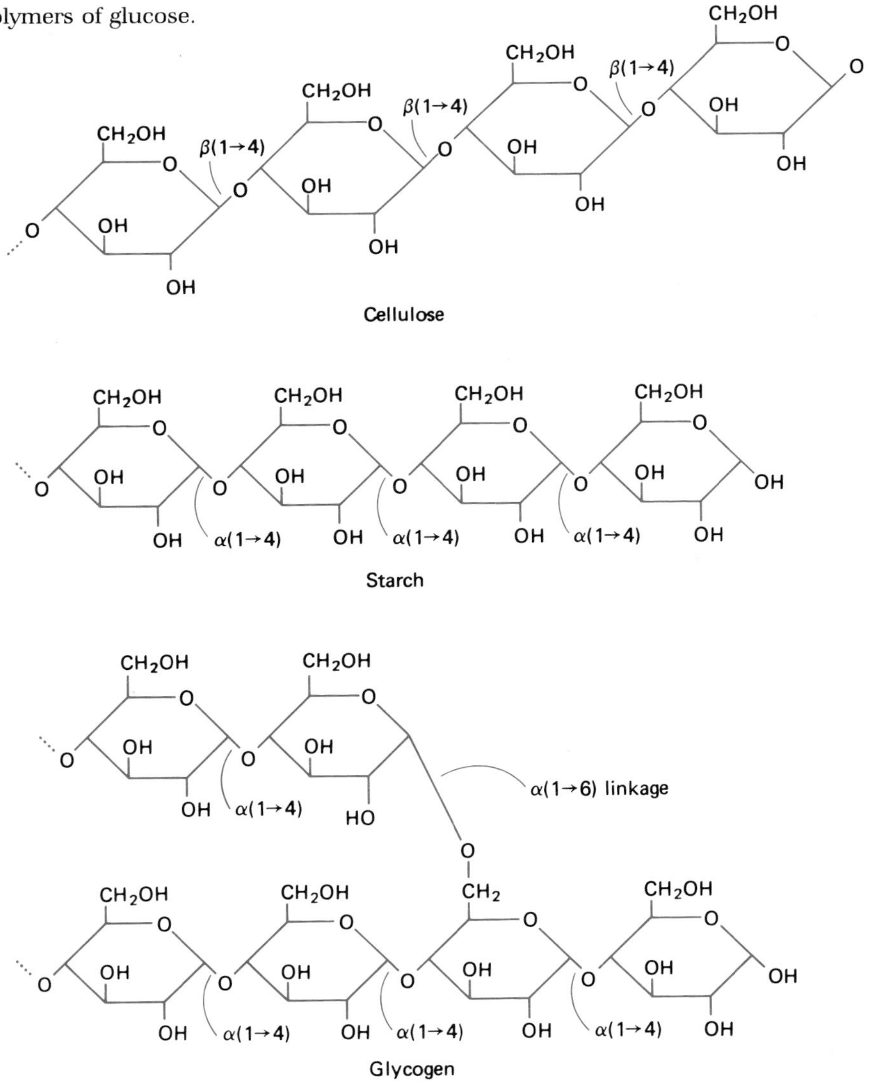
β(1→4)
Cellulose
α(1→4)
Starch
α(1→6) linkage
Glycogen

reactive groups of both monosaccharides are linked. The same monosaccharides can form various disaccharides and polysaccharides by forming glycosidic bonds with different orientations; glycosidic bonds are covalent bonds formed by the elimination of water when the alcohol group of one carbohydrate molecule reacts with the aldehyde group of another carbohydrate. For example, maltose is formed by an α-1-4 linkage of two glucose units, and cellobiose is formed by a β-1-4 linkage of two glucose units (Figure A.16). The polysaccharide cellulose represents a polymeric chain of glucose monosaccharides that are linked with β-1-4-glycosidic linkages (Figure A.17). The 1-4 indicates that the bond is between the number 1 carbon atom of one of the glucose molecules and the number 4 carbon atom of the other glucose molecule; the α and β designate the three-dimensional orientation of the bond.

Lipids

Lipids are important energy storage molecules. Lipids are water-insoluble biochemicals that are soluble in nonpolar solvents, such as chloroform. Many organisms are unable to store large amounts of carbohydrates and convert carbohydrates to lipids—fats and oils—for energy storage. Fats are even more energy-rich than carbohydrates because they have a higher proportion of the C—H bonds; C—H bonds store more energy than the C—O bonds, which are more abundant in carbohydrates than lipids. Lipids are a major constituent of biological membranes, and the solubility properties of lipid molecules enable them to function as the components of these semipermeable barriers.

Types of Lipids. There are several types of lipids, and these lipids can be divided into two major classes: complex lipids, which normally are composed of fatty acids bonded to an alcohol, and simple lipids, such as steroids. The most abundant complex lipids are the fats that are combinations of fatty acids bonded to glycerol. The linkage of the fatty acid to the alcohol is by an ester bond formed between the carboxyl group of the acid and the alcohol group of the glycerol molecule; an ester bond is a covalent bond that is the product of the elimination of water in a reaction between an acid and an alcohol molecule. In triglycerides three fatty acids, which may be the same or different, are linked to the three carbons of the glycerol molecule. The nonpolar nature of the relatively long-chain fatty acids of these lipids causes them to exhibit hydrophobic interactions with water.

Phospholipids, the principal lipids of biological membranes, are derivatives of glycerol in which fatty acids are bonded to only two of the carbons of the glycerol molecule (Figure A.18). The third carbon is linked by an ester bond to phosphate. The negatively charged phosphate acts as a hydrophilic head of the lipid molecule and is attracted to water, and the hydrocarbon portion of the fatty acid acts as a hydrophobic tail. The phosphate may further be linked by a second ester bond to another alcohol, forming more complex phospholipid molecules. Lipid molecules may combine with carbohydrates to form glycolipids or with proteins to form glycoproteins.

In contrast to the complex lipids, which typically contain saturated fatty acids (all single bonds between carbons), most of the simple lipids are

Figure A.18 The structural formula of a phospholipid, showing both the hydrophilic and hydrophobic nature of this class of compound.

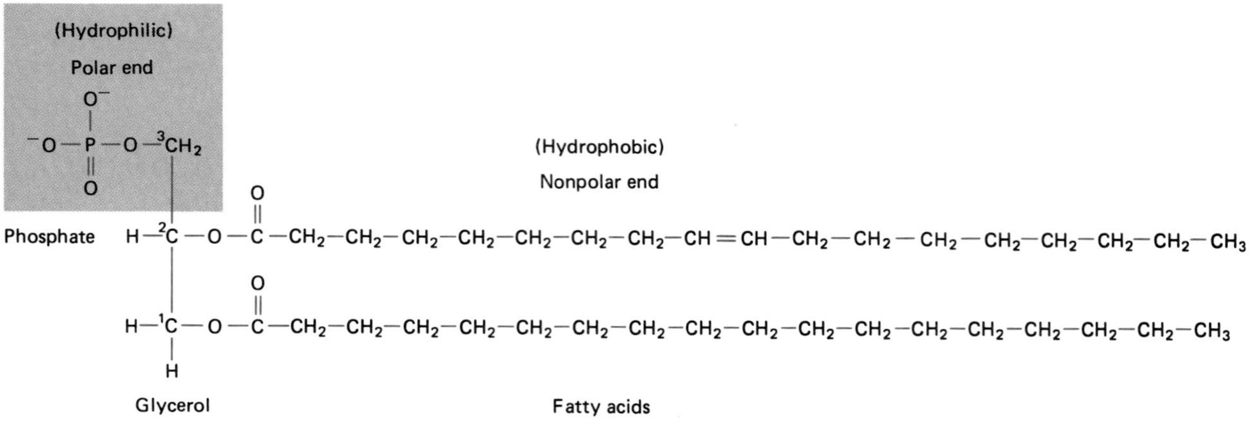

(Hydrophilic)
Polar end

$^{-}O-P(=O)(O^{-})-O-{}^{3}CH_2$

(Hydrophobic)
Nonpolar end

Phosphate

$H-{}^{2}C-O-C(=O)-CH_2-CH_2-CH_2-CH_2-CH_2-CH_2-CH_2-CH=CH-CH_2-CH_2-CH_2-CH_2-CH_2-CH_2-CH_2-CH_3$

$H-{}^{1}C(H)-O-C(=O)-CH_2-CH_2-CH_2-CH_2-CH_2-CH_2-CH_2-CH_2-CH_2-CH_2-CH_2-CH_2-CH_2-CH_2-CH_2-CH_2-CH_3$

Glycerol

Fatty acids

unsaturated hydrocarbon molecules (double bonds between some of the carbon atoms). Steroid molecules contain four fused rings with double bonds (Figure A.19).

Proteins

Proteins are large macromolecules with molecular weights of 6000 to nearly a million. These molecules serve as structural, regulatory, and catalytic components of the cell. Proteins are composed of long chains of amino acids linked by peptide bonds and are extremely important molecules in biological systems, often composing 50 percent of the cell's dry weight. The amino acid subunits of proteins possess two functional groups; an amino group and a carboxylic acid group. Only 20 amino acids normally are found in the protein macromolecules. All have an L configuration, and in each of these amino acids both the amino group and the carboxylic acid group are linked to the same α-carbon atom (Figure A.20). Each of the amino acids also has an additional chemical group bonded to the α-carbon atom that is designated as an R group. These R groups distinguish the amino acids and establish the chemical properties of the protein molecule.

The bonding of the carboxyl group of one amino acid to the amino group of another amino acid establishes the peptide bond that characterizes the protein molecule (Figure A.21). The formation of the peptide bond can be considered as a condensation reaction between the amino group of one amino acid and the carboxyl group of another amino acid, with the elimination of water. Two amino acids linked together by a peptide bond are termed a dipeptide; three amino acids so linked, a tripeptide; and the macromolecules formed by linking many amino acids by peptide bonds are called polypeptides. A protein may be composed of one or more polypeptide chains of amino acids. In the middle of the polypeptide chain, the carboxyl and amino groups are linked to each other. When a polypeptide forms, one end of the molecule will have a free carboxylic acid group, and the other end of the molecule will have a free amino group (Figure A.22). The amino free end is termed the amino or N-terminal and the carboxyl free end, the carboxyl or C-terminal. These two ends of a peptide chain are biochemically distinguishable and impart an ordered direction to the protein molecule. The ability of biochemical molecules to exhibit directionality is important in their functioning in biological systems.

The protein molecule also has a three-dimensional structure that determines its biological properties (Figure A.23). The sequence of amino acids in the polypeptide forms the primary structure of a protein. It is the number and order of specific amino acids within the peptide chains of the protein—that is, the primary structure—that establish the properties of the protein macromolecule. All proteins have primary structure regardless of whether they also have secondary, tertiary, or quaternary structure.

The specific conformational pattern of a protein, stabilized by hydrogen bonding, is known as the secondary structure. The secondary structure is a folded or twisted configuration of part of the primary sequence that is common to many proteins. Often the long polypeptide chain of a protein molecule is twisted so that it forms a helix. The nature of the peptide linkage and the conformation of the amino acids result in the spontaneous formation of an alpha helix or a beta sheet for most polypeptides.

Some protein molecules also exhibit folding of the polypeptide chains that establishes tertiary structure. The tertiary structure is primarily determined by the R groups of the various amino acids of the polypeptide chain. Interactions between the R groups causes the protein to assume a particular shape. This occurs because when a polypeptide chain forms, the interactions between amino acids that are quite distant from each other along the polypeptide sequence stablize the molecule, establishing the highly complex and coiled structure of

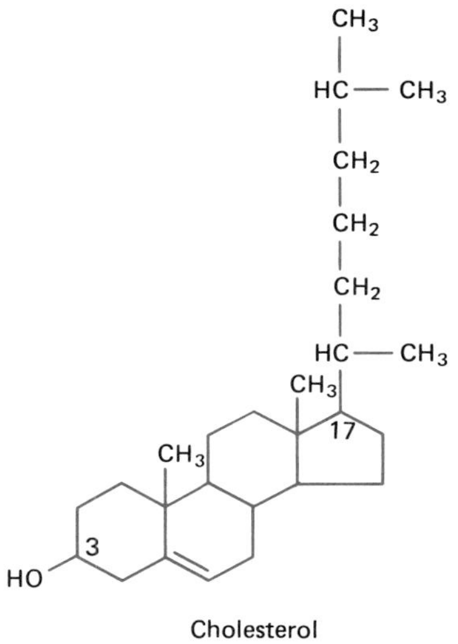

Figure A.19 The structural formula of cholesterol, which is an example of a steroid structure.

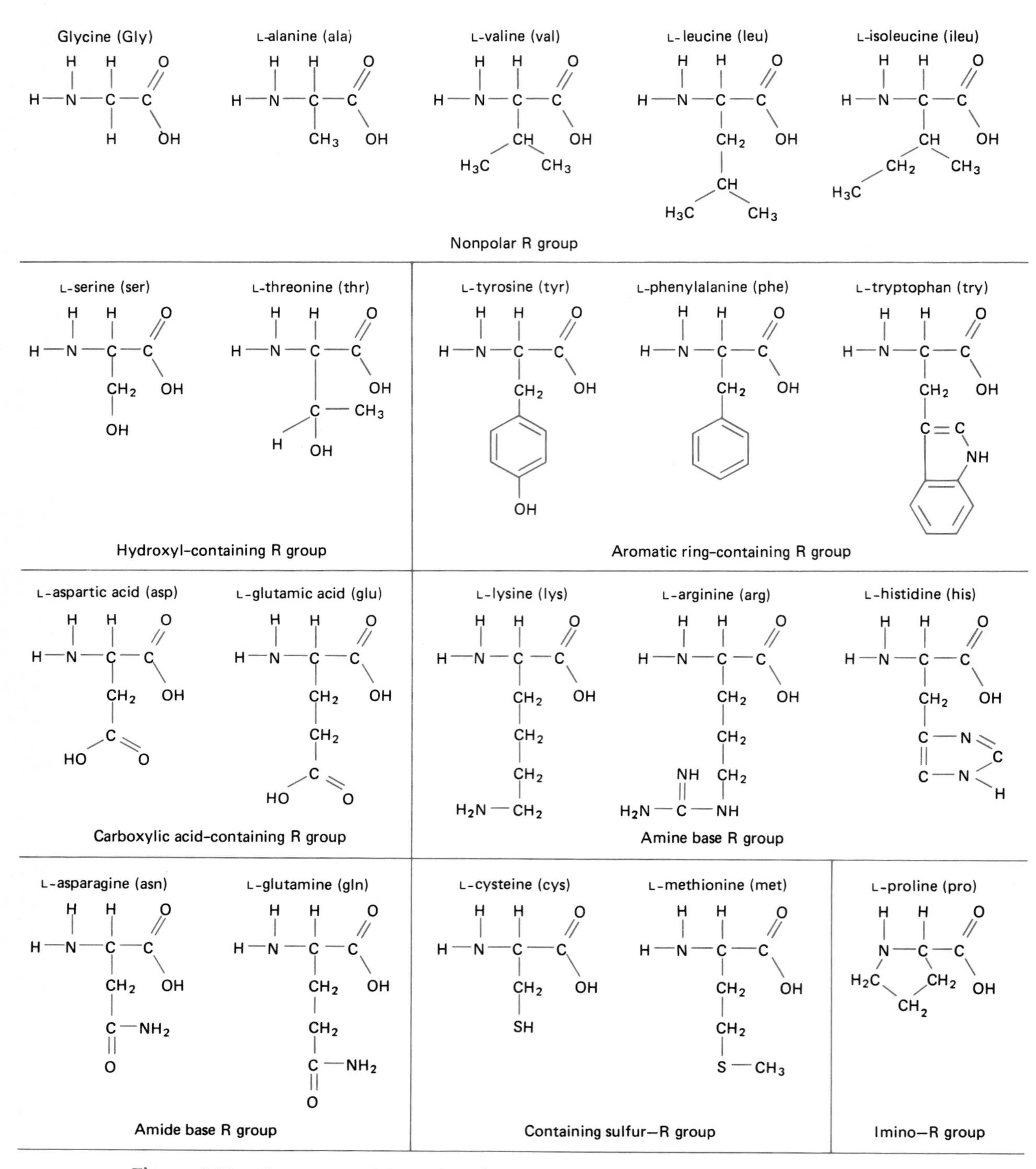

Figure A.20 The structural formulas of 20 common amino acids. Each is an L-α-amino acid. The structures differ in the other constituents.

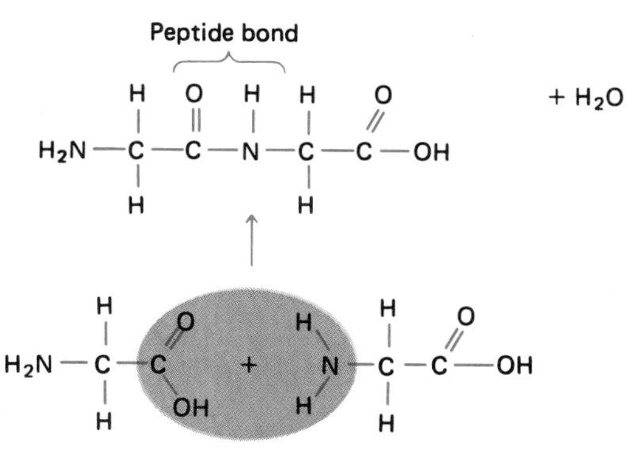

Figure A.21 The peptide bond links amino acids into polymeric units, binding together the units of proteins.

the protein. Folding of the polypeptide molecule is also stabilized by interactions of the sulfhydryl groups of the sulfur-containing amino acids. The interactions of sulfhydryl groups of the amino acid cysteine form disulfide bridges that contribute to the protein molecule's tertiary structure. The tertiary conformation is very important in determining the properties of a protein, and if altered by heat, changes in pH, or changes in ionic concentration, the properties of the protein can change drastically. Some proteins are composed of more than one polypeptide molecule. The quarterary structure of a protein reflects the interactions of two or more polypeptide chains and describes how the peptide chains are arranged or clustered in space.

The high specificity exhibited by enzymes is based on their biochemical composition, which establishes their proper three-dimensional configuration, and even slight changes, such as the change in the order of relatively few amino acids in a long polypeptide chain, can alter the activity of an enzyme. Almost all enzymes are proteins. Active enzyme molecules must have the proper three-dimensional conformation to exhibit catalytic activity. Denaturation of the protein molecule, that is, disruption of the three-dimensional shape, eliminates the activity of an enzyme molecule. Similarly immunoglobulins (antibodies) are proteins whose properties depend upon their quaternary structure.

Nucleic Acids

Nucleic acids are large macromolecules composed of nucleotides linked together to form a polymer. The nucleotide subunits of nucleic acids are themselves composed of smaller units. A nucleotide is composed of three covalently linked individual units: a nitrogenous base (a heterocyclic ring structure containing nitrogen as well as carbon, sometimes referred to as a nucleic acid base or simply as the base), a five-carbon monosaccharide (either ribose or deoxyribose), and a phosphate group (Figure A.24). The nitrogenous base is linked to carbon-1 of the monosaccharide by an N-glycosidic bond. The nitrogenous base plus the monosaccharide, without the phosphate group, is called a nucleoside. The phosphate group forms a diester bond between the carbon-3 of one monosaccharide and the carbon-5 of another monosaccharide unit, and it is this bond that links the molecule together (Figure A.25).

Two types of nucleic acids occur in biological

Figure A.22 A polypeptide has a free amino end and a free carboxyl end.

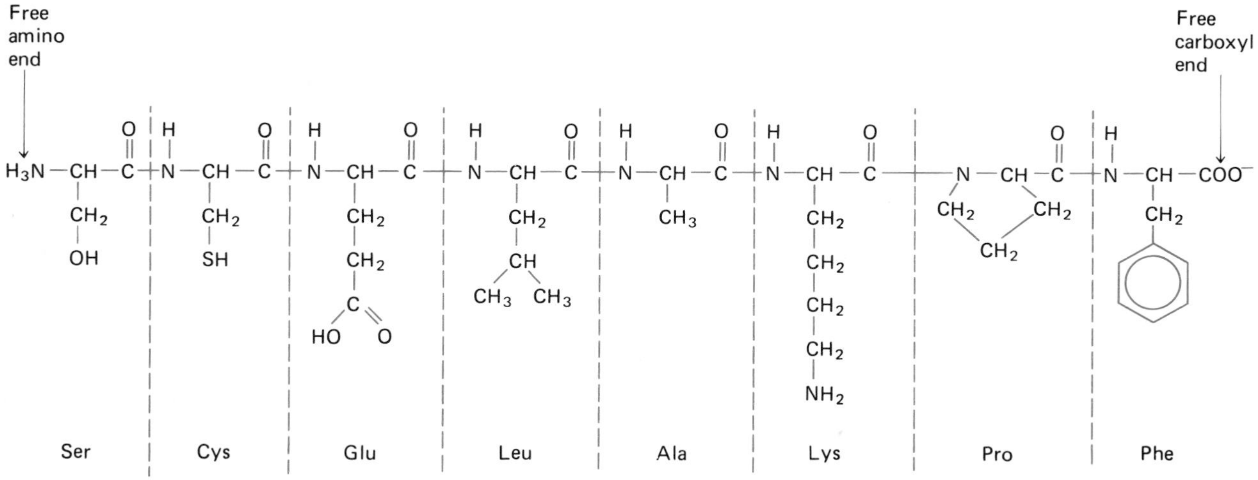

Free amino end

Free carboxyl end

Ser | Cys | Glu | Leu | Ala | Lys | Pro | Phe

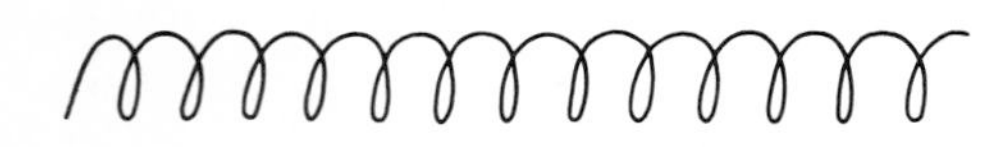

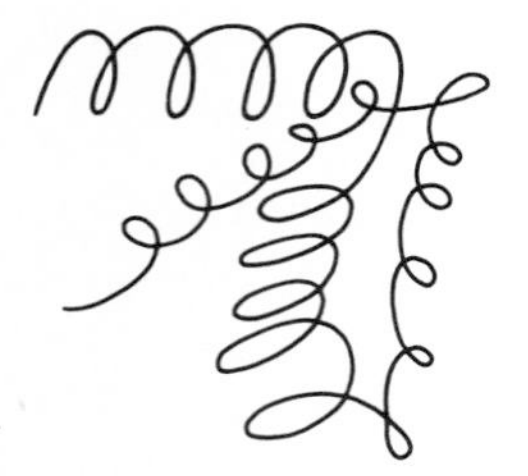

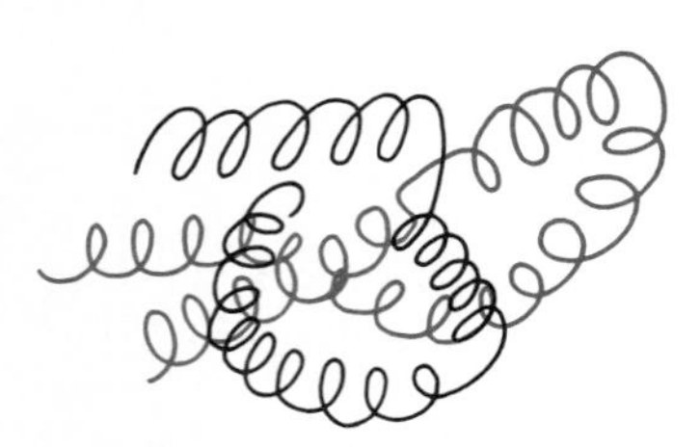

Figure A.23 Proteins have primary, secondary, tertiary, and quaternary structures.

Figure A.24 The structural formula of a nucleotide, the basic unit of the informational macromolecules of living systems.

Base + sugar
nucleoside
Base
Sugar
Nucleoside + phosphate group
nucleotide

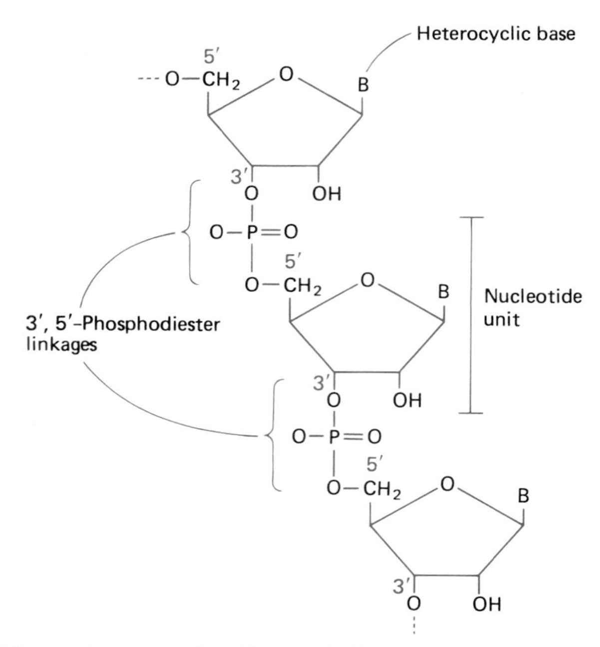

Figure A.25 Nucleotides are linked by phosphate diester bonds to form dimers and polymeric units.

systems: deoxyribonucleic acid (DNA) and ribonucleic acid (RNA). In DNA the monosaccharide that occurs is deoxyribose; ribose is the monosaccharide found in RNA. Four nitrogenous bases occur in DNA: adenine, guanine, cytosine, and thymine. Adenine (A) and guanine (G) are substituted purine bases that are two-ringed structures. Cytosine (C) and thymine (T) are substituted pyrimidine bases and have only one ring. The pyrimidine base thymine is not found in RNA; rather, another pyrimidine, uracil (U), is used in the RNA molecule. The bases adenine, guanine, and cytosine also occur in RNA.

In both RNA and DNA the nucleotides are linked by 3′-5′phosphate diester linkages (Figure A.26). Consequently, at one end of the nucleic acid molecule, there is no phosphate diester bond to the carbon-3 of the monosaccharide; thus there is a free hydroxyl group at the carbon-3 position (3′-OH free end). At the other end of the molecule, the carbon-5 is not involved in forming a phosphate diester linkage, and there is a free hydroxyl group at the carbon-5 position (5′-OH free end). The fact that the ends of the nucleic acid macromolecule differ permits directional recognition at the biochemical level in the same sense that we can recognize left and right.

The DNA double helix is composed of two primary polynucleotide chains held together by hy-

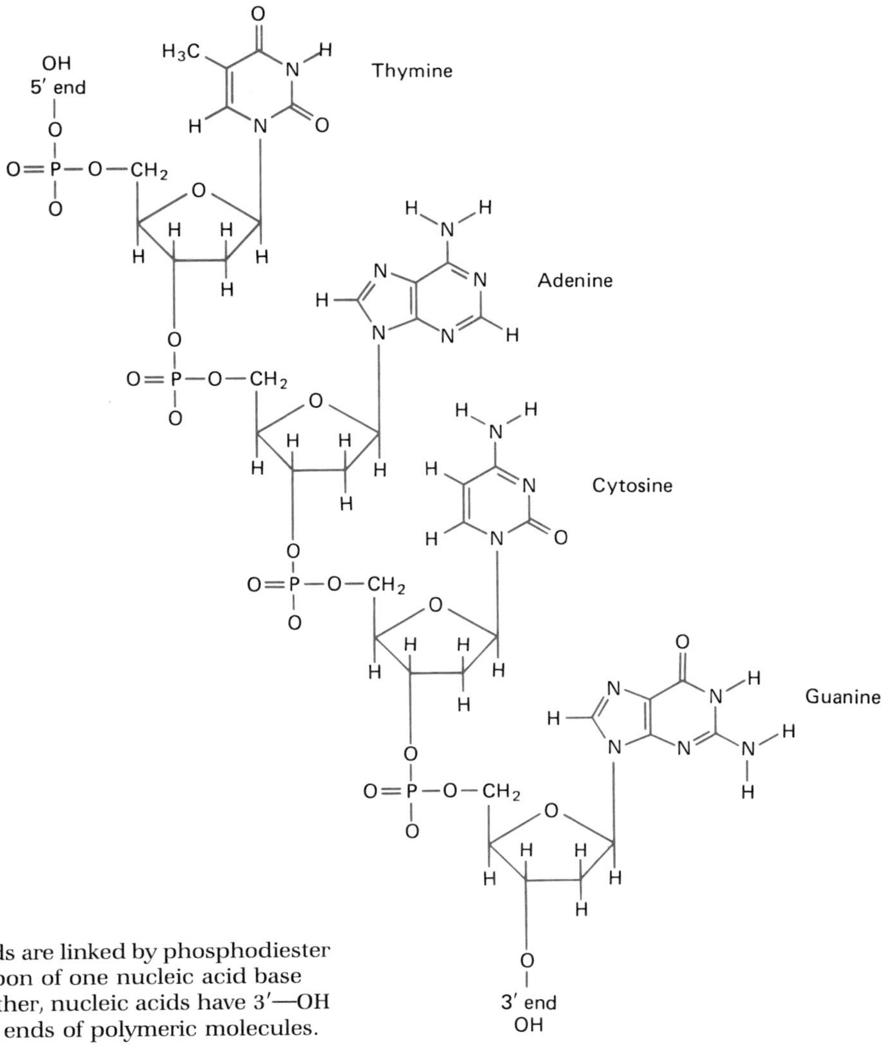

Figure A.26 Because nucleic acids are linked by phosphodiester bonds between the number 3 carbon of one nucleic acid base and the number 5 carbon of the other, nucleic acids have 3′—OH and 5′—OH free ends at opposite ends of polymeric molecules.

drogen bonding between nucleotide bases (Figure A.27). Within the double helical DNA, adenine always pairs with thymine, and guanine always pairs with cytosine and these pairs of bases are therefore said to be complementary. There are two hydrogen bonds established between the adenine–thymine base pairs and three hydrogen bonds established between the guanine–cytosine base pairs. The sequence of nucleotides within the DNA molecule codes for the sequence of bases in RNA and ultimately the sequence of amino acids in proteins. Reading the sequence in the appropriate order is essential for converting stored genetic information into the functional activities of the organism.

Several nucleotides form derivatives that have extremely important functions in living systems. Adenosine triphosphate (ATP) is a molecule produced in cells that is used to move energy through biological systems and to drive various energy-requiring biochemical reactions. Nicotinamide adenine dinucleotide (NAD) is another nucleotide derivative that plays a critical role in metabolism. NAD and related compounds act as coenzymes and are involved in oxidation–reduction reactions.

Chemical Equations

Chemical reactions involve the formation and breakage of chemical bonds, and it is through this process that change occurs. In a chemical reaction the bonds holding the atoms of a molecule together

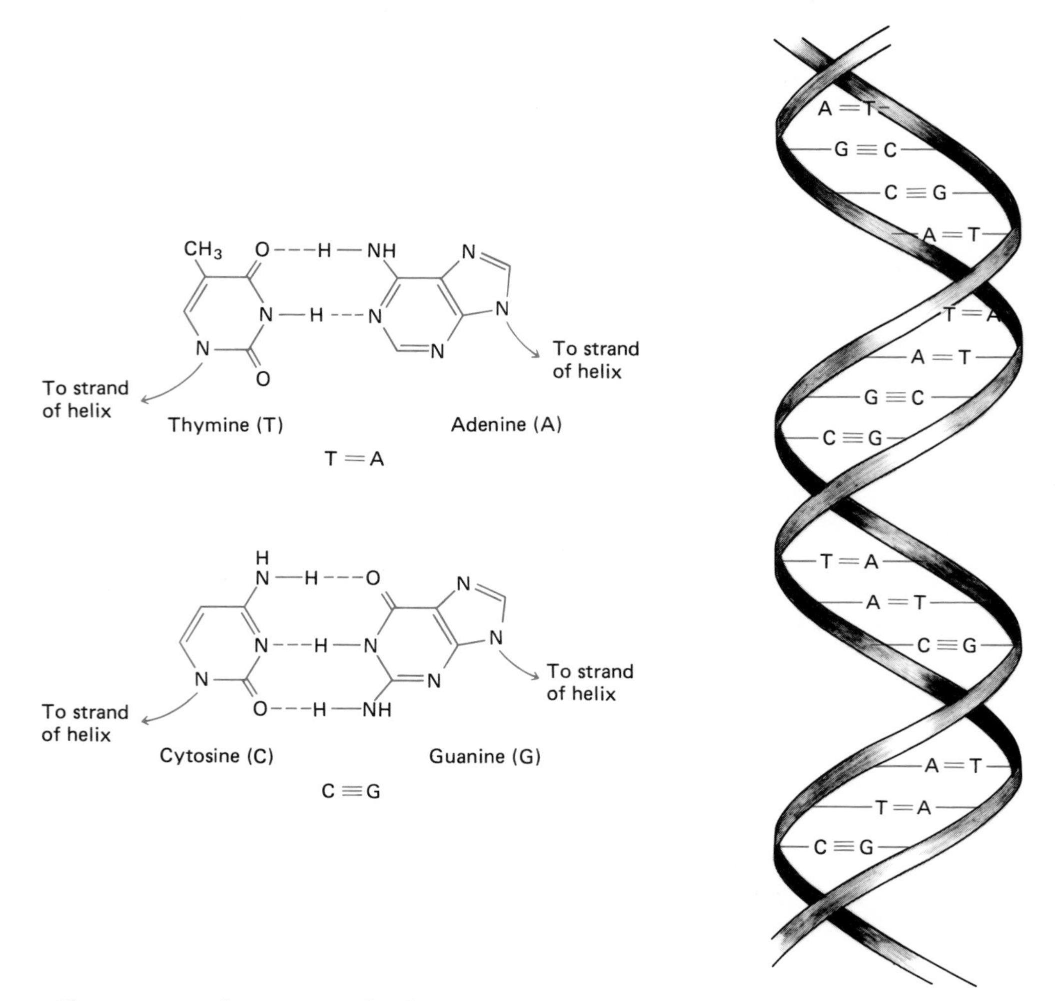

Figure A.27 The two strands of DNA are held together by hydrogen bonding between complementary base pairs.

may break to form new combinations (products), and molecules may interact to form new bonds (new molecules). The total number of each type of atom remains unchanged in a chemical reaction; atoms are neither formed nor destroyed in the chemical reactions found in living systems. Only the electrons in the outer valence orbitals are involved in biochemical reactions. Because there is a conservation of matter and energy in chemical reactions, there must be a balance between what goes into a chemical reaction (reactants) and what is produced by that reaction (products).

The balanced chemical equation shows the relationship between the reactants (left side of the equation) and the products (right side of the equations) of the reaction. The chemical equation identifies the reactants and products by name and formula and shows the balance of the equation so that the number of each kind of element in the reactants equals the number of that same kind of element in the products. For example, the equation for the dissociation of sodium chloride (NaCl) in water to form sodium (Na^+) and chloride (Cl^-) ions is written

$$NaCl \rightarrow Na^+ + Cl^-$$

The numbers of sodium and chlorine atoms on both sides of the equation are the same, and therefore the equation is properly balanced. Note, however, that although water is required for this reaction to occur, it is not shown in the equation because it is not transformed in the reaction; when substances act as solvents and do not participate directly in the reaction, they are not shown within the equation; sometimes, however, solvents are indicated above the arrow to show that their presence is necessary for the reaction.

$$NaCl \xrightarrow{\text{water}} Na^+ + Cl^-$$

As another example, the equation for the reaction of the carbohydrate glucose ($C_6H_{12}O_6$) with oxygen (O_2) to form water (H_2O) and carbon dioxide (CO_2) is

$$C_6H_{12}O_6 + 6\ O_2 \rightarrow 6\ H_2O + 6\ CO_2$$

This equation shows us that six water and six carbon dioxide molecules are formed from one glucose and six oxygen molecules; there is an exact balance of atoms on each side of the equation—6 carbon, 12 hydrogen, and 18 oxygen atoms are found in both the reactants and products.

Chemical Equilibrium

Chemical reactions are reversible; that is, under differing conditions the reactants become the products and the products the reactants. Therefore equations generally are written with arrows pointing in both directions to indicate the reversible nature of chemical reaction.

$$C_6H_{12}O_6 + 6\ O_2 \rightleftharpoons 6\ H_2O + 6\ CO_2$$

If a chemical reaction is allowed to proceed long enough, it will reach a point of equilibrium where there is no net change in the concentrations of products and reactants.

The actual direction in which the chemical reaction proceeds depends upon a variety of conditions, including the initial concentrations of the reactants and products and the energy supplied to the reaction. All reactions proceed toward a state of equilibrium. In order to know in what direction a reaction will go, we need to know the initial concentrations of reactants and products and the concentrations of reactants and products at equilibrium. To describe the concentrations of reactants and products at equilibium we define an equilibrium constant (K_{eq}) as the ratio of the concentrations of products divided by the concentrations of reactants at equilibrium. For the generalized reaction

$$A + B \rightarrow C + D$$

the equilibrium constant is defined as

$$K_{eq}\ [C][D]/[A][B]$$

Brackets indicate concentration. If the K_{eq} is greater than 1, the concentration of products is greater than the concentration of reactants at equilibrium; that is, the reaction proceeds to the right side of the equation, favoring product formation. The reverse is true if K_{eq} is less than 1. The net flow of the reaction can shift, either towards the products or towards the reactants, depending on the concentrations of the products and the reactants and on the temperature; addition or removal of reactants or products alters the direction of the reaction.

The more collisions between reacting molecules, the greater the potential for product formation; therefore, all other things being equal and assuming colliding molecules have sufficient energy to bring about the reaction, the rate of reaction is proportional to the concentrations of the reactants—high relative concentrations of reactants can lead to the rapid formation of products as the reaction moves toward a state of equilibrium. By adding reactants and/or removing products, the reaction can be driven in a particular direction. Enzymes can speed up the *rates* of reactions but do not affect the *equilibria* of the reactions.

Types of Chemical Reactions

Acid–Base Reactions

One of the most important types of chemical reactions is the acid–base reaction. Acids are substances that when added to water produce or increase the concentration of hydrogen (H^+) ions. Bases are substances that when added to water produce or increase the concentration of hydroxyl (OH^-) ions.

If the complete dissociation of a molecule in water produces a large amount of hydrogen ions, the compound is called a strong acid. Thus, hydrochloric acid—an acid found in the stomach—is a strong acid that completely dissociates in water.

$$HCl \rightarrow H^+ + Cl^-$$

The same reasoning applies to substances that behave as bases. Sodium hydroxide (lye) is a strong base because it completely dissociates in water.

$$NaOH \rightarrow Na^+ + OH^-$$

Alternatively, if a basic substance only partially dissociates in water, it will contribute only a limited amount of hydroxyl ions and therefore will be a weak base. For example, ammonium hydroxide is such a weak base.

$$NH_3 + H_2O \rightleftharpoons NH_4{}^+ + OH^-$$

Similarly, when acidic substances only partially dissociate in water, the concentration of hydrogen ions will be small; the substance is therefore a weak acid. An example of a weak acid is acetic acid (vinegar).

$$CH_3COOH \rightleftharpoons CH_3COO^- + H^+$$

Carbon dioxide (CO_2) behaves as a weak acid even though it does not contain hydrogen, because aqueous solutions of carbon dioxide become slightly acidic. When CO_2 dissolves, it initially forms carbonic acid (H_2CO_3)

$$CO_2 + H_2O \rightarrow H_2CO_3$$

which partially dissociates to bicarbonate and hydrogen ions

$$H_2CO_3 \rightarrow HCO_3{}^- + H^+$$

Subsequently, the bicarbonate dissociates in a second reaction so that the overall equation for the dissolution of carbon dioxide in water can be written

$$CO_2 + H_2O \leftarrow H_3O^+ + CO_3{}^{2-}$$

In addition to the abilities of weak acids and bases to regulate pH through their buffering capacity, that is, through shifts in the equilibrium of acid/base dissociation reactions, acids react directly with bases to regulate pH. When an acid reacts with a base, there is a reaction between the hydrogen ions produced by the acid and the hydroxyl ions produced by the base, resulting in the formation of water and a salt; this is known as a neutralization reaction because it balances the hydrogen ion and hydroxyl ion concentrations and thus achieves a neutral pH of 7. For example, when hydrochloric acid reacts with sodium hydroxide, the products are sodium chloride and water.

$$HCl + NaOH \rightarrow NaCl + H_2O$$
$$H^+ + Cl^- + Na^+ + OH^- \rightarrow Na^+ + Cl^- + H_2O$$

If the amounts of acid and base are balanced, all the free hydrogen ions react with all the free hydroxyl ions, and the result is a solution of the salt that has a neutral hydrogen ion concentration. The buffering capacities of weak acids and bases are important in preventing radical changes in the pH of physiological fluids when hydrogen or hydroxide ions are added.

Oxidation–Reduction Reactions

Oxidation–reduction reactions are based on the exchange of electrons between molecules. Whenever electrons are exchanged between atoms, there is a change in the oxidation state of the atoms. Oxidation is the process whereby an atom or molecule loses one or more electrons; reduction is the gain of one or more electrons by a molecule or atom. For example, when ferrous iron (Fe^{2+}) is converted to ferric iron (Fe^{3+}), it is oxidized because it loses an electron; conversely when ferric iron (Fe^{3+}) is converted to ferrous iron (Fe^{2+}), it gains an electron and thus is reduced. Similarly when acetate is converted to ethanol, it gains an electron and thus is reduced. Such oxidation and reduction reactions can not occur alone—the oxidation of one substance must always be coupled with the reduction of another; this requirement for coupling is a consequence of the laws governing the conservation of energy and matter.

In many reactions the oxidation of one substance is coupled with the reduction of oxygen, and in most biological systems oxygen is the molecule of choice to act as an oxidizing agent. For example, during cellular respiration, the oxidation of glucose is coupled with the reduction of oxygen. In such oxidation reactions, large amounts of energy are released. Living organisms use the energy released from such oxidation reactions for maintaining the myriad life processes. In contrast to oxidation reactions, reduction reactions require the input of energy. Reduction reactions are used by living systems to store energy. Thus when carbon dioxide is reduced to glucose in photosynthesis, energy is stored within the organic molecules of the organism. This stored energy can later be released when organisms oxidize sugars during cellular respiration. The energy released by such oxidation reac-

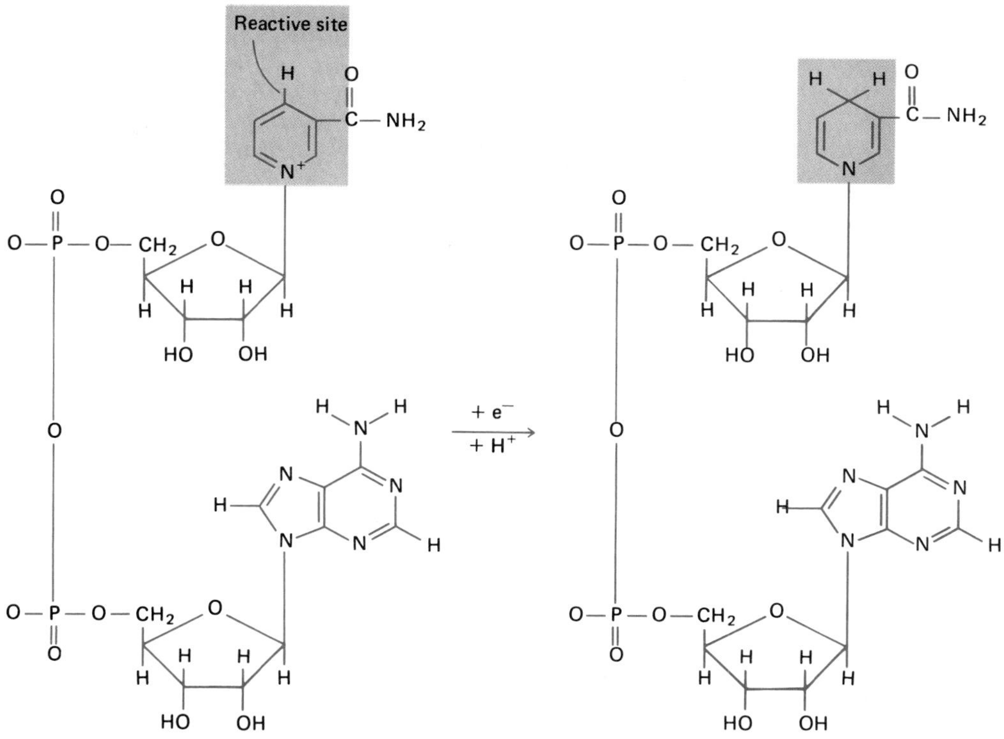

Reactive site

NAD → NADH ($+ e^-$, $+ H^+$)

$$NAD^+ + 2e^- + 2H^+ \rightarrow NADH + H^+$$

Oxidized coenzyme NAD ⟶ Reduced coenzyme NADH

Figure A.28 The reduction of NAD^+ to $NADH + H^+$ is a critical reaction that often is coupled with the oxidation of substrates within a cell. This reaction can be written several ways; the abbreviated form NAD → NADH is employed throughout this book.

tions can be coupled with energy-requiring biosynthetic reactions. Living systems must generate enough energy from oxidation reactions to meet the energy demands of the biosynthetic reactions involved in growth and reproduction.

In many metabolic reactions, the oxidation of an energy-rich substrate is coupled with the reduction of a molecule called a coenzyme. The coenzyme acts as an accessory to the reaction by temporarily accepting the electron. An example of such a reaction is the conversion of the coenzyme NAD (nicotinamide adenine dinucleotide) to its reduced form NADH (Figure A.28). NAD is a major oxidizing agent within all biological systems, and this coenzyme plays a critical role in many of the energy-generating reactions of organisms. Reduced coenzymes play the opposite role, serving as reducing agents in biosynthetic reactions. In such reactions a reduced coenzyme can donate an electron, thus becoming oxidized, such as occurs when the reduced coenzyme NADPH (nicotinamide adenine dinucleotide phosphate) is oxidized to NADP in a reaction that is coupled with the reduction of a substrate. Biosynthesis is a reductive process, and cells must generate reduced coenzymes to provide the reducing power needed to carry out biosynthetic reactions.

Bioenergetics

There are three types of work that cells can perform; mechanical work (movement), biochemical work (making new cellular components), and concentration (maintaining the internal environment of the cell different from the external). In all chemical reactions, including those that occur in biological systems, the flow of energy through the system is governed by the laws of thermodynamics; these laws prescribe the nature of chemical reactions based solely upon the initial and final energy states of a system. According to the first law of thermodynamics, energy is conserved; that is, chemical reactions neither create nor destroy energy. It is important to keep this in mind when considering the transformations of chemicals and the flow of energy through a biological system.

In a chemical reaction there must be a net balance between the energy required to break chemical bonds, the energy released by the new bonds that are formed, and the energy—such as heat energy—that is exchanged with the surroundings. The change in the stored energy between the amount contained in the bonds of the reactants and those of the products of a chemical reaction is described as the ΔH (enthalpy) of the reaction, the change in heat content of the molecules (Figure A.29). Reactions that absorb heat have a positive ΔH and are termed endothermic reactions; chemical reactions which release heat have a negative ΔH and are termed exothermic reactions.

In addition to the heat content of molecules, energy is needed to maintain the arrangement of molecules in a highly ordered state. This energy is greater than the energy present in a random distribution of molecules (state of disorder). According to the second law of thermodynamics, all processes proceed in the direction that increases the total entropy of the system and the surroundings, that is, in the direction of maximum randomness or disorder. Equilibrium is the least ordered state of a system, and therefore is the state where entropy is the largest. The change in energy that occurs when a system goes from a highly ordered state to a highly disordered state is not available to living organisms for performing work. Whether a chemical reaction occurs or not depends upon the relative state of order of the products and reactants as well as on the energy stored within the reactant and product molecules. Living organisms are highly ordered systems and they must continually expend energy to maintain this high degree of order. The only way that a biological organism can maintain its high degree of order (a state of low entropy) is by increasing the disorder (increasing the entropy) of the surroundngs. Order can increase within a biological system only if disorder increases elsewhere in the universe.

Figure A.29 Energy diagrams showing ΔH for exothermic and endothermic reactions.

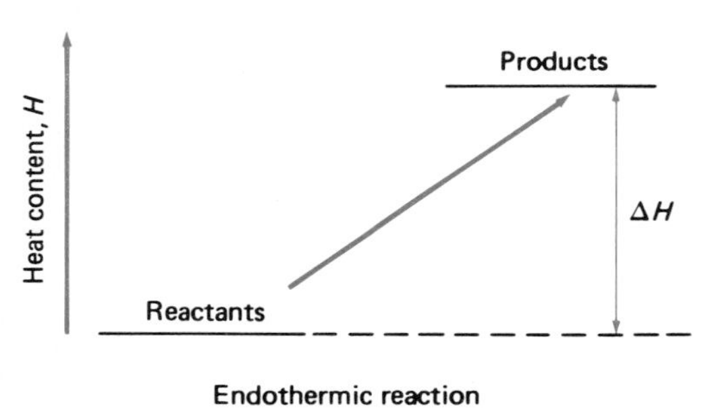

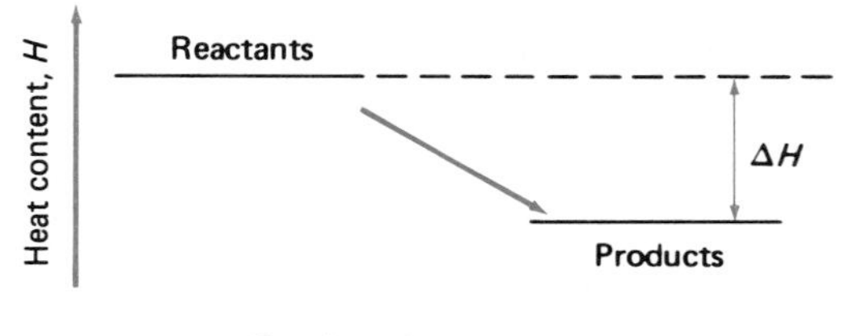

In predicting whether or not a chemical reaction can spontaneously occur, the change in free energy (ΔG) of the reaction must be considered; the free energy describes the energy that is available for doing work in a chemical reaction. Biological systems make use of free energy to perform the work necessary to transform substances into the macromolecules of the organism and to maintain the organizational integrity of the system. The concept of free energy takes into consideration the degree of order, as well as the stored energy. The change in free energy of a reaction (ΔG) is related to the change of the heat of reaction (ΔH), the temperature (T, absolute temperature in Kelvin), and the change in the state of order or entropy (ΔS) between products and reactants.

$$\Delta G = \Delta H - T\Delta S$$

Reactions that release free energy have a negative ΔG and are termed exergonic; reactions that do not result in enough energy to do work have a positive ΔG and are termed endergonic. Viewed in another way, some chemical reactions require an input of energy from another source to drive them uphill (endergonic reactions), and other chemical reactions release energy as they run downhill (exergonic reactions) (Figure A.30). Fortunately, endergonic and exergonic reactions can be coupled so that as one reaction runs downhill, it pulls another uphill; that is, the free energy released in exergonic

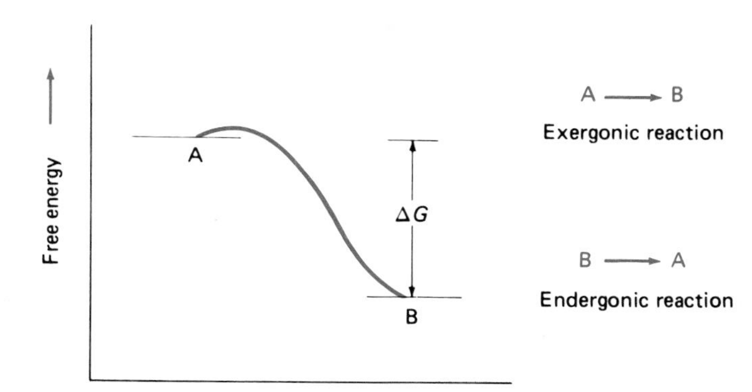

Figure A.30 This graph shows the changes in free energy during exergonic and endergonic reactions.

reactions can be used to drive endergonic reactions. When reactions are coupled to achieve a favorable ΔG (negative ΔG), reactions occur spontaneously, the reaction favors the formation of products, and there is a decrease in the free energy of the system.

ATP and Biochemical Reactions

The free energy (ΔG°) for an overall reaction will be the same regardless of the number of steps or the pathway required to go from reactants to products. Consequently, it is possible to couple individual reactions to achieve a favorable overall ΔG° that will allow the complete process to occur. In particular, the formation of ATP (adenosine triphosphate), an endergonic reaction, can be driven by various exergonic metabolic reactions, and the hydrolysis (breaking a bond by the addition of water across that bond) of ATP can be coupled with endergonic reactions. ATP is called an energy-rich or high-energy compound because it contains chemical bonds that upon hydrolysis release a large amount of energy. The hydrolysis of ATP to form ADP (adenosine diphosphate) and inorganic phosphate (P_i) is highly exergonic, releasing approximately 7.3 kcal (kilocalories) per mole of ATP hydrolyzed (Figure A.31).

The same amount of energy (7.3 kcal) is also required for the formation of ATP from ADP + P_i. Many of an organism's metabolic reactions that break down organic compounds are involved in generating ATP from ADP; the energy captured in the ATP molecule can be used to drive energy-requiring reactions, including those involved in biosynthesis, movement, and transport. ATP serves almost universally in biological systems as the energy source for energy-requiring reactions; most endergonic biochemical reactions are coupled with the exergonic hydrolysis of ATP. As such, ATP can be termed the "universal currency of energy" in biological systems; cellular processes requiring energy most likely depend upon the use of ATP.

Reaction Kinetics

The rate of a chemical reaction depends on the temperature and the relative concentrations of reactants and products. At elevated temperatures, molecular motion is increased; thus, reaction rates

Figure A.31 The hydrolysis of ATP to ADP is used to drive endergonic reactions.

$$\text{ATP} + H_2O \longrightarrow \text{ADP} + O^- {-}PO_2^-{-}O^- + H^+ + \approx 7.5 \text{ kcal/mole}$$

ATP ADP P_i

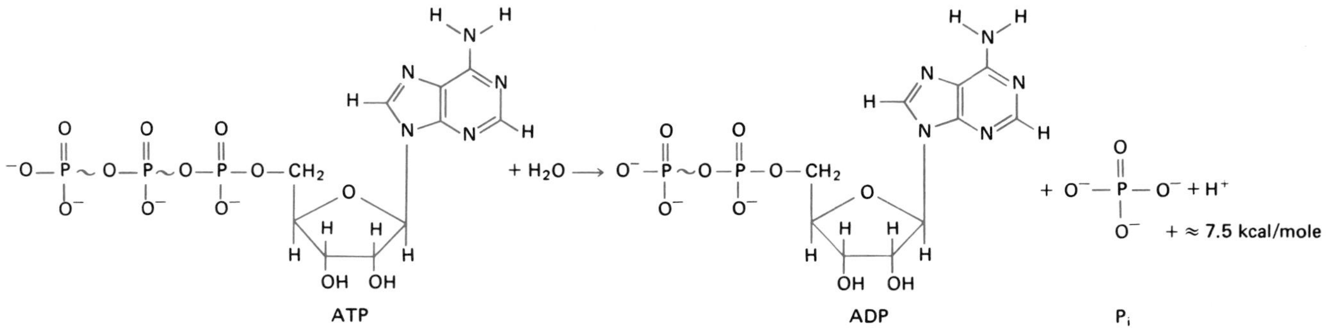

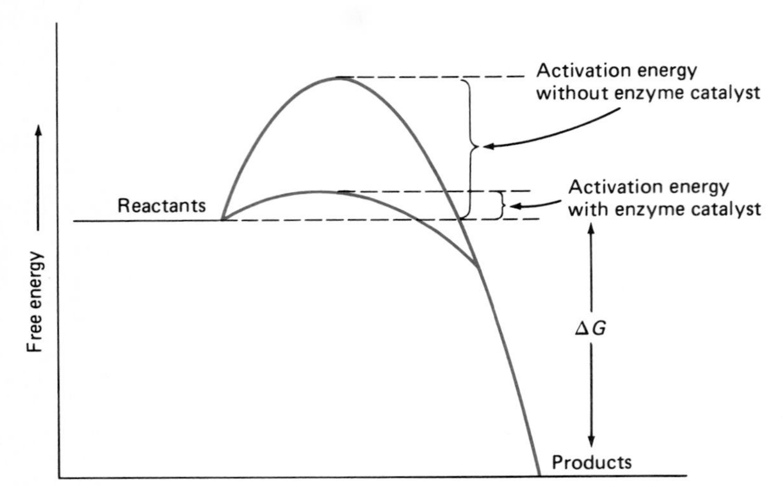

Figure A.32 This graph illustrates the activation energy required for a chemical reaction to start and that an enzyme effectively lowers this energy of activation.

can be increased by heating, because it results in more collisions between reactant molecules, which increases the rate of product-formation. However, macromolecules are disrupted at elevated temperatures and therefore, although chemical reactions can be greatly accelerated in a test tube using a bunsen burner, they cannot be increased greatly in biological systems by increasing the temperature of the reaction. Reaction rates also can be speeded up by increasing the concentrations of reactants. Although changes in concentrations of reactants do affect the rates of chemical reactions, such changes do not explain the high rates of biochemical reactions whose rates are higher than can be accounted for by simply increasing the concentrations of reactants.

Enzymes

The rapid rates by which biochemical reactions in living systems occur are possible because of the role of enzymes as biological catalysts. For a chemical reaction to occur, the reactant molecules must collide with sufficient energy to bring about the reaction, that is, the two colliding molecules must have a certain minimal amount of energy—called the activation energy—in order to react (Figure A.32). The energy of activation must be achieved even if the $\Delta G°$ of the reaction is negative; that is, the en-

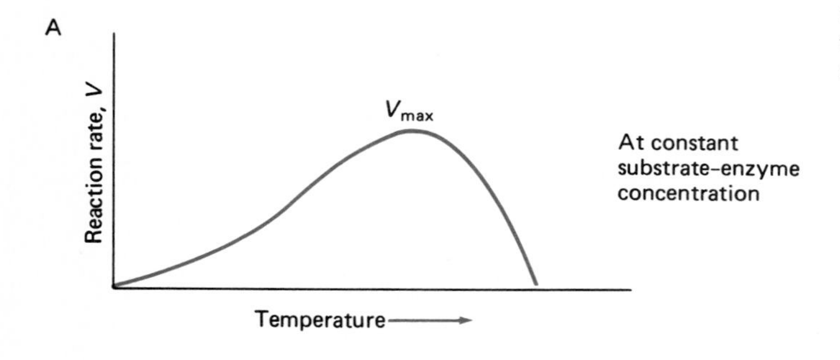

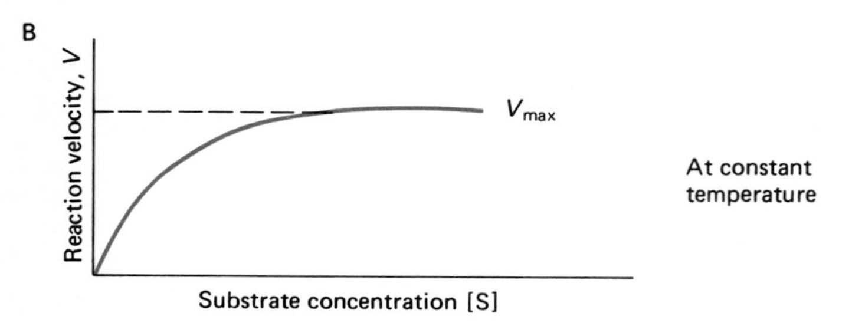

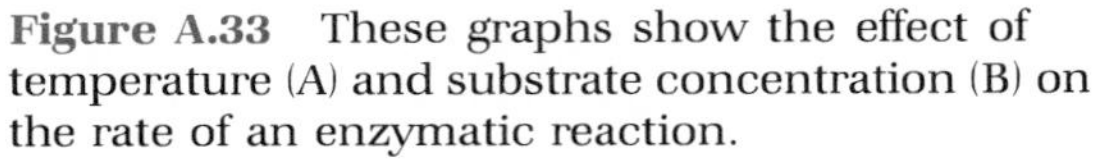
Figure A.33 These graphs show the effect of temperature (A) and substrate concentration (B) on the rate of an enzymatic reaction.

ergy must be elevated to a higher energy state before the barrier to the reaction can be overcome and the reaction can occur. At the relatively low temperatures of biological systems, few collisions between molecules have the energy required to overcome the activation energy and to react spontaneously; therefore, biological reactions are dependent upon the presence of protein catalysts (enzymes). A catalyst is a substance that lowers the energy of activation required to bring about a reaction, and therefore at a given temperature, a catalyzed reaction proceeds more rapidly than an uncatalyzed reaction; within certain limits, the rate of an enzymatically catalyzed reaction doubles for every 10°C rise in temperature (Figure A.33). An essential property of a catalyst is that it is not consumed in the reaction, and thus enzymes theoretically may continue to function indefinitely.

A particular enzyme catalyzes only a specific reaction. There are six major classes of enzymes based on the type of reaction catalyzed (Table A.5). Within each class, enzymes are named according to the substrate of the reaction; for example, succinate dehydrogenase catalyses the dehydrogenation of succinate to form fumarate. Most macromolecules are formed through the actions of hydrolases and ligases. For example, the glycosidic linkage, peptide linkage, and ester linkage are formed by hydrolases; the phosphodiester linkages of the nucleic acids are formed by ligases.

The degree of substrate specificity exhibited by enzymes reflects the fact that the enzyme and the substrate must fit together in a specific way for the enzyme to lower the activation energy. The fit has been likened to a lock and key, indicating the precise match that exists between an enzyme and its substrate (Figure A.34). Unlike the rigid lock and key model however, the binding of the substrate molecule to the enzyme may alter the three-dimensional configuration of the enzyme to establish the proper fit. The enzyme has a specific active site at which the substrate binds. When all the active sites of an organism's enzyme molecules are occupied, saturation occurs, and the reaction proceeds at the maximal rate. The binding of the enzyme to the substrate involves the formation of weak bonds that are sufficient to place a strain on the substrate molecule, lowering the activation energy and permitting the reaction to proceed. In the case of reactions involving multiple reactant molecules, the enzyme effectively positions the reactants in space so that they are brought together with the right

Table A.5 International Classification of Enzymes—Class Names, Code Numbers, and Types of Reactions Catalyzed

1. Oxido-reductases (oxidation–reduction reactions)
   - 1.1 Acting on CH—OH
   - 1.2 Acting on C═O
   - 1.3 Acting on C═CH
   - 1.4 Acting on CH—NH
   - 1.5 Acting on CH—NH^2
   - 1.6 Acting on NADH; NADPH
2. Transferases (transfer of functional groups)
   - 2.1 One-carbon groups
   - 2.2 Aldehydic or ketonic groups
   - 2.3 Acyl groups
   - 2.4 Glycosyl groups
   - 2.7 Phosphate groups
   - 2.8 Sulfur-containing groups
3. Hydrolases (hydrolysis reactions)
   - 3.1 Esters
   - 3.2 Glycosidic bonds
   - 3.3 Peptide bonds
   - 3.5 Other C—N bonds
   - 3.6 Acid anhydrides
4. Lyases (addition to double bonds)
   - 4.1 To C═C
   - 4.2 To C═O
   - 4.3 To C═N—
5. Isomerases (isomerization reactions)
   - 5.1 Racemases
6. Ligases (formation of bonds with ATP cleavage)
   - 6.1 C—O
   - 6.2 C—S
   - 6.3 C—N
   - 6.4 C—C

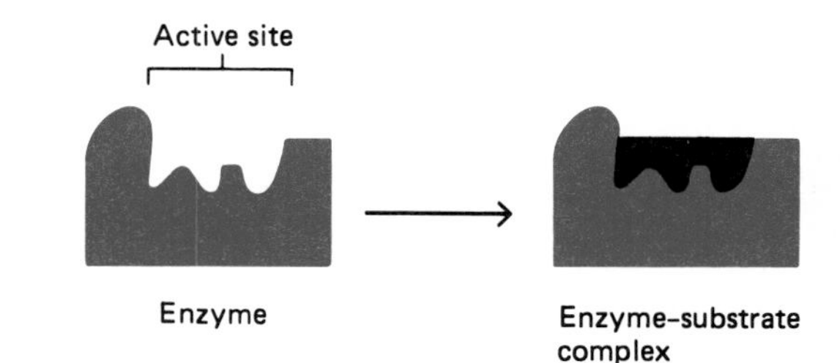

Figure A.34 The fit between enzyme and substrate has been likened to a lock and key. Actually, the interaction of substrate with enzyme modifies the three-dimensional structure of the enzyme. The precision of fit is responsible for the high degree of specificity of enzymes for particular substrates.

orientation to bring about the biochemical reaction. The precision of fit between enzyme and substrate molecule permits the establishment of exactly the right spatial orientation so that the numerous biochemical reactions of an organism can occur with great rapidity.

The specificity of the configuration needed to form the enzyme–substrate complex, however, renders enzymes quite susceptible to inhibition. The activities of an enzyme can be completely inhibited by denaturing the enzyme, that is, by radically distorting its three-dimensional shape, such as occurs at high temperatures. The activities of an enzyme can also be inhibited by the binding of substances other than the specified substrate to the enzyme (Figure A.35). If an inhibitory substance exhibits an affinity for the active site of the enzyme because of its similarity to the normal substrate, it can act as a competitive inhibitor, competing with the specified substrate for the active site of the enzyme. Competitive inhibition can be overcome by increasing the substrate concentration. In contrast, in-

Figure A.35 Enzymes are subject to various types of inhibition. (A) Competitive inhibition occurs when an inhibitory substance competes with the natural substrate for the active and/or binding sites. (B) Noncompetitive inhibition occurs when an inhibitory substance reacts with the enzyme at a site other than the active site, decreasing the activity of the enzyme.

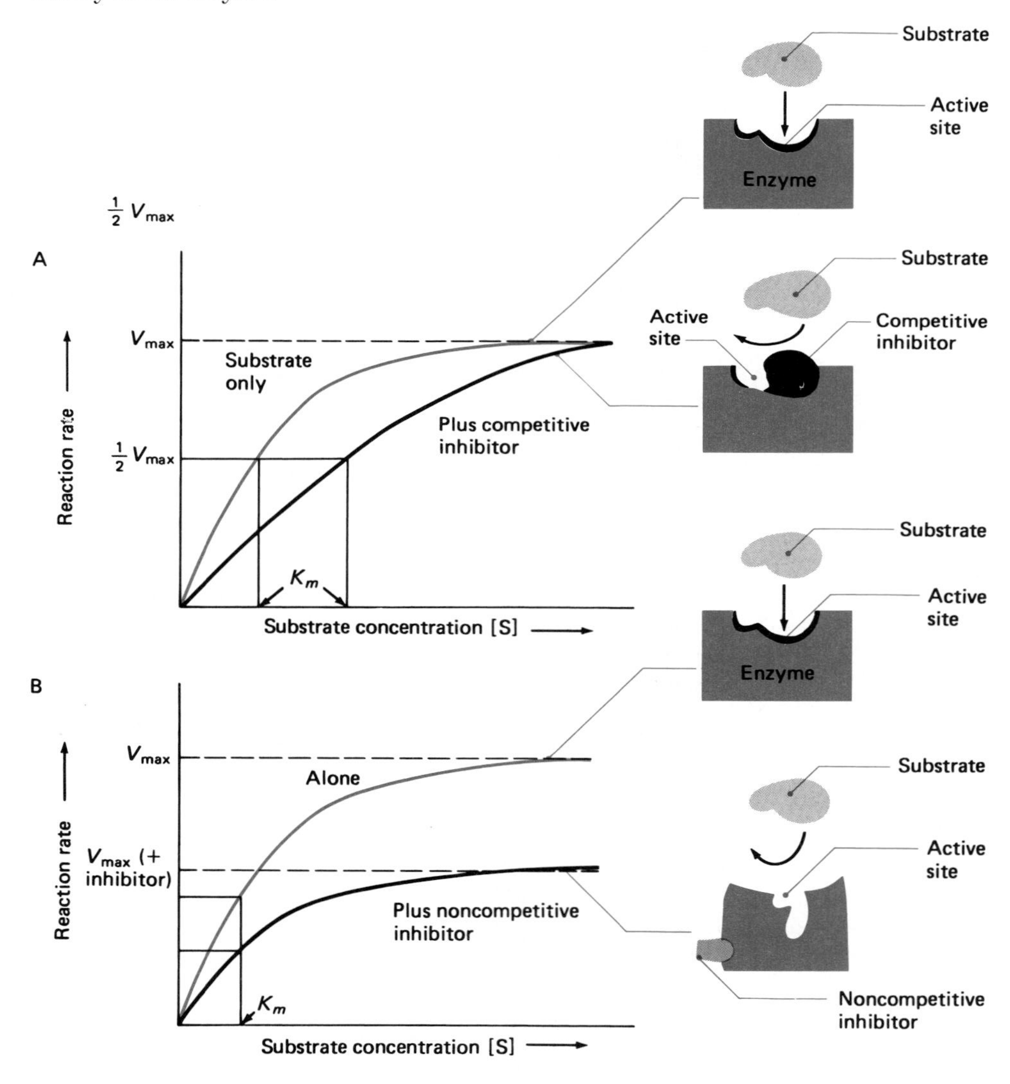

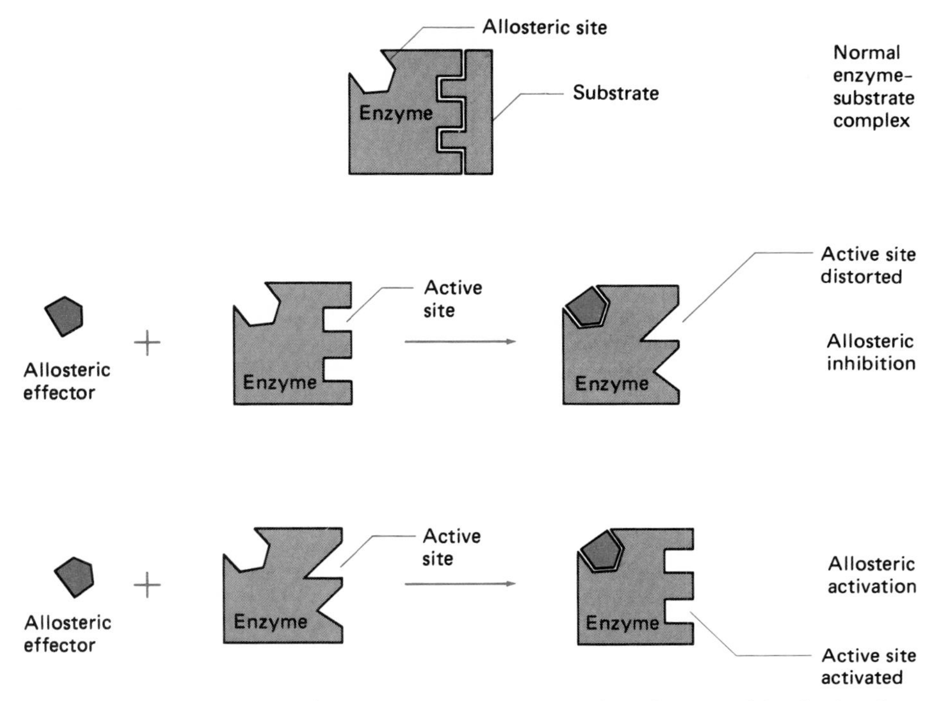

Figure A.36 The activities of enzymes can be increased or decreased by the binding of substance other than the substrate to allosteric effector sites.

creasing substrate concentration does not overcome noncompetitive inhibition. Noncompetitive inhibitors do not compete with the specified substrate, but they may bind irreversibly at the active site or may distort the enzyme by binding at a site on the enzyme other than the active site.

The ability to modify the three-dimensional shape of the enzyme molecule provides a basis for the regulation of the rates of the enzymatic activities within an organism. Substances may bind to the enzyme at sites removed from the active site and still alter the properties of the enzyme molecule. Additional binding sites of enzyme molecules that are involved in regulating the activities of enzymes are called allosteric effector sites (Figure A.36). Allosteric effector sites may be involved either in the inhibition or the activation of an enzyme. Just as with the reaction between enzyme and substrate, the affinity of a allosteric inhibitor for an allosteric effector site normally shows a great deal of specificity. In some cases, the end product of an enzyme reaction sequence may act as an allosteric inhibitor by reversibly binding with an enzyme within that sequence. Such a system is self-regulating because there is a feedback mechanism through which the reaction is shut off when excessive product is formed. This type of regulation by allosteric inhibition is called feedback or end product inhibition. In other cases, the reversible binding of a substance to an allosteric site may activate the enzyme, increasing its activity (feedback activation). Both feedback inhibition and feedback activation are important processes that regulate the activities of enzymes and thus the rates of metabolic reactions.

Glossary

α-Hemolysis partial hemolysis of red blood cells as evidenced by the formation of a zone of partial clearing (greening) around certain bacterial colonies growing on blood agar.

Abiotic referring to the absence of living organisms.

Abrasion an area denuded of skin, mucous membrane, or superficial epithelium caused by rubbing or scraping.

Abscess a localized accumulation of pus.

Absorption the uptake, drinking in, or imbibing of a substance; the movement of substances into a cell; the transfer of substances from one medium to another, e.g., the dissolution of a gas in a liquid; the transfer of energy from electromagnetic waves to chemical bond and/ or kinetic energy, e.g, the transfer of light energy to chlorophyll.

Accessory pigments pigments that harvest light energy and transfer it to the primary photosynthetic reaction centers.

Acellular lacking cellular organization; not having a delimiting cytoplasmic membrane.

Acetobacter bacterial genus—cells ellipsoidal to rod-shaped; motile by peritrichous flagella or nonmotile; endospores not formed; young cells Gram negative, in older cultures some strains become Gram variable; metabolism respiratory, never fermentative; oxidize ethanol to acetic acid at neutral and acid reactions (pH 4.5); strict aerobes; optimal growth ca. 30°C; G + C 55–64 mole%.

Acetyl a two carbon organic radical containing a methyl group and a carbonyl group.

Acetyl CoA acetylcoenzyme A; a condensation product of coenzyme A and acetic acid; an intermediate in the transfer of two-carbon fragments, notably in its entrance into the tricarboxylic acid cycle.

Achievable serum levels maximal (peak) concentrations of an antimicrobial agent that can occur in serum.

Achromatic lens an objective lens in which chromatic aberration has been corrected for two colors and spherical aberration for one color.

Acid fast the property of those bacteria that retain their initial stain and do not decolorize after washing with dilute acid–alcohol.

Acid foods foods with a pH value less than 4.5.

Acidic a compound that releases hydrogen (H^+) ions when dissolved in water; a compound that yields positive ions upon dissolution; a solution with a pH value less than 7.0.

Acidic stains a stain with a positively charged chromophore (colored portion of the dye) that is attracted to negatively charged cells.

Acidophiles microorganisms showing a preference for growth at low pH, e.g., bacteria that grow only at very low pH values, ca. 2.0.

Acinetobacter bacterial genus—rods, usually very short and plump, approaching coccus shape in stationary phase, predominantly in pairs and short chains; no spores formed; flagella not present; Gram negative; oxidative metabolism; oxidase negative, catalase positive; optimal growth 30–32°C; G + C 40–47 mole%.

Acne an inflammatory disease involving the oil glands and hair follicles of the skin, found chiefly in adolescents and marked by papules or pustules, especially about the face.

Acquired immune deficiency disease (AIDS) an infectious disease caused by HTLV-3 virus, characterized by the loss of normal immune response system followed by various opportunistic infections.

Acquired immunity the ability of an individual to produce specific antibodies in response to antigens to which the body has been exposed based upon the development of a memory response.

Actinomyces bacterial genus—Gram positive, irregularly staining bacteria; nonacid-fast, nonspore-forming and nonmotile; filaments with true branching may predominate and are particularly evident in 18–48 hour microcolonies; carbohydrates are fermented with the production of acid but no gas; some species may show greening or complete lysis of rabbit red blood cells; facultative anaerobes; most are preferentially anaerobic, and one species grows well aerobically; carbon dioxide is required for maximum growth.

Actinomycetes members of an order of bacteria in which species are characterized by the formation of branching filaments and/or true filaments.

Activated sludge the solid portion of sewage containing active populations of aerobic microorganisms.

Activated sludge process an aerobic secondary sewage treatment process using sewage sludge containing active complex populations of aerobic microorganisms to break down organic matter in sewage.

Activation energy the energy in excess of ground state that must be added to a molecular system to allow a chemical reaction to start.

Active immunity immunity acquired as a result of the individual's own reactions to pathogenic microorganisms or their antigens; attributable to the presence of antibody or immune lymphoid cells formed in response to an antigenic stimulus.

Active transport the movement of materials across cell membranes from regions of lower to higher concentration requiring the expenditure of metabolic energy.

Acute referring to a disease of rapid onset, short duration, and pronounced symptoms.

Acute stage stage of disease when symptoms appear and disease is most severe.

Adaptive enzymes enzymes produced by an organism in response to the presence of a substrate or a related substance; also called an inducible enzyme.

Adenine a purine base component of nucleotides, nucleosides, and nucleic acids.

Adenosine a mononucleoside consisting of adenine and D-ribose.

Adenosine diphosphate (ADP) a high-energy derivative of adenosine containing two phosphate groups, one phosphate group less than ATP; formed upon hydrolysis of ATP.

Adenosine monophosphate (AMP) a compound composed of adenosine and one phosphate group formed upon the hydrolysis of ADP.

Adenosine triphosphatase an enzyme that catalyzes the reversible hydrolysis of ATP; the membrane-bound form of this enzyme is important in catalyzing the formation of ATP from ADP and inorganic phosphate.

Adenosine triphosphate (ATP) a major carrier of phosphate and energy in biological systems, composed of adenosine and three phosphate groups; the free energy released from the hydrolysis of ATP is used to drive many energy-requiring reactions in biological systems.

Adhesins substances involved in the attachment of microorganisms to solid surfaces; factors that increase adsorption.

Adhesion factors see adhesins.

Adjuncts starchy substrates, such as corn, wheat, and rice, that provide carbohydrates for ethanol production and are added to malt during the mashing process in the production of beer.

Adjuvants substances that increase the immunological response to a vaccine; can be added to vaccines to slow down absorption and increase effectiveness; substances that enhance the action of a drug or antigen.

ADP see adenosine diphosphate.

Adrenaline hormone secreted by the adrenal medulla in response to stress, causes a rise in blood pressure; used as a heart stimulant.

Adsorption a surface phenomenon involving the retention of solid, liquid, or gaseous molecules at an interface.

Aer- combining form meaning air or atmosphere.

Aerated piles method of composting for decomposition of organic waste materials where the wastes are heaped in separate piles.

Aerial mycelia a filamentous mass of hyphae occurring above the surface of a substrate.

Aerobes microorganisms whose growth requires the presence of air or free oxygen.

Aerobic having molecular oxygen present; growing in the presence of air.

Aerobic bacteria bacteria requiring oxygen for growth.

Aerosol a fine suspension of particles or liquid droplets sprayed into the air.

Aflatoxin a carcinogenic poison produced by some strains of the fungus *Aspergillus flavus.*

African sleeping sickness also known as African trypanosomiasis; a protozoan disease that affects the nervous system, caused by *Trypanosoma*, a flagellate that is injected by the bite of the tsetse fly.

Agar a dried polysaccharide extract of red algae used as a solidifying agent in various microbiological media.

Agglutin an antibody capable of causing the clumping or agglutination of bacteria or other cells.

Agglutination the visible clumping or aggregation of cells or particles due to the reaction of surface-bound antigens with homologous antibodies.

Agrobacterium bacterial genus—motile Gram negative rods; metabolism respiratory; optimal growth 25–30°C; G + C 59.6–62.8 mole%.

AIDS see acquired immune deficiency syndrome.

Alcaligenes bacterial genus—cells rods, coccal rods, or cocci usually occurring singly; motile with peritrichous flagella; Gram negative; metabolism respiratory, never fermentative; do not fix gaseous nitrogen; oxidase positive; optimal growth 20–37°C; G + C 57.9–70 mole%.

Alcohols organic compounds characterized by one or more —OH (hydroxyl) groups.

Aldehydes a class of substances derived by oxidation from primary alcohols and characterized by the presence of a —CHO group.

Algae a heterogeneous group of eukaryotic, photosynthetic, unicellular, and multicellular organisms lacking true tissue differentiation.

Algicides chemical agents that kill algae.

Alkaline a condition in which hydroxyl (OH^-) ions are in abundance; solutions with a pH of greater than 7.0 are alkaline or basic.

Allele any one or more alternative forms of a given gene concerned with the same trait or characteristic; one of a pair or multiple forms of a gene located at the same locus of homologous chromosomes.

Allelopathic substances substances produced by an organism that adversely affect another organism.

Allergen an antigen that induces an allergic response, i.e., a hypersensitivity reaction.

Allergy an immunological hypersensitivity reaction; a type of antigen–antibody reaction marked by an exaggerated physiological response to a substance in sensitive individuals.

Allosteric enzymes enzymes with a binding and catalytic site for the substrate and a different site where a modulator (allosteric effector) acts.

Allotypes antigenically different forms of a given type of immunoglobulin that occur in different individuals of the same species.

Alpha-hemolysis see α-hemolysis.

Amastigotes rounded protozoan cells lacking flagella; a form assumed by some species of Trypanosomatidae, e.g., *Plasmodium*, during a particular stage of development.

Amebic dysentery an inflammation of the colon caused by *Entamoeba histolytica*; also known as amebiasis.

Amino an $—NH_2$ group.

Amino acids a class of organic compounds containing an amino (NH_2) group and a carboxyl (COOH) group; the subunits of proteins.

Aminoacyl site site on a ribosome where a tRNA molecule attached to a single amino acid initially binds during translation.

Amino end the end of a peptide chain or protein with a free amino group, i.e., an alpha-amino group not involved in forming the peptide bond.

Aminoglycoside antibiotics broad-spectrum antibiotics containing an aminosugar, an amino- or guanidoinositol ring, and residues of other sugars that inhibit protein synthesis, e.g., kanamycin, neomycin, and streptomycin.

Ammonification the release of ammonia from nitrogenous organic matter by microbial action.

Ammonium ion the cation NH_4^+.

AMP see adenosine monophosphate.

Amphibolic pathway a metabolic pathway that has both catabolic and anabolic functions.

Amylases enzymes that hydrolyze starch.

Anabolism the process of synthesis of cell constituents from simpler molecules, usually requiring the expenditure of energy; biosynthesis.

Anaerobes organisms that grow in the absence of air or oxygen; organisms that do not use molecular oxygen in respiration.

Anaerobic referring to the absence of oxygen; not requiring oxygen for growth.

Anaerobic digester a secondary sewage treatment facility used for the degradation of sludge and solid waste.

Anaerobic respiration metabolism using inorganic electron acceptors other than oxygen as terminal electron acceptors for energy-yielding oxidative metabolism.

Anamnestic response a heightened immunological response in persons or animals to the second or subsequent administration of a particular antigen given some time after the initial administration; a secondary or memory immune response, i.e., the rapid reappearance of antibody in the blood following the administration of an antigen to which the subject had previously developed a primary immune response.

Anaphylactic hypersensitivity an exaggerated immune response reaction of an organism to foreign protein or other substances involving degranulation of mast and or basophil cells and the release of histamine.

Anaphylactic shock a physiological condition of shock resulting from an anaphylactic hypersensitivity reaction, e.g., from a hypersensitive reaction to penicillin or bee bites, and in severe cases death can result within minutes.

Anemia a condition characterized by having less than the normal amount of hemoglobin, reflecting a depressed number of circulating red blood cells.

Animal virus a virus that multiplies in animal cells.

Anions negatively charged ions.

Anisogametes gametes differing in shape, size, and/or behavior.

Anode the positive terminal of an electrolytic cell.

Anorexia absence of appetite.

Anoxic absence of oxygen; anaerobic.

Anoxygenic photosynthesis photosynthesis that takes place in the absence of oxygen and during which oxygen is not produced; photosynthesis that does not split water and evolve oxygen.

Anoxyphotobacteria bacteria that can carry out only anoxygenic photosynthesis.

Antagonism the inhibition, injury, or killing of one species of microorganism by another.

Anthrax an infectious disease of animals, including humans, cattle, sheep, and pigs, caused by *Bacillus anthracis*.

Anti- combining form meaning opposing in effect or activity.

Antibacterial agents agents that kill or inhibit the growth of bacteria, e.g., antibiotics, antiseptics, and disinfectants.

Antibiotic resistance the natural or acquired ability of a microorganism to overcome the inhibitory effects of an antibiotic.

Antibiotics substances of microbial origin that in very small amounts have antimicrobial activity; current usage of the term extends to synthetic and semisynthetic substances that are closely related to naturally occurring antibiotics and have antimicrobial activity.

Antibodies glycoprotein molecules produced in the body in response to the introduction of an antigen that can specifically react with that antigen; also known as immunoglobulins, which are part of the serum fraction of the blood formed in response to antigenic stimulation and which react with antigens with great specificity.

Antibody-dependent cytotoxic hypersensitivities type 2 hypersensitivity, reaction in which an antigen present on the surface of the cell combines with an antibody, resulting in the death of that cell by stimulating phagocytic attack or initiating the complement pathway; examples include blood transfusions between incompatible types and Rh incompatibility.

Antibody-mediated immunity immunity produced by the activation of the B-lymphocyte population, leading to the production of several classes of immunoglobulins.

Anticodon a sequence of three nucleotides in a tRNA molecule that is complementary to the codon triplet in mRNA.

Antifungal agents agents that kill or inhibit the growth and reproduction of fungi; may be fungicidal or fungistatic.

Antigen any agent that initiates antibody formation and/or induces a state of active immunological hypersensitivity and that can react with the immunoglobulins formed.

Antigen–antibody referring to the specific reaction between antigen and complementary antibody molecules.

Antihistamines compounds used for treating allergic reactions and cold symptoms by inactivating histamine, which is released as part of the immune response.

Antihuman gamma globulin antibodies that react specifically with human antibodies.

Antimicrobial agents chemical or biological agents that kill or inhibit the growth of microorganisms.

Antimicrobial susceptibility testing laboratory tests designed to determine the resistance of disease-causing microorganisms to antimicrobial agents.

Antimicrobics chemicals that inhibit or kill microorganisms (antimicrobial agent) and can be safely introduced into the human body; include synthetic and naturally produced antibiotics and generally is used instead of the term antibiotic.

Antiseptics chemical agents used to treat human or animal tissues, usually skin, in order to kill or inactivate those microorganisms capable of causing infection; not considered safe for internal consumption.

Antisera blood or blood fraction containing antibodies; serum fraction of blood containing immunoglobulins.

Antitoxin antibody to a toxin capable of reacting with that poison and neutralizing the specific toxin.

Antiviral agents substances capable of destroying or inhibiting the reproduction of viruses.

Appendaged bacteria those bacteria that form extensions or protrusions from their cells.

Appendicitis inflammation of the appendix, generally caused by excessive growth of indigenous microbes and characterized by abdominal pain, nausea, vomiting, and elevated white blood cell count.

Apochromatic lens an objective microscope lens in which chromatic aberration has been corrected for three colors and spherical aberration for two colors.

Aquatic growing in, or living in, or frequenting water; a habitat composed primarily of water.

Aqueous of, relating to, or resembling water; made from, with, or by water; solutions in which water is the solvent.

Archaebacteria prokaryotes with cell walls that lack murein, having ether bonds in their membrane phospholipids; analysis of rRNA indicates that the archaebacteria represent a primary biological kingdom related to both eubacteria and eukaryotes; considered to be a primitive group of organisms that were among the earliest living forms on earth.

Arthrobacter bacterial genus—cells that in complex media undergo a marked change in form during the growth cycle; older cultures (generally 2–7 days) are composed entirely or largely of coccoid cells; on transfer to fresh complex medium growth occurs by enlargement (swelling) of the coccoid cells followed by elongation; do not form endospores; Gram positive, but the rods may be readily decolorized and may show only Gram positive granules in otherwise Gram negative cells; not acid fast; metabolism respiratory, never fermentative; strict aerobes; optimal growth 20–30°C; G + C 60–72 mole%.

Arthropods animals of the invertebrate phylum Arthropoda, many of which are capable of acting as vectors of infectious diseases.

Arthrospores spores formed by the fragmentation of hyphae of certain fungi, algae, and cyanobacteria.

Arthus reaction immune-complex reaction, which occurs because there is an excess of antibody, characterized by tissue destruction and inflammation.

Artificially acquired active immunity the production of antibodies and development of a memory response by the body as a result of administration of a vaccine (vaccination).

Artificially acquired passive immunity the transfer of humoral antibodies formed by one individual to a susceptible individual, accomplished by injection of antiserum.

Ascomycetes members of a class of fungi distinguished by the presence of an ascus, a sac-like structure containing sexually produced ascospores.

Ascospores sexual spores characteristic of ascomycetes, produced in the ascus after the union of two nuclei.

Ascus the sporangium or spore case of fungi consisting of a single terminal cell.

-ase suffix denoting an enzyme.

Asepsis state in which potentially harmful microorganisms are absent; free of pathogens.

Aseptic meningitis any infection of the meninges not caused by bacteria or fungi, a viral cause is suspected.

Aseptic techniques precautionary measures taken in microbiological work and clinical practice to prevent the contamination of cultures, sterile media, etc., and /or infection of persons, animals, or plants by extraneous microorganisms.

Aseptic transfer technique the movement of microbial cultures in such a way as to minimize or eliminate exposure to the atmosphere.

Asexual lacking sex or functional sexual organs.

Asexual reproduction reproduction without union of gametes; formation of new individuals from a single individual.

Asexual spores spores produced asexually that often are involved in survival or dissemination of microorganisms.

Aspirate to draw or remove fluid by suction.

Assay analysis to determine presence, absence, or quantity of one or more substances.

Assimilation the incorporation of nutrients into the biomass of an organism.

Asymptomatic carriers persons who harbor infectious agents, who are not themselves ill, but who can infect others.

Ataxia the inability to coordinate muscular action.

Atmosphere the whole mass of air surrounding the earth; a unit of pressure approximating 1×10^6 dynes/cm^2.

Atopic allergies localized expression of type 1 hypersensitivity reaction; examples include hay fever and food allergies.

ATP see adenosine triphosphate.

ATPase see adenosine triphosphatase.

Attentuated pathogen a pathogen whose virulence has been diminished.

Attenuation any procedure in which the pathogenicity of a given organism is reduced or abolished; reduction in the virulence of a pathogen; control of protein synthesis involving the translation process.

Atypical pneumonia pneumonia that is not secondary to another acute infectious disease; pneumonia not treatable with penicillin; pneumonia caused by a bacterial pathogen that is not culturable on routine defined media.

Aureomycin 7-Chlorotetracycline; an antibiotic obtained from *Streptomyces aureofaciens*.

Autochthonous microorganisms and/or substances indigenous to a given ecosystem; the true inhabitants of an ecosystem; often used to refer to the common microbiota of the body or those species of soil microorganisms that tend to remain constant despite fluctuations in the quantity of fermentable organic matter in the soil.

Autoclave apparatus in which objects or materials may be sterilized by air-free saturated steam under pressure at temperatures in excess of 100°C.

Autoimmunity immunity or hypersensitivity to some constituent in one's own body; immune reactions with self-antigens.

Autolysins endogenous enzymes involved in the breakdown of certain structural components of the cell during particular phases of cellular growth and development.

Autolysis the breakdown of the components of a cell or tissues by endogenous enzymes, usually after the death of the cell or tissue; enzymatic self-destruction of a cell.

Autotrophic capabable of synthesizing organic components from inorganic sources and generating ATP from light or the oxidation of inorganic compounds.

Autotrophic metabolism metabolism that does not require organic carbon as source of carbon and energy.

Autotrophs organisms whose growth and reproduction are independent of external sources of organic compounds, the requirement for cellular carbon being accomplished by the reduction of carbon dioxide and the need for cellular energy being met by the conversion of light energy to ATP or the oxidation of inorganic compounds to supply the free energy for the formation of ATP.

Avirulent lacking virulence; a microorganism lacking the properties that normally promote the ability to cause disease.

Axopodia semipermanent pseudopodia (false-feet extensions), e.g., the pseudopodia that emanate radially from the spherical cells of heliozoans and some radiolarian species.

Azotobacter bacterial genus—large ovoid cells with marked pleomorphism; Gram negative, with marked variability; fix atmospheric nitrogen; grow well aerobically but can also grow under reduced oxygen tension; optimal growth between 20 and 30°C; G + C 63–66 mole%.

β-Galactosidase an enzyme catalyzing the hydrolysis of β-linked galactose within dimers or polymers.

β-Hemolysis complete lysis of red blood cells as shown by the presence of a sharply defined zone of clearing surrounding certain bacterial colonies growing on blood agar.

β-Lactamase an enzyme that attacks a β-lactam ring, such as a penicillinase that attacks the lactam ring in the penicillin class antimicrobials, inactivating such antibiotics.

β-Oxidation metabolic pathway for the oxidation of fatty acids resulting in the formation of acetate and a new fatty acid two carbon atoms shorter than the parent fatty acid.

Babesiosis disease caused by the protozoan *Babesia microti*, which is transmitted by ticks and endemic to Nantuckett Island and coastal areas of New England.

Bacillus bacterial genus—cells rod-shaped, straight or nearly so; majority motile; heat-resistant endospores formed; not more than one in a sporangial cell; Gram reaction is positive, or positive only in early stages of growth; metabolism strictly respiratory, fermentative, or both respiratory and fermentative, using various substrates; strict aerobes or facultative anaerobes; G + C 32–62 mole%.

Bacteremia condition in which viable bacteria are present in the blood.

Bacteria (singular, **bacterium**) members of a group of diverse and ubiquitous prokaryotic, single-celled organisms.

Bacterial chromosome circular DNA macromolecule that stores the genome of a bacterial cell; lacks histones associated with eukaryotic chromosomes.

Bacterial dysentery dysentery caused by *Shigella* infections; also called shigellosis.

Bacterial meningitis inflammation of the meninges caused by a bacterial infection of the spinal column.

Bactericidal any physical or chemical agent able to kill some types of bacteria.

Bacteriocides chemical agents that kill bacteria.

Bacteriological filter a filter with pores small enough to trap bacteria, about 0.45 μm or smaller, used to sterilize solutions by removing microorganisms during filtration.

Bacteriology the science dealing with bacteria, including their relation to medicine, industry, and agriculture.

Bacteriophage a virus whose host is a bacterium.

Bacteriostatic an agent that inhibits the growth and reproduction of some types of bacteria but need not kill the bacteria.

Bacteroides bacterial genus—Gram negative, nonspore-forming rods; metabolize carbohydrates or peptone; fermentation products of sugar-utilizing species include combinations of succinic, lactic, acetic, formic, or propionic acids, sometimes with short-chained alcohols; obligately anaerobic; optimal growth 37°C; G + C 40–55 mole%.

Bacteroids irregularly shaped forms that some bacteria can assume under certain conditions, e.g., *Rhizobium* in root nodules.

Balantidiasis a protozoan disease of the digestive system also known as balantidial dysentery, caused by *Balantidium coli*, typically a mild disease consisting of abdominal pain, nausea, vomiting, diarrhea, and weight loss.

Bartonella bacterial genus—in stained blood films the organisms appear as rounded or ellipsoidal forms or as slender, straight, curved, or bent rods, occurring either singly or in groups within erythrocytes; within tissues they are situated within the cytoplasm of endothelial cells as isolated elements or are grouped in rounded masses; Gram negative; not acid fast; stain poorly or not at all with many aniline dyes but satisfactorily with Romanowsky's and Giemsa's stains.

Bartonellosis Carrioń's disease; a bacterial infection of humans endemic to the Andes caused by *Bartonella bacilliformis*, which attacks red blood cells.

Basic stain a dye whose active staining part consists of a cationic, negatively charged group that may be combined with an acid, usually inorganic, that has affinity for nucleic acids.

Basidiocarps the fruiting bodies of basidiomycetes.

Basidiomycetes a group of fungi distinguished by the formation of sexual basidiospores on a basidium.

Basidiospores sexual spores formed on basiodcarps by basidiomycetes.

Basidium (plural, **basidia**) club-like structure of basidiomycetes on which basidiospores are borne.

Basophils white blood cells containing granules (granulocyctes) that readily take up basic dye.

Batch culture bacteria grown by inoculation on fresh media; models situations such as occur when a canned food is contaminated with a bacterium.

Batch process common simple form of culture in which a fixed volume of liquid medium is inoculated and incubated for an appropriate period of time; cells grown this way are exposed to a continually changing environment; when used in industrial processes, the culture and products are harvested as a "batch" at appropriate times.

Bauer–Kirby test a standardized antibiotic susceptibility procedure using the agar diffusion method, whereby the concentration gradient established by diffusion of the antibiotic into the agar reveals the sensitivity of the tested microorganism to the specific antibiotic.

B cell a differentiated lymphocyte involved in antibody-mediated immunity.

BCG vaccine vaccine of bacillus of Calmette–Gurein; active vaccine prepared from *Mycobacterium bovis* that is used to immunize against tuberculosis.

Bdellovibrio bacterial genus—cells are single, small, curved, motile rods in the parasitic state; motile; Gram negative; parasitic strains attach to and penetrate into bacterial host cells; metabolism respiratory, not fermentative; optimal growth 30°C; G + C 45.5–51.3 mole%.

Beer alcoholic beverage produced via microbial fermentation and brewing of cereal grains.

Bergey's Manual reference book describing the established status of bacterial taxonomy; describes bacterial taxa and provides keys and tables for their identification.

Beta-galactosidase see β-galactosidase.

Beta-hemolysis see β-hemolysis.

Beta-lactamase see β-lactamase.

Beta-oxidation see β-oxidation.

Bifidobacterium bacterial genus—rods highly variable in appearance; Gram positive; nonacid-fast; nonspore-forming; nonmotile; anaerobic; optimal growth 36–38°C; G + C 57.2–64.5 mole%.

Binary fission a process in which two similarly sized and shaped cells are formed by the division of one cell.

Binomial nomenclature the scientific method of naming plants, animals, and microorganisms, using two names consisting of the species epithet and genus name.

Bioassay the use of a living organism to determine the amount of a substance based on the extent of growth or activity of the test organism under controlled conditions.

Biochemicals substances produced by and/or involved in the metabolic reactions of living organisms.

Biocides chemical agents that kill organisms including microbes and can be used to sterilize materials.

Biodegradation the process of chemical breakdown of a substance to smaller products caused by microorganisms or their enzymes.

Biodisc see biodisc system.

Biodisc system a secondary sewage treatment system employing a film of active microorganisms rotated on a disc through sewage; also known as rotating biological contactor.

Bioenergetics the transfer of energy through living systems; energy transformations in living systems.

Biogeochemical cycling the biologically mediated transformations of elements that result in their global cycling, including transfer between the atmosphere, hydrosphere, and lithosphere.

Biological control the deliberate use of one species of organism to control or eliminate populations of other organisms; used in the control of pest populations.

Biological oxygen demand the amount of dissolved oxygen required by aerobic and facultative microorganisms to stabilize organic matter in sewage or water; also known as biochemical oxygen demand or BOD.

Bioluminescence the generation of light by certain microorganisms.

Biomagnification the phenomena of the increase in the concentration of a chemical substance, such as a pesticide, as the substance is passed to higher members of a food chain.

Biomass the dry weight, volume, or other quantitative estimation of organisms; the total mass of living organisms in an ecosystem.

Biosphere the part of the earth in which life can exist; all living things together with their environment.

Biosynthesis the production (synthesis) of chemical substances by the metabolic activities of living organisms; also called anabolism.

Biotic of or relating to living organisms; caused by living things.

Biotype a variant form of a given species or serotype, distinguishable by biochemical or other means.

Blastomycosis a chronic mycosis (fungal infection) caused by *Blastomyces* in which lesions develop, e.g., in the lungs, bones, and skin.

Blood–brain barrier cell membranes that control the passage of substances from the blood to the brain.

Blood plasma the fluid portion of the blood minus all blood corpuscles.

Blood serum the fluid expressed from clotted blood or clotted blood plasma.

Bloodstream the flowing blood in the circulatory system.

Blood type an immunologically distinct, genetically determined set of antigens on the surfaces of erythrocytes (red blood cells), defined as A, B, AB, and O.

Bloom a visible abundance of microorganisms, generally referring to the excessive growth of algae or cyanobacteria at the surface of a body of water.

B lymphocytes white blood cells that are able to produce specific immunoglobulins; their surfaces carry specific immunoglobulin antigen-binding receptor sites.

B memory cells specifically stimulated B lymphocytes not actively multiplying but capable of multiplication and production of plasma cells on subsequent antigenic stimuli.

BOD see biological oxygen demand.

Booster vaccines vaccine antigens administered to elicit an anamnestic response and to maintain extended active immunity.

Bordetella bacterial genus—minute coccobacilli arranged singly or in pairs, more rarely in short chains; Gram negative, bipolar; colonies on potato–glycerol–blood agar medium, smooth, convex, pearly, glistening, nearly transparent, surrounded by zone of hemolysis without definite periphery; metabolism respiratory, never fermentative; strict aerobes; optimal growth 35–37°C.

Borrelia bacterial genus—cells helical, with 3–10 or more coarse, uneven, often irregular coils; Gram negative; fermentative metabolism; strict anaerobes; optimal growth 28–30°C.

Botulism food intoxication or poisoning that is severe and often fatal, caused by ingestion of toxins produced by *Clostridium botulinum*.

Bradykinin a peptide formed by enzymatic modification of an α-globulin (kininogen) acting as a vasodilator and playing a role in neural recognition of noxious stimuli.

Brain abscesses free or encapsulated collections of pus in the brain, secondary to some other infection usually caused by pyogenic bacteria.

Broad spectrum antibiotics antibiotics capable of effective action against a wide range of pathogenic microorganisms.

Bronchitis inflammation occurring in the mucous

membranes of the bronchi often caused by *Streptococcus pneumoniae*, *Haemophilus influenzae*, and certain viruses.

Brown algae members of the division Phaeophycophyta that produce xanthophylls.

Brucella bacterial genus—coccobacilli or short rods; mammalian parasites and pathogens; no capsules; nonmotile; do not form endospores; Gram negative; metabolism respiratory; strict aerobes; some require 5–10 percent added carbon dioxide for growth especially on initial isolation; optimal growth 20–40°C; G + C 56–58 mole%.

Brucellosis a remittent febrile disease caused by infection with bacteria of the genus *Brucella*.

Bruton's congenital agammaglobulinemia chromosomal linked B-cell immumodeficiency disease affecting males in which all immunologic classes are totally or partially absent.

Buboes enlarged regional lymph nodes usually associated with cases of bubonic plague.

Budding a form of asexual reproduction that occurs in many yeasts and some bacteria, in which a daughter cell develops from a small outgrowth or protrusion of the parent cell, the daughter cell being smaller than the parental cell; continuous slow release of viruses from an infected host cell by exocytosis, in which the viruses acquire a covering of host cell membrane—budding viruses do not kill the host cell and therefore such infections persist.

Budding bacteria bacteria that reproduce by budding.

Buffer a solution that tends to resist the change in pH when acid or alkali is added.

Burkitt's lymphoma malignant tumorous growth on jaw or abdomen caused by the Epstein–Barr virus, usually affecting children.

Burst size the average number of infectious viral units released from a single cell.

Calvin cycle the primary pathway for carbon dioxide fixation in photoautotrophs and chemolithotrophs.

Campylobacter bacterial genus—slender non-spore-forming spirally curved rods; motile; respiratory; methyl red and Voges–Proskauer negative; found in the reproductive organs, intestinal tract, and oral cavity of animals and humans; G + C 30–35 mole%.

Cancer a malignant, invasive cellular tumor that can spread throughout the body.

Candidiasis mild, superficial fungal infection caused by *Candida* sp., infecting the skin, nails, or mucous membranes.

Canning method for the preservation of foodstuffs where suitably prepared foods are placed in metal containers that are heated, exhausted, and hermetically sealed.

Capillary one of a network of tiny hair-like blood vessels connecting the arteries to the veins.

Capsid a protein coat of a virus enclosing the naked nucleic acid.

Capsomere the individual protein units that form the capsid of a virus.

Capsule a mucoid envelope composed of polypeptides and/or carbohydrates surrounding certain microorganisms; a gelatinous or slimy layer external to the bacterial cell wall.

Carbohydrates a class of organic compounds consisting of many hydroxyl (—OH) groups and containing either a ketone or aldehyde.

Carbolic acid phenol.

Carbon cycle the biogeochemical cycle by which carbon is recycled by natural processes; carbon, in the form of the gas carbon dioxide, is converted into food by producer organisms, consumer organisms obtain their carbon by eating these producers, carbon leaves producers and consumers in the form of wastes and dead material, which are broken down by decomposers.

Carboxyl end the terminus of a polypeptide chain with a free alpha-carboxyl group not involved in forming a peptide linkage; also known as the C-terminal end.

Carboxylic acid an organic chemical having a —COOH functional group.

Carcinogen cancer-causing agent.

Cardiovascular system the system that circulates blood throughout the body; includes the heart, blood vessels, arteries, veins, and capillaries.

Carditis inflammation of the heart.

Caries bone or tooth decay with formation of ulceration; also known as dental caries.

Carotenoid pigments a class of pigments usually yellow, orange, red, or purple, that are widely distributed among microorganisms.

Carriers individuals who harbor pathogens but do not exhibit any signs of illness.

Catabolism reactions involving the enzymatic degradation of organic compounds to simpler organic or inorganic compounds with the release of free energy.

Catabolite repression repression, by glucose or other readily utilizable carbon sources, of the transcription of genes coding for certain inducible enzyme systems.

Catalases enzymes that catalyze the decomposition of hydrogen peroxide (H_2O_2) into water and oxygen and the oxidation of alcohols to aldehydes by hydrogen peroxide.

Catalyst any substance that accelerates a chemical reaction but itself remains unaltered in form and amount.

Catalyze to subject to modification, especially an increase in the rate of chemical reaction.

Catheterization insertion of a hollow tubular device (a catheter) into a cavity, duct, or vessel to permit injection or withdrawal of fluids.

Catheters narrow, hollow tubes that are inserted into a body cavity in order to inject or to withdraw fluids.

Cathode the electrode at which reduction takes place in an electrolytic cell; a negatively charged electrode.

Cations positively charged ions.

Cell the functional and structural subunit of living organisms separated from its surroundings by a delimiting membrane.

Cell-mediated hypersensitivities type 4, or delayed hypersensitivities, involve T lymphocytes and occur 24–72 hours after exposure to the antigen; contact dermatitis, including poison ivy, is an example.

Cell-mediated immunity specific acquired immunity involving T cells, primarily responsible for resistance to infectious diseases caused by certain bacteria and viruses that reproduce within host cells.

Cellulase an extracellular enzyme that hydrolyzes cellulose.

Cellulose a linear polysaccharide of β-D-glucose.

Cell wall structure outside of and protecting the cell membrane, generally containing murein in prokaryotes and composed chiefly of various other polymeric substances, e.g., cellulose or chitin, in eukaryotic microorganisms.

Centers for Disease Control branch of the U. S. Public Health Service in Atlanta, Georgia, responsible for the generation, collection, and dissemination of epidemiological information; works to identify the causes of major public health problems; also collects and maintains samples of microbiological cultures.

Central nervous system that part of the nervous system contained within the brain and spinal cord.

Centrifuge an apparatus using centrifugal force to separate by sedimentation particulate matter suspended in a liquid.

Cephalosporins a heterogeneous group of natural and semisynthetic antibiotics active against a range of Gram positive and negative bacteria, by inhibiting the formation of crosslinks in peptidoglycan.

Cerebrospinal fluid the fluid contained within the four ventricles of the brain, the subarachnoid space, and the central canal of the spinal chord.

Cestodes tapeworms; endoparasites, adult forms of which live in the intestines of vertebrate hosts.

Chagas' disease also known as South American trypanosomiasis; caused by the protozoan *Trypanosoma cruzi*, which is carried by the conenose bug that lives in wood or mud houses.

Chanchre the lesion formed at the site of primary inoculation by an infecting microorganism, usually an ulcer.

Chancroid a lesion produced by an infection with *Haemophilus ducreyi* involving the genitalia; a sexually transmitted disease caused by *H. ducreyi*.

Cheese product of the microbial fermentation of milk by lactic acid bacteria.

Chemical preservative those chemical substances specifically added to prevent the spoilage of a food or the biodeterioration of any substance, by inhibiting microbial growth and/or activity.

Chemiosmosis see chemiosmotic hypothesis.

Chemiosmotic hypothesis the theory that the living cell establishes a proton and electrical gradient across a membrane and that by controlled reentry of protons into the region contained by that membrane, the energy to carry out several different types of endergonic processes may be obtained, including the ability to drive the formation of ATP.

Chemoautotrophs bacteria that derive their energy source from the oxidation of inorganic compounds; organisms that obtain energy through chemical oxidation and use inorganic compounds as electron donors, also known as chemolithotrophs.

Chemolithotrophic bacteria see chemoautotrophs.

Chemolithotrophs see chemoautotrophs.

Chemostat an apparatus used for continuous-flow culture to maintain bacterial cultures in the log phase of growth; based on maintaining a continuous supply of a solution containing a nutrient in limiting quantities that controls the growth rate of the culture.

Chemotaxis a type of locomotive response in which the stimulus is a chemical concentration gradient; movement of microorganisms toward or away from a chemical stimulus.

Chemotherapy the use of chemical agents for the

treatment of disease, including the use of antibiotics to eliminate infecting agents.

Chicken pox common, acute, and highly contagious infection caused by the herpesvirus varicella–zoster, occurring most frequently in children and producing a distinctive rash.

Childbed fever see puerperal fever.

"Chinese letter" formation see snapping division.

Chitin a polysaccharide composed of repeating N-acetylglucosamine residues, abundant in nature in arthropod exoskeletons and fungal cell walls.

Chlamydia bacterial genus—nonmotile, spheroidal cells; obligately intracellular growth cycle; the principal developmental stages are (1) the elementary body, which is a small, electron-dense spherule, containing a nucleus and numerous ribosomes surrounded by a multilaminated wall; (2) the initial body, which is a large, thin walled, reticulated spheroid, containing nuclear fibrils and ribosomal elements; and (3) an intermediate body representative of a transitional stage between the initial body and elementary body; the elementary body is the infectious form of the organism; the initial body is the vegetative form that divides by fission intracellularly but is apparently noninfectious when separated from the host cells; Gram negative; because they are unable to synthesize their own high energy compounds, they have been described as "energy parasites;" optimal growth 33–41°C; G + C 39–45 mole%.

Chlamydiae members of the bacterial genus *Chlamydia,* all of which are obligate intracellular inhabitants.

Chlamydospores thick-walled, typically spherical or ovoid resting spores produced asexually by certain types of fungi from cells of the somatic hyphae.

Chlor- combining form meaning having chlorine as a substitute for hydrogen.

Chlorination the process of treating with chlorine as for disinfecting drinking water or sewage.

Chlorophyll the green pigment responsible for absorption of light energy during photosynthesis in plants; the primary photosynthetic pigment of algae and cyanobacteria.

Chloroplasts membrane-bound organelles of photosynthetic eukaryotes where the biochemical conversion of light energy to ATP occurs; the sites of photosynthesis in eukaryotic organisms.

Cholera an acute infectious disease caused by *Vibrio cholerae,* characterized by severe diarrhea, delirium, stupor, and coma.

Chromatic aberration an optical lens defect causing distortion of the image because light of differing wavelengths is focused at differing points instead of at a single focal point.

Chromatids fibrils formed from a eukaryotic chromosome when it replicates prior to meiosis or mitosis.

Chromatin the deoxyribonucleic acid–protein complex that constitutes a chromosome; the readily stainable protoplasmic substance in the nuclei of cells.

Chromosomes structures that contain the nuclear DNA of a cell.

Chronic disease any slowly developing illness that is likely to continue or recur for a long period of time.

-cide suffix signifying a killer or destroyer.

CIE see countercurrent immunoelectrophoresis.

Cilia thread-like appendages, having a 9 + 2 arrangement of microtubules, occurring as projections from certain cells, which beat rhythmically, causing locomotion or propelling fluid over surfaces.

Ciliates members of the protozoan phylum Ciliata that use cilia for locomotion.

Ciliophora members of one subphylum of protozoa that possess simple to compound ciliary organelles in at least one stage of their life history; protozoa motile by means of cilia.

Circulatory system the vessels and organs comprising the cardiovascular and lymphatic systems of animals.

Cistron the functional unit of genetic inheritance; a segment of genetic nucleic acid that codes for a specific polypeptide chain; synonym for gene.

Citric acid cycle see Krebs cycle.

Citrobacter bacterial genus—motile, peritrichously flagellated rods, not encapsulated; Gram negative; can use citrate as sole carbon source.

Clamp cell connections hyphal structures in many basidiomycetes formed during cell division by dikaryotic hyphal cells, i.e., formed by hyphal cells containing two nuclei of different mating types.

Classification the systematic arrangement of organisms into groups or categories according to established criteria.

Clinical immunologist a certified individual in charge of a diagnostic immunology laboratory, who is responsible for the submission of diagnostic reports to physicians.

Clinical microbiologist a certified individual in charge of a diagnostic microbiology laboratory, who is responsible for the submission of diagnostic reports to physicians.

Clonal selection theory a theory to account for

antibody formation; supposes that each B lymphocyte in the body is capable of responding to a specific nonself antigen with which the body may or may not have had previous contact.

Clone a population of cells derived asexually from a single cell, often assumed to be genetically homologous; a population of genetically identical individuals.

Clostridium bacterial genus—rods, usually motile by means of peritrichous flagella; form endospores; Gram positive but may appear Gram negative in the late stages of growth; fermentative; most strains are strictly anaerobic; G + C 23–43 mole%.

Club fungi basidiomycete-producing fungi.

Coagglutination an enhanced agglutination reaction based on using antibody molecules whose Fc fragments are attached to cells so that a larger matrix is formed when the Fab portion reacts with other cells for which the antibody is specific.

Coagulase an enzyme produced by pathogenic staphylococci, causing coagulation of blood plasma.

Cocci spherical or near-spherical bacterial cells, varying in size and sometimes occurring singly, in pairs, in regular groups of four or more, in chains, or irregular clusters.

Coccidioidomycosis a disease of humans and domestic animals that generally is self-limiting caused by *Coccidioides immitis,* usually occurring via the respiratory tract; also known as Valley Fever and San Joaquin Valley Fever.

Coccobacilli bacteria that are very short oval-shaped rods.

Codon a triplet of adjacent bases in a polynucleotide chain of an mRNA molecule that codes for a specific amino acid; the basic unit of the genetic code specifying an amino acid for incorporation into a polypeptide chain.

Coenocytic referring to any multinucleate cell, structure, or organism formed by the division of an existing multinucleate entity or formed when nuclear divisions are not accompanied by the formation of dividing walls or septa; multinucleate hyphae.

Coenocytic hyphae fungal filaments that lack septa and so do not divide into uninucleate cell-like units.

Coenzymes the nonprotein portions of enzymes; biochemicals that act as acceptors or donors of electrons or functional groups during enzymatic reactions.

Cofactors see coenzymes.

Cold sores also known as fever blisters; lesions that erupt on the mouth or face, caused by a recurrent infection with herpes simplex 1 virus.

Coliforms Gram negative, lactose-fermenting, rod-shaped bacteria that generally inhabit the gastrointestinal tract.

Colinear two related linear information sequences, arranged so that the unit may be taken from the other without rearrangement.

Colonization the establishment of a site of microbial reproduction on a material, animal, or person without necessarily resulting in tissue invasion or damage.

Colony the macroscopically visible growth of microorganisms on a solid culture media.

Colorado tick fever the only recognized tick-borne viral disease in the United States; an acute febrile disease characterized by sudden onset of fever, headache, and severe muscle pain.

Coma a state of unconsciousness.

Cometabolism the gratuitous metabolic transformation of a substance by a microorganism growing on another substrate; the cometabolized substance is not incorporated into an organism's biomass, and the organism does not derive energy from the transformation of that substance.

Commensalism an interactive association between two populations of different species living together, in which one population benefits from the association and the other is not affected.

Common cold an acute self-limiting inflammation of the upper respiratory tract due to a viral infection.

Common source outbreak disease outbreak characterized by a sharp rise and rapid decline.

Communicable diseases those diseases that can be spread from one host to another.

Competent bacteria bacteria capable of taking up a DNA fragment from a bacterial donor cell.

Competition an interactive association between two species both of which need some limited environmental factor for growth and thus grow at suboptimal rates because they must share the growth limiting resource.

Competitive inhibition the inhibition of enzyme activity caused by the competition of an inhibitor with a substrate for the active (catalytic) site on the enzyme; impairment of function of an enzyme due to its reaction with a substance chemically related to its normal substrate.

Complement group of proteins normally present in plasma and tissue fluids that participates in antigen–antibody reactions, allowing reactions such as cell lysis to occur and enhancing phagocytosis by polymorphs and macrophages.

Complement fixation the binding of complement to an antigen–antibody complex so that the complement is unavailable for subsequent reactions.

Complementary strands designation used to indicate that one strand of a DNA macromolecule is the antithesis of the other strand—the juxtaposed nucleotides are complementary bases (adenine opposite thymine and guanine opposite cytidine), one stand runs from the 3′-hydroxyl free end to the 5′-hydroxyl free end and the other runs in the opposite direction.

Completed test the third stage in the standard test for fecal contamination of water; if it is positive, gas is formed when colonies showing a green metallic sheen on EMB are cultured in lactose broth and incubated at 35°C.

Complex-mediated hypersensitivities type 3 hypersensitivity reactions occur when excess antigens are produced during a normal inflammatory response and antigen–antibody–complement complexes (immune complexes) are deposited in tissues.

Composting the decomposition of organic matter in a heap by microorganisms; a method of solid waste disposal.

Computer-assisted identification identification system that is widely used in clinical laboratories, in which a computer is employed to compare the results of a series of phenotypic tests of an unknown isolate against the test results of organisms that have already been classified as belonging to a particular taxonomic group; many comuter-aided commercial systems provide identificaions within hours.

Condenser lenses the lenses on a microscope used for focusing or directing light from the light source onto the object.

Confirmed test the second stage of the standard test for fecal contamination of water, colonies are grown on EMB agar.

Congenital diseases those illnesses present at birth that are the result of some in utero condition.

Conidia thin-walled, asexually derived spores, borne singly or in groups or clusters in specialized hyphae.

Conidiophores branches of mycelia-bearing conidia.

Conjugation the process in which genetic material is transferred from one microorganism to another, involving a physical connection or union between the two cells; a parasexual form of reproduction, sometimes referred to synonymously as mating.

Conjunctivitis inflammation of the mucous membranes covering the eye, the conjunctiva.

Constipation a condition in which the bowels are evacuated at long intervals or with difficulty; the passage of hard, dry stools.

Constitutive enzymes enzymes whose synthesis is not altered in response to changes in the environment but rather are continuously synthesized.

Contagion the process by which disease spreads from one individual to another.

Contagious disease an infectious disease that is communicable to healthy, susceptible individuals by physical contact with one suffering from that disease, contact with bodily discharges from that individual, or contact with inanimate objects contaminated by that individual.

Contamination the process of allowing uncontrolled addition of microorganisms to an area or substance.

Continuous culture a method for growing microorganisms without interruption by continual addition of substrates and recovery of products.

Continuous feed reactors method of composting, using a reactor that establishes environmental parameters that maximize the degradation process.

Continuous process an industrial process for growing microorganisms without interruption by continual addition of substrates and recovery of products.

Continuous strand the strand of DNA that can be synthesized continuously because it runs in the appropriate direction for the continuous addition of new free nucleotide bases; also referred to as the leading strand.

Contractile vacuoles pulsating vacuoles found in certain protozoa, used for the excretion of wastes and the exclusion of water for the maintenance of proper osmotic balance.

Convalescence recovery period of a disease during which signs and symptoms disappear.

Cortex a layer of a bacterial endospore important in conferring heat resistance to that structure.

Corticosteroids derivatives of the steroid hormones released by the adrenal glands used to treat inflammatory diseases.

Cortisone hormone that relieves arthritis pain; derivatives alleviate allergy symptoms and other inflammatory respones; can be chemically synthesized.

Corynebacterium bacterial genus—straight to

slightly curved rods with irregularly stained segments and sometimes granules; frequently show club-shaped swellings; snapping division produces angular and palisade (picket fence) arrangements of cells; generally nonmotile; Gram positive, although some species (e.g., *C. diphtheriae*) decolorize easily, especially in old cultures; not acid fast; carbohydrate metabolism fermentative and respiratory; grow best aerobically; G + C 52–68 mole%.

Coryza an inflammation of the mucous membranes of the nose, usually marked by sneezing and the discharge of watery mucus.

Counter current immunoelectrophoresis a technique based on immunological detection of substances relying on the movement of antibody and antigen toward each other in an electric field resulting in the rapid formation of a detectable antigen–antibody precipitate.

Counter immunoelectrophoresis see countercurrent immonoelectrophoresis.

Covalent bond a strong chemical bond formed by the sharing of electrons.

Cowpox a mild, self-limiting disease caused by a vaccinia virus, involving the formation of vesicular lesions in the hands and arms of man and the udders of cows.

Coxiella bacterial genus—short rods resembling organisms of the genus *Rickettsia* in staining properties, dependence on host cells for growth and close natural association with arthropod and vertebrate hosts; grows preferentially in the vacuoles of the host cell rather than in cytoplasm or nucleus as do the species of *Rickettsia*.

CPE see cytopathic effects.

Creutzfeldt–Jakob disease rapidly progressive disease that results in degeneration of the central nervous system (dementia) caused by a prion; this disease normally affects adults between ages 40 and 65.

Critical point drying a method for removal of liquids from a microbiological specimen by adjusting temperature and pressure so that the liquid and gas phases of the liquid are in equilibrium with each other; used to minimize disruption of biological structures for viewing by scanning electron microscopy.

Crossing over the process in which a break occurs in effect in each of two adjacent DNA strands and the exposed 5′-hydroxyl and 3′-hydroxyl ends unite with the exposed 5′-hydroxyl and 3′-hydroxyl ends of the adjacent strands so that there is an exchange of homologous regions of DNA.

Culture to encourage the growth of particular microorganisms under controlled conditions; a growth of particular types of microorganisms on or within a medium as a result of inoculation and incubation.

Curvature of field distortion of a microscopic field of view in which specimens in the center of the field are in clear focus while the peripheral region of the field of view is out of focus.

Cutaneous pertaining to the skin.

Cyanobacteria prokaryotic, photosynthetic organisms containing chlorophyll a, capable of evolving oxygen by the splitting of water; formerly known as blue-green algae.

Cyst a dormant form assumed by some microorganisms during specific stages in their life cycles, or assumed as a response to particular environmental conditions, in which the organism becomes enclosed in a thin- or thick-walled membranous structure, the function of which is either protective or reproductive; a normal or pathological sac with a distinct wall containing fluid.

Cysticercosis disease caused by ingestion of pork tapeworm larvae that occurs when eggs or larvae move from the infested gastrointestinal tract and develop in the liver, muscles, heart, or brain.

Cystitis an inflammation of the urinary bladder.

Cytochromes reversible oxidation–reduction carriers in respiration.

Cytomegalovirus inclusion disease severe, often fatal disease of newborns caused by the cytomegalovirus, usually affecting the salivary glands, kidneys, lungs, and liver.

Cytomegaly characteristic cellular response that results in formation of oversized cells in response to the presence of cytomegaloviruses.

Cytopathic effects (CPE) generalized degenerative changes or abnormalities in the cells of a monolayer tissue culture due to infection by a virus.

Cytoplasm the living substance of a cell, exclusive of the nucleus.

Cytoplasmic membrane the selectively permeable membrane that forms the outer boundary limit of the cell; in bacteria this cell boundary is normally bordered externally by the cell wall.

Cytosine a pyrimidine base found in nucleic acid.

Cytosis the movement of materials into or out of a cell, involving engulfment and formation of a membrane-bound structure rather than by passage through a membrane.

Cytoskeleton protein fibers and tubules composing the structural support framework of a eukaryotic cell.

Cytostomes mouth-like openings of some protozoa, particularly ciliates.

Cytotoxic T cells specialized class of T lymphocytes that are able to kill cells, including infected host cells, as part of the cell-mediated immune response.

Cytotoxins substances capable of injuring certain cells without lysis.

Darkfield microscope a microscope where the only light seen in the field of view is reflected from the object under examination, resulting in a light object on a dark background.

Deaminase an enzyme involved in the removal of an amino group from a molecule, liberating ammonia.

Deamination the removal of an amino group from a molecule, especially an amino acid.

Death permanent cessation of life functions; at the microbial level death occurs when the microbe loses its ability to reproduce and carry out metabolism needed to maintain cellular organization; at the human level death is often interpreted as complete loss of brain or independent cardiovascular functions.

Death phase the part of the normal growth curve that occurs when the organisms can no longer reproduce.

Decarboxylase an enzyme that liberates carbon dioxide from the carboxyl group of a molecule by hydrolysis.

Decarboxylation the splitting off of one or more molecules of carbon dioxide from organic acids, especially amino acids.

Decimal reduction time the time required at a given temperature to heat-inactivate or kill 90 percent of a given population of cells or spores; the time required at a given temperature to reduce the number of viable microorganisms by one order of magnitude.

Decomposers organisms, often bacteria or fungi, in a community that convert dead organic matter into inorganic nutrients.

Defined medium the material supporting microbial growth in which all the constituents, including trace substances, are quantitatively known; a mixture of known composition used for culturing microorganisms.

Definitive hosts organisms that harbor the adult, sexually mature form of a helminthic parasite.

Dehydrogenase an enzyme that catalyzes the oxidation of a substrate by removal of hydrogen.

Dementia severe impairment or loss of intellectual capacity and personality integration.

Denaturation the alteration in the characteristics of an organic substance, especially a protein, by physical or chemical action; the loss of enzymatic activity due to modification of the tertiary protein structure.

Dengue fever a human disease caused by togavirus and transmitted by mosquitos, characterized by fever, rash, and severe pain in joints and muscles.

Denitrification the formation of gaseous nitrogen or gaseous nitrogen oxides from nitrate or nitrite.

Dental caries tooth decay; erosion of dental enamel and development of a cavity due to growth of lactic acid–producing bacteria.

Dental plaque the combination of bacterial cells, dextran, and debris that adheres to the teeth and is removed by brushing and flossing.

Deoxyribonucleic acid (DNA) the carrier of genetic information; macromolecule housing the genetic information of a cell; a type of nucleic acid occurring in cells, containing adenine, guanine, cytosine, and thymine, and D-2-deoxyribose linked by phosphodiester bonds.

Deoxyribose a five-carbon sugar having one oxygen less than the parent sugar ribose; a component of DNA.

Dermatitis an inflammation of the skin.

Dermatophytes see dermatophytic fungi.

Dermatophytic fungi fungi characterized by their ability to metabolize keratin and capable of growing on the skin surface, causing diseases of the skin, nails, and hair called ringworm or tinea.

Dermis the lower layer of the skin, containing specialized nerve endings.

Desensitization prevention of inflammatory allergic responses usually by a programmed course of allergen injections.

Desert a region of low rainfall; a dry region; a region of low biological productivity.

Desiccation the removal of water; method of food preservation against microbial attack.

Desulfovibrio bacterial genus—curved rods; motile; Gram negative; obtain energy by anaerobic respiration reducing sulfates or other reducible sulfur compounds to hydrogen sulfide strict anaerobes; optimal growth 25–30°C.

Detergent a synthetic cleaning agent containing surface active agents that do not precipitate in hard water; a surface active agent having a hydrophilic and hydrophobic portion.

Deuteromycetes fungi with no known sexual stage; also known as Fungi Imperfecti.

Diagnostic table a table of distinguishing features used as an aid in the identification of unknown organisms.

Diapedesis the process by which leukocytes move out of blood vessels.

Diarrhea a common symptom of gastrointestinal disease, characterized by increased frequency and fluid consistency of stools.

Diatomaceous earth a silicaceous material composed largely of fossil diatoms, used in microbiological filters and industrial processes.

Diatoms unicellular algae having a cell wall composed of silica, the skeleton of which persists after the death of the organism.

Diauxie the phenomena in which, given two carbon sources, an organism preferentially metabolizes one completely before utilizing the other.

Dichotomous key a key for the identification of organisms, using steps with opposing choices until a final identification is achieved.

Differential blood count procedure for determining the ratios of various types of blood cells, used to determine the relative numbers of white blood cells as a diagnostic indication of an infectious process.

Differential media bacteriological media on which growth of specific types of organisms leads to readily visible changes in the appearance of the media so that the presence of specific types of microorganisms can be determined.

Differential staining the procedure wherein the use of a stain distinguishes microbial strains on the basis of different reactions to the staining procedure.

Diffraction the breaking up of a beam of light into bands of differing wavelengths due to interference.

Diffusion the movement of ions or molecules across a concentration gradient from an area of higher concentration to an area of lower concentration.

DiGeorge syndrome immune disorder caused by the partial or total absence of cellular immunity, resulting from a deficiency of T lymphocytes because of incomplete fetal development of the thymus.

Dikaryotes cells with two different nuclei as a result of the fusion of two cells.

Dimorphism the property of existing in two distinct structural forms, e.g., fungi that occur in filamentous and yeast-like forms under different conditions.

Dinoflagellates algae of the class Pyrrhophycophyta, primarily unicellular marine organisms, possessing flagella, including algae that cause red tide.

Diphtheria an acute, communicable human disease caused by *Corynebacterium diphtheriae*.

Diplococci spherical bacterial cells (cocci) occurring in pairs.

Diploid having double the haploid number of chromosomes; having a duplicity of genes.

Dipole moment the polarity resulting from the separation of electric charges, in chemical bonds resulting from an unequal distribution of electrons.

Direct fluorescent antibody staining method used to detect the presence of an antigen by staining with a specific antibody linked with a fluorescent dye—the conjugated fluorescent antibody reacts directly with the antigens.

Disaccharides carbohydrates formed by the condensation of two monosaccharide sugars.

Discharge emission.

Discontinuous strand the strand of DNA that lags behind the replication of the continuous strand because DNA polymerases can only add nucleotides in one direction and therefore synthesis of this strand can only begin after some unwinding of the double helix has occurred and occurs via the synthesis of short segments that run in the opposite direction to the overall direction of synthesis; also referred to as the lagging strand.

Disease syndrome stages in the course of a disease.

Disinfectants chemical agents used for disinfection.

Disinfection the destruction, inactivation, or removal of those microorganisms likely to cause infection or give rise to other undesirable effects.

Dispersal breaking up and spreading in various directions, e.g., the spread of microorganisms from one place to another.

Dissemination the scattering or dispersion of microorganisms or disease, e.g., the spread of disease associated with the dispersal of pathogens.

Dissociation separation of a molecule into two or more stable fragments; a change in colony form often occurring in a new environment, associated with modified growth or virulence.

Distilled liquors commercial product of alcoholic fermentation and chemical distillation, typically beverages with higher alcohol contents than could be obtained by fermentation alone.

DNA see deoxyribonucleic acid.

DNA homology the degree of similarity of base sequences in DNA from different organisms.

DNA polymerases enzymes that catalyze the phosphodiester bonds in the formation of DNA.

Dolipore septa the thick internal transverse opening between cell walls of basidiomycetes.

Donor any cell that contributes genetic information to another.

Donor strain bacterial cell line that donates DNA during mating.

Dormant an organism or spore that exhibits minimal physical and chemical change over an extended period of time but remains alive.

Double bag method disposal method for materials potentially contaminated with infectious microorganisms in which a bag containing the contaminated materials is sealed within a second bag for safety.

Double-stranded DNA virus virus with a genone consisting of double-stranded DNA.

Double-stranded RNA virus virus with a genone consisting of double-stranded RNA.

Doubling time see generation time.

DPT vaccine a single vaccine used to provide active immunity against diphtheria, tetanus, and pertusssis; contains diphtheria and tetanus toxoids and killed *Bordetella pertussis* cells.

Drugs substances used in medicine for the treatment of disease.

Dust cells macrophage cells fixed in the alveolar lining of the lungs.

***D* value** see decimal reduction time.

Dysentery gastrointestinal disease characterized by severe diarrhea.

Dysfunctional immunity an immune response that produces an undesirable physiological state, e.g., an allergic reaction, or the lack of an immune response resulting in a failure to protect the body against infectious or toxic agents.

Eclipse period the time during viral replication when complete, infective virions are not present.

Ecology the study of the interrelationships between organisms and their environments.

Ecosystem a functional self-supporting system that includes the organisms in a natural community and their environment.

Edema the abnormal accumulation of fluid in body parts or tissues that causes swelling.

Effluent the liquid discharge from sewage treatment and industrial plants.

Eikenella bacterial genus—rods to coccobacilli; Gram negative; acid and gas not produced from carbohydrates; facultative anaerobes; G + C 56 mole%.

Electron a negatively charged subatomic particle that orbits the positively charged nucleus of an atom.

Electron acceptors substances that gain electrons during oxidation–reduction reactions.

Electron donors substances that give up electrons during oxidation–reduction reactions.

Electron transport chain a series of oxidation–reduction reactions in which electrons are transported from a substrate through a series of intermediate electron carriers to a final acceptor, establishing an electrochemical gradient across a membrane that results in the formation of ATP.

Electrophoresis the movement of charged particles suspended in a liquid under the influence of an applied electron field.

Elephantiasis caused by the nematode *Wuchereria bancrofti*, which breeds in the tissues of the circulatory system damaging the lymphatic vessels, causing swelling and distortion (gross enlargement) of the legs.

ELISA see enzyme-linked immunosorbant assay.

EMB agar see eosin methylene blue agar.

Embden–Meyerhof pathway a specific glycolytic pathway; a sequence of reactions in which glucose is broken down to pyruvate.

Embryonated eggs hen or duck eggs containing live embryos used for culturing viruses and preparing tissue cultures.

Encephalitis an inflammation of the brain.

End- combining form indicating within.

Endemic peculiar to a certain region, e.g., a disease that occurs regularly in an area.

Endemic typhus also known as murine, rat, or flea typhus, this milder form of epdidemic typhus is caused by *Ricksettsia typhi* and is transmitted to humans by the bites of infected fleas or lice; produces fever, rash, headache, cough and muscle aches; prevention is through rat control.

Endo- combining form indicating within.

Endocarditis infection of the endocardium or heart valves caused by bacteria or, in the cases of intravenous drug abusers, fungi.

Endocardium endothelial cells lining the cavities of the heart; the membrane lining the interior of the heart.

Endocytosis the movement of materials into a cell by cytosis.

Endogenous produced within; due to internal causes; pertaining to the metabolism of internal reserve materials.

Endonuclease an enzyme that catalyzes the cleavage of DNA, normally cutting the DNA at specific sites.

Endoplasmic reticulum the extensive array of internal membranes in a eukaryotic cell involved in coordinating protein synthesis.

Endospore-producing bacteria bacteria that produce endospores; group of bacteria that includes the genera *Clostridium* and *Bacillus*; particularly important bacterial group because of the heat resistance of the endospores that are produced.

Endospores complex spores formed within bacterial cells that are heat resistant.

Endosymbionts bacterial genera that live within the cells or tissues of other organisms without adversely affecting the other organism.

Endosymbiotic a symbiotic association in which one organism penetrates and lives within the cells or tissues of another organism.

Endothelial a single layer of thin cells lining internal body cavities; the inner layer of the seed coat of some plants.

Endotoxin toxic substances found as part of some bacterial cells; the lipopolysaccharide component of the cell envelope of Gram negative bacteria, also known as LPS.

End product the chemical compound that is the final product in a particular metabolic pathway.

Energy the capacity to do work.

Energy transfer the movement of energy from one chemical compound to another.

Enrichment culture any form of culture in a liquid medium that results in an increase in the numbers of a given type of organism while minimizing the growth of any other organism present.

Enrichment media any culture medium that favors the growth of a particular microorganism.

Enter- combining form meaning the intestine.

Enteric of or pertaining to the intestines.

Enterotoxins toxins specific for cells of the intestine, causing intestinal inflammation and producing the symptoms of food poisoning.

Envelope the outer covering surrounding the capsid of some viruses.

Enzymatic reactions chemical reactions catalyzed by enzymes.

Enzyme-linked immunosorbant assay (ELISA) a technique used for detecting and quantifying specific serum antibodies.

Enzymes proteins that function as efficient biological catalysts, increasing the rate of a reaction, without altering the equilibrium constant, by lowering the energy of activation.

Eosin methylene blue agar a medium used for the detection of coliform bacteria; the growth of Gram positive bacteria is inhibited on this medium, and lactose fermenters produce colonies with a green-metallic sheen.

Eosinophils white blood cells that have an affinity for eosin or any acid stain.

Epi- prefix meaning upon, beside, among, above, or outside.

Epidemic an outbreak of infectious disease among a human population in which for a limited time a high proportion of the population exhibits overt disease symptoms.

Epidemic hemorrhagic conjunctivitis an infectious disease of the eye caused by enteroviruses and characterized by subconjunctival hemorrhages.

Epidemic keratoconjunctivitis acute, self-limiting adenoviral infection of the eyes characterized by redness, edema, swelling, and discomfort.

Epidemic typhus also known as European, classic, or louse-borne typhus; caused by *Ricksettsia prowazekii* transmitted by the bite of infected body lice; diagnosis is by the Weil–Felix reaction and treatment is with antibiotics.

Epidemiology the study of the factors and mechanisms that govern the spread of disease within a population, including the interrelationships between a given pathogenic organism, the environment, and populations of relevant hosts.

Epidermis the outer layers of the skin.

Epifluorescence microscopy form of microscopy employing stains that fluoresce when excited by light of a given wavelength, emitting light of a different wavelength; exciter filters are used to produce the proper excitation wavelength, and barrier filters are used so that only fluorescing specimens are visible.

Epinephrine see adrenaline.

Epizootic an epidemic outbreak of infectious disease among animals other than humans.

Equilibrium a state of balance, a condition in which opposing forces equalize one another so that no movement occurs; in a chemical reaction, the condition where forward and reverse reactions occur at equal rates so that no net change occurs; when a reaction is at equilibrium the amounts of reactants and products remain constant.

Equilibrium constant describes the relationship among concentrations of the substances within an equilibrium system regardless of how the equilibrium condition is achieved.

Ergotism the condition of intoxication that results from the ingestion of grain contaminated by ergot alkaloids produced by *Claviceps purpurea*, characterized by neurological symptoms including hallucinations.

Erwinia bacterial genus—cells predominantly single, straight rods; motile by peritrichous flagella; Gram negative; fementative; facultative anaerobes; pathogenic for plants; optimal growth 27–30°C; G + C 50–58 mole%.

Erythema abnormal reddening of the skin due to local congestion, symptomatic of inflammation.

Erythrocytes red blood cells.

Erythromycin an antibiotic produced by a strain of *Streptomyces* that inhibits bacterial protein synthesis.

Escherichia bacterial genus—straight rods; Gram negative; motile by peritrichous flagella or nonmotile; citrate cannot be used as sole carbon source; glucose and other carbohydrates are fermented with production of lactic, acetic, and formic acids; the formic acid is split into equal amounts of carbon dioxide and hydrogen; lactose is fermented by most strains; facultative anaerobes; indole positive; no hydrogen sulfide produced from TSI agar; methyl red positive; Voges–Proskauer negative; G + C 50–51 mole%.

Ethanolic fermentation a type of fermentation where glucose is converted to ethanol and carbon dioxide.

Etiology the study of the causation of disease.

Eubacteria prokaryotes other than archaebacteria.

Euglenoids members of the algal division Euglenophycophyta; unicellular organisms surrounded by a pellicle.

Eukaryotes cellular organisms having a membrane-bound nucleus within which the genome of the cell is stored as chromosomes composed of DNA; eukaryotic organisms include algae, fungi, protozoa, plants, and animals.

Eukaryotic cell a cell with a true nucleus.

Eutrophication the enrichment of natural waters with inorganic materials, especially nitrogen and phosphorus compounds, that support the excessive growth of photosynthetic organisms.

Evolution the directional process of change of organisms by which descendants become distinct in form and/or function from their ancestors.

Exo- prefix indicating outside, outside layer, or out of.

Exogenous due to an external cause; not arising from within the organism.

Exon the region of a eukaryotic genome that is known to encode the information for protein or RNA macromolecules or the regulation of gene expression; a segment of eukaryotic DNA that codes for a region of RNA that is not excised during posttranscriptional processing.

Exotoxins protein toxins secreted by living microorganisms into the surrounding medium.

Exponential growth see exponential phase.

Exponential phase that period during the growth cycle of a population when growth is maximal and constant and there is a logarithmic increase in population size.

Extracellular external to the cells of an organism.

Exudate viscous fluid containing blood cells and debris that accumulate at the site of an inflammation or lesion.

FAB (fluorescent antibody staining) see direct fluorescent antibody staining.

Fab fragment either of two identical fragments produced when an immunoglobulin is cleaved by papain; the antigen binding portion of an antibody, including the hypervariable region.

Facilitated diffusion the movement across a plasma membrane of a substance from an area of higher concentration to an area of lower concentration mediated by permease carrier proteins.

Facultative anaerobes microorganisms capable of growth under either aerobic or anaerobic conditions; bacteria capable of both fermentative and respiratory metabolism.

FAD see flavin adenine dinucleotide.

$FADH_2$ reduced flavin adenine dinucleotide.

Family a taxonomic group; the principal division of an order; the classification group above a genus.

Fastidious an organism difficult to isolate or culture on ordinary media because of its need for special nutritional factors; an organism having stringent physiological requirements for growth and survival.

Fatty acids straight chains of carbon atoms, with a COOH at one end, in which most of the carbons are attached to hydrogen atoms.

Fc fragment the crystallizable portion of an immunoglobulin molecule containing the constant region; the end of an immunoglobulin that binds with complement.

Feedback inhibition a cellular control mechanism by which the end product of a series of metabolic reactions inhibits the activity of an earlier enzyme in the sequence of metabolic transformations, and thus, when the end product accumulates, its further production ceases.

Fermentation a mode of energy-yielding metabolism that involves a sequence of oxidation–reduction reactions in which an organic substrate and the organic compounds derived from that substrate serve as the primary electron donor and the terminal electron acceptor; in contrast to respiration, there is no requirement for an external electron acceptor to terminate the metabolic sequence.

Fermenter an organism that carries out fermentation.

Fermentor a reaction chamber in which a fermentation reaction is carried out; a reaction chamber for growing microorganisms used in industry for a batch process.

Fertility fruitfulness; the reproductive rate of a population; the ability to support life; the ability to reproduce.

Fertility pilus F or sex pilus involved in bacterial mating and found only on donor cells.

Fertility plasmid see F plasmid.

Fetus embryo after the third month of gestation in the womb.

Fever the elevation of the body temperature above normal.

Fever blisters see cold sores.

Fibrin the insoluble protein formed from fibrogen by the proteolytic action of thrombin during normal blood clotting.

Fibrinogen a protein in human plasma synthesized in the liver that is the precursor of fibrin; used to increase the coagulability of blood.

Fibrolysin a proteolytic enzyme capable of dissolving or preventing the formation of a fibrin clot.

Filament any elongated thread-like bacterial cell.

Filamentous fungi fungi that develop hyphae and mycelia; also called molds.

Filterable virus an obsolete term used to describe infectious agents that were able to pass through bacteriological filters.

Filtration the separation of microorganisms from the medium in which they are suspended by passage of a fluid through a filter with pores small enough to trap the microbes.

Fire algae members of the algal division Pyrrophycophyta; unicellular, biflagellate, store starch or oils and have xanthophyll pigments.

Fission a type of asexual reproduction in which a cell divides to form two or more daughter cells.

Flagella flexible, relatively long appendages on cells used for locomotion.

Flagella antigens see H antigens.

Flagellates organisms having flagella; one of the major divisions of protozoans, characterized by the presence of flagella.

Flavin adenine dinucleotide a coenzyme used in oxidation–reduction reactions; of lower energy than NAD.

Flatfield objective a microscope lens that provides an image in which all parts of the field are simultaneously in focus; an objective lens with minimal curvature of field.

Floc a mass of microorganisms caught together in a slime produced by certain bacteria, usually found in waste treatment plants.

Flukes flatworms belonging to the class Trematoda.

Fluorescence the emission of light by certain substances upon absorption of an exciting radiation; the emitted light being of a different wavelength than the wavelength of the excitation radiation.

Fluorescence microscope a microscope in which the microorganisms are stained with some form of fluorescent dye and observed by illumination with short wavelength light, e.g., ultraviolet light.

Fluorescent antibodies antibodies capable of giving off light of one color when exposed to light of another color.

Fluorescent antibody staining (FAB) see direct fluorescent antibody staining.

Folliculitis bacterial infection of hair follicles that causes the formation of a pustule.

Fomes inanimate objects that can act as carriers of infectious agents.

Fomites inanimate objects and materials that have been associated with infected persons or animals and that potentially harbor pathogenic microorganisms.

Food additive a substance or mixture of substances other than the basic food stuff, which is intentionally present in food as a result of any aspect production, processing, storage, or packaging.

Food infection disease resulting from the ingestion of food or water containing viable pathogens that can establish an infectious disease, e.g., gastroenteritis from ingestion of food containing *Salmonella*.

Food intoxication disease resulting from the ingestion of toxins produced by microorganisms that have grown in a food.

Food poisoning a general term applied to all stomach or intestinal disorders due to food contaminated with certain microorganisms, their toxins, chemicals or poisonous plant materials; disease resulting from the ingestion of toxins produced by microorganisms that have grown in a food.

Food preservation the prevention or delay of microbial decomposition or self-decomposition of food and prevention of damage because of insects, animals, mechanical causes, etc.; the delay or prevention of food spoilage.

Food spoilage the deterioration of a food that lessens its nutritional value or desirability, often due to the growth of microorganisms that alter the taste, smell, or appearance of the food, or the safety of ingesting that food.

Food web an interrelationship among organisms by which energy is transferred from one organism to another, where each organism consumes the preceding one and in turn is eaten by the following member of the sequence.

Foraminiferans marine members of the protzoan class Sarcodina that form silicaceous tests.

Formalin a 40 percent solution of formaldehyde, a pungent-smelling, colorless gas used for fixation and preservation of biological specimens and as a disinfectant.

Forssman antigen a heat-stable glycolipid; a heterophile antigen, an immunologically related antigen found in unrelated species.

F pilus see fertility pilus.

F plasmid fertility plasmid coding for donor strain, includes genes for the formation of the F pilus.

F^- strain bacterial strain that lacks the fertility factor; it acts as the receptor strain in conjugation.

F^+ strain bacterial strain that has the fertility factor as a plasmid in the cytoplasm.

Frame shift mutation a type of mutation that causes a change in the three base sequences read as codons, i.e., a change in the phase of transcription arising from the addition or deletion of nucleotides in numbers other than three or multiples of three.

Francisella bacterial genus—very small coccoid to ellipsoidal pleomorphic rods; nonmotile; Gram negative; strictly aerobic; optimal growth 37°C.

Freeze drying the removal of water from fozen foods under high vacuum.

Freeze etching a technique used to examine the topography of a surface exposed by fracturing or cutting a deep-frozen cell, making a replica, and removing the biological material; used in transmission electron microscopy.

Fruiting body a specialized fungal structure that bears sexually or asexually derived spores.

Frustules the silica containing cell walls of a diatom.

Fungi (singular, **fungus**) diverse and widespread unicellular and multicellular eukaryotic organisms, lacking chlorophyll and usually bearing spores and often filamentous.

Fungicides agents that kill fungi.

Fungi Imperfecti fungi with septate hyphae that reproduce only by means of conidia, lacking a known sexual stage; Deuteromycetes.

Fungus see fungi.

Furuncles boils; painful, inflammatory sores around a central core resulting from folliculitis.

γ-Globulin a specific class of serum proteins with antibody activity; serum fraction of blood containing antibodies; also called immunoglobulin G (IgG) or gammaglobulin.

γ-Hemolysis test result when no zone of clearing arises from the inoculation of a microorganisms on blood agar; no hemolysis has occurred.

Galls abnormal plant structures formed in response to parasitic attack by certain insects or microorganisms; a tumor-like growth of plants in response to an infection.

Gametes haploid reproductive cells or nuclei, the fusion of which during fertilization leads to formation of a zygote.

Gammaglobulin see γ-globulin

Gamma-hemolysis see γ-hemolysis.

Gamma radiation short wavelength (10^{-8} to 10^{-1}) electromagnetic radiation that has high penetration power and can kill microorganisms by inducing or forming toxic free radicals.

Gas gangrene a disease condition involving tissue death developing when certain species of toxin-producing bacteria grow in anaerobic wounds or necrotic tissues.

Gasohol a mixture of gasoline and ethanol used as a fuel.

Gastroenteritis an inflammation of the stomach and intestine.

Gastroenterocolitis an inflammation of the gastrointestinal tract accompanied by the formation of pus and blood in the stools.

Gastrointestinal syndrome gastroenteritis associated with nausea, vomiting, and/or diarrhea.

Gastrointestinal tract the stomach, intestines, and accessory organs.

Gas vacuoles membrane-limited, gas-filled vacuoles that occur commonly in groups in the cells of a number of cyanobacteria and certain other bacteria.

Gelatin a protein obtained from skin, hair, bones, tendons, etc., used in culture media for the determination of a specific proteolytic activity of microorganisms.

Gelatinase a hydrolytic enzyme capable of liquifying gelatin.

Gene a sequence of nucleotides that specifies a particular polypeptide chain.

Generation time the time required for the cell population or biomass to double.

Generic drug name commonly used name of drug not protected by trademark registration.

Genetic engineering the deliberate modification of the genetic properties of an organism either through the selection of desirable traits, the introduction of new information into DNA, or both; the application of recombinant DNA technology.

Genetics the science dealing with heredity.

Genital herpes a sexually transmitted disease caused by a herpes virus; an infection by herpes simplex virus marked by the eruption of groups of vesicles often in the genital region.

Genital warts disease characterized by benign tumor development caused by human papilloma virus transmitted by sexual contact; warts most commonly develop on the moist areas of the genitalia.

Genitourinary tract the combined urinary and genital systems; the combined reproductive system and urine excretion system, including the kidneys, ureters, urinary bladder, urethra, penis, prostrate, testes, vagina, fallopian tubes, and uterus.

Genome the complete set of genetic information, as contained in a haploid set of chromosomes.

Genotype the genetic information contained in the entire complement of alleles.

Genus a taxonomic group, next above species and forming the principal subdivisions of the family.

Geometric isomers the formation of nonequivalent structures based on the particular positions at which ligands are attached to a central atom.

Germ a disease-causing microorganism; a pathogenic microbe.

German measles rubella, an acute systemic infectious disease of humans caused by rubella viruses invading via the mouth or nose, characterized by a rash.

Germ free animal an animal with no normal microbiota; all its surfaces and tissues are sterile, and it is maintained in that condition by being housed and fed in a sterile environment.

Germicide a microbicidal disinfectant.

Germination a degradative process by which an activated spore becomes metabolically active, involving hydrolysis and depolymerization.

Giardiasis disease cauased by the protozoan *Giardia* when it infects the human intestine.

Gingivitis inflammation of the gums.

Gliding bacteria bacteria that exhibit gliding motility.

Gliding motility movement when some bacteria are in contact with solid surfaces.

Globular protein the general name for a group of water soluble proteins.

Glomerulonephritis an inflammation of the filtration region of the kidneys.

Glucose the monosaccharide sugar $C_6H_{12}O_6$.

Glycocalyx a structure of bacteria that contains polysaccharides and lies outside of the cell wall.

Glycogen a nonreducing polysaccharide of glucose found in many tissues and stored in the liver where it is converted when needed into sugar.

Glycolysis an anaerobic process of glucose breakdown by a sequence of enzyme-catalyzed reactions to a pyruvic acid.

Glycoproteins a group of conjugated proteins that upon decomposition yield a protein and a carbohydrate.

Glycosidic bonds bonds in disaccharides and polysaccharides formed by the elimination of water.

Glyoxylate cycle a metabolic shunt within the tricarboxylic cycle involving the intermediate glyoxylate.

Golgi apparatus a membranous organelle of eukaryotic organisms involved with the formation of secretory vesicles and the synthesis of complex polysaccharides.

Gonorrhea a sexually transmitted disease caused by *Neisseria gonorrhoeae*; specific infectious inflammation of the mucous membrane of the urethra and adjacent cavities caused by *N. gonorrhoeae*.

Gram stain differential staining procedure by which bacteria are classified as Gram negative or positive depending on whether they retain or lose the primary stain when subject to treatment with a decolorizing agent; the staining procedure reflects underlying structural differences in the cell walls of Gram negative and Gram positive bacteria.

Grana a membranous unit formed by stacks of thylakoids within a chloroplast.

Granules small intracellular particles, usually staining selectively.

Granuloma an inflammatory growth composed of granulation tissue (normal and scar tissue).

Granuloma inguinale the chronic destructive ulceration of external genitalia due to *Donovania granulomatis*.

Grazers organisms that prey upon primary producers; protozoan predators that consume bacteria nondiscriminately; filter-feeding zooplankton.

Green algae members of the algal division Chlorophycophyta found in aquatic ecosystems.

Groundwater all subsurface water.

Growth any increase in the amount of actively metabolic protoplasm accompanied by an increase in cell number, cell size, or both.

Growth curve a curve obtained by plotting increase in size or number of microorganisms against elapsed time.

Growth factors any compound, other than the carbon and energy source, that an organism requires and cannot synthesize.

Growth rate increase in the number of microorganisms per unit time.

Growth temperature range established by the maximum and minimum temperatures at which a microorganism can grow.

Guanine a purine base that occurs naturally as a fundamental component of nucleic acids.

Guillain-Barré syndrome acute febrile polyneuritis; a diffuse neuron paresis that results from infection with a prion.

Habitat a location where living organisms occur.

Haemophilus bacterial genus—coccobacillary; nonmotile; Gram negative; strict parasites, requiring growth factors present in blood; aerobic to facultatively anaerobic; optimal growth 37°C; G + C 38–42 mole%.

Halophiles organisms requiring NaCl for growth; extreme halophiles growing in concentrated brines.

H antigen a type of flagella antigen found in certain bacteria.

Haploid a single set of homologous chromosomes; having half the normal diploid number of chromosomes.

Hapten a substance that elicits antibody formation only when combined with other molecules or particles but that can react with preformed antibodies; a molecule that can react with antibody but cannot itself elicit antibody formation.

Helix a spiral structure.

Hemagglutination the agglutination or clumping of red blood cells.

Hemagglutination inhibition the prevention of hemagglutination, usually by means of specific immunoglobulins or enzymes.

Heme an iron-containing porphyrin ring occurring in hemoglobin.

Hemocytometer a counting chamber used for estimating the number of blood cells.

Hemoglobin the iron-containing, oxygen-carrying molecule of red blood cells containing four polypeptides in heme group.

Hemolysin a substance that causes lysis of erythrocytes.

Hemolysis the lytic destruction of red blood cells and the resultant release of hemoglobin.

Hemolytic disease of the newborn disease that stems from an incompatibility of fetal (Rh positive) and maternal (Rh negative) blood resulting in maternal antibody activity against fetal blood cells; also known as erythroblastosis fetalis.

Hemorrhagic showing evidence of bleeding, the tissue becomes reddened by the accumulation of blood that has escaped from capillaries into the tissue.

Hepatitis inflammation of the liver; disease caused by hepatitis viruses that involves the liver.

Hepatitis A hepatitis caused by type A hepatitis virus, usually transmitted by fecal-oral route.

Hepatitis B hepatitis caused by type B hepatitis virus usually transmitted via blood.

Hepatitis nonA–nonB hepatitis caused by hepatitis virus lacking both A and B antigens; commonly transmitted by contaminated blood transfusions.

Herbicides chemicals used to kill weeds.

Herd immunity concept that an entire population is protected against a particular pathogen when 70% of the population is immune to that pathogen.

Heritable any characteristic that is genetically transmissible.

Herpes encephalitis form of encephalitis (inflammation of the brain) caused by Herpes simplex virus; treatable with acyclovir.

Herpes simplex infections localized blistery skin rash caused by herpes simplex virus, usually on the lip or the genitalia.

Hetero- combining form meaning other, other than usual, different.

Heterocysts cells that occur in the trichomes of some filamentous cyanobacteria that are the sites of nitrogen fixation.

Heterogeneous composed of different substances; not homologous.

Heterogeneous RNA (hnRNA) high molecular weight RNA formed by direct transcription in eukaryotes that is then processed enzymatically to form mRNA.

Heterolactic fermentation fermentation of glucose that produces lactic acid, acetic acid, and/or ethanol, and carbon dioxide, carried out by *Leuconostoc* and some *Lactobacillus* species.

Heterophile antibody antibody that reacts with heterophile antigens, commonly found in sera of individuals with infectious mononucleosis.

Heterophile antigens immunologically related antigens found in unrelated species.

Heterotrophic incapable of utilizing carbon dioxide as sole carbon source; requiring one or more organic compounds for nutrition.

Heterotrophic metabolism metabolism that requires organic carbon as source of carbon and energy.

Heterotrophs organisms requiring organic compounds for growth and reproduction, the organic compounds serving as sources of carbon and energy.

Hfr see high frequency recombinant.

High frequency recombinant a bacterial strain that exhibits a high rate of gene transfer and recombination during mating; the F plasmid is integrated into the bacterial chromosome.

Histamine a physiologically active amine that plays a role in the inflammatory response.

Histiocytes macrophages that are located at a fixed site in a certain organ or tissue.

Histocompatibility antigens genetically determined isoantigens present on the lipoprotein membranes of nucleated cells of most tissues that incite an immune response when grafted onto a genetically disparate individual and thus determine the compatibility of tissues in transplantation.

Histones basic proteins rich in arginine and lysine that occur in close association with the nuclear DNA of most eukaryotic organisms.

Histoplasmin test skin test designed to detect antibodies against *Histoplasma*.

Histoplasmosis a disease of humans and animals caused by the fungus *Histoplasma capsulatum*, characterized by fever, anemia, leukopenia, and emaciation, primarily involving the reticuloendothelial system.

HnRNA see heterogeneous RNA.

Homo- combining form denoting like, common, or same.

Homolactic fermentation the fermentation of glucose that produces lactic acid as the sole fermentation product, carried out by many species of *Lactobacillus* and *Streptococcus*.

Homologous pertaining to the structural relation between parts of different organisms due to evolutionary development of the same or corresponding part; a substance of identical form or function.

Homology genetic relatedness.

Hookworms roundworms that cause infestations of the small intestine.

Hospital institution in which sick and injured people receive medical and surgical treatment.

Host a cell or organism that acts as the habitat for the growth of another organism; the cell or organism upon or in which parasitic organisms live.

HTST process high temperature–short time pasteurization process at a temperature of at least 71.5°C for 15 seconds; the most widely used commercial pasteurization.

Humoral referring to the body fluids

Humoral immune defense system see antibody-mediated immunity.

Hyaluronidase enzyme that catalyzes breakdown of hyaluronic acid; spreading factor.

Hybridomas cells formed by fusion of lymphocytes (antibody precursors) with myeloma (tumor) cells that produce rapidly growing cells that secrete monoclonal antibodies.

Hydatid cysts larval stages of the tapeworm *Echinocococcus granulosus*.

Hydr- combining form meaning water.

Hydrocarbons compounds composed only of hydrogen and carbon.

Hydrogen bond a weak attraction between an atom that has a strong attraction for electrons and a hydrogen atom that is covalently bonded to another atom that attracts the electron of the hydrogen atom.

Hydrogen cycle the biogeochemical movement of hydrogen, usually in conjunction with carbon and oxygen; the principle transformations involve conversion between water and organic carbon compounds.

Hydrolysis the chemical process of decomposition involving the splitting of a bond and the addition of the elements of water.

Hydrolyze see hydrolysis.

Hydrophilic a substance having an affinity for water.

Hydrophobia fear of water, one of the symptoms of rabies.

Hydrophobic a substance lacking an affinity for water, not soluble in water.

3′-hydroxyl end (3′-OH) the end of a nucleic acid macromolecule (DNA or RNA) at which carbon-3 of the carbohydrate is not involved in forming the phosphate diester linkage that bonds the macromolecule; the lack of bonding to carbon-3 makes this end biochemically recognizable and confers directionality upon the nucleic acid macromolecule.

5′-hydroxyl end (5′-OH) the end of a nucleic acid macromolecule (DNA or RNA) at which carbon-5 of the carbohydrate is not involved in forming the phosphate diester linkage that bonds the macromolecule; the lack of bonding to carbon-5 makes this end biochemically recognizable and confers directionality upon the nucleic acid macromolecule.

Hyperchromatic shift the change in absorption

of light exhibited by DNA when it is melted, forming two strands from the double helix.

Hyperplasia the abnormal proliferation of tissue cells resulting in the formation of a tumor or gall.

Hypersensitivity the state of an exaggerated immunological response upon reexposure to a specific antigen.

Hypertonic a solution whose osmotic pressure is greater than that of a standard solution.

Hypertrophy an increase in the size of an organ, independent of natural growth, due to enlargement or multiplication of its constituent cells.

Hypervariable region a region of immunoglobulins that accounts for the specificity of antigen–antibody reactions; genetically specified terminal regions of the Fab fragments.

Hyphae branched or unbranched filaments that constitute the vegetative form of an organism, occurring in filamentous fungi, algae, and bacteria.

Hypotonic a solution whose osmotic pressure is less than that of a standard solution.

Icosahedral virus a virus having cubical symmetry and a complex 20-sided capsid structure.

Icosahedron a solid figure contained by 20 plane faces.

Identification the process of determining the greatest affinity of an unknown organism to a group that has already been defined.

Identification key a series of questions that leads to the unambiguous identification of an organism.

IgA immunoglobulin A; a class of immunoglobulin molecules found in body secretions and on mucous membranes.

IgD immunoglobulin D; one of five classes of immunoglobulins.

IgE immunoglobulin E; a class of immunoglobulins important in hypersensitivity reactions.

IgG immunoglobulin G; the main class of circulating immunoglobulins that is very important in protection against systemic infections.

IgM immunoglobulin M; a class of immunoglobulins formed early in the immune response to an infection.

Immune the condition following initial contact with a given antigen in which antibodies specific for that antigen are present in the body; the innate or acquired resistance to disease.

Immune-complex disease illness caused by the formation of antibodies against antigen–antibody complexes.

Immune defense network the integrated defense system of the body that protects against infection; the combined and interactive system that includes the B cell (antibody–mediated) and T cell (cell–mediated) immune responses.

Immune interferon type of lymphokine having antiviral properties secreted by lymphocytes in response to a specific antigen to which they have been sensitized.

Immune response system the mechanism for response to the invasion of the body of particular pathogenic microorganisms and other foreign substances; it is characterized by specificity, memory, and the acquired ability to detect foreign substances.

Immunity the relative unsusceptibility of a person or animal to active infection by pathogenic microorganisms or the harmful effects of certain toxins; the condition of a living organism whereby it resists disease.

Immunization procedure in which an antigen is introduced into the body in order to produce a specific acquired immune response.

Immunodeficiency the lack of an adequate immune response due to inadequate B or T cell recognition and/or response to foreign antigens; a lack of antibody production.

Immunoelectrophoresis a two-stage procedure used for the analysis of materials containing mixtures of distinguishable proteins, e.g., serum using electrophoretic separation and immunological detection.

Immunofluorescence any of a variety of techniques used to detect a specific antigen or antibody by means of homologous antibodies or antigens that have been conjugated with a fluorescent dye.

Immunogenicity the ability of a substance to elicit an immune response.

Immunoglobulin A see IgA.

Immunoglobulin D see IgD.

Immunoglobulin E see IgE.

Immunoglobulin G see IgG.

Immunoglobulin M see IgM.

Immunoglobulins specific classes of glycoproteins found in plasma and other body fluids; also known as antibodies; the antibody-containing fraction of serum.

Immunological referring to the immune response.

Immunology the study of immunity.

Immunosuppressant a drug that depresses the immune response.

Immunosuppression depression of the immune response.

Impetigo an acute inflammatory skin disease caused by bacteria (often *Streptococcus* and

Staphylococcus), characterized by small blisters, weeping fluid, and crusts.

IMViC tests a group of tests (indole, methyl red, Voges–Proskauer, citrate) used in the identification of bacteria of the Enterobacteriaceae family.

Inclusion conjunctivitis chlamydial infection affecting mostly newborns, causing an acute ocular inflammation, characterized by reddening of the eyelids and a thick purulent discharge; treatment is with tetracycline, erythromycin, or sulfonamide eyedrops.

Incubation the maintenance of controlled conditions to achieve the optimal growth of microorganisms.

Incubation period the period of time between the establishment of an infection and the onset of disease symptoms.

Incubators controlled temperature chambers.

Indicator organism an organism used to indicate a particular condition, commonly applied to coliform bacteria, e.g., *Escherichia coli* or *Streptococcus faecalis,* when their presence is used to indicate the degree of water pollution due to fecal contamination.

Indigenous native to a particular habitat.

Indirect fluorescent antibody a fluorescent antibody test to detect the presence of specific antigens associated with a microorganism, in which an immunoglobulin molecule is first reacted with the antigens to form a complex and then a conjugated fluorescent antibody dye is added that reacts with the first unlabelled immunoglobulin that was used to form the first antigen–antibody complex; the test is indirect because the fluorescent-labelled dye actually reacts with the unlabelled immunoglobulin that was used to react with the antigens and not directly with the antigens.

Inducers substances responsible for activating certain genes, resulting in the synthesis of new proteins.

Inducible enzymes enzymes that are synthesized only in response to a particular substance in the environment.

Induction an increase in the rate of synthesis of an enzyme; the turning on of enzyme synthesis in response to environmental conditions.

Infant botulism caused by toxin produced by the infection of an infant's gastrointestinal tract with *Clostridium botulinum*.

Infection a condition in which pathogenic microorganisms have become established in the tissues of a host organism.

Infectious a disease that can be transmitted from one person, animal, or plant to another; an individual with an infectious disease; a pathogen that can establish an infection.

Infectious diseases diseases caused by pathogens.

Infectious dose the number of pathogens that are needed to overwhelm host defense mechanisms and establish an infection.

Infectious hepatitis hepatitis A caused by ingestion of hepatitis A viruses in water or food.

Infectious mononucleosis glandular fever, an acute infectious disease that primarily affects the lymphoid tissues; caused by Epstein–Barr virus, characterized by formation of abnormal white blood cells.

Inflammation the reaction of tissues to injury or infection, characterized by local heat, swelling, redness, and pain.

Inflammatory exudate pussy material from blood vessels deposited in tissues or on tissue surfaces as a defensive response to injury or irritation.

Inflammatory response see inflammation.

Influenza an acute, highly communicable disease tending to occur in epidemic form; caused by influenza viruses, and characterized by malaise, headache, and fever.

Inhibition the prevention of growth or multiplication of microorganisms; the reduction in the rate of enzymatic activity; the repression of chemical or physical activity.

Inhibitors substances that repress or stop a chemical action.

Inoculate to deposit material, an inoculum, onto medium to initiate a culture, carried out with an aseptic technique; to introduce microorganisms into an environment that will support their growth.

Inoculum the material containing viable microorganisms used to inoculate a medium.

Insecticides substances destructive to insects; chemicals used to control insect populations.

In situ in the natural location or environment.

Interferons glycoproteins produced by animal cells that act to prevent the replication of a wide range of viruses by inducing resistance; also play an important role in the regulation of the immune defense network.

Intermediary metabolism intermediate steps in the cellular synthesis and breakdown of substances.

Intermediate hosts organisms that harbor the larval stage of a helminth.

Intermediately sensitive (I) one of the standardized zones of inhibition used to determine the degree to which a microorganism is sensitive to a particular antibiotic.

Intoxication poisoning as by drug, serum, alcohol, or any poison (toxin).

Intracellular within a cell.

Intradermal within the skin.

Intramuscular within the substance of a muscle.

Intravenous within or into the veins.

Intron an intervening region of the DNA of eukaryotes not coding for a known protein nor regulatory function.

Invasiveness the ability of a pathogen to spread through a host's tissues.

In vitro in glass; a process or reaction carried out in a culture dish or test tube.

In vivo within the living organism.

Ion an atom that has lost or gained one or more orbital electrons and is thus capable of conducting electricity.

Ionic bond a chemical bond resulting from the transfer of electrons between metal and nonmetal atoms; positive and negative ions are formed and held together by electrostatic attraction.

Ionization the process that produces ions.

Ionizing radiation radiation, such as gamma and X radiation, that induces or forms toxic free radicals, which cause chemical reactions disruptive to the biochemical organization of microorganisms.

Iso- combining form meaning for or from different individuals of the same species.

Isolation any procedure in which a given species of organism present in a sample is obtained in pure culture.

Isomer one of two or more compounds having the same chemical composition but differing in the relative positions of the atoms within the molecules.

Isotope an element that has the same atomic number as another but a different atomic weight.

-itis suffix denoting a disease, specifically an inflammatory disease of a specified part.

Jaundice yellowness of the skin, mucous membranes, and secretions resulting from liver malfunction.

Juvenile dysentery dysentery of infants often caused by *Campylobacter fetus*.

Kala-azar disease disease caused by the flagellate protozoan *Leishmania donovani*, generally transmitted to humans by the bite of a sandfly that has fed on an infected dog or rat; also known as visceral leishmaniasis, dumdum fever, and black fever.

Kappa particles bacterial particles that occur in the cytoplasm of certain strains of *Paramecium aurelia*, such strains have a competitive advantage with other strains of *Paramecium* and are known as killer strains.

Kelps brown algae having vegetative structures consisting of a holdfast, stem, and blade; can form large macroscopic structures.

Keratin a highly insoluble protein that occurs in hair, wool, horn, and skin.

Keratoconjunctivitis disease caused by adenoviruses, characterized by inflammation of the eyes accompanied by redness, swelling, and discomfort.

Ketone an organic compound derived by oxidation from a secondary alcohol, containing a characteristic $-\overset{\displaystyle O}{\overset{\|}{C}}-$ group.

Killer T cells see cytotoxic T cells.

Kingdom a major taxonomic category consisting of several phyla or divisions; the primary divisions of living organisms.

Klebsiella bacterial genus—nonmotile, encapsulated rods; can use citrate and glucose as sole carbon source; glucose is fermented with production of acid and gas; Voges–Proskauer positive; methyl red negative; hydrogen sulfide not produced from TSI; catalase positive; oxidase negative; optimal growth 35–37°C; G + C 52–56 mole%.

Koch's postulates a process for elucidating the etiologic agent of an infectious disease.

Koplik's spots small red spots surrounded by white areas occurring on the mucous membranes of the mouth during the early stages of measles.

Krebs cycle the tricarboxylic acid cycle; the citric acid cycle; the metabolic pathway in which acetate derived from pyruvic acid is converted to carbon dioxide and reduced coenzymes are produced.

Kupffer cells macrophages lining the sinusoids of the liver.

Kuru disease caused by a prion affecting the central nervous system observed among cannibals in New Guinea.

-labile unstable, readily changed by physical, chemical, or biological processes.

Lac-operon inducible enzyme system of *Escherichia coli* for the utilization of lactose.

Lactam an organic compound containing a —NH—CO— group in ring form that occurs in penicillins and cephalosporins.

Lactamase an enzyme that breaks a lactam ring.

Lactic acid fermentation fermentation that produces lactic acid as the primary product.

Lactobacillus Bacterial genus—rods; chain formation common; do not produce spores; Gram positive but becoming Gram negative with increasing age; metabolism fermentative even though growth generally occurs in the presence of air; some are strict anaerobes; lactic acid major end product of fermentation; optimal growth 30–40°C; G + C 33.3–53.9 mole%.

Lactoferrin an iron-containing compound that binds the iron necessary for microbial growth resulting in a slight antimicrobial action.

Lactose a disaccharide in milk; when hydrolyzed it yields glucose and galactose.

Lagging strand see discontinuous strand.

Lag phase a period following inoculation of a medium during which numbers of microorganisms do not increase.

Laminar flow the flow of air currents in which streams do not intermingle; the air moves along parallel flow lines; used in laminar flow hood to provide air free of microbes over a work area.

Latent potential; not manifest; present but not visible or active.

Latent period the period of time following infection of a cell by a virus before new viruses are assembled.

Late onset hypogammaglobulinemia immunodeficiency disorder where there is a shortage of circulating B cells and/or B cells with IgG surface receptors.

Leach to wash or extract soluble constituents from insoluble materials.

Leading strand see continuous strand.

Leavening substance used to produce fermentation in dough or liquid; the production of carbon dioxide that results in the rising of dough.

Lecithinases extracellular phospholipid-splitting enzymes.

Legionella bacterial genus—cells rod–shaped; Gram negative; weakly oxidase positive; catalase positive; nonmotile; fastidious, with narrow optimal temperature and pH ranges; do not utilize carbohydrates; urea not utilized; aerobic; G + C 39 mole%.

Legionnaire's disease a form of pneumonia caused by *Legionella pneumophilia*.

Leishmaniasis disease caused by protozoa of the genus *Leishmania*.

Leprosy a chronic contagious disease affecting humans and armadillos, caused by *Mycobacterium leprae*; also known as Hansen's disease.

Leptospirosis disease of humans or animals caused by *Leptospira*.

Lesion a region of tissue mechanically damaged or altered by any pathological process.

Lethal dose the amount of a toxin that results in the death of an organism.

Leukemia type of cancer characterized by the malignant proliferation of abnormally high numbers of leukocytes; treatment is with chemotherapy.

Leukocidin an extracellular bacterial product that can kill leukocytes.

Leukocyte a type of white blood cell, characterized by a beaded, elongated nucleus.

Leukocytosis an increase above the normal upper limits of the leukocyte count.

Leukopenia a decrease below the normal number of leukocyctes in the blood.

Life a state that characterizes living systems, encompassing the complex series of physicochemical processes essential for maintaining the organization of the system and the ability to reproduce that organization.

Ligases a group of enzymes that catalyze reactions in which a bond is formed between two substrate molecules using energy obtained from the cleavage of a pyrophosphate bond.

Lipases fat-splitting enzymes.

Lipids fats or fat-like substances that are insoluble in water and are soluble in nonpolar solvents.

Lipophilic preferentially soluble in lipids or nonpolar solvents.

Lipopolysaccharides (LPS) molecules consisting of covalently linked lipids and polysaccharides.

Lipopolysaccharide toxin an endotoxin.

Listeria bacterial genus—small, coccoid, Gram positive rods; do not produce spores or capsules; not acid fast; motile by peritrichous flagella; acid but no gas from glucose and several other carbohydrates; esculin is hydrolyzed; aerobic to microaerophilic; optimal growth 20–30°C; G + C 38 mole%, except one species with G + C 56 mole%.

Liter a metric unit of volume equal to 1000 milliliters.

Litmus plant extract dye used as an indicator of pH and oxidation or reduction.

Living system a system separated from its surroundings by a semipermeable barrier, composed of macromolecules, including proteins and nucleic acids, having lower entropy than its surroundings, and thus requiring inputs of energy to maintain the high degree of organization; capable of self-replication, and normally based on cells as the primary functional and structural unit.

Lockjaw see tetanus.

Logarithmic phase see exponential phase.

Long-term immunity acquired immunity that establishes a bank of memory cells that persist within the body and permit recognition of specific antigens to which the body has been previously exposed thereby insuring long-term protection (immunity) against disease.

Low acid food food with pH above 4.5.

LPS see lipopolysaccharides.

LTH process low temperature–hold pasteurization process, e.g., 63°C for 30 minutes.

Lumbar puncture the removal of cerebrospinal fluid from the vertebral canal.

Luminescence the emission of light without production of heat sufficient to cause incandescence; produced by physiological processes, friction, chemical, or electrical action.

Ly-, lys-, lyt- combining forms meaning loosen or dissolve.

Lyme disease an infectious disease that produces arthritis caused by a spirochete and transmitted by a tick.

Lymph a plasma filtrate that circulates through the body.

Lymphatic system the widely spread system of capillaries, nodes, and ducts that collects, filters, and returns tissue fluid, including protein molecules, to the blood.

Lymph nodes an aggregation of lymphoid tissues surrounded by a fibrous capsule found along the course of the lymphatic system.

Lymphocytes lymph cells.

Lymphogranuloma venereum a sexually transmitted disease caused by a *Chlamydia*, characterized by an initial lesion, usually on the genitalia, followed by regional lymph node enlargement and systemic involvement.

Lymphokines a varied group of biologically active extracellular proteins formed by activated T lymphocytes involved in cell-mediated immunity.

Lymphomas cancers of the lymph glands and other lymphoid tissues.

Lyophilization the process of rapidly freezing a substance at low temperature, then dehydrating the frozen mass in a high vacuum.

Lysins antibodies or other entities that under appropriate conditions are capable of causing the lysis of cells.

Lysis the rupture of cells.

Lysogeny the nondisruptive infection of a bacterium by a bacteriophage.

Lysosomes an organelle containing hydrolytic enzymes involved in autolytic and digestive processes.

Lysozymes enzymes that hydrolyze peptidoglycan, acting as bactericidal agents when they degrade the bacterial cell walls.

Lytic of or relating to lysis or a lysin; viruses that cause lysis of cells within which they reproduce.

Lytic phage bacterial viruses capable of bursting a cell by the destruction of its membrane.

MacConkey's agar a solid medium used for the growth of enteric bacteria.

Macro- combining form meaning long or large.

Macromolecules very large molecules having polymeric chain structures, as in proteins, polysaccharides, and other natural and synthetic polymers.

Macrophages mononuclear phagocytes; large actively phagocytic cells found in spleen, liver, lymph nodes, and blood, important factors in nonspecific immunity.

Macroscopic of a size visible to the naked eye.

Macular rash small red dots on the skin.

"Magic bullets" term used to describe early synthetic drug compounds, particularly those from Paul Ehrlich's laboratory, that were protrayed as being able to seek out and destroy disease-causing pathogens.

Magnetosomes structures within bacterial cells that contain iron granules and act as magnetic compasses, permitting bacteria to move in response to the earth's magnetic field.

Magnetotaxis motility directed by a geomagnetic field.

Magnification the extent to which the image of an object is larger than the object itself.

Major histocompatibility complex the genetic region in human beings that controls not only tissue compatibility but also the development and activation of part of the immune response.

Malaise a general feeling of illness, accompanied by restlessness and discomfort.

MBC see minimum bactericidal concentration.

Measles an acute, contagious systemic human disease caused by a paramyxovirus that enters via the oral and nasal route, characterized by the presence of Koplik spots.

Media plural form of medium; see medium.

Medical technologist an allied health professional trained and certified to perform tests used in the diagnosis of disease.

Medium (plural, **media**) the material that supports the growth/reproduction of microorganisms.

Meiosis cell division that results in a reduction of the state of ploidy, normally from diploid to haploid, during the formation of the germ cells.

Membrane filter a cellulose-ester membrane used for microbiological filtrations.

Meninges the membranes covering the brain and spinal cord.

Meningitis inflammation of the membranes of the brain or spinal cord.

Merozoites vegetative forms of the protozoan *Plasmodium* found in red blood cells.

Mesophiles organisms whose optimum growth is in the temperature range of 20–45°C.

Mesosomes intracellular membranous structures found in the infoldings of bacterial cell membranes; their function is as yet unknown.

Messenger RNA (mRNA) the RNA that specifies the amino acid sequence for a particular polypeptide chain.

Metabolic pathway a sequence of biochemical reactions that transforms a substrate into a useful product for carbon assimilation or energy transfer.

Metabolism the sum total of all chemical reactions by which energy is provided for the vital processes and new cell substances are assimilated.

Metabolites chemicals participating in metabolism; nutrients.

Metabolize to transform by means of metabolism.

Metachromatic granules cytoplasmic granules of polyphosphate occurring in the cells of certain bacteria that stain intensely with basic dyes but appear a different color; also known as volutin.

Metastasis the spread of cancer from a primary tumor to other parts of the body.

Methanogenic bacteria see methanogens.

Methanogens methane-producing prokaryotes; a group of archaebacteria capable of reducing carbon dioxide or low molecular weight fatty acids to methane.

Methylation the process of substituting a methyl group for a hydrogen atom.

MIC see minimum inhibitory concentration.

Micro- combining form meaning small.

Microaerophiles aerobic organisms that grow best in an environment with less than atmospheric oxygen levels.

Microbes microscopic organisms; microorganisms.

Microbicidal any agent capable of destroying microbes.

Microbiology the study of microorganisms and their interactions with other organisms and the environment.

Microbiostatic agents that inhibit the growth of microorganisms but do not kill them.

Microbiota the microorganisms normally associated with a given environment; the microorganisms associated with a particular tissue.

Microbodies collections of functionally related enzymes contained within a membranous envelope in eukaryotic organisms.

Micrococcus bacterial genus—cells spherical, occurring singly, in pairs and characteristically dividing in more than one plane to form a regular cluster, tetrads, or cubical packets; no resting stages known; Gram positive; metabolism strictly respiratory; aerobes; optimal growth 25en30°C; G + C 66–75 mole%.

Microflora see microbiota.

Microglia macrophages of the central nervous system.

Micrometer one millionth (10^{-6}) part of a meter; 10^{-3} of a millimeter.

Microorganisms microscopic organisms, including algae, bacteria, fungi, protozoa, and viruses.

Microscope an optical or electronic instrument for viewing objects too small to be visible to the naked eye.

Microtome an instrument used for cutting thin sheets or sections of tissues or individual cells for examination by light or electron microscopy.

Microtubules cylindrical protein tubes that occur within all eukaryotic organisms and aid in maintaining cell shape, comprise the structure of organelles of cilia and flagella, and serve as spindle fibers in mitosis.

Microwave radiation long wavelength radiation having poor penetrating power that apparently is unable to kill microorganisms directly.

Mildew any of a variety of plant diseases in which the mycelium of the parasitic fungus is visible on the affected plant; biodeterioration of a fabric due to fungal growth.

Millipore filter a specific commercial brand of membrane filters.

Mineralization the microbial breakdown of organic materials into inorganic materials brought about mainly by microorganisms.

Miniaturized commercial identification systems small devices containing multicompartmentalized chambers that each perform separate biochemical tests used for the identification of bacterial species.

Minimum bactericidal concentration (MBC) the lowest concentration of a chemotherapeutic agent that will prevent the growth of a particular microorganism.

Minimum inhibitory concentration (**MIC**) the lowest concentration of a particular antimicrobial drug necessary to inhibit the growth of a particular strain of microorganism.

Mitochondria (singular, **mitochondrion**) semiautonomous organelles found in eukaryotic cells; the site of respiration and other cellular processes; consisting of an outer membrane and an inner one that is convoluted.

Mitosis the sequence of events resulting in the division of the nucleus into two genetically identical cells during asexual cell division, each of the daughter nuclei having the same number of chromosomes as the parent cell.

Mixed acid fermentation a type of fermentation carried out by members of the Enterobacteriaceae, converting glucose to acetic, lactic, succinic, and formic acids.

MMR vaccine a single vaccine designed to provide immunity against measles, mumps, and rubella.

Moiety a part of a molecule having a characteristic chemical property.

Mold a type of fungus having a filamentous structure.

Mole% G + C the proportion of guanine and cytosine in a DNA macromolecule.

Mollicutes a class of prokaryotic organisms that do not form cell walls, e.g., *Mycoplasma*.

Monera prokaryotic protists with a unicellular and simple colonial organization; kingdom of the bacteria.

Mono- combining form meaning single, one, or alone.

Monoclonal antibody an antibody produced from a clone of cells making only that specific antibody.

Monocytes amoeboid agranular phagocytic white blood cells derived from the bone marrow.

Mononuclear having only one nucleus.

Mononuclear phagocyte system the macrophage system of the body, including all phagocytic white blood cells except granular white blood cells; also known as the reticuloendothelial system.

Monosaccharide any carbohydrate whose molecule cannot be split into simpler carbohydrates; a simple sugar.

Moraxella bacterial genus—rods, usually very short and plump (coccobacilli); Gram negative; oxidative metabolism; a limited number of organic acids, alcohols, and amino acids serve as carbon and energy sources; carbohydrates not utilized; oxidase positive; catalase usually positive; hydrogen sulfide not produced; strict aerobes; optimal growth 32–35°C; G + C 40–46 mole%.

Morbidity the state of being diseased; the ratio of the number of sick individuals to the total population of the community; the conditions inducing disease.

Mordant a substance that fixes the dyes used in staining tissues or bacteria; a substance that increases the affinity of a stain for a biological specimen.

Morphogenesis morphological changes, including growth and differentiation of cells and tissues during development; the transformations involved in the growth and differentiation of cells and tissues.

Morphology the shape and structure of organisms, including microorganisms, their cells and multicellular organizations.

Mortality death; the proportion of deaths to population.

Mortality rate death rate; number of deaths per unit population per unit time.

Most probable number (**MPN**) the statistical estimate of a bacterial population through the use of dilution and multiple tube inoculations.

Motility the capacity for independent locomotion.

MPN see most probable number.

mRNA see messenger RNA.

Muco-ciliary escalator system defense system that lines the upper respiratory tract and protects the respiratory tract against pathogens; the system consists of mucous membranes and cilia—mucus secretions trap microbes and cilia beat with an upward wave-like motion to expel microbes from the respiratory tract.

Mucosa mucous membranes, the linings of body cavities that communicate to the exterior.

Mucous membrane the type of membrane lining cavities and canals that have communication with air.

Mucus a viscid fluid secreted by mucous glands consisting of mucin, water, inorganic salts, epithelial cells, and leukocytes.

Multilateral budding budding that occurs all around the mother cell in some yeasts.

Multiple antibiotic resistance the ability to resist the effects of two or more unrelated antibiotics by bacterial strains generally containing R plasmids.

Mumps an acute infectious disease caused by a virus, characterized by swelling of the salivary glands.

Murein peptidoglycan; the repeating polysaccharide unit comprising the backbone of the cell walls of eubacteria.

Mushroom poisoning food poisoning caused by the ingestion of toxin-producing fungi in which the toxin accumulates in the fruiting body (mushroom) that is eaten; often mushroom poisoning effects the central nervous system and in many cases is fatal; *Amanita* mushrooms are often called the death-angel mushrooms because of their beauty and production of deadly toxins.

Must the fluid extracted from crushed grapes; the ingredients, e.g., fruit pulp or juice, used as substrate for fermentation in wine making.

Mutagen any chemical or physical agent that promotes the occurrence of mutation; a substance that increases the rate of mutation above the spontaneous rate.

Mutant any organism that differs from the naturally occurring type because its base DNA has been modified, resulting in an altered protein that gives the cell different properties than its parent.

Mutation a stable heritable change in the nucleotide sequence of the genetic nucleic acid, resulting in an alteration in the products coded by the gene.

Myc- combining form meaning fungus.

Mycelium (plural, **mycelia**) the interwoven mass of discrete fungal hyphae.

Mycetoma a chronic infection usually involving the foot, characterized by the presence of pussy nodules and caused by a wide variety of fungi or bacteria; also known as madura foot.

Mycobacterium bacterial genus—slightly curved or straight rods; filamentous or mycelium-like growth may occur; acid–alcohol fast at some stage of growth; Gram positive, but not readily stained by Gram's method; nonmotile; no endospores, conidia, or capsules; growth slow to very slow; optimal growth at about 40°C; G + C 62–70 mole%.

Mycolic acid fatty acids found in the cell walls of *Mycobacterium* and several other bacteria related to the actinomycetes.

Mycology the study of fungi.

Mycoplasmas members of the group of bacteria composed of cells lacking cell walls and exhibiting a variety of shapes.

Mycorrhizae a stable, symbiotic association between a fungus and the root of a plant; the term also refers to the root-fungus structure itself.

Mycosis any disease in which the causal agent is a fungus.

Mycotoxins toxic substances produced by fungi, including aflatoxin, amatoxin, and ergot alkaloids.

Myocarditis infection of the myocardium; can result from viral, bacterial, helminthic, or parasitic infections, hypersensitivity immune reactions, radiation therapy, or chemical poisoning,

Myocardium the muscular tissue of the heart wall.

Myx- combining form meaning mucus.

Myxamoebae nonflagellated amoeboid cells that occur in the life cycle of the slime molds and are members of the Plasmodiophorales.

NAD see nicotinamide adenine dinucleotide.

NADH reduced nicotinamide adenine dinucleotide.

NADP see nicotinamide adenine dinucleotide phosphate.

NADPH reduced nicotinamide adenine dinucleotide phosphate.

Narrow spectrum antibiotics antibiotics that are highly selective with respect to the species that they can inhibit; such an antibitotic must be targeted at a particular pathogen.

Nasopharyngeal swabs culture taken from the nasopharynx by means of a polyester attached to and wrapped around a thin stick.

Nasopharynx the upper part of the pharynx continuous with the nasal passages.

Necrosis the pathologic death of a cell or group of cells in contact with living cells.

Negative stain the treatment of cells with dye so that the background, rather than the cell itself, is made opaque; used to demonstrate bacterial capsules or the presence of parasitic cysts in fecal samples; a stain with a positively charged chromophore.

Negri bodies acidophilic, intracytoplasmic inclusion bodies that develop in cells of the central nervous system in cases of rabies.

Neisseria bacterial genus—cocci, occurring singly but often in pairs with adjacent sides flattened; endospores not produced; nonmotile; capsules may be present; Gram negative; complex growth requirements; few carbohydrates utilized; aerobic or facultatively anaerobic; catalase positive; oxidase positive; optimal growth ca. 37°C; G + C 47.0–52.0 mole%.

Nematodes worms of the class Nematoda.

Neoplasm the result of the abnormal and excessive proliferation of the cells of a tissue; if the progeny cells remain localized, the resulting mass is called a tumor.

Nephrons the microscopic functional units of the kidneys that control the concentration and volume of blood by removing and adding selected amounts of water and solutes and excreting wastes.

Neurotoxin a toxin capable of destroying nerve tissue or interfering with neural transmission.

Neutropenia a decrease below the normal standard in the number of neutrophils in the peripheral blood.

Neutrophils a large granular leukocyte with a highly variable nucleus consisting of 3-5 lobes and cytoplasmic granules that stain with neutral dyes and eosin.

NGU see nongonococcal urethritis.

Niche the functional role of an organism within an ecosystem; the combined description of the physical habitat, functional role, and interactions of the microorganisms occurring at a given location.

Nicotinamide adenine dinucleotide (NAD) a coenzyme used as an electron acceptor in oxidation–reduction reactions.

Nicotinamide adenine dinucleotide phosphate the phosphorylated form of NAD formed when NADPH serves as an electron donor in oxidation–reduction reactions.

Nitrate a salt of nitric acid, NO_3^-.

Nitrate reduction the reduction of nitrate to reduced forms, e.g., when under anaerobic and microaerophilic conditions, bacteria use nitrate as a terminal electron acceptor for respiratory metabolism.

Nitrification the process in which ammonia is oxidized to nitrite and nitrite to nitrate; a process primarily carried out by the strictly aerobic, chemolithotrophic bacteria of the family Nitrobacteraceae.

Nitrifying bacteria Nitrobacteraceae; Gram negative, obligately aerobic, chemolithotrophic bacteria occurring in fresh and marine waters and in soil that oxidize ammonia to nitrite or nitrite to nitrate.

Nitrite a salt of nitrous acid, NO_2- nitrites of sodium and potassium are used as food additives and preservatives.

Nitrobacter bacterial genus—cells short rods; Gram negative; chemolithotrophs that oxidize nitrite to nitrate and fix carbon dioxide; strictly aerobic; temperature range for growth 5–40°C; G + C 60.7–61.7 mole%.

Nitrofurantoin a synthetic, broad-spectrum antibacterial agent widely used in the treatment of urinary tract infections.

Nitrogenase the enzyme that catalyzes biological nitrogen fixation.

Nitrogen cycle the biogeochemical cycle by which atmospheric nitrogen gas is converted into a usable form for plants by microbial enzymatic reactions and back to nitrogen gas again.

Nitrogen fixation the reduction of gaseous nitrogen to ammonia, carried out by certain prokaryotes.

Nitrogenous containing nitrogen.

Nitrosomonas bacterial genus—cells ellipsoidal or short rods, motile or nonmotile, occurring singly, in pairs, or short chains; Gram negative; chemolithotrophic, oxidize ammonia to nitrite and fix carbon dioxide; strictly aerobic; temperature range for growth 5–30°C; G + C 47.4–51 mole%.

Nocardia Bacterial genus—produce true mycelium, but mycelium production may be rudimentary; Gram positive; some species acid fast to partially acid fast; obligate aerobes; nonmotile; pigments are produced by several species; G + C 60–72 mole%.

Nodules tumor-like growths formed by plants in response to infections with specific bacteria, within which the infecting bacteria fix atmospheric nitrogen.

Nomenclature the naming of organisms, a function of taxonomy governed by codes, rules, and priorities laid down by committees.

NonA–nonB hepatitis see hepatitis nonA–nonB.

Nongonococcal urethritis (NGU) any inflammation of the urethra not caused by *Neisseria gonorrhoeae*.

Nonself-antigens foreign antigens; antigens not found as part of the normal cells and tissues of the body.

Nonsense codon a codon that does not specify an amino acid but acts as a punctuator of mRNA.

Nonsense mutation a mutation in which a codon specifying an amino acid is altered to a nonsense codon.

Nonspecific defense system host resistance that tends to afford protection against various pathogens that is innate.

Nonspecific urethritis (NSU) see nongonococcal urethritis.

Normal growth curve characteristic growth curve exhibited by bacteria when inoculated on fresh media, obtained by plotting increases in the numbers of micoorganisms against elapsed time; consisting of the lag, log, stationary, and death phases.

Nosocomial infection an infection acquired while in a hospital.

NSU see nongonococcal urethritis.

Nuclease an enzyme capable of splitting nucleic

acids to nucleotides, nucleosides, or their components.

Nucleic acid a macromolecule containing phosphoric acid, sugar, and purine and pyrimidine bases; the nucleotide polymers RNA and DNA.

Nucleoid region the region of a prokaryotic cell in which the genome occurs.

Nucleolus an RNA-rich intranuclear body not bounded by a limiting membrane that is the site of rRNA synthesis in eukaryotes.

Nucleoprotein a conjugated protein closely associated with nucleic acid.

Nucleoside a class of compound in which a purine or pyrimidine base is linked to a pentose sugar.

Nucleosome the fundamental structural unit of DNA in eukaryotes having approximately 190 base pairs folded and held together by histones.

Nucleotide the combinination of a purine or pyrimidine base with a sugar and phosphoric acid; the basic structural unit of nucleic acid.

Nucleus an organelle of eukaryotes in which the cell's genome occurs; the differentiated protoplasm of a cell surrounded by a membrane that is rich in nucleic acids.

Numerical aperature the property of an objective lens that describes the widest cone of light that can enter that microscope lens from the specimen; important in determining resolution.

Numerical profile used in commercial systems for the identification of clinical isolates; calculates and compares the test pattern of an unknown with that of a defined group to determine the probability that the test results could represent a member of that taxon.

Nurse person trained to take care of the ill and infirm.

Nutrient a growth supporting substance.

Nutrition requirments of living organisms for growth and sustenance.

O antigens lipopolysaccharide–protein antigens occurring in the cells of Gram negative bacteria.

Objective lens the microscope lens nearest the object.

Obligate anaerobes organisms that grow only under anaerobic conditions, i.e., in the absence of air or oxygen; organisms that cannot carry out respiratory metabolism.

Occluded closed or shut up.

Ocular lens the eyepiece of a microscope.

-oid combining form meaning resembling.

Oil immersion lens a high power microscope objective designed to work with the space between the objective and the specimen filled with oil to enhance resolution.

Oligodynamic action the ability of a small amount of a heavy metal compound to act as an antimicrobial agent.

Oncogenes genes that can lead to malignant transformations.

Oncogenic viruses viruses capable of inducing tumor formation.

Operator region a section of an operon involved in the control of the synthesis of the gene products encoded within that region of DNA; a regulatory gene that binds with a regulatory protein to turn on and off transcription of a specified region of DNA.

Operon a group or cluster of structural genes whose coordinated expression is controlled by a regulator gene.

Opportunistic pathogens organisms that exist as part of the normal body microbiota but that may become pathogenic under certain conditions, e.g., when the normal antimicrobial body defense mechanisms have been impaired; organisms not normally considered pathogens but that cause disease under some conditions.

Opsonization the process by which a cell becomes susceptible to phagocytosis and lytic digestion by combination of a surface antigen with an antibody and/or other serum component.

Optical isomers compounds having the same number and kind of atoms and grouping of atoms but differing in their configurations or arrangements in space; specifically their structures are not superimposable.

Optimal growth temperature the temperature at which cells exhibit the maximal growth rate.

Oral cavity mouth.

Orchitis inflammation of the testes.

Organelle a membrane-bound structure that forms part of a microorganism and performs a specialized function.

-ose combining form denoting a sugar.

-osis combining form meaning disease of.

Osmophiles organisms that grow best or only in or on media of relatively high osmotic pressure.

Osmosis the passage of a solvent through a membrane from a dilute solution into a more concentrated one.

Osmotic pressure the force resulting from differences in solute concentrations on opposite sides of a semipermeable membrane.

Osmotic shock any disturbance or disruption in a cell or subcellular organelle that occurs when it is transferred to a significantly hypertonic or hypotonic medium, with lysis of cells resulting from osmotic pressure.

Osmotolerant organisms that can withstand high osmotic pressures and grow in solutions of high solute concentrations.

Otitis externa also known as swimmer's ear, inflammation of the skin of the external ear canal and auricle.

Otitis media inflammation of the inner ear.

Oxidase an oxidoreductase that catalyzes a reaction in which electrons removed from a substrate are donated directly to molecular oxygen.

Oxidation an increase in the positive valence or decrease in the negative valence of an element, resulting from the loss of electrons that are taken on by some other element.

Oxidation pond a method of aerobic waste disposal employing biodegradation by aerobic and facultative microorganisms growing in a standing water body.

Oxidation–reduction potential a measure of the tendency of a given oxidation–reduction system to donate elections, i.e., to behave as a reducing agent, or to accept electrons, i.e., to act as an oxidizing agent; determined by measuring the electrical potential difference between the given system and a standard system.

Oxidative phosphorylation a metabolic sequence of reactions occurring within a membrane in which an electron is transferred from a reduced coenzyme through a series of electron carriers establishing an electrochemical gradient across the membrane that drives the formation of ATP from ADP and inorganic phosphate.

Oxidative photophosphorylation a metabolic sequence of reactions occurring within a membrane in which light initiates the transfer of an electron through a series of electron carriers establishing an electrochemical gradient across the membrane that drives the formation of ATP from ADP and inorganic phosphate.

Oxidize to produce an increase in positive valence through the loss of electrons.

Oxygen cycle the biogeochemical cycle by which oxygen is exhanged and distributed throughout the biosphere.

Oxyphotobacteria bacteria capable of evolving oxygen during photosynthesis.

Ozonation the killing of microorganisms by exposure to ozone.

Palindrome a word reading the same backward and forward; a base sequence the complement of which has the same sequence; a nucleotide sequence that reads the same when read in the antiparallel direction.

Pandemic an outbreak of disease that affects large numbers of people in a major geographical region or that has reached epidemic proportions simultaneously in different parts of the world.

Papular rash a skin rash characterized by raised spots.

Paralytic shellfish poisoning caused by toxins produced by the dinoflagellates *Gonyaulax*, which concentrate in shellfish such as oysters and clams.

Parasites organisms that live on or within the tissues of another living organism, the host, from which it derives its nutrients.

Parasitism an interactive relationship between two organisms or populations in which one is harmed and the other benefits; generally, the population benefiting, the parasite, is smaller than the population that is harmed.

Parenteral route route of infection when microorganisms are deposited directly into tissues beneath the skin and mucous membranes.

Parfocal pertaining to microscopical oculars and objectives that are so constructed or so mounted that in changing from one to another the image will remain in focus.

Parotitis inflammation of the parotid gland, as in mumps.

Passive agglutination a procedure in which the combination of antibody with a soluble antigen is made readily detectable by the prior adsorption of the antigen to erythrocytes or to minute particles of organic or inorganic materials.

Passive immunity short-term immunity brought about by the transfer of preformed antibody from an immune subject to a nonimmune subject.

Pasteurella bacterial genus—cells ovoid or rod-shaped; nonmotile; do not produce endospores; Gram negative, but bipolar staining is common; metabolism fermentative; methyl red negative; Voges–Proskauer negative; aerobic to facultatively anaerobic; optimal growth 37°C; G + C 36.5–43.0 mole%.

Pasteurization the reduction in numbers of microorganisms by exposure to elevated temperatures but not necessarily the killing of all microorganisms in a sample; a form of heat treatment that is lethal for the causal agents of a number of milk transferable diseases as well as for a proportion of normal milk microbiota, which also inactivates certain bacterial enzymes that may cause deterioration in milk.

Pathogens organisms capable of causing disease in animals, plants, or microorganisms.

Pathology the study of the nature of disease through the study of its causes, processes, and effects, along with the associated alterations of structure and function.

Pellicle a thin protective membrane occurring around some protozoa, also known as a periplast; a continuous or fragmentary film that sometimes forms at the surface of a liquid culture, which may consist entirely of cells or may be largely extracellular products of the cultured organisms.

Pelvic inflammatory disease (PID) any acute, subacute, recurrent, or chronic infection of the oviducts and ovaries, with adjacent tissue involvement, most commonly caused by *Neisseria gonorrhoeae*.

Penicillinase a β-lactamase which hydrolyzes the beta-lactam linkage of many penicillins, rendering it ineffective as an antibiotic.

Penicillins a group of natural and semisynthetic antibiotics, having a β-lactam ring, that are active against Gram positive bacteria inhibiting the formation of crosslinks in the peptidoglycan of growing bacteria.

Pentose a class of carbohydrates containing five atoms of carbon.

Pepsin a proteolytic enzyme.

Peptidase an enzyme that splits peptides into amino acids.

Peptide bond a bond in which the carboxyl group of one amino acid is condensed with the amino group of another amino acid.

Peptides compounds of two or more amino acids containing one or more peptide bonds.

Peptidoglycan the rigid component of the cell wall in most bacteria.

Peptidyl site the site on the ribosome where the growing peptide chain is moved during protein synthesis.

Peptones a water soluble mixture of peptides and amino acids produced by the hydrolysis of natural proteins either by an enzyme or an acid.

Perfringens food poisoning food poisoning by ingestion of *Clostridium perfringens* type A, a self-limiting condition characterized by abdominal pain and diarrhea.

Periferal nervous system that part of the nervous system outside the brain and spinal cord; the ganglia and the cranial and spinal nerves.

Period of illness the acute phase of an infection during which the patient experiences characteristic symptoms.

Period of infection the time period during which viable disease-causing microbes are present in the body.

Periodontal disease disease of the tissues surrounding the teeth.

Periodontal pockets holes in the gums deepened by periodontal disease.

Periodontitis inflammation of the periodontium, the tissues surrounding a tooth.

Peritrichous flagella flagella that are uniformly distributed over the surface of the cell.

Permeability the property of cell membranes that permits transit of molecules and ions in solution across the membrane.

Permease an enzyme that increases the rate of transport of a substance across a membrane.

Peroxidase an oxidoreductase that catalyzes a reaction in which electrons removed from a substrate are donated to hydrogen peroxide.

Peroxide the anion O_2^{2-} or HO_2^-, or a compound containing one of these anions.

Peroxisomes microbodies that contain D-amino acid oxidase, α-hydroxy acid oxidase, catalase, and other enzymes, found in yeasts and certain protozoans.

Person-to-person epidemic epidemiological disease pattern characterized by a relatively slow and prolonged rise and decline in the number of cases.

Person-to-person transmission spread of infectious disease from one person to another.

Pest a population that is an annoyance for economic, health, or aesthetic reasons.

Pesticides substances destructive to pests, especially insects.

Petri dish a round, shallow, flat-bottomed dish with a vertical edge, together with a similar, slightly larger structure that forms a loosely fitting lid, made of glass or plastic, widely used as receptacles for various types of solid media.

pH expression of the hydrogen ion concentration; the logarithm to the base 10 of the reciprocal of the hydrogen ion concentration; $-\log [H^+]$.

Phage see bacteriophage.

Phagocytes any of a variety of cells that ingest cells and particulate matter; blood cells that can ingest and digest microorganisms.

Phagocytosis the process in which particulate matter is ingested by a cell, involving the engulfment of that matter by the cell's membrane.

Phagosomes membrane-bound vesicles in phagocytes formed by the invagination of the cell membrane and the phagocytized material.

Phagotrophic referring to the ingestion of nutrients in particulate form by phagocytosis.

Pharmaceutical a drug used in the treatment of disease.

Pharyngitis inflammation of the pharynx.

Phase contrast microscope a microscope that achieves enhanced contrast of the specimen by altering the phase of light that passes through the specimen relative to light that passes through the background, eliminating the need for staining in order to view microorganisms and making the viewing of live specimens possible.

Phenetic pertaining to the physical characteristic of an individual without consideration for its genetic makeup; in taxonomy, a classification system that does not take evolutionary relationships into consideration.

Phenol coefficient a measure of the antibacterial power of a substance relative to the disinfectant phenol; the ratio of the disinfecting power of a substance relative to phenol.

Phenolics class of antiseptics and disinfectants derived from carbolic acid.

Phenotype the totality of observable structural and functional characteristics of an individual organism, determined jointly by its genotype and environment.

Phenotypic characteristics the observable qualities of an organism.

-phile combining form meaning like, or having an affinity for.

-phobic combining form meaning having an aversion to or lacking affinity for.

Phosphate diester linkage bond that links nucleotides of DNA and RNA, consisting of a phosphate group bonded via ester linkages between carbon-3 of one carbohydrate and carbon-5 of the next carbohydrate portion of the macromolecule; the bonding of two moieties through a phosphate group in which each moiety is held to the phosphate by an ester linkage.

Phosphodiester bond see phosphate diester linkage.

Phospholipid a type of lipid compound that is an ester of phosphoric acid and also contains one or two molecules of fatty acid, an alcohol, and sometimes a nitrogenous base.

Phosphorylation the esterification of compounds with phosphoric acid; the conversion of an organic compound into an organic phosphate.

Photo- combining form meaning light.

Photoautotrophs organisms whose source of energy is light and whose source of carbon is carbon dioxide, characteristic of algae and some prokaryotes.

Photoheterotrophs organisms that obtain energy from light but require exogenous organic compounds for growth.

Photophosphorylation a metabolic sequence by which light energy is trapped and converted to chemical energy with the formation of ATP.

Photoreactivation a mechanism whereby the effects of ultraviolet radiation on DNA may be reversed by exposure to radiation of wavelengths in the range 320–500 nm; an enzymatic repair mechanism of DNA present in many microorganisms.

Photosynthesis the process in which radiant energy is absorbed by specialized pigments of a cell and is subsequently converted to chemical energy; the ATP formed in the light reactions is used to drive the fixation of carbon dioxide with the production of organic matter.

Phototrophic bacteria see phototrophs.

Phototrophs organisms whose sole or principal primary source of energy is light; organisms capable of photosynthesis.

Phycobilisomes granules found in cyanobacteria and some algae on the surface of their thylakoids.

Phycology the study of algae.

Phycomycete a group of true fungi, lacking regularly spaced septae in the actively growing portions of the fungus and producing sporangiospores by cleavage as the primary method of asexual reproduction.

Phylogenetic referring to the evolution of a species from the simpler forms; in taxonomy a classification based on evolutionary relationships.

Phylum a taxonomic group composed of groups of related classes.

Physician doctor of medicine.

Physiology the science that deals with the study of the functions of living organisms and their physicochemical parts and metabolic reactions.

Phytoplankton passively floating or weakly motile photosynthetic aquatic organisms, primarily cyanobacteria and algae.

PID see pelvic inflammatory disease.

Pili filamentous appendages that project from the cell surface of certain Gram negative bacteria, apparently involved in adsorption phenomena.

Pink eye infection of the eye caused by *Haemophilus aegyptius*, characterized by swelling of the eyelids, discharge from the eye, and bleeding within the conjunctiva as well as redness and itching characteristic of many eye inflammations.

Pinworm disease caused by the nematode *Enterobius vermicularis* which spend their life cycles in human hosts, particularly the large intestine.

Pinworms *Enterobius vermicularis* nematodes.

Pitching the inoculation of mash with yeast during production of alcoholic beverages; the inoculation of a substrate with microorganisms.

Plague a contagious disease often occurring as an epidemic; an acute infectious disease of humans and other animals, especially rodents, caused by *Yersinia pestis* transmitted by fleas.

Planapochromatic lens a flat field apochromatic objective microscope lens.

Plankton collectively, all microorganisms that passively drift in the pelagic zone of lakes and other bodies of water, chiefly microalgae and protozoans.

Plant viruses viruses that multiply in plant cells.

Plaque a clearing in an area of bacterial growth due to lysis by phages; the accumulation of bacterial cells within a polysaccharide matrix on the surfaces of teeth; also known as dental plaque.

Plasma membrane see cytoplasmic membrane.

Plasmids extrachromosomal genetic structures that can replicate independently within a bacterial cell.

Plasmodium malaria-causing protozoans; the life stage of acellular slime molds characterized by a motile multinucleate body.

Plate counts method of estimating numbers of microorganisms by diluting samples, culturing on solid media, and counting the colonies that develop to estimate viable microorganisms in the sample.

Pleomorphic exhibiting pleomorphism.

Pleomorphism the variation in size and form between individual cells in a clone or a pure culture.

Ploidy the number of complete sets of chromosomes of a eukaryotic nucleus or cell.

PMNs see polymorphs.

Pneumonia inflammation of the lungs, often of bacterial or viral etiology; an infection of the lower respiratory tract that produces inflammation of the lungs.

Polar located at an end.

Polar budding budding of yeast cells restricted to one or both polar ends of a mother cell.

Polar flagella flagella emanating from one or both polar ends of a cell.

Poliomyelitis inflammation of the gray matter of the spinal cord, caused by a picornavirus.

Poly-β-hydroxybutyric acid a polymeric storage product formed by some bacteria.

Polyenes antimicrobial agents capable of altering sterols in eukaryotic plasma membranes.

Polymerase an enzyme that catalyzes the formation of a polymer.

Polymers the products of the combination of two or more molecules of the same substance.

Polymorphonuclear having a nucleus connected together by thin strands of nuclear substance.

Polymorphonuclear leukocytes see polymorphs.

Polymorphs leukocytes having granules in the cytoplasm, also known as polymorphonuclear leukocytes or PMNs.

Polymyxins antibiotics whose effectiveness is based on their ability to cause the disintegration of phospholipids.

Polypeptide a chain of amino acids linked together by peptide bonds but of lower molecular weight than a protein.

Polysaccharides carbohydrates formed by the condensation of several to many monosaccharides, e.g., starch and cellulose; having multiple monosaccharide subunits; macromolecules composed of sugar subunits.

Portals of entry the sites through which a pathogen can gain access and entry to the body.

Positional isomers molecules with identical molecular formulas but differening in the locations of substituents; isomers with functional groups located at different positions.

Positive stain a stain with a positively charged chromophore; a stain that is attracted to negatively charged cells.

Postpartum sepsis see puerperal fever.

Potable fit to drink.

Pour plate a method of culture in which the inoculum is dispersed uniformly in molten agar or other medium in a petri dish, the medium is allowed to set and is then incubated.

Precipitin reaction a serological test in which the interaction of antibodies with soluble antigens is detected by the formation of a precipitate.

Predation a mode of life in which food is primarily obtained by killing and consuming animals; an interaction between organisms in which one benefits and one is harmed, based on the ingestion of the smaller–sized organism, the prey, by the larger organism, the predator.

Predators see predation.

Preemptive colonization phenomenon that occurs when pioneer organisms alter environmental conditions in such a way that discourages further colonization by other microorganisms.

Preservation the use of physical and/or chemical means to kill or retard the growth of those microorganisms that cause spoilage.

Preservatives chemicals used for preservation.

Presumptive diagnosis preliminary diagnosis of

a disease based on signs and symptoms reported to the physician.

Presumptive tests tests that point toward a probable diagnosis but do not definitively identify the etiology of a disease—generally used to aid in the preliminary diagnosis so that treatment can be initiated while additional tests are run to define the best and most specific treatment.

Primary atypical pneumonia pneumonia caused by *Mycoplasma pneumoniae*.

Primary immune response the first immune response to a particular antigen; has a characteristically long lag period and relatively low titer of antibody production.

Primary producers those organisms capable of converting carbon dioxide to organic carbon, including photoautotrophs and chemoautotrophs.

Primary sewage treatment the removal of suspended solids from sewage by physical settling in tanks or basins.

Prions infectious proteins; substances that are infectious and reproduce within living systems but appear to be proteinaceous, based on degradation by proteases, and to lack nucleic acids, based on resistance to digestion by nucleases.

Probabilistic identification matrix characterizes large numbers of strains of a taxonomic group to establish the variability within a group for a particular feature; used to allow organisms of unknown affiliation to be identified as members of established taxa.

Processed cheeses blends of various cheeses.

Prodromal stage time period in the infectious process following incubation when the symptoms of the illness begin to appear.

Progeny offspring.

Projector lens the lens of an electron microscope that focuses the beam onto the film or viewing screen.

Prokaryotes cells whose genomes are not contained within a nucleus; the bacteria.

Prokaryotic cells cells lacking true nuclei; bacterial cells.

Promastigote an elongated and flagellated form assumed by many species of the Trypanosomatidae during a particular stage of development.

Promoter specific initiation site of DNA where the RNA polymerase enzyme binds for transcription on the DNA.

Promoter region starting point on DNA for the transcription of RNA by RNA polyermerase.

Propagated transmission see person-to-person transmission.

Prophage the integrated phage genome formed when the phage genome becomes integrated with the host's chromosome and is replicated as part of the bacterial chromosome during subsequent cell division.

Prophylaxis the measures taken to prevent the occurrence of disease.

Propionibacterium bacterial genus—Gram positive, nonspore-forming, nonmotile rods; usually pleomorphic, diphtheroid, or club-shaped; metabolize carbohydrates, peptone, pyruvate, or lactate; fermentation products include combinations of propionic and acetic acids; anaerobic to aerotolerant; optimal growth 30–37°C; G + C 59–66 mole%.

Prostaglandins naturally occurring fatty acids that circulate in the blood, stimulate the contraction of smooth muscles, and have the ability to lower blood pressure and affect the action of certain hormones.

Prostatitis inflammation of the prostate gland, usually caused by a Gram negative bacterial infection.

Prosthecae a cell wall-limited appendage forming a narrow extension of a prokaryotic cell.

Proteases exoenzymes that break proteins down into their component amino acids.

Proteinase one of the subgroups of proteases or proteolytic enzymes that act directly on native proteins in the first step of their conversion to simpler substances.

Proteins a class of high molecular weight polymers composed of amino acids joined by peptide linkages.

Protein toxins proteins secreted by bacteria that act as poisons.

Proteolytic enzymes enzymes that break down proteins.

Proteus bacterial genus—straight rods ; motile by peritrichous flagella; nonpigmented; acid produced from glucose; methyl red test usually positive; Gram negative; optimal growth 37°C; G + C 38–42 mole%.

Protista in one proposed classification system, a kingdom of organisms lacking true tissue differention, i.e., the microbes; in another classification system, a kingdom that includes many of the algae and protozoa.

Protoplasm the viscid material constituting the essential substance of living cells upon which all the vital functions of nutrition, secretion, growth, reproduction, irritability, and locomotion depend.

Protoplasts spherical, osmotically sensitive structures formed when cells are suspended in an isotonic medium and their cell walls are completely removed; a bacterial protoplast consists of an intact cell membrane and the cytoplasm contained within.

Prototrophs parental strains of microorganisms that give rise to nutritional mutants known as auxotrophs.

Protozoa diverse, eukaryotic, typically unicellular nonphotosynthetic microorganisms generally lacking a rigid cell wall.

Protozoology the study of protozoa.

Pseudomonas bacterial genus—cells single, straight, or curved rods; motile by polar flagella; monotrichous or multitrichous; no resting stages known; Gram negative; metabolism respiratory, never fermentative; some are facultative chemolithotrophs; G + C 58–70 mole%.

Pseudoplasmodium in cellular slime molds, an aggregation of cells formed before the production of a fruiting body; in net slime molds, a network of cells linked by slime filaments; in Myxobacterales, a number of individual cells imbedded in a slime matrix.

Pseudopodia false feet formed by protoplasmic streaming in protozoa and used for locomotion and the capture of food.

Psittacosis an infectious disease of parrots, other birds, and humans, caused by *Chlamydia psittaci*; also called parrot fever.

Psychro- combining form meaning cold.

Psychrophile an organism that has an optimum growth temperature below 20°C.

Psychrotroph a mesophile that can grow at low temperatures.

Puerperal fever an acute febrile condition following childbirth caused by infection of the uterus and/or adjacent regions by streptococci.

Pure culture a culture that contains cells of one kind; the progeny of a single cell.

Purine a cyclic nitrogenous compound, the parent of several nucleic acid bases.

Purulent full of pus; containing or discharging pus.

Pus a semifluid, creamy yellow, or greenish yellow product of inflammation composed mainly of leukocytes, serum, and cellular debris.

Pustule a small elevation of the skin containing pus.

Putrefaction the microbial breakdown of protein under anaerobic conditions.

Pyelonephritis inflammation of the kidneys.

Pyknosis the condition of having a contracted nucleus.

Pyoderma a pus-producing skin lesion.

Pyogenic pus producing.

Pyrimidine a six–membered cyclic compound containing four carbon and two nitrogen atoms in a ring, the parent compound of several nucleotide bases.

Pyrite a common mineral containing iron disulfite.

Pyrogenic fever producing.

Pyrogens fever-producing substances.

Q fever an acute disease in humans characterized by sudden onset of headache, malaise, fever, and muscular pain, caused by *Coxiella burnetii*; the reservoirs of infection are cattle, sheep, and ticks.

Quarantine the isolation of persons or animals suffering from an infectious disease in order to prevent transmission of the disease to others.

Quaternary ammonium compounds group of cationic detergents that disrupt bacterial cell membranes used as antiseptics and disinfectants.

Quats see quaternary ammonium compounds.

Rabies an acute and usually fatal disease of humans, dogs, cats, bats, and other animals, caused by the rabies virus and commonly transmitted in saliva by the bite of a rabid animal.

Radappertization the reduction in the number of microorganisms by exposure to ionizing radiation.

Radiation the process in which energy is emitted in particles or waves.

Radioisotopes radioactive isotopes; forms of elements that spontaneously decompose with the emission of radiation.

Radioimmunoassay a highly sensitive serological technique used for assaying specific antibodies or antigens employing a radioactive label.

Radiolarians free-living protozoa occurring almost exclusively in marine habitats, having axopodia, with a skeleton of silicon or strontium sulfate.

Radurization sterilization by exposure to ionizing radiation.

Rancid having the characteristic odor of decomposing fat, chiefly due to the liberation of butyric and other volatile fatty acids.

rDNA recombinant DNA.

Reagins a group of antibodies in serum that react with the allergens responsible for the specific manifestations of human hypersensitivity; a heterophile antibody formed during syphilis infections.

Recalcitrant a chemical that is totally resistant to microbial attack.

Recipient strain any strain that receives genetic information from another.

Recombinant any organism whose genotype has arisen as a result of recombination; also any nucleic acid that has arisen as a result of recombination.

Recombinant DNA technology see genetic engineering.

Recombination the exchange and incorporation of genetic information into a single genome, resulting in the formation of new combinations of alleles.

Red algae members of the Rhodophycophyta containing phycoerythrin pigment, which gives it the characteristic color, occuring in marine habitats.

Red tide caused by toxic blooms of *Gonyaulax* and other dinoflagellates that color the water and kill invertebrate organisms—the toxins concentrate in the tissues of filter-feeding molluscs causing food poisoning.

Reduced flavin adenine dinucleotide the reduced form of the coenzyme flavin adenine dinucleotide; $FADH_2$.

Reducing power the capacity to bring about reduction.

Reduction an increase in the negative valence or a decrease in the positive valence of an element resulting from the gain of electrons.

Refraction the deviation of a ray of light from a straight line in passing obliquely from one transparent medium to another of different density.

Refractive index an index of the change in velocity of light when it passes through a substance causing a deviation in the path of the light.

Relapsing fever a human disease characterized by recurrent fever, caused by a *Borrelia* sp. and transmitted by ticks and lice.

Renin an enzyme obtained from a calf's stomach that can hydrolyze proteins.

Replica plating a technique by which various types of mutants can be isolated from a population of bacteria grown under nonselective conditions, based on plating cells from each colony onto multiple plates of selective and/or differential media and noting the positions of inoculation.

Replication multiplication of a microorganism; duplication of a nucleic acid from a template; the formation of a mold for viewing by electron microscopy.

Replication fork the Y-shaped region of a chromosome that is the growing point during replication of DNA.

Repression the blockage of gene expression.

Repressor protein a protein that binds to the operator and inhibits the transcription of structural genes.

Reproduction a fundamental property of living systems by which organisms give rise to other organisms of the same kind.

Reservoirs the constant sources of infectious agents found in nature.

Resistant (R) in antimicrobial sensitivity testing, the standardized zone of inhibition shows that the antibiotic disc has little or no effect on the microorganism.

Resolution the fineness of detail observable in the image of a specimen.

Resolving power a quantitative measure of the closest distance between two points that can still be seen as distinct points when viewed in a microscope field; depends largely on the characteristics of its objective lens and the optimal illumination of the specimen; approximately equal to the wavelength of light divided by two times the numerical aperature of the objective lens.

Respiration a mode of energy-yielding metabolism requiring a terminal electron acceptor for substrate oxidation, with oxygen frequently used as the terminal electron acceptor; breathing; inhalation of air.

Respiratory therapist an allied health professional who aids patients with respiratory problems and in alleviating or preventing pneumonias.

Respiratory tract the structures and passages involved with the intake of oxygen and expulsion of carbon dioxide in animals.

Restriction enzymes enzymes capable of cutting DNA macromolecules.

Reticuloendothelial system see mononuclear phagocyte system.

Retroviruses family of enveloped RNA animal viruses that use reverse transcriptase to form a DNA macromolecule needed for their replication.

Reverse transcriptase an enzyme that synthesizes a complementary DNA from an RNA template.

Reye's syndrome a neurological disease that sometimes occurs after a viral infection.

Rh antigen (Rhesus antigen) an antigen that occurs on the surfaces of some red blood cells; individuals with the antigen have Rh-positive blood and those lacking this antigen have Rh-negative blood.

Rheumatic fever a febrile disease, characterized by painful migratory arthritis and a predilection to heart damage leading to chronic valvular disease, that results from the systemic spread of *Streptococcus pyogenes* toxins.

Rhizobium bacterial genus—rods, commonly pleomorphic; Gram negative; metabolism respiratory; characteristically able to invade root hairs of leguminous plants and initiate production of root nodules; within nodules bacteria are pleomorphic (bacteroids); nodule bacteroids characteristically involved in fixing molecular nitrogen; optimal growth 25–30°C; G + C 59.1–65.5 mole%.

Ribonucleic acid (RNA) a linear polymer of ribonucleotides in which the ribose residues are linked by 3′,5′-phosphodiester bonds; the nitrogenous bases attached to each ribose residue may be adenine, guanine, uracil, or cytosine.

Ribosomal RNA (rRNA) RNA of various sizes that make up part of the ribosomes, constituting up to 90 percent of the total RNA of a cell; single-stranded RNA but with helical regions formed by base-pairing between complementary regions within the strand.

Ribosomes intracellular structures composed of rRNA and protein; the sites where protein synthesis occurs.

Rickettsia bacterial genus—parasitic bacteria occurring intracellularly or intimately in association with tissue cells other than erythrocytes or with certain organs in arthropods; growth generally occurs in the cytoplasm of host cells; transmitted by arthropod vectors; short rods; no flagella; Gram negative; G + C 30–32.5 mole%.

Rickettsialpox a disease caused by *Rickettsia akari*, characterized by enlargement of the lymph nodes, fever, chills, headache, secondary rash, and leukopenia following an initial papule at the locale of the bite of a mite.

Rickettsias members of the family Rickettsiaceae; Gram negative bacterial parasites or pathogens of vertebrates and arthropods that reproduce intracellularly.

Ringworm any mycosis of the skin, hair, or nails in humans or other animals in which the causal agent is a dermatophyte; also called tinea.

Ripened cheeses cheeses that have undergone additional microbial growth beyond a single fermentation step to achieve a characteristic taste, texture, or aroma.

RNA see ribonucleic acid.

RNA polymerase enzyme that catalyzes the formation of RNA macromolecules.

Rocky Mountain spotted fever a tick-borne human disease caused by *Rickettsia rickettsii* that occurs in parts of North America.

Roundworm disease any infestation of the body by roundworms.

Roundworms members of the phyla Aschelminthes, commonly called helminths; parasitic animals that spend part or all of their life cycle in human hosts.

R plasmid a plasmid encoding for antibiotic resistance.

rRNA see ribosomal ribonucleic acid.

Rubella see German measles.

Sac fungi the ascomycetes, fungi that produce sexual spores within an ascus.

Salmonella bacterial genus—rods, usually motile by peritrichous flagella; can utilize citrate as carbon source; Gram negative; indole negative; hydrogen sulfide produced on TSI; methyl red positive; Voges–Proskauer negative; catalase positive; oxidase negative; nitrate reduced; optimal growth 37°C; G + C 50–53 mole%.

Salmonellosis any disease of humans or animals in which the causal agent is a species of *Salmonella*, including typhoid and paratyphoid fever, but most frequently referring to a gastroenteritis.

Salt tolerant organisms that can grow at elevated salt concentrations (greater than 5 percent) but that do not require added sodium chloride for growth.

Sanitary landfill a method for disposal of solid waste in low-lying areas, with wastes covered with a layer of soil each day.

Sanitize to make sanitary as by cleaning or sterilizing; to remove microorganisms and/or substances that support microbial growth.

Sanitizers substances capable of sanitizing.

Saprophytes organisms, e.g., bacteria and fungi, whose nutrients are obtained from dead and decaying plant or animal matter in the form of organic compounds in solution.

Sarcodina a major taxonomic group of protozoans characterized by the formation of pseudopodia.

Sarcomastigophora phylum of protozoans, that includes the Sarcodina and the Mastigophora; members are motile by means of flagella, pseudopodia, or both; reproduction is by syngamy.

Scanning electron microscope (SEM) a type of electron microscope in which a beam of electrons systematically sweeps over the specimen, the intensity of secondary electrons generated at the specimen's surface where the beam impacts is measured, and the resulting signal is used to determine the intensity of a signal viewed on a cathode ray tube that is scanned in synchrony with the scanning of the specimen.

Scanning electron microscopy a form of electron microscopy in which the image is formed by a beam of electrons that has been reflected from the surface of a specimen.

Scarlet fever infection caused by *Streptococcus pyogenes* transmitted by inhalation and direct contact, most common in children 2–10, characterized by sore throat, nausea, vomiting, fever, rash, and strawberry tongue; treated with penicillin or erythromycin.

Scar tissue tissue that forms after a wound, composed of normal functional and supportive tissues.

Schistosomiasis slowly progressive disease caused by trematodes that infect the urinary or intestinal tracts.

Schizogamy a form of asexual reproduction characteristic of certain groups of sporozoan protozoa; coincident with cell growth, nuclear division occurs several or numerous times, producing a schiziont that then further segments into other cells.

Schlicter test a direct method for determining the antibacterial activity of serum of patients receiving antimicrobial drugs, that is, the dilution of a patient's serum that kills the infecting organism is determined, also known as the serum bactericidal test.

Scolex the head of a tapeworm, containing suckers and sometimes hooks.

Scrub typhus acute infection caused by *Rickettsia tsutsugamushi* and transmitted to humans by mite larvae; also known as tsutsugamushi fever.

Secondary immune response the response of an individual to the second or subsequent contact with a specific antigen, characterized by a short lag period and the production of a high antibody titer.

Secondary sewage treatment the treatment of the liquid portion of sewage containing dissolved organic matter, using microorganisms to degrade the organic matter that is mineralized or converted to removable solids.

Secretory pertaining to the act of exporting a fluid from a cell or organism.

Sedimentation the process of settling, commonly of solid particles from a liquid.

Selective culture medium an inhibitory medium or one designed to encourage the growth of certain types of microorganisms in preference to any others that may be present.

Selective toxicity the property of some antimicrobial agents to have a toxic effect on some microorganisms and none on others.

Self-antigens antigens associated with the normal cells and tissues of the body; one's own antigens, in contrast to nonself, or foreign, antigens.

Self-limiting that characteristic of a disease that runs its own definite and limited course and is not influenced by outside factors.

Semiconservative replication the production of double-stranded DNA containing one new strand and one parental strand.

Sensitive (S) in antimicrobial sensitivity testing, the standardized zone of inhibition shows that the antibiotic disc is effective on the microorganism.

Septate separated by crosswalls.

Septicemia a condition where an infectious agent is distributed through the body in the bloodstream; blood poisoning, the condition attended by severe symptoms in which the blood contains large numbers of bacteria.

Septic tank a simple anaerobic treatment system for waste water where residual solids settle to the bottom of the tank and the clarified effluent is distributed over a leaching field.

Septum in bacteria the partition or crosswall formed during cell division that divides the parent cell into two daughter cells; in filamentous organisms, e.g., fungi, one of a number of internal transverse crosswalls that occur at intervals of length within each hyphae.

Serology the in vitro study of antigens and antibodies and their interactions.

Serotypes the antigenically distinguishable members of a single species.

Serotyping tests to identify microorganisms based upon serological procedures that detect the presence of specific characteristic antigens.

Serratia bacterial genus—motile, peritrichously flagellated rods; many strains produce pink, red, or magenta pigments; glucose is fermented with or without production of a small volume of gas; methyl red negative; Voges–Proskauer positive; Gram negative; facultatively anaerobic; catalase positive; oxidase negative; G + C 53–59 mole%.

Serum the fluid fraction of coagulated blood.

Serum hepatitis a form of viral hepatitis transmitted by the parenteral injection of human blood or blood products contaminated by hepatitis viruses; hepatitis B and hepatitis nonA–nonB.

Serum killing power the antimicrobial activity of the serum of a patient receiving antibiotics; an in vivo measure of antibiotic activity.

Serum sickness a hypersensitivity reaction that occurs 8–12 days after exposure to a foreign antigen; symptoms caused by the formation of immune complexes include a rash, joint pains, and fever.

Severe combined deficiency a genetically determined type of immunodeficiency caused by the failure of the lyphoid tissues to develop.

Sewage the refuse liquids or waste matter carried by sewers.

Sewage fungus *Sphaerotilus natans* found in river water receiving sewage effluent.

Sewage treatment the treatment of sewage to reduce its biological oxygen demand and to inactivate the pathogenic microorganisms present.

Sexually transmitted diseases diseases whose transmission occurs primarily or exclusively by direct contact during sexual intercourse; also known as venereal diseases.

Sexual reproduction reproduction involving the union of gametes from two individuals.

Sexual spore a spore resulting from the conjugation of gametes or nuclei from individuals of different mating type or sex.

Sheath a tubular structure formed around a filament or around a bundle of filaments, occurring in some bacteria.

Sheathed bacteria bacterial group characterized by the presence of sheaths.

Shelf life the period of time during which a stored product remains effective, useful, or suitable for consumption.

Shigella bacterial genus—nonmotile rods; not encapsulated; cannot use citrate or malonate as sole carbon source; hydrogen sulfide is not produced; glucose and other carbohydrates are fermented with the production of acid but not gas; Gram negative; generally catalase positive; oxidase negative.

Shigellosis bacillary dysentery caused by bacteria of the genus *Shigella*.

Shingles an acute inflammation of the periferal nerves caused by reactivation of an infection with the herpes varicella virus, which has remained latent since causing chicken pox in that individual, characterized by painful small, red nodular skin lesions; also known as herpes zoster.

Shipyard eye see epidemic keratoconjunctivitis.

Short-term immunity immunity that only lasts a relatively short time because it does not result in the formation of long–lasting B and T memory cells; type of immunity that can be conferred by passive immunization and transfer of immunoglobulins from mother to fetus.

Signs observable and measurable changes in a patient caused by a disease.

Silent mutations changes in the nuleotide sequence that do not change the amino acid specified by the codon; are possible because the genetic code is degenerate.

Simple stain procedure in which bacteria are stained with a single basic dye before viewing under a brightfield microscope.

Single-stranded DNA viruses viruses with genomes consisting of single-stranded DNA.

Single-stranded RNA viruses viruses with genomes consisting of single-stranded RNA.

Skin the external covering of the body, consisting of the dermis and epidermis.

Slime layer a capsular layer surrounding microbial cells composed of diffuse secretions that adhere loosely to the cell surface.

Slime molds Gymnomycota, fungal group that borders on protozoa; their vegetative cells lack cell walls; their nutrition is phagotropic.

Sludge the solid portion of sewage.

Smallpox an extinct disease caused by the variola virus that in humans caused an acute, highly communicable disease characterized by cutaneous lesions on the face and limbs.

Snapping division exhibited by *Corynebacterium*, when after binary fission cells do not separate completely and appear to form groups resembling "Chinese letters."

Sneeze a sudden, noisy, spasmodic expiration through the nose, caused by the irritation of nasal nerves.

Sodium a metallic metal of the alkali group (Na).

Somatic antigens antigens that form part of the main body of a cell, usually at the cell surface and distinguishable from antigens that occur on the flagella or capsule.

Somatic cells any cell of the body of an organism except the specialized reproductive germ cell.

Sore throat inflammation of the mucous membrane lining the throat that produces pain and redness; pharyngitis.

Sp. species singular.

Species a taxonomic category ranking just below a genus and including individuals that display a high degree of mutual similarity and that actually or potentially interbreed.

Specificity the restrictiveness of interaction; of an antibody, refers to the range of antigens with which an antibody may combine; of an enzyme, refers to the substrate that is acted upon by that enzyme; of a pathogen or parasite, refers to the range of hosts.

Spectrophotometer an instrument that measures the transmission of light as a function of wavelength, allowing quantitative measure of intensity of two sources or wavelengths.

Spectrum a range, e.g., of frequencies within which radiation has some specified characteristic, such as the visible light spectrum.

Spectrum of action the range of bacteria against which an antibiotic may be targeted, may be narrow or broad.

Sphaerotilus bacterial genus—straight rods, occurring in chains, with a sheath of uniform width, which may be attached by means of a holdfast; Gram negative; metabolism respiratory, never fermentative; aerobic but can grow at reduced oxygen concentrations; optimal growth 25–30°C; G + C 65.5–70.5 mole%.

Spherical aberration a form of distortion of a microscope lens based on the differential refraction of light passing through the thick central portion of a convex–convex lens and the light passing through the thin peripheral regions of the lens.

Spheroplasts spherical structures formed from bacteria, yeasts, and other cells by weakening or partially removing the rigid component of the cell wall.

Spiral bacteria bacterial group characterized by the presence of helically curved rods, motile by means of polar flagella.

Spirillum bacterial genus—rigid, helical cells; motile by means of polar multitrichous flagella; generally exhibit bipolar flagellation; Gram negative; strictly respiratory metabolism; aerobic-microaerophilic; optimal growth 30°C; G + C 38–65 mole%.

Spirochetes bacterial group characterized by the presence of helically coiled rods wound around one or more central axial filaments; motile by a flexing motion of the cell.

Split genes genes coded for by noncontiguous segements of the DNA so that the mRNA and the DNA for the protein product of that gene are not colinear.

Spontaneous generation theory that held that certain forms of life could arise naturally from nonliving matter.

Sporangiospores an asexual fungal spore formed within a sporangium.

Sporangium a sac-like structure within which numbers of motile or nonmotile asexually derived spores are formed.

Spore an asexual reproductive or resting body that is resistant to unfavorable environmental conditions, capable of generating viable vegetative cells when conditions are favorable; resistant and/or disseminative forms produced asexually by certain types of bacteria in a process that involves differentiation of vegetative cells or structures, characteristically formed in response to adverse environmental conditions.

Sporocidal capable of killing bacterial spores; any agent capable of killing bacterial spores.

Sporotrichosis chronic disease caused by the fungus *Sporotrix schenckii* that usually enters the body through breaks in the skin producing characteristic skin lesions on the fingers and hands; treated with application of potassium iodide solution.

Sporozoa a subphylum of parasitic protozoa in which mature organisms lack cilia and flagella, characterized by the formation of spores.

Sporozoite the cells produced by the division of the zygote of a sporozoan.

Sporulation the process of spore formation.

Spp. species plural.

Spreading factor see hyaluronidase.

Spread plate a method of microbial inoculation in which a small volume of liquid inoculum is dispersed with a glass spreader over the entire surface of an agar plate; also the plate so inoculated.

Sputum the material discharged from the surface of the air passages, throat or mouth, consisting of saliva, mucus, pus, microorganisms, fibrin, and/or blood.

Stab cell an immature lymphocyte.

Staining coloring cells or tissues with dyes.

Stains substances use to treat cells or tissues to enhance contrast so that specimens and their details may be detected by microscopy.

Stalked bacteria bacteria with appendages (stalks) that are relatively wide; the stalks can attach to a substrate or to other cells, or may serve to increase the efficiency of nutrient acquisition.

Staphylococcal food poisoning an acute, nonfebrile condition caused by the enterotoxins of certain strains of *Staphylococcus*.

Staphylococcal scalded skin syndrome (SSSS) severe skin disorder most often in infants caused by *Staphylococcus aureus* marked by erythema, peeling, and necrosis that give the skin a scalded appearance.

Staphylococcus bacterial genus—cells spherical occurring singly, in pairs, and characteristically dividing in more than one plane to form irregular clusters; nonmotile; no resting stages known; Gram positive; metabolism respiratory and fermentative; catalase positive; a wide range of carbohydrates may be utilized with the production of acid; facultative anaerobes; optimal growth 35–40°C; G + C 30–40 mole%.

Starter cultures a pure or mixed culture of microorganisms used to initiate a desired fermentation process; used in the microbial production of food.

Stationary phase a growth phase during which the death rate equals the rate of reproduction, resulting in zero growth rate in batch cultures.

Stem cell a formative cell; a blood cell capable of giving rise to various differentiated types of blood cells.

Sterilization the process that yields a condition totally free of microorganisms and all other living forms.

Sterilizing agents substances capable of yielding a sterile condition.

Steroids lipid molecules with four fused carbon rings that include cholesterol and hormones.

Stock culture a culture that is maintained as a source for authentic subcultures; a culture whose purity is ensured and from which working cultures are derived.

Strain a cell or population of cells that has the general characteristics of a given type of organism, e.g., bacterium or fungus, or of a particular genus, species, and serotype.

Streak plate method of microbial inoculation whereby a loopful of culture is scratched across the surface of a solid culture medium so that single cells are deposited at a given location.

Strep throat disease caused by *Streptococcus pyogenes* transmitted by direct contact and inhalation, characterized by sore throat with pain and difficulty swallowing; treated with penicillin or erythromycin, rest, and isolation; also known as streptococcal pharyngitis.

Streptococcus bacterial genus—cells spherical, occurring in pairs or chains; Gram positive; metabolism fermentative; predominant end product of glucose fermentation is lactic acid; catalase negative; facultative anaerobes; minimal nutritional requirements generally complex; optimal growth ca. 37°C; G + C 33–42 mole%.

Streptomyces bacterial genus—produce true mycelia; slender hyphae; the aerial mycelium at maturity forms chains of three to many spores; reproduction by germination of the aerial spores, sometimes by growth of fragments of the vegetative mycelium; Gram positive; produce a wide variety of pigments; highly oxidative heterotrophs; aerobes; optimal growth 25–35°C; G + C 69–73 mole%.

Streptomycin an aminoglycoside antibiotic produced by *Streptomyces griseus*, affecting protein synthesis by inhibiting polypeptide chain initiation at the 70S ribosomes.

Structural gene a gene whose product is an enzyme, structural protein, tRNA, or rRNA, as opposed to a regulator gene whose product regulates the transcription of structural genes.

Subcutaneous beneath the skin.

Substrate a substance upon which an enzyme acts.

Succession the replacement of populations by other populations better adapted to fill the ecological niche.

Sulfide a compound of sulfur with an element or basic radical.

Superoxide dismutase an enzyme that catalyzes the reaction between superoxide anions and protons, the products being hydrogen peroxide and oxygen.

Superoxide radical a toxic oxygen free radical.

Suppressor mutation a mutation that alleviates the effects of an earlier mutation at a different locus.

Surfactant a surface-active agent.

Susceptibility a characteristic reflecting the likelihood that an individual will acquire a disease if exposed to the causative agent.

Svedberg unit the unit in which the sedimentation coefficient of a particle is commonly quoted.

Swarm cells flagellated cells of Myxomycetes that swarm together to form true plasmodia.

Swimmer's ear see otitis externa.

Symbiosis an obligatory interactive association between two populations producing a stable condition in which the two different organisms live together in close physical association to their mutual advantage.

Symptomology the symptoms of disease taken together.

Symptoms physiological disorders that result in detectable deviations from the normal healthy state and as reflected in complaints from a patient.

Synergism an interactive but nonobligatory association between two populations in which each population benefits.

Synergy an additive interaction between two drugs; in antibiotic action, when two or more antibiotics are acting together, the production of inhibitory effects on a given organism that are greater than the additive effects of those antibiotics acting independently.

Syngamy the union of gametes to form a zygote.

Syntrophism a phenomenon in which the growth or improved growth of an organism is dependent on the provision of one or more metabolic factors or nutrients by another organism growing in the vicinity.

Syphilis a sexually transmitted disease of humans caused by *Treponema pallidum* that is characterized by a variety of lesions and stages.

Systematics the range of theoretical and practical studies involved in the classification of organisms; a synonym of taxonomy.

Systemic infections infections that are disseminated throughout the body via the circulatory system.

T aggressor cells see cytotoxic T cells.

Taxis a directional locomotive response to a given stimulus exhibited by certain motile organisms or cells.

Taxon a taxonomic group, e.g., genus, family , order, etc.

Taxonomic hierarchy organizational levels used to group living things; levels are kingdom, phylum, class, order, family, genera, and species.

Taxonomy the science of biological classification; the grouping of organisms according to their mutual affinities or similarities.

T cells lymphocyte cells that are differentiated in the thymus and are important in cell-mediated immunity, as well as modulation of antibody-mediated immunity.

Teichoic acids polymers of ribitol or glycerol phosphate found in the cell walls of some bacteria.

Temperate phage bacteriophage with the ability to form a stable, nondisruptive relationship within a bacterium; a prophage.

Template a pattern that acts as a guide for directing the synthesis of new macromolecules.

Tertiary sewage treatment a sewage treatment process beyond secondary ones, aimed at the removal of nonbiodegradable organic pollutants, mineral nutrients, and microorganisms.

Tests the outer protective coverings or shells formed by some protozoa.

Tetanus a disease of humans and other animals in which the symptoms are due to a powerful neurotoxin formed by the causal agent, *Clostridium tetani*, present in an anaerobic wound or other lesion, characterized by sustained involuntary contraction of the muscles of the jaw and neck; also known as lockjaw.

Tetracyclines a group of natural and semisynthetic antibiotics that have a common structure and that inhibit protein synthesis in a wide range of bacteria.

Theca a layer of flattened membranous vesicles beneath the external membrane of a dinoflagellate; an open or perforated shell-like structure that houses part or all of a cell.

T helper cells a class of T cells that enhance the activities of B cells in antibody-mediated immunity.

Therapeutic value a measure of the usefulness of an antimicrobial agent for use in treating disease based upon the degree of effectiveness and selectivity against an infecting pathogen compared to the toxicity toward the host organism.

Thermal death time the time required at a given temperature for the thermal inactivation or killing of a specified number of microorganisms.

Thermophiles organisms having an optimum growth temperature above 45°C.

Thiobacillus bacterial genus—small rod-shaped cells; motile by means of a single polar flagellum; no resting stages known; Gram negative; energy derived from the oxidation of one or more reduced or partially reduced sulfur compounds; final oxidation product is sulfate; obligate aerobes; optimal growth 28–30°C; G + C 50–68 mole%.

Thrush candidiasis of the mucous membranes of the mouth of infants, characterized by the formation of whitish spots.

Thylakoids flattened, membranous vesicles that occur in the photosynthetic apparatus of cyanobacteria and algae; the thylakoid membrane contains chlorophylls, accessory pigments, and electron carriers and is the site of light reaction in photosynthesis.

Thymine a pyrimidine component of DNA.

Tinea a disease of the skin, hair, or nails caused by dermatophytic fungi; also known as ringworm.

Tissue culture the maintenance or culture of isolated tissues and plant or animal cell lines in vitro.

Titer the concentration in a solution of a dissolved substance.

T killer cells see cytotoxic T cells.

T lymphocytes lymphocytes that function in the cell-mediated immune response.

Tonsillitis inflammation of the tonsils, commonly caused by *Streptococcus pyogenes*.

Toxic enzymes enzymes produced by microorganisms that adversely effect human cells or tissues; the production of toxic enzymes contributes to the virulence of some pathogens.

Toxic shock syndrome a disease caused by the release of toxins from *Staphylococcus*, resulting in a physiological state of shock—major outbreaks of the disease have been associated with the use of tampons during menstruation.

Toxicity the quality of being toxic; the kind and quantity of a poison produced by a microorganism or possessed by a nonbiological chemical.

Toxin any organic microbial product or substance that is harmful or lethal for cells, tissue cultures, or organisms; a poison.

Toxinogenicity the ability to produce toxins; the production of factors that cause disease.

Toxoid a modified protein exotoxin that has lost its toxcity but has retained its specific antigenicity.

Toxoplasmosis an acute or chronic disease of humans and other animals, caused by the intracellular pathogen *Toxoplasma gondii*; transmission is by ingestion of insufficiently cooked meats containing tissue cysts.

Trachoma a communicable disease of the eye, caused by *Chlamydia trachomatis*.

Transcription the synthesis of hnRNA, mRNA, rRNA, and tRNA from a DNA template.

Transduction the transfer of bacterial genes from one bacterium to another by bacteriophage.

Transferrin serum beta-globulin that binds and transports iron.

Transfer RNA (tRNA) a type of RNA involved in carrying amino acids to the ribosomes during translation; for each amino acid there is one or more corresponding tRNAs that can bind it specifically.

Transformation a mode of genetic transfer in which a naked DNA fragment derived from one bacterial cell is taken up by another and subsequently undergoes recombination with the recipient's chromosome; in tissue culture, the conversion of normal cells to cells that exhibit some or all of the properties typical of tumor cells; morphological and other changes that occur in both B and T lymphocytes on exposure to antigens to which they are specifically reactive.

Transfusion incompatibility because blood contains red cells with surface antigens and antibodies to the antigens that are not present, transfused blood must be compatible between donor and recipient to prevent deleterious antigen–antibody reactions; the compatility of blood types is primarily determined by the presence or absence of A and B antigens and anti-A and anti-B antibodies.

Translation the process of synthesis of polypeptide chains with mRNA serving as the template that occurs at the ribosomes.

Transmission electron microscope (TEM) a type of electron microscope in which the specimen transmits an electron beam focused on it; image contrasts are formed by the scattering of electrons out of the beam, and various magnetic lenses perform functions analogous to those of ordinary lenses in a light microscope.

Transplant to graft tissues from the same body or another or move organs from one body to another.

Transposable genetic elements see transposons.

Transposons translocatable genetic elements that can move from one location to another within a chromosome with relatively little homology between the movable DNA segments.

Transtracheal aspiration procedure used to obtain samples from lower respiratory tract free of upper respiratory tract microbes by collecting fluid via a tube passed through the trachea.

Traveler's diarrhea see gastroenteritis.

Trematodes see flukes.

Trench mouth also known as Vincent's angina and necrotizing ulcerative gingivitis; caused by fusiform bacillus or spirochete infection, characterized by painful, superficial bleeding gingival ulcers; treated with antibiotics, analgesics, and possibly removal of tissue.

Treponema bacterial genus—unicellular, helical rods, with tight regular or irregular spirals; cells have one or more axial fibrils inserted at each end of the protoplasmic cylinder; motile; Gram negative; metabolism fermentative, using amino acids and/or carbohydrates; strict anaerobes; G + C 32–50 mole%.

Treponemes small spirochetes, such as *Treponema pallidum*, the etiologic agent of syphilis.

Tricarboxylic acid cycle see Krebs cycle.

Trichinosis *Trichinella spiralis* infection acquired by eating encysted larvae in pork or other meat.

Trichome a chain or filament of cells that may or may not include one or more resting spores.

Trickling filter a simple, film-flow type, aerobic sewage treatment system; the sewage is distributed over a porous bed coated with bacterial growth that mineralizes the dissolved organic nutrients.

tRNA see transfer RNA.

-troph combining form indicating relation to nutrition or to nourishment.

Trophic level steps in the transfer of energy stored in organic compounds from one organism to another.

Trophozoite a vegetative or feeding stage in the life cycle of certain protozoans.

Trypanosomiasis any of a number of human and

animal diseases in which the causal organism is a member of the genus *Trypanosoma*.

T suppressor cells a class of T cells that depresses the activities of B cells in antibody-mediated immunity.

Tuberculosis an infectious disease of humans and other animals, caused in humans by *Mycobacterium tuberculosis* and *M. bovis* and may affect any organ or tissue of the body, most usually the lungs.

Tularemia an acute or chronic systemic disease characterized by malaise, fever, and an ulcerative granuloma at the site of infection, caused by *Francisella tularensis*.

Turbidity cloudiness or opacity of a solution.

Tyndallization a sterilization process aimed at the elimination of endospore formers in which the material is heated to 80–100°C for several minutes on each of three successive days and incubated at 37°C for the intervening periods.

Typhoid fever an acute, infectious disease of humans caused by *Salmonella typhi* that invades via the oral route; the symptoms include fever and skin, intestinal, and lymphoid lesions.

Typhus fever an acute infectious disease of humans characterized by a rash, high fever, and marked nervous symptoms, caused by rickettsia and transmitted by vectors.

Ultracentrifuge a high-speed centrifuge that will produce centrifugal fields up to several hundred thousand times the force of gravity; used for the study of proteins and viruses, for the sedimentation of macromolecules, and for the determination of molecular weights.

Ultraviolet light short wavelength electomagnetic radiation in the range 4–400 nm.

Unicellular having the form and characteristics of a single cell.

Unripened cheeses cheeses produced by a single-step fermentation.

Uracil a pyrimidine base, a component of nucleic acids.

Urea $CO(NH_2)_2$, a product of protein degradation.

Ureaplasma bacterial genus—cells coccoid; lacks a true cell wall; cells bounded by a triple-layered membrane; colonies exhibit cauliflower head or fried egg appearance; urea hydrolyzed with simultaneous production of carbon dioxide and ammonia; catalase negative; glucose fermented aerobically and anaerobically; optimal growth 35–37°C; G + C 26.9–29.8 mole%.

Ureases enzymes that split urea into carbon dioxide and ammonia.

Urethra the canal through which urine is discharged.

Urethritis inflammation of the urethra.

Urinary tract the system that functions in the elaboration and excretion of urine.

Urkaryote the proposed progenitor of prokaryotic and eukaryotic cells; the primordial living cell.

Use-dilution method test in disinfectant evaluation that determines how much a chemical can be diluted and still remain effective.

UV see ultraviolet light.

Vaccination the process of introducing a vaccine into a living organism in order to confer immunity.

Vaccine any antigenic preparation administered with the object of stimulating the recipient's specific immune defense mechanisms with respect to given pathogens or toxic agents.

Vacuoles a membrane-bound cavity within a cell that may function in digestion, secretion, storage, or excretion.

Vaginal tract a region of the female genital tract; the canal that leads from the uterus to the external orifice of the genital canal.

Vaginitis inflammation of the vagina.

Variant a strain that differs in some way from a particular, named organism.

Vasodilators substances that cause the dilation or enlargement of blood vessels.

Vectors organisms that act as carriers of pathogens and are involved in the spread of disease from one individual to another.

Vegetative cells cells that are engaged in nutrition and growth, not acting as specialized reproductive or dormant forms.

Vegetative growth production of a new organism from a portion of an existing organism exclusive of sexual reproduction.

Veillonella bacterial genus—small cocci; nonmotile; Gram negative; carbohydrates and polyols not fermented; produce acetate, propionate, carbon dioxide and hydrogen from lactate; oxidase negative; complex nutritional requirements; carbon dioxide required; anaerobic; optimal growth 30–37°C; G + C 40–44 moles %.

Venous puncture medical procedure in which a hypodermic needle is passed through the skin and into a vein; used for intravenous injection of drugs and for obtaining blood samples.

Viability the ability to grow and reproduce.

Vibrio bacterial genus—short rods, axis curved or straight; motile by a single polar flagellum; Gram negative; not acid fast; endospores not produced;

nonencapsulated; metabolism both respiratory and fermentative; fermentation of carbohydrates produces mixed products but no carbon dioxide or hydrogen; oxidase positive; facultative anaerobes; G + C 40–50 mole%.

Vinegar a condiment prepared by the microbial oxidation of ethanol to acetic acid.

Viral of or pertaining to a virus.

Viral gastroenteritis gastroenteritis caused by adeno-, echo-, or coxsackieviruses.

Virion a single, structurally complete, mature virus.

Viroids the causal agents of certain diseases, resembling viruses in many ways but different in the apparent lack of a virus-like structural organization and their resistance to a wide variety of treatments to which viruses are sensitive; naked infective RNA.

Virology the study of viruses and virus diseases.

Virucides chemicals capable of killing viruses.

Virulence the capacity of a pathogen to cause disease, broadly defined in terms of the severity of the disease in the host.

Virulence factors special properties of disease-causing microorganisms that enhance their pathogenicity and allow them to invade human tissue and disrupt normal body functions.

Virulent pathogen an organism with specialized properties that enhance its ability to cause disease.

Virus a noncellular entity that consists minimally of protein and nucleic acid and that can replicate only after entry into specific types of living cells; it has no intrinsic metabolism and replication is dependent on the direction of cellular metabolism by the viral genome; within the host cell viral components are synthesized separately and are assembled intracellularly to form mature, infectious viruses.

Vitamins a group of unrelated organic compounds some or all of which are necessary in small quantities for the normal metabolism and growth of microorganisms.

Volutin see metachromatic granules.

Vomiting the forcible ejection of the contents of the stomach through the mouth.

Vulvovaginitis inflammation of the vulva and vagina, usually caused by *Candida albicans*, herpes viruses, *Trichomonas vaginalis*, or *Neisseria gonorrhoeae*.

Wandering cells cells capable of ameboid movement, include free macrophages, lymphocytes, mast cells, and plasma cells.

Wandering macrophage see wandering cells.

Warts small benign tumors of the skin, caused in humans by the human papilloma virus.

Wasserman test a classic complement fixation test for the diagnosis of syphilis that has been replaced by more accurate methods.

Water activity (Aw) a measure of the amount of reactive water available, equivalent to relative humidity; the percent water saturation of the atmosphere.

Weil–Felix test a serological test used for the diagnosis of typhus fever.

Whey the fluid portion of milk that separates from the curd.

Whipworms nematode *Trichuris truchiura*, which infest the first part of the human large intestine.

Whooping cough an acute, respiratory-tract disease mainly in children, caused by *Bordetella pertussis*,characterized by paroxysms of coughing that usually end in loud whooping inspirations; also known as pertussis.

Widal test an agglutination test used for the diagnosis of typhoid fever.

Windrow a slow composting process that requires turning and covering with soil or compost.

Wine the product of the microbial fermentation of grapes and other fruits.

Wobble hypothesis hypothesis that accounts for the observed pattern of degeneracy in the third base of a codon and says that the third base of the codon can undergo unusual base pairing with the corresponding first base in the anticodon.

Wort in brewing, the liquor that results from the mixture of mash and water held at 40–65°C for 1–2 hours, during which the starch is broken down by amylases to glucose, maltose, and dextrins, and proteins are degraded to amino acids and polypeptides.

Wound botulism infection of a wound with *Clostridium botulinum*, diffusions of toxins from the site of infection can affect the nervous system.

X rays type of ionizing radiation with wavelengths from 10^{-3} to 10^{2} nm that may be used for sterilization.

Xanthophyll a pigment containing oxygen and derived from carotenes.

Xenobiotic a man-made synthetic product not formed by natural biosynthetic processes; a foreign substance or poison.

Xerophiles organisms that grow under conditions of low water activity.

Yeasts a category of fungi defined in terms of morphological and physiological criteria; typically a unicellular, saprophytic organism that characteristically ferments a range of carbohydrates and in which asexual reproduction occurs by budding.

Yellow fever an acute, systemic disease that affects humans and other primates, caused by a togavirus, transmitted to humans by mosquitos.

Yersinia bacterial genus—cells ovoid or rods; nonmotile or motile with peritrichous flagella; nonencapsulated; various carbohydrates fermented without production of gas; methyl red positive; Voges–Proskauer negative; optimal growth 30–37°C; G + C 45.8–46.8 mole%.

Yersiniosis infection caused by *Yersinia enterolytica* in which symptoms resemble those of appendicitis.

Yogurt the product of the fermentation of milk by *Lactobacillus bulgaricus* and *Streptococcus thermophilus*.

Zone of inhibition the area of no bacterial growth around an antimicrobial agent in an agar diffusion test for antimicrobial sensitivity.

Zoonoses diseases of lower animals.

Zoospores motile, flagellated spores.

Zygospore thick-walled resting spores formed subsequent to gametangial fusion by members of the zygomycetes.

Zygote a single, diploid cell formed from two haploid parental cells during fertilization.

Index

Italicized numbers indicate figure or table

B

C

E

F

H

I

M

N

O

P

Q

R

W

X

Y

Z